Advances in Intelligent Systems and Computing

Volume 1278

The series "Advances in Intelligent Systems and Computing" contains publications on theory, applications, and design methods of Intelligent Systems and Intelligent Computing. Virtually all disciplines such as engineering, natural sciences, computer and information science, ICT, economics, business, e-commerce, environment, healthcare, life science are covered. The list of topics spans all the areas of modern intelligent systems and computing such as: computational intelligence, soft computing including neural networks, fuzzy systems, evolutionary computing and the fusion of these paradigms, social intelligence, ambient intelligence, computational neuroscience, artificial life, virtual worlds and society, cognitive science and systems, Perception and Vision, DNA and immune based systems, self-organizing and adaptive systems, e-Learning and teaching, human-centered and human-centric computing, recommender systems, intelligent control, robotics and mechatronics including human-machine teaming, knowledge-based paradigms, learning paradigms, machine ethics, intelligent data analysis, knowledge management, intelligent agents, intelligent decision making and support, intelligent network security, trust management, interactive entertainment, Web intelligence and multimedia.

The publications within "Advances in Intelligent Systems and Computing" are primarily proceedings of important conferences, symposia and congresses. They cover significant recent developments in the field, both of a foundational and applicable character. An important characteristic feature of the series is the short publication time and world-wide distribution. This permits a rapid and broad dissemination of research results.

**** Indexing: The books of this series are submitted to ISI Proceedings, EI-Compendex, DBLP, SCOPUS, Google Scholar and Springerlink ****

More information about this series at http://www.springer.com/series/11156

Eftihia G. Nathanail · Giannis Adamos ·
Ioannis Karakikes
Editors

Advances in Mobility-as-a-Service Systems

Proceedings of 5th Conference on Sustainable Urban Mobility, Virtual CSUM2020, June 17–19, 2020, Greece

Springer

Editors
Eftihia G. Nathanail
Department of Civil Engineering
University of Thessaly
Volos, Greece

Giannis Adamos
Department of Civil Engineering
University of Thessaly
Volos, Greece

Ioannis Karakikes
Department of Civil Engineering
University of Thessaly
Volos, Greece

ISSN 2194-5357 ISSN 2194-5365 (electronic)
Advances in Intelligent Systems and Computing
ISBN 978-3-030-61074-6 ISBN 978-3-030-61075-3 (eBook)
https://doi.org/10.1007/978-3-030-61075-3

This Springer imprint is published by the registered company Springer Nature Switzerland AG
The registered company address is: Gewerbestrasse 11, 6330 Cham, Switzerland

Preface

The 5th edition of the Conference on Sustainable Urban Mobility (CSUM) 2020, held virtually from Greece on June 17–19, 2020, offered the floor to scientists, researchers, engineers and decision makers from twenty-five countries in Europe, Asia, America and Africa, to share their newest inventions, methods, findings and experience on sustainability and urban mobility. Special focus was given to the main theme of the Conference "Advances in Mobility-as-a-Service Systems," which is nowadays one of the main challenges of transportation and the society.

This book, which gathers the proceedings of CSUM 2020, presents 110 works covering a wide range of topics on public transport and demand-responsive systems, connected and autonomous vehicles and fleets, electromobility, infrastructure resilience, data sharing and digitalization and addresses emerging needs for reshaping transport modeling, transformational technologies and governance and business models for accelerating deployment.

The rapid introduction of shared mobility, along with shared economy and services, set the urge to seek for technological and organizational solutions, methodological frameworks and models, which depict accurately and timely the effectiveness and impacts of the new mobility options.

The dynamic changes in the demand and supply and their proper coupling appear to be the main objective of the developers and operators. Under the umbrella of "Smart Cities," societies demonstrate new behaviors for their traveling needs and adopt active and micro-mobility modes, with or without the combination of public transit, shared vehicles and shared rides. The flexibility to plan, book and pay for their trips through an integrated coordinated information and communication system, reinforced by the proper infrastructure and equipment, proves to be notably attractive to people and is expected to provide high service level, reduced environmental and societal impacts and optimized space usage.

This book is intended to embrace the readers' expectations for innovative thinking, intelligent and effective mobility and up-to-date sustainable solutions. As the most recent advances of the domain are discussed here, we anticipate that the book will constitute a comprehensive source of knowledge. With a sincere acknowledgment to the authors and reviewers, who contributed to the high quality of the papers, we wish you an enjoyable and thoughtful reading.

August 2020

Eftihia Nathanail
Giannis Adamos
Ioannis Karakikes
Guest Editors

Organization

Organizing Committee Chair

Eftihia Nathanail	Department of Civil Engineering, University of Thessaly, Volos, Greece

Organizing Committee

Giannis Adamos	Department of Civil Engineering, University of Thessaly, Volos, Greece
Ioannis Karakikes	Department of Civil Engineering, University of Thessaly, Volos, Greece
Maria Karatsoli	Department of Civil Engineering, University of Thessaly, Volos, Greece

Scientific Committee Members

Joao Abreu	University of Lisbon, Portugal
Giannis Adamos	University of Thessaly, Greece
Constantinos Antoniou	Technical University of Munich, Germany
Georgia Ayfadopoulou	Hellenic Institute of Transport, Greece
Sokratis Basbas	Aristotle University of Thessaloniki, Greece
Evangelos Bekiaris	Hellenic Institute of Transport, Greece
Maria Boile	University of Piraeus, Greece
Oded Cats	TU Delft, The Netherlands
Floridea Di Ciommo	CENIT (Center for Innovation in Transport), Spain
Wouter Dewulf	University of Antwerp, Belgium
Nikolaos Eliou	University of Thessaly, Greece
Sonja Forward	VTI Swedish Road and Transport Research Institute, Sweden

Contents

Accelerating Deployment: Trials, Pilots and Case Studies

System Performance and Assessment

Public Transport and Demand Responsive Systems

Gender Impact on Transit Quality of Service Importance and Performance Assessment

Maria Tsami and Eftihia Nathanail(✉)

Traffic, Transportation and Logistics Laboratory – TTLog, Department of Civil Engineering, University of Thessaly, Pedion Areos, 38334 Volos, Greece
marttsami@gmail.com, enath@uth.gr

Abstract. This paper examines the impact of gender on transit travelers' quality assessment considering 26 indicators' importance and performance ratings. Through an online survey and using a 5 point Likert scale, transit travelers were asked to rate both the importance and the performance they recognize for the examined quality indicators. The online survey was distributed to a list of contacts, organizations, businesses, universities and operations' websites in three Greek cities (Athens, Thessaloniki and Volos), aiming to cover all modes and quality conditions someone may face in Greek transit operations. In total, 211 PT users (96 male and 116 female transit travelers) provided their feedback that was further analyzed, aiming to underline the impact of their gender on transit quality importance and performance assessments. Research results, showed that women attribute a higher importance than men on route and service related characteristics, cleanliness and safety and security related indicators, along with the availability of shelter and benches at stops, the ease of purchasing ticket and the use of ecological vehicles. Similarly, women recognize a higher performance for the ease of purchasing ticket. Research results explicitly analyze the gender impact on transit quality assessment, providing useful knowledge and insights for decision makers planning and operations.

Keywords: Transit quality of service · Gender analysis · Importance · Performance

1 Introduction

Gender differences in travel behavior have been studied by a number of researchers, justifying that gender analysis is essential to be considered for assessing performance, importance and satisfaction ratings, on one hand, and explain travel behavior relative choices, on the other [1–6]. It is also well known, that the level of satisfaction is affected by the gap between perceptions and expectations from the service [2, 3] and consequently this satisfaction is linked with travel behaviors/choices [1, 4, 5, 6].

Women are believed to be more critical and thus their assessments usually guide service promotion/integration strategies [4, 5, 6]. Aiming to increase knowledge on how women evaluate the service, Rojo et al. [4], developed an ordered logit model to assess gender differences on perceived interurban bus quality. Based on their model, quality and price had the strongest relationship, while bus conditions and service

E. G. Nathanail et al. (Eds.): CSUM2020, AISC 1278, pp. 3–10, 2021.
https://doi.org/10.1007/978-3-030-61075-3_1

frequency were also important factors for their analysis [4]. Similarly, Tsami and Nathanail [5] compared female travelers' perceptions with operators' views and performance ratings, while Silveira et al. [6] developed a linear regression model proving that female travelers tend to be more critical on expected and delivered QoS (women recognized a higher importance than men in the majority of quality indicators tested). Based on this study, male travelers, attributed higher importance ratings in all service performance related indicators (i.e. reliability, frequency, travel time) while females in comfort and safety related ones (i.e. safety and security, crowding, vehicle conditions). Finally, female travelers of that study stated lower satisfaction for vehicle crowding conditions and safety [6].

Based on such findings, this research comes to examine the impact of the gender on travellers perceived quality importance and performance, considering a wide range of transit indicators that were assessed by transit users coming from three Greek cities.

This paper is organized in 4 Sections. Following the introductory part (Sect. 1), the research methodology is presented in Sect. 2, including indicator selection and presenting sample characteristics. Section 3 presents research analysis and results, while Sect. 4 presents research conclusions and discussion.

2 Methodology

Based on a literature review on Quality of Service indicators (QoS), a list of 26 indicators was selected to be used for this research. Following, an online survey dedicated to Public Transport (PT) users was conducted, organized in three parts. The first one collected socioeconomic characteristics of respondents, the second captured their travel characteristics and the third their importance and performance assessment from the selected QoS indicators by using a 5 point Likert scale.

The survey link was distributed to a list of contacts, organizations, businesses, universities and operations' websites in three Greek cities (Athens, Thessaloniki and Volos). The city selection was made based on the criterion of their representativeness in the public transport service offered in Greek cities. Two main categories were defined; one, included Athens, which is the only city with a combined and dense transit network, including tram, trolley and metro; the second, included cities serviced only by bus. Two cities were selected, to provide a representativeness on the bus network density and ownership model; Thessaloniki, being a city serviced only by bus, organized as a frequency-based model, with a dense network operated by a Public Transport Operator; Volos, being also serviced only by bus, however, organized as a schedule-based model, and operated on a sparse network by a private service company.

2.1 Indicator Selection

Having reviewed a number of research papers [10–18], it was evident that different QoS dimensions have been assessed by researchers, based on the research focus. According to dell' Olio et al. [7, 8] travel time, waiting time, cleanliness and comfort are among the important parameters in transit quality assessment, while de Ona et al. [9] mentioned that frequency and speed matter more. Still, comfort, safety and travel time are commonly accepted to be among the crucial parameters by the majority of researchers [10–18].

Considering the list of studies regarding quality determinants in public transportation the aspects mainly characterizing bus services are: service availability, service reliability, comfort, cleanliness, safety and security, fare, information, customer care and environmental impacts [5].

Based on the QUATTRO project outcomes and the developed Public Transport Quality Matrix there are 8 main categories of indicators and a list of a number of sub-indicators grouped under the main indicators [17]. This project's outcomes were the basis for the European Standard EN 13816 Quality of service in passenger transport services [18]. Following the list of the 8 main categories of QUATTRO, and considering the findings from the quality indicators review [4–18], this research indicators' selection was made by considering the most used and relevant to the service indicators, covering at the same time all QUATTRO main categories. The final main indicators used in current research followed the 11 indicator categories of Eboli and Mazzula (2007;2012) [12, 13], while a list of 26 sub-indicators were selected based on their suggestions and grouped under each main indicator category, thus: route characteristics, service characteristics, service reliability, comfort, cleanliness, fare, information, safety and security, personnel, customer services and environmental protection.

2.2 Sample Characteristics

In total, 211 PT users provided their feedback and respondents socioeconomic and travel characteristics are represented in Table 1.

Table 1. General characteristics of transit users' sample.

Indicator	Attributes	N	%
Gender	Male	96	45,5
	Female	115	54,5
Age	<=20	21	10
	21–40	139	65,9
	41–65	49	23,2
	>65	2	0,9
Occupation	Private sector employee	28	13,3
	Public sector employee	50	23,7
	Self employed	50	23,7
	Student	64	30,3
	Pensioner	5	2,4
	Unemployed	14	6,6
Mode usually used	Urban bus	152	72
	Trolley	3	1,4
	Metro	41	19,4
	Suburban rail	13	6,2
	Tram	2	0,9

(continued)

Table 1. (*continued*)

Indicator	Attributes	N	%
Purpose of travel	Work	76	36
	Study	42	19,9
	Markets/Shopping	20	9,5
	Entertainment	26	12,3
	Doctor/Hospital	3	1,4
	Personal issues	31	14,7
	Other	13	6,2
Travel Frequency with Public Transport	Daily (>=5 days/week)	56	26,5
	Many times in a week (3–4 times/week)	49	23,2
	Some days in a week (1–2 times/week)	50	23,7
	Occasionally (1–3 times/ month)	36	17,1
	Rarely (<1 time/ month)	20	9,5

As indicated in Table 1, the sample of travelers consisted of 96 men and 115 women, while the majority of them (66%) were among 21–40 year old. Almost 30% were students, 24% work for the public sector employees, whilst another 24% were self-employed. The urban bus was the mode mostly used by survey respondents (72%) as it is the only mean of transportation that operates in the cities of Thessaloniki and Volos. Next most frequent mode was the metro, used by 19% of respondents located in the city of Athens. The majority of the sample (36%) used their usual mode of travel to go to work, while 20% for education purpose, 15% for personal issues and 12% for entertainment. 73% of these trips are being made either daily (26,5%), many times in a week (23%) or weekly (24%).

Responses were further examined through a list of statistical analysis tools that were applied to further investigate any relations between genders that influence user's importance and performance assessments.

3 Research Analysis and Results

The analysis aimed at investigating the gender impact on user's perceptions by measuring gender differences on the importance and performance ratings of transit travelers. Mean importance and performance values were calculated for all transit users (both males and females) as indicated in Table 2 showing that all QoS indicators tested had a higher mean rating on their importance than on their performance assessment, with one exception, the cleanliness of bus exterior, that performs better than needed. Additionally, the difference between Importance and Performance mean ratings was above 1 point for indicators related with service characteristics and service reliability, fare and information, proving the need for them to perform better.

Table 2. QoS indicators tested and mean values of Importance and Performance Assessment of all PT users.

Indicator	Sub-indicator	Code	Importance		Performance	
			mean	Std. Dev	mean	Std. Dev
Route characteristics	Path	1.1	3,94	0,964	3,51	0,938
	Number of stops and distance between stops	1.2	3,91	0,908	3,66	0,848
	Bus stop location	1.3	3,68	0,995	3,52	0,917
Service characteristics	Service frequency	2.1	4,45	0,677	3,28	1,135
	Daily service time	2.2	4,34	0,837	3,15	1,217
Service reliability	Reliability of runs that come on schedule	3.1	4,57	0,729	3,18	1,221
	Punctuality (runs that come on time)	3.2	4,45	0,776	3,26	1,185
Comfort	Bus crowding	4.1	4,22	0,936	3,01	1,177
	Comfort of seats on bus	4.2	3,51	0,968	3,32	1,019
	Air conditioning on bus	4.3	4,16	0,885	3,42	1,128
	Levels of noise and vibrations on bus	4.4	3,61	0,986	3,13	1,103
	Availability of shelter and benches at stop	4.5	3,77	1,085	3,05	1,194
Cleanliness	Cleanliness of bus interior, seats and windows	5.1	4,1	0,853	3,29	1,073
	Cleanliness of bus exterior	5.2	3	1,047	3,16	1,066
Fare	Ticket cost	6.1	4,49	0,807	2,63	1,294
Information	Availability of schedule/maps on bus, and announcements	7.1	3,99	0,983	2,64	1,324
	Availability of schedule/maps at stops	7.2	4,27	0,831	2,66	1,286
	Availability of information by phone, mail	7.3	4,08	0,978	2,89	1,262
Safety and security	Safety and competence of drivers	8.1	4,23	0,903	3,64	0,977
	Security against crimes on bus	8.2	4,21	0,955	3,42	1,103
	Security against crimes at bus stops	8.3	4,14	1,026	3,34	1,089
Personnel	Personnel appearance	9.1	3,51	0,968	3,08	0,965
	Personnel helpfulness	9.2	3,71	1,072	3,18	1,086
Customer services	Ease of purchasing the ticket	10.1	4,25	0,892	3,49	1,169
	Administration of complaints	10.2	3,36	1,148	2,65	1,086
Environmental protection	Use of ecological vehicles	11.1	3,85	1,063	2,93	1,161

3.1 Gender Impact on QoS Indicator Importance

To analyze the impact of the gender on the perceived importance, a Mann-Whitney test for two independent samples was applied, with dependent variable the importance rating of each indicator and independent (factor) the Gender. Based on the test output, the group of females had the highest mean rank values and therefore the highest importance evaluations in all 26 indicators. Table 3 presents the statistically significant comparisons based on the Mann-Whitney test output, applied for two independent samples to identify whether importance ratings in the 26 quality indicators differ between male and female travelers.

Table 3. Impact of gender on importance of QoS indicators.

Indicators	Gender	Mean rank	Mann-Whitney U	Asymp. Sig. (2-tailed)
1.1	Male	90.94	4074.5	0.001
	Female	118.57		
1.2	Male	93.01	4273	0.003
	Female	116.84		
1.3	Male	95.89	4549	0.022
	Female	114.44		
2.1	Male	93.31	4302	0.002
	Female	116.59		
2.2	Male	96.57	4615	0.023
	Female	113.87		
4.5	Male	95.77	4537.5	0.021
	Female	114.54		
5.1	Male	93.8	4349	0.005
	Female	116.18		
5.2	Male	97.3	4684.5	0.049
	Female	113.27		
8.1	Male	92.19	4194	0.001
	Female	117.53		
8.2	Male	94.74	4439.5	0.008
	Female	115.4		
8.3	Male	92.92	4264	0.002
	Female	116.92		
10.1	Male	96.92	4648	0.032
	Female	113.58		
11.1	Male	90.2	4003	0
	Female	119.19		

As we can notice from Table 3, it can be concluded that importance evaluation in the female group was statistically significantly higher than the male group for indicators 1.1- Path (U = 4074.5, p = 0.001), 1.2- Number of stops and distance between stops (U = 4273, p = 0.003), 1.3- Bus stop location (U = 4549, P = 0.022), 2.1- Service frequency (U = 4302, p = 0.002), 2.2- Daily service time (U = 4615, p = 0.023), 4.5- Availability of shelter and benches at stop (U = 4537.5, p = 0.021), 5.1- Cleanliness of bus interior, seats and windows (U = 4349, p = 0.005), 5.2-

Cleanliness of bus exterior (U = 4684.5, p = 0.049), 8.1- Safety and competence of drivers (U = 41,94, p = 0.001), 8.2- Security against crimes on bus (U = 4439.5, p = 0.008), 8.3- Security against crimes at bus stops (U = 4264, p = 0.002), 10.1- Ease of purchasing the ticket (U = 4648, p = 0.032) and 11.1- Use of ecological vehicles (U = 4003, p = 0.000).

Therefore, females draw a higher importance on route (1.1,1.2,1.3) and service (2.1,2.2) related characteristics, cleanliness (5.1, 5.2) and Safety and Security indicators (8.1,8.2,8.3), along with the availability of shelter and benches at stops (4.5), the ease of purchasing ticket (10.1) and the use of ecological vehicles (11.1).

3.2 Gender Impact on QoS Indicator Performance

Similarly, a Mann-Whitney test for two independent samples was applied, with dependent variable the performance rating of each indicator and independent (factor) the Gender. Based on the test output only the indicator 10.1- Ease of purchasing ticket had a statistically significant output ($p = 0.019 < 0.05$) meaning that performance evaluation is statistically significant higher in female travelers than in male (U = 4519.5) for this indicator performance, as indicated in (Table 4).

Table 4. Impact of gender on performance of QoS indicators.

Indicators	Gender	Mean rank	Mann-Whitney U	Asymp. Sig. (2-tailed)
10.1	Male	95,58	4519,5	0,019
	Female	114,7		

4 Discussion and Conclusions

The present research validates that gender importance and performance evaluation on transit QoS are different for a number of QoS indicators. Females draw higher importance on route and service related characteristics, cleanliness and safety and security issues. Additionally, they matter more for the availability of shelter and benches at stops, the ease of purchasing ticket and the use of ecological vehicles than males, while they recognize a higher performance for the ease of purchasing ticket. In general, this research proved that females' assessments, as being more critical, could support decision making of transit operators in order to increase the quality of their services and females satisfaction.

The authors are aware of the limitations of the current survey, considering the sample size and potential bias. However, this research was a preliminary feedback collection research that was about to provide researchers with initial insights for the transit system quality assessment, in order to build on the next steps of research that assessed in great detail PT quality in the respective cities.

Findings from this survey were used to develop a larger research experiment, that was conducted in the city of Volos, Greece, focusing on the assessment of the local PT system quality assessment. That research reached more than 1000 transit travelers, using a face-to-face interview questionnaire approach, covering the whole city travellers.

Future research, suggests applying such analysis in a greater number of transit travelers including more cities and comparing results with operators' views and assessments.

References

1. Simićević, J., Milosavljević, N., Djoric, V.: Gender differences in travel behaviour and willingness to adopt sustainable behaviour. Transp. Plan. Technol. **39**(5), 527–537 (2016). https://doi.org/10.1080/03081060.2016.1174367
2. Oliver, R.L.: A cognitive model of the antecedents and consequences of satisfaction decisions. J. Mark. Res. **17**(November), 460–469 (1980). https://doi.org/10.2307/3150499
3. Parasuraman, A., Zeithaml, V.A., Berry, L.L.: A conceptual model of service quality and its implications for future research. J. Mark. **49**(Fall), 41–50 (1985). https://doi.org/10.2307/1251430
4. Rojo, M., Gonzalo-Orden, H., dell'Olio, L., Ibeas, A.: Modelling gender perception of quality in interurban bus services. Proc. Inst. Civil Engineers Transp. **164**(1), 43–53 (2011a)
5. Tsami, M., Nathanail, E.: Opening ground to female transit movements. Women's vs operator's perspective in transit quality of service. In: 5th International Conference on Women's Issues in Transportation - Bridging the Gap, Paris - La Défense, France, 14–16 April 2014 (2014)
6. Silveira, T.C., Dos, R., Krainer, C.W.M., Krainer, J.A., Romano, C.A., Matoski, A., Santo, A.L., eMoreira, A.R.: Does gender influence travel satisfaction with public transport? Revista ESPACIOS **40**(25) (2019). ISSN 0798 1015
7. dell'Olio, L., Ibeas, A., Cecín, P.: Modelling user perception of bus transit quality. Transp. Policy **17**(6), 388–397 (2010)
8. dell'Olio, L., Ibeas, A., Cecin, P.: The quality of service desired by public transport users. Transp. Policy **18**(1), 217–227 (2011)
9. de Oña, J., de Oña, R., Eboli, L., Mazzulla, G.: Perceived service quality in bus transit service: a structural equation approach. Transp. Policy **29**, 219–226 (2013)
10. Friman, M.: Implementing quality improvements in public transport. J. Public Transp. **7**(4), 49–65 (2004)
11. Ceder, A.: Public Transit Planning And Operation: Theory, Modelling And Practice. Elsevier, Amsterdam (2007)
12. Eboli, L., Mazzulla, G.: Service quality attributes affecting customer satisfaction for bus transit. J. Public Transp. **10**(3), 21–34 (2007)
13. Eboli, L., Mazzulla, G.: Performance indicators for an objective measure of public transport service quality. Eur. Transp. **51**, 1–21 (2012)
14. Nathanail, E.: Measuring the quality of service for passengers on the hellenic railways. Transp. Res. Part A Policy Pract. **42**(1), 48–66 (2008)
15. Litman, T.: Developing indicators for comprehensive and sustainable transport planning. Transp. Res. Rec. J. Transp. Res. Board **2017**, 10–15 (2007)
16. Rojo, M., Gonzalo-Orden, H., dell'Olio, L., Ibeas, A.: Modelling gender perception of quality in interurban bus services. Transport **164**(1), 43–53 (2011b)
17. European Commission. Quality in transportation (QUATTRO) study. Brussels: European Commission (2000)
18. European Standard EN 13816 Quality of service in passenger transport services (2002)

Investigation of Minibus Public Transport Service Characteristics in an Urban Area Through the Use of a Stated and Revealed Preference Survey

Maria Akrioti[1], Socrates Basbas[1], Georgios Georgiadis[2](✉), and Eftihia Nathanail[3]

[1] School of Rural and Surveying Engineering, Faculty of Engineering, Aristotle University of Thessaloniki, 54124 Thessaloniki, Greece

[2] School of Civil Engineering, Faculty of Engineering, Aristotle University of Thessaloniki, 54124 Thessaloniki, Greece
ggeorgiadis@civil.auth.gr

[3] School of Civil Engineering, Faculty of Engineering, University of Thessaly, Pedion Areos, 38334 Volos, Greece

Abstract. The operation of minibuses as public transport modes worldwide is considered as a both efficient and effective way to meet the mobility needs of special population groups in urban areas. Minibus services are more flexible in terms of routing and scheduling as well as more affordable compared to regular bus public transport. This paper investigates the potential introduction and the characteristics of a minibus transit service in the Municipality of Kalamaria (~90,000 pop.) in Thessaloniki Greater Area, Greece. The research is based on a joint revealed and stated preference survey, which took place in 2015 and was addressed to the citizens of the Municipality, where such a minibus service was formerly active. The revealed preference survey mainly explored the current travel behavior of citizens and their level of acceptance for such a service in case of its reopening. The stated preference survey investigated the intentions and travel preferences of the respondents under different scenarios of travel time, service frequency and travel cost. A total of 140 questionnaires was collected and analyzed using descriptive and inferential statistics. Causal analysis was conducted to reveal interrelations of respondents' willingness to use the service depending on the above attributes. According to the research findings, the perceived "quality of service" is strongly associated with safety, reliability and comfort of the transport mode used. Additionally, travel time is considered more important than cost. Although respondents seem to prefer a free of charge service, it was found that Willingness-to-Pay for minibus services ranges around 0.5 euros.

Keywords: Minibus · Public transport · Willingness-to-pay · Travel time · Travel cost · Urban mobility

E. G. Nathanail et al. (Eds.): CSUM2020, AISC 1278, pp. 11–20, 2021.
https://doi.org/10.1007/978-3-030-61075-3_2

1 Introduction

The achievement of sustainable urban mobility goals is largely dependent on well-functioning Public Transport (PT) systems [1]. Worldwide, cities are struggling to improve and modernize their current PT networks by implementing intelligent transport system technologies, optimizing PT routes and schedules in response to rapidly changing lifestyles and travel habits as well as by introducing innovations and investing in PT fleet and infrastructure. Such initiatives are gradually being considered as required steps towards successfully compete with private car use and result in greater PT ridership figures.

In this context, a lot of cities are currently dealing with the introduction or formalization of minibuses as a special-purpose or complementary PT mode. Minibuses are small buses, typically capable of carrying 20 passengers or less [2]. They are already extensively used as PT modes in Asia, Latin America, eastern European countries [3–5] and USA, where minibuses' fleet has been substantially increased during the previous decade [6]. Their further introduction continues to be studied in other parts of Europe and Asia [7–10]. Minibuses are most often used for making short trips [11], delivering demand-responsive transport services for special user groups, such as students and people with disabilities [7, 11], realizing tailored PT services in low-density areas [7] or historical urban districts [12], providing additional supply to regular PT services [7, 9] as well as for accommodating last-mile and feeder transport services to main PT routes [10]. Past research findings indicated that minibuses, due to their greater flexibility and higher speed they achieve compared to standard buses, can be profitable, even at lower occupancy rates and higher fares, and less detrimental to traffic congestion and environmental conditions [5, 13, 14]. Recent findings highlight comparatively greater customer satisfaction rates from minibus services than standard bus ones [8].

In any case, however, transport authorities need to be aware of both local travel needs and conditions for the establishment of such minibus PT systems in order to ensure their public acceptance. This paper investigates the introduction of a minibus PT system which would operate as an alternative travel option for trips with origin and destination points within a single urban municipality (*internal trips*). The study area is the municipality of Kalamaria, which belongs to the functional urban area of Thessaloniki, Greece. The specific objective is to highlight the main service characteristics of a potential minibus PT system which would be compliant with local travel preferences, needs and expectations. For this reason, a customized Revealed Preference (RP) and Stated Preference (SP) questionnaire survey was conducted with Kalamaria's citizens and a corresponding analysis setting was developed to provide the necessary research findings.

The remainder of the paper is as follows. Next section gives a short profile of the study area in terms of its demographic and transport related characteristics. The questionnaire survey method, the collected data and their analysis are explained in Sect. 3. Results are presented and discussed in Sect. 4. Section 5 summarizes the main conclusions.

2 Background

Municipality of Kalamaria is located in the southeastern part of Thessaloniki's functional urban area. The latter has almost 1 Mio. Residents while the municipality of Kalamaria has a so far increasing population of 91,258 inhabitants and its corresponding population density is approximately 14,258 inh/Km^2 [15]. Within the boundaries of Kalamaria there are critical points of interest, such as hospitals, commercial buildings and sports related facilities, that attract visitors from all over Thessaloniki's greater area.

Currently, PT in Kalamaria, and in Thessaloniki in general, is provided only by buses. There is only one PT operator in the city which is named UTOT (Urban Transport Organization of Thessaloniki). UTOT schedules standard and articulated diesel buses except for the Thessaloniki's historical district of Ano Poli where minibuses are deployed [16, 17]. UTOT PT lines connect Kalamaria to Thessaloniki's city center with high and medium service frequencies which are dependent on the specific line and time period. PT stops for UTOT lines are placed along the main arterial roads that pass through Kalamaria's central areas. Therefore, though the existing PT services are considered as satisfactory for commuting to/from Thessaloniki's city center, they fail to properly support internal trips, since no PT lines and stops are assigned to collector and local roads. In addition to this, PT coverage inadequacies are observed in the municipality's recently developed areas, where residential land uses are mostly hosted.

In response to this, municipality of Kalamaria had before long launched municipal PT services with minibuses in 1991 with the aim to: (a) increasing the existing PT coverage of the area under its responsibility by providing additional routes and services to those delivered by UTOT, (b) stimulating economic and leisure activities in the municipality and (c) reducing private car use along with tackling congestion and parking problems in municipality's central districts. The municipal PT fleet was comprised of two (2) 19-seat and one (1) 33-seat minibuses. The network was built onto two (2) circular fixed PT lines which connected locations within municipality and had bus stops which enabled access to all important places of interest, such as public sector services, healthcare facilities, educational establishments and commercial and recreation areas. Timetable frequencies were approximately 20–30 min, span of service was 12 h (06:00–21:00) during weekdays only and there was no charge for minibus PT passengers. Averagely speaking, the highest number of minibuses' passengers was observed during morning periods (i.e. approximately 50 passengers per itinerary) and the lowest figures in the afternoon (i.e. approximately 10 or fewer passengers per itinerary). Due to severe financial difficulties, the municipality permanently stopped minibus PT services in 2009. Limited resources already had affected PT system's operation and reliability before closure since the frequent cancellation of itineraries and the incapacity to modernize fleet have disappointed its users. Along with these shortcomings, certain service design drawbacks were also affecting minibuses' performance and patronage since the circular routes often resulted in high travel times between locations that, in practice, were within walking distance to each other. The quality of minibus PT services was further deteriorated by the fact that no serious investment was ever made for creating and delivering either static or real-time information to passengers. Due to all these issues, there were PT users who continued to prefer UTOT services over minibuses even when they had to make internal trips. Additionally, private car use for internal trips was never significantly threatened because of minibuses' operation.

3 Material and Methods

3.1 Survey Method

The necessary data for meeting the study's objective were collected via an ad-hoc questionnaire survey which included face-to-face interviews with a representative sample of Kalamaria's citizens. The interviews were based in a structured questionnaire form which combined both RP and SP survey methods and included four parts. The first part included questions on the socioeconomic characteristics of survey participants, such as their age, gender, household income and use of household's car (if any). In the second part, respondents were asked to provide details, i.e. trip purpose, cost, duration, origin-destination etc., on their most recent internal trip as well as to share their experiences from the former minibus local PT service, in case they have used it. Third part examined, in a general sense, the willingness to replace current travel mode with a potential minibus PT service. It also explored the relative importance of travel quality attributes such as comfort, cost, duration, safety and reliability.

Table 1. Alternative travel options and attribute levels for card sets A, B and C of SP survey.

Attribute	Card set	Travel option 1 Bus (A), Car (B), Walking (C)	Travel option 2 Minibus (All sets)
Travel time (%)	A	As today	−30, Equal, +20
	B	As today	−20, Equal, +20
	C	As today	−30, Equal, +30
Frequency (min)	All sets	As today	10, 20, 30
Travel cost (€)	All sets	As today	Free, 0.4, 0.8

For the fourth part of the questionnaire (SP survey), respondents were clustered into three (3) groups according to the travel mode they used for their most recent internal trip within Kalamaria (A: PT bus; B: Private car; C: Walking). Then, they were shown a set of nine (9) cards in line with the group they belonged to. In each set of cards two (2) alternative travel options for their most recent trip within Kalamaria were described. These options were the actual mode they previously used (A, B or C) and a hypothetical minibus PT service with routes inside Kalamaria and vehicle sizes of 15 persons. Each alternative was further defined by specific attributes, i.e. percentage difference of travel time between the two (2) alternative travel options, service frequency and travel cost. Attribute levels varied across cards so that each card would present a specific combination of travel options (Table 1). Respondents were asked to select one (1) travel option from each card that better expressed their personal preferences compared to the other available ones in the same card. All cards also contained a third "don't know/ don't answer" option.

3.2 Data Collection and Process

The questionnaire survey was performed in March 2015. Interviews were conducted, within the municipality of Kalamaria, with citizens at bus PT stops of UTOT and at specific locations, where important activities (e.g. commercial, employment, leisure etc.) along with pedestrian flows are concentrated. A total of 140 valid interviews were completed.

Table 2 presents the general characteristics of the survey participants and shows that our sample was well distributed against gender, age and monthly household income categories. Most of the respondents lived in households with cars and they were also their main users. Almost half of the most recent internal trips, which were reported by respondents, was made on foot and the other half was equally shared between private car and PT bus. This modal split probably explains the relatively low travel time and cost figures of Table 2. Thought that all trip purpose main categories appear in respondents' answers, leisure and commuting are the most popular ones. Regarding the former municipal minibus PT system, Table 2 indicates that more than half of the survey's participants had no experience from using it, but those who had were satisfied from its services.

Table 2. Personal and trip characteristics of the respondents (n = 140).

Characteristics	Statistics
Gender	Male: 45.7%; Female: 54.3%
Age group	19–24: 15.7%, 25–39: 37.9%; 40–54: 25.7%; 55–64: 15.7%; > 64: 5.0%
Monthly household income (€)	0–800: 21.4%; 801–1,600: 46.4%; 1,601–2,400: 7.9%; > 2,400: 7.1; No answer: 17.2%
Possession of household car	Yes: 88.6%; No: 11.4%
Most frequent user of household's car	Yes: 59.7%; No: 40.3%
Trip mode	PT bus: 28.6%; Private car: 28.6%; Walking: 42.8%
Trip purpose	Commuting: 27.8%; Shopping: 17.1%; Leisure: 36.4%; Business: 11.4%; Other: 7.3%
Travel time (minutes)	3–5: 16.4%, 7–10: 42.8%; 11–15: 23.5%; 20–30: 15.7%; > 30: 1.6%
Travel cost (€)	0: 55%, 1: 22.9%; 2: 17.9%; 3–5: 4.2%
User of former minibus PT service	Yes: 37.9; No: 62.1%
Satisfied from former minibus PT service	Yes: 92.3%, No: 7.7%

In order to investigate citizens' preferences and opinions towards a potential minibus PT system for Kalamaria, we further analyzed the collected data. We used descriptive statistics to underline the factors that affect travel experience, to investigate whether willingness-to-use minibus PT is affected by travel time, service frequencies and travel cost as well as to highlight the fares that would be considered as affordable for such a PT service.

Furthermore, we employed inferential statistics methods to investigate whether current mode choice and willingness to use a potential minibus PT mode are dependent on personal characteristics. In order to examine personal characteristics' variables with two (2) categories, i.e. gender etc., the Mann-Whitney U test was performed and in all other cases the Kruskal-Wallis test was applied. For all test statistics, the null hypothesis (H_0) we adopted is that there is no relationship between personal characteristics and (current and future) mode choice. The alternative hypothesis (H_1) is that there is a statistically significant relationship between them. We rejected the null hypothesis for p-values lower than 0.05. All calculations were performed with SPSS software [18].

4 Results and Discussion

Descriptive analysis of RP results showed that almost all respondents would be interested into trying minibus PT services for their internal trips, since Fig. 1 demonstrates that the willingness-to-shift rate is over 90%. Figure 1 also shows that many of them (43%) expect that such service would be free of charge. In fact, willingness-to-pay results imply that the maximum acceptable fare for travelling by PT minibuses would be 0.5 euros.

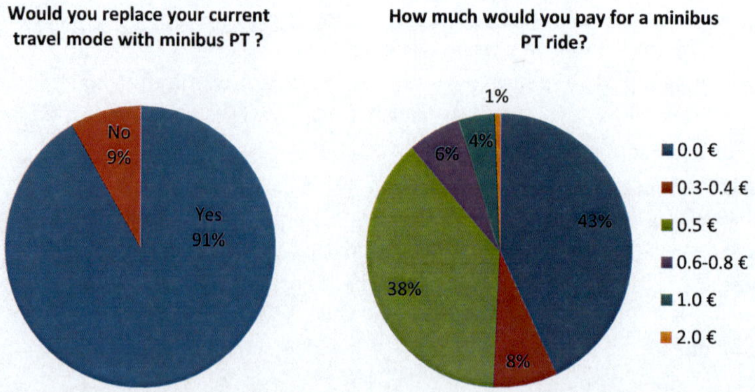

Fig. 1. Willingness-to-shift to (left) and willingness-to-pay for (right) a potential minibus PT service.

SP results, however, indicate that the willingness-to-shift rates may greatly vary against minibus PT service characteristics. Travel time appears to be a key decision factor for preferring minibuses over current travel modes. Specifically, Table 3 reports that the highest acceptance rates for minibuses were observed when travel time was reduced, and routes were scheduled every 10 min. Willingness-to-shift rates significantly dropped when travel time was increased, and service frequencies were over 20 min. Travelers seem to accept a fare that does not exceed 0.4 euros, considering that all tested combinations which included a fare of 0.8 euros are associated with very low selection rates in favor of minibuses. This finding reconfirms Fig. 1 willingness-to-pay results.

Table 3. SP results: Selection of minibus as travel option.

Hypothetical Minibus PT trip attributes				Willingness to shift to minibus PT service
# Card	Travel time*	Service frequency (min)	Travel cost (€)	
1	Reduced	10	0	94.3%
2	Increased	20	0	45.0%
3	Equal	30	0	45.0%
4	Reduced	20	0.4	76.4%
5	Equal	10	0.4	90.7%
6	Increased	30	0.4	23.60%
7	Increased	10	0.8	18.60%
8	Reduced	30	0.8	35.70%
9	Equal	20	0.8	14.30%

*compared to current travel option (all modes considered)

Table 4. Respondents' opinions on travel experience determinants.

Travel experience statements	Level of agreement 1: Fully disagree; 5: Fully agree
Travel Safety is a key quality of service attribute	4.34
Travel Comfort is a key quality of service attribute	4.16
Travel Reliability is a key quality of service attribute	4.12
Travel Cost is more important than Travel Time	2.97

The higher importance of travel time over travel cost is also confirmed by the RP results. Table 4 shows the level of respondents' agreement, using a five-point likert scale, on certain statements which dealt with specific trip quality attributes. Table 4 results make evident that the majority of respondents thinks that travel time is more important than travel cost. In Table 4 it is also evident that safety is the most important travel aspect amongst those examined. Reliability and comfort are equally appreciated by local travelers. Both of these factors are assigned with relatively high importance scores which are similar to safety ones.

Table 5 reports the p-value results obtained from the application of inferential statistical techniques on selected variables of RP survey. The techniques followed were explained in Sect. 3.2. According to Table 5, both the selection of current mode for internal trips and the willingness-to-shift to a potential minibus PT service do not present any statistically significant correlations with gender, age and monthly income. This indicates that service characteristics of minibuses are more critical to personal ones when it comes to the acceptance of such a new PT service. Table 5 also shows that statistically significant relationships are found only for the mode choices of the most frequent users of households' cars (p-values < 0.05). This group of travelers, as expected, mostly use private car for their internal trips. They are also associated with statistically significantly lower willingness to use a hypothetical minibus PT system compared to the other respondents.

Table 5. Personal characteristics and mode choice: Inferential statistics results (p-values).

Personal characteristics	Mode choice		Test statistic
	Trip mode currently used	Willingness to shift to minibus PT service	
Gender	0.252	0.132	Mann-Whitney U
Age	0.998	0.571	Kruskal-Wallis H
Monthly income	0.721	0.396	
Most frequent user of household's car	0.000	0.017	Mann-Whitney U

5 Conclusion

Minibuses are widely considered as a promising PT mode that due to its greater flexibility and lower cost can be utilized for a variety of PT purposes. Currently, minibuses are mainly used for short trips that pertain to demand responsive transport systems and PT lines which provide additional supply to the core PT networks.

In this paper we investigated the introduction of a potential minibus PT system that would accommodate internal trips within the urban municipality of Kalamaria in Thessaloniki's greater area, Greece. We accordingly designed and performed a joint RP and SP survey to collect data on personal characteristics, current travel preferences and opinions of Kalamaria's citizens regarding the future introduction of such a minibus PT service in their local community. We employed both descriptive and inferential statistics methods to determine those minibus PT service characteristics which would be compatible with citizens' needs and expectations.

Overall, we may conclude that any new minibus PT system for Kalamaria should have far better service characteristics compared to the former system which functioned ten years ago in the same area. Kalamaria's citizens would welcome the launch of a minibus PT network provided that its operation would realize tangible travel time and quality of service benefits. More specifically, according to survey participants' views, an ideal minibus PT system for internal trips within their local area should: (a) provide shortened total travel times compared to the current ones, (b) schedule minibus vehicles every 10 min, (c) charge no fare or maximum 0.5 euros per ride and (d) meet the quality criteria of reliability, safety and comfort. This set of service characteristics is commonly agreed by all respondents regardless to their personal attributes. We also found that the frequent private car drivers were the most reluctant group of travelers to shift to minibuses. These findings indicate that the introduction and operation of a future minibus PT system should also be accompanied by a tailored marketing strategy which would both target specific groups of trip makers and promote the quality of service attributes of minibuses in order to gain considerable modal shares and most importantly reduce private car use.

A limitation of this study is the fact that the survey was conducted six (6) years after the termination of the local minibus PT service. Therefore, there may have been cases where participants did not adequately remember their previous travel experience on minibuses while they were providing their answers. Another limitation of the study is the sample size of respondents which should be attributed to the inherent difficulties of conducting face-to-face interviews with people who, during survey periods, may have time restrictions for participating. Web-based surveys and focus group discussions could be additional tools for obtaining a larger sample.

References

1. European Commission: COM 2011 (144): White Paper. Roadmap to a Single European Transport Area – Towards a competitive and resource efficient transport system, Brussels (2011)
2. American Association of State Highway and Transportation Officials: AASHTO Transportation Glossary. 4th edn. AASHTO, USA (2009)
3. Vandevelde, D.: A new regulation for the European public transport. Res, Transp. Econ. **22** (1), 78–84 (2008)
4. International Association of Public Transport (UITP): Organisation and major players of short-distance public transport. New developments in the European Union. UITP, Brussels (2010)
5. Gwilliam, K.: A review of issues in transit economics. Res. Transp. Econ. **23**(1), 4–22 (2008)
6. Federal Transit Administration: National Transit Summaries and Trends. National Transit Database 2010 Report Year. U.S. Department of Transportation, Washington, D.C. (2011)
7. de Jong, W., Vogels, J., van Wijk, K., Cazemier, O.: The key factors for providing successful public transport in low-density areas in the Netherlands. Res. Transp. Bus. Manag. **2**, 65–73 (2011)
8. Noor, H.M., Nasrudin, N., Foo, J.: Determinants of customer satisfaction of service quality: city bus service in Kota Kinabalu, Malaysia. Procedia Soc. Behav. Sci. **153**, 595–605 (2014)
9. Eiró, T., Viegas, J.M., Martínez, L.M.: A new optimisation procedure to design minibus services: an application for the Lisbon Metropolitan Area. Procedia Soc. Behav. Sci. **20**, 856–865 (2011)

10. Martínez, L.M., Eiró, T.: An optimization procedure to design a minibus feeder service: an application to the Sintra Rail Line. Procedia Soc. Behav. Sci. **54**, 525–536 (2012)
11. The Transit Cooperative Research Program (TCRP).: Report 165: Transit Capacity and Quality of Service Manual, Third Edition. Transportation Research Board of the National Academies, Washington, D.C. (2013)
12. Pitsiava-Latinopoulou, M., Basbas, S., Nikolaou, K., Toskas, G.: The use of environment friendly public transport system in historical urban areas. In: Sucharov, L.J., Brebbia, C.A. (eds.) 6th International Conference on Urban Transport and the Environment in the 21st Century – Urban Transport VI 2000, International Series: Advances in Transport, vol. 6, pp. 359–368. WITpress (2000)
13. Glaister, S.: Competition on an urban bus route. J. Transp. Econ. Policy XIX **1**, 217–244 (1985)
14. Walters, A.A.: The benefits of minibuses. J. Transp. Econ. Policy XII **3**, 320–337 (1979)
15. Hellenic Statistical Authority: Population and Housing census (2011). https://www.statistics.gr/en/statistics/pop. Accessed 07 Mar 2020
16. Basbas, S., Nikolaou, K., Toskas, G.: Relative contribution of various diesel vehicle classes to the Thessaloniki urban air pollution. Water Air Soil Pollut. Focus **4**, 55–63 (2004)
17. Basbas, S., Nikolaou, K., Toskas, G.: Environmental impacts of bus traffic in the Thessaloniki Metropolitan Area. J. Environ. Prot. Ecol. **2**(3), 567–574 (2001)
18. IBM Corp. Released 2017. IBM SPSS Statistics for Windows, Version 25.0. IBM Corp., Armonk (2017)

Effect of Self-driving Buses on Vehicle Scheduling

Viktor Nagy(✉) and Balázs Horváth

Széchenyi István University, Győr, Hungary
nagy.viktor@sze.hu

Abstract. We are more and more close to the time when a higher number of autonomous vehicles are appearing in road traffic. The number of unanswered questions does not diminish but grow. One such issue is the role of autonomous vehicles in public transport. When talking about autonomous vehicles we often think of only cars and we think less about self-driving buses. But the economic potential inherent in autonomous buses is huge. In the Hungarian vehicle and crew scheduling practice (and also in other countries) the one driver-one vehicle control is typical. This method closely links the vehicles and the drivers. Vehicles should therefore adapt to the rest time of the crew and the employment rules. Unused reserves are generated in the system. Autonomous vehicles can release this overcapacity. Thanks to that, fewer vehicles can carry out public transport tasks and we can save extra rides. It also provides a solution to the lack of drivers, which is a basic problem in many countries. In our study we show the reserves that can be recovered from the system in the case of three Hungarian cities (Eger, Dunaújváros, Győr). We show how much savings can be achieved by running autonomous buses in European cities with a population of 45 000, 54 000 and 130 000 inhabitants. The results are promising. In smaller cities we could achieve about 20% of economical savings but in bigger cities 40% is also realistic. Our statements are based on only rough calculations and they try to help in preparation for the future.

Keywords: Public transport · Autonomous vehicles · Vehicle scheduling · Transport modelling

1 Introduction

Nowadays, there are more and more words about autonomous vehicles. Many researchers and examinations discuss the chances and opportunities of the age of self-driving vehicles, which is expected to come in the next 10–15 year. We can read a lot about autonomous vehicles and their potential. Some researchers state that with autonomous vehicles the traffic will be lower than now [1]. Others predict the opposite of that [2]. Some examination expects more and more free public spaces, other researches state that they need additional capacity. Despite these studies, there are only a few research studies which examine what self-driving vehicles can do in public transport despite the fact that this technical solution is not the future but rather the present as Sion, Vienna and Helsinki are already operating their autonomous buses

E. G. Nathanail et al. (Eds.): CSUM2020, AISC 1278, pp. 21–29, 2021.
https://doi.org/10.1007/978-3-030-61075-3_3

with test runs. Our study is a smaller slice of a larger scale research, where we examine the vehicle scheduling operation opportunites, costs and line route optimisation of autonomus public transport service, based on the passenger's traveling behaviour. In this paper, we try to give some answers to the following question: what kind of consequences could the appearance of self-driving vehicles have in the schedule-based Hungarian public transport service? We examined the public transport of three Hungarian cities, with different population, historical development and geographical aptitude. In this study we focus on work organization, we mention the other aspects only briefly.

We can observe this question in two aspects. We can examine the system as work organization (in this study we focus on this), on the other hand we can talk about the effect of autonomous vehicles on passengers. The new service could alter user preferences, since it will be more reliable and accurate but also the journey will lose the human side without the bus driver. Just like in the case when ticket machines displaced the ticket offices.

2 Autonomous Vehicles in Public Transport

We can examine the public transport systems usually in three ways: user (passenger), customer (usually the local government), operator (transport company). These two latter aspects are often fused, which is not a big mistake as these two create the system as a whole service. According to this idea we are approaching the issue of self-driving vehicles from two sides in this paper [3].

What kind of differences can passengers expect when using autonomous vehicles? In the case of a self-driving vehicle service, the passengers could expect to receive a more accurate service compared to the current service (i.e. an autonomous vehicle will not hurry home after the last run). The service may be much more standardized as there are no different driving styles nor good or bad drivers. The vehicles always drive in the same way. On the other hand, this service is machine-only, there is no personal attachment. There is no driver, no one can oversee the passengers. This could disturb some passengers, even if the distant oversight of the vehicles is solved. There is less emotion on the operator side, rather the cost of implementing the service.

But which cost elements will change?

The most important factor is the wage cost of bus drivers which is practically eliminated when autonomous vehicles are used. However, it does not cease completely as it may require smaller movements and occasional runs in case of failure. However, this is only a small part of the original costs and it will appear in service costs. Another important element is the fuel cost which will not disappear but it is changing as according to today's trends the autonomous vehicles usually have electric engines. It also raises the possibility that the operator can even think about power generation (thinking about solar panels). The third important issue is the amortisation. The price of self-driving vehicles will be much higher than the price of conventional buses. Also the maintenance will require more sophisticated solutions.

Another important question is the scheduling system. With ceasing the constraints of the crew and the vehicles (which is very tight in Hungarian practice), we can use our vehicles more efficiently and we can save costs. In this paper, we focus on this aspect of the appearance of autonomous vehicles.

3 Limitations of Vehicle and Crew Scheduling in the Hungarian Practice

The vehicles are maybe the most expensive parts of the system but according to Hungarian practice the vehicle and the driver are connected to each other in most cases. If the driver decides to rest, then the vehicle rests too. It is a luxury as it is the most expensive item of the system as mentioned before. The efficient use of vehicles would require that the vehicles always have to do useful activities but this does not happen. The vehicle scheduling utilisation does not exceed 70% but in some cases it does not reach 60% either. It means that the vehicle does not generate income in one third of its working time simply because the crew is spending its obligatory rest period.

If we recall the definition of vehicle scheduling: “a daily program of a vehicle” there is not a word about the driver, so it is possible to organize vehicles and drivers separately which means more efficient use of vehicles. The goal is not 100% efficiency obviously, as it is still necessary to use spare time for technology and traffic reasons but it is much more efficient than today’s practice. Keeping in mind the basic principles of vehicle scheduling, we can also create a driverless program for the vehicle where we only need to pay attention to the technological times and to spare times (to compensate delays caused by traffic).

4 The Impact of Autonomous Buses on Vehicle Scheduling Through an Example

In addition to the theoretical approach, we also dealt with practical cases. In our previous study, we examined the city of Eger (Hungary) [4]. In this paper, we observed the vehicle scheduling system of another two Hungarian cities (Dunaújváros and Győr). From earlier, we knew the vehicle schedules of the buses and the timetable was also known in these cities. The data table contains all of the vehicle schedules, the boarding and alighting numbers in each bus stop and we also know the number of passengers between the bus stops. We used Microsoft Excel to examine the datasets since the task was not complicated. We sorted the data by vehicle schedule ID and by time and we got the results. We also examined the utilisation of each of the vehicle schedules.

4.1 City of Eger

In the city of Eger, a review of the local public transport system took place in 2011. At that time, the city had the following line network as seen in Fig. 1 [5]. The city has a population of about 54 000 inhabitants and the street network is based on historical roots. The transport of the city is highly influenced by the geographical relief.

In 2011, the city's bus network and timetable was served by 40 blocks. That means that there were 40 buses in the system. However, due to some time overlaps and part-time work, 37 buses were physically sufficient to provide the service. (The question of reserves is not discussed here.) These 37 vehicles complete 5005 useful kilometres with 130 empty km (useless extra km).

Based on our previous papers [6], in our research we were interested in the effects self-driving buses could have on this network. Autonomous buses were considered as not requiring inter-job breaks and can be used from the start to the end of the service time. This does not appear in current Hungarian practice as the vehicle is typically set aside during the rest time of the driver.

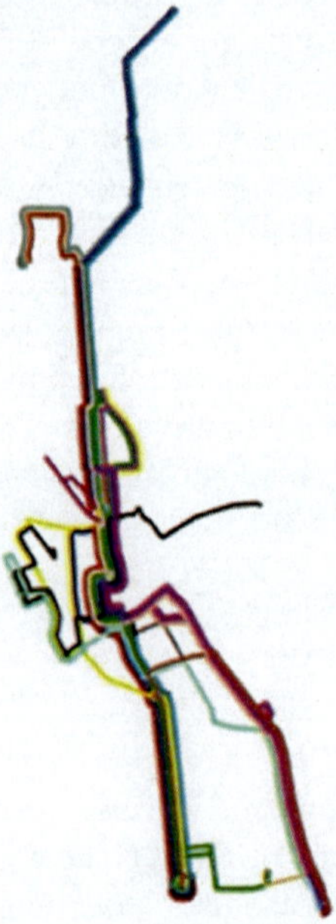

Fig. 1. Public transport lines of Eger in 2011.

Our calculations were made using the software PTV VISUM which is able to create block systems under specified conditions [5]. Before we did our calculations, we examined the theoretical vehicle minimum with the current timetable. It was 33 vehicles which means that having fewer than 33 vehicles, the city is physically unable to solve the timetable. Assuming that the autonomous vehicles can operate continuously, we have the result that 35 vehicles are enough to provide the service. Figure 2 shows a sample of the block versions.

In addition to significant savings in human resource (about 90 drivers), this intervention saves two vehicles resulting significant investment and lower operating costs. In contrast, the useful km is the same (5005) but the additional runs are growing from 130 km to 1590 km.

Examining the timetable, we came to the conclusion that the current minimum 33 vehicles can be reduced with minor changes to the schedule. If we change the timetable with a few minutes (up to 5 min) in the case of some runs then we can reduce the number of vehicles to 32. However, it is important to emphasize that the timetable has to be changed for this but in this case, we could reduce the number of additional runs to 1323 km.

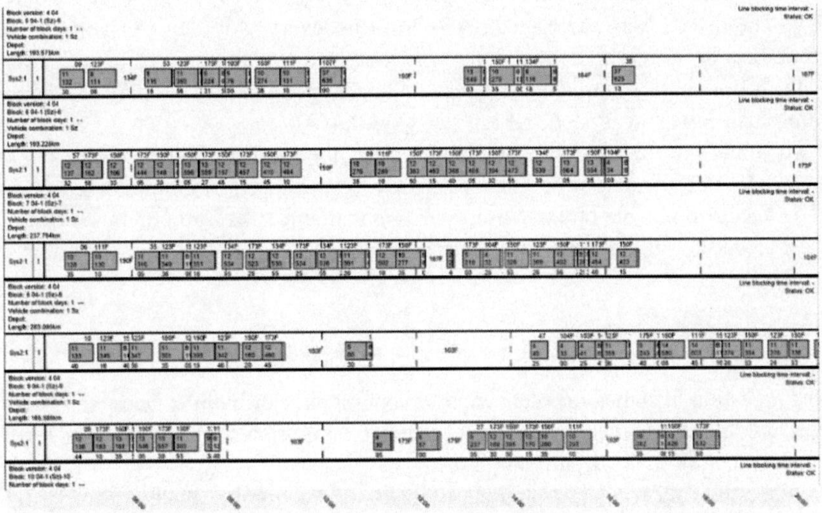

Fig. 2. Block versions with autonomous vehicles (sample).

4.2 City of Dunaújváros

In the Hungarian city of Dunaújváros, the population is about 45 000. The former country town significantly transformed in the 50's to an industrial town. It lies next to the Danube river. The public transport system was examined in 2008. This investigation was our data source, which showed 89 stops on 43 lines to serve forty thousand inhabitants of the city. That time the service needed 27 vehicles with a performance of 5466 km on every workday.

If we imagine a self-driving bus fleet – as described previously – 17 vehicles would be enough to serve the city. On the other hand, it would lead to a drastic improvement in empty runs (from 172 km/day to 1519 km/day). Although it is a huge amount, the range is similar to the city of Eger and here the decline in the number of vehicles is much higher.

Similar to the previous city, we investigated the optimised timetable. It would lead to saving an extra vehicle, so just 16 buses would be needed. However, it is very interesting that the amount of empty runs would also decrease from 1519 km to 1340 km.

4.3 City of Győr

As the third city, we studied Győr which is a lowland town. The traffic is influenced by the city's rivers and bridges. This town is on another scale, it has a mass transit sytem (not "just" public service) since this is a city with 130 000 inhabitants (officially) or more. It means that Győr is an economically very important city that is 3 times bigger than the cities investigated previously. We had a dataset from 2012, it was our starting point.

In 2012, the city had 46 lines with 451 stops. On this network, 87 vehicles served the city. These 87 buses performed 14 787 km on every workday. As you can see, these numbers between this city and the other ones are also three times bigger. Due to the fact that this service is much more systematic and periodic, the proportion of empty runs are much lower: it is just 353 km per workday.

If we would have a similar self-driving system, the needed number of buses would decrease to 78 and the amount of empty runs would increase just a little bit to 405 km.

The effect of timetable optimisation is also remarkable: it is further 3 buses (total 75 vehicles needed), and the empty runs increase to 471 km per workday.

5 Financing Questions

The big question is whether such an intervention is worthwhile since self-driving vehicles are more expensive so their amortisation is also higher. We made some economic calculations to decide this issue.

We used the following assumptions to make the economic calculations. The basis of these values were the cost level of the local public transport operator. The running cost were calculated from the cost of the drivers (projected value based on the salary of the drivers and on the running distances); from the average fuel consumption and fuel cost in Hungary; and from the local company other costs, projected to one km.

Due to the high purchase cost of autonomous vehicles, we count amortisation separately.

- total run cost: 2 €/km*
  - o driver: 0.8 €/km (projected cost)
  - o fuel: 0.35 €/km
  - o other (direct and indirect): 0.85 €/km
- total running cost in future: 1.05 €/km*
  - o driver: 0 €/km (self-driving vehicle)
  - o fuel: 0.2 €/km
  - o other (direct and indirect): 0.85 €/km

* without amortisation of the vehicles.

One element is missing from this list: the amortization of the buses. We used linear amortization for ten years. The estimated price for a traditional bus is 200 000 €, while the price for a self-driving bus is as high as 400 000 €. The cost elements for each version are listed in the tables below (Tables 1, 2 and 3).

Although we have at every case extra cost for amortisation, the savings in running costs and staff costs are much higher, so the total costs of the system were lower than in the original situation in all three cities. It is very important to note: all calculations are

Table 1. Cost elements in Eger [5].

	Current system	Self-driving buses (same timetable)	Self-driving buses (optimised timetable)
Number of vehicles	37	35	32
Vehicles [EUR]	7 400 000	14 000 000	12 800 000
Loaded runs [km/day]	5 004,90	5 004,90	5 004,90
Empty runs [km/day]	129,8	1 589,20	1 322,70
Running cost [EUR/km]	2	1,05	1,05
Daily operating cost [EUR/day]	10 269	6 924	6 644
Yearly operating cost [EUR/year]	3 080 820	2 077 142	1 993 194
Difference in yearly operating cost [EUR/year]		−1 003 679	−1 087 626
Difference in amortisation [EUR/year]		+660 000	+540 000
Difference of yearly costs [EUR/year]		−343 679	−547 626

Table 2. Cost elements in Dunaújváros.

	Current system	Self-driving buses (same timetable)	Self-driving buses (optimised timetable)
Number of vehicles	27	17	16
Vehicles [EUR]	5 400 000	6 800 000	6 400 000
Loaded runs [km/day]	5 466,10	5 466,10	5 466,10
Empty runs [km/day]	172,1	1 519,00	1 339,70
Running cost [EUR/km]	2	1,05	1,05
Daily operating cost [EUR/day]	11 276	7 334	7 146
Yearly operating cost [EUR/year]	3 382 920	2 200 307	2 143 827
Difference in yearly operating cost [EUR/year]		−1 182 614	−1 239 093
Difference in amortisation [EUR/year]		+140 000	+100 000
Difference of yearly costs [EUR/year]		−1 042 614	−1 139 093

Table 3. Cost elements in Győr.

	Current system	Self-driving buses (same timetable)	Self-driving buses (optimised timetable)
Number of vehicles	87	78	75
Vehicles [EUR]	17 400 000	31 200 000	30 000 000
Loaded runs [km/day]	14 787,10	14 787,10	14 787,10
Empty runs [km/day]	352,6	405,10	470,80
Running cost [EUR/km]	2	1,05	1,05
Daily operating cost [EUR/day]	30 279	15 952	16 021
Yearly operating cost [EUR/year]	9 083 820	4 785 543	4 806 239
Difference in yearly operating cost [EUR/year]		−4 298 277	−4 277 582
Difference in amortisation [EUR/year]		+1 380 000	+1 260 000
Difference of yearly costs [EUR/year]		−2 918 277	−3 017 582

based on estimated and average cost elements. The calcualtions are simplified, e.g. there is no cost for charging stations and special service stations for these kind of new buses. The maintenance staff's cost could be higher as well.

Despite these uncertainties, we still believe that the estimated cost savings of 20,9% (Eger), 36,2% (Dunaújváros) and 39,5% (Győr) are not unrealistic, although we know that in reality, it would be a bit lower than calculated, due to the listed uncertainty factors. Let us hope the range of the savings is true, then the extra costs of the new vehicles in the first year would come back in 3–5 year.

6 Conclusions

The appearance of self-driving vehicles is expected to have a major impact not only on private transport but also on public transport. In our paper, we pointed out that the use of autonomous buses may lead to concrete savings compared to current Hungarian practice as savings can be achieved both in vehicle numbers and in the number of the drivers, which was shown in three cities of Hungary.

We can say that based on these three cities, the results are promising. The different typle of cities behave differently. The smaller city's have less common mass traffic. Therefore, these cities have less potential on the field of savings. Both of the cities have opportunities. Maybe the original system has too much waste, which could be corrected. On the other hand, the bigger cities have more to save, as can be seen in the case of Győr. In smaller towns, the empty runs increased drastically but in Győr this change is not considerable.

In the future, we would like to do the same calculation for other Hungarian cities, as well as for other European cities. It would lead to more general statements based on more examples.

Acknowledgements. The research shown in this paper was supported by the project EFOP-3.6.2–16-2017–00016 "Dynamics and Control of autonomous vehicles".

References

1. Fagnant, D.J., Kockelman, K.: Preparing a nation for autonomous vehicles: opportunities, barriers and policy recommendations. Transp. Res. Part A Policy Pract. **77**, 167–181 (2015)
2. Martinez, L.M., Viegas, J.M.: Assessing the impacts of deploying a shared self-driving urban mobility system: an agent-based model applied to the city of Lisbon. Portugal Int. J. Transp. Sci. Technol. **6**, 13–27 (2017)
3. Horváth, B., Horváth, R., Nagy, V.: Önvezető járművek hatása a közforgalmú közlekedés szervezésére. II. Magyar Közlekedési Konferencia (43. Útügyi Napok, IX. Nemzeti Közlekedési Napok) Eger, 3–15 November 2018 (2018)
4. Horváth, B., et al.: Eger megyei jogú város helyi közforgalmú közlekedési vonalhálózatának és menetrendjének megújítása. Universitas-Győr Nonprofit Kft (2011)
5. PTV AG: PTV VISUM 17 Manual, Karlsruhe (2018)
6. Nagy, V., Horváth, B.: The effects of autonomous buses to vehicle scheduling system. Procedia Comput. Sci. **170**, 235–240 (2020)

Check for updates

Sustainability of Public Transport in Nottinghamshire: A Look at Bus Service Quality

Agnes Boscoe-Wallace(✉) and Sunday Chizoba Okafor

Nottingham Trent University, 50 Shakespeare Street, Nottingham NG1 4FQ, UK
Agnes.Boscoe-Wallace@ntu.ac.uk

Abstract. Sustainable urban development entails decreasing reliance on private cars as a means of transportation and endorsing public transport use. Nevertheless, consistent evaluation and enhancement of quality of service is needed if public transport is to be made more appealing particularly to car owners. A comprehension of customers perceived and expected quality of service can highly benefit in knowing areas of incapability needing precedence for upgrading.

This paper uses the SERVQUAL model to analyze quality of service of public bus transport by evaluating perceptions and expectations of the quality of service to customers, and its influence on general customer contentment about Nottinghamshire's public bus transport service.

An online survey resulted in 1306 responses, which was deemed statistically suitable for evaluation and it constituted 42.8% males and 57.2% females.

Findings from the study showed the existence of significant differences between customers perceived and expected quality of service ensuing in a level of dissatisfaction with the bus service. The most disapproval from respondents were in relation to the courteousness of drivers, fare and the punctuality of the service having −1.18, − 2.07 and −1.17 gap scores respectively. Of prime concern to respondents were the responsiveness of the service and reliability. Customers' perceptions of quality of service was found to be more pivotal at improving the general contentment in the service. Perceptions of the frequency of the service and timeliness had notable association with general contentment.

Limitations of the research, suggestions to policy makers and for further work are indicated.

Keywords: Sustainability · Public bus transport · Quality of service

1 Introduction

Carbon emissions in proportion to fuel combustion in the United Kingdom (UK) increased from 23.6% in 2000 to 28.5% in 2014 with private vehicles causing these [1]. A vital approach to contribute to enhanced choices in mobility for everyone particularly with respect to the underprivileged users of transport and increase support is by enhancing urban public transport conditions [2]. The use of customers' expectations as a basis for evaluating quality of service will advise excellence in decision making and

E. G. Nathanail et al. (Eds.): CSUM2020, AISC 1278, pp. 30–41, 2021.
https://doi.org/10.1007/978-3-030-61075-3_4

make certain the most advantageous distribution of funds and resources [3] for refining bus services. The basis for evaluating bus quality of service is laid out in the European Standard on Transportation [4] and discussed in the European Union (EU) transport research vision [5].

This paper evaluates quality of service for public bus transport in Nottinghamshire, UK from the viewpoint of the stakeholder. The objective is to assess this through a survey, with the gap between customers' perceptions and expectations assessed. The results will help recommend procedures for enhancing public transport quality of service. The multi-scale tool SERVQUAL was used for the evaluation since it is simple for service providers to comprehend outcomes and execute [6]. Its capability to assess both customers perceived and expected quality of service [7] was the basis for adopting it.

The SERVQUAL model was first developed by Parasuraman et al. [8] with the initial study assessing 10 aspects of quality of service. A re-evaluation of the model to refine its measurement accuracy and ease of use resulted in a 22-item scale centered on five quality of service dimensions of empathy, assurance, reliability, responsiveness and tangibles, [9]. Empathy entails providing individualized attention to the customers. Assurance includes the hospitable attributes of the service providers mainly the drivers who have more direct association with the customers. Reliability is the potential of service providers to carry out contracted services steadily, reliably and correctly. Responsiveness is the preparedness of service providers to assist customers and give punctual service. Tangibles constitutes the visible proof of the service likely to be visibly assessed and perceived [10].

There are three realizable scenarios to assess quality of service using SERVQUAL, i) When perceptions of customers of the quality of service is lesser than expectations; ii) When perceptions of customers of the quality of service is equal to their expectations; and iii) When expectations of customers of the quality of service is exceeded. In this situation of delivering a service, expectation is the quality of service the customer desires out of the service provider. Judgement of the provided service by the customer is perception. Quality of service is thus the gap between service perception and expectation of customers. The variation in service contexts entreats a modification to SERVQUAL with the original scale serving as a primary framework making it a versatile application in various circumstances [9, 11]. There is however no confirmed gap value where taking action is not required. All service attributes with negative gap scores require a level of improvement but with consideration to resource allocation [12]. Generally, in the Perception-Expectation model, service improvement will ideally be unrequired in cases where perception equals expectation (i.e. gap value is zero), but this is rare. Outcomes from the SERVQUAL model is primarily to aid practitioners and policy makers decide on service attributes to be prioritized for improvement but does not state what value of negative gap score should taking no action be required. However, for accuracy it is recommended to include 'Tolerance Zone Factor' in the evaluation process of quality of service using SERVQUAL [12]. The technique establishes the difference between desired (or expected) and adequate quality of service and is reported [12] to have been developed by Parasuraman [13]. In the interim, comparing adequate quality of service with perceived quality of service can provide better comprehension of where service improvement may not be required.

SERVQUAL was revised in this study to a 12-item scale taking into account four of the five service dimensions with the aim of measuring quality of service of public bus transport. As in other studies [6, 11, 14], 'Empathy' was omitted from the scale since in view of the number of passengers using the bus daily it is not practical in the situation of public transport to offer customers individual attention.

2 Methodology

The study area is Nottinghamshire, a county within East Midlands of England, UK. The traditional county town is Nottingham. The approximated population of the county in the middle of 2018 was 823,100 having circa 40% being city inhabitants [15]. Nottingham City Transport (NCT), which is the bus company with the most proportion of customer participation in the survey shuttle in and out of the city.

Twelve attributes established encompassing four quality dimensions measured on a 7-point Likert-scale commencing with 1 being 'strongly disagree' to 7 being 'strongly agree' are shown in Table 1. The statements assessed participants' perceptions and expectations of public bus services with a questionnaire used as the method to obtain information about customers. Questionnaires enable survey of a large population to be

Table 1. Revised SERVQUAL dimensions and attributes for perception subscale.

Dimension and attribute statement	item
Tangibility	
Bus stops are conveniently located	1
Buses are clean and conducive	2
Seats are available on most trips	3
Responsiveness	
Bus fares are good value for money	4
Bus information is easily obtainable	5
Buses are frequent	6
Reliability	
Buses are punctual	7
Journey time is relatively short and acceptable	8
Bus information are correct and updated	9
Service is stable and does not change regularly	10
Assurance	
Drivers are courteous and helpful	11
I feel safe and secure during my bus journey	12

effectively done in a short period at a relatively low cost and eliminate bias in responses in comparison to methods like interviews [16]. The 'practical approach' adopted for obtaining the sample size allows a lot of participants to take part with schedule and price allowing. Quota, convenience and snowball sampling methods were used such that; Major bus companies operating in the city were approached to disseminate the survey web page to their customers and circulate to their social network. Contacted Nottinghamshire County Council and Nottingham City Council to help forward the web page to suitable participants. Emails to students and staff and asked to forward to friends and family in the category for eligible participants. Finally used the university and student union social networking web pages for participants. A pilot survey was done to assure questions were phrased free of ambiguity with the main survey launched for 2 months on 24th June 2019. Ethical approval was obtained and compliance with the UK Data Protection Act 2018 ensured.

2.1 Data Analysis and Scale Reliability

Descriptive and inferential statistical methods was applied to the 'cleaned' data. Statistical results such as means, standard deviation, variances etc. was obtained from descriptive methods. Confirmation from the sample assembled to establish a deduction regarding accepting or rejecting a null hypothesis was obtained from inferential methods. Statistical tests conducted on the quantitative data was to a 95% degree of confidence and 0.05 level of significance. The tests were the paired t-test and Bivariate (Pearson) Correlation. A scale reliability test was performed on the four SERVQUAL dimensions used, to establish its internal consistency and reliability for quality of service evaluation. The Cronbach alpha measures reliability and a value of 0.7 or greater, is deemed allowable. The Cronbach alpha representing the dimension for perception and expectation subscales were found to be high as shown in Table 2 with the mean Cronbach alpha for perception and expectation being 0.76 and 0.79 respectively suggesting the four dependent variables are dependable in being used to assess the quality of service.

Table 2. Reliability coefficients for the four SERVQUAL dimensions [10]

Dependent variable	No. of items	Cronbach alpha	
		Perception subscale	Expectation subscale
Tangibility	3	0.72	0.73
Responsiveness	3	0.70	0.75
Reliability	4	0.85	0.79
Assurance	2	0.78	0.84

3 Results

3.1 Profile of Respondents

Respondents for the Nottingham City Transport (NCT) analysis was made up of 57.2% females, 95.9% are inhabitants of Nottinghamshire and 66% have a driving license. Table 3 shows the socio-demographic characteristics of the 1306 respondents. There were 50.4% respondents aged 45 year and above and 57.3% have a higher education with only 1.5% having no qualification. Majority of respondents are employed representing 67.2% and retirees represented 13.5%. Respondents who use the bus three or more times a week constituted 61% with those who have been using it for 10 year or more being 65.5%. The gross annual income with the highest number of respondents being 31.5%, were those earning 15,000 to 24,999 British pound sterling.

Table 3. Socio-demographic attributes of respondents

Socio-demographic factors	n	%
Gender		
Male	559	42.8
Female	747	57.2
Place of habitation		
In Nottinghamshire	1253	95.9
Out of Nottinghamshire	53	4.1
Driving license		
Respondents with driving license	862	66.0
Respondents without driving license	444	34.0
Age group		
16–24	187	14.3
25–34	226	17.3
35–44	235	18.0
45–54	291	22.3
55–64	239	18.3
65 and above	128	9.8
Level of education		
No qualification	20	1.5
Secondary school	189	14.5
College/sixth form	349	26.7
Higher education (University)	748	57.3

(continued)

Table 3. *(continued)*

Socio-demographic factors	n	%
Economic status		
Education	128	9.8
Employed	878	67.2
Unemployed and seeking work	46	3.5
Looking after home or family	47	3.6
Permanently sick or disabled	33	2.5
Retired	176	13.5
Usage frequency		
Once in a while	187	14.3
Once a week	149	11.4
Twice a week	174	13.3
Three or more times a week	797	61.0
Usage duration		
Less than 1 year	71	5.4
1–5 years	242	18.5
5–10 years	140	10.7
10 year or more	855	65.5
Gross annual income (GBP)		
4,999 or less	157	12.0
5,000–14,999	272	20.8
15,000–24,999	411	31.5
25,000–34,999	223	17.1
35,000–44,999	118	9.0
45,000 or above	125	9.6

3.2 Quality of Service - Dimension Level

The outcome of paired t-test statistics for the presence of significant difference between means of respondents perceived and expected quality of service at dimension level is given in Table 4. For every dimension, the perceived quality of service was found to be significantly different compared to the expected quality of service. The least perceived quality of service M value of 5.26 was noted for 'Responsiveness' along with 'Reliability' with an M value of 5.46. Contrarily, the greatest perceived quality of service with M value 5.65 was noted for 'Tangibility' with the least service gap M value of −0.73, t (1305) = −24.14, p = 0.000. 'Responsiveness' and 'Assurance' having mean gap scores M values of −1.22 and −1.07 respectively were noted as the two service dimensions that badly performed. For all dimensions, the greatest service gap was obtained for 'Assurance'.

Table 4. Comparing quality of service perception and expectation - Dimension Level).

Service dimension	N	Perception (P)		Expectation (E)		Gap score (P-E)		95% Confidence interval of the difference		t	df	Sig. (2-tailed)
		M	SD	M	SD	M	SD	Lower	Upper			
Tangibility	1306	5.65	1.02	6.38	0.67	−0.73	1.09	−0.78	−0.67	−24.14	1305	0.000
Responsiveness	1306	5.26	1.21	6.47	0.63	−1.22	1.29	−1.29	−1.15	−34.18	1305	0.000
Reliability	1306	5.46	1.11	6.28	0.68	−0.81	1.18	−0.88	−0.75	−25.02	1305	0.000
Assurance	1306	5.57	1.24	6.64	0.60	−1.07	1.30	−1.14	−1.00	−29.77	1305	0.000

3.3 Quality of Service - Attribute Level

In Table 5 customers' perceptions were notably lesser in comparison to expectations for all the service attributes. The largest and least perceived quality of service were noted for 'Bus stops are conveniently located' having M value of 5.89 and 'Bus fare is good value for money' having M value of 4.52. 'Bus fare is good value for money' having gap score M value of −2.06, t (1305) = −38.08, p = 0.000 was noted as the most poorly performing service. 'Drivers are courteous and helpful' having gap score M value of −1.18, t (1305) = −26.91, p = 0.000 had the second worst level of performance and 'Buses are punctual' having a gap score M value of −1.17, t (1305) = −25.79, p = 0.000 had the third worst level of performance. Poor degrees of performance to different extents was observed for all the service attributes and 'Journey time is relatively short and acceptable' was noted to have the minimum negative service gap M value of −0.29. Even though 'Journey time is relatively short and acceptable' had a high perceived quality of service M value of 5.63, it also recorded the smallest expected quality of service M value of 5.92. The greatest expected quality of service was noted for 'I feel safe and secure during my bus journey' with an M value of 6.71 having a corresponding mean service gap score M value of −0.97, t (1305) = −27.10, p = 0.000.

'Journey time is relatively short and acceptable', 'Bus stops are conveniently located' and 'Seats are available on most trips' were the three foremost achieving attributes. 'Bus fare is good value for money', 'Drivers are courteous and helpful' and 'Buses are punctual' were the three least achieving attributes.

Table 5. Comparison of quality of service perception and expectation - attribute level.

Service dimension	Attributes	N	Perception (P)			Expectation (E)			Gap score (P − E)		95% Confidence interval of the difference		t	df	Sig (2-tailed)
			M	Correlation	SD	M	Correlation	SD	M	SD	Lower	Upper			
Tangibility	Bus stops are conveniently located	1306	5.89	0.431**	1.16	6.46	0.085**	0.82	−0.56	1.24	−0.63	−0.49	−16.34	1305	0.000
	Buses are clean and conducive	1306	5.46	0.505**	1.31	6.49	0.050	0.75	−1.03	1.44	−1.11	−0.95	−25.83	1305	0.000
	Seats are available on most trips	1306	5.60	0.457**	1.34	6.18	−0.002	0.92	−0.59	1.56	−0.67	−0.50	−13.60	1305	0.000
Responsiveness	Bus fares are good value for money	1306	4.52	0.522**	1.81	6.58	0.031	0.76	−2.06	1.96	−2.17	−1.96	−38.08	1305	0.000
	Bus information is easily obtainable	1306	5.80	0.476**	1.25	6.57	0.056*	0.67	−0.77	1.28	−0.84	−0.70	−21.88	1305	0.000
	Buses are frequent	1306	5.46	0.561**	1.49	6.27	−0.044	0.87	−0.81	1.70	−0.90	−0.72	−17.25	1305	0.000
Reliability	Buses are punctual	1306	5.16	0.552**	1.47	6.32	0.015	0.81	−1.17	1.63	−1.25	−1.08	−25.79	1305	0.000
	Journey time is relatively short and acceptable	1306	5.63	0.498**	1.21	5.92	0.032	1.02	−0.29	1.47	−0.37	−0.21	−7.07	1305	0.000
	Bus information are correct and updated	1306	5.41	0.481**	1.33	6.58	0.052	0.68	−1.17	1.38	−1.25	−1.10	−30.55	1305	0.000
	Service is stable and does not change regularly	1306	5.65	0.531**	1.32	6.28	0.004	0.91	−0.63	1.51	−0.72	−0.55	−15.21	1305	0.000
Assurance	Drivers are courteous and helpful	1306	5.40	0.471**	1.47	6.58	0.002	0.68	−1.18	1.58	−1.26	−1.09	−26.91	1305	0.000
	I feel safe and secure during my bus journey	1306	5.74	0.475**	1.25	6.71	0.014	0.61	−0.97	1.29	−1.04	−0.90	−27.10	1305	0.000

** Correlation is significant at the 0.01 level (2 – tailed) * Correlation is significant at the 0.05 level (2-tailed)

4 Discussion

This paper evaluated customers' viewpoint of the quality of service of public bus transport with the SERVQUAL model. There were 1306 useful responses made up of 42.8% male and 57.2% female. Expectations regarding all service attributes for the quality of service surpassed the perceptions of customers', ensuing in negative service gap. The presence of negative service gaps indicates customers' expectation of a top-level quality of service in correspondence to perceptions of the service delivered [17]. Attributes with the foremost negative service gap was linked to the fare succeeded by the courteousness of drivers and punctuality of the service. This evidences that respondents prioritize these attributes and look forward to services being delivered at more economical prices accompanied by timely services and courteousness of drivers. A related study representing the 'education' category noted the frequentness of the service provided, timeliness and current details about the services to have the biggest negative service gap [11]. Sam et al. [18] with a study sample of university students noted the attractiveness of the fare, solace and security in the bus to be considerably essential. The minimum service gap was associated with the time of the trip, succeeded by bus stop position. The small negative service gap derived for trip duration suggests contentment with the prevailing trip duration by most respondents.

From the bivariate analysis, the association connecting the perceived quality of service of customers and that of the general contentment in service provision was noted to be compelling. This indicates that the key to growing customers' contentment about public bus transport is by tackling service perception, which should also be taken into account when setting out to enhance the service [6].

For service dimensions, the least service perception of 5.26 was observed for responsiveness with reliability having a service perception of 5.46 coming next. In opposite to this, in Too and Earl [11], the minimum perception by customers was noted for service reliability and after that, service responsiveness. The variability in the findings probably results from the sample formation. Too and Earl [11], had more than sixty percent of respondents being in education, with this study, having about seventy percent respondents being employed and fewer than ten percent being in education.

Concerning quality of service, with a value of −1.22, responsiveness recorded the greatest negative service gap and assurance succeeding with a value of −1.07. Notwithstanding a minimal perceived quality of service for service reliability in comparison with service assurance, the rise in respondents' expectation at a value of 6.64 outcome out of 7 resulted from assurance having an elevated negative service gap. It suggests that security on buses and courteousness of drivers are hoped for by customers, contradicting that by Sam et al. [14] and Too and Earl [11] but concurring with Barabino et al. [6]. The variation in outcomes of quality of service performance in research evidences its strong nature being dependent on the study formation populace, and total attributes used with individual users normally prioritizing various facets of the service [14].

5 Conclusion and Recommendations

This study evaluated the quality of service of public bus transport in Nottinghamshire, UK from customers' perspective to give useful information to service providers and policy makers for improving the service sustainably, to persuade growth in usage and decrease reliance on the car [5]. The service characteristics of each bus route and other service providers was not discussed. Data used was exclusively from Nottingham City Transport customers owing to poor participation from other bus operators.

The alternative hypotheses tested was that 'a significant difference exists between customers' perception and expectation of the quality of service of buses'. Findings indicate significant differences exist between the respondents' perceived and expected quality of service of the bus ensuing in degrees of discontentment in the service.

In line with the requirements in BS EN 13816 [4] and the EU transport research vision [5], these service improvements are recommended: i) Encourage occasional public transport promotional offers in addition to consistent quality service to attract private car users. ii) Deliver economical and equitable fare bus services with prioritized reliability of the service. iii) Prioritize service frequency and punctuality in real time iv) Enhance the provision of protected shelters and current details about buses at bus stops with research indicating that non-regular users can be conscious of service details like timetables at bus stops. v) Implement use of bus lanes effectually and restrict unlawful parking along them. vi) Ensure suitable implementation of laws in case of disorderly passenger conduct likely to impact customers security in the bus and to guarantee the attentiveness of drivers. vii) Implement cash-free payment, provide warmth in the cold months and air conditioning or improve ventilation in summer on buses and viii) Provide training for drivers to improve interaction with customers regarding being courteous and helpful.

Based on the findings and limitation, subsequent investigations should consider; comparing the quality of service from all service providers by adopting an approach that guarantees a rise in customer involvement. Appraise the quality of service on a specific route to find service areas needing prioritizing for upgrading. Increase the number of attributes evaluated to take into account other service areas that were not examined to give strong clarification about the service.

Acknowledgments. Our sincere appreciation to Anthony Carver-Smith of Nottingham City Transport for his contribution especially with regard to data collection.

References

1. World Bank. Carbon emission indicator from transport for United Kingdom, World bank: IEA (2014)
2. Barabino, B., Deiana, E., Mozzoni, S.: The quality of public transport service: the 13816 standard and a methodological approach to an Italian CASE. Ing. Ferrov **68**, 475–499 (2013)
3. Eboli, L., Mazzulla, G.: Performance indicators for an objective measure of public transport service quality. Eur. Transp. **51**, 1–21 (2012)

4. Institution, B.S.: BS EN 13816:2002 Transportation. Logistics and Services. Public Passenger Transport. Service Quality definition, targeting and measurement. British Standards Institution (2002)
5. Woodcock, A., Tovey, M.: Designing whole journey, multimodal transport provision. Des. J. **23**(1), 91–112 (2020)
6. Barabino, B., Deiana, E., Tilocca, P.: Measuring service quality in urban bus transport: a modified SERVQUAL approach. Int. J. Qual. Serv. Sci. **4**(3), 238–252 (2012)
7. Morton, C., Caulfield, B., Anable, J.: Customer perceptions of quality of service in public transport: evidence of bus transit in Scotland. Case Stud. Transp. Policy **4**, 199–207 (2016)
8. Parasuraman, A., Zeithaml, V.A., Berry, L.L.: A conceptual model of service quality and its implications for future research. J. Mark. **49**, 41–50 (1985)
9. Parasuraman, A., Berry, L.L., Zeithaml, V.A.: Refinement and reassessment of the SERVQUAL scale. J. Retail. **67**(4), 420–450 (1991)
10. Okafor, S.C.: Sustainability of Public Transport: A Stakeholder Perspective on Bus Service Quality in Nottinghamshire. Nottingham Trent University, Nottingham (2019)
11. Too, L., Earl, G.: Public transport service quality and sustainable development: a community stakeholder perspective. Sustain. Dev. **18**(1), 51–61 (2010)
12. Susniene, D.: Quality approach to the sustainability of public transport. Transport **27**(1), 102–110 (2012)
13. Parasuraman, A.Z.V.A., Berry, L.L.: Reassessment of expectations as a comparison standard in measuring service quality: implications for further research. J. Mark. **58**(1), 111–124 (1994)
14. Sam, E.F., Hamidu, O., Daniels, S.: SERVQUAL analysis of public bus transport services in Kumasi metropolis, Ghana: core user perspectives. Case Stud. Transp. Policy **6**, 25–31 (2018)
15. Nottinghamshire County Council. Nottinghamshire County Council (2019). https://www.nottinghamshire.gov.uk/business-community/economic-data/population-estimates. Accessed 17 July 2019
16. Saris, W.E., Gallhofer, I.N.: Design, Evaluation and Analysis of Questionnaires for survey research. A John Wiley and Sons, Inc., Hoboken (2007)
17. Fellesson, M., Friman, M.: Perceived satisfaction with public transport service in nine European cities. J. Transp. Res. Forum **47**(3), 93–104 (2008)
18. Sam, E., Adu-Boahen, K., Kissah-Korsah, K.: Assessing the factors that influence public transport mode preference and patronage: perspectives of students of university of cape coast (UCC) Ghana. Int. J. Dev. Sustain. **3**(2), 323–336 (2014)

Demand Responsive Public Transport System in Airport Travel: Case Study of Delhi

Sujata Savant[1(✉)], Neeraj Sharma[1], and Amit Singh Baghel[2]

[1] RITES Ltd., Plot 1, Sector 29, Gurgaon 122001, Haryana, India
sujatas@rites.com, neerajsharma.084@gmail.com
[2] VERCOS, C2, Sector 1, Noida 201301, UP, India
amitsinghbaghel@vercos.in

Abstract. Air travel due to its inherent advantage over other modes, is on the path of rapid growth in India. All over the world, airport travel is predominantly by private modes and Delhi is no exception. With 66% of trips by cars/taxis, the airport has started experiencing congestion whereas metro connectivity to Delhi's IGI airport suffered due to ridership issues.

Literature study suggests that all over the world, there is a gradual shift to public transport (PT) in airport travel. This paper, realizing the need to increase public transport share in airport travel, attempts to understand the user behavior, reasons for preference to private transport, status of existing PT connectivity etc. and suggests feasible solution. A data intensive study was conducted at IGI. Field surveys including OD cum opinion survey with sample of 11000 passengers was done at all airport terminals of IGI, metro stations, Public Transport Center and parking lots to identify the issues that will facilitate movement from private to public transport.

The paper identifies issues with existing metro connectivity and suggests demand responsive bus system with dynamic routing to fill the current PT void at IGI airport. The service, Satellite Airport Service (SAS), is proposed to connect all the terminals of IGI with major origin destination points in Delhi. The SAS service will be linked with a web-based journey planner to plan access/dispersal trips at the time of ticket booking. These measures will help in increasing share of public transport in airport travel in Delhi.

Keywords: Demand responsive transport · Airport access · Public transport system

1 Introduction

Air travel due to its inherent advantage over other modes, is on the path of rapid growth. India is already the world's third-largest domestic and overall civil aviation market and it is estimated that the number of passengers travelling through Indian airports would reach 450 million by the end of year 2020. Indira Gandhi International (IGI) Airport, the only international airport of Delhi (the capital city of India), has witnessed a surge in air travel and catered to about 69.2 million passengers in 2019, up by about 69% within 5 year duration (Fig. 1).

E. G. Nathanail et al. (Eds.): CSUM2020, AISC 1278, pp. 42–55, 2021.
https://doi.org/10.1007/978-3-030-61075-3_5

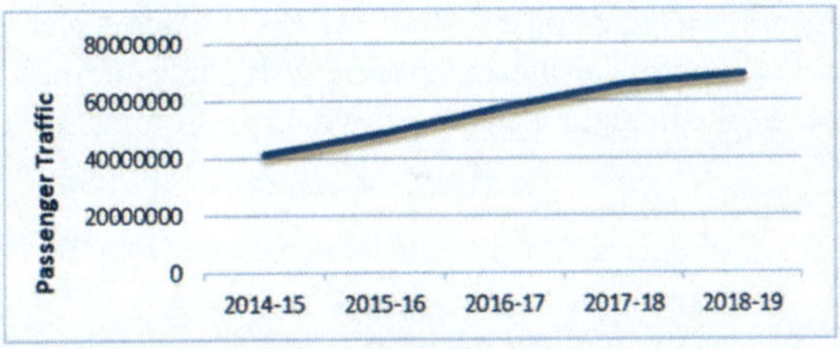

Fig. 1. Passenger traffic growth at IGI airport, Delhi

The increasing trend in air travel is likely to continue in coming years. Like most of the airports around the world, IGI has also observed the dominance of private modes in airport access dispersal activity. The unprecedented growth in air travel coupled with high share of private modes in access/dispersal has triggered congestion at the kerb front, choking of the external road network around the airport. To cater to increasing private vehicles in airport activities, airport authorities are providing more and more parking space within the airport premise. An urgent need to bring a paradigm shift in airport travel was felt to arrest the situation from getting worse. This led Airport Authority of India, a national level agency under Ministry of Civil Aviation of India, responsible for development and operation of about 130 airports across the country, to conduct a study focused towards multimodal integration at airports with an aim to increase share of mass transport in access/dispersal activities at the airports.

A data intensive study was conducted at all the 3 terminals of IGI including conduct of traffic surveys at terminal kerb-front, origin-destination (OD) cum opinion surveys (all terminals, metro stations, Public Transport Center and parking lots) and parking demand surveys to study various aspects related to airport travel. The main focus of the surveys was to understand the present travel preferences of the airport passengers and factors affecting their mode choice. The status of existing public transport (PT) connectivity, its role and relevance in airport travel was also studied [1].

This paper presents the outcomes of this comprehensive study giving an insight into the airport travel pattern, user behavior and limitations of existing public transport systems catering to the busiest and most important airport of India. The paper assesses the travel characteristics of users and their willingness to shift to a new system. It also envisages DRT as a potential new PT system for airport travel and assesses the impact it would create if implemented at IGI.

2 Study Methodology

The objectives of the study included development of a conceptual landside multi-modal transport integration plans for passenger movement with focus on public transport and formulate proposals to meet the traffic requirements for the ultimate capacity of the airport. The methodology developed to achieve the desired objectives was divided into three broad stages namely study approach, data analysis and proposal formulation. At first stage, the need and objectives of the study were detailed to plan for the requisite database generation and undertake literature review to study the best practices being followed around the world including need to switch to sustainable modes in airport travel (Fig. 2).

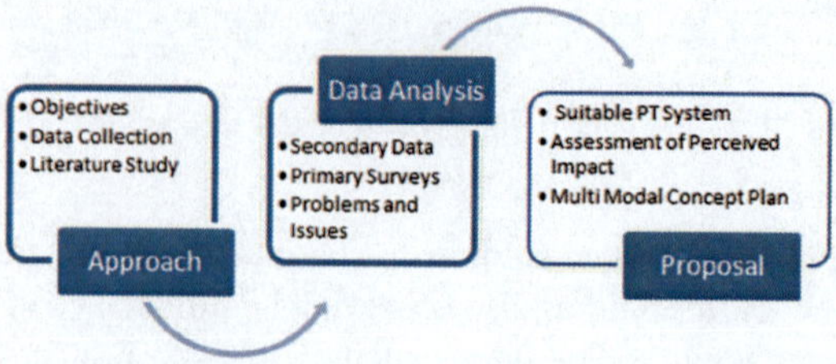

Fig. 2. Study methodology

The second stage involved the analysis of secondary and primary data collected as part of the study. The primary data included traffic volumes, passenger volumes, parking study etc. To arrive at a feasible public transport option, passenger interviews regarding origin, destination, opinion on existing modes and willingness to shift to a new mode were taken as the part of OD Surveys of airport users. This user opinion data formed the basis of arriving at solutions to address the issues reported by airport users with the focus on promoting public transport in airport travel. The perceived impact of PT measures was also assessed based on two key parameters of PT i.e.waiting time and access-dispersal time.

3 Understanding IGI Airport

The Indira Gandhi International (IGI) Airport, Delhi, is the most important airport in North India for both domestic and international travel. The immediate influence area of IGI is Delhi National Capital Region (NCR), an area spanning about 55,000 sq. km including Delhi and its satellite cities. For international trips the influence area extends to cities as far as 500 km and beyond (Fig. 3). Study started with understanding the IGI precinct and existing transport infrastructure at the terminal in detail. Field surveys were carried out to assess the existing traffic and travel characteristics of the airport users along with their travel preferences. The surveys included mid-block traffic counts at entry/exit points of airport and access roads for passenger terminals, parking survey, Boarding/Alighting and OD survey at Public Transport Centre (PTC), PT/IPT stops and air passenger terminals [1].

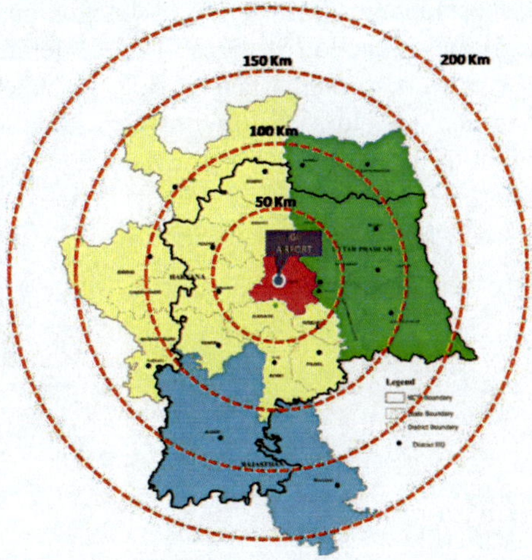

Fig. 3. Map showing influence area of IGI airport

3.1 Master Plan of IGI

The IGI sprawls on 2100 ha land and was upgraded in 2010 on PPP basis. At present, it is operated by a consortium of GMR Group (54%), Fraport (10%), Malaysia Airports (10%) with 26% stake of Airports Authority of India (AAI). The Master Plan for IGI (Fig. 4) envisages ultimate handling capacity of 109 million passengers. At present, there are three passenger terminals i.e. T1, T2 and T3 along with a cargo terminal that operate from IGI. T1 and T2 are used for domestic operations while T3 is an international terminal with limited domestic operations. The airport handled about 70 million passengers with an average 950 daily flights operated from IGI in FY 2018–19. The airport is expected to be saturated by year 2026–27 considering the traffic growth trends of the past years [2].

Terminal T3 is located at about 7.2 km and 2.2 km from T1 and T2 respectively. At present, inter-terminal journeys are facilitated by a shuttle bus providing free services to airport passengers. An Automated People Mover (APM) has been envisaged in the Master Plan to connect the passenger terminals, cargo terminal and real estate development within airport premise (Aero City) to facilitate inter terminal transfer of passengers, employees and visitors. Surface parking facilities for private vehicles are presently available at T1 and T2 terminals whereas T3 has a Multilevel Car Parking (MLCP) with capacity of about 4,300 cars.

3.2 Public Transport Connectivity

There is a dedicated express metro connectivity to IGI airport known as Airport Express Line connecting T3 terminal directly to New Delhi Railway Station (NDRS) with route length of 23 km. The services of Airport Express Metro are operational from 04.45 A.M to 11.30 P.M with a frequency of 10 min. Being a standalone system with high fares, the patronage of Airport metro was dwindling since its inception. However, the ridership increased substantially after its recent integration with city's MRTS network (at Dhaula Kuan) and reduction in fares. The T1 terminal is also recently connected to city's MRTS network through Magenta line (Janakpuri West to Botanical Garden). Terminal T3 is also in close vicinity of Bijwasan Railway Station, an important railway station on Delhi's rail network and being developed as a world class station by Indian Railways but it lacks integration with the airport terminal. Figure 5 shows the map of MRTS connectivity to IGI.

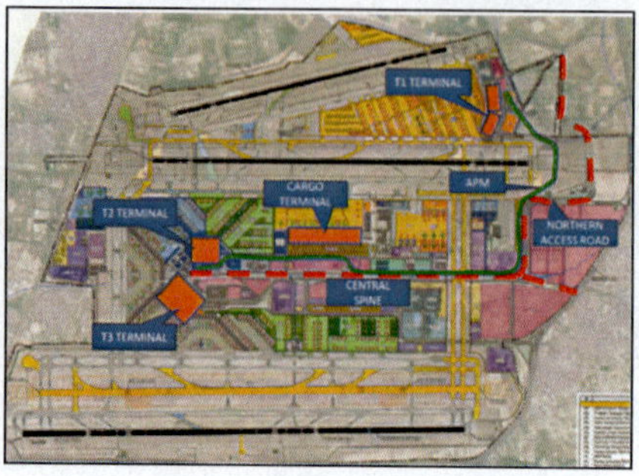

Fig. 4. Master plan of IGI

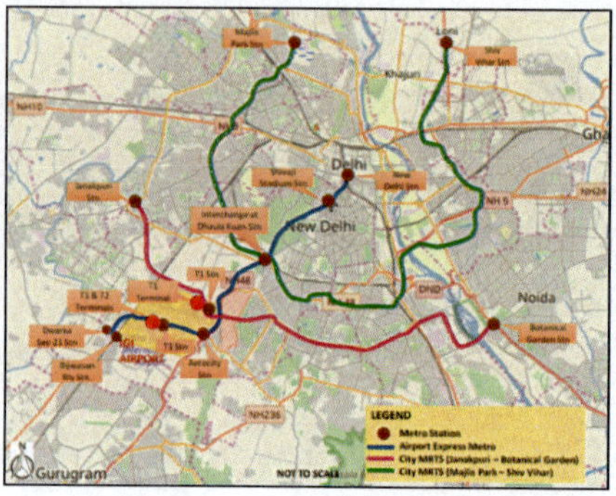

Fig. 5. MRTS connectivity to IGI airport

The Public Transport Centre (PTC), envisaged as a hub integrating different public transport modes serving IGI, is located at considerable distance from all the terminals with limited functioning. It houses separate pick up drop off zones for buses, autos & shuttle buses along with waiting lounge for passengers. Delhi Transport Corporation (DTC), the city bus transport corporation of Delhi, provides bus services from different parts of Delhi to PTC with limited operational hours i.e. from 5 am to 11 pm. Except chartered buses, all the passengers travelling to airport by city buses and auto rickshaw are required to deboard at PTC and board the airport shuttle service.

3.3 Characteristics of Travel to IGI: Results of Field Surveys

Based on primary surveys, daily traffic at terminal kerb-front was found to be about 82,000 PCUs with 98% of private vehicles. Cars (47%) and taxis (50%) were the dominant mode of travel. The share of buses was negligible at 1%. Hourly distribution of daily traffic suggests that about 26% of total traffic accessed the kerb-front between 11 pm to 6 am, outside the operational hours of the existing public transport system and is entirely dependent on private vehicles for access/dispersal from airport.

The boarding/alighting data of public transport users collected for 24 h at total 7 locations (Table 1) including Public Transport Centre, Metro stations and PT/IPT stops inside the airport precinct, suggests that public transport has a share of 33% with metro (28.6%) followed by bus (4.4%). Utilization of PTC was found to be mere 1.4% of daily passengers, indicating serious intermodal integration issues at the PTC.

Table 1. Daily boarding/alighting at PT/IPT stops at IGI airport

S. No	Location name	Daily passengers (Nos.)		
		Boarding	Alighting	Total
1	T-3 Metro station	6558	6447	13006
2	Aero city Metro station	2401	2196	4596
3	T1 Metro station	1491	1275	2766
4	Public transport centre	1499	1090	2590
5	T-3 Bus stop	1307	1480	2787
6	T-2 Bus stop	986	1273	2259
7	T-1 Bus stop	1739	2321	4060
	Total	**15981**	**16082**	**32064**

The origin - destination (OD) survey conducted with the sample size of 11000 passengers at airport terminals, PTC, Metro Stations and Bus stops suggests that trips originating from or destined to areas outside Delhi city was about 21%, indicating a regional influence of IGI, spanning up to 250 km. The internal trips (within Delhi) constituted 79% of the share with maximum trips from West Delhi (23%) along with major share from Gurgaon (4.7%) and NOIDA (4.8%) in Delhi NCR.

Distance based analysis of the trips suggests the trips up to 50 km were mainly intra city trips spanning across Delhi NCR. The trips above 50 km were intercity trips, originating from/destined to areas outside NCR. The OD data indicates that most trips to airport (51%) had their origin within distance of 20–30 km and share of Intercity trips is also quite significant with share of about 13%. The average trip length of intra city trips was observed to be 24 km and 152 km for intercity trips. Travel cost analysis indicates that majority (70%–80%) of the passenger trips to airport have travel cost in between INR 50 to INR 500. The average travel cost for passengers travelling to airport is about Rs. 268. The higher average travel cost indicates the high use of private vehicles/taxis in commute to/from airport.

The user opinion survey revealed various issues faced by them with current PT system (Fig. 6). User preferences data indicates that majority of passengers are not using private cars and taxis by choice but are doing so in lieu of an efficient public transport system. Also, as the existing public transport is not in sync with the 24 × 7 airport operations, it became irrelevant to majority of international trips being performed during night. For preference to PT system, the users gave maximum weightage to 24 × 7 availability of public transport followed by provision of last mile connectivity, less travel time, cost effectiveness and comfort with space for luggage (Fig. 7). About 90% of passengers showed willingness to pay additional Rs.100 (100% increase) on the present highest public transport fare, for a safe, fast, comfortable and reliable public transport service. Similar behavior was also observed in case of airport employees with 85% dependent on private modes for their commute to airport. The employee opinion survey also revealed similar issues with PT service. Poor reliability and low frequency were the other major concerns for employees.

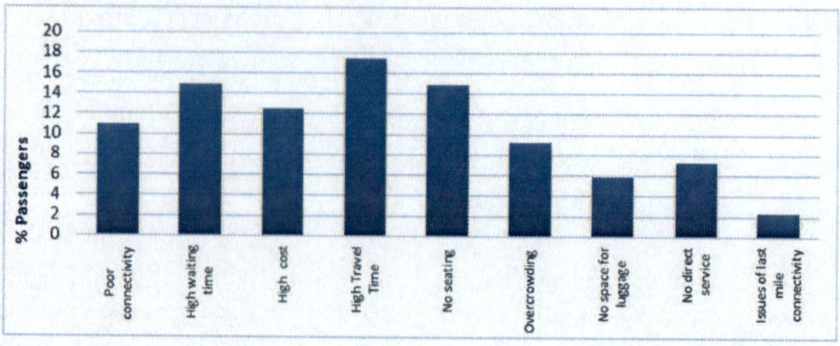

Fig. 6. User opinion on public transport service at IGI airport

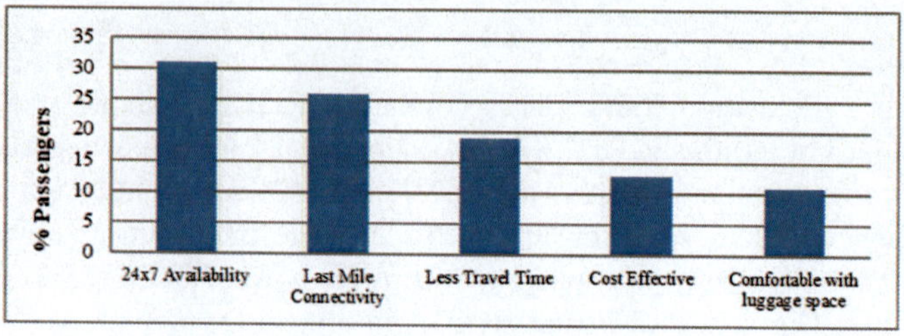

Fig. 7. User preferences in new PT system

3.4 Issues in Current Public Transport Services to IGI Airport

Discussions were held with Delhi Metro Airport Express Line operator and DTC to sync the operation of services with 24 × 7 nature of airport operations but both operators expressed major constraints to operate their services 24 × 7 round the clock. The traffic study concluded that the available public transport infrastructure is not able to serve the purpose majorly due to:

- Lack of coverage: Airport Metro Line connects only T3 and is not connected with other terminals and PTC
- Lack of integration: Airport Metro Line is poorly integrated with city's Metro system. Bus connectivity to IGI is limited till PTC, forcing passengers for an unnecessary interchange
- Lack of comfort: City's bus system and metro system are overcrowded and do not offer the level of comfort expected by an airport passenger
- Lack of priority: The shuttle bus stop is located at the terminal end, with prime areas allocated for boarding/alighting of private cars and taxis. The shuttle bus itself is dilapidated and is operated by DTC as a city bus service with dilapidated bus condition and poor service quality
- Lack of reliability: Limited operational hours of the Metro and Bus system with long waiting time for bus

A comparative study of landside transport plan with respect to public transport connectivity in other cities of India suggests that in Bangalore, BMTC, the city transport corporation operates 24 × 7 special bus service from different parts of the city to Kempegowda International Airport (KIA) with the headway of 30 min. About 25% of the passengers arriving/departing daily at KIA use these services. It has a fare which is almost one fifth of the taxi fare. Also, the Karnataka State Road Transport Corporation (KSRTC) runs a premium bus service called "Fly Bus" from KIA to Mysore city (about 200 km). About 54% of the Mysore passengers use the services costing Rs. 1,000/passenger as against Rs. 2,500/- charged by taxi. The terminals for these services are in front of the airport terminal and are accessible by walk. In case of Rajiv Gandhi International Airport (RGIA), Hyderabad, Pushpak Airport Liner bus service provides public transport connectivity from RGIA to important destinations in the city. Presently 39% passengers are using this service for accessing the airport. The bus stop for Pushpak buses is in front of the terminal building. The service is available for Rs. 250/- for journeys above 34 km length, costing about Rs. 800/- for the same journey by taxi. These examples suggest that dedicated bus services to airport can garner considerable ridership despite being costly than a normal bus service. Though the above services can be pre-booked from the website of respective transport corporations, high frequency of these services allows passengers to buy tickets after they board these services.

4 Proposals to Promote Public Transport Share

Considering the acceptability of bus based public transport in airport travel in other cities of India, a concept plan was prepared targeting at least 10% to 15% increment in the existing modal share of PT in airport travel. Measures such as PT prioritization at terminal kerb front, PT connectivity for 24 × 7 and disincentives to private modes by levying high parking charges were included as the major components of concept plan. The concept plan included the conversion of existing PTC into major transport hub accommodating pick/drop and parking of cars/taxis with connectivity of terminals with PTC along with remote check in facility. Moreover, an efficient external connectivity to the airport was also viewed as a major factor.

5 Bus Based Public Transport for IGI- Satellite Airport Service (SAS)

Based on the inferences (Sect. 3.4) drawn from traffic study and passenger and employee opinion, Demand Responsive Transport (DRT) was identified as the potential public transport system which could address the issues of PT in airport travel to IGI. Accordingly, DRT was studied in detail to understand its nature and its characteristics. DRT is a transportation option that falls between private car and conventional public bus services. It was previously considered to be an option only for less developed countries and for niches like elderly and disabled people [3]. Lately, DRT has emerged as an intermediate form of transport, somewhere between the bus and taxi, covering a wide range of transport services ranging from less formal community transport through to area-wide service networks [4]. To explore the applicability of DRT in airport travel, the intrinsic characteristics of the DRT system was studied from various literature and excerpts. The literature study suggests that five aspects of the DRT system are [5]:

- Coverage and routing: Defined by the operating area and the degree of flexibility in the operation. The latter can range from a fully flexible door-to- door service, to more rigid configurations in which only partial deviations from a planned route are possible.
- Operating hours: Equivalent to the service span considered in fixed transit.
- Vehicle characteristics: DRT fleets often consist of minibus vehicles, with previous research suggesting that a vehicle capacity of eight is the preferred vehicle size [6].
- Booking system: The system can allow for instant bookings in real-time, require advanced bookings or allow for both options. Traditionally, dial-a- ride schemes relied on telephone reservations, but large-scale real-time DRT systems require internet connection to perform the bookings in an efficient manner.
- Request acceptance criteria: Time required to the pick-up point or vehicle availability are the often-used criteria in deciding whether a request is accepted or rejected.

As the DRT intrinsic characteristics matched the requirements of airport travel, the concept of Satellite Airport Service (SAS) based on DRT was proposed to be adopted to provide 24×7 PT connectivity to IGI. The concept envisaged dedicated bus system to fill in the operational gap of Airport Express Metro and conventional bus system. Though the vehicle was accepted to be a minibus type vehicle in line with DRT systems, it was envisaged that larger buses may also be deployed based on concentration of demand from a specific area. In any case, the vehicle must allow comfortable seats and dedicated space for luggage. To ease the operation of these buses, it was proposed that the booking for SAS shall be prompted at the time of booking the flight ticket to allow the decision-making process for access/dispersal way before passenger reaches the kerb-front.

The SAS was envisaged to be operating from major origin destination points as identified from the OD survey results. For identifying SAS stops within Delhi city, it was divided into 5 zones i.e. North, South, East, West and Central Delhi. From each zone, the nearest rail terminal/bus terminal/metro station of the major OD point identified was taken as SAS stop. Similarly, SAS stops were identified from the satellite cities in Delhi NCR. The terminals/metro stations have been used so that to begin with, their infrastructure can be used for SAS. The map showing identified SAS stops for IGI airport is given in Fig. 8.

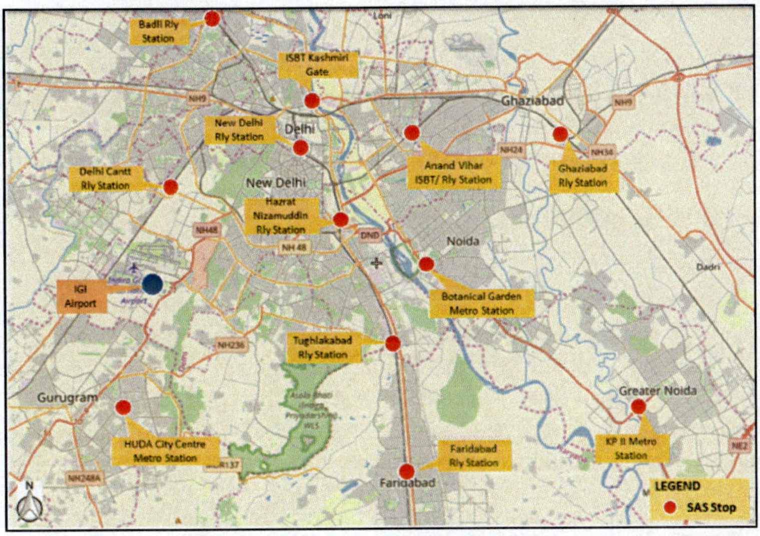

Fig. 8. Identified SAS stops for IGI airport

It was envisaged that the success of SAS would depend on the implementation of the complete environment that could provide the comfort comparable to that of taxi as far as possible and the following was included:

- Booking and payment of SAS ride (seat and luggage) during online booking of air tickets
- Information on passenger volume and schedule goes into SAS central command system located at PTC
- Command system sorts the demand based on computer algorithms and sends data to SAS terminals
- The SAS starts from the identified SAS terminal and moves on the route as suggested by the system, picking passengers one by one near their origin points allowing guaranteed seat and space for luggage
- SAS is prioritized at terminal forecourts with straight access to the terminal

The Study proposals have been discussed with Airport Authority of India and they are in discussion with airport operators to implement the study proposals. Precedence are available in Indian cities (KIA, Bangalore and RGIA, Hyderabad) suggesting that bus-based transport system can compete with taxi services for airport travel, given they provide similar level of service, albeit for a higher cost than that of a normal bus service. In line with this, DRT system has been proposed to address to most of the issues highlighted by airport passengers at IGI. Once implemented in Delhi, KIA and RGIA will be asked to operate their bus services on DRT basis to ensure more efficiency in the system.

5.1 Perceived Impact of SAS

Considering that DRT is a new concept in airport travel, it was felt imperative to statistically assess the impact of DRT in travel to/from IGI. The OD survey and the user opinion data formed the basis for assessing DRT impact. The methodology adopted included identification of the potential users and the importance attached to the key parameters of waiting time and access/dispersal time.

Potential Users of DRT

•As the SAS is proposed to connect IGI with all major terminals in the city (rail and bus), it is assumed that SAS is applicable to intracity as well intercity passengers
•DRT would address all concerns expressed by IGI users in their opinion survey (Sect. 3.4) except issues of high cost and last mile connectivity. Accordingly, it was estimated that 85% of airport passengers shall be the potential users of DRT considering that 2% of passengers facing issues of last mile connectivity and 13% passengers finding existing public transport fares to be high will not find DRT attractive.

DRT Acceptability and Sensitivity Towards Waiting Time

- 15% of the passengers indicated high sensitivity towards waiting time
- 100% indicated acceptability of 5-min waiting time for public transport system
- About 64% passengers indicated waiting time of 10 min for taxi
- Passenger sensitivity towards waiting time lowers with increase in distance from the airport
- Overall, passengers indicated that waiting time needs to be less than 20% of the total journey time
- Accordingly, a range of 5 to 10 min as a nominal range for waiting time for DRT was assessed

Based on above, using the data related to distance from airport, trunk journey time, with additional time of access & dispersal and waiting time, potential share of passengers of DRT was assessed considering a cap of waiting time to be less than 20% of total journey time (Table 2).

Above analysis indicates that passengers in less than 10 km radius of IGI may not find DRT to be that attractive and may continue to use car/taxi. The % share of trips originating from <10 km region around the airport is about 10%. Accordingly, potential users of DRT based on waiting time criteria would be about 90% passengers. Thus, based on above two criterions, about 76% of IGI passengers would be the potential users of DRT.

Table 2. Applicability to passengers based on waiting time

Distance from airport	Trunk journey time (min)	Assumption on access/dispersal time (min)	Proposed waiting time (min)	Overall journey time (min)	Share of waiting time in overall journey time (%)
5	10	10	7	27	26%
10	20	10	7	37	**19%**
15	30	10	7	47	**15%**
20	40	10	7	57	**12%**
25	50	10	7	67	**10%**
30	60	10	7	77	**9%**
35	70	10	7	87	**8%**
40	80	10	7	97	**7%**
45	90	10	7	107	**7%**
50	100	10	7	117	**6%**
55	110	10	7	127	**6%**
60	120	10	7	137	**5%**
65	130	10	7	147	**5%**
70	140	10	7	157	**4%**
75	150	10	7	167	**4%**

Shift from Other PT Modes

About 33% passenger trips are presently catered by public transport (metro and bus). Of this 33%, about 5% trips are catered by bus system. Based on users perception of present systems, it is expected that 20% of the trips catered by bus and about 10% of the existing ridership of metro at present shall shift to SAS i.e. about 1% of total bus trips and 2.8% of total trips.

Shift from Private Transport

Scenario based approach has been followed to assess the shift from private modes/taxi to DRT. The approach considers saving on cost and lower convenience with DRT as compared to using a private mode and taxi. Following scenarios for modal shift have been envisaged:

- Optimistic Scenario: 20% shift from private cars (private and taxi) i.e. 12.8% of total trips
- Pessimistic Scenario: 10% shift from cars (private and taxi) i.e. 6.4% of total trips

Considering all of the above, the modal share of SAS under optimistic and pessimistic scenario (Table 3) will be:

Table 3. Modal share of SAS in different scenarios

Potential user of DRT	Optimistic scenario	Pessimistic scenario
Based on reported existing issues in PT systems	85%	
Based on sensitivity towards waiting time	90% of 85%	
Overall	76%	
Shift from other PT Modes	3.8%	
Shift from private modes/Taxi	12.8%	6.4%
Overall shift	16.6%	10%
Impact in number of trips (Annual)	11.6 million trips	7.1 million trips
Impact in number of trips (Daily)	32,161	19,717

Thus, introduction of SAS can have tremendous effect not only on passenger convenience, on airport authorities in managing traffic but also in curbing air pollution.

6 Conclusion

Volume of airport passengers is way less in an Indian city in comparison to people accessing other transport nodes such as railway and bus terminals. But at the same time, it is high enough to create congestion at the airport if all the passengers travel by private modes. Also, airport trips originate from/are destined to various parts of the city and cannot be served by a fixed dedicated system. Planning a dedicated rail-based system for allowing passenger connectivity at airports may not be viable financially always as well as from accessibility point of view. On the other hand, a typical city bus system may have issues such as lack of comfort, space for luggage and transit delays. As such, demand responsive transport is found to be best suited system for airport travel in the case city. The success of this system will depend on the availability of the associated infrastructure and the role of private sector for the same can be explored. To allow a paradigm shift in airport travel in India, the SAS model need to be adopted in all major cities in phased manner. The detailed methodology adopted here can be replicated for other cities as well.

To take the study forward, Use of DRT as a premium public transport service for daily commute in Indian cities can be taken up.

References

1. Multimodal Transport Integration Study at 6 Metropolitan Airports in India: RITES Ltd., India (2016)
2. Master Plan, Delhi International Airport Limited (2016)
3. Bakker, P.: Large scale demand responsive transit systems-a local suburban transport solution for the next millennium. In: Proceedings of the Public Transport Planning and Management, Proceedings of Seminar E Held at the Aet European Transport Conference, Robinson College, Cambridge, UK, 27–29 September 1999, p. 433 (1999)

4. Marcus, E., Stephen, P., Graham, P., Marker, S.: INTERMODE: innovations in demand responsive transport. Department for Transport and Greater Manchester Passenger Transport Executiv, p. 32 (2004)
5. Alonso-Gonza´lez, M.J., Liu, T., Cats, T., Van Oort, N., Hoogendoorn, S.: The potential of demand-responsive transport as a complement to public transport: an assessment framework and an empirical evaluation. Transp. Res. Rec. **2672**(8), 879–889 (2018)
6. Khattak, A.J., Yim, Y.: Traveler response to innovative personalized demand-responsive transit in the San Francisco bay area. J. Urban Plann. Dev. **130**(1), 42–55 (2003)

Sustainable Mobility and Public Transportation Systems in Medium-Sized Cities

Elias Papastavrinidis(✉), George Kollaros, Antonia Athanasopoulou, and Vasiliki Kollarou

Department of Civil Engineer, Democritus University of Thrace, 67100 Xanthi, Greece
eliaspapastavrinidis@gmail.com

Abstract. The subject of this project is the elaboration of a sustainable public transportation systems plan in a mid-sized city. The study aims to maximize the benefits of an urban area in order to improve the quality of life for both residents and visitors, with participatory planning and the principles of sustainable development. Sustainable mobility is defined as the movement of people and goods both in the urban and suburban axes of cities, which is based on three main pillars: environment, society and economy. These pillars also outline the key objectives of a Sustainable Urban Mobility Plan, which are none other than providing urban mobility for all, providing safety while adopting environmental transport standards. Furthermore, the financial element is also taken into account, contributing to increased efficiency. For the reasons mentioned above, sustainable mobility aims to create ideal public transport conditions, contributing to improving the quality of city life. This research demonstrates the necessity of adopting a Sustainable Urban Transport Plan by the Municipality of Xanthi, in northern Greece, as evidenced by the recording of the current situation. Therefore, the main scope of the study is to formulate policies and implement measures that will gradually strengthen the use of public transport changing the mentality of citizens who move around using the car.

Keywords: Sustainable mobility · Transport policy · Urban transport · Buses · Passengers

1 Introduction: Public Transportation

Despite the impressive development of technological innovations, contemporary medium-sized cities continue to encounter significant problems with regards to their public transport. Travel within the city comprises the majority of the daily commute, something that is justified by the movement of a large part of the population for work reasons. Intra-city movement as a terminology refers not only to the movement of residents living in a city permanently, but also to those who move there for their own personal reasons [1].

The huge traffic load has always been a matter of concern for both Greece and Europe in general. Thus, with the aim of reducing the use of private vehicles for inter-

E. G. Nathanail et al. (Eds.): CSUM2020, AISC 1278, pp. 56–64, 2021.
https://doi.org/10.1007/978-3-030-61075-3_6

city transportation, the focus has shifted to promoting the use of public transport. A wider and more intensive use of public transport brings not only a better quality of life but also benefits the economy and the environment [2].

More specifically, transport in the area of Xanthi consists of the urban transport which performs radial routes. The routes of the itineraries have been designed based on the needs of service and the connection of the city with its surroundings. However, it is found that there are shortages both in areas of specialized demand (Hospital Area, Train Station, etc.) and in movements of greater distance within the city itself.

2 Transportation in the Urban Area of Xanthi

2.1 Public Transport Itineraries

The city of Xanthi is the capital of the homonymous Regional Unit and administratively belongs to the Region of Eastern Macedonia and Thrace. The permanent residents in Xanthi are 65,133 according to the 2011 census. Xanthi is connected to the regions of Macedonia and Epirus via the National Road, which is located just a few kilometers outside the city. The train station is also in the outskirts of the city, where the route Thessaloniki - Alexandroupolis is operated daily.

Xanthi serves a large number of passengers daily with the bus being the main means of public transport. In addition, both the location of the starting point and the routes of the city buses affect the whole physiognomy of the city, as they affect both its social and economic characteristics [3].

The Municipality of Xanthi is administratively composed of two spatial units: regional area of Xanthi and regional area of Stavroupoli. Municipal unit of Xanthi is consisted of the urban area of the main city and residential areas which are at a relatively short distance. This spatial unit is the area under consideration, as it is served by public transport (Table 1).

Table 1. Itineraries of urban buses in Xanthi.

Line	Route
1	Xanthi – Evmoiro – Xanthi
2	Xanthi – Kalithea – Xanthi
3	Xanthi – Kimmeria – Xanthi
4	Xanthi – Magiko – Xanthi
5	Xanthi – Campus – Xanthi
6	Xanthi – Pigadia – Xanthi
7	Xanthi – Feloni – Xanthi
8	Xanthi – Chrysa – Xanthi

2.2 Suggestions from Public Transport Passengers

For the purposes of this study, a survey was conducted on the mapping of traffic conditions in the area of Xanthi and the reasons for the use of public transport. The aim of this process was to draw conclusions about the frequency of use of public transport, user preferences and customer satisfaction. The survey was conducted by compiling and a questionnaire and collecting responses from a total of 200 passengers. The following conclusions are drawn through the questionnaires with on-site collection of answers and through personal experience acquired by acting both as a user and observer of each route separately [4]:

Existence of intense traffic congestion in the city centre due to the radial routes of the buses resulting in the accumulation of a large number of vehicles. This does not help the overall picture of the city and interrupts pedestrian flows, thereby endangering their safety. The phenomenon is becoming increasingly concerning due to the fact that the starting point of urban transport is near the central square of Xanthi.

There are several population gathering areas that are not being served by bus services (e.g. Indoor Sports Centre). These specific points could easily be accommodated and included in the existing routes through a slight detour of the bus.

Lack of a special ramp at the bus stops to improve accessibility for people such as the elderly and the disabled. This makes urban transport unattractive to a portion of the population and as a result they travel by taxi without the possibility of choice.

Lack of routes at traction points where transport is necessary to combine means of transport (e.g. to the train station). For example, the movement from the Xanthi railway station with luggage is currently performed only by taxi as no other means of transport is available.

Inability to obtain a ticket in non-central areas. Consequently, urban transport users have to purchase tickets in advance.

Many existing parking spaces do not have any signage or shelter, therefore only local resident residence is aware the existence of these public transport stops.

Disrespect of users in public property, since the phenomenon of vandalism is observed in some stops with evidence of dirty or damaged - broken shelters.

Small width and poor condition of sidewalks that prevent the smooth movement of pedestrians and people with disabilities.

3 Critical Issues for Promoting Public Transport in the City of Xanthi

3.1 Redesign of Bus Itineraries in the Urban Area

The improvement of the urban transport network aims to limit the use of the car in the city. The change in the existing radial structure of the routes in the city of Xanthi with a logic of serving diametrically opposed destinations with single routes can con-tribute positively to addressing several problems, such as reducing traffic congestion and reducing congestion in central areas, or the time of movement and development of the routes (Fig. 1).

In the context of the above proposal for the operation of bus lines must be considered: data on the origin of the destination, passenger traffic, travel times, frequencies, various economic and technical data (number of staff, line operating costs, ticket costs, vehicle maintenance costs, etc.), alternative routes due to geometric constraints, positioning based on geometric and traffic data [4].

Fig. 1. Starting point of buses in the city centre of Xanthi.

3.2 Deterioration of Bus Stops due to Incomplete Coverage of Points

According to the Canadian Urban Transport Association, bus stops in residential areas are recommended to stand 200–400 m apart, which corresponds to about 5 min of walking. However, the distance between the stops should not be less than 200 m as the bus does not manage to develop speed [5].

In the city of Xanthi there are population gathering points which are not served by a bus stop at a distance of less than 400 m, which is the maximum acceptable walking time (e.g. Indoor Sports Centre). Therefore, an additional number of stops could be installed at such "dead" points in order to better meet the needs for urban transport in the city (Fig. 2).

3.3 Upgrade Bus Station Shelters

Urban transport reflects the image of a city and so does the urban equipment used as infrastructure for the waiting of the passenger public in the existing stops. In some places, waiting shelters and urban equipment could be placed semantically so that the stop would be in the form of a sculpture, with the aim of aesthetically upgrading the wider area.

Another alternative is to install shelters with photovoltaic panels for energy autonomy. These panels can be equipped with USB device charging ports, a touch screen for local information, LED lighting and free internet connection. An energy shelter can make the stay of urban transport users a very pleasant experience.

Finally, new shelters will have to be installed in non-existent stations and the existing damaged ones should be replaced. The installation of shelters is also important as it clearly marks the area where passengers can safely wait for the bus [5].

Fig. 2. Bus stops in the city of Xanthi.

3.4 Flexible Transport System with Small Buses

Flexible transport systems complement the existing urban transport. Flexible transport systems can consist of small buses to cover special local routes. These special itineraries can be carried out to not only serve the city's shopping centre, but also to cover travel attractions such as the railway station in the southern part of Xanthi or the airport in Kavala.

This particular transportation system can serve groups of five or more people, who need more than one car and move to areas that are not served by existing public transport. People belonging to vulnerable social groups, such as people with disabilities, people with limited mobility and children can also be served.

In terms of costing, each route may be proportional to the number of people served in relation to the mileage. This makes it possible to streamline the cost of providing transportation services (Fig. 3).

Fig. 3. Small bus serving Xanthi's shopping centre.

3.5 Installation of Information Boards and "Smart Software"

In Xanthi, the hours of public transport routes are recorded only in some of the kiosks across the city. However, it is very important for an urban transport user who is waiting at a stop to know exactly how long it takes for the bus to arrive. Modern tables are updated immediately, and the user is notified in real time via a monitor for the exact time of departure.

In fact, the next step of this system is to create a corresponding application for smart devices that informs the user about the exact location of the line vehicle he has chosen. Also, the passenger can be informed in real time how long will take for the bus to arrive at the selected stop.

In addition, the software will enable the user to be informed about various announcements concerning urban transport, therefore planning of the movement will become a much easier task [6].

3.6 Development of Ticketing Systems in the Form of a Single and Intelligent Fare System

Urban transport of Xanthi presents problems in the faring system as the points of issuance of tickets are limited and the price is moderately satisfactory. The real problem is that users have to buy tickets from the city centre even if they want to use the bus from a more remote stop (Fig. 4).

Fig. 4. Simple paper tickets for buses in the city of Xanthi.

The solution to this problem comes from the single and intelligent fare system in the form of an electronic ticket. This system comes to replace the paper ticket and has a number of advantages such as charges depending on the distances traveled or discounts depending on how often the user moves. The smart ticket consists of smart cards or applications on "smart devices" and special machines with sensors inside the public transport to activate the ticket. These cards are recharged either via the inter-net or from a bank account or via "smart devices" for security and convenience.

This particular fare system can be gradually applied beyond public transport, such as on intercity routes or on rented vehicles. It can also be used for transactions in parking lots or in the rental of shared bicycles. In total, each system has its own Control Centre and the corresponding equipment. This practice makes public transport more attractive in the long run and contributes substantially to environmental up-grading and reduced waiting time when buying tickets [6].

3.7 Ensuring Road Safety for Vulnerable Users

In the area of Xanthi, the existing bus stops do not have the infrastructure to serve the disabled. However, it is recommended that special care be taken to facilitate the movement of this portion of the population to the boarding and disembarking ramps.

Specifically, the term obstructed persons refer to the elderly, the disabled, cyclists or generally persons with a mobility problem. Therefore, ensuring roadside assistance through appropriate infrastructure at stops and the wider road area is very crucial for their service. This has a positive effect on these people in the sense that they feel safe and use public transport more frequently and with increased comfort. After all, when a bus approaches a bus stop, the driver is required to use the special ramp for even greater safety [7].

All the above have to be applied in the city of Xanthi, as every person has the right to travel by public transport. After all, respect for every human being is expressed through the infrastructure that a city offers to its residents.

3.8 Developing a Tram Network

The tram as a means of public transport promotes maximum green and clean mobility, as it has almost zero emissions of emissions and carbon dioxide, while providing comfort, safety and easy transfer to users, all the while reducing the daily use of cars in the city.

In the city of Xanthi, it is proposed that a fixed orbit of mean of transport, i.e. tram, should be developed in accordance with the General Urban Plan. The tram line is designed to connect the city centre with the campus and the General Hospital creating a central artery. The development of a fixed orbit in Xanthi can significantly up-grade the urban area and travel in the city [8].

4 Conclusions and Recommendations

The traffic problem in the region of Xanthi is recognized as important. As a result, it is necessary to take action to formulate policies and implement measures that will gradually strengthen the use of public transport and alternative means of transport changing the mentality of citizens who move around using the car.

Urban transport is the main pillar for achieving the goal of sustainable mobility and urban development. Reducing the external costs of daily commuting, especially in urban areas, creating more jobs by increasing the share of public transport and enabling citizens to live in a more humane society are just some of the benefits related to the use of public transport [9].

Through the effective cooperation of all stakeholders involved in a Sustainable Urban Mobility Plan, policies are being formulated to implement important priorities for the promotion of public transport, such as the creation of a tram network, the development and implementation of intelligent systems such as electronic tickets and informing and raising public awareness.

The design of the appropriate urban transport network in the area of Xanthi can have beneficial results both in terms of transport service and in terms of environmental and urban regeneration, contributing to the reduction of space currently used by cars.

References

1. Wefering, F., Rupprecht, S., Bührmann, S., Böhler-Baedeker, S.: Guidelines - Developing and implementing a sustainable urban mobility plan. Directorate-General for Mobility and Transport, European Commission (2014)
2. European Metropolitan Transport Authorities: EMTA Barometer 2019 of public transport in European metropolitan areas based on 2017 data (2019)
3. Technical Program 2020: Municipality of Xanthi 2019/11/27 (in Greek)
4. Kollaros G.: Transportation compatible land uses and bus-stop location. In: Transactions on the Built Environment, vol. 41, pp. 459–468. WIT Press (1999)
5. Association, C.U.T.: Canadian Transit Handbook, 3rd edn. Canadian Urban Transit Association, Toronto (1993)
6. Goldman, T., Gordham, R.: Sustainable urban transport: Four innovative directions. Technol. Soc. **28**, 261–273 (2006)
7. Rudolph, F., Black, C., Glensor, C., Hüging, H., Lah, O., Mcgeever, J., Mingardo, G., Parkhurst, G., Plevnik, A., Shergold, I., Streng, M.: Decision-making in sustainable urban mobility planning: common practice and future directions. World Transp. Policy Pract. **21**(3), 54–64 (2015)
8. General Urban Plan 2013: Municipality of Xanthi 2013/05/22 (in Greek)
9. Duportail, V., Meerschaert, V.: Final ADVANCE audit scheme and guidelines. TRAJECT, European Commission (2013)

Case Studies in the Emilia Romagna Region in Support of Intermodality and Accessibility of Public Transport

Margherita Pazzini[1(✉)], Claudio Lantieri[1], Valeria Vignali[1], Andrea Simone[1], Giulio Dondi[1], Giuseppe Luppino[2], and Denis Grasso[2]

[1] DICAM Department, School of Engineering and Architecture, University of Bologna, Viale Risorgimento 2, 40136 Bologna, Italy
margherita.pazzini2@unibo.it

[2] Istituto sui trasporti e la logistica, Via dei Mille 21, 40121 Bologna, Italy

Abstract. The concept of intermodality is becoming increasingly important for the development of global communication and environmental sustainability. Creating an intermodal network capable of expanding connections between inland and coastal areas of the same country or between different countries, is fundamental in order to improve mobility and the level of service offered. In Italy numerous projects have been presented to increase national intermodality. An example is the Emilia-Romagna Region which, thanks to the Inter-Connect project, aims of enhancing the accessibility of passengers between the city of Bologna and the coastal areas, by promoting the use of collective transport. In this study, an analysis was carried out on the connectivity between Bologna and Ravenna to reduce the journey time on the railway line and increase the use of collective transport. The analysis included the redesign of the railway stops for a faster and more efficient service, as well as improving user satisfaction. In addition, an integrated bus-train ticket was created to promote the use of public transport. The results showed that this new solution led to have less delays on the line, although the reorganization of the stops involved the cancellation of some of them. Moreover, numerous surveys have shown that the use of an integrated bus-train ticket helps to improve intermodality within cities. The study has also highlighted the need for more efficient ticket advertising campaigns to better inform users.

Keywords: Intermodality · Sustainable mobility · Integrated bus-train tickets · Environmental improvements

1 Introduction

An intermodal mobility behavior is defined as the flexible combination of different means of transport during a single trip (Chlond (2013); Jones et al. (2000)). Intermodality is integral part of sustainable mobility and its enhancement is of great importance, especially in congested urban areas (Pitsiava-Latinopoulou and Iordanopoulos (2012); Infas (2010); Yeh (2008)). An improvement of intermodal

E. G. Nathanail et al. (Eds.): CSUM2020, AISC 1278, pp. 65–74, 2021.
https://doi.org/10.1007/978-3-030-61075-3_7

connectivity will be useful both to promote travels and communication within a country (Ryosuke and Hironori (2019)) and to reduce congestion (Allard and Moura (2018)) with the aim of boosting economy and local tourism.

An inadequate planning of travel solutions leads to a reduction of the level of public transport resulting in a shift of user choices towards different means of transport, mainly private cars (Pitsiava-Latinopoulou and Iordanopoulos (2012); Allard and Moura (2018)).

The study on mobility shows how the level of service offered to passengers affects their behaviour about the transport mode (Pitsiava-Latinopoulou and Iordanopoulos (2012)). One of the most important indicators when evaluating the quality of the urban public transport service offered to passengers is how long it took the users to travel (Kopylova et al. (2018)). The total amount of time taken by users for a single move is equal to the sum of the travel time, of the journey time and of the waiting time at interchange points (Vande Walle and Steenberghen (2006); Xumei et al. (2011)). Oostendorp and Gebhardt (2018) showed that time efficiency is a very important aspect for intermodal users and underlined how binding it was for most of them when choosing the means of transport. Lu et al. (2018) showed how socio-demographic characteristics of passengers and their travel options have an impact on the value of waiting time and of the time on board. The better users are informed about the different modes of travel the shorter the waiting time is perceived. Therefore, user needs shall be taken into account in urban planning whenever aiming at getting the best advantages from intermodal transport combinations for an efficient urban transport system. Monzon et al. (2017) stated that the weakest point when managing intermodal travel solutions is the lack of internal coordination which negatively affects the quality of information given to passengers. In their study about intermodal transport of passengers, Allard and Moura (2018) identified three key factors relevant to intermodal transport: connectivity, cooperation and choice.

Another aspect to consider in the intermodal travel solution is the higher ticket fee due to the use of different means of transport just in one journey. On one side travellers don't like long waiting times involved in interchange solutions, but on the other side they are not willing to accept a higher cost for intermodal transfers (Yen et al. (2017)). The union of complementary or cooperative network structures and the integration of fares and information provision are defined as the integration of different transport solutions (Clifton and Mulley (2018)).

The purpose of this paper is to provide alternative solutions to enhance the use of public transport by improving connection between inland and coastal areas thus increasing the level of service offered to users. Special solutions have been sought not involving large investments and yet, offering valid options to private transport without too high costs. This study is divided in two parts analysing connections for short and long distances: in the first case the focus will be on reducing delays of trains at stations, in the second case bus-train integrated solutions will be studied.

2 The Case of the Emilia-Romagna Region in the Inter-connect Project

With its high-quality industrial base producing wealth and creating employment and a coastline that during the summer season is an attractive destination for tourists, Emilia-Romagna is one of the biggest and wealthiest regions in Italy.

The main transport planning tool of the Region is called “PRIT – Piano Regionale Integrato dei Trasporti” (Transport Integrated Regional Plan) and its purposes are:

- To guarantee passenger and freight accessibility in the region;
- To reduce transport-related energy consumption;
- To reduce transport-related emission (pollutants and greenhouse gases).

With regards to passenger transport, PRIT aims to promote an integrated system built upon a core regional transit network (trains, trams, buses). Figure 1 shows the transport modal distribution for all kinds of journeys within single municipalities. Intra-municipality trips (i.e. urban trips) seem to be more sustainable with 40% of people using public transport, bike or walking. When considering all kinds of journey, as per Fig. 1, a 10% decrease is to be pointed out. However, public transports are increasing, from 7,2% urban to 8,3% for all trips.

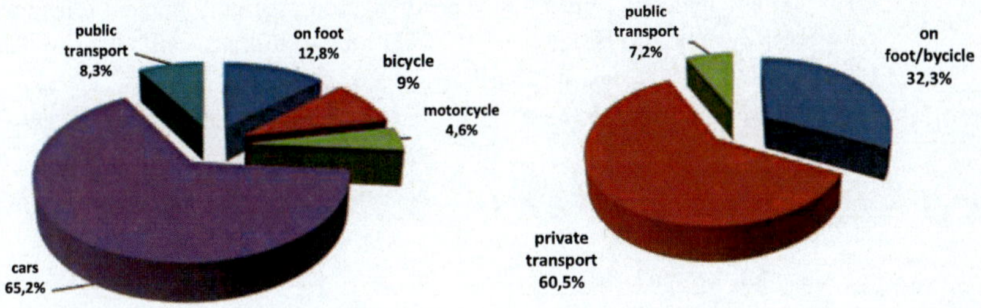

Fig. 1. Emilia-Romagna mode share transport modal distribution in 2018. On the left all trips, on the right only intra-municipalities trips.

Emilia Romagna Region aims at improving its railway transport for passengers within Bologna and Romagna Area to enhance accessibility of coastal areas (Romagna – provinces of Ravenna, Forlì-Cesena and Rimini) and connections to inland hubs (Bologna). The rail passenger transport should be improved with a network which will connect different areas within Emilia Romagna. Although transport of passengers is highly seasonal, especially in the coastal part, both “summer” flows, mainly of tourists, and “winter” flows of commuters have been investigated in order to achieve the best and more convenient railway transport for both types of users. The comparative analysis has shown a delay in improving the main inter-urban links despite the increasing demand. The connection between Bologna and Ravenna, for example, hasn't been improved for 10 years. Among the main destinations from Bologna, Ravenna is the closest in terms of travel distance, but the farthest in terms of travel time. While the connections with other towns have faced a reduction of travel time, the connections with Ravenna only partially have. In fact, only 4 services per day are faster than in 2008 and 16 take a longer time than in 2008.

This case study will look at how to improve travel time without major infrastructure investment but focusing on re-designing train stops to have a faster and more satisfactory service. As for the railway line along Ravenna-Rimini coastal line, given the tourist vocation of this area, in recent years local authorities (i.e. Rimini municipality) have emphasised the need to promote public transport for tourists as a measure meant to reduce negative environmental externalities like traffic congestion, environmental impact, road safety, etc. The analysis is therefore divided in two different study cases, A and B, described in Table 1, with the same purpose, namely enhancing and fostering intermodal transport within the region.

Table 1. Emilia-Romagna case study aims and solutions.

	Case study A	Case study B
Aims	• improve the quality of life of Ravenna commuters • create an efficient rail connection between Ravenna and Bologna • reduce delays on the railway line	• attract more tourists not travelling by car • offer affordable and efficient transport solutions to tourists without using car
Solutions	• reduce travelling time between Bologna and Ravenna • reduce the number of train stops ensure new bus services	• introduce integrated bus-train tickets • create 3 or 7 days integrated tickets for tourists

3 Case Studies

3.1 Case Study A: A Fast Rail Connection Between Bologna and Ravenna

The aim of case study A is to improve and revitalize the passenger railway service between the towns of Ravenna, Rimini and Bologna reducing the travel time to reach the destination. The railway lines connecting Rimini to Ravenna and Ravenna to Castel Bolognese are single track lines as shown in Fig. 2.

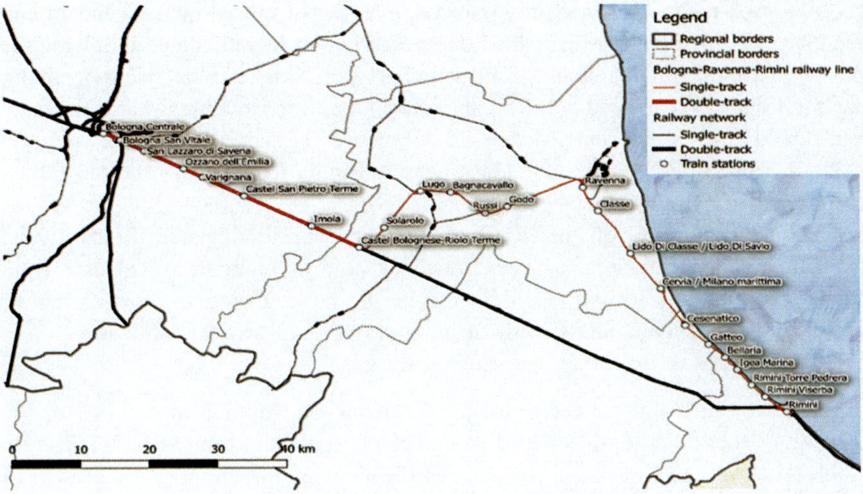

Fig. 2. Rail infrastructures in the pilot area.

An infrastructural improvement of these lines would involve large investments and long time. So, in order to get an improvement without high costs, this study will focus mainly on definition and testing of solutions capable of reducing train travel time.

The number of passengers on the Bologna-Rimini-Ancona line significantly increased in 2018 compared to the previous years. During the winter months the rise in the number of passengers was especially at the stations in Bologna, Imola and at the intermediate stations between the two towns. In summer the number was higher between Bologna and Rimini and at the other stations along the Adriatic Coast of Romagna.

In order to size the rail network properly according to the users' needs, an analysis has been carried out using data from on-board train surveys made during those times of the year considered to represent the passenger flow at best, usually July and November. The data collected in different days can be averaged to obtain the most significant day. The same data can also be extended, according to some balancing standards, to similar periods in order to represent the year total value. The stations served by regional trains in Emilia Romagna are 258 and, according to the recognition acquired, passengers daily getting on a train in 2018, on an "average working day" in winter in the region were about 159,100 (in 2017 they were about 149,300). The average number of passengers throughout the year is nearly 47,100,000. The data show an increase in passengers in 2018 compared to the previous years, both during winter and summer times. After some years of stability of data, in year 2018 a sharp increase in the winter data was pointed out (+8% compared to 2017) as well as a rise during the summer (+4% compared to 2017). The increasing in the number of passengers catching a train to move within the region resulted in an increase in delays on lines.

The analysis of relevant data shows a worsening of travel times (negative trend has gone on for three years): after a positive result in 2015, 2016, 2017 and above all in 2018 negative results occurred as for punctuality, back to levels registered in

2013/2014. Apart from June and July, punctuality was always below the standard and worse than in 2017. Furthermore it should be taken in consideration that, following the extension of the railway line with the introduction of high speed rail services operating in the last decade (Vignali et al. (1001)), the travel time from Bologna towards the main Italian capital cities significantly decreased between 2008 and 2018. On the contrary, this didn't happen between the different towns in Emilia Romagna as the rail infrastructure remained almost unchanged.

After studying delays on the line and the users' needs compared to size of the railway line, a new solution has been found capable of reducing travel time from Bologna, Ravenna and Rimini in order to make the railway services more attractive, both for tourists and commuters, without modifying the railway infrastructure.

The main goals of the study were as follows:

- Reduce travelling time between Bologna, Ravenna and Rimini from 81"–83" to 69" in both directions. The objective is to reduce the travelling time of 12"–14";
- Reduce journey time between Ravenna and Rimini serving all the coastal towns of Romagna at best. The objective is to reduce the journey between Ravenna and Rimini and vice versa to less than an hour;
- A 3% increase of direct trips by train (12,475 to 12,888);
- Generate benefits on the whole line of regional trains.

The best technical solution provided by the analysis of train stops along the railway between Bologna and Ravenna (Fig. 3) in order to shorten travelling time consisted in reducing the number of stops. Some train stops at stations with the lowest numbers of passengers every day were removed accordingly. In the present case, some stops were cancelled at Russi and Classe for some trains (see picture below). In the case of Russi stops of the most used trains have been granted every day, while all stops in Classe would be cancelled.

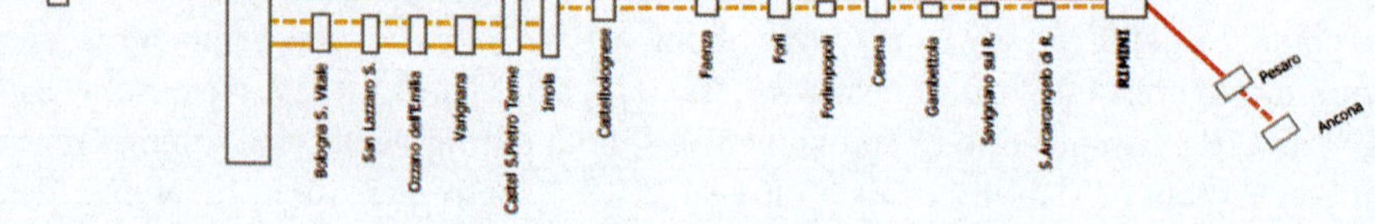

Fig. 3. Railway stations between Bologna and Ravenna.

Here below data of the two railway stations of interest:

- Godo: 110 was the average number of passengers a day. They were mainly commuters with a season ticket. Train stops would be reduced from 34 to 12 per day;
- Russi: 70 was the average number of passengers a day. They were mainly commuters with a season ticket here too.

Since the reduction of the train stops gave rise to significant difficulties for passengers, new services of public buses were introduced linking the two above stations to the nearest railway station with active stops. Moreover, the connection of buses to the rail stations allowed the development of intermodal public transports highly interesting both for tourists and workers. In fact, some train times were coordinated to the times of two buses serving industrial areas and the city centre. This was possible thanks to an increase in the train service in both directions during the peak hours (7 to 9 in the morning and 4 to 6 in the afternoon).

3.2 Case Study B: The Survey in Rimini

The case study B deals with the bus-train connection, within the whole of costal area, thanks to a single ticket. The increasing number of tourists getting to Romagna during the summer from different countries has elicit the request of a new survey about their needs as far as the intermodality of public transport in that region is concerned. A survey was conducted in Rimini, the main hub town with an airport, a train station and several bus stations, in the period August-September 2018, the one with the highest peak of tourists, on four working days. A sample group of 163 people was interviewed, 141 interviews took place at the Rimini railway station, at the Riccione bus station in Piazzale Curiel and on buses number 4,8,11. The 42 remaining interviews were held at Rimini's "Federico Fellini" Airport. The 53% of the sample interviewed was made up of foreign tourists (mainly from Russia), while the rest were Italian tourists. The survey aimed to collect data on the travel experience of tourists in Romagna as a whole, assessing whether they were aware of the existence of an integrated bus ticket (Romagna Smart Pass), mapping their interest in a new type of integrated ticket (bus +train) and stating how far they are willing to pay for it. Since the interviews were conducted at the airport or railway station, the fact that most of the tourists interviewed had arrived in Rimini by train or plane should not be considered a significant event. The analysis on the means of transport chosen by tourists to move to Romagna during their holidays seems to be more relevant.

From the results obtained it was observed that the average holiday lasts 7.5 days. Foreign tourists spend longer holidays on average than Italian tourists - 8.8 days against 6.2 days. Each person interviewed was on vacation with 2.3 people, on average, regardless of nationality.

Since the interviews were conducted at the airport or train station, the fact that most of the tourists interviewed arrived in Rimini by train or plane should not be considered a significant event. More relevant seems to be the analysis on the mean's of transport chosen by tourists to move to Romagna during their holidays. 44% of tourists use the bus, while 22% move on foot. Moreover, the main tourist destination in Romagna is Rimini followed by Riccione.

The cities of Romagna are the main holiday destinations for Italian tourists while foreign tourists still prefer the major Italian attractions such as Venice, Rome, Florence. For these tourists, Rimini is only a hub from which to start their journey through Italy. Independent travelers mainly use high-speed trains, while others join organized trips and travel by bus. In both cases, San Marino is the main tourist attraction for a trip to Romagna. Although interest in an integrated bus-train flat-rate ticket valid within the Romagna region turned out to be quite high, only a few tourists (18%), no matter if foreigners or Italians, knew about Romagna Smart Pass. The availability to pay an integrated nus-train flat-rate ticket valid in Romagna area was evaluated according to different costs tourists were willing to pay for it. The results showed that foreign tourists (57%) are willing to spend more on an integrated ticket, both for 3 and 7 days, compared to Italian tourists (48%). More than 60% of the tourists interviewed were interested in visiting and/or had visited at least one of the tourist attractions using a Romagna Smart Pass (discounts on tickets and services). This percentage shows the importance of these kinds of agreements in order to better promote Romagna Smart Pass among tourists. The highest request is for a tourist integrated flat rate ticket lasting 7 days (40%) or 3 days (33%). For each kind of ticket, a range of three different prices were suggested. Prices were set in collaboration with Emilia-Romagna Region and Rimini municipality on basis of the prices of 2018 Romagna Smart Pass (11€ for a 3-day ticket, 22€ for a 7-day ticket) and of the price of the ticket for the longest train line within Romagna (Riccione-Faenza, 6.65€ for a one-way ticket by local trains).

Figure 4 shows the percentage of tourists available for the purchase of an integrated 3 or 7 day flat-rate ticket for buses and trains valid in the Romagna region. Tourists are willing to pay less than €27 for a 3-day ticket in free circulation and less than €55 for a 7-day one. Furthermore, the results showed that foreign tourists are willing to spend more on both integrated tickets than Italian tourists.

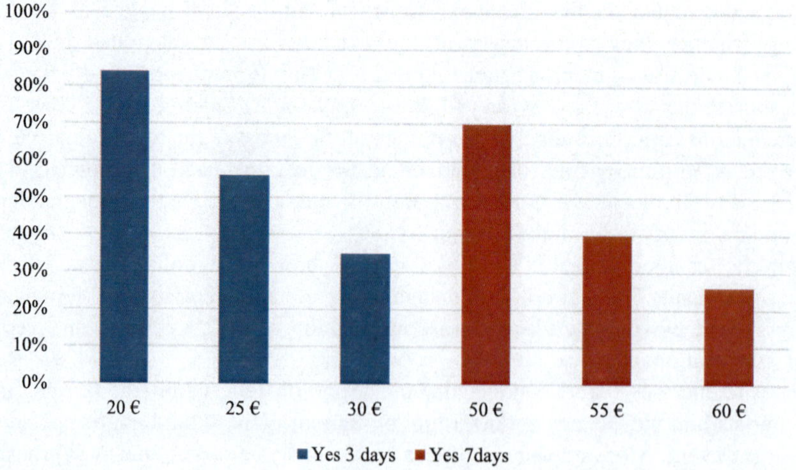

Fig. 4. Availability to pay for the integrated ticket lasting 3 days or 7 days (% of responses).

4 Conclusion

In this study, an analysis was carried out on the connectivity of collective transport services in order to encourage their use. Has shown by Kopylova et al. (2018) the time taken by the journey considerably influences the choice of public means of transport by users. The redesign of the railway stops with the elimination of some stations suggested by this study, has the consequence of having a faster and more efficient service, improving user satisfaction. Oostendorp and Gebhardt (2018) showed that time efficiency is the main topic for intermodal users. To meet the ever increasing demand of efficient public transport, this study proposes the elimination of the stations with few users as the most convenient solution to have no delays on the line, managing to provide a better service as well. Furthermore, the connection of bus services with railway stations has made it possible to develop intermodal public transport solutions available to users who used transport in the eliminated stations. In particular, it was possible to coordinate train timetables with two bus services serving industrial areas and the city center. Monzon et al. (2017) underline the important of internal coordination among the different intermodal means of transport. This study focuses on the creation of an integrated bus-train ticket to offer efficient and economic intermodal solutions capable of attracting users who move from central areas to coastal areas. The main results of this study show that both foreign and Italian tourists are interested in an integrated flat-rate ticket valid in Romagna. In general, it is possible to see as foreign tourists are usually more interested than Italian tourists in a flat rate integrated ticket and they are even willing to spend more for such a ticket. In fact, the interest for an integrated ticket is higher among foreign tourists (57%) compared to Italian tourists (48%). As for the duration of the flat rate integrated ticket, the data collected shows as the most requested duration for an integrated ticket is 7 days (40%) or 3 days (33%). In relation to the costs of the future Smart Pass tickets, the analysis carried out shows as, on average, tourists are willing to pay less than 27€ for a 3-day flat rate integrated ticket and less than 55€ for a 7-day flat rate integrated ticket. These data on tourists' availability to pay suggest maintaining both the existing integrated Smart Pass bus tickets and the new integrated bus-train tickets (case A) since higher prices for the integrated bus-train tickets may reduce the number of people using these tickets.

However, the analysis found that only 18% of the tourists interviewed knew about the existence of an integrated ticket already available in the areas examined. Moreover, the qualitative observations collected by the interviewers show that the ticket is not properly promoted at the main tourist transport hubs. Therefore, in order to encourage the use of public transport by users, effective promotional campaigns aimed at launching the integrated bus-train ticket should be implemented and supported at the main tourist transit hubs.

Acknowledgments. The authors thank all the participants in the Inter-Connect project and in particular Mr. Alessandro Meggiato, Mr. Andrea Normanno and Mr. Cesare Sgarzi of the Emilia-Romagna Region for their help and support in this study.

References

Ahrens, G.-A., Ließke, F., Hubrich, S., Wittwer, R.: Datenaufbereitung der Verkehrserhebung 'Mobilität in Städten – SrV, Dresden (2010)

Allard, R.F., Moura, F.: Effect of transport transfer quality on intercity passenger mode choice. Transp. Res. Part A **109**, 89–107 (2018)

Chlond, B.: Mulitmodalität und Intermodalität. In: Beckmann, K.J., Klein-Hitpaß, A. (eds.) Nicht weniger unterwegs, sondern intelligenter? Neue Mobilitätskonzepte, pp. 271–293, Berlin (2013)

Clifton, G., Mulley, C.: Barriers and facilitators of integration between buses with a higher level of service and rail: an Australian case study. Res. Transp. Econ. **69**, 39–50 (2018)

Infas, D.L.R.: Mobilität in Deutschland 2008. Ergebnisbericht, Bonn (2010)

Jones, W.B., Cassady, C.R., Bowden, R.O.: Developing a standard definition of intermodal transportation. Transp. Law J. **27**, 345 (2000)

Kopylova, T., Mikhailov, A., Shesterov, E.: A Level-of-Service concept regarding intermodal hubs of urban public passenger transport. Transp. Res. Procedia **36**, 303–307 (2018)

Lu, H., Burge, P., Heywood, C., Sheldon, R., Lee, P., Barber, K., Phillips, A.: The impact of real-time information on passengers' value of bus waiting time. Transp. Res. Procedia **31**, 18–34 (2018)

Monzon, A., Alonso, A., Lopez-Lambas, M.: Joint analysis of intermodal long distance-last mile trips using urban interchanges in EU cities. Transp. Res. Procedia **27**, 1074–1079 (2017)

Oostendorp, R., Gebhardt, L.: Combining means of transport as a users' strategy to optimize traveling in an urban context: empirical results on intermodal travel behavior from a survey in Berlin. J. Transp. Geogr. **71**, 72–83 (2018)

Pitsiava-Latinopoulou, M., Iordanopoulos, P.: Intermodal passengers terminals: design standards for better level of service. Procedia Soc. Behav. Sci. **48**, 3297–3306 (2012)

Ryosuke, Y., Hironori, K.: Success factors in the introduction of an intermodal passenger transportation system connecting high-speed rail with intercity bus services. Case Stud. Transp. Policy **7**, 708–717 (2019)

Vande Walle, S., Steenberghen, T.: Space and time related determinants of public transport use in trip chains. Transp. Res. Part A **40**, 151–162 (2006)

Vignali, V., Cuppi, F., Lantieri, C., Dimola, N., Galasso, T., Rapagnà, L.: A methodology for the design of sections block length on ETCS L2 railway networks. J. Rail Transp. Plan. Manage. **13**, 100160 (2019)

Xumei, C., Qiaoxian, L., Guang, D.: Estimation of travel time values for urban public transport passengers based on SP survey. J. Transp. Syst. Eng. Inf. Technol. **11**, 77–84 (2011)

Yeh, C.-F.: A study on feasibility of passenger intermodal transport in city of the developing world. In: Conference Proceedings Codatu XIII. Ho-Chi-Minh-Stadt (2008)

Yen, B.T.H., Tseng, W., Mulley, C., Chiou, Y., Burke, M.: Assessing interchange effects in public transport: a case study of South East Queensland. Transp. Res. Procedia **25**, 4019–4037 (2017)

Forecasting of Urban Public Transport Demand Based on Weather Conditions

Ricardo Correia[1,2], Tânia Fontes[2(✉)], and José Luís Borges[1,2]

[1] Faculty of Engineering, University of Porto, Rua Dr. Roberto Frias, 4200-465 Porto, Portugal
[2] INESC TEC, Rua Dr. Roberto Frias, 4200-465 Porto, Portugal
tania.d.fontes@inesctec.pt

Abstract. Weather conditions have a major impact on citizens' daily mobility. Depending on weather conditions trips may be delayed, demand may be changed as well as the modal shift. These variations have a major impact on the use and operation of public transport, particularly in transport systems that operate close to capacity. However, the influence of weather conditions on transport demand is difficult to predict and quantify. For this purpose, an artificial neural network model – the Multilayer Perceptron – is used as a regression model to estimate the demand of urban public transport buses based on weather conditions. Transit bus ridership and weather conditions were collected along a year from a medium-size European metropolitan area (Oporto, Portugal) and linked under the assumption that individuals choose the travel mode based on the weather conditions that are observed during the departure hour, the hour before and two hours before. The transit ridership data were also labelled according to the hour, day of the week, month, and whether there was a strike and/or holiday or not. The results demonstrate that it is possible to predict the demand of public transport buses using the weather conditions observed two hours before with low error for the entire network (MAE = 143 and RMSE = 322). The use of weather conditions allow to decreases the error of the prediction by $\sim$8% for the entire network.

Keywords: Urban public transport · Buses demand · Weather conditions · Neural networks

1 Introduction

With climate change, the influence of meteorology on public transport has been generated growing interest. Several works have been conducted last years in Portugal [11], Netherlands [21], Spain [12], Australia [1], China [3, 4], and United States [2, 6, 13]. Buses [1, 2] and metro [3, 4] have been the means of transport studied. In these studies several techniques were applied as multivariate regressions [4, 12], linear regression [21], Quasi-Poisson regression [22], negative binomial distribution [23], minimum square method [2, 5] and moved average [3].

Several studies explain that the use of public transport is directly influenced by weather conditions variations [e.g. 6]. Temperature is one of the variables that most affect users, unlike atmospheric pressure that does not appear to cause any negative

E. G. Nathanail et al. (Eds.): CSUM2020, AISC 1278, pp. 75–84, 2021.
https://doi.org/10.1007/978-3-030-61075-3_8

effect [11]. Several studies highlight that precipitation and high wind speed, contributed negatively to the demand for public transport [4–8]. Other variables as snow [5] and relative humidity [4, 6] can also play an import role on public transport demand.

Meteorology, as it is one of the main influencing factors of mobility, intervenes in users' decisions and habits. Meteorological conditions tend to determine the behaviour of users when choosing to travel or not, as well as when choosing a destination, means of transport and route to follow [9]. Several studies have demonstrated the existence of a direct correlation between the variation of meteorological conditions and the demand generated in transport services [5, 6]. Meteorological conditions affect differently each of the means of public transport [10] and adverse weather conditions have a greater impact on occasional travel [9].

Adverse conditions induce a greater variation in the demand for public transport on weekends [3] which can be due to the purpose of the trips (mostly for leisure purpose). Pendular movements are the least affected by meteorological changes, mainly when it comes to cancelling the trip [9]. The effects of weather conditions tend to vary if there are major events such as national holidays and vacations. The season only plays a small role in the impact of meteorological conditions in the demand for public transport, and a given meteorological event causes similar responses in different seasons [13].

Several factors contribute to the choice of the type of transport to be used, be it private or public. The distance of travel, gender, age and whether or not have a driving license are some of these factors [14]. This thinking more focused on socio-economic variables triggered in some authors the need to study the influence of weather conditions in different social strata and ages [15]. Adverse events have a greater impact on senior passengers, over the age of 75 years, in winter compared to summer [16]. This fact occur due to the reduced mobility of senior passengers and their greater fragility to environmental conditions.

Depending on the weather conditions, trips can be delayed, and demand and modal split can be significantly changed, since under certain conditions some groups of users, such as the elderly, the sick and children, tend not to travel or decrease their travel frequency. The main goal of this work is to understand if the meteorological conditions could improve the accuracy of the forecast demand of buses.

2 Materials and Methods

This section describes the methodology followed to develop a forecasting model of public transport as function of temporal variables and meteorological conditions. It is intended that the model predict the total demand for buses on the overall metropolitan network. Several scenarios were analyzed in order to select the most robust input variables of the model. Figure 1 shows the methodology followed.

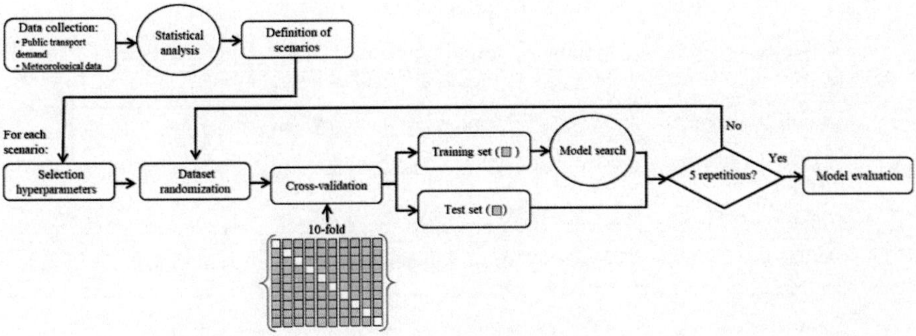

Fig. 1. Methodology overview.

2.1 Data Collection

Data on demand for public transport and weather conditions were collected. These data were obtained for the Metropolitan Area of Porto (AMP) for the period from 1 January 2013 to 31 December 2013.

The demand (*D*) for public transport was obtained from the validations of the smartcard Andante. The data collected (N = 77,213 million records) refer to trips made on buses, using the main company operating in Porto (STCP). Each record contains information related to a trip, namely information on the card ID, the stop and the line, the service operator, the time of validation, the user's profile and the zones to which the user has access.

The meteorological data were collected for an automatic meteorological station managed by the Meteorological Institute (Pedras Rubras). The station is located in the airport of Porto, which allow to have representative data for the whole study area. The collected meteorological variables were: temperature (*T*), wind speed (*WS*), relative humidity (*RH*), dew point (*DP*), atmospheric pressure (*P*), and cloudiness (*CL*). Data were collected each half an hour.

To obtain a uniform set, the data were pre-processed and normalized. First, records with empty variables were removed. Then, the meteorological data were normalized using the z-score method. This normalization was applied because the lower and upper limits are different for each of the variables, which can negatively influence the model's learning.

In addition to the demand and meteorological variables, four variables were also analyzed: hour (*H*), day of the week (*DS*), month (*M*), strike (*S*), and holiday (*O*). These variables were considered as they usually tend to influence the variation in demand for public transport. Table 1 shows the variables collected.

Table 1. Data collected.

Variables	Acronym	Description/Unit	Periodicity
Hour	*H*	01 h:00 min; 01 h:30 min…	30 min
Day of the week	*DS*	Monday: 1; Tuesday: 2…	–
Month	*M*	January: 1; February: 2…	–
Strike	*S*	Yes: 1; No: 0	–
Holiday	*O*	Yes: 1; No: 0	–
Demand	*D*	nº of validations	30 min
Temperature	*T*	°C	30 min
Wind speed	*WS*	km/h	30 min
Relative humidity	*RH*	%	30 min
Dew point	*DT*	°C	30 min
Atmospheric pressure	*P*	Mb	30 min
Cloudiness	*CL*	Clean; Some Clouds; Rain…	30 min

2.2 Groups of Users

Different types of users tend to react differently to changes in weather conditions, which can influence their predisposition to travel [15, 16]. Thus, to better understand the impacts of meteorological variables on different types of users, they were segmented into four groups. For this purpose, only the data from travel passes validations were used (66.7% of total journeys). The different groups were created taking into account the age group, occupation and the existence of social supports:

- G1 (students): includes students aged between 4 and 23 years old (6%);
- G2 (normal): includes workers in general and validations carried out with single agents (62%);
- G3 (elderly): includes retired users and users over 65 years old (18%);
- G4 (social): includes users who are given social support (14%).

2.3 Input Scenarios

In real world situations, it is usually difficult to collect a wide range of variables, as in the case study presented here. Thus, it is important to understand the impact on the model accuracy when different/reduced subset of variables are used. In order to select the variables with the greatest influence on the demand forecasting capacity, seven scenarios were defined considering different sets of inputs.

To define the scenarios, a statistical analysis of the input variables was performed. Correlation between bus passenger demand (*D*) and weather conditions variables (*T*, *WS*, *RH*, *DT*, *P* and *CL*) was analysed. The importance and direction of the association between these variables was analysed to check if the weather conditions influences the occurrence of a trip to be taken immediately, in the next hour or in the next two hours. Meteorological variables without significance correlation ($p > 0.05$) with demand are candidates to be discarded. Spearman's correlation (r_s) was used as the Kolmogorov-Smirnov test shows no evidence that the variables follow the normal distribution.

Table 2 presents the Spearman's correlation coefficient (r_s) between demand for public transport and the temporal and meteorological variables for each group of users (G1, G2, G3 and G4) for a trip taken immediately and in the next two hours. Statistical significance ($p < 0.05$ or $p < 0.001$) values of correlation are appropriately marked (*5% level; **1% level). In general, low correlation are found with high statistical significance ($p < 0.001$). Temperature (*T*), wind speed (*WS*) and relative humidity (*RH*) are the variables with the greatest significant correlation with the demand for buses. On the other hand, atmospheric pressure ($r_{s(G1)} = 0.026$ and $p < 0.01$; $r_{s(G2)} = -0.003$; $r_{s(G3)} = 0.012$; $r_{s(G4)} = 0.002$) does not appear to have a direct correlation with demand. With regard to studies where there is a delay of one hour (results now show on Table 3) or a delay of two hours between demand and meteorology, it appears that the correlations of variables with demand tend to decrease, except for the cloudiness where the correlation is significant ($r_s < 0.01$) in the Normal (G2) and Social (G4) user groups.

Based on the results provided in Table 2, seven scenarios (C1–7) were defined based on different combinations of the input variables. C1, C2 and C3 use all the variables available, while C4, C5 and C6 only use the variables that have a significant correlation with the demand for all the groups of users (G1-4). C7 uses only time variables. C1 and C4 uses the meteorological variables measured at the same time as the trips take place, while for C2 and C5 the demand estimates are made based on the meteorology observed in the previous hour and for C3 and C6 the demand estimates are made based on the meteorology observed in the previous two hours. Table 3 shows a summary of the variables considered by these scenarios.

Table 2. Spearman correlation.

	Demand vs. Meteorology				*Demand vs Weather conditions from previous 2h*			
Variable	*G1*	*G2*	*G3*	*G4*	*G1*	*G2*	*G3*	*G4*
T	0.206**	0.253**	0.324**	0.253**	0.162**	0.183**	0.228**	0.176**
WS	0.249**	0.221**	0.272**	0.225**	0.230**	0.177**	0.192**	0.173**
RH	-0.305**	0.275**	-0.372**	-0.276**	0.265**	-0.206**	-0.259**	-0.197**
DT	-0.004	0.056**	0.055**	0.053**	0.039**	0.015	0.020**	0.012
P	0.026**	-0.003	0.012	0.002	0.020*	-0.011	0.008	-0.007
CL	0.009	0.018*	0.007	0.018*	0.014	0.026**	0.008	0.025**

Notes: Dark values point out significant values of correlation at *5% level ; and ** 1% level .

Table 3. Set of input variables for each scenarios.

Scenarios	Temporal variables					Meteorological variables					
	H	DS	M	S	O	T	WS	RH	DT	P	CL
C1	x	x	x	x	x	x	x	x	x	x	x
C2 (With meteorology of 1 h ago)	x	x	x	x	x	x	x	x	x	x	x
C3 (With meteorology of 2 h ago)	x	x	x	x	x	x	x	x	x	x	x
C4	x	x		x	x	x	x	x			
C5 (With meteorology of 1 h ago)	x	x		x	x	x	x	x			
C6 (With meteorology of 2 h ago)	x	x		x	x	x	x	x			
C7	x	x	x	x	x						

2.4 Modelling

Artificial neural networks (ANN) are statistical models inspired by the biology of brain networks. These models have been used successfully in several areas of knowledge, from engineering to the social sciences [17–19].

ANNs were used due its ability to generalize and to model complex nonlinear relationships. Unlike other forecasting techniques, ANNs do not impose restrictions on input variables and are able to model even with data with heteroscedasticity.

The model was implemented in Python using the KERAS library [20]. Due to its modular structure, this library allows for rapid development of ANNs, with a greater number of iterations and intuitive model creation using modules that generate neuron layers, activation functions, cost functions, initialization schemes, regularization schemes and optimizers.

This section describes the model used, how it was optimized and how it was applied to predict demand for the use of public buses in a metropolitan area.

Model Structure

We consider a Multilayer Percepron (MLP) a feedforward artificial neural network that allows mapping a series of input variables and obtaining a set of outputs through a composition of non-linear functions. The MLP architecture can be seen as a graph composed of a series of layers, which in turn are composed of a series of processing units (neurons). The layers between the entrances and exits are called hidden layers. For each of the scenarios defined, an ANN architecture was defined based on a base architecture that can be represented in the following form: d: n_{hid}: n_{hid} : n_{hid} : 4, where d represents the number of input variables (and which depends on the complexity of each scenario defined), n_{hid} is the number of neurons in each hidden layer and 4 output neurons to estimate the demand for each user group previously defined (G1–4). A previous analysis shows that the problem requires a complex model structure in order to predict buses demand accurately.

Learning Optimization

To search for the optimal set of weights, the batch backpropagation algorithm was used.

To define the ANN architectures for each scenario, first a set of preliminary experiments was carried out to define the number of training seasons (iterations of the optimization algorithm), the number of neurons in the hidden layers (n_{hid}), the number of samples per training lot (batch size) and the dropout value between layers of each ANN. Random search was used as it is more appropriate because not all hyperparameters are important for the proper functioning of ANN. This research was carried out with different ranges for each of the hyperparameters, namely:

- *Number of neurons*: ranging between 50 and 500 neurons in each hidden layer;
- *Epochs*: ranging between 100 and 600;
- *Batch size*: ranging between 8 and 100 records;
- *Dropout per layer*: ranging between 0 and 0.6 on each hidden layer.

For each set, 5 repetitions of stratified 10-fold cross-validation was perform. For each repetition, the whole dataset was randomized, and each training fold was normalized (Z-score), to have inputs with zero means and unit standard deviation.

The Leaky rectified linear activation function (Leaky ReLU) was used. To increase, the model's ability to adjust or train from the data, a slope (a = 0.01) was considered to the negative part of the ReLU activation function domain as proposed by Maas and Ng [24] avoiding in this way the conversion of negative values into zeros.

The optimization was carried using the algorithm HYPERAS (version 0.4.1) during 100 iterations. This algorithm evaluates several configurations at random, taking into account the parameters provided, in order to generate a good configuration of the hyperparameters. The model used 70% of the data set for training, 15% of the data set for validation and 15% of the data set for testing.

3 Results

Imbalance can compromise the performance of learning algorithms that are not usually prepared for such imbalance class distribution. Also, traditionally accuracy measures are distribution dependent and do not provide a clear picture of the classifier's functionality. Several strategies have been assessed to solve (or at least minimize) the effects of class imbalance, either by adapting the traditional classifiers or by considering different measures of performance. We follow the latter approach. For that purpose we consider three distinct measures of performance: the Mean Absolute Error (MAE), the Root Mean Square Error (RMSE) and the coefficient of determination (R2).

Table 4 show the hyperparameters used in each scenario, while Table 5 presents the results obtained. An analysis of these data show that the results of the different performance measures for each of the models are quite similar for C1, C2 and C3, except for the RMSE value which is much lower in the C3 (case in point that there is a two-hour advance of the values of the meteorological variables). Observing the deviation patterns of the prediction errors, we can verify once again that the C3 is the one that obtains the best results. Also, an analysis of the scenarios using only the meteorological variables with the significant correlation (C4, C5 and C6), show worsen results when compared to the first ones (C1, C2 and C3).

In general, the models that used all meteorological variables as input variables (C1, C2 and C3), performed better than when these variables were not used (C7) and when the variables used are only those that have a greater correlation with demand (C3, C4 and C5). The best results of the demand forecast were obtained when the models used the meteorological variables registered two hours before the trip took place (C3) as input variables. When all the meteorological variables are used, the RMSE decreases by 8.2% and the MAE decreases by 8.8% (comparison of C3 with C7).

Table 4. Architecture, number of epochs, bach size and dropout values of each scenario.

Scenario	Architecture	Epochs	Bach size	Dropout (1; 2; 3)
C1	11 : 230 : 230 : 230 : 4	450	50	0.15; 0.2; 0.3
C2	11 : 300 : 300 : 300 : 4	350	70	0.1; 0.1; 0.1
C3	11 : 250 : 250 : 250 : 4	250	80	0.15; 0.15; 0.35
C4	7 : 240 : 240 : 240 : 4	200	80	0; 0.25; 0.2
C5	7 : 150 : 150 : 150 : 4	300	60	0.15; 0.27; 0.10
C6	6 : 200 : 200 : 200 : 4	250	70	0.05; 0.15; 0.10
C7	5 : 150 : 150 : 150 : 4	400	60	0.126; 0.12; 0.11

Table 5. Results.

Scenario	MAE		RMSE	R2
	X	SD		
C1	151.7	332.3	343.5	0.86
C2	154.1	343.9	356.0	0.83
C3	143.4	312.1	322.7	0.89
C4	183.4	355.5	373.5	0.85
C5	165.1	355.6	366.9	0.86
C6	170.5	362.7	375.7	0.86
C7	157.3	348.8	351.4	0.84

X: average; SD: Standard Deviation.

4 Conclusions

A multilayer perceptron with three hidden layers was applied to forecast the demand for buses across a metropolitan transport network taking into account the influence of meteorological factors. Demand was forecast for four distinct groups of users: students (G1), workers (G2), elderly people (G3), and users who are given social support (G4). Seven scenarios were defined based on different combinations of the input variables. In these scenarios the meteorology observed in the same hour, in the previous hour and in the previous two hours of the demand were analyzed.

It was found that the use of meteorological variables allows to substantially increase the forecasting capacity of the model without meteorological variables. At the same time, it appears that when a two-hour delay in the demand is applied to the meteorological variables, the forecast results improve, with a decrease in the RMSE of 6% for total demand (comparison of C1 with C3). However, a high complex model (with 150 to 300 hidden neurons) is required to obtain MAE ranging from 143 and 170.

The present work demonstrate that such a model can help decision makers to foster buses demand, in particular in periods with fluctuations of meteorology. This allow to transport providers to manage more efficiently the resources available, particularly vehicles and drivers.

Acknowledgements. This work is financed by the ERDF - European Regional Development Fund through the Operational Programme for Competitiveness and Internationalisation - COMPETE 2020 Programme and by National Funds through the Portuguese funding agency, FCT - Fundação para a Ciência e a Tecnologia within project PTDC/ECI-TRA/32053/2017 - POCI-01-0145-FEDER-032053. Tânia Fontes also thanks FCT for the Post-Doctoral scholarship SFRH/BPD/109426/2015.

References

1. Tao, S., Corcoran, J., Hickman, M., Stimson, R.: The influence of weather on local geographical patterns of bus usage. J. Transp. Geogr. **54**, 66–80 (2016)
2. Stover, V., McCormack, E.: The impact of weather on bus ridership in Pierce County Washington. J. Public Transp. **15**(1), 95–110 (2015)
3. Li, J., Li, X., Chen, D., Godding, L.: Assessment of metro ridership fluctuation caused by weather conditions in Asian context: using archived weather and ridership data in Nanjing. J. Transp. Geogr. **66**(35), 356–368 (2018)
4. Zhou, M., Wang, D., Li, Q., Yue, Y., Tu, W., Cao, R.: Impacts of weather on public transport ridership: results from mining data from different sources. Transp. Res. Part C Emerg. Technol. **75**, 17–29 (2017)
5. Singhal, A., Kamga, C., Yazici, A.: Impact of weather on urban transit ridership. Transp. Res. Part A Policy Pract. **69**, 379–391 (2014)
6. Guo, Z., Wilson, N., Rahbee, A.: Impact of weather on transit ridership in Chicago, Illinois. Transp. Res. Rec. J. Transp. Res. Board **2034**, 3–10 (2008)
7. Guo, Z., Wilson, N. H., Rahbee, A.: The impact of weather on transit ridership Chicago. In: TRB Annual Meeting, pp. 3–10 (2007)
8. Kalkstein, A.J., Kuby, M., Gerrity, D., Clancy, J.J.: An analysis of air mass effects on rail ridership in three US cities. J. Transp. Geogr. **17**(3), 198–207 (2009)
9. Cools, M., Moons, E., Creemers, L., Wets, G.: Changes in travel behavior in response to weather conditions. Transp. Res. Rec. J. Transp. Res. Board **2157**, 22–28 (2010)
10. Sabir, M., van Ommeren, J., Koetse, M.J., Rietveld, P.: Impact of weather on daily travel demand (2010)
11. Costa, V., Fontes, T., Borges, J., Dias, T.: Impacts of weather conditions in urban public transport: understanding the effects of climatic changes using big data. Transp. Res. Board, Washington D.C (2017)
12. Arana, P., Cabezudo, S., Peñalba, M.: Influence of weather conditions on transit ridership: a statistical study using data from Smartcards. Transp. Res. Part A Policy Pract. **59**, 1–12 (2014)
13. Kalkstein, A.J., Kuby, M., Gerrity, D., Clancy, J.J.: An analysis of air mass effects on rail ridership in three US cities. J. Transp. Geogr. **17**(3), 198–207 (2009)

14. Yagi, S., Mohammadian, A.: Policy simulation for new BRT and area pricing alternatives using an opinion survey in Jakarta. Transp. Plan. Technol. **31**(5), 589–612 (2008)
15. Mahrsi, M.K.E.: A novel method of multi-information acquisition for electromagnetic flow meters. p. 9 (2014)
16. Hjorthol, R.: Winter weather – an obstacle to older people's activities? J. Transp. Geogr. **28**, 186–191 (2013)
17. Zupan, J., Gasteiger, J.: Neural Networks for Chemists: An Introduction. Wiley, Cambridge (1993)
18. Cochocki, A., Unbehauen, R.: Neural Networks for Optimization and Signal Processing, 1st edn. John Wiley Sons Inc, Chichester (1993)
19. Rowland, Z., Vrbka, J.: Using artificial neural networks for prediction of key indicators of a company in global world. In: Klieštik, T. (ed.) Globalization and its Socio-Economic Consequences, 16th International Scientific Conference Proceedings. PTS I-V, Žilina, Slovensko, pp. 1896–1903 (2016)
20. Chollet, F.: keras. https://github.com/fchollet/keras (2015)
21. Creemers, L., Wets, G., Cools, M.: Meteorological variation in daily travel behaviour: evidence from revealed preference data from the Netherlands. Theo. Appl. Climatol. **120**(1–2), 183–194 (2015)
22. de Montigny, L., Ling, R., Zacharias. J.: The effects of weather on walking rates in nine cities. Environ. Behav. **44**(6), 821–840 (2011)
23. Schmiedeskamp, P., Zhao, W.: Estimating daily bicycle counts in Seattle, Washington, from seasonal and weather factors. Transp. Res. Rec. J. Transp. Res. Board **2593**(1), 94–102 (2016)
24. Maas, A.L., Ng. A.Y.: Rectifier non linearities improve neural network acoustic models, vol. 28 (2013)

Reshaping Transport Modelling

Attitudes of E-Scooter Non-users Towards Users

Athanasia Kostareli[1], Socrates Basbas[1], Nikiforos Stamatiadis[2], and Andreas Nikiforiadis[1(✉)]

[1] School of Rural and Surveying Engineering, Aristotle University of Thessaloniki, 54124 Thessaloniki, Greece
anikiforiadis@topo.auth.gr
[2] Department of Civil Engineering, University of Kentucky, Lexington, KY, USA

Abstract. Micromobility aims to provide environmentally friendly options for trips that cannot be accomplished with public transport and reduce automobile dependence. The most recent transport mode for accomplishing such trips is the electric scooter (e-scooter) and they have flooded several countries and cities taking a significant share of trips. However, there is little guidance regarding their operations and initial observations demonstrate public disdain. This article deals with the evaluation of the public opinion in regard to e-scooters based on those who do not use them. To solicit input, an in-person and an electronic questionnaire were used in Thessaloniki, Greece. Statistical analysis of the collected data was undertaken to establish possible pattern of public opinion and identify situations of e-scooters that cause problems in pedestrian and motor vehicle traffic. Furthermore, the lack of infrastructure, which is a deterrent to e-scooter use, and the absence of a legislative framework that sets out the rules for traffic seem to be main concerns. Finally, the surveys noted that e-scooters will be promoted and integrated into traffic by designing and building adequate infrastructure, training users in Highway Code and imposing fines on offenders.

Keywords: Micromobility · E-Scooters · Urban mobility

1 Introduction

The population of urban areas is growing rapidly, and it is estimated that by 2050 it will account for 2/3 of the world population [1]. The increased urbanization poses new challenges for the transportation systems and networks. Satisfying mobility needs with the use of public transport is often difficult due to lack of extended network coverage and frequency of service resulting in forcing travelers to use their own vehicles or, in the worst case scenario, to not be able to complete their mobility needs to address employment opportunities, health care, etc. [2]. Furthermore, the deterioration of air quality and environmental conditions along with efforts to increase energy saving [3] have provided the opportunity to develop a new form of mobility, i.e., micromobility [4]. Micromobility services include bicycles (conventional or electric) and e-scooters. These transport services are shared and they are used by many different people, many

E. G. Nathanail et al. (Eds.): CSUM2020, AISC 1278, pp. 87–96, 2021.
https://doi.org/10.1007/978-3-030-61075-3_9

times a day on an as-needed basis. Micromobility services are able to fill the "first and last mile' gaps that public transport cannot cover, offering more options to commuters [4]. The length of urban trips is mostly short, and it is estimated that micromobility can replace more than 4 trillion miles worldwide [2].

E-scooters have affected most countries from the first moment they appeared because of their low travel costs, fulfilling mobility needs for areas without an adequate public transport system [5], but also because they are a fun way to travel for adults [6]. Another factor contributing to their rapid and massive acceptance is the tendency of young people who do not to want to acquire assets by buying private means of transportation, but to choose personalized mobility tools at a cost per use [7].

The rapid spread of e-scooters has found governments around the world unprepared as there are no clear guidelines and legal frameworks defining their permitted areas of operation, allowable maximum speed limit, authorized parking spaces, and other operational rules [8]. As a result, existing infrastructure for other transport modes, such as bicycle lanes and sidewalks, are used. However, these facilities do not have the required dimensions to allow an e-scooter to safely operate simultaneously with them. The absence of an organized and unified network of bicycle infrastructure often results in the driving of e-scooters on sidewalks. This behavior elicits strong reactions of pedestrians and people with disabilities, who complain that their movement is blocked and feel unsafe even in pedestrian-only areas [2]. Others express dissatisfaction with the new means of transportation, causing damage to equipment and even throwing it into the sea [9]. At the same time, there have been cases of injuries of varying severity [10] to death [11] due to a lack of driving safety measures. Many local administrations have implemented bans and strict regulations on the movement of e-scooters due to the problems arisen from their operation [2].

The present study aims to examine the attitudes of e-scooter non-users in the city of Thessaloniki, Greece in order to understand their issues associated with the use of e-scooters and define potential areas where efforts should be concentrated to address their use. The findings of this work are expected to address an important aspect of the continued use of e-scooters in Greece and provide a basis for developing possible policies for addressing their efficient and safe operation.

2 Literature Review

Manufacturing of scooters began in the early 1900s, with the first petrol-powered scooter appearing in 1919 and the 1940s characterized by their increased use [12]. In 1990, a folding aluminum scooter was introduced, which dominated Japan and the US, and it was the forefather of today's e-scooters [13]. The addition of an electric motor in 2003 turned the scooters into e-scooters, but they failed to dominate due to the high use of bicycle-share systems. However, in the last four years, $5.7 billion has been invested by small startups, while larger ones have invested $2 billion to advance e-scooter use [14].

E-scooter share systems rely on applications that users can install on their mobile phones. Through these applications, the user is able to locate e-scooters on a map, identify their charges, and select one for their trip. To start the e-scooter, users scan the code on the steering wheel with their mobile phone and repeats this at the end of the trip to lock the e-scooter. Charges are based on an initial fee per use with additional mileage charges. Overnight, companies locate their e-scooters through their positioning system to charge and place them again for use [15].

The rapid spread of e-scooters is a challenge for governments around the world, as the lack of regulations on their use has caused a backlash in society due to problems encountered in both vehicular and pedestrian traffic [2]. For this reason, many countries have found it necessary to inform and educate users of e-scooters in order to prevent negative opinions from greater usage and improvement of their operations. Such efforts result in distribution of leaflets and posters that are posted on public transport stations and in parking lots and display rules for safe use [16]. Other efforts include the development of web pages where citizens can pose questions and seek answers to frequently asked questions [17].

Some countries have amended traffic laws to ensure the safe integration of e-scooters into traffic. For example, in the US, the state of California with AB Bill 2989, passed September 18, 2018, sets a maximum speed limit of 15 mph, allowing e-scooters to be driven alongside motorized traffic on highways with speeds up to 25 mph, while on roads with higher speeds it forces them to be traveling on bicycle lanes. It also requires a driving license or a driving license for an e-scooter and the use of a helmet [18]. Germany in June 20, 2019 set 20 km/h as the maximum speed limit and allowed e-scooters to travel on bicycle lanes and on the road (when there is no bicycle lane), prohibited driving on sidewalks, set a user age limit 14 yr and required liability insurance from both e-scooter sharing companies and privately owned e-scooters. Violators will be punished with monetary penalties [3]. The United Kingdom, under the Motorways Act 1835, prohibits driving on roads and sidewalks, with a fine of £300 and the removal of 6 diploma points (out of 12). In London, e-scooter driving is only permitted at Queen Elizabeth's Olympic Park [19].

There are several issues that create a negative overall attitude towards the expansion of e-scooter usage. One of the major complaints is the use of sidewalks interfering with pedestrian traffic and especially that of handicapped persons. The large number of rental companies and uncontrolled number of e-scooters that often appear with no prior local approval is another complaint. E-scooters parked at random locations present an untidy urban scene. A recent study also concluded that ecologically, e-scooter travel has a larger negative environmental impact than other transport modes. This is due to the overall requirements of the system that requires their production, continuous maintenance, daily collection and recharging and relocation at high-demand spots [20].

The literature review shows that there is a somewhat uncontrolled operation of e-scooters and cities and national governments are racing to establish a set of guidelines that would define their efficient and safe operation. This study will provide a meaningful contribution to understanding the attitudes of the non-users in the Greek environment and aid in the development of a set of rules that could address efficient use of e-scooters.

3 Study Area and Data Collection

E-scooters appeared first in Thessaloniki in December 2018. Local authorities have argued that the use of e-scooters will help improve traffic quality and offload the road network [21]. Thessaloniki is the largest city in Northern Greece and second largest in population after the capital city of Athens. According to the last census conducted in 2011 its population is 1,110,551 [22] and it also hosts a large number of students. The only means of public transport available in the city are the city buses of the Thessaloniki Urban Transport Organization. It also has a 12 km long bike lanes network [23] and a bike-sharing system [24]. The acceptance of e-scooters by many users was rapid as an alternative to overcrowded traffic.

The purpose of this study is to evaluate the operation of e-scooters as non-users perceive it and identify their opinion and attitudes towards them. To this end, a set of questions were developed to collect the required information. The sample size was initially set at 300 questionnaires to allow for obtaining a representative sample of the population.

The questionnaire was developed based on reviews of the Greek and international literature and consisted of sections that identified the socioeconomic characteristics of the respondents, their most common means of transport, perceived situations of e-scooters affecting safety of both users and non-users, and their opinion on certain interventions that will improve e-scooter use.

To conduct the research, two forms of collection were employed: in-person and electronic surveys. For the in-person survey, the interview technique was followed to provide the interviewers with the necessary clarification when needed and thus to avoid any misinterpretation when completing the questionnaire. A total of 140 questionnaires were obtained during the September-October 2019 period with most of the interviews conducted in the afternoon and evening. The electronic questionnaire was posted on a website and within a week in October 2019 an additional 167 questionnaires were completed. It is noted that there are three additional questions in the online questionnaire that indicate the profession, the level of education and the area of residence of the respondents. The total sample number is 307 questionnaires.

4 Analysis

4.1 Descriptive Statistics

Demographics

Respondents' demographic data shows that there is a slightly larger number of female (54.1%) than male respondents. Half of the respondents, specifically 50.8%, are between 18–27 yr old, while the other age groups showed lower representation with increasing age. Also, most of the respondents (24.1%) have annual family income 7.000€–12.000€, while only 7.5% exceed 36.000€. The majority of respondents (50.1% only from electronic questionnaire) are employees and university graduates (37.7%).

E-Scooter Usage

The main reason for not using an e-scooter is that 48.5% of respondents prefer to walk and 18.6% use a public bus. Ownership of a private vehicle is high (42%), most identify as pedestrians (36.2%) and a large percentage use car (27%). The main reason for not using an e-scooter is that they prefer to walk (48.5%) or use a public bus (18.6%). It should be noted that a large percentage (64.3%) of those who have not used an e-scooter so far are interested in using them in the future.

E-Scooter Issue

The next set of questions attempts to identify concerns and issues from e-scooter usage. Referring to pedestrian blockage, it is mainly due to the fact that e-scooters drive on the sidewalk (53.1%) and jeopardize vulnerable users (elderly, disabled) (43.1%). Furthermore, the majority of respondents (74.6%) consider that the movement of vehicles is hindered by users of e-scooters who do not comply with the Highway Code, while the least important issue is their lower than automobile speeds (56.4%). The safety of users of e-scooters is considered by most (57.3%) to be low because they do not take protective measures; the use of underaged persons was also noted as important (41.7%).

Means to Improve E-Scooter Use

The findings on the various aspects that non-users currently see as important for improving the use of e-scooters were also examined. The majority of participants (71.7%) identified the need for training for e-scooter users as means of improving operations and safety, while the use of fines for traffic violations was considered also important (33.9%). Reducing e-scooter speeds was considered less important (36.8%). Finally, the costs for using them is a fairly important reason to avoid their use (34.5%), but the lack of infrastructure is the main reason why a large percentage of people do not use e-scooters (60.8%). The construction of bicycle infrastructure will be a key factor for the majority (75.9%) in choosing e-scooters.

4.2 Comparative Analysis by Gender and Age

This section presents an analysis based on gender and age groups in order to examine the respondents' attitudes regarding e-scooter use and suggestions to improve their operations. Males and females rank similarly the different issues for the pedestrians. Both male and female respondents consider that driving e-scooters on sidewalks as their most important negative aspect since it could obstruct pedestrian flow while parking them on sidewalks was also viewed as an important aspect (Table 1). Also, males and females totally agree that e-scooter users in many cases do not comply to the Highway Code resulting in important issues for their co-existence with motorized vehicles. Concerning their co-existence with pedestrians, both males and females state that there are issues of safety for the elderly and these issues become greater due to the high speed of the e-scooters. There is also an agreement between the two genders regarding the issues that setting e-scooter users' safety in danger. More specifically, they have the opinion that not wearing helmet and the fact that they are used by minors can pose safety issues for the riders.

Table 1. Gender based e-scooters issues (means).

	Male	Female
Issues for pedestrians		
They move on the sidewalk	3.07	3.22
They park in the sidewalk	2.59	2.7
Many parked scooters accumulate on the sidewalks	2.07	1.91
Interfere with pedestrian crossing	2.27	2.16
Issues for vehicles		
Do not comply to the Highway Code	2.61	2.67
Illegally parked on the road	1.84	1.73
Move at slower speeds than vehicles	1.55	1.59
Issues for pedestrians safety		
Move at high speed	2.17	2.16
Are silent	1.65	1.63
Endangering vulnerable users such as elderly	2.20	2.22
Issues for e-scooter users safety		
They do not take protective measures (e.g. Helmet)	3.28	3.36
They are no different that's riding a motorcycle	2	1.81
They are no different that's riding a bicycle	1.82	1.87
They are used by minors (regardless of regulation)	2.91	2.96

The attitudes of males are also to a large extent similar regarding the needed improvements. Both males and females believe that fines for improper usage could be the most effective means to improve safety and operations of e-scooters while reducing their operating speeds is not viewed as an important incentive for improving their operations (Table 2). Lack of adequate infrastructure is considered to be an essential factor for not using e-scooters, both for males and females; however females tend to assign greater importance to this factor. On the other hand, males tend to assign higher importance to the e-scooter usage cost.

The next analysis examined the opinions provided as they relate to the respondents' age groups (Table 3). Respondents in all age groups identified as the most important issue the use of e-scooters on sidewalks resulting in interference with pedestrian flow. For respondents in the 55+ age group the usage of e-scooter on sidewalks seems to be a great issue, while the least important issue is the interfere with pedestrians on cross-walks. All three age groups identified the non-compliant behavior of e-scooter users as the most important issue while the lower speeds of e-scooters were viewed as the least important, especially for older people. Likewise, all three groups consider equally important their operating speeds and the fact that e-scooters could pose safety concerns for handicapped persons on sidewalks, while their silent operating mode is not considered important. Finally, all age groups believe that user safety is reduced because users do not use any protective measures and due to the fact that the e-scooters are used by minors. Older people give a greater importance, comparing to the other age groups, to the use of e-scooters by children.

Table 2. Gender based e-scooters improvements (means).

	Male	Female
Compliance with Highway Code		
Warning for improper parking/driving on sidewalks	2.55	2.45
Imposing fines	2.82	2.75
Suspending user account for document incidents	2.57	2.53
Speed reduction	2.06	2.27
Reasons that prevent from using e-scooter		
Lack of infrastructure	4.01	4.27
Limited number of available e-scooters	2.16	2.08
Cost of transportation	3.69	3.33
Mandatory use of credit card	2.95	3.05
Mandatory use of smartphone	2.18	2.28

Table 3. Age based e-scooters issues (means).

	18–36	37–54	55+
Issues for pedestrians			
They move on the sidewalk	3.05	3.43	4
They park in the sidewalk	2.63	2.72	2.71
Many parked scooters accumulate on the sidewalks	2.06	1.69	2.14
Interfere with pedestrian crossing	2.26	2.15	1.14
Issues for vehicles			
Do not comply to the Highway Code	2.59	2.82	2.86
Illegally parked on the road	1.77	1.77	2.14
Move at slower speeds than vehicles	1.63	1.42	1
Issues for pedestrians safety			
Move at high speed	2.13	2.28	2.29
Are silent	1.70	1.46	1.43
Endangering vulnerable users such as elderly	2.19	2.26	2.29
Issues for e-scooter users safety			
They do not take protective measures (e.g. Helmet)	3.35	3.22	3.29
They are no different that's riding a motorcycle	1.83	2.11	2
They are no different that's riding a bicycle	1.89	1.71	1.57
They are used by minors (regardless of regulation)	2.92	2.97	3.14

In contrast with people between 18 and 54 yr old, who believe that imposing fines is the most effective means for the users to obey the Highway Code, older respondents consider the warning for improper parking or use and the cancellation of users' account as equally important (Table 4). Lack of infrastructure is the most important reason for the under 55 respondents for not using e-scooters, while the comparative analysis highlights the diffidence of those over 55 for using a credit card, which may constitutes a significant barrier for their use.

Table 4. Age based e-scooters improvements (means).

	18–36	37–54	55+
Compliance with Highway Code			
Warning for improper parking or driving on sidewalks	2.51	2.40	2.71
Imposing fines	2.72	3.03	2.71
Suspending user account for document incidents	2.57	2.45	2.71
Speed reduction	2.19	2.12	1.86
Reasons that prevent from using e-scooter			
Lack of infrastructure	4.05	4.54	3.86
Limited number of available e-scooters	2.14	2.12	1.43
Cost of transportation	3.55	3.35	2.86
Mandatory use of credit card	3.06	2.69	4
Mandatory use of smartphone	2.20	2.29	2.86

5 Discussion

This paper presents the results of a survey conducted in Thessaloniki, Greece that aimed to gauge public opinion regarding the use of e-scooters targeting non-users. The study tried to identify the issues that could result from their use as well as to seek potential measures resulting in safe operations and inclusion of e-scooters in traffic.

The study findings support the casual observations of the high use of private automobiles for addressing mobility needs in the city which is mainly attributed to the high automobile ownership rates and limited, ineffective public transport network.

The results showed that the main reason for the respondents to not select e-scooters as a travel mode is the lack of infrastructure that would allow for proper operation of e-scooters. This is due to perceived lack of safety when moving on the roadway along with vehicular traffic as well as when they ride on the sidewalks, since they pose safety concerns for the pedestrians and other vulnerable users. The necessity of using a credit card also constitutes a barrier, especially for older people. However, a large percentage of respondents (64%) are interested in using e-scooter in the future and this finding is in agreement with a survey in Portland, Oregon, which showed that the respondents expressed a positive attitude towards future e-scooter usage [25].

The use of e-scooter by minors and the fact that the users do not take protective measures are considered very important safety issues by all respondents, regardless of their gender and age. The obstructions of e-scooters to other vehicles as well as where they are allowed to operate underscore the need for education to not only address these issues but to also allow the safer and more proper incorporation of e-scooters in the traffic stream. Traffic fines related to improper operation could be viewed as a deterrent and means for protecting pedestrians and would align proper e-scooter usage and operation.

Future efforts should consider collecting a larger number of questionnaires that would have a greater age representation, since in this research the largest position of the respondents belongs to the 18–27 age group. In general, public opinion agrees that there is a need for infrastructure development and possibly bicycle infrastructure to allow for a safer and more efficient e-scooter mobility. This would result in elimination of improper sidewalk usage and thus improve pedestrian safety. User training regarding the existing Highway Code and overall proper use of e-scooters is also central to creating a safer environment for all users. Finally, respective regulations need to be designed to allow for a more complete inclusion of e-scooters to traffic streams.

References

1. United Nations: World Urbanization Prospects, 2014 revision (2015)
2. Deloitte. Insights: https://www2.deloitte.com/us/en/insights/focus/future-of-mobility/micro-mobility-is-the-future-of-urban-transportation.html. Accessed 8 Nov 2019
3. Gubman, J., Jung, A., Kiel, T., Strehmann, J., Horn, B., Dr. Hebes, P.: Shared e-scooters: paving the road ahead policy recommendations for local governmen. AgoraVerkehrswende (2019)
4. DuPuis, N., Griess, J., Klein, C.: Micromobility in Cities: A History and Policy Overview. National league of cities (2019)
5. Lee, M., Chow, J.Y.J., Yoon, G., He, B.Y.: Forecasting e-scooter competition with direct and access trips by mode and distance in New York City. Cornell University. https://arxiv.org/ftp/arxiv/papers/1908/1908.08127.pdf. Accessed 4 Feb 2020
6. Boston Consulting Group: https://www.bcg.com/publications/2019/promise-pitfalls-e-scooter-sharing.aspx. Accessed 4 Feb 2020
7. Dataconomy: https://dataconomy.com/2019/08/micromobility-what-does-it-mean-for-the-future-of-transportation/. Accessed 4 Feb 2020
8. Cnn.Business: https://money.cnn.com/2018/02/15/technology/electric-scooters-bird/index.html. Accessed 2 Feb 2020
9. iefimerida: https://www.iefimerida.gr/ellada/ilektrika-patinia-sto-eleos-ton-hoyligkan. Accessed 12 Feb 2020
10. Shaheen, S., Cohen, A.: Shared Micromoblity Policy Toolkit: Docked and Dockless Bike and Scooter Sharing. Transportation Sustainability Research Center, UC Berkeley (2019). https://doi.org/10.7922/G2TH8JW7
11. Cnn.Travel: https://edition.cnn.com/travel/article/electric-scooter-bans-world/index.html. Accessed 4 Feb 2020
12. Unagiscooters: https://www.unagiscooters.com/blogs/features/the-scooter-a-brief-history. Accessed 8 Nov 2019
13. Citylab: https://www.citylab.com/transportation/2018/09/man-behind-urban-scooter-revolution/570109/. Accessed 8 Nov 2019
14. Forbes: https://www.forbes.com/sites/carltonreid/2019/03/18/bicycling-take-a-hike-the-micromobility-revolution-will-be-motorized/#5c09777c135d. Accessed 10 Nov 2019
15. American Association of Motor Vehicle Administrators: https://www.aamv.org. Accessed 3 Feb 2020
16. City of Santa Monica: https://www.santamonica.gov/press/2018/08/22/santa-monica-launches-public-education-campaign-on-e-scooter-safe-rules-of-the-road. Accessed 6 Feb 2020

17. Portland Bureau of Transportation: https://www.portlandoregon.gov/transportation/77294. Accessed 12 Feb 2020
18. California Legislative Information: https://leginfo.legislature.ca.gov/faces/billTextClient.xhtml?bill_id=201720180AB2989. Accessed 3 Feb 2020
19. Forbes: https://www.forbes.com/sites/alexledsom/2019/09/10/e-scooter-havoc-across-french-cities-is-a-crackdown-needed/#769822b03038. Accessed 5 Feb 2020
20. The globe and mail: https://www.theglobeandmail.com/opinion/article-why-e-scooters-have-been-on-a-bumpy-ride/. Accessed 25 Feb 2020
21. Insider: https://www.insider.gr/aytokinito/ellada/102683/thessaloniki-metakiniseis-me-ilektrika-patinia-meso-kinitoy. Accessed 5 Feb 2020
22. Elstat: https://www.statistics.gr/el/statistics/-/publication/SAM03/-. Accessed 5 Feb 2020
23. Nikiforiadis, A., Basbas, S.: Can pedestrians and cyclists share the same space? The case of a city with low cycling levels and experience. Sustain. Cities Soc. **46**, 101453 (2019). https://doi.org/10.1016/j.scs.2019.101453
24. Boufidis, N., Nikiforiadis, A., Chrysostomou, K., Aifadopoulou, G.: Development of a station-level demand prediction and visualization tool to support bike-sharing systems' operators. Transp. Res. Procedia **47**, 51–58 (2020)
25. Portland Bureau of Transportation: E-scooter Findings Report. Portland, Oregon (2018)

Evaluating the Performance of Reinforcement Learning Signalling Strategies for Sustainable Urban Road Networks

Haris Ballis(✉) and Loukas Dimitriou

University of Cyprus, Nicosia, Cyprus
lucdimit@ucy.ac.cy

Abstract. Smart Cities promise to their residents, quick journeys in a clean and sustainable environment. Despite, the benefits accrued by the introduction of traffic management solutions (e.g. improved travel times, maximisation of throughput, etc.), these solutions usually fall short on ensuring the environmental sustainability around the implementation areas. This is because the environmental dimension (e.g. vehicle emissions) is usually absent from the optimisation methodologies adopted for traffic management strategies. Nonetheless, since environmental performance corresponds as a primary goal of contemporary mobility planning, solutions that can guarantee air quality are significant. This study presents an advanced Artificial Intelligence-based (AI) signal control framework, able to incorporate environmental considerations into the core of signal optimisation processes. More specifically, a highly flexible Reinforcement Learning (RL) algorithm has been developed in order to identify efficient but -more importantly- environmentally friendly signal control strategies. The methodology is deployed on a large-scale micro-simulation environment able to realistically represent urban traffic conditions. Alternative signal control strategies are designed, applied, and evaluated against their achieved traffic efficiency and environmental footprint. Based on the results obtained from the application of the methodology on a core part of the road urban network of Nicosia, Cyprus the best strategy achieved a 4.8% increase of the network throughput, 17.7% decrease of the average queue length and a remarkable 34.2% decrease of delay while considerably reduced the CO emissions by 8.1%. The encouraging results showcase ability of RL-based traffic signal controlling to ensure improved air-quality conditions for the residents of dense urban areas.

Keywords: Reinforcement learning · Traffic signal control · Traffic management · Air quality · Large-scale micro-simulation

1 Introduction

Efficient traffic demand management has been the epicentre of many researchers' attention [1]. During the last decades multiple traffic signal control strategies ranging from fixed time to demand responsive (adaptive) have been suggested and evaluated [1]. Nonetheless, methodologies stemming from the Artificial Intelligence (AI) field

E. G. Nathanail et al. (Eds.): CSUM2020, AISC 1278, pp. 97–105, 2021.
https://doi.org/10.1007/978-3-030-61075-3_10

have only recently emerged [2]. Between the AI-based methodologies, Reinforcement Learning (RL) stands as one of the most suitable approaches to address the traffic signalling optimisation issue, since it is has proven very effective at optimising highly dynamic and stochastic systems. RL attempts to identify optimum policies (i.e. strategies) in complex environments which require actions by one or more agents, taking place during multiple episodes. At the end of each episode, the agents, evaluate the state of the system and get rewarded depending on the results of their actions.

With regards to the application of RL to traffic optimisation, signal controllers are usually perceived as the agents who make decisions regarding the state of the governed intersection. The agents adapt their strategies depending on the volume of the reward they receive at the end of the episode (e.g. evaluation time interval). In the various presented examples [2–4], the reward has been often expressed among many others as the vehicle throughput, the average delay, the number of stops, etc. Based on the previous, it becomes apparent that the observability of the simulated system is a pre-requisite for the application of the methodology. Although the required level of observability is difficult to be achieved on the ground since numerous traffic detectors or even more sophisticated equipment would be required, nonetheless the vehicles of the future (CAVs) will be able to beacon the relevant information and thus allow for the full tractability of the system's state.

The presented study contributes by evaluating multiple reward mechanisms under fully realistic demand conditions as manifested in the large-scale road network of Nicosia, Cyprus. The following sections describe the design of a RL-based signal control optimisation methodology (Sect. 2), implemented for the urban road network of Nicosia, Cyprus (Sect. 3) as well as the summary of the paper (Sect. 4).

2 Methodology

2.1 Reinforcement Learning Components

Agents. The agents constitute the RL controlled traffic signals ($n \in N$)

Actions. Agents (N) take decisions (AS_t^n) at each time interval (t) whether to keep the current traffic signal phase ($AS_t^n = 0$) or to progress to the next ($AS_t^n = 1$).

Reward. The reward (r_{t-1}^n) is a measure of the success for each action. Rewards are calculated for all agents ($n \in N$) and refer to the previous time-interval ($t - 1$).

State. The state represents the conditions of the simulated environment while its exact definition depends on the reward that will be used. The state must include all the required information to calculate the reward (r_{t-1}^n) for each agent (n) at the previous time-interval ($t - 1$). Regardless of the reward type, the state must include the active stage $\left(AS_t^n\right)$ of all agents (N) at all time-intervals (T). Based on this, the state can be defined as $S = (AS_t^n, r_{t-1}^n) \forall n \in N, t \in T$.

3 Reinforcement Learning for Traffic Signal Control

The next section describes the main elements defining the RL paradigm in the context of traffic signal control. Traffic signals ($n \in N$) constitute the agents of the system who decide at each time interval ($t \in T$) to take the action ($a_t \in [0, 1]$) of switching to the next stage or not. Note that for the purposes of this study stages are assumed to take place cyclically, as usually recommended for safety reasons [5]. The micro-simulation framework is responsible for updating the state (S_t) of the environment as well as for the calculation of the reward (r_t^n) that will be provided to each agent at the end of each interval. The objective of each agent is to identify the set of actions (policy) which results to the maximisation of his reward. The components of the RL design are presented below. With the successive completion of multiple simulation runs, agents improve their decisions by learning through their action-based rewards. For the purposes of this study, the learning of agents is accomplished by the widely used Q-Learning approach. In Q-Learning, the Q values based on which agents make their decisions are updated as:

$$Q_{t+1}(s_t, a_t) = Q_t(s_t, a_t) + alpha * (r_t^n + gamma * Q_{max_a}(s_{t+1}, a) - Q_t(s_t, a_t)) \quad (1)$$

where $alpha \in [0, 1]$ is an input parameter controlling the learning rate of agents while the discount factor $gamma \in [0, 1]$ represents the way agents regard future rewards. Low *gamma* values result in myopic agents accounting mostly for the immediate rewards while larger values result in agents who value future rewards more.

An important element of RL-based methodologies is the balancing between the exploration-exploitation trade-off. Agents can potentially learn more efficient policies by exploring the available search space and consequently evaluating a wider range of possibilities. On the other hand, the continuous exploration of the available search space can prevent agents from deciding and eventually following the most beneficial policy. For this study the commonly used ε-greedy algorithm [6] has been used to achieve the exploration-exploitation balance.

4 Experimental Results

4.1 Network Description

The abovementioned methodology was evaluated on the large-scale urban road network of Nicosia, which covers an area of almost 20 km^2 and includes the majority of Nicosia's metropolitan area (Fig. 1). The network consists of 3,365 links and more than 200 junctions out of which 19 are signalised. For the simplification of the analysis, the RL signal control system was implemented only on the three signalised junctions presented in the shaded area of Fig. 1. The implementation is more clearly depicted in Fig. 2. With regards to traffic demand, the centre of Nicosia faces significant congestion since more than 16,000 trips are executed in the during the morning peak hour (07:00–08:00). To the best of the authors' knowledge, this is one of the very few examples evaluating an RL signal control optimisation framework on an urban road network under real traffic conditions.

Fig. 1. The simulation network of Nicosia, Cyprus and the Reinforcement Learning signal-controlled area (under green shade).

5 Evaluation Scenarios

The evaluated scenarios explore the effect of different reward types to the network performance. More precisely, the evaluated reward mechanisms are the vehicle throughput (A-Vehicles), the average queue length (B-Queue) and the volume of CO emissions (C-Emissions). This research uses the widely used traffic simulator VISSIM to mimic the urban traffic conditions. A specifically designed module developed in the programming language Python, enabled the communication of rewards between the agents (traffic signal controllers) and the environment through the Component Object Model (COM) interface. Additionally, the Python module was responsible to complete the transition between traffic stages when that was decided by the agents. At this point it should be noted that the transition between stages respected the required amber time as dictated in the fixed timed timings. Finally, the TRANSYT-7F model [7] was used for the estimation of the environmental impact.

All the evaluation scenarios were executed for 2,000 iterations during the AM peak hour (07:00–08:00) with the following configuration (Table 1). The scenarios were

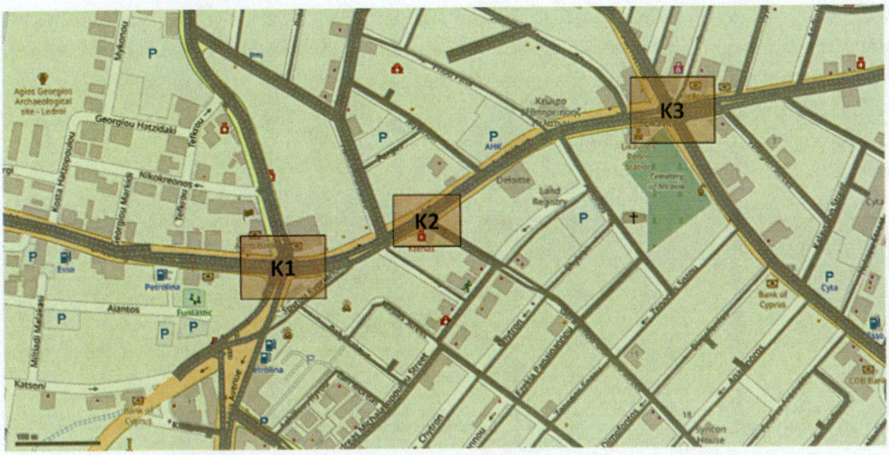

Fig. 2. Reinforcement Learning signal-controlled junctions.

executed simultaneously and were completed in approximately four days using an Intel™ Xeon E3–1230@3.40 GHz CPU, 32 GB RAM computer.

Table 1. Parameter settings for reinforcement learning.

Parameter	Value
alpha	0.7
gamma	0.9
epsilon (ε-greedy)	0.7
Time-interval duration	10 s

Finally, the efficiency of the previously mentioned implementation strategies is compared against the currently implemented (fixed time) signal control program for the Nicosia's network. The metrics to be evaluated are namely:

- The total number of vehicles crossing the evaluated junctions (Vehicles)
- The average delay at junctions (Avg. Delay)
- The average queue length at junctions (Avg. Queue length)
- The volume of CO emissions

6 Discussion of Results

The next section presents the results of the application of the suggested RL methodology for the four evaluated reward mechanisms.

Based on the results presented in Fig. 3, the vehicle throughput (a) reward mechanism presents the most potential. Using the vehicle throughput (A-Vehicles) as reward mechanism leads to a 4.8% increase in total network flow and significant reductions to queue lengths (−17.7%), delays (−34.2%) and CO emissions (−8.1%). On the other hand, the delay (C-Delay) and emissions (D-Emissions) reward mechanisms deteriorate the overall performance.

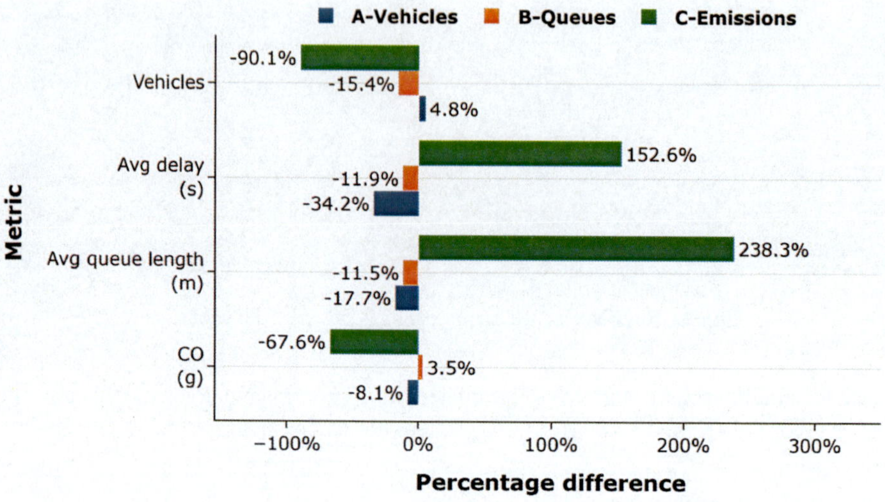

Fig. 3. Comparison of the evaluated reward mechanisms. Differences are calculated relatively to the base scenario.

The evaluation of the executed scenarios is also presented in Fig. 4. Each evaluation metric in the graph has been normalised based on the maximum value across all scenarios. As it can be noticed, the A-Vehicles reward mechanism outperforms all the evaluated reward mechanisms as well as the current signal control strategy (Base).

The currently presented results emphatically showcase the effect of the reward mechanism on the solution. For example, contradictory conclusions would have been drawn if the methodology had been evaluated using only a single reward mechanism. If for instance, any other than the vehicle throughput (A-Vehicles) reward mechanism had been used, then the potential of the methodology would have been considerably underestimated. Moreover, it becomes apparent that ensuring the environmental quality and the sustainability of urban areas is not a trivial task. As it can be observed, attempting to optimise the system towards the minimisation of emissions can have a considerable negative effect on the traffic flow conditions of the transport network. Nonetheless, RL based traffic demand management approaches seem capable of simultaneously improving the environmental as well as the network conditions (i.e. flows, delays, queues, etc.) around the implementation area.

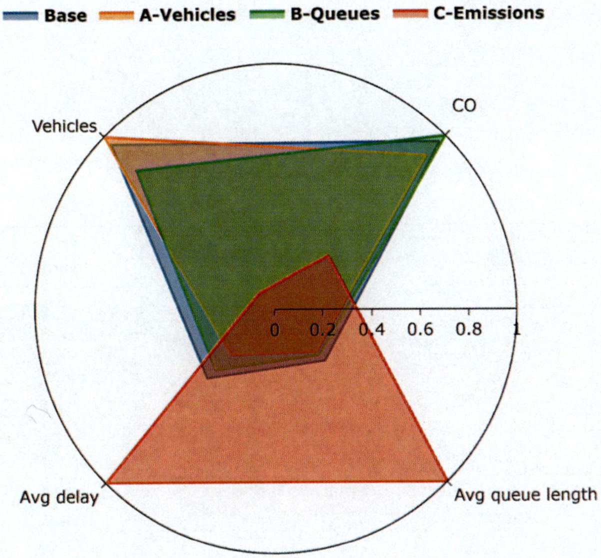

Fig. 4. Evaluation of the Reinforcement Learning reward mechanism. Values are normalized according to the maximum value for each metric.

7 Further Evaluation of the Recommended Reward Mechanism

The current section delves in the analysis of the best reward mechanism as previously identified (i.e. A-Vehicles). Fig. 5 presents the dynamic evaluation of the four suggested evaluation metrics between 07:00 and 08:00. As it can be observed, the RL optimized traffic signal control area outperforms the currently implemented (Base), fixed-time signal control strategy throughout the evaluated period. Vehicle throughput is consistently higher in the case of RL controlled signals while the opposite holds true for the average queue length, the average delay and the CO emissions. These results showcase the significant potential of emerging RL signal optimization approaches to improve environmental conditions without disrupting traffic operations.

Despite the previous encouraging results and the potential of the suggested RL-based methodology for the addressing of efficient traffic management, some aspects require further studying. Firstly, the methodology should be evaluated over a larger implementation area (preferably at city level) in order to assess its efficiency at improving traffic conditions at a large-scale. Secondly, modifications of the methodology allowing the "communication" between the agents should be evaluated since it is very likely that the isolation between agents can deteriorate the performance of the methodology.

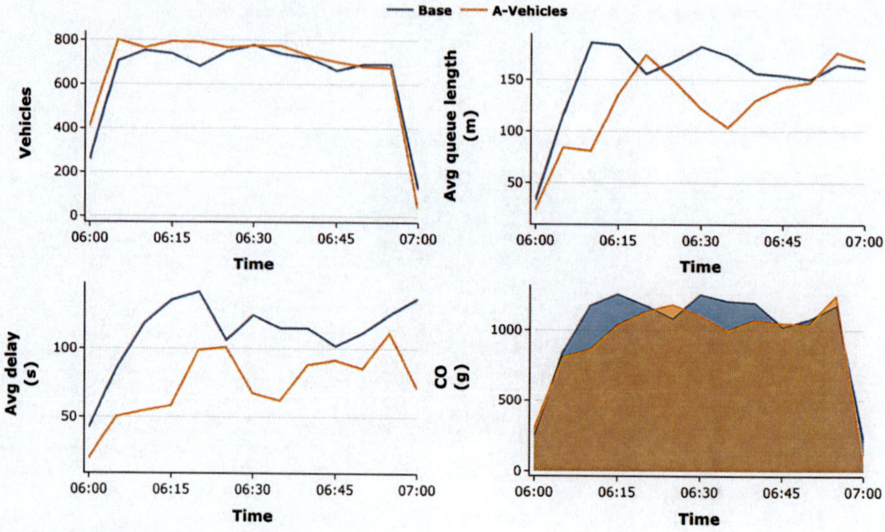

Fig. 5. Dynamic comparison between the Base and the best reward mechanism (A-Vehicles) for the suggested evaluation metrics.

8 Conclusions

Modern traffic management systems in the era of sustainable mobility face the challenge of achieving optimum traffic conditions while maintaining high-level of environmental quality. In this study, an advanced Reinforcement Learning signal control optimisation framework is presented and evaluated. According to the preliminary results obtained from the large-scale urban network of Nicosia, Cyprus, RL driven signal controls, prove highly suitable for the efficient traffic demand management. In detail, the application of multiple implementation scenarios to the road network of Nicosia (19 signalised junctions) resulted in the best-case scenario to significantly improved conditions compared to the currently implemented traffic signalling system. Specifically, the total network throughput was increased by 5% while at the same time significant reductions were achieved for queue lengths (18%), delays at junctions (35%) and CO emissions (8%). Although, further assessment is required, RL based methodologies seem very promising at ensuring clean environmental conditions without disrupting the transport network's performance.

References

1. Papageorgiou, M.: Overview of road traffic control strategies. In: IFAC Proceedings Volumes (IFAC-PapersOnline). IFAC Secretariat, pp. 29–40 (2004)
2. Mannion, P., Duggan, J., Howley, E.: An Experimental Review of Reinforcement Learning Algorithms for Adaptive Traffic Signal Control. In: McCluskey, T., Kotsialos, A., Müller, J., Klügl, F., Rana, O., Schumann, R. (eds.) Autonomic Road Transport Support Systems, pp. 47–66. Springer International Publishing, Cham (2016)

3. Bakker, B., Whiteson, S., Kester, L., Groen, F.C.A.: Traffic light control by multiagent reinforcement learning systems. Stud. Comput. Intell. **281**, 475–510 (2010). https://doi.org/10.1007/978-3-642-11688-9_18
4. Zhong, D., Boukerche, A.: Traffic signal control using deep reinforcement learning with multiple resources of rewards. In: Proceedings of the 16th ACM International Symposium on Performance Evaluation of Wireless Ad Hoc, Sensor, & Ubiquitous Networks - PE-WASUN 2019, pp. 23–28. ACM Press, New York (2019)
5. Urbanik, T., Tanaka, A., Lozner, B., et al.: Signal Timing Manual, 2nd edn. Transportation Research Board (2015)
6. Buşoniu, L., Babuška, R., De Schutter, B.: A comprehensive survey of multiagent reinforcement learning. . IEEE Trans Syst. Man Cybern. Part C Appl. Rev. **38**, 156–172 (2008)
7. Penic, M.A., Upchurch, J.: TRANSYT-7F: enhancement for fuel consumption, pollution emissions, and user costs. Transp. Res. Rec. 104–111 (1992)

Community Participation Towards Sustainability Enhancement of Transportation Sector for Baghdad City

Firas Alrawi[1](✉), Khalid Alwani[2], Hamid Alacash[2], and Seda Mesrop[3]

[1] Urban and Regional Planning Center, University of Baghdad, Baghdad, Iraq
dr.firas@uobaghdad.edu.iq
[2] College of Engineering, University of Anbar, Ramadi, Iraq
{khalid.hardan, hamid.awad}@uoanbar.edu.iq
[3] Department of Civil Engineering, Institute of Technology, Baghdad, Iraq
seda.S.T.C@mtu.edu.iq

Abstract. Population opinions and preferences are one of the requirements that city planners strive to meet. Regulating the vital sectors in the city, including the transportation sector, without paying attention to its residents' viewpoints, may lead to the failure of this organization process.

This research represents an analysis of many aspects of the use of mass transit and automobiles in the city of Baghdad, depending on Baghdad's residents' point of view. In order to determine the most critical aspects that transport planners and traffic engineers should focus on, to developing the transport sector in the city and achieving sustainable urban mobility. To achieve the research goal, 200 questionnaires were distributed to a group of Baghdad residents who had even a slight knowledge of the characteristics of the city's transportation sector. A set of relationships was built based on distributed forms highlighting the essential aspects that city residents want to develop in the transport sector.

Keywords: Community participation · Transport planners · Mass transit and automobile · Sustainable urban mobility

1 Introduction

Planning is a mechanism for organizing and developing societies. The primary intention of the plan is to improve the livelihoods of the population and raise the community's economic, social, environmental, and so. The transportation sector is one of the most vital aspects of cities that contribute to linking its various activities. Since the local communities are the most familiar with their requirements, the quality of services they need, and the main problems in society, the community's involvement in the planning process has become an essential matter.

E. G. Nathanail et al. (Eds.): CSUM2020, AISC 1278, pp. 106–115, 2021.
https://doi.org/10.1007/978-3-030-61075-3_11

1.1 Planning and Community Participation

Generally, the participation of citizens in the decision-making process in contemporary planning is considered fundamental [1]. The population and users of our cities are the beneficiaries of accurate urban planning. Good urban plans are established on an in-depth knowledge of local culture, and requirements. The real participation of the community allows people to be part of urban plans decision-making processes. This kind of partnership strengthens the feeling of owning the places which have had a hand in planning [2]. Sometimes changing the planning procedures is not suitable for everyone, or maybe it conflicts with the short-term interests of a part of society; therefore, it is impossible to judge the behavior from above; it can only be achieved through persuasion [3]. The planner has extensive practical and theoretical experience in planning problems such as transport problems. They need a broader understanding of the local problems that are unique to a particular community. In other words, professional planners need more in-depth insight into the problems of society. Statistical data for noise pollution and various planning problems do not necessarily reflect how residents feel about these difficulties or their proposals to address them [3].

Theoretically, the more the public is participating in the planning process, the possibility to implement planning decisions increases; however, the proper level of public involvement may vary depending on project aims, goals, and available resources from one case to another [1]. Popular participation received significant attention from countries and governments, and as a result of this increased interest was the emergence of the concept of participatory planning, or what is known as bottom-up planning, to distinguish it from planning without participation [4]. Planning and development processes must reflect the needs and desires of the local target group, and imported development models should not be applied unless adapted to the prevailing economic, social, political, and cultural conditions [4].

1.2 Transportation Sector and Sustainability

As the concern for the effects of transportation on quality of life and the environment grew, broader approaches to transportation planning were being developed. This concern was being expressed in the worldwide [5].

During the past two decades, many attempts within the transportation planning, policy, and research communities to use new models to urge population to use the sustainable transportation.

Some view it as an anti automobile or anti highway, but this is not what sustainable transport truly is. The sustainable transport recognizes that until now the automobile is the most common mode for transport in the worldwide. Its most ardent admirers also recognize that this vehicle has many problems. In the transportation system the highway recognized as the main part in this system and its need to improve and rebuilt as needed. However, some of the transportation problems not necessarily solved by increasing the numbers of highway. Thus, the all transport research, planning, and policymaking today attempted to solve the problems of transportation sectors through the perspective of sustainable transportation [6]. The significant increase in the number of vehicles used in the world has a negative impact on the environment since the

Second World War. The car was the reason for switching from standard pedestrian roads to wide roads and thus losing green spaces and increasing socioeconomic costs for road construction and maintenance. The aims for the sustainability of transportation to reduce the need for cars as a source of pollution and loss of social relations [7].

The Problems in Transport Sector. The biggest problem facing the transportation sector is traffic congestion. It's described by Ministers of Transport for European Conference as a "traffic condition in which vehicles are constantly stopping and starting and in which vehicle concentration is high while flow speeds are low". Fig. 1 illustrates the relationship between vehicle density and speed of traffic [6, 8].

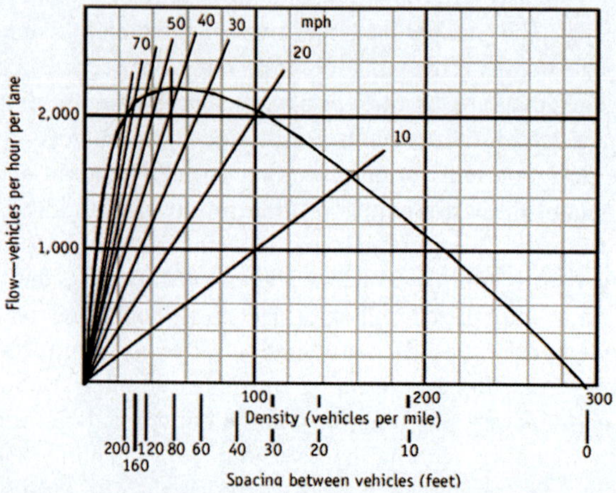

Fig. 1. A highway capacity curve for uninterrupted flow.

The main barrier to sustainable transportation is congestion, apparently, because the resulting impacts are very diverse. Congestion decreased traffic speed, which leads to both lower fuel efficiency and increased emissions detrimental to human health Fig. 2 represents an increase in carbon emissions with a decrease in speed [6].

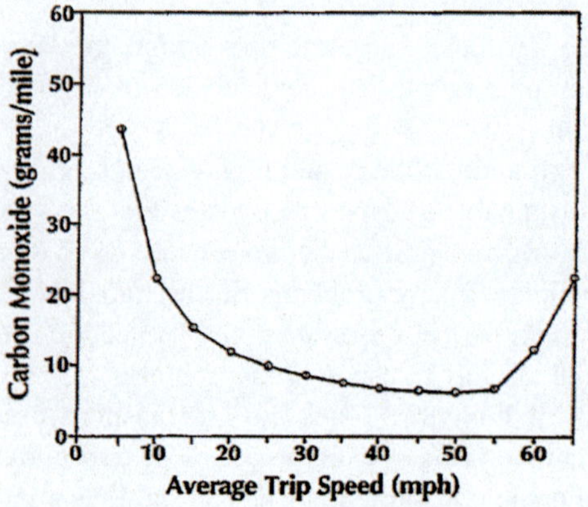

Fig. 2. Generalized carbon monoxide emissions as a function of speed.

Transportation and oil go hand in hand. In the United States the transportation sector consumed more than two-thirds of the petroleum products, while it's consumed about more than half of world oil Fig. 3 [9].

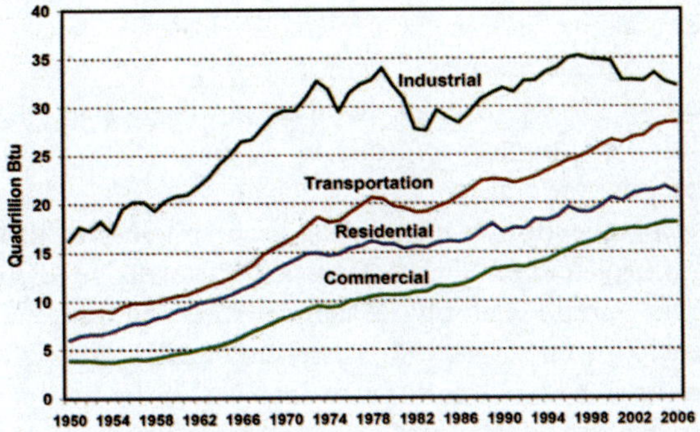

Fig. 3. USA Energy Consumption by Sector, 1950–2006.

Range of Possible Solutions. Many solutions to transport sector problems, some of which are covered by transport planning and traffic engineering, others through the operations of transport and traffic management, and the formulation of various policies, such as congestion pricing, reorganizing traffic on the roads, improving the characteristics of the transportation modes, etc. Land use planning does not directly address significant transport problems sustainability, but if it can manage the activity space over which driving occurs, it can substantially affect emission, fuel consumption, and traffic amount [6]. Solutions like what so-called travel demand management involves a wide range of programs that run the gamut from encouraging greater use of bicycles and more walking by individuals to large-scale global solutions to some of the problems, one specific set of policies like speed and speed limits [9].

The main aim for transportation planning is to provide safe and smooth movement for traffic flow in the network system. The traditional approach for transportation planners to provide a smooth traffic flow is including widening existing roads and building new roads. However, this approach will increase the traffic flow in the network system, especially in situations where traffic demand controls are minimal or absent [10].

The major transportation facilities are already built, and the problem faced by transportation engineers is to operate these systems at their most productive and efficient level. The term used to describe this operational planning process is transportation systems management (TSM) [11]. In many cities, the approach of incrementally adding supply to meet increasing demand is close to reaching its physical limits.

The primary objective of TSM is to create more efficient use of existing facilities through improved management and operation of vehicles and the roadway.

The TSM approach entails planning for the future by managing demand more effectively on the existing road network rather than constructing new road links [12].

2 Analysis of the Transportation Sector in Baghdad

In the city of Baghdad after 2003, complete collapse occurred in the public transportation system, which is still evident through the inability to cover the demand for daily urban trips. Reliance on automobiles has resulted in severe traffic congestion, especially in the city center and sector centers, an increase in taxi hours, and the emergence of many problems associated with it.

The study relied on the data for the population who live in Baghdad city, Iraq collected through 200 questionnaire forms, to determine the priorities they deem appropriate in developing the transport sector for the city. The specializations of the surveyed persons were spread in several fields, such as transport planning, transport engineering, and urban planning, who had extensive knowledge of the transport network in Baghdad. The workplaces for these samples were distributed among university instructors and municipal workers, as well as students in those specializations. Several indicators have been used to relate to the development of the transport sector, which focuses on public transport and transport management methods.

The indicators related to the public transportation sector were analyzed by linking them to the characteristics of the socioeconomic surveyors, while the indicators related to transport management were analyzed according to the importance of their application in the city from the point of view of the study sample.

3 Result of Study

One of the main goals of the transportation planners is to introduce a safe and comfortable transport system for the citizens, therefore, it's important to support the citizens and their community role inside Baghdad city to be part of the planning conception. Thus, we delivered some application forms to many sections of the community to find out there point of views in these problems and solutions to reach the sustainable concept in transport sector by applying the public transit and TSM concept in the city through using several indicators.

3.1 The Terms Related to Public Transit

The form consisted of nine indicators related to transportation currently used by individuals automobile or mass transit (A-T), or for which they may convert in the future after developing the public transport sector, depending on the improvement in the characteristics of public transport for each of these indicators:

X1: The mode of transportation that you use to go to work (A-T)?
X2: The mode of transportation that you use to go shopping and recreation (A-T)?
X3: Your favorite transportation means (A-T)?

If you use a private car, what are the means you will use (A-T)? If:

X4: Improve public transit and increase number of buses to satisfy Baghdad's needs.

X5: Improve reliability of buses arrive at its stations to decrease waiting time.
X6: Creation of a rapid transit system between home and work with an increase in payment more than usual.
X7: Providing air-conditioned buses with higher ride fees than usual.
X8: Provide affordable means.
X9: Providing special transportation systems such as the metro and monorail, which need a distance to walk to and from their stations in the origins and destinations.

We find the relationship between these indicators and several socioeconomic characteristics for individuals and household which was:

Sex: There is a strong relationship between sex and transport means that people prefer as shown in the Table 1 some categories preferred public transit on work trips especially males, and prefer private cars in shopping. In the case of enhancing public transportation, the study indicates an increase in the use of public transportation. But not up to the required level due to the poorly understood concept of public transportation. When we asked about preferred means if public service improved the study indicate public transit.

Table 1. Relation between sex and study prefer for (A-T).

Sex		X1	X2	X3	X4	X5	X6	X7	X8	X9
Males	A	37	58	68	37	18	34	24	29	45
	T	63	42	32	63	82	66	76	71	55
Females	A	47	80	100	47	27	47	76	71	55
	T	53	20	0	53	73	53	24	29	45
Total	A	40	64	77	40	21	38	32	32	36
	T	60	36	23	60	79	62	68	68	64

Age category: In the Table 2 there is preferred to the public transit for the young category and that varies when they got older, which includes work, shopping, and recreation trips.

Table 2. Relation between age category and study prefer for (A-T).

Category		X1	X2	X3	X4	X5	X6	X7	X8	X9
−20	A	40	73	80	53	33	60	53	33	20
	T	60	27	20	47	67	40	47	67	80
21–30	A	21	47	68	26	0	26	21	16	32
	T	79	53	32	74	100	74	79	84	68
31–40	A	46	56	100	46	36	46	36	36	46
	T	54	44	0	54	64	54	64	64	54
41–50	A	80	0	40	80	80	100	100	40	60
	T	20	100	60	20	20	0	0	60	40
51-	A	67	100	67	67	33	33	33	67	100
	T	33	0	33	33	67	67	67	33	0

Household size: As in Table 3 there is a strong relationship between household size and transit use for all purposes trips, especially if we enhance public transit.

Table 3. Relation between household size and study prefer for (A-T).

Household size		X1	X2	X3	X4	X5	X6	X7	X8	X9
−4	A	45	59	73	45	27	41	32	36	55
	T	55	41	27	55	73	59	68	64	45
5–7	A	43	70	83	39	17	35	35	39	26
	T	57	30	17	61	83	65	65	61	74
8-	A	13	63	75	25	13	38	25	0	13
	T	87	37	25	75	87	62	75	100	87

Number of a worker per household: The possibility of using public transit increase in all study indicators where the number of worker per household increased, especially in the work trip. As in Table 4.

Table 4. Relation between number of a worker per household and study prefer for (A-T).

N. of worker		X1	X2	X3	X4	X5	X6	X7	X8	X9
1	A	37	58	74	47	26	47	37	37	37
	T	63	42	26	53	74	53	63	63	63
2	A	43	74	74	43	22	43	35	35	43
	T	57	26	26	57	78	57	65	65	57
3-	A	36	55	91	18	91	91	19	19	19
	T	64	45	9	82	9	9	81	81	81

Average income: A less average income to household meaning more dependent on public transit for all trip purposes. As in Table 5.

Table 5. Relation between average income and study prefer for (A-T).

Average income *1,000,000 ID		X1	X2	X3	X4	X5	X6	X7	X8	X9
−1.5	A	20	48	68	40	16	60	36	24	40
	T	80	52	32	60	84	40	64	76	60
2–2.5	A	57	76	86	38	29	19	29	38	33
	T	43	24	14	62	71	81	71	62	67
3-	A	57	86	86	43	14	14	29	43	29
	T	43	14	14	57	86	86	71	57	71

Car owned: As in Table 6 increasing car ownership leads to depending on private cars with the possibility to use public transit if it enhanced.

Table 6. Relation between car owned and study prefer for (A-T).

Car owned		X1	X2	X3	X4	X5	X6	X7	X8	X9
0	A	0	8	83	33	25	33	17	8	42
	T	100	92	17	67	75	67	83	92	58
1	A	48	76	70	39	15	42	36	36	36
	T	52	24	30	61	85	58	64	64	64
2-	A	62	100	100	50	38	25	38	50	13
	T	38	0	0	50	62	75	62	50	87

3.2 The Terms Related to TSM

The questions in the questionnaire form were "what is your opinion for applying the following TSM policy? And are you supporting this policy?

X10: Controlling traffic for freeways, including changeable message signs.
X11: Reserving the curb lane of arterial roads for buses only.
X12: Reserving an entire street, usually in the CBD, for the exclusive use of buses and/ or pedestrian.
X13: Reserving a lane in the opposite direction of traffic (contraflow lane) when traffic is heavy in one direction and light in the other direction.
X14: Setting up reversible freeway lanes for inbound morning heavy traffic and outbound evening traffic.
X15: Providing separate lanes on freeways for use by buses or high-occupancy vehicles.
X16: Peak hour pricing in which tolls are charged on vehicles that use the area during congested hours.
X17: Automobiles restricted areas that limit the use of cars within the CBD. (one technique is to divide the CBD into zones and restrict travel across zone boundaries. Transit is permitted into the CBD using street rights of way that define the zones).
X18: Vehicle restrictions in residential areas, including stop signs, closed streets, speed bumps, etc.
X19: Curb parking restrictions to reduce the amount of on-street parking.
X20: Off-street parking restrictions in the CBD, such as pricing differentials to discourage all-day parking.
X21: Provide parking spaces for high-occupancy vehicles, to encourage ride-sharing.
X22: Parking rate charges designed to encourage ride-sharing or to limit vehicular traffic.
X23: Parking information for motorists regarding space availability.

X24: Adjusting hours of work schedules for the purpose of dispersing peak hours.
X25: Flextime that permits an employee to begin and end the workday within a flexible time range.
X26: A four-day workweek, based on 10 h/day.
X27: TSM strategies encourage nonmotorized travel by making walking and bicycling safer and more pleasant. Among the most effective means are widening side-walks; providing lighting, benches, and pedestrian lanes; building grade separations such as underpasses or overpasses to avoid conflicts at intersections, building bikeways, and installing pedestrian controls at intersections.

The Table 7 shows that the greatest category from the study understands the importance of applying TSM policy, 71.7% from the study sample for all strategies were choosing Yes to applying this policy, 20.5% picking on No, and 7.8% choosing don't know. If we divided these policies into a category (1) Most wanted. (2) Not wanted. (3) wanted. Depending on the opinions of the questionnaire in the previous table. We can notes that all (X10, X11, X12, X15, X21, X23, X24, X25, X26, and X27) policies variables concentrate on Improving the reality of public transportation, introducing policies related to car-pooling, ride-sharing, developing flexible work schedules, and allocating spaces for pedestrians and bicyclists. While the study sample did not interact positively with policies related to restricting and pricing the movement in the city center and other sectoral centers, shown by variables (X16, X20).

Table 7. The result of TSM indicators application form.

TSM. No	X10	X11	X12	X13	X14	X15	X16	X17	X18
Yes	73%	85%	97%	68%	67%	79%	40%	66%	61%
No	12%	11%	3%	11%	13%	9%	50%	23%	33%
Don't know	15%	4%	0%	21%	20%	12%	10%	11%	6%
TSM. No	X19	X20	X21	X22	X23	X24	X25	X26	X27
Yes	54%	37%	71%	64%	91%	87%	92%	83%	76%
No	46%	59%	21%	23%	3%	9%	5%	14%	23%
Don't know	0%	4%	8%	13%	6%	4%	3%	3%	1%

4 Conclusions

Involving the population in the planning process helps planner to gain a deeper understanding of the problems in the urban environment. The vision for the beneficiaries of the planning process is of great importance to the planner, which contributes to facilitating his mission in developing alternatives and formulating solutions and proposals for that community. The transportation sector is one of the most vital sectors in the city and is closely related to the population and their daily life. This sector is affected by the behavior of the population, which made it difficult for the planner to know the system in which users of the transportation system operate. Especially with all this complexity of the city's transportation system. This study was an attempt to involve Baghdad residents in formulating proposals to organize the transportation sector in the city. The opinions of the residents were consistent with the planning point of view in many aspects. 71.7% of the study sample indicated the importance of applying new models for the urban transport sector. Most of the population supported methods such as the priority of public transit. While they did not support policies related to restricting movement and imposing fees on the right of passage.

Acknowledgments. To institute of international education IIE.

References

1. Kamaci, E.: A novel discussion on urban planning practice: citizen participation. ICONARP Int. J. Archit. Plan. **2**(1), 1–19 (2014)
2. MENZ Homepage. https://www.mfe.govt.nz/publications/towns-and-cities/urban-design-toolkit-third-edition. Accessed 23 Nov 2019
3. Enyedi, G.: Public Participation in Socially Sustainable Urban Development, 1st edn. Hungarian Academy of Sciences, Pécs (2004)
4. Qararih, M.: The mechanism of awareness activation & popular participation in constructional planning at West Bank. Master thesis. An-Najah University, Nablus (2004)
5. Weiner, E.: Urban Transportation Planning in the United States: An Historical Overview, 1st edn. Greenwood Publishing Group, Inc., Westport (1999)
6. William, R.: Sustainable Transportation Problems and Solutions. Guilford Press, New York (2010)
7. Kunstler, J.: Home from Nowhere. Simone and Schuster, New York (1996)
8. William, L., Jerry, D.: Tomorrow's Transportation: Changing Cities, Economies, and Lives, 1st edn. Artech House, Inc., New York (2000)
9. Randolph, J., Gilbert, M.: Energy for Sustainability Technology, Planning, Policy. Island Press, Washington, D.C. (2008)
10. Fwa, T.: The Handbook of Highway Engineering. Taylor & Francis Group, New York (2006)
11. Nicholas, J., Lester, A.: Traffic & Highway Engineering, 3rd edn. Thomson, New York (2002)
12. Rogers, M.: Highway Engineering, 1st edn. Blackwell Publishing Ltd., Malden (2003)

Impact of Congestion Pricing Policies in Round-Trip and Free-Floating Carsharing Systems

Carolina Cisterna(✉), Giulio Giorgione, and Francesco Viti

University of Luxembourg, 2, avenue de l'Université, 4365 Esch-sur-Alzette, Luxembourg
{carolina.cisterna,giulio.giorgione, francesco.viti}@uni.lu

Abstract. Carsharing is a short-period car rental system where being member is the first step to have access to its services. In a previous study a membership choice model of the city of Berlin was estimated using the agent-based model MATSim and discrete choice theory. Results have shown two distinct member profiles: two-way members use the system for primarily pre-planned long trips while free-floating members usually employ the service as a substitute to private cars in pursuing daily activities. Starting from the developed membership model this study investigates how congestion pricing policies affect carsharing mode choice by implementing different price ranges and applying them in MATSim on the synthetic population of Berlin in order to assess the system behaviour during an ordinary day. Results show an increase of carsharing choice with different impact and a decrease of car mode usage. Free-floating service customers decreased their travel distance, as congestion pricing impacts their choices more than travel cost in a daily user's plan, instead they do not change their willingness to access to the service. Two-way customers are not affected by congestion pricing as the service is used as a substitute of both public transport and private car.

Keywords: Membership choice · Carsharing · Congestion pricing

1 Introduction

Carsharing has the potential to satisfy individualized transportation demand in a sustainable and socially beneficial way, increasing social cohesion amongst sharers [1].

The first commercial station-based carsharing in Europe was launched in the 40's in Switzerland, while the service was later developed in Germany in the 80's [2]. Nowdays, carsharing is employed in more than 1100 cities around the world, 26 countries and on 5 continents [3]. Currently there are three distinct types of car sharing systems: round-trip (or two-way), free floating and one-way. The first system has a station-based nature and the users must return the vehicle into the same picking up area. The second type is more flexible, since the users can pick up the car and drop it off wherever they want inside a pre-defined zone. One-way is an intermediate type of

E. G. Nathanail et al. (Eds.): CSUM2020, AISC 1278, pp. 116–126, 2021.
https://doi.org/10.1007/978-3-030-61075-3_12

service between the two, as customers can pick up the car in a station, but they could return it at any other station.

According to the German federal carsharing association, in 2018 round-trip carsharing systems have increased of 17.6% and free-floating systems of about 25%, with respectively 535.000 and 1.575.000 users in Germany [4]. The city of Berlin, for instance, it is possible to find both round-trip and free-floating services.

Previous studies have shown how carsharing can be a reasonable alternative amongst mobility systems by connecting a missing link, reducing private car usage, and reducing emissions [5]. For this reason, policies on different cities have been implemented to disincentivize the use of private cars, and favor shared mobility options. Carsharing is therefore being included as mobility option in municipality policy planning, with carrot solutions like free and/or guaranteed parking, access to limited access zones, exception or discounts of tolls, etc. [6]. It is therefore important to gain insight into the impact of policies to carsharing in order to understand how this service could be used as alternative to the private cars. For instance, a combination of exemptions to congestion pricing and access restrictions to combustion engine cars has been found to be strong a determinant for the carsharing business in the city of Milan. It was reported that, after the implementation of the 'Area C', the number of owned cars in the municipality decreased in favor of an increased number of carsharing members [7].

With the goal of better capturing the impact of sustainable policies to carsharing systems, this study aims to investigate how congestion pricing strategies can affect carsharing mode usage in an ordinary day, starting from typical members' profile for both free floating and two-way systems. We deal with these systems as independent services, since their members have different characteristics.

By simulating daily users' plans using the agent-based software MATSim [8] the authors capture the different members' behaviour. The remainder of the paper is structured as follows: Sect. 2 provides a review of the state of art, in Sect. 3 the methodology is presented, in Sect. 4 the results are analyzed and finally in Sect. 5 the discussion and conclusions are drawn.

2 State of the Art

Several pricing policies have been studied in the literature, mainly using stated preference surveys and travel behavior analysis and modeling.

By a stated preference survey, three different pricing components have been analysed in Switzerland: fuel cost, tolling and parking, and how they affect mode choice [9]. Three different scenarios have been implemented and linked to route, mode and departure time; results have shown how travelers' evaluation of cost and travel time attributes have no linear correlation. Travelers seemed to expect more travel time savings from toll or parking cost because they have not internalized this cost in their budget constraint. Pricing was found to be mostly associated to travel time saving and to be dependent on household income [10].

Dynamic congestion effects on users' car utility has been simulated in a daily users' plan in MATSim, to evaluate how it could affect travel choices [11]. By converting the daily time and queuing in monetary costs and adding these into the utility of the agent, results show a reduction of congestion and a very large increase of efficiency within a car mode by a decrease of the number of trips and the overall travel time.

Moreover, different pricing schemes have been analyzed for three modes: private car, public transit and station-based shared vehicles in [12]. For private cars, the impact of parking fees in the destination areas has been evaluated in relation to value of time and vehicle cost operations linked to travelled distances. Instead for shared vehicles utility, four different pricing strategies have been studied: simple distance-based, origin-based, destination-based, and a combination of the previous plans. By employing a multinomial logit mode choice model in an agent-based framework and combining pricing schemes, results showed that there is a global significant shift from private vehicle use to shared modes. Furthermore, shared modes substitute short trips done by public transit, due to their long waiting time and transit access which have decreased the utility function for this type of trips. The performance metrics also revealed two main impacts: travelers were observed to reduce waiting times and pricing at the expense of longer-distance trips.

Dynamic pricing has been studied in a two-way carsharing system using a synthetic population of the city of Berlin and simulating it by agent-based software MATSim in [13]. The adopted pricing model is dependent on vehicle availability in booking stations and on users' value of time. Results show how dynamic pricing does not affect users with a high value of time and generally population with an average value of time tends to reduce reservation times to keep rental costs low.

There are multiple decisional variables linked to the user's utility function, the most studied in the literature being travel time and distance-based costs [5], but there are also tolls and fares, in other words, fixed costs, which do not depend on any attributes related to trip characteristics. The evaluation of these pricing schemes on users' mode choice for carsharing systems has not been deeply investigated yet. Taking into account a typical profile for carsharing members and considering both congestion prices as fixed cost and travel cost in the private car utility function, this study aims to evaluate the impact of these costs on carsharing modal split, considering two inherently different carsharing service types.

3 Methodology

In this study we adopt an agent-based simulation approach to study the impact of pricing policies to mode choice, in particular focusing on the shift from car and public transport to carsharing. We adopted an agent-based model since a trip-based model (e.g. the 4-stage model) does not allow to provide a temporal and spatial carsharing vehicle locations, and because carsharing is a mode that individuals adopt for specific activity-travel patterns [14].

Typical daily mobility has been simulated using the agent-based software MATSim on the network of city of Berlin [15]. MATSim is an open source program based on Java programming language, designed to implement large-scale transport agent-based simulations. This open source software is capable of simulating daily mobility patterns at a microscopic level using information about daily schedules. The software executes the agent plan and calculates a scoring value according to the utility function of each transport mode. The score usually increases if the utility of time spent by users performing their activities increases whereas the time spent travelling decreases the utility. Hence, the time spent in traffic is a disutility because the agent is not able to perform her/his activities. Agents choose the preferred mode(s) for reaching their activities, and they re-evaluate their choices based on the outcomes of the previous days in a day-to-day learning process.

This paper starts from a synthetic population and a congestion pricing policy as input and runs a daily simulation to reach stochastic equilibrium using MATSim, which generates carsharing performance indicators such as travel time, accessibility, mode choice and trip purpose (see Fig. 1).

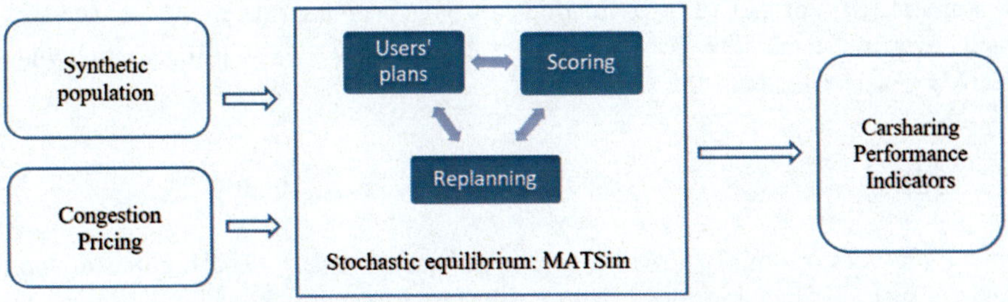

Fig. 1. Flow diagram of the adopted methodology.

This study uses a synthetic population representing 10% of the population of the city of Berlin with homogeneous characteristics previously estimated and available on web [16]. The dataset contains the users' daily activities plan, their socio-demographic attributes and their car availability. Moreover, every agent is assumed to have accessibility to all transport modes, in particular every user is a member of both the carsharing systems, no restrictions have been implemented, but 1000 m has been set as threshold for the willingness to search for a car sharing station.

The simulation has been run with 500 iterations in order to reach a Stochastic User Equilibrium, i.e. the users' score does not change anymore in the last iterations.

Congestion pricing has been implemented by considering the whole city of Berlin as a congestion pricing area following a realistic variation of a real congestion pricing scheme employed in Europe (i.e. in Milan and London [17]): citizens have to pay a fixed daily toll if they want to use their own car in a specific area. This congestion price has been considered in the private car utility as a constant together with the travel cost already implemented in MATSim [17], since the daily fee impacts private car choice while carsharing systems are assumed in this study to be exempted by this policy. By adding this daily congestion price to car mode we aim to model the increase in the disutility of private car usage and analyze how this can impact the carsharing usage.

In this study 3 scenarios have been implemented as it is shown in Table 1.

Table 1. Scenarios.

Scenario	Daily toll
0	0 €/day
2	2 €/day
8	8 €/day

The congestion pricing range has been imposed considering the current monthly prices to access low-emission zones in the city of Berlin according to the "Senate Department of the Environment, Transport and Climate Protection", and unifying these fees in a daily fee for all the categories of private vehicles [18].

Furthermore, 61 stations have been simulated with two cars per each station according to [19], instead 121 free floating cars have been considered in a service area which covers more than the city center area of Berlin, according to [20] and an unique price has been set for both the services.

4 Results

Table 2 provides a summary of the results per each scenario. The number of total bookings seems to increase more significantly in the scenario 8, while the impact in scenario 2 is not evident. This applies to both free floating and two-way services: the

Table 2. Summary of results.

Variable	Scenario 0	Scenario 2	Scenario 8
Population	273232	273232	273232
Number of carsharing stations	61	61	61
Number of free-floating vehicles	106	106	106
Number of two-way vehicles	122	122	122
Numbers of users:			
Total bookings	907	940	1304
Two-way bookings	62	81	122
Free floating bookings	845	859	1181
Two-way system:			
Number of trips	167	227	367
Average trip duration (hours)	1.70	1.22	1.27
Total travel time (hours)	105	97	153
Total distance (km)	2763	3462	5520
Free-floating system:			
Number of trips	845	859	1181
Average trip duration (hours)	0.87	0.68	0.51
Total travel time (hours)	734	582	603
Total distance (km)	15890	15300	16849

number of trips made by two-way users is almost three times greater than the number of bookings, whereas free-floating bookings are equal to the number of trips.

The average trip duration for both services decreases with increasing fees, while the total travel distance does not vary significantly in the scenario 2 for free-floating, instead it significantly increases in scenario 8, whereas travel distance in two-way systems increases steadily.

The simulation is considered well representative of a typical day since the results of the two-way are comparable with data provided by a carsharing company operating in Berlin, for which the typical booking numbers are 65 with about 150 cars [21], instead for the free floating service we do not have specific data to compare.

A more specific evaluation of performance indicators has been performed: travel time and accessibility cumulative distribution for both services, mode choice and trip purpose.

Analyzing the differentials between scenarios 8 and 0 of the cumulative function of distances traveled an increase of about 10% of the probability of travelling less than 30 km has emerged, whereas between scenarios 0 and 2 there not such type of difference (see Fig. 2). Free-floating users have increased their usage, but they have reduced their traveled distances in scenario 8. Instead, congestion policies seem not to impact two-way traveled distances, since there is no evident raise among scenarios (see Fig. 2).

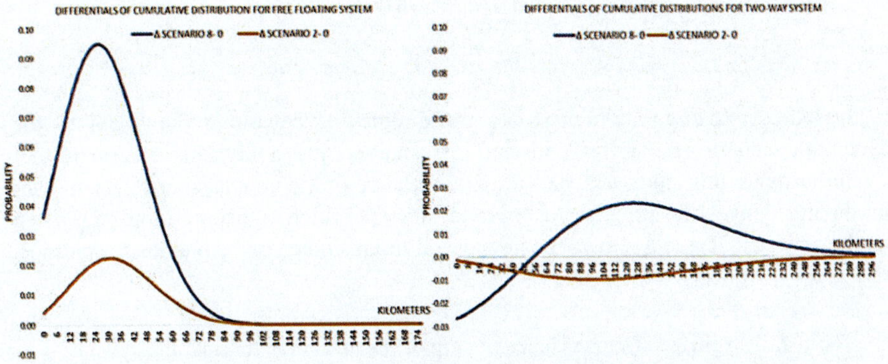

Fig. 2. Differentials of cumulative distributions of travel distance for both services.

Looking at the differentials of in terms of accessibility, measured as the time between reservation time and when users physically access the car, either free-floating and two-way users are found not be apt to walking more to access the service cars and the stations (see Fig. 3). Concerning mode choice, Table 3 provides the percentage of mode choice of the population for each scenario. Car mode is globally decreased among scenarios 2 and 8, instead all the other transport modes are increased after introducing congestion pricing.

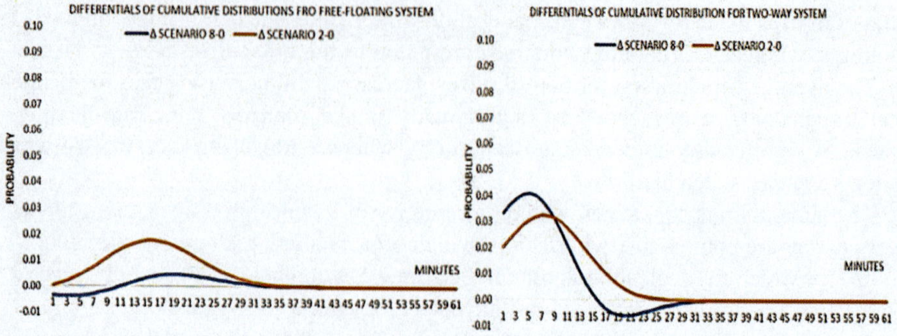

Fig. 3. Differentials of cumulative distributions of accessibility for both s services.

Table 3. Global mode choice per trip.

Mode choice (%)	Scenario 0	Scenario 2	Scenario 8
Car	44.0	38.7	28.0
Pt	29.5	30.4	40.2
Walk	12.6	15.0	16.6
Bike	13.8	15.8	19.0
Carsharing:	0.10	0.10	0.20
Two-way	0.02	0.03	0.04
Free-floating	0.08	0.09	0.13

Overall, carsharing mode increases the double in scenario 8 compared to the benchmark scenario (scenario 0), instead in scenario 2 just a slight increase is found.

Furthermore, this study analyses the mode choice of the new carsharing users after introducing congestion pricing, in order to analyze which transport mode has been mainly impacted, Table 4 shows the new users' mode choice per trip among scenarios.

Table 4. Percentage of mode choice per trip of the new users for both the systems.

Scenario	Car	Pt	Bike	Walk	Free-floating
0	3	61	3	4	30
2	3	59	2	6	30
8	4	59	3	6	29
Scenario	Car	Pt	Bike	Walk	Two-way
0	20	14	15	5	46
2	16	14	11	11	48
8	15	17	12	8	48

Free-floating mode is employed about 30% of the daily trip in each scenario together with public transport which has more than 60% of usage followed by walk, car and bike. These latter shares indicate that these modes are the most impacted by congestion pricing among new free-floating users. On the other hand, we cannot identify an incisive difference per scenario except for two-way system which is chosen almost 50% in every scenario, these percentages confirm how two-way customers are willing to keep the car for more than one trip or longer than other modes. The number of both carsharing members has not been reported because it has a no significant change.

In addition, we analyze which transport mode has been substituted by carsharing (see Fig. 4). We extracted the data relative to new users (i.e. those agents that did not use previously the carsharing mode) to capture how their mode choice is affected by congestion pricing. Private car choice is the mode which has been reduced most in both scenarios, that indicates free-floating users have employed the service as private car substitute, instead bike and public transport are not affected by congestion pricing. Whereas, two-way system ha substituted private car but also public transport mode in both scenarios, that indicates that the service captures two different kind of customers, because it is employed by different types of travelers with different travel patterns.

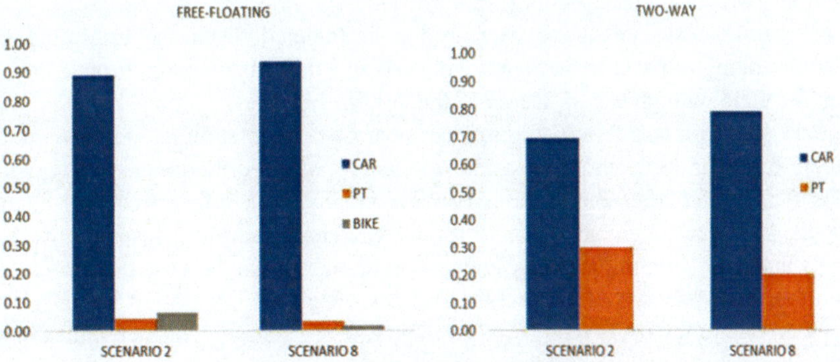

Fig. 4. Mode substituted by carsharing.

Furthermore, car sharing trip purpose for both systems has been analyzed. We observed that congestion pricing does not seem to affect users' trip purpose. The two-way system final destination and origin is more than 50% home, that means users mainly start their trip from home, indicating that this carsharing service is not used for commuting. The free-floating service seems to be used for all purposes but more than 45% for leisure and 40% for returning back home.

5 Discussion and Outlook

This study investigates how a congestion pricing policy can affect carsharing mode choice in a daily simulation for both two-way and free-floating services and by using two pricing values. This paper takes as assumption the typical profile of members estimated in a previous study [22].

The round-trip members have been found to use the service for a long daily trip with at least two intermediate destinations and when they do not have always their own car available. Whereas, free-floating members seem to have always their own car available and they mostly have a point to point trips with travel time of at least 15 min. They are apt to use the service as a substitutional of private car.

Since free-floating customers use the service as a substitute of their own car, it seems reasonable that they are more positively impacted by congestion pricing policies since the free-floating service allows avoiding paying the fixed daily fare, whereas round-trip customers usually seem not prone to increase their carsharing usage after including pricing policies. Moreover, they already know their daily budgets since they have planned their trips and their costs. Looking at Table 2, the results confirm the assumption that the number of the trips made by two-way users is higher than bookings, which means that users have at least one intermediate destination during their trips, instead the number of booking and trips with free floating are as much as reservations, which indicates point-to-point trips.

Free-floating customers have decreased their traveled distances, indicating that congestion pricing impacts more than travel costs in a daily user's plan, instead they do not change their willingness to access to the service.

The free-floating users' behavior reflects a car ownership attitude, i.e. they want to find their own car as close as possible to their origin. Furthermore, mode choice analysis before and after introducing congestion policies confirm that free-floating customers substitute their own car with the carsharing service. Instead, two-way members have not been impacted by congestion pricing, since they use the service as a substitute of both private car and public transport. A global increase of the usage has been shown due to the users who shifts from private car, but performance indicators of the service have not experienced by new policies.

This study has confirmed the members' behavior of both systems previously estimated. The main limitations of the current study are the assumed unconditional accessibility of the services, which has been granted to everyone, the number of cars per station has been set equal to 2 without considering a possible supply-demand unbalance. Moreover, congestion pricing has been taken as fixed cost in the whole city of Berlin and it has not been diversified per zones. Considering specific zones and applying a congestion pricing policy, free-floating can strongly reduce the private car ownership in a long-term horizon. Furthermore, considering the worldwide evolution of electric carsharing systems, the policy applied in this paper could also contribute to the reduction of private car purchasing [23] in an electric carsharing scenario. Policymakers and carsharing system can definitely contribute to find a rebalance of the number of drivers and cars in congested urban areas.

Considering a system as Mobility-as-a-Service (MaaS), which means gathering shared transport modes in just one membership, with the massive interest that is arising [24], the long-term goal of this research will be to implement and study a MaaS membership choice model. In this research direction, future steps will investigate users' attributes of different shared transport modes to identify the common variables to individualize the most suitable MaaS package to offer.

References

1. Münzel, K., Boon, W., Frenken, K., Vaskelainen, T.: Carsharing business models in Germany: characteristics, success and future prospects. ISeB **16**(2), 271–291 (2018)
2. Heilig, M., Mallig, N., Schröder, O., Kagerbauer, M., Vortisch, P.: Implementation of free-floating and station-based carsharing in an agent-based travel demand model. Travel Behav. Soc. **12**, 151–158 (2018)
3. Shaheen, S.A., Cohen, A.P.: Carsharing and personal vehicle services: worldwide market developments and emerging trends. Int. J. Sustain. Transp. **7**(1), 5–34 (2013)
4. bcs Bundesverband CarSharing Homepage. https://carsharing.de/alles-ueber-carsharing/carsharing-zahlen/carsharing-staedteranking-2017. Accessed 3 Jan 2019
5. Shaheen, S.A., Cohen, A.P.: Worldwide carsharing growth: an international comparison. In: 94th Conference in Transportation and Research Board, Washington, D.C. (2007)
6. Guideline for municipalities and governments Homapage. https://itcarsharing.it. Accessed 27 Sept 2019
7. Laurino, A., Grimaldi, R.: The Italian way to carsharing. TeMA J. Land Use Mob. Environ. **5**(3), 77–90 (2012)
8. MATSim Homapage. https://matsim.org/. Accessed 27 Jan 2020
9. Hilgert, T., Kagerbauer, M., Schuster, T., Becker, C.: Optimization of individual travel behavior through customized mobility services and their effects on travel demand and transportation systems. Transp. Res. Procedia **19**, 58–69 (2016)
10. Vrtic, M., Schüssler, N., Erath, A., Axhausen, K.W.: Route, mode and departure time choice behaviour in the presence of mobility pricing. In: 86th Conference in Transportation and Research Board, Washington, D.C. (2007)
11. Kaddoura, I., Kickhofer, B.: Towards an agent-based marginal social cost approach. J. Transp. Econ. Policy **49**(Part 2), 200–218 (2015)
12. Chen, T.D., Kockelman, K.M.: Management of shared, autonomous, electric vehicke fleet: implications of pricing schemes. In: 94th Conference in Transportation and Research Board, Washington, D.C. (2015)
13. Giorgione, G., Ciari, F., Viti, F.: Availability-based dynamic pricing on a round-trip carsharing service: an explorative analysis using agent-based simulation. Procedia Comput. Sci. **151**, 248–255 (2019)
14. Balac, M., Ciari, F., Axhausen, K.W.: Carsharing demand estimation: Zurich, Switzerland, area case study. Transp. Res. Rec. J. Transp. Res. Board **2536**, 10–18 (2015)
15. Horni, A., Nagel, K., Axhausen, K.W.: Introducing MATSim. In: The Multi-Agent Transport Simulation MATSim, pp. 3–8. Ubiquity Press, London (2016)
16. Ziemke, D., Kaddoura, I., Nagel, K.: The MATSim Open Berlin Scenario: an openly available agent-based transport simulation scenario based on synthetic demand modeling and Open Data. In: 8th International Workshop on Agent-based Mobility, Traffic and Transportation Models, Leuvens. Elsevier (2019)

17. Rotaris, L., Danielis, R., Marcucci, E., Massiani, J.: The urban road pricing scheme to curb pollution in Milan, Italy: description, impacts and preliminary cost–benefit analysis assessment. Transp. Res. Part A Policy Pract. **44**(5), 359–375 (2010). https://doi.org/10.1016/j.tra.2010.03.008
18. Senate Department for the Environment, Transport and Climate Protection Homapage. https://www.berlin.de/sen/uvk/en/. Accessed 16 Mar 2020
19. Stadtmobil Berlin Homepage. https://berlin.stadtmobil.de/. Accessed 16 Mar 2020
20. car2go Homepage. https://www.car2go.com/. Accessed 16 Mar 2020
21. Oply Homepage. https://www.oply.com/. Accessed 16 Mar 2020
22. Cisterna, C., Giorgione, G., Cipriani, E., Viti, F.: Supply characteristics and membership choice in round-trip and free-floating carsharing systems. In: 6th Conference in Models and Technologies for ITS, Krakow. IEEE (2019)
23. Firnkorn, J., Müller, M.: Free-floating electric carsharing-fleets in smart cities: the dawning of a post-private car era in urban environments? Environ. Sci. Policy **45**, 30–40 (2015)
24. Utriainen, R., Pöllänen, M.: Review on mobility as a service in scientific publications. Res. Transp. Bus. Manag. **27**, 15–23 (2018)

Spatiotemporal Diversifications of Urban Activities and Travels in Egaleo Municipality, Attica Region

D. G. Perperidou(✉) and M. Sfakianaki

University of West Attica, Egaleo, Greece
dgperper@uniwa.gr

Abstract. This paper presents diversifications of systematic activities and correlated systematic travels, within 12 years a period. As systematic activities and correlated ssystematic travels are defined those urban activities and travels that are frequently repeated, have same start and end time, travels have same origin and destination, using same transport means and activities take place in the same spatial location. Data derive from qualitative questionnaire survey, Systematic Activities – Travels (SAT) Survey, specifically designed for easy and fast implementation, resulting to large volumes of data on systematic activities and correlated systematic travels characteristics. As 2004–2005 initial SAT Survey covered the Greater Athens Metropolitan Area and for the 2018 survey Egaleo was selected as a low income, high population density. Municipality facing the outcomes of Greece's recent economic crisis. Egaleo is also a typical example of Metropolitan's Area deprived western region. Research results indicate significant changes in transport mode choices, travels' length and duration, as well as changes to activities' participation, due to the economic recession. At the same time, the Metro Line 3 expansion and the operation of the new Egaleo Metro Station has a significantly positive impact in systematic activities participation choices, especially in the economic recession period.

Keywords: Systematic urban travels · Systematic urban activities · Diversifications

1 Introduction

The complexity of cities' structure and functions affects urban activities participation and urban travels. The choice of an individual to participate in an activity, which takes place in a specific location at a specific time, affects transport mode choice and travel duration. The activity – based approach launched in 1954 by Mitchell and Rapkin's landmark study established the link between activities participation proscess and urban travel characteristics [1]. In 1970 Hägerstrand described constraints that influence and determine individual selections and participation in activities in time and space [2]. Chapin identified particular spatiotemporal activity – travel patterns [3], while Jones argued that the need for urban travels derives from the need to participate in activities

E. G. Nathanail et al. (Eds.): CSUM2020, AISC 1278, pp. 127–137, 2021.
https://doi.org/10.1007/978-3-030-61075-3_13

[4]. According to Jones et al. travel is an essential component of activities participation and decision on urban travels originating from an activity participation sequence [5].

Daily participation in activities is part of an overall way of life, thus activity and urban travel planning has a specific weekly frequency [6, 7]. Short term activity scheduling and rescheduling process, affects travel choices [8, 9], while individuals should be asked for their typical activity programs [10]. The need for more field data to identify activities – travels inter-actions is always a necessity [11, 12].

Public transport choice can be affected by residence change and free public transport tickets offer [13]. While there are no significant diversifications on personal, business and free time activities, men's journey distance to work is greater than women's, while women spent more time for shopping than men [14]. Non-work activities for women are more intense than those of men within 8 km from home [15]. Walking for educational activities is dominant among youngsters, while young women walk more for shopping activities and older men walk more for entertainment [16]. Besides gender and activities type (work and non-work activities) various socio-economic characteristics like income, residence area, number of available cars per household, affect transport mode selection and travel distance[17–21].

2 Method

2.1 Systematic Activities and Travels

Urban activities have a specific scheduling process and daily, weekly or monthly – systematic - repetition frequency, as part of citizen' way of life. Based on the methodology developed for the record and analysis of systematic urban activities and correlated travels by Perperidou [21], two travels correspond to a frequently repeated, systematic, urban activity: one for going to activity's location and one for returning from it.

By the term systematic activities and correlated systematic travels are defined those urban activities and travels that: i) are frequently repeated within the week, ii) have same start and end time, iii) travels have same origin and destination, using same transport modes and iv) activities take place in the same spatial location and residence is the departure and return destination for the majority of systematic travels, to and from an activity's location.

Systematic activities and correlated systematic travels are affected by various parameters and city functions, such as: residence's spatial allocation and its distance from the location of individual's activities, spatial distribution of activities locations both for work and non-work activities, activities start and end time (day-time), transport networks' availability and provided services, selected transport modes for systematic travels, travel day-time and travel duration, working timetables, household-members relationships (especially in cases of protected members, like children and elderly people) and household socioeconomic background (income, availability of private car, motorcycle or bicycle).

Systematic activities and correlated systematic travels are divided into:

- Programmed – compulsory: work and education, which are repeated usually 5 times per week and take place mainly during working days, Monday to Friday;
- By choice (non – work activities): shopping, sports, entertainment, visiting friends or social gathering, public or private services transactions activities (tax office, banks etc.) and others activities like person accompany, church etc. Compered to programmed – compulsory their repetition frequency varies from 1 to 3 times per week and they take place not only during week days but also in weekends.

2.2 Survey Method

A specialized qualitative questionnaire survey, based on interviews, the Systematic Activities Travels Survey (SAT Survey) is designed for systematic activities and correlated systematic travels data acquisition [21]. SAT Survey's advantages are:

- Simultaneous record of systematic activities and correlated systematic travels characteristics, such as: start and end time, activity type and location, used transport mean (or means), etc.
- Systematic activities – travels frequency record,
- Easy and fast Questionnaire fill in.

To facilitate data collection, process and analysis and in order to minimize interview time, systematic activities – correlated systematic travels are divided into 9 categories and sub-categories (Table 1).

Table 1. Systematic activities–correlated Travels Categories/Sub-categories.

Systematic activities-Travels category	Systematic activities–Travels sub - categories
Work	(No sub-categories)
Shopping	Daily shopping: supermarkets, kiosks, street market, groceries, etc.
	Long term shopping: Clothing, books, electronics, cosmetics, household equipment, etc.
Education	Nursery – kindergarten, primary school, high school, technical high school, learning center, professional training institute, technological institute, university
Health	Visit to doctor, pharmacy, rest home, etc.
Entertainment	Coffee shops, cinema, theater, concert, restaurants/tavern, night club/bar, sport event (spectator)
Social	Visit to friend's house, cultural events, private club, participation to political party
Sports/Leisure	Walk in park/open spaces, playground, exercise sports, gym
Transactions with Public Services/Banks	Bank, tax office/public pay office, ministry/municipality/prefecture services
Other	Dog walk, hairdresser's/beauty salon, church, person accompany (in case of children and elderly)

The SAT Questionnaire Survey Method, combines and uses elements from travel – based, activity – based and time-use surveys, so thus this combination facilitates large collected data volumes regarding usual citizens' activities and travels patterns, in respect to qualitative social research principles and classification [22].

For each systematic activity and its correlated systematic travels their exact start & end time is recorded (in day-time basis), along with their repetition frequency. The end time of a travel to an activity's spatial location is activity's start time, while this location corresponds to travel's destination. Furthermore, activity's end time indicates return travel's start time and activity's location corresponds to return travel's origin. Residence's spatial location is origin and destination of an activity's forth and return travels respectively, unless otherwise stated. Any outdoor time spent, during participation to activities, is recorded. For travels, walking time from/to car or public transport stations is also recorded. Participants' demographic and social characteristics (age & gender, marital status, residence address, education level, professional status and annual income) and participant's household characteristics (number of members, household members' age & gender, education level, professional status and financial level, residence status - owned or rented, cars' ownership - number, type & CC, motorcycle ownership - number, type & CC) are also recorded for further statistical analysis and sample classification.

Implementing SAT Survey, systematic activities – travels characteristics are easily collected and analyzed. Intercorralations between systematic activities & correlated travels and participants sociodemographic characteristics are also determined, thus is facilitated understanding of citizens activities, travel pattern (or patterns) and their overall way of life [16, 23]. Survey's implementation to smaller or larger urban areas or to smaller or larger population groups, generates volumes of information, that can be used to evaluate transport modes in various periods, travel choices, or even exposure to outdoor air pollution [16, 21, 23].

3 Results

The data used herein were collected during 2004–2005 SAT Survey [21] and 2018 SAT Survey, Sfakianaki [24], conducted in Egaleo Municipality.

Egaleo Municipality is located four kilometers west of Athens city center, in Greater Athens West Regional Unit. The West Regional Unit is characterized as low living standard Unit. Egaleo area is 650 HA, out of which 112 Hα form the Eleonas industrial Area and another 134 Ha the Baroutadiko Park and is crossed by five major urban high ways (Fig. 1). Both campuses of the University of West Attica (with over 10.000 students) are located within municipality's administrative boundaries. The Egaleo Metro Station began its operation in May 2007 and is part of Metro Line 3 extension to Greater Athens West Regional Unit. Prior to Metro Line 3 extension the West Regional Unit was not properly served by public transport and in combination with the poor road network, faced heavy road traffic. Egaleo is characterized as low financial level (annual income/household €6,000 to €20,000), high population density (106 residents/hectare in total area, 404 residents/hectare in urban area) municipality.

Its population dropped to 69400 residents in 2011 [24] from 74046 residents in 2001 [25], probably as a result of Greece's recent economic recession and domestic or international migration.

Fig. 1. Map of Greater Athens Metropolitan and Egaleo Municipality

The 2004–2005 SAT Survey covered Greater Athens Metropolitan Area and was conducted right after 2004 Athens Olympic Games, when Greece's economy was strong and growing. The 2008 Global Financial Crisis led to Greece's economic recession, which affected low living standards areas, like West Regional Unit (of Athens Metropolitan area), further deteriorating living conditions in such urban areas. In this context for the 2018 SAT Survey, Egaleo Municipality was selected in order to investigate changes in activities and transport habits prior to and within the economic crisis, as well as the impact of Metro Line 3 and Egaleo Metro Station operation, especially in the era of the economic recession.

Both 2004–2005 and 2018 SAT surveys, include detailed data on systematic activities and correlated systematic travels. Activities locations and travels origins and destinations were geocoded. Precise travels distances were calculated, along with precise calculation of travels duration and activities participation duration correlated on day-time basis.

From 2004–2005 SAT Survey 291 systematic activities corresponding to 582 correlated systematic travels referring to Egaleo were extracted, from 54 participants (out of 5456 systematic activities corresponding to 10912 systematic travels from 1081 participants), while in 2018 Survey 480 systematic activities and 960 correlated systematic travels were recorded, 65 participants, classified according demographic and socioeconomic characteristics Table 2.

Table 2. SAT Survey 2004–2005/2018 participants' demographic characteristics.

Variable	SAT Survey 2004–2005	SAT Survey 2018
Participant characteristics		
Gender: Male/Female	54.3%/45.7%	52.3%/47.7%%
Marital Status: Single/Married	55%/45%	70.7%/30.3%
Age: 0–19/20–34/35–64/65+	15.8%/45%/32.3%/ 6.9%	14%/46.9%/30.4%/8.7%
Household characteristic		
No household members: 1/2/3 4/5/6	17.5%/8.9%/24.4% 45.7%/3.4%/0%	7.7%/21.5%/33.8% 30.8%/3.1%/3.1%
No of cars: 0/1/2/3	14.4%/41.9%/34%/ 9.6%/	26.15%/56.92%/13.85%/ 3.08%
No of motorcycles: 0/1	90%/10%	37.7%/64.3%
Residence ownership status: rented/owned	15%/85%	45.25%/54.75%

Research results, revealed major diversifications in systematic activities and correlated systematic travels between 2004–2005 and 2018, as well as significant changes in social and demographic characteristics of participants' households. There has been a huge decrease in car ownership, while motorcycle ownership increased, indicating efforts to minimize systematic travels costs and expenditures. 2004–2005 SAT Survey systematic activities/correlated systematic travels mean weekly repetition frequency is 2.51 (2.64 men/2.41 women), while in 2018 SAT Survey repetition frequency is 2.65 (2.48 men/2.83 women), Table 3.

Table 3. Weekly frequency per systematic activity and gender.

Activity	SAT Survey 2004–2005		SAT Survey 2018	
	Men	Women	Men	Women
Work:	3.93	4.27	5.17	4.92
Shopping:	3.00	2.23	2.52	2.33
Education:	3.93	3.77	3,50	4,20
Health:	1.00	0.00	1,00	1,00
Entertainment:	2.12	2.06	1,52	1,33
Social:	1.67	2.12	1,39	1,71
Sports/Leisure:	2.43	1.00	2,29	2,25
Transactions Public Services/Banks:	1.00	1.33	1,14	1,64
Other:	1.20	1.44	4,80	6,00

Significant diversifications both for compulsory and non-compulsory systematic activities between 2004–2005 and 2018 are also revealed. Shopping (daily and other) is the dominant systematic activity, along with entertainment (Table 4). Systematic activities for work, shopping and health were increased, while those for entertainment and social gathering were dramatically decreased.

Table 4. Distribution of systematic activities.

Variable	SAT Survey 2004–2005	SAT Survey 2018
Work	8.93%	19.45%
Shopping Daily	13.06%	17.77%
Shopping other	9.62%	10.14%
Education:	18.56%	16.68%
Health	0.69%	2.45%
Entertainment	23.02%	14.70%
Social	15.81%	4.14%
Sports/Leisure	6.19%	8.03%
Transactions Public Services/Banks	2.06%	1.97%
Other	2.06%	4.67%

The need to support family income resulted to systematic activities for work increase, while participation in entertainment and social activities has decreased due to their expenditures (travel and service costs). Furthermore, in the case of systematic entertainment activities, there is a considerable drop of their repetition frequency. The drop of educational systematic activities could be attributed to inadequate household financial means, especial for non-school and personal training. In the case of systematic activities for health purposes there has been a notable increase due health care services development, within a 500 m radius from Egaleo Metro Station. The increase of sports/leisure and other (which mainly includes dog walk) systematic activities indicates a growing trend for participation to low cost (or zero cost) non-work systematic activities, which at the same time offers the opportunity for exercise either in organized sports facilities (mainly outdoor) or in open spaces. The 134 hectares Baroutadiko Park is a characteristic recreational attraction pole, not only local but also regional, providing the opportunity for low cost outdoor exercise and walk in frequent basis.

According to 2004–2005 Survey, travelling on foot and travelling by car (driver & passenger) were the two dominant transport modes. The situation is completely reversed on 2018 Survey, travelling on foot and Metro are the two dominant transport modes (Table 5).

Table 5. Systematic travels modal split

Systematic travels modal split	SATS 2004–2005	SATS 2018
Travelling on foot	36.42%	46.77%
Bicycle	0.68%	0.00%
Motorcycle	2.10%	5.23%
Motorcycle/Metro	0.00%	0.50%
Private car passenger	32.60%	8.55%
Private car driver	10.30%	4.28%
Private car driver/Metro	0.70%	1.66%
METRO	0.70%	19.71%
METRO/Public buses	3.40%	1.43%
Public buses	11.00%	11.40%
Taxi	2.10%	0.48%

Due to Greece's economic recession there has been a tremendous drop of private car modal split between 2004–2005 and 2018. From 42%, in total, for both private car driver and passenger, in the 2004–2005 Survey, private car use dropped to 15% in 2018 Survey. On the other hand operation of Egaleo Metro Station and metro line, resulted to sharp increase of Metro modal split and to new modes Motorcycle/Metro combination, which was recorded on 2018 Survey. The expansion and operation of Metro network at Greater Athens West Regional Unit, facilitated inexpensive and longer distance travels for low income residential areas.

Mean travel distance for all transport modes was increased for the majority of motorized modes among 2004–2005 and 2018 Surveys (Table 6), mean travel time duration was decreased (Table 6). The operation of Egaleo Metro Station, after 2007, provided quick and easy access to low cost and long distance travels, while the

Table 6. Systematic travels mean distance & mean duration

	SATS 2004–2005	SATS 2018	SATS 2004–2005	SATS 2018
	Systematic travels mean distance (m)		Systematic travels mean duration (min)	
Travelling on foot	817.5	688.2	8.3	11.3
Bicycle	2591.0	0.0	15.0	0.0
Motorcycle	5979.3	4372.4	25.0	12.5
Motorcycle/Metro	0.0	6341.2	0.0	30.0
Private car driver	5247.6	6595.7	22.4	17.3
Private car pass	2531.0	7621.2	35.0	25.0
P.Car driver/Metro	5233.7	4950.1	40.0	36.1
Metro	4284.7	8532.8	35.0	29.4
Metro/PB	6374.5	6250.6	65.5	44.9
Public buses	5038.4	5554.3	39.4	27.8
Taxi	3530.3	6119.5	20.0	16.6

decrease of private car use, resulted to less traffic along road network and thus faster trips. There was no significant change in public buses modal split between 2004–2005 and 2018. Systematic travels with public buses distance was increased by 10%, while their mean travel time was drop by 25%, indicating that the decrease of car use has a positive impact on public buses travel time.

Table 7. Systematic travels mean walking time per mode

Variable	SAT Survey 2004–2005	SAT Survey 2018
Systematic travels walking time per mode		
Bicycle	2.00	0.00
Motorcycle	3.33	0.00
Motorcycle/Metro	0.00	5.00
Private car driver	3.85	0.19
Private car passenger	2.00	1.00
Private car driver/Metro	15.00	8.21
METRO	15.00	10.50
METRO/Public buses	12.50	9.67
Public buses	8.44	7.28
Taxi	5.00	1.25
Total mean	6.71	5.38

Walking time to and from motorized means decreases in 2018 compared to 2004–2005 Survey. In the case of public transport, walking time decrease indicates that the expansion of Metro lines, combined with local public buses routes rescheduling and increase, facilitates the fast and easy accessibility to public transport stations. In the case of private cars and motorcycle, ownership reduction provides easier access to parking place closer to both departure (mainly residence) and destination locations (Table 7).

4 Conclusions

The majority of urban activities has a specific repetition frequency, thus are systematic. Systematic activities are correlated to systematic travels and have specific characteristics, which are recorded by specialized quantitative questionnaire survey, the Systematic Activities – Travels Survey (SAT Survey). This research presents the outcomes of SAT Survey in a low income, high populated municipality in two periods: 2004–2005 right after 2004 Athens Olympic Games and prior to economic recession and 2018 within the economic crisis as well as the impact by Egaleo Metro Station and metro line operation.

Results revealed that in economic recession periods activities' participation is limited only to the absolute necessary or low/non costs activities (mainly sports and recreation), affecting also correlated systematic travels. The dramatic reduction of private car ownership and subsequently use indicates that in low living standards regions car use is prohibitive due to its cost. The operation of Metro proved to be a valuable alternative for easy, fast and mainly low cost travels, enhancing the opportunities for longer distance travels e.g. for work. The decrease of private cars use, results to faster travel times for public buses routes. Public transport network expansion eases and facilitates participation to activities, as individuals have access to low cost and extended spatial coverage transport network.

Further analysis of the research data could reveal correlations between systematic activities – travels patterns, geographic diversifications between activities locations and selected transport mode.

References

1. Mitchell, R., Rapkin, C.: Urban Traffic: A Function of Land Use. Columbia University Press, New York (1954)
2. Hägerstrand, T.: What about people in regional science? Pap. Reg. Sci. Assoc. Pap. **24**, 7–21 (1970)
3. Chapin Jr., F.S.: Human Activity Patterns in the City. Wiley, New York (1974)
4. Jones, P.M.: New approaches to understanding travel behavior: the human activity approach. In: Hensher, D.A., Stopher, P.R. (eds.) Behavioral Travel Modelling, pp. 55–80. Redwood Burn, London (1979)
5. Jones, P.M., Dix, M.C., Clarke, M.I., Heggie, I.G.: Understanding Travel Behavior. Gower, Aldershot (1983)
6. Hirsh, M., Prashker, J.N., Ben-Akiva, M.: Dynamic model of weekly activity pattern. Transp. Sci. **20**(1), 24–36 (1986)
7. Pas, E.I.: Intrapersonal variability and model of goodness-of-fit. Transp. Res. A **21**(A), 431–438 (1987)
8. Axhausen, K., Gärling, T.: Activity-based approaches to travel analysis: conceptual frameworks, models, and research problems. Transp. Rev. **12**(4), 323–341 (1992)
9. Doherty, S.T.: An activity scheduling process approach to understanding travel behavior. Presented at 79th Annual Meeting of the Transportation Research Board, Washington, D.C., 9–13 January 2000 (2000)
10. Lee, M.S., McNally, M.G.: On the structure of weekly activity/travel patterns. Transp. Res. Part A Policy Pract. **37**(10), 823–839 (2003)
11. Lee-Gosselin, M.: Scope and potential of interactive stated response data collection methods. In: Household Travel Surveys: New concepts and Research Needs, Irvine, California, 12–15 March 1995, pp. 115–133. Transportation Research Board Conference Proceedings 10 (1996)
12. Bowman, J.L., Ben-Akiva, M.: Activity-based disaggregate travel demand model system with activity schedules. Transp. Res. Part A **35**, 1–28 (2001)
13. Gärling, T., Axhausen, K.: Introduction: habitual travel choice. Transportation **30**, 1–11 (2003)
14. Niemeier, D.A., Morita, J.G.: Duration of trip-making activities by men and women. Transportation **23**, 353–371 (1996)
15. Kwan, M.P., Lee, J.: Geovisualization of human activity patterns using 3D GIS: a time-geographic approach. In: Goodchild, M.F., Janelle, D.G. (eds.) Spatially Integrated Social Science: Examples in Best Practice, Chap. 3. Oxford University Press, Oxford (2003)

16. Vlastos, Th., Perperidou, D.G.: Planning for walking. Tech. Chronicle Sci. J. TGG **I**(3), 33–43 (2007)
17. Anastasopoulos, P.Ch., Islam, M.B., Perperidou, D., Karlaftis, M.G.: Hazard-based analysis of travel distance in urban environments: longitudinal data approach. J. Urban Plan. Dev. **138**(1), 53–61 (2012)
18. Chen, J., Shaw, S.L., Yu, H., Lu, F., Yanwei, C., Lia, Q.: Exploratory data analysis of activity diary data: a space–time GIS approach. J. Transp. Geogr. **19**, 394–404 (2011)
19. Haustein, S., Hunecke, M.: Identifying target groups for environmentally sustainable transport: assessment of different segmentation approaches. Curr. Opin. Environ. Sustain. **5** (2), 197–204 (2013)
20. Große, J., Olafson, A.S., Carstensen, T.A., Fertner, C.: Exploring the role of daily "modality styles" and urban structure in holidays and longer weekend trips: travel behaviour of urban and peri-urban residents in Greater Copenhagen. J. Transp. Geogr. **69**, 138–149 (2018)
21. Perperidou, D.G.: Development of methodology for the record & analysis of systematic activities & travels with use of geostatistical methods – contribution in the estimation of the exposure to air pollution. Ph.D. Dissertation, National Technical University of Athens, Greece (2010)
22. Patton, M.Q.: Qualitative Evaluation and Research Methods, 2nd edn. Sage, Newbury Park (1990)
23. Perperidou, D.G.: Systematic activities and travels record and analysis as tool for evaluation and monitoring transport networks. Presented at 5th International Congress on Transport Research in Greece, 27–28 September 2010, Volos, Greece (2010)
24. Sfakianaki, M.: Urban add traffic changes in Egaleo Metro Station determination and land use. Thesis, University of West Attica, Greece (2019)
25. Hellenic Statistical Authority, Population – Housing Census, Athens, Greece (2001 & 2011)

Travellers' Propensity to Cycle: The Case of Dublin and Athens

Konstantinos Tsepenta[1](✉), Ioanna Spyropoulou[1], and Aoife Ahern[2]

[1] National Technical University of Athens, 9 Iroon Polytechniou str., 15780 Zografou-Athens, Greece
kos_tsep@outlook.com, iospyrop@central.ntua.gr
[2] School of Civil Engineering, University College Dublin, Belfield, Dublin 4, Ireland
aoife.ahern@ucd.ie

Abstract. The aim of this study is to identify the factors that affect a person's propensity to cycle in urban areas. A survey was conducted in two European cities: Dublin and Athens. Dublin boasts a substantial increase in cycling during the past years as a result of the implementation of targeted measures promoting cycling. While, in Athens the design of dedicated cycling infrastructure has commenced only recently, and Athenians' attitudes towards cycling are still rather negative. Cycling propensity was investigated through the design of a stated preference questionnaire, in which participants were asked to state their willingness to cycle in specific scenarios, with trip purpose, trip distance and infrastructure quality being the parameters defining those scenarios. Probit models were designed and results highlighted both similarities and differences between the two sub-populations. This isolation of specific parameters defining the cycling propensity both regionally and internationally could prove important to design a future suitable transport system for each city, focused on more sustainable transport modes.

Keywords: Cycling · Infrastructure · Travel behavior · Traveller preferences

1 Introduction

Cycling comprises an environmentally friendly, cheap and sustainable way of transport, and its integration in the transport systems is a prerequisite towards achieving sustainable mobility in urban areas [1]. Thus, a greater understanding of the factors affecting the propensity to cycle in urban areas is crucial for the design of future transportation systems. Research in different countries shows that a number of factors influence the willingness to cycle in a population. These factors include the built and natural environment, trip characteristics and socioeconomics, amongst others [2–4]. The risk associated with cycling and high levels of traffic in urban areas can discourage cycling [5]. However, the provision of good cycling infrastructure can change perceptions of risk and can create safer cycling environments, thereby increasing willingness to cycle, [6, 7]. Research has found that other factors such as road surface

E. G. Nathanail et al. (Eds.): CSUM2020, AISC 1278, pp. 138–147, 2021.
https://doi.org/10.1007/978-3-030-61075-3_14

quality, lane width, provision of good facilities at destination, and proximity to public transport can all play a role in increasing levels of cycling in a population [8, 9, 10, 11]. It is also important to consider the characteristics of the trip, such as distance and purpose. For example, for very short distances (less than 2 km) walking is preferred to cycling. For longer journeys, cycling is not a preferred option as exposure to uncomfortable and unsafe travelling conditions is higher [12–14]. Traveller characteristics, such as age and gender, also impact upon cycling rates. Several studies indicate that cycling propensity declines with age [8, 15–18], although some studies show that age does not really have a significant effect on cycling [19, 20]. Furthermore, gender is an additional contributing factor, with men riding more than women [17, 18, 21]. At the same time, for specific gender groups (working population) riding propensity has been observed to higher for women [3, 20]. Other contributory parameters include income, status, profession, weather conditions and topology [16, 17, 22, 23].

The aim of this study is to identify the factors that affect a person's propensity to cycle in urban areas. The study considers two different cities which exhibit both different characteristics, as well as different traveller behaviour. Thus, the aim is twofold. First to indentify global contributory factors and second to define local factors which may explain the differences between the two populations. The city of Dublin and the city of Athens have been selected as the case studies areas, with the first being a city where a 10% of trips are made by the bicycle and the second a city exhibiting cycling proportions close to zero [24].

2 The Two Cities

Athens, capital of Greece is located in the Attika region. When we refer to Athens as an urban area, we consider the greater Athens and Piraeus area covering an area of 412 sq. km. Its population is 3.827.624 people (census 2011), but based on indicators, this number has risen to probably over 4 million people in 2019. According to Eurostat in 2011, the functional urban area of Athens was the 9th most densely populated in the EU, with a population density of 17,040 per square kilometer. Athens has a Mediterranean climate and is listed as the warmest capital city of Europe, enjoying 300 days of sunshine, and only about 397 mm | 15.6 inch of precipitation falls annually. Considering dedicated cycling infrastructure, the total length of existing cycleways (all being segregated ones) is about 55 km and there are plans to expand the network with the design of a 27 km cycling route crossing the city in the next years. Relatively recently, public bike rental stations have been implemented providing around 1000 bicycles. Still, lack of maintenance and use of cycle lanes from motorcycles as traffic lanes or from cars as parking areas hinder their use. Last, Athens is a city with major elevation points, resulting in cycling trips requiring more effort compared to other cities.

Dublin is the capital city of Ireland. By European standards, it is a small city. It is the most populous city on the island, with approximately 1.9 million people, living in the area called the Greater Dublin Area (GDA) [24]. The GDA includes the 4 Dublin local authorities of Dun Laoghaire Rathdown, Fingal, South County Dublin and Dublin

City, plus the 3 counties of Wicklow, Meath and Kildare. Dublin has a temperate climate, with cool summers and mild winters and rainfalls of 30.2/767 mm inches per year. Dublin is a city with low residential density, and residents tend to live in houses in suburbs rather than apartments in the city centre, which leads to congested traffic conditions. Cycling rates in Ireland have increased in the last 10 years with the last census in 2016 reporting numbers cycling to work had risen by 43% since 2011, with over 56,000 cycling to work. The same census reports that 66% of those who cycle to work are in Dublin, where 7.6% (or 38,870) of commuters cycle, and that most cyclists are young, male and professional [24]. This increase in cycling may be due to investment in cycling infrastructure that has taken place in the same period.

3 Field Survey and Data Collection

3.1 Field Survey Design

To achieve the aim of the study, a field survey was conducted through the design of a stated preference questionnaire, which was distributed to travellers in Dublin and Athens. Data was collected employing face to face, on-site interviews. The sampling procedure was random.

The study follows the same format as work conducted by Konstantinidou and Spyropoulou [4] in a study of cycling propensity in Thessaloniki. The distributed questionnaire consisted of four sections as follows: The first part of the questionnaire collected information on trip characteristics and preferences, in the third part factors encouraging and discouraging cycling propensity were explored, while the fourth part involved socioeconomic characteristics. The second part comprised the stated preference experiment, under which the participant had to state his/her probability to cycle under specific hypothetical scenarios. The propensity to cycle in each scenario was stated using a 5-point Likert scale ranging from definitely not to definitely yes. Each scenario was described by 3 parameters with 3 levels each: trip purpose (work, shopping-weekly and social-evening), trip distance (<2 km, 2–8 km, >8 km) and available cycling infrastructure (no dedicated infrastructure, some bike lanes exist but not continuous, segregated bike lanes everywhere-fully continuous network). Thus a total of 27 different scenarios were designed. These were divided into 3 blocks of 9 scenarios per block, based on the criteria of orthogonality, which ensures that the characteristics presented are statistically independent from each other [2].

The questionnaires were distributed at different days of the week, different times of the day and in different locations, to increase population representativeness. The average time of completing the questionnaire ranged between 13–18 min. The questionnaires were distributed in July 2019 in Dublin and in August 2019 in Athens.

3.2 Sample Characteristics

The total sample collected consists of 300 respondents, 150 from Athens and 150 from Dublin. 45% were men 54% women and 1% identified themselves as "other" gender. Considering age groups, 3% was younger than 18 years old, 23% was 18–24 years old,

32% 25–34, 15% 35–44, 15% 45–54, 8% 55–64 and 4% was over 64 years old. According to participants' answers, 15% of the participants have completed only secondary education or lower education levels, 17% have attended university or finished an institute of vocational training (IEK), 43% have a bachelor's degree or a diploma from a university or technological education institute (TEI), 23% have completed a master's degree and 2% have a Phd. In Athens 70% of the participants own at least one motorized vehicle, while in Dublin this percentage is rather lower, and is 40%. On the other hand, in Athens only 9% of the participants own a bicycle, whereas in Dublin the respective proportion is 32%. 39% of the participants use their car as their primary transport mode, while 34% use the public transport, 11% cycle and 11% walk, 4% use a motorcycle and only 1% use a taxi.

The participants rated the cycling infrastructure of their city using a 5-point Likert scale, with 1 representing "non-existent" and 5 representing "excellent" infrastructure. Dubliners rated the cycling infrastructure of Dublin as 2.5, while Athenians as 1.5. Considering the factors that travellers perceive as encouraging and discouraging cycling respectively (a -5 point Likert scale is employed with 1 representing "strongly disagree" and 5 "strongly agree"), all 5 variables concerning cycling infrastructure have received high ratings, along with the more respectful/careful drivers variable. It is important to note that all variables in both cities were rated above average. Road safety and the presence of big road junctions seem to be the factors that discourage travellers the most, especially in Athens. Elevation, bad weather conditions and the transportation of goods/people are also rated above average in both cities. More specifically, people in Dublin rated bad weather conditions as the most discouraging factor considering cycling. Results indicated that travellers in Dublin rate encouraging factors higher than travellers in Athens, whereas the exact opposite is observed considering discouraging factors. This is an indication of the general relative attitudes considering cycling between these two cities, which is also depicted in the actual cycling proportions.

4 Cycling Propensity

4.1 Methodology

The methodology employed in this research is discrete choice models analysis, as it describes more accurately travellers' choice mechanism and has been widely used in transport studies [2, 4]. The design of probit models is preferred over logit models as it is more general. In addition, "random effects" is selected to represent the correlation between the responses of the same individual, as its participant provides answers for 9 scenarios. Ordered probit analysis is performed as travellers' choice is represented in an ordered scale, with the possible answers describing the probability of a participant performing the trip by bicycle being "definitely not", "probably not", "probably", "probably yes" and "definitely yes".

4.2 Probit Models

The R programming language was used for the statistical analysis, and more specifically the pglm package was used for the estimation of the ordered probit models. Three models were designed to capture cycling propensity considering the whole sample population, Athens and Dublin travellers. The model considering the whole sample population (both Athens and Dublin travellers) was designed to capture global factors affecting traveller cycling propensity (more information can be found in [25]). Results indicated that infrastructure and distance are the variables affecting the respondents in the most consistent manner as they presented the highest t-values. Distance is a discouraging factor, reducing greatly the propensity, while infrastructure increased the propensity to cycle significantly. Results concerning trip purpose indicated that travellers exhibited the highest propensity to cycle for commuting. Trips concerning shopping exhibited a reduced propensity compared to work trips, while socializing trips exhibited the lowest propensity. The impact of trip purpose on cycling has not been explored in significant detail by other researchers but this is similar to findings made in a study by one of the authors of cycling in Thessaloniki [4].

Considering socioeconomic factors, education, income, age and profession were found to influence the propensity to cycle both in Athens and Dublin. In particular, highly educated travellers, demonstrated a much higher propensity to cycle compared to those who had not continued their studies after school. Travellers with higher personal income also demonstrated a slightly higher propensity to cycle, which is interesting as many studies find that those on the highest and lowest incomes are less likely to cycle [4]. Concerning profession, the model indicated that students in both countries had a higher propensity to cycle compared to people with other professions. Finally, age was also found to influence cycling propensity, with travellers aged 18–24 years being less likely to cycle compared to people younger than 18 years old.

Another subgroup of factors included in this model involves transport preferences. The model indicated that people who chose walking as their main transport mode had a reduced propensity to cycle, while people riding motorcycles as their main transport mode had an increased cycling propensity, both compared to travellers whose primary transport mode was their private cars: perhaps this is because the shift from motorcycle to cycling is not that great and those who use motorcycles are already less risk-adverse than non-cyclists. In addition, the more people could not afford a private vehicle the higher the propensity to cycle, Lastly, people who experienced delays in their daily trips had a higher propensity to cycle, perhaps seeing cycling as a mode that would enable them to overcome those delays. The model included two cycling discouraging factors, fatigue and transport speed, and two encouraging factors, a more extended cycling network and better environmental conditions. Last, the model demonstrated that Athenians have a significantly lower propensity to cycle than people living in Dublin city.

This study presents the results of the probit model for the two different explored cities (Table 1).

Table 1. Probit model considering the population of Athens and Dublin.

Variable	Estimate value	t-Value	Estimate value	t-Value
	Athens		Dublin	
Intercept	−0.841	−1.396	0.170	0.267
Infrastructure [some bike lanes exist but not continuous]	0.86418	9 .521	0.4112	4.711
Infrastructure [segregated bike lanes everywhere, fully continuous network]	1.23628	13.331	0.961	10.774
Distance [2–8 km]	0.71217	−8.329	−0.813	−9.435
Distance [>8 km]	−1.12595	−12.457	−1.447	−15.490
Purpose [social-evening]	−0.93677	−10.419	−0.492	−5.630
Purpose [shopping-weekly]	−0.32291	−3.840	−0.482	−5.523
Socioeconomic factors				
Income [personal income group 6]	−0.53503	−2.561	0.412	1.690
Education [=PhD]	1.65601	5.772		
Education [>=attending university-college]			1.308	6.766
Age [18–24]	−0.61106	−2.313		
Age [25–34]	−0.38960	−1.682		
Age [35–44]	−1.31731	−4.717		
Age [45–54]	−0.61641	−2.644		
Age [55–64]	−0.67489	−2.775	−0.544	−2.308
Family status [divorced]	−0.93476	−4.697		
Profession [retired]	−0.73119	−2.945		
Profession [student]			1.241	5.157
Gender [female]	−0.32860	−3.256		
Transport preferences				
Main transport mode [public transport]	0.28179	2.363	−1.052	−8.839
Main transport mode [pedestrian]			−1.230	−6.631
Congestion delays my travels [>=disagree]	1.60900	4.955		
Cannot afford private vehicle [disagree & neither agree nor disagree]			1.542	6.744
Cannot afford private vehicle [agree]			2.136	8.468
Discouraging factors				
Transportation speed [strongly agree]	−1.69451	−6.098		
Too many pedestrians [>=disagree]	−0.49372	−2.734		
Too many pedestrians [>agree]			−0.498	−3.808
Fatigue [>neither agree nor disagree]	−0.61284	−5.533		
Fatigue [strongly agree]			−1.913	−9.495
Road safety [>=disagree]			−2.194	−5.481

(*continued*)

Table 1. (*continued*)

Variable	Estimate value	t-Value	Estimate value	t-Value
Encouraging factors				
Less traffic [disagree & neither agree nor disagree]	1.24787	2.780		
Less traffic [>=agree]	1.61249	3.667		
Bigger cycling network [strongly agree]	0.39983	3.189		
Better air quality [>disagree]			1.625	3.446

Models' AIC were estimated to be 2665.95 for the Athens model, and 2690.37 for the Dublin model. It should be noted that the intercept for the Athens model is negative, whereas for the Dublin mode is positive. All three explored scenario parameters affect cycling propensity in more or less the same manner. In particular, the increase of infrastructure quality results in higher cycling propensity, and the increase of trip distance results in lower cycling propensity. Considering trip purpose, working is the purpose demonstrating the highest cycling propensity, followed by shopping. Social purposes exhibit a substantially lower propensity compared to the other two purposes for trips in Athens, and a somewhat similar propensity to shopping trips for trips in Dublin.

Five socioeconomic factors were included in this model. The most significant one (which was not significant in the general model) is gender. Results demonstrate that women in Athens have a significantly lower propensity to cycle than men, whereas gender does not affect cycling propensity in Dublin. Furthermore, in Athens age constitutes a more important factor, with people older than 18 years old presenting reduced propensity to cycle compared to people younger than 18 years old. The reduced cycling propensity is only evident for the older population in Dublin (55–64 years old). Just like in the general model, highly educated people exhibit a higher propensity to cycle. In addition, high personal income results in reduced propensity in Athens and in increased propensity in Dublin.

Considering trip characteristics, transport habits relevant to the selected primary mode influence cycling propensity. People using the public transport as their main transport mode in Athens exhibit a higher cycling propensity and a lower cycling propensity in Dublin compared to those using passenger cars. Furthermore, travellers in Dublin whose primary transport mode is walking also exhibit reduced cycling propensity. In addition, the more a person in Dublin cannot afford a private vehicle, the higher his/her propensity is to cycle. While in Athens, travellers who experience congestion in their everyday trips exhibit a higher cycling propensity. Last, considering discouraging and encouraging factors, Athenians' propensity to cycle is affected considerably by fatigue, while this effect is smaller for Dubliners, reflecting the size of the two cities and the different topologies. Furthermore, Athenians who consider cycling being a slow mode are less likely to cycle, while for Dubliners concerns on road safety reduce their cycling propensity. Lastly, the presence of many pedestrians discourages both Athenians and Dubliners from cycling. The two most important changes that would increase cycling propensity of travellers in the city of Athens are the construction of a more extended cycling network and the reduction of traffic on the streets, while in Dublin the only encouraging factor identified in the model is environmental conditions and better air quality.

5 Discussion

This study explored travellers propensity to cycle in two cities: Athens and Dublin, which present distinct differences considering both environmental conditions (infrastructure, topology, weather conditions) but also travellers' habits and preferences. Conclusions should be generalized with care, as due to the sample size and data collection method, representativeness of the population is not fully ensured. Furthermore as results indicated, for specific parameters travellers in the different cities exhibited completely different behaviour. Thus, specific findings might not be applicable to all cities. Trip distance and infrastructure presented the highest t-values indicating that the attitude of travellers towards them is greatly consistent. Furthermore, the greater the trip distance, the lower the cycling propensity, while better infrastructure was found to increase cycling propensity in all the models. These results come in accordance with findings from similar studies. Last, in all three models, trips to work exhibited the highest cycling propensity, followed by shopping and socializing trips. These results partly explain the great difference of the cycling proportions between the two cities. This was also depicted in the general population model results with travellers in Dublin exhibiting higher cycling propensity compared to Athenians. The Dublin cycling network is much more extended compared to that of Athens, while at the same time the city of Athens is much more dispersed which entails longer trip distances.

The highest cycling propensity was associated with work trips in both cities, and the majority of working trips in Athens involve trips from the suburbs to areas in or close to the city centre, involving average distances greater than 8 km, usually between 8 and 15 km, which are somewhat longer to those in the city of Dublin. Thus, to promote cycling in the city of Athens, dedicated infrastructure, which was rated from the survey respondents as being rather poor, needs to be substantially improved. Still, the cycling proportions will probably still not increase as much due to the long trip distances involved. On the other hand, in a survey exploring traveller preferences in Greece and Germany, results indicated that Greek travellers were more willing to accept trips involving longer travel times, if improved infrastructure was offered [26].

Socioeconomic factors including age and gender affected Athenians more compared to Dubliners. More specifically, cycling propensity was similar between men and women in Dublin in the examined sample, although census statistics indicate increased cycling for the male population, while Athenian women were found to be less willing to cycle. Furthermore, fatigue, was found to be a discouraging factor affecting more Athenians compared to Dubliners, as the majority of cycling trips in Athens include inclined sections. Equipping the city with bike rental stations including e-bikes and smart applications that ease their use may increase cycling for all the population and also for women. Considering age, in Dublin only travellers aged over 55 years old exhibited a significantly lower cycling propensity compared to youngsters. In Athens on the other hand, all age groups exhibited a lower cycling propensity compared to youngsters. This indicates that cycling is not part of the everyday life of the adult population, which might be partly associated with less free time, lifestyle status and fatigue but mainly for older age groups (or women). To change this, promotional campaigns mainly through social media and smart applications need to be designed, mainly to affect young age groups, as the effect of social influence and social interactions amongst peers in selecting alternative transport modes is significant [27].

An interesting finding is that although higher education level resulted in higher cycling propensity in both cities, the effect of income differed between them. Higher income is associated with higher cycling propensity in Dublin and with lower cycling propensity in Athens. Although this may be related with the trip duration and the value of time associated with the two study populations, as in the model describing cycling propensity considering Athenians the low speed associated with cycling was included as a contributory factor, it might also be an indication of status mentality. In a previous study in the second biggest city of Greece, Thessaloniki [4], travellers who stated that the private car serves as a symbol of status and demonstrated lower cycling propensity.

Cycling constitutes a modern, alternative transport mode. The investigation of the differences between a city with moderately good practices, Dublin, and a city where only a small proportion of the population cycles to their destination, Athens, can help design effective measures to promote cycling. Still, further insights are required. Towards this direction a revealed preference study in the city of Athens would be very relevant, especially now, as walking and cycling exhibited a substantial increase during the recent Covid-19 lockdown.

References

1. Handy, S., van Wee, B., Kroesen, M.: Promoting cycling for transport: research needs and challenges. Transp. Rev. **34**(1), 4–24 (2014)
2. Hensher, D.A.: Stated preference analysis of travel choices: the state of practice. Transportation **21**, 107–133 (1994)
3. Wiltox, F., Tindemans, H.: Evaluating bicycle-car transport mode competitiveness in an urban environment: an activity-based approach. World Transp. Policy Pract. **10**(4), 32–42 (2004)
4. Konstantinidou, M., Spyropoulou, I.: Factors affecting the propensity to cycle – the case of Thessaloniki. Transp. Res. Procedia **24**, 123–130 (2017)
5. Rissel, C.: Have helmet laws put the skids on Australia's bike share scheme? https://theconversation.com/have-helmet-laws-put-the-skids-onaustralias-bike-share-scheme-2703. Accessed 04 Apr 2020
6. Pucher, J., Buehler, R.: Making cycling irresistible: lessons from the Netherlands, Denmark and Germany. Transp. Rev **28**, 495–528 (2008)
7. Pucher, J., Dill, J., Handy, S.: Infrastructure, programs, and policies to increase bicycling: and international review. Prev. Med. **50**, 106–125 (2010)
8. Sener, I.N., Fluru, N., Bhat, C.R.: An analysis of bicyclists and bicycling characteristics: who, why and how much are they bicycling? In: 88th Annual Meeting of the Transportation Research Board, Washington, D.C. (2009)

9. Li, Z., Wang, W., Liu, P., Ragland, R.D.: Physical environments influencing bicyclists perception of comfort on separated and on-street bicycle facilities. Transp. Res. Part D **17**, 256–261 (2012)
10. Noland, R., Kunreuther, H.: Short-run and long-run policies for increasing bicycle transportation for daily commuter trips. Transp. Policy **2**(1), 67–79 (1995)
11. Hunt, J.D., Abraham, J.E.: Influences on bicycle use. Transportation **34**, 453–470 (2007)
12. Xing, Y., Handy, S.: Factors associated with proportions and miles of bicycle rides for transportation and recreation in 6 small US cities. In: 88th Annual Meeting of the Transportation Research Board, Washington, D.C. (2009)
13. Kejer, M.J.N., Rietveld, P.: How do people get to the railway station? The Dutch experience. Transp. Plan. Technol. **23**, 215–235 (2000)
14. Van Wee, B., Rietveld, P., Meurs, H.: Is average daily travel time expenditure constant? In search of explanations for an increase in average travel time. J. Transp. Geogr. **14**, 107–122 (2006)
15. Shafizadeh, K., Niemeier, D.: Bicycle journey-to-work: travel behavior characteristics and spatial attributes. Transp. Res. Rec. **1578**, 8490 (1997)
16. Pucher, J., Komanoff, C., Schimek, P.: Bicycling renaissance in North America? Recent trends and alternative policies to promote bicycling. Transp. Res. Part A Policy Pract. **33**, 625–654 (1999)
17. Moudon, A.V., Lee, C., Cheadle, A.D., Collier, C.W., Johnson, D., Schmid, T.L., Weather, R.D.: Cycling and the built environment: a US perspective. Transp. Res. Part D **10**, 245–261 (2005)
18. Dill, J., Voros, K.: Factors affecting bicycling demand: initial survey findings from the Portland, Oregon Region. Transp. Res. Rec. **2031**(1), 9–17 (2007)
19. deGeus, B.: Cycling to work: psychosocial and environmental factors associated with cycling and the effect of cycling on fitness and health indexes in an untrained working population. Doctoral dissertation, Department of Human Physiology and Sports Medicine, Vrije Universiteit Brussel (2007)
20. Wardman, M.R., Tight, M.R., Page, M.: Factors influencing the propensity to cycle to work. Transp. Res. A **41**(4), 339–350 (2007)
21. Banister, C., Gallant, N.: Sustainable commuting: a contradiction in terms? Reg. Stud. J. Reg. Stud. Assoc. **33**(3), 274–280 (1999)
22. Dieleman, F.M., Dijst, M., Burghouwt, G.: Urban form and travel behavior: microlevel household attributes and residential context. Urban Stud. **39**(3), 507–527 (2002)
23. Buehler, R.: Trip-end facilities at work and bicycle commuting in the Washington, DC. In: 92nd Annual Meeting of the Transportation Research Board, Washington, D.C. (2013)
24. Central Statistics Office 2016 – Census of Population: Commuting in Ireland. https://www.cso.ie/en/releasesandpublications/ep/p-cp6ci/p6cii/p6mtw/
25. Tsepenta, K.: Factors affecting cycling and walking propensity: the case of Dublin and Athens. Diploma thesis, National Technical University of Athens (2020)
26. Hardinghaus, M., Papantoniou, P.: Evaluating cyclists' route preferences with respect to infrastructure. Sustainability **12**, 3375 (2020)
27. Manca, F., Sicakumar, A., Polak, J.W.: The effect of social influence and social interactions on the adoption of a new technology: the use of bike sharing in a student population. Transp. Res. Part C **105**, 611–625 (2019)

The Role of Transport in Urban Planning in Greece: How to Integrate Sustainable Mobility Planning in Local Spatial Planning?

Efthimios Bakogiannis[1], Vasilios Eleftheriou[1], Charalampos Kyriakidis[1](✉), and Ioannis Chatziioannou[2]

[1] National Technical University of Athens (NTUA), 9, Ir. Polytechneiou Str., 10682 Athens, Greece
kyriakidisharry@gmail.com

[2] Instituto de Ingeniería (UNAM), Circuito Escolar, Ciudad Universitaria, 04510 Mexico City, Mexico

Abstract. Urban planning in Greece was an unclear and multifaceted procedure, failing to produce clear results. Such an example is related to various traffic problems recorded in several Greek cities. Indeed, urban planning, which was mainly applied through the implementation of General Development Plans (G.D.Ps.), has failed to examine the aspects of transportation. To face this problem, Sustainable Urban Mobility Plans (S.U.M.Ps.) are on the forefront of urban planning process and many Greek cities tend to implement such plans, in the near future. Meanwhile, urban planning legislative framework has been modified and G.D.Ps. have been replaced by another similar type of tool: Local Spatial Plans (L.S.Ps.). This change reveals an opportunity, as a holistic-comprehensive planning approach is emerged. In that context, measures and solutions, whose viability will not be questioned, in practical terms, over a few years, can be proposed in order for L.S.Ps. and S.U.M.Ps. to function effectively. This is the main topic of this paper that tries to approach the potential problems that may arise at legal and practical level from the coexistence of S.U.M.Ps with L.S.Ps which are new planning tools of urban design. This knowledge is derived by analyzing the way in which S.U.M.Ps. have already been implemented in Greece and the problems emerged by the previous institutionalized plans. Concerning the institutional provision for S.U.M.Ps and L.S.Ps, we attempt to identify some cases where problems may arise during the planning procedure, in order to give additional directions to the studiers of both plans, during the study process.

Keywords: Local Spatial Plans (L.S.Ps.) · Sustainable Urban Mobility Plans (S.U.M.Ps.) · Planning legislation

E. G. Nathanail et al. (Eds.): CSUM2020, AISC 1278, pp. 148–157, 2021.
https://doi.org/10.1007/978-3-030-61075-3_15

1 Introduction: How Close is Greece to a Comprehensive Urban and Traffic Planning Model?

Urban planning plays an important role in urban studies [1]. The goal of this field is related to the study and the organization of urban space. On the other hand, mobility is considered as an essential condition for social interaction and, consequently, for cities formation and function [2–3]. This is a reality in many countries worldwide, including Greece; however, although both urban and mobility planning types should have a supplementary role, in practice, they are implemented on an independent way and thus such a differentiation is mentioned. The problem is rooted in different issues. In Greece it is due to the (a) brief history of urban planning which has its roots in the regulation of building activity [4] and moves in parallel to the short history of the autonomous Greek state [officially Hellenic Republic], (b) historical established existence of informal construction system - a fact which is reinforced through increased refugee flows during the period of "Asia Minor Disaster" [5] which over the years had an increasing trend and functioned as a means of developing many settlements and cities in the Greek Territory, (c) intensive use of the car as a dominant means of transportation in cities following an open–city development model, (d) economic structure of the country and the cities which didn't follow the growth rate of central and north European cities, and the (e) strong role of professional guilds in organizing and operating the planning system.

The lack of practical synergy between these two aforementioned planning types, while the urbanization intensified during the second half of the 20th Century, centered in Athens [6], has resulted in increased demand for commuting and, consequently burdening cities on environmental, social and economic levels [7–8]. During the last years, decision makers and transport planners try to tackle such challenges by establishing a mobility system that meets society's social, economic and environmental needs, according to the EU agenda [9]. The combination of various means of transport, the development of mild traffic zones, the development of cycling and pedestrian networks are parts of the puzzle of proposed interventions in order to modernize the infrastructure of cities [10] so that they can cope with the emerging needs of citizens who are invited to use more sustainable solutions in their everyday lives in order to circulate through their city. The above and some additional actions are being considered and proposed, where appropriate, in the context of implementation of Sustainable Urban Mobility Plans (SUMPs); SUMPs aim is to promote compact city that seems to be a more sustainable and economic efficient model [10].

Taking all the above into consideration, a query is raised: To what extend can SUMPs be well integrated into the existing traffic and urban planning system in Greece? By approaching the above question more broadly, it could be determined that the purpose of the research is to investigate whether the incorporation of a new type of design into the country's urban planning contributes substantially to the development of a comprehensive design, with two aspects that complement each other. Are we talking about a reconciliation of the two above planning types or a persistence in adversity? Such topics are approached in this paper in order for a dialogue to start among planners and decision makers; through that, ideal solutions can be provided in order for potential problems to be faced, on time.

Section 2 briefly summarizes the historical evaluation of the institutional framework for urban planning in Greece. Two important planning tools are presented in that context: (a) General Development Plans (GDPs), which were the earliest major urban planning tools that have been in the spotlight for 31 years, b) Local Spatial Plans (LSPs) which are the new tools that are currently active. Section 3 focuses on the aim and the design approach that results through SUMPs' procedure. In this section, a comparative analysis is also attempted between SUMPs and conventional traffic studies that have been the main traffic planning tool in Greece until recently. Section 4 studies, on a comparative way, SUMPs and LSPs. More specifically, the degree in which both of those types can satisfactorily keep up in practice is examined. Moreover, their institutional role as well as their position in the planning system is also mentioned. Finally, Sect. 5 summarizes what it is mentioned above. Provisions are also made for the way those plans will operate.

2 The Planning Legislation in Greece: A Brief Overview

In Greece, urban planning is usually identified as building control regulation due to the fact that building activity is an expression of economic and social life of people [4]. The outcome of such a perception is the identification of cities with their structured surface, severely limiting the urban – rural relationship, for decades. In fact, by examining the planning history of the Greek state, and particularly its early period (1828–1923) where the development of the first street (plot) plans of the cities across the newly established Greek state is observed, it is found that planning procedure did not follow specific legislation [4]. The first attempt of state intervention in the field of urban planning was made in 1923, with the issuance of the Legislative Decree (L.D.) 17.07.1923 "On the plans and cities, towns and settlements of the state and their construction" (Government Gazette A` 228/16.08.1923). This Decree was consisting the guide for the reconstruction of most of the Greek cities. Its aim was, on the one hand, to meet people needs during that period (hygiene, aesthetics, security, technological modernization and economy) and, on the other hand, to gain the homogenization of the multicultural and multinational population of the country [11]. This goal was based on the adaption of traditional and pre-industrial structures of the Greek cities, to the standards of Western Europe, in terms of economy and urban development [12].

For almost half a century, the above L.D. has consisted the only institutional document for town planning.

In 1979, the L.D. No. 947/1979 "On Residential Areas" (Government Gazette A`169/28.07.1979) attempted to encourage a new impetus in urban planning, modernizing the established regime [13], which showed significant tolerance to informal housing (Institute of Local Government, 2006 please comply with reference standard). However, even that Decree did not remain active for a long time due to intense social criticism and condemnation [14]. Therefore, it was replaced by the Act No. 1338/1983 on "Expansion of Urban Plans, Residential and Related Arrangements" (Government

Gazette A`33 /14.03. 1983), which although initially seemed to be transitional (it was voted to meet emergency requirements and was expected to be limited in duration), it has had a significant impact on the Greek planning system until today [15–16], signaling the modernization of the country's institutional planning framework [17].

One of the most important regulations promoted by the Act No. 1337/1983 was the organization of town planning for settlements of more than 2000 residents on two levels [18]: (a) General Development Plan (GDPs), and b) Urban Planning Study. Through this act, another sub-level is added; Implementation Plans [16] that apply the urban space and street plans of a settlement [18]. Planning proposals derived by the GDPs -they are also combined with a perimetric Residential Control Zone (ZOE) [16, 19]- are realized at the urban and peri –urban space. They relate to an integrated framework of guidelines that evaluates, in addition to urban-planning characteristics, the characteristics of the natural environment [20].

GDPs (or Open-City Spatial and Housing Organization Plans – SHOOAPs- for less than 2000 residents) are key-elements that have also been proposed by the Act No. 2508/1997 for the "Sustainable Urban Development of Cities and Settlements of the Country and Other Provisions", which was the next legislative act. However, GDPs proposed by the Act No. 2508/1997 differ from the previous GDPs in their spatial reference and, consequently, with their objects. In that way, the transformation of urban planning thought in Greece is expressed; after this act, urban planning deals with a whole municipality [15]. In this sense, the new plans work as a synthesis of heterogeneous perceptions and situations, taking into consideration specific spatial, quantitative and qualitative properties [21]. To the above parameters are also identified the characteristics of transport that are associated with the development of activities within a geographic area. GDPs and SHOOAPs should be consistent with the overarching plans such as the Regulatory Plans (RPs) (RPs have been established by the L.D. No. 1262/1972 on "Regulatory Plans of Urban Areas", but they were first implemented after the adoption of Act No. 1337/1983″). However, it is crucial that through the Act No. 2508/1997 a better hierarchy of these two planning levels has been achieved, in proportion to the planning system in the countries of the European Union (Great Britain, Germany and France) [4].

In order for the urban planning to be implemented in more sustainable terms, Act No. 4269/2014 on "Spatial Planning and Sustainable Reform- Sustainable Development" (Government Gazette 142/A/28-06-14) amends the provisions of the Act No. 2508/1997. Therefore, GDPs are replaced by LSPs.

In the foreground are also mentioned the Special Spatial Plans (SSPs), that are hierarchically aligned to the same level as the LSPs and refer to urban regeneration programs, environmental protection schemes or initiatives related to the neutralization of the consequences of natural disasters. In this sense, the latest urban planning law follows the same logic. Because the Act No. 4447/2016 on "Spatial Planning –Sustainable Development and Other Provisions" retains the aforementioned planning tools. However, a modification is suggested regarding the plans of implementation.

It is noteworthy that both LSPs, in accordance with the requirements of Government Gazette B 1975/ 6-6-2017, as well as the GDPs are characterized by a comprehensive planning, that takes into account several issues, such as transport. In that context, SUMPs, that came to the forefront a few years ago, promise a different design approach which has to be combined with the guidelines of LSPs. Are any problems expected when both of these plans are going to be used in a combined way? Before answering this question, Sect. 3 summarizes the framework for the implementation of SUMPs in Greek cities and their relationship with the existing traffic planning framework.

3 SUMPs in the Modern Greek Planning Reality

In order to reverse the situation and alleviate congestion and environmental degradation problems, European Union has been urging its Member States to implement policies in favour of the compact city [6]. Their goal is to address transport related problems in a sustainable way [22] and thus, to improve the quality of life in cities by ensuring a reliable and effective environment-friendly urban transport system [2]. SUMPs propose policies that encourage a shift towards more sustainable transport modes, reduce travel speeds within the cities and neighborhoods, upgrade the aesthetics of the road environment and provide suitable mobility conditions for vulnerable road users (pedestrians, cyclists, disabled, etc.) [23]. In order to be more specific, a SUMP consists of a strategic planning tool that contribute in transforming the current situation in a city by changing the mobility choices of citizens without leaving their mobility needs unsatisfied [24]. It is the result of a structured process that comprises inter-disciplinary status analysis, building vision, objective and target setting, policy and measure selection, active communication-participation, monitoring and evaluation.

Concerning the SUMP's area of intervention, this is defined within the administrative boundaries of the elaborator body of each project. The elaborator body is a Local Government Organization (LGO) of first or second degree. The above geographic area may extend beyond the administrative boundaries of the elaborator body, if necessary. In this case, the elaborator body is obliged to invite all the LGOs of the intervention area in order to participate in the operator's network. Apart from LGOs of first or second degree, operator's network may include the competent regional state bodies, transport service providers, chambers of commerce, trade and professional associations as well as environmental associations.

Regarding the procedure of approval of a SUMP, the competent departments of the elaborator body (in each case) should approve it. After the text composition of a SUMP, and once the consultation procedures have been completed, the plan is disclosed, prior to its approval, to the: (a) competent department of the Ministry of Infrastructure and Transport that is responsible for its designation as well as its rejection if the SUMP is incomplete (b) competent departments of Ministries of Interior, Environment and Energy, and Infrastructure and Transport for the purpose of expressing an opinion and (c) competent regional bodies of the State that are responsible for interventions within each region as well as to all the authorities and the wider public sector affected by the implementation of each SUMP or the ones that will be subsequently asked to adopt the measures proposed by the SUMP. It is obvious that such plans are characterized by a low level of institutionalizing, something that is not the case for plans included in the official spatial planning system. Section 4 focus on the issues arising from this incompatibility of the institutional processes.

4 GDPs and LSPs VS SUMPs: Problems and Divisions in the Context of a Comprehensive Planning Procedure

As presented so far, it is found that urban planning in Greece is, more or less, instable. During the last decades, a range of institutional documents that modify the intended spatial planning tools or differentiate their application scale, the objects they examine and their character, in general, have been produced. At the same time, experience has shown that the completion and the approval of various urban plans (i.e. GDPs) has proved to be a complicated and time consuming process. As a result, in many municipalities across the country, there are not approved plans in order to organize (peri-) urban space. In such settlements, where the population is less than 2000 residents, a special transitional regime of land use is established, so that, a regulator of the situation to be in place, until planning to be completed. The above are noted because they should be already known when a new plan (such as SUMPs) is adopted.

Below are presented some cases where allegedly problems are expected after the implementation of SUMPs. More specifically:

- Conducting of a SUMP in a municipality for which (or a part of it) a GDP is not available: As noted above, one of the differences between the GDPs established by the Act. No. 1337/1983 and the ones institutionalized by the Act. No. 2508/1997 is related to the planning scale and the scale of the proposed interventions. Many kallikratean municipalities (since the 2011 Kallikratis reform, 325 municipalities have been emerged by unifying the former municipalities -municipalities that have been emerged through the 1997 Kapodistrias reform- that were smaller in size), have not updated urban planning in their areas by conducting new GDPs for the whole municipality, according to the Act No. 2508/1997. Such an example, although it is not the only one, is the Municipality of Platanias located in the Regional Unit of Chania in Crete. In that case, there is only a GDP which was conducted according to the Act No. 1337/1983 and it concerns the municipal unit of Platanias. As a result, there is a large part of the municipality for which an urban planning document is not available. In such cases, implementation of a SUMP raises queries about the directions that this plan should follow, until a LSP is going to be institutionalized. Taking into account that such planning directions are not available, there are two possible scenarios about the future of planning in those cases: (a) LSPs are required to converge to the proposals of SUMPs or (b) a divergence in terms of objectives and proposals may be observed between the two plans.
- Conducting of a SUMP in a municipality for which a GDP is available but a LSP is pending. Most of the municipalities across Greece fall into this category. This is observed due to the fact that LSPs have not been implemented, so far. However, SUMPs have already started or, even, completed in many municipalities. Such cases are Thessaloniki, Larisa, Kozani, Rethymno, Drama, Kallithea and Zografos. Since then, the Green Fund has announced a funding program in order for 180 Greek municipalities to implement a SUMP. Nevertheless, these municipalities but also others may face dilemmas related to the queries raised on the previous case, where there is not a GDP, since: (a) LSPs seem to follow the strategy of the previous

GDPs because this strategy has also been acted as the basis in order for SUMPs to be conducted, or (b) LSPs will bring to the fore a different objective targeting from the one presented on the previous GDPs and which was the basis in order for SUMPs to be conducted. As a result, LSP's strategy will be different from the SUMP's one.

- Conducting of a SUMP in an inter-municipal level, where the intervention area is part of more than one municipality for which (or some of them) GDP is not available. The reasons why this may be the case have been discussed previously. The problems arising from the preparation of a SUMP under these circumstances, include those presented in the first case (a) and the possibility that the plans of the neighboring municipalities do not converge, at a policy level. This is likely to happen during the implementation of a SUMP; especially in areas where spatial planning presents gaps (i.e. in touristic areas, taking into account that the Special Spatial Framework for Tourism has been canceled).
- Conducting of a SUMP in an inter-municipal level, where the intervention area is part of more than one municipality for which (or some of them) GDP is available. Concerns raised in that case are similar to the ones presented in the previous cases, as well.

The problems arising in the previous cases can be categorized in two types:

- Institutional problems related to the validity of planning proposals: As has been noted previously, SUMPs approaches are institutionalized at a low-level procedure. More specifically, the General Secretary of the Decentralized Administration is the one who is going to approve a SUMP. Then, the decision have to be published in the Government Gazette in order for the plan to be expressly approved. Contrary to that procedure, a LSP is approved by a Ministerial Decision. It should be mentioned that in case of a SUMP which is conducted in an inter-municipal level, it is necessary to be related to the RP -if exists- that consists of an Act. Moreover, when RPs and LSPs are conducted, Strategic Environmental Impact Studies are also necessary. There is no similar provision of environmental assessment when a SUMP is implemented. Such variations in the way various types of plans are adopted and approved proves that urban planning is "more important" than SUMPs. The only way to avoid canceling a SUMP is to be based on the existing RP or LSP. But what about cases where SUMPs conducted by taking into account the directions of existing DGPs (i.e. in case of road network hierarchy) rather than the LSPs, the implementation of which is scheduled? Institutional problems also include what may arise when applying the guidelines of a SUMP to areas where existing (or new) plans approved by a Presidential Decree.
- Practical problems related to the scale of the plans and the proposed interventions: Experience resulting from the implementation of SUMPs that are in progress -one such example is the SUMP for Rethymno- demonstrates that the study area differs from the municipal territory. This is evident both in cities located not only in the countryside but also in metropolitan areas, such as in the Athens Metropolitan Areas (AMA), where many residents regularly move from one municipality to another, on a daily basis. In small municipalities, this is more intense, as many services such as (schools of all levels, health care, etc.) are not provided. Contrary to that, the study

area of a GDP is focused on the municipal territory, according to the Act No. 2508/1997. This study area is mainly used in case of LSPs. However, they can also be implemented in a smaller geographic unit, like the one of municipal unit. As a result, there are differences between the study areas as well as the intervention areas of various plans. Hence, it is possible to find out discrepancy in terms of targeting and directions.

5 Conclusions

This paper tries to approach the potential problems that may arise at legal and practical level from the coexistence of SUMPs with LSPs which are new planning tools of urban planning. This concern arose from the fact that both plans are relatively new. After all, any new plan integrated in the planning system brings to the fore a number of issues, especially in the case of Greek urban planning system, which is characterized by complexity, due to the numerous institutional documents and planning tools. The summary of the legislation and the instability of the tools used (Sect. 2) support this hypothesis.

The above speculation is related to the nature and the institutional role of SUMPs (Sect. 3) as strategic and operational plans adopted at the regional level. This seems to make them "inferior" to the plans that are part of the official planning system (RP, LSPs, GDPs, Street -Plot- Plans, Urban Planning Studies) that are enacted through Acts, Ministerial Decisions and Presidential Decrees, respectively.

Taking into account the institutional provision for both planning types (transport and urban planning), it has been attempted to identify some cases where problems may arise during the planning procedure. The reason why assumptions are mentioned instead of empirical analysis has to do with the fact that: (a) only a few SUMPs have implemented and completed, so far. Only in Larisa, constructions have started to be implemented. (b) no LSPs have been conducted, so far.

Due to the low level of institutionalization of the SUMPs, it is believed that most problems are likely to arise due to the fact that SUMPs have been implemented before the implementation of LSPs. Indeed, in case a different approach adopted by researchers, practitioners and local authorities during the implementation of a LSP may result in contradiction to the proposals of the SUMP. In fact, in the case of inter-municipal SUMPs, problems are more likely to be detected as there will be two or more LSPs (after all, LSPs can be conducted even for municipal units), whose proposals do not imply convergence with prior urban planning, if any, or in the directions of the existing SUMP.

Another issue has to do with the lack of existing urban planning in Greek territory. Such a (permanent) omission has allowed for increased informal construction, which has not been eliminated, in practice, but only by regulation, due to addressing the issue through voter-centered and economically–centered legislation. The absence of organized planning through specific directions implies the development of the SUMPs only under guidelines of overhead plans (i.e. spatial plans).

In any case, institutional and practical problems that may arise during the conduction of SUMPs and LSPs should not discourage us about the future of planning process. The reason why all of those issues are mentioned is to prevent possible problems and provide solutions in order to alleviate them. In that way, planning process may be accelerated. Moreover, it should be mentioned that if we can address all the issues presented above this means that transport and urban planning can be applied in a complementary way.

References

1. Papagrigoriou, V.: Urban Planning, Introduction, Institutions, Policy. Sakkoulas Publications, Athens (2007)
2. Okraszewska, R., Romanowska, A., Wolek, M., Oskarbiski, J., Birr, K., Jamroz, K.: Integration of a multilevel transport system model into sustainable urban mobility planning. Sustainability **10**(2), 479–499 (2018)
3. Gavanas, N., Papaiannou, P., Pitsiava-Latinopoulou, M., Politis, I.: Urban Transport Networks and Mobility Management. Kallipos Publications, Athens (2015)
4. Siolas, A., Vassi, A., Vlastos, Th., Kyriakidis, C., Bakogiannis, E., Siti, M.: Methods, Tools and Urban Planning Applications. Kallipos Publications, Athens (2015)
5. Bakogiannis, E., Kyriakidis, C., Milioni, T.: The struggle for integration of refugees and immigrants: examples of integration of refugees in European Cities and the Case Study of Athens. Eur. J. Eng. Formal Sci. **3**(1), 18–27 (2019)
6. Bakogiannis, E., Kyriakidis, C., Siti, M., Karolemeas, C., Christopoulos, K.: Issues on social and technical transition concerning the participatory planning process in the context of sustainable urban mobility plans. In: 11th International Conference of the Hellenic Geographical Society, Laurio, Greece (2018)
7. Arnoykati, E., Maraš, V., Bugarinović, M., Paladini, S.: Paving the way to sustainable mobility. In: Proceedings of the International Conference on Traffic and Transport Engineering, Belgrade, Serbia (2016)
8. Nikitas, A., Wallgren, P., Rexfelt, O.: The paradox of public acceptance of bike sharing in Gothenburg. Proc. Inst. Civ. Eng. Eng. Sustain. **169**(3), 101–113 (2015)
9. European Commission (EC): A sustainable future for transport, towards an integrated, techologial-led and user friendly system. COM (2009) 279 final edn. Brussels (2009)
10. Skayannis, P., Goudas, M., Rodakinias, P.: Sustainable mobility and physical activity: a meaningful marriage. Transp. Res. Procedia **24**, 81–88 (2017)
11. Institute of Local Development: Urban planning: Application problems and reformation proposals – Scientific Committee Report (2006)
12. Karadimou-Gerolympou, A.: Reforming concepts in organizing urban spaces. 1917–1920. Contemp. Top. **39**, 55–60 (1987)
13. Tzika-Chatzopoulou, A.: Urban Planning Legislation. NTUA Publishing, Athens (2003)
14. Mimis-Stergiakis, D.: The Regulatory Plan for Larisa-Volos dipole – Und. Thesis, University of Thessaly, Volos (2004)
15. Aravantinos, A.: Urban Planning: For the Sustainable Development of Urban Space. Symmetria Publications, Athens (2007)
16. Economou, D.: Valuation of Law 1337/1983 after 25 years. In: Petrakos, G., Beriatos, H., Kokkosis, H., Kotzamanis, V. (eds.) 25 Papers for Planning and Development of Space – Collective Volume for the 20 Years of the Department of Planning and Regional Development (PRD), pp. 241–262. University of Thessaly Press, Volos (2009)

17. Giannakou, A.: Urban plans for Thessaloniki: ideology and practice during the 20th century. In: Labrianidis, L., Papamichos, N. (eds.) The Moving Frontier: The Changing Geography of Production in Labour Intensive Industries, pp. 447–487. Kritiki Publications, Athens (2008)
18. Koudouni, A.: Legislation framework of General Development Plans (GDP) and Open-City Spatial and Housing Organization Plan (SHOOAP) – Coursework for the Course: "Dynamics of Spatial Structures and Land Use and Contemporary Design Practices", School of Architecture, National Technical University of Athens (NTUA), Athens (2006). Case studies and sustainable urban mobility schemes: A communication channel among researchers and interdisciplinary community groups. International Journal of Service Science
19. Economou, D.: Spatial planning systems: the Greek reality and international experience. Soc. Res. Rev. **101**(101–102), 3–57 (2000)
20. Urban Reconstruction Company (EPA) 1982–1984: General Development Plan: Standards. In: Kousidonis, C., Lalenis, K., Economou, D. (eds.) Urban Planning Studio II: Programming – GDPs. University of Thessaly Press, Volos (2005)
21. Lalenis, K., Kyriazis, A.: Institutional conflicts and planning deadlocked: urban environment and forest in conflict with the general development plan for Kavala. In: Gospodini, A., Kotzamanis, V., Beriatos, H., Economou, D., Christopoulou, O., Gousios, D., Kougoulos, A.,Duquenne, M.N., Skayannis, P. (eds.) Proceedings of the 3rd Pan-hellenic Conference on Planning and Regional Development, Part I, pp. 312–317 (2009)
22. Papaioannou, P., Politis, I., Nikolaidou, A.: Steps towards sustaining a SUMP network in Greece. Transp. Res. Proceedia **14**, 945–954 (2016)
23. Bakogiannis, E., Kyriakidis, C., Siti, M., Floropoulou, E.: Reconsidering sustainable mobility patterns in cultural route planning: Andreas Syngrou Avenue, Greece. Heritage **2**, 1702–1723 (2019)
24. Arsenio, E., Martens, K., Di Ciommo, F.: Sustainable urban mobility plans: bridging climate change and equity targets? Res. Transp. Econ. **55**, 30–39 (2016)

A Vision for Urban Micromobility

From Current Streetscape to City of the Future

Shengwei Tan and Ken Tamminga(✉)

The Pennsylvania State University, University Park, PA 16802, USA
krtl@psu.edu

Abstract. As urban transport technology accelerates, various novel modes of electric-assisted personal transportation are emerging. These create both opportunities and constraints for transportation engineers and urban designers. Our research suggests that it is becoming increasingly clear that traditional road designs and public transportation infrastructures are struggling to accommodate the challenges. Micromobility (MM), including e-bikes, e-scooters, e-skateboards, Segways and hoverboards, is becoming more popular and acceptable by people in the urban environment. Benefits include portability, ease of use, and affordability through shared services. Yet questions abound: How can the increased presence of MM be part of the necessary mixed streaming on urban streets? How can existing infrastructure and spatial allocations be more accommodating of MM, while not unduly disadvantaging other transport forms? Using a case study from the core of Washington, DC, we model the possibilities for adaptable road features that might be implemented for MM based on different traffic loads and infrastructure configurations. We conclude with a brief examination of how micromobility accommodation is poised to leverage urban transformation more broadly, including as it relates to sustainable green infrastructure and stormwater management opportunities.

Keywords: Micromobility · Urban streetscape · PEV infrastructure · Autonomous vehicle · Urban transformation

1 Introduction

Micromobility Newsletter defines micromobility (MM) as any transportation tool that is less than 500 kg. Secondary traits include products with electrification ability and utility uses for commuting or public transporting. MM provides opportunities for both shared and private owned vehicles [1]. This quickly emerging technology is evidently increasing urban mobility with new degrees of convenience and affordability. Electric-powered products such as e-scooters, e-skateboards, hoverboards, Segways, and unicycles are changing the lifestyle and behaviors of human-beings in the city. However, these emerging micromobility products also pose social and functional issues because they are so new that current road infrastructures do not yet fully support their operational requirements [2]. For example, in many North American cities group-shared e-scooters are allowed to ride on bike lanes but are restricted from occupying space traditionally reserved for pedestrians: the sidewalk. However, because of a lack of

E. G. Nathanail et al. (Eds.): CSUM2020, AISC 1278, pp. 158–167, 2021.
https://doi.org/10.1007/978-3-030-61075-3_16

well-designed bike or multi-purpose lanes in almost all US cities, many e-scooter users ride on pedestrian pathways. Though illegal, this is also understandable considering that circulating in traffic dominated by larger cars and buses can be a risky proposition.

2 Problem Statement

Current street corridor and streetscape design in US cities do not effectively accommodate Personal Electronic Vehicles (PEV). E-scooters, Segways, moped-motorcycles, e-skateboards and other emerging micromobility forms currently compete for space along traffic lanes, bicycle lanes and transit corridors, and frequently interact with pedestrians, delivery vehicles and other modes—not always with positive results. PEV users are often confused about where they belong, and chaotic situations emerge as unsafe behaviors proliferate in the current urban environment.

The development of emerging personal transit technologies has exceeded the capacity of traditional public roadway infrastructures to safely and civilly accommodate them. In fact, traditional roadway designs limit the implementation of new transportation technologies because they are mainly designed for cars. Urban street spaces need to be reconsidered and rearranged in order to adapt emerging personal electrical vehicles and future autonomous vehicles. Traditional thinking on urban road engineering, design and management needs to be replaced by re-thinking circulation patterns and conventional infrastructure solutions.

3 Literature Review

3.1 Where to Ride and Park?

Scooters are allowed to run on bike lanes since in most US cities they are considered vaguely bicycle-like in comparison with other transportation types. At the same time, current road infrastructure provides a mostly inappropriate environment for scooters to operate. Generally, thus, riders are confused about where they should operate scooters and other micromobility modes. Regulations are either minimal or inconsistent, and poorly designed street corridors negatively influence the behavior of the riders, resulting in dangerous situations for riders, pedestrians and drivers. "Because of these problems and more, San Francisco, Nashville, Denver, Scottsdale, and Charlotte, among other cities, have written cease and desist letters to scooter vendors operating in their respective cities." [9]. Overall, the limitation of current urban streets in accommodating PEVs is mostly due to lack of multi-modal bike lanes, or otherwise poor connections and lack of safety protections and enforcement associated with existing bike lanes.

Scooter parking is also a key challenge for developing urban micromobility systems due to the lack of accommodation by conventional streetscape design. "There are a few ways PEV fleet suppliers and municipalities have been experimenting to create better parking, including requiring vendors to apply for a space within a "furniture zone" and to apply for a 'corral' in a traditional car parking spot" [9]. However, the storage for scooters cannot practically be designed for single units; rather, they need to be connected with many other transportation modes. Designated storage infrastructure should be considered as an essential ancillary infrastructure; such storage might also be able to integrate with other functions such as self-supply charging service stations, communication base stations, visitor guidance kiosks, green infrastructure provisions, and even public shelters. The goal for scooter parking should be the provision of a range choices for city inhabitants as they consider their needs use based on the circumstances at play.

3.2 Bike Lane Design Conventions

Based on design guidance from the National Association of City Transportation Officials (NACTO) the minimum width of a one-way urban bike lane should be 91 cm (36 inches, with 150 to 183 cm (5 to 6 feet) desirable [10]. If the bike lane is placed adjacent to a parking lane, the width of the bike lane should be at least 150 cm (5 ft.), which means there has to be a 61 cm (24 inch) buffer between the riding area and the parking area. The 61 cm buffer is also useful as the buffer distance from any other physical barriers.

NACTO recommends that bike lane iconography (e.g. words, symbols, arrow markings) be placed outside of the motor vehicle tread path at intersections, driveways, and merging areas in order to minimize wear from the motor vehicle path and enhance safety for all modes. A solid white lane line marking is recommended to separate motor vehicle travel lanes from the bike lane. Most jurisdictions use a 1.83 to 2.44 m (6 to 8 ft.) wide painted line [10].

3.3 Autonomous Urbanism with Micromobility

Autonomous urbanism is defined by NACTO as an environment that distributes the benefits of the public data-driven operations equally for citizens by designing or allocating city streetscapes in full consideration of emerging autonomous vehicle, smart city, and micromobility technologies [6]. One major factor of autonomous urbanism is the development of Autonomous Vehicles (AVs). Some experts believe that AVs are going to change the blueprint of the streetscape, and that this huge technology shift will benefit micromobility because the implementation of autonomous vehicle culture could open up space for other modes along the roadway cross-section [6]. For example, the width of the road could be narrowed because AVs have the ability to promote "road softening": the initial 3.05 to 3.66 m (10 to 12 ft.) traffic lane width could be reduced to 2.74 to 3.05 m (9 to 10 ft.) since AVs will be able to avoid most of the typical dangers associated with conventional vehicles. Conventional parking might also be transformed because the act of dropping-off would replace the act of parking; individual parking spaces could be removed for other potential benefits, but drop-off design for autonomous vehicles might be introduced to future road design [7].

The concept of road cross-sectional rights-of-way that are adaptive to real-time demands also presents possibilities for MM to thrive as part of the future autonomous urbanism scenario [8]. In the context of "big data", city streets could adapt during certain circumstances, thus changing the function of the road. For example, if online, real-time traffic information for a particular street shows that there are few cars driving on the street during the weekend, then this street may provide its street space for community events or other activities during the weekend.

4 Case Study: Washington DC

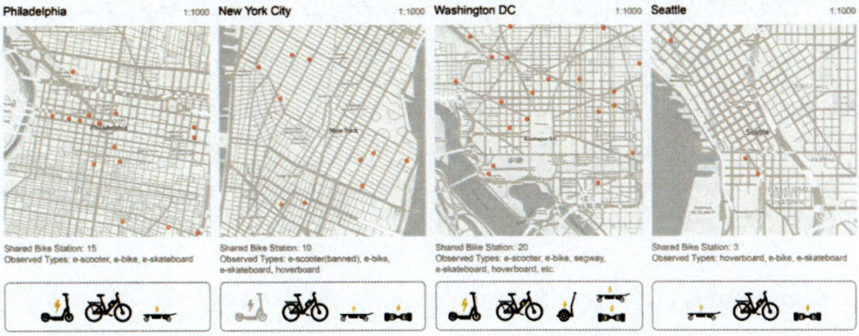

Fig. 1. Site selection process: identify existing MM-related infrastructures and observe variety of MM types in use.

Cities that accept shared e-scooters on the U.S. East coast or mid-West that are a practical distance for researchers at Penn State University to visit (see Fig. 1) include Philadelphia, New York City, and Washington DC. Seattle was also briefly assessed. These four cities were receptive to pioneering PEV fleet purveyors, providing sufficient services from several e-scooter/e-bike firms, a relatively flat geography, and a generally supportive municipal regulatory context. After visiting each city, Washington DC emerged as the most suitable venue for case study. The research areas were chosen based on key commercial precincts and important residential neighborhoods.

Detailed site selection was based on suitability analysis using the ESRI-based geographic information system (GIS). Data compiled included building height, land use, street width, landmarks, and green spaces. Along with extensive reconnaissance in the summer of 2019, this information helped to identify which areas might suitable as case study areas. The three selected study areas are shown on Fig. 2. Due to article length constraints, we will mostly address Area B, Dupont Circle, a major intersection in the DC core. Analysis and design details will also briefly address Area A, the Georgetown residential area, and Area C, K Street, a commercial intersection in downtown DC.

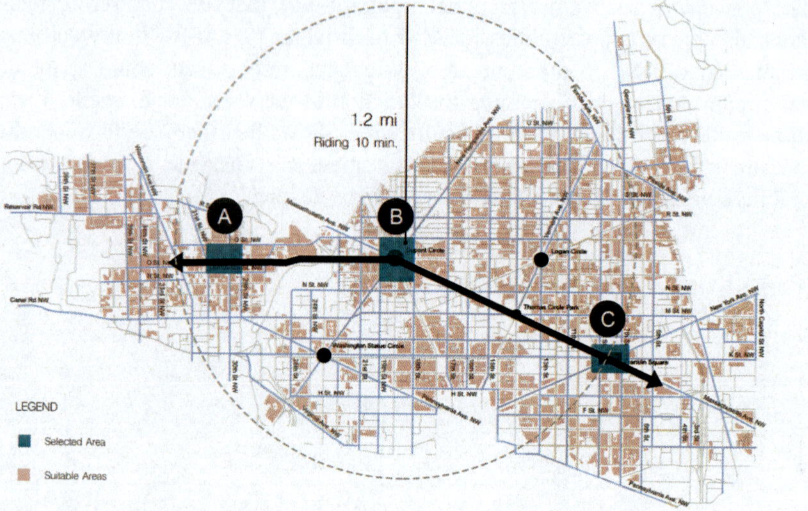

Fig. 2. Suitable research areas in light red, with selected case study sites shown in cyan, based on GIS suitability analysis.

4.1 Design Strategy

The design strategy that follows shows a range of possible solutions in addressing problems summarized from the design principles that are framed below. Because design conventions that created the traditional urban streetscape have resulted in a maze of standards and requirements for engineers and designers, design prototypes based on broad principles grounded in realistic case study present an opportunity to develop a whole new design for roadways in the 21st century city. Therefore, the researchers provide design prototypes applied to test areas in Washington, DC as an experiment with prototype effectiveness in terms of spatial, functional and management aspects.

4.2 Design Principles: Safety, Connectivity, Experience

Design for micromobility requires carving out space within the street and creating a spatial identity, or presence, for PEVs. Heightened recognition and spatial separation from other transport modes—especially larger vehicles—is essential if PEVs are to become an integral part of future urban mobility systems.

Connectivity has two levels of meaning. The first is the continuous movement and ease of circulation PEV movement, without too many obstacles caused by other vehicles, confrontational behaviors, or inappropriate physical barriers. The second meaning has to do with the transition from other types of transport modes to PEVs. Transitions should be direct and convenient; this means that the location of PEV infrastructures should be integrated with urban transit stations such as metro stations and bus stops. People should also have the ability to carry their PEVs onto public transportation such as trains, metros and buses. Finally, lane allocation and design, and implications for both MM and complementary modes, requires re-thinking.

In terms of individual human experience, people experience less visual and other sensory stimuli when they are in a car or a bus because they have limited access to sights, smells and sounds along the street corridor. PEV riders, however, have much more direct contact with the city landscape. A comprehensive micromobility ecosystem should be able to integrate with progressive approaches to urban green infrastructure and sustainable stormwater facilities.

4.3 Design Inquiry

Our first design exploration re-allocates on-street parking by taking advantage of curbside space. Concurrently, space for the PEV lane is created through substituting curbside space for street parking space (see Fig. 3). The physical barriers in the roadway median can be reduced, generating spaces for personal electric vehicles—as long as there is at least a 91 cm (3 ft.) width which can be deducted from the barrier. For lanes that cannot generate PEV space based on those two methods, PEV users must share lanes with larger vehicles, and speed limits would need to be limited to 40 kph (25 mph).

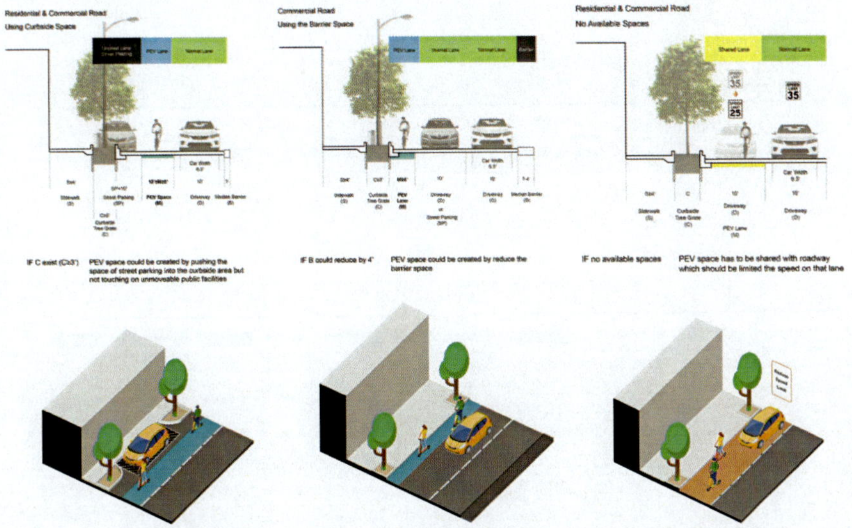

Fig. 3. This concept reconfigures street parking and median barriers to generate space for PEVs.

Prototypical design concepts can be transformed into many other different forms based on the specific situation of each area. The Georgetown case area demonstrates how the design prototype will be applied at P Street NW (see Fig. 4) within a 10-year timeline. A key intervention is that street parking is pushed from the roadway edge to the curbside area between each tree grate.

Fig. 4. Pushing street parking into the curbside area to create a PEV lane.

In the future, PEV lanes could also be considered as an adaptive lane for riders to use during off-hours which could maximize the usage of the road. This adaptive lane can be also designated as a temporary fast lane for the more vulnerable PEVs (speed could be over 20 mph) such as e-bikes, fast unicycles, moped e-bikes, etc. (see Fig. 5).

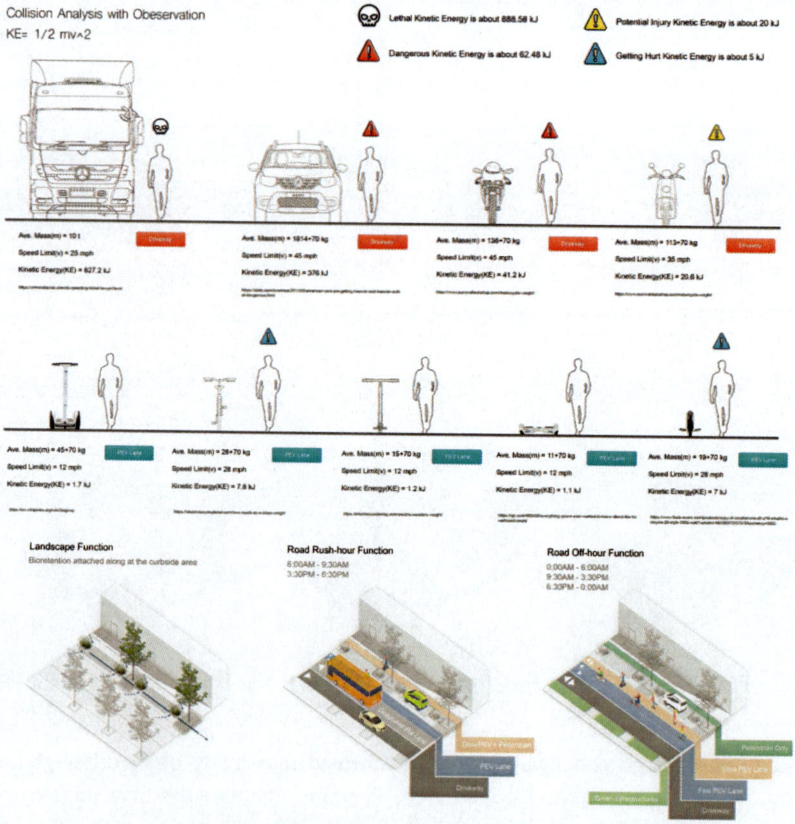

Fig. 5. Adaptive PEV lane with Fast & Slow function based on the collision analysis. Products with higher potential damage should be placed away from the pedestrian.

Another example is Area B, Dupont Circle. This iconic intersection requires a much more complex analysis in order to reduce negative interactions between pedestrian, PEVs and vehicles. Multi-modal interference analysis helped researchers to be aware of conflict points where PEV users might engage with vehicles. A painted dashed PEV line indicates the conflict potential, and PEV speed is slowed by integrating dashed lines with high-resistance/textured surface materials (see Fig. 6).

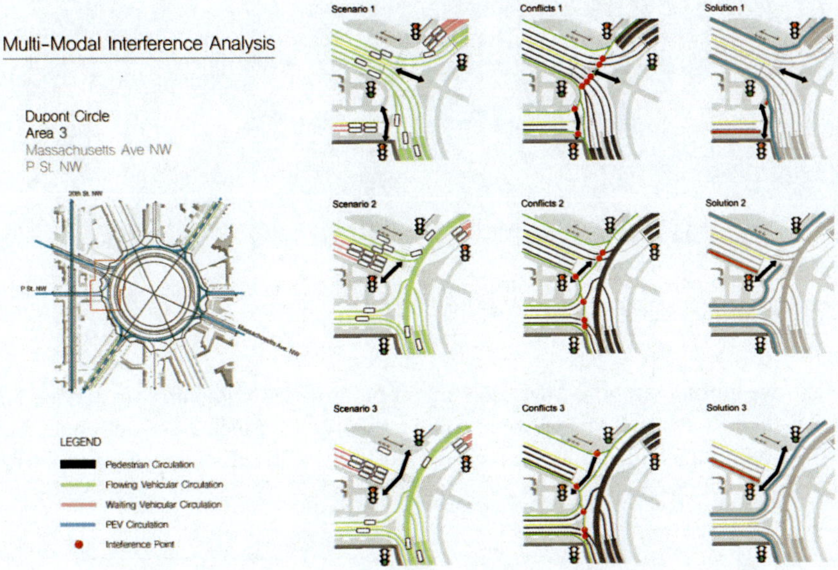

Fig. 6. Analysis of potential inter-modal conflicts to determine how PEV users may proceed under various traffic signal combinations.

When level 5 autonomous (i.e. completely self-driving) vehicles are introduced, the street pattern should fully adapt to accommodate micromobility as well. By removing vehicular on-street parking and replacing one lane of the roadway with a protected PEV lane with an 8 cm (3") raised vertical separation, micromobility can become a fully viable transport choice for future citizens to move through the city more safely and enjoyably (see Fig. 7).

Fig. 7. The future Mixed-streaming urban streetscape includes an equal space division between vehicles and Micromobility.

Next, we considered how MM users might be provided with carry-on service and easy transitions between modes. People could easily carry on their e-scooter on the bus, metro or even into a building as we transition to a city with self-supplied green energy technology (see Fig. 8).

Fig. 8. Carry-on PEVs on e-bus, with smaller PEVs stashed in recharging/locker rooms located in the foyers of large buildings.

5 Conclusion

Micromobility is poised to transform the way people move through the city and the way they inhabit the streetscape. Our scenarios envision that traditional urbanism that favors low-efficiency vehicles which have long dominated roadways will shift to a future urbanism that promotes high-efficiency, multi-modal mixed-streaming along the city street corridor.

In addition, our scenarios still require further studies about practical application in the regional disparity in terms of weather influences and municipal codes and engineering standards. But the research and design inquiry herein suggest that MM could be helpful to support a more equitable and dynamic roadway ecosystem that achieves high efficiency on the street and creates a good infrastructural foundation for the introduction of autonomous vehicles in the future. Micromobility could also help achieve a more sustainable city, particularly considering that by the 2050s fossil fuels are expected to run low [4]. Localized, renewable energy could integrate with the development of micromobility because PEVs have relatively lower energy footprint compared to larger vehicles. As an entire category of cleaner, quieter and safer vehicles is in the offer in cities worldwide, PEV users must be provided the space to interact positively with other city inhabitants, thus enhancing the vitality and sustainability of the city.

References

1. Stubblefield, C.: The micromobility landscape. In: Micromobility Industries (2019). https://micromobility.io/blog/the-micromobility-landscape. Accessed 4 Mar 2020.
2. Field, K.: What is the ultimate personal electric vehicle? In: CleanTechnica (2018). https://cleantechnica.com/2018/08/05/what-is-the-ultimate-personal-electric-vehicle/. Accessed 11 Feb 2020.
3. Habersham, R.: New safety measures set for Atlanta's e-scooter riders. In: ajc (2019). https://www.ajc.com/news/local/atlanta-beltline-new-safety-measures-target-scooter-riders/Ol90WYClUZn2jRVapMPnLN/. Accessed 26 Feb 2020.
4. Holloway, J.: Hot, crowded, and running out of fuel: Earth of 2050 a scary place. In: Ars Technica (2012). https://arstechnica.com/science/2012/03/hot-crowded-and-running-out-of-fuel-earth-of-2050-a-scary-place/. Accessed 10 Mar 2020.
5. Ojeda, A.: Mother, daughter seriously injured in boardwalk scooter crash. In: NBC 7 San Diego (2018). https://www.nbcsandiego.com/news/local/serious-injury-scooter-accident-boardwalk-mission-beach/49448/. Accessed 26 Feb 2020.
6. Ink S NACTO Blueprint for Autonomous Urbanism. https://nacto.org/publication/bau2/. Accessed 11 Feb 2020
7. Goodyear, U.K.: The Goodyear Eagle-360 concept tyre (2016). https://www.youtube.com/watch?v=oSFYwDDVgac. Accessed 11 Feb 2020
8. Yadan, L.: From transportation infrastructure to green infrastructure: adaptable future roads in autonomous urbanism. Landscape Architect. Front. **7**, 92 (2019)
9. Kirstin, A., Brandon, B., Riley, O., Christopher, S.: Governing Micromobility: A Nationwide Assessment of Electric Scooter Regulations. TRB 2019 Annual Meeting (2019)
10. NACTO: Bike Lanes. Urban Bikeway Design Guide, pp. 1–26 (2014). https://doi.org/10.5822/978-1-61091-582-3_1

Transformational Technologies

Deep Bidirectional and Unidirectional LSTM Neural Networks in Traffic Flow Forecasting from Environmental Factors

Georgios N. Kouziokas(✉)

School of Engineering, University of Thessaly, Volos, Greece
gekouzio@uth.gr

Abstract. The application of deep learning techniques in several forecasting problems has been increased the last years, in many scientific fields. In this research, a deep learning structure is proposed, composed mainly of double Bidirectional Long Short-Term Memory (Bi-LSTM) Network layers, for the prediction of the traffic flow in the study area. Also, traffic flow-related environmental factors were taken into consideration in order to construct the deep learning forecasting model. The final results have showed an increased accuracy of the proposed deep learning Bi-LSTM – based model compared to other machine learning models that were tested such as unidirectional LSTM networks, Support Vector Machines and Feedforward Neural Networks.

Keywords: Artificial neural networks · Bidirectional LSTM · Support Vector Machines · Environmental factors · Traffic flow forecasting · Public management

1 Introduction

The development of information technology has led to the adoption of new technology-based management systems and techniques in public management [1, 2, 3, 4] and also to the development of new intelligent transportation systems in urban environment for facilitating decision making in transportation [5–6]. The adoption of artificial intelligence – based methods as prediction techniques has rapidly increased the last years in several scientific fields, such as smart cities [7–9], sustainable public management [10–11], safety and security in smart cities [12–13], socioeconomic factors [14–15], transportation safety and management [16–18]. The applicability of the deep bidirectional and unidirectional LSTM neural networks has been increased the last years due to the improved prediction accuracy compared to other state of the art prediction techniques in many scientific problems [19–21]. This research investigates the application of deep bidirectional and unidirectional long short-term memory neural networks in traffic flow forecasting in order to improve the urban management strategies and apply better proactive measures in traffic related problems.

Several researchers have studied forecasting techniques based on machine learning or conventional forecasting techniques in order to predict the traffic flow and improve urban transportation management strategies [22–26]. Hu et al. [22], have proposed a

E. G. Nathanail et al. (Eds.): CSUM2020, AISC 1278, pp. 171–180, 2021.
https://doi.org/10.1007/978-3-030-61075-3_17

hybrid model in traffic flow forecasting by implementing particle swarm optimization and support vector machines. The final results of the experiments have shown that the proposed technique achieve more accurate prediction results compared to the other forecasting methodologies that were developed such as, ARIMA (Autoregressive Integrated Moving Average) and BPNN (Back Propagation Neural Network). Lippi et al. [23], have studied the application of several prediction models in traffic flow prediction such as: ARIMA (Seasonal Autoregressive Integrated Moving Average) coupled with a Kalman filter, support vector machines with an RBF kernel, artificial neural networks. According to the experimental results, the most accurate model compared to the other forecasting techniques that were implemented, was the ARIMA model coupled with a Kalman filter. Moretti et al. [24], have applied a hybrid neural network bagging ensemble model. The hybrid model combines Artificial Neural Networks (ANNs) and a statistical approach to forecast the urban traffic flow rates. The final results have shown that the hybrid statistical and bagging ensemble model had an improved prediction accuracy compared to the artificial neural network models, the basic ensemble model, and the bagging ensemble model. Hong et al. [25], has proposed hybrid evolutionary algorithms in traffic flow prediction. The authors have proposed a hybrid algorithm based on genetic algorithm, simulated annealing algorithm and support vector regression. The final results have shown that the proposed model produce better prediction results than the other tested models: SARIMA (Seasonal Autoregressive Integrated Moving Average), Holt–Winters, back-propagation neural network, and seasonal Holt–Winters techniques. Yang et al. [26], have proposed a deep learning approach in traffic flow forecasting. The authors have implemented a stacked autoencoder Levenberg–Marquardt (SAE-LM) neural network model with one hidden layer. The results have shown that the proposed model had better prediction accuracy than the other tested models such as: particle swarm optimized neural network (PSONN) and the RBF neural network.

In this research, a novel deep learning double Bidirectional Long Short-Term Memory (Bi-LSTM) Network layer model is proposed for the first time according to the literature, in order to predict the traffic flow and advance the transportation safety and management strategies by taking into consideration environmental factors. In the next sections the proposed methodology, the experimental results and the discussion are presented.

2 Preliminaries

2.1 Deep Bidirectional and Unidirectional LSTM Networks

The last years deep unidirectional Long Short – Term Memory (LSTM) networks have been used successfully in several scientific problems as a prediction technique. The LSTM networks has a structure similar to the Recurrent Neural Networks (RNNs). The main advantage of LSTM networks over the normal RNNs is that they overcome the vanishing gradients problem which is a major drawback of the regular RNNs [27].

The deep LSTM network structure consists of four types of gates-components: the input gate, the forget gate, the cell candidate and the output gate. The following mathematical equations express the functions of the gates-components at time step *t*.

The corresponding equation for the input gate:

$$i_t = \sigma_g(W_i \mathbf{x}_t + R_i \mathbf{h}_{t-1} + b_i) \tag{1}$$

The corresponding equation for the forget gate:

$$f_t = \sigma_g(W_f \mathbf{x}_t + R_f \mathbf{h}_{t-1} + b_f) \tag{2}$$

The corresponding equation for the cell candidate:

$$g_t = \sigma_c(W_g \mathbf{x}_t + R_g \mathbf{h}_{t-1} + b_g) \tag{3}$$

The corresponding equation for the output gate:

$$o_t = \sigma_g(W_o \mathbf{x}_t + R_o \mathbf{h}_{t-1} + b_o) \tag{4}$$

where the parameters *f, i, g, o* represent the forget gate, the input gate, the cell candidate, and the output gate, respectively. The parameter *R* represents the recurrent weights the parameter *W* the input weights, the b factor denotes the bias, h_t the hidden state at the time step *t*, σ_g the gate activation function and σ_c the state activation LSTM function.

2.2 Support Vector Machines

The Support Vector Machines were applied in this research as proposed by Vapnik [28]. In regression problems Support Vector Machines implement an ε-insensitive loss function. The function that Support Vector Machines apply in regression problems is expressed by the following mathematical formula:

$$f(x) = (w \cdot \Phi(x)) + b) \tag{5}$$

Where the conditions are: $w \subset R^n$, the $b \subset R$. The Φ represents the SVM non-linear transformation from the R^n to a high dimensional space.

2.3 Artificial Neural Networks

The Artificial Neural Networks (ANNs) are computing systems trying to simulate the complex structure of the brain system. A neural network processes information received from the input parameters. The data traverses via connections of the network layers to produce an output according to the processed input parameters [29]. The Artificial neural networks were implemented in this study to predict the traffic flow, since they are universal approximators and they can model non-linear relationships. A Feedforward Multilayer Perceptron was implemented, since it is considered as the most suitable in time series prediction problems [30].

3 Proposed Methodology

The proposed deep learning model focuses on the development of a double Bidirectional LSTM layer model. The proposed deep learning structure consists of an input layer, two Bi-LSTM layers, two dropout layers, a fully connected layer and an output regression layer. The dropout layers were added in the network to the feed-forward LSTM connections in order to avoid the overfitting problem [31].

The research methodology includes the following basic stages: data collection and preparation, data cleaning and normalization, machine learning prediction model creation, LSTM, Bi-LSTM, SVM, ANN models development, and finally the comparison of the developed forecasting models in the selected study area.

According to the literature, some environmental factors such as precipitation and visibility affect the traffic flow [32–33]. These factors and the historical data of the traffic flow were selected as input factors. The one-time step forecasting method was used which means that the input factors were used to feed the machine learning models in order to predict the value of the traffic flow (number of vehicles) of the next time step (The next five minutes). In the first stage, the input data were collected and prepared regarding the traffic flow, the precipitation and the visibility. In the second stage, the dataset was examined for duplicates and then was normalized by using the min-max normalization method in order to be used as inputs into the machine learning models. In the third stage, the Bi-LSTM, LSTM, SVM, ANN prediction models were developed by setting up the experimental parameters. In the final stage, comparison and evaluation of the final machine learning models were implemented in order to discover the optimal forecasting model in the study area (Fig. 1).

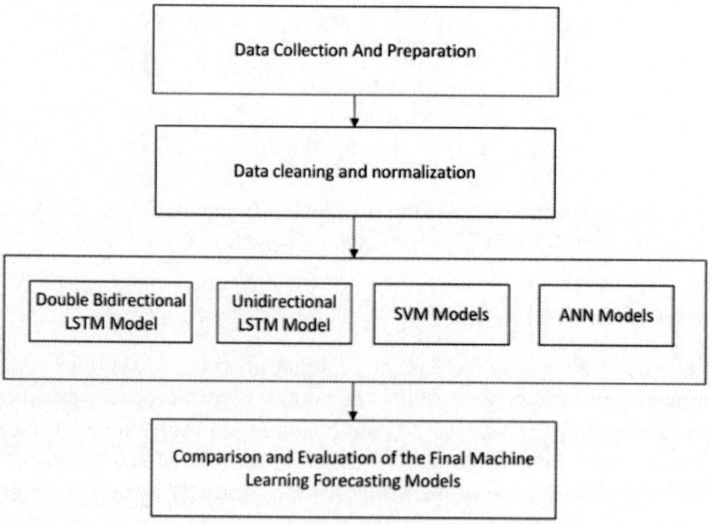

Fig. 1. Methodology overview.

4 Results and Discussion

4.1 Data Collection and Normalization and Preparation

The Traffic Flow data were collected from the Caltrans Performance Measurement System (PeMS)[1] in a five-minute interval for highway segments in the 4th district in the Alameda Bay Area in Oakland for the time period 04/01/2016 to 31/03/2016. The precipitation data (inch/h) and visibility data (mile) were collected from the Automated Surface Observing System (ASOS) of the National Centers for Environmental Information[2] for the study area. The collected dataset was prepared and pre-processed for duplicates or other incoherencies. The min-max normalization method was implemented to normalize the dataset values. The 10-Fold cross validation was implemented to overcome the overfitting problem in all the machine learning models.

4.2 The Proposed Bi-LSTM Deep Learning Model

The proposed Bi-LSTM deep learning model was developed by adding an input layer, two Bi-LSTM layers, two dropout layers, a fully connected layer and an output regression layer. The dropout layers were added to feed-forward LSTM connections in order to avoid the overfitting problem [31]. The adaptive moment estimation training algorithm was used. 250 hidden units in every Bi-LSTM layer that produced the optimal results. The 10-fold cross validation was implemented during training. The initial learning rate was 0.001 and the gradient threshold was the L1 norm. The hyperbolic tangent function (tanh) was used as the state activation function. The sigmoid function was implemented as the gate activation function.

4.3 Unidirectional LSTM Model

Also, a deep unidirectional LSTM model was constructed to be compared with the double Bi-LSTM layer deep learning model. An LSTM layer, a dropout layer, a fully connected layer and an output regression layer were applied. The number of hidden units of the unidirectional LSTM was set to the 250 hidden units. The 10-fold cross validation was implemented during training. The initial learning rate was set to 0.001, and the L1 norm was implemented as gradient threshold method. The state activation function was the hyperbolic tangent function and the sigmoid function as the gate activation function. The adaptive moment estimation algorithm was implemented as the optimization algorithm.

4.4 SVM Models

Several kinds of SVM kernels were tested by using 10-fold cross validation: linear, Gaussian, Radial Basis Function (RBF) and the polynomial SVM. Several hyperparameters were investigated: The C constraint, the kernel scale (γ), the ε (epsilon)

---

[1] https://pems.dot.ca.gov/

[2] https://www.ncdc.noaa.gov/

parameter. The best prediction results were performed by the RBF kernel. The following table summarize the results (Table 1).

Table 1. Comparison table of the SVM kernels

SVM kernel	RMSE
Linear	29.0814
RBF	12.3996
Gaussian	14.7101
Polynomial	16.8425

4.5 Artificial Neural Network Models

The Artificial Neural Network models were constructed by using a feedforward multilayer perceptron and 10-fold cross validation. Several network topologies were tested regarding the hidden layer neurons, and also the types of the hidden layer transfer functions such as the linear transfer function, the tanh-sigmoid transfer function, and also the log-sigmoid transfer function. The Levenberg-Marquardt (LM) algorithm was implemented as the training algorithm. The structure of the optimal model was the one with 16 neurons and the tan-sigmoid transfer function in the hidden layer. The RMSE error of the optimal model was 15.4646.

4.6 Model Comparison

The experimental results have shown that the proposed double bidirectional dropout Bi-LSTM layer model had better prediction accuracy than the other tested machine learning models: Unidirectional LSTM networks, Support Vector Machines and Feedforward Artificial Neural Networks. The experimental results are illustrated in the comparison Table 2.

Table 2. Comparison table of the developed forecasting models

Method	RMSE
Unidirectional Dropout LSTM	9.3159
Double Bidirectional Dropout Bi-LSTM	6.5278
SVM	12.3996
ANN	15.4646

The Root Mean Squared Error (RMSE) of the proposed model was found 6.5278. Figure 2 shows the error distribution of the Bidirectional LSTM model. Figure 3. Illustrates the error variances of the proposed double Bi-LSTM model. Figure 4 shows the comparison chart of the Bidirectional LSTM model for the target and the predicted traffic flow values of the test set. Figure 5 shows the comparison diagram of the true and the predicted values for all the developed Bi-LSTM, LSTM, SVM, ANN models.

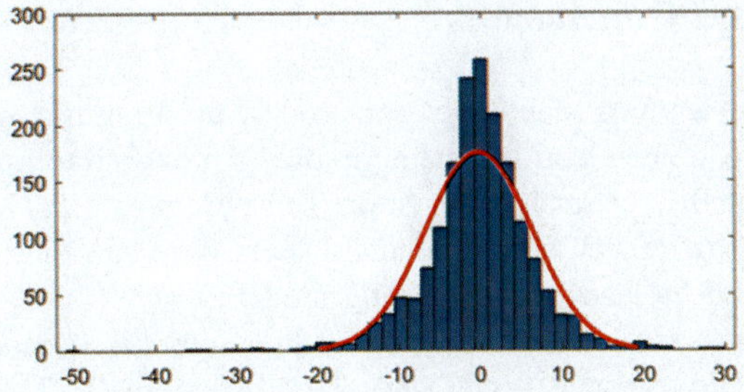

Fig. 2. Error distribution of the proposed double Bi-LSTM model.

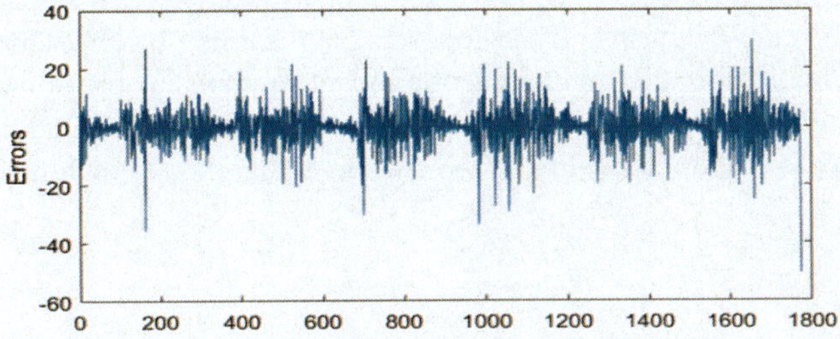

Fig. 3. Error variances of the proposed double Bi-LSTM model for the last 1800 time steps.

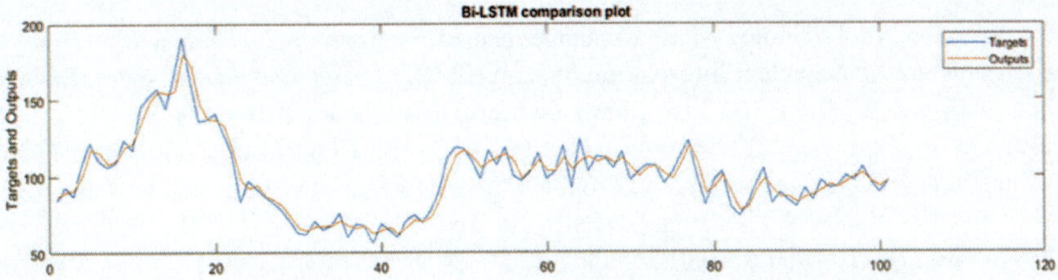

Fig. 4. Comparison chart of the Bi-LSTM model for the last 100 targets of the test set.

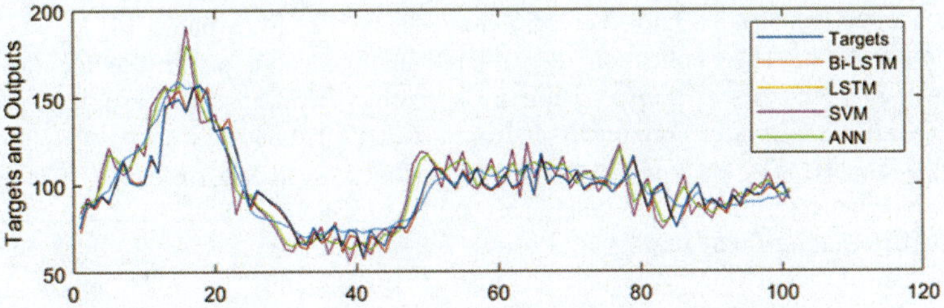

Fig. 5. Comparison chart for the developed models for the last 100 targets of the test set.

5 Discussion and Conclusions

The adoption of deep learning techniques can be very useful in transportation decision making. In this study, a deep learning methodology is proposed to forecast traffic flow by adopting a double Bidirectional Long Short-Term Memory (LSTM) Network layer model. Also, traffic flow-related environmental factors were taken into consideration in order to develop the deep learning forecasting model.

Compared to other researches [22–27] of the literature this research utilizes for the first time a double Bidirectional Long Short-Term Memory (LSTM) model for predicting the traffic flow. The experimental results of this study have shown an improved forecasting accuracy of the proposed double Bidirectional LSTM deep learning model in the study area, compared to the other machine learning techniques that were implemented: unidirectional LSTM networks, Support Vector Machines and Feedforward Artificial Neural Networks. The proposed deep learning-based methodology for developing the optimal traffic flow prediction model can be valuable in transportation management, urban planning and public management, and also in adopting better proactive strategies facilitate the life of the citizens in the urban environment.

References

1. Rezende, D.A.: Digital city projects: information and public services offered by Chicago (USA) and Curitiba (Brazil). In: Open Government: Concepts, Methodologies, Tools, and Applications, pp. 1452–1468. IGI Global (2020). https://doi.org/10.4018/978-1-5225-9860-2.ch068
2. Kouziokas, G.N.: Technology-based management of environmental organizations using an Environmental Management Information System (EMIS): design and development. Environ. Technol. Innov. **5**, 106–116 (2016). https://doi.org/10.1016/j.eti.2016.01.006
3. Batunova, E., Popovich, T., Smirnova, O., Truhachev, S.: Concept of intelligent decision-making support system for city environment management. In: Information Fusion and Intelligent Geographic Information Systems, pp. 167–178. Springer (2020). https://doi.org/ https://doi.org/10.1007/978-3-030-31608-2_12
4. Kouziokas, G.N.: SVM kernel based on particle swarm optimized vector and Bayesian optimized SVM in atmospheric particulate matter forecasting. Appl. Soft Comput. **93**, 106410 (2020a). https://doi.org/10.1016/j.asoc.2020.106410
5. Grant-Muller, S., Usher, M.: Intelligent transport systems: the propensity for environmental and economic benefits. Technol. Forecast. Soc. Change **82**, 149–166 (2014)
6. Kouziokas, G.N.: The application of artificial intelligence in public administration for forecasting high crime risk transportation areas in urban environment Transp. Res. Procedia **24**, 467–473 (2017). https://doi.org/10.1016/j.trpro.2017.05.083
7. Li, D., Deng, L., Cai, Z.: Intelligent vehicle network system and smart city management based on genetic algorithms and image perception. MSSP **141**, 106623 (2020). https://doi.org/10.1016/j.ymssp.2020.106623
8. Chui, K.T., Lytras, M.D., Visvizi, A.: Energy sustainability in smart cities: artificial intelligence, smart monitoring, and optimization of energy consumption. Energies **11**, 2869 (2018)

9. Agarwal, P.K., Gurjar, J., Agarwal, A.K., Birla, R.: Application of artificial intelligence for development of intelligent transport system in smart cities. J. Traffic Transp. Eng. **1**, 20–30 (2015)
10. Wirtz, B.W., Müller, W.M.: An integrated artificial intelligence framework for public management. Public Manage. Rev. **21**, 1076–1100 (2019). https://doi.org/10.1080/14719037.2018.1549268
11. Kouziokas, G.N., Perakis, K.: Decision Support System Based on Artificial Intelligence, GIS and Remote Sensing for Sustainable Public and Judicial Management. Europ. J. Sustain. Dev. **6**(3), 397–404 (2017). https://dx.doi.org/https://doi.org/10.14207/ejsd.2017.v6n3p397.
12. Srivastava, S., Bisht, A., Narayan, N.: Safety and security in smart cities using artificial intelligence—a review. In: 2017 7th International Conference on Cloud Computing, Data Science & Engineering-Confluence, pp. 130–133. IEEE (2017)
13. Nigon, J., Glize, E., Dupas, D., Crasnier, F., Boes, J.: Use cases of pervasive artificial intelligence for smart cities challenges. In: 2016 International IEEE Conferences on Ubiquitous Intelligence & Computing, Advanced and Trusted Computing, Scalable Computing and Communications, Cloud and Big Data Computing, Internet of People, and Smart World Congress (UIC/ATC/ScalCom/CBDCom/IoP/SmartWorld), pp. 1021–1027. IEEE (2016). https://doi.org/https://doi.org/10.1109/UIC-ATC-ScalCom-CBDCom-IoP-SmartWorld.2016.0159
14. Kouziokas, G.N.: A new W-SVM kernel combining PSO-neural network transformed vector and bayesian optimized SVM in GDP forecasting. Eng. Appl. Artif. Intell. **92**, 103650 (2020b). https://doi.org/10.1016/j.engappai.2020.103650
15. Wagner, D.N.: Economic patterns in a world with artificial intelligence. Evol. Instit. Econ. Rev. **17**, 111–131 (2020). https://doi.org/10.1007/s40844-019-00157-x
16. Kouziokas, G.: Artificial intelligence and crime prediction in public management of transportation safety in urban environment. In: Proceedings of the 3rd Conference on Sustainable Urban Mobility, pp. 534–539 (2016)
17. Gura, D., Dubenko, Y., Shevchenko, G., Dyshkant, E., Khusht, N.: Three-dimensional laser scanning for safety of transport infrastructure with application of neural network Algorithms and methods of artificial intelligence. Transportation Soil Engineering in Cold Regions, vol. 2, pp. 185–190. Springer (2020). doi:https://doi.org/https://doi.org/10.1007/978-981-15-0454-9_19
18. Kouziokas, G.N.: Neural network-based road accident forecasting in transportation and public management. In: Nathanail, E.G., Karakikes, I.D. (eds.) Advances in Intelligent Systems and Computing, Data Analytics: Paving the Way to Sustainable Urban Mobility, vol. 879, pp. 98–103. Springer International Publishing, Cham (2019). https://doi.org/https://doi.org/10.1007/978-3-030-02305-8_12
19. Hu, F., Zhu, Y., Liu, J., Li, L.: An efficient long short-term memory model based on laplacian eigenmap in artificial neural networks. Appl. Soft Comput. **106218** (2020). https://doi.org/https://doi.org/10.1016/j.asoc.2020.106218
20. Liu, L., Finch, A., Utiyama, M., Sumita, E.: Agreement on target-bidirectional recurrent neural networks for sequence-to-sequence learning. J. Artif. Intell. Res. **67**, 581–606 (2020)
21. Kouziokas, G.N.: Long Short-Term Memory (LSTM) deep neural networks in energy appliances prediction. In: Proceedings of the 2019 IEEE Panhellenic Conference on Electronics & Telecommunications (PACET), pp. 1–5 (2019). https://dx.doi.org/https://doi.org/10.1109/PACET48583.2019.8956252
22. Hu, W., Yan, L., Liu, K., Wang, H.: A short-term traffic flow forecasting method based on the hybrid PSO-SVR. Neural Process. Lett. **43**, 155–172 (2016)

23. Lippi, M., Bertini, M., Frasconi, P.: Short-term traffic flow forecasting: an experimental comparison of time-series analysis and supervised learning. IEEE Trans. Intell. Transp. Syst. **14**, 871–882 (2013)
24. Moretti, F., Pizzuti, S., Panzieri, S., Annunziato, M.: Urban traffic flow forecasting through statistical and neural network bagging ensemble hybrid modeling. Neurocomputing **167**, 3–7 (2015)
25. Hong, W.C., Dong, Y., Zheng, F., Wei, S.Y.: Hybrid evolutionary algorithms in a SVR traffic flow forecasting model. Appl. Math. Comput. **217**, 6733–6747 (2011)
26. Yang, H.F., Dillon, T.S., Chen, Y.P.P.: Optimized structure of the traffic flow forecasting model with a deep learning approach. IEEE Trans. Neural Netw. Learn. Syst. **28**, 2371–2381 (2016)
27. Cortez, B., Carrera, B., Kim, Y.J., Jung, J.Y.: An architecture for emergency event prediction using LSTM recurrent neural networks. Expert Syst. Appl. **97**, 315–324 (2018)
28. Vapnik, V.: The Nature of Statistical Learning Theory Springer, New York (1995)
29. Svozil, D., Kvasnicka, V., Pospichal, J.: Introduction to multi-layer feed-forward neural networks. Chemom. Intell. Lab. Syst. **39**(1), 43–62 (1997)
30. Hornik, K.: Approximation capabilities of multilayer feedforward networks. Neural Netw. **4** (2), 251–257 (1991)
31. Pham, V., Bluche, T., Kermorvant, C., Louradour, J.: Dropout improves recurrent neural networks for handwriting recognition. In: 2014 14th International Conference on Frontiers in Handwriting Recognition, pp. 285–290. IEEE (2014).
32. Wang, C., Dai, Y., Zhou, W., Geng, Y.: A vision-based video crash detection framework for mixed traffic flow environment considering low-visibility condition. J. Adv. Transport. **2020,** 1–11 (2020)
33. Rakha, H., Farzaneh, M., Arafeh, M., Sterzin, E.: Inclement weather impacts on freeway traffic stream behavior. Transp. Res. Rec. **2071**, 8–18 (2008)

Accelerating the Deployment of Electric Light Vehicles for Sustainable Urban Mobility: A Harmonized Pilot Demonstration Methodology

Anna Antonakopoulou[1(✉)], Evangelia Portouli[1], Nikolaos Tousert[1], Maria Krommyda[1], Angelos Amditis[1], Maria Pia Fanti[2], Alessandro Rinaldi[2], and Bartolomeo Silvestri[2]

[1] Institute of Communication and Computer Systems, I-SENSE Group, Athens, Greece
{anna.antonakopoulou, v.portouli, nikos.tousert, maria.krommyda, a.amditis}@iccs.gr

[2] Department of Electrical and Information Engineering, Politecnico di Bari, Bari, Italy
{mariapia.fanti, alessandro.rinaldi, bartolomeo.silvestri}@poliba.it

Abstract. The development of sustainable, smart mobility has been accelerated by the arrival of innovative technologies. With the paradigm shift towards transport electrification, Electric Light Vehicles (EL-Vs) represent a very promising pathway to smart urban mobility. Still, the current market penetration of EL-Vs is relatively low compared to that of conventional vehicles. Via one-year long demonstrations of such vehicles in six different European cities (Rome, Genoa, Bari, Málaga, Trikala and Berlin) ELVITEN EU funded project proposes a holistic approach to boost the EL-Vs usage by addressing all of the issues hindering the wide market penetration of EL-Vs, which are the Users' Low Awareness, the Consumers' Concerns and the Inadequate Mobility Planning for EL-Vs. This paper presents the harmonized, user-centered, controlled, step-by-step methodology that has been followed in order to prepare and set up the pilot in each City, to create awareness among users and to collect and analyse data from the different Cities so that the findings can be comparable and thus being able to derive recommendations and guidelines to accelerate the deployment of EL-Vs in complex and demanding urban environments. The demonstration activities target the collection of appropriate sizes of various types of data, based on the methodological triangulation concept and a blended qualitative and quantitative study approach, in order to increase the credibility and validity of the results.

Keywords: Pilots · Methodology · Deployment · Sustainable urban mobility · EL-Vs · Data collection · Users

E. G. Nathanail et al. (Eds.): CSUM2020, AISC 1278, pp. 181–191, 2021.
https://doi.org/10.1007/978-3-030-61075-3_18

1 Introduction

EL-Vs are a further step towards an even more sustainable urban mobility. The Users' Low Awareness about the EL-Vs performance and functionalities has been acknowledged as one reason hindering the EL-Vs market penetration and it is partially due to the limited direct experience of such vehicles and the consumers' concerns about the relatively high cost of such vehicles and the feeling of uncertainty as regards the possible need to re-charge the vehicle during the trip. Additionally, there is a lack of consistent knowledge and information needed by planning authorities to prepare an adequate traffic and charge infrastructure to achieve their integration in the transport and electricity networks. For example, little is known on how such vehicles are used and how much basic infrastructure is required in terms of parking spaces and type and location of charge facilities [1]. ELVITEN project [2] organizes long demonstrations of EL-Vs in 6 EU cities to demonstrate the potential market penetration of EL-Vs. The main objectives are to enhance users' awareness about EL-Vs and to collect data in order to generate guidelines for manufacturers and authorities for the better integration of such vehicles in the transportation and electricity networks. In order to enhance the credibility and validity of the upcoming analysis results the methodological simultaneous triangulation concept has been exploited [3], [4] where qualitative and quantitative methods for data gathering are used concurrently and the findings complement one another at the end of the analysis. Following this methodological framework, the consortium partners collect and analyze, using inductive and deductive approaches [5], various types of city transport data, data from the real trips done by people with such vehicles and data relevant to the usage of the related services and tools. Opinions and attitudes of people towards the EL-Vs are also collected via specific questionnaires administered via related services and apps as well as via wide online surveys and interviews. The data will be processed within the framework of the project during the evaluation phase in order to analyze mobility behaviour, actual usage of EL-Vs, charging behavior, experiences and attitudes towards EL-Vs and the associated services that will be evaluated against specific KPIs. In order to accomplish the data collection, ELVITEN utilizes a harmonized, user-centred process for demonstrations set-up and operation to ensure also that the needs of the citizens who use the system services and tools are fully taken under consideration, thus motivating them also to use the EL-Vs during the pilot phase. Apart from the data collection process that is of utmost importance, the analysis of the mobility situation in the demonstration cities and consequently the definition and deployment of the usage schemes, the identification of users types as well as their recruitment constitute very important activities to the demonstration methodology. Furthermore, the design, implementation and verification of the ICT tools and apps are focused on the individual users and are a mix of automated data logging via loggers installed on the EL-Vs, via user-friendly apps and services and via direct surveys through questionnaires and interviews. A design of the appropriate means (what type of survey, what questions to ask) to collect meaningful feedback from users has been conducted. Finally, the usage and data collection monitoring of the demonstrations is done via custom designed visualization tools (Fig. 1).

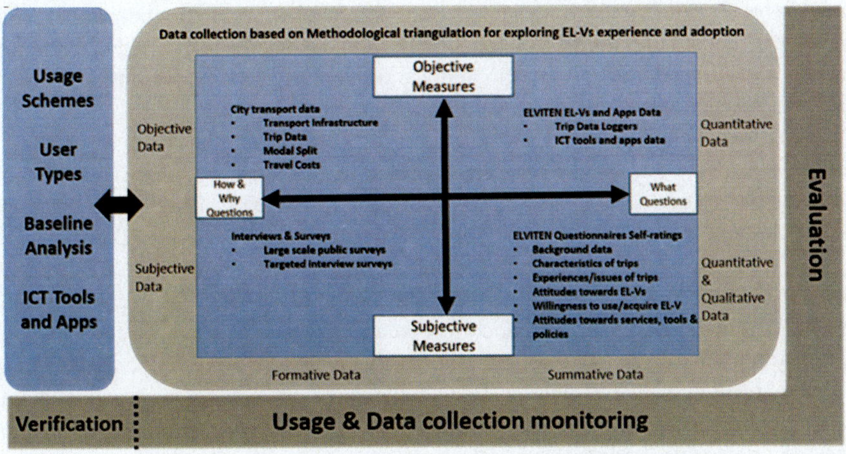

Fig. 1. Demonstration methodology and data collection.

The description of the demonstration methodology followed in ELVITEN project constitutes the focus of this paper. The remainder of this paper is structured as follows: Sect. 2 presents the methodology for the analysis of the current situation in each city, to be used as reference baseline in the analysis and evaluation phase, Sect. 3 describes the procedures to involve and engage users and stakeholders, to get users' consent to the processing of their data and to sign agreements with owners and long-term sharers of such kind of vehicles while Sect. 4 describes the demonstration set-up activities as well as the data collection and demonstration monitoring tools. Finally, Sect. 5 gives the conclusions.

2 Baseline Analysis Per City

For assessing changes and impact of ELVITEN, as it establishes a basis for comparing the situation before, during and after the ELVITEN project, a baseline analysis carried out to be a useful benchmark for examining what level of change is triggered by the project and can be also a crucial element in research and planning, besides in any monitoring and evaluation activities [6]. The data used in the baseline calculation is based on information provided by cities relevant to the current status of mobility: number of vehicles, typology, and other. The baseline values can be considered as benchmarks/starting values for the Agreed KPIs and can be used to analyse changes over time, by measuring and/or comparing values during and after the demonstrations. The Agreed KPIs are selected by asking the six cities to assign a numerical score, based on the level of importance of the indicator (with a scale from 0 to 10; 0 not important and 10 necessary). An initial screening included the selection of Agreed KPIs with scores greater than 8. For each Study Question (SQ) the number of KPIs that satisfied this parameter are identified; if no KPIs satisfied the first selection criterion for the Study Question, the KPI with the highest score is selected. Subsequently, the average values of the score assigned by each city are evaluated in order to choose the same KPIs for the cities. Afterwards, the baseline mobility situation for the Demonstration cities has been investigated in order to analyse the mobility conditions, infrastructures and EL-Vs present before the demonstration phase.

The last step of the adopted methodology for the baseline analysis was the calculation of baseline values of the Agreed KPIs. In the literature, different methods described how to set a baseline value. Sensible and useful baseline values can be obtained through enough performance measure values calculated [8]. Other studies estimate the baseline values using other techniques [8]. The methods to set the baseline values are related to the maturity of the KPIs: in some cases, it is possible to use historical data, in other it is needed to collect data in advance in order to set it or to use data estimation techniques. The common scenarios to set a baseline value for KPIs are (i) a brand-spanking new KPI with no data yet, (ii) a mature KPI with a lot of historical data and (iii) a mature KPI with a seasonal or cyclical pattern. In this study, as the selected Agreed KPIs are indicators without historical data, so they belong to the first category. More in detail, the calculation methodologies for each KPI are based on input data provided by users and EL-Vs during the demonstrations. The input data to calculate the baseline values are collected from the surveys conducted at the beginning of the project and information provided by the cities and the Agreed KPIs are the same for all the cities.

3 Users' and Stakeholders Involvement

3.1 Usage Schemes and User Types

One of the main targets of ELVITEN project is to develop and demonstrate usage schemes for EL-Vs in six demonstration cities. At first stage a study related with the mobility situation in each city has been conducted and proposed the most promising usage schemes and target user clusters for the best integration of EL-Vs into the existing transport and mobility network in each city. Mobility demand, user needs, opportunities and barriers to EL-Vs use were gathered as follows:

- Statistical data regarding the transport networks and mobility usage in each city (geographical/socio-economic characteristics, transport infrastructure, trip data, modal split, travel costs, etc.)
- Data from a large scale public questionnaire survey (approximately 7000 responses) aimed at all citizens in the six cities (and also answered by citizens in other European cities) covering current travel patterns, whether they would potentially use different types of EL-V, and perceptions of advantages and barriers to using EL-Vs.
- Data from a more limited set of targeted interview surveys of fleet operators and professional drivers (mostly focused on delivery companies) on their current fleet use and the potential scope of switching some or all of their ICE vehicles to EL-Vs (including perceptions of advantages and barriers).

Based on these data analysis the following different usage schemes were identified: (i) Public short-term sharing scheme, (ii) Private (individual) ownership scheme, (iii) Delivery sharing & ownership scheme, (iv) Corporate sharing & ownership scheme.

Therefore the target user clusters involved in ELVITEN demonstration activities were identified in general as owners and sharers while the types of users related with the functional usage of EL-Vs during the demonstration activities taking place at each city were identified as Short-term sharers of EL-Vs, Long-term sharers of EL-V and Owners of EL-Vs. The involvement of stakeholders is also considered very important for the acceptance of EL-Vs, regarding their willingness to use fleets of EL-Vs for their personnel and/or to offer the organisation's parking facilities for charging private EL-Vs. The type of stakeholders are light delivery fleet owners, rental and sharing companies, tourism service providers, planning and public authorities.

3.2 User's Involvement and Recruitment Procedure

ELVITEN uses several ways to involve the participants. In more detail, ELVITEN uses the partners' networks and their national members, local associations of users' groups and established Regional Support Groups (RSGs) in all the pilot Cities. It also undertakes wide awareness campaigns in each City, starting well in advance the start of the demonstrations period, to attract the interest by a large number of users and stakeholders. Such campaigns are being undertaken via local press and media announcements, announcements in relevant local and national associations and municipalities, web sites and social networks. Posters at key locations in each City and leaflets are also employed for this purpose as well as advertisement of the incentives given by the Municipalities from the usage of the vehicles and ELVITEN ICT tools and services.

3.3 Getting Users' Consent and Ethical Procedures

In order to get consent from the users, an Information Sheet has been prepared and ELVITEN users are asked to read it before giving their consent to be involved in the demonstrations and the main procedures implemented are the following:

- All personal data collected will be treated as confidential and no actions will be undertaken unless the partners have the consent by the users (data subjects). The data handling procedures strictly adhere to the Regulation (EU) 2016/679 [9].
- Users are asked to read the information sheet and sign beforehand a consent form on the terms of data collection, treatment and further use.
- Users are provided with access to their data when requested and ultimately have the right to force the deletion of said data or withdraw their consent.
- Personal data will not be in any case shared with or disclosed to anyone outside the research team.
- Activities only include adult research participants who are healthy and fully able and capable to provide informed consent to their participation and who sign the informed consent form before any such activity.
- Participation of users is strictly voluntary and they have the right to withdraw themselves and their data from any ELVITEN activity.

3.4 Agreements with EL-V Owners and Long-Term Sharers

Apart from the consent form, two more types of agreements are signed by some owners of EL-Vs and long-term sharers before joining the demonstrations. The scope of the agreements is to define the terms under which these users commit to the demonstrations. For owners, this entails the acceptance of using a trip data logger and the conditions when ELVITEN has the right to request it back, for example in case of inactivity, so that it can be given to another user who would be more active and thus could contribute more data. Similarly, for long-term sharers the agreement entails terms relevant to the provision of the equipped shared EL-V and the conditions when ELVITEN has the right to request it back before the end of the agreement period. In more detail, the agreements define the following issues: (i) Equipment that is given to the user, (ii) Duration of the agreement & Conditions to terminate the agreement, (iii) User's and ELVITEN partners' responsibilities & Handling of disputes.

4 Demonstrations Set-up for Data Collection and Monitoring

This chapter describes the steps that the demonstration cities and the ELVITEN consortium partners follow in order to collect the necessary data for the analyses and evaluations that will be conducted at a later stage of the project.

4.1 Verification Methodology

The first demonstration activities were the short pilot dry-run tests that have been organized in each City with limited numbers of participants, to verify that all services and the ICT tools operate smoothly and to report any issues for corrective actions before the demonstration start. The usage schemes tests gave quite focused and detailed user feedback in order to refine the schemes, the services and tools as needed, before the start of the open demonstrations. A template for dry-run tests was developed, organizing the verification activities into use cases to cover all aspects involved in the verification. The dry-run tests use cases covered the following aspects:

- Preparation: included verification whether the City was prepared for the tests, including the availability of test users and EL-Vs, the presence of ICT providers, etc.
- Inventory Management: to verify the existing EL-Vs inventory and the installation of inventory-management oriented applications.
- User registration: to verify the coherence and storage of data for the registration of users in order to give them access to the ELVITEN services and to verify the information provided to users.
- Booking & Brokering Services: to verify different aspects of the booking and brokering of resources (EL-Vs, Parking Spaces and Charging Points) by users focusing on real availability as well as ICT tools that support these services and the data exchange between the middleware and the ELVITEN Data Warehouse.

- Operations: to verify the complete process for long/short term sharing of resources to a user, including questionnaires administration and tracking of data exchanges from cities, users and vehicles through the middleware up to the Data Warehouse.

4.2 Tools for Quantitative Data Collection During the Pilot Phase

The trip data (such as location, speed, acceleration, trip duration, etc.) are collected via the trip data loggers (EL-V Live Data) which are installed on the EL-Vs and via the ELVITEN ICT applications from where also the questionnaires are administered. In general the ICT tools deployed include registration services, a Brokering and a Booking service for EL-Vs, parking and charging points, an EL-V fleet monitoring tool and an Eco-Drive app, a Serious Game app and an Incentive Management Smart Card. All the data are transmitted and stored in the Data Warehouse via the ELVITEN middleware. The detailed description of the ELVITEN ICT platform is out of scope for this paper however its schema has been provided in Fig. 2.

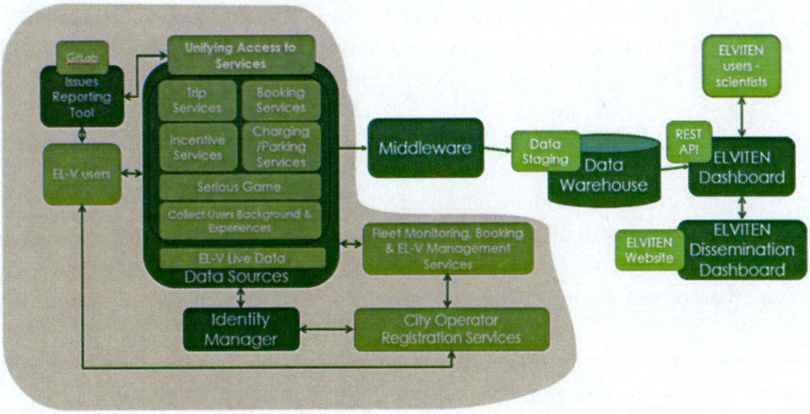

Fig. 2. The ELVITEN ICT platform and data repository.

4.3 Tools for Qualitative Data Collection During the Pilot Phase

Qualitative data (reports on experienced problems, acceptance of EL-Vs, services and ICT tools, attitudes, willingness to use, etc.) are collected via the questionnaires which are administered by the ELVITEN apps. The questionnaires contain yes/no and ranking questions (5-point Likert scale about user's satisfaction), the user's unique ID, as well as open comment sections and their content is related with the following aspects:

- Background data related to demographics, mobility habits and EL-V experience.
- Characteristics of the last trip with an EL-V regarding the ride, parking and charging.
- Experiences / issues encountered during the last trip.
- Attitudes towards EL-Vs & Willingness to use/acquire an EL-V.

- Attitudes and opinions towards each service, ICT tool and policy in place in the reference city.

Users are asked to fill in the questionnaires at a predefined time plan according to the usage scheme and submit the questionnaires immediately after their completion. If these questionnaires are not submitted in time, the users are reminded to complete and submit them via each tool and application. For all types of users the Background Data questionnaire is filled in and submitted before the first trip along with the registration process. The Experiences / issues questionnaire is filled in and submitted after each trip, if an issue has occurred during the last trip. The rest of questionnaires (mobility change and attitudes towards EL-Vs and towards the ELVITEN services and apps) for short-term sharers and owners are being administered in four months intervals while for long-term sharers they are administered after the end of the sharing period (e.g. week). Additionally the service operators report on a daily basis on i) the local weather and ii) how many of the shared El-Vs are operational. Furthermore, a wide online survey will be conducted to the middle of demonstration period related with the problems the users' are facing as well as to charging patterns in order to support the data collection process. Finally, a citizens' wide attitudes survey will be conducted after the end of the demonstrations, similar to the one carried out at the beginning of the project, to analyse the differences in the wide citizens population attitudes and opinions regarding EL-Vs before and after ELVITEN demonstrations.

4.4 Tools for Demonstrations Monitoring

During the demonstrations, mechanisms to effectively monitor usage such as number or registrations, number of submitted questionnaires, timely submission of questionnaires etc. have been implemented. For this scope, the monitoring indicators include: (i) Total user count (sharers, owners), (ii) Number of inactive users as inverse to usage of the vehicles and applications during a specific timeframe, (iii) Number of trips and questionnaires submitted per city on average on a timeframe. Other measures related to user activity monitoring constitute the reminders, e.g. mail or automated notifications via the apps, to users to take care of the questionnaires they are expected to fill in at different times during the whole demonstration phase.

4.5 The ELVITEN Dashboard

The usage of EL-Vs, trip data and questionnaires collection is bi-weekly monitored during the demonstration phase via a custom implemented interface, the "ELVITEN Dashboard", a web application used for visualizing the data stored in the Data Warehouse through pertinent tables, charts and labels. It visualizes data from various sources including the trip data loggers on the EL-Vs, data from the ICT tools and the users' responses to questionnaires and of the KPIs. The inspection of the Dashboard content sheds light also on reported problems and on the number of trips and questionnaires completed in each city (e.g. small number of trips or poor numbers in the submitted questionnaires can be easily spotted through the Dashboard, in order to timely address the cause of the problems and alarm partners to intensify their actions). Moreover, through the presentation of the historical and current views of operations, it can also serve as an alert mechanism by timely indicating possible integration problems between ELVITEN components. The various views of the Dashboard are the following (Fig. 2):

- Cities Data Collection Overview: presenting statistics and counters for each city related with the number of registered users per city, trips, submitted questionnaires and problems encountered. Pie and bar charts have been incorporated in order to visualize the proportion of questionnaires and the proportion of ICT tools usage.
- KPIs Overview: Visualization of the KPIs calculated at the beginning of each month.
- Trip Data Tables View: exposes the content of the Data Warehouse related with the data coming from the trip data logger of the vehicles in the form of tables.
- ICT Tools Data Tables View: exposes the content of the Data Warehouse related with the data coming from the ICT Tools and apps in the form of tables
- Questionnaires Data View: exposes the content of the Data Warehouse related with the data coming from the questionnaires in the form of tables.
- Dashboard SQL Explorer: A user-friendly interface in order to quickly write and share SQL queries through a simple and usable SQL editor.

Finally, the Cities Data Collection Overview is integrated to the ELVITEN website, namely "ELVITEN Dissemination Dashboard" to let users and other interested parties to be informed about the pilot activities and their progress (Fig. 3).

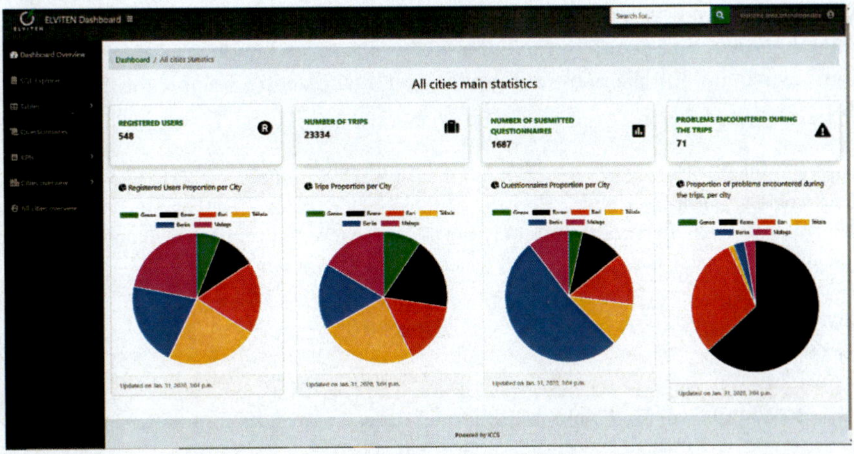

Fig. 3. The ELVITEN dashboardall cities overview.

4.6 Issues Monitoring from Demonstrations

To ensure continuous and effective technical support to the demonstrations, the development of a dedicated issue-reporting functionality was implemented and included in the ELVITEN Unified application which integrates all the ELVITEN apps.

The issue-reporting functionality allows the ELVITEN users to report any kind of issues (i.e., concerning vehicles, infrastructures, or applications) that occur while using the ELVITEN services; in addition, users can report their suggestions about the ELVITEN services in general. However, major emergencies such as vehicles out-of-service and road accidents, can be reported to the local help desk by phone call directly via the Unifying App's issue-reporting functionality. The Unifying App, by posing direct questions to the user, can evaluate whether the issue requires a direct phone call, and thus present the user with a phone call button or not. The GitLab [10] issue tracker has been chosen to send the issues on the basis of a decision tree.

5 Conclusions

This paper focused on the harmonized user-centered pilot methodology framework designed for setting the fundamental procedures which have been followed in the ELVITEN project, although its basic concepts can be used also in other technological domains, in order to involve and recruit users, get their consent, establishing the terms of their data processing and the agreements for the involvement of the EL-V owners and long-term EL-V sharers during the demonstrations, identify the current mobility situation of a pilot city and thus implement and deploy the necessary measurement, data collection and monitoring tools, so the targets set regarding creation of a big data bank can be met and consequently serve the analysis and evaluation procedures. Finally, it is worth to note that currently the data collected regarding the EL-Vs usage and users' involvement, corresponding to about 34.000 trips, 400 vehicles in operation and 700 registered users, prove the efficiency of the ELVITEN demonstration methodology and will properly serve the analysis process, guidelines and recommendations extraction for planning authorities and manufacturers upon the end of the project.

Acknowledgment. This work is a part of the ELVITEN project. ELVITEN has received funding from the European Union's Horizon 2020 research & innovation programme under grant agreement no 769926. Content reflects only the authors' view and European Commission is not responsible for any use that may be made of the information it contains.

References

1. Santucci, M., Pieve, M., Pierini, M.: Electric L-category vehicles for smart urban mobility. Transportation Research Procedia **14**, 3651–3660 (2016)
2. EU H2020 ELVITEN Project homepage, https://www.elviten-project.eu/en/. Accessed 03 Feb 2020.
3. Denzin, N.: Sociological Methods: A Sourcebook. 5th edn. Aldine Transaction (2006).
4. Rothbauer, P: Triangulation. The SAGE Encyclopedia of Qualitative Research Methods, pp. 892–894. In Given, Lisa edn. Sage Publications (2008).
5. Wiklund-Engblom, A.: Exploring conative constructs and self-regulation of e-learners: A mixed methods approach. In: Proceedings ascilite Sydney (2010).
6. Silvestri, B., Rinaldi, A., Roccotelli, M., Fanti, M.P.: Innovative baseline estimation methodology for key performance indicators in the electro-mobility sector. CoDIT **2019**, 1367–1372 (2019)

7. Barr, S.: How to set a KPI Baseline to Monitor Improvement (2018). https://www.staceybarr.com/measure-up/set-kpi-baseline-monitor-improvement/. Accessed 05 Sep 2018
8. Proforecast "How to Set KPI Baseline?" (2018). https://proforecast.com/blog/how-to-set-kpi-baseline/. Accessed 05 Sep 2018
9. Regulation (EU) 2016/679 of the European Parliament and of the Council of 27 April 2016 on the protection of natural persons with regard to the processing of personal data and on the free movement of such data, and repealing Directive 95/46/EC
10. GitLab Homepage. https://about.gitlab.com/. Accessed 03 Feb 2020

Investigating the Impacts of Additive Manufacturing on Supply Chains

Vissarion Manginas, Eftihia Nathanail(✉), and Ioannis Karakikes

Traffic, Transportation and Logistics Laboratory, University of Thessaly, Pedion Areos, 38334 Volos, Greece
amagginas@yahoo.gr

Abstract. Additive manufacturing is an emerging technology that is gaining more ground in various production areas, like medicine, industrial production, and consumer product printing. Aim of this paper is to analyze the applications of this technology on supply chains, investigate how it affects the structure and characteristics of the traditional supply chain, and finally, develop a business model for its optimal use, through the case study of the LEAP-1A jet engine fuel nozzle supply chain. This was achieved in three steps. First, current literature was reviewed, to fully understand the technology, its applications and the production processes it affects. Second, three different supply chain scenarios, i.e. conventional supply chain, centralized 3D printing and decentralized 3D printing were developed and elaborated, so as to allow the easier identification of the differences of the supply chain processes among the scenarios. To select the most effective scenario, an online survey with the participation of experts was conducted, to attribute weights to indicators identified through the literature review. Finally, a business model for the chosen scenario was developed using the Business Model Canvas. The analysis and evaluation of the indicators revealed that the most effective supply chain model is a decentralized additive manufacturing model. Based on the overall results it can be concluded that additive manufacturing simplifies the supply chain, increases flexibility in design and production and reduces transport and logistics costs.

Keywords: 3D printing · Supply chain · SCOR · Impact assessment · Transport

1 Introduction

Production, and subsequently supply chains today, are mostly focused on economies of scale. For this reason, production is especially centralized in locations, usually found in developing countries, taking advantage of inexpensive labor and potentially less strict regulations. This fact increases the need for shipping the resulted products from the manufacturing countries to the market and imposes a high share of the overall internal and external costs to transportation (Boon and van Wee 2017).

Additive manufacturing seems to have the potential to change this practice, due to the flexibility it can provide to a producer. Additive manufacturing is generally defined as "*the process of joining materials to make objectsfrom three-dimensional model data,*

E. G. Nathanail et al. (Eds.): CSUM2020, AISC 1278, pp. 192–201, 2021.
https://doi.org/10.1007/978-3-030-61075-3_19

usually layer-by-layer, as opposed to subtracting manufacturing methodologies" (Piazza and Alexander 2015).

Additive manufacturing can have many effects on production, like encouraging mass customization, improving resource management, decentralizing production, reducing complexity, optimizing logistics and inventory processes, improving product design and prototyping and also effecting changes on legal and regulatory frameworks (Mohr and Khan 2015).

Additive manufacturing can offer many advantages to the production process, mainly related to design and production flexibility, reduced environmental waste and reduced transportation loads and costs.

Its disadvantages have mostly to do with regulatory and legal issues, product quality and consistency, higher production costs, especially for mass production and difficulty of change in production methods.

In the future, additive manufacturing is expected to be used more extensively, both for industrial and domestic uses, making production more flexible, and being particularly helpful in the prosthetic limbs industry and medicine in general (Campbell et al. 2012; Negi et al. 2013).

This paper investigated the impacts that additive manufacturing technologies can have on production and, more specifically, the supply chain and a business model was developed, using the Business Model Canvas (Osterwalder et al. 2010).

2 Additive Manufacturing

2.1 About Additive Manufacturing

Additive manufacturing has been used in a lot of different production processes, usually for products that are very specific, have high value, are produced in small numbers and require a lot of customization. Some examples include medicine (especially prosthetic limbs and bone transplants), industrial uses (mainly prototyping, but also in the automotive and the aerospace industry) and miscellaneous uses, like architectural models and home printing (Campbell et al. 2012; Wong and Hernandez 2012; Negi et al. 2013).

As additive manufacturing is a general term, it encompasses many slightly different technologies. (Wong and Hernandez 2012).

2.2 Additive Manufacturing and Supply Chain

There are four main pillars that can optimize logistics of supply chains. These are improved resource management, faster respond to demand, overall logistics enhancement and reduction of the environmental footprint (Attaran 2017).

Additive manufacturing can affect supply chain, as mentioned before. Generally, a wide use of additive manufacturing could lead to a total overhaul of the supply chain and logistics, mainly due to the fact that production would be decentralized and moved closer to the end user, while the supply chain would become shorter and more simplified. This could be a result of the production flexibility that additive manufacturing can offer, by being able to produce products in smaller scale without increasing the average cost (Silva and Rezende 2013; Marchese et al. 2015; Mohr and Khan 2015; Stapleton and Pande 2016; Boggers et al. 2016; Attaran 2017; Boon and van Wee 2017).

Likewise, additive manufacturing may have multiple impacts on the economy. The main impact would be the change in the dynamics between economies of scale and scope, with economies of scope gaining significant ground, and even gaining the advantage in some products with certain characteristics (Petrick and Simpson 2013). Another impact is the shift of the supply chain towards digitalization, with digital designs and their distribution playing a more important role than the supply chain itself (Jiang et al. 2017). One economic aspect where the effect of additive manufacturing cannot be determined is production cost per unit.

Additive manufacturing could also have a positive effect on reducing the overall environmental footprint (Mashhadi et al. 2015). The decentralization of production is going to help reduce emissions and energy expended due to goods transport (Boon and van Wee 2017).

In terms of the social impacts additive manufacturing may have, it is the shift of the workforce towards digital industries, due to the increase in digital design that will accompany a broad adoption of additive manufacturing (Ratto and Ree 2012), as well as the empowerment of the consumer and developing countries, since additive manufacturing can allow users to produce their own products (Thiesse et al. 2015), leads slowly into a merger between producers, distributors and end users (Chen 2017).

Finally, the supply chain business models are expected to experience changes due to a potential shift towards additive manufacturing as a means of production. Businesses will have to focus on finding new suppliers for the new raw materials that will be used and specifying a way to harmonize the old and the new models (Hartmann and Lebherz 2017; Bogers et al. 2016). There are four possible models that could take advantage of additive manufacturing's potential. These are a leagile model, a completely agile model, a digital business model, and finally a service model, where businesses will function only as service providers, providing the users with the option of printing their own designs, either bought or self-made (Mashhadi et al. 2015).

3 Methodology

3.1 Scenario Development

To better illustrate the impacts additive manufacturing can have on the supply chain, the case study of the supply chain of the fuel nozzles of the LEAP-1A jet engine was used. The current supply chain was analyzed in Scenario 1, and then two additional scenarios were tested. Scenario 2 regarded centralized production with additive manufacturing techniques and that of decentralized additive manufacturing. It assumed more decentralized production in comparison with the conventional one, but the products are still produced in some central locations, albeit closer to the end user, due to the flexibility the technology allows. Scenario 3 was based on decentralized additive manufacturing and assumed complete decentralization with the products being produced on site. Scenarios 2 and 3 were then compared to Scenario 1, considering economic, environmental, social impacts, impacts on logistics and inventory and supply chain flexibility and structure.

3.2 Criteria and Indicators

In each scenario, the structure of the supply chain was determined and all the scenarios were evaluated in terms of their economic, environmental and social impact, as well as their impact on logistics.

A hierarchical structure was followed. Based on this structure, its criteria category was further subdivided in sub-criteria and respective indicators. All evaluation components were based on literature review on current supply chains, aerospace parts supply chains and additive manufacturing supply chains (European Commission 2016; Nathanail et al. 2016; Boon and van Wee 2017; Nguyen 2017; Ghadge et al. 2018) (Table 1).

Table 1. Categories, subcategories and indicators.

Criteria	Sub-criteria	Indicators
Economy	Risk	Demand uncertainty
		Deficiency risk
	Cost	Transport cost
		Raw material cost
		Investment cost
		Manufacturing cost per unit
	Production effectiveness	Time to market
		Production process reliability
		Quality
Environment	Pollution	Waste
		Energy consumption
		Emissions
Society	Framework	Legal framework
	Familiarization and uptake	Employee familiarization
		Manufacturer's specialization
		Ease and effectiveness of design
		Customization
Transport and Logistics	Transport	Response time
		Tons
		Ton-kilometers
	Inventory	Inventory
	Supply and distribution	Raw material availability

3.3 Weights

The weighted sum method was implemented. Weights were attributed based on the Analytical Hierarchy Process (AHP) (Saaty 1990). First, criteria, sub-criteria and indicators were subjected to pair-wise comparisons. For this reason, an online questionnaire survey was developed. An invitation to participate in the survey was sent out to 200 experts, selected based on their appearance in literature of similar topics as the one of the present paper. Of all invitees, 24 accepted to complete the survey. The experts were asked to rate each pair of criteria, sub-criteria and indicators on a 1–9 scale, with 5 representing equal importance between the two items. The experts were asked to rate these items based on their relative importance, with 9 representing the maximum degree to which the second item is considered more important than the first and 1 representing the opposite. This modification of Saaty's scale was made to better align with the other scales used in this research, especially the ones used during the assessment of the scenarios. After the responses were gathered, the mean for each pair-wise comparison was calculated, based on which the Eigen value vector was calculated (Saaty 1990). An example of the average pair-wise comparison values and the relative weights of the considered evaluation elements is given in Table 2.

Table 2. Example of pair-wise comparison and relative weights.

Sub-criterion	Pair-wise comparison index		Relative weight
	Familiarization and uptake	Framework	
Familiarization and uptake	5,00	6,29	0,71
Framework	0,79	5,00	0,29

The final weight of each indicator is the product of the relative weights of the indicator, its sub-criterion and criterion. Table 3 displays all relative weights and the final weights of the indicators.

Table 3. Criteria, sub-criteria and indicator weight.

Criteria	Relative criterion weights	Sub-criteria	Relative sub-criterion weights	Indicators	Relative indicator weights	Final indicator Weights
Economy	0,42	Cost	0,51	Manufacturing cost	0,42	0,09
				Investment cost	0,27	0,06
				Raw material cost	0,18	0,04
				Transport cost	0,11	0,02
		Production Effectiveness	0,31	Quality	0,52	0,07
				Time to market	0,30	0,04
				Reliability	0,17	0,02
		Risk	0,16	Demand uncertainty	0,68	0,04
				Deficiency risk	0,31	0,02

(continued)

Table 3. (*continued*)

Criteria	Relative criterion weights	Sub-criteria	Relative sub-criterion weights	Indicators	Relative indicator weights	Final indicator Weights
Environment	0,17	Pollution	1	Waste	0,51	0,09
				Energy consumption	0,30	0,05
				Emissions	0,18	0,03
Society	0,12	Familiarization and uptake	0,71	Customization	0,42	0,03
				Ease and efficiency of design	0,28	0,02
				Manufacturer's specialization	0,17	0,01
				Employee familiarization	0,11	0,01
		Framework	0,28	Legal framework	1	0,03
Transport and Logistics	0,27	Supply and distribution	0,51	Raw material availability	0,68	0,09
				Reverse logistics	0,31	0,04
		Transport	0,31	Response time	0,57	0,04
				Tons	0,26	0,02
				Ton-kilometers	0,16	0,01
		Inventory	0,16	Inventory	1	0,04

4 Scenarios Assessment

The next step was the evaluation of each indicator, for each of the three scenarios. There were three types of indicators, those for which quantitative data for each scenario were available, those for which data was available in the form of percentage change between the scenarios and qualitative indicators, which were evaluated with the use of quantification through text coding (Sandelowski 2009). For the first type of indicators, the actual values were used, for the second, the percentages and for the third, a five level Likert scale was assumed to interpret the indicator assessment. For the second type of indicators, values were attributed to Scenario 1 by taking into account the percentage change, in each scenario. The scenario with the largest positive change was attributed with the maximum value (100%), and the other scenarios were rated based on that. Furthermore, normalization was conducted to result in the final indicators scores. For this purpose, a common 0–10 scale was used. The normalization formula was

$$y = c + (x - a) * (d - c)/(b - a) \tag{1}$$

In this formula, y represents the final normalized value of each indicator, x the initial value of each indicator, a and b the lower and upper limit of each indicator's initial scale for each type of indicator and c and d the lower and upper limit of the final common scale. For the first indicator type, the initial scale varies, based on the lower and higher value between the three scenarios, for the second indicator type the initial scale is 0%-100% and for the third indicator type the initial scale is a 1–5 Likert scale. Table 4 illustrates the final normalized values of the indicators on the common 0–10 scale for all three scenarios. Detailed analysis of the calculations may be found in (Manginas 2020).

Table 4. Normalized indicator values.

Indicator	Scenario 1: Conventional manufacturing	Scenario 2: Centralized additive manufacturing	Scenario 3: De-centralized additive manufacturing
Manufacturing cost	7	10	10
Investment cost	10	0	10
Raw material cost	10	1	1
Transport cost	0	8,5	10
Quality	6	4	4
Time to market	3,6	10	10
Reliability	6	4	4
Demand uncertainty	6	8	10
Deficiency risk	6	4	4
Waste	1	10	10
Energy consumption	9,6	10	10
Emissions	9,45	10	10
Customization	6	10	10
Ease and efficiency of design	6	10	10
Manufacturer's specialization	10	2	2
Employee familiarization	6	4	4
Legal framework	6	4	4
Raw material availability	6	4	4
Reverse logistics	6	8	10
Response time	1	10	10
Tons	1	10	10
Ton-kilometers	3,5	10	10
Inventory	0	9,5	10

The final score for each scenario was calculated by multiplying the value of each indicator by its weight, adding all products and dividing by the number of indicators (23).

On Table 5 can be found the final score for each of the three scenarios.

Table 5. Final scenario scores.

Scenario 1: Conventional manufacturing	Scenario 2: Centralized additive manufacturing	Scenario 3: Decentralized additive manufacturing
0,236740125	0,302192731	0,337266873

From Table 5 it can be deduced, that the scenario of decentralized additive manufacturing is the most suitable for the jet engine fuel nozzle supply chain. Therefore, this scenario was further analyzed in a business model, which was developed based on the Business Model Canvas (Osterwalder et al. 2010).

5 Business Model

A business model was formulated based on the Business Model Canvas. The model analyzes a business model in terms of nine main components: Key Partners, Key Activities, Key Resources, Value Propositions, Customer Relationships, Channels, Customer Segments, Cost Structure and Revenue Streams.

Regarding Key Partners, these are the raw material suppliers and processors, the additive manufacturing printers' manufacturers, logistics service providers and storage providers for digital files.

Key Activities would be the design and printing of products, digital storage and the provision of logistics services.

Key Resources are the digital designs, the printers, the raw materials, the infrastructure and the human resources.

Value Propositions are quality improvement, less waste, design and production flexibility, reduced cost, reduced demand for logistics services and reduced environmental footprint.

In terms of Customer Relationships, there should be constant communication on a personal level between the manufacturers and their customers, in order for the customers to be as informed as possible at all times.

Regarding Channels, these are the internet, specialized expositions and conferences, scientific publications and one on one presentations to potential customers.

Customer Segments are aerospace manufacturers and independent competitors that lack the capital and technology to compete effectively.

In terms of Cost Structure, the main costs would be personnel cost, the cost of equipment and infrastructure, the cost of raw materials and research and development and the cost of promotion and logistics.

Finally, the Revenue Streams would be the sale of parts, direct investment, loans, tech development, digital designs and the provision of logistics services.

6 Conclusions

Based on all of the above, it can be concluded, that additive manufacturing can have a significant effect on supply chains and affect their economic, social, environmental and logistics impact.

Regarding fuel nozzle supply chain in particular, it was concluded that additive manufacturing can help produce parts in small production runs or even individually. In addition, the capability of production on location that the decentralized model can offer, can also help reduce transportation costs.

This new technology is expected to have a significant contribution in the domain of manufacturing and logistics. However, it is still in an early stage and the floor is wide open for scientific teams.

References

Piazza, M., Alexander, S.: Additive Manufacturing: A Summary of the Literature. Urban Publications, Cleveland (2015)

Mohr, S., Khan, O.: 3D printing and its disruptive impacts on supply chains of the future. Technol. Innov. Manage. Rev. **5**(11), 20–25 (2015)

Campbell, I., Bourell, D., Gibson, I.: Additive manufacturing: rapid prototyping comes of age. Rapid Prototyping J. **18**(4), 255–258 (2012)

Wong, K.V., Hernandez, A.: A review of additive manufacturing. ISRN Mech. Eng. **4** (2012)

Negi, S., Dhiman, S., Sharma, R.J.: Basics, applications and future of additive manufacturing technologies: a review. J. Manuf. Technol. Res. **5**(1–2), 75–96 (2013)

Nathanail, E., Mitropoulos, L., Karakikes, I., Adamos, G.: Sustainability for assessing urban freight transportation measures. Logistics Sustain. Transp. **9**(2), 16–36 (2018)

Osterwalder, A., Pigneur, Y., Clark, T.: Business Model Generation: A Handbook for Visionaries, Game Changers, and Challengers. Wiley, Hoboken (2010)

Pirjan, D., Petrosanu, D.-M.: The impact of 3D printing technology on the society and economy. Rom. Econ. Bus. Rev. **7**(2), 360–370 (2013)

Attaran, M.: Additive manufacturing: the most promising technology to alter the supply chain and logistics. J. Serv. Sci. Manage. **10**, 189–205 (2017)

Silva, J.V.L., Rezende, R.A.: Additive manufacturing and its future impact on logistics. In: 6th IFAC Conference on Management and Control of Production and Logistics, Fortaleza, Brazil (2013)

Marchese, K., Crane, J., Haley, C.: 3D Opportunity for the Supply Chain: Additive Manufacturing Delivers. Deloitte University Press (2015)

Stapleton, D., Pande, V.: Evaluating additive manufacturing as a disruptive technology in transportation & logistics. In: POMS 27th Conference Innovative Operations in an Information and Analytics Driven Economy, Orlando, USA (2016)

Bogers, M., Hadar, R., Bilberg, A.: Additive manufacturing for consumer-centric business models: implications for supply chains in consumer goods manufacturing. Technol. Forecast. Soc. Chang. **102**, 225–239 (2016)

Boon, W., van Wee, B.: Influence of 3D printing on transport: a theory and experts judgement based conceptual model. Transp. Rev. **38**(5), 556–575 (2017)

Petrick, I.J., Simpson, T.W.: 3D printing disrupts manufacturing: how economies of one create new rules of competition. Res. Technol. Manage. **56**(6), 12–16 (2013)

Jiang, R., Kleer, R., Piller, F.T.: Predicting the future of additive manufacturing: a Delphi study on economic and societal implications of 3D printing for 2030. Technol. Forecast. Soc. Chang. **117**, 84–97 (2017)

Weller, C., Kleer, R., Piller, F.T.: Economic implications of 3D printing: market structure models in light of additive manufacturing revisited. Int. J. Prod. Econ. **164**, 43–56 (2015)

Mashhadi, A.R., Esmaeilian, B., Behdad, S.: Impact of additive manufacturing adoption on future of supply chains. In: ASME 2015 International and Manufacturing Science and Engineering Conference, Charlotte, USA (2015)

Ratto, M., Ree, R.: Materializing information: 3D printing and social change. First Monday. **17** (7) (2012)

Thiesse, F., Wirth, M., Kemper, H.-G., Moisa, M., Morar, D., Lasi, H., Piller, F., Buxmann, P., Mortara, L., Ford, S., Minshall, T.: Economic implications of additive manufacturing and the contribution of MIS. Bus. Inf. Syst. Eng. **57**(2), 139–148 (2015)

Chen, Z.: The service-oriented manufacturing mode based on 3D printing: a case of personalized toy. Procedia Eng. **174**, 1315–1322 (2017)

Rayna, T., Striukova, L.: From rapid prototyping to home fabrication: how 3D printing is changing business model innovation. Technol. Forecast. Soc. Chang. **102**, 214–224 (2015)

Hartmann, J., Lebherz, M.: Commercializing Additive Manufacturing Technologies: A Business Model Innovation Approach to Shift from Traditional to Additive Manufacturing, Master's thesis, Halmstad University, Halmstad, Sweden (2017)

European Commission: Identifying Current and Future Application Areas, Existing Industrial Value Chains and Missing Competencies in the EU, in the Area of Additive Manufacturing (3D Printing), EASME, Brussels, Belgium (2016)

Nathanail, E., Adamos, G., Gogas, M.: A novel framework for assessing sustainable urban logistics. In: 14th World Conference on Transport Research, Shanghai, China (2016)

Nguyen, Q.T.: Environmental Impact of Additive Manufacturing on the Global Supply Chain, Bachelor thesis, Helsinki Metropolia University of Applied Sciences, Helsinki, Finland (2017)

Ghadge, A., Karantoni, G., Chaudhuri, A., Srinivasan, A.: Impact of additive manufacturing on aircraft supply chain performance: a system dynamics approach. J. Manuf. Technol. Manage. **29**(5), 846–865 (2018)

Saaty, T.: How to make a decision: the analytic hierarchy process. Eur. J. Oper. Res. **48**, 9–26 (1990)

Manginas, V.: Investigating the impact of additive manufacturing on supply chains, Master thesis, University of Thessaly, Volos, Greece (2020)

Modelling MaaS Plans and Commitment Length: Experience from Two European Cities

Athena Tsirimpa(✉), Ioannis Tsouros, Ioanna Pagoni, and Amalia Polydoropoulou

Department of Shipping, Trade and Transport, Transportation and Decision-Making Laboratory, University of the Aegean, Mitilini, Greece
atsirimpa@aegean.gr

Abstract. Mobility –as –a Service (MaaS) is a novel concept in transportation, which is received with enthusiasm and its rising popularity is met by a growing number of studies. Exploring potential user and early adopters' preferences is an imperative step towards the formulation of realistic, sensibly priced MaaS plans. This paper explores user preferences of MaaS plans, focusing on the duration of the plan (user commitment); adjusting pricing for duration. This type of research provides insight into the decision-making process, especially the trade-off between longer commitment and unit prices of included travel modes in the plans. To perform the described research, data from over 500 individuals from Budapest, Hungary and Manchester, UK was utilized. Collected data includes socio-demographic characteristics, attitudinal data, habitual travel patterns and stated preferences towards various MaaS plans with different commitment duration. Model estimation results indicate a general preference for MaaS packages instead of pay as you go. Additionally, the paper explores user heterogeneity and identifies different segments of the sample which have different reactions to a longer commitment. Results indicate certain willingness-to-accept values for longer commitments which may drive stakeholders (for example a MaaS operator) decisions on pricing packages of longer duration.

Keywords: Mobility as a service · User preferences · Stated preferences · Demand model

1 Introduction

Mobility as a Service (MaaS) offers seamless mobility through an integrated application that allow users to access a number of mobility and non-mobility services offered either as part of a bundled package or via pay as you go. In this context, MaaS promises to change mobility as we currently know it, by encouraging the switch from vehicle ownership to usership [**Error! Reference source not found.**]. Due to its potential benefits, MaaS has attracted interest between researchers, policymakers, transport service providers and city planners.

In addition, numerous real-life applications of MaaS currently exist in Finland, Germany, Netherlands, Austria, Italy and Sweden [2], as well as in other parts of the world (i.e. US, Australia, UK), leading many researchers to explore demand regarding

E. G. Nathanail et al. (Eds.): CSUM2020, AISC 1278, pp. 202–209, 2021.
https://doi.org/10.1007/978-3-030-61075-3_20

MaaS and compare the different developments between countries [3]. Although there is a growing body of the literature that focuses on MaaS, studies exploring demand for MaaS are rather limited, with the majority of them focusing on users' preferences towards different MaaS offerings [4, 5, 6, 7], as well as on individuals' attitudes towards MaaS. Matyas and Kamargianni [**Error! Reference source not found.**] through a mixed-method approach explored individual preferences towards MaaS in Greater London. Their analysis indicated that apart from public transport that was the mode mostly preferred, other modes such as car sharing, bike-sharing and taxi were not preferred by the respondents. On the contrary, Guidon et al. [8] and Reck and Axhausen [9] through stated preference experiments, found that individuals are more interested in car sharing if it is part of a MaaS package compared to a stand-alone service. Caiati et al. [10] provides a detailed conceptual modelling framework for MaaS adoption. Esztergár-Kiss and Kerényi [11] developed a methodology for identifying the services that should be included in a MaaS package based on a number of parameters (such as demographics, weather conditions, environmental consciousness) that can be found at a city level. The developed approach was applied in 15 European cities; however individual preferences were not taken into account.

The current study seeks to explore preferences and stated choices of individuals regarding the purchase of MaaS plans, focusing on the duration of the plan (user commitment); adjusting pricing for duration. For this purpose, an online questionnaire survey was conducted in two European cities: Greater Manchester (United Kingdom) and Budapest (Hungary) collecting both revealed and stated preference data. These data were then used to develop an econometric discrete choice model for MaaS plans.

The rest of the paper is organized as follows: a short description of the survey and descriptive statistics of the sample are presented in Sect. 2. Section 3 presents the modeling framework. Section 4 presents the model estimation results, while Sect. 5 concludes the paper.

2 Survey and Sample Descriptive Statistics

This research was conducted as part of MaaS4EU, a Research and Development project funded by the European Horizon2020. In order to be able to address the objectives of the MaaS4EU project and understand the behavior of MaaS potential end-users, an online survey was designed and conducted by the authors of the paper and colleagues from UCL-MaaSLab [Error! Reference source not found.]. The survey consisted of two parts: a section with socioeconomic, current travel patterns and attitudinal data and a section with stated preference experiments. The target population for our sampling strategy were identified during the focus groups that were previously organized in the two study areas [14]. Based on these findings, the target customer segments, in both cities, are younger than the actual population average.

Overall, 574 respondents from both Manchester and Budapest, replied to the online questionnaire and their responses were used for the development of the model presented in this paper. The average age of the sample is 39 yr old and 42% are females. The majority of the respondents are full-time employees (28%), followed by 7% students and 6% pensioners. Approximately one third of the sample holds a university degree (a bit higher than the EU average), while 50% have flexible working hours. Concerning their current travel patterns, the respondents stated that they conduct on average 7 trips per week by car, 5 trips per week with public transport and 7 trips per week on foot. In addition, 7 trips per week are conducted with bicycle, either owned by the respondent (4 trips per week) or as part of a bike-sharing scheme (3 trips per week). More than two thirds of the sample have a car license and 34% has a public transport pass.

3 Modeling Framework

This section presents the modeling framework and the specification of the model developed for MaaS plan choice. The participants of our survey, were presented with three different MaaS alternatives, as part of the stated preference (SP) experiments, resulted in repeated choices. Hence, the observations generated by each participant are dependent and correlated (the assumption of identically independent distributed (iid) error components is violated). In Fig. 1, the rectangles at the top represent the explanatory variables of the model. The solid arrow links the explanatory variables to the decision utility (illustrated by an ellipsis), while the dashed arrow links the utility to the choice of the MaaS plan. The utility for each MaaS plan alternative, and subsequently the probabilities of choosing it, is assumed to be affected by the MaaS plans related attributes (i.e. subscription cost, cost per ride, length of commitment), individuals' socioeconomic characteristics and habitual travel behavior. The model developed, is an error component multinomial logit capturing: (a) the correlation between alternatives with common characteristics and (b) the correlation among the observations of the same individual. These correlations are captured through the inclusion of an appropriate random error term [12, 13].

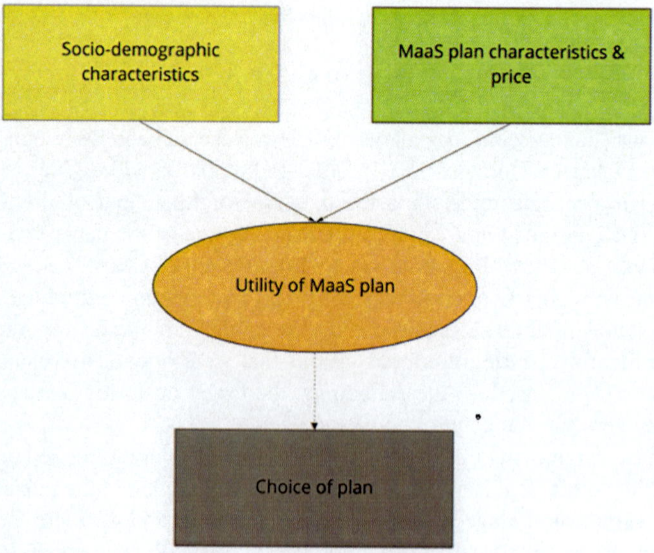

Fig. 1. Modeling framework.

In the error component model, the utility *Unj* of an individual *n* derived from alternative *j* is given by Eq. (1):

$$U_{nj} = \boldsymbol{\beta}' x_{nj} + \varepsilon_{nj}, \tag{1}$$

Where, $\boldsymbol{\beta}'$ is a vector of fixed coefficients, x_{nj} are observable variables with regards to alternative j and ε_{nj} is the stochastic term of the utility function. In particular, x_{nj} concern the respondent's characteristics that can be observed (i.e. socio-demographics), as well as the attributes of the MaaS plan alternatives. The probability of an individual n to choose alternative j is defined in Eq. (2):

$$P_{nj} = \int L_{nj}(\boldsymbol{\beta}) f(\boldsymbol{\beta}) d\boldsymbol{\beta} \tag{2}$$

where $L_{nj}(\beta)$ is the logit probability evaluated at parameters β (see Eq. (3)) and $f(\beta)$ is a density function.

$$L_{nj}(\boldsymbol{\beta}) = \frac{e^{a_n + \beta_n x_{nj}}}{\sum_{j=1}^{J} e^{a_n + \beta_n x_{nj}}} \tag{3}$$

A non-zero error component z_{nj} is included in some of the utility functions so that to capture the correlation among different alternatives, thus the stochastic term of the utility is expressed as $\varepsilon_{nj} = \mu' z_{nj} + e_{nj}$, where μ is a vector of random terms with zero mean and e_{nj} is an iid extreme value.

4 Model Estimation Results

This section presents the model estimation results. A description of the explanatory variables of each model is provided and the statistical significance of each variable is indicated in Table 1. The presented model was selected in the basis of statistical goodness-of-fit and parsimony. To ensure the model's identification, a sufficient number of draws (1000, 5000) was used to ensure the stability of the estimated parameters.This section presents the model estimation results of the error component logit model developed, containing pay-as-you-go, weekly and monthly plans. The model is identified for three different alternatives, the pay-as-you-go option, weekly plan and monthly plan.

Overall the model performs well the parameters have the expected signs and most of them are statistically significant at the 95% confidence interval level. Ceteris paribus, the pay-as-you-go option seems to be the least preferable among respondents. This is reflected in the signs of the alternative specific coefficients (ASC), where the pay as you go option is the base. The variables associated with contracts that end automatically after the scheduled period or have a longer period commitment, when interacted with the price coefficient they indicate individual's willingness to commit. Coefficients that measure the effect of the monthly cost of restricted and unrestricted public transport

Table 1. Model estimation results.

Variable	Coefficient	t-test
Cost of plan pay as you go	−0.0794	−3.88
Cost of taxi in pay as you go	−0.0124	−2.45
Cost of car sharing in pay as you go	−0.00304	−0.72
Alternative Specific Constant (ASC) Weekly plan	0.325	0.39
Weekly Cost Restricted Public Transport	−0.0416	−2.31
Weekly Cost Unlimited Public Transport	−0.0544	−2.02
Presence of Carsharing in Weekly plans	1.34	2.72
Alternative Specific Constant (ASC) Monthly Plan	0.321	0.38
Monthly Cost for Unlimited Public Transport	−0.0186	−2.17
Monthly Cost for Restricted Public Transport (PT)	−0.0118	−2.03
Presence of Carsharing in Monthly plan	0.979	1.97
Contract Duration – Contract automatically ends after the scheduled period have a longer period commitment (Generic to Weekly and Monthly plans)	0.278	1.79
Contracts have a longer period commitment (Generic to Weekly and Monthly plans)	0.256	1.60
SIGMA_MaaS	−2.64	−7.56

access either in the weekly or the monthly plans are statistically significant and have the expected sign (negative). Thus, as the cost of public transport increases the probability of selecting the respective plan decreases. In addition, it appears that individuals are more price sensitive when it comes to weekly packages when considering public transport modes. The presence of car sharing positively affects both weekly and monthly plans. A finding similar to the findings of Guidon [8] et al. and Reck and Axhausen [9]. The pay-as-you-go plan price coefficient is negative and statistically significant as expected, while taxi cost coefficient in the pay-as-you-go plan is statistically significant whereas the car sharing coefficient is not. The estimated sigma for MaaS plans (monthly and weekly) is significantly different from zero, capturing the correlation between MaaS plans, as well as the correlation among the observations of the same individual.

5 Willingness-To-Commit

This section presents the willingness of individual to commit by comparing the coefficients of the pay-as-you-go, weekly and monthly plans, as well as, the coefficients of public transport presence (both limited and unlimited) in the case of automatically stopping contract and contract with longer commitment.

Table 2 presents the estimated willingness-to-commit prices for different MaaS plans. As it can be seen, individuals are willing to pay 6.3 € per week more in an unlimited public transport MaaS plan to avoid an automatically stopping contract and 6.6 per week more to avoid a longer commitment contract. This means that in order to convince the users to sign up for a contract (subscription package) the MaaS operator should offer a discount compared to the pay-as-you-go option that is at least of this size. Similar results are also obtained in MaaS packages containing limited (restricted) public transport options. In the monthly commitment plans the overhead becomes 22 € for automatically stopping contracts and 23 € for contracts of longer commitment for unlimited public transport plans and 14.7 €/15.5€ for the limited (restricted) PT plans.

Table 2. Willingness-to-commit.

	Automatically stopping contract		Longer commitment contract	
Plan length	Unlimited Public Transport	Restricted Public Transport	Unlimited Public Transport	Restricted Public Transport
Weekly (€/week)	6.31	4.82	6.64	5.07
Monthly (€/month)	22.08	14.72	23.25	15.5

6 Conclusions

This paper explores consumer preferences and potential demand regarding the purchase of MaaS plans. The study uses revealed and stated preference data collected into cities, Budapest (Hungary) and Manchester (UK), through an online survey within MaaS4EU (H2020 funded project). An error component logit model was estimated, indicating individuals' preferences regarding different MaaS plans. The dependent variable of the model includes plans that vary in the length of the commitment (pay-as-you-go, weekly, monthly) and the model estimated provides useful insight about the price elasticity in the different time periods of commitment, as well as, the willingness-to-commit to either automatically ending contracts or longer period contracts. In addition, the model captures the shared heterogeneity between respondents that prefer MaaS plans as opposed to those that choose the pay as you go alternative. Results reveal that user are willing-to-pay around 6€ per week in order to avoid commitment (either in the form of automatically stopping contract or longer duration contract), a price that can reach more than 20€ in monthly settings if concerns the purchase of unlimited public transport plans.

A study limitation can be identified to the lack of explored heterogeneity between the survey respondents in order to identify user segments which have different commitment, preferences and price elasticities. This exploration could be conducted by utilizing an ICLV (integrated choice and latent variable) model or latent class segmentation techniques and is an interesting direction for future research. Identifying user segments is important both for constructing personalized services and for designing meaningful MaaS plans, combining popular alternatives in optimal commitment form and length. Given that the number of papers in the literature investigating the demand for MaaS are still scarce, we consider that the findings of this paper contribute significantly to the existing literature. The presented work provides useful insight on aspects that have not been yet explored in the literature, such as preferred duration of user commitment and type of contract. This information can be quite useful to MaaS operators while developing their MaaS plans.

Acknowledgments. This research is part of the Project "MaaS4EU" (End-to-End Approach for Mobility-as-a-Service tools, business models, enabling framework and evidence for European seamless mobility). This project has received funding from the European Union's Horizon 2020 research and innovation programme under grant agreement No 723176. This publication only reflects the authors' view and the European Union is not liable for any use that may be made of the information contained therein.

References

1. Matyas, M., Kamargianni, M.: The potential of mobility as a service bundles as a mobility management tool. Transportation **45**, 1951–1968 (2019)
2. Durand, A., Harms, L.: Mobility-as-a-Service and changes in travel preferences and travel behaviour: a systematic literature review, Bijdrage aan het Colloquium Vervoersplanologisch Speurwerk (2018)
3. Smith, G., Sochor, J., Sarasini, S.: Mobility as a service: comparing developments in Sweden and Finland. Res. Transp. Bus. Manage. **27**, 36–45 (2018)
4. Kamargianni, M., Li, W., Matyas, M., Schäfer, A.: A critical review of new mobility services for urban transport. Transp. Res. Procedia **14**, 3294–3303 (2016)
5. Ho, C.Q., Hensher, D., Mulley, C.: Wong, Y: Potential uptake and willingness-to-pay for Mobility as a Service (MaaS): a stated choice study. Transp. Res. Part A Policy Practice **117**, 302–318 (2018)
6. Kamargianni, M., Matyas, M., Li, W., Muscat, J.: Londoners' attitudes towards car-ownership and Mobility-as-a-Service: Impact assessment and opportunities that lie ahead, UCL Energy Institute's MaaSLab report prepared for Transport for London (2018). MaaSLab homepage https://www.maaslab.org/copy-of-maas-publications. Accessed 20 May 2020
7. Hensher, D.A.: Future bus transport contracts under a mobility as a service (MaaS) regime in the digital age: are they likely to change? Transp. Res. Part A Policy Practice **98**, 86–96 (2017)
8. Guidon, S., Wicki, M., Bernauer, T., Axhausen, K.: Transportation service bundling - For whose benefit? consumer valuation of pure bundling in the passenger transportation market. Transp. Res. Part A Policy Practice **131**, 91–106 (2020)
9. Reck, D.J., Axhausen, K. W.: Understanding long-term multimodal mobility demand to inform MaaS service bundling. In: 19th Swiss Transport Research Conference, Monte Verita/Ascona, 15–17 May (2019).
10. Caiati, V., Feneri, A.M., Rasouli, S. Timmermans, H.J.P.: Innovations in urban mobility and travel demand analysis: Mobility as a Service context. In: Proceedings BIVEC-GIBET Transport Research Days, pp. 492–503 (2017).
11. Esztergár-Kiss, D., Kerényi, T.: Creation of Mobility Packages Based on the MaaS Concept. Travel Behav. Soc.**21**, 307–317 (2019)
12. Train, K.E.: Discrete choice methods with simulation. Cambridge University Press, Discrete Choice Methods with Simulation (2002)

13. Hess, S., Daly, A., Batley, R.: Revisiting consistency with random utility maximisation: theory and implications for practical work. Theory and Decision. Springer US, **84**(2), 181–204 (2018)
14. Polydoropoulou, A., I. Pagoni, Tsirimpa, A.: Ready for mobility as a service? insights from stakeholders and end-users. Travel Behaviour and Society (2018)

A Regional Competence Centre for SUMPs in Central Macedonia, Responding to the Identified Local Needs

Maria Chatziathanasiou[1(✉)], Maria Morfoulaki[1], Konstantia Mpessa[2], and Lambrini Tsoli[3]

[1] Centre for Research and Technology Hellas, Hellenic Institute of Transport, Thessaloniki, Greece
{mariacha,marmor}@certh.gr
[2] Region of Central Macedonia, Thessaloniki, Greece
K.Mpessa@pkm.gov.gr
[3] Regional Development Fund of Central Macedonia, Thessaloniki, Greece
l.tsoli@rdfcm.gr

Abstract. The municipalities of the Region of Central Macedonia (RCM) still face significant problems in their SUMP launch and implementation. Unclear jurisdiction on sustainable urban mobility within the city departments, incomplete knowledge and/or limited in-house capacities for SUMP development, ‘silo’ approaches in the political level of SUMP planning and limited synergies with inter-municipal/ regional strategies are some of the issues that hinder the wide deployment of SUMPs in RCM. The Hellenic Institute of Transport, the Regional Development Fund of Central Macedonia and RCM have been working closely during the last 3 years through the Interreg Europe REFORM project cooperation in order to involve regional municipalities in an exchange of experience and capacity building process regarding SUMPs. Through this synergy, they have gained insight into the real problems that the city staff members face (through dedicated learning events and surveys) and they launched a Competence Centre for supporting the technical implementation of SUMPs and providing a ‘stage’ for dialogue between municipalities, regional authorities and scientific/knowledge experts. The Competence Centre of RCM was built as an online tool providing up-to-date technical guidance, based on the official EU SUMP guidelines, information on the recent developments in SUMP in the region and beyond, a library of useful documents and Greek good practices, a forum for exchange of experience between staff members of RCM’s municipalities and the opportunity for tailor-made SUMP training courses. Its overall aim is to align the local SUMPs into a wider, regional planning.

Keywords: SUMP · Competence center · Capacity building

E. G. Nathanail et al. (Eds.): CSUM2020, AISC 1278, pp. 210–219, 2021.
https://doi.org/10.1007/978-3-030-61075-3_21

1 Introduction

Over the last decade Sustainable Urban Mobility Plans (SUMPs) have been actively supported by the European Commission. The origin of SUMPs can be found in the Thematic Strategy on the Urban Environment, where "publication of guidelines for sustainable urban transport plans" is proposed [1]. Since then, the concept of SUMPs has been specifically formulated through the Urban Mobility Package in 2013 [2] and the Eltis Guidelines in 2013 [3] and 2019 [4].

SUMPs have been recognized as an important tool to promote healthier and safer mobility systems [5, 6] and several good practices have already demonstrated successful results for cities and regions [4]. Nonetheless, the adoption of SUMPs in Europe still faces significant challenges, as the 2017 SUMP assessment needs survey, indicated that only a 37% of the responding cities have implemented a SUMP, with Greece holding one of the last positions in the list (6%) [7].

Although the wide adoption of SUMPs is a key target for the European and national policies, their philosophy requires an integrated planning, translated into a meaningful cooperation "across institutional barriers" [4]. This requires complementarity in a cross-sectoral level (i.e. transport, land use, societal planning, environment, etc.) and cross-administrative level (i.e. city, region, country).

Bringing the SUMP development and management outside the limits of a local (city) SUMP, several SUMP governmental structures apply [8], from informal/soft coordination (as the case of RCM, in Greece), where, usually SUMPs are developed in the city level and the region or metropolitan unit does not have a mandate to supervise them, to supra-municipal authorities (such as Greater Manchester, in UK), which are responsible for delivering and managing the mobility policies.

Regardless of their role in SUMP, regions and metropolitan units are responsible for the large scale territorial mobility planning, managing funds of important pillars of SUMPs, such as public transport services and cross-municipal infrastructure developments. The role of the regions in framing and serving the wider mobility strategy is also recognized by the Commission, as it highlights the importance of [9]: developing an approach to urban mobility which ensures coordinated and mutually reinforcing action at national, regional and local level; ensuring that SUMPs are developed and implemented in their urban areas and that they are integrated into a wider urban or territorial development strategy; reviewing and amending where necessary the technical, policy based, legal, financial, and other tools at the disposal of local planning authorities.

2 Current Status of SUMPs in RCM

RCM is one of the thirteen administrative regions of Greece. It has a total population of around 1,9 million inhabitants, which places it to the second most populous regions in Greece [10]. The most important sector, regarding its contribution to the GDP is the tertiary sector and the region has a highly popular touristic profile [11].

At the same time, the region's mobility heavily depends on the use of private cars, as their modal share in the metropolitan area of Thessaloniki reaches the 41% [12].

There is a greater need to focus on sustainable modes of transport within local SUMPs, as well as to coordinate relevant actions at regional level, thus scaling them up to the territorial planning.

RCM is characterized by a relatively low percentage of SUMP implementation, as, at the end of 2019, only 10 (out of the 38) municipalities of RCM have concluded or started the process of developing a SUMP, while 5 were in the phase of public procurement, facing severe delays with the process.

As in the case of the majority of Greek municipalities, cities in RCM often suffer from limited resources and know-how for the SUMP definition and they highly need [13]: guidance, expertise and training in SUMPs; support in financing SUMP development and measures; support in the institutional (responsibilities and requirements for cooperation) framework of SUMPs; support in the legal framework for mobility planning and the legal framework for integrating mobility and land use planning.

Even the allocation of tasks to external experts, which is the common practice for Greek municipalities, requires comprehensiveness of the SUMP philosophy and technical and administrative requirements. At the same time, a sustainable mobility policy requires that cities, especially the ones belonging to a wider metropolitan area, abandon their 'stand-alone culture' and participate into a constructive open dialogue with other local authorities, governments, stakeholders and citizens. 'Silo' approaches in the political level of local sustainable mobility planning are 'transferred' in the technical departments who are eventually asked to implement or supervise the implementation of local SUMPs. Staff members are often straggling with legislative and administrative issues that have might have been solved, though, in other municipalities.

It should be noted, that the current national framework of SUMPs is reflected through Article 22 of Law 4599/2019 [14], where the definition of SUMPs and guidance for its implementation are provided, nonetheless, more detailed specifications are still expected. Furthermore, to the knowledge of the author and as reflected from other research [7], there is currently no up-to-date national/regional/local websites dedicated to SUMPs. It should be mentioned, thought, some supporting initiatives for creating Greek SUMP networks have been undertaken, mostly in the context of European project such as EPPOM or ENDURANCE [15, 16].

3 Overall Methodological Framework

The overall approach towards the implementation of RCMs' Competence Center (CC) for SUMPs initiated from a structured exchange of experience process that took place as part of an Interreg Europe Programme cooperation: the REFORM project [17].

The REFORM initiative (2017–2020) has set the goal of enhancing the role of the regions in the SUMP development and implementation. One of its main pillars was the continuous exchange of experience process, involving consortium technical meetings, Good Practices (GPs) transferability, regional learning events and stakeholders' involvement in various levels (European, regional, local). Through a structured technical approach and a solid internal and external communication strategy, REFORM succeeded in inspiring four regional Actions Plans towards the definition of actions that can address the needs, gaps and priorities of the regions in sustainable urban mobility planning.

The project's methodology for the Action Plans' development and implementation involved a multi-stakeholder participatory approach for generating intra- and inter-regional knowledge and strengthening the links between the regional authorities and the stakeholders involved in mobility planning (local authorities, citizens, transport operators, experts, scientists, etc.) (see. Fig. 1).

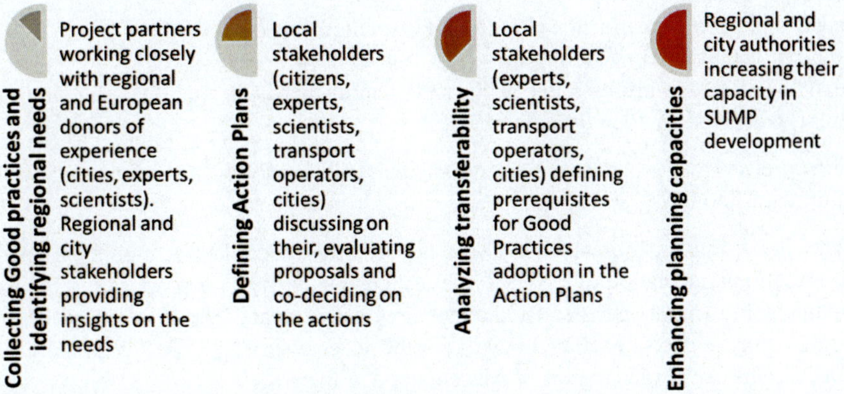

Fig. 1. Involvement of stakeholders in the main exchange of experience activities of REFORM.

Through a specific approach, RCM co-defined (with the cities) the needs in sustainable mobility planning and the support that the region should establish for the wider – but integrated to the regional strategy – SUMP adoption. This, eventually, constituted the main input for the development of the operational architecture of a regional SUMP support mechanism (RCM's CC). The step-by-step implementation of the specific approach is presented in the next sections.

4 Defining the Feasibility and the Operational Architecture of the SUMP Competence Centre

As a first step, a critical review of the current situation on SUMP development in RCM has taken place, in order to validate the need for developing a SUMP supporting mechanism. Afterwards, the collection and analysis of the specific needs, barriers and problems of the technical staff, who is responsible for SUMP implementation and/or management, took place, giving a clear understanding of the content that this mechanism should provide. The relevant information was collected through a semi-structured interview process and a questionnaire survey, both presented below in more details.

4.1 Interviewing Regional Authorities and Cities on the SUMP Status and Administration

A semi-structured interview process took place as part of RCM Action Plan preparation activities, in order to describe the [18]:

- regulatory framework and technical instruments of the regional policy about SUMPs
- strengths and weaknesses of the national, regional and municipal regulations on mobility planning
- costs and procedures for the drafting and development of a SUMP
- state of the SUMPs in the region
- existing in-house professional skills and training needs in mobility planning, as these perceived by the city authorities.

The interviews involved regional and municipal staff members and indicated the following main findings [18, 19]:

- The Greek implementation of SUMPs is currently undertaken at municipal level. For the implementation of a SUMP the responsible body is (or proposed to be) the municipality (more specifically, the technical department). The legal check of the administrative decisions that are taken for the acceptance of a SUMP is done by the Decentralized Administration of Macedonia – Thrace.
- Main weaknesses of the national, regional and municipal regulations on mobility planning are mostly related to the fragmentation of responsibilities of the bodies for the planning and a luck of a systematic approach at national level.
- There is currently no funding or any support for SUMPs included in the regional policy instruments of RCM, due to the fact that there have been no relevant requests from the cities' administrations. Nonetheless, a significant opportunity for the wide adoption of SUMPs was provided through the funding mechanism of the Green Fund of the Ministry of Environment, which was launched in 2016 and provided financial support to 153 Greek municipalities (24 from RCM) to develop their SUMPs.
- Training on SUMPs is considered necessary for the majority of the city authorities' interviewees, as relevant experience regards mostly traffic studies. Currently, all city authorities develop their SUMPs with the support of external experts.

4.2 Questioning City Authorities About Their Requirements from a Regional SUMP Support Mechanism

After the first round of discussions with the city and regional authorities, RCM confirmed the feasibility of establishing a regional mechanism (Competence Centre) to support its 38 municipalities in developing, adopting and eventually implementing and monitoring their SUMPs.

For that purpose, a second round of discussion with the city authorities took place (taking the form of a survey), in order to define the specific content to be delivered by the mechanism. The survey was conducted through an electronic questionnaire. The questionnaire included three groups of questions [20]:

1. General information of the participant and the municipality he/she represents
2. Closed and Likert scale questions reflecting on the characteristics of the municipality concerning the management of SUMP, such as: Are responsibilities in sustainable urban mobility assigned to specific department? How do you perceive the level of cooperation with neighboring municipalities and the regional administration? What is the level of in-house knowledge regarding SUMP development and citizens' and stakeholders' engagement?
3. Scaled questions that prioritize the needs of each municipality in the management SUMPs, regarding both the SUMP procurement phase and the steps of the SUMP development.

The questionnaires were completed by 25 representatives of 18 municipalities of RCM. Main findings of the analysis include:

- In the majority (89%) of the participating municipalities, there are no clear responsibilities for sustainable urban planning and sustainable mobility and the knowledge about SUMP is characterized as incomplete (low by 52% of the respondents and moderate by 32% of the respondents). Nonetheless, the few municipalities that operate a dedicated department for sustainable mobility have staff members of high experience and knowledge regarding both the SUMPs and the consultation procedures that are required.
- A high percentage of the respondents (44%) consider that they do not develop at all the synergies with the neighboring municipalities for the planning of urban mobility, while a similar percentage (40%) considers that synergies are accomplished in a moderate level.
- There are several respondents (40%) who consider that the cooperation between the municipalities and RCM is good. However, there is a remarkable number of respondents (56%), representing 12 municipalities, who consider that there is no or moderate cooperation.
- Concerning the cooperation within the city departments, a 20% replied that it does not exist at all, a 36% that it exists to a moderate level and a 44% that it is very good or excellent.
- The experience concerning the stakeholders' consultation processes seems to be considered non-existent (12%), low (36%) or moderate (36%) by the majority of the respondents.
- Regarding the hierarchy of the needs of the municipalities: as far as the planning is concerned (drafting of a SUMP tender), the description of the data needed and its collection methodologies is of outmost important (see Fig. 2). As far as the implementation of SUMPs is concerned, the stage of defining visions, objectives and priorities, but also the tasks of overseeing those steps that require data collection and analysis and the use of the traffic model are considered the topics for which the municipalities would mostly welcome support (see Fig. 3).

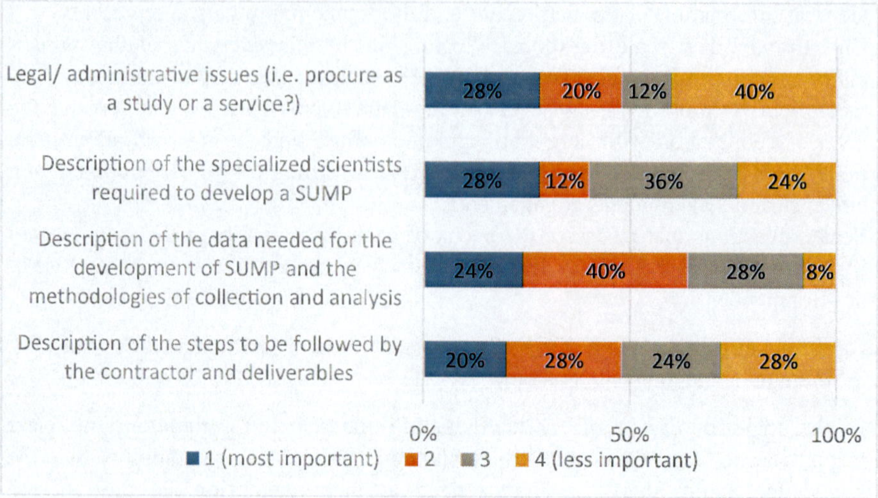

Fig. 2. Hierarchy of the municipal needs for the drafting of a SUMP tender.

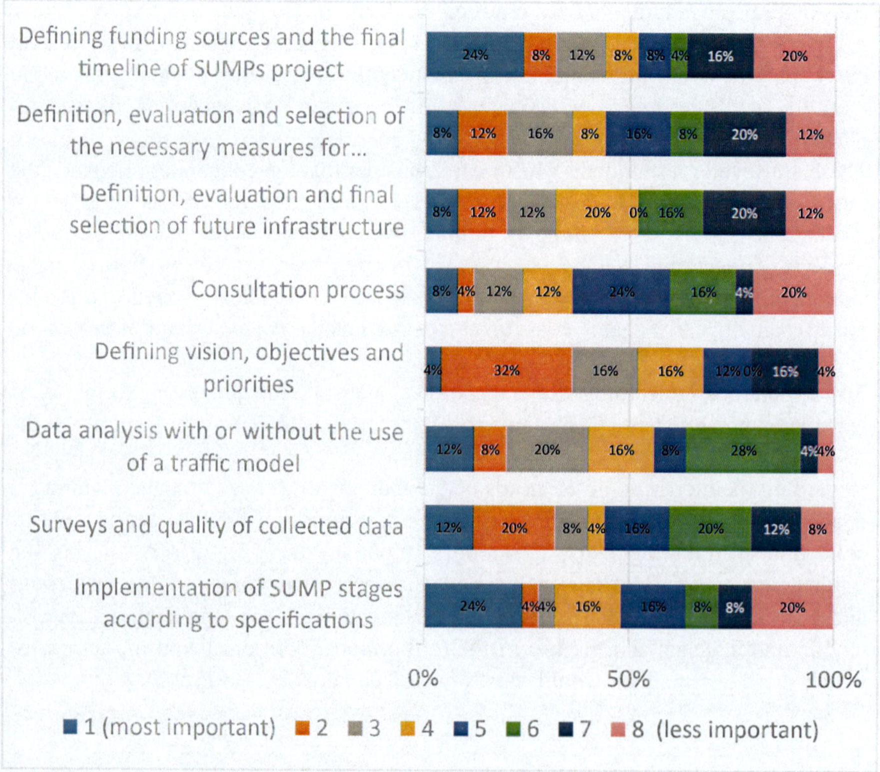

Fig. 3. Hierarchy of the municipal needs related to the steps of the SUMP cycle.

5 Development of RCM's SUMP Competence Centre

Capitalizing the identified barriers and needs of the cities' technical staff for supervising and/or implementing SUMPs, the Hellenic Institute of Transport (HIT), the Regional Development Fund of Central Macedonia and RCM cooperated in the development and operation of the Competence Centre (CC) which constitutes a 'stage' for dialogue between municipalities, regional authorities and scientific/knowledge experts.

The CC of RCM was built as an online tool [21] (see Fig. 4), aiming at providing technical support to the municipalities for implementing their SUMPs. The main provisions of the CC are listed in Table 1 and matched to the needs they address.

Table 1. Main components of the CC and needs they address.

Component	Need addressed
Getting acquainted with SUMPs, the European actions and the Greek legislation	Familiarize with the new approach in urban mobility
Up-to-date technical guidance, delivering the SUMP cycle in an interactive – user friendly mode (one-click for entering each phase or step of the cycle)	Increase knowledge on SUMPs > facilitate SUMP development and/or management
Library of useful documents (including examples of public procurement documents for SUMP development)	Facilitate administrative procedures related to SUMP planning and/or management; facilitate SUMP development and/or management
Frequently Asked Questions (FAQs) for each sub step of the SUMP cycle (32 in total)	Increase knowledge on SUMPs > facilitate SUMP development and/or management
Recent developments in SUMP in RCM and beyond	Support SUMP community feeling; increase SUMP visibility; develop synergies
Online forum for staff member of municipalities	Develop synergies between cities; facilitate administrative procedures
Online communication form with the support team	Respond to new requests > update content
Training material	Build capacity in SUMP planning and management

Further to the above, a special online "space" is reserved in the CC for setting up a future interoperability with the "Urban Mobility Observatory". The Observatory is a tool, currently under implementation by RCM, which will collect, analyze and set up data of the local SUMPs.

The CC is supported by a team of specialized personnel from HIT and RCM, who will provide technical advice within the context of the official European and national guidelines and specifications.

The CC's content will be constantly tailor-made to any further requests of the cities, as follow-up consultations with the city authorities will take place, keeping pace with the progress of SUMPs in the region. To this context, consultation, work, seminars, workshops, meetings, etc. will be constantly organized by the support team. It is envisaged that the CC will be eventually developed to a living network for sustainable mobility in the region and be part of a new cooperative structure, data driven by a "Regional Observatory", which will monitor local SUMP implementation, thus allowing RCM a better knowledge of the regional planning needs in sustainable mobility.

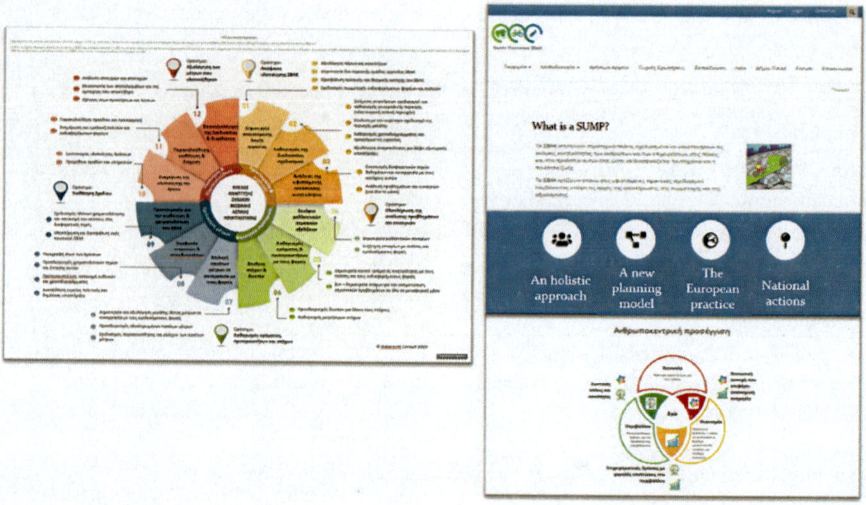

Fig. 4. Example of a content of the CC: familiarizing with the new approach in urban mobility.

6 Conclusions

SUMPs are recognized as a valuable tool for addressing urban mobility challenges and tackling the negative impacts of transport in health, safety and living conditions. At the same time, they require integration of policies, both between different sectors and between different administrative levels. It is still evident, though, that cities haven't yet reached the stage of a wide SUMP deployment and that they face significant challenges in engaging their SUMPs into wider territorial policies. To that end, the role of the regions can be crucial.

This paper follows the set-up of a Competence Center for supporting SUMPs development and implementation in the region of Central Macedonia. To achieve that, RCM has followed a structured exchange of experience process within the REFORM project in order to engage the city authorities in co-defining their needs in SUMP development and implementation and co-designing the content of the mechanism that will support them on that matter. As a result, a regional Competence Center was established and currently operates in an online open space, providing: technical guidance on SUMP development; useful documents; answers to FAQs; news on SUMP development in RCM and a forum for the city staff members. Consultation and training will be further tailored-made to the needs of the cities, transforming the Competence Center to a "live" one-stop shop for SUMP in RCM and beyond.

References

1. COM (2005) 718: Communication from the Commission to the Council and the European Parliament on Thematic Strategy on the Urban Environment, Brussels (2006)
2. European Commission: Communication from the Commission to the European Parliament, The Council, The European Economic and Social Committee and the Committee of the Regions: Together towards competitive and resource-efficient urban mobility', Brussels, 913 final (2013)
3. Rupprecht Consult: Guidelines for Developing and Implementing a Sustainable Urban Mobility Plan, First Edition (2013)
4. Rupprecht Consult: Guidelines for Developing and Implementing a Sustainable Urban Mobility Plan, Second Edition (2019)
5. European Commission, Joint Research Center: Quantifying the effects of sustainable urban mobility plans. JRC Technical reports (2013)
6. Pisoni, E., Christidis, P., Thunis, P., Tombetti, M.: Evaluating the impact of "Sustainable Urban Mobility Plans" on urban background air quality. J. Environ. Manage. **231**, 249–255 (2019)
7. Durlin, T.: Status of SUMP in European member states, SUMPS-UP project (2018)
8. Chinellato, M., Morfoulaki, M.: Sustainable urban mobility planning in metropolitan regions. Sustainable urban mobility planning and governance models in EU metropolitan regions (2019)
9. COM (2013) 913: Annex. A Concept for Sustainable Urban Mobility Plans to the Communication from the Commission to the European Parliament, the Council, the European Economic and Social Committee and the Committee of the Regions, Brussels (2013)
10. Association of Greek Regions. https://www.enpe.gr/en/. Accessed 04 Apr 2020
11. European Commission. Region of Kentriki Makedonia. https://ec.europa.eu/growth/tools-databases/regional-innovation-monitor/base-profile/region-kentrikimakedonia. Accessed 07 Jan 2020
12. Municipality of Thessaloniki: Sustainable Urban Mobility Plan (2019)
13. Chnellato, M., Staalens, P., Wennberg, H., Sundberg, R., Bohler, S., Brand, L., Adams, R., Dragutescu, A.: Users' need analysis on SUMP take up, SUMPS-UP project (2017)
14. Greek Parliament: Law 4599/2019 (2019)
15. Papaioannou, P., Politis, I., Nikolaidou, A.: Steps towards sustaining a SUMP Network in Greece. Tansp. Res. Procedia **14**, 945–954 (2016)
16. SUMPnet. https://sumpnet.gr/. Accessed 27 Mar 2020
17. REFORM project. https://www.interregeurope.eu/reform/. Accessed 04 Apr 2020
18. Crowther, M., Dolce, C.: The selected good practices in the REFORM project regions. REFORM project (2018)
19. Pantazi, K., Nikopoulou, A., Mihailidis, K.: The State of Development of SUMPs in RCM Definition of Regional Needs and Priorities. REFORM project internal document (2017)
20. REFORM: Authorities' survey. Assessment of the needs to establish a regional mechanism for the support of local SUMPs. REFORM project internal report (2018)
21. RCM's Competence Center. https://www.keyp-svak-rcm.imet.gr/. Accessed 04 Apr 2020

Mobility as a Service (MaaS): Past and Present Challenges and Future Opportunities

António Amaral[1], Luís Barreto[2(✉)], Sara Baltazar[2], and Teresa Pereira[2]

[1] CIICESI, Escola Superior de Tecnologia e Gestão, Instituto Politécnico do Porto, Felgueiras, Portugal
ama@estg.ipp.pt

[2] Escola Superior de Ciências Empresariais, Instituto Politécnico de Viana do Castelo, Valença, Portugal
lbarreto@esce.ipvc.pt

Abstract. Recently, Mobility as a Service (MaaS) concept and its main theoretical approaches have been under discussion, to positively influence the future of mobility. Namely, by contextualizing MaaS's role in modern societies explaining its main functions, characteristics, and attributes, as well as identifying all the stakeholders involved in this comprehensive challenge towards ensuring its widespread implementation. The environmental, societal, technological and cultural changes needed to ensure a sustainable mobility ecosystem are an utmost challenge that requires an intense effort and involvement of all different types of stakeholders within their perspectives, roles, responsibilities and contributions to the mobility system overall behavior and performance. Notwithstanding, the global tendency of digital transformation, also referred as digitization, in society and businesses are upbringing a new technological evolution that will lead to a new mobility paradigm bringing together MaaS and the internet of Mobility (IoM), thus creating what we call the Internet of Mobility as a Service (IoMaaS). The future trends of mobility will have to be 'human-centric', to properly balance the amount of technology requested into the ecosystem to ensure the whole system's universality, to be inclusive, as well as developing the appropriate amount of technology, accordingly to the different users' technological skills. Furthermore, different types of incentives and penalties need to be included in supporting a broad cultural shift regarding citizen's mobility routines habits. This will be of great importance to ensure the sustainability of this new mobility paradigm as well as of the ability to attain all its benefits.

Keywords: Mobility as a Service (MaaS) · Mobility opportunities · Sustainable mobility · Integration · Technology · Internet of Mobility as a Service (IoMaaS)

E. G. Nathanail et al. (Eds.): CSUM2020, AISC 1278, pp. 220–229, 2021.
https://doi.org/10.1007/978-3-030-61075-3_22

1 Introduction

The urban population growth, due to the increase level of urbanization, combined with an inadequate and misaligned transportation systems, along with shortfall investment in public infrastructure and the transportation challenges such as congestion, inefficiency and pollution have been producing and introducing deep effects on the citizen's life. Most of the cities in the world are developing strong efforts to meet the accessibility and mobility needs of their citizens. The accessibility and the mobility concepts are part of the actual societies, representing together a challenge and a concern regarding the achievement of a more sustainable and a more inclusive society in order to promote the well-being development. It is recognized the importance of roads' infrastructure development to respond to the increase demand for mobility. The improve of the main entry nodes, in large urban centers, contributed to the migration of residence places to areas more distant from workplaces. As a result, the planned transport systems, mainly focused on car use, contributes to the control of the mobility market, which in many areas can offer superior level of service concerning effort and journey/travel times, despite the high fixed costs, compared to other alternative mobility services use [1]. Nevertheless, it is noticed a change in urban mobility, namely by the decreasing usage of the private car to the growing use of public transportation (PT) and active modes – e.g. car sharing, walking and bicycle use. Actually, the global challenge of more sustainable mobility is to change the paradigm centered on the availability of the car, to a paradigm more centered on the citizen and his daily activities. Naturally, this challenge implies behavioral and cultural changes and, therefore, it is essential to find ways to change behaviors, mobility routines and achieve users' engagement. To address these challenges, it should be experiencing a thoughtful restructuring of transportation, with the surely technology as a main driving force, in order to promote the integration of sustainable and smart cities' concepts [2]. This gradual change in mobility can be supported and driven by the Mobility as a Service (MaaS) approach.

In the following subsections are presented the MaaS overview and concepts.

1.1 Mobility Concepts Towards MaaS

As previously mentioned, mobility is related with diverse concepts. There is an interaction between all the concept regarding mobility, from which MaaS as also emerged. In Fig. 1 are depicted some mobility concepts and its evolution.

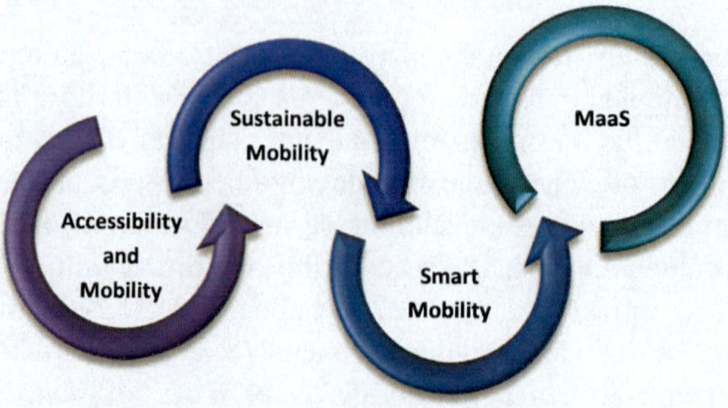

Fig. 1. Mobility system evolving concept scheme.

Accessibility can be connected to the relationship between the land use and the transportation system used [3]. Regarding the **mobility** itself, foresees the movements measured by number of trips, distance and speed, among other characteristics [4]. As a concept which need to evolve without compromising the future generations to fulfill their needs, it appears that **sustainable mobility** can be seen as a promoter of equal opportunities for travelling and at the same time as a promoter of energy consumption reduction of transport, contributing to improve a better efficiency of the resources invested in transport and to reduce environmental pollution [5]. In the actual society and facing the technological gadgets advances, widely used, contributes to the emergence of **smart mobility**, as suggested in Fig. 1. This concept aims to embrace a mobility system where all users can access seamlessly and on-demand mobility, contributing to a sustainable mobility [6]. Lastly, the **MaaS** concept, which is a relatively recent concept where transportation converge, accordingly to users' needs and monthly trip packages [7], plays a significant and promising perspective to replace private car use, with the use of responsive on-demand services [8]. In practice, mobility needs a shared digital framework allowing aggregation of new data sources from connected infrastructures, vehicles, smartphones and more. The MaaS paradigm' evolvement involves the integration with other services (e.g. museums, gym, concerts, cinema or restaurant bookings) guaranteeing the transport service to the end-users, facilitating their access [9]. This means, the last concept – MaaS – should be viewed as an inclusive, sustainable and smart mobility overall approach.

1.2 MaaS Concept

The MaaS concept can be associated with two main drivers: a) the transport approach, in order to provide a seamlessly transport integrating (means of transport, associated information and purchasing) response; and b) the computing component combines the "mobility" and the advertised "as-a-Service" concepts, transforming the transport sector, in accordance with the data processing capabilities [10]. The idea of an integrated mobility system fully aligned and combined with extra services was first named Mobility-as-a-Service – MaaS, by Heikkilä [11] in her Master Thesis, presented in

2014. In addition, the Maas Lab [12] defines MaaS concept as a "*user-centric, intelligent mobility management and distribution system, in which an integrator brings together offerings of multiple mobility service providers, and provides end-users access to them through a digital interface, allowing them to seamlessly plan and pay for mobility*". This emerging concept can be seen as a solution to meet the present customer needs and tackling future mobility opportunities challenges [13], assisting modern societies in shifting their mobility solutions, while creating a more inclusive and efficient society, in a sustainable and environmentally friendly manner [14].

2 MaaS Description

In this section, it is introduced the MaaS role in modern societies, its main functions, features and attributes, Information and Communications Systems (ICT) role as well as identifying all the stakeholders involved in this comprehensive challenge towards ensuring its widespread implementation.

2.1 MaaS Features

In order to provide users with seamless intermodal mobility, MaaS is based on three main elements, specified in the respective situations [15]:

1. Ticket & Payment integration: by using one smart card or ticket which allows access to all the modes taking part in the different types of services, as well as providing, in one unique account, a unified bill by the use of those services;
2. Mobility Package: customers could have different combinations of payment options (in time or distance) for their mobility services; and
3. ICT integration: a single application or online interface to enable the information access about all the transport modes.

Therefore, MaaS deeply relies on a digital platform that is able to integrate end-to-end trip planning throughout all modes of (public or private) transportation, as well as assuring booking, electronic ticketing and payment services capabilities. Moreover, encompasses integrated payments and ticketing, which means the capability to do payments for an entire multimodal trip with a unique transaction. This unified payment encourages the use of multi-modal transportation [16]. In this context, it is critical the cooperation, collaboration and engagement of all the involved (end users, electricity providers, electronic ticketing providers, payment service providers, private mobility service providers, public transit operators, regulatory and policies, et cetera), extended to a global transportation system designed to work in a cooperative, integrated and interconnected manner (in which are included services, infrastructure, information, and payment) to serve the public interest [17, 18].

2.2 MaaS Stakeholders

The environmental, societal, technological and cultural changes needed to ensure a sustainable mobility ecosystem are an utmost challenge that requires an intense effort and involvement of all different types of stakeholders. Their own perspectives, roles, responsibilities and contributions are strategic to the mobility system' overall behavior and performance.

Citizens and stakeholders' involvement and cooperation are fundamental requirements for a successful mobility improvement, as it requires the integration of real-time data from diverse smart infrastructures and a high degree of support and acceptance to share their individual data [19]. In the scope of MaaS, a number of umbrella bodies have been applied to bring together all stakeholders [20].

Table 1 presents the MaaS stakeholders (primary or beneficiaries and some secondary) and their roles and responsibilities. It should be notice, the roles presented may vary between countries, accordingly the context of specific country [13] as well as between environments' typologies [21].

Table 1. Roles and responsibilities of MaaS stakeholders. Adapted from [13].

	Stakeholders	Roles and responsibilities
Primary stakeholders	National authorities	National incentives and subsidies; legislation processes; transport policy and strategies; infrastructure investments; funding; implementation of transport policy, strategy and investments; open transport data; permits; licenses; collaboration; road network operations; tolling
	Local and regional authorities	Regional strategic planning and development; transportation and traffic planning design: redefining the locations and positions of stations and stops; statutory transportation
	MaaS operators	Combine existing transport services on the "one-stop-shop" principle; creating a unique customer service level experience; Offer the adequate services supported by the "right" business models
	Users	User interactions mechanisms to assess their needs and obtain its prompt feedback; providing key data for services design ("prosuming")
	Service providers (transport, logistics and digital)	Providing real-time information of Schedules; vehicle information; locations and availability; drivers and rides platform; management of material; key enabling technology solutions, services and applications to MaaS operators and service providers which ensures booking; ticketing
Secondary stakeholders	Funding agencies	Creation of financial instruments to provide support and risk capital availability to stimuli the MaaS ecosystem
	Academia	Evaluation and impact assessment; new trends and knowledge; recommendations
	Original Equipment Manufacturers (OEMs) and Resellers	Fleet; technology; servitization
	Insurance companies	New insurances for MaaS

From the above table, MaaS operators are the organizations that have to integrate and manage the Mobility Service Providers (MSPs)' offerings, towards designing suitable MaaS Products, in order to sell them to users. In sum, they work as an intermediate between MSPs and users [12]. The involvement and the commitment of all stakeholders, as part of a MaaS system are vital. Nevertheless, it seems that the hyped expectations and optimistic timeframes of a fully integrated MaaS' offerings bring benefits for the society and users, although the income generated from the MaaS operators are changing towards the difficulties faced [1].

2.3 MaaS Technology

The integration of transportation modes, real-time information and instant communication and dispatch is redefining private automobility [22]. ICT has an important role in enabling and promoting MaaS, once it is supported by a digital transport service platform, should enable users to get real-time information on a range of public and private transport options, and to access and pay for it [10]. Thus, it can be perceived as a one-stop online ICT interface including [10]:

- An intermodal journey planner, operating in real-time, combining different transport modes, e.g. car, ride and bike-sharing, car rental, taxi, metro, rail, bus, and other transportation modes;
- A single payment portal, whereby users can pay "as they go" or buy in advance a "service bundle";
- A booking system integrating the whole end-to-end journey stages.

Nowadays, MaaS is accessed through digital platforms - either a multimodal trip-planning app - or webpages and, instead of making a fierce competition to these platforms [18], a common and integrated base should be promoted and delivered.

3 Internet of Mobility as a Service (IoMaaS), the Next Step

The current global tendency of digital transformation, also referred as digitization, in society and businesses are upbringing a new technological evolution, leading to a new mobility paradigm named the Internet of Mobility as a Service (IoMaaS), which result from MaaS and the Internet of Mobility (IoM) concepts.

3.1 Internet of Mobility (IoM)

New challenges related with the digitization era have arisen, bringing broad implications within our way of living [23]. Nowadays, user behavior is changing, younger generations – highly technological – show environmental concerns as well as an increase interest to access circular economy services. Moreover, it is noticing a decrease enthusiasm on having a private car [11]. In fact, as it has been evidenced, consumers are becoming aware of the societal and environmental impacts of their mobility choices and they are beginning to favor options which are more convenient and more sustainable [16].

On the socio-economic perspective new trends have been emerging. The land use and transport system policy favor modal shift from car to mobility services [1]. It can also be highlighted, technology-enabled transportation services, have the potential to reduce driving and car ownership [24].

Thus, a technological (r)evolution can lead to the evolvement of the IoM – an open-source, decentralized mobility aggregation marketplace. IoM entails a global network aimed to empower the full benefits of MaaS, focusing on providing individualized user-centric services [25].

3.2 Future Trends

The future trends of mobility will have to be "human centric", in order to properly balance the amount of technology requested into the ecosystem to ensure the whole system's universality, to be inclusive, as well as developing the appropriate amount of technology, accordingly to the different users' technological skills. The converging technological and social trends, e.g. the rapid growth of car and ride sharing possibilities, the increasing evolvement of electric power alternatives within the political agenda of governments and transnational entities, new lightweight materials, and ultimately, the increasing level of autonomous vehicle driving will, without any kind of doubt, challenge the way people and goods will travel [18].

Figure 2 depicts MaaS integration levels, according to the available services, and the corresponding user' cognitive effort, as stated in [26–28]. On one hand, the low

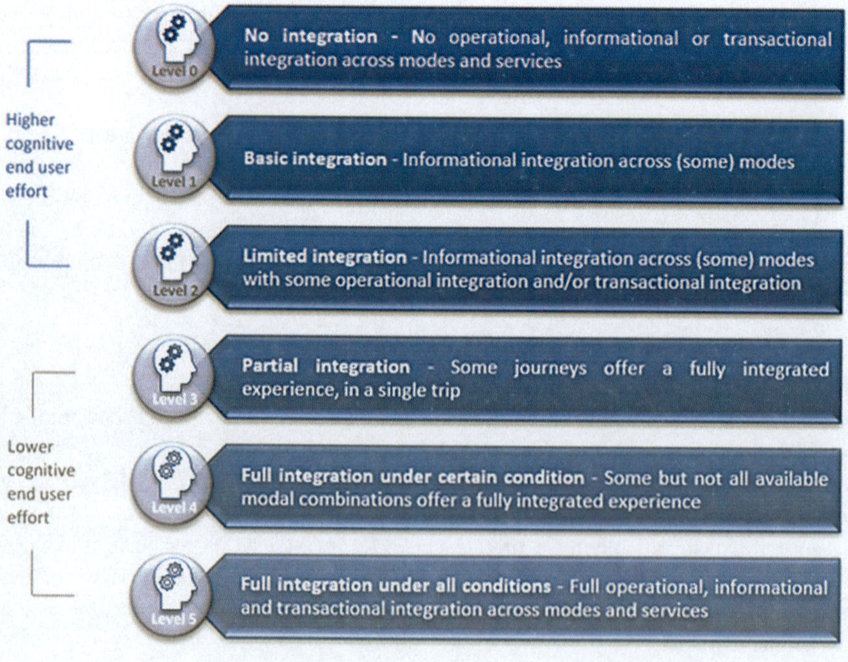

Fig. 2. MaaS integration levels and cognitive user effort. Source: [23].

integration level corresponds to a high cognitive effort matured by the user, when choosing its mobility; on the other hand, high integration level corresponds to a minimum cognitive user effort when using mobility services.

The levels 0–2 of MaaS integration imply a high cognitive effort by the end user, and the levels 3–5 imply a low cognitive effort by the end user. Therefore, in the last three MaaS integration levels, the user can enjoy its mobility plans and choose without having to expend time and cognitive effort, which means, in a "natural" and easy way. The aim is to achieve a mobility system which embraces high integration levels with low cognitive effort. The increase pressure on the availability of mobility systems has increased the need of innovative solutions, in order to increase its efficiency [15]. Gamified systems, for example, when exploited to launch MaaS-based applications can contribute to engage users and retain them on the long-term [29]. ICT systems are a crucial part in the implementation and development of new mobility systems [23], since they ensure the integration of all information, as a dynamic and interactive process, and enable the convergence between all stakeholders and the services' connection. Furthermore, it can be materialized by the IoMaaS systems, which can improve geographical mobility, as well as to learn from the end user' experiences and actions, promoting a reduction effortlessness in using a sustainable mobility [23]. The next step evolution of MaaS if integrated with IoM, will certainly potentiate a cultural change in the mobility user's behavior, that can replicate attitudes and routines which can have repercussions in all citizens' well-being. These new and upcoming IoMaaS challenges introduce cybersecurity concerns, which are crucial for guarantying the commitment regarding the system safety along with its usability and confidence – reducing the cognitive end user effort and its requirements to adapt to new mobility features. The development and dissemination of standard mechanisms to connect the different platforms and integrate all systems will play a vital role, providing also a systematic evaluation, with risk analysis and its impact in terms of security. Hence, every platform associated to IoMaaS system should guaranty a high level of security as well as the maturity levels.

4 Conclusion and Future Work

Based on the current MaaS development stage and integrating the potentialities of IoM, IoMaaS or similar future smart mobility solution is looming. The development of strategies to reduce the technological illiteracy, through the augmented reality assistants attained to some mobility systems members, can easily promote the usability of an IoM systems.

The spread of population and its low-density especially in rural areas, arises the challenge to provide transport networks to cover long distances. Additionally, the employment in the primary industry, reducing the permanence of young people on such regions will create economic and social problems.

The introduction of gamification can be a strategy to promote awareness on the citizens' current mobility choices, encouraging a more sustainable, efficient, or newly introduced mobility services. However, it must be noticed, gamification comes with some disadvantages, as it blurs the boundaries between virtual versus reality and resources or budget to be used to improve other resources.

It is undeniable the advantages of smart mobility systems. However, the interconnected systems, the data sharing and integration are invitations to fraud, theft, terrorism, et cetera. In addition to these security problems, there are many other emergent attacks that will continue to escalate, requiring, as future work, further studies regarding such threats. On the other hand, it is also important to develop further studies regarding the users' perception and feelings about smart mobility systems and how these systems can improve their well-being and be more inclusive in the society. Moreover, the definition of a MaaS roadmap and a work methodology, including focus groups, workshops, interviews, surveys and meetings with transport authorities, entities and academia should be implemented, to validate and define future strategies.

References

1. Liimatainen, H.: Implications of Mobility as a Service (MaaS) in Urban and Rural Environments: Emerging Research and Opportunities. IGI Global, Hershey (2020)
2. Gonçalves, L., Silva, J., Baltazar, S., Barreto, L., Amaral, A.: Implications of Mobility as a Service (MaaS) in Urban and Rural Environments: Emerging Research and Opportunities. IGI Global, Hershey (2020)
3. Nuzzolo, A., Coppola, P., Papa, E.: Marginal activity access cost (MAAC): a new indicator for sustainable land use/transport (LUT) planning. In: EWGT2013 – 16th Meeting of the EURO Working Group on Transportation, Porto, vol. 111, pp. 450–459 (2014)
4. Litman, T.: Evaluating Accessibility for Transportation Planning - Measuring People' s Ability to Reach Desired Goods and Activities. Victoria Transport Policy Institute, Canada (2015)
5. Campos, V., Ramos, R.: Proposta de Indicadores de Mobilidade Urbana Sustentável Relacionando Transporte e uso do Solo. In: 1st Congresso Luzo-Brasileiro para o Planejamento Urbano, Regional, Integrado e Sustentável (PLURIS), Brazil, pp. 86–91 (2005)
6. Docherty, I., Marsden, G., Anable, J.: The governance of smart mobility. Transp. Res. Part A Policy Pract. **115**, 114–125 (2018)
7. Hietanen, S., Sahala, S.: Mobility as a Service - Can it be even better than owning a car? ITS, Canada (2014)
8. Jittrapirom, P., Caiati, V., Feneri, A.-M., Ebrahimigharehbaghi, S., González, M.J.A., Narayan, J.: Mobility as a service: a critical review of definitions, assessments of schemes, and key challenges. Urban Plan. **2**(2), 13–25 (2017)
9. Pangbourne, K., Stead, D., Mladenovic, M., Milakis, D.: Governance of the Smart Mobility Transition. Emerald Publishing, Bingley (2018)
10. Enoch, M.: Mobility as a Service (MaaS) in the UK: change and its implications. Government Office for Science, London (2018)
11. Heikkilä, S.: Mobility as a service - a proposal for action for the public administration case Helsinki. Master thesis, Aalto University, Finland (2014)
12. MaaSLab. https://www.maaslab.org/. Accessed 11 Feb 2020
13. Eckhardt, J., Aapaoja, A., Haapasalo, H.: Implications of Mobility as a Service (MaaS) in Urban and Rural Environments: Emerging Research and Opportunities. IGI Global, Hershey (2020)

14. Amaral, A., Barreto, L., Baltazar, S., Silva, J.P., Gonçalves, L.: Implications of Mobility as a Service (MaaS) in Urban and Rural Environments: Emerging Research and Opportunities. IGI Global, Hershey (2020)
15. Kamargianni, M., Li, W., Matyas, M., Schäfer, A.: A critical review of new mobility services for urban transport. Transp. Res. Procedia **14**, 3294–3303 (2016)
16. World Business Council for Sustainable Development. https://www.wbcsd.org/Programs/Cities-and-Mobility. Accessed 01 Mar 2020
17. Karlsson, I.C.M., Sochor, J., Strömberg, H.: Developing the "service" in mobility as a service: experiences from a field trial of an innovative travel brokerage. Transp. Res. Procedia **14**, 3265–3273 (2016). 6th Transport Research Arena, Poland
18. Goodall, W., Dovey Fishman, T., Bornstein, J., Bonthron, B.: The rise of mobility as a service-reshaping how urbanites get around. Delotitte Rev. **20**, 111–130 (2017)
19. Gogola, M., Sitányiová, D.: Implications of Mobility as a Service (MaaS) in Urban and Rural Environments: Emerging Research and Opportunities. IGI Global, Hershey (2020)
20. Pangbourne, K.: Implications of Mobility as a Service (MaaS) in Urban and Rural Environments: Emerging Research and Opportunities. IGI Global, Hershey (2020)
21. Shaheen, S., Cohen, A., Yelchuru, B., Sarkhili, S.: Mobility on Demand Operational Concept Report (No. FHWA-JPO-18–611). United States Department of Transportation, USA (2017)
22. Shaheen, S.A., Cohen, A., Farrar, E.: Implications of Mobility as a Service (MaaS) in Urban and Rural Environments: Emerging Research and Opportunities. IGI Global, Hershey (2020)
23. Baltazar, S., Amaral, A., Barreto, L., Silva, J., Gonçalves, L.: Implications of Mobility as a Service (MaaS) in Urban and Rural Environments: Emerging Research and Opportunities. IGI Global, Hershey (2020)
24. Hallock, L., Inglis, J.: The innovative transportation index: the cities where new technologies and tools can reduce your need to own a car. Frontier Group, USA (2015)
25. Travelspirit Foundation "Open Internet of Mobility". https://travelspirit.foundation/internet-of-mobility. Accessed 04 Feb 2020
26. Durand, A., Harms, L., Hoogendoorn-Lanser, S., Zijlstra, T.: Mobility-as-a-Service and changes in travel preferences and travel behaviour: a literature review. KiM Netherlands Institute for Transport Policy Analysis, Netherlands (2018)
27. Lyons, G., Hammond, P., Mackay, K.: The importance of user perspective in the evolution of MaaS. Transp. Res. Part A Policy Pract. **121**, 22–36 (2019)
28. Sochor, J., Arby, H., Karlsson, M., Sarasini, S.: A topological approach to Mobility as a Service: a proposed tool for understanding requirements and effects, and for aiding the integration of societal goals. Res. Transp. Bus. Manag. **27**, 3–14 (2018)
29. Marconi, A., Loria, E.: Implications of Mobility as a Service (MaaS) in Urban and Rural Environments: Emerging Research and Opportunities. IGI Global, Hershey (2020)

Planning the Urban Shift to Electromobility Using a Cost-Benefit-Analysis Optimization Framework: The Case of Nicosia Cyprus

Filippos Alogdianakis and Loukas Dimitriou(✉)

Laboratory for Transport Engineering, Department of Civil and Environmental Engineering, University of Cyprus, 75 Kallipoleos Street, P.O. Box 20537, 1678 Nicosia, Cyprus
lucdimit@ucy.ac.cy

Abstract. Electromobility is considered the foreseeable future of road transportation and a significant shift to vehicles' consumer behaviour. Electric Vehicles (EV) adoption has environmental benefits linked to lower emissions and noise levels, that improve the quality of life, especially in urban areas. Nevertheless, to reap these benefits, both the energy and the transport sector have to be reformed. However, the combination of both sectors provides a complex system within which electromobility adoption has to be planned. The work herein proposes a framework to plan the shift to electromobility on the urban level. The proposed framework uses system theory and the Cost-Benefit-Analysis (CBA) tool to plan the shift to electromobility. An application of the framework is provided for the case of Cyprus's capital, Nicosia, within a 20-year scope. The results of the do-nothing scenario presented herein, indicate that increased socio-environmental benefits can be anticipated by embracing electromobility, providing room for optimal policies.

Keywords: Electromobility · Feasibility study · Renewable energy sources · Future consumer behaviour

1 Introduction

Electromobility is considered the foreseeable future of road transportation and a significant shift to vehicles' consumer behaviour. The adoption of advanced technology Electric vehicles (EVs) has environmental benefits linked to lower emissions and noise levels which improve the quality of life, especially in urban areas. However, electromobility is highly dependent on the energy sector for the electricity consumed by EVs. Therefore, electromobility adoption can provide an attractive opportunity for transforming both emission-intensive sectors simultaneously, especially in a time where country-specific goals have been set to limit (Green House Gas) GHG emissions and the use of carbon fuels. These country-specific goals are subjected to international agreements that have set gradual decarbonization goals that lead to a net-zero emissions target within the second half of the 21st century. However, this goal is achievable, if only both sectors are transformed from the urban/municipal level within the concept of sustainable urban mobility.

E. G. Nathanail et al. (Eds.): CSUM2020, AISC 1278, pp. 230–240, 2021.
https://doi.org/10.1007/978-3-030-61075-3_23

This study provides a unified framework as part of a Sustainable Urban Mobility Plan (SUMP) of electromobility adoption for urban areas. Initially, factors of the EV ecosystem are recognized, and the system's boundaries are set. Then, using the concepts of Cost-Benefit Analysis (CBA) and system theory, the system of both transport and energy sectors is defined. Decision variables are incorporated, where policy planning can be performed, altering both sectors and generating the system's costs and benefits. A case study of Cyprus's capital, Nicosia, is provided within a 20-year framework where electromobility's feasibility sustainability is studied under the do-nothing scenario. For this scenario, no changes in policies are planned that alter electromobility's adoption pace or the energy sector's production. The framework developed herein is fully parametric and applicable to countries/cities aiming to embrace electromobility.

2 Literature Review

Electromobility adoption is a multidisciplinary study ranging from consumer behaviour modelling to effects in urban mobility and a country's economy. An essential aspect of EV adoption is modelling the market's EV penetration rate. Common factors used in the modelling regard the purchase price, energy prices, operating costs, charging infrastructure, policy measures and EV range (Choi et al. 2018; Qiao et al. 2017; Zhao and Heywood 2017). Moreover, modelling approaches followed for EV adoption rate relate to consumer choice models, agent-based models, diffusion rate and time series models. However, the prediction outcomes of these models vary substantially, and as a result, they are used to understand the external influences of market diffusion.

Policy measures successfully used by countries to promote electromobility comprise another research topic. Among the most commonly imposed policies are monetary incentives either with direct subsidies for EV purchase or indirect subsidies such as road tax exemptions. In a study to assess policy effectiveness, Lieven (2015) used conjoint analysis and k-means clustering to analyse data of 20 countries across five continents. The results showed a similar public's appreciation to either high monetary direct incentives or lower direct monetary incentives that also included investments in charging infrastructure.

Assessing the sustainability of electromobility adoption has also been the focus of other studies where within the context of Life Cycle Analysis (LCA) to economic, environmental and social impacts have been assessed (Choi et al. 2018; Onat et al. 2016; Zhao and Heywood 2017). However, a holistic system-based approach was presented for the case of the US, where socioeconomic and environmental impacts were assessed (Onat et al. 2016). The importance of following a holistic scientific approach to obtain optimal policies was outlined in their study, and a similar approach is followed in the present work.

3 System Description and Methods Used

The framework of the study includes four stages. In the first stage, subsystems of the electromobility system are recognized along with relationships linking them. These relationships are used to form equations and determine variables and define the costs and benefits of the system. In the second stage, the CBA tool is used to assess the feasibility of EV adoption. The third stage incorporates the system's stochasticities, and a sensitivity analysis is performed. Lastly, in the fourth stage, optimization methods are incorporated to achieve optimal planning based on sustainability criteria. However, for brevity purposes in the present paper, the first two stages are provided.

3.1 System Description

Figure 1 depicts the basic parts of the EV adoption system with arrows indicating relationships connecting the system's parts. Moreover, the Figure is separated into four distinct regions, indicated with dotted rectangles. The first two are located in the central area (Fig. 1) and show the sectors of energy and transport. The other two, relate to the costs and benefits generated from electromobility adoption. Systems, subsystems, the costs and benefits, are represented with ovals which are placed within the designated parts. In this study, the starting point of the system is a country's economy indicated by the Gross Domestic Product (GDP), which affects both the consumption of energy but also the registration of new vehicles. EV penetration is depicted with red colour (Fig. 1) while decision variables affecting policies are depicted with dashed lines and grey colour (Fig. 1). The decision variables are located in both sectors, and their alteration affects EV penetration as well as the costs and benefits generated. In the following section, the links of the parts involved are described and the system's equations are formed.

Transport Sector

Each year the urban fleet is enriched with new vehicles, whose purchasers withdraw their old cars. In this study the consumer's choices are limited to Internal Combustion Vehicles (ICV) and EVs which for this study regard battery electric vehicles. Incentive policies alter the consumer's choice and increase EV penetration rate to the market. Transportation is dependent on a country's economy thus, GDP can be considered to affect the new vehicle registrations. With a given growth rate function, an estimation of future new registrations can be obtained at each point in future time. For the case study, a linear regression approach is used to obtain the yearly new registrations based on GDP from Cyprus' specific data.

$$G_{DP,t} = G_{DP_0} + \left(\sum_{t=1}^{t=n} G_{DP,t} r_t\right) \tag{1}$$

$$X_t = a_1 G_{DP,t} + a_2 \tag{2}$$

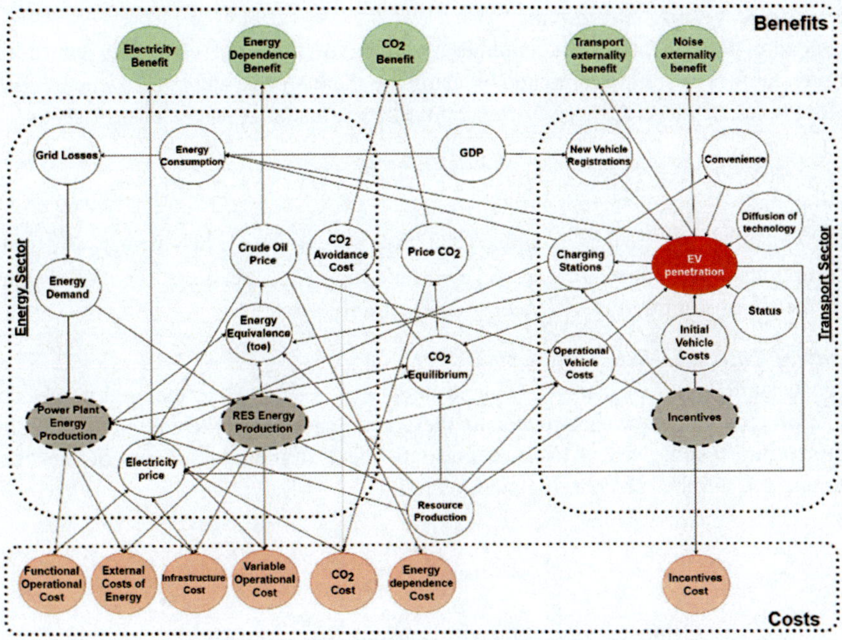

Fig. 1. EV system basic parts with their connections linked to the costs and benefits considered in this framework, resulting from policy related variables denoted by dashed grey ovals.

where, $G_{DP,t}$: GDP in currency units in time t in years, G_{DP_0}: last year of GDP before the analysis, r_t: growth rate unitless if $r_t > 1$ growth if $r_t < 1$ recession, X_t: new vehicle registrations of year t, a_1 and a_2: linear regression coefficients.

Vehicle Choice Model

A binomial logit model-BLM is used for modelling the vehicles market organization, through vehicle type choices among ICV and EV for each year based on: acquiring vehicle cost, operational vehicle costs, refuelling/charging convenience, the status criteria valuation and on the diffusion of innovation curve.

Incentive Policies

Incentive policies can alter the yearly penetration rate of EVs. Two types of incentive policies are used in this work. The first type relates to a limited number of subsidies (i.e. funding per EV or/and funding per old vehicle withdrawal) provided to the public. The public's demand for these incentive policies is modelled using a Gompertz function. The second type of incentives relates to taxation and infrastructure construction (i.e. charging stations) that affect the vehicle choice model.

Costs and Benefits of the Transport Sector

Benefits of the transport sector relate to vehicle registrations (i.e. sales taxes) and travel (i.e. fuel taxes, electricity price, road tax), however additional costs and benefits have to be included in order to assess the sustainability of EV adoption. In this part, all costs and benefits considered in the framework are provided.

Benefits from Vehicle Circulation

To estimate these benefits the commercial price of fuel and electricity are used. Commercial fuel prices are estimated by using the market price and adding import costs and fuel taxes, while electricity prices should be estimated from the energy sector.

$$B_U^T, t = T_f^T(N_{ICV,t}\, m_{v,t} f_{ICV,t}^{mean}\, p_{f,t}^T) + T_e^T(N_{EV,t}\, m_{v,t} f_{EV,t}^{mean}\, p_{e,t}) \quad (3)$$

where, $T_{f,t}^T$: taxation of fuel, $p_{f,t}^T$: commercial price of fuel, $N_{ICV,t}$: ICV fleet, $f_{ICV,t}^{mean}$: mean consumption of ICV fleet, $m_{v,t}$: average distance travelled per year, $N_{EV,t}$: EV fleet, $f_{EV,t}^{mean}$: mean consumption of EV fleet, $p_{e,t}$: electricity price.

Incentives Costs and Benefits

The costs and benefits of incentive policies can be either related to the total number of vehicles in circulation and thus related to operational costs, or to the number of vehicles registered in a specific year. Costs are also dependent on the total amount of subsidies given and the number of charging stations built.

$$C_{In,t}^T = n_{EV,t}\, s_{c1,t}^{EV}\, (T_{S,t}^{EV}) + N_{EV,t} \sum_{i_2=1}^{k_2} \left(c_{o,t}^{In}\right) + s_{e,t}^{In}\, c_{CS,t}^{In} + \sum_{k=1}^{z} \left(G_{k,t}^{In}\, c_{k,t}\right) \quad (4)$$

$$B_{In,t}^T = n_{ICV,t}\, T_{S,t}^{ICV}\, s_{c1,t}^{ICV} + N_{ICV,t} \sum_{i_2=1}^{k_2} \left(c_{o,t}^{In}\right) \quad (5)$$

where, $C_{In,t}$: cost of incentive policies, $n_{EV,t}$: the number of new EV registrations, $T_{S,t}^V$: sales tax (where $V \in \{EV, ICV\}$), $s_{c1,t}^{EV}$: EV purchase price, $c_{o,t}^{In}$: operational costs or benefits per vehicle, $s_{e,t}^{In}$: number of charging stations, c_{CS}^{In}: cost of fast public charging station, $G_{k,t}^{In}$: number of k subsidies for EV purchase, $c_{k,t}$: cost of subsidy type and $B_{In,t}$: benefit of incentive policies.

Externalities Costs and Benefits

In this work the externalities considered relate to the external costs of noise (c_N), air pollution (c_A) habitat loss (c_H). These costs can be obtained from literature or from country specific studies and depend on the distance travelled ($m_{v,t}$). Moreover, since research is ongoing reduction factors have been used for external costs which have not yet been determined such as the case of noise pollution.

$$C_{N,t}^T = m_{v,t}\left[N_{EV,t}(c_A^{EV} + c_H^{EV} + c_N^{EV} b_{N,t}) + N_{ICV,t}(c_A^{ICV} + c_H^{ICV} + c_N^{ICV})\right] \quad (6)$$

$$B_{N,t}^T = m_{v,t}\, N_{EV,t}\left[(c_A^{ICV} + c_H^{ICV} + c_N^{ICV}) - (c_A^{EV} + c_H^{EV} + c_N^{EV} b_{N,t})\right] \quad (7)$$

where, $C_{N,t}^T$: cost of transport externalities, c_N^V: noise externality cost for combustion vehicles (where $V \in \{EV, ICV\}$), $b_{N,t}$: reduction factor for EVs in the urban environment, c_A^V: air pollution externality cost, c_H^V: habitat loss externality cost and $B_{N,t}^T$: externality benefits.

Climate Change Costs and Benefits

In this study avoidance costs are used to estimate the costs and benefits of climate change external costs. Avoidance costs are estimated from the expenditure made for avoiding climate change effects and it implies a general international commitment (i.e. Paris agreement). Avoidance costs are estimated per tonne of CO_2 equivalent and vary in time with lower and upper limits. For the transport sector climate change costs are generated from ICVs relative to the CO_2 while benefits are generated from EVs which are mitigated by the CO_2 generation of the energy sector of the next section.

$$C^T_{GHG,t} = N_{ICV,t} f^{mean}_{ICV,t}\, m_{v,t}\, b_{CO_2,t}\, c_{GHG,t} \tag{8}$$

$$B^T_{GHG,t} = m_{v,t}\, b_{CO_2,t}\, c_{GHG,t}\left[N_{EV,0} f^{mean}_{ICV,0} + \sum\nolimits_{t=t_0}^{t} (n_{EV,t} f_{ICV,t})\right] \tag{9}$$

where, $C^T_{GHG,t}$: climate change avoidance cost, $b_{CO_2,t}$: emissions factor providing the weight of CO_2 equivalent emissions per volume of fuel burnt, $c_{GHG,t}$: avoidance cost per CO_2e and $B^T_{GHG,t}$: benefit of climate change avoidance costs.

Energy Dependence Costs and Benefits

In this framework, energy dependence is considered by the cost of fuel imports needed, to cover the transport sector's consumption. Thus, ICVs that use carbon fuels generate costs based on their consumption and the global price of fuel. However, depending on a country's endogenous oil resources and their use to cover ICV demand, costs and benefits can be generated. On the other hand, EVs generate benefits as they do not need fuels from the transport sector. However, these benefits are mitigated based on the energy dependence of the energy sector.

$$C^T_{D,t} = P^T_{f,t} N_{ICV,t} f^{mean}_{ICV,t}\, m_{v,t}\, p^D_{f,t} \tag{10}$$

$$\begin{aligned} B^T_{D,t} =& [(1 - P^T_{f,t}) N_{ICV,t} f^{mean}_{ICV,t} + N_{EV,0} f^{mean}_{ICV,0} \\ &+ \sum\nolimits_{t=t_0}^{t} (n_{EV,t} f_{ICV,t})] m_{v,t} (p^D_{f,t} - p^I_{f,t}) \end{aligned} \tag{11}$$

where, $C^T_{D,t}$: energy dependence cost, $P^T_{f,t}$: portion of ICV supply covered by imported fuels (%), $B^T_{D,t}$:energy dependence benefit, $p^D_{f,t}$: imported fuel price and $p^I_{f,t}$: endogenous resource fuel price.

Energy Sector

The energy sector is vital for the sustainability and feasibility of EV adoption as lower externalities, and lower energy dependence costs will follow from widespread RES adoption, to obtain a country's energy and emission goals. Following a similar approach to the transport sector, the electricity demand is estimated using GDP. Future energy consumption can be obtained at each point in future time using a linear regression approach to country-specific data. Since this approach cannot incorporate societal advances such as EV adoption and circulation, which can seriously alter energy demand, EV consumption has to be considered separately. Each year's energy supply is estimated by using a variable to include network losses.

$$D_t = (1 + l_t)(D_{0,t} + N_{EV,t}\, m_{v,t} f^{mean}_{EV,t}) \quad (12)$$

where, D_t: electricity demand, $D_{0,t}$: electricity consumption in (kWh), $f^{mean}_{EV,t}$: electricity consumption of EV per km and l_t: losses percentage (%).

Decision Variables

Energy demand is covered by a selection of RES and power plants based on a country's available (re)sources. Although, electricity generation is one of the main sources of GHG emissions, depending on the resources used, emissions can be decreased. In this work in order to mathematically model the energy policy, two decision variables are used to indicate the electricity production ratio of different available sources. These decision variables, are in the form of utilization factors, one indicating the ratio of ($i = \{1 : k\}$) different power plants ($P_{PP,t,i}$) and a second ratio for the demand covered of ($j = \{1 : l\}$) different RES ($P_{RES,t,j}$). The sum of the decision variables equals to one (=1) providing the macroscopic energy policy, where different combinations of RES and power plants are utilized to achieve emission targets.

Installation, Operational and System Costs

Installation costs occur when current energy infrastructure cannot cover the demand of the policy defined by the decision variables. The new units assigned may regard both powerplants and RES. Operational costs are also considered and relate to fixed operational costs for power plants and variable operational costs for RES, which for the case of powerplants consider fuel consumption. Additionally, for both RES and power plants, system costs are considered that include grid costs, balancing costs and profile costs.

Externalities Costs and Benefits

The external costs of energy relate to impacts on other sectors such as human health, crop yield loss, and other important issues that have not been taken to consideration up to this point. However, it should be mentioned that externalities related to the production of certain goods (i.e. photovoltaic panels) impact a non-producing country by import costs. Externalities can be sought from country-specific studies or from literature regarding general sectors (z = {1…w}) affected by energy production. A general form is used for these costs in the following equations related costs while externality benefits are estimated based on the relative benefit of each considered energy production option compared to all others.

$$C^e_{Ex,t} = \sum\nolimits_{i=1}^{k} P_{PP,t,i}\, D_t\left(\sum\nolimits_{z=1}^{w} c^{PP_i}_{E,z}\right) + \sum\nolimits_{j=1}^{l} P_{RES,t,i}\, D_t\left(\sum\nolimits_{z=1}^{w} c^{RES_j}_{E,z}\right) \quad (13)$$

$$B^e_{Ex,t} = \sum\nolimits_{i=1}^{k} P_{PP,t,i}\, D_t\left(\sum\nolimits_{z=1}^{w} c^{PP_i}_{R,z}\right) + \sum\nolimits_{j=1}^{l} P_{RES,t,i}\, D_t\left(\sum\nolimits_{z=1}^{w} c^{RES_j}_{R,z}\right) \quad (14)$$

where $C^e_{Ex,t}$: external costs of energy, $c^{PP_i}_{E,z}$: external cost of energy to (z) considered sectors (e.g. Health, Crops yield losses, forestry etc.) from power plant energy production (i), $c^{RES_j}_{E,z}$: external cost of energy to (z) considered sectors from power plant energy production (j), $B^e_{Ex,t}$: relative benefits of power production, $c^{PP_i}_{R,z}$ and $c^{RES_j}_{R,z}$:

relative costs of each power plant and RES considered weighted by the decision variable of each plant or RES respectively.

Climate Change Costs

Climate change costs of the energy sector are estimated using avoidance costs based on the volume of CO_2e generated by energy production. The volume of CO_2 can be estimated from the decision variables indicating the use of each power plant, the power plant's efficiency and a corresponding CO_2 emissions factor based on the fuel used by the respective plant. On the other hand, CO_2 benefits are obtained from RES power generation and can be estimated by using the mean emissions per unit of energy produced.

$$C^e_{GHG,t} = \sum_{i=1}^{k} \left[\frac{Q_{PP,t,i}}{H_{PP,t,i}} P_{PP,t,i}\, e_{CO_2,i} \right] D_t\, c_{GHG,t} \tag{15}$$

$$B^e_{GHG,t} = \frac{1}{k(1-P_{RES})} \sum_{i=1}^{k} \left[P_{PP,t,i} \frac{Q_{PP,t,i}}{H_{PP,t,i}} e_{CO_2,i} \right] \left[\sum_{j=1}^{l} \left(P_{RES,t,j}\right) D_t c_{GHG,t} \right] \tag{16}$$

where, $C^e_{GHG,t}$: climate change cost of energy sector, $Q_{PP,t,i}$: mean capacity of powerplant (i), $H_{PP,t,i}$: mean efficiency of powerplant (i), $e_{CO_2,i}$: emission factor indicating the generated CO_2 from the power plant (i) and B^e_{GHG}: climate change benefit of the energy sector.

Energy Dependence Costs and Benefits from the Energy Sector

In this study, only energy dependence benefits are taken into consideration for the energy sector; fuel costs have been calculated in operational costs of power plants. Energy dependence benefits are obtained both from RES and from powerplants utilizing endogenous resources. For the case of RES, no fuels are used for electricity generation thus the benefit is calculated from the mean efficiency of each power plant, the price of the corresponding fuel ($p^D_{r,t,i}$) and its transport cost ($c^D_{R,t,i}$) as if they were imported. On the other hand, for the case of endogenous fuel sources ($i_I = \{1\ldots k_I\}$), the mean efficiency of each power plant (i_I) is multiplied by the benefit derived by subtracting the endogenous fuel price (p_{r,t,i_I}) and transportation cost (c_{R,t,i_I}), from the import price (p^D_{r,t,i_D}) and the transport cost ($c^D_{R,t,i}$) if that fuel was imported.

$$\begin{aligned} B^e_{D,t} = {} & \sum_{j=1}^{l} \left(P_{RES,t,j}\right) D_t \frac{1}{k} \sum_{i=1}^{k} \left[\left(\frac{Q_{PP,t,i}}{H_{PP,t,i}} \right) p^D_{r,t,i} \left(1 + c^D_{R,t,i}\right) \right] \\ & + \sum_{i_I=1}^{k_I} \left[\left(P_{PP,t,i_I} D_t \frac{Q_{PP,t,i}}{H_{PP,t,i}} \right) \left(p^D_{r,t,i_I} \left(1 + c^D_{R,t,i_I}\right) - p_{r,t,i_I}\left(1 + c_{R,t,i_I}\right) \right) \right] \end{aligned} \tag{17}$$

where, $B^e_{D,t}$: energy dependence benefit, $p^D_{r,t,i}$: price of imported resource, $c^D_{R,t,i}$: transport cost of imported resource as percentage, p^D_{r,t,i_I}: price of endogenous (i_I) resource if it was imported, c^D_{r,t,i_I}: transport cost of endogenous (i_I) resource if it was imported, p_{r,t,i_I}: endogenous resource (i_I) cost and c_{r,t,i_I}: endogenous resource transport cost as (%).

3.2 Cost Benefit Analysis

The obtained costs and benefits are used within a CBA analysis framework to estimate the feasibility of EV adoption. First, the time window of 20 years is selected for the years (t) depicted Eqs. (1–17). Moreover, a discount rate (4%) is determined to estimate the Net Present Value (NPV) of each year's costs and benefits. Then, by summing all yearly costs (C_t) and benefits (B_t), the (B_t/C_t) ratio can be used to indicate the feasibility of the project. To estimate these two parameters the cost and benefit functions of the transport sector are used as is, while the energy sector's functions are used based on the electricity consumption of the transport sector that incorporates also EV consumption by considering the factor J_t. In this way costs and benefits of the two sectors are combined to study the feasibility of EV adoption.

$$C_t = C_{In,t} + C^T_{N,t} + C^T_{GHG,t} + C^T_{D,t} + J_t\left(C^e_{I,t} + C^e_{Op,t} + C^e_{S,t} + C^e_{Ex,t} + C^e_{GHG,t}\right) \tag{18}$$

$$B_t = B^T_U, t + B_{In,t} + B^T_{N,t} + B^T_{GHG,t} + B^T_{D,t} + J_t\left(B^e_{Ex,t} + B^e_{GHG,t} + B^e_{D,t}\right) \tag{19}$$

where, C_t: total costs generated at time t, J_t: factor indicating the ratio of electricity consumed by the transport sector, $C^e_{I,t}$: energy sectors installation costs, $C^e_{Op,t}$: operational costs of energy sector, $C^e_{S,t}$: system costs of energy sector and B_t: total benefits generated at time (t).

4 Results and Discussion

This part provides a case study of Cyprus's capital Nicosia and the do-nothing scenario where 13% of the electricity demand is produced by RES and the natural gas plant is delivered in 2028. The importance of the Cyprus' case study is based on the facts that: (a) the country's capital has been ranked first in noise pollution (Peris 2020), (b) Cyprus has a large vehicle per person ratio, (c) both the transport and energy sectors are highly dependent on carbon fuel, making Cyprus the most highly dependent member State of the EU which is conflicting with emission goals of the common European goals. Decarbonization of both sectors is of crucial importance to achieve the country's future emission goals. For decarbonizing the energy sector both solar and wind have high energy potential (Zachariadis and Taibi 2015), while on the other hand Natural Gas will be soon an available endogenous resource. For all mentioned reasons the study's framework is applied to Nicosia to assess the feasibility of EV adoption. For the case study of Cyprus national and regional statistics and official documents were used to calibrate the models and perform the analysis. The resulting energy mix is depicted in Fig. 2(a) while the CO_2 production and the national limits are shown in Fig. 2b. The vehicle choice model (Fig. 2c) based on country data indicates that in 2027 people will equally prefer EVs to ICVs. The effect of the monetary subsidy currently provided is also shown in Fig. 2 where 200 direct subsidies have been provided as incentive to attract EVs. Moreover, Nicosia's fleet composition is presented in Fig. 2(e), where it can be observed that EVs have replaced 80% of the ICV fleet by 2040. The results of the CBA are shown in Fig. 2(f) and discussed in the next section.

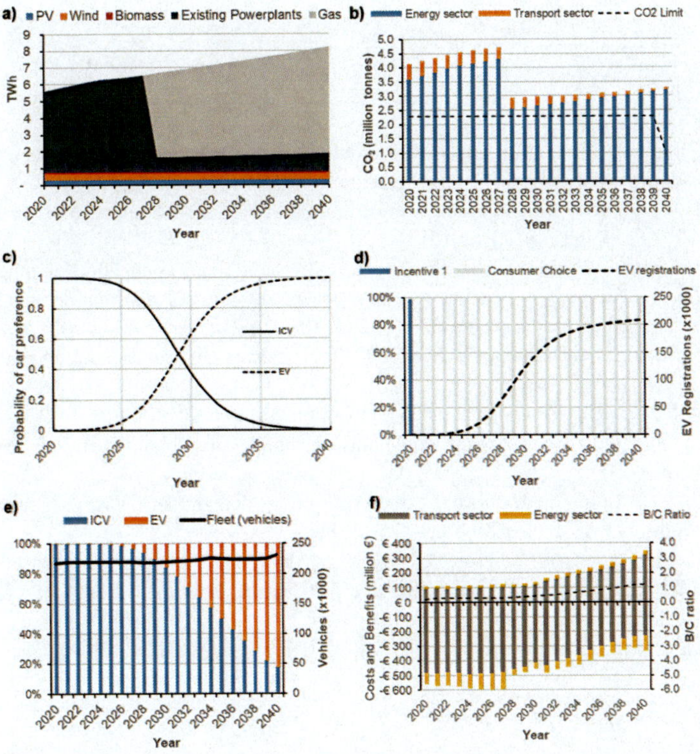

Fig. 2. Results of the described framework related to: (a) electricity demand and energy supply from different energy sources for do-nothing scenario, (b) CO_2 generation per sector and CO_2 emission limit (c) vehicle choice model showing the probability of car preference, (d) new registered EVs and existent incentives (e) Nicosia's fleet consistency and (f) CBA and B/C ratio.

5 Conclusions

This paper concerns the development of an evaluation framework for the introduction of EVs in cities or countries. A unified framework is provided for the SUMP of electromobility adoption. Initially, factors of the EV ecosystem are recognized allowing the boundaries of the system to be set. An analysis based on benefit cost ratio has been developed to evaluate the monetarized short and long-term impacts and the cost of decisions of a country's energy goal timeframe. To illustrate the framework, the case of Cyprus's capital, Nicosia has been used and the do-nothing scenario has been presented. The results indicate that EV adoption provides future benefits that exceed externality and other costs provided that cleaner energy sources are used. Since, benefits exceed costs in future years, incentive policies should be used to embrace electromobility earlier when EV penetration rate is low. Moreover, additional benefits can be reaped by reducing CO_2 emissions by increasing the RES share. From the above findings it is apparent that an optimization framework is needed to obtain optimal policies.

References

1. Choi, H., Shin, J., Woo, J.R.: Effect of electricity generation mix on battery electric vehicle adoption and its environmental impact. Energy Policy **121**, 13–24 (2018)
2. Lieven, T.: Policy measures to promote electric mobility - a global perspective. Transp. Res. Part A Policy Pract. **82**, 78–93 (2015)
3. Onat, N.C., Kucukvar, M., Tatari, O., Zheng, Q.P.: Combined application of multi-criteria optimization and life-cycle sustainability assessment for optimal distribution of alternative passenger cars in U.S. J. Clean. Prod. **112**, 291–307 (2016)
4. Peris, E.: Environmental noise in Europe - 2020. European Environment Agency (Issue 22/2019) (2020)
5. Qiao, Q., Zhao, F., Liu, Z., Jiang, S., Hao, H.: Comparative study on life cycle CO_2 emissions from the production of electric and conventional vehicles in China. Energy Procedia **105**, 3584–3595 (2017)
6. Zachariadis, T., Taibi, E.: Exploring drivers of energy demand in Cyprus - Scenarios and policy options. Energy Policy **86**, 166–175 (2015)
7. Zhao, S.J., Heywood, J.B.: Projected pathways and environmental impact of China's electrified passenger vehicles. Transp. Res. Part D Transp. Environ. **53**, 334–353 (2017)

Connected and Autonomous Vehicles and Fleets

Ex-Post Evaluation of an In-Vehicle Warning System for Rail-Road Level Crossings: The Case of Taxi Drivers

Anastasios Skoufas[1(✉)], Neofytos Boufidis[2], Josep Maria Salanova Grau[2], Georgia Ayfantopoulou[2], and Socrates Basbas[1]

[1] Department of Transportation and Hydraulic Engineering, School of Rural and Surveying Engineering, Aristotle University of Thessaloniki, 541 24 Thessaloniki, Greece
{askoufas, smpasmpa}@topo.auth.gr

[2] Centre for Research and Technology Hellas, Hellenic Institute of Transport, Charilaou-Thermi Rd, P.O. Box 60361, 57001 Thermi, Thessaloniki, Greece
{boufidisneo, jose, gea}@certh.gr

Abstract. Connected and autonomous mobility are of great interest in transport research. Furthermore, Rail-Road Level Crossings represent high-risk locations of the network and the accidents that take place at them are considered as one of the most significant accident categories that occur at rail infrastructure. Hence, the evaluation of cooperative systems with the aim of increasing safety at Rail-Road Level Crossings is a crucial issue especially towards the evaluation of the system' objectives as well as decision making for investments regarding in-vehicle warning systems. However, there are many barriers regarding the ex-post evaluation of these systems such as difficulty in collecting and analyzing quantitative data as well as GPS low accuracy. The present research examines the ex-post evaluation of an in-vehicle warning system for Rail-Road Level Crossings developed within the Horizon 2020 project "SAFER-LC" and tested in the city of Thessaloniki, Greece. The evaluation made with a questionnaire-based survey which was carried out in August-October 2019. Statistical analysis revealed numerous interesting findings between drivers' socioeconomic attributes and the way they assess the in-vehicle warning system, indicating the high level of acceptance towards the tested driver assistance system, by a demanding professional drivers' group.

Keywords: C-ITS ex-post evaluation · In-vehicle warning systems · Rail-Road Level Crossings · Taxi drivers · Sociodemographic attributes

1 Introduction

Transport is one of the European Union' (EU') leading sectors regarding its high external costs with main components the environmental, accident and congestion costs [26]. Information Communication Technologies (ICT) such as Cooperative Intelligent Transport Systems (C-ITS) are technologies that aim in the reduction of the

E. G. Nathanail et al. (Eds.): CSUM2020, AISC 1278, pp. 243–252, 2021.
https://doi.org/10.1007/978-3-030-61075-3_24

environmental impacts and in the enhancement of safety and operational efficiency [7]. EU invests largely on the deployment of C-ITS. Specifically, Connecting Europe Facility (CEF) Transport and Horizon 2020 allocated €23.7 and €6.3 billion respectively [8]. Limited research has been conducted regarding C-ITS services whose goal is the increase of safety at RRLCs as well as their evaluation by the end users. RRLC accidents are one of the most significant accident categories that take place in the EU since one person is killed and close to one is seriously injured, on average daily basis, at European RRLCs [9]. Furthermore, accidents' external costs concerning delays, material and environmental damages are far very high [9]. However, the ex-post evaluation is rarely required. In some countries such as France, UK and USA a lot of initiatives are taken regarding transport projects' ex-post evaluation [10].

This paper focuses on the ex-post evaluation of the in-vehicle warning system which was developed by the Hellenic Institute of Transport of the Centre for Research and Technology Hellas (HIT/CERTH) within the Horizon 2020 research project "SAFER-LC" in the city of Thessaloniki, Greece. The objective of this paper is to investigate how the socioeconomic characteristics of the taxi drivers who used the service affect the way that they assess its effect in terms of safety. Section two provides an overview regarding the evaluation of technological solutions whose goal is the increase of safety at RRLCs. Survey' data collection (Sect. 3) and statistical analysis subsequently (Sect. 4) are the two main steps of the research' methodological approach. The paper concludes with the discussion and the future research directions.

2 Literature Review: Evaluation of Current Technological Systems Towards Safer Rail-Road Level Crossings (RRLCs)

RRLCs include the active and the passive RRLCs where automatic warning devices (e.g. bells, boom gates, flashing lights, audible warnings) and static warning signs (e.g. stop sign, St. Andrews cross) are used respectively. Passive RRLCs are proportionally riskier in terms of fatal accidents than the active [12] while crashes that take place at active RRLCs are less severe than the passive ones [18]. Several studies highlight that the impacts of different warning devices on drivers' behavior differ. Lenne et al. [16] simulation study reports that vehicle mean approaching speed was decreased more rapidly in RRLCs equipped with flashing lights than to RRLCs equipped with traffic signals. Unintentional human errors in RRLC accidents include failure of warnings detection or failure of warnings' meaning apprehension during RRLC approach [29] as well as inattention and distraction caused by different reasons [22]. Young et al. [35] point out the impact of texting on drivers' behavior in a sample of 28 participants. Texting doubled the time that drivers looked off road and contributed to drivers' loss of situation awareness (SA). Endsley [6] defined SA as the internal conceptualization of the current situation while Bolstad [3] study reveals that SA is higher to younger (16–25 years old) and middle-aged adults (40–50 years old) than the older adults (65–80 years old). Stanton et al. [28] point out that reduced SA is also a key factor for unintentional noncompliance at RRLCs. Intentional errors concern risky behavior (e.g. beat the train) despite drivers full warnings' awareness [32].

Alternative low-cost technologies are being used today so that the operational and maintenance cost of the RRLCs' warning equipment can be reduced. Some examples include sensors of different types (audio, video, radar, laser) as well as new services based on ICT that enable the communication between trains, car drivers, RRLCs and control centres [11]. However, it should be highlighted that new technologies should not be concerned as subsequent to the traditional existing warning devices rather than an additional control for RRLC safety. These technologies, as a part of RRLC active protection, inform drivers about RRLC presence or about train' arrival at RRLC. The 2 main approaches that these technologies serve are: a) Vehicle to Vehicle Communication (V2V) with the direct communication between trains and vehicles, b) communication between vehicles with the infrastructure (Vehicle to Infrastructure – V2I). Messages are transmitted through antennas, transmitters and receivers, radar, microwave technology, Global Navigation Satellite Systems (GNSS), short-range communication devices and closed-circuit television (CCTV) while cars should be properly equipped in order to receive the warning messages [31]. In-Vehicle warning systems are the most prominent Intelligent Transport Systems (ITS) approach for RRLC safety while numerous pilot tests took place internationally [14]. In the context of the project "ITS to improve safety at Level Crossings" the cooperative service' alerts occurred in the right time as described by the 82% of the drivers [27]. Furthermore, the perceived effectiveness of a similar system deployed at Illinois was described as "high" or "very high" by the 43% of the drivers while the 25% of the drivers described it as "low" or "very low" in a sample of 300 drivers [1, 2, 19]. In Minnesota, the in-vehicle alert system alerts were described as valuable by the 80% of the drivers [30]. In Finland, the in-vehicle warning system developed in the context of the project "Junavaro" was described by 56.1% of reliability and by 82.9% of precision [21].

However, limited research has been conducted for the evaluation of the effect of these systems on drivers' behavior, especially on systems' failure modes as well as on human factor consequences of the unreliable warnings [34]. Landry et al. [13] research on the effect of the in-vehicle auditory alerts (IVAAs) in a sample of 20 participants and 4 RRLCs with different protection equipment revealed that IVAAs had a longer lasting effect at passive RRLCs while lower drivers' compliance scores were noticed at active RRLCs. 16 out of the 20 participants are willing to use the system in their private cars as well especially out of familiar areas. However, 60% of the participants stated that IVAAs can become annoying after long exposure which implicates that further research should be conducted towards the making of more pleasant sounds for IVAAs. Larue and Wullems [15] simulation study in a sample of 15 participants familiarized with the in-vehicle audio system showed that system' use neither led in the increase of the RRLC approaching speed nor to drivers' distraction. Nonetheless, the system' use led in the decrease of their stopping compliance to 53.4% compared to the presence of signage with stopping probability of 69.8%. Furthermore, drivers' positive feedback about the system focused on the potential drivers' increased awareness as well as on the reduction of drivers' inattention levels due to fatigue driving conditions. On the other hand, negative feedback concerns potential drivers' complacency and potential system' failures while drivers highlight the need for system' failures in a safe manner. Patterson

[23] highlights the legal liability issues regarding the adoption of non-fail-safe alternative technologies that are responsible for accident cause while the use of failsafe technology would probably prevent the realization of the same accident. Wullems [33] notes that legal liability issues as well as the required reliability levels for active RRLC protection systems are overriding obstacles for the adoption of low-cost RRLC technology.

3 The In-Vehicle Warning System Tested in Thessaloniki, Greece

3.1 Description of the In-Vehicle Warning System

The in-vehicle warning system pilot test is one of the first at European level regarding multimodal cooperative services. The service was tested at 29 RRLCs in the Thessaloniki Greater Area ($\sim$1 million inhabitants) by the fleet of the "TaxiWay" taxi association which consists of up to 1000 taxis. The successful deployment of many ITS and C-ITS services in cooperation with "TaxiWay" by HIT/CERTH (COMPASS4D, C-MOBILE) [20] and the taxi drivers' professionality are two main reasons for the selection of the user group of taxi drivers. The local ecosystem was created by services' deployment, facilitated the SAFER-LC pilot [24]. Additionally, TRAINOSE, the main railway operator in Greece, equipped 25–30 trains with Galileo-enabled devices so that the latter could be real-time monitored [24]. System' warnings are based on mobile communication technology and aim to increase safety at RRLCs by providing warning alerts to the taxi drivers in the surrounding area of the RRLCs. The main elements of the in-vehicle system are the following: a) Location tracking devices and monitoring system, b) Detection system, c) Alert system/Human – Machine Interface (HMI) device. Each RRLC is traced out by 2 predefined polygons of the road and rail network. Polygons are unique for every RRLC because the polygons are adopted to the RRLC' topology. Polygons' design is based on 2 principles: a) They should include all road sections heading to the RRLC within a 80-m radius from the rail, to ensure that alerts will be generated for all test vehicles heading to RRLC well before they reach the dangerous area., b) They should exclude all nearby road sections not heading to RRLC, to avoid irrelevant alerts for vehicles not heading to the level crossing) (Fig. 1).

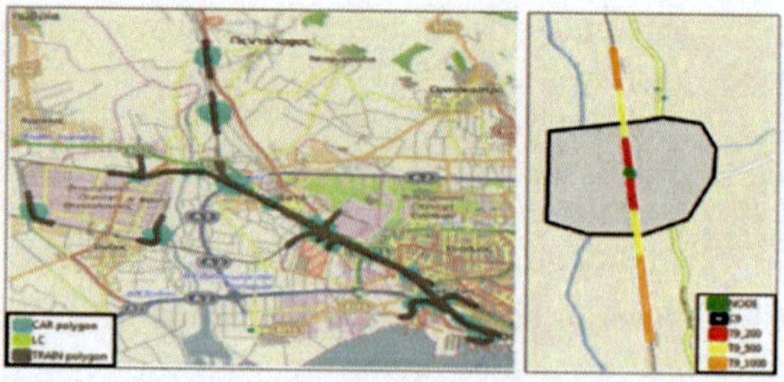

Fig. 1. Polygons of the alert system implemented in Thessaloniki [24].

The in-vehicle system provides static and dynamic warnings. The static one informs the driver about the RRLC presence when the driver enters the polygons while the dynamic one informs the driver about train' estimated time of arrival (ETA) when a train enters the polygon simultaneously [25]. ETA is calculated with machine learning algorithms [4] using train' position and speed [5]. The alert is provided through a pop-up window appearing on the taxi' navigation device while the advices are generated at a distance of 1000, 500 and 200 m between the train and the RRLC (Fig. 2).

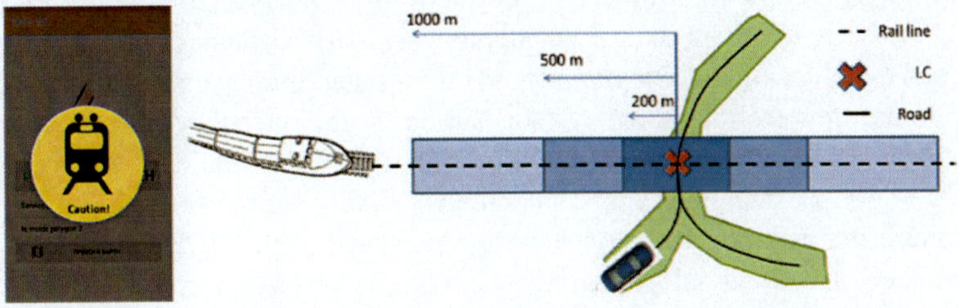

Fig. 2. Warning (static) message when approaching the RRLC (left) and theoretical representation of rail and road polygons for the alert system in Thessaloniki, Greece (right) [24].

3.2 Questionnaire Survey

In order to conduct the ex-post evaluation of the in-vehicle warning system, a face to face stated-revealed preference questionnaire survey was carried out. The questionnaire consists of 14 questions and the survey took place between August 1 to October 9, 2019. 82 taxi drivers of the "TaxiWay" taxi association participated in the survey and hence 82 valid questionnaires were completed. The first part of the questionnaire includes 8 questions concerning the socioeconomic characteristics of the taxi drivers and their involvement in different types of traffic accidents. The second part includes 6 questions and it concerns the evaluation of the in-vehicle warning system in terms of perceived safety feeling and system' warnings positive effect along with the drivers' perception regarding its hypothetical use from other road users such as new drivers and vulnerable road users. Last, in this part, the drivers' attitude regarding the hypothetical integration of such a system in navigation systems as well as the encouragement extent of the system' use by drivers' relatives (in case that it is free) are examined. The five-point Likert scale is used in the second part [17].

4 Descriptive and Inferential Statistics

4.1 Descriptive Statistics

SPSS v25.0 software was used for the realization of the whole statistical analysis. 58.2% of the drivers used the in-vehicle warning system while the rest 41.8% did not use the system. The overriding majority is male with 97.8% while only 2.2% is female. The age of the respondents lies between 27–60 years old with the mean age being 48.

Then, sample' mean driving experience is 16 years with the most experienced drivers reaching 35 driving years. Concerning the drivers' household annual family income, the majority (29.3%) earns < €7000 while only 9.7% lies between €24001–36000. 24.4% of the sample preferred not to reveal their annual income. The vast majority (69.8%) stated that they are High School graduates followed by University graduates (20.9%). Additionally, 2.3% are University students (undergraduate, postgraduate) and 7% are University graduates (with Master).

The overwhelming majority (73.2%) stated that it has never had an accident with injuries followed by drivers who had 1 accident (24.4%) and 2.4% that they had 2 accidents. 74.4% is involved to 1–5 accidents with material damages. The rest stated that they had 6–15 accidents. The majority (81.4%) stated that owns a private car and it is willing to use new technological systems similar to the ones they have in their taxi. 9.3% responded that either they do not own a private car or that, although they own one, they will not use such systems. The perceived safety feeling from the system' use for the significant majority was considerably enhanced with 47.7%. 21.7% of them stated that their perceived safety feeling was completely enhanced. 40% were completely affected and 33.3% were considerably affected on the way they approach RRLCs.

Drivers were asked to evaluate the effect of the warnings if the system was used by other road users (new drivers, drivers who did not know about the RRLC) and Vulnerable Road Users (VRUs). It should be highlighted that the sample revealed that warnings would completely affect the way that other road users and VRUs approach RRLCs with 42.2% and 71.1% respectively. As far as the hypothetical integration of a similar safety system in the navigation system of modern cars to increase safety, 95.6% supports positively this integration. Most drivers (88.9%) would encourage their relatives to use the safety system considering it is free of charge (Fig. 3).

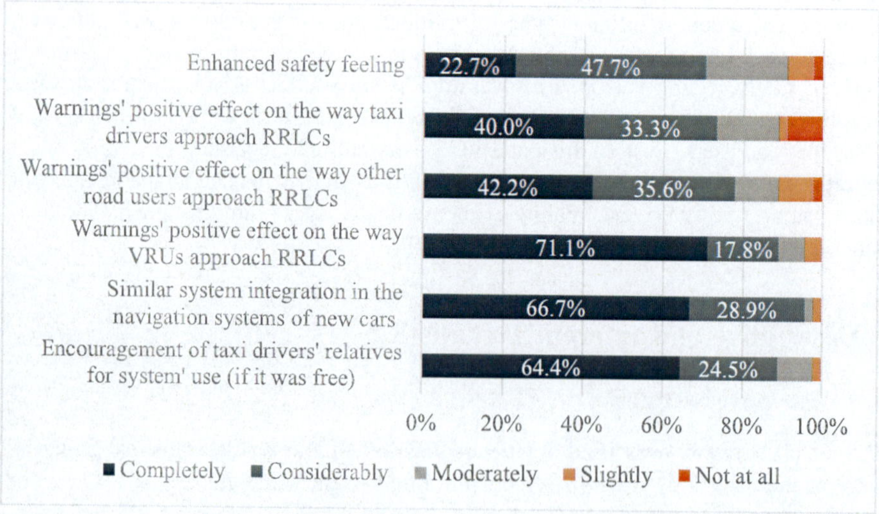

Fig. 3. Descriptive statistics regarding the evaluation of the in-vehicle warning system.

4.2 Inferential Statistics

The level of the drivers' annual income found to correlate with the drivers' opinion regarding system' warnings positive effect on the way they approach rail-road level crossings ($r = 0.354$, $p < 0.05$). Drivers' private car ownership status is significantly correlated with the extent of encouragement of the system use by drivers' relatives in case it was free. According to the Kruskal-Wallis H test there is a statistically significant difference in the extent of encouragement system use between the different private car ownership status, $\chi^2(2) = 10.456$, $p = 0.005$, with a mean rank extent of encouragement of 6.38 for no private car ownership, 24.54 for private car ownership and willingness to use new technological systems and 20.25 for private car ownership and no willingness to use new technological systems.

5 Discussion and Future Directions

The sample size of drivers' who used the in-vehicle system (46 drivers) can be considered as small. Few statistically significant correlations were found. Similar research should be conducted towards a model estimation regarding the quantification of the effect of specific drivers' attributes on a variable related with the system' evaluation in term of user acceptance and compliance (e.g. enhanced feeling of safety, system' positive effect on drivers' way of approaching RRLCs). Alongside, questionnaires were completed only by the taxi drivers who used the in-vehicle system. It would be more interesting if the sample consisted of drivers who were and were not use the in-vehicle warning system. Hence, enlightening comparative results would be achieved regarding the in-vehicle system evaluation from the two different groups.

A rather limited amount of research has been carried out for the evaluation of C-ITS regarding RRLC safety. The in-vehicle warning system developed within the research project "SAFER-LC" is one of the first at European level. However, within many European research projects (e.g. COSMO, DRIVE C2X, FREILOT, Compass4D) questionnaire surveys were carried out for the evaluation of C-ITS services such as Red Light Violation Warning, Road Hazard Warning, Energy Efficient Intersection Service, Approaching Emergency Vehicle Warning in terms of driver acceptance and compliance. Further research should be undertaken for the testing and evaluation of such C-ITS. Directive 2010/40/EU entitled "Ex post evaluation of the Intelligent Transport Systems" refers to monitoring ITS' benefits and to investigating factors hold the achievement of the policy goals back. Ex-post evaluation' results can benefit both policy makers as well as the communication with the public [10].

6 Conclusions

Rather limited research has been conducted for the evaluation of C-ITS for RRLC safety but with encouraging results. This survey was designed to evaluate the in-vehicle warning system which was developed within the research project "SAFER-LC" taking into account the sociodemographic attributes of the taxi drivers. Furthermore, additional features that affect the perceived usefulness of the in-vehicle warning system such as the level of drivers' fatigue, weather conditions (daylight or night) and familiarity of the drivers with the road network should be investigated in relevant research. Thus, even more interesting and valuable findings can be arisen. Results can contribute to the scarce literature review regarding the ex post evaluation of such systems as well as professional drivers' perceptions regarding C-ITS solutions for RRLCs. The in-vehicle warning system led to an enhanced feeling of safety for the taxi drivers when approaching RRLCs. Furthermore, results reveal the drivers' positive belief for the in-vehicle warning system effect on the way they approach RRLCs and the systems' potential positive effect on the way other road users (new drivers, drivers who do not know about the RRLC presence) as well as VRUs approach RRLCs. The potential integration of a similar system in drivers' private cars and the potential encouragement of the system' free use by drivers' relatives as stated by the overriding majority is an important finding. Finally, this research reveals 2 interesting correlations between the taxi drivers' socioeconomic characteristics with the way drivers assess the in-vehicle warning system.

Acknowledgements. The questionnaire survey has been conceded in the framework of "SAFER-LC" project by HIT/CERTH. The project has received funding from the EU's Horizon 2020 research and innovation programme under grant agreement No 723205.

References

1. Benekohal, R.F., Aycin, M.F.: Analyses of drivers' opinions about railroad grade crossings traffic control devices and safety: background survey (2004)
2. Benekohal, R.F., Rawls, C.G.: Analysis of drivers' responses to in-vehicle receiver (IVR) after experiencing two modes of operation (2004)
3. Bolstad, C.A.: Situation awareness: does it change with age? In: Proceedings of the Human Factors and Ergonomics Society Annual Meeting, vol. 45, no. 4, pp. 272–276 (2001). https://doi.org/10.1177/154193120104500401
4. Boufidis, N., Aifandopoulou, G., Salanova Grau, J.M., Tzenos, P.: Estimating train arrival time with supervised machine learning algorithms. Paper presented at the 13th ITS European Congress, Brainport, Netherlands (2019)
5. Boufidis, N., Salanova Grau, J.M., Tzenos, P., Aifandopoulou, G.: Estimating time of arrival of trains at level crossings for the provision of multimodal cooperative services. Paper presented at the 22nd EURO Working Group on Transportation Meeting, Barcelona (2019)
6. Endsley, M.R.: The role of situation awareness in naturalistic decision making. In: Zsambok, C.E., Klein, G. (eds.) Expertise: Research and Applications. Naturalistic Decision Making, pp. 269–283. Lawrence Erlbaum Associates (1997)
7. European Commission: Cooperative, connected and automated mobility (CCAM). https://ec.europa.eu/transport/themes/its/c-its_en. Accessed 20 Jan 2020
8. European Commission: Intelligent Transport Systems (ITS) for Road (2019). https://ec.europa.eu/inea/sites/inea/files/cefpub/cef_transport_its-2019-web.pdf
9. European Union Agency for Railways, ERA: Railway Safety Performance in the European Union (2016). https://op.europa.eu/en/publication-detail/-/publication/313a2c56-7277-11e7-b2f2-01aa75ed71a1/language-en

10. ITF: Ex-Post Assessment of Transport Investments and Policy Interventions, ITF Roundtable Reports. OECD Publishing, Paris (2017). https://doi.org/https://doi.org/10.1787/9789282108154-en
11. Khoudour, L., Ghazel, M., Boukour, F., Heddebaut, M., El-Koursi, E.: Towards safer level crossings: existing recommendations, new applicable technologies and a proposed simulation model. Eur. Transp. Res. Rev. **1**(1), 35–45 (2009). https://doi.org/10.1007/s12544-008-0004-z
12. Laapotti, S.: Comparison of fatal motor vehicle accidents at passive and active railway level crossings in Finland. IATSS Res. **40**(1), 1–6 (2016). https://doi.org/10.1016/j.iatssr.2015.12.003
13. Landry, S., Jeon, M., Lautala, P., Nelson, D.: Design and assessment of in-vehicle auditory alerts for highway-rail grade crossings. Transp. Res. Part F Traffic Psychol. Behav. **62**, 228–245 (2019). https://doi.org/10.1016/j.trf.2018.12.024
14. Larue, G., Soole, D., Rakotonirainy, A.: Intelligent Transport Systems for Safer Level Crossings Literature Review - Background Research, CRC for Rail Innovation. Project No: R2.111 (2010)
15. Larue, G.S., Wullems, C.: Driving simulator evaluation of the failure of an audio in-vehicle warning for railway level crossings. Urban Rail Transit **1**(3), 139–148 (2015). https://doi.org/10.1007/s40864-015-0018-5
16. Lenné, M.G., Rudin-Brown, C.M., Navarro, J., Edquist, J., Trotter, M., Tomasevic, N.: Driver behaviour at rail level crossings: responses to flashing lights, traffic signals and stop signs in simulated rural driving. Appl. Ergon. **42**(4), 548–554 (2011). https://doi.org/10.1016/j.apergo.2010.08.011
17. Likert, R.: A technique for the measurement of attitudes. Arch. Psychol. **140**, 1–55 (1932)
18. Liu, J., Khattak, A.J., Richards, S., Nambisan, S.: What are the differences in driver injury outcomes at highway-rail grade crossings? Untangling the role of pre-crash behaviors. Accid. Anal. Prev. **85**, 157–169 (2015). https://doi.org/10.1016/j.aap.2015.09.004
19. Medina, J., Benekohal, R.: Analysis of drivers' opinions on the effectiveness of traffic and in-vehicle receiver control devices at rail-highway crossings. In: 85th Annual Meeting of the Transportation Research Board, Washington, DC, USA (2006)
20. Mitsakis, E., Stamos, I., Salanova Grau, J.M., Chrysohoou, E., Iordanopoulos, P., Aifadopoulou, G.: Urban mobility indicators for Thessaloniki. Journal Traffic Logistics Engineering 1(2), 148–152 (2013). https://doi.org/10.12720/jtle.1.2.148-152
21. Öörni, R.: Reliability of an in-vehicle warning system for railway level crossings a user-oriented analysis. IET Intell. Transp. Syst. **8**(1), 9–20 (2014). https://doi.org/10.1049/iet-its.2012.0129
22. Parnell, K.J., Stanton, N.A., Plant, K.L.: Exploring the mechanisms of distraction from in-vehicle technology: the development of the PARRC model. Saf. Sci. **87**, 25–37 (2016). https://doi.org/10.1016/j.ssci.2016.03.014
23. Patterson, J.: Fail-Safe Forum: Legal Issues, Parliamentary Road Safety Committee (2008)
24. SAFER-LC – Deliverables & Publications: Implementation guidelines. SAFER Level Crossing by integrating and optimizing road-rail infrastructure management and design (2018). https://saferlc.eu/IMG/pdf/saferlc_20180430_d41_v08_dlr_implementation_guidelines.pdf
25. Salanova Grau, J.M.: Multimodal cooperative safety services: a location-based alert system at level crossings. Presented at ILCAD 2019, International Level Crossing Awareness Day, Act Safety at Level Crossings, Amersfoort, Netherlands (2019)

26. Schroten, A., van Essen, H., van Wijngaarden, L., Sutter, D., Andrew, E.: Sustainable transport infrastructure charging and internalisation of transport externalities: Executive summary. Publications Office of the European Union, Luxembourg (2019). https://doi.org/10.2832/246834
27. Singh, J., Desai, A., Acker, F., Ding, S., Prakasamul, S., Rachide, A., Bentley, K., Nelson-Furnell, P.: Cooperative intelligent transport systems to improve safety at level crossing. In: Level Crossing 2012 London, London, UK (2012). https://pdfs.semanticscholar.org/0dfb/a031899f467e908e0d3624e36783ab97005c.pdf
28. Stanton, N.A., Salmon, P.M., Walker, G.H., Salas, E., Hancock, P.A.: State-of-science: situation awareness in individuals, teams and systems. Ergonomics **60**(4), 449–466 (2017). https://doi.org/10.1080/00140139.2017.1278796
29. Tenkink, E., Van der Horst, R.: Car driver behavior at flashing light railroad grade crossings. Accid. Anal. Prev. **22**(3), 229–239 (1990). https://doi.org/10.1016/0001-4575(90)90015-D
30. United States Department of Transportation: Intelligent transportation systems at highway-rail intersections, a cross-cutting study: Improving safety and mobility at highway-rail grade crossings. United States Department of Transportation, Washington, DC, USA (2001)
31. Victoria State Government Towards zero: A strategy for improved level crossing safety in Victoria. Victoria State Government, State of Victoria, Australia. https://www.towardszero.vic.gov.au/__data/assets/pdf_file/0010/183556/STU_0206_RS_STRATEGY_2016_web.pdf
32. Witte, K., Donohue, W.A.: Preventing vehicle crashes with trains at grade crossings: the risk seeker challenge. Accid. Anal. Prev. **32**(1), 127–139 (2000). https://doi.org/10.1016/S0001-4575(99)00061-5
33. Wullems, C.: Towards the adoption of low-cost rail level crossing warning devices in regional areas of Australia: a review of current technologies and reliability issues. Saf. Sci. **49**(8–9), 1059–1073 (2011). https://doi.org/10.1016/j.ssci.2011.04.006
34. Wullems, C., Wayth, R., Galea, V., Nelson-Furnell, P.: In-vehicle railway level crossing warning systems: can intelligent transport systems deliver? In: Conference on Railway Excellence, Adelaide, Australia (2014)
35. Young, K.L., Lenné, M.G., Salmon, P.M., Stanton, N.A.: The impact of texting on driver behaviour at rail level crossings. Accid. Anal. Prev. **118**, 269–276 (2018). https://doi.org/10.1016/j.aap.2018.05.002

A Conceptual Model for the Simulation of the Next Generation Bike-Sharing System with Self-driving Cargo-Bikes

Imen Haj Salah[1(✉)], Vasu Dev Mukku[1], Stephan Schmidt[1], and Tom Assmann[1,2]

[1] Otto von Guericke University Magdeburg, Universitätsplatz 2, 39106 Magdeburg, Germany
imen.hajsalah@ovgu.de

[2] Fraunhofer Institute for Factory Operation and Automation IFF, Sandtorstraße 22, 39106 Magdeburg, Germany

Abstract. Car- and Bike-sharing systems have dramatically altered mobility services in metropolitan areas in recent years. They improve flexibility and convenience for non-private car-based mobility. This transformation seems to be further revolutionized by the introduction of self-driving vehicles. Indeed, Autonomous Mobility-on-Demand (AMoD) systems are gaining a huge research interest as they allow to address conventional system problems such as fleet imbalance. In this context, we present our AuRa (Autonomous Rad) project which aims to develop an on-demand shared-use self-driving bikes service (OSABS). Unlike the AMoD systems with car fleets, the AuRa concept highly reduces space occupancy, energy consumption, and air pollution by leveraging bikes. It can be further integrated with public transportation. This work describes the AuRa system design from a logistical perspective. Furthermore, we represent this system with a conceptual model that can be used to develop an agent-based simulation model in Anylogic software.

Keywords: Autonomous vehicles · Autonomous bikes · Bike-sharing system · Mobility-on-Demand · Conceptual model · Agent-based simulation

1 Introduction

1.1 Motivation

The societal mission of mitigating air pollution and greenhouse gas emissions are forcing urban agglomerations worldwide to strongly green their urban transportation systems. Shared autonomous car fleets are heavily discussed and put forward as a promising solution to reduce emissions and congestion by allowing more efficient use of resources. However, these positive effects are challenged by some recent studies demonstrating that such mobility services increase the

E. G. Nathanail et al. (Eds.): CSUM2020, AISC 1278, pp. 253–262, 2021.
https://doi.org/10.1007/978-3-030-61075-3_25

mileage traveled in urban agglomerations and subsequently increase congestion and energy consumption [1,2].

To solve these and other collateral problems caused by car traffic, the move towards environmentally sustainable transport has become a priority in urban areas. Bike-sharing systems are identified as one potential alternative to serve short-distance trips and enhance connectivity to public transportation networks [3]. Nevertheless, one prevalent issue of conventional systems is the fleet unbalance, particularly during peak hours. This generally leads to service unavailability and customer dissatisfaction.

Taking those backsides of shared autonomous car fleets and conventional bike-sharing systems, we propose our novel project "AuRa" which aims at contributing to sustainable mobility through the development of a self-driving bike as a means for an on-demand shared-use autonomous bike service (OSABS). As autonomous bikes ride on specific cycling ways, OSABS helps to decrease congestion on roads and consequently reduces pollution. Through leveraging the autonomous driving, our system can redistribute bikes dynamically to meet future demand. Self-driving cargo-bikes can make the bike-sharing more attractive to customers as they can (i) reduce service unavailability, which occurs when a manually driven system is unbalanced (ii) provide a sustainable and environmentally-friendly demand-responsive system.

As a new innovative idea, we still need to prove its economic viability. Throughout the project, we aspire to provide fundamental knowledge about OSABS and explore its potential for sustainable urban mobility. In this first work, we present a conceptual model that captures the main elements of our system. The definition of such a conceptual model lays the ground for the development of a realistic simulation model that will help to evaluate different strategies and allows a better understanding of the system.

1.2 Literature Study and Scientific Contribution

During the last years, extensive studies have been conducted on shared autonomous vehicle (SAV) services. Different aspects of SAVs were studied, mainly fleet management aspects such as vehicle assignment e.g. [4,5], vehicle rebalancing [6–9] and fleet dimensioning [7,10]. For a taxonomy on SAV fleet management problems, the reader may refer to [11].

Other studies have focused on the business aspects [12,13] where the authors try to evaluate different pricing schemes. Vehicle charging [14,15] and station distribution [16,17] were also covered in the literature in order to assess different parking strategies and charging technologies. In all the above papers, a car fleet system is modeled using mostly an agent-based simulation. For an exhaustive literature review, we recommend reading [18]. The paper provides a comprehensive outlook on relevant studies in the field of SAV.

In our project, we aim to cover all those mentioned aspects while considering our new service specifications. OSABS can be compared to autonomous car fleets at the management level, mainly as a part of AMoD services. However, it is utterly distinct at the business and operations levels. Bike utilization implies a different price scheme, distinct routing network, particular parking requirements, specific charging and energy supply options and different demand patterns. Integrating these business and operation specifications to the management aspects can also create new fleet management problems that have not yet been covered.

In this paper, we provide several contributions. In Sect. 2, we briefly present the design of our new system and describe its key elements. In Sect. 3, we raise the open questions that need to be answered throughout the project. In Sect. 4, we present a conceptual model for the simulation. Finally, conclusions and future research directions are highlighted in Sect. 5.

2 System Description

2.1 System Design

The system design of the next-generation bike-sharing is a modification of conventional bike-sharing designs [19,20]. At its core is an electric cargo-bike which drives autonomously with no user sitting on it and turns into a manual mode when under control of a user. A customer can, therefore, call a bike to its point of demand and will receive it within a given service time. Next, the manual ride, with or without goods, leads directly to the final destination. The management will secure this functionality by firstly assigning the bikes to demands placed over time and space and secondly redistributing bikes after usage to either station in areas with prospective demand or directly to the next user. Thus, it also becomes possible to secure intermodal, seamless trip chains by providing a pre-booked bike at the public transport stop for the last mile.

Initially, this system will be based within the city of Magdeburg, Germany, in suburban areas. The public transport integration will be limited to trips between a stop and final destinations without considering complete trip chains since we consider only on-demand bookings. We assume a service time of 10 min, which is derived from first surveys with students of Magdeburg University.

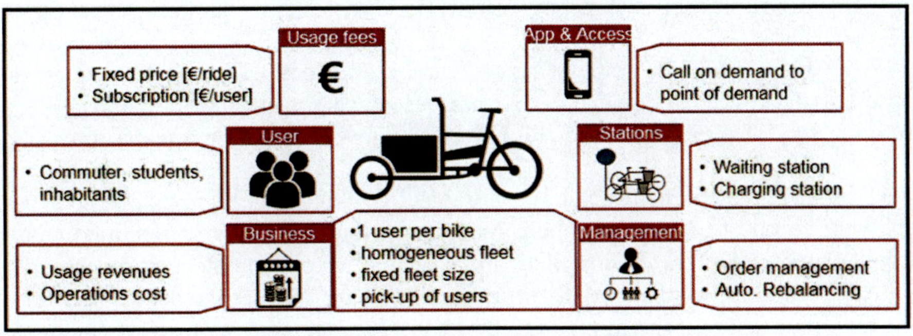

Fig. 1. System design of on-demand shared-use autonomous bike-sharing

As a first simulation-based feasibility study, we will limit our system under investigation, as shown in Figure 1. The key elements of OSABS are:

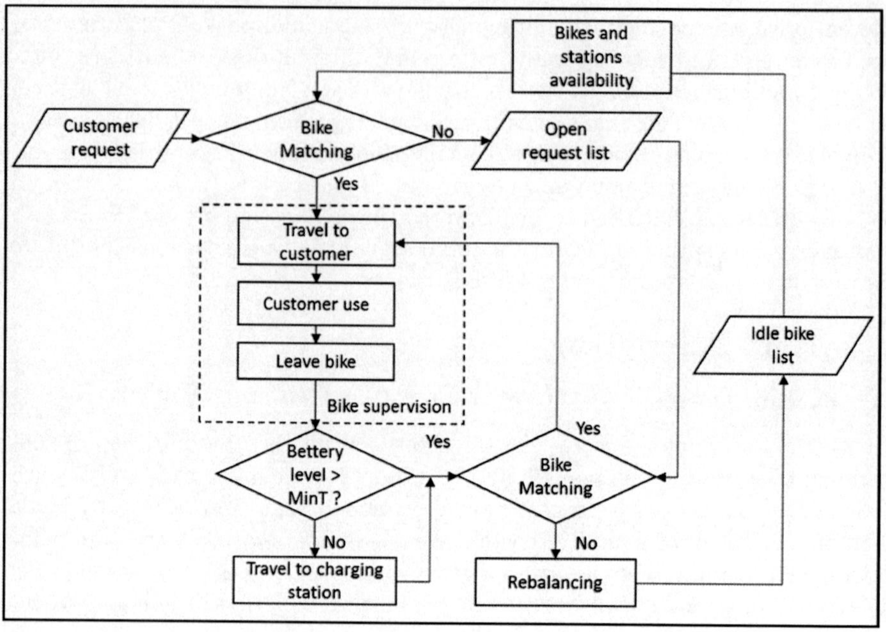

Fig. 2. Order management process (MinT: minimum battery threshold)

Bikes: The bikes are tilting three-wheeled electric-assisted cargo bikes which need charging. This configuration allows stable autonomous rides as well as dynamic manual rides. We consider an homogenous fixed-size fleet.

Users: The user, their usage potential, mobility needs and actual mobility patterns will be investigated via structured, representative surveys with citizens of Magdeburg and an analysis of the transportation demand model of the city, representing origin-destination between statistic district per mode.

Stations: They can be either waiting or charging stations. Waiting stations need no infrastructure and enable bikes to park close to prospective demand areas. Charging stations allow refilling the battery and require major infrastructure.

Order Management: The fleet of autonomous bikes needs to be managed continuously to handle the dynamic environment efficiently. Since it is assumed that we will track and monitor our bikes in real-time, intelligent fleet management strategies could be implemented to minimize waiting times, increase the service rate and reduce operational costs. The order management algorithms include two main sections. One is the matching strategy which allows assigning available bikes to open customer requests while reducing the operating costs and staying within the 10-minute constraint. The second is the rebalancing strategy which consists of distributing idle bikes during low demand times based on future demand forecasting. Its goal is to determine a dynamic rebalancing scheme that guarantees our service rate while minimizing the operational cost of bike relocation. In our framework, we prioritize serving current demand before dispatching bikes for future demand. This prioritization is meaningful as the main objective is to satisfy the present demand and the forecasting information is not as accurate (we deal with high stochastic demand). The two sections use a routing algorithm in order to calculate both route cost and duration. The decision process of the order management strategy is described in Fig. 2.

Business Model: The next generation bike sharing will operate either as functional extended bike sharing provider, offering real door-to-door mobility, or can furthermore become a part of mobility-as-a-service packages. The primary revenue streams can be assumed to stay within usage related fees (more diversification possible), public subsidies and advertisement. The big advantage lies in the autonomous redistribution. This function cuts between 30 to 80% of operational costs [21] currently spent for moving bikes between stations by vans and furthermore decreases the number of bikes needed per user.

3 Research Questions and Intended Experiments

The new system raises different open questions that need to be investigated. A system simulation will help to evaluate different strategies and provide responses. We are mainly focusing on four aspects which are:

Station Distribution Strategies: To identify the best station distribution strategy for the AuRa bikes across the city, the following questions need to be answered:

- How should waiting stations be distributed in order to guarantee customer request satisfaction in less than 10 min?
- What are the optimal number and location for charging stations considering cost and infrastructure constraints?

Energy Supply Technologies: The aim of this experiment is to identify the suitable energy supply for the AuRa bike. We investigate the possible energy supply technologies for autonomous vehicles available in the market and evaluate the cost, benefits and drawbacks of each one. A comparison study allows concluding the most economically-viable technology for the bike.

Order Management: The main fleet management issues are:

- What is the best assignment strategy?
- How can the rebalancing and charging strategy be jointly optimized to meet the demand, especially during peak hours?
- What is the minimum fleet size required to meet the target service rate?
- How does the waiting time affect the assignment strategy performance?
- Which rebalancing strategy is the best for OSABS?

By comparing the revenue and service rates of different matching and rebalancing strategies, we can determine the ones who achieve the best performance.

Business Model: Against the autonomous rebalancing benefits stand high investment cost for the autonomous bikes and the charging stations. Open questions on the business concepts are:

- Do reduced efforts for rebalancing and fewer bikes per user needed compensate for the higher investment per bike?
- Which additional user potential can be accessed through on-demand service, seamless trip chains and the possibility to transport goods and people?
- Can flexible and value-added pricing schemes be implemented and what other revenue streams are possible?

The simulation will enable the revenue calculation for different demand and pricing scenarios. It further determines the best service rate for economic viability.

4 Conceptual Model

The simulation model provides a platform for identifying suitable solutions for the research questions and intended experiments. Therefore, a modular simulation model which integrates the main activities of OSABS using the agent-based paradigm will be implemented. We aim to run it for a year with a time step of one minute. In this section, a conceptual model for its implementation is presented. Figure 3 shows the concept of the modular simulation model with its interfaces. We structure it into four layers and stand-alone modules with defined functionalities.

Model Input Layer: This layer allows the introduction of different business and demand scenarios as input to the model through the business and demand modules. The job server simulates the interface between the customer and the system. It generates job requests based on the demand data and obtains the matching result from the central information processing unit (CIPU). Table 1 illustrates the interfaces and functionalities.

Table 1. Modules of model input layer

Module name	Interfaces	Functionality
Business	Output and Evaluation	It imports different business cases to the model
Demand	Job Server	It provides different demand scenarios
Job Server	Demand, CIPU and Forecasting	It translates the demand data into job requests and collects job status from the CIPU

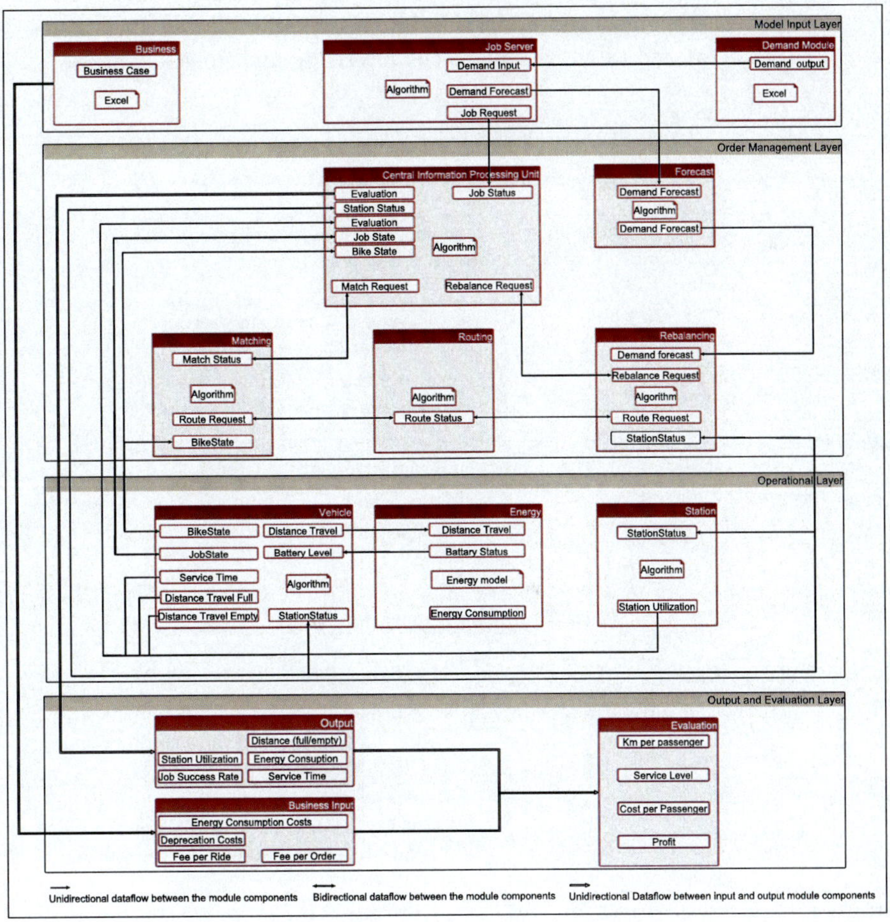

Fig. 3. Bird view of modular simulation model

Order Management Layer: This layer contains four modules which perform fleet management operations. The CIPU allows the implementation of the decision process and define the priorities between the matching and rebalancing modules. Once receiving a job request, it forwards it to the matching module, which checks for bike availability through the vehicle module and calculates the best route between the bike and the request location using the routing module. In the absence of requests, the rebalancing module is responsible for bike redistribution in time and space based on the demand forecasting module (Table 2).

Operational Layer: This layer includes the vehicle, energy and station modules. The vehicle component simulates and provides the status of the bikes as a basis for the system operation and evaluation. The energy part informs about the energy consumed and the battery level of the bikes. The station module provides the station status such as station capacity, number of idle bikes and free spaces available in the station. Table 3 explains the interfaces and functionalities.

Table 2. Modules of order management layer

Module name	Interfaces	Functionality
CIPU	Job Server, Matching, Rebalancing, Vehicle, Station and Output	It monitors the activities of the different modules in the model
Matching	CIPU, Routing, Vehicle	It assigns bikes to customers considering the routing information
Routing	Matching, Rebalancing	This provides the route information of bike to the matching and rebalancing module
Rebalancing	Routing, Station, CIPU and Forecast	It distributes bikes in space based on future demand and idle bikes distribution
Forecasting	Job Server, Rebalancing	Provides demand forecast information to the rebalancing module

Table 3. Modules of operational layer

Module Name	Interfaces	Functionality
Vehicle	Energy, Station, Matching, CIPU	It provides information about bike state and battery level to the matching and CIPU modules
Energy	Vehicle	It provides the battery level, utilization of energy based on vehicle parameters (speed, route, distance) to the vehicle module
Station	Vehicle, Rebalancing, CIPU	It manages Inventory, type of station, station ID, monitoring station utilization

Output and Evaluation Layer: This layer represents output and evaluation. With data gathered from CIPU and the business module, the evaluation module depicts the statistics of the system and aggregates it KPIs. We assess the service level, distance travelled by the bikes, the number of customers travelled, costs and revenues to evaluate the economic viability of the system (Table 4).

Table 4. Modules of output and evaluation layer

Module Name	Interfaces	Functionality
Output	CIPU	Collects the primary output
Business Input	Business module	It provides costs and fees to the evaluation module
Evaluation	Business input and Output	It provides the statistic evaluation of the overall system

5 Conclusion and Outlook

The main results of this work are the introduction of a new concept (OSABS) in the mobility services and the development of a conceptual modular simulation model which allows a holistic representation of a real-world system. Our conceptual model is designed with several layers to investigate different aspects of OSABS and examine the interaction between the different modules. The CIPU module in the order management layer enables central control of the system. With centralized information on the fleet and demand, we can optimize resource use more efficiently and implement integrated management strategies. It also coordinates between the matching and rebalancing algorithms enabling to test various management policies.

This first model could be extended. Future scope leans towards adding more complexity by considering flexible prices, in-advance booking, a heterogeneous fleet and intermodal trips. Furthermore, we will investigate the viability with respect to other cities and different urban forms.

Next steps are implementing an agent-based simulation model in Anylogic, relaxing the assumptions and hopefully setting up a small scale living lab demonstrator by the projects end.

Acknowledgments. This project is funded by the federal state of Saxony-Anhalt and the European Regional Development Funds (EFRE, 2014–2020), project number 19-15003/004.

References

1. Fraedrich, E., Kröger, L., Bahamonde-Birke, F.J., Frenzel, I., Liedtke, G., Trommer, S., Lenz, B., Heinrichs, D.: Automatisiertes Fahren im Personen- und Güterverkehr. Auswirkungen auf den Modal-Split, das Verkehrssystem und die Siedlungsstrukturen. e-mobil BW Landesagentur für Elektromobilität und Brennstoffzellentechnologie Baden-Württemberg GmbH (2017)
2. Hörl, S., Becker, F., Dubernet, T.J.P., Axhausen, K.W.: Induzierter Verkehr durch autonome Fahrzeuge: Eine Abschätzung. Technical report, ETH Zurich (2019)
3. Becker, S., Rudolf, C.: Exploring the potential of free cargo-bikesharing for sustainable mobility. GAIA-Ecol. Perspect. Sci. Soc. **27**(1), 156–164 (2018)
4. Fagnant, D.J., Kockelman, K.M., Bansal, P.: Operations of shared autonomous vehicle fleet for Austin, Texas, market. Transp. Res. Rec. **2563**(1), 98–106 (2015)
5. Hyland, M., Mahmassani, H.S.: Dynamic autonomous vehicle fleet operations: optimization-based strategies to assign AVs to immediate traveler demand requests. Transp. Res. Part C Emerg. Technol. **92**, 278–297 (2018)
6. Babicheva, T., Burghout, W., Andreasson, I., Faul, N.: The matching problem of empty vehicle redistribution in autonomous taxi systems. Procedia Comput. Sci. **130**, 119–125 (2018)
7. Fagnant, D.J., Kockelman, K.M.: Dynamic ride-sharing and fleet sizing for a system of shared autonomous vehicles in Austin, Texas. Transportation **45**(1), 143–158 (2018)

8. Marczuk, K.A., Soh, H.S.H., Azevedo, C.M.L., Lee, D.-H., Frazzoli, E.: Simulation framework for rebalancing of autonomous mobility on demand systems. MATEC Web Conf. **81**, 1005 (2016)
9. Rossi, F., Zhang, R., Hindy, Y., Pavone, M.: Routing autonomous vehicles in congested transportation networks: structural properties and coordination algorithms. Auton. Robots **42**, 1427–1442 (2018)
10. Ma, J., Li, X., Zhou, F., Hao, W.: Designing optimal autonomous vehicle sharing and reservation systems: a linear programming approach. Transp. Res. Part C Emerg. Technol. **84**, 124–141 (2017)
11. Hyland, M.F., Mahmassani, H.S.: Taxonomy of shared autonomous vehicle fleet management problems to inform future transportation mobility. Transp. Res. Rec. **2653**(1), 26–34 (2017)
12. Gurumurthy, K.M., Kockelman, K.M., Simoni, M.D.: Benefits and costs of ride-sharing in shared automated vehicles across Austin, Texas: opportunities for congestion pricing. Transp. Res. Rec. **2673**(6), 548–556 (2019)
13. Salazar, M., Rossi, F., Schiffer, M., Onder, C.H., Pavone, M.: On the interaction between autonomous mobility-on-demand and public transportation systems. In: 2018 21st International Conference on Intelligent Transportation Systems (ITSC), pp. 2262–2269 (2018)
14. Chen, T.D., Kockelman, K.M., Hanna, J.P.: Operations of a shared, autonomous, electric vehicle fleet: implications of vehicle & charging infrastructure decisions. Transp. Res. Part A Policy Pract. **94**, 243–254 (2016)
15. Iacobucci, R., McLellan, B., Tezuka, T.: Optimization of shared autonomous electric vehicles operations with charge scheduling and vehicle-to-grid. Transp. Res. Part C Emerg. Technol. **100**, 34–52 (2019)
16. Azevedo, C.L., Marczuk, K., Raveau, S., Soh, H., Adnan, M., Basak, K., Loganathan, H., Deshmunkh, N., Lee, D.H., Frazzoli, E., et al.: Microsimulation of demand and supply of autonomous mobility on demand. Transp. Res. Rec. **2564**(1), 21–30 (2016)
17. Kondor, D., Zhang, H., Tachet, R., Santi, P., Ratti, C.: Estimating savings in parking demand using shared vehicles for home-work commuting. IEEE Trans. Intell. Transp. Syst. **20**(8), 2903–2912 (2018)
18. Narayanan, S., Chaniotakis, E., Antoniou, C.: Shared autonomous vehicle services: a comprehensive review. Transp. Res. Part C Emerg. Technol. **111**, 255–293 (2020)
19. Büttner, J., Mlasowsky, H., Birkholz, T., et al.: Optimising bike sharing in European cities. Recuperado el **28**, 2017 (2011)
20. Institute for Transportation and Development Policy (New York, NY), Gauthier, A.: The bike-share planning guide. ITDP Institute for Planning & Development Policy (2013)
21. Koska, T., Friedrich, M., Rabenstein, B., Bracher, T., Hertel, M.: Innovative öffentliche Fahrradverleihsysteme: Ergebnisse der Evaluation und Empfehlungen aus den Modellprojekten. Bundesministerium für Verkehr und digitale Infrastruktur (2015)

An Image-Based Approach for Classification of Driving Behaviour Using CNNs

Evaggelos Spyrou[1(✉)], Ioannis Vernikos[1], Michalis Savelonas[2], and Stavros Karkanis[2]

[1] Department of Computer Science and Telecommunications, University of Thessaly, Lamia, Greece
{espyrou,ivernikos}@uth.gr

[2] General Department, University of Thessaly, Lamia, Greece
{msavelonas,sk}@uth.gr

Abstract. In this work we present an approach for the classification of driving behaviour using Convolutional Neural Networks (CNNs), based on measurements that have been obtained by the internal CAN-bus of the vehicle. As is the case with different driving behaviours, CAN-bus sensor data reflect the driving patterns associated with different types of vehicles. The experimental evaluation is performed on a real-life dataset composed by measuring 27 attributes, for 4 different car types, namely vacuum, car, truck and garbage truck. These features are processed to form pseudocolored images, capturing both temporal and qualitative features of parts of routes. For classification, we use a deep CNN architecture. Results indicated an accuracy of 91% and increased performance compared to other neural network-based approaches.

Keywords: Convolutional neural networks · Deep learning · Driving behaviour · CAN-bus measurements

1 Introduction

Controller area network (CAN)-bus standards using on-board diagnostics (OBD) ports, as well as telematics sensors, have been shown to provide a signature of driving behaviour by Meiring and Myburgh [1]. Several intelligent approaches analysing such data have been proposed in the last decade.

In [2], statistical features of various driving events have been derived and used in the context of either unsupervised learning (k-means clustering) or supervised learning (SVMs), so as to distinguish different drivers. In [3], several statistical features of 3-axis accelerometer signals where calculated and used to classify driving behaviour as normal or aggressive. These features include central and dispersion characteristics, histogram moments, Kendall's tau rank correlation coefficient, covariance between each pair of acceleration signals etc. Using sequential

E. G. Nathanail et al. (Eds.): CSUM2020, AISC 1278, pp. 263–271, 2021.
https://doi.org/10.1007/978-3-030-61075-3_26

forward feature selection and k-nearest neighbours (k-NN) classification, they maintained 7 features and achieved 100% accuracy. However, the authors note that the driving samples comprising their dataset might be too easily separable. In [4], thresholds associated with events such as acceleration, braking and turning where defined and used for event detection, aiming at classifying driving behaviour on data acquired by mobile phone sensors. The same group introduced UAH dataset [5], acquired by means of mobile phone sensors. This dataset comprises route samples of normal, aggressive and drowsy driving behaviour, under various conditions (motorway or secondary road, six different drivers etc.). The UAH dataset facilitates research in driving behaviour, although mobile phone sensors are less accurate than standard telematics.

In [6], speed and acceleration measurements where discretized into a finite risk space. A rule-based approach was employed to assess person specific risk. In [7], standard speed and acceleration measurements where used and models where defined to reflect personalized characteristics and road information. Experimental comparisons between various classifiers where presented to support random forests with Bayesian optimization. In, [8], various measurements where used, including speed and acceleration. The authors reduced data size by defining numerical domains and employed either probabilistic automata or labelled directed graphs, in order to define a formal model of driving behaviour. Overall, the above referenced works where based on driving patterns in large time frames. Recently, some works employed recurrent neural networks (RNNs) [9] without defining hand-engineered features. In [10], RNNs where applied on the UAH dataset, whereas in [11], two RNN variants: long short-term memory networks (LSTMs) [12] and gate recurrent units (GRUs) [13], where applied on a dataset comprising signals acquired by means of telematics sensors. These works where based on the assumption that driving behaviour manifests itself in limited time frames.

We should note that the majority of the aforementioned works is based on the extraction of handcrafted features, calculated from raw measurements. However, as it is well-known, handcrafted features are "manually" engineered, while their performance may vary, depending on the task at hand. Also, in many cases they have a significant computational cost, making them impractical on their application on big amounts of data. On the other hand, as we shall see in Sect. 3, the deep learning approach does not require such a feature extraction step. Instead, features are "learned" by the network. This way, they are tailored to fit the needs of each specific task. In this work, we aim to take advantage of the latter approach, More specifically, we propose a simple, yet effective methodology for transforming raw measurements that have been collected by the CAN-bus of vehicles into images, in such a way that temporal and qualitative properties of routes are preserved. Since CAN-bus sensor data reflect the driving patterns associated with different types of vehicles, our problem ends up to the classification of vehicle type. Thus, the aforementioned images are then fed to deep convolutional neural networks (CNNs), which classify them based on the type of the vehicle. We also compare our approach to other neural network-based methodologies and demonstrate its superior results over them.

2 Convolutional Neural Networks

Since the previous decade, the broad research area of machine learning has been vastly dominated by deep learning approaches, which have been able to achieve breakthrough results in several domains of interest. Through those years many different approaches and deep architectures have been proposed. Within the area of computer vision, the most dominant one is undoubtedly the CNN [14]. CNNs may be regarded as a significant alteration of traditional neural networks (NNs). A traditional NN, consists of layers of connected, non-linear neurons; the output of each one is computed by imposing a non-linear function on the sum of its inputs. Connections of neurons typically have a weight. The goal is to learn the set of weights of the entire network, upon a learning process. The input of a NN is usually a vector, which may consist of e.g., raw measurements or from the extraction of a set of derived features. Note that a change of input typically affects the whole set of weights.

On the other hand, in case of CNNs, the input consists of a raw image, i.e., of its pixel values. Moreover, the goal is to learn a set of convolutional filters that operate to the input, upon optimization of the classification error at the output of the network. As with every other NN, CNNs may consist of an arbitrary number of layers. However, neurons in each layer, are affected only by a sub-region of the previous one and not by the entire image. In a sense, the aforementioned filters are used to "learn" how to extract features in multiple levels of abstraction. CNNs are considered as a very powerful model, able to provide better generalization, while keeping the number of free parameters relatively low, when compared to traditional approaches. Also, they do not require any kind of prior knowledge in the domain of applications, since they omit the feature extraction step. The process of training a CNN strongly resembles the one of a typical NN; it consists of a forward propagation of data, followed by a backward propagation of the error, so as to tune the set of weights. Common approaches used in the context of the learning process are gradient-based variations. The key component of a CNN is the set of the *convolutional* layers, yet layers may also fall into three other categories, namely pooling, normalizing and fully-connected layers. More specifically:

Convolutional Layers: Within each convolutional layer, neurons are grouped so as to form a rectangular/squared grid, which performs a convolution within a given part of the image. This type of layers is ideal for image processing, where input data are rectangular. However, 1-D convolutional matrices may also be an alternative option in certain cases. Thus, these layers convolve with the part of the image that consists their input, passing the result to the next layer. Note that this process is biologically inspired; it has similarities with the response of a neuron of the visual cortex to a stimulus.

Pooling Layers: These layers are usually placed upon a convolutional layer or a set of convolutional layers. They operate by subsampling rectangular blocks from the previous convolutional layer and producing a single output from each one, e.g., the max or average value; this way the outputs of a given group of neurons in the previous layer is combined to a single neuron in the next one. This way, they enable the extraction of complex features from various resolutions. Moreover, it has been proven that they consist a strong mechanism towards the prevention of over-fitting which is the main cause for lack of generalization in unseen data.

Normalization Layers: The role of these intermediate layers is to normalize data, resulting to (minor) improvements in classification accuracy. More specifically, they perform normalization over local input regions; this provides an increased discrimination ability over their neighboring regions.

Fully-Connected Layers (Dense): These layers are the top-level layers of every CNN architecture, performing the high-level reasoning of the entire network, i.e., they combine features detected from image regions. They are responsible for the transformation of the output of their previous layer ultimately into a N-dimensional vector, where N denotes the number of classes. However, typically more than one consecutive fully-connected layers are needed. One may argue that the fully-connected layers are the actual network, while the previous ones (convolutional, pooling, normalization) are feature extractors. Note that these layers are not spatially located anymore, while the last one is tied to a loss function, used to provide an estimation of the classification error, which is then used for weight update during the aforementioned back propagation process.

The rest of this paper is organized as follows: Sect. 3 presents the CNNs, which are used in this work and also the proposed methodology for using CAN-bus measurements with 2D CNNs. The dataset used and the experiments performed are presented in Sect. 4. Finally, conclusions are drawn in Sect. 5, where plans for future work are also presented.

3 Proposed Methodology

In this section we present the main contribution of our work. More specifically, in Subsect. 3.1 we describe the procedure we follow for creating the input of our network. Then, in Subsect. 3.2 we present the proposed CNN architecture.

3.1 Network Input

Our work has been partially inspired by Jiang and Yin [15], who concatenated raw signals captured from sensors of smart mobile phones and created 2D images, in order to recognize human behaviour. Similarly, our work is based on raw measurements that have been obtained by the internal CAN-bus of vehicles. Each column of a given image corresponds to a temporal sequence of measurements of a given sensor. Unlike the aforementioned work, we omit the Fast Fourier Transformation of the image and instead we map corresponding signal values to colors, using an RGB color map. As it will be described in Sect. 4, the input of the network is a color image, which captures both temporal and qualitative properties of a small part of a route. In Fig. 1, we present typical examples of the created images for all 4 types of cars.

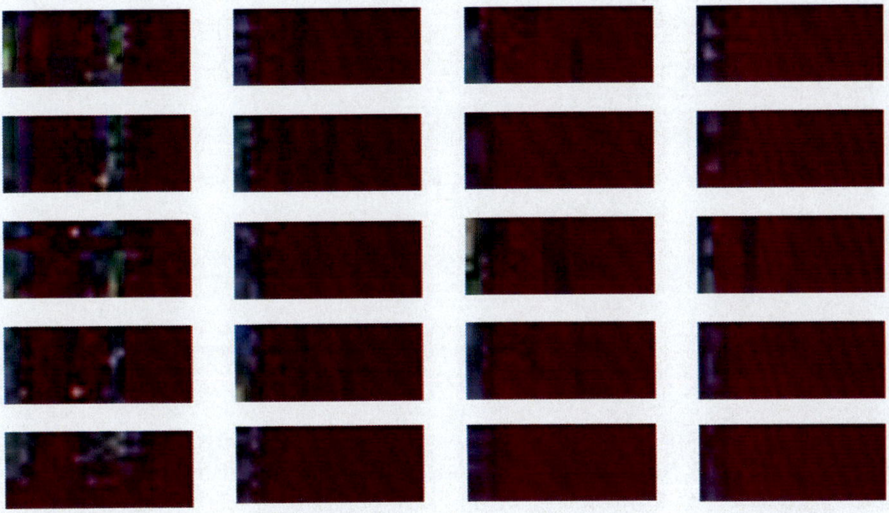

Fig. 1. Representative examples of training images for the 4 classes, created by the proposed methodology.

3.2 Network Architecture

We should herein note that one of the major drawbacks of almost all deep learning approaches is that they are prone to overfitting in case of limited available training data. This is the major cause for lack of generalization. It is well-known that in most real-life scenarios this is the case, since data collection processes are costly and tedious. Good practices have shown that in order to avoid overfitting, tens of thousands of training samples are usually adequate, however this is closely connected to the problem at hand. In an effort to surpass the aforementioned limitation, and allow the network to learn discriminating features, we adopt the "dropout technique" proposed by Srivastava et al. [16]. The main idea of dropout is to randomly choose neurons during the training phase, which are ignored (i.e., not updated) and reinserted into the network at a next stage.

The deep architecture that we have used throughout our experiments is presented in detail in Fig. 2. It is based on our previous work in human action recognition using sensor measurements from depth cameras [17]. In order to allow the extraction of features at several scales, we scaled the image by a factor of 10. This way, the first convolutional layer filters the 100×270 resized input image with 8 kernels of size 3×3. The first pooling layer uses "max-pooling" to perform 2×2 sub-sampling. Then, the second convolutional layer filters the 49×134 resulting image with 16 kernels of size 3×3. A third convolutional layer filters the 47×132 resulting image with 16 kernels of size 3×3. A second pooling layer also uses "max-pooling" to perform 2×2 sub-sampling which is then used as input to a fourth convolutional layer, which filters the 21×64 resulting image

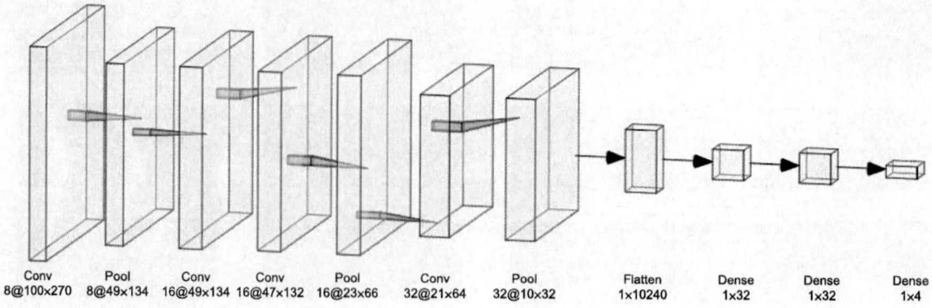

Fig. 2. The CNN architecture used for classification of the activity images.

with 32 kernels of size 3×3, using the dropout technique. A third pooling layer uses "max-pooling" to perform 2×2 sub-sampling, followed by a flatten layer that transforms the output image of the last pooling to a vector, which is then used as input to a dense layer using dropout. Then we use two dense layers with 32 units each. Finally, a fourth dense layer produces the output of the network.

Convolutional and dense layers used the "relu" activation function, except from the last dense layer that used the "sigmoid" activation function. Moreover, we used the adam optimizer and the "categorical cross-entropy" as the loss function.

4 Experiments

The experimental evaluation is performed on the real-life dataset that we have created for the needs of our previous work [11]. It has been collected from standard OBD ports of four types of vehicles. More specifically, we have used a private car, a waste collection vehicle (garbage truck), a normal truck and a sweeper vehicle (vacuum). Note that measurements have been collected in real-time and under real conditions; i.e., vehicles were driven while performing their everyday tasks.

Each measurement (feature) vector is composed by measuring 27 attributes, i.e., maximum positive acceleration, maximum negative acceleration, latitude, longitude, speed, maximum transverse acceleration and a 21-bin acceleration histogram with values ranging from -0.5 g to $+0.5$ g, for 4 different car types, namely vacuum, car, truck and garbage truck. Min-max normalization is applied to derive feature values ranging from 0 to 1. In this work, each sample is a $10 \times 27 \times 3$ pseudocolored image, that has been created upon concatenation of 10 consecutive feature vectors and mapping corresponding values to colors using an RGB color map, ending up with a dataset comprising of 14288 samples. Therefore, we may argue that the created images capture both temporal and qualitative features of parts of routes.

The experiments were performed on a personal workstation with an Intel™i7 5820K 12 core processor on 3.30 GHz and 16 GB RAM, using NVIDIA™Geforce

GTX 2060 GPU with 8 GB RAM and Ubuntu 18.04 (64 bit). The deep CNN architecture has been implemented in Python, using Keras 2.2.4 [18] with the Tensorflow 1.12 [19] backend.

We set the batch size equal to 50 and ran the aforementioned CNN model for 170 epochs, and 171 steps per epoch. We used 60% of the data for training, 30% for validation and the remaining 10% for testing. We repeated the experiment 10 times and managed to achieve approx. 91% accuracy. The confusion matrix indicating the performance of our model per class is depicted in Table 1. Moreover, precision, recall and F1-scores for each class are depicted in Table 2. Finally, in Table 3 we compare the results of this work, to the findings of [11]. More specifically, we compare the 2D CNN presented herein to a) a long short-term memory (LSTM) network [12]; b) a gated recurrent unit (GRU) [13]; and c) an 1D CNN. As it may be easily observed, the proposed methodology clearly outperforms the remaining three in terms of accuracy.

Table 1. Normalized confusion matrix for the 4 classes.

	Garbage	Truck	Car	Vacuum
Garbage	0.77	0.23	0	0
Truck	0.06	0.94	0	0
Car	0	0.10	0.90	0
Vacuum	0	0	0	1

Table 2. Precision, recall and F1-score for each class.

	Precision	Recall	F1-Score
Garbage	0.95	0.77	0.85
Truck	0.75	0.94	0.83
Car	1	0.90	0.95
Vacuum	1	1	1

Table 3. Comparisons of the accuracy achieved to those of [11]. Best result indicated by bold.

	2D CNN	LSTM[11]	GRU[11]	1D CNN[11]
Accuracy	**0.91**	0.78	0.84	0.78

5 Conclusions and Future Work

The increasing need for daily mobility in developing countries causes a higher need for safer vehicle traffic. In addition to the development of infrastructures, the contribution of driving behaviour plays an important role to the field of human behaviour in safety traffic. Understanding the relationship between driving behavior and driving safety can be based on the classification of the driving in predetermined driving modes or profiles. In this work we presented an approach for the classification of driving behaviour using CNNs, based real data from the 4 different types of vehicles. The data have been acquired from the CAN-bus protocol of each vehicle and demonstrated superior results over other deep approaches.

We need to go one step further to combine driving style information in different situations and driving phases to build a unique “fingerprint driver behaviour” among different types of driving style. Additionally, AI techniques could introduce “intelligent” mobility criteria for safer and smarter mobility in modern cities with sustainable mobility management.

Acknowledgment. This research has been co-financed by the European Regional Development Fund of the European Union and Greek national funds through the Operational Program Competitiveness, Entrepreneurship and Innovation, under the call RESEARCH-CREATE-INNOVATE (project code: T1EDK-03459).

References

1. Meiring, G.A.M., Myburgh, H.: CA review of intelligent driving style analysis systems and related artificial intelligence algorithms. Sensors **15**(12), 30653–30682 (2015)
2. Van Ly, M., Martin, S., Trivedi, M.M.: Driver classification and driving style recognition using inertial sensors. In: IEEE Intelligent Vehicles Symposium (IV), pp. 1040–1045. IEEE (2013)
3. Vaitkus, V., Lengvenis, P., Žylius, G.: Driving style classification using long-term accelerometer information. In: 19th International Conference on Methods and Models in Automation and Robotics (MMAR), pp. 641–644. IEEE, September 2014
4. Bergasa, L.M., Almeria, D., Almazán, J., Yebes, J.J., Arroyo, R.: DriveSafe: an app for alerting inattentive drivers and scoring driving behaviors. In: IEEE Intelligent Vehicles Symposium Proceedings, pp. 240–245. IEEE (2014)
5. Romera, E., Bergasa, L.M., Arroyo, R.: Need data for driver behaviour analysis? Presenting the public UAH-DriveSet. In: IEEE 19th International Conference on Intelligent Transportation Systems (ITSC), pp. 387–392. IEEE (2016)
6. Joubert, J.W., De Beer, D., De Koker, N.: Combining accelerometer data and contextual variables to evaluate the risk of driver behaviour. Transp. Res. Part F Traffic Psychol. Behavi. **41**, 80–96 (2016)
7. Yi, D., Su, J., Liu, C., Quddus, M., Chen, W.H.: A machine learning based personalized system for driving state recognition. Transp. Res. Part C Emerging Technol. **105**, 241–261 (2019)

8. Bouhoute, A., Oucheikh, R., Boubouh, K., Berrada, I.: Advanced driving behavior analytics for an improved safety assessment and driver fingerprinting. IEEE Trans. Intell. Transp. Syst. **20**(6), 2171–2184 (2018)
9. Giles, C.L., Kuhn, G.M., Williams, R.J.: Dynamic recurrent neural networks: theory and applications. IEEE Trans. Neural Networks **5**(2), 153–156 (1994)
10. Saleh, K., Hossny, M., Nahavandi, S.: Driving behavior classification based on sensor data fusion using LSTM recurrent neural networks. In: IEEE 20th International Conference on Intelligent Transportation Systems (ITSC), pp. 1–6. IEEE (2017)
11. Mantzekis, D., Savelonas, M., Karkanis, S., Spyrou, E.: RNNs for classification of driving behaviour. In: 10th International Conference on Information, Intelligence, Systems and Applications (IISA), pp. 1–2. IEEE (2017)
12. Hochreiter, S., Schmidhuber, J.: Long short-term memory. Neural Comput. **9**(8), 1735–1780 (1997)
13. Chung, J., Gulcehre, C., Cho, K., Bengio, Y.: Empirical evaluation of gated recurrent neural networks on sequence modeling. arXiv preprint arXiv:1412.3555 (2014)
14. LeCun, Y., Bottou, L., Bengio, Y., Haffner, P.: Gradient-based learning applied to document recognition. Proc. IEEE **86**(11), 2278–2324 (1998)
15. Jiang, W., Yin, Z.: Human activity recognition using wearable sensors by deep convolutional neural networks. In: Proceedings of the 23rd ACM International Conference on Multimedia, pp. 1307–1310. ACM (1998)
16. Srivastava, N., Hinton, G., Krizhevsky, A., Sutskever, I., Salakhutdinov, R.: DropOut: a simple way to prevent neural networks from overfitting. J. Mach. Learn. Res. **15**(1), 1929–1958 (2014)
17. Papadakis, A., Mathe, E., Vernikos, I., Maniatis, A., Spyrou, E., Mylonas, P.: Recognizing human actions using 3D skeletal information and CNNs. In: International Conference on Engineering Applications of Neural Networks, pp. 511–521. Springer, Cham (2019)
18. Chollet, F.: Keras-team/keras. GitHub. https://github.com/fchollet/keras. Accessed 16 Par 2020
19. Abadi, M., et al.: TensorFlow: a system for large-scale machine learning. In: Proceedings of 12th USENIX Symposium on Operating Systems Design and Implementation (OSDI 16) (2016)

Introducing Automated Shuttles in the Public Transport of European Cities: The Case of the AVENUE Project

Eliane Horschutz Nemoto[1,2(✉)], Ines Jaroudi[1,2], and Guy Fournier[1,2]

[1] Institute for Industrial Ecology, Pforzheim University of Applied Sciences, Tiefenbronner Str. 65, 75175 Pforzheim, Germany
eliane.horschutz-nemoto@hs-pforzheim.de
[2] Laboratory of Industrial Engineering, Sustainable Economy Group, École CentraleSupélec Université Paris-Saclay, 91190 Gif-sur-Yvette, France

Abstract. Our current mobility paradigm has reached a tipping point. Individual mobility, based on cheap fossil fuel and high CO_2 emissions no longer meet the needs posed by a globally increasing demand for passenger mobility, neither corresponds to the climate agenda.

In this regard, innovations and technologies play an important role to shape the future mobility and provide solutions for more efficient, affordable, accessible, and sustainable mobility in cities.

This paper aims to explore how innovations on mobility, such as shared automated electric vehicles (SAEV) can contribute to a positive change in the mobility paradigm and sustainable mobility, and to this end, which are the current obstacles to be overcome and the key factors related to SAEV's deployment. Thereby, it presents the case of the Autonomous Vehicles to Evolve to a New Urban Experience - 'AVENUE', a European project that has implemented pilot trials to test automated shuttles within the public transport of Lyon, Geneva, Luxembourg, and Copenhagen.

Based on primary data from the project and secondary data from AVENUE public reports, the study reports on the project implementation in the four cities and first learnings through obstacles and key factors to accelerate the deployment of automated shuttles in cities. It contributes to the discussion on technical & operational, social, and legal obstacles as well as key elements in the deployment of automated shuttles.

Keywords: Automated vehicles · Shared mobility · Public transport · Sustainable mobility

1 Introduction

The nature of mobility is changing. These changes are triggered by the rise of new technologies, new sharing economy models, and the consumer's preference for convenient and flexible mobility without relying on individual cars (Attias 2017). Such elements call in question the current car-based mobility model.

E. G. Nathanail et al. (Eds.): CSUM2020, AISC 1278, pp. 272–285, 2021.
https://doi.org/10.1007/978-3-030-61075-3_27

These transformations have to be aligned and could reinforce the so-called Sustainable Urban Mobility Plans (SUMPs). SUMPs are a new approach to urban mobility planning, that focuses on "sustainable and integrative planning processes", taking into account the adoption of new modes of transport based on micro-mobility, Mobility as a Service (MaaS) and shared transport (Eltis 2020).

This paper sheds light on the potential changes and transitions on the mobility paradigm triggered by the integration of shared automated electric vehicles (SAEV's) in the public transport of cities, coupled with the discussion on the current obstacles and key factors related to SAEV's deployment (Banister 2011; Merriman 2020; Sheller and Urry 2016). In this regard, it seeks to answer the following research question: what are the main limitations and key factors for the deployment of SAEV towards sustainable mobility transitions in cities?

To answer this question, the paper relies on the case of the AVENUE, a project operating pilot tests in four European cities - Lyon, Geneva, Luxembourg, and Copenhagen. Key learnings of the first 18 months of the project are now reported on. These reports address obstacles and key factors for the deployment of automated shuttles in European cities and are the basis of this article.

The study combines theoretical research with practical experience from the pilot projects. The methodology applied is qualitative, with an exploratory-descriptive nature (Gil 2002). Primary data from the project as well as secondary data from public reports of the project were analysed, summarized, presented, and discussed.

Main results and contributions of this study address the technical & operational, social, and legal obstacles, and respective key factors in the deployment of automated shuttles in cities, as well as, the relation between theoretical reflection and the field experience.

The paper is structured as follows. Section 2 presents the theoretical background, addressing concepts on the mobility paradigm, and transitions to a new and sustainable mobility, as well as concepts on shared, electric and automated mobility, substantiate how they can contribute to future sustainable mobility. Section 3 describes the applied methodology. Section 4 presents the case of the AVENUE project and the test sites. Section 5 presents then the main findings concerning obstacles and key factors to deploy automated shuttles in cities. Section 6 contains the discussions and reflections connecting the theoretical background and results, and Sect. 7, the concluding remarks.

2 Theoretical Background

2.1 The Mobility Paradigm – Concepts and the Transition to Sustainable Mobility

The discussion addressing the mobility paradigm has raised attention in the last decades, due to the increasing externalities in the transport sector. Indeed, externalities of the transportation sector vary from accidents, congestion, pollution, etc. These costs borne by societies represent a catalyst for a modal shift towards more sustainable and public means of transit (Storchmann 2003). In the EU for instance, the transport sector is responsible for 27% of the total greenhouse gas emissions (GHG) in 2017 (European

Environment Agency 2019). Globally, the transport sector represents 14% of the total anthropogenic GHG emissions (IPCC 2014).

The current mobility paradigm is based on individual mobility, characterized by car ownership, individual rides, cheap fossil fuel, and consequently high CO2 emissions (Fournier 2017). This model has influenced how cities and streets are planned, infrastructure is built, and policies are designed.

Merriman (2020) describes a 'new mobility paradigm', which, first, addresses "the significance of mobility and mobilities in the modern world; and second, but less commonly, the emergence of new forms of mobile practice and technology which are reconfiguring social and economic life".

Sheller and Urry (2016) state that the new mobility paradigm presents intersections among public policy, planning, and applied research, and it plays a role in recognizing the complexity embedded in mobility systems.

Further, Sheller and Urry (2016) refer to a shift towards a "new mobility paradigm" that is occurring nowadays lead by sharing, connectivity, and accessibility. These are socio-technical transitions aiming to decrease the use of automobiles and preparing for a post-auto mobility transition.

Regarding this transition, Banister (2008, 2011) sustains that sustainable mobility provides an alternative paradigm to investigate the complexity of cities. He proposes four types of actions towards a 'sustainable mobility paradigm', being: i) Reducing the need to travel - substitution; ii) Transport policy measures - modal shift; iii) Land-use policy measures - distance reduction; iv) Technological innovation - efficiency increase. The author also reinforces the importance of a transition process that involves the support of all stakeholders.

In this paper, the focus is given to the technological, social, and legal aspects of SAEV's, in order to contribute to a transition to a new mobility paradigm.

2.2 Shared Automated Electric Vehicles (SAEV) - Technologies for the Transition to a New Mobility Paradigm

Shared automated electric vehicles represent a disruptive potential on mobility, by combining shared mobility, electrification, and automation (Sprei 2017). Such a technology combination has the potential to tackle mobility externalities and contribute to a transition for a sustainable mobility paradigm.

Automated vehicles are vehicles equipped with automated driving systems (ADS) to support or replace human driving, with six levels of driving automation, from Level 0 - no driving automation to Level 5 – full driving automation (SAE 2018). In this article when referring to 'automated vehicles', it refers to Level 3, Level 4, and Level 5 of driving automation as it was defined by the On-Road Automated Driving committee, 2018.

Automated driving has been seen beneficial to improve mobility safety and avoid accidents, improve the traffic flow and reduce traffic jams, and improve accessibility for instance for elderly people and people with reduced mobility (Litman 2019). Its benefits can be broadened when combined with shared and electric mobility.

Shared mobility services, such as car-sharing, ride-sharing, and on-demand ride services, have had a transformative impact on mobility in cities. It enables flexibility, convenience, mobility service customization, and cost savings on mobility (Narayanan et al. 2020). Besides its contribution to reducing private vehicle usage, and the modal shift from vehicle ownership to shared mobility services (Shaheen and Chan 2015).

Hand (2017) explores the potential of automated vehicles and shared mobility to reshape our cities. In the sense that such combination can contribute to eliminate parking spaces in cities, prioritize space for walking, biking and shared mobility, and alleviate pressures on the built environment.

The study from Jones and Leibowicz (2019) investigated the benefits from the diffusion of shared automated vehicles associated with electrification, proving to be beneficial to reduce mobility greenhouse emissions and to boost renewable electricity generation.

The study from Antonialli (2019) focuses more specifically on shared automated electric shuttles that provide services on public transportation. The study points to 92 experimentations worldwide providing first and last mile itineraries and micro transit commute, with shuttles running in closed or controlled areas or mixed traffic routes. Such experimentations present legal and technological constraints, nonetheless they represent a step forward to understand shared mobility, electrification, and automation can contribute to sustainable mobility in cities.

Further, as pointed out by Sprei (2017), besides new technologies, suitable regulations and policies are crucial for the transition to sustainable mobility, as well as to avoid potential rebound effects.

Shared automated vehicles belong to an array of new technologies that address mobility gaps and provide Mobility-as-a-Service solutions. Thus, SAEV's present an opportunity to reduce automobile use and to reduce pollution. In order to further advance this technology, it is crucial to acknowledge the difficulties encountered in the field tests and provide targeted solutions.

3 Methodology

This study combines theoretical research with the practical experience from pilot projects deploying automated shuttles in the public transport of European cities. Initially, a literature review provided concepts on the mobility paradigm, and the transitions to a new mobility paradigm and sustainability, as well as concepts on shared automated electric vehicles, its deployment and potential contributions to sustainable mobility in cities.

In order to answer the research question - what are the main limitations and key factors for the deployment of SAEV towards sustainable mobility transition in cities? – the paper relies on the AVENUE project. Primary data were collected based on the learnings of the first 18 months of the project. These learnings address obstacles and key factors to accelerate the implementation of automated shuttles, focusing on technical & operational, social, and legal issues. In addition, secondary data from public reports of the project were analysed, summarized, presented, and discussed.

Studies on environmental and economic impacts, as well and surveys to assess the social impacts of automated shuttles integration on public transport are in the scope of AVENUE project and ongoing. However, due to their early stage, their results are not available yet and not in the scope of this paper.

4 The Case of the AVENUE Project

The focal point of this paper relies on the case of AVENUE - Autonomous Vehicles to Evolve to a New Urban Experience, a Horizon 2020 EU project. It aims to promote public transport services in urban and suburban areas through the deployment of new technologies. The project deploys a fleet of automated electric shuttles in four European cities: Lyon, Copenhagen, Luxembourg, and Geneva (see Table 1). The full-scale demonstrations will run for 4 years, from 2018 to 2022.

AVENUE focuses on door-to-door, MaaS solutions, and personalized mobility services, targeting a shift from individually-owned vehicles to sustainable transportation. Such a project presents a plethora of opportunities to test and enhance the services of AV's within the urban public transport network.

Table 1 summarizes information about the different sites and environments where the automated shuttles have been deployed and operational details.

Table 1. The AVENUE pilots test

City	Pilot	Route - site	Type of passenger
Geneva	Meyrin area	Fixed circular line (route 2,1 km) Open road, mixed traffic	Resident of the area
	Belle-Idée	Hospital/Private area Speed limit 30 km/h Mixed traffic, to be on demand	Visitors of the hospital, patients
Lyon	Groupama Stadium	Public, open road (route 1,3 km) connects the tramway line with Groupama stadium	Different passengers, people with reduced mobility
Copenhagen	Nordhaven	Industrial port	Not applicable
Luxembourg	Contern	Industrial area (route 2,2 km) Speed limit 50 km/h Open road, mixed traffic	Employees working at Campus Contern
	Pfaffenthal	Urban area (route 1,2 km) Speed limit 30 km/h Open road, mixed traffic	Workers, tourists, residents, and visitors of Luxembourg city

The automated shuttles have a capacity of 15 passengers, run in urban and suburban areas, and most of them operate in open roads with mixed traffic. The ride is free of charge, and the speed varies between 11–18 km/h. Currently, all automated shuttles follow a specific itinerary with fixed routes, although, one of the biggest ambitions of the project is to offer mobility on-demand on a specific area, as the example of Belle Idée in Geneva.

The automated shuttle service provides first and last mile connections on journeys not covered by public transport, or complementing the service intervals offered by public transport, as occurs in Groupama Stadium, in Lyon. The pilots are diverse as they run in different settings. For instance, Contern and Nordhaven are industrial areas whereas Belle idée is a hospital area. The user public varies between employees, tourists, and local residents.

5 Findings and Results

This section presents the learnings of the AVENUE project by addressing the obstacles and pointing key factors to accelerate the implementation and deployment of automated vehicles in the public transport of European cities.

The findings focus on four domains: technical & operational, social, and legal. It is important to notice that the findings on technical & operational obstacles present the Public Transport Operators' (PTO's) experience. These reports describe the operational management and the difficulties perceived, therefore, these obstacles are context-based-cases.

5.1 Technical and Operational Obstacles and Key Factors for Deployment

The main obstacles concerning the technical capabilities of the shuttle and its operational performance to provide first and last-mile journeys within the AVENUE test trials are presented in Table 2. They inhibit a full level 4 automation. To better understand the obstacles, they are classified based on the classic '*Sense – Plan – Act*' design: a robotics and automation method. *To sense* means the shuttle is capable of determining its location, perceiving relevant static and dynamic objects, and predicting the future behavior of relevant objects in its environment. *To plan* means to ensure a collision-free and accurate driving that respects regulations. *To act* means to correctly execute the driving plan and to communicate with other road users (Wood et al. 2019; Haag 2018). Moreover, the infrastructure and consequent operational restrictions faced by the Public Transport Operators are presented as well.

Subsequently, the key factors to overcome such obstacles are addressed in the text. The following findings and information relied on AVENUE public reports.

Table 2. Technical and resulting operational obstacles

Technical and resulting operational obstacles
Sense • Sensors range limited to certain distances • Sensors cannot detect obstacles less than 3.5 cm in height • Lack of optimized detection systems to distinguish between static and dynamic obstacles: e.g. people versus animals, snowflakes versus big raindrops, etc. • Sensor limited with bad weather conditions • Spatial Resolution of LIDAR Sensors limits the detection of far or moving objects Plan • Lack of capacity to modify the route to avoid and overpass obstacles on the road • The shuttle is pre-synchronized with the traffic regulations, but it cannot on its own comply to traffic rules • Incapacity to overcome road obstacles without the interference of an operator • Lack of fleet orchestration capability Act • Unnatural driving behavior, e.g. hard braking • Interoperability: e.g. lack of communication system with other road users and other automated systems • Lack of capacity to manoeuvre in bad weather conditions • Operation is limited to pre-mapped and predefined routes and pre-defined stops • Automated shuttle capacity to operate in different road elevations
Vehicle and Infrastructure related • Charging and de-charging problems in cold weather (0 degrees Celsius or lower) • Low battery capacity to sustain faster driving, due to the shuttle being equipped with more sensors and hardware for automated driving technology • Slow battery charging • Infrastructure limitations: reference points, - no building in rural areas, (buildings are used as a reference point for positioning); roadway grades - problems to surpass passages where the grade of the street goes above 12° • Digital infrastructure limitations: Shuttle may stop due to signal loss • Charging infrastructure: Lack of green electricity source to fuel the electric shuttles; Lack of management capabilities of recharging infrastructure
Resulting operational obstacles for the Public Transport Operators (PTO's) • Mapping and routes: operation limited to predefined routes with involvement from the shuttle provider; Low speeds affect the other road users; Weather limits the operation of the shuttles; • Maintenance: the provider is responsible for maintaining the shuttles which lead to prolonged delays to operate the shuttles back • Transport planning: automated shuttles are not integrated into the public transport ecosystem; Lack of means/algorithms to count passengers to estimate occupancy and capacity of passengers inside the shuttle

Source: Elaborated by the authors, based on Bürkle 2019; Zinckernagel et al., 2018; Zinckernagel et al. 2019; Beukers 2019; Guldmann et al., 2019; Reisch 2019; Zuttre 2019.

Currently, such obstacles hinder the full operation and full performance of the automated shuttles. Key factors to overcome the automation-related obstacles described in the 'Sense-Plan-Act' are related to improvements on Hardware and Software components, in more details:

i) Improvements of Hardware - such as cameras, radars, sensors resolution, LiDAR, and GPS, for a better capacity to read the environment
ii) Improvements of Software - developing algorithms for perception and object analysis, in order to detect and classify objects and obstacles in the environment; mechanisms for prediction of movements and actions by other motorized and non-motorized moving objects; greater data power processing.

Such improvements would boost the capacity of the automated shuttle to read and interact with the environment, drive faster and more fluidly, distinguish obstacles on the road and bypass it without the manual control, and operate on bad weather conditions. However, the research and development of these new technologies take time and are costly. For instance, the hardware components are limited by a few suppliers, and the technology required is not available on the market or is too expensive (Bürkle 2019). And a more powerful powertrain would enable the shuttle to operate in hilly conditions and higher road elevations.

Regarding the 'Vehicle and Infrastructure related' obstacles, improvements on the battery capacity and endurance will be necessary in the future for longer performance and to provide mobility on demand.

In the future, it will be key factors 5G Networks for data transmission and the interoperability of the automated shuttles with other automated systems, as well as with infrastructure and non-automated systems (Zinckernagel et al. 2019).

The support and investment from the cities that host the pilot projects are also key points for coordination and investments on infrastructure to provide and build together connected electric and green mobility.

Most of the operational obstacles faced by the PTO's in the field would be solved by tackling the previous obstacles addressed. In addition, deploying a fleet orchestration platform will be also a crucial tool, when considering the services management, coordination, and maintenance of automated shuttle fleets. For mobility on demand, developing algorithms to estimate the shuttles occupancy and capacity left out is an important point.

According to Bestmile, a partner in the project, fleet orchestration here refers to a broader meaning than fleet management. In the sense that fleet orchestration is not limited just to telematics and preventative maintenance. It is also about coordinating a demand-responsive transit, aiming the operator efficiency, passenger convenience, and benefits for the city traffic (Mellano 2019).

5.2 Social Obstacles and Key Factors for Deployment

Social acceptance towards new technologies, human behavior, and interaction with these technologies are also major points for automated driving deployment and diffusion. These social obstacles are here classified based on accessibility and People with Reduced Mobility (PRM) accessibility, safety, security, and users' perceptions and acceptance.

Thus, it is discussed here below the social barriers and obstacles facing the deployment of the automated shuttles. The information here presented is based on AVENUE public reports (see summary in Table 3).

Table 3. Social obstacles

Social obstacles
Accessibility and PRM accessibility • Barriers related to the use of technology can exclude certain population groups, e.g.: elderly people or people not familiar with smartphones and applications • Accessibility and usability for disabled people, e.g. adjust on the vehicle are needed for people on wheelchair
Safety • Passengers want to talk to a driver and rely on a driver to help them • Stories about accidents with automated vehicles • Worries that other road users will not be able to anticipate the behavior of automated vehicles • Road users are not used to automated shuttles on the road • The automated shuttle is overtaken by other road users (cars, buses, trucks)
Security • Risk of cyber-attacks: hackers could take control of AV's;
Users' Perceptions and Acceptance • The traffic situation is very complex to be handled by technology • Doubts that the technology is mature enough to be trusted • Recent news about accidents with automated vehicles • Automated vehicles will lead to more delays and failures and more traffic jams for other road users • Passengers do not like the idea that there could be no operator aboard the automated shuttle, (e.g.: to perform first aid, to have an authority figure present, risk of vandalism, robberies or assaults, no information or support of a supervisor during the trip if required, no support to reach a connection, no support to get on and off the shuttle)

Source: Elaborated by the authors, based on Dubielzig et al. 2018; Mathé et al. 2019; Zinckernagel et al. 2019.

To tackle part of the social obstacles mentioned, a key factor is the presence of an operator on board the shuttle. For the shuttles' deployment, it is required by law, but also according to the AVENUE findings, the operator plays a role in acceptance. It is a major point for people to feel safe, to explain the technology, to provide information, to have a supervising element, and to help elderly people and people with reduced mobility to use the vehicle (Mathé et al. 2019).

Trust-building measures and the creation of awareness about automated driving technology are necessary (Dubielzig et al. 2018). In this direction, it is recommended to make use of videos and flyers to explain how the technology works and to report real data from the pilot project for the general public. For example about the number of passengers that have travelled on the shuttles and/or number kilometers driven to demonstrate the progress of the technology.

Also, it is important to create awareness of road users about the pilot test and automated vehicles, in order to respect the speed of the local roads and the shuttles' speed. In addition, to implement adjustments inside the vehicle to meet the needs of the different people with reduced mobility is a major point to improve the accessibility of the shuttles.

5.3 Legal Obstacles and Key Factors for Deployment

To advance the deployment of automated shuttles, the legal framework has to adapt accordingly. The insights gathered in this section were collected from the AVENUE public reports (see Table 4):

Table 4. Legal obstacles

Legal obstacles
• Homologation (technical inspections and documentation)
• Legal guidelines: there are no dedicated laws designated for automated driving projects (liability, interoperability, pollution restrictions, data safety)
• Institutional development and innovation: Fragmented European collusion and laws concerning AV projects

Source: Elaborated by the authors, based on Beukers 2019; Guldmann et al. 2019; Reisch 2019; Zuttre 2019; Bürkle 2019; Attias et al. 2018.

The territorial jurisdictions vary between France, Switzerland, Luxembourg, and Denmark. The legal procedures depend on the political will to promote new innovative mobility solutions. For instance, Swiss and Luxembourgish authorities are supportive of future transport modes (Reisch 2019; Beukers 2019). Regarding the technical inspections and documentation- "the homologation phase" -, the pilot administrators needed to apply for different permissions, especially, if there are alterations needed for the test sites. In Copenhagen, the process requires a third party (an engineering firm) to weigh in on the feasibility and risks of usual transportation projects (Guldmann et al. 2019).

Another obstacle refer to the fact that these projects are pioneers. Therefore, there are no legal guidelines designated for them. So, the project has to adhere to other transportation and infrastructure requirements that are not suitable for the technological and experimental level. This could lead to discordant development. Due to the lack of legal precedents, there are no specific liability clauses about whom takes responsibility in case of an accident, e.g. the need for a driver's license is an open question (Bürkle 2019). For now, the operators onboard have to possess a valid driving license.

Interoperability defines a struggle also for the lack of precedent of connected systems. Another hurdle that presented itself is data safety and the European restrictions on the matter. As stated in the First report on regulatory requirements: "European action on the development of automated and connected vehicles is particularly important… in the field of data access and exchange, to ensure the interoperability of services and interfaces, ensuring security and privacy" (Couzineau et al. 2018).

Partners presented key factors to deploy the shuttles successfully in their project deliverables. Keolis emphasized the need for starting new projects of AV's in order to advance the technology and garner political support. They also pointed out that public transport operators have to provide technical transparency communication and admit the challenges facing AV's projects (Zuttre 2019).

Moreover, there is a need for uniform and specific guidelines for AV's pilots (Guldmann et al. 2018). This could be agreed upon through a joint EU-based AV law department that, centrally, can ensure state-of-the-art knowledge regarding type approvals. In this case, not only type approvals of the vehicles but also approvals of routes and software across EU borders - one legal framework to accommodate all approvals necessary to implement AV's. Most importantly, one department with the ability to constantly update the legal framework to ensure safety and agile development of AV.

6 Discussion

The results, although from an incipient phase of the project, show the current stage of SAEV's performance and services within the public transport of European cities. The technological and operational obstacles require continuous improvements, and for this, regular feedback from the field is valuable to guide the advances in technology and services. To a certain extent, the performance of the SAEV's may furthermore influence social acceptance and legal developments. For social acceptance, the pilot projects are a good start for users to get familiar and build trust towards autonomous driving, and evaluate the usability of SAEV's. The project contributes this way to the diffusion of innovation towards sustainable urban mobility. In addition, the pilot projects provide relevant legal recommendations envisioning to build a common background for standards, regulations, and one legal framework for AV's in Europe.

Except for some technological improvements, many of the obstacles and key factors pointed in this paper concern consequently a process of medium and long-term goals for SAEV's to be ready to be deployed on-demand and on a broader scale in cities. Furthermore, it may be a long-term process for AV's to bring real impacts to the current mobility paradigm as stated by Jones and Leibowicz (2019) and Shaheen and Chan (2015), by replacing journeys with individual cars, favoring shared and electric powertrain, and reducing mobility externalities.

However, as suggested by Banister (2008, 2011), the AVENUE project contributes to introducing innovating transport systems to better satisfy the transport needs of passengers to a transition towards sustainable mobility and a new mobility paradigm.

It is also important to note that these new trends add more complexity to mobility and that technology per se will not solve the problem. Therefore, future mobility requires more common efforts to strengthen the intersections among mobility planning, governance, policies, and regulations that contextualize the use of technology, as pointed by Sheller and Urry (2016). In addition, SAEV's may be deployed in consistency with other soft modes of transport, considering for instance that more walkable spaces and cycling lines are aspirated in cities, therefore, as stated by Hand (2017), AV's technologies could be an enabler to reshape cities and alleviate pressures on the built environment.

7 Conclusion

Based on the case of the AVENUE project, this paper presents how the pilot projects have performed within diverse contexts testing automated shuttles in four European cities. The AVENUE project reveals that automated shuttle services can provide solutions to mobility gaps on a local scale where there is no public transport (or still weak public transport service). Further, once deployed on a broader scale, SAEV's could also contribute to changes in the mobility paradigm by replacing journeys with individual cars with shared mobility, fostering electric powertrain, and tackling mobility externalities for sustainable mobility in cities.

The obstacles and key factors in the deployment of SAEV presented and discussed may entail a process of medium and long-term for technology improvements and tests, social acceptance and familiarization with the technology, and legal support. In addition, the transition to a new and sustainable mobility paradigm will require to take into perspective the complexity of mobility and all its intersections.

In this regard, projects like AVENUE play an important role in setting new mobility ecosystems and the ground for manufacturers, software developers, and transport operators to work together to diversify and innovate the mobility system to provide more convenient and sustainable public transport.

It raises awareness among citizens and road users on a city scale level about automated driving, and it provides relevant findings of the citizens' willingness and positioning towards this technology. This sort of project pushes the boundaries of existing legal frameworks. It requires collective action that lies down solid regulatory foundations that promote sustainable, accessible, and affordable mobility. By drafting adequate regulations, mobility innovations could thrive while meeting the community's and the city's needs.

To conclude, the pilot tests can be seen as a preparation for the transition for the mobilities of the future, aiming to contribute to a more sustainable mobility paradigm, based on shared, electric, and automated mobility.

Disclaimer. The research leading to these results has received funding from the European Union's Horizon 2020 research and innovation program under grant agreement No. 769033, for the AVENUE project.

References

Attias, D.: The automobile world in a state of change: from the automobile to the concept of auto-mobility. In: The Automobile Revolution. Springer International Publishing (2017). https://doi.org/10.1007/978-3-319-45838-0_2

Banister, D.: Cities, mobility and climate change. J. Transp. Geogr. **19**(6), 1538–1546 (2011). https://doi.org/10.1016/j.jtrangeo.2011.03.009

Beukers, J.: First Iteration Geneva Large Scale Pilot Use Case Demonstration Report: AVENUE project (2019)

Bürkle, T.: Autonomous Shuttles: Technical Obstacles for the Diffusion, Implementation and the Deployment (2019)

Denmark, D.: The outsiders: planning and transport disadvantage. J. Plann. Educ. Res. **17**(3), 231–245 (1998). https://doi.org/10.1177/0739456X9801700304

Dubielzig, M., Reisch, M., Panou, M.: First Passenger needs analysis and specifications: AVENUE project (2018)

Eltis: The SUMP Concept and Guidelines (2020). https://www.eltis.org/mobility-plans/sump-concept

European Environment Agency: Greenhouse gas emissions from transport in Europe (2019). https://www.eea.europa.eu/downloads/7009a89effc04dbea8b5f94ff0a39912/1576584662/transport-emissions-of-greenhouse-gases-12.pdf

Fournier, G.: The new mobility paradigm. transformation of value chain and value proposition through innovations. In: Attias, D. (eds.) The Automobile Revolution. Springer https://doi.org/10.1007/978-3-319-45838-0_3

Guldmann, S., Zinckernagel, C., Lytzen, P.: First iteration Copenhagen Large Scale Pilot Use Case Demonstration report: AVENUE project (2019)

Haag, A.: State of the Art in Autonomous Driving (2018). https://events.techcast.com/btd/2018/Sydney_1005/

Hand, A.: Redefining urban mobility: four ways shared autonomous vehicles will reshape our cities. Real Estate Issues, 49–52 (2017)

Harper, D.: Paradigm: origin and meaning of paradigm (2020). https://www.etymonline.com/word/paradigm

IPCC: Climate Change: Mitigation of Climate Change Working Group III Contribution to the Fifth Assessment Report of the Intergovernmental Panel on Climate Change. Cambridge University Press, New York (2014)

Jones, E.C., Leibowicz, B.D.: Contributions of shared autonomous vehicles to climate change mitigation. Transp. Res. Part D Transp. Environ. **72**, 279–298 (2019). https://doi.org/10.1016/j.trd.2019.05.005

Litman, T.: Autonomous Vehicle Implementation Predictions Implications for Transport Planning (2019)

Mathé, L., Dubielzig, M., Lindemann, M.: Second Passenger needs analysis and specifications: AVENUE project (2019)

McKay, J.: Transport or Mobility: What's the difference and why does it matter? (2019). https://www.forumforthefuture.org/blog/transport-or-mobility

Mellano, A.: Fleet orchestration: keys to successful AV mobility service (2019). https://www.autonomousvehicletech.com/articles/1879-fleet-orchestration-keys-to-successful-av-mobility-service

Merriman, P.: Mobility. Int. Encycl. Hum. Geogr., 141–148 (2020). https://doi.org/10.1016/B978-0-08-102295-5.10292-6

Narayanan, S., Chaniotakis, E., Antoniou, C.: Shared autonomous vehicle services: a comprehensive review. Transp. Res. Part C Emerg. Technol. **111**, 255–293 (2020). https://doi.org/10.1016/j.trc.2019.12.008

Reisch, M.: First Iteration Luxembourg Large Scale Pilot Use Case Demonstration Report: AVENUE project (2019)

SAE: Taxonomy and Definitions for Terms Related to Driving Automation Systems for On-Road Motor Vehicles (2018). https://saemobilus.sae.org/content/J3016_201806

Shaheen, S., Chan, N.: Mobility and the sharing economy: impacts synopsis (2015). http://innovativemobility.org/wp-content/uploads/Innovative-Mobility-Industry-Outlook_SM-Spring-2015.pdf

Shannon, T.: The ethics of mobility: a framework for assessing mobility paradigms, pp. 569–578 (2006). https://doi.org/10.2495/SC060541

Sheller, M., Urry, J.: Mobilizing the new mobilities paradigm. Appl. Mobilities **1**(1), 10–25 (2016). https://doi.org/10.1080/23800127.2016.1151216

Sprei, F.: Disrupting mobility. Energy Res. Soc. Sci. **37**, 238–242 (2017). https://doi.org/10.1016/j.erss.2017.10.029

Wood, M., et al.: Safety first for automated driving (2019). https://www.aptiv.com/docs/default-source/white-papers/safety-first-for-automated-driving-aptiv-white-paper.pdf

Zinckernagel, C., Guldmann, S., Lytzen, P.: First Gap analysis and recommendations on autonomous vehicles for public service: AVENUE project (2018)

Zinckernagel, C., Lutgens, E., Guldmann, S., Lytzen, P.: Second Gap analysis and recommendations on autonomous vehicles for public service: AVENUE project (2019)

Zuttre, Q.: First iteration Keolis Lyon Large Scale Pilot Use Case Demonstration Report: AVENUE project (2019)

Strategic Planning for Urban Air Mobility: Perceptions of Citizens and Potential Users on Autonomous Flying Vehicles

Tomás Ferreira[1] and Sofia Kalakou[2(✉)]

[1] Instituto Universitário de Lisboa (ISCTE-IUL), Avenida das Forças Armadas, 1649–026 Lisbon, Portugal
tlmfa@iscte-iul.pt

[2] Business Research Unit (BRU-IUL), Instituto Universitário de Lisboa (ISCTE-IUL), Avenida das Forças Armadas, 1649–026 Lisbon, Portugal
sofia.kalakou@iscte-iul.pt

Abstract. World's current mobility systems are often inefficient and unsustainable, therefore the need for new schemes to satisfy mobility needs appears. This quest has given the impetus to the industry to invest in new technologies such as autonomous systems enabling self-driving vehicles. In this context, the concept of Urban Air Mobility (UAM), a term used for short-distance, on-demand, highly automated, passenger or cargo-carrying air mobility services, has arisen. This paper presents the introduction phase of strategic planning for the era of urban air mobility focusing on user and citizen acceptance of the system required for its operation. A survey is designed to capture the perception of citizens and potential users on aspects such as safety, security, well-being of the society (including issues of aesthetics, quality of life, social impacts), driving behaviour, mobility behaviour, expected benefits and their impact on the acceptance and the intention to use these systems. The acceptance of citizens and potential users (considered as two different groups) is analysed in terms of its potential uses (e.g. health emergencies, leisure, connectivity to remote regions). The survey is applied to the Metropolitan area of Lisbon and 207 responses were gathered. The collected data was analysed through correlation analysis and non-parametric tests. Conclusions are made on perceptions of citizens over different adoption and embracement levels.

Keywords: Unmanned aerial vehicles · Urban air mobility · Adoption · Technology

1 Introduction

Population growth is today one of the major concerns that societies need to deal with. Currently, the earth is populated by 7 Billion people and that figure is expected to increase [1]. According to the United Nations, the number of inhabitants will grow by 10% until 2030 (reaching 8,9 Billion) and 26% by 2050, making a total of 9,7 Billion people, more 2 Billion that our present record [2]. This growth will affect various aspects of human life, one of them being urban transport. Cities are becoming more and

E. G. Nathanail et al. (Eds.): CSUM2020, AISC 1278, pp. 286–295, 2021.
https://doi.org/10.1007/978-3-030-61075-3_28

more crowded, aggravating the problem of urban mobility. The future is not looking bright either, as every year more than 77 million people move from rural to urban areas [3]. In this context, for many cities moving by Public Transport (PT) is not a viable option [4]. Most of PT systems face unstable demand during weekdays, peak hours with crowded vehicles creating discomfort among the passengers and on low ridership periods it becomes difficult for PT companies to cope with high costs. Also, PT systems have a high infrastructure maintenance cost making it harder to maintain the service level and keep up with the demand.

This inefficiency of PT in conjunction with other aspects leads passengers towards private transport choices, mainly private cars. However, neither cars have proven to be efficient as a mode. Increased volumes of cars at streets result to congestion and increasing commuting times. Studies have shown that the average American citizen spends a total of 26.9 min per day in commute [5], and in India, the average commuter spends over 2 h of their day on the road commuting to work or home [6]. Overall transportation in urban areas is characterized by extreme traffic congestion, long commuting time, air pollution and inadequate public transportation [7].

World's current mobility systems are clearly insufficient and unsustainable, therefore the demand for new mobility services has arisen. Noticing this demand, multiple companies started to develop new mobility services such as car-sharing (e.g. ZipCar, Car2Go) and ride-hailing services (i.e. Uber, Lyft) that offer more personalized services. Companies' business based on Sharing Economy began to thrive, creating opportunities including new business models such as Mobility as a Service (MaaS).

MaaS combines services from different modes of transport to provide customised mobility services, all in one interface [8], giving the user more flexibility by providing all the transportation means needed to commute. Such platforms combine different transport modes and give the users the possibility of choosing the ones they prefer and the ticket option they wish to use as well (e.g. pay-as-you-go or mobility packages).

This quest for the next big step in mobility has made companies invest heavily in research and development (R&D) of new technologies. One technology that has seen great growth in the last years is the autonomous systems enabling self-driving vehicles. Multiple companies started to invest and later test this system in cars and began to study the application of such technologies on other modes of transport, such as the self-driving aircrafts. Uber is one of the companies focused on developing this technology. In 2019, the ride hailing service company invested $457 Million in R&D of autonomous vehicles [9] indicating their vision of future urban mobility.

Gradually other companies, institutions and policymakers are analysing the possibilities of the urban mobility in the vertical dimension forming the concept of Urban Air Mobility (UAM) that expresses on-demand, highly automated (pilotless), passenger or cargo-carrying air transportation services [10]. This concept relies on short distance vertical take-off and landing aircraft (VTOL), therefore giving the flexibility needed to operate this aircraft. There are three main uses of UAM: last-mile delivery, air metro and air taxi. Last mile deliveries aim to transport goods from the distribution hub to the final delivery transportation. The Air Metro is an autonomously operated aircraft, that can accommodate 2 to 5 passengers at a time; it resembles public transport (PT) services since it has fixed routes, schedules and stops. Air Taxi, much like the Air Metro aircraft, is autonomous and can carry multiple passengers, the difference relies on the fact that this transportation mode has no fixed route, no schedule and no predetermined stops, it works only on-demand and requires multiple possible stops so that the service can truly become door-to-door [11]. Vertical take-off and landing aircrafts (VTOL) can be the solution to mobility problems, providing a fast, clean and ubiquitous alternative.

The current work analyses the aspects that can affect user adoption in the future and the embracement of this new mode from the societies. The next section presents background work that provides insights in the development of the current work. Then the design of the survey that reveals the attitude and perspectives of the public towards passenger aerial vehicles is presented and the preliminary results of the collected data are illustrated and discussed. The last section includes the conclusion and future work.

2 Literature Review

The Urban Air Mobility (UAM) concept aims to enable a world where people or goods can be transported in the urban environment in minutes rather than hours, always on demand. UAM can be realised in the form of air taxis and shared or owned vehicles, creating an on-demand flying service network. Currently, manufacturers are already on the prototype flying-test phase and are getting ready to reach market availability in 2023 [12]. These unmanned aerial vehicles (UAV) will be VTOLs, an all-electric aircraft that has the capability to vertically take-off and land, therefore does not require any runways [13]. According to Holden et al., UAM will add the third dimension, which increases the accessibility between suburbs and cities and, ultimately within urban areas [12]. In resemblance to the automated vehicles (AV), VTOLs will be fuelled by electricity and produce zero emissions whilst producing a much lower noise than a traditional helicopter [13]. The major manufacturers at the moment are Airbus with their model Vahana and CityAirbus; Boeing with Aurora (partnership made with Aurora flight sciences); and Volocopter with the Volocopter. These VTOLs are expected to have a range of 50 km and a top speed of 120 km/h.

The availability of the technology generates opportunities for the study of the future of mobility especially in urban and suburban areas with the respective requirements in infrastructure and service operation. Al Haddad et al. performed a study to observe which factors affect the adoption and use of UAVs and found that safety plays a crucial role in early and late adoption. Other factors such as affinity with automation, data and ethical concerns were also found to have an impact on adoption [14]. These findings were coherent with a NASA study, where respondents reported that safety, costs and environmental aspects were determinants of adoption and the majority of respondents (over 70%) stated they would be comfortable with other people using air taxis services regardless of them using it or not [11].

Eker et al. found that women are more concerned with safety than men (e.g. safe interactions between UAVs) meaning that this safety concerns may prevent women from being early adopters [15]. This finding corroborates with conclusions of Al Haddad et al. in which women expressed a lower interest in UAM, lower trust in automation, greater security and safety concerns. Moreover, women had a higher desire of having extra safety measures such as an operator on the ground and in-vehicle safety cameras [14]. Income and education background can also be indicators of the likelihood of adopting this service. Castle found that having a higher income and a greater degree of education would translate into a higher willingness to use UAM [16]. Furthermore, there is evidence that young respondents are more likely early adopters which might be explained due to the largely unknown capabilities of UAVs (Eker et al.). Additionally, it was suggested that informational campaigns should be designed and implemented to increase awareness [15].

Much like in Autonomous Vehicles, data privacy and ethical concerns can also have influential negative impacts on the early adoption of UAVs. Regulations should be created to establish standards for liability, security and data privacy [14]. Furthermore, environmental concerns were also observed as crucial points in the adoption of UAM service. NASA observed that the environmental impact was the third highest concern on their study with over 2,500 responses [11]. Al Haddad stated that there is the need of policymaking regarding in the area of noise and visual impact. Regulating these areas could lead to a higher public acceptance [14].

NASA also studied what key actions that policymakers and constructors could undertake that would increase the public comfort with air taxis. The actions were related with safety, environmental concerns, legal issues and with noise impact. The respondents showed a higher desire for actions such as proven lower accident rates than cars, successful human demonstrations of their safety and successful trials in other cities. The fourth most highlighted action was related with the environment, where respondents stated that they would feel more comfortable with air taxis if they are less harmful to the environment than regular cars. The least picked action was related with noise showing that respondents have a lack of concern with UAM noise [11].

Although AVs and UAVs are not the same to the user's eyes, they share strong commonalities not only on the constituents' technology [12] but also on the challenges they face as a new way of transportation and the expected benefits. It is expected that automation in mobility will potentially improve the societies' quality of life by contributing against the traffic and the environmental crisis, improving users' productivity, reducing car crashes and hence increase safety [17]. Social benefits will arise as well, as AVs would be the possibility to solve the mobility problems of the elderly, people with disabilities or even children [18] and could enable higher independent mobility for the non-drivers whilst increasing road capacity and reducing traffic congestion [19]. AVs can offer last-mile solutions and fill the transportation needs in places with less frequently used routes. When compared to Public Transportation (PT), AVs offer more privacy, comfort and intimacy, seating availability would be guaranteed and walking time would be significantly reduced [20].

3 Methodology

As technology is rapidly evolving, the industry is developing prototypes of passenger aerial vehicles and some of them have already performed thousands of test flights (e.g. Airbus Vahana). With this rapid growth the need of research that assesses people's perception towards UAVs and their intention to use them for mobility needs arises. In this study it is proposed that the integration of UAVs in the future transport systems is decomposed to two dimensions, the intention of citizens to use them and the embracement of this vehicle type from the society which indicates the voluntary inclusion of the mode in the transport system. UAM has various benefits for the cities that adopt this transport mode. However, it is important to ensure that the UAM does not decrease the quality of life by congesting the sky and increasing the noise pollution. Therefore, having the public involved and co-creating becomes crucial for the adoption of UAM [21]. The rate of adoption is suggested that depends on the trust people have on technology and the tendency towards new technologies' adoption, the perception of the people over the expected benefits and safety of UAVs, concerns over cybersecurity, people's travel well-being, mobility and driving behaviour, their environmental concerns and their sociodemographic characteristics. Hence, a conceptual model is developed to research the impact of these aspects on the people's embracement and intention to use UAVs (Fig. 1).

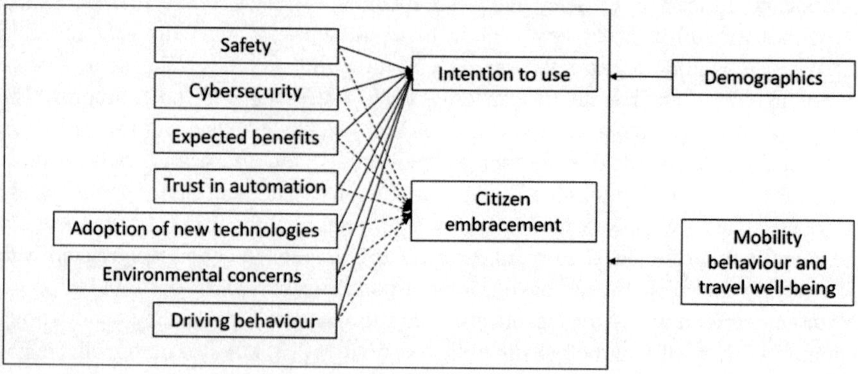

Fig. 1. Conceptual model of UAVs adoption

3.1 Presentation of Hypothesis to Test

In order to access the citizens embracement and intention to use UAVs, the following hypothesis will be tested according with the Table 1 below.

3.2 Survey Design

To collect the data that will assess the conceptual model, a survey was designed consisting of four parts, with a total of 49 questions. The collected data had the form of categorical (in a 7-point Likert scale), continuous and ordinal variables. The first part of the survey is composed by questions that reflect the trust of the respondents in automation and their attitude towards the adoption of new technologies. Statements were presented to the respondents and they were asked to express the level of agreement as well as the level of adopting a new technology or mobility innovation. Respondents were also asked their view on shared mobility services and other mobility innovations.

At the second part, the participants were introduced to the UAVs and were presented some of the aircraft and service characteristics. This introduction helped the

Table 1. Hypothesis over adoption and embracement levels

#	Hypothesis
H1	Men intend to use UAVs earlier than women
H2	Safety is perceived in a different way among the public embracement levels
H3	Young people are willing to adopt UAVs earlier than older
H4	Familiarity with shared mobility services has an impact on adoption and embracement
H5	Public embracement levels vary across adoption levels
H6	Income levels don't vary across adoption levels.
H7	The expected benefits are differently perceived among the public embracement levels
H8	Accident history vary across adoption levels
H9	Cybersecurity is perceived in a different way across public embracement groups
H10	People satisfied with ride hailing services are willing to embrace this mode early

respondents to get more familiarized with the subject before expressing perceptions on it. Then the level of agreement to statements related to expected benefits, safety and cybersecurity of UAVs were measured. To assess the respondent's view on expected benefits, their perception regarding UAVs impact on road traffic, mobility behaviour, safety and independence on mobility (e.g. disabled people mobility) was captured. Safety concerns were reflected through the measurement of the respondent's agreement with situations that may occur such as flying under poor weather conditions or their fear of having a mid-air collision. The respondents were also asked to state if they would feel safer if the UAV service had some characteristics such as a pilot on ground ready to take over the aircraft if needed, the possibility to speak with an operator at any time or security cameras inside the VTOL cabin. The participants' view on cybersecurity was assessed through the measurement of the degree of concern they had towards critical points of cybersecurity such as data privacy, user tracking, loss of privacy and loss of control. The intention to use UAVs was measured using the Technology Adoption Life Cycle (from innovators to laggards) and the purpose of use was also reported. The embracement of the new mode as a citizen (not necessarily as a user) was also included; statements towards the level of comfort if UAVs are available in their city, the availability of UAVs to everyone, and the possible purposes of use of UAVs were evaluated.

The next part of this survey consisted of questions about the respondents' mobility (mode of transport, travel time, transfers) and driving behaviour (e.g. enjoying driving, driving after drinking, involvement in accidents) and environmental concerns. To finish the survey participants provided some socio-demographic information (e.g. age, gender, income, residence).

4 Results

A total of 207 random replies to the online survey were collected. The survey was created on Lime Survey and shared through WhatsApp, Facebook and E-mail. It was found that there is reliable relationship between the measured dimensions and the measured aspects of intention and embracement (alpha Cronbach). Among the respondents, 55% are male and 45% female, 28% are younger than 25, 34% belong to the age interval 25–34, 6% to 35–44, 11% to 45–54, 20% to 55–65 and 1% the rest. Regarding the type of residence area, 38% live in a big city (1 million - 10 million inhabitants), 45% in a city with less than 1 million inhabitants, 11% in a town and 5% at a village. The composition of the sample's employment state is the following: 47% are employed full-time, 7% part-time, 10% self-employed, 3% currently unemployed, 20% students at the university and the rest are retired and volunteers. When analysing the statistics of the trip purposes UAVs should be used for, it is found that the replies of the respondents vary significantly in the "strongly agree" level of the replies where it is seen that 20% believe they should be used for healthcare service, 7% for social activities, 6% for leisure and 3% for work. More than 50% of the respondents disagree (at any level) with the use of UAVs for work trips, 23% for leisure and 32% for social activities.

The embracement of the new vehicle type was analysed in its relationship to safety aspects, cybersecurity, expected benefits, trust in automation and adoption of new technologies through the Pearson correlation indicator; when appropriate the Spearman correlation indicator was applied. It was found that there is a statistically significant (up to 0.05 level) correlation among the perception of some aspects and embracement. Specifically, the concern of respondents about possible collisions of the first vehicles appears to affect the embracement aspect with the highest correlation to be among the possible collisions and the generated to the public (0.452). The respondents indicated that whether they would feel comfortable living in a city that adopts this mode is related to the risks of terrorism, possible collisions and technology readiness the first years. The expected benefits to mobility independence and better conditions of traffic were positively related to the aspects of embracement. Familiarity with new mobility services was also positively related to embracement. Contrary to what was expected, the availability of connection to ground operations did not affect significantly all the embracement aspects apart from the perception of UAVs being beneficial for the society that had a low (0.263) and positive relationship with this service option. Weather conditions were not found to influence embracement. Finally, noise and visual pollution had a negative impact on the adoption of UAVs and their perception as acceptance means of transport in a city.

Regarding the relationship of intention to use to safety and cybersecurity, the fear to fly, lack of communication with on-ground services and fear of possible collisions had an impact on the prevention of people from using the new mode. On the other hand, the adoption of new existing mobility solutions, the perception of gains in road congestion and travel time had a positive correlation to the adoption of UAVs.

In order to analyse variations of perceptions among different adoption and embracement levels, a Shapiro test was performed to assess the applicability of ANOVA analysis. It was shown that the data doesn't follow a normal distribution and hence, non-parametric tests were applied. The results are presented below separately for the intention to use and the perception of UAVs as an acceptable transport mode. No differences were found among the adoption and embracement levels when analysing Affinity to Automation, Mobility Technologies, perceptions on Cybersecurity and Environmental issues while Safety perceptions varies only among different embracement levels.

4.1 Variations in Intention to Use

A Mann-Whitney *U* Test indicated that the adoption levels of male respondents are higher that the adoption level of female respondents, indicating that men are willing to adopt this mode earlier than women. When considering the respondents' accident history, the Mann-Whitney *U* Test indicated that the adoption levels of respondents who had at least one accident as a driver exceeded those of the respondents who have never had a accident as a driver. Kruskal-Wallis tests were performed to explore differences in other socioeconomic aspects across adoption levels. Contrary to what expected, considering age, there were no differences in the distribution of the groups. Also, no differences were found when considering monthly income of respondents. Tests on mobility behaviour indicated that respondents with higher adoption levels had higher satisfaction levels with ride hailing services but the frequency of use did not vary.

Differences on the perception of expected benefits across Laggards and the other adoption levels were found when considering the perception of the aspects "The use of Air Vehicles will make my travel time more productive across intention to use", "Air vehicles will offer a less stressful mobility experience" and "Air vehicles will increase the trips people will make". The aspect "Air vehicles will offer a safe and fast mean of transportation" received lower rates by Laggards and Late Majority compared to the other adoption levels. In the driving behaviour of the respondents, their feeling of safety when driving the car on their own was higher in the cases of Early Adopters compared to Early and Late Majority. Differences were also tested among some aspects of public embracement. Considering the perception of the aspect "Air Vehicles will increase the quality of life in the cities that offers this transport mode" among the adoption categories, it is found that Laggards rated lower this statement compared to the other adoption categories. The same difference was found on the analysis of the improvement of transport accessibility for all citizens, the equality of plane's and UAV's safety, and the sense of safety and comfort that UAVs inspire. They also have a higher rate of perceived risk for the public, stress and fear they cause. Regarding the visual and noise pollution aspects, the Kruskal-Wallis test indicated that Early Adopters tend to have lower concerns over these aspects compared to the other categories.

4.2 Variations in Embracement

The analysis of the socioeconomic aspects was similar to adoption's, only gender differences were found on the perception of "Air Vehicles are an acceptable means of transport". Kruskal-Wallis tests were also performed with multiple groups. The tests revealed that when considering the aspects "The use of Air Vehicles will reduce road congestion" and "The use of Air Vehicles will reduce accident on roads", respondents that accept more UAVs as a means of transport tend to believe that UAVs will reduce road congestion. The analysis of differences in the aspects "The use of Air Vehicles will make my travel time more productive" and "AVs will significantly reduce travel time" across the public embracement, showed that people more open to UAVs tend to believe more on travel time and time productivity benefits generated by UAVs compared to those that are less receptive on UAVs as a mode in the city. In terms of urban planning and accessibility to transport networks, perceptions over the facilitation of connecting remote areas to bigger cities and multimodal nodes, the release of free space in the urban environment for other facilities (such as pedestrian zones), the mobility opportunities to people with reduced mobility independence and the seamless mobility of police and healthcare staff were higher at the groups that find UAVs an acceptable transport mode. The same conclusions were made for the reduction of CO_2 emissions and the belief that UAVs will offer a safe and fast transport mode and a less stressful mobility experience. Differences among high and average levels of public embracement were observed among respondents that "Somewhat agree" with the statement "I'm concerned that the first Air Vehicles available will be unsafe due to possible vehicle collisions in the air above cities" on the public embracement tend to differ with the respondents "Strongly agree" and have a higher perception of the safety of UAVs. Low embracement levels also demonstrated higher propensity to drive after having drunk alcohol.

5 Conclusion and Future Work

There is a need for a new, safer, faster, and greener solution of transportation. UAVs can be a new opportunity in urban mobility. This study presented the dimensions to be studied in the process of the introduction of UAVs in mobility systems in a way that they can serve citizens and potential users. The intention to use and the embracement of UAVs are both important for the successful planning and implementation of the new service in the third dimension of urban space. A survey was conducted in Lisbon area to collect data on citizen's perceptions over aspects related to the introduction of UAVs in the transport system of a city. Through the analysis of 207 random replies and the conduction of correlation and non-parametric tests, six out of ten proposed hypotheses were validated (H1, H2, H5, H7, H8 and H10) indicating differences among the perceptions of men and women, variations in the adoption and embracement levels according to the safety perceptions, expected benefits and satisfaction with shared mobility services. Further analysis will be conducted through the development of discrete choice models that will model the intention to use and public embracement levels.

References

1. Worldometers. https://www.worldometers.info/world-population/? Accessed 26 Sept 2019
2. United Nations: World Population Prospects 2019: Highlights. Department of Economic and Social Affairs, ((ST/ESA/SER.A/423)), pp. 1–46 (2019). http://www.ncbi.nlm.nih.gov/pubmed/12283219

3. UN-Habitat: From Habitat II to Habitat III: Twenty Years of Urban Development. World Cities Report 2016, pp. 1–26 (2016). http://wcr.unhabitat.org/wp-content/uploads/sites/16/2016/05/Chapter-1-WCR-2016.pdf
4. Rodrigue, J.: (2017) https://transportgeography.org/?page_id=462. Accessed 26 Sept 2019
5. Saldivia, G.: (2018). https://www.npr.org/2018/09/20/650061560/stuck-in-traffic-youre-not-alone-new-data-show-american-commute-times-are-longer. Accessed 30 Sept 2019
6. TomTom (2019). https://www.tomtom.com/en_gb/traffic-index/ranking/. Accessed 30 Sept 2019
7. Taiyab, N.: Transportation in mega-cities : a local issue, a global question. The Frederick S. Pardee Center for the Study of the Longer-Range Future - Issues in Brief, pp. 1–8 (2008)
8. Jittrapirom, P., Marchau, V., van der Heijden, R., Meurs, H.: Future implementation of mobility as a service (MaaS): results of an international Delphi study. Travel Behaviour and Society, (November), pp. 1–14 (2018). https://doi.org/10.1016/j.tbs.2018.12.004
9. Korosec, K.: (2019). https://techcrunch.com/2019/04/11/uber-spent-457-million-on-self-driving-and-flying-car-rd-last-year/. Accessed 3 Oct 2019
10. Wright, K.: (2018). https://www.mitre.org/publications/project-stories/urban-air-mobility-adds-a-new-dimension-to-travel. Accessed 4 Oct 2019
11. Urban Air Mobility: November 2018. Uam-Market-Study-Executive-Summary-Pr.Pdf. https://www.nasa.gov/sites/default/files/atoms/files/uam-market-study-executive-summary-pr.pdf
12. Holden, J., Goel, N.: Fast-forwarding to a future of on-demand urban air transportation, pp. 1–98 (2016). https://www.uber.com/elevate.pdf
13. Lineberger, R., Hussain, A., Mehra, S., Pankratz, D.: Passenger drones and flying cars (2018). https://www2.deloitte.com/content/dam/insights/us/articles/4339_Elevating-the-future-of-mobility/DI_Elevating-the-future-of-mobility.pdf
14. Al Haddad, C., Chaniotakis, E., Straubinger, A., Plötner, K., Antoniou, C.: Factors affecting the adoption and use of urban air mobility. Transp. Res. Part A Policy Pract. (2020). https://doi.org/10.1016/j.tra.2019.12.020
15. Eker, U., Fountas, G., Anastasopoulos, P.C., Still, S.E.: An exploratory investigation of public perceptions towards key benefits and concerns from the future use of flying cars. Travel Behav. Soc. (2020). https://doi.org/10.1016/j.tbs.2019.07.003
16. Castle, J., Fornaro, C., Genovesi, D., Lin, E., Strauss, D.E., Waldewitz, T., Edridge, D.: Flying solo – how far are we down the path towards pilotless planes? UBS Glob. Res. **53** (2017). https://neo.ubs.com/shared/d1ssGmLAVeEB/
17. Shabanpour, R., Golshani, N., Shamshiripour, A., Mohammadian, A.(Kouros): Eliciting preferences for adoption of fully automated vehicles using best-worst analysis. Transp. Res. Part C Emerg. Technol. **93**(June), 463–478 (2018). https://doi.org/10.1016/j.trc.2018.06.014
18. König, M., Neumayr, L.: Users' resistance towards radical innovations: the case of the self-driving car. Transp. Res. Part F Traffic Psychol. Behav. **44**, 42–52 (2017). https://doi.org/10.1016/j.trf.2016.10.013
19. Begg, D.: Vision for London : a 2050 Vision for London (2014). https://www.transporttimes.co.uk/Admin/uploads/64165-transport-times_a-2050-vision-for-london_aw-web-ready.pdf
20. Krueger, R., Rashidi, T.H., Rose, J.M.: Preferences for shared autonomous vehicles. Transp. Res. Part C Emerg. Technol. **69**, 343–355 (2016). https://doi.org/10.1016/j.trc.2016.06.015
21. Agouridas, V.: Saving the planet. Gov. Eur. Q. (3) (2019). http://edition.pagesuite-professional.co.uk/html5/reader/production/default.aspx?pubname=&edid=028a000f-2738-4a6c-bed0-b963dd885266

How Autonomous Vehicles May Affect Vehicle Emissions on Motorways

Panagiotis Papantoniou(✉), V. Kalliga, and Constantinos Antoniou

Technical University of Munich, Munich, Germany
p.papantoniou@tum.de

Abstract. The objective of the present research is to investigate the vehicle emissions that may be produced in mixed traffic conditions of autonomous vehicles and human drivers on motorways. For this purpose, simulation scenarios will be developed in a specific part of Attiki Odos motorway, a modern motorway extending along 70 km, which constitutes the ring road of the greater metropolitan area of Athens and the backbone of the road network of the whole Attica prefecture. Attiki Odos is an urban motorway, with two separate directional carriageways, each consisting of 3 lanes and an emergency lane. For the purpose of the present research, peak hour traffic demand is estimated from 7:00 to 9:00, while both congested, as well as uncongested conditions will be simulated. To achieve this objective, five simulation scenarios are developed, including different percentages of automated and human driven vehicles (0%, 25%, 50%, 75% and 100% of AVs) while NOx and CO emissions are investigated in each scenario. Results indicate that Autonomous Vehicles have the potential to increase the emissions on the motorway. Additionally, the specific increase of emissions is estimated in all different scenarios of autonomous vehicles' percentages in the mixed traffic scenarios.

Keywords: Autonomous vehicles · Emissions · Motorway · Traffic simulation

1 Background and Objectives

Transport is Europe's biggest source of carbon emissions, contributing 27% to the EU's total CO_2 emissions, with cars and vans representing more than two thirds of these, according to the European Environment Agency (EEA 2017). Transport is the only sector in which emissions have grown since 1990, contributing to the increase in the EU's overall emissions in 2015 (EEA 2016). Moreover, transport related emissions further increased in 2016 and in 2017 EU oil consumption – a good proxy for transport CO_2 – increased at its fastest pace since 2001 (IEA 2018).

Several methods are used to measure vehicle emissions, such as on-board emission measurements, remote sensing, near-road air quality measurements, tunnel studies, simulation studies (Smit et al. 2008). All methods have their own strengths and weaknesses, and the weaknesses in particular must be clearly understood and explicitly considered in any subsequent use of the emissions data, e.g., model development and validation (Smit et al. 2010).

E. G. Nathanail et al. (Eds.): CSUM2020, AISC 1278, pp. 296–304, 2021.
https://doi.org/10.1007/978-3-030-61075-3_29

Furthermore, Autonomous Vehicles operations are inherently different from human driven vehicles and have the potential to offer several important benefits. Autonomous vehicles can be programmed to not break traffic laws, have superior reaction times, do not drink and drive etc. (Fagnant and Kockelman 2014). On the other hand, autonomous vehicles encounter several limitations. One barrier is the cost, as in the first years they are likely to be more expensive, and therefore many people may not be able to afford them. Another important barrier is electronic security due to computer hackers which may target autonomous vehicles and intelligent transportation systems more generally, causing collisions and traffic disruptions (Davidson and Spinoulas 2015).

However, the effect of Autonomous cars on emissions is still not clear. A study conducted in USA indicated that with AVs having a 90% share in traffic, delays on motorways will decrease by 60% and fuel consumption by 25% (ENO 2013). On the other hand, Makridis et al. (2020), in a traffic simulation study found that AVs have the highest fuel consumption per km travelled among both connected autonomous vehicles and human drivers.

Based on the above, the objective of the present research is to investigate the vehicle emissions that are will be produced in mixed traffic conditions of autonomous vehicles and hu-man drivers on motorways. For this purpose, five simulation scenarios will be developed in a specific part of a motorway including different percentages of automated and human driven vehicles.

The paper is structured as follows. In the next section, the methodological approach is presented including details regarding the implementation of the simulation scenarios, characteristics of the motorway as well as the theoretical background of the analysis. The results are presented in the third section whilst general conclusions are stated alongside with proposals for further research.

2 Methodological Approach

2.1 Motorway Case Study

For the purpose of the present research, simulation scenarios are developed in a specific part of Attiki Odos motorway. Attikes Diadromes SA, also known as Attica Tollway Operations Authority, is the Operating and Maintenance Company of the Attiki Odos Motorway (Attica Tollway) in Athens, Greece. The Tollway is a 70 km-long urban Motorway, fully access-controlled through 39 toll barriers (6 mainline barriers at the extremities plus 33 entry ramps). The Company activities cover full operation and maintenance including the 12.60 km of tunnels and cut-and-covers with the longest twin-bore tunnel having a length of about 1 km (total unidirectional length).

Attika tollway is the Athens Ring Road, providing free-flow traffic conditions in the city centre periphery, linking the downtown areas by radial connections and main arteries. It provides a link between the national motorway network to the south and to the north of the Greek Capital and connects the city and its suburbs with the new Athens International Airport. It is part of the general transportation plan for the development of the Greater Athens Transportation System. It also provides connections with mass transport mode facilities (Metro, Suburban Rail and Buses) (Fig. 1).

Fig. 1. Motorway case study.

Attica Tollway is an urban motorway with tolls paid at all entry points to the Tollway (a flat toll of €2,80 for category 2 is charged at all 39 Toll Stations, no matter the distance driven), with two directionally-separated carriageways, each consisting of 3 lanes and an emergency lane (hard shoulder). The suburban railway of Athens has been constructed in the central reservation of the motorway. Being a closed motorway, it is fully-controlled at its access points and consists of two sections, which are perpendicular to one another:

2.2 Simulation Characteristics

2.2.1 Programming Tool

SUMO (Simulation of Urban Mobility) is an open source, highly portable, microscopic and continuous traffic simulation package designed to handle large road networks. It is mainly developed by employees of the Institute of Transportation Systems at the German Aerospace Center (Lopez et al. 2018).

SUMO road networks represent real world network graphs, where nodes are intersections and roads are represented by edges. Intersections consist of a position, a shape, and a right of way rules, which may be overwritten by a traffic light. Edges are unidirectional connections between two nodes and contain a fixed number of lanes. SUMO road networks can be imported as a digital road map using *netconvert* which converts networks from either other simulations or from other sources like *OpenStreetMap*.

2.2.2 Vehicles and Routes

Two types of vehicles are used in the current simulation network: Human-driven vehicles and AVs, where the human driven cars are not responsible for regaining control at any point of their trip. Below there is a small description of the models used for the simulation of the above-mentioned vehicle types.

- Human-driven Vehicles: For the simulation of manually driven vehicles, the default option from SUMO is a modified Krauss car-following model (Krauß 1998)
- Autonomous vehicles: For no and full automation vehicles, the deceleration and the emergency deceleration remained the same, considering the safety. The emergency deceleration was set to 8 m = s2. This value was based on the study of Kudarauskas (2007). While the mingap, acceleration and time headway were taken from Atkins Ltd (2016) (Table 1).

Table 1. Simulator parameters.

Vehicle type	MinGap	Accel	Decel	Max decel	Sigma	Tau
Manual	2.5	3.5	4.5	8.0	0.5	1
AV	0.5	3.8	4.5	8.0	0	0.6

- Mingap: the offset to the leading vehicle when standing in a jam (m).
- Accel: the acceleration ability of vehicles of this type (m/s2).
- Decel: the deceleration ability of vehicles of this type (m/s2).
- Emergency Decel: the maximum deceleration ability of vehicles of this type in case of emergency (in m/s2).
- Sigma: the driver imperfection (between 0 and 1).
- Tau: the driver's desired (minimum) time headway (reaction time) (in s).

Demand is setting up by the use of some available applications by utilizing different source of information. For large-scale scenarios the so-called Origin-Destination (O/D) matrices are used. For the specific case study, since a small part of Attiki Odos is simulated data from detectors were used. Based on the fact that, most of highways are well equipped with induction loops, which measures the number of vehicles entering and leaving the motorway, it is assumed that flows are known in the motorway. The algorithm namely *DFROUTER*, provided by SUMO, uses the information collected from induction loops to build the vehicle amount and routes.

SUMO has the possibility to produce the emission is each simulation step. Values for each vehicle are calculated and will be recorded and can be further researched. For the specific case study, this output is quite useful since it is possible to compare the benefits of a more sustainable traffic consisting of autonomous/electric cars.

2.2.3 Scenarios

Five different scenarios have been developed in the present research in order to estimate the impact of automation on the motorway. For the mixture of vehicles, 5 different cases were studied, creating all possible combinations of penetrations of each vehicle type ranging from 0 to 100% with intervals of 25% as shown in the table below. Each scenario was running for two hours with the second hour being the network's peak hour (Table 2).

Table 2. Autonomous Vehicles and Human Drivers percentages.

Penetration rate of AVs						
Penetration rate of Human drivers		0%	25%	50%	75%	100%
	0%					X
	25%				X	
	50%			X		
	75%		X			
	100%	X				

2.3 Theoretical Background

While a histogram is an approximate representation of the distribution of numerical or categorical data, a box plot is a method for graphically depicting groups of numerical data through their quartiles. Box plots may also have lines extending from the boxes (whiskers) indicating variability outside the upper and lower quartiles, hence the terms box-and-whisker plot and box-and-whisker diagram (Williamson et al., 1989).

With regard to the interpretation of boxplots, it should be noted that the spacing between the different parts of the box plot indicates the degree of dispersion (spread) and skewness in the data and identifies outliers. More specifically:

- The line in the middle of the boxes is the median
- The bottom of the box indicates the 25th percentile.
- The top of the box represents the 75th percentile.

3 Results

Within the framework of the present research the following emissions types are further analysed:

- NOx: The amount of NOX emitted by the vehicle in the actual simulation step
- CO: The amount of CO emitted by the vehicle in the actual simulation step

In Table 3 the average values regarding all deferent scenarios are presented both for NOx and CO emissions.

Table 3. Total CO_2 emissions in mg/s.

Emissions	100% AV	75% AV 25% HD	50% AV 50% HD	25% AV 75% HD	100% HD
NOx	65	60.8	58	56.2	55
CO	0.22	0.23	0.23	0.22	0.22

3.1 NOx

Nitrogen oxides (NOx) are produced when fuel is combusted in the engine in the presence of air. NOx comprises a mixture of nitric oxide (NO) and nitrogen dioxide (NO_2). NO is not harmful to health at the concentrations typically found in the atmosphere. However, in contrast, NO2 is associated with a range of environmental and health problems (EEA 2016).

In the first step of the analysis the boxplot regarding all deferent scenarios is presented.

Results of Table 3 and Fig. 2 indicate that the average value of NOx emission is higher in the scenario of 100% Autonomous Vehicles while in all the other scenarios, results are very similar. It is also notable that mixed traffic results in slightly higher NOx emissions than fully human driving cars. In the next figure, the five different histograms are presented aiming to explore the NOx emissions individually in each scenario (Fig. 3).

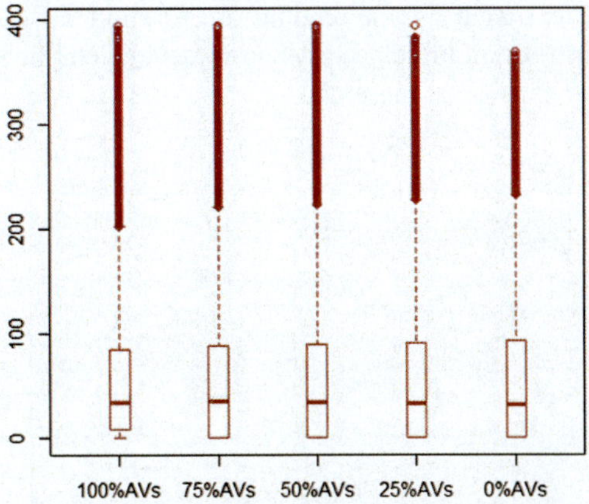

Fig. 2. Boxplot of NOx emissions per scenario.

Results indicate that in the scenario of 100% Autonomous vehicles, the lowest frequency of emissions in the 0–20 emissions gap is found. On the other hand, in the scenario of 100% human drivers, the same emissions gap has the highest frequency which leads to lower overall NOx emissions.

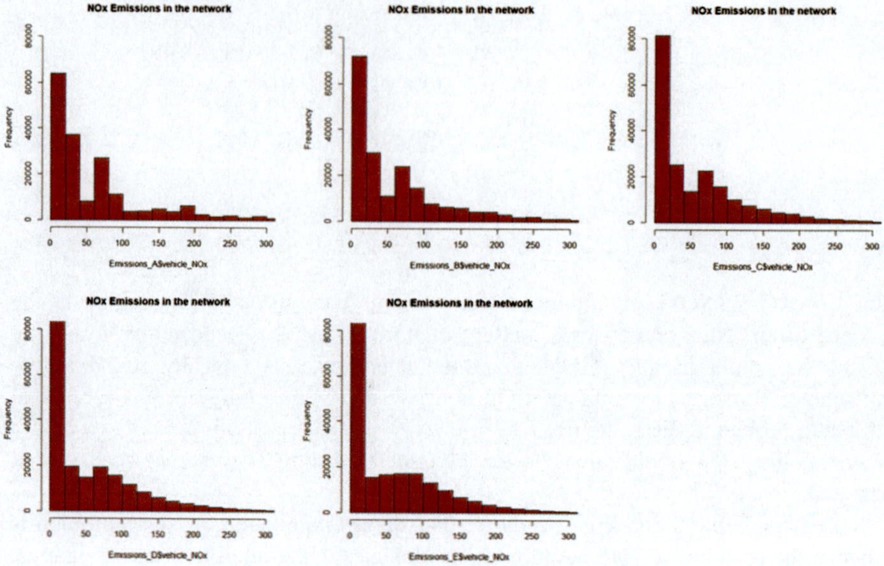

Fig. 3. Histogram of NOx emissions per scenario.

3.2 CO

Carbon monoxide (CO) is a temporary atmospheric pollutant in some urban areas, chiefly from the exhaust of internal combustion engines (including vehicles). In the second part of the present analysis, the amount of CO emitted by the vehicles in the actual simulation is estimated for the 5 scenarios developed. In Fig. 4 the boxplot regarding all deferent scenarios is presented.

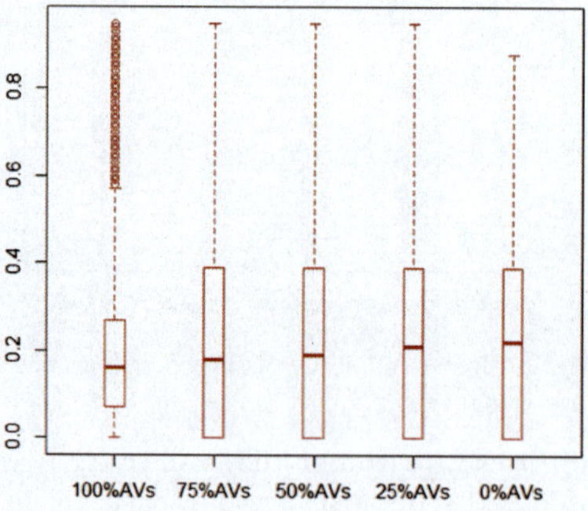

Fig. 4. Boxplot of CO emissions per scenario

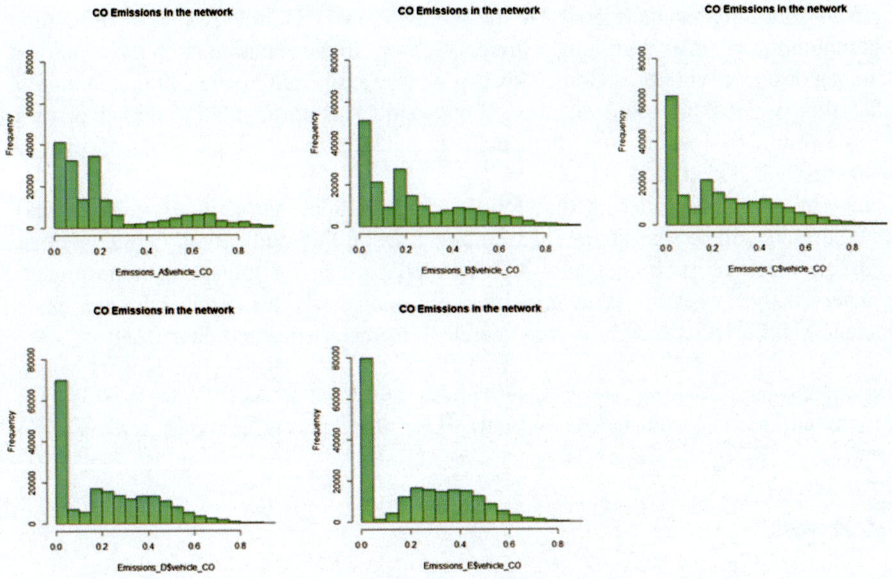

Fig. 5. NOx emissions per scenario

Results from Table 3 and Fig. 4 indicate that the average value of CO emissions remain the same in the scenarios of full autonomous vehicles and human drivers while in mixed traffic conditions the production of such emissions is slightly higher. In the next figure, 5 different histograms are presented aiming to explore the CO emissions individually in each scenario.

Results indicate that in the scenario of 100% Autonomous vehicles, the lowest frequency of emissions in the 0–0,05 emissions gap is found. On the other hand, in the scenario of 100% human drivers, the same emissions gap has the highest frequency (Fig. 5).

4 Conclusions

The present study relied on two key objectives. The first was to simulate five scenarios of mixed traffic conditions with Autonomous Vehicle and human drivers on a realistic motorway transport network. Subsequently, the second objective was to extract and analyse the different emissions of NOx and CO in each scenario in order to compare and provide significant qualitative conclusions.

Results indicate that Autonomous Vehicles both in the mixed traffic as well as in the scenario of their unique existence in the motorway produce more NOx emissions in comparison with the scenario of 100% human drivers. This output is in line with a recent microsimulation study that investigated the impact of vehicle automation and connectivity in terms of traffic flow and emissions on a realistic highway transport network, the ring road of Antwerp, Belgium (Makridis et al. 2020). Makridis et al. (2020) indicated that AVs have the poorest performance in terms of average speed and flow and generate the highest emissions values per kilometer.

In addition, the results referring to the production of CO indicate that mixed traffic of autonomous vehicles and human drivers leads to higher emissions than the unique existence of one driving behaviour. This is probably explained by the fact that in mixed traffic, the "perfect" behaviour of autonomous vehicles is not familiar to human drivers and as a result leads to more conflicts and consequently more vehicle emissions due to the aggressiveness on the conflict events.

In conclusion, considering that for the next decades autonomous vehicles and human drivers will be sharing the roads, the findings of this study highlight the fact that the different driving performance of AVs will have a negative impact on environment. As a result, the target of reducing vehicle emissions, which is crucial for the environment, should focus mainly on the vehicles' characteristics and clean fuels.

Acknowledgement. The project has received funding from the European Union's Horizon 2020 research and innovation programme under the Marie Skłodowska-Curie grant agreement No 754462.

References

Atkins Ltd.: Research on the Impacts of Connected and Autonomous Vehicles (CAVs) on Traffic Flow, Technical Report. Department for Transport (2016)

Davidson, P., Spinoulas, A.: Autonomous vehicles – what could this mean for the future of transport. In: AITPM 2015 National Conference (2015)

Eno: Preparing a Nation for Autonomous Vehicles: Opportunities, Barriers and Policy Recommendations, ENO Center for Transportation, pp. 8–10 (2013)

European Environment Agency: Explaining road transport emissions A non-technical guide. Publications Office of the European Union, Luxembourg (2016)

European Environment Agency: Greenhouse Gas – Data Viewer. Publications Office of the European Union, Luxembourg (2017)

Fagnant, D., Kockelman, K.: The travel and environmental implications of shared autonomous vehicles, using agent-based model scenarios. Transp. Res. Part C Emerg. Technol. **40**, 1–13 (2014)

International Energy Agency: Global Energy and CO_2 status report 2017 (2018)

Krauß, S.: Microscopic modeling of traffic flow: investigation of collision free vehicle dynamics. Ph.D. dissertation, Universitat zu Koln (1998)

Kudarauskas, N.: Analysis of emergency braking of a vehicle. Transport **22**(3), 154–159 (2007)

Lopez, P.A., Behrisch, M., Bieker-Walz, L., Erdmann, J., Flötteröd, Y., Hilbrich, R., Lücken, L., Rummel, J., Wagner, P., Wießner E.: Microscopic traffic simulation using SUMO. In: IEEE Intelligent Transportation Systems Conference (2018)

Makridis, M., Mattas, K., Mogno, C., Ciuffo, B., Fontaras, G.: The impact of automation and connectivity on traffic flow and CO_2 emissions. A detailed microsimulation study. Atmos. Environ. **226** (2020)

Smit, R., Brown, A., Chan, Y.: Do air pollution emissions and fuel consumption models for roadways include the effects of congestion in the roadway traffic flow? Environ. Model Softw. **23**, 1262–1270 (2008)

Smit, R., Ntziachristos, L., Boulter, P.: Validation of road vehicle and traffic emission models—a review and meta-analysis. Atmos. Environ. **44**, 2943–2953 (2010)

Williamson, D., Parker, R., Kendrick, J.: The box plot: a simple visual method to interpret data. Ann. Intern. Med. **110**(11), 916–921 (1989)

A Taxonomy of Skills and Knowledge for Efficient Autonomous Vehicle Operation

Foteini Orfanou(✉), Eleni Vlahogianni, and George Yannis

National Technical University of Athens,
5 Iroon Polytechniou, 15773 Athens, Greece
{forfanou,elenivl,geyannis}@central.ntua.gr

Abstract. The autonomous vehicles are expected to bring unprecedented changes in the labor sector and the workforce. Traditional jobs will be alleviated, new will be created while people involved in the autonomous vehicle operation should be qualified with additional skills and knowledge in order to be able to deal with the new technology and the various systems. Furthermore, the impact on the role of the 'driver' is anticipated to be significant in all transportation modes. The purpose of the present research is to identify the skills and knowledge required for an efficient and proper operation of any autonomous vehicle. Both professional and private operators and all transportation sectors (road, rail, maritime, aviation) and autonomous levels will be considered as each one has different requirements.

Keywords: Autonomous vehicles · Taxonomy · Skills · Knowledge

1 Introduction

Autonomous or driverless vehicles are equipped with various systems and sensors for assisting driver during the driving task or fully substituting him in higher levels of automation. Automation is expected to increase road capacity, minimize or alleviate accident occurrence and congested phenomena, reduce traffic violations and improve safety levels as they react faster and more appropriate than human drivers [1]. The advent of autonomous vehicles will also bring major changes to the labor sector and the workforce as many jobs will be alleviated, new jobs will be created while workers may need to be reskilled and upskilled. People involved in the autonomous vehicle operation should be qualified with additional skills and knowledge in order to be able to deal with the new technology and its various functions. Additionally, the impact on the role of the 'driver' is anticipated to be significant in all transportation modes as the driver (professional or private) will have the opportunity to abstain from the driving process and be focused on secondary tasks or even remotely control and operate the vehicle.

The scope of the present research is to identify skills and knowledge required for an efficient operation of any autonomous vehicle considering all transportation sectors and automaton levels. Both professional and private drivers, various categories of the labor sector and professions involved in the autonomous vehicle construction and operation will be analyzed in order to reveal the new needs arisen from the advent of automation.

E. G. Nathanail et al. (Eds.): CSUM2020, AISC 1278, pp. 305–315, 2021.
https://doi.org/10.1007/978-3-030-61075-3_30

Project deliverables, reports, studies, articles, scientific papers, websites were reviewed and experts were contacted for identifying the requirements for all workers and drivers. The taxonomy of skills and knowledge formulated can be related to HMI development in order to meet anticipated skills of the operator and his new training needs.

2 Road Sector

Each level of automation requires different systems and sensors inside the vehicle and therefore additional driver skills and knowledge as automation level increases. In levels 2 and 3 the driver continuously cooperates with the vehicle on the driving task [2] and he should monitor the systems in terms of supervising their status, performance and appropriate operation. The intense of monitoring decreases in level 4 and 5 and requires driver concentration maintenance [2]. The driver should be familiar with all sensors and systems in each automation level and be aware of their location, how they work, the principles governing them, their capabilities and limitations, understand their decisions and actions and recognize errors [3, 4]. It is necessary, for the driver of levels 1–4 to be able to take over the vehicle control when necessary either due to system limitations or failure. The control take over must be achieved within a short period and therefore driver keeps high levels of situational awareness and concentration [5]. The road sector is considered to be very complex and the safe interaction and communication between all different (vulnerable) road users is of major importance. People with programming skills in machine learning and artificial intelligent are needed for the algorithms and software development so that the autonomous vehicles can understand the surroundings, detect any (physical) object around and (re)act safely. Since big amount of data will be recorded continuously from the AV's various sensors, backend software engineers are necessary for data storage services, design of APIs for proper communication and data transmission. Along with the software engineers, robotics and electrical engineers will be involved in the design of hardware, electrical and communication systems, while the automotive engineering skills will be upgraded to meet future cars characteristics. Existence of communication models and wireless networks should enable the information and data transmission and exchange between the vehicle, the infrastructure as well as the traffic management centre (TMC). In automation levels 3 and 4 the vehicle communicates with the infrastructure and its various units (e.g. lane marking, traffic lights) and as a result all infrastructure characteristics should be appropriately designed (e.g. road surface quality) [6]. People working in the TMC should know how to recognize the data received and the skills to process and analyse it.

Concerning public transport (PT), the driver should be trained for operating an autonomous vehicle and skilled for monitoring its operation remotely ensuring safety for the passengers and the other interacting road users. The passengers of an autonomous bus should be familiar with its operation principles, recognize its actions and know who they should contact or how to evacuate the bus in case of emergency [7]. Programming and engineering skills are required for developing the systems necessary for the automated bus operation Due to the absence of driver, there will be need for high quality in-vehicle means of communication [8]. Logistic operators should also be skilled and have knowledge in remotely controlling the process and managing the delivered cargo if driverless vehicles are used in terms of dispatching, scheduling and routing and monitor the process [9] while intervention is possible via an operation center using wireless transmission. The very critical issue of cyberattack risk requires experts with continuous education on these issues for ensuring cybersecurity and encryption protection. Law specialists are required for formulating and establishing suitable regulations and framework for the operation of autonomous vehicles as well as solve liability issues in case of failure or incident occurrence.

Table 1. Skills and Knowledge for AV operation in the road sector.

Skills	Description
Social Skills	Communication, Team working, organization, problem-solving
Programming and Computer Skills	Artificial Intelligence, Algorithms, software development, backend/frontend skills, machine learning, higher-order skills in big data analysis, Cybersecurity and encryption protection, security systems for protecting external communication for AVs, data protection
Engineering/Technical Skills	Sensors and systems development, hardware development, Robotics, electrical engineering, automotive engineering, digital road map database access, firmware, Smart Traffic Light controller system, smart signs, advisory road marking, Testing and Simulation Skills
Driver Skills and Knowledge	Cooperation-collaboration with the vehicle, Efficiently monitoring and supervising the system, Concentration maintenance, Familiarity with all electronic devices and sensors, Knowledge of their limitations and capabilities, Understand the information and warnings from the systems based on the surroundings, Knowledge of differences among different levels of automation, Situational awareness and transition of control skills, Capability of recognizing errors -malfunctions and act properly
Remote operation (PT)	Skills and knowledge for efficient remote monitoring and operation of the PT vehicle
Communication skills	V2I and V2V communication model, Wireless communication, ad hoc network, DSRC Multi-Channel Test Tool
Traffic management center	Collection and processing skills from the data transmitted from the infrastructure and the vehicles
Law skills	Legal framework and standards for the autonomous vehicle operation Liability issues in case of incident occurrence, Data generated by V2X infrastructures to be compliant with national or international law

3 Rail Sector

The automatic train protection is already installed in GoA1 (Grade of Automation) for ensuring the automatic activation of the brakes in case of speeding or other risky situations while the automated train operation is introduced from GoA2 for safe movement and control. Continuous monitoring, track supervision, communication, operational knowledge, selective attention, high levels of situational awareness, skills of critical diagnosis and other social, perceptual and cognitive skills are considered important for the train driver of GoA1 and 2 also in case of intervention [10].

Table 2. Skills and Knowledge for AV operation in the rail sector.

Skills	Description
Social Skills	Communication, Team working, organization, skills in timetable management, problem-solving, slit-second decision making, Knowledge in human factors for passengers and workers safety
Programming and Computer Skills	Same as in the road sector (Table 1)
Engineering/Technical Skills and Knowledge	Same as in the road sector (Table 1) Systems for driverless and unattended train operation, automatic train protection and automatic train operation, diagnostics, Signaling technologies, new signaling and position technologies, Knowledge of the European Train Control System (ETCS) and wireless delivery of mission-critical rail communications, digital interlocking system
Driver/Crew Skills and Knowledge	Same as for road sector (Table 1) Maintenance of on route driving skills, knowledge of new on board systems, Monitoring of the passenger exchange, detection and accomplishment of emergency conditions, supervision of the train's state.
Communication skills	V2I communication model, Wireless communication, ad hoc network, Wireless interface/connection and components, data transmission
Law skills	Legal framework and standards for the autonomous vehicle operation Liability issues in case of incident occurrence, Data generated by V2X infrastructures to be compliant with national or international law
Skills for workers in front line and network control, train driving	Rail vehicle setup and deconstruction skills and knowledge for a safe and efficient pre-journey, in journey and post journey autonomous train operation, Skilled rail network controllers
Safety management skills	Preparing for emergencies related to both safety and environmental protection, fatigue management
Remote Control Skills	Off site and remote fault support skills, Incident recovery procedures for autonomous trains and rail vehicles, including fault identification and rectification, remote operations, processing of large amount of data
Signaler	Knowledge of all new signaling technologies and systems, ready to intervene efficiently any time

Since the first two GoAs require the presence of the driver, he should have the skills and knowledge of a driver of an automated vehicle (Table 1) while network controllers should ensure safety by providing the right real time information to the driver [11]. Concerning GoA3, the driver becomes a remote control operator in strong cooperation and communication with the train attendant who is still in the train supervising passengers exchange and detecting emergencies [6]. Due to the large amount and complexity of data transmitted to the control center, remote operators are required to have high skills in big data analytics and problem-solving for maintaining high operation and reliability levels [12]. Manual intervention is still necessary in cases of any emergency situation occurred and the train operator takes over the train control remotely [13]. Remote operators and drivers should also have safety management skills for incident recovery including fault identification and fault rectification while on route driving skills should be maintained in case of emergency [11]. Technical and engineering skills [14] are necessary for GoA3 and GoA4 for enabling driverless and unattended train operation respectively in terms of track and passenger transfer supervision as well as the train operation in event of disruption, (physical) obstacle detection for collision avoidance, existence of other trains on the route [15]. Furthermore, wireless signaling, sensors and communication technologies are also need to be designed and developed for enabling the data capture and transmission between the train, track and signals [6, 11]. Since rail safety work is considered to be dangerous enough, skills and knowledge in human factors for ensuring health and safety of the workers and the passengers are also important [16]. Artificial intelligence and software and hardware skills are also required for efficient visual perception in case of driverless trains while skills in software and hardware assessment are necessary for ensuring safety. Augmented and virtual reality and simulation skills are a prerequisite for developing and testing a rail infrastructure and control or maintenance operations [11, 16]. Regulations and guidelines should be also established in the rail sector in case of operation of trains of different levels of automation and if both semi and fully automated trains are using the same track [12]. Liability issues for driverless trains are necessary to be defined in terms of responsibility in case of incident or failure. The role of signaler will also be affected as they are required to have deep knowledge of all the new signaling systems, continuously monitoring the decisions the systems take and intervene when necessary [12].

4 Maritime Sector

The professional driver of an autonomous vessel is required to have technical and engineering skills for dealing with any malfunction or failure of the hull structure, the machinery and other systems [17]. In case of remote vessel control, skilled personnel in the shore control centre should understand and interpret the pertinent data transmitted from the various vessel sensors to the shore-based facility [6] as well as navigate it. Operation monitoring, emergency situations handling, autonomous ship surveillance and additional safety related tasks are performed in the shore control center. Data should be monitored and controlled via maritime broadband radio, and satellite communication [18]. Human intervention may be required at any time and under various conditions and as a result seafarers on the bridge or in the shore based center should maintain high levels of situational awareness [18]. Due to the fact that the centre may serve both autonomous and conventional vessels as well as different types of vessels, people involved should have interoperability skills and be able to distinguish the different principles governing each type [17]. The seafarer of the shore based center should have apart from maritime knowledge, digital and software engineering skills as well as data fluency and programming skills for interpreting large amount of data [18]. He would need to deal with various parties (i.e. shipyards, port authorities), in a way that differs significantly with the ship type and therefore communication skills are considered among the most important competencies [19]. One of the limitations and major concerns is the new legislation that has to be established and liability issues that should be solved as it is considered that existing legislation issues do not favour the existence and increase of the number of the autonomous vessels [20]. It is apparent, that AV operators need to acquire knowledge about the legal framework associated with the autonomous vessels [19]. Programming, engineering and technical skills are also required for developing all the systems, sensors and technologies the autonomous vessel should be equipped with so that it can navigate itself on a specific route, detect obstacle or avoid collisions [21]. Fully autonomous ships will be equipped with automatic mooring and unmooring systems or with detachable bridge. For auto mooring the required infrastructure and communication infrastructure should be developed [21]. Besides, V2V and V2I connectivity and communication should also be enabled by developing sensors, platforms and systems such as maritime broadband radio or Global System for Mobile Communications (GSM) [6] (Table 3).

Table 3. Skills and Knowledge for AV operation in the maritime sector.

Skills	Description
Social Skills	Same as in the rail sector (Table 2), onboard and shore-based personnel
Programming and Computer Skills	Same as in the road sector (Table 1), augmented and virtual reality skills and knowledge
Engineering/Technical Skills	Same as road sector (Table 1), obstacle detection, surroundings mapping, mooring and unmooring systems, HD Maps of the relevant port transport infrastructure, naval engineer, Testing and Simulation Skills, Airborne or underwater drones for hazardous inspection and maintenance tasks, either by remote control or autonomously
Driver/Crew Skills and Knowledge	Same as in the road sector (Table 1) Knowledge of new on board systems, Interoperability Skills, Docking skills, Coast water crews inner-port navigation the mooring skills, Monitoring of the passenger exchange, detection and accomplishment of emergency conditions, supervision of the vessel's state.

(*continued*)

Table 3. (*continued*)

Skills	Description
Communication skills	Satellite communication capacity and the bandwidth, advanced data transmission technology systems, communication network, V2V and V2I communication
Law skills	Same as in the road sector (Table 1)
Safety management	Preparing for emergencies related to safety/environmental protection
Remote Control Skills	Understand and interpret the pertinent data transmitted from the vessel to the shore-based facility in case of a machinery/equipment/hull damage event and any other case concerning safety, Distinguish the different principles governing each type -Interoperability skills, Mooring and unmooring operation skills, Complex engines and machinery aboard monitoring, Data analytic experts and system controllers

Maritime operations and ship maintenance can be performed remotely using robots creating safer conditions for the coastal workers. Data transmission networks, sensors, augmented and virtual reality and drone technology [22] can enable surface and under water communication, remote services and maintenance work. Cybersecurity is extremely important also in the maritime for supporting safe and secure shipping. Safety management skills are also required for an autonomous vessel operation "*related to both safety and environment protection*" [23]. Communication and team working skills are essential for the successful accomplishment of tasks by on-board crew depending on adequate instructions or assistance by a remote control centre [24]. Finally, quay cranes driver will be upskilled with general knowledge on electronics and mechanics along with control panel handling skills [18]. Since each port or terminal may have different operational processes a docker is required to have deep knowledge of them and skills such as efficient planning, equipment dispatching or remote control of ship handling [18].

5 Aviation Sector

Monitoring the automation systems is a skill required also in the aviation sector when the state of autonomy is level 3 or 4. The pilot should be capable of detecting any system malfunction or suspicious performance as well as be ready to react properly in case of failure in order to avoid air crashes. In contrast to the car drivers, who can recall their driving skills even if automated systems are used, the basic flying skills of a pilot are not retained due to the complexity of the pilot tasks. Additionally, the pilot of airplanes of these autonomous levels is required to have a deep knowledge of the systems and appropriately distinguish the various kind of information received from them and take the right decisions accordingly. In the case of autonomous airplanes the pilot should have all the necessary and required skills and knowledge in order to efficiently and safely supervising remotely the airplane. Due to the fact that remote control is more difficult and demanding than the on board control and supervision, he should be completely aware of every system installed in the plane as well as its level of autonomation, its capabilities and limitations [25]. It is necessary that he is capable of promptly detecting any suspicious activities of the systems and any abnormal behavior of the plane and be prepared for handling any situation. Monitoring tasks are among the basic skills of a UAS operator including instrument monitoring, navigation, route and long term monitoring so that safety is ensured [25]. Additionally, in the future, one pilot may have to supervise more than one unmanned airplanes simultaneously and it is obvious that he should be high skilled and well trained. Different types of aircrafts have different types of systems [25], they follow different routes and they are flying on different airways and therefore the remote controller should ensure their safe operation and journey and conduct a preflight check of the systems of the plane Social skills are also considered important for an UAS operator such as decision making, risk assessment skills or team leadership and communication skills [25–27]. Similarly to the other transportation sectors, high levels of situation awareness should be preserved. Apart from the technical part, the operation of UAS needs the establishment of standards, regulations and operational rules. The Federal Aviation Administration [28] formulated rules for small unmanned aircraft systems in 2016 including operational limitations, Remote Pilot in Command Certification and Responsibilities, Aircraft Requirements and model aircraft. Towards the full automation of bigger airplanes, it is necessary that people involved in the technical and legal sector should be cooperated in order to formulate the appropriate legal and operation framework as well as rules for the safe operation of UAS. Drones are nowadays used for short range surveillance purposes and their operators should be able to identify obstacles and modify the drone route accordingly. In the future, drones will also be used for transporting people and goods within urban environments and therefore they will cover longer distances (Table 4).

Table 4. Skills and Knowledge for AV operation in the aviation sector.

Skills	Description
Social Skills	Communication, Team working, organization, skills in timetable management, problem-solving, slit-second decision making
Programming and Computer Skills	Same as in the road sector (Table 1)
Engineering/Technical Skills	Sensors and systems development, hardware development, Robotics, electrical engineering, aeronautics, automotive engineering, safe navigation systems development, Testing and Simulation Skills, Airborne drones can perform potentially hazardous inspection and maintenance tasks, either by remote control or autonomously (in cooperation with programming and computer skills)
Driver/Crew Skills and Knowledge	Same as for road sector (Table 1) Knowledge of new on board systems, Interoperability Skills, Monitoring of the passenger exchange, detection and accomplishment of emergency conditions.

(*continued*)

Table 4. (*continued*)

Skills	Description
Communication skills	Satellite communication capacity and the bandwidth, advanced data transmission technology systems, communication network
Law skills	Legal framework and standards for the autonomous vehicle operation, liability issues in case of incident occurrence, data generated by V2X infrastructures to be compliant with national or international law
Safety management	Preparing for emergencies
Remote Control Skills	Supervision, Detection of suspicious activities or abnormal behavior of the plane, Simultaneously monitoring and supervision of more than one unmanned airplanes, Knowledge of characteristics of different types of aircraft, the routes they follow, Preflight Check
Urban Environment Operation	Engineering/Technical/Programming Skills (Landing and take-off without a runway, obstacle detection and avoidance)

According to [29], the safe operation and performance of these aircrafts are based on three functions: the landing and taking off process will be executed without a runway, the aircrafts should be able to detect, see and avoid obstacles, like buildings and vehicles as they fly in low altitudes and last but not least the efficient management of emergency situations such as weather conditions. Similarly to all the other transport modes, engineering and programming skills are required for sensors (lidar, radar, cameras) and systems development, simulation modelling, software and hardware development and testing so that the autonomous aircraft will optimum perform the tasks of perception, decision/planning and execution. People involved in these processes should ensure that information is received from all systems and sensors and that this information is accurate and correct for achieving a safe flight. Finally, regulations and legislation should be established for their safe operation and efficient management of the airspace [30, 31]. The guidelines should also include the characteristics and specific requirements of the drones. Finally, unmanned traffic management including sensors, communication systems and servers will be introduced for coordinating and monitoring the large number of drones in an urban or rural environment as well as for receiving data and providing real time information [30].

6 Conclusions

The automated vehicles will severely affect the labor sector. Workers involved in the design, manufacturing and in the various stages of operation are required to be upskilled or reskilled. Drivers will have a new role based on the automation level. This research presented a taxonomy of skills and knowledge drivers, workers and other people involved in the field of autonomous vehicles. Road, rail, maritime and aviation sector were analyzed and all levels of automation were considered. The research revealed similarities in the need of social, programming, engineering and communication skills, as in all different transport modes the design of the vehicle, the development of its sensors and systems enabling the automated operation and the proper communication and information exchange among the vehicles and the infrastructure are required. Regulations should also be established while liability issues are necessary to be clarified. Among the differences are the various systems each transport mode needs for automated operation, as well as the remote control skills and knowledge required.

Acknowledgement. The analysis is conducted within the framework of Drive2theFuture project (Needs, wants and behavior of "Drivers" and automated vehicle users today and into the future" funded by European Commission under the MG-3.3.2018: "Driver" behavior and acceptance of connected, cooperative and automated transport; Research and Innovation Action (RIA).

References

1. Litman, T.: Autonomous vehicle implementation predictions: implications for transport planning. Transp. Res. Board Ann. Meet., 36–42 (2014)
2. Alonso Raposo, M., Grosso, M., Després, J., Fernández Macías, E., Galassi, C., Krasenbrink, A., Krause, J., Levati, L.M., Mourtzouchou, A., Saveyn, B., Thiel, C., Ciuffo, B.: An analysis of possible socio-economic effects of a Cooperative, Connected and Automated Mobility (CCAM) in Europe. European Union (2018)
3. Manser, M., Noble, A., Ghanipoor Machiani, S., Shortz, A., Klauer, S., Higgins, L. Ahmadi, A.: Driver training research and guidelines for automated vehicle technology (2019)
4. Marinik, A., Bishop, R., Fitchett, V., Morgan, J.F., Trimble, T.E., Blanco, M.: Human factors evaluation of level 2 and level 3 automated driving concepts: concepts of operation. National Highway Traffic Safety Administration, Washington, DC (2014)
5. Hutschins, C.: Autonomous Vehicles and Driver Capability (2018)
6. Fiedler, R., Bosse, C., Gehlken, D., Brummerstedt, K., Burmeister, H.C.: Autonomous vehicles' impact on port infrastructure requirements (2019)
7. Lundquist, M.: Autonomous bus passenger experience. Ph.D. dissertation (2018)
8. Mirnig, A., Gärtner, M., Wallner, V., Trösterer, S., Meschtscherjakov, A., Tscheligi, M.: Where does it go? A study on visual on-screen designs for exit management in an automated shuttle bus. In: Proceedings of the 11th International Conference on Automotive User Interfaces and Interactive Vehicular Applications (AutomotiveUI 2019), pp. 233–243. Association for Computing Machinery, New York (2019)
9. Flämig, H.: Autonomous Vehicles and Autonomous Driving in Freight Transport (2016)
10. Brandenburger, N., Hörmann, H.J., Stelling, D., Naumann, A.: Tasks, skills, and competencies of future high-speed train drivers. Proc. Inst. Mech. Eng. Part F J. Rail Rapid Trans. (2016)
11. Australian Industry Standards – Rail IRC Skills Forecast (2018)
12. Brandenburger, N., Naumann, A., Friedrich, B., Grippenkoven, J.: Automation in railway operations: effects on signaller and train driver workload (2018)
13. Brandenburger, N., Naumann, A.: From in-cabin driving to remote interventions – train driver tasks change with railway automation. In: Poster at HFES Europe Chapter Annual Meeting (2018). https://www.hfes-europe.org/posters-2018/

14. Shift2Rail: Bridging the Skills Gap for the Rail Sector - Analysis of Six Measures and Recommendations (2019)
15. Bienfait, B., Zoetardt, P., Barnard, B.: Automatic train operation: the mandatory improvement for ETCS applications. Aspect Irse, pp. 1–10 (2012)
16. Australian Industry Standards – Rail Sector Overview (2018)
17. Infante, V., Sousa, D.: Deliverable D3.1: proposed future training curricula and courses for the transport sector. The skilful project (2018)
18. World Maritime University: Transport 2040: Autonomous ships: A new paradigm for Norwegian shipping - Technology and transformation (2019). Reports. 59
19. International Chamber of Shipping: Seafarers and digital disruption - The effect of autonomous ships on the work at sea, the role of seafarers and the shipping industry (2018)
20. IMOConvention on the International Regulations for Preventing Collisions at Sea (COLREGs) (1972)
21. Veritas, B.: Guidelines for autonomous shipping. guidance note NI 641 DT R00 E (2017), checked on 11 June 2018
22. Fraunhofer CML: RoboVaaS - Robotic Vessels as-a-Service (2019). https://www.cml.fraunhofer.de/en/researchprojects/current-projects.html
23. Vartdal, B.J., Skjong, R., Clair, A.L.S.: Remote-Controlled and Autonomous Ships (2018)
24. Carey, L.: Report on BIMCO autonomous ships seminar (2019). https://law.nus.edu.sg/cml/pdfs/reports/CML-R1901.pdf
25. Pavlas, D., Burke, S., Fiore, S., Salas, E., Jensen, R., Fu, D.: Enhancing unmanned aerial system training: a taxonomy of knowledge, skills, attitudes, and methods. Hum. Fact. Ergon. Soc. Ann. Meet. Proc. **53**, 1903–1907 (2009)
26. Tvaryanas, A.P., Thompson, W.T., Constable, S.H.: Human factors in remotely piloted aircraft operations: HFACS analysis of 221 mishaps over 10 years. Aviat. Space Environ. Med. **77**(7), 724–732 (2007)
27. Wilson, J.R.: UAVs and the human factor. Aerospace Am. **40**, 54–57 (2002)
28. Federal Aviation Administration: Summary of Small Unmanned Aircraft Rule (Part 107) (2016). https://www.faa.gov/uas/media/Part_107_Summary.pdf
29. Colombo, P.: Self-Flying planes are here; autonomous aircraft are the future (2019). https://www.ansys.com/blog/self-flying-planes-vs-autonomous-aircraft
30. Duval, T., Green, A., Langstaff, M., Miele, K.: Air-mobility solutions: what they'll need to take off (2019). https://www.mckinsey.com/industries/capital-projects-and-infrastructure/our-insights/air-mobility-solutions-what-theyll-need-to-take-off
31. European Union Aviation Safety Agency: Concept of Operations for Drones A risk based approach to regulation of unmanned aircraft (2015)

Towards the Adoption of Corporate Mobility as a Service (CMaaS): A Case Study

António Amaral[1], Luís Barreto[2(✉)], Teresa Pereira[2], and Sara Baltazar[2]

[1] CIICESI, Escola Superior de Tecnologia E Gestão, Instituto Politécnico do Porto, Felgueiras, Portugal
ama@estg.ipp.pt
[2] Escola Superior de Ciências Empresariais, Instituto Politécnico de Viana do Castelo, Valença, Portugal
lbarreto@esce.ipvc.pt

Abstract. The increasing level of awareness gained, by citizens in general and companies in particular, around the sustainability issues and of the climate change are producing changes in how organizations are dealing and projecting their future vision. Therefore, new managerial approaches are being embraced towards adopting a set of a strategies fully aligned with the reduction of the greenhouse gas emissions. Due to this increase evidence of sensitivity, organizations are embracing their role as stakeholders that need to contribute, throughout its corporate social agenda, to a responsible and smart policy promoting the implementation of strategies that could endeavor the cultural shift of their workers, clients, suppliers, among others, towards effectively contributing to sustainability and social responsibility. The case study of a medium size company reported is related to a structural change in how the organization foresees its mobility behavior and how it intends to follow the concepts of Corporate Mobility as a Service (CMaaS). This case study discloses the strategies that have been implemented and the Information and Communication Technologies (ICT) platform that has been developed towards having a broader view about the impacts of the mobility requested by all the organization. In addition, it is presented a group of Key Performance Indicator (KPI) that point the benefits attained with this effort as well as projecting the following steps that will support the CMaaS roadmap implementation in the future.

Keywords: Corporate Mobility as a Service (CMaaS) · Carpooling · Sustainability · Corporate social responsibility · Cultural change

1 Introduction

Companies and citizens are gaining an increased awareness regarding the importance of being sustainable, as well as of its corporate impacts, especially in citizens' quality of life within the context of smart cities development [1, 2]. The mobility theme is gaining enhanced visibility and importance within the corporate and social responsibility agendas definition among companies, intending to adopt new managerial approaches to use, for example, their car fleet in a far more smart and sustainable way

E. G. Nathanail et al. (Eds.): CSUM2020, AISC 1278, pp. 316–325, 2021.
https://doi.org/10.1007/978-3-030-61075-3_31

[3]. Nevertheless, this motivation for re-adjusting their management approaches towards the corporate mobility design is, somehow, related with the fiscal incentives that the Portuguese government is introducing to stimulate a paradigm shift. Due to this increase evidence of sensitivity, organizations are embracing their role as stakeholders that need to contribute to a responsible and smart policy promoting the implementation of strategies that could endeavor the cultural shift of their workers, clients, suppliers, among others, towards effectively contributing to sustainability and social responsibility. Some of the sustainability challenges could be tackled through using smart mobility approaches, improving mobility for the individual user and through the implementation of the Mobility as a Service (MaaS) concept. For that reason, the reduction of private car commuting is essential towards ensuring cities' decarbonization and the transition of behavior towards more sustainable practices and approaches, which will favor the appearance of shared mobility services as a viable alternative for daily commuting [4]. Corporate mobility deals with how companies enable their employees to use the company's car fleet in a sustainable manner. Full corporate mobility can be achieved by complementing how employees use transports in commuting travels [5]. Thus, Corporate Mobility as a Service (CMaaS) can then be easily described as the use of the MaaS concept applied to the context of corporate mobility, knowing that MaaS can be supported using shared mobility service integration like carpools, car rentals, city bikes and ride sharing services, using also well-developed public transport systems [6–8]. The CMaaS adoption could start by introducing travel plans (adopted since 1998 in the UK [9]) supported in a digital platform. Such strategy can use a broad range of measures that could contribute to increase employees' quality of life and to more favorable working conditions.

The main goal of this paper is to present a case study of a Portuguese medium sized company that foresees a structural change in its corporate mobility behavior. The case study discloses the strategies that have been implemented, the Information and Communication Technologies (ICT) platform that has been developed towards having a broader view about the corporate mobility changes and impacts. In addition, a group of Key Performance Indicator (KPI) are pointed along with the benefits attained, as well as projecting the following steps that will support the CMaaS roadmap implementation in the future, pointing out the major outcomes and benefits expected through this new managerial vision. The paper is organized as follows: the next section, Sect. 2, presents and introduces some concepts about mobility, and their progress. The Sect. 3 presents the developed features performed by a Portuguese construction company, that seeks to have a full operational CMaaS system. Finally, Sect. 4 presents the conclusion and limitations of the case study.

2 The Evolution of Mobility Solutions

The car's ownership and the amount of vehicular travels have increased significantly throughout time, in the urban areas [10] along with its mischievous environmental impacts, especially when it concerns to the developing countries [11]. However, in the developed countries, the new generations – also called millennial generation – are not so interested in car ownership and as a consequence of that the driver's license rate is decreasing in European countries and in the US [12]. The technological advances in the ICT can improve and propel the need for new mobility paradigms [13]. Nevertheless, these new mobility systems request an adequate amount of involvement between a substantial number of stakeholders that have critical roles on ensuring the mobility system balance according to MaaS paradigm and its own governability structure [14]. In order to guarantee the MaaS' promises about being efficient, universal and also plural to enable different possibilities of choice to its users it will require the inclusion of a public entity legally mandated to intervene into the system towards being able to ensure its efficiency and equity [15]. Using ICT as a mainstream in the optimization process of any mobility activity towards being able to improve the quality of life as well as to reduce the greenhouse gas emissions [16]. Thus, following are presented some established and emerging mobility solutions and services that can be needed to contribute for a smart and sustainable mobility [1].

2.1 Mobility as a Service (MaaS)

MaaS is a mobility concept that can integrate different transport modes, enabling users to be aware of all different possibilities in a single service, allowing users to purchase tickets and plan trips and book door-to-door trips from different suppliers in one place [17]. According to Jittrapirom et al. [18] and Gonçalves et al. [19] the main goal of MaaS is to create a seamless experience, with higher level of flexibility and a more customized transport solution, supported in a form of a digital mobile application. Therefore, services like carpools, car rentals, city bikes and ride and car sharing services, and of course public transport systems and services, from different stakeholders with their own platforms, contracts and payment solutions, can be bundled by a MaaS service provider, that can offer a more user friendly single interface for booking, travel's planning and ticket purchasing. According to Laine et al. (2018) [20] a MaaS service could on one hand reduce the need of users having their own car, but on the other hand not decrease the amount of vehicular travel. Notwithstanding, it still lacks enough evidences and conceptual clarity about its overall potential and acceptance level by the general public [21]. However, there are some examples reported in literature of MaaS applications that can help clarifying the first developments attained: a) In the study performed by Schikofsky et al. (2020) [22], some insights from Germany are presented in a structural equation model about the motivational mechanisms behind the intention to adopt MaaS. The identification of the latent acceptance factors towards the MaaS adoption are very important to practice and for supporting the definition of suitable public policies; b) Wright et al. (2020) [23] presented the results obtained from four pilot tests performed with the RideMyRoute App in different European sites – the App was able to suggest solutions which included carpool options in 20% of the planning solutions and between those 85% were solutions that involved a mix between carpool services and public transport; c) Storme et al. (2020) [21] studied, in Ghent (Belgium), the relationship between MaaS use and private car ownership. One of the main conclusions obtained in this exploratory pilot study was that the majority of participants were willing to explore MaaS services (especially public transport and car sharing services). Nevertheless, an evident reduction of the private car use was difficult to attain, especially in a real-life setting or even to leisure trips; d) In the research study developed by Ho et al. (2020) [8] in Sydney (Australia) and in Tyneside (UK), some insights were gained about MaaS demand towards identifying different subscription models and willingness to pay for mobility services. Simultaneously, it was tried to assess the MaaS demand by the following groups: tourists, businesses, and households.

2.2 Corporate Mobility

Corporate mobility can be defined as a mean of transportation provided by employers or by other external companies, normally in the form of buses (shuttle services). This type of mobility can also be induced by the change of attitudes, specially by the younger employees (millennials) that are not interested in having or using own cars in commuting trips [12]. Therefore, companies and organizations that are aware of this attitude shift, need to provide mobility services to be able to retain and capture qualified working people. For example, in the US, Amazon and Microsoft companies, as a consequence of not having a well-defined public transport system and as a managerial option towards increasing social responsibility awareness of their employees related to the unsustainable use of private cars in commuting trips, are offering employee shuttles in Seattle [24]. In Europe, IKEA also provided, during the summer of 2017, employee shuttles, due to some public transport problems during that season [25]. Corporate mobility is not only about transporting people to their workplace, but also ensuring that employees have good mobility within the working hours, for meetings, lunch, etcetera. Another example of corporate mobility is when a Higher Education Institution promotes different forms of mobility for their students and employees. One such example is the "Bus Académico" of the Polytechnic Institute of Viana do Castelo (IPVC), in Northern Portugal. IPVC provides shuttle services that connects 17 locations to the IPVC Schools, thus allowing both students and employees from the Alto Minho Region to travel to and from any of IPVC's schools.

2.3 Corporate Mobility as a Service (CMaaS)

CMaaS can be viewed as the application of the MaaS concept on corporate mobility, allowing to transfer some of the proclaimed benefits from MaaS into corporations. Thus, CMaaS implies that an employer can provide MaaS for its employees, as a complimentary service or as complement benefit, allowing the support and management of a variety of different transport modes using a simple interface, but aligned with the scope of corporate mobility [7]. CMaaS is a very novel concept of MaaS and there are scarce real-life examples. Some CMaaS initiatives are implemented in Europe, especially in Central and Northern Europe, and one of their common features is the use of a carpooling, car-sharing or ride-sharing platforms. One example is from the study performed by Hesselgren et al. (2019) [7] where empirical results are presented from a study of CMaaS at a company with 14,000 employees in Sweden. CMaaS are socio-technical systems, therefore several perspectives need to be properly handled and integrated like user applications functionalities and attributes, transport modes, technology illiteracy, travel habits, lifestyle choices, and employer relations, among others.

Throughout the adoption of CMaaS and sustainable transport three barriers were identified: (1) inadequate integration of the internal transport system with external transport systems; (2) corporate policies, culture and norms that are not aligned with the usage of other services; and (3) system limitations due to particular regulations.

3 Fostering CMaaS - A Portuguese Case Study

This section presents the steps given by the DST [26] company towards the goal of fully implementing a CMaaS solution. DST is one of the biggest construction companies in Portugal created in the late forties of the past century. It has several construction sites in Portugal and in other countries, in addition to engineering and construction, they have now businesses associated with the environment, renewable energies, telecommunications, real estate and capital ventures. In DST fleet cars are assigned to employees of a certain hierarchy level and can be used for business trips but also private trips, which can result in large and inefficient car fleets, and in most cases the fleet cars are often over dimensioned since all mobility needs have to be covered with just one car [27]. This type of corporate mobility management can also promote that other employees are neglected in their mobility necessities due to the lack of efficiency in managing the corporate car fleets. The increase visibility and importance of sustainability, climate change, social responsibility, and social equity awareness within DST, the company managers decided to approve an internal project with the goal to promote a cultural shift within all workers towards the adoption of several mobility practices that can enhance the corporate sustainability. Those were the main reasons that led them towards the definition of DST's CMaaS strategy. The first development resulted in the "dst group carsharing" digital platform (Fig. 1) available since October 2017, which main function was to promote sharing of trips among

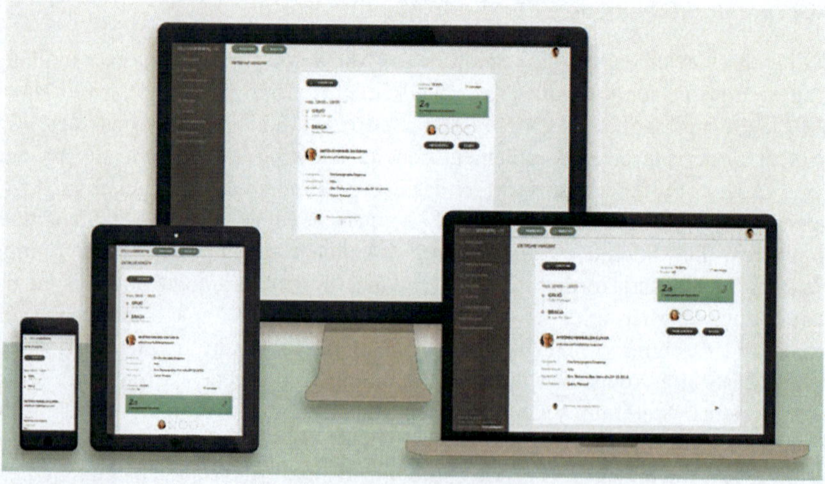

Fig. 1. The DST group digital carsharing platform.

employees. Employees (users) can share leisure trips, work trips or work commute trips. The platform also enables to share parcel transport when requested. Its main functionalities are the following: to register trips with origin and destination, if there is already one travel with a similar route on that date the user is warned; to search trips based on its origin, destination and date trip, reporting suggested trips registered in the route or in a route with a slight deviation; email notifications or through the platform regarding trip changes in which the employee is involved; possibility to create alerts to notify when a new trip is entered with some predefined parameters; and the listing of trips and parcel delivery. The platform discloses a dashboard (Fig. 2) in which its presented some general information regarding the economy that the user effort meant for the company, the number of kilometers shared, the number of trips shared, the number of parcels that were transported, and the percentage of employees that shared at least one trip and the total kilograms of CO_2 saved with the travel shares.

Fig. 2. DST's group car sharing platform dashboard.

An innovative functionality of this platform is the use of gamification [28], to stimulate car sharing to reward employees. The reward can be obtained such as concert, cinema or theater tickets or as consumption in companies' vending machines. The main goal of the car sharing game is to reduce the number of trips made by DST's employees, both in transporting other employees or parcels, thus benefiting the DST's sustainability commitment. The challenge is to use car sharing and to transport parcels whenever possible, and the greater the number of kilometers and shared loads of parcels, the greater the reward. For keeping this gamification updated the platform sends notification to all users informing the total number of shared kilometers, the total savings and the CO_2 emission reduction obtained by all users complemented with information regarding the three best employees (Fig. 3). During the last two years the company registered a set of Key Performance Indicator (KPI): CO_2 kilograms reduction; number of saved kilometers traveled; and percentage of employees that use the digital platform.

Fig. 3. Information regarding the kilometers shared by the three best employees and the overall kilometers shared by all users. Note: Due to GDPR regulations, employee's names are hidden.

According to the data presented in Table 1 the introduction of the platform has meant a significant decrease of CO_2 emissions, with a reduction of 44.417 kg in 2018 and 63.381 kg in 2019, meaning that in 2019 there was a reduction of 42,70% when compared to 2018. Table 1 also shows that in 2018 the number of saved kilometers by the company car fleet was 186.704 and in 2019 was 283.017 thus, 2019 had an increase of more than 53% in the saved kilometers when compared to 2018. This also means that there was a reduction of costs in terms of fuel consumption and fleet car maintenance, however this is not still a significant value but clearly will have a significant impact in the next years. Finally, it is important to refer that the users/employees that use the platform increased more than 7% in 2019, which clearly states that sustainable mobility and social responsibility awareness is increasing among the company's employees.

Table 1. Key Performance Indicators (KPI).

KPI	2018	2019	Δ% to 2018
CO_2 reduction (kg)	44.417	63.381	+42,70
Saved kilometers (km)	186.704	283.017	+53,73
% of participation	16.57	24.21	+7.64

3.1 Next Steps Towards CMaaS

The previous section has presented, from the perspective of a Portuguese construction company, the first step that the company has realized to pursuit a fully integrated corporate mobility system, also called Corporate Mobility as a Service (CMaaS). The company is aware that more steps are needed towards fulfilling CMaaS requirements.

In that sense, some initiatives are suggested/advised that can be planned for the years to come. These initiatives have in mind a more flexible and more sustainable mobility service for the company's employees. Next are pointed some of the initiatives that can be developed: a) **Renew the car fleet** - In order to have a more sustainable car fleet, it is advised that the company should increasingly use electricity powered vehicles such as fully Electric Vehicles (EVs) and plug-in hybrid electric vehicles (PHEVs). This, of course, is a medium-long term initiative, and can be supported by the European Union financial incentives and taxes; b) **Integrate alternative mobility offers/solutions** - The concept of CMaaS is not limited to vehicular solutions and to only company owned solutions. It is advisable that other modes of transportation must be considered as well, including, public transportation, electric bicycles and/or taxis and also the integration of other active modes of transportation in order to combat employee's inactivity habits. Such transparent integration of those mobility solutions, will clearly advance the use of low emissions within the company context; c) **Key-less or code driven access** - To allow a more manageable, ease and effective access to the car's fleets, or other mobility offer, it is proposed the integration of a key-less access solution. This means that each employee (or person with access rights) has access to, for example, an RFID chip card or a smartphone application that allows access to the mobility solutions, as well as the right to start and stop the vehicle. This can also be used to in real time update the state of the mobility solution in the digital platform; d) **Motivational strategies** - The company has already started to create motivational strategies through its gamification strategy. However, more motivational strategies can be used in order to speed up the needed mind shift, the use of economic incentives, motivational strategies can address emotional factors of the relevant user groups and must be used transversely within the company, from the upper management, to the accounting department of the company, to the fleet management division, to an anonymous employee; e) **Definition of new KPIs** - The company is also aware that new KPIs can be used to gauge the progress of the employee's and the company vision in terms of mobility habits and sustainability. Some examples of new KPIs are: the reduction of the number of parking slots used in company's sites towards being used in the future to other value-added activities; the integration of other active modes of transportation in order to combat employee's inactivity habits; the establishment of new partnerships with different stakeholders towards involving them in this ecosystem and towards being possible to exchange the credits gained through the mobility to exchange into other services or products; the reduction of the amount of traffic and noise within the company's sites and the region; among other possibilities.

4 Conclusion and Future Developments

This paper presents the first solution developed by a Portuguese company concerning its internal mobility habits. A car sharing platform was used, with an innovative gamification strategy to encourage changes in the employees' mobility habits. Due to the increased level of awareness gained by the company, concerning sustainability and climate change issues, it has embraced a new managerial approach that seeks to have a full CMaaS system implemented. The primary purpose of the CMaaS is to provide the company employees with accessible and reliable corporate mobility, and also a reliable commuter mobility option to and from the workplace. The car sharing platform presented has some innovative functionalities, like the possibility to share parcel transport when requested, the use of a dashboard with personal and overall information and a gamification strategy that promotes competitiveness for reaching more sustainable and environmental friendly attitudes among the employees. It is also introduced a set of features that should be introduced in order to strive the CMaaS system as is the purpose of the company. The main limitation of the actual system is having a unique provider (the company itself), which owns and operates all transport modes and there is only one group of users, employees. For a CMaaS setup it is fundamental that the company can offer more transportation modes and the public transport can be included as a collaborator, creating a fully integrator mobility system.

As future developments are thought to study the inclusion of new partners towards assessing the right business model for collaboration, as well as it is under discussion to have other pilot test in an industrial park with multiple companies, most of them in the automotive sector, to adopt the sustainable mobility paradigm through CMaaS.

References

1. Yigitcanlar, T., Kamruzzaman, M., Foth, M., Sabatini-Marques, J., da Costa, E., Ioppolo, G.: Can cities become smart without being sustainable? A systematic review of the literature. Sustain. Cities Soc. **45**(June 2018), 348–365 (2019)
2. Silva, B.N., Khan, M., Han, K.: Towards sustainable smart cities: A review of trends, architectures, components, and open challenges in smart cities. Sustain. Cities Soc. **38** (August 2017), 697–713 (2018)
3. Lyons, G.: Getting smart about urban mobility Aligning the paradigms of smart and sustainable. Transp. Res. Part A: Policy Pract. **115**, 4–14 (2018)
4. Sopjani, L., Stier, J.J., Hesselgren, M., Ritzén, S.: Shared mobility services versus private car: implications of changes in everyday life. J. Clean. Prod. **259**, 8 (2020)
5. Lindblad, S., Nygårds, S.: Degree project in the field of technology: corporate mobility as a valuable service? An exploratory study from a business model. Master of Science thesis, KTH Industrial Engineering and Management (2018)
6. Merkert, R., Bushell, J., Beck, M.J.: Collaboration as a service (CaaS) to fully integrate public transportation Lessons from long distance travel to reimagine mobility as a service. Transp. Res. Part A Policy Pract. **131**(September 2019), 267–282 (2020)
7. Hesselgren, M., Sjöman, M., Pernestål, A.: Understanding user practices in mobility service systems: Results from studying large scale corporate MaaS in practice. Travel Behav. Soc. **21**(December), 318–327 (2019)
8. Ho, C.Q., Mulley, C., Hensher, D.A.: Public preferences for mobility as a service: insights from stated preference surveys. Transp. Res. Part A Policy Pract. **131**(September 2019), 70–90 (2020)
9. Rye, T.: Travel plans: do they work? Transp. Policy **9**(4), 287–298 (2002)
10. Contreras, G., Platania, F.: Economic and policy uncertainty in climate change mitigation: The London Smart City case scenario. Technol. Forecast. Soc. Change **142**(December 2017), 384–393 (2019)
11. Moriarty, P., Honnery, D.: Low-mobility: the future of transport. Futures **40**(10), 865–872 (2008)

12. Klein, N.J., Smart, M.J.: Millennials and car ownership: less money, fewer cars. Transp. Policy **53**, 20–29 (2017)
13. Gairal, R., Golightly, D., Laing, K., Palacin, R., Todd, L.: Children, Young people and Mobility as a Service: opportunities and barriers for future mobility. Transp. Res. Interdisc. Perspect. 100–107, 6–8 (2020)
14. Docherty, I., Marsden, G., Anable, J.: The governance of smart mobility. Transp. Res. Part A Policy Pract. **115**(October 2017), 114–125 (2018)
15. Pangbourne, K., Mladenovič, M.N., Stead, D., Milakis, D.: Questioning mobility as a service: Unanticipated implications for society and governance. Transp. Res. Part A Policy Pract. **131**(October 2019), 35–49 (2020)
16. Barreto, L., Amaral, A., Baltazar, S.: Flexible on Demand Transport Services (FDTS): the future of mobility systems. In: Proceedings - 2019 IEEE International Conference on Engineering, Technology and Innovation, ICE/ITMC (2019)
17. Goodall, B.W., Dovey, T., Bornstein, J.: The rise of mobility as a service. Deloitte, Technical report 20 (2017)
18. Jittrapirom, P., Caiati, V., Feneri, A.M., Ebrahimigharehbaghi, S., AlonsoGonzález, M.J., Narayan, J.: Mobility as a service: a critical review of definitions, assessments of schemes, and key challenges. Urban Plann. **2**(2), 13–25 (2017)
19. Gonçalves, L., Silva, J.P., Baltazar, S., Barreto, L., Amaral, A.: Challenges and implications of Mobility as a Service (MaaS). In: Amaral, A., Barreto, L., Baltazar, S., Siva, J., Gonçalves, L. (eds.) Implications of Mobility as a Service (MaaS) in Urban and Rural Environments: Emerging Research and Opportunities, pp. 1–20. IGI Global, Hershey (2020)
20. Laine, A., Lampikoski, T., Rautiainen, T., Bröckl, M., Bang, C., Stokkendal Poulsen, N., Kofoed-Wiuff, A.: Mobility as a Service and Greener Transportation Systems in a Nordic context (2018)
21. Storme, T., De Vos, J., De Paepe, L., Witlox, F.: Limitations to the car-substitution effect of MaaS. Findings from a Belgian pilot study. Transp. Res. Part A Policy Pract. **131**(March 2019), 196–205 (2020)
22. Schikofsky, J., Dannewald, T., Kowald, M.: Exploring motivational mechanisms behind the intention to adopt mobility as a service (MaaS): insights from Germany. Transp. Res. Part A Policy Pract. **131**(October 2019), 296–312 (2020)
23. Wright, S., Nelson, J.D., Cottrill, C.D.: MaaS for the suburban market: Incorporating carpooling in the mix. Transp. Res. Part A Policy Pract. **131**(April 2019), 206–218 (2020)
24. Soper, T.: Seattle traffic disappears as Amazon, Microsoft, others enforce remote work policies, March 2020
25. Sustainable life at home – transportation. https://www.ikea.com/ca/en/this-is-ikea/sustainable-everyday/transportation-pub1982d56e
26. DST Group. https://www.dstsgps.com/?locale=en_US
27. Bates, J., Leibling, D.: Spaced Out Perspectives on parking policy. RAC Foundation, no. July, p. 132 (2012)
28. Marconi, A., Loria, E.: Key role of gamified motivational systems for citizens engagement and behavior change. In: Amaral, A., Barreto, L., Baltazar, S., Silva, J.P., Gonçalves, L. (eds). Implications of Mobility as a Service (MaaS) in Urban and Rural Environments: Emerging Research and Opportunities, pp. 211–246. IGI Global, Hershey (2020)

Accelerating Deployment: Governance and Business Models

Creating Smart(er) Cities by Accelerating Innovation in Transport Small and Medium Sized Enterprises (SMEs): The Case of West Midlands Region

Eleni Anoyrkati(✉) and Alba Avarello

Coventry University Enterprises Ltd., Puma Way, Coventry CV1 2TT, UK
e.anoyrkati@coventry.ac.uk,
aavarello@cad.coventry.ac.uk

Abstract. Small and Medium sized Enterprises (SMEs) at present employ 55% of the European workforce in transport and their essential role in the value chain will increase [1].

The rigid value chain of the transport sector is preventing innovation. Tier 2 SMEs generally find it difficult to interact with vehicle manufacturers as they tend to have short-term supply contracts to Tier 1 companies. They have no collective voice or influence at European level and the EU is not supporting innovation in these companies [2].

As part of a wider research project, this paper investigates this market failure at regional level and focuses on the opportunities for innovative and proactive transport SMEs. The main objective is to identify the innovation process of transport SMEs and the way to improve their capacity and capability to further develop and grow. Based on desk research of relevant literature and interviews with SMEs and Business Support organisations in the transport sector in the West Midlands region of England, this paper draws conclusions on the key future success factors that can boost transport innovation.

Keywords: Transport · Innovation · Smes

1 Introduction

This research paper aims to equip the policy makers with the necessary evidence and a solid knowledge base to better channel funds for Research and Innovation. The main objective was to perform a transport innovation mapping exercise on the West Midlands region of the UK. The analysis focused on the transport SME competitiveness support in the region (status quo) by employing a systemic approach in order to analyze the SME innovation environment. Further research was conducted to identify the opportunities arising from the new forms of transport such as low carbon vehicles, smart mobility, light vehicles etc. and investigate what is the role of the SMEs into the innovation chain and how they can benefit from the arising opportunities.

E. G. Nathanail et al. (Eds.): CSUM2020, AISC 1278, pp. 329–334, 2021.
https://doi.org/10.1007/978-3-030-61075-3_32

2 Methodology

The main research questions answered in this paper included:

- What is the status quo of the regional transport SME competitiveness level.
- What are the key strengths that the region should capitalize on and the main opportunities that should take advantage of
- What are the main obstacles and threats that should be overcome to offer a more competitive transport SME sector

To address these research objectives, two main methodologies were used. A secondary research on scientific and grey literature about the transport SMEs status on the region. Also, regional policy papers and current funding mechanisms and measures were investigated. To validate the results identified and also capture fields that were not covered in the literature, ten interviews were conducted. The following framework conditions were investigated: geography and governance, economy, stakeholders, existing regional policy. A template (semi-structured) questionnaire was used for the interviews that were conducted with the following key stakeholders:

Business Support organizations: They deal with a great number of SMEs on daily basis and therefore they have a very good picture of the transport SMEs position in the region.

SMEs: As the main target group, they were in a good position to state their perspective on advantages and disadvantages of the innovation support framework that currently exists on a bottom up level.

Policy makers: The regional policy makers provided a top down perspective of the support provided as well as the innovation levels.

The interviews aimed at identifying the following: challenges of developing and commercialization products and services, opportunities and obstacles, how transport innovation products are financed. Based on the literature review and the answers provided during the interviews, a SWOT analysis was conducted.

3 West Midlands Region

The UK is located off mainland Europe, North-West of France and separated by the English Channel. Although made up of disparate islands, the UK is mainly comprised of the island of Great Britain and a portion of the island of Ireland, namely Northern Ireland in the North-East. Therefore, the constituent countries of Scotland, England and Wales located in Great Britain, and the constituent country of Northern Ireland are together part of the sovereign nation of the UK [3, 4].

The SWOT analysis of the region revealed the following [5–10] (Table 1):

Table 1. SWOT analysis of the West Midlands Region.

Strengths	Weaknesses
One of the most accessible locations in the UK Upgraded transport network and infrastructure, improving connectivity, productivity and movement of goods Historic and cultural significance of transport sector and related industries in the region, birthing ancillary services and products Strong and stable economy, growth potential through strategy and policies Political support for the industry and innovation, nationally and regionally Very strong research clusters, manufacturers (OEM's and Tier 1 suppliers) Specialisms and trade bodies across major modes of transport - planes, trains and automobiles (aviation, rail and road) Potential for cross modal transport initiatives and concepts, integrated transport, Mobility as a Service models Home to world leading universities, science parks and incubators Combined authority for spreading of funds and resources across wards and districts Joined up policy approach through Transport for West Midlands Manufacturing remains the largest industry in the region	Huge congestion problem on key route network (comprising 7% of entire network yet accounting for 50% of traffic volume), impact on traffic flows and public transport Midlands Engine, Midlands Connect skewed focus across regions, East vs West Skills shortages to fulfil innovation potential Gaps in business support provision Skills gaps and knowledge Some transport modes concentrated in one or two localities, e.g. Metro WM commercial innovation activity and investment is moderate by UK standards and lower than international competitors Innovation gaps are generally larger for Greater Birmingham and Solihull Local Enterprise Partnership and the Black Country Local Enterprise Partnership than for Coventry and Warwickshire Stagnant growth for SMEs SMEs time constrained, many not accessing the support available or unsure how Some gaps in broadband and energy provision R&D can be concentrated in larger manufacturers Complex and complicated funding streams for SMEs, the less experienced or less networked in companies may struggle to access the resources they need
Opportunities	*Threats*
There is scope for an increased role for sustainable travel Hub/cluster for intelligent mobility, reputation for transport innovation and inward investment Home of 'Future Mobility' built from successful partnership working Pioneer of driverless car technology and 5G testing Different and innovative funding and procurement models post-Brexit Emerging products, services and sectors as a result of and preparing for environmental concerns (Low Carbon/Circular economies) WM firms are the most successful exporters in UK and the only region exporting mainly to China. There are also a number of planned infrastructure projects that will improve the region The Midlands Engine is a strategy launched in 2016 to enhance productivity in the region. This includes investment in transport to reduce congestion, a new university to increase graduates in Engineer and funding to support the growth of small businesses The HS2 rail link will boost Birmingham's economy by £1.4 billion by reducing connection times with London and the north	Brexit, disruption to manufacturing, supply chains, investment and movement of goods and services Businesses may not be ready to trade post Brexit HS2 Delays in delivery, over budget Competition for 'smart city', 'living lab' status with other regions in the country, namely London, Bristol Pace of societal change, innovation and competition Emergence of disruptive industries and technologies such as Robotics, AI, VR and 3D Printing Global supply chain disruption, risk of global recession, geopolitical instability and uncertainty Delivering on a low carbon future, cutting emissions through transport and innovation on time to government and international recommendations

4 Results

The West Midlands is the centre of transport innovation in UK, leading smart low-carbon movement of people and goods and creating new opportunities. It is home to renowned manufacturers (like Jaguar Land Rover, Aston Martin and JCB) that support an extensive supply chain of small firms. It also has a great connectivity, being well-served by major road and rail networks that help to support its entrepreneurialism. The service sector also plays a big role in the region, employing almost half of the working population.

The next ten years will be a time of massive changes for the mobility of people and goods and consumers behaviors. The main priorities and interventions relevant to SME competitiveness reflect the several local economic contexts across the region, but a number of themes are common. In particular:

- The need to create employment and GVA (gross value added);
- Supply chain development to support growth especially in advanced manufacturing and engineering;
- The full spectrum of Innovation and R&D activities, from general process innovation and productivity improvement to more intensive and tech focused activities to generate spin out companies;
- Strong innovation and technology flavour in Enterprise and start-ups, although some areas also prioritise the more general development of an entrepreneurial culture;
- Focus on priority sectors to drive growth;
- The advanced manufacturing sector is crucial, with automotive, aerospace and aeronautical sectors as key strengths.

  The future prosperity of the WM lies in the adaptability to long-term trends in mobility:
- Creating new markets like electric and connected autonomous vehicles (CAV) and mobility as a service since CAV is worth between £50 and £100 billion to the UK economy;
- Stimulating further innovation in key areas for the region like battery research and manufacturing, 5G and dat;
- Taking advantage of growing global markets in very light rail, digital rail and electric and autonomous flight;
- Continuing to develop a clean, integrated transport network, maximising the opportunities presented by HS2, optimising the value of the Transforming Cities Fund and other locally led investments and working smartly with Midlands Connect (an integrated transport network and arrival of HS2 could add £4 billion to the WM economy, driving major centres of growth such as UK Central Solihull).

The region will monitor its progress on a network of over 50 miles of roads in Coventry, Birmingham and Solihull, a globally leading 'real world' testbed for developing the next generation CAVs after a recent public and private investment of over £50 millions.

The funding landscape is improving with increasing amounts of business expenditure on R&D and venture capital investment.

The interviews revealed that the main barriers to innovation the SMEs face are money, skills shortage for people with technical knowledge, skills shortage in low carbon technologies especially and lack of knowledge/uptake in hydrogen fuel cell technology.

University degrees in the region are popular – students' study e.g. Environment, Engineering or Energy course, then if there are no jobs locally, they go abroad to Europe or International. However, there is a knowledge gap in the industry and amongst students/general public which creates a need for a public demonstration to showcase newer innovations, how to get to a low carbon future fast.

Barriers to partnering for SMEs include time constraints in making applications and awaiting a decision. A big barrier is all the form filling just to find out if they are viable or eligible for a particular programme. Other transport modes that the government invests in always proves to be too late, strategizing period and then the implementation is too late or haphazard, moreover the uncertainty and delays around high-speed rail are ongoing.

However, SMEs admitted that there are support services that can help them bring their innovations to market if they know where to look. For example, CWLEP have helped services come to market: The Innovation programme gives them non-financial support and a grant to help launch to the market. The Green Business programme funds energy grants. From idea generalisation to commercialisation, the Innovation Programme (CWLEP) supports companies at varying stages, some are very early concept, help them to prototype through to launching. The LEP also supports a number of companies through Innovate UK and the universities through venture capital, partnerships, signposting to other services through the Growth Hub.

The SMEs have also stated that they are missing out on opportunities and there is a need to invest more to support, with the onset of Brexit –financially and non-financially.

In regards to accessing new markets, there are support services that can help SMEs such as the international trade advisors appointed by the Department of International Trade and they help companies regulate the logistics of moving their goods and services, from border control and documentation.

The Chambers of Commerce offer generic business support programmes via local authority contracts. Not targeted at transport sector but can include transport business that fall within SME definition and are seeking generic business support.

5 Conclusions

In addition to a strong research base, West Midlands shows a well-developed network of science parks, associated innovation, incubator centers and accelerator programmes.

The region benefits from strong physical assets and 'hard' infrastructures to support transport SMEs' growth, development and expansion.

West Midlands faces multiple skills challenges: very few people have high-level qualifications and too many have no qualifications. This is a major issue for the development of innovative businesses.

There is a remarkable set of local networks, as well as institutions at regional, national and international levels developing industrial and academic collaborations and partnerships. This can offer the opportunity to actually bridge the skill gap by creating university-industry collaborations that could support the creation of innovative projects.

The business ecosystem is strengthened by a robust local policy focus on driving-up levels of innovation and maximising the potential of key assets, including major research-intensive universities, RTOs and R&D active firms.

Acknowledgements. The results presented in this paper have received funding from the European Regional Development Fund under the Interreg Europe programme PGI05275 (project acronym: RECREATE).

References

1. Muller, P., Julius, J., Herr, D., Koch, L., Peycheva, V., McKiernan, S.: Annual Report on European SMEs 2016/2017. European Commission, Brussels (2017)
2. Brooks, R., Maher, S., Morris, D., Davalli, C., Adams, N., Pickering, C.: SMEs-acquiring new technology in different regions in Europe, Innovative Transport SME Action-INTRASME (2015)
3. Constituent state. https://en.wikipedia.org/wiki/Constituent_state. Accessed 07 Jan 2020
4. Definitions dictionary. https://www.definitions.net/definition/constituent%20country. Accessed 07 Jan 2020
5. Coventry University. https://www.coventry.ac.uk/research/areas-of-research/institute-for-future-transport-and-cities/our-facilities/. Accessed 04 Feb 2020
6. Department for Business Innovation and Skills, SME journey towards raising external finance, London (2013)
7. Innovation Alliance for West Midlands. https://innovationwm.co.uk/2018/04/24/west-midlands-finance-for-small-business-spring-2018-update/. Accessed 07 Feb 2020
8. West Midlands Combined Authority: West Midlands Traver Trends 2017, Birmingham (2017)
9. Department for Transport: Future of Mobility-Urban Strategy, Industrial strategy, London (2019)
10. European Investment Bank: Using Financial Instruments for SMEs in England in the 2014–2020 programming period, London (2015)

Policy Directions for Enhancing Transport Innovation Infrastructure for Smarter Regions

Tessa Lukehurst and Eleni Anoyrkati(✉)

Coventry University Enterprises Ltd., Puma Way, Coventry, UK
tlukehurst@cad.coventry.ac.uk,
e.anoyrkati@coventry.ac.uk

Abstract. The European transport companies are leading innovators. Transport innovation can contribute to tackle major societal challenges, which are becoming more urgent by the day. However, budgets for transport are usually spent on maintaining or improving current infrastructure as statutory obligations drive the policy agendas for investment. There are opportunities offered by the rapid development of digital technology in 5G communications, connected and autonomous vehicles, alternative power sources and interconnected devices means that the transport sector is on the cusp of stochastic change. There are numerous issues around access to the technology as it becomes available and how to maximise the benefit to a region. This paper investigates the competitive advantage of five regions and explores the innovation potential along with the success factors needed in tackling any innovation obstacles. Policy suggestions are put forward in creating a funding strategy that builds upon existing and potential areas of competitive advantage, avoiding fragmentation and insularity and linking and leveraging the assets in new and different ways.

Keywords: Transport · Innovation · Policy

1 Introduction

This research paper aims to assist Policy makers in developing economic development activity and Policy Support for innovation in transport. Starting from a review of the transport industry and research activity in the West Midlands of the UK and taking account of best practices reviewed there and in four other European regions it makes general and specific recommendations for improvement.

2 Methodology

The analysis was conducted using desk-based research of relevant literature on Transport innovation policies and transport sector innovation systems as well as interviews with Policy Makers and leading Researchers. The main objective of the investigation was to identify the innovation process of transport oriented Small to Medium Enterprises (SMEs) in the partner regions.

E. G. Nathanail et al. (Eds.): CSUM2020, AISC 1278, pp. 335–346, 2021.
https://doi.org/10.1007/978-3-030-61075-3_33

The following framework conditions were investigated: geography and governance, economy, stakeholders, existing regional policy. The interviews aimed at identifying the following: challenges of developing and commercialization products and services, opportunities and obstacles, how transport innovation products are financed.

3 The West Midlands Transport Research Environment

3.1 The West Midlands

The West Midlands is one of 12 regions of the UK. Situated centrally in England it covers an area of about 13,000 km^2 and consists of a number of disparate areas. The population of the region was estimated at 5.8 million in 2016. Of these 2.9 million live in the West Midlands Combined Authority including 532,911 in Coventry. The combined authority covers an area of less than 1,000 km^2 which means that 50% of the population lives on 7% of the land area. The majority of the region's heavy industry is located here.

The region historically underperforms the south of the country, but has been able to show consistent growth over the past twenty years. Figure 1 shows that in local terms Coventry has grown at a slower rate than the rest of the region overall.

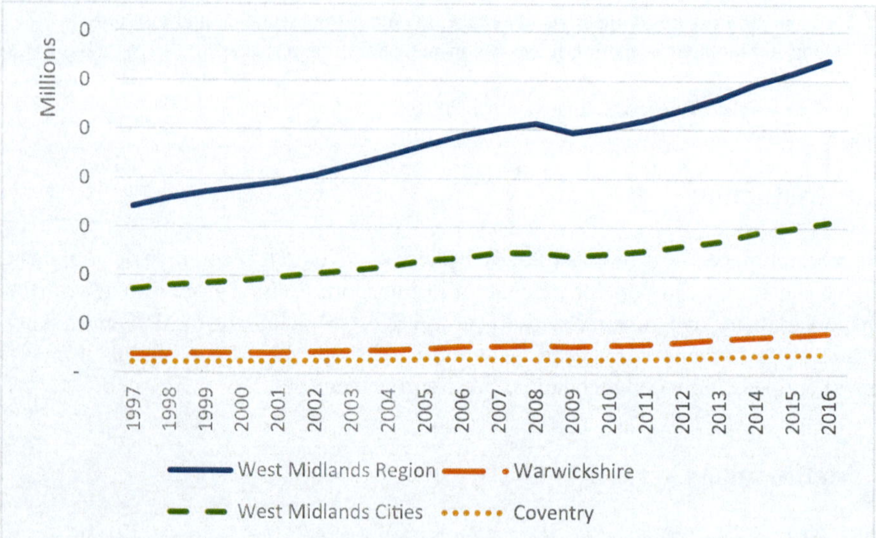

Fig. 1. GVA Growth in the West Midlands Region [1]

The region is the central nexus for Road and Rail communication throughout the UK.

The motorway circle of M6 and M42 around Birmingham is one of the busiest road networks in the UK and acts as a distribution node centrally connecting motorway links to all parts of the country. Birmingham's New Street Station is a central hub for the rail network and is the busiest rail station outside of London. This makes the area very attractive to transport businesses with many local warehousing and organisational hubs and makes transport one of the most important issues in the region.

As shown in Fig. 2, employment in this sector declined over many years, but has recently risen from below the national average to above, as the value of the central hub is realised.

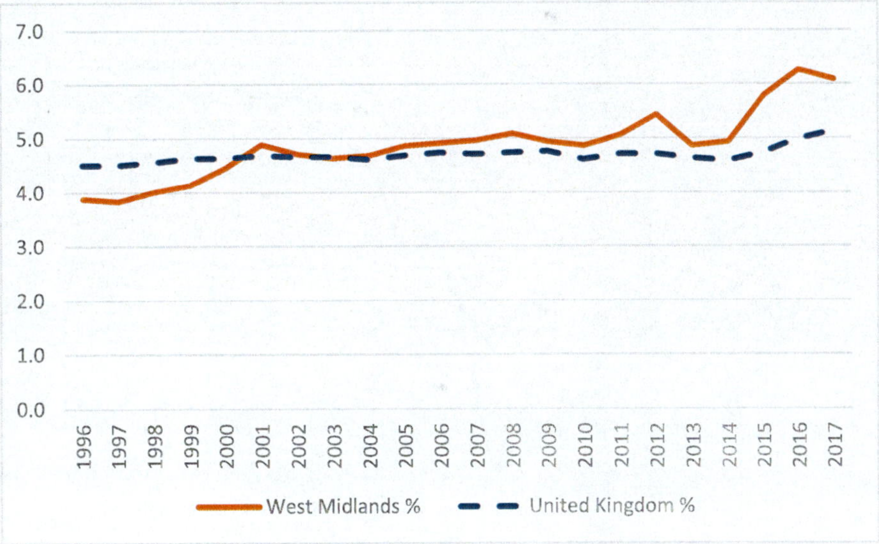

Fig. 2. Transport and Storage Employment as a percentage of Total [2]

3.2 Research in the West Midlands

The region has a number of high quality universities, with 3 currently in the top 30 universities in the UK, according to The Times Higher Education Magazines combined rankings table. (Warwick, Birmingham, Coventry) with three more in the top 50, (Harper Adams, Keele and Aston). Two of these universities are located in Coventry. Two others are close by in Birmingham. This creates a potentially powerful and influential research hub. Additionally there are major centres of transport research at MIRA and the Manufacturing Technology Centre in Coventry. The region overlaps the northern half of the Motor Industry centre in the UK and a number of major automotive companies have research units in the area, including: Jaguar Land Rover, Tata Motors, Ricardo, ZF (formerly TRW Konekt).

Traditionally a region which has been at the centre of innovation in transport the West Midlands has been in decline for decades. An unsustainable government run giant automotive business slowly collapsed through the latter half of the last century, dragging the region's productivity and innovation with it. The region began a remarkable turnaround in 2008 when locally based Jaguar Land Rover was sold to the Tata Group. Figure 3 demonstrates that this investment has begun a change in fortunes that has seen a tripling of value added in the region and which is leading a regional recovery in both transport innovation and manufacturing.

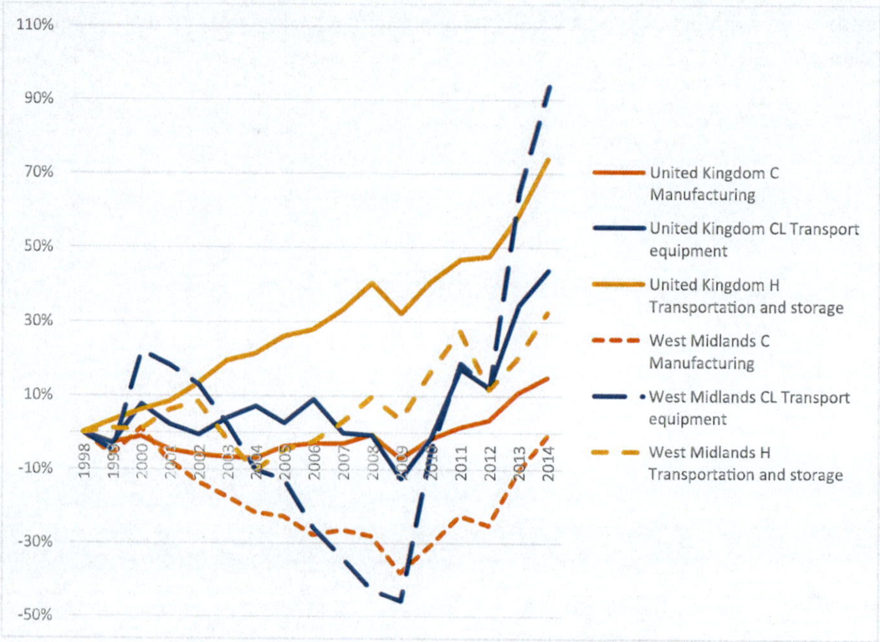

Fig. 3. Increased Value from Transport Manufacturing in the West Midlands 1998–2014 [3]

Although spending on research in the region has traditionally lagged the rest of the country, the resurgence of the automotive industry has driven major change. Figure 4 shows that 2010 growth of research in the region in the region has been driven by the arrival of new transport and related manufacturing research centres.

3.3 Academic Innovators

For Coventry the two principle academic sources of Innovation in Transport are Warwick and Coventry Universities.

Warwick University
Warwick University does not have a research department specifically dedicated to transport research, but it does have facilities with strong transport focus. In particular the Warwick Manufacturing Group, a long-established research department in the University has a number of research areas that are focused on transport. In particular:

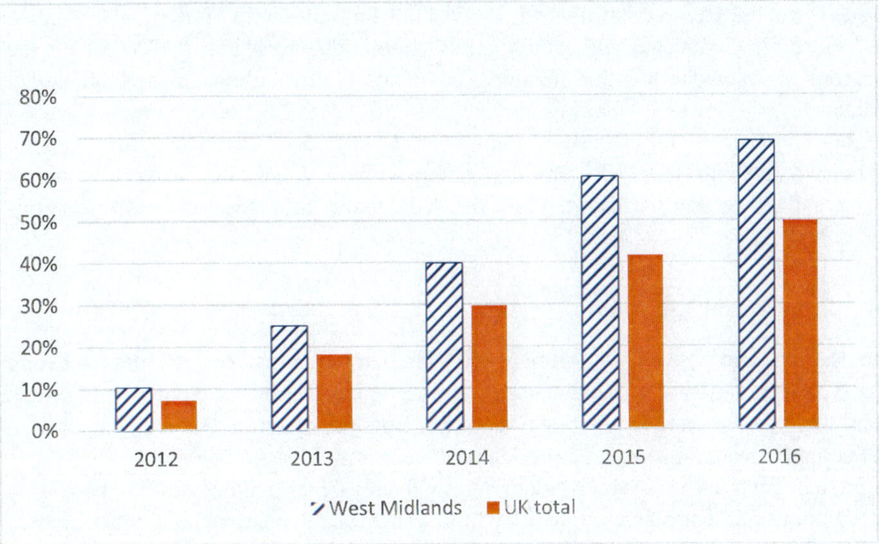

Fig. 4. RTD Growth over 2011 base [4]

- The Very Light Rail Innovation Centre
- Revolution Very Light Rail Project

The very light rail sector offers a significant opportunity for UK companies to develop new solutions embracing technology transferred from the automotive sector, leading to the growth of a new industry supplying UK and international rail schemes.

The Warwick Manufacturing Group have a number of transport Innovation related projects in development. Tis includes the *National Automotive Innovation Centre*, which opened in summer 2018. It focusses on breakthrough technologies and increasing the flow of skilled engineers into the workplace.

Coventry University

Coventry University houses the UKs largest research department working on Transport. The Centre for Future Transport and Cities has almost 300 staff working on Innovation in mobility for the future.

Included in this centre are:

The National Transport Design Centre (NTDC) is a State-Of-The-Art Clustering and Hub Facility.

Opened in May 2017, the NTDC is an incubation facility designed to explore new areas of transport design research and find new ways to use existing equipment, as well as creating new technologies. It's both provocative and surprising, ambitious and disruptive, creating real transferable outcomes and impact. The NTDC functions in a truly cross-disciplinary way, bringing designers, technologists, coders, together with artists, gamers, material specialists and fashion experts.

Last year the Centre also opened a new joint initiative with Horiba MIRA called the Centre for Connected and Autonomous Automotive Research (CCAAR)[1] at the forefront of research into the changing nature of highly automated and intelligent vehicles.

The University is currently building the Centre for Advanced Low Carbon Propulsion Systems C-ALPS[2]. In cooperation with FEV, one of the world's largest Power and Powertrain companies this will help place the region at the front of this research.

3.4 Commercial Innovators

The Region has a strong assortment of Commercial Innovators. Having a strong Automotive Manufacturer headquartered in the region creates an environment which attracts an infrastructure led by tier 1 suppliers. Both the Manufacturer and a number of these suppliers have research in the region.

Jaguar Land Rover have headquarters and are a major innovator in the region through development centres at Whitley in Coventry and at nearby Gaydon, which also hosts a development centre for Aston Martin. Tata Motors also have a development centre based which works with both Coventry and Warwick Universities. Supporting them locally are tier 1 suppliers with development centres including ZF, Ricardo and MIRA Horiba.

Smaller innovators include companies focussed on low carbon transport, like London EV Company who build electric taxis and Microcab. Others are focussed on autonomous vehicles, like Aurrigo, Conigital and Westfield.

4 The Policy Environment

4.1 National Policy

The Region does not have a specific RIS3 strategy, but refers to the national one. The national policy situation is highly influenced by a government view that we should be taking a position as leaders in some aspects of transport innovation – in particular Smart Mobility and driverless vehicles.

The national strategy notes that the UK [4] has the potential to be a world leader in innovation. It also stresses that the UK needs to strengthen the ability to commercialise this innovation and defines the role of Government as an enabler through improving the interface between Universities, business and finance. It also recognises that competition is an essential part of the drive to innovate and that multi-partner collaborations can provide synergies that are more than the sum of their parts.

1 https://www.coventry.ac.uk/primary-news/horiba-mira-and-coventry-university-launch-autonomous-vehicle-research-centre/

2 https://www.coventry.ac.uk/research/areas-of-research/institute-for-future-transport-and-cities/our-facilities/fev-low-carbon-centre/

This strategy has identified a number of key sectors. Of particular interest to innovation in transport are the identification of Aerospace and Automotive among ten key Industrial Sector strategies.

Britain is a leader in the Aerospace industry in Europe. It is responsible for about 4% of GVA and the growth in Global air traffic means that this is set to grow. The RIS3 strategy recognizes that, as incumbent leader, this position is at risk as changes in next generation aircraft means that product and manufacturing technologies will change. In response to this the Aerospace industrial Strategy is focusing investment in the four key, high value, highly complex areas of modern aircraft – wings, engines, aero structures and advanced systems.

The UK is the fourth largest vehicle manufacturer in Europe. Whilst it is a smaller sector for the country than Aerospace, the strategy identifies that it provides more manufacturing jobs than aerospace and has momentum in growing. The RIS3 strategy identifies that there is a major shift coming in demand for ultra-low emission vehicles and focusses on creating an environment which will nurture and invest in the SMEs which are essential increasing the country's share of the supply chain.

The UK government recognises that many of the innovations in transport at present are disruptive, often to the extent of needing changes in laws to make their adoption possible. In response to this they have launched the Foresight project with the following aims:

"Understanding what the future of transport could hold is important for society and government.

The Foresight Future of Mobility project will help policy makers to think about the future of transport by providing the latest scientific evidence and tools.

The project is developing evidence in the following areas:

1. the interaction between people, technology and data
2. new transport business models
3. alternative transport futures

4.2 Local Policy

Advanced manufacturing and engineering are recognised as a key driver of economic growth in the region. According to the regional ESIFstrategy it represents 10% of all employment in the region. This is 57% greater than the overall average and 500% greater than the UK average for the automotive industry. The strategy therefore places AME at the front of every strategic priority in the region. In terms of transport the strategy specifically states: "The Coventry and Warwickshire LEP area is a core part of the UK High Value Manufacturing Catapult through both WMG and the Manufacturing Technology Centre. The Catapult provides UK business with a gateway to access the best manufacturing talent and facilities in the country. It also acts as a conduit for funding from both the public and private sectors for projects and initiatives with due merit. In addition the focus on AME supports the Governments strategy for both the automotive and aerospace industries and linkage to the eight great technologies including; Advanced Materials, Robotics and Autonomous Systems and Energy Storage." (Coventry and Warwickshire LEP, 2016) [5].

The ESIF policy goes on to identify that Intelligent Mobility, Niche Vehicles aerospace and transport equipment present a significant opportunity for Coventry and Warwickshire to become a leading centre in the UK, utilising local strengths in innovation. By 2025 the strategy projects that the sector will be worth £2.4 billion and 12% of the sub-region's output. It also predicts that the automotive sector will increase in importance with the location quotient rising from five to seven (employment in the sector 7 times more concentrated than the average).

The policy identifies market failures in three main areas.

- Technology or knowledge spillovers: the difficulty that SMEs have in funding research and development frequently means that common interest in the wider benefits of the results of the work is not realised.
- Coordination and Network failures: the difficulty that small firms often encounter engaging in and managing collaboration
- Imperfect and asymmetric information: a market failure that makes it difficult for innovative SMEs to find the investment they need to develop this innovation.

To deal with this situation the policy proposes specialist technology support programmes, and demonstrator programmes around integrated transport systems, driverless vehicles, large data sets and automated vehicle control systems, all of which are very relevant to the sector. Programmes to stimulate collaborative research and knowledge transfer are intended to stimulate commercialisation of innovation from both public and private sectors.

Despite this declared focus and reliance on the sector there are more Research and Innovation projects focused on Life sciences including healthcare and Food and Agriculture than Transport in both the wider West Midlands region and in the Coventry region. Most projects in this area are loosely focused in ways which will pick up Transport Innovation, but do not place an emphasis on the sector.

Sources of Innovation Funding

Most of the funding allocated to transport is for infrastructure. Development funding comes from a number of sources. The Department of Transport provides specific funding schemes for innovation which are advertised on a national basis and the region has been successful in gaining some of these funds. Additionally there is funding available both from the UK government Department for trade and Industry through Innovate UK and from EU Structural Funds. This funding is directed at business support and is targeted at regional growth. Whilst almost all of the funding is generic for businesses, the size of the transport industry in the area means that some of it is used for transport innovation.

Use of Innovation Funding

It is difficult to identify the region's specific record on patents in transport. Information released by the UK Patent Office shows that number of patents both applied for and issued in the west midlands is a rising trend, but perhaps the most significant factor is that in 2017 West Midlands based Jaguar Land Rover were the UKs third largest applicant for patents and received more granted patents than any other person or organization in the Country [6].

5 Barriers and Enablers of Transport Innovation

The region is affected by the general drivers and barriers to innovation in transport as described by Weisenthal et al. in their 2015 review carried out for EU Joint Research Centre [7]. In respect of the general drivers the region is fortunate in having the headquarters, design and some manufacture of one major automotive manufacturer and the research departments of a number of niche manufacturers. This capacity attracts other supplier design functions and a number of tier one manufacturers. Figure 5 demonstrates that most income from innovation is demonstrated by the automotive sector (Weisenthal).

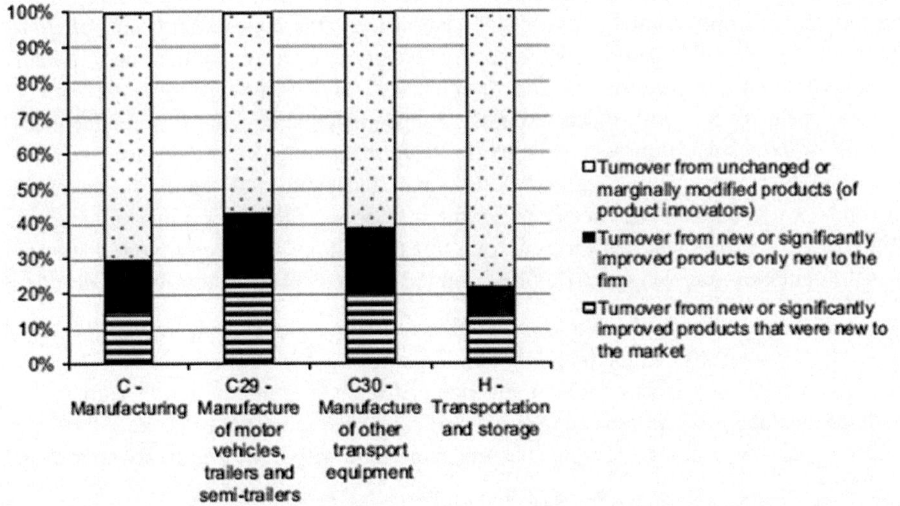

Fig. 5. Contribution of innovative products (new to the market; new to the firm) to the turnover of companies in transport-related sectors. Data source: Eurostat CIS survey 2010 (based on NACER2 sectors). (taken from Weisenthal et al.) [7]

As suggested above commercial research in the region is highly concentrated in the automotive sector with much less carried out in smaller companies.

Driving and barrier factors in the region can be summarized in the diagram below, constructed from the evidence gained in stakeholder interviews.

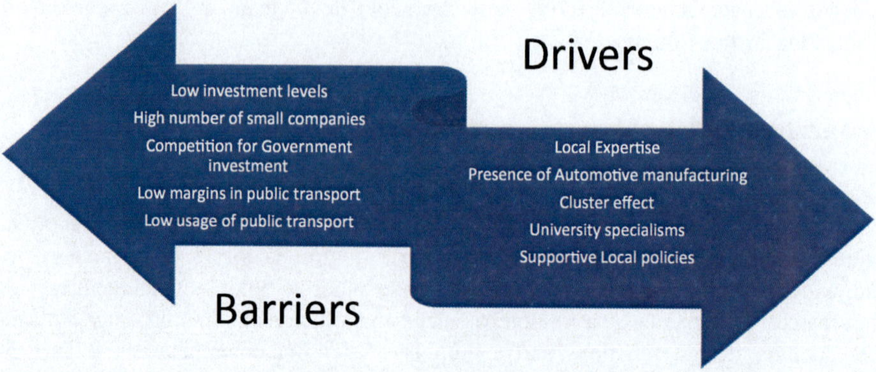

In particular the region's ESIF strategy highlights the difficulty of gaining investment for smaller businesses in the region, particularly in Coventry and Warwickshire. It highlights that 94% of commercial R&D expenditure in the region is accounted for by the top 5% of firms, compared with 90% nationally. This concentration on research coming from a few companies has the positive effect of ensuring that the majority is in the automotive sector, a strong driver of innovation. This creates a good environment for the cluster effect, with much expertise in the region and a good potential local market for the results. Both local universities have strong engineering departments and host initiatives in engineering, advanced manufacturing and transport research. Coventry City Council has been successful in gaining funding from a number of initiatives to create a test bed for new transport technology in the city, with a special focus on autonomous vehicles. This has resulted in a powerful vision for the city to.

- Reduce congestion and pollution via effective traffic management
- A Global centre for testbeds, pilots and trials of new technologies
- Demonstrate operational CAV in advance of the 2022 Commonwealth games
- Introduce attractive ridesharing service
- Meet new demands for responsive transport that will link seamlessly to public transport
- Giver the public a reliable, safe and value for money alternative to private car ownership
- Cluster of top universities attracting talent in manufacturing, engineering, science and tech

The focus on a single major manufacturer and the Tier one and two suppliers that support them is a barrier to development. Having one primary large company in the area means that investment outside of this focus is difficult to find. The region has a high number of small companies who, according to the ESIF policy have a low potential for innovation. Many of those businesses are dependent on the income that Jaguar Land Rover bring to the area. Government investment in innovation in transport is based on competitive tendering for funding. This leads to fragmentation of effort with organisations reluctant to collaborate and, despite the strength in Transport Innovation this region, it has led to the UK Catapult for Transport research being based 50 miles away in Milton Keynes.

There is an apparent disconnect between the strategy and the implementation of funding under ESIF in that there are very few support projects which directly fund Transport Innovation. Universities and research organisations are clearly very active in research in Transport, as evidenced by the list of projects above. Recently a bid to set up a Transport Innovation Accelerator failed to receive funding so there is an opportunity in this area.

6 Conclusions and Recommendations

It is clear from the work carried out that the region has many of the elements of a potential centre of excellence in transport innovation, however without better coordination there is a risk of losing that advantage. The failures of the market in not supporting transport innovation are defined in the policy, but there is nothing in the applied policy that addresses this. Having identified that there is potential competitive advantage in encouraging the smaller local innovators who support or are spinning out of the major players we identified some actions which can be taken to support them. Additionally the region is dependent on a single major original manufacturer who can dictate focus or whose loss could cripple the local transport innovation infrastructure.

This has led to our making two important recommendations.

1) Given the importance of Transport to the region, increase the emphasis on Transport Innovation in all economic development projects, ensuring that more innovators are aware of support available
2) Increase the exchange of knowledge and needs between transport departments, innovators and small businesses to create collaboration opportunities and increase the relevance ability to build on previous relevant ideas.

The objective is to create a better opportunity for transport innovation in the region.

It will create a network of transport-oriented innovators solving the ongoing issues between the current state of the art and future multimodal and autonomous transport systems. The network will work cooperatively to gain funding and viable business opportunities in innovative transport projects creating employment and investment in the region. It will have the strength to work in directions other than driven by one major player.

It will provide direction and help to put together cutting-edge innovation projects that the region needs.

Adopting these measures will create sustainable future economic development in transport for the region and more opportunities to increase multi-modal transport systems.

Acknowledgements. The results presented in this paper have received funding from the European Regional Development Fund under the Interreg Europe programme PGI02182 (project acronym: INNOTRANS).

References

1. Office Of National Statistics: Regional gross value added (income approach) reference tables (2017)
2. Office Of National Statistics: Workforce Jobs by Industry and Region (2018)
3. Office of National Statistics: UK Gross Domestic Expenditure on Research and Development (2018)
4. Department for Business Innovation and Skills: Smart Specialisation in England (2014)
5. Coventry and Warwickshire LEP: European Structural and Investment Funds 2014–2020 (2106)
6. Intellectual Property Office: Historical data of patents issued in the UK (2016)
7. Wiesenthal, T., Condeço-Melhorado, A., Leduc, G.: Innovation in the European transport sector: a review, no.1 (2015)

Energy Consumption and Perspectives on Alternative Fuels for the Transport Sector: A National Energy Policy for Greece

Alkiviadis Tromaras[1], Dimitris Margaritis[1(✉)], and Tatiana Moschovou[2]

[1] Centre for Research and Technology, Hellas/Hellenic Institute of Transport, 6th km Charilaou – Thermi, 57001 Thessaloniki, Greece
dmarg@certh.gr

[2] Department of Transportation Planning and Engineering, National Technical University of Athens (NTUA), 5, Iroon Polytechniou Street, 15773 Athens, Greece

Abstract. During the last years, gasoline consumption, mainly for road transport, has dropped by more than 30% in Greece, while diesel consumption has seen a generally upward trend after 2013. In contrary, consumption of alternative fuels follows a positive trend. However, in 2016, the national use of Renewable Energy Sources in transport was at 1.4%, though the EU average was at 7,1%. The implementation of two National Action sub-Plans for Greece, one regarding the Renewable Energy Sources and the second about Energy Efficiency of Vehicles will boost the use of the alternative fuels and consequently the national production and distribution effort. This paper reviews the energy consumption over a period of years, defines the main pillars and describes the context of a rational energy policy plan for the transport sector.

Keywords: Transport · Energy reduction · Alternative fuels · Fuel consumption · Strategic plan · Energy policy

1 Introduction

Under the Renewable Energy Directive (European Commission 2018), Member States have to ensure that at least 10% of their transport fuels come from renewable resources. During the last five years, gasoline consumption, mostly used in road transport, has dropped by more than 30% in Greece, while diesel consumption has faced a generally upward trend after 2013. For railways, a 22% of the network is electrified placing Greece at the lowest European position of electrified railways. For maritime transport, an increasing use of crude oil over diesel is observed, which is the main maritime fuel for the rest of the EU (YPEKA 2016).

This paper is concerned with the potential for energy savings and the use of alternative fuels in Greece. It aims to define the main pillars of a national strategic plan for a transport energy policy and to describe a wider context by presenting actual cases.

E. G. Nathanail et al. (Eds.): CSUM2020, AISC 1278, pp. 347–356, 2021.
https://doi.org/10.1007/978-3-030-61075-3_34

2 Current Status of Alternative Fuels in Greece

Greece produces conventional fuels (gasoline, diesel, kerosene and fuel oil) for transport use and imports the remaining quantities to satisfy the demand. Production facilities also exist for the alternative fuels:

1. *Biodiesel.* The biodiesel production network in the Greek market consists of 16 producers and 7 importers (Government Paper of the Hellenic Republic 2016).
2. *Compressed Natural Gas (CNG).* Natural Gas has been introduced to Greece by DEPA (2019) while FISIKON is the main CNG distributor, offering a network of 14 CNG refuelling stations mainly in large cities with another 7 planned to become operational in 2020 (FISIKON 2020).
3. *Liquefied Petroleum Gas (LPG)* is the most widely spread gas fuel used in transport in the country with more than 1000 available refuelling stations all over Greece.
4. *Electric energy.* Currently about 140 charging points are located in several spots in Greece supported either by FORTISIS, or Blink Europe (FORTISIS 2019; Blink Europe 2020), or in the form of pilot installations by petroleum companies. According to the Hellenic Institute of Electric Vehicles (2020), several spots existing along the national highways, but also in a number of public open area and parking spaces (e.g. Polis Park).
5. Biogas. According to the European Biogas Association (2018), Greece has 37 biogas production plants with all the volume being used for generating electric energy.
6. *Liquefied Natural Gas (LNG).* An LNG storage facility is already in operation on Revithousa island for hosting imported volumes of gas. Two more installations are also planned and under construction in Alexandroupolis area. Nevertheless, due to the absence of LNG refuelling stations as well as the absence of LNG vehicles, LNG is not used as a fuel for transport activities in Greece.

Table 1 presents the total consumption and production of conventional and alternative fuels in the Greek transport sector for year 2010 and 2014. According to the data, the consumption of conventional fuels has an overall negative trend with fuel oil having the biggest drop of −42%. The only conventional fuel with a positive change of +4.3% in consumption was kerosene. Gasoline, has a negative change of −32% and diesel, predominantly used in Greek road, maritime and rail sector, faced a reduction of −22%. On the contrary, consumption of alternative fuels shows a positive trend with LPG having a 374% increase. On an annual basis, between 2013–2014, LPG usage presented a positive 6.9% trend. The usage of biodiesel also had positive increase by 8% with a yearly constant increase rate. This data reveals the effect of the last years' financial crisis. The decrease of the country's GDP as well as the establishment of high taxation on conventional fuels, led people to seek cheaper resources such as alternative fuels for transportation.

Table 2 presents an overall picture of conventional and alternative fuels' consumption. It is evident that the share of alternative fuels still remains small, compared to conventional fuels. Thus, alternative fuels in the Greek market contribute to only a very small share of 4 to 6% to the overall usage of fuels. It is worth stating that the figures for alternative fuels for year 2010 and 2012 might be due to inaccurate data reported by official sources to Eurostat.

Table 1. Comparison of production and consumption figures of conventional and biofuels for Greece for the years 2010 and 2014 (ktoe).

	Fuel type	2010		2014		2018		% change	
		Production	*Consumption*	*Production*	*Consumption*	*Production*	*Consumption*	*Production*	*Consumption*
Conventional	*Gasoline*	4715	3867	4974	2643		2403	+5	−32.0
	Kerosene	–	234	–	178	–	220	–	+4.3
	Diesel	4642	2730	8145	2155		2504	+75	−22.0
	Fuel oil	–	441	–	258	–	340	–	−42.0
Alternative	*LPG*	771	46	831	222		242	+8	+374.0
	CNG	–	14.1	–	14		15	–	−0.9
	Biodiesel	112	124	142	134	151	159	+27	+8.0

Elaboration from: (ΥΡΕΚΑ 2016), and (Eurostat 2016)

Table 2. Consumption of conventional and alternative fuels for Greece (2010–2014).

	Total consumption (ktoe)				
	2010	2011	2012	2013	2014
Conventional fuels	8056	7114	6153	5970	6061
Alternative fuels	188	320	178	349	383
Share of alternative fuels	2.3%	4.3%	2.8%	5.5%	5.9%

Elaborated from: Eurostat (2016)

3 Fuel Consumption Per Transport Mode

3.1 Road Transport

The main conventional fuels used in road transport are gasoline and diesel, while biodiesel, CNG, LPG and electric energy constitute alternative fuels used in Greece. Figure 1 presents a comparison of fuel consumption for diesel and gasoline between Greece and EU 28. Gasoline consumption in Greece seems to have gradual declined by −46.2% from 2010 (3.847 ktoe) to 2018 (2.403 ktoe). Although, road diesel consumption seems to decline after 2010 from 2.437 ktoe to 1.623 ktoe in 2012, in the following year consumption seemed to begin recovering again. Despite the fact that road diesel consumption for Greece and EU 28 share a similar pattern, the reduction in consumption for Greece is greater to that for Europe. The fluctuation of gasoline and diesel consumption seems to have been affected by the financial state of the country as well as the imposed taxes on fuels. Hence, based on the presented data, gasoline is the predominant fuel used for road transport in Greece while diesel comes second, which is the exact opposite to the European figures where diesel is the main fuel consumed. The recovery of diesel in 2012 is a direct result of the policy regarding diesel usage that Greece has followed over the years for not allowing diesel passenger cars in urban areas due to national smoke standards. In late 2011 the Greek government has allowed the usage of diesel vehicles that meet Euro 5 and 6 regulations.

Another interesting fact is that according to Table 1 production figures of gasoline are much higher than the consumption from the road transport sector. This means that the country can export gasoline, something that is also verified by Eurostat (2018) statistical data of the country's exports.

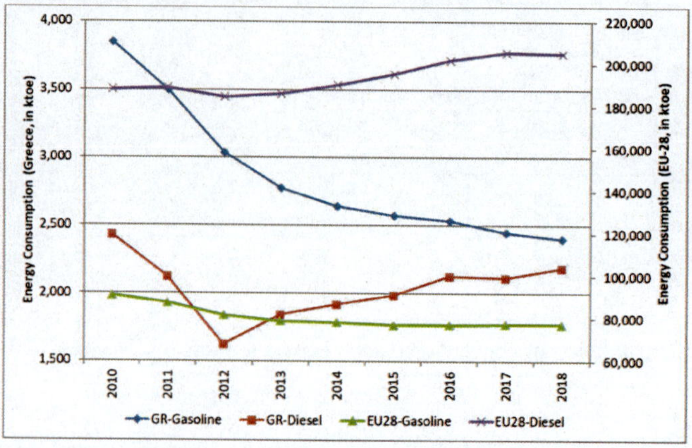

Fig. 1. System architecture Energy consumption for diesel and gas in the road transport sector during 2010–2018 for Greece and EU28 (in ktoe). Source: Elaboration on data from Eurostat (2018).

Figure 2 presents a comparison on fuel consumption for alternative fuels between Greece and EU 28. Overall, in both cases consumption of alternative fuels shows a positive growth. LPG seems to be an exception to this rule especially for years 2010 and 2012 where significant dropdowns have occurred, possibly as mentioned earlier, due to inaccurate reported data from official sources of Eurostat. CNG consumption is very low with figures of 13–15 ktoe, except of 2016 when a significant growth happened (19 ktoe). Low consumption figures are also shown for the EU28 with both pattern lines sharing similar characteristics. Finally, values for electricity with the exception of 2 ktoe reported in 2014 (Eurostat 2018).

3.2 Rail Transport

Diesel and electric energy are the main types of fuels used in the Greek rail sector with diesel being the predominant fuel. The highest diesel consumption figures of 41 ktoe are reported from 2014 with the remaining years showing fluctuations (Fig. 3, left side). Consumption of biodiesel started on year 2013. On the contrary, electric energy comes at the first place with a large difference in terms of consumption for the EU28 member states (Fig. 3, right side). This trend could become reality in Greece as well when the railway network along Athens-Thessaloniki-Eidomeni corridor will be electrified.

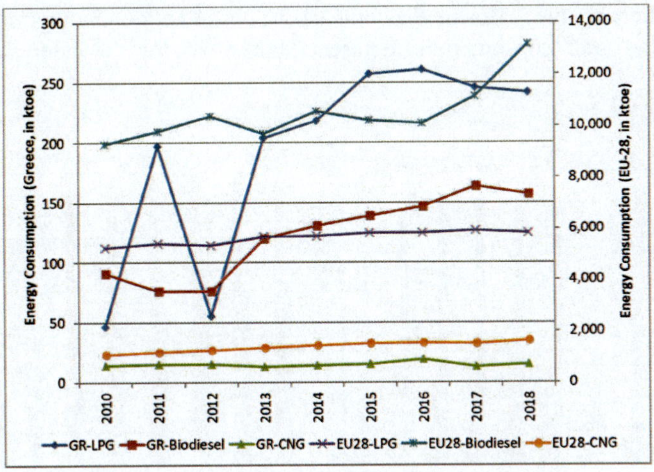

Fig. 2. Energy consumption of alternative fuels in the road transport sector for Greece and EU28 for 2010–2018 (in ktoe). Source: elaboration on Eurostat (2018).

Greece remains one of the few countries in the EU without electrification in the entire rail network. Therefore, biodiesel should be seen as an alternative until further electrification of the entire network. However, biodiesel in the rail sector is low even at an EU28 level (Fig. 3 right side).

3.3 Air Transport

The main fuels used in aviation in the Greek market are gasoline and kerosene (for piston driven propeller aircrafts and turbine driven aircrafts, respectively). According to Eurostat (2018) use of aviation gasoline in Greece is very low compared to kerosene and is only used in domestic aviation with quantities of 2–3 ktoe per annum. According to Fig. 4, the consumption of aviation kerosene for domestic flights shows a 6% decrease between 2010 (234 ktoe) and 2018 (220 ktoe).

3.4 Maritime Transport

Maritime diesel and fuel oil are the main fuels used in the domestic Greek maritime sector according Fig. 5 (left part). Figure 5 (right part) presents the respective figures for the EU28. After 2011 the consumption of maritime diesel in the Greek market shows a decrease of - 31% in 2014, topped by a −42% decrease in fuel oil usage in 2014 from a maximum value of 440 ktoe in 2010. Maritime diesel and fuel oil usage seem to be decreasing for both Greece and the EU 28. A significant difference between the compared markets is that in the case of Greece, fuel oil is predominantly used for domestic maritime transport compared to the EU28 figures where the opposite occurs. In the latter case maritime diesel comes in first position in terms of consumption with almost doubled figures compared to fuel oil. An interesting fact worth noting is that the consumption of fuel oil for the Greek domestic maritime transport represents the 27–

30% (depending on the year) of the total EU28 consumption. Furthermore, Greek maritime diesel fuel consumption represents the 6–7% of the equivalent EU28 consumption.

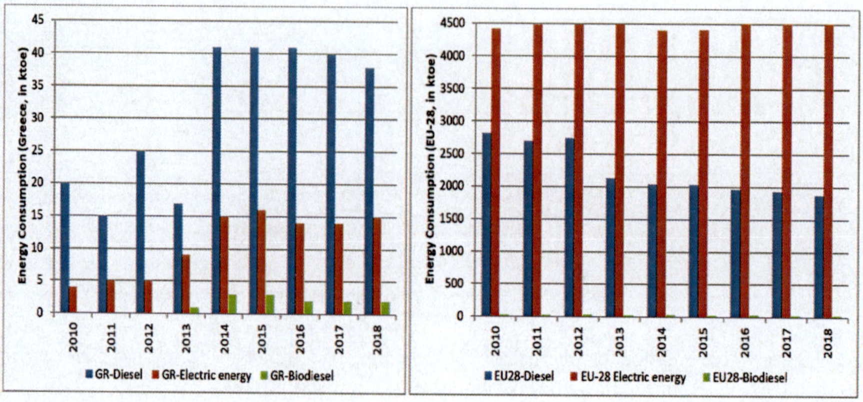

Fig. 3. Consumption of conventional and alternative fuels used in the Greek rail sector (left part) and in the EU 28 rail sector (right part) in the period 2010–2018 (ktoe). Source: Elaboration on data from Eurostat (2018).

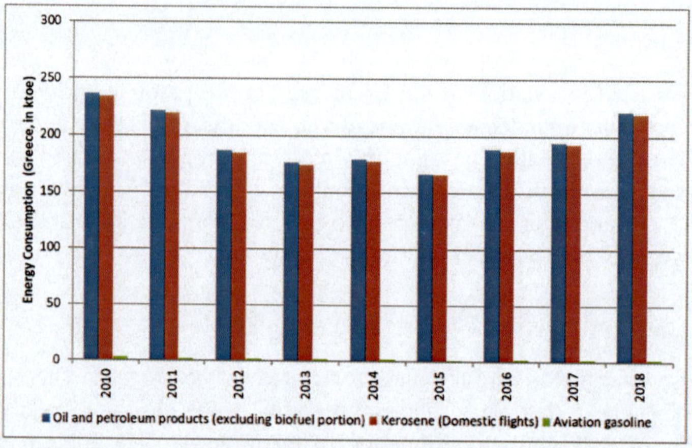

Fig. 4. Consumption of conventional fuels in Greek air transport 2010–2018 (ktoe). Source: elaboration on Eurostat (2018).

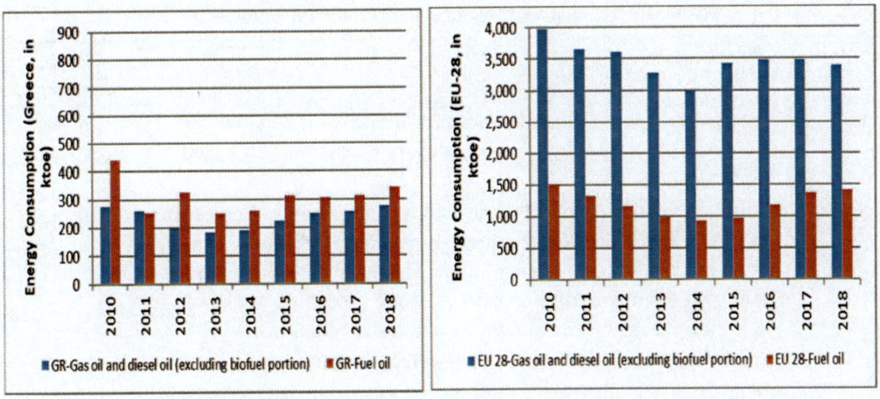

Fig. 5. Consumption of conventional fuels (ktoe) in the domestic maritime transport sector in Greece (left part) and in the EU28 (right part) for the period 2010–2018. Source: elaboration on data from Eurostat (2018).

4 Towards a Rational National Energy Policy for the Greek Transport Sector

4.1 Parameters to Be Taken into Account for a National Transport Energy Policy

The first and foremost element of a National Transport Energy Strategic Plan would be the delineation of the objectives and goals of the national policy in relation to a number of parameters that will define the energy mix for the transport sector in the coming 30 years or so (until 2050). These parameters include:

A. The national targets for the reduction of transport related GHG emissions in all transport sectors. For the EU member countries in the region, these targets would be largely set by the EU's policies and legislation already set in place as mentioned in the beginning. For the other countries in the region each government should by 2022 define these targets within the so-called *Nationally Determined Contributions – NDCs* that have to be defined as part of the COP21 Paris Agreement of 2015.
B. National targets for specific energy mix to be used by each transport mode.
C. The targets and objectives regarding the fleet mix i.e. the types of vehicles that will be allowed to circulate in the country. A number of European countries are already in the process of proposing the banning of vehicles with internal combustion engines in urban areas after 2040.

4.2 Supporting the Production and Use of Renewable and Alternative Fuels

In Greece, a wide range of prospects and potentials exist for the development and utilization of renewable and alternative fuels. Table 3 presents alternative fuels currently present in the Greek market including those with future potential and their relation to the previous list of parameters.

Table 3. Alternative fuels and their potential in Greece.

Fuel type	Potentials	Parameter
LPG	The production and distribution infrastructures of LPG is continuously expanding and has already reached the so called "critical mass"	A, B
Biofuels	Their future use is modest while its production seems - at present - to be low. The production of second-generation *biodiesel* coming from fried oils could be another solution. This prospect is quite important and with the appropriate support from the government, it could lead to a future innovative solution for alternative fuels	A
	Biomethane derived from upgrading biogas could become an alternative fuel although it will require extensive capital investment the creation of upgrade plants	A
	Ammonia derived from biomass, energy crops, natural gas or renewable energies can be used as fuel directly on internal combustion engines or fuel cells and has attracted a lot of attention in terms of its usage as a future maritime fuel (Lloyd's register 2017). Greece could potentially become a producer of this fuel	A, B
Natural Gas	A relatively new transport fuel in Greece that has not yet reached critical mass in terms of distribution stations although there are good prospects of its use. Therefore, establishing a pricing policy in the first years of expansion will be crucial for attracting attraction of new users	A, B
Hydrogen	Production and distribution units for hydrogen do not exist in Greece. Nevertheless, pilot production units exist at the Centre for Research and Technology Hellas (CERTH) and the Center for Renewable Energy Sources and Savings (CRES). Hydrogen is considered to be the most promising fuel for future transport systems. It can be produced either from water by electrolysis provided that electricity from renewable sources is used or from more innovative techniques, such as Hydrosol a method developed by Centre for Research and Technology Hellas (CERTH), where hydrogen is produced by CO2 and water under the impact of sun energy (Pagliaro et al. 2010). Aspects of the production, provision and transport of hydrogen in Greece have to be considered with particular attention while investment issues have to be promptly planned by state and industry	A, B, C
Biomass	The available reserves of agricultural residues that could be used to produce biomass in Greece are significant (Christou 2013). Available agricultural and forestry residues that can be utilized, equal to 3–4 billion tons of oil per year and correspond to the 30–40% of oil amount that is annually consumed in Greece. Such figures show impressive potentials that cannot be ignored	A, B

4.3 Promotion of Energy Friendly Modes and Clean Vehicles

Energy "friendly" or "clean" transport modes, are defined as those that consume energy produced from "energy friendly" ways and do not intensify the problem of climate change. Such modes are: i. all electrified rail or road transport means since the electricity they use is produced from hydroelectric or solar energy or other "energy friendly" sources, ii. the vehicles that use renewable or alternative fuels and iii. transport on foot or by bicycle. Also, all public transport means using hydrocarbon fuels could be considered as energy friendly transport modes concerning their high level of passengers' service and their consequent low energy consumption per passenger. However, they should be the first to be considered for conversion to using alternative energy fuels in the future.

The promotion of energy "friendly" or "clean" transport modes must form a systematic and permanent aspect of transport and energy policies for successive Greek governments. Incentives, within the context of this policies, could include: a. reduction of car tax for "clean vehicles", b. reduction of tax or public subsidies for purchase of friendly vehicles, c. reduction of tax of renewable fuels, d. ease of licensing and launching of stations for the supply of environmentally friendly fuels through granting procedures, e. update of the compensatory measures for drivers of "friendly vehicles" in urban areas (e.g. free access in areas with traffic restrictions, special parking places or free parking, free parking places for electrical cars). In addition, vehicle importers in Greece will need to show more willingness and offer additional electric vehicle models in the market compared to the limited choices that currently exist.

5 Conclusions

Greece must make an intensive effort for the improvement of energy resources in the transport sector and the larger use of renewable and alternative fuels following the corresponding EU Guidelines.

It is of importance now to integrate, and plan renewable and alternative fuels sources within a National Transport Energy Plan that will be in line with the country's general Transport Policy and Energy Policy as well as to all associated policies for dealing with climate change. This Plan, should, among others, include: (i) Quantified targets in relation to the fuels in use in transport sector in the future (target years 2030, 2050). (ii) Specific measures to be taken for increased production and use of renewable and alternative fuels in Greece incorporating emissions from field to tank in order to estimate alternative fuel pathways that are suitable for the country. (iii) Specific measures for increased use of energy friendly means of transport and clean vehicles (special emphasis in electric cars). (iv) Targets for improvement of the Greek energy mix with larger penetration of renewable energies. (v) Specific measures for creating production facilities for alternative fuels using renewable energies. (vi) Collection and study of statistical data regarding the capacity that the different regions in Greece have in terms of renewable energies. (vii) Insight to social behavioral issues and user acceptance of clean sources of energy in transport.

Boosting the national production and use of "clean" fuels in transport is also very compatible regarding the techno-economic opportunities of the country as well as the responsibilities towards the European and international treaties. Such a prospect will give the country a powerful development perspective. Although the energy intensity of Greece has almost reached the European average, there is still significant room for energy savings and rational use of energy in transport.

Acknowledgements. We would like to thank Prof. G. Giannopoulos for his valuable contribution, by providing us relevant literature for Greece.

References

Christou: Available reserves of agricultural residues in Greece, Energy and Biomass, CRES, Athens, October 2013 (2013)

DEPA: Company site (2019). http://www.depa.gr/. Accessed 12 Jan 2019

European Commission: Directive (EU) 2018/2001 of the European Parliament and of the Council of 11 December 2018 on the promotion of the use of energy from renewable sources (2018)

Eurostat: Energy Balances (2018). http://ec.europa.eu/eurostat/web/energy/data/energy-balances

Eurostat: Energy Balances (2016). http://ec.europa.eu/eurostat/web/energy/data/energy-balances

European Biogas Association: (2018). http://european-biogas.eu/. Accessed 19 Feb 2020

FISIKON: (2020). Company website. http://www.fisikon.gr/. Accessed 29 Mar 2020

FORTISIS: (2019). Company site. https://www.fortisis.eu/. Online 12 Jan 2019

Government Paper of the Hellenic Republic: Volume 2, Paper No. 1417, 19 May 2016 (2016). http://www.ypeka.gr/LinkClick.aspx?fileticket=7C1ZcYKhf4s%3D&tabid=292&language=el-GR. Accessed 2 Nov 2016. (in Greek)

Hellenic Institute of Electric Vehicles: (2020). https://www.heliev.gr/. Accessed 31 Mar 2020

Lloyd's register: Zero-Emission Vessels 2030. How do we get there (2017). http://www.lrs.or.jp/news/pdf/LR_Zero_Emission_Vessels_2030.pdf. Accessed 23 Jan 2019

Pagliaro, M., Konstandopoulos, G.A., Ciriminna, R., Palmisano, G.: Solar hydrogen: fuel of the near future. Energy Environ. Sci. **3**(3), 279–287 (2010)

YPEKA (Ministry of Environment, Energy and Climate Change). Data collected after contact with the Hydrocarbons Department of the General Secretariat for Energy and extracted resources of YPEKA, April 2016 (2016)

Building Capacity of Small-Medium Cities' Local Authorities to Implement MaaS and Other Innovative Transport Schemes

Anastasia Founta[1(✉)], Olympia Papadopoulou[1], Sofia Kalakou[2,4], and Georgios Georgiadis[3]

[1] Lever Development Consultants S.A., 26th Octobriou str. 43, 54627 Thessaloniki, Greece
anasfou@gmail.com

[2] Instituto Universitário de Lisboa (ISCTE-IUL) - Business Research Unit (BRU-IUL), Avenida das Forças Armadas, 1649-026 Lisbon, Portugal

[3] Transport Engineering Laboratory, Department of Civil Engineering, Faculty of Engineering, Aristotle University of Thessaloniki, 54124 Thessaloniki, Greece

[4] VTM, Ed. Central Plaza - Av. 25 de Abril de 1974, 23 - 2ºA, 2795-197 Linda-a-Velha, Portugal

Abstract. Sustainable development requirements in combination with extreme technological evolution have changed the way mobility is considered, creating challenges to Local Authorities (LAs) both in planning and implementation phases of mobility solutions. This paper focuses on the capacity building of LAs to deliver Mobility-as-a-Service (MaaS) and other innovative transport schemes as part of Sustainable Urban Mobility Plans (SUMPs). It presents a methodological approach for the design and employment of an integrated learning tool that intends to increase the adoption rates of measures' packages through LAs knowledge strengthening. The learning tool consists of a detailed facilitator guide to run an one-day classroom course along with the conceptual background and necessary training material. The methodological approach consists of a multilevel and multicriteria process that integrates the results/outcomes of the assessment of the cities capacity to implement SUMP through an evaluation framework. The classroom course has been structured in order to clarify the value of MaaS and other innovative measures for small-medium cities, analyze successful case studies under the spectrum of overcoming challenges efficiently, present tools and guidelines supporting collaboration between team members. The results of this work have been validated through the pilot application to six LAs. The overall evaluation of the pilots showed that content's accuracy and achievement of workshop's objectives was more than satisfactory (more than 50% of the participants gave the highest rate) and participants became more engaged with SUMP measures implementation.

Keywords: Capacity building · Sustainable mobility · Innovative mobility measures · MaaS · SUMPs

E. G. Nathanail et al. (Eds.): CSUM2020, AISC 1278, pp. 357–367, 2021.
https://doi.org/10.1007/978-3-030-61075-3_35

1 Introduction

Sustainable development requirements in combination with extreme technological evolution have changed the way mobility is considered, creating challenges to Local Authorities (LAs) both in planning and implementing mobility solutions, as they are the main actors for the development and implementation of Sustainable Urban Mobility Plans (SUMPs) [1]. This paper focuses on capacity building for LAs to deliver Mobility-as-a-Service (MaaS) and other innovative transport schemes, as part of SUMPs measures. In the case of Small-Medium (S-M) cities[1], where resources and expertise are more restricted than in larger ones, a capacity building program can support LAs to become more independent and resilient to changes.

So far, capacity building programs in the sustainable mobility sector, that address/target organizational level, include webinars, e-learning courses, guidelines on designing and implementation procedures produced either in the framework of projects funded by European programs also available in [2] and [3] or by other organizations [4, 5]. This paper introduces a Capacity Building Program (CBP) developed in the framework of SUITS Project funded by Horizon 2020 program, focusing on the training module "Building capacity of S-M cities' Local Authorities to implement MaaS and other innovative transport schemes".

SUITS CBP, overall ambition is the transformation of LAs into learning organizations, as a response to the intensively changing ecosystem of mobility. In learning organizations, people continually expand their capacity to create the results they desire when new and expensive patterns of thinking are nurtured while learning how to learn together [6]. In this context and taking into account that capacity development in organizational level is also a process of optimizing the utilization of human and financial resources [7], SUITS CBP has adopted the "train the trainer" approach, an approach applied in different sectors to achieve long-term capacity development [8], while raising awareness [9]. In SUITS case, CBP incorporates instruments more concerted to small target groups (in-house workshops and guides on how to run a workshop - facilitator's guides) next to mass training approaches (i.e. webinars and guidelines). These practices aim to optimize LAs resources, and enhance leadership, management and technical skills of LAs staff. Moreover, this approach raises awareness on the value of collaboration between LAs' staff and knowledge transfer when dealing with integrated mobility planning.

By presenting SUITS CBP formulation, this paper aims to contribute to CBPs addressed to LAs, focusing on mobility solutions implementation, by updating the knowledge on current capacity issues, proposing the in-house workshop approach along with facilitator's guide and highlighting the need of capacity development regarding innovative transport schemes such as MaaS.

[1] 50,000–250,000 residents in cities' urban centers.

2 Methods

A training program for capacity building of S-M cities has been developed taking into account the following elements: (a) cities' specified needs in the sector of transport and mobility (sample from the participant cities at SUITS project) and (b) the new mobility concepts, methodologies and tools deriving from the combined expertise of SUITS consortium and CIVITAS network, as well as from the results of related EU projects. The process is described by three (3) steps (Fig. 1).

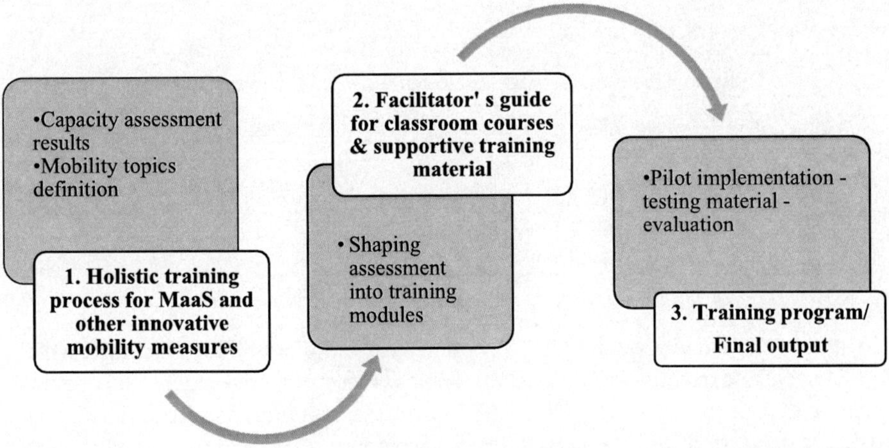

Fig. 1. Proposed methodology.

The first step includes topic definition and capacity assessment. A multicriteria analysis was carried out to identify the more interesting mobility topics. MaaS, among other innovative mobility concepts, is revealed as one of the topics to consider in the CBP. At this step, a baseline capacity assessment also defined the cities' specific capacity needs (financial, organizational, technical etc.) related to SUMP measures implementation. The capacity assessment results, both aggregated (all cities) and dis-aggregated (S-M cities), reveal the need of a holistic approach. The second step refers to the transformation process of the findings from step one into training module. The output of this process is the facilitator's guide along with the supportive material. "Building capacity of S-M cities' Local Authorities to implement MaaS and other innovative transport schemes" training module is further analyzed. The third step validates the output based on pilot training program implementation on participant cities at SUITS project. Pilots contributed to outputs amelioration and prepared the ground for post capacity assessment.

2.1 Innovative Transport Schemes and MaaS as Part of SUITS CBP

The selection of module topics began by considering LAs' critical feedback regarding the content of SUITS CPB as well as the accumulated experience and expertise of

SUITS academic and consulting partners. Selection of topics considered also the Urban transport priorities of S-M CIVITAS cities and SUMP v2.0 directions. An initial list of topics included forty-two (42) mobility measures and aspects, derived from relevant EU project database [10]. A more detailed selection was decided to be made in order to determine the exact content of each CBP module and to avoid overlaps with other training material already available from other relevant projects or sources. To meet this objective, a multicriteria-based approach of four (4) stages, equivalent to four (4) criteria, was followed to rank the initially selected topics according to their importance. Weights assigned to criteria were based on the subjective weighting of direct rating method [11] (Table 1).

Table 1. Criteria and weights selected for multi-criteria analysis.

Criteria (c_i)	Criteria description	Weights (w_i)	Weights value
c_1	Stage 1 ranking	w_1	1
c_2	Stage 2 ranking	w_2	2
c_3	Stage 3 ranking	w_3	3
c_4	Stage 4 ranking	w_4	4

The first stage reviewed the CIVITAS database and the mobility measures which CIVITAS cities currently focus on [12]. Topics were compared against this review considering their relevance to CIVITAS network's mobility priorities. They were assigned a score from 0 (no relevance) to 1 (maximum relevance) and weighted with w_1. Scorings were provided by the SUITS partner leading the CBP development.

The second stage included a doodle pool and communication process with other research projects which at that time or previously were also dealing with CBPs for SUMP development in European cities to ensure the originality of the work. In this respect, representatives of the projects were asked to provide a score (0 for very high overlapping to 1 for no overlapping) to up to 20 topics, after revising the overall list of the 42 topics. The scores were weighted with w_2.

The third stage (w_3) examined whether the initial topics adequately fulfill the overall CBP ambition. Scores from 0 (no relevance) to 1 (maximum relevance) were provided from SUITS academic and mobility consulting partners to align the final decision with SUITS main objectives.

The fourth stage was considered as the most determining one (w_4) considering the actual mobility needs, priorities, challenges and objectives of 7 cities (partners of SUITS project), three big cities, and four S-M ones. An ad-hoc workshop was made for this purpose so that SUITS cities were able to provide their importance scores (0 for very low and 1 for very high importance) against the initially selected topics. The total scores per topic have been calculated by summing up the four stages' results (1).

$$s = \sum_{i=1}^{4} w_i * c_i \qquad (1)$$

where $s =$ *total score* per topic, $0 \leq s \leq 10, 0 \leq c_i \leq 1, for\ c_i = 0, 0.2, 0.4\ldots1$

The top sixteen (16) topics are displayed in Table 2, as part of the overall ranking of the 42 topics. MaaS solution and innovative procedures, methods or tools (innovative mobility data gathering methods, innovative procurement and financing, bankable projects, crowdsourcing etc.) are, among other topics, highly rated, raising the need to be included in SUITS CBP training modules.

Table 2. Sixteen (16) top results of multicriteria analysis. Topics' nature has been added in order to be considered to modules formulation.

	Topics under consideration	Score	Topics nature
1	Electric mobility and clean fuels	10	Mobility solution
2	Low emission zones	10	Mobility solution
3	Mobility as a Service (MaaS)	10	Mobility solution
4	Innovative mobility data gathering methods	9	Tool
5	Innovative Procurement	9	Procedure
6	Parking management	9	Mobility solution
7	Demand Management Strategies	9	Mobility solution
8	Traffic calming measures	9	Mobility solution
9	Vehicular traffic management	9	Mobility solution
10	Safety and security	8.5	Mobility solution
11	Engagement of people and stakeholders	8.5	Procedure
12	Financing for environmentally friendly transport systems.	8	Procedure
13	Innovative and sustainable financing	8	Procedure
14	Development of bankable projects and partnerships	8	Method
15	Cycling	8	Measure type
16	Crowdsourcing	7.5	Tool

Specifically, MaaS was one of the topics with the highest rate. From cities' perspective, MaaS is considered as a topic with moderate importance (0.74 in average) in comparison to bike sharing (0.92 average score), carpooling and car sharing (0.8). MaaS has been more highly evaluated by large cities (0.8 average score). Nevertheless, experts' opinion and current availability of training tools and material addressed to local authorities, prioritized MaaS.

2.2 Capacity Assessment Results and the Correlation with the Development of Training Material

After prioritizing and grouping candidate topics for CBP, the application of a capacity assessment framework provided insights for the content, the structure and the form of the modules. The evaluation aimed to assess the effectiveness of LAs to plan and implement mobility measures and identify key indicators to be improved in order to increase LA capacity. Workshops and interviews with mobility planners of the LAs and transport operators of the SUITS cities provided insights. Each city's mobility measures were associated with challenges and pertinent indicators and according to the results of the evaluation, capacity targets were set which could be evaluated after the training to measure the impact of the CBP in the future (Fig. 2). In particular, the set of indicators was composed in order to assess the capacity to face challenges and reveal possible inefficiencies in all the elements that form the capacity [13]. The indicators assess the current operations of the cities in 4 main areas (organizational, political, legal and societal) and 4 subareas (communicational, financial, managerial and technical) related to the environment in which an authority exists and operates.

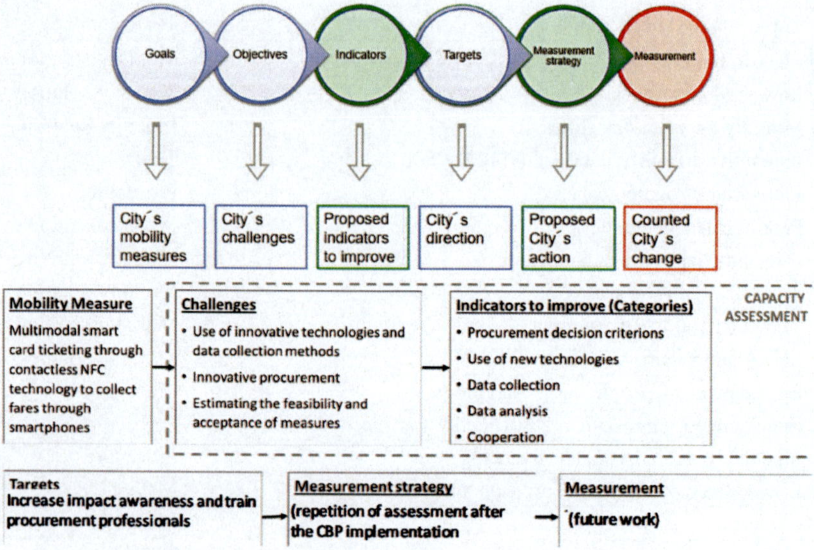

Fig. 2. Example of capacity assessment results for a city.

The results are described in detail in [14]. Fifteen (15) challenges that cities cope with when shaping sustainable urban mobility plans were derived (Tables 3, 4, 5). At an aggregated level, the capacity assessment of the cities demonstrated that the priority areas for interventions are project monitoring, innovative financing and training, regular self-assessment, cover staff's needs, coordination and cooperation between sectors, understanding of legal and regulatory framework and acquisition of power delegation. Each city's mobility measures were associated with challenges and pertinent indicators and, according to the results of the evaluation capacity, targets were set which could be evaluated after the training to measure the impact of the CBP.

2.3 Shaping Assessment into Training Module for MaaS and Innovative Transport Schemes

The afore-presented processes provided insights for the content and training format. Conclusions focused on elements that raise the need of capacity development on MaaS and innovative schemes implementation, and determine training objectives, are analyzed. Insights contributed to the formulation of the integrated SUITS CBP [15].

In this framework, it is concluded that measure-oriented modules, rather than procedures or method-oriented modules, allow capturing cities interest directly. A measure is usually the starting point from which LAs staff launch their effort to deal with implementation barriers and search for capacity building solutions (Fig. 2). Resuming Sect. 2.1 results, MaaS, is a new concept of mobility solutions, attracting the interest of both S-M and large cities while few integrated CBPs are available. It embeds also mobility topics such as "Development of bankable projects and partnerships", "Innovative mobility data gathering methods", and all available and future transport systems/schemes such as shared mobility. Moreover, LAs staff struggle to demonstrate measures value in order to enhance political commitment and sustainable thinking. Therefore, better understanding of measure requirements and its wider context of implementation could allow to overcome corresponding challenges (Table 3).

Table 3. Challenges associated to specific training objectives supporting module's structure.

A. Module's structure	
Challenges	Objectives of module (identification number)
Understanding political interests and affecting political decisions	Highlight how measures serve relevant European goals. (1)
	Highlight the value of measures for the society in local and national level. (2)
Sustainability Thinking	Explain sustainability positive impact from a more social and economic aspect. (3)
	Highlight negative impact of current mobility habits. (4)
Effective project management and monitoring	Explain monitoring procedures and tools. (5)
	Explain implementation steps to assist project management giving real - world examples. (6)

Classroom course/workshop is considered the appropriate form to help cities facing three (3) capacity challenges (Table 4). Through exercises, teamwork, and discussion, LAs could recognize weaknesses and strengths, make collective commitments and get to know requirements and available resources to perform sustainable urban mobility measures more efficiently. In this respect, facilitator's guide approach replies potentially to knowledge management and systematic staff development.

Table 4. Challenges associated to specific training objectives supporting module's form.

B. Module's form	
Challenges	Objectives of CBP
Institutional cooperation	Introduce classroom courses instead of only e-learning courses and webinars. (7)
Knowledge management/knowledge transfer	Introduce facilitator's guide to enable LAs' staff to manage and transfer knowledge running course in-house. (8) Develop teamwork exercises rather than individual tests/workshop. Knowledge transfer between LAs to be available. (9)
Systematic staff deployment and development	Introduce facilitator's guide providing insights of the course, sources & references to keep staff updated after course completion. Enable knowledge transfer to new staff. (10)

Procedures such as procurement, financing and data management are important for all types of mobility measures (Table 5). Indeed, innovative data collection methods, innovative financing, procurement and partnerships have been highlighted also by capacity baseline assessment results. Therefore, introducing these topics into one inclusive module entitled "Building capacity of S-M cities LAs to implement MaaS and other innovative transport schemes" has been considered more valuable for the cities. Hence, decision makers, political representatives and staff of the procurement department are considered as potential participants along with technical staff.

Table 5. Challenges associated to specific training objectives supporting module's range of topics.

C. Module's inclusive topics	
Challenges	Objectives of CBP
Use of innovative technologies and data collection methods	Inform about innovative technologies and data collection methods. (11) Develop modules addressed to innovative technologies. (12)
Understanding and applying innovative financing methods	Present innovative financing and innovative procurement methods and familiarize with the processes required. (13)
Innovative procurement	
Interaction and cooperation with business partners	Present partnerships concepts already applied. (14)

3 Results

"Building capacity of S-M cities LAs to implement MaaS and other innovative transport schemes" module is comprised of two main elements: "facilitator's guide" and "supportive material" such as workbook to be distributed to course participants, presentation for use in the classroom and handouts (exercises). Educational strategies and methods from training programs already available in various fields are adopted to formulate the facilitator's guide. Its components aim to provide all necessary information to run the workshop without requiring previous experience as a facilitator. The content of the training is structured into chapters including team-work activities based on the results of the applied methodology and in particular on final conclusions of Sect. 2.3. In Table 6, the course flow is shown with a short description of the provided content, developed activities and correspondence with training objectives. For each chapter the facilitator obtains: (a) a condensed version of the content with reference to the respective workbook pages, where the content is further employed, (b) an estimation of its duration, (c) instructions on how to run each training section including relevant information (team-work activities, a reference to the supporting PowerPoint slides etc.). References and additional sources are provided to enhance knowledge on the topic. The classroom course/workshop is designed to be conducted within a single day and with representatives from a single authority. However, the course flow is formed in a way that can be split into more days and allows participation of more than one local authority to better serve local needs and expectations.

Following the module development, three (3) pilot workshops were carried out to identify course drawbacks in terms of content, teaching processes, distributed material and to evaluate its effectiveness regarding the wider ambition of SUITS CBP.

Table 6. Course flow of the module. Main content, activities and correspondence with training objectives.

No	Chapters title	Description (training objectives)	Activities
1	Introduction	Course/workshop objectives Why sustainable mobility?	Ice-braker activity
2	Description of measures	MaaS, shared mobility	
3	Value for the cities	Measures benefits, stakeholders, participatory methods, relevant EU strategies, Social Impact Assessment (1, 2, 3, 4)	Exercise A: Analyzing benefits and views of stakeholders on MaaS and shared mobility
4	Successful case studies or Best practices of SUITS cities	Free floating bike sharing in Turin (SUITS city) MaaS "Whim app" in Helsinki (2, 3, 5, 6, 9)	Discussion on any similarities or differences with the city
5	Innovative financing, procurement and partnerships	Recommended for MaaS and shared mobility Reference to relative guidelines (13, 14)	EXERCISE B: Matching funding mechanisms and partnership schemes with implementation components
6	Business model canvases for MaaS and shared mobility	Presentation of methodology and exercise on MaaS business model canvas (2, 6, 11, 12, 13, 14)	EXERCISE C: Filling out business model canvas for MaaS and shared mobility schemes
7	Process and implementation aspects	Required data, available data collection methods, evaluation procedure (5, 6, 11, 12)	Discussion on implementation activities/steps tailored to the city
8	Available tools and guidelines	Reference to available tools to support implementation steps (10)	EXERCISE D: Performing CIVITAS ECCENTRIC tool for the city

One pilot focused only on MaaS and shared mobility and other two on combined modules (all innovative mobility solutions). The participants were from different departments of selected LAs (policy and strategy, project management, procurement, technical) and with different roles (head of department, transport graduates etc.).

The aggregated evaluation results showed that participants were either fully satisfied or almost fully satisfied with most of the issues under evaluation. More than 74% of the participants gave positive rate with regard to content accuracy, exercises structure, course structure, gaining knowledge/ideas/skills and finding solutions to their problems. More than 80% of the participants awarded the two highest rates regarding confidence in applying the learnings from the workshop and believed that the workshop achieved its objectives. Successful case studies have been proved one of the most interesting part. The workbook was characterized useful and user-friendly. Finally, alternative funding mechanisms and exercise on business model canvases have also been highlighted as important by the participants.

4 Discussion

The overall purpose of this course is to increase understanding about the value of innovative measures such as MaaS and shared mobility solutions. The course is designed to offer concrete practical tools and guidance to better implement these measures and to build specific skills regarding how success of these measures can be ensured by convincing stakeholders and by overcoming financial, administrative and technical barriers. The course aims to strengthen cooperation between LA's staff on different levels, from policy makers to junior engineers, through the conduction of interactive exercises and eventually to advance local priorities on innovation such as MaaS and other emerging mobility solutions. It can be applicable to other S-M cities. The parameters to be considered when willing to enhance implementation of innovative mobility services and raise their positive impact, exceed beyond technical issues and pertain to other capacity gaps (cooperation issues, financial issues, etc.) and challenges such as behavioral change. It requires LAs not only to overcome the implementation barriers but become highly resilient and responsive to new challenges and changes. Future work includes the capacity assessment of LAs after application of the capacity program and the evaluation of their progress as learning organizations.

References

1. Zoeteman, B.C.J.: What's behind the leadership sustainable development from politicians to CEOs? Environ. Dev. **8**(1), 113–130 (2013)
2. ELTIS training: https://www.eltis.org/resources/training. Accessed 18 Mar 2020
3. CIVITAS learning centre: https://civitas.eu/learning-centre. Accessed 18 Mar 2020
4. MAAS-Alliance. https://maas-alliance.eu/. Accessed 16 May 2020
5. ERTICO ITS Europe. https://ertico.com/. Accessed 16 May 2020
6. Senge, P.M.: The Fifth Discipline: The Art and Practice of the Learning Organization. Doubleday/Currency, New York (1990)

7. Austin, M.J., Packard, T.R., Brody, R.: Managing the Challenges in Human Service Organizations: A Casebook. Sage Publications, Thousand Oaks (2009)
8. Kenny, S., Clarke, M.: Challenging Capacity Building: Comparative Perspectives. Palgrave Macmillan, London (2010)
9. Pawar, M.S.: Social and Community Development Practice. Sage Publications, New Delhi (2014)
10. KONSULT. http://www.konsult.leeds.ac.uk. Accessed 25 Mar 2020
11. Odu, G.O.: Weighting methods for multi-criteria decision making technique. J. Appl. Sci. Environ. Manag. **23**(8), 1449–1457 (2019)
12. CIVITAS cities. https://civitas.eu/cities/civitas-cities. Accessed 25 Mar 2020
13. Martins, S., Kalakou, S., Pimenta I.: Capacity Building Requirements-Evaluation Framework, cities. Deliverable D.2.2 of SUITS Horizon 2020 Project (2017)
14. Kalakou, S., Spundflasch, S., Diaz, A., Pirra, M.: Contextualisation of Project cities. Deliverable D.2.1 of SUITS Horizon 2020 Project (2018)
15. Papadopoulou, O., Georgiadis, G., Founta, A.: Integrated Subject Module and facilitator's guide. Deliverable D.5.1 of SUITS Horizon 2020 Project (2018)

Mapping and Analyzing the Transport Innovation Framework of the Region of Central Macedonia, Greece

Evangelos Genitsaris(✉), Vasiliki Amprasi, Aristotelis Naniopoulos, and Dimitrios Nalmpantis

School of Civil Engineering, Faculty of Engineering, Aristotle University of Thessaloniki, PO Box 452, 541 24 Thessaloniki, Greece
egenitsa@civil.auth.gr

Abstract. The regions of the European Union (EU) countries could play a significant role in implementing European and national policies towards the enhancement of innovation and competitiveness of the transport-related ecosystem of key stakeholders and actors. Regional policies may cover a wide spectrum of fields, such as investments in enhancing main innovation capacities, funding schemes for the expansion of transport-related entrepreneurship, institutional changes, etc. Thus, understanding the regional transport innovation framework through a structured and systematic approach is required in order to inform the relevant decision-making process at a regional level. In this paper, we explore the transport innovation framework of the Region of Central Macedonia (RCM) in Greece. Based on desk research, we identify and list the main transport innovation capacity-related cases that have already been, are ongoing to be, or are planned to be implemented in the future. Then, we analyze the transport innovation context by investigating the Strengths, Weaknesses, Opportunities, and Threats (SWOT) analysis, through which the main existing and potential areas of competitive advantage, as well as the barriers and enablers of transport innovation in the RCM emerge. For this purpose, we apply a qualitative research method. In particular, we conducted ten (10) in-depth personal interviews with key individuals, coming from various bodies of the "triple helix." The approach adopted and followed provides a framework, based on which more targeted regional policy interventions could be promoted, aiming to leverage the transport innovation potential at the regional level.

Keywords: Regional policy · Transport systems · Entrepreneurship · Research and Development (R&D) · SWOT · Interviews

1 Introduction

In this paper, we attempt to understand the transport innovation "framework" of the Region of Central Macedonia (RCM), Greece, through a structured and systematic approach. Based on desk research, we identify and list key transport innovation capacity-related cases that have been, are ongoing to be, or planned to be implemented in the future.

E. G. Nathanail et al. (Eds.): CSUM2020, AISC 1278, pp. 368–378, 2021.
https://doi.org/10.1007/978-3-030-61075-3_36

This way, we distinguish and interpret in an ad-hoc way, three main different conceptual abstracts, in particular, the innovation mechanisms, the innovative solutions, and the innovation providers. We provide a comprehensive Strengths, Weakness, Opportunities, and Threats (SWOT) analysis for the RCM in relation to the prospects of the transport innovation enhancement. This qualitative analysis resulted through the consideration of (i) the information coming from the individual/personal in-depth interviews with representatives of the key local actors, as well as (ii) the interpretation and filtering of this information by the authors' viewpoint, as it is formed through their personal experiences, expertise, and perspectives. Little information from secondary sources, such as reports, is complementing the analysis. The paper builds on the research process followed in the frame of the INTERREG Europe "Supporting regions innovation capacity in transport" (INNOTRANS) project's activities in the RCM [1].

We build on a definition of the term innovation as *"the implementation of a new or significantly improved product, or process, a new marketing method, or a new organisational method in business practice, workplace organisation or external relations"* [2]. By expanding to the whole spectrum of transportation a past description of innovation, initially used within the public transport field, we adopt a broad interpretation of it as "*every idea coming from other fields, and it is currently not applied in [public] transport or even any idea that, while it has been conceived and may already been implemented in some areas of the world, it has not been yet spread or adopted in a certain local or/and national context*" [3–8].

2 Research Methodology

We used a method of qualitative research, i.e., personal in-depth interviews with individuals coming from various transport-related and innovation-oriented local stakeholders and actors. The interviewees do not necessarily express the opinion(s) of the organization in which they work or are being involved, as the strategy and views of any legal body are reflected through and expressed only by its Board of Directors and its legal representatives. Interviews were conducted following a semi-structured style. For the facilitation of the personal interviews, we used three questionnaire guides that were slightly differentiated in order to be in line with the characteristics of the type of organization (Public Authorities, Businesses, and Academia), with which each respondent was affiliated.

The questionnaires were developed in the frame of the INNOTRANS project to be used in the frame of local research activities, in our case in the RCM. The questionnaire guides we used covered several aspects of the topic under exploration, such as organization profile, information about product/technology development background and capacity, drivers and barriers relevant to the development of Research and Development (R&D) activities, experiences and views related to the innovation initiatives within the region, preferences related to the type of support by the RCM, experiences from past participation in R&D activities, etc. The interviews were not recorded, as a recording is better to be avoided whenever interviewing within the "industry" sphere. Besides, we were not interested in the way interviewees were expressing themselves.

However, in order not to miss any information and facilitate the whole process efficiently, two researchers attended most of the interviews, complementing one another in the tasks of the interviewer and rapporteur. A detailed report was written no more than one day after the completion of each interview so that the memory of the dialogue to be still fresh.

The interviews took place during the second semester of 2018. Ten (10) people were interviewed in total. In most of the cases, two researchers were simultaneously participating in the interviewing process, complementing one another in asking questions and taking notes. Out of the ten interviewees: (i) with regard to their gender, seven (7) are men and three (3) women; (ii) in terms of their educational level, they are separated equally to PhD and MSc holders; and at last (iii) regarding their age, half of them belong to the age group of 55–65, four (4) are between 45–55, and one (1) is over 65 years old. Although the people interviewed were ten (10), some of them had two affiliations by the time the research was conducted, both of which were relevant to the survey study. In Table 1, we present the respondents' affiliations and the broad category of their organization. Considering the above data, we can group the organizations based on the "Triple Helix" rationale as follows (inside the curly brackets the number of bodies is indicated): (a) Governance {3} (public authorities {3}); (b) Business sector {5} (transport services/operators {3}, technology company {1}, chamber {1}); and (c) Academia {3} (academic research units {3}). There are also four bodies of mixed type, which could be characterized as "hybrid entities" {4} (business research associations {2}, technology park {1}, training and research unit {1}).

Table 1. Respondents' affiliations and the corresponding type of organization.

	Category of organization	Affiliation
1	Logistics company (warehouse operation & freight transport)	Director
2	Municipal authority	Director
3	Innovative company on Intelligent Transportation Systems (ITS)	Founder/Chairman
4	Public Transport Authority (PTA)	Transport planner
5	Municipal authority	Officer
6	Vocational training & research center	Chairman
7	University research laboratory	Director
8	Port operator	Chairman
9	Transport research institute	Research Director
10	Logistics research association	Chairman
11	Chamber	Vice-Chairman
12	University research laboratory	Director
13	Freight transport company	CEO
14	International transport research association	Vice-chairman
15	Technology park	CEO

3 Mapping Transport Innovation Capacity of the RCM

In order to understand the current transport innovation capacity of the RCM, we identified key cases that are relevant to or have the potential to contribute to its increase. Based on this identification, three generic concepts of higher abstraction level emerged: in particular, the innovation mechanism, the innovative solution, and the innovation provider. In this perspective, any identified case is approached as a way of enhancing the innovation capacity of the transport ecosystem, as follows:

1. *Innovation mechanisms:* by definition, they comprise ways to enhance and promote innovation. Some of them are particularly specialized in the transport domain, while others concern all sectors of the economy. As a mechanism, we mean any structure, service or agency that acts as a facilitator of innovation and is established for the support of any stage of innovation production.
2. *Innovative solution (technology and/or management system):* they result to a more efficient and effective transport and mobility system overall, encouraging the further establishment of innovations that can be built on them, producing this way multiplying effects. As an innovative solution, we imply any outcome of the intellectual and creative activity taking, e.g., the form of a system, product, management procedure, business model, service or scheme that fulfills the criteria of "innovation." An innovative solution may refer either: (i) to a real application of it by any actor, or (ii) to a marketable and exportable product or know-how.
3. *Innovation providers:* they provide innovative products or innovative know-how on which other businesses can build on. They constitute an essential and integral part of a dynamic, active, and innovative transport ecosystem.

The cases that concern the transport innovation capacity can be grouped based on specific criteria, such as (i) innovation case (e.g., mechanism, solution, provider); (ii) implementation phase; and (iii) spatial/geography scale (Table 2).

Table 2. Key indicative cases reflecting the RCM's transport innovation capacity.

Short name	Innovation case	Description	Implementation Phase	Spatial scale
OK!Thess	Mechanism	Pre-incubator for the support of potential startups including the transport-related ones	In operation	Urban level
One Stop Liaison Office	Mechanism	"One-stop-shop" service of RCM for the support of businesses	Launched recently	Regional level
MobiThess	Solution	Website developed by a partnership of local actors for traffic information & sustainable mobility awareness	In operation & constant updating	Urban level
Easybike	Solution	A Greek bike-sharing system (technology and business model)	In operation	Urban level

4 Qualitative SWOT Analysis

Based on the findings resulted from the ten (10) in-depth personal interviews (Interview Number: IN1, IN2, …, IN10) conducted on the individual level, we summarize and aggregate our results by leveraging the widely used SWOT analysis. SWOT analysis comprises a rather comprehensive and straightforward qualitative method of identifying the Strengths, Weaknesses, Opportunities, and Threats from a particular viewpoint, e.g., of an organization, a region, etc. The output resulted from the interviews is organized by the authors in specific themes that are grouped under each one of the four main categories of the SWOT analysis. The number of interviewee(s) is indicated as the source of any position mentioned, or any information provided is cited. Wherever it was possible, we considered issues being to our knowledge that are relevant to transport and innovation. This way we enriched interviewees' claims. Although some opinions expressed by the individuals in some cases may contradict one another, we choose to cite all of them in the analysis below, indicating explicitly the interview number (e.g., IN2) in almost all the cases and in particular, whenever a claim seems to be a subjective opinion, rather than an objective fact. The SWOT analysis attempts to capture significant transport and innovation aspects.

- Strengths

  1. *Co-existence and spatial proximity of a "critical mass" of academic actors:* There are many academic, educational, and research institutions, which form a significant human capital of young graduates and researchers (IN8, IN5 [9]).
  2. *Geographical location and its impact on logistics:* The position of the RCM favors its evolvement to a transport-logistics hub serving the freight flows between Europe, Asia, and Africa. In particular, the Port of Thessaloniki comprises the physical gate to the Balkan area and through it to Central Europe, encouraging the launching and development of logistics (transport, warehousing, and forwarding) services and the implementation of innovative transport systems.
  3. *The distance of RCM from the national capital center and implications on innovation culture:* Although being located far from the decision-making capital may be considered as a weakness at first glance, the barrier to the easy access of public funds encourages the development of more sustainable research activities in economic terms, urging for international competitiveness and innovation (IN4).

- Weaknesses

  1. *Absence of an outward-looking business attitude and action:* Although there are few "model" examples of transport/logistics services' companies and technology providers in the RCM, it is noticed that some of them are still not much active in participating in international activities, e.g., research, exhibitions, etc. (IN2, IN5). There is not any widely adopted business "culture" (IN4) that favors the development of a well-structured marketing and sales strategy, which promotes, among others, active participation in international events, exhibitions, etc. (IN2, IN5, IN4), securing and attracting this way additional funding.

2. *Absence of an innovation-favoring culture:* The potential of transport innovation promotion is hindered by the unwillingness of the companies to invest in R&D and by the unawareness of the future trends in the market by the academic actors (IN3).
3. *Limited cooperation among academia and business sector dependent on public funds:* The cooperation of the academia with the industry is usually not facilitated directly by establishing links with companies. On the contrary, it is usually implemented through European Union (EU) funded or national R&D programs (IN3).
4. *Lack of cooperation among the business actors:* Cooperation is not a mainstream way of working, as there is much fear and concern associated with it, although it could be a win-win choice that could reduce the operational costs (IN3).
5. *Poor governance of innovation at the regional level*: According to the Regional Innovation Scoreboard 2017, the RCM is ranked as "Moderate Innovator." The actual impact of several structures/agencies or intermediary services, which were launched in the last few years for the promotion of innovation (e.g., the Technology Park, the Alexander's Zone of Innovation, etc.), is still limited (IN3). Most of the actors of the (transport) innovation eco-system are providing either design/planning support for innovation, or soft activities (e.g., promotion of software development, support for the creation of start-ups, etc.). This way, the industrial process is not promoted adequately, despite the fact that industry could play the role of a real multiplier towards innovation generation (IN8). There is no Technology Park that allows businesses (e.g., spin-offs) to be located in the same space enabling knowledge and information exchange (IN4). Funding provided by the RCM for the support of entrepreneurship is not focusing or concentrating on a few and carefully selected thematic areas that could promote growth for all (IN3).
6. *Overlapping and unclear responsibilities and regulations*: In some cases, several distinct authorities are being involved in the decision making in the mobility and logistics field (such as urban logistics, traffic signals management, etc.), having overlapping competences and responsibilities, and increasing this way the complexity towards the overcoming of problems (IN9).
7. *Lack of communication and information flow both externally and internally*: This is identified not only between different sectors (e.g., business vs academia), or within a sector itself (e.g., among competitive companies) but also within the distinct administrative units of the same organization or even between an authority and other structures under its direct control (IN9).
8. *Fragmented and isolated research*: Research activity is concentrated within university laboratories and is fragmented among numerous small units without any particular clear goal or connection [9].
9. *Scattered and "individual" innovation support activities*: There is a fragmentation of innovation support actions and a lack of coordination at the regional level [9]. An integrated monitoring and evaluation system of innovation supporting activities that could provide useful feedback for policy planning is currently absent [10].

10. *Intellectual property rights and patenting expertise*: The inadequate consideration of the intellectual property rights and the ambiguity that this may create, could comprise a discouraging parameter that can limit motivation towards working on a more commerce-oriented way after the end of a research project (IN6).
11. *Miscellaneous weaknesses*: (i) limited funding sources [10]; (ii) limited business self-financing capacity [9]; (iii) high rates of unemployment that have led to a significant brain-drain trend (migration of high level educated persons); (iv) poor standardization in the logistics field (IN6); (v) absence of useful technology expertise within the business consulting sector (IN3); (vi) absence of a Centre of Excellence on Transport that could promote the cooperation between academia and businesses (IN5); (vii) bureaucracy and taxing system, as the current taxation system leads many companies to migrate in neighboring countries (IN4); the bureaucracy and the absence of a culture favoring business initiatives in Greece hinder innovation (IN4); the tax rates and the limited access to the banking system could also comprise factors discouraging innovation (IN6).

- Opportunities

1. *Privatization of the current transport modes/terminals and the anticipated establishment of new ones*: The Port and Airport at Thessaloniki, as well as TRAINOSE (the main interurban rail operator) were privatized since almost two years ago, shaping a new context that could favor innovation initiatives. For example, the Port of Thessaloniki intends to invest in the development of a Port Community System; the adoption of a new terminal operating system; the monitoring of specific Key Performance Indicators (KPI) (e.g., per commodity); the exploitation of the Internet of Things (IoT); the decision taking supporting technologies, etc. (IN7). In addition, the application of the European Regulation 1370/2007 in the bus sector may boost innovation (IN1, IN5). The introduction of new transport modes, e.g., metro, sea-boat service, etc., could potentially lead to new fields for innovation (IN1).
2. *Innovation driven by clients' requests*: Business clients of companies may request the incorporation of specific state-of-the-art business approaches or systems driving in some cases, innovation promotion (IN2).
3. *Main market trends and transport/logistics innovation potential*: The identification of the leading market trends (IN4, IN5) that will drive innovation during the next decades could be a fundamental prerequisite for the innovation promotion as it is emerged by the few "model" transport innovation research or business "islands." Nowadays, there is a greater potential for development and progress in the RCM, within a short-term period in specific R&D fields, e.g.:
   i. *logistics*: green logistics, agro-food logistics, city logistics, routing, network design/intermodality, autonomous/robotic vehicles used in goods' handling, organizational models (IN3), truck appointment systems (IN7);
   ii. *mobility*: on-demand services, services' customization, cooperative schemes and services (e.g., Mobility as a Service [MaaS] or Logistics as a Service [LaaS]), behavior change, traffic management, etc. (IN8);

   iii. *Information and Communication Technology (ICT) in transport*: incorporation of 5G, Galileo, and LORA technologies in the telecommunication capabilities of the systems (IN5); and
   iv. *alternative fuels and sustainability*: readiness for Liquefied Natural Gas (LNG) usage in the road transport sector in future (IN6), air pollution monitoring systems (IN7), and impact evaluation framework of port activity on traffic (IN7).
4. *Establishment of communication links and structured forms of cooperation among business and research actors*: Direct and informal/formal relations between the companies/businesses can work for their benefit and be very effective (IN3). The existence of a "Cluster" or an "exhibition space" (IN3) or a "Centre of Excellence on Transport" (IN5), which could promote the cooperation between academia and industry, providing space for the demonstration of technological systems to be developed, is currently missing. The cooperation of the companies/businesses with the academic sector can produce exciting results in relation to the promotion of transport innovation.
5. *Funding priorities and potential funding instruments for the RCM*: The RCM should fund mainly sectors related to food and metals, the two main exporting products of the RCM, resulting in a good rate of return, and spreading the benefits to a wider spectrum of actors (IN3). Among several EU funding sources for R&D, Knowledge and Innovation Communities (KIC) is an instrument for huge research infrastructure, promoting the cooperation of the industry with academia (IN8).
6. *Supporting the small academic research groups*: Special Accounts for Research Funds (SARF) that supervise the activity of the independent research groups within each public university, could implement specific horizontal actions, such as marketing, infrastructure sharing, etc. for the benefit of all teams of researchers (IN3).
7. *Enhancing the mobility of researchers and high-educated employees*: The enhancement of researchers' mobility, the exploitation of Greeks that gained experiences abroad, and the encouragement of involvement in international fora and activities could help in innovation promotion (IN4).
8. *Repatriation of migrated high educated Greeks*: Reversing the brain-drain phenomenon by attracting back in the RCM expatriated experts and scientists from more innovative regions of Europe, is a significant opportunity that could lead to the launching of innovative businesses, including the transport sector (IN4, IN5).
9. *Public, Private and Mixed Partnership initiatives towards innovation capacity enhancement*: There are many planned and ongoing initiatives during the last five years, undertaken by actors of the public or private sector, either on an individual level or in cooperation, to develop tangible, intangible or hybrid innovation mechanisms (such as incubators and clusters) [9]. In particular:
   i. The *Thess INTEC initiative* that is steered by the Thessaloniki's Technological Park (IN4, IN8) whose intention is to be evolved in a 4th generation Tech Park with important research infrastructure, accommodating

Mega Projects, among which two will be related to transport (the "Future Mobility Applications" project and the "Competence Centre for Business and Logistics Challenges" that will try to co-locate industry and academia actors) (IN8). The aim is to promote spatial proximity and cooperation of businesses with the research sector (IN8, IN4).

ii. Several Expressions of Interest (EoI) for the creation in the near future of *innovation clusters or competences centers* by public or private funds, such as the: (i) Innovation Hub by the Port of Thessaloniki, (ii) International Centre of Pfizer for digital technologies, (iii) International Cooperative Centre of Innovation and Digital Skills by CISCO, and the (iv) Clusters' creation funded by the Regional Operational Programme of Central Macedonia.

iii. The *One Stop Liaison Office of RCM* (2019-…), an agency under the "umbrella" of the RCM for the support of innovation.

10. *Opportunities related to the national economic situation*: Great need to increase exports may encourage innovation-driven business [10].

- Threats

1. *Mobility policy implementation risks:* due to overlapping roles among various authorities (IM1).
2. *The crucial importance of personal relations among individuals*: The role of individuals/persons in key positions is crucial, as they can become the catalysts or the barriers of change (IM3).
3. *Threats related to the conditions of the national economy and the State's administration*: (i) bureaucracy of public initiatives for the support of innovation and entrepreneurship [9]; (ii) unclear and constantly changing taxation system; (iii) regularly changing framework of research activities monitoring at the academia sector; (iv) social challenges related to aging of the population, high poverty, and exclusion [10]; (v) absence of the capability so far, to restructure the economy towards the production of goods and services of added-value [10].

5 Discussion and Concluding Remarks

It is worthwhile noticing that there are few good examples of innovative business and research "islands" in the RCM providing transport innovation, which have managed to create an international portfolio and client base (e.g., Laboratory of Applied Thermodynamics (LAT) of the Aristotle University of Thessaloniki [AUTh], LINK Technologies, KENOTOM, BETA CAE, and few AUTh's spin-offs).

All of them share two common attributes; in particular: (i) the professional mobility experience of their founders in other countries abroad, and (ii) the strong linkages with the local academia sector, for instance, by recruiting well-educated employees with much more competitive personnel cost than in Central and Northern Europe. Although the number of these successful and innovative "structures" (companies and research teams) on transport is low, their examples seem to be rather encouraging for the development of the sector. Their presence could attract similar and competitive big actors that would like to benefit from the same advantages that these actors enjoy, mainly the availability of well-educated personnel that costs less than in Central and Northern Europe.

Overall, through the study of the RCM's transport innovation framework, it was drawn that: (a) transport innovation initiatives by the local businesses are rather scattered and not shared among them, preventing the establishment of possible synergies; (b) information flow channels between the transport innovation stakeholders are missing, hindering networking and cooperation; (c) local authorities are not aware of the real transport innovation capacity of the Region, leading to a non-straightforward policy-making process; (d) although many businesses are innovating in ICT in the RCM, only few of them have focused on transport-related fields.

The Regional Authority can support transport innovation stakeholders in order to come closer to each other creating synergies. In addition, it can encourage and mobilize businesses towards innovating on the transport sector by allocating grants for the establishment of special innovation mechanisms, the adoption of innovative solutions, and the production of novel systems and know-how to be leveraged by the transport industry and sector.

References

1. INNOTRANS Homepage. https://www.interregeurope.eu/innotrans/. Accessed 31 March 2020
2. Organisation for Economic Co-operation and Development: Oslo Manual: Guidelines for collecting and interpreting innovation data. 3rd edn. OECD Publishing, Paris (2005). http://www.oecd-ilibrary.org/science-and-technology/oslo-manual_9789264013100-en
3. CIPTEC project: D2.1: Guidelines for field research design and relevant material on existing innovative practices. CIPTEC project, Thessaloniki (2016). http://ciptec.eu/deliverables
4. Genitsaris, E., Roukouni, A., Stamelou, A., Nalmpantis, D., Naniopoulos, A.: Co-creating innovative concepts to address crucial trends and challenges that public transport faces in Thessaloniki. In: Ketikidis, P., Solomon, A. (eds.) University-Industry Links: Coproducing Knowledge, Innovation & Growth: Proceeding of the 10th International Conference for Entrepreneurship, Innovation and Regional Development (ICEIRD 2017), pp. 159–166. ICEIRD, Thessaloniki, Greece (2017). http://iceird.eu/2017/conference-proceedings
5. Genitsaris, E., Stamelou, A., Nalmpantis, D., Trochidis, I., Naniopoulos, A.: Crowdsourcing innovative ideas by public transport users and non-users of the city of Thessaloniki. In: Proceedings of the 8th International Congress on Transportation Research (ICTR 2017), pp. 28–29. Thessaloniki, Greece, September 2017
6. Nalmpantis, D., Roukouni, A., Genitsaris, E., Stamelou, A., Naniopoulos, A.: Evaluation of innovative ideas for public transport proposed by citizens using multi-criteria decision analysis (MCDA). Eur. Transp. Res. Rev. **11**(1), 22 (2019). https://doi.org/10.1186/s12544-019-0356-6
7. Naniopoulos, A., Genitsaris, E., Stamelou, A., Kostopoulos, I., Aissati el., H., Schmitz, W.: Co-creating innovation: concepts and ideas for public transport resulted through participatory processes applied in four different urban areas of Europe. In: Proceedings of the 7th Transport Research Arena (TRA 2018), pp. 16–19. Vienna, Austria, April 2018

8. Tsafarakis, S., Gkorezis, P., Nalmpantis, D., Genitsaris, E., Andronikidis, A., Altsitsiadis, E.: Investigating the preferences of individuals on public transport innovations using the maximum difference scaling method. Eur. Transp. Res. Rev. **11**(1), 3 (2019). https://doi.org/10.1186/s12544-018-0340-6
9. Reid, A., Komninos, N., Jorge, A., Sanchez, P., Tsanakas, P.: RIS3 Assessment: Central Macedonia (2012). https://www.urenio.org/wp-content/uploads/2013/04/RIS3-review-report-Central-Macedonia-final-edited-2012.pdf
10. Region of Central Macedonia: Stratigiki proothisis tis koinonikis entaxis, katapolemisis tis ftocheias kai kathe morfis diakriseon stin Perifereia Kentrikis Makedonias [Strategy for the promotion of social inclusion, fight against poverty and all forms of discrimination in the Region of Central Macedonia]. Region of Central Macedonia, Thessaloniki, Greece (2015). http://www.pepkm.gr/attachments/stratigikes/PESKE-Kentrikis_makedonias.pdf

Integrated Parking Management Plan in Medium-sized Cities

Elias Papastavrinidis(✉), George Kollaros, Antonia Athanasopoulou, and Vasiliki Kollarou

Department of Civil Engineering, Democritus University of Thrace, 67100 Xanthi, Greece
eliaspapastavrinidis@gmail.com

Abstract. The expected benefits of implementing an integrated plan with controlled parking zones in a mid-sized city centre will be manifold and readily visible to both traffic and pedestrian's traffic and safety. More specifically, the pricing of on-street parking spots and the imposition of time constraints will discourage the unnecessary use of private vehicles and free up parking spaces while increasing their use, that is, the number of parking spaces serviced. In addition, it is proposed to organize and deploy special parking spaces exclusively for taxis, tour buses and trucks in the urban area, to serve the needs arising from existing uses and to improve citizens' accessibility. Restricted parking both at home and at the trip destination combines to low possibilities of car use. Parking restrictions will have the greatest effect in compact cities. In Xanthi, Thrace, parking management has often remained a domain untouched by the authorities, unless parking problems have spiraled out of control and the city wants to gain financial revenue. This has led to a merely reactive and operational way of dealing with parking, mainly only responding when a specific problem pops up (at a certain location), and using an isolated approach, further facilitating car use. Thus, a "predict and provide" mechanism – often focusing on infrastructure – has dominated parking policy in the city for many years. An effort is made to face the problem with the aid of an analysis based on counts and questionnaires in different locations of Xanthi.

Keywords: Sustainable mobility · Parking · Parking management plan · Controlled parking area · Parking spaces

1 Introduction: Management of Parking Spaces

The availability and cost of a parking lot is a crucial determinant of whether or not people prefer to drive to a specific destination, and also whether or not they prefer to own a car.

Local authorities have direct control over the utilization of kerb space (other than on national roads) in their areas, and thus of the availability and price of on-street parking. Many authorities own public off-street car parks, over whose use and price they even have control [1].

Car parking constitutes a problem of significance both at local and at strategic level of designing. Parking policy plays a serious role within the management of

E. G. Nathanail et al. (Eds.): CSUM2020, AISC 1278, pp. 379–387, 2021.
https://doi.org/10.1007/978-3-030-61075-3_37

transportation systems in dense urban areas. So as for parking policy decisions to be founded, the analysis of parking behaviour and therefore the effect of parking policies should be fully integrated with the other elements of the transport planning process [2].

Today, illegal parking of cars on streets in the city centre of Xanthi has increased dramatically, resulting in serious violation in the quality of inhabitants' life and severe degradation of urban space. The main reasons for this situation are the lack of awareness of drivers in the area and the inability of authorities to control the ban on parking vehicles. A driver who wants to park in the city centre of Xanthi is looking for one of the sparse vacancies (burdening traffic and consuming extra fuel), until he finally parks his car.

2 Controlled Parking Zones in Medium-Sized Cities

Controlled parking zones has the following goals [3]:

- Implementation of an extensive system of controlled parking on the road, with separation of seats, in places of visitors (short-term or long-term) and in places of exclusive use of residents.
- Implementation of the system of strict and systematic policing of the places where the prohibition of parking is proposed, with special emphasis on the critical places, such as the accesses of the junctions, the stops and the turns of buses, etc.
- Creation of off-road parking spaces, in order to cover, in addition to the previous measures, the parking needs of both the visitors and the residents of the areas in which they will be implemented.

The expected benefits of implementing a comprehensive plan with controlled parking zones are multiple and directly visible both in vehicle traffic and in pedestrian traffic and safety. Specifically, the pricing of the places on the road and the imposition of time restrictions, discourage the unnecessary use of the car and release parking spaces, while at the same time increasing the degree of their use, i.e. the number of serviced parking slots [4].

Increasing the rotation rate in conjunction with the organization and delimitation of parking spaces help to comply with the rules and restrictions of the system through policing. A framework is created with defined rules that apply and at the same time are accepted by all users. As a result, it is estimated that illegal parking will be drastically reduced, with positive effects on walking and road network traffic.

Payment of parking fees using a mobile phone can be made by credit/debit bank cards. The controlled parking system contacts with the driver through short text messages. The driver's communication with the system is interactive, since for each action he receives a corresponding response from the system, which either confirms the successful execution of his request or guides him for the required actions or even warns him of errors or emergencies.

The system allows the controller to check the legality or not of parking a vehicle by checking its license plate and parking code. The system cross-checks the database to see if the vehicle is legally or illegally parked if the owner of the vehicle has paid for it via text message. The system also recognizes whether the vehicle belongs to an owner who has joined one of the special groups of citizens who have the right to free parking such as the disabled, regular hospital visitors, etc.

3 Parking Infrastructure in the City of Xanthi

3.1 Parking Spaces for Taxis

Taxis are located in the city centre and in urban areas where land use attracts a large number of people (Table 1) (Fig. 1).

Table 1. Capacity of taxi parking spaces.

Number	Site	Capacity
1	M. Vogdou St.	10
2	Pesonton Iroon St.	6
3	Vas. Sofias St.	10
4	28 Oktovriou St.	10
5	Plateia Eleutherias	15
6	Mprokoumi St.	8
7	Pavlou Mela St.	7
8	Miaouli St.	7
9	Hospital	7
10	Train Station	10

Fig. 1. Taxi parking spaces in the city of Xanthi.

3.2 Parking Spaces for Tourist Buses

The places for stop and parking of tourist buses in the city of Xanthi are determined by regulatory decisions of the Municipal Council (Table 2) (Fig. 2).

Table 2. Capacity of tourist bus stops and parking spaces.

Number	Site	Capacity
1	28 Oktovriou St. – Xanthippion Hotel	1
2	28 Oktovriou St. – Plateia Eleutherias	2
3	28 Oktovriou St. – Democritus Hotel	1
4	Plateia Dimokratias – Military Club	1
5	Plateia Dimokratias – M. Karaoli St.	2
6	M. Karaoli St. – Orfeas Hotel	1
7	Pesonton Iroon St.	2
8	Andrea Papandreou St.	5
9	Vas. Sofias St. – AOX Stadium	7

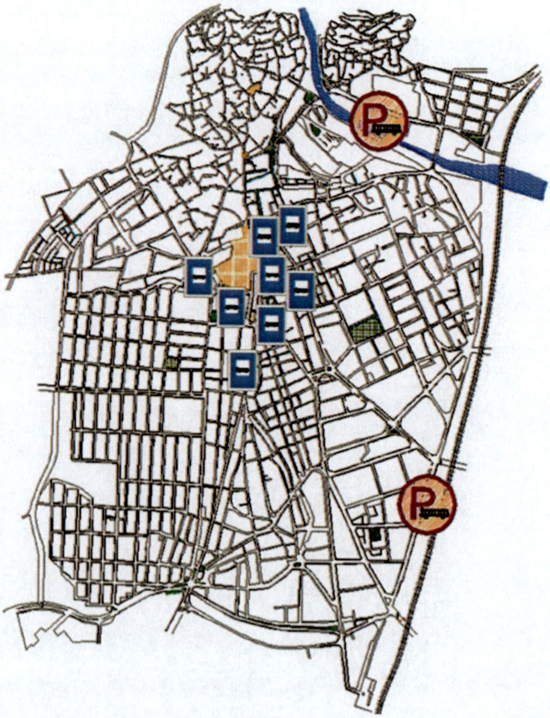

Fig. 2. Tourist bus stops and parking spaces in the city of Xanthi.

To park the buses along the road, a width of 3 m is required, according to the dimensions of the bus. In the above sections, the width of the road is sufficient for the parallel parking of buses and the necessary width for the traffic of vehicles.

3.3 Parking Spaces for Loading and Unloading - Operational Needs

Functional parking spaces are required to service vehicles that are regularly and compulsorily required for the operation of the dependent land use.

The required parking space for operational needs must be on the same plot as the land use. For the location of these places, the relevant regulations are taken into account in order to meet the parking needs of the individual developments and to ensure the prevention of illegal parking on the road (Fig. 3).

Fig. 3. Influence of existing loading and unloading parking spaces in the city of Xanthi.

The location based on standards is therefore in the competence of the service of the Municipality to be granted where the corresponding license is needed [5].

3.4 Parking Spaces for Private Cars

The operation of parking spaces is directly related to the development of awareness among citizens for the cost of parking in urban areas and the strong policing to maintain the effectiveness of the measure. The available public and private parking spaces provide a satisfactory level of service for the downtown area (Table 3) (Fig. 4).

Table 3. Capacity of public and private parking spaces for vehicles.

Number	Site	Ownership	Capacity
1	Pazari	Public	270
2	Limnio	Public	260
3	Municipal Parking	Public	220
4	Old Town	Public	25
5	Plateia Galaxias – Karaoli St.	Private	231
6	Cosmos Centre – Karaoli St.	Private	220
7	Post Office – Miltiadi Georgiou St.	Private	17
8	Tsimiski St.	Private	55
9	Platonos St.	Private	30
		TOTAL	1328

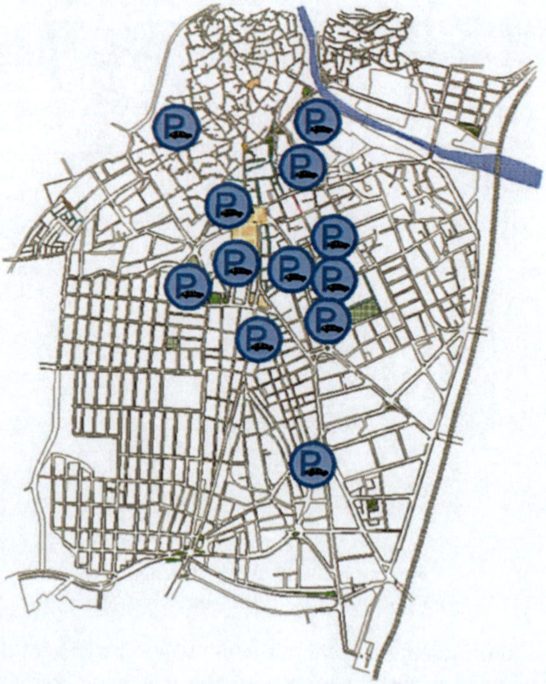

Fig. 4. Public and private parking spaces in the city of Xanthi.

In the city centre of Xanthi, some places can be configured and function as parking spaces (Table 4).

Table 4. Estimated capacity of proposed new parking for vehicles.

Number	Site	Ownership	Capacity
1	Municipal Parking (expansion)	Public	80
2	Old Town (expansion)	Public	200
3	Parking of Regional Services	Public	100
4	Indoors Sports Centre	Public	120
5	28 Oktovriou St.	Public	80
		TOTAL	580

4 Controlled Parking System in the City of Xanthi

4.1 Parking Time

From the analysis of the data and the calculation of the parking demand balance in the study area, it appears that in the early morning hours from 07:30 to 09:30 the balance is positive. However, with the opening of commercial stores and the local market, the balance is turning negative with the lack of parking spaces showing its maximum value between 12:30–13:30. Then the deficit shows a decreasing trend and even shows a peak in the period 14:30–15:30 during which the employees in both the public and the private sector leave the area. As the local market reopens, demand is rising but "marginally" this is being met by the number of parking spaces on offer.

Alternation measurements show that the majority of vehicles park for up to an hour. The percentage of vehicles that park for two and three hours is also important. The duration of legally parked vehicles ranges from three to twelve hours, while illegally parked vehicles for obvious reasons take less time to park. The largest percentage of legal parking is for short-term parking, while the lowest percentage is for long-term parking. The illegal parking of vehicles is due to the fact that users have access to specific land uses in order to meet their needs, with any results in the operation of the road network.

The controlled parking system allows the parking of visitors' vehicles for a limited period of time with a fee, which will be determined by the local authorities, while ensuring the necessary and sufficient number of parking spaces to meet the needs of the residents. The main criterion for the selection of the application methods of the measures mentioned above is the coverage of the city centre without creating points of excessive demand for parking [6].

4.2 General Characteristics of the System

During the morning operation of the controlled parking system, residents will move their vehicles to meet their needs. Therefore, all users will have the right to park in the parking spaces if they pay the corresponding fee. As for the amount of the fare, this will be determined by the Municipal Authority. For the visitors there will be a time limit regarding their stay in controlled parking spaces and it will be applied in a different pricing system which will be defined by the Municipal Authority. Special parking spaces for the service of specific users (disabled, professional carriers, banks, TAXI, etc.) will be excluded from the controlled parking system and will continue to exist in the same status.

Regarding the residents of the operating area of the controlled parking system, they will be provided with a special signal from the municipal authority and parking for them will be free. Residents will park at designated locations, for any period of time they wish without any financial burden. It will be possible to accredit a vehicle per household [7].

Although the proposed system seems complicated, it is pointed out that the implementation of a controlled parking system based on new technologies regarding the detection of the data of the parked vehicles will simplify the control procedures.

In addition to the installation of the Controlled Parking System and in particular the technology that will support its operation, its viability will depend on the observance of the rules that will be set. Systematic and reliable policing to minimize the effects of illegal parking or non-payment should be ensured. At the same time, policing should be done in such a way that it requires a short time. In addition, policing should be continuous and cover the entire area [8].

5 Conclusions and Recommendations

The Controlled Parking System for the city of Xanthi aims at the better management of the offered parking infrastructure, in order to increase the level of parking. The largest number of vehicles can be served by the same number of parking spaces on offer while reducing demand. This eliminates the phenomenon of illegal parking on the road.

The problem of offering parking spaces is a priority for the operation and organization of land uses in the city. Any possibility of reduction of parking spaces should be considered from the shutdown of a place, as long as such a possibility cannot be avoided. If substantial problems are observed, the Controlled Parking System is recommended in a "high" importance category. The operation of a Controlled Parking System can have positive effects on the available load capacity of the examined area.

References

1. Young, W., Thompson, R.G., Taylor, M.A.: A review of urban car parking models. Transp. Rev. **11**(1), 63–84 (1991)
2. Coombe, D., Guest, P., Bates, J., Le Masurier, P.: Study of parking and traffic demand. Traffic Eng. Control **38**(2), 62–75 (1997)
3. Pickett, M.: Special Parking Areas (Network Management Notes). Transport Research Laboratory (TRL), London (2000)

4. Rye, T., Koglin, T.: Parking Management. In: Ison, St., Mulley, C.: Parking Issues and Policies (Transport and Sustainability vol. 5). Emerald Group Publishing Limited (2014)
5. Technical Program 2020: Municipality of Xanthi 2019/11/27 (in Greek)
6. Still, B., Simmonds, D.: Parking restraint policy and urban vitality. Transp. Rev. **20**(3), 291–316 (2000)
7. Rye, T.: Parking Management and Pricing (STEER Program COMPETENCE). Napier University, Edinburgh (2006)
8. Khalid, M., Ibad, U.R., Muhammad, A., Mohammad, T.K., Muhammad, R.N.: Development of android controlled arduino based intelligent car parking system. Int. J. Wireless Microw. Technol. (IJWMT) **10**(1), 48–61 (2020)

Factors Affecting the Adoption of New Technologies: The Case of a New Sharing Economy Application in the Transport Sector of Thessaloniki

Maria Natalia Konstantinidou[1(✉)] and Erifili Christina Chatzopoulou[2]

[1] Hellenic Open University/Centre for Research and Technology Hellas-Hellenic Institute of Transport, Thessaloniki, Greece
mariakon@certh.gr

[2] Hellenic Open University, Patras, Greece
chatzopoulou.erifili@ac.eap.gr

Abstract. Societies today invest in transport sustainability by developing and promoting smarter and greener transport solutions through targeted strategies. Among these solutions, sharing economy applications are increasingly gaining ground. In Greece, however, the use of such applications is still at a very early stage of adoption. Under this light, the present study aims to investigate citizens' predisposition towards the adoption of a sharing economy application in light of the development of a new transport solution in the city of Thessaloniki, the second largest city in Greece. For the purposes of the study, a questionnaire survey was developed investigating current trip patterns and characteristics of the respondents, opportunities and barriers for adopting the new sharing system as well as a stated preference experiment. Discrete choice analysis was performed, and a model was developed describing the determining factors for the integration of a sharing economy concept in the existing transport system amongst the population. The results of this study aim to contribute to the identification of the target market as well as the maximization of the benefits for the users. Furthermore, the results of the study are expected to be useful for the design of appropriate promotion campaigns for the new transport sharing application, providing citizens/users of Thessaloniki with an opportunity for a quicker, less stressful way to arrive to their destinations, as opposed to the services currently provided (e.g. intra city buses).

Keywords: Sharing economy · Factors · Transport · Model · Thessaloniki

1 Introduction

Daily trips are an integral part of everyday life and society in general. Nowadays, the concept of sustainability has become more and more common in urban planning. In the transport sector, this concept known as "sustainable mobility" [1] encompasses the use of more environmental friendly transport modes and the creation of safer conditions for urban trips within the city. Among the greener transport solutions, sharing economy applications are increasingly gaining ground in the formulation of transport policies

E. G. Nathanail et al. (Eds.): CSUM2020, AISC 1278, pp. 388–397, 2021.
https://doi.org/10.1007/978-3-030-61075-3_38

that implement the principles of sustainable mobility. Although the contribution of mobility sharing systems to the implementation of a more sustainable transport sector has been extensively studied in the literature, the factors affecting the propensity to use such sharing systems and their impact in different social contexts and on travellers' behaviour remain largely unexplored [2].

The aim of the present paper is to contribute to this gap in the literature by investigating the factors that may prompt the citizens of Thessaloniki to choose a new sharing economy based transport service for their daily trips. Section 2 reviews the factors affecting the transport choice and the adoption of new technologies based on the literature. Section 3 presents the methodological components that were used in the research, including the experimental design and the analysis. The results which include the model estimation results and the factors that may encourage or discourage the use of the taxi-sharing service are outlined in Sect. 4 and discussed in Sect. 5. Finally, the main conclusions as well as suggestions for further research are provided in Sect. 6.

2 Literature Review

Sharing economy developments in the transport sector have important implications on the formulation of transport policies as they influence the future demand and are predicted to be one of the largest sectors in terms of revenues in many countries. Ride-sharing, car-pooling and car-sharing are the most famous types of passengers transport sharing applications.

To identify the factors affecting the propensity to use a mobility sharing application, we build on knowledge from two distinct streams of research. The first one concerns the factors that influence the choice of transport mode in general. But, since the ride-sharing services are typically performed through an online/mobile platform, the adoption of a new technology is also required. Thus, the second stream of research concerns the factors that influence the level of adoption of new technologies.

The factors affecting the choice of transport mode can be categorized in five groups [3]: socio-economic and demographic characteristics, spatial development patterns, policies directly or indirectly affecting travel behaviour, national cultures or individual factors and other factors including the cost, the duration and the security of the trip.

Regarding the socio-economic and demographic characteristics, the income and the car ownership are considered to be two of the most determinant factors identified in the majority of studies for explaining the mode choice. Factors such as household composition, gender and age seem also to be crucial for the determination of the transport mode choice in wealthy countries [3]. Some studies argue that there is higher probability for the females to shift to public transport. This implies stronger preference among males to drive as they are less likely to use public transport [4]. Car-sharing can have positive implications in the reduction of vehicle ownership and consequently in the reductions of the average vehicle miles travelled and the emissions produced [5], and thus it may have positive effects on the use of public transport [6]. According to Dickinson et al. [7], the influence of age is not significant on the decision to use a car-pooling service and thus, people of all ages can benefit of these schemes.

Although many researchers confirm the mutual influence between the spatial development patterns and the travel behaviour, their relationship has not yet been established. The dense cities are characterized by mixed land uses and consequently shorter distances and thus the urban residents are more likely to choose active transport modes such as the bicycle or walking as means of transport than the suburban residents [8].

In terms of policies affecting travel behaviour, the high operating costs for the car (sales taxes, tolls etc.) reduce the likelihood of choosing car as a travel mode and result in higher percentages of trips by public transport, walking, or cycling [3]. Similarly, the increase in the fares of public transport leads to a decrease in its use [9]. Therefore, cost savings is one of the main incentives for the mobility sharing services since the trip's cost is shared among the driver who usually owns the car and the passengers [10]. Governments also influence the existence of the suitable transport infrastructure [11] that can facilitate the trips by car making them faster and more convenient. Lastly, a land use planning characterized by high densities supports public transport, walking, and cycling [3].

The differences in travel behaviour are often explained by the differences in culture and social attitudes. Dissimilarities in travel may be attributed to different government systems and market economies. Thus, countries with a strong tradition in the car manufacturing such as Germany have high rates of motorization and consequently high percentages of car use. Ecological beliefs are also likely to affect the choice of transport mode. People with deep environmental beliefs make more frequent use of environmental friendly modes such as the bicycle and have a more positive view of new services that reduce the environmental impacts [12].

The time and the purpose of a trip can also affect the choice of transport mode. Individuals have different expectations regarding the transport mode they use to go for work or studies than the one they use to go for leisure activities. The unavailability of a sufficient number of public transport vehicles and the unfixed timetables lead travellers to not prefer public transport for their trips to work [9]. Other encouraging factors for using mobility sharing systems found in the literature include the building of stronger community structures since people learn to be cooperative rather than individualistic [10]. Although sharing space and social engagement are referred in the majority of the studies as positive motivations for participating in car-pooling schemes, they are reported also as discouraging factors by a part of travellers [13].

On the basis of the above, it thus expected that the age, the family status, the education, the income, the area where someone's destination (work) is located, the trip cost and purpose, the ecological beliefs, the good's ownership and the status that reveals the chosen transport mode affects the propensity to use the taxi-sharing service.

The taxi sharing concept is based on a new technology system which clusters users' origins and destinations based on the starting time of the trips and after the analysis of offer/demand each cluster is assigned to one taxi vehicle. Thus, except of the factors affecting the choice of transport mode, it is also important to understand the factors responsible for technological innovations adoption. Over the last 20 years a variety of models have emerged such as the Social Cognitive Theory by Compeau and Higgins and the Infusion Model by Cooper and Zmund [14]. The more recent model that tries to combine findings of eight different models creating a unified model is the Unified

Theory of Acceptance and Use of Technology (UTAUT) model (Patel and Connolly, 2007). The eight acceptance models are: 1) Theory of Reasoned Action (TRA); 2) Technology Acceptance model (TAM); 3) Motivational model (MM); 4) Theory of Planned Behavior (TPB); 5) Combined TAM and TPB (C-TAM-TPB); 6) Model of PC Utilization (MPCU); 7) Model based on Innovation Diffusion Theory (IDT); 8) Model based on Social Cognitive Theory (SCT) [13]. After the combination of the eight above models four factors are identified as determinants: performance expectancy, effort expectancy, social influence and facilitating conditions.

Parameters that cover these four factors are expected to positively affect the propensity to use the taxi-sharing service and thus their inclusion in our model does not contribute to the creation of knowledge. Nevertheless, they are included in the third part of the questionnaire as encouraging or discouraging factors (e.g. service's reliability, existence of mobile app, friend's recommendation and rewarding system) and are analysed through descriptive statistics.

3 Methodology

3.1 Experimental Design

The approach undertaken in this study involves a stated preference questionnaire survey distributed to inhabitants of Thessaloniki. The questionnaire consists of four distinct parts including questions based on the factors that have been found in the literature as affecting the choice of transport mode.

The first part of the questionnaire involves general questions on participant moving preferences. In the second part of the questionnaire nine hypothetical scenarios of a trip in the city of Thessaloniki are presented to the respondent. Each hypothetical scenario is constructed combining three parameters: trip purpose, trip duration and trip cost. These three factors are recorded in the literature among the main ones that affect the choice of transport mode [10]. Trip purpose involved: (i) work (ii) leisure or (iii) shopping. Trip duration involved: (i) short duration (up to 15 min) (ii) medium duration (16–30 min) or (iii) long duration (31–45 min). Last, trip cost involved: (i) low cost (up to 2.00€) (ii) medium cost (2.10–5.00€) or (iii) high cost (5.10–8.00€). The use of these levels in a full factorial design resulted in 27 ($3 \times 3 \times 3$) different scenarios. To avoid respondents' fatigue from the unreasonable sets of scenarios, three blocks of scenarios were created, each of which contained 9 scenarios. The approach proposed by Rizzi and Ortúzar [16] is applied in the block design methodology. Responses considering traveller preference were recorded on a five-point Likert scale [17], ranging from 1 'Definitely not' to 5 'Definitely'. In the third part of the questionnaire, the encouraging and discouraging factors for the use of taxi sharing service are explored. Finally the fourth part investigates the socioeconomic characteristics of the participants.

As far as the data collection method is concerned, the Internet-base survey was chosen as the easiest way to reach inaccessible areas and create a more representative sample of the population. This method was considered to be the most appropriate for collecting raw data as it has high response rates and allows the sample to be flexible and controllable. Then the sampling pattern had to be selected. As the aim of the present study is to determine the parameters that affect the level of adoption of a new sharing economy service, the sample may consist of anyone who resides and moves in the city of Thessaloniki and therefore no particular part of the population had to be included or excluded from the study. The data collection period was from March until April 2019.

The sample survey comprised 153 participants (70 male and 83 female) with their age distribution being 6% for 18–24, 53% for 25–34, 25% for 35–44, 16% for 45–54 and no participant for the groups 55–64 and over 64 years old. The modal split of the sample concerning their main mode of transport is 50% private car, 20% public transport, 4% motorcycle, 2% bicycle, 20% foot and 4% taxi.

3.2 Discrete Choice Analysis

Respondents in surveys are often asked to express their preferences in an ordering scale. The first step designing an ordered logit model is the determination of the utility associated with each alternative. A more appropriate approach is that the researcher believes that the respondent knows the utility level related to the subject of the question and answers the questions, based on this knowledge. This utility consists of one observed and one not observed factor:

$$U = \beta' x + \varepsilon \tag{1}$$

In the current model implementation, the dependent variable is the ordered response which has five possible choices ("Definitely not", "Probably not", "Maybe", "Probably" and "Definitely") based on the participants' answers to the hypothetical scenario question of the second part of the questionnaire. Discrete choice analysis, through the design of appropriate ordered logit model with random effects, is performed to identify the contributing factors for taxi sharing use. Analysis results are presented in the following section.

4 Results

4.1 Model Estimation

The resulting model is presented in Table 1. The base level for all the independent variables that are presented in Table 1 is the last level (5) and the utility differences for the rest levels are calculated. All variables are statistically significant with a confidence level greater than 95%.

Table 1. Model estimation results.

Independent Variables/Levels	Estimate	P-value
Work area		
Centre	0.470	0.000
East	–	–
Trip purpose		
Work	−0.505	0.000
Shopping	−0.338	0.004
Leisure	–	–
Trip cost		
Up to 2.00€	1.379	0.000
2.10–5.00€	0.821	0.000
5.10–8.00€	–	–
People have the right to change the natural environment according to their needs		
Disagree	0.606	0.000
The ownership of goods is not a trend anymore		
Disagree	−0.462	0.000
The so-called "ecological crisis" is exaggerated		
Disagree	0.369	0.013
The transport mode reveals our status		
Strongly Disagree	−0.593	0.039
Age		
25–34 years	0.384	0.029
Family Status		
Married	−1.166	0.000
Single	−0.874	0.004
Education		
Primary, Secondary, High School	−0.601	0.007
Income (€)		
901–1300	−0.800	0.000
1701–2100	−0.794	0.000

4.2 Descriptive Statistics

In the third part of the questionnaire, the possible factors that would discourage or encourage the taxi sharing use were sought and they are analysed through descriptive statistics (Figs. 1 and 2).

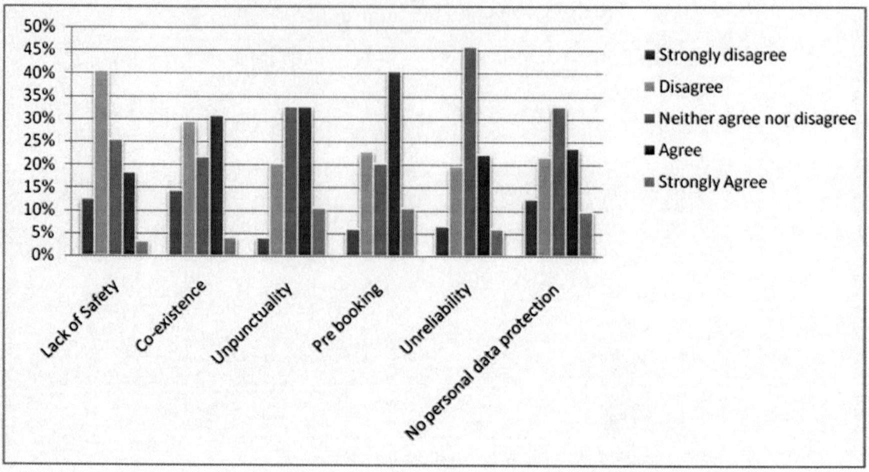

Fig. 1. Factors discouraging the use of the taxi-sharing service.

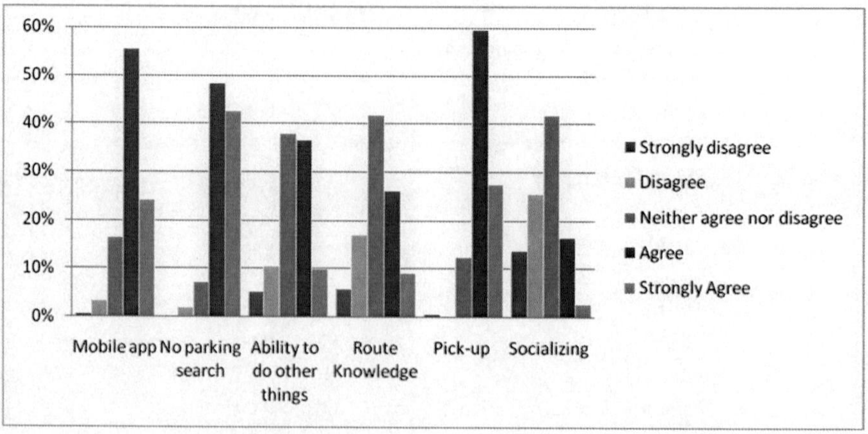

Fig. 2. Factors encouraging the use of the taxi-sharing service.

5 Discussion

The results indicate that trip characteristics such as the purpose and the cost affect the use of the taxi sharing service. Socio-economic characteristics such as age, family status, education and income also affect the likelihood of taxi-sharing use, with the results of the survey being similar to those of the international literature. More specifically, our findings corroborate the results of Buehler [3], who found that factors such as household composition, age and income are crucial for the determination of the transport mode choice. Lastly, parameters related to the general mindset of the travelling population as reflected in the question of what a means of transport should provide affect the use of the service.

More specifically, two of the selected parameters, trip purpose and trip cost, used for the stated preference design, influence traveller propensity to use the new sharing economy service on transport. In particular, participants are more likely to use the service for recreational or shopping trips than for work. Polat [9] also noted that people desiring to be consistent in their working timetable often prefer transport modes such as private cars for work trips which provide them more autonomy. Considering trip cost, travellers' propensity to use the taxi-sharing mode declines as the cost of the trip increases; this finding is in line with the findings of prior studies demonstrating that although cost saving is one of the main incentives for the mobility sharing services since the trip's cost is shared among the passengers [10], high operating costs for the car [3] or increase in the fares of public transport [9] reduce the likelihood of choosing car or public transport respectively as a travel mode. More specifically, those who are asked to pay up to 2.00€ are more likely to use the new transport service in Thessaloniki. The latter can be considered in the design and implementation of marketing strategies for the services' providers in order to promote sharing economy applications in Thessaloniki. Lastly, the trip duration which is the third factor used in the stated preference design does not play an important role in travellers' tendency to use the service. This contradicts the findings of prior research that highlighted time savings among the benefits of car-sharing services. A possible explanation is that as the transport sharing services, and especially taxi-sharing service, have only been recently introduced to Thessaloniki (and Greece), it is not yet linked to time savings in travellers' mind.

A set of parameters that reflected travellers' attitudes towards the chosen mode of transport was investigated and specific attributes were found to affect the use of taxi-sharing service. In particular, those whose work place is in the central part of the city are more likely to use the taxi sharing service than those who move to the western or eastern part of Thessaloniki. Due to the high level of traffic congestion in the city centre during the working times, people prefer to use a quick transport mode avoiding the time consuming process of searching for a parking place.

In addition, participants who believe that they don't have the right to change the natural environment according to their needs or the so called "ecological crisis" is not exaggerated are more likely to turn to transport modes that reduce the environmental impacts. Individuals often report environmental effects such as the lower emissions due to the fewer cars on the road as an encouraging factor for choosing a mobility sharing service [10]. Also, those who believe that the ownership of goods is not a trend anymore are more likely to use the taxi-sharing service. Although the environmental consciousness of the population seems to be increasing, the car use can still serve as a status proof. Participant attitudes on whether the transport mode indicates the user's status affect taxi-sharing use. Participants who strongly agree with this statement are more probable to use the taxi sharing as it still includes the car use instead of public transport.

In terms of respondents' characteristics, age was found to affect the service use, while gender was not found to have a significant effect. Younger travellers (aged 25–34 years) are more probable to use the taxi sharing service than the elderly. This can be explained by the fact that younger people are more open-minded in testing new transport modes. There is a correlation between taxi-sharing use and the family status.

In particular, the propensity to use the service is found to be higher for single individuals than married ones and especially for those who have kids, who may have more obligations and thus, prefer a more autonomous transport mode. Individuals who have a Bachelor's, Master's or PhD degree seem to be more positive for using the taxi-sharing service than those who have completed primary, secondary or high school. The fact that they have a wider range of knowledge may be the reason for not being hesitant with the use of such mobility schemes. Lastly, higher income implies higher likelihood to select the taxi-sharing service since individuals with lower income may prefer more economical transport modes. This contradicts Buehler's [3] statement that people with high incomes strongly prefer the use of private vehicles since car's ownership and maintenance is more feasible for them.

Regarding the factors that discourage the use of the taxi-sharing service use, the majority of the participants (about 45%) strongly disagrees or disagrees that the coexisting with other passengers is a discouraging factor. The building of stronger community structures is also noted by Belk [10] as an encouraging factor for using mobility sharing systems. Also the parking search, the avoidance of the obligation to know the route, the pick-up of the passenger from the desired origin point and the existence of a mobile application are also found to be among the factors affecting the propensity to taxi-sharing use. The majority of the participants placed the above factors among the encouraging ones, with the two last factors (pick up and mobile application) having a percentage higher than 50%. Regarding the factors that would lead someone to use more the taxi sharing service, individuals stated agreement or strongly agreement with the ability to choose the driver and their co passengers' number and profile. Also a positive similar experience in the past and the implementation of a rewarding system would also lead individuals to use the service more.

6 Conclusions

This study serves as a first attempt to capture the preferences of travellers in Thessaloniki regarding the use of a new mobility sharing scheme, in order to use this knowledge and integrate the findings in the business model and the exploitation plan of the taxi sharing service, aiming to maximize the benefits for the users.

The present research could be complemented with a revealed preference survey which could provide more accurate and detailed results for citizens' habits and preferences regarding the taxi-sharing use. Further research could include the investigation of traveller preferences for specific sub-groups of the population (private car users vs. public transport users, and so on). Performing a similar research for other cities in Greece may also reveal additional factors and provide a new topic for discussion in the research field, since the living conditions in each city shape differences in the mindset and the habits of their inhabitants.

References

1. Cohen, B., Kietzmann, J.: Ride On! mobility business models for the sharing economy. Organ. Environ. **27**(3), 279–296 (2014)
2. Neoh, J.G., Chipulu, M., Marshall, A.: What encourages people to carpool an evaluation of factors with meta-analysis. Transportation **44**(2), 423–447 (2017)
3. Buehler, R.: Determinants of transport mode choice: a comparison of Germany and the USA. J. Transp. Geogr. **19**, 644–657 (2011)
4. Chee, W.L., Fernandez, J.L.: Factors that influence the choice of mode of transport in penang: a preliminary analysis. Soc. Behav. Sci. **91**, 120–127 (2013)
5. Circella, G., et al.: What effects U.S. passenger travel. current trends and future perspectives. California: California Department of Transportation and National Centre for Sustainable Travel, pp. 1–76 (2016)
6. Standing, C., Standing, S., Biermann, S.: The implications of the sharing economy for transport. Transp. Rev. **39**(2), 226–242 (2018)
7. Dickinson, J.E., Hibbert, J.F., Filimonau, V., Cherrett, T., Davies, N., Norgate, S., Speed, C., Winstanley, C.: Implementing smartphone enabled collaborative travel: routes to success in the tourism domain. J. Transp. Geogr. **59**, 100–110 (2017)
8. Witlox, F., Tindermans, H.: Evaluating bicycle-car transport mode competitiveness in an urban environment. An activity-based approach. World Transp. Policy Pract. **10**(4), 32–42 (2004)
9. Polat, C.: The demand determinants for urban public transport services: a review of the literature. J. Appl. Sci. **12**(12), 1211–1231 (2012)
10. Belk, R.: You are what you can access: sharing and collaborative consumption online. J. Bus. Res. **67**(8), 1595–1600 (2014)
11. Banister, D.: The sustainability mobility paradigm. Transp. Policy **15**, 73–80 (2008)
12. Ascher, F.: Métapolis ou l'avenir des villes. Editions Odile Jacobs, Paris (1995)
13. Nielsen, J.R., Hovmøller, H., Blyth, P.L., Sovacool, B.K.: Of "white crows" and "cash savers": a qualitative study of travel behavior and perceptions of ridesharing in Denmark. Transp. Res. Part A **78**, 113–123 (2015)
14. Röcker, C.: Why traditional technology acceptance models won't work for future information technologies. World Acad. Sci. Eng. Technol. **65**, 237–243 (2010)
15. Patel, H., Connolly, R.: Factors influencing technology adoption: a review. In: Proceedings of the 8th International Business Information Management Association Conference Information Management in the Networked Economy: Issues & Solutions, pp. 20–22. Dublin, Ireland, June 2007
16. Rizzi, L.I., de Ortúzar, J.D.: Stated preference in the valuation of interurban road safety. Accid. Anal. Prev. **35**(1), 9–22 (2003)
17. Likert, R.: A technique for the measurement of attitudes. Arch. Psychol. **140**, 1–55 (1932)

Carsharing in Greece: Current Situation and Expansion Opportunities

Alexandra Boutla, Chrysanthi Sfyri, Georgios Palantzas, Evangelos Genitsaris, Aristotelis Naniopoulos, and Dimitrios Nalmpantis(✉)

School of Civil Engineering, Faculty of Engineering, Aristotle University of Thessaloniki, PO Box 452, 54124 Thessaloniki, Greece
dnalba@civil.auth.gr

Abstract. This paper aims to explore the potential of carsharing in Greece and to estimate its level of acceptance by the Greeks. The growth of carsharing services as a new and more sustainable way of transport is shifting mobility from private ownership to shared use. Cities and the lives of people are transformed by the sharing economy and collaborative consumption, affecting the transportation industry as well. Carsharing offers an ecological and economical alternative to private car ownership, providing significant benefits. In real-time, from anywhere, anytime, someone can book a car by a phone call or via an application that enables him/her to choose the closest vehicle. In order to evaluate the acceptance of carsharing by citizens, we conducted a survey in which 100 respondents participated. The participants responded to a structured questionnaire, and the results revealed the citizens' daily mode of transport preferences, as well as the level at which they know and accept the new alternative of carsharing. Our results show that the Greeks think that carsharing is an economical, ecological, and attractive way to travel, ranked in descending order. They would use carsharing mainly for commuting, and they would prefer small-engine cars, an Internet application for booking, and charging per kilometer. No special preference was found for free-floating or station-based schemes. The findings also suggest other preferences of people who would probably make them choose carsharing in their daily lives and the fact that the Greek market seems to be ready for carsharing.

Keywords: Sustainable mobility · Sharing economy · Disruptive · Acceptability

1 Introduction

In recent years, an emerging and alternative economy has emerged, i.e., the sharing economy. Sharing economy "is a system that modifies traditional product and service market structures, affecting existing business models for production, distribution, and consumption" [1, 2]. The sharing economy is expanding into more and more areas and is increasingly used by a large part of the population. A typical industry where this phenomenon has been observed is transportation.

E. G. Nathanail et al. (Eds.): CSUM2020, AISC 1278, pp. 398–407, 2021.
https://doi.org/10.1007/978-3-030-61075-3_39

In particular, the urbanization of the last decades increased transport, created congestion problems, and ultimately created a strong need to share these resources. At the same time, the rising awareness regarding environmental issues and better utilization of resources created the need for collective use and sustainable transport that led to new transport schemes, such as the concept of "shared cars," or else carsharing, introducing, thus, sharing economy in the field of transportation.

In Greece, there is not much research on carsharing implementation or acceptability [3]. The aim of this paper is to estimate the level of acceptance of carsharing by the Greeks and to identify the reasons for their preferences.

1.1 The Benefits of Carsharing

Carsharing affects the whole society and each person individually. Firstly, one main advantage of this mode of transportation is the reduced cost. Carsharing enables its users to avoid fixed ownership costs, such as insurance, roadside assistance, parking, etc. [4]. Users are only charged according to the duration that they traveled with the shared vehicle. Moreover, carsharing gives access to each person regardless of their financial situation, and, moreover, they can use a car when needed, paying in total smaller amounts of money compared to the cost of ownership. Thus, carsharing provides the opportunity for more and more people to sell their own car, or not buy one, and use a shared vehicle for their journeys.

Apart from individual cost benefit, there are also many social benefits, mainly due to the prospect of a lower car ownership rate. According to studies, one shared vehicle can replace approximately 13 private cars [5]. Such a benefit is less traffic congestion [4]. More specifically, better road traffic is expected to be achieved with shared cars and also a lower car ownership rate. There will be less congestion at rush hours, with fewer delays for each resident, avoiding annoyance, and economic and social costs. The environment also will be positively affected by carsharing. Shared vehicles are, on average, newer than privately owned ones, meaning that any advances in engine technology, fuel efficiency, and emission levels are integrated faster in the carsharing fleet than in privately owned cars' fleet [6]. Furthermore, reducing the number of cars, through carsharing, reduces the pressure and time to find a parking space in neighborhoods. This fact enables the conversion of public urban spaces from parking lots to green spaces that are an indicator of urban sustainability while, at the same time, upgrading the quality of the existing open public spaces [6, 7], promoting thus walkability [8]. All these advantages could increase public support for developing better infrastructure for shared modes of transport. Indeed, carsharing users tend to be more eco-conscious and, therefore, prefer to walk, cycle, and use public transport more often, in conjunction with shared cars [6], since most of them do not own a car.

1.2 How Does Carsharing Work in Practice?

Carsharing is a growing trend, being an innovative and disruptive scheme of car rental. What distinguishes it from traditional car hire is that it is designed to be an attractive service for those customers who wish to rent cars for short periods.

The new service of carsharing can be rented for even a few hours, and the driver only pays for its use, depending on how long a car is used and the distance it travels [9]. Another difference with a traditional car rental service that makes carsharing more practical is that carsharing gives access to cars without the need to own one, i.e., customers can book a car at any time of the day or night, by phone or Internet [9].

In addition, shared cars are located in dedicated parking spaces in the city, or everywhere in free-floating schemes, so there is a high possibility that a shared car will be parked near the user's house. Customers pick up the car, and they simply return it to any station, or they park it on-road in free-floating schemes, after its use. There are no keys to unlock the car; just a smartcard or a Personal Identification Number (PIN) through an app can give access to the user. The most common charging modes are per hour of use of the vehicle or per kilometer traveled. However, there are variations depending on the company, and the policies followed [9].

1.3 Carsharing Schemes

There are various types of carsharing schemes. In some cases, a single carsharing operator provides more than one scheme. Thus, understanding the distinctions between the different types of carsharing services is essential for the user to choose each time the best scheme according to his/her needs [10].

Round-Trip Carsharing
This type of carsharing is the best established commercial carsharing scheme. Users generally book a car when they want to use it, via mobile apps or special websites. In most cases, but not all, the user must determine the time at which he/she prefers to start the reservation and its duration. The duration of its use is completed when the customer returns the car to the same place where he/she received it, and the charging of cost is valid from the time he/she accesses the vehicle until the time he/she returns it at the end of the reservation. The fleet of vehicles is either owned by a company or leased by another competent body. Cars are available in exclusive parking lots, which in some cases are on-road and, in other cases, are off-road [10]. Zipcar is the world's largest provider of round-trip carsharing services [10, 11].

Peer-To-Peer Carsharing
This scheme is based on the round-trip carsharing scheme, with the main difference that the fleet of vehicles is not supplied by a particular company but owned by individuals. People who choose to make their private cars available for use by others receive payment when rented. In some cases, the vehicles are equipped with telematics devices allowing vehicle tenants to access the vehicle via a smartcard; otherwise, the owner of the vehicle has to carry the keys to the renter. This carsharing scheme usually has a wider variety of vehicles for selection. The primary role of the operator is to create an online electronic vehicle platform. The operator, as part of the business model, provides an adapted insurance product that protects the owner of the vehicle and collects a percentage of each lease made through their online purchase [10].

Point-To-Point Free-Floating Carsharing
Point-to-point free-floating carsharing allows single-sided routes within a specific geographic area, unlike the above carsharing schemes. Use is usually spontaneous, for example, no reservation is made at all, or the vehicle is booked a few minutes before its use. The system administrator owns fleets. An agreement with the entity that manages the street parking is necessary, which will oblige the payment of an agreed amount in exchange for customers' right to park in any legal parking lot on the way [10]. Although this type of carsharing allows one-sided routes, round trips are also possible. The largest operator of this carsharing scheme is Car2Go [12, 13].

Point-To-Point Station-Based Carsharing
Some point-to-point carsharing services are based on stations, which means the user takes a car from one parking station and returns it to another [10]. Stable infrastructure serving carsharing can be located in parking lots, such as electric vehicle charging points and customer service kiosks. The French company Autolib [14] is the largest one in the point-to-point station-based carsharing industry [10], with a plan to expand globally while piloting in Boston by Zipcar [10, 11]. The management of this model seems to be less demanding than free-floating carsharing, but it is necessary to have exclusive parking for use.

2 Carsharing Expansion Opportunities in Greece

While carsharing is widespread in foreign countries, in Greece, it remains an untapped opportunity to improve the quality of life [6].

However, the idea of a carsharing service has started in 2016 in the Greek market as well. Carsharing was brought to Greece by Carky, an ambitious startup company that enabled individuals to register their car for rent from other users. According to the company's CEO, Akis Stark, "this is a widespread and highly successful idea abroad, based on the concept of 'sharing economy' and has been adapted to the Greek market. The significant presence of car rental companies and tourism create a very lucrative territory for the development of this model, so we decided to start Carky in Greece" [15, 16]. The placement of Carky in the Greek market was promising, but it seems that it changed its orientation as its website is no longer available [17]; it redirects the user to the ride-hailing app "Aegean Taxis" [18].

Interestingly, the big company "Avis" also has development plans in the Greek market. Leasing prospects are high in the Greek market, as today, the proportion of vehicles purchased by leasing in Europe stands at 78%, compared to just 26% in Greece, which means there is a potential for growth. However, according to the company's CEO Mr. Andreas Taprantzis, to develop carsharing, the company has to collaborate with the Municipality of Athens or other big cities in Greece to find suitably parking spots needed throughout the city. Currently, Avis runs a pilot application of a carsharing service in Athens with a fleet of 50 vehicles [19–22].

3 Methodology

This research is an attempt to investigate the acceptance of carsharing by the Greeks and the reasons that determine its levels. The main tool of the research was a structured questionnaire we developed. The questionnaire was formed according to the overview of the behavioral characteristics of a population subset, i.e., the sample, examining its travel habits, in terms of mode and frequency, as well as preferences and intentions to use a carsharing service.

The questionnaire was developed using the Google Forms platform, as a synthesis of numerous questions utilized in former tools, adapted to the Greek reality, in coherence with the characteristics of the Greek urban environment. The questionnaire consists of two sections. The first one included questions about demographics and some other characteristics of the respondents, and the second one included carsharing specific questions. The answers to all the questions are included in the following chapter, and, therefore, there is no need to describe it further here. The questions were posed in the form of multiple-choice questions as well as Likert-scale questions, to make it quick to complete and easy to read.

The collection of the answers was realized through the Internet. The questionnaire was published and disseminated through social media, mainly Facebook, and by asking respondents to spread it even further, using, thus, the snowball sampling technique. It remained open for answers from Monday 15 April 2019 until Sunday 21 April 2019, after reaching 100 answers of balanced gender representation, so that conclusions could be drawn about the likely attitudes of both men and women towards carsharing. At the same time, more representation was sought from people who hold a driver's license and possibly own their own car so that they could give a more informed view of the costs of owning a car. Probably most respondents are residents of Thessaloniki, Greece, as its dissemination started there. After the implementation of the snowball sampling technique, we did not had control over our sample.

4 Results and Discussion

4.1 Demographics

The sample consisted of 100 participants, 47 females and 53 males, whom age had to be more than 18 years old to answer the questions. Most of the participants were 18–24 years old (73%), a few were 24–34 years old (26%), and only one (1) was 34 years old. The majority had a driver's license (75%), although only 38% of the sample owned a car. Most of the participants reported that the mode of transport that they use on a daily basis is public transportation, which was more than those who prefer the private car, i.e., 38% and 27%, respectively. The rest of them chose walking or cycling (33%), and barely 2% taxi. However, despite the significant percentage of using public transportation, the participants seemed to be disappointed by its use, since it could not satisfy their transport needs and the desired number of trips during the day.

4.2 Carsharing

The results of seven (7) carsharing related questions are analyzed in this section.

Concerning the question of whether the respondents would prefer a shared car instead of a car that they owned so they can avoid paying fixed ownership costs, 51% answered "Yes," and 49% answered "No."

In regard to the question of why they would use a carsharing service, it seems that the respondents would use carsharing mainly because it is an economical way to travel, followed by that it is an ecological way to travel, and that it is an attractive way to travel (Fig. 1).

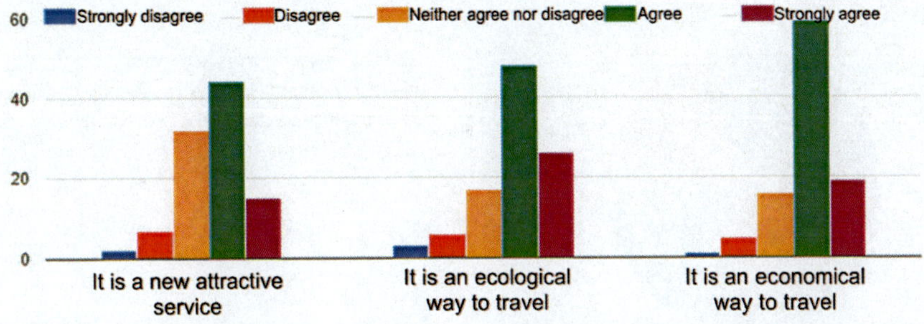

Fig. 1. Reasons to use carsharing: a) economy, b) eco-friendliness, and c) attractiveness.

The most common purpose of traveling with a shared vehicle would be commuting (61%), followed by intercity travel (25%), and periurban recreational travel (14%) (Fig. 2). This result can also explain the preference of the vast majority of the respondents about the preferable type of shared vehicle; small-engine vehicles like ordinary cars will be the most likely chosen vehicles (79%), in contrast to medium-engine vehicles like jeeps (16%), and large-engine vehicles like minivans (5%) (Fig. 3).

Concerning the booking method, an online Internet application was found to be the best option (78%). The use of phone calls or Short Message Service (SMS) was the preference of 13%, while 9% said that an interactive info-kiosk is their preferred option (Fig. 4).

60% of the respondents stated that charging per kilometer is the most affordable method. The rest of the respondents preferred charging per time: 24% per hour and 16% per month (Fig. 5).

Regarding the question for the way of return of the shared vehicle, the respondents chose equally the two possible options, which they were pick up and delivery from/to specific stations in the city (50%), and from/to random parking spots, i.e., free-floating fleet (50%) (Fig. 6).

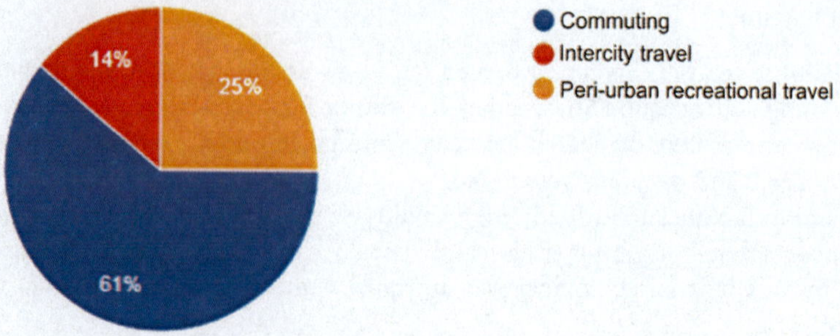

Fig. 2. Reasons to travel: a) commuting, b) intercity travel, and c) periurban recreational travel.

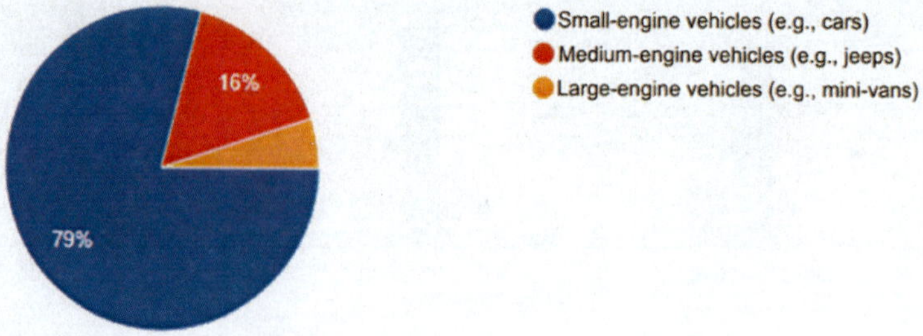

Fig. 3. Preferred type of shared vehicle: a) small-, b) medium-, and c) large-engine vehicle.

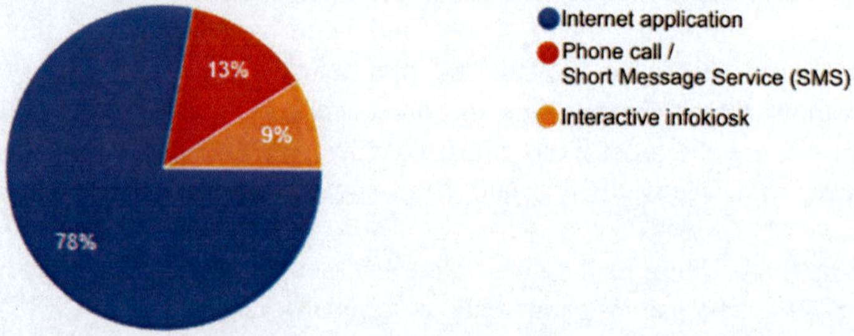

Fig. 4. Booking options: a) Internet application, b) phone call or SMS, and c) info kiosk.

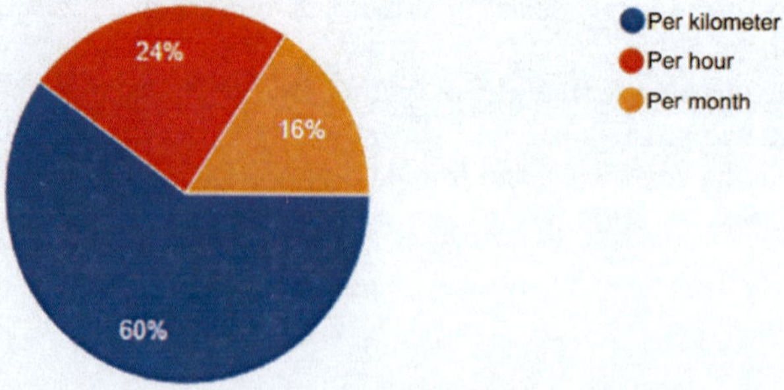

Fig. 5. Carsharing charging schemes: a) per kilometer, b) per hour, and c) per month.

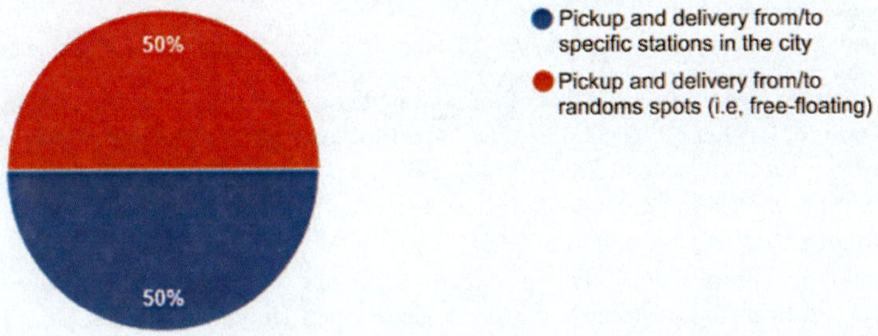

Fig. 6. Pick up and delivery of the shared vehicle: a) from/to stations, and b) free-floating.

5 Conclusions and Recommendations for Further Research

Carsharing could be a better option compared to car ownership in Greece for many people as a more economical alternative. Although there is much research in Greece on how to increase public transportation's modal share using innovation and public consultation [23–26], it seems that even more disruptive approaches should be followed to decrease the fleet of privately owned cars. Thus, carsharing, apart from being just a fancy new disrupting mode of transport, has a more critical role to play in the future [27].

The public sector, universities, research institutes, and other stakeholders, could play an important role, initially by informing citizens about this mode of transport and how it operates, as well as about its various benefits so that everyone is properly informed. At the same time, another proposal is for incentives, such as tax deductions, subsidies, and privileges (e.g., they could be in the form of special traffic lanes), both for businesspersons to invest and for individuals to decide to use shared vehicles. All of the above, if implemented, could form the basis for proper development of carsharing in Greece, with as much universal acceptance and adoption as possible.

One recommendation for future research concerns the cost of the service, as the results of the present study show that the importance of the cost factor has gained more considerable significance, compared to that of the attractiveness of the service and the eco-friendliness.

Besides, saving money is perhaps the most important motivation for changing a habit [6]. Therefore, investigating the cost of the service and the exact way of pricing in Greece is necessary.

At the same time, an extensive study is needed to determine what factors might influence the perception as to which is the most advantageous choice of commuting. Providing a detailed realistic comparison of the actual cost of privately owned vehicles with that of shared vehicles is the key to make the desired shift from car ownership to carsharing.

The sample size of our survey was not large enough to allow for clustering and inferential statistics. A future relevant study, with a larger sample size, could allow for clustering per demographic and other characteristics, and inferential statistics with statistically significant results, that could lead to much better findings and conclusions.

References

1. Grant Thornton: Operation and effect of the sharing economy in the hotel industry in Greece. Grant Thornton, Athens (2015). https://www.tornosnews.gr/files.php?force&file=Sharing_economy_Impact_Assessment_Study_354423404.pdf
2. Sarros, I.: Sharing economy. Diploma thesis, Technical University of Crete, Chania (2018). https://doi.org/10.26233/heallink.tuc.77414
3. Papanaoum, D.: From car ownership to carpooling and carsharing: trends' investigation using the Delphi method. Master's thesis, Hellenic Open University, Patras (2018). https://apothesis.eap.gr/handle/repo/40511
4. University of Kentucky Webpage: Car Sharing (Zipcar). https://www.uky.edu/transportation/rideshare/zipcar. Accessed 30 Mar 2020
5. Shaheen, S., Cohen, A., Chan, N., Bansal, A.: Sharing strategies: carsharing, shared micromobility (bikesharing and scooter sharing), transportation network companies, microtransit, and other innovative mobility modes. In: Deakin, E. (ed.) Transportation, Land Use, and Environmental Planning, pp. 237–262. Elsevier, Amsterdam (2020). https://doi.org/10.1016/B978-0-12-815167-9.00013-X
6. Lalioti, K.M.: Carsharing: the study of an innovative and ecological way of transport. Master's thesis, University of Piraeus, Piraeus (2012). http://dione.lib.unipi.gr/xmlui/handle/unipi/5807
7. Nalmpantis, D., Lampou, S.C., Naniopoulos, A.: The concept of woonerf zone applied in university campuses: the case of the campus of the aristotle university of thessaloniki. Transp. Res. Procedia **24**, 450–458 (2017). https://doi.org/10.1016/j.trpro.2017.05.071
8. Gkavra, R., Nalmpantis, D., Genitsaris, E., Naniopoulos, A.: The walkability of Thessaloniki: citizens' perceptions. In: Nathanail, E., Karakikes, I. (eds.) Data Analytics: Paving the Way to Sustainable Urban Mobility. CSUM 2018. Advances in Intelligent Systems and Computing, vol. 879, pp. 191–198. Springer, Cham (2019). https://doi.org/10.1007/978–3-030-02305-8_23
9. Souris, C., Theofilatos, A., Giannis, G.: Investigation of the acceptance of autonomous vehicles by Greek drivers. In: Proceedings of the 8th International Congress of Transportation Research in Greece (ICTR2017), pp. 28–29. Thessaloniki, Greece, September 2017. https://www.nrso.ntua.gr/geyannis/wp-content/uploads/geyannis-pc263.pdf
10. Le Vine, S., Zolfaghari, A., Polak, J.: Carsharing: evolution, challenges and opportunities. European Automobile Manufacturers Association, Brussels (2104). https://www.acea.be/uploads/publications/SAG_Report_-_Car_Sharing.pdf
11. Zipcar Homepage. https://www.zipcar.com/. Accessed 30 Mar 2020
12. Car2Go: Free-floating carsharing made easy: how car2go works (Press release). Car2Go, Stuttgart (2018). https://www.car2go.com/media/data/germany/microsite-press/files/2018_car2go-en.pdf
13. Car2Go Homepage. https://www.car2go.com/. Accessed 30 Mar 2020
14. Autolib Homepage. https://www.autolib.eu/. Accessed 30 Mar 2020
15. Mavraganis, K.: The startup that brings "car sharing" to Greece: Carsharing by Carky. (2016, June 25). HuffPost Webpage. https://www.huffingtonpost.gr/2016/06/25/car-sharing-carky_n_10629578.html. Accessed 30 Mar 2020
16. LiFO Webpage: Carky The Airbnb of cars, now in Greece (2016, July 18). https://www.lifo.gr/articles/choice/107790. Accessed 30 Mar 2020
17. Carky Homepage. http://www.carky.gr. Accessed 17 Sept 2019
18. Aegean Taxi Homepage. https://aegeantaxi.com/. Accessed 30 Mar 2020

19. Insider.gr Wepage: Andreas Taprantzis (Avis): Leasing in Greece has enormous potential (2016, January 21). https://www.insider.gr/epiheiriseis/aytokinito/105675/andreas-taprantzis-avis-leasing-stin-ellada-ehei-terastia-dynamiki. Accessed 30 Mar 2020
20. Protonotariou, M.: Andreas Taprantzis (Avis): Investments of 200 million in 2019 – we start pilot sharing car sharing (2016, June 25). Mononews Webpage. https://www.mononews.gr/business/andreas-taprantzis-avis-ependisis-200-ekat-to-2019-xekiname-pilotika-to-car-sharing. Accessed 30 Mar 2020
21. Engnews 24h Webpage: Andreas Taprantzis: Avis brings car-sharing to Greece (2020, March 7). https://engnews24h.com/andreas-taprantzis-avis-brings-car-sharing-to-greece/. Accessed 30 Mar 2020
22. Protonotariou, M.: Taprantzis (Avis): We will bring carsharing to Athens - Investments over 200 million euros in 2019 (2019, January 30). Mononews Webpage. https://www.mononews.gr/business/taprantzis-avis-tha-feroume-to-car-sharing-stin-athina-ependisis-pano-apo-200-ekat-evro-to-2019. Accessed 30 Mar 2020
23. Tsafarakis, S., Gkorezis, P., Nalmpantis, D., Genitsaris, E., Andronikidis, A., Altsitsiadis, E.: Investigating the preferences of individuals on public transport innovations using the maximum difference scaling method. Eur. Transp. Res. Rev. **11**(1), 3 (2019). https://doi.org/10.1186/s12544-018-0340-6
24. Nalmpantis, D., Roukouni, A., Genitsaris, E., Stamelou, A., Naniopoulos, A.: Evaluation of innovative ideas for public transport proposed by citizens using multi-criteria decision analysis (MCDA). Eur. Transp. Res. Rev. **11**(1), 22 (2019). https://doi.org/10.1186/s12544-019-0356-6
25. Stamelou, A., Genitsaris, E., Nalmpantis, D., Naniopoulos, A.: Investigating potential synergies among social entrepreneurship and public transport through experts' consultation in Greece. In: Nathanail, E., Karakikes, I. (eds.) Data Analytics: Paving the Way to Sustainable Urban Mobility. CSUM 2018. Advances in Intelligent Systems and Computing, vol. 879, pp. 496–503. Springer, Cham (2019). https://doi.org/10.1007/978–3-030-02305-8_60
26. Papadima, G., Genitsaris, E., Karagiotas, I., Naniopoulos, A., Nalmpantis, D.: Investigation of acceptance of driverless buses in the city of trikala and optimization of the service using conjoint analysis. Utilities Policy **62**, 100994 (2020). https://doi.org/10.1016/j.jup.2019.100994
27. Papanaoum, D., Palantzas, G., Chrysanidis, T., Nalmpantis, D.: The impact of megatrends on the transition from car-ownership to carsharing: a Delphi method approach. In: Nathanail, E., Adamos, G., Karakikes, I. (eds.) Advances in Mobility as a Service Systems. CSUM 2020. Advances in Intelligent Systems and Computing, vol. 1278, pp. xx–yy. Springer, Cham (2019). https://doi.org/10.1007/978-3-030-61075-3_51

Accelerating Deployment: Trials, Pilots and Case Studies

Good Practice for Student Mobility in University of Pavia

Davide Barbieri(✉), Michele Rostan, and Andrea Zatti

University of Pavia, Pavia, Italy
barbieri@unipv.it

Abstract. Pavia is a city with 70,000 inhabitants 40 km south of Milan, the regional capital of Lombardy, one of the more populated and rich Italian regions. The University of Pavia was established in 1361 and until the 20th century has been the only university in the area of Milan and in Lombardy. Todays within the region there are 13 universities, 7 of which are located in Milan. Seven institutions, including Pavia University, are state universities.

Currently, some 24,000 students study at the University of Pavia. About 21,500 students attend short first cycle courses (55%), that is undergraduate or Bachelors' programmes, long first cycle (28%) and second cycle (16%) courses, that is graduate programmes equivalent to Masters'. The rest are doctoral students and students attending advanced specialised courses.

Less than 10% of the students are from Pavia; about 55% of them come from other places within Lombardy, while 35% come from outside the region.

Within Pavia, almost all the students walk around the city. More than 60% of them travel by bus and a considerable minority travel by car. About one out of four students travel by bicycle [1].

As urban mobility is one of the most important aspect of student life, since 2003 the University of Pavia has developed a protocol with the municipality and the Urban Public Transport society for the student mobility at discounted tickets (20 Euros per year for students; 175 Euros for specializing and majoring students). About 50% of university students use it.

Keywords: Urban mobility · Public transport · University students

1 The Context. Pavia as a University Town

The University of Pavia, founded in 1361, has approximately 25 thousand students, of whom two thirds come from outside the province, and 2 thousand employees. The educational and research activities of its eighteen departments - some of which form the medical and engineering faculties - take place in three locations. The major part of them are carried out, of course, in the city of Pavia, while the city of Cremona houses the Department of Musicology and Cultural Heritage and that of Voghera the Sports Sciences course. In addition, the University Institute of Higher Studies (IUSS) is also located in Pavia. Next to the university ones - and closely connected with these - there are other important scientific and educational institutions: the Policlinico San Matteo Hospital, two other scientific hospitalization and care institutions, the local offices of

E. G. Nathanail et al. (Eds.): CSUM2020, AISC 1278, pp. 411–417, 2021.
https://doi.org/10.1007/978-3-030-61075-3_40

national scientific institutes and 16 university colleges", that is special institutions providing - independently or in collaboration with the University and the IUSS - study support services, educational, training and sports activities.

The University is made up of about fifty buildings organized in poles distributed along an axis of about ten kilometres crossing the city from the east to the west.

Part of the poles are located in the city centre. Another part is located in an intermediate area between the railway station and the suburbs, near the hospital. Finally, the Polo Cravino is located on the extreme western outskirts of the city, beyond the ring road (see Fig. 1). The colleges are also distributed in the urban fabric, partly in the central area, partly in the peripheral one.

Pavia has about 70 thousand inhabitants. The structure of the city is characterized by the presence of three waterways, the Ticino river, the Naviglio Pavese and the Navigliaccio, which "close" part of the urban fabric in a triangle that contains the "old" city in its many historical stratifications, from the Lombard one to the Visconti one to the neoclassical one. The map shows two *caesura* that "cut" the urban fabric in the western part of the city: the Milan - Genoa railway and the ring road.

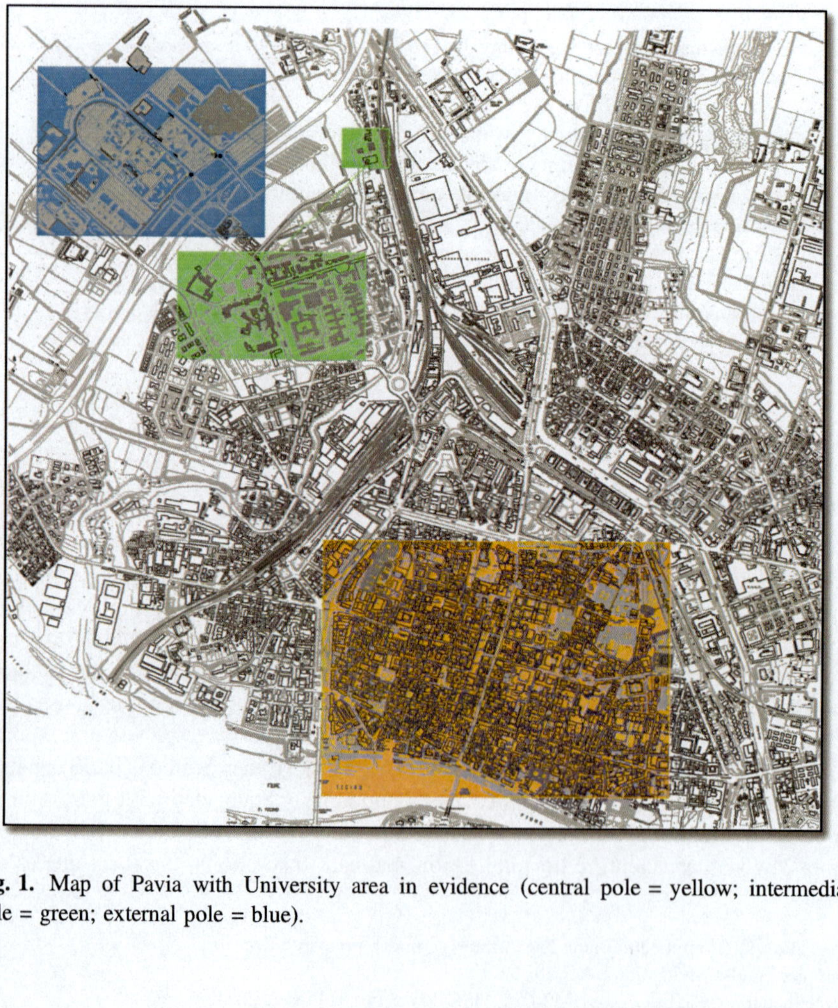

Fig. 1. Map of Pavia with University area in evidence (central pole = yellow; intermediate pole = green; external pole = blue).

The morphology of the city and the distribution of the poles and colleges that make up the local higher education system result in one of the main problems that the University and the city have to face is that of urban student mobility. In fact, when classes are in session, thousands of students travel to the city every day to reach their places of study. From the train and bus stations, commuting students must reach various university buildings. Long term and short term "living-in" students move around the city.

The university poles must be connected to each other and to the colleges. Local public transport for students was a first field of collaboration between the University, the Municipality and the local transport company. Subsequently, thanks also to the appointment of the University Mobility Manager in 2012, the collaboration between different subjects expanded to other fields through specific projects aimed at sustainable urban mobility.

2 The UNIPASSBUS Project

In order to encourage the use of public transport within the urban area by university students, since the academic year 2003/2004 an agreement issued by the University of Pavia, the Municipality of Pavia and the company of public transport, allows the free movement of university students using public transportation at reduced rates [2].

The current agreement has two types of recipients. The first includes regularly enrolled students, doctoral students, Erasmus students and volunteers from the national civil service. The second includes graduating and post-graduates students.

The former can use public transportation paying € 20 per capita per year; the latter can do the same at a cost of € 175 per capita per year.

In view of the right to move freely granted to students, the Municipality of Pavia contributes to the costs incurred by the public transport company by paying an annual sum of € 33,500 while the University pays to the same company a fixed annual lumpsum (equal to € 876,000 in 2019/2020).

2.1 Data Analysis

Table 1 shows the data on the use of the UNIPASSBUS service by students since 2007, the starting year of the systematic data collection by the local public transportation company. Data are updated to the 2019/2020 academic year. After an initial increase, the average percentage of students who use local transportation thanks to the UNIPASSBUS Project is around 42% of the overall student population.

Thanks to the availability of the list of students with the UNIPASSBUS travel document provided by the Public Transportation Company for the academic year 2018/2019 and the administrative data on students enrolled in the same year, it is possible to analyse some characteristics of the students who use the "free movement" agreement.

Table 1. Data on the use of the Public Urban transport by students.

	2007	2008	2009	2010	2011	2012	2013	2014	2015	2016	2017	2018	2019
Subscribers Students s.l.	8.085	8.620	8.777	8.943	9.134	9.996	10.001	9.516	[redacted]	[redacted]	[redacted]	9.914	10.005
Subscribers Specializing and undergraduates				23	28	42	57	61	[redacted]	[redacted]	[redacted]	70	114
Total subscribers	8.085	8.620	8.777	8.966	9.162	10.038	10.058	9.577	[redacted]	[redacted]	[redacted]	9.984	10.119
Total students enrolled and students in mobility (Erasmus)	23.099	23.181	23.514	23.382	22.792	22.768	22.487	22.201	22.145	22.570	23.708	24.655	23.287*
% subscribers students	35.00	37.19	37.33	38.25	40.08	43.90	44.47	42.86	-	-	-	40.21	42.96*

*unconsolidated data

The results of the analysis show that some characteristics of the students appear to influence the decision to "subscribe" and to use the local public transport service. Female students are more likely to "subscribe" than male students (Table 2) and students enrolled in single-cycle master's degree programs and international "incoming" students are more likely to get the pass than others (Table 3).

Table 2. Students enrolled in possession of Unipassbus by gender (percentage values).

	Subscribers enrolled	Not subscribers enrolled	Total
Female	43,1	56,9	100,0
Male	34,2	65,8	100,0
Total	40,2	59,8	100,0

What matters most in the decision to take advantage of the "free movement "agreement, however, is, on the one hand, the condition defined by residence, and, on the other, the place where lessons are attended.

Table 3. Students enrolled with or without Unipassbus by type of degree course (percentage values).

	Subscribers enrolled	Not subscribers enrolled	Total
Bachelor degree courses	36,6	63,4	100,0
Master degree courses	38,4	61,6	100,0
Single cycle degree courses	50,6	49,4	100,0
Post-graduate courses	16,3	83,7	100,0
Mobility: exchange programs	61,4	38,6	100,0
Total	40,2	59,8	100,0

Thanks to the administrative data on the residence of students enrolled in first and second level and single cycle study courses, students can be divided into three groups. The first group is made up of those who live in the Municipality of Pavia. We consider these as "resident" students. The second group is composed of those who reside in other municipalities of the Province of Pavia and in the provinces of Milan, Lodi, and Alessandria, that is - roughly - within a range of about 50 km from Pavia. These students are considered daily commuters. The third group is made up of those who reside in other provinces of Lombardy, in other regions of the North West, North East, Central, South and Islands, and abroad. They are considered as long term and short term "living-in" students. Data show that UNIPASSBUS is taken much more by resident and "off-site" students than daily commuters (Table 4).

Finally, students can be divided according to the location of the department in which they study.

Assuming - to simplify - that the lessons of a course of study offered by a department are held at its headquarters, it is possible to place students in two large areas. The central area includes the departments of Law, Political and Social Sciences, Humanities, Economic and Business Sciences and Brain and Behavioural Sciences. 43.2% of enrolled students attend lessons in this area. The peripheral area - where 55.3% of the students study - includes the Departments of Chemistry, Physics, Internal Medicine and Medical Therapy, Molecular Medicine, Public Health, Experimental and Forensic Medicine, Clinical-Surgical, Diagnostic and Paediatric Sciences, Pharmaceutical Science, Biology and Biotechnology, Civil Engineering and Architecture, Industrial and Information Engineering, Mathematics and Earth and Environmental Sciences. The location of the place of study in the city centre or in the suburbs also influences the propensity to use urban public transport. In fact, enrolled students who attend lessons in the suburbs are much more likely to get the UNIPASSBUS than those who study in the centre (Table 5).

Table 4. Students enrolled with or without Unipassbus by type of condition with respect to residence (percentage values).

	Subscribers enrolled	Not subscribers enrolled	Total
Resident	47,7	52,3	100,0
Daily traveller	26,0	74,0	100,0
Offsite	54,9	45,1	100,0
Total	40,2	59,8	100,0

This is due to the fact that the two poles are both less accessible on foot or by bicycle.

Table 5. Students enrolled with or without a Unipassbus by study area (percentage values).

	Subscribers enrolled	Not subscribers enrolled	Total
Centre city	29,6	70,4	100,0
Periphery	52,1	47,9	100,0
Cremona town	0,6	99,4	100,0
Total	40,2	59,8	100,0

2.2 Results

We can conclude that about half of the students - about ten thousand people - require the pass to move freely using the urban public transport. This service therefore represents an important element of student welfare.

The characteristics of the students most associated with the decision to get the pass are gender, the type of course attended, the condition defined on the basis of the residence and the place where the lessons are attended. Female students, students enrolled in longer courses, "resident" and long term and short term "living-in" students and students studying in the suburbs are more likely to subscribe.

In 2018, the daytime urban transport service was integrated without additional fees with night runs (the so-called Morpheus service), in order to allow students to enjoy the city of Pavia also for out-of-class activities and facilitate the transfer of weekly commuter students from the station to their residence.

3 Conclusions

Generally, the UNIPASSBUS Project is considered displaying some strengths: its popularity among students, 10 thousand of them, almost 50% of the total, use it; the "good practice" of collaboration between the University and the Municipality in terms of hospitality, attractiveness, promotion of public transport, fight against pollution, promotion of the image of both the city and the University, etc.; the good collaboration between the local transport company, the municipality and the University in promoting and managing this special "service"; the significant contribution to contrasting the practice of traveling without a ticket and to smoothing out conflicts between students - largely non-residents of Pavia - and the city. Next to these, however, there are also some weaknesses: a very large financial commitment for the University, with a cost of the agreement that grows annually; a service that does not meet all needs: crowded buses, lack of connection with colleges, especially in the evening, colleges or didactical poles that are insufficiently or not served; the presence of an advantage for students who use urban public transport, while little or nothing is done for those who use - even or only - extra urban means of transport: buses and trains.

References

1. Anzivino, M., Rostan, M.: University student participation in out-of-class activities: consequences on study career and academic achievement. In: Deem, R., Eggins, H. (eds.) University as a Critical Institution?, Rotterdam, p. 185. Sense Publisher (2017)
2. Barbieri, D., Rostan, M.: Pavia: buone pratiche di collaborazione per una mobilità sostenibile. In: UniTown, Città universitaria. Dalle buone pratiche all'identità, Ferrara, Faust Edizioni, pp. 157–170 (2015)

Willingness of Cruise Tourists to Use & Pay for Shared and Upgraded Sustainable Mobility Solutions: The Case of Corfu

Maria Morfoulaki[1], Michail Agathos[2], Glykeria Myrovali[1(✉)], and Maria Natalia Konstantinidou[1]

[1] Hellenic Institute of Transport-Centre for Research and Technology Hellas, Thessaloniki, Greece
{marmor,myrovali,nconstantinidou}@certh.gr
[2] Port of Corfu, Region of Ionian Islands, 49100 Corfu, Greece
agathos@olke.gr

Abstract. Tourism and cruise tourism is an important economic activity in most Ionian Islands, such as Corfu, which is located at the northern part of the Ionian Sea at the entrance of the Adriatic Sea. All inhabitants of Corfu directly or indirectly are occupied with tourism which of course supports the economy of the island but also brings a lot of difficulties in the everyday life of the local community, such as traffic congestion, delays, noise, environmental pollution etc. The intense of these problems is even larger in the historical center, since it is the main destination of the island especially for the cruise visitors who have very limited time for sight-seeing and usually prefer a tour inside the city or in the nearby monuments and beaches. The current work is attempting to shed light on different aspects of the cruise tourism in Corfu as regards mainly the demographical characteristics of the cruise visitors, their most popular destinations, the mode that they usually use but also their willingness to use and to pay for a range of sustainable mobility solutions-that is to say a Bike and Car sharing system and an upgraded and reliable Bus-Based transport system. The analysis and results is based on a questionnaire survey that was realized in the Port of Corfu and was addressed to several cruise visitors during the summer period of 2019.

Keywords: Willingness to pay · Cruise tourism · Bike–sharing · Car-sharing · Reliable bus-based system

1 Introduction

Since the 20th century, tourism has developed as one of the foremost sources of income for numerous urban and inter-urban agglomerations. The relationship between tourism and transportation growth as strictly related aspects has been recognized very early [1], while a balanced progress of both sectors affects the local economy as well as the nationwide and international competitiveness [2]. The special sector of tourism-cruise tourism, experienced great development recently, at a global level as well as in the Mediterranean Sea and it is strategically linked to local operations and GDP growth

E. G. Nathanail et al. (Eds.): CSUM2020, AISC 1278, pp. 418–427, 2021.
https://doi.org/10.1007/978-3-030-61075-3_41

[3, 4]. Around USD 15.000 billion was the cruise related impact from both passengers and crew spending at port cities at global level in 2013–average spending in home-ports reaches USD 316.3 per passenger while transit passengers are estimated to spend around USD 92 per visit (this number regarding crew is smaller reaching the USD 56.7 per visit) [5].

Numerous Greek islands are considered very important cruise tourism destinations [8] with the Ionian Islands to be in the top of the visitors preferences. Tourism and cruise tourism is an important economic activity in most of the Ionian Islands and mainly in Corfu [8]. According to recent data, cruise spending per passenger for 2018 reaches the 140euro, however when considering just home ports this amount is considerably higher while for transit passengers is around 60 euro. Spending is linked with food, souvenirs, transportation and visit of points of interest and touristic attractions [6, 7].

Within this frame, Corfu is presenting an intense touristic attractiveness that is however associated with robust and seasonal demand variations for passengers' transportation. Direct externalities of the intense seasonality are the cumulative growing problems, such as traffic congestion, environmental pollution, and increased travel times [4, 9]. Being cruise tourism an important economic activity in Corfu [10], questions related to the solution of the severe traffic problems affecting both residents and cruise tourists arising. Cruise tourists arriving to Corfu, have a limited period to visit the main attractions of the island using mainly walking or public transport buses for the close to the port destinations and/or private touristic buses and rented cars for points of interests that are placed farther away. As most of the cruise visitors claim, they usually confront severe delays due to the traffic problems in the road network of Corfu. From the other hand the buses and cars that are used for serving the people who are disembarking from a cruise ship, aggravate more the already congested network.

The current work, aims to identify the current characteristics and the mobility profile of the cruise tourists in Corfu and also the alternative new mobility solutions that are willing to use for visiting the island's monuments, protecting also its sustainability especially in the historic center-which consists a UNESCO World Heritage Site. Additionally, their "willingness to pay" for using these modes is also identified. With the term new mobility solutions we refer to non-existed in Corfu mobility services such as bike sharing, car sharing or a more advanced service of Public Buses [11].

The present study was part of Corfu Port Authority activities in the framework of INTER-PASS Interreg ADRION 2014-2020 Project. In the general context of the project for improving the intermodal connections of ports and airports in the Adriatic–Ionian Region in order to better serve tourists and local communities ('global' overview), Corfu pilot case concentrated on the collection of 'global' needs for the development of a platform that will facilitate the tourism planning while simultaneously acting as information provision single point for tourists.

2 Corfu Island–Description and Assessment of the Current Situation of the Cruise Visitors' Mobility

Corfu Island is located at the entrance of the Adriatic Sea, at the northern point of the Ionian Sea, in the Mediterranean basin. To its east, Corfu borders with the coast of Epirus and to its south-west with Albania, while it faces southern Italy to the west. The island is 592 sq2 and is inhabited by 100,000 individuals. The two core entrances/exits of the island are the Port of Corfu and the international Airport "Ioannis Kapodistrias" Corfu port manages an average of 1.3 million passengers, 536000 vehicles and 485 cruise ships per year. The recent years, cruise tourism, is becoming one of the most active and fast-growing branch of the touristic sector in Corfu, as well. More detailed in the year 2017, 408 cruise ships arrived at Corfu Port with 630.000 travelers [12, 13].

Nevertheless, the cruise touristic growth in Corfu, worsened even more the related traffic problems, as the existing road network and the infrastructure are not adequate to satisfy this huge demand, especially during summer. This big number of cruise tourists, who disembark in Corfu have specific alternative modes for commuting towards the desired destinations. These modes are buses (public transport system, hop on hop off, touristic coaches), taxis, walking and rented cars.

The main characteristics of the above-mentioned alternatives are:

- Current urban P.T routes seem to cover most of the travelers' needs, but frequently problems related to the reliability of the provided services occur, due to traffic congestion and illegal parking of private vehicles.
- The main traffic problems and delays in the network of the city center occur due to the large volumes of touristic buses that are serving the cruise and the other visitors' needs.
- The practice of new technologies (i.e. mobility planning services, information provision, advanced mobility solutions) and eco-friendly vehicles is not widespread.
- Even if the port is very close to the city center and also to most of the island monuments, there is no bike sharing system offered to the island visitors.

Corfu cruise tourists are supposed to remain in ports approximately 9.9 h [8]. Understanding their profile and needs and offering effective, cost–efficient and environmentally friendly mobility solutions for their trips can be a useful step towards sustainability of the area. Trying to cover this need, a questionnaire survey has been implemented, in order to identify the cruise tourists' sociodemographic characteristics, mobility preferences and for revealing their willingness to pay and willingness to use new mobility solutions. The survey was targeted to the Cruise tourists disembarked at the port of Corfu (O.L.K.E. SA) during the summer period of 2019.

1045 cruise tourists participated in the questionnaire survey. The answers were inserted in an excel database, and its validity and reliability was checked, in order to eliminate possible random and systematic errors. More detailed, missing value analysis was conducted; data cleaning from irrational responds (irrational pricing of a transport system or numeric value errors).

The majority of the cruise tourists belong to the age group of 60 and over (42%), followed by the age groups of 18–40 and 40–60 years (31% and 27% respectively). The high percentages of cruise tourists under 60 gives a first indication for the potentials of a bike shared system. From the other side, the group over 60 may require even more dedicated and on demand services. The sample of the cruise tourists belong to different nationalities with visitors from USA, UK and Canada to consist the 60% of the sample (Table 1).

Table 1. Summary statistics for observed variables.

Variable	Description	Percentage
Age Group	18–40	31%
	40–60	27%
	60+	42%
Origin Country	USA	32%
	U.K	13%
	Canada	13%
	Philippines	4%
	Germany	4%
	Spain	2%
	Italy	2%
	Indonesia	2%
	Russia	1%
	Other	27%

The most popular (68%) tourist destination is Corfu's Old Town (2.3 km distance from Port of Corfu). The rest Corfu sites that are sharing low rates in the cruise visitors' preferences (2–8%), are Achillion Palace (10.6 km distance from Port of Corfu), Old Fortress (3 km distance from Port of Corfu), Kanoni and Mouse Island (6.6 km distance from Port of Corfu), Saint Spyridon (3.2 km distance from Port of Corfu), New Fortress (1.7 km distance from Port of Corfu) and City Museums (2.7 km distance from Port of Corfu).

The mode that the cruise tourists use for visiting the above-mentioned destinations is bus. Even for the Old Town which is only 2.3 kms from the port, visitors stated that they use bus (41%), foot (33%), rented car (9%) while only 3% stated that they use bike (maybe due to the fact that no bike sharing system is currently offered).

An analysis of the modes used by each group category shows also that bus is the most popular mode for the people between 18–40 (Table 2).

A review of the local published articles, which refer to the traffic problems of the city, highlighted the fact that the bus which is the main mode of transport used by cruise visitors, consists the main obstacle for the proper operation of the traffic network, in the city center. The hop on hop off buses as well as the touristic coaches are too long and large for the specific road capacity, their frequencies-in order to serve the cruise demand-are too high, causing huge delays, and traffic jams.

Table 2. Modal split per age group.

Age Group	Car renting	Bus	Bike	Foot	Combining models (bus, bike, walking)
18–40	10%	38%	7%	31%	14%
40–60	10%	47%	3%	27%	13%
60plus	2%	51%	0%	37%	10%

Alternative new mobility solutions such as a bike sharing system for the short distance trips seem that they could be very efficient solutions for the case of Corfu. The willingness of the visitors to use and to pay for such solutions is presented in the next section.

3 Willingness of Cruse Visitors to Use & Pay for New Mobility Solutions in Corfu

Willingness to pay refers to the supreme value for money an individual would provide in exchange for a service or product. It is a significant factor for planning and implementation of new services or products [14]. Research on willingness to pay, can offer valuable insights for measuring the value of new mobility solutions [15] in transport, namely Bike sharing, Car sharing, reliable Bus - Based systems (e.g. PT on demand). In this case, the research question rising, is related to individuals' willingness to use and pay for transport advancements, acknowledging that the above mentioned mobility solutions are still new and the potential willingness to use or pay, is remaining under investigation.

The literature so far has focused on behavioral aspects of cruise tourism and transport mode choice in the destinations, as well as planning and designing issues of Bike sharing for connecting the port areas with a city center/touristic location [16, 17]. The willingness to pay has been scarcely approached. Two case studies where identified through the study of the literature in Ravenna, Italy and in Valencia, Spain. Both case studies intended to calculate willingness to pay of a Bike sharing service. Although, the implementation of these 2 Bike sharing services and the calculation of the willingness to pay are providing with useful insights, the specific cases are characterized by the authors unsuccessful. The third case presented below, though, do not include a willingness to pay analysis, is a useful paradigm of a successful cruise tourism and Bike sharing combination, providing with insights for Bike sharing pricing.

The Bike sharing service implemented in Ravenna demonstrated that Bike sharing cruise tourists users in Ravenna, are willing to pay a service fare in the range of €10–20 per day. This price range is considered by the authors too high and they suggest treating the results with caution. In addition to that, users also revealed a greater tendency to use the Bike sharing service to reach the historic city center [16]. A pilot project in Valencia, provided tourists with an alternative transport mode to reach the city center. During the six-month duration of the pilot project, the Port Authority provided a bike depot with 10 electric bikes that could be rented by cruise tourists. The pricing of the service was: 9 euros up to 1 h, 18 euros up to 2 h and 24 euros up to 1 day. Even though the service was efficient and well-advertised, the results at the end of the pilot were unsatisfactory, since only 8 individuals rented a bike [18].

"Quikbyke" is a spin-off related to e-bikes, promoting their use in tourist locations by means of a solar-powered rental point, which can be removed and transferred to another location, to satisfy the seasonal travelers' flows. The first trials were installed in Omaha, Nebraska, United States. The solar-powered rental point can be run while recharging up to six e-bikes at the same time, with no external energy inputs. The standard size and self-contained design of the box makes the handling easy by ship or truck. Vehicles are equipped with lithium batteries that allow a 40–50 km range. The fare is $5 for 30 min and the attempt was successful, planning to expand in the Caribbean area a famous cruise tourist destination [18].

The introduction of new mobility solutions in the case of Corfu can be the answer to the traffic problems of the city. Due to this fact, an investigation of the willingness of the cruise visitors to use them as well as the amount that they are willing to pay for them, is an important element in order to ensure their possible viability.

The analysis of the relative data in the Questionnaire survey demonstrated that the 23% of the cruise tourists are willing to use a bike sharing system and 31% a car sharing system. In addition, a bus based more reliable intra - city and inter– City accordingly, attract the interest of the travelers involved in the survey to a greater extent in relation to the Bike and Car sharing mobility options (49 and 45% accordingly).

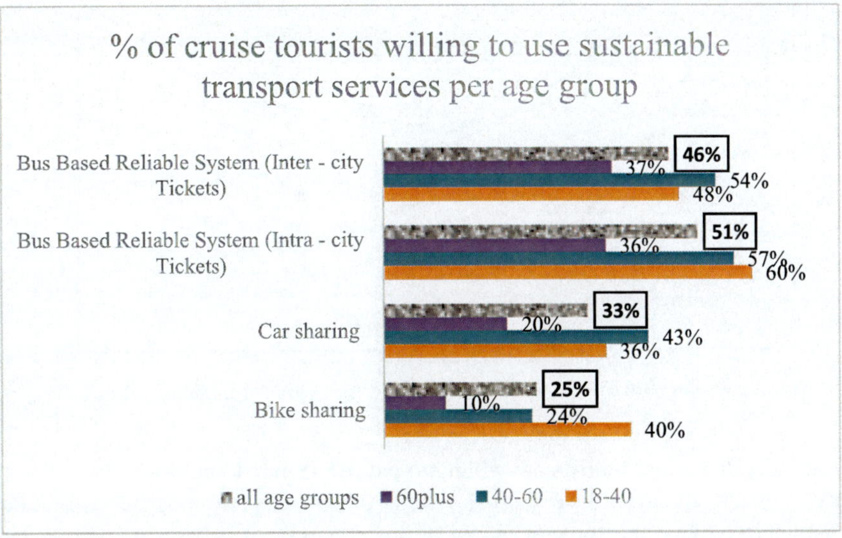

Fig. 1. Willingness of cruise tourists to use sustainable mobility options–average and per age group.

The above-mentioned data are assigned per age group as it is presented below (Fig. 1).

According the replies of the cruise visitors almost half of the younger group are willing to use a bike sharing system for their short distance trips in the city monuments as well as the a more advanced intercity bus.

The eldest people are not so willing to use a bike sharing system even for short distance trips. The respective percentages are 24% for the 40–60 group and only 10% for the 60+ group.

The age group of 40–60 years old, prefer the introduction of an advanced bus service for their trip while almost half of them are willing also to use a car-sharing system for visiting their destination.

The elder people are those who are not willing to use a sharing system not a bike nor a car and it seems that they prefer to use the current solution of a touristic coach than an advance public transport bus service.

The cruise visitors of Corfu stated more precisely and directly the amount of money they would be willing to pay in order to use new shared mobility solutions. Figure 2 demonstrates the willingness to pay graph, according to the cruise tourist's responds.

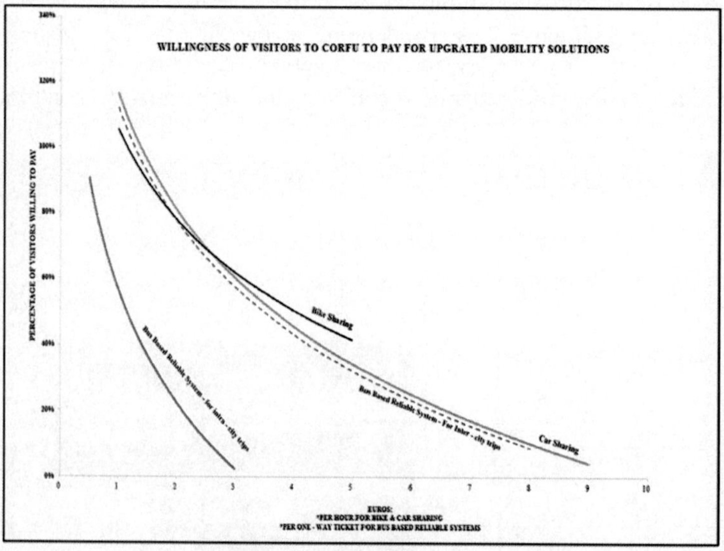

Fig. 2. Willingness of visitors to Corfu to pay for upgraded mobility solutions.

Therefore, the cruise tourists are willing to pay 1,5–5 euros/hour for a Bike sharing service, 2–9 euros/hour for a car sharing system, 0.5–3 euros/one–way ticket for a bus based reliable system for Intra–city and 2–8 euros/one–way a bus based reliable system for Inner–city one-way ticket service.

The average prices per group category shows that even if there are big differentiations in the willingness of the different age categories to use a mode, the amount that they are willing to pay for these services are almost the same, as it is presented for the case of bike sharing, in the following table (Table 3).

Table 3. Willingness of Cruise tourists to pay per mobility solutions per age group.

Age group	Bike sharing average WTP (euros)
60plus	2.5
18–40	2.8
40–60	2.7

4 Discussion and Conclusions

The present study aims to provide some insights of Corfu cruise tourists' profile and their willingness to use and willingness to pay for upgraded or shared mobility solutions, namely Bike sharing, Car sharing and Reliable Bus Based system.

The questionnaire survey conducted, and the data collection can benefit both the tourists and the port/city authorities. The port and relevant city planners could gather useful evidence for their future planning in order to provide sustainable transport alternative modes, services and systems that will answer to the needs of the tourists and will upgrade the traffic and environmental conditions of the island.

To this extend the data analysis for the case of Corfu, demonstrated that:

- The younger age groups of 18–40 and 40–60 years, share a significant percentage and
- The most popular tourist destination is Corfu's old town (small distance from the starting point-Port of Corfu).

Almost half of the younger group are willing to use a bike sharing system for visiting city center and other close to the port destinations. The development of such a system it will minimize the volumes to/from the old city center of Corfu which is currently suffering from traffic volumes and delays.

The car sharing system seems that it can become an alternative mode choice for the age group of 40–60 years old, in order to visit long distance destinations. The use of an e-car share system can also become an optimum solution for small group of cruise visitors who wish to visit monuments outside the city of Corfu as it minimizes the environmental and noise emissions.

Both the above-mentioned age group categories, prefer to use an advance bus inter and/or intra city bus system. In order this system to upgrade the traffic and environmental problems of the city, should have specific characteristic. It should be operated on demand, it should be adjustable to the demand (use of small vehicles for low demand) and also use low emission fleet (e-buses or hybrid).

Finally, the current research work went deeper in order to find the WTP for the proposed mobility solutions. According to the results, it seems that for the short distance destinations such as the historical city center, visitors are willing to pay an average cost of 3 euros/hour for a bike sharing system (1euro/hour is the current price in places such as Thessaloniki) and an average of 1,75 euros/one–way for a reliable bus service. At the same time, for longer distance destination such as Achillion or Mouse island, visitors are willing to pay an average of 5,5 euros/one–way ticket for a car sharing service, or 5 euros/one–way ticket for a reliable and comfortable bus service.

This conclusion has a potentially significant application at policy level, steering the ports authorities, the municipality and the stakeholders to foster shared or upgraded mobility solutions, for connecting the port and the city center or the locations with touristic interest. The primary findings may shed the light on further discuss and analysis for the integration of new shared mobility options not only to the transportation system of Corfu but also to all the port-cities that accommodate high volumes of cruise ships and visitors. The introduction of these solutions will not only upgrade the offered mobility services to the cruise tourists but it will also support sustainability of the cities reducing the environmental effects of the traffic.

References

1. Chew, J.: Transport and tourism in the year 2000. Tourism Manage. **8**(2), 83–85 (1987). ISSN 0261-5177, https://doi.org/10.1016/0261-5177(87)90003-3
2. Kaplan, S., Manca, F., Sick Nielsen, T.A., Prato, C.G.: Intentions to use bike-sharing for holiday cycling: an application of the theory of planned behavior. Tour. Manage. **47**, 34–46 (2015)
3. Gui, L., Russo A-P.: Cruise ports: a strategic nexus between regions and global lines–evidence from the mediterranean, pp. 129–150 (2011). https://doi.org/10.1080/03088839.2011.556678
4. Carić, H., Mackelworth, P.: Cruise tourism environmental impacts–the perspective from the Adriatic Sea. Ocean Coast. Manage. **102**, Part A, pp. 350–363 (2014). https://doi.org/10.1016/j.ocecoaman.2014.09.008
5. BREA: The global economic contribution of cruise tourism 2013, Study prepared for Cruise Lines International Association, September 2014, USA (2014)
6. ELSTAT, elaborated by Greek Tourism Confederation (SETE) Intelligence. https://www.maritimes.gr/news/uploads/uploads/2019_SymvolhTourismou-2018.pdf
7. Triantafyllopoulos, C.: Cruise market and the position of Greece, University of Piraeus, Thesis (2017)
8. Maragkogianni, A., Papaefthimiou, S.: Evaluating the social cost of cruise ships air emissions in major ports of Greece. Transp. Res. Part D: Trans. Environ. **36**, 10–17 (2015). https://doi.org/10.1016/j.trd.2015.02.014
9. Morfoulaki, M., Kotoula, K., Mirovali, G., Chrysostomou, K., Stathacopoulos, A., Batsoulis, A.: Investigating the implementation of potential strategies for enhancing urban mobility and a city logistics system on the Island of Corfu. WIT Trans. Ecol. Environ. **191**, 15–26 (2014). https://doi.org/10.2495/SC140021
10. Apostolou, A.: All-inclusive resorts, cruises and sustainability. The perspective of the Corfu community. J. Tourism Cult. Territorial Dev. **9**(18), 73–103 (2018)

11. Burrieza, J., Djukic, T., Aifadopoulou, G., Grau, J.M.S., Masegosa, D.A., Rojo, M.J., Ripa, F., Frederix, R., Pápics, P., Himpe, W., Antoniou, C., Narayanan, S., Bousse, Y., Fabianski, C.: D2.1 new mobility options and urban mobility: challenges and opportunities for transport planning and modelling. Deliverable Report, pp. 123–129 (2019)
12. Corfu Port Authority: The Action Plan–Corfu Port Authority, Intermodal Passengers Connectivity between Ports and Airports (INTER-PASS), Interreg ADRION (2018)
13. Myrovali, G., Morfoulaki, M., Agathos, M.: Crowd-learning, target setting and intermodality investing, Corfu leading by example areas of high seasonal demand. In: ICTR 2019, Athens (2019)
14. Foreit, J.R., Foreit, K.G.F.: Willingness to Pay Surveys for Setting Prices for Reproductive Health Products and Services: A User's Manual. The Population Council, USA (2014)
15. Liu, P., Guo, Q., Ren, F., Wang, L., Xu, Z.: Willingness to pay for self-driving vehicles: Influences of demographic and psychological factors. Transp. Res. Part C Emerg. Technol. **100**, 306–317 (2019). https://doi.org/10.1016/j.trc.2019.01.022
16. Bardi, A., Mantecchini, L., Grasso, D., Paganelli, F., Malandri, C.: Flexible mobile hub for e-bike sharing and cruise tourism: a case study. Sustainability **11**(19), 5462 (2019). https://doi.org/10.3390/su11195462
17. Bakogiannis, E., Vassi, A., Christodoulopoulou, G., Siti, M.: Bike sharing systems as a tool to increase sustainable coastal and maritime tourism. The case of Piraeus. Reg. Sci. Inquiry **10**(3), 57–70 (2018). Hellenic Association of Regional Scientists
18. From shared resources to joint solutions: Report on the technical and economic sustainability of the Moses "Mobile Depot" tested in Ravenna Port, Business model for the replicability of the Moses Ravenna pilot, Report 2018, Interreg Italy–Croatia, European Union (2019)

Road Safety for School Zones in Medium-Sized Cities

Elias Papastavrinidis(✉), George Kollaros, Ioannis Karamanlis, Antonia Athanasopoulou, and Vasiliki Kollarou

Department of Civil Engineer, Democritus University of Thrace, 67100 Xanthi, Greece
eliaspapastavrinidis@gmail.com

Abstract. Schools are social utility units where many people are gathered every day. The best for such organizations is to develop in campuses. However, in the urban environment of mid-sized cities, this is very rarely accomplished. School travel systematically addresses barriers for walking to school. The term "school travel planning" could be thought as a community-based model for implementing active school travel conditions. Students, parents and teachers could leave their cars at home and encouraged to increase walking, cycling, use of buses and carpooling. Also, school related trip times can be differentiated by for example picking up children later. Finally, it's a great help for taking different route to avoid congested roads. Its primary goal could cost-effectiveness of the way to get more kids walking and wheeling to school. Xanthi is a medium-sized city in Greece with school units scattered in various areas. Most parents move kids to school by car, because they believe that pupils, on their own, cannot circulate safely. That creates a need for the establishment of safe routes that will encourage pupils to use methods and means of alternative transport to schools. This research includes two sections about the city of Xanthi: a) an analysis of the school's educational, building, urban and traffic data, identifying road safety problems, and b) proposals for specific projects and practices that can improve the conditions and contribute to the sustainable operation of the school. The development of a school travel plan is expected to produce long-term benefits for the city of Xanthi.

Keywords: Sustainable mobility · Road safety · Pupils · School · Travel time

1 Introduction: School Travel Planning

One of the most significant objectives of sustainability in urban environment is the safe movement of pedestrians. School complexes where the concentration of people is increased due to specific uses are a special category as their function attracts people from various social groups such as teachers, parents, children and students. Criteria to be considered are:

Street type at the entrance of the school unit.
Site position of the school unit.

E. G. Nathanail et al. (Eds.): CSUM2020, AISC 1278, pp. 428–434, 2021.
https://doi.org/10.1007/978-3-030-61075-3_42

Characteristics of the pavement at the entrance of the school or school complex.
School complex educational level.

It also takes into account how pupils come and go as part of the school's working hours as well as the presence of additional cultural or sports activities for the students or other social groups in the neighborhood [1].

In any case, the existing situation is examined in detail and the necessary measures for the formation of the school ring are proposed. Where possible, additional interventions are foreseen on the roadway and the pavement.

A School Travel Plan is a package of initiatives designed to enhance safety, reduce car use and support sustainable travel for the school journey. While some are led by the school, others–such as changes to the highway–can only be implemented by the local authorities [2].

2 Walking to School

The promotional strategies which schools may follow vary depending on age. Walking reward programs, walking buses, and 'Walk to School' events can be common options in primary schools. Schemes often may run alongside each other and work well in combination. Different factors affect in encouraging secondary school students to walk [3].

2.1 Walking Incentive Schemes

Primary schools implemented programs to award children with flags, patches, awards, and rewards for walking to school. It can be very effective in facilitating walking travel. Many local authorities are now supporting such initiatives, though the details vary. For certain schemes each child is given marks. For other schemes the whole class gains points, and a prize is awarded to the student with the maximum amount of walking. Many programs incorporate rewards for people as well as for schools.

Incentives can be applied to include certain modes of transport, with points for children arriving by bicycle or car sharing. They often gear to 'park and walk' tactics and points are rewarded to travel a minimum distance. Schemes may run every day or once a week and certain schemes are limited to the summer term. Parents are told of the scheme by letter to home and often is required to give written permission to children in order to participate [4].

2.2 Promoting Walking in Secondary Schools

Generally, the project becomes more effective when secure route steps are paired with constructive participation of the students. The perceived safety of the road system is likely to be especially relevant as children make independent journeys and the implementation of a secure route scheme represents a tremendous opportunity to increase knowledge about travel choices [4].

Some of the above walking programs may theoretically be tailored to suit older children. One secondary school, for example, can encourage its younger pupils to play for a walking trophy. On top of this, site planning and other plans, mentioned below, are likely to be relevant in promoting school walking.

2.3 Pedestrian Friendly Sites and Arrangements

The urban environment conveys a number of information regarding how we want to travel. Meagre footways and accommodating driveways put the car at the top of the hierarchy of transports. Schools ought to think twice about how to design their sites to welcome those on foot and by bicycle.

Opening an additional entrance to the site or having a main footpath along a preferred pedestrian line may make walking quicker and more comfortable, which could enable certain children to make safer journeys on quieter roads. A separate pedestrian entry renders it difficult for children to mingle with traffic. A broad route that directly leads to the main doors reinforces the priority is provided to pedestrians. Improving lighting and curtailing excessive vegetation can help to deal with safety [4].

2.4 School Crossing Patrols

Many schools find important areas where a school crossing patrol can assist children undertake individual journeys. Recruitment may be a problem. Where local authorities have deliberately sought recruiting and have failed, options such as more regulated crossings should be looked at. Small facilities, such as traffic islands, may also help make the job safer and more attractive [5].

3 Reducing Road Danger

The phase of the School Travel Plan is structured to predict potential road danger, including close misses and real injuries. In the basis of their transportation knowledge, children, parents and others are essentially expected to predict when injuries are likely to arise even though they have not yet occurred. It is critical these concerns to be responded.

Although addressing injury hot spots should be a goal, delivering a step change in home to school transport involves a comprehensive solution, with the long-term aim of developing high quality walking and bicycle paths across school catchments. An authority that dismisses the safety concerns of parents and children, without looking ways to deal with these–whether through engineering or other solutions–risks demoralizing everyone involved within the plan [5].

4 Managing Car Use

The school run is a major contributor to congestion particularly in urban environments, with unregulated parking outside the school gates making the problem worse. The percentage of children who go to school by vehicle has almost increased in the past years. The patterns indicate increased ownership of the car and a related decrease in the amount and period of adult journeys on foot or by bicycle.

A school that keeps encouraging parents or students to park outside on the road may possibly find it difficult to limit vehicle use. Restrictions regulation is always a concern, and certain schools launch educational programs to facilitate conformity and inspire safe actions, usually working in partnership with the police.

In certain instances, police make regular visits as a deterrent to inconsiderate parking. For some, the identification numbers of offending vehicles are gathered by parents and neighbors to be sent to authorities, who then issue a note. Issuing a warning letter ahead of a possible fine can be seen as a diplomatic step. Nonetheless, motivating parents to check on one another in this manner can cause resentment, and not everyone is comfortable with this approach. Less controversially, schools should draw up a Code of Conduct for drivers which can be available to parents and inserted into the home-school agreement. Many schools prohibit parents from parking on school grounds and limit their usage to staff and visitor car parks. Exceptions are made in special cases such as for parents picking up diseased children.

Parking appears in school travel plans as one of the steps included in the proposals to increase protection at the school entrance. Parents, for example, may park in a convenient place further from the school and walk the rest of the way with their children rather than dropping them off at the school gate. This may be achieved individually or as part of a walking shuttle, whereby children are gathered at an agreed time to a prearranged place and accompanied to school [6].

5 School Urban and Traffic Data in the City of Xanthi

5.1 Traffic Conditions Around School Units

The case study presents the interventions which are proposed for four school complexes in the city of Xanthi (Fig. 1). The criteria set for their selection are proximity to high traffic volumes, pupils' number and recording of accidents [7].

School Complex A is located next to a park, which features a landscaped recreation area and sports facilities. Pupils most of the time enter to school facilities from entrances on the pedestrian street within the building block. The various uses that have been developed are separated by pedestrian crossings. The streets around the building block are asphalt-paved, single-directional with relatively low traffic load [8].

Intervention proposals: pavement widening; pedestrian crossing; traffic signs; improving accessibility; bicycle path network; parking facilities.

School Complex B is located in the city centre (Fig. 2). Pupils get to school from two entrances. At the beginning/ending of school activities, school traffic wardens coordinate vehicles on the main street contributing to the safe passage of students, as the traffic is booming especially during peak hours [8].

Intervention proposals: pavement widening; pedestrian crossing; low traffic routes; traffic signs; improving accessibility; parking facilities.

School Complex C is located in the west side of the city. The streets around the school are asphalt paved and categorized as low traffic. A road has an obstacle separating the opposite traffic currents and generating more favorable traffic conditions while the three others have a single carriageway. The main route allows the development of alternative approaching the city centre [8].

Intervention proposals: pedestrian widening; pedestrian landscaping; signs installation; improving access for people with disabilities; connecting to a bicycle network; arranging parking spaces.

Fig. 1. School complexes in the city of Xanthi.

School Complex D is located in the south side of the city, next to a park. In the northwestern part of the block there is a church and the Mental Health Centre. The school complex has three entrances. The main entrance/exit of the students is on Zalongou St. Zaloggou-Thyateiron and Grigoriou E' streets are local one-way streets, while Ilioupolis St. is a two-way collection road, which serves traffic loads to the Primary and Secondary Road Connecting the superior network by creating alter-native routes [8].

Intervention proposals: pedestrian crossing layout; mild traffic road layout/cycling paths; signs installation; accessibility improvement for the disabled; parking space configuration.

Fig. 2. Primary school in 28 Oktovriou street, Xanthi.

5.2 Intervention Proposals Improving Infrastructures Around School Units

It is proposed to narrow the road, which is achieved by widening the sidewalk for the formation of soft traffic roads near schools. At the same time, an effort is being made to preserve the parking lot, where this is possible by forming appropriate recess-es on the sidewalk.

The sidewalks around the school complexes are in some cases recently constructed and are wide enough. Otherwise, it is proposed to widen them, reducing the width of the roadway at the expense of the parking lane, in order to place the necessary urban equipment in this area.

In cases of roads where there is not the necessary total width for the formation of a traffic lane and sidewalks on both sides of the road, it is proposed to create a road with a smooth road with a single deck

An additional means of implementing mild traffic measures near schools is the construction of elevated pedestrian crossings. At the points of intersection, the elevation may occupy the entire surface of the intersection (cruciform type). In any case, the disabled must be serviced. Before the vehicle's ascent ramp at the elevated crossing, a signal is placed to stop the speed of passing vehicles. In case it is not possible to apply the type of elevated cross-sectional crossing, a simple crossing can be constructed.

In any case, cross-sections must be given special attention to their maintenance. In the case of elevated lanes, the damage they will suffer over time will on the one hand reduce their efficiency, on the other hand they will be dangerous points for two-wheeler drivers. In the case of simple crossings, special care must be taken to maintain the road marking. Particular attention should be paid to signage and electric lighting, so that the provisions of the passages are visible during the night.

School signs can be equipped with bright information lights for students, which are a good way to stimulate the attention of drivers. The principle of reliability of signage must be followed and therefore the operation of traffic lights must be stopped during the period when students are not passing, such as in summer [9].

6 Conclusions and Recommendations

The goal of this project, is to encourage pupils and their parents to travel to school in a more active or sustainable way to help reduce the congestion caused by short (single occupancy car trips) on the 'school run'. The results include changes in safety and wellness, and also a favorable effect on air quality. The study also explicitly ties successful school trips with enhanced academic achievement.

Pupils can develop a greater appreciation in safety, economic and social advantages of walking and cycling as alternatives to needles automobile journeys. This will allow them to establish more positive travel patterns that can keep in their adult life.

Participation of pupils is crucial to the implementation of an effective School Travel Plan, as it is their travel conduct that will ideally be affected and changed by this program. It is important that pupils' opinions are sought and brought under consideration.

Finally, pupil participation can reach delivering projects and monitoring progress through the curriculum, school council or out-of-school-hours activities [10].

References

1. Brighton and Hove City Council: School travel plans-A guide for schools and developers (2013). http://www.brighton-hove.gov.uk/schooltravelplans/. Accessed 30 Apr 2020
2. School Travel Plans: Guidance and information pack for primary schools. http://www.bridgend.gov.uk/. Accessed 29 Feb 2020
3. Newson, C., Cairns, S., Davis, A.: Making school travel plans work: experience from English case studies. Trans. Qual. Life, 20–53 (2010)
4. School Travel Plans – What's so special. https://tfl.gov.uk/. Accessed 29 Feb 2020
5. Surrey County Council's Sustainable Transport Team: Road safety outside schools (2014). http://www.surreycc.gov.uk/. Accessed 30 Apr 2020
6. Parking its role in workplace and school travel planning. http://www.britishparking.co.uk/. Accessed 29 Feb 2020
7. Technical Program 2020: Municipality of Xanthi, 27 Nov 2019. (in Greek)
8. General Urban Plan 2013: Municipality of Xanthi, 22 May 2013. (in Greek)
9. Goodman, M., Lewis, M.: A guide to school travel plans. Surrey county council (2018). http://www.surreycc.gov.uk/. Accessed 30 Apr 2020
10. Lay, J., Kennedy, J.: The costs and benefits of school travel planning projects in Ontario, Canada. Metrolinx Green Communities Canada, Toronto (2014)

A Multi-criteria-Based Methodology for Assessing Alternative Bicycle Lane Implementation Solutions in Urban Networks

Ioannis Politis, Efthymis Papadopoulos(✉), Ioannis Fyrogenis, and Zoi Fytsili

Laboratory of Transportation Engineering, Department of Civil Engineering, School of Technology, Aristotle University of Thessaloniki, 541 24 Thessaloniki, Greece
{pol, efthympg, fyrogeni, zfytsili}@civil.auth.gr

Abstract. In order for the bicycle usage to be increased, cities need to provide users, inter alia, with high-quality infrastructure. Yet, there is not a generally accepted approach to optimize placemaking of such infrastructure in the urban environment, especially on the micro-level. This paper presents a methodological approach for assessing alternative bicycle lane implementations in urban networks, from a micro perspective. Specifically, a multicriteria analysis-based methodology is proposed, which accounts for specific criteria, including geometric characteristics of road network (i.e. road width and slope), effect on on-street parking supply, public acceptance, land-use and built environment characteristics, etc. These criteria are weighted, in line with their relative importance, by micromobility and sustainable urban mobility experts. This methodology is applied to the city of Karditsa, Greece, where two (2) alternative streets were examined in terms of hosting a different type of bicycle lane that would complement the existing cycle route network. The MCA results suggested that safety forms the major factor that most heavily affect the final placemaking and type of the bicycle lane decision. The proposed methodology could contribute towards successful bicycle routing, forming a useful tool for traffic engineers and local authorities.

Keywords: Bicycle route planning · Sustainable urban mobility · Multi-criteria analysis · Safety

1 Introduction and Brief Literature Review

Mainly focusing on "Active Mobility", the promotion of bicycle usage is one of the major priorities of Sustainable Urban Mobility Plans (SUMPs) [1]. Among other aspects, the development of appropriate infrastructure, as well as its integration with the existing cycle route network, are broadly believed to play a key role towards the desired goal of bicycling rates increase [2, 3].

In order for an integrated and coherent cycling network to be developed, various factors should be considered, mostly related to topography, land use and built environment, traffic, spatial distribution of destination points, etc. In this context, cycleway

E. G. Nathanail et al. (Eds.): CSUM2020, AISC 1278, pp. 435–444, 2021.
https://doi.org/10.1007/978-3-030-61075-3_43

selection entails a planning and engineering process, which usually follows a deductive sequencing of actions: after the identification of the major destination points and the recognition of the desired facility to be provided, the aforementioned process leads to specific corridors based on real-world restrictions, such as road width availability, budget, etc. [4].

Bicycle route planning forms a complex and multi-dimensional problem and thus several studies have already sought to develop a methodology for the evaluation of the alternative cycleway routes. To this end, various decision-making techniques have been applied - along with other GIS-based or algorithmic methods - with the multi-criteria analysis (MCA) being among the most widely used techniques [2, 5–7]. Being broadly applied for the evaluation process of transport-related projects [8], the MCA approach is usually employed when the appraisal of alternative options aiming to serve a specific goal is based on different criteria [9]. Following a deductive approach, all the above-mentioned studies performed different MCA procedures, with the overall aim of planning an integrated cycling network in urban areas.

Contrary to what is mentioned above, the process of cycleway route planning in Greece often follows an inductive rather than a deductive approach: usually, newly constructed buildings generate the need for new individual bicycle lanes, with the major restriction being their proximity to the relevant point of interest. It follows from the above that while most of the aforementioned MCA applications were utilized on a macro level, which is usually the most fitting approach, certain cases-especially to the Greek reality-would benefit more from a micro-level variant.

In this paper, a multi-criteria-based methodology for the assessment of alternative bicycle lane implementation solutions on a micro-level is proposed and applied as a case study in the city of Karditsa, Greece. Moreover, different criteria are considered and weighted against each other, with the aim of capturing the influence of those factors on the final placemaking and type of the bicycle lane decision.

2 Case Study Area

Karditsa is a city in mainland Greece and the capital of the Karditsa regional unit, with a population of almost 40,000. The city is characterized by a pedestrian-friendly urban center and an extensive network of radial cycleways leading to it. The bicycle is traditionally a widely accepted and used mode of transport and the city has an extensive - for Greek standards - infrastructure to support it. The existing cycleway network's total length is 7.3 km, when including only the urban cycleways [10].

As already mentioned above, the deductive approach regarding the bicycle route planning process is not the norm in Greece, where new individual bicycle lanes are usually proposed to be constructed, with the principal aim of meeting the increased transport demand stemming from newly emerging land uses. This is the case in Karditsa, where during the last few years a new facility in the southern part of the city-planned to pool multiple administration services - has been under construction. Considering that this facility is expected to be a significant trip attractor, new bicycle infrastructure is needed to accommodate the expected increased transport demand. While the existing cycleway network extends towards the general area of the new building, the expected increase in terms of transport demand and the integration in the existing radial design, justify a direct connection to the city center.

From the above, it is clear that regarding the study area, the problem of bicycle route planning needs to be examined from a micro perspective, in terms of identifying which road is the most appropriate to host a new bicycle lane, in an already quite extensive cycle route network. Although a new bicycle lane rather than a whole cycle network needs to be constructed, the problem remains complex, since there is not a well-defined framework for the assessment and selection of the most appropriate road and type of cycleway.

3 Materials and Methods

To address our research goal of identifying-among the alternatives-the most appropriate road and type of cycleway to be constructed, a multi-criteria-based analysis was performed, while several on-field measurements, questionnaire surveys and GIS-based calculations were chosen as the data gathering methods. Overall, the methodological steps taken in the current research are comparable to those of a conventional multi-criteria analysis application:

After having defined the problem, a set of alternatives was developed. In this context, a number of different roads were identified, with the primary restrictions being their ability to host a new bicycle lane and their proximity to the aforementioned point of interest. While considering the spatial distribution of the city's cycle route network in its entirety and using our engineering judgement, we identified two (2) alternative streets that were considered to meet the aforementioned requirements. Each alternative street necessitates a different bikeway type:

a) ***Alt_1***: two one-way bicycle lanes on either side of Koumoundourou str. (from Thessaliotidos str. to Ipsilantou str.).
b) ***Alt_2***: one two-way bicycle lane on the left side of Ir. Politechniou str. (from Thessaliotidos str. to Ipsilantou str.).

Various criteria categories and sub-criteria that suit the decision problem were then identified, with the purpose of structuring an appropriate hierarchical criteria tree for the forthcoming analysis and thus, developing the means by which the set of alternatives will be assessed and compared. In the first stage, a wide range of criteria categories and sub-criteria was recognized, largely based on the pertinent literature, as well as the experience gained so far by relevant projects. In the second stage, a group brainstorming session was carried out, with a dozen of micromobility and sustainable urban mobility experts from the local community attending. This way, the criteria categories and the sub-criteria resulted from the first stage were further assessed and a final set of seventeen (17) sub-criteria and eight (8) parent criteria categories was developed, reflecting the major factors that affect the placemaking and type of the bicycle lane decision. An overview of the criteria categories and the respective sub-criteria, is provided in Table 1.

Subsequently, a spreadsheet was developed that included all the above-mentioned sub-criteria, as well as their parent criteria categories. The spreadsheet was forwarded

Table 1. Overview of bicycle lane implementation criteria categories and sub-criteria.

CC Code	CC	SC Code	SC
CC1	Road geometric characteristics	SC1.1	Road width
		SC1.2	Road slope
CC2	Road operational features	SC2.1	Traffic volume (PCUs)
		SC2.2	Pedestrian volume (Persons)
CC3	On-street parking supply	SC3.1	Num. of parking spaces lost (Cars)
		SC3.2	Num. of parking spaces lost (2-wheel)
		SC3.3	Parking turnover rate (avg)
CC4	Public acceptance	SC4.1	Public acceptance (bicycle users)
		SC4.2	Public acceptance (citizens, car users)
CC5	Land use & built environment characteristics	SC5.1	Proximity to new administrative building
		SC5.2	Proximity to recreational land uses (parks, public sport facilities, etc.)
		SC5.3	Proximity to schools
CC6	Ability to meet future demand	SC6.1	Ability for further expansion
		SC6.2	Ability for integration with existing network
CC7	Safety	SC7.1	Num. of conflicts between car/bicycle users
		SC7.2	Num. of conflicts between bicycle users/pedestrians (at pedestrian crossings)
CC8	Cost	SC8.1	Construction cost

CC: Criteria Category, SC: Sub-Criterion

to fifteen (15) micromobility and sustainable urban mobility national experts, who were asked to assign weights for each of the criteria categories and sub-criteria that would reflect their relative importance to the decision. Criteria categories were assigned a relative weight, ranging from 0 to 100. The more important the criteria category, the higher its weight. The total weight of all criteria categories should sum up to 100. Similarly, the weights assigned to sub-criteria ranged between 0 and 100 and should sum up to 100, for each individual criteria category.

The experts' responses were gathered in a single spreadsheet, where the average weights of the criteria categories and the sub-criteria were calculated. In order for the MCA analysis to be performed, different scores also needed to be calculated for each of the alternatives, in a way to reflect the performance of the latter against each distinct sub-criterion. The computation of the appropriate scores, required two (2) individual processes: the calculation of a set of indicators and the application of a normalisation process, in order for the values of those indicators to be converted into scores.

For each sub-criterion, two (2) indicators - reflecting each alternative - were calculated. In order to conclude to the values of those indicators, different on-field measurements and questionnaire surveys were conducted, while various GIS-based spatial calculations were also performed. Table 2 summarizes the set of indicators, that were taken into account for the calculation of the relative scores.

Table 2. Overview of sub-criteria-related indicators for relative scores calculation.

SC	Alt_1	Alt_2	Source of Data
Road width	7,2m	6,7m	Google Earth, Arc-GIS Earth, GIS-based calculations
Road slope	0,22%	0,24%	
Traffic volume (PCUs)	227 (avg)[1]	287 (avg)[1]	Hourly traffic and pedestrian volume measurements (09:00-10:00)
Pedestrian volume (Persons)	283 (avg)[1]	233 (avg)[1]	
Num. of parking spaces lost (Cars)	115[2]	64[2]	Parking demand and supply survey
Num. of parking spaces lost (2-wheel)	13[2]	12[2]	
Parking turnover rate (avg)	0,9 (avg)[3]	0,86[3]	Parking turnover survey (10:00-13:00, per 30min intervals)
Public acceptance (bicycle users)	58%[4]	68%[4]	Questionnaire survey on bicycle and car users
Public acceptance (citizens, car users)	34%[4]	56%[4]	
Proximity to new administrative building	300m[5]	360m[5]	Google Maps, GIS-based calculations
Proximity to recreational land uses (parks, public sport facilities, etc.)	390m (avg)[5]	360m (avg)[5]	
Proximity to schools	190m (avg)[5]	190m (avg)[5]	
Ability for further expansion	High	Low	Estimations, Engineering Judgement
Ability for integration with existing net-work	High	High	
Num. of conflicts between car/bicycle users	31	14	OpenStreetMap, GIS-based calculations
Num. of conflicts between bicycle users/ pedestrians (at pedestrian crossings)	6	2	
Construction cost (€)	22.500	19.700	Estimations based on relevant projects

[1]Measurements in two different segments on each Koumoundourou and Ir. Politechniou str: 176 PCUs and 448 persons (Koumoundourou/Ipsilantou), 277 PCUs and 117 persons (Koumoundourou/Dim. Lappa), 283 PCUs and 333 persons (Ir. Politechniou/Ipsilantou), 290 PCUs and 132 persons (Ir. Politechniou/Dim. Lappa)

[2]Koumoundourou str: 57/13 legal car/2-wheel vehicle parking spaces lost (left side), 58 legal car parking spaces lost (right side), Ir. Politechniou str: 64/12 legal car/2-wheel vehicle parking spaces lost (left side)

[3]Koumoundourou str: parking turnover rate = 0,89 (right side) and 0,92 (left side), Ir. Politechniou str: parking turnover rate = 0,86 (left side)

[4]100 questionnaires (50 bicycle and 50 car users): 58%/34% of bicycle/car users assessed Alt1 (Koumoundourou str) as "good" or "very good", 68%/56% of bicycle/car users assessed Alt2 (Ir. Politechniou str) as "good" or "very good"

[5]Proximity to newly constr. building: network distance, proximity to recr. land uses: average euclidian distance, 5 such land uses taken into account, proximity to schools: average network distance, 2 schools taken into account

SC: Sub-Criterion, Alt: Alternative

In order for the values of the above-mentioned indicators to be converted into different scores per alternative, a normalisation process should be applied. The normalisation is achieved using a function relating the arithmetic scores to the indicator values measured on the respective scale. Among the different existing methods [11], min-max was the normalisation method used for the purpose of the current research, while the scores were calculated on an arithmetic scale 0–100. The min-max indicator values did not reflect values of specific alternatives, but lower-higher values, in order for the scores to be representative for each alternative. The described normalisation process was applied for all the indicators presented in Table 2, either expressed in measurable form, or in qualitive terms. For the latter, a Likert scale from 1 to 5 was used (1: none, 2: low, 3: moderate, 4: high, 5: very high). The normalisation process used in the context of the current research, is illustrated in the Eq. 1 below:

$$z_i = \frac{x_i - min(x)}{max(x) - min(x)} \tag{1}$$

Where:

z_i is the normalised indicator value
x_i is the actual indicator value
min(x) is the minimum actual indicator value (lower for representative purposes)
max(x) is the maximum actual indicator value (higher for representative purposes)

At this point, it is pertinent to note that while the weighting of the criteria categories and the sub-criteria was handled by the experts, the scores per alternative were calculated by the authors of the current research.

As the final step in this process, each sub-criterion average weight was multiplied by the average weight of its relevant, parent criteria category and thus, the cumulative weight for each sub-criterion was calculated. Each sub-criterion cumulative weight was then multiplied by its corresponding score. Following the same approach for all criteria categories and sub-criteria and adding the products of the cumulative weights by the scores, the overall MCA score for each of the alternatives was derived, on an arithmetic scale 0 to 100. The overall score per alternative was given by the following Eq. 2:

$$S_c = \sum_{c=1}^{k} w_c \left(\sum_{j=1}^{n} w_i s_{ij} \right) \tag{2}$$

Where:
w_c is the weight of the criteria category c, for c = 1, k
$\left(\sum_{j=1}^{n} w_i s_{ij}\right)$ is the sum of the weighted scores of all sub-criteria per criteria category.

4 Results and Further Discussion

Based on the MCA results, Ir. Politechniou (Alt_2) was proven to be the most appropriate road to host the new bicycle infrastructure, since its overall score was calculated to higher than the corresponding score of Koumoundourou str. (Alt_1). More precisely, the overall score of Ir. Politechniou str. was calculated to be 73, 90, compared to 62, 12 that was the overall score calculated for Koumoundourou str. In view of the above, the construction of a two-way bicycle lane on the left side of Ir. Politechniou str. is proposed.

Apart from the overall scores per alternative, a deeper look at the MCA results is needed in order for a better understanding to be gained. Figure 1 summarizes the final MCA scores for those criteria categories and sub-criteria of which the aforementioned scores appeared to be significantly different between the two alternatives.

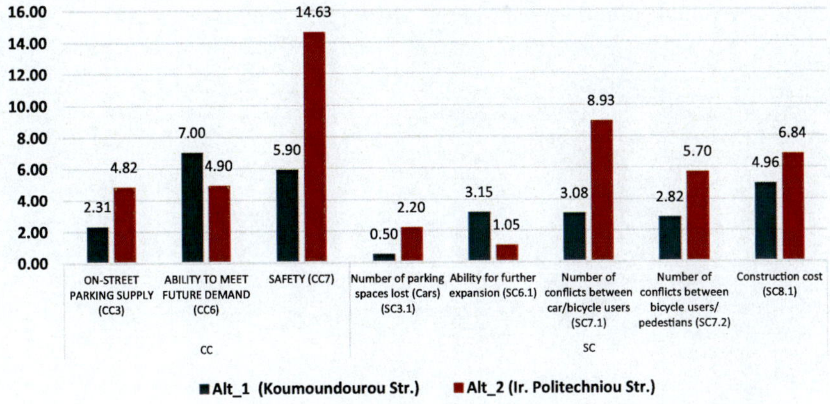

Fig. 1. Significant differences in final MCA scores per Criteria Category (CC) and Sub-Criterion (SC).

In broad terms, both of the alternatives scored higher in half of the criteria categories taken into account. However, the differences between the two alternatives' scores in those criteria categories that Alt_1 scored higher, appeared to be minor, with the biggest difference being detected to the criteria category "ability to meet future demand" (Alt_1: 7,00, Alt_2: 4,90). On the other hand, Alt_2 scored significantly higher in half of the criteria categories, with the major differences being detected to the criteria categories "safety" (Alt_1: 5,90, Alt_2: 14,63) and "on-street parking supply" (Alt_1: 2,31, Alt_2: 4,82). It could therefore be argued that "safety" formed that criteria category, which most heavily turned the final decision in favour of Alt_2.

Moving to the analysis of the MCA results in a more disaggregated level, a further examination of the final MCA scores per sub-criterion is needed, regarding both of the alternatives. When compared to Alt_2, Alt_1 scored slightly higher in five (5) out of the total of the seventeen (17) sub-criteria taken into account in the analysis, with the greater difference being detected to the sub-criterion "ability for further expansion" (Alt_1: 3,15, Alt_2: 1,05). On the contrary, Alt_2 scored higher in ten (10) out of the total of the seventeen (17) sub-criteria, with the most significant differences referring to the sub-criteria "number of conflicts between car and bicycle users" (Alt_1: 3,08, Alt_2: 8,93), "number of conflicts between bicycle users and pedestrians" (Alt_1: 2,82, Alt_2: 5,70), "cost" (Alt_1: 4,96, Alt_2: 6,84) and "number of legal parking spaces lost (cars)" (Alt_1: 0,50, Alt_2: 2,20). It follows from the above that in terms of the final MCA scores per sub-criterion, the final decision about the placemaking of the new bicycle infrastructure was mostly affected by the two sibling sub-criteria of the criteria category "safety".

As already mentioned, except for the final decision resulting from the MCA application, the current research also seeks to detect the deciding factors (criteria categories and sub-criteria) affecting the final placemaking and type of the bicycle lane decision. 7 of the criteria categories and the sub-criteria, which was handled by the experts, is required. Figure 2 presents the cumulative weights (the products of each sub-criterion average weight by the average weight of its relevant criteria category) for all the sub-criteria taken into account in the MCA performed.

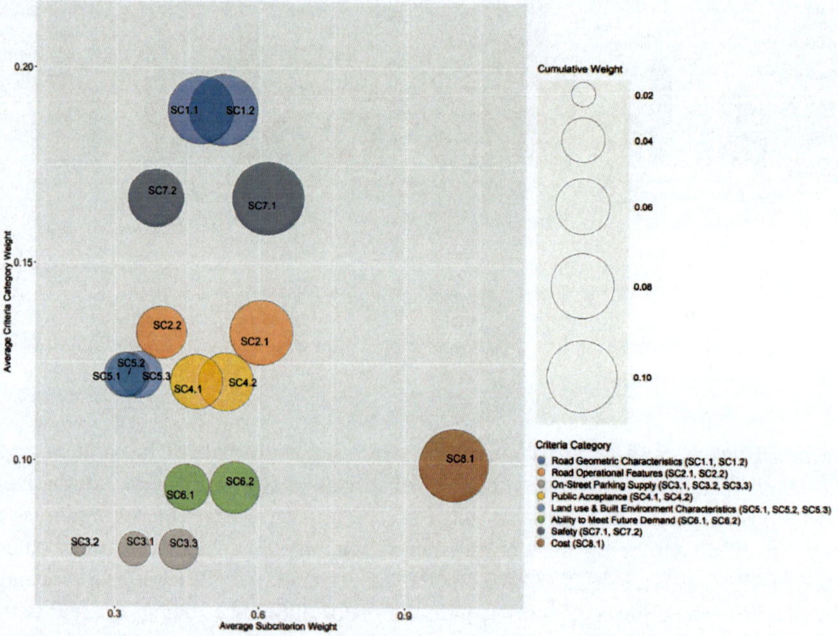

Fig. 2. Cumulative weights per Sub-Criterion (SC).

Among the criteria categories taken into account in the MCA, "road geometric characteristics" and "safety" were assigned the highest weights (with their average criteria category weight being calculated to be 0, 1892 and 0,1667 respectively), followed by "road operational features", "land use and built environment characteristics" and "public acceptance", the average weight of which was calculated to be 0,1325, 0,1217 and 0,1200 respectively. Lastly, "cost", "ability to meet future demand" and "on-street parking supply" formed the least heavily weighted criteria categories, with their average weight ranging between 0, 0992 and 0, 0775.

Regarding the sub-criteria weighting, "number of conflicts between car and bicycle users" and "traffic volume (PCUs)" formed those sub-criteria that were most heavily weighted by the experts (with their average sub-criterion weight being calculated to be 0,6158 and 0,6033 respectively). On the other hand, "number of legal parking spaces lost (2-wheel vehicles)" was the least heavily weighted sub-criterion, since its average weight was calculated to be 0,2250.

From the above it could be concluded that, based on the experts' weighting, "road geometric characteristics" and "safety"/"number of conflicts between car and bicycle users" and "traffic volume (PCUs)" form those upper-level/lower lever factors that most heavily affect the final placemaking and type of the bicycle lane decision.

At this point, it should be noted that a high cumulative weight does not necessarily imply a strong degree of influence on the final decision. For example, while the highest average cumulative weights were calculated for the sub-criteria "road slope" (0,0990) and "number of conflicts between car and bicycle users" (0,1027), only the latter appeared to have a great influence on the final MCA decision. The fact that this is the case is largely down to the scores calculated by the research team on the basis of the relevant indicators, which-regarding the sub-criteria "road slope"-were more or less the same for both of the alternatives. On the contrary, that is not the case for the sub-criterion "number of conflicts between car and bicycle users", the scores of which were calculated to differ considerably regarding the two alternative roads.

5 Conclusions and Recommendations for Future Research

In the overall context, the final decisions resulting from MCA applications, depend on multiple factors (criteria categories weights, sub-criteria weights, scores per alternative), which interact with-and in certain cases offset-one another. That is certainly reflected in the way that the overall MCA scores for each alternative are calculated.

The MCA application suggested that, when compared to Alt_1 (Koumoundourou str.), Ir. Politechniou str. (Alt_2) is the most appropriate road to host the new bicycle lane. This is mainly due to the fact that high average cumulative weights were calculated for the two individual sub-criteria of the criteria category "safety" and at the same time, the scores of the aforementioned sub-criteria-calculated on the basis of the relevant indicators-appeared to be much higher for Alt_2. Overall, the criteria category "safety" and its sibling sub-criteria ("number of conflicts between car and bicycle users", "number of conflicts between bicycle users and pedestrians") appeared to have the greater influence on the final MCA decision, in favour of Alt_2.

Among all the criteria categories/sub-criteria taken into account in the MCA performed, "road geometric characteristics" and "safety"/"number of conflicts between car and bicycle users" and "traffic volume (PCUs)" were the most heavily weighted by the experts. While the highest average cumulative weights were calculated for the sub-criteria "number of conflicts between car and bicycle users" and "road slope", the latter did not finally affect the final decision, because of its relatively close scores calculated for the two alternatives.

The vast majority of the related studies have examined the problem of bicycle route planning on the macro-level, with a view to planning an integrated and coherent cycling network in urban and suburban areas. However, the case study area has already a quite extensive-for Greek standards-cycle network and an expansion of the latter is needed to accommodate the expected increased transport demand resulting from the newly constructed administrative building. To that end, the current research examines the problem of bicycle route planning from a micro perspective.

So far, we have performed a multi-criteria-based analysis for assessing alternative bicycle lane implementation solutions in urban networks. Our next step in the research is to perform a complementary Sensitivity Analysis in order for the robustness of the MCA results to be tested (check how much the MCA results will be altered if the different weights and/or scores change by a one percentage unit) and also to apply different decision-making techniques in order to provide comparisons between their results.

References

1. European Commission Homepage. https://ec.europa.eu/transport/themes/urban/guidance-cycling-projects-eu/policy-development-and-evaluation-tools/sumps-and-cycling_en. Accessed 3 Mar 2020
2. Hsu, T.P., Lin, Y.T.: A model for planning a bicycle network with multi-criteria suitability evaluation using GIS. WIT Trans. Ecol. Environ. **148**, 243–252 (2011)
3. Pucher, J., Dill, J., Handy, S.: Infrastructure, programs and policies to increase bicycling: an international review. Prev. Med. **50**, 106–125 (2010)
4. Schultheiss, B., Goodman, D., Blackburn, L., Wood, A., Reed, D., Elbech, M.: Bikeway Selection Guide. U.S., Department of Transportation, Federal Highway Administration (2019
5. Grisé, E., El-Geneidy, A.: If we build it, who will benefit? A multi-criteria approach for the prioritization of new bicycle lanes in Quebec City, Canada. J. Transp. Land Use **11**(1), 217–235 (2018)
6. Ana, S., Pinto, I., Ribeiro, D., Delgado, J.: Multicriteria analysis for evaluation of bike lane routes integrated to public transportation. Procedia-Soc. Behav. Sci. **162**, 388–397 (2014)
7. Rybarczyk, G., Changshan, W.: Bicycle facility planning using GIS and multi-criteria decision analysis. Appl. Geogr. **30**, 282–293 (2010)
8. Basbas, S., Makridakis, C.M.: A review of the contribution of multi-criteria analysis to the evaluation process of transportation projects. Int. J. Sustain. Dev. Plann. **2**(4), 387–407 (2007)
9. Dodgson, J.S., Spackman, M., Pearman, A., Phillips, L.D.: Multi-Criteria Analysis: A Manual, Department of Communities and Local Government, London (2009)
10. Vlastos, T., Bakogiannis, E.: Towards A Greece With Fewer Cars. Spatial Planning and Sustainable Urban Mobility Strategies Against Climate Change, Grigori, Athens (2019)
11. E C. Joint Research Centre., Org. for Economic Co-operation and Development: Handbook on constructing composite indicators: methodology and user guide, OECD (2008)

Examination of the Level of Service of the 2K Bus Line in Thessaloniki, Greece, and Proposed Improvements

Christos Braziotis, Ioanna-Eirini Tsali, Evangelos Genitsaris, Aristotelis Naniopoulos, and Dimitrios Nalmpantis(✉)

School of Civil Engineering, Faculty of Engineering, Aristotle University of Thessaloniki, PO Box 452, 541 24 Thessaloniki, Greece
dnalba@civil.auth.gr

Abstract. The purpose of this paper is to examine the level of service of the 2K bus line in Thessaloniki, Greece, and to make proposals for its improvement. The paper is divided into four parts. In the first part, a questionnaire survey that was answered online and on-field by 137 people is used to determine the current situation and level of service from the users' perspective. According to the satisfaction score of users regarding certain factors (timeliness, security, comfort, and reliability) and how these affect their travel experience, the overall satisfaction score of traveling on the 2K bus line is calculated. In the second part, real data provided by the public transport operator are analyzed. In the third part, the problems based on the results of the questionnaire and the real conditions of the bus line are identified, and in the fourth and final part, possible improvements are proposed. The results show an overall satisfaction score of 26.7%, with the lowest satisfaction rate being related to the comfort factor and the highest one with reliability. The real data analysis leads to several conclusions such as low average speeds, deviations between the estimated and the actual travel times, and the fact that one out of three bus itineraries is not executed at all. The methodology followed in this paper can be used to examine the level of service of different bus lines or whole bus line networks, as it is possible to be adjusted to the specific needs of each research.

Keywords: Bus line · Level of Service (LoS) · Public transport · Thessaloniki

1 Introduction

Problems of public transport in Thessaloniki, Greece, are increasing and overcrowding and poor transport conditions are constantly observed. The existing network and infrastructure are considered unable to serve the city sufficiently, and several improvements need to be made. In this research, we examine the operating conditions of a specific bus line of Thessaloniki, Greece, i.e., the 2K bus line. This line is of high interest as it crosses the city from one end to the other in a 17 km route, it serves many people, and its operation started about one year ago by a Public Transport Operator (PTO) that was recently nationalized, an exception in European practices that needs to be examined.

E. G. Nathanail et al. (Eds.): CSUM2020, AISC 1278, pp. 445–454, 2021.
https://doi.org/10.1007/978-3-030-61075-3_44

The following of this paper is structured as follows: in Sect. 2 the methodology followed to collect and analyze the required data is described, in Sect. 3 the results are displayed, in Sect. 4 the problems are presented and some improvements are suggested, and in Sect. 5 some conclusions are drawn.

2 Methodology

2.1 Questionnaire Survey

In order to examine the Level of Service (LoS) of the 2K bus line, a questionnaire survey was carried out. Thessaloniki's PTO is known as "OASTh", an acronym derived from its Greek name, which is translated in English as "Organisation of Urban Transportation of Thessaloniki" [1]. Questionnaires were distributed to passengers of OASTh from the 8th to the 17th of December 2019, and a total of 137 answers were collected, 108 of which were online and 29 on the field.

The questionnaire was divided into three (3) parts. In the first part, questions about general information of the users, such as demographics and travel habits, were included. During the second part, the satisfaction of passengers about the bus service quality was investigated through questions divided into four (4) categories (travel time, security, comfort, and reliability). Since the satisfaction level of passengers regarding these factors represents their viewpoint for the offered service, it could assess the quality of service of the examined bus line. In order to improve the evaluation process and discover the crucial problems, each of the above categories was divided into two subcategories. Users of the 2K bus line were asked to rate their satisfaction of each subcategory on a 5-point Likert scale, from 1, which means "not at all satisfied", to 5, which means "fully satisfied". The third part consisted of questions concerning general trip characteristics, such as the frequency and reasons of fare evasion, the rate of inability to board due to full occupancy, and the opinion of the passengers about the most significant factors that could improve their trip experience. The methodology used is based on relevant published literature [2].

2.2 Actual Data

For a better view of the LoS, it was considered necessary to analyze the actual operating conditions of the 2K bus line. Real data about average speeds and actual and estimated journey times from a period between November 1 and November 30, 2019, were collected directly from OASTh's archives. Based on this information, an attempt to compare actual and theoretical operating conditions of the 2K bus line was made.

3 Results

3.1 Questionnaire

The majority of the respondents consists of users aged from 18 to 25 years old (75.2%), and users who travel mostly for school or study purposes (53.3%).

More than ⅔ of the respondents (69.3%) stated than they cannot use an alternative means of transport instead of the 2K bus line. The detailed distribution of the questionnaire samples is shown in Table 1.

Table 1. Sample distribution of the questionnaire.

Characteristics		Frequency	Percentage
Type of questionnaire	Online	108	78.8%
	On-field	29	21.2%
Age	<18	3	2.2%
	18–25	103	75.2%
	26–40	11	8.0%
	41–65	15	10.9%
	<65	5	3.6%
Trip frequency	Daily	36	26.3%
	4–5 times per week	33	24.1%
	2–3 times per week	40	29.2%
	Rarely	28	20.4%
Trip purpose	School/Study	73	53.3%
	Work	19	13.9%
	Entertainment	21	15.3%
	Other needs/businesses	21	15.3%
	Other	3	2.2%
Trip length (number of stops)	Less than 5	36	26.3%
	About 10	62	45.3%
	About 20	32	23.4%
	More than 30	7	5.1%
Possibility of traveling with an alternative mean of transport	Yes	22	16.1%
	No	95	69.3%
	Sometimes	20	14.6%

As mentioned earlier, in the second part of the questionnaire, the respondents were asked to rate their satisfaction with respect to eight (8) subcategories. To calculate the average Satisfaction Rate (*SR*) of each category and subcategory of the 2K bus line from the users' perspective, Eq. (1) is used:

$$SR = \sum\nolimits_{1}^{5} \frac{n_i \cdot i}{n_{tot.}} \qquad (1)$$

where i is the satisfaction rate (from 1 to 5), n_i is the number of answers for each rate, and $n_{tot.} = 137$ is the number of total answers. The results are displayed in Table 2.

Table 2. The average satisfaction rate of each category and subcategory.

Category	Subcategory	Average Satisfaction Rate (SR)	
1. Travel time	1.1 Waiting time at the bus stop	1.985	2.183
	1.2 Travel time in the bus	2.380	
2. Security	2.1 Security of waiting and boarding	2.547	2.226
	2.2 Security inside the bus	1.905	
3. Comfort	3.1 Travel comfort	1.292	1.475
	3.2 Quality of service	1.657	
4. Reliability	4.1 Timetable punctuality	2.044	2.573
	4.2 Telematics accuracy	3.102	

Each category has a different impact on the quality of customer service, and, thus, the Weight Factor (*WF*) of each category needs to be estimated. Question 17 contributed to the estimation since the respondents were asked to prioritize the categories using a 5-point Liker scale from 1, "most important" to be improved, to 5, "less important" to be improved. Given the answers, the *WF* was calculated from Eq. (2).

$$WF = 5 - \sum\nolimits_1^5 \frac{n_i \cdot i}{n_{tot.}} \quad (2)$$

Given the average *SR* of each category and the *WF*, the Weighted Average satisfaction rate (*W.Avg.*) is calculated from Eq. (3).

$$W.Avg. = \frac{\sum_1^4 SR \cdot WF}{\sum_1^4 WF} \quad (3)$$

The final percentage of the Service Quality (*SQ*) is given by Eq. (4), which is finally estimated to be equal to 26.68%.

$$SQ = \frac{(W.avg.) - 1}{5 - 1} \quad (4)$$

Figure 1 shows the results of Question 17, while in Table 3, the results of the calculations mentioned are displayed.

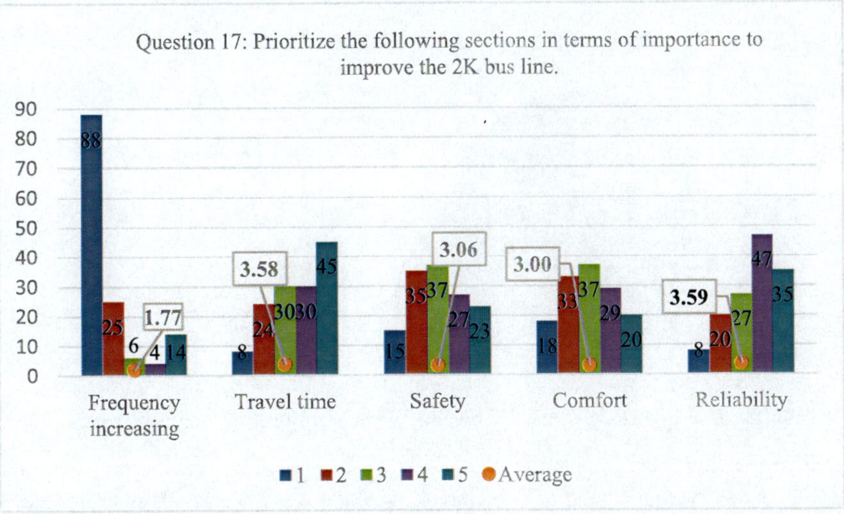

Fig. 1. Prioritization of different categories in terms of importance for improvement.

Table 3. Service quality of the 2K bus line.

Category	Average Satisfaction Rate (*SR*)	Weight Factor (*WF*)	Service Quality (*SQ*)
Travel time	2.183	1.42	29.58%
Security	2.226	1.94	30.65%
Comfort	1.475	2.00	11.88%
Reliability	2.573	1.41	39.33%
Weighted average satisfaction rate (*W.Avg.*)	2.067		26.68%

The answers in the third part of the questionnaire have shown that more than ⅓ (37.2%) of the respondents commute most of the times by the 2K bus line without a ticket and only 36.5% always buy one (Fig. 2)! The survey shows that the reasons for fare evasion are mostly a) the fact that the quality of service does not correspond to the ticket price (57%), and b) the lack of ticket validation due to overcrowding (54.8%). Another critical factor is the lack of ticket controls (34.4%) and other reasons with a lower percentage (14%). Moreover, overcrowding seems to be a critical problem since respondents said that almost 4 out of 10 times (41%) they could not board on the bus due to it.

3.2 Actual Data

Based on the data derived directly from OASTh, important conclusions about the average speed and the actual and estimated travel time of the 2K bus line were drawn.

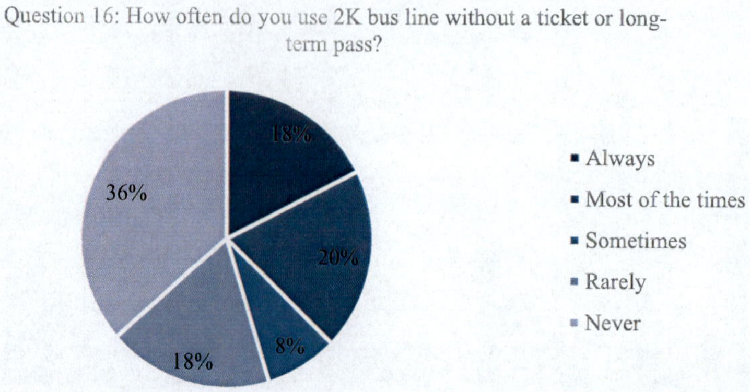

Fig. 2. Use frequency of the 2K bus line without a ticket or long-term pass.

In Fig. 3 the average speed of vehicles on weekdays and weekends is presented. It is concluded that for most of the day, the vehicles' average speed on weekdays is lower than the vehicles' average speed on weekends, which was expected due to the increased traffic volume on weekdays (due to education purposes, commuting, etc.). More specifically, the average speed of buses is 14.99 km/hr on weekdays, 16.33 km/hr on Saturdays, and 18.41 km/hr on Sundays. Nevertheless, the average speed of vehicles on the evening hours is lower on weekends than that on weekdays. This is possibly due to the fact that people usually choose to go out late hours for entertainment during weekends. In addition, it is concluded that from Monday to Friday the lowest average speed is observed around 17:00 and 18:00 whereas the highest one at 00:00. The lowest

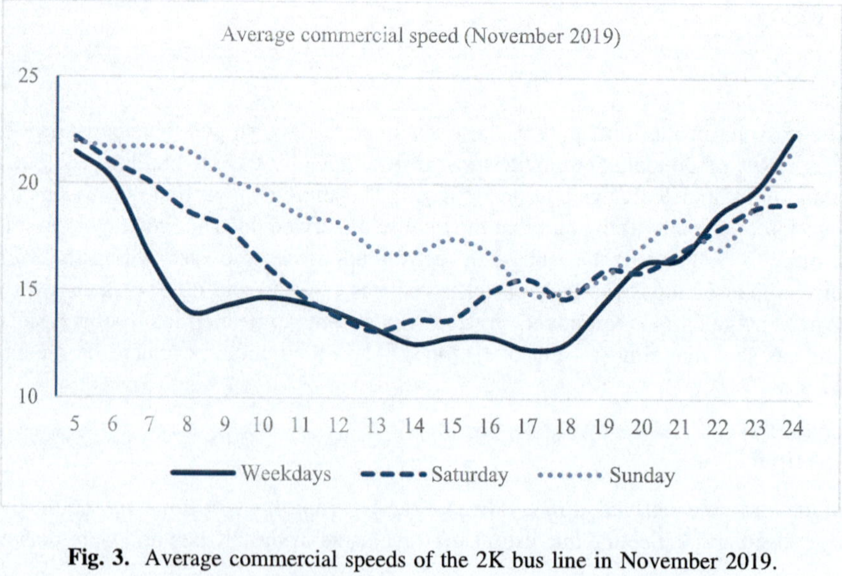

Fig. 3. Average commercial speeds of the 2K bus line in November 2019.

average speed on Saturday is between 12:00 and 13:00 and the highest one at 5:00. Regarding Sunday, the lowest average speed is observed between 17:00 and 18:00 and the highest one at 5:00 and 00:00. Generally, on weekdays the most hours of the day (8:00–18:00) vehicles' speed is too low, and the average speed is even lower than 15 km/h. As will be mentioned below, travel time is about 78′ at outbound trips and 83′ at inbound trips because of the high traffic volume and crowdedness.

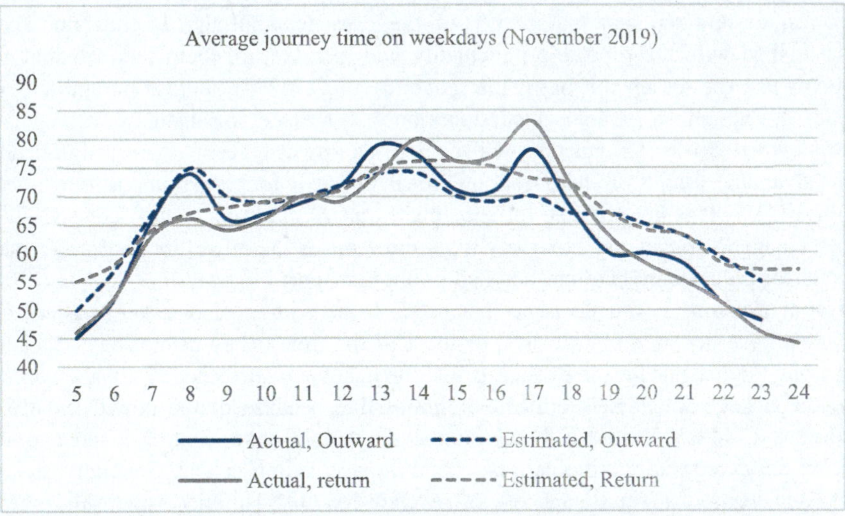

Fig. 4. Average journey time of the 2K bus line on weekdays in November 2019.

In Fig. 4 the estimated travel time and the actual travel time are presented. In outbound trips, it is evident that the estimated travel time is longer than the actual one until 12:00. On the other side, later by 18:00, the actual travel time is longer than the expected time. In general, there are no significant discrepancies between the actual and estimated time, except for evening hours that the maximum deviation is 13′.

4 Discussion

4.1 Problems

The results of the questionnaire survey, as well as the analysis of real data, indicate the following problems related to the operation of the 2K bus line:

- **Comfort:** As shown in Table 2, comfort has the lowest satisfaction rate between the four (4) categories, despite being really important for the users' travel experience (Fig. 1). Specifically, 75.9% of the respondents claimed that they were not satisfied with travel comfort, while only 0.7% said they were satisfied or very satisfied.
- **Travel time:** Travel time is the category with the second lowest service quality rate. Based on the results of the questionnaire, the reason is mainly due to the long

waiting time at the bus stop, where 70.1% of the respondents said they were not at all satisfied or low satisfied, while none claimed to be very satisfied. Better, but still with a low satisfaction rate, is ride time in the bus, where real data analysis shows an average commercial speed of 15 km/hr on weekdays. At peak times, this speed reaches only 12 km/hr, resulting in a journey time of more than 80 min.

- **Security:** Users' safety is also very low. It results that safety at the bus stop is higher than inside the bus, where 73.7% of the users claimed not to be satisfied.
- **Reliability:** Reliability is on a higher level compared to the other four (4) categories, despite the fact that 68.6% of the passengers of the 2K bus line feel unsatisfied with the timetable punctuality and just 11% of them feel satisfied. It seems that the reason for that is the reliability of OASTh's application and telematics, for which 46.7% feel satisfied and only 33.6% feel unsatisfied.
- **Bus crowdedness:** Crowdedness of the 2K bus line also seems to be a significant disadvantage. Based on the questionnaire survey, it is found that about four in ten times (41%) it is not possible for someone to board due to bus crowdedness. This percentage is considered extremely high and reduces the reliability and preference level for the specific bus line but also public transport in general.
- **Ticket evasion:** A crucial issue that needs to be addressed is ticket evasion, as 37.2% have declared that always, or most of the times, they do not buy a ticket. Among the reasons for the evasion is that the ticket is overpriced (57%), and access to the ticket validators is difficult or impossible because of the crowd (54.8%). Moreover, 34.4% attributed the evasion to the lack of ticket control, whereas only 5.4% answered that the problem is to find a place selling tickets. Generally, ticket evasion leads to a significant loss of revenue for OASTh, and, as a result, strict measures should be taken regarding the main reasons for fare evasion.
- **Route frequency:** Based on OASTh's data, between January 1, to November 30, 2019, only 65.6% of the scheduled routes were realized. This percentage is considered low, and passenger traffic is not served sufficiently, resulting in congestion and a low level of reliability. This problem is also reflected in the questionnaire survey, where routes' increasing is the most important factor in which 2K bus line needs improvement according to 64.2% of the respondents.
- **Route planning:** The comparison between the scheduled and the actually executed routes reveals a problem in the proper route planning. Scheduled and actual routes differ in a variety of areas, e.g., the frequency, departure times, and travel duration.

4.2 Improvement Suggestions

Survey findings, and the problems mentioned above, indicate the need for improvements towards a more orderly and efficient operation of the 2K bus line. Some of the suggested improvements are mention below. Prioritization has been made, so those that affect more the line operation, according to the users, are placed on top of the list:

- **More frequent service:** The fact that one out of three of the scheduled routes is not realized results in high occupancies and long waiting times. More frequent service will solve these problems, and the comfort and reliability will be improved.

- **More frequent ticket inspections:** Almost one fourth (23.4%) of the respondents claimed that they use the 2K bus line without a ticket due to lack of inspections. Concerning the fact that 45.3% of them often do not use a ticket, this leads to the need for more ticket inspections. This measure will increase OASTh's revenue.
- **Routes restructuring:** The deviations between estimated and actual journey times and failure to observe departure times result in reliability problems. A restructuring of the timetable based on the fleet capabilities and new data is necessary.
- **Service quality improvement:** According to the results, the respondents are not satisfied with the service quality. Cleanliness, buses and seats' quality, ticketing systems, etc. can easily get better with significant LoS improvement.
- **Security measures:** Most passengers do not feel safe while waiting or traveling by the 2K bus line. Thus, measures, such as the installation of Closed-Circuit Tele-Vision (CCTV) and better lighting at bus stops, should be taken.
- **Innovative applications:** The current economic conjuncture should not be an excuse for OASTh to avoid implementing innovative solutions and applications since many of them are of low cost and increase the service quality significantly. Recently, much research on this field has been realized in Thessaloniki [3–9].

5 Conclusions

The main goal of this research was to determine the service quality of the 2K bus line of OASTh as perceived by passengers, and the following conclusions were drawn:

Regarding the passengers' profile, it is concluded that:

- Almost 70% of the 2K bus line users are not able to use another means of transport.
- Almost half of the respondents use the 2K bus line at least four (4) times a week.
- Almost 37.2% use the 2K bus line without validating a ticket.

Regarding the operation of the 2K bus line, it is concluded that:

- Almost 35% of the scheduled bus routes are not realized.
- The average commercial speed is relatively low, at about 15 km/hr.

Regarding the service quality of the 2K bus line, it is concluded that:

- About 4 out of 10 times (41%) boarding is not possible due to bus crowdedness.
- Survey respondents ranked the attributes of the 2K bus line from the highest to the lowest level of satisfaction as follows: comfort, travel time, safety, and reliability.
- Passengers ranked in priority order the following as having the greatest need for improvement: increased frequency, comfort, safety, travel time, and reliability.
- The final service quality percentage of the 2K bus line is only 26.68%.

For the improvement of the 2K bus line the following are proposed:

- Timetable restructuring and increasing the frequency of the 2K bus line iterations.
- Performing more frequent ticket controls to eliminate ticket evasion.
- Improvement of services and safety in the bus and the infrastructure.

In this research, the percentage of 2K bus line service quality was estimated using a questionnaire survey, and its actual operating conditions were analyzed. Subsequently, issues were detected, and solutions were proposed.

The methodology of this research can be also used to evaluate the status of any line of all urban transportation modes, and it has the potential to extend to the overall transport network of a city. The questionnaire can be modified properly to include more or fewer categories and subcategory levels, depending on the survey's purpose.

In conclusion, the methodology used in this research is a simple and easy method to estimate the service quality and identify the problems of one or more transport lines, and, with further study, it can provide comparisons between different lines or networks.

References

1. Organisation of Urban Transportation of Thessaloniki Homepage. http://oasth.gr/. Accessed 01 Jan 2020
2. Weng, J., Di, X., Wang, C., Wang, J., Mao, L.: A bus service evaluation method from passenger's perspective based on satisfaction surveys: a case study of Beijing. China. Sustain. **10**(8), 2723 (2018). https://doi.org/10.3390/su10082723
3. Tsafarakis, S., Gkorezis, P., Nalmpantis, D., Genitsaris, E., Andronikidis, A., Altsitsiadis, E.: Investigating the preferences of individuals on public transport innovations using the maximum difference scaling method. Eur. Transp. Res. Rev. **11**(1), 3 (2019). https://doi.org/10.1186/s12544-018-0340-6
4. Nalmpantis, D., Roukouni, A., Genitsaris, E., Stamelou, A., Naniopoulos, A.: Evaluation of innovative ideas for public transport proposed by citizens using multi-criteria decision analysis (MCDA). Eur. Transp. Res. Rev. **11**(1), 22 (2019). https://doi.org/10.1186/s12544-019-0356-6
5. Stamelou, A., Genitsaris, E., Nalmpantis, D., Naniopoulos, A.: Investigating potential synergies among social entrepreneurship and public transport through experts' consultation in Greece. In: Nathanail, E.G., Karakikes, I.D. (eds.) CSUM 2018. Advances in Intelligent Systems and Computing, vol. 879, pp. 496–503. Springer, Cham (2019). https://doi.org/10.1007/978-3-030-02305-8_60
6. Genitsaris, E., Stamelou, A., Nalmpantis, D., Naniopoulos, A.: A criteria-based evaluation framework for assessing public transport related concepts resulted from collective intelligence approaches. In: Nathanail, E.G., Karakikes, I.D. (eds.) CSUM 2018. Advances in Intelligent Systems and Computing, vol. 879, pp. 529–537. Springer, Cham (2019). https://doi.org/10.1007/978-3-030-02305-8_64
7. Giannaka, D., Malliaris, S., Genitsaris, E., Karagiotas, I., Naniopoulos, A., Nalmpantis, D.: Investigating user preferences in night bus lines operation using Conjoint Analysis in the city of Thessaloniki, Greece. In: Proceedings of the 8th Transport Research Arena (TRA 2020), Helsinki, Finland, 27–30 April 2020. https://doi.org/10.5281/zenodo.3708354
8. Tsoukanelis, A., Genitsaris, E., Nalmpantis, D., Naniopoulos, A.: Conjoint Analysis for the optimization of a potential flexible transport service (FTS) in the region of Zagori, Greece. In: Nathanail, E.G., Karakikes, I.D. (eds.) CSUM 2018. Advances in Intelligent Systems and Computing, vol. 879, pp. 478–486. Springer, Cham (2019). https://doi.org/10.1007/978-3-030-02305-8_58
9. Papadima, G., Genitsaris, E., Karagiotas, I., Naniopoulos, A., Nalmpantis, D.: Investigation of acceptance of driverless buses in the city of Trikala and optimization of the service using Conjoint Analysis. Util. Policy **62**, 100994 (2020). https://doi.org/10.1016/j.jup.2019.100994

Using Alternative Fuel Vehicles in Medium-Sized Cities

Elias Papastavrinidis(✉), Vasiliki Kollarou, Antonia Athanasopoulou, and George Kollaros

Department of Civil Engineer, Democritus University of Thrace, 67100 Xanthi, Greece
eliaspapastavrinidis@gmail.com

Abstract. Today many people are interested in vehicles that use alternative fuels or advanced technologies that can help reduce operating costs, meet emissions requirements, support sustainability and energy independence. Vehicle manufacturers are responding to this interest with a wide range of options across a steadily growing number of vehicle applications. The EU wants to phase out conventionally fueled vehicles by 2050 and move to-wards carbon neutral urban logistics by 2030. To achieve this ambitious objective, it developed legislation and provided financial incentives to encourage cities to use alternatively fueled vehicles and create related infrastructure. Nowadays many medium-sized cities in EU started to adopt policies promoting the use of alternative fuel vehicles. In Xanthi, northern Greece, local authorities recently developed policies promoting the use of electric vehicles as they upgraded their fleet and started using them in public services. However, there are additional actions that can be implemented for vehicle use with non-conventional and advanced fuels from residents and visitors. This project aims in an effort to face the problem with the aid of an analysis based on counts and questionnaires in different locations of Xanthi.

Keywords: Sustainable mobility · Vehicles · Alternative fuels · Non-conventional fuels · Energy independence

1 Clean Vehicles and Alternative Fuels as Part of an Integrated Sustainable Urban Mobility Strategy

It is broadly recognized that European cities have a key role to play when it comes to the deployment of clean vehicles and the use of alternative fuels. In order to implement an effective strategy towards a wider uptake of the alternative-fuel market at the city-level, an integrated approach to sustainable mobility is needed. The Sustainable Energy Action Plans (SEAPs) and the Sustainable Urban Mobility Plans (SUMPs) are the two major tools that can help cities to achieve an effective clean vehicle strategy [1, 2].

Both SEAP and SUMPs foresee the implementation of key measures to be developed in cooperation with different policy areas and sectors, different levels of and administration and stakeholders. In particular for SUMPs, integrating alternative fuels and clean vehicles should result in a sub-plan or sub-strategy that strengthens the role

E. G. Nathanail et al. (Eds.): CSUM2020, AISC 1278, pp. 455–461, 2021.
https://doi.org/10.1007/978-3-030-61075-3_45

of these vehicles as part of a multimodal system. When it comes to the intro-duction of electric vehicles, for example, SEAPs and SUMPs constitute local planning strategies that clearly have some aspects in common. Cities, therefore, need to develop an effective package of measures where clean vehicles are part of a broader mobility system [1, 2].

2 Contribution of Local Authorities to the Implementation of Alternative Fuels Infrastructure

The National Policy Frameworks which are based within the context of the Alternative Fuels Infrastructure Directive represent an important milestone. The EC liaise with EU Member States regarding these NPFs and then report on their assessment to the EU.

The EC is in charge of monitoring the implementation of the Directive and suggesting an Action Plan to support Member States. The EC can revise if necessary, the AFI Directive by the end of 2020 [3].

To further the uptake of the alternative fuels and vehicle infrastructures, also the local level can adopt a set of policy measures to facilitate the roll-out of clean vehicles and support the implementation of the Directive. Some of the policy measures mentioned in the Directive are within the remit of cities as well [3]:

- Direct incentives for the acquisition of means of transport using alternative fuels or for building the infrastructure.
- Tax incentives to market means of transport using alternative fuels and therefore the relevant infrastructure.
- Use of public procurement including joint procurement.
- Demand-side non-financial incentives like access to restricted areas, parking policy and dedicated lanes.

In fact, many cities already implement such incentives and procurement initiatives, thus paving the way for the further roll-out of alternative fuels and vehicles and related infrastructures.

3 Policies Deploying Alternatively Fueled Vehicles

Nowadays, local authorities demonstrate a clear political will on the local level to deploy alternatively fuelled vehicles. At the same time, there are significant contextual differences across cities and their capacity to undertake substantial financial investments to cope with their ambitions to switch from fossil fuel vehicles to hybrids, alternative or electric vehicles can vary substantially as well.

This is for example linked to the national frameworks, subsidies or incentives in place. Europe's main capitals have already made good progress and clearly take the lead in clean vehicles deployment, while medium-sized cities are rather looking at such transformations within a 2020–2025 time-horizon.

Cities also express a clear wish to further exchange with their counterparts on how to best developing and tailor their own local policies in such a way that they could boost electromobility both for the freight and the public or private passenger segment. Particularly, cities would like to learn about the trajectories and forecasts for decarbonizing transport by sector that are being developed in other member states [4].

3.1 Vehicle Data Collection

One major barrier to collecting vehicle data relates to the difficulty for cities of keeping track of private vehicles that are registered in another Member State. While the free movement of goods and workers in the EU constitutes a fundamental principle embedded in the EU Treaty, it also prevents cities from accurately assessing the real number of electric cars circulating within their geographical area.

Similarly, cities noted that they still lack a clear picture of the number of private cars operated by taxi fleets. Both elements should be taken into account in all preparatory work at the EU level, for example regarding a pan-European register log for vehicles or by strengthening data communication between the national and the local level.

Some cities also face the fact that vehicles do not necessarily need to be registered in their respective Member States, which in turn does not allow for accurate information on this category for vehicle data provision [5].

3.2 Charging Infrastructure Data Collection

Some cities, have already privatised their charging network. While this privatisation substantially speeds up the deployment of charging infrastructure, it also changes the way information is shared. In that perspective, precise figures for charging infrastructure coverage are unavailable.

Cities are also bound to confidentiality clauses in some of the major projects currently under development. This could hamper data collection at the national level and hence create difficulties in the near future to assess the fulfilment of the EU tar-gets set in terms of charging points per vehicle on the national market.

Some additional guidelines in Green Public Procurement and overall, within the Clean Vehicle Directive revision should therefore specify a requirement for sharing such information in future tenders or public procurement templates. Additionally, tracking the use of charging points in cities is a challenging issue. Firstly, because some of the city fleets or taxis use rented parking lots, with few data available and therefore only allowing a rough estimation of use of the charging infrastructure rather than exact data. While some local authorities have already set up an interoperable scheme which keeps track of the use of the charging points and also ensures an interoperable billing system when charging points belong to subcontracted third parties, most cities have not yet deployed a comprehensive system that would allow to measure in real time the use of charging points [5].

The role of the EU would be crucial in this respect, by further supporting research and exchange of good practices while also exploring the possibility of comprehensive yet nonbinding guidelines to address this challenge.

3.3 Bus and Infrastructure Data

Cities express the need for more exchange at the EU level on bus depots or maintenance and storage facilities related to electric fleets. This could either take the form of strengthening the current EU project addressing this issue or could be explored within the context of new projects [6].

Some elements of concern were also raised regarding charging plugs and enroute recharging systems' standards. Both will be needed in the future to fully deploy electric buses in cities while ensuring at the same time the interoperability of charging systems for domestic or international coaches accessing the urban area of the city centre.

Local authorities finally pointed out that more evidence is needed on LNG/CNG and comparisons with different vehicle engine types to economically understand what type could best be deployed locally. This could be supported by creating an EU public/private pool with industrial partners to help with policy setting/investment decisions in this area.

4 Use of Alternative Fuel Vehicles in Greek Cities

The Ministry of Environment and Energy in Greece has established various financial incentives for the promotion of electric vehicles. Specifically, electric cars are exempt from the obligation to pay traffic fees, registration fees and luxury tax, while hybrid cars are exempt from 50% of the prescribed registration fee [7].

In addition, it is emphasized that integrated interventions are planned at the regulatory level in order to launch all the conditions for the sustainable development of electric propulsion in Greece. Emphasis are going to be placed on categories of vehicles with high transport work (e.g. taxis, buses, courier companies, etc.) and thus potentially large energy and environmental benefits.

According to the objectives included in the National Energy and Climate Plan [7]:

Especially for electric vehicles in road transport, the results refer to the participation of 10% of electric passenger vehicles in relation to the total fleet of passenger vehicles that will be in circulation in the year 2030.

The gradual renewal of the fleet of passenger vehicles by modern ones with higher energy efficiency per unit of transport work, like electric vehicles and more efficient internal combustion vehicles, will cause a reduction within the final energy consumption of passenger vehicles whose absolute number will amount to 293 ktoe for the period 2016–2030.

The electrification of road passenger transport results in a significant improvement in energy efficiency per unit of transport work, with unit consumption falling to 24 ktoe/kpkm in 2030 compared to 32 ktoe/kpkm in 2016.

The electrification of the fixed orbit media will lead to a reduction of the final energy consumption of the sub-sector by 26%.

The electrification of the transport sector is going to be achieved mainly through railways, while within the subsectors of passenger vehicles and buses within the year 2030, the contribution of electricity is predicted to rise to 5% and 3% respectively [7].

5 Policies Promoting the Use of Alternative Fuel Vehicles in the City of Xanthi

Municipality of Xanthi recently procured three electric cars and officially entered the world of electric vehicles. These are essentially ecological city vehicles that run exclusively on electricity (eco cars), which are used to provide services to citizens in order to serve them more efficiently (Fig. 1).

Fig. 1. Electric vehicle in the Municipality of Xanthi, Greece.

Saving resources with new technologies is a big bet that the Municipal Authority has made a priority from the first moment, which is gradually becoming a reality. The new electric vehicles enhance the green environmental footprint of Xanthi's Municipality. In addition, they provide zero maintenance costs, zero traffic charges and zero fuel costs, as their consumption is limited to € 1 per 100 km.

Recently, local authorities decided to place recharging points at the vehicle service facilities in publicly accessible recharging points along the urban road network as well as in the parking spaces of public buildings [8].

Moreover, the Municipality of Xanthi organizes and implements events for the promotion of electric mobility. The main goal is to raise awareness about environmental protection and to inform the public about the management of renewable energy sources for the benefit of citizens, as well as to protect public health from the effects of internal combustion engines - especially in urban centres.

Emphasis is also placed on the more rational management of conventional fuels, which are imported goods, so that the national economy is not burdened.

6 Definition of a Clean Vehicle and Further Reliability on Euro Standard

Implementing Low Emission Zones or local air quality targets requires a reliable ranking system for clean vehicles and clean vehicle procurement. Cities raise the issue of the lack of a clear and common definition of a clean vehicle at the European level. Furthermore, recent developments also showed a lack of transparency and reliability of the Euro standards, which constitute the basis for implementing access restrictions and/or Low Emission Zones locally.

From this perspective, cities pointed out that an understandable, and easy-to-apply- and -communicate common definition of a clean vehicle is needed, based on the combination of harmful emissions and well-to-wheel CO2 emissions produced, as well as the energy efficiency of the car. In turn, such definition would allow a global ranking system similar to the one used under the Energy consumption labelling scheme (A to G) and the set-up of EU objectives based on thresholds that would evolve over time.

Cities would also welcome a revamped Clean Vehicle Portal in that respect, either as a stand-alone website or by merging it with a one-stop-shop portal such as the EAFO website. On the consumer/user end, cities propose that within the upcoming revision of the EU Directive on the labelling system for CO2 emissions from cars, the European Commission would take the opportunity to also include emissions from the transport sector described under the National Emission Ceiling and the Ambient Air Quality Directives. It should be noted that these issues are currently under discussion within the Sustainable Transport Forum's City subgroup, a multi-stakeholder expert group which aims to support the EC with the application of the Clean Power for Transport strategy, to facilitate the implementation of the AF Directive and implement EU activities and programs aimed at fostering the deployment of AF infrastructure (energy and climate goals) [9].

It offers a platform for structural dialogue, exchange of technical knowledge, co-operation and coordination and as such provides advice and technical expertise on the development and implementation of legislation, policies, projects and programs, delivers opinions, submit reports, and develops and proposes innovative solutions.

References

1. Sustainable Energy and Climate Action Plan: Municipality of Xanthi. Accessed 20 May 2016, (in Greek)
2. Wefering, F., Rupprecht, S., Bührmann, S., Böhler-Baedeker, S.: Guidelines - Developing and implementing a sustainable urban mobility plan. European Commission: Directorate-General for Mobility and Transport (2014)
3. Bauen, A., Gomez, I., OudeNijeweme, D., Paraschiv, M.: Alternative fuels expert group report - Studies and reports. European Commission (2017)
4. Winkel, R., Hamelinck, C., Bardout, M., Bucquet, C., Ping, S., Cuijpers, M., Artuso, D., Bonafede, S.: Alternative fuels and infrastructure in seven non-EU markets. European Commission (2016)

5. Mattson, J.: Use of alternative fuels and hybrid vehicles by small urban and rural transit systems. North Dakota State University, Fargo (2012)
6. U.S. Department of Energy: Clean cities guide to alternative fuel and advanced medium and heavy-duty vehicles. Office of Energy Efficiency and Renewable Energy (2011)
7. National Energy and Climate Plan of Greece 2019: Hellenic Republic, Ministry of the Environment and Energy. Accessed 30 Dec 2019, (in Greek)
8. Technical Program 2020: Municipality of Xanthi. Accessed 27 Nov 2019, (in Greek)
9. D' Appolonia, S.P.A., Ramboll, T.M.: Leuven: Clean Transport - Support to the Member States for the implementation of the directive on the deployment of alternative fuels infrastructure - Good practice examples. European Commission (2016)

Investigating the Athens – Thessaloniki Door-to-Door Intercity Transport Connection by All Means from the Students' Point of View

Anna-Angeliki Sarantakou, Evangelos Genitsaris, Aristotelis Naniopoulos, and Dimitrios Nalmpantis(✉)

School of Civil Engineering, Faculty of Engineering, Aristotle University of Thessaloniki, PO Box 452, 541 24 Thessaloniki, Greece
dnalba@civil.auth.gr

Abstract. In Greece, Athens and Thessaloniki are the two largest cities with the largest universities and, therefore, with a large student population. Students are one of the groups with the higher, if not the highest, use of public transport, as most of them do not own private cars and need to travel long distances both in the city they study and to and from the city they came. Moreover, it is still very common for students of Athens and Thessaloniki to travel to and from Thessaloniki and Athens, respectively, both for recreational reasons and because the parents and friends of many of them live in the other city. This paper presents a general overview of the different transport means available in Athens and Thessaloniki and the ways the two cities are interconnected, followed by a comparison of the several ways the intercity trips can be realized. Emphasis is placed on how students choose to travel, which are their criteria, and how much time they make from the moment they leave their home to the point of arrival to the other city (door-to-door). For this reason, a questionnaire survey was conducted among students of Athens and Thessaloniki. The questionnaire was answered by 155 students from Athens and 166 students from Thessaloniki. A summary table presents the cost, time, number of means, and comfort for all the possible door-to-door intercity routes. As expected, considering the current financial conjuncture, the most important factor was found to be the cost of the trip.

Keywords: Criteria · Cost · Comfort · Time · Airplane · Train · Bus · Car · Greece

1 Introduction

Public transportation is essential for a city as it serves the transport needs of its citizens for activities of daily living (e.g., work, shopping, leisure, etc.). Public transportation is important for all citizens, including vulnerable users with all the difficulties they face (e.g., People with Disability [PwD], the elderly, etc.), and is mostly needed by the weaker social groups that often do not own a private car. In large cities where universities are located, there are also large students' populations.

E. G. Nathanail et al. (Eds.): CSUM2020, AISC 1278, pp. 462–471, 2021.
https://doi.org/10.1007/978-3-030-61075-3_46

In Greece, Athens and Thessaloniki are the two largest cities with the largest universities and, consequently, with the largest student populations. Students are a group of people who mainly use public transportation and taxis for their transport, as most of them do not own a private car and need to travel long distances. It is common for Athens' students to travel to Thessaloniki and vice versa, as both Thessaloniki and Athens are attractive inland tourist destinations, and to visit their families as for many of them, their family lives in the other city than the one they study. Therefore, for Athenian students who study in Thessaloniki and for Thessalonian students who study in Athens, the investigation of several door-to-door intercity travel options that there are available is of great importance. This was the main reason for the survey we conducted and will be presented in this paper.

As far as we know, there is not much published research on the topic. We used the Scopus indexing service with the following keywords: a) Athens, b) Thessaloniki, and c) travel, in the title, abstract, and keywords, and we found only six (6) papers. The five (5) of them were irrelevant to the topic, and the other one is an old publication of 1983 about safety perceptions of intercity travel modes [1]. Although there is much research in Greece about urban public transport in big cities [2, 3], and also in smaller towns [4], it seems that there is a lack of research regarding the whole door-to-door intercity travel chain between Athens and Thessaloniki.

2 Methodology

The transportation connection between Athens and Thessaloniki is one of the most important connections in Greece as these are the two most significant urban centers that are connected by almost any means of intercity transportation. Using data such as travel time, ticket fare, travel conditions, etc. could lead to some results. Nevertheless, it was essential to compare the door-to-door transportation connections between the two cities, also considering the subjective view of the students themselves.

For this reason, we developed an online questionnaire using the Google Forms platform. The questionnaire was disseminated to students in Athens and Thessaloniki in three rounds: a) through a personal Facebook page to close contacts who were students in Athens and Thessaloniki, b) through a story published at an Instagram personal account of around 550 followers, and c) through a Facebook group of 406 members who were students of civil engineering at the National Technical University of Athens (NTUA) to gather more answers from Athens. The respondents were asked to spread the questionnaire even further, using, thus, the snowball sampling technique.

Regarding the structure of the questionnaire, after the demographic questions, the questionnaire was initially divided per departure city into two sections, in which they were asked how they are served by the urban transport system, and further on into other two sections depending on the availability or not of urban public transportation from their home to the intercity departure stations. Following, they were asked to choose their preferred way of traveling from one city to the other, and to rank the following criteria: a) total cost, b) comfort, c) total time, d) time to/from stations, and e) transport means to/from stations, using a 5-point Likert scale (Fig. 1).

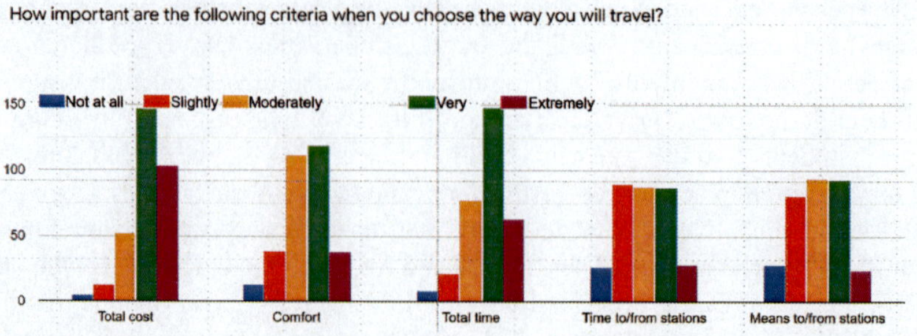

Fig. 1. Importance of the criteria when choosing the way to travel.

At the end of the survey, a fairly large amount of responses had been collected, giving an adequate sample for our survey: 321 responses, with 155 (48.3%) students being in Athens and 166 (51.7%) students in Thessaloniki. 162 (50.5%) respondents were men, 156 (48.6%) were women, and 3 (0.9%) did not know or did not want to answer. 224 (69.8%) of the respondents did not work, and 97 (30.2%) worked. By averaging the answers, the time and cost of each scheme of door-to-door intercity journey were calculated, and some interesting conclusions were drawn regarding the results of the questionnaire.

3 Results and Discussion

3.1 Overview of Public Transportation in Athens and Thessaloniki

Athens Public Transportation

The Athens transportation network is extensive and supported by many means. Thus, in many areas of Athens, public transportation buses, which in the other cities of Greece have a leading role, have a purely supportive role and serve mainly the commute that takes place within an area. The public transportation means available in Athens are the following: a) the urban public transport buses, b) the trolleys, c) the metro, d) the electric train (metro line 1), and e) the tram. There is also the suburban railway that serves some distant areas [5].

City buses and trolleys cover areas not only inside but also outside the city of Athens. More specifically, there are 277 city buses and trolleys, of which five lines (X14, X93, X95, X96, and X97) are express lines that serve travelers to and from the "Eleftherios Venizelos" Athens International Airport without intermediate stops. There are also public transport buses and trolleys that serve transfers to all intercity stations, including the "Kifissos" Intercity Bus Station. The ticket price for students is €0.60 for a single-use using a personalized card with the exception of the ticket for the aforementioned express bus lines, which costs €3.00 for students [5].

The Athens Metro is one of the most convenient means of public transportation in the city as it is easy to understand and use, it is independent of the road traffic, and it is the most reliable on arrival time, which makes it convenient for anyone to use. The city metro is currently operating two lines, and a third one will be built soon. It will have a total length of 38.2 km, including 35 stations in total. Line 2 has a length of 39 km and 20 stations, while Line 3 is 39.1 km long with 21 stations. In terms of city intercity stations, it serves the public for the Athens International Airport and the "Larissa" Railway Station, and it connects to the Suburban Rail Station in many locations throughout the city. The ticket price for students is €0.60 for a single-use using a personalized card with the exception that of the ticket to and from the Athens International Airport that costs €5.00 for the students. The electric train now has the name metro Line 1. It is a ground-based electric train with a length of 25.7 km and 24 stations. It has lower speeds than the metro and joins one of the most southwestern areas, Piraeus, with one of the northernmost areas, Kifissia. It serves travelers to the Port of Piraeus and is connected to the other two metro lines. The ticket price for students is €0.60 for a single-use using a personalized card [5].

Finally, the Tram has a total system length of 27 km and has three (3) lines with a total of 48 stops. The tram does not serve so many travelers or inhabitants for commuting, but it plays mainly a supportive role for the city buses in these areas. The ticket price for students is €0.60 for a single-use using a personalized card [5].

Questionnaire Results About Public Transport in Athens

Of the 155 students in Athens who responded to the question of whether they were connected to their departure stations by public transportation, 129 (83.2%) answered "Yes," and 26 (16.8%) answered "No."

Of those who answered "Yes," 78 (60.5%) responded that they were concerned about their connection to urban public transportation when choosing the way they would travel, while 51 (39.5%) said "No."

They were also asked about how many different means of transport they need to reach each departure station, i.e., the Athens International Airport, the Railway Station, and the Intercity Bus Station, separately. The mean of the answers to the three questions is as follows: 0 different means: 2.58%; 1 different means: 30.49%; 2 different means: 50.13%; 3 different means 15.24%; and more than 3 different means: 1.56%.

Finally, there was a question about the means of public transport they mostly use to reach their departure station, and the results are the following: 26.4% urban bus, 58.1% metro, 7% electric train, 0.8% trolley, 1.6% tram, and 6.2% none.

Thessaloniki's Public Transportation

Thessaloniki's transportation network is well-structured and easy to use since everyone can easily navigate in the city and the road network. Nevertheless, its only public transport mode is urban public transportation buses. The area it serves is covered by 79 bus lines, of which 18 are suburban. There are four traveler-specific lines: X1, N1, 45, and 79, with the 45 bus line being express having fewer stops than the rest. The lines mentioned above serve the following intercity stations to Athens: a) the "Makedonia" Thessaloniki Airport, b) the "New" Thessaloniki Railway Station, and c) the "Macedonia" Intercity Bus Station.

Moreover, the public transportation system of Thessaloniki offers the possibility of crossing through the city and connecting the intercity stations with a single line, making it suitable for intercity travelers. The urban public bus ticket fare for students is

€0.50 for single-use. For lines 50 (Cultural Line), 01X (KTEL-Airport), and 01 N (KTEL-Airport), the ticket price is €1.00 for students.

Questionnaire Results About Public Transport in Thessaloniki

Of the 166 students studying in Thessaloniki who responded to the question of whether they were connected to their departure stations by public transportation, 150 (90.4%) answered "Yes," and 16 (9.6%) answered "No."

Of those who answered "Yes," 84 (56%) said that they were concerned about their connection to urban public transportation when choosing how to travel, while 66 (44%) answered "No."

3.2 Connection Routes Between Athens and Thessaloniki

Traveling between Athens and Thessaloniki and vice versa is one of the most popular trips in Greece and is done in the most variable ways with different criteria and amenities each, which means it can be the ideal domestic trip for anyone. In this section, we will analyze how the two cities are linked and what each way offers individually in regard to time, cost, security, and comfort. The travel modes are as follows: a) airplane, b) train, c) intercity bus, and d) private vehicle (i.e., car).

Travel by Airplane

The 502 km distance between the two largest cities in Greece may be considered a lot for a road trip, but they are not so much for an air flight. The usual flight time between Athens and Thessaloniki is just one hour with very frequent daily departures making it easy to save a lot of time. Additionally, times of arrival and departure are more precise than the other means of transport because the human factor interferes less, and delays are less common. It is important to note, however, that traveling by airplane also takes a long time for the door-to-door scale because the airports are farther from the cities' centers compared to any other station of intercity transport.

Specifically, according to the respondents, the average time for the students in Athens to reach the Athens International Airport is 45′–60′, which coincides with the predominant answer (41.9%), while for the Thessaloniki Airport the average and dominant answer is 30′–45′ (31.3%). Also, worth noting is the fact that the airport is the only departure station requiring a long arrival time before departure. In particular, depending on the flight and the luggage, the passenger is expected to be 60′–90′ earlier at the airport. To summarize, the total door-to-door journey time by airplane from Athens to Thessaloniki and vice versa is estimated to be 3.5 to 4.5 h.

Traveling by airplane is considered luxurious for students, mainly due to the price of the ticket, which is volatile and somehow unpredictable, because it is constantly changing and usually increasing as the day of departure approaches. This means that for someone who does not know the exact date he/she wishes to travel a reasonable amount of time before that date, in order to find an affordable ticket price, he/she will probably avoid this way of travel unless it is necessary. Still, due to the price volatility, it is difficult to predict the cost of travel. After an airline survey, we found that the cost of a ticket for a flight between Athens and Thessaloniki can range from €45 to €200. The questionnaire also shows that the average cost of traveling from Athens to the Athens International Airport is €6–12, with the most frequent answers being €3–6

(38%) and €6–12 (37.2%). The average cost of traveling from Thessaloniki to the airport is €6–12, with the most frequent answer being €0–3 (61.3%). With all this in mind, traveling by airplane can cost in total €60–215.

Traveling by airplane is considered one of the safest and most comfortable ways to travel. In terms of comfort, it is important to note that waiting at the airport as well as continuous checks, queues, access to the airplane, and the cramped space are its major disadvantages, while in terms of the time of the flight and the experience it becomes enjoyable.

Another result of the questionnaire shows that the majority of students who responded need to get one (1) (41.9%) or two (2) (45%) different means of transportation to get to the Athens International Airport. From that, it follows that traveling to and from the Athens International Airport can cause fatigue, which also affects travelers' comfort experience.

Travel by Train

The railway line between Athens and Thessaloniki is 518.18 km. The train journey time can be 4 h and 10 min, 4 h and 23 min, and 5 h and 30 min depending on the itinerary, with six (6) to seven (7) daily routes from Athens to Thessaloniki and vice versa, depending on the day and the demand. Arrival and departure times are precise, and there are very few delays. Train stations in Athens and Thessaloniki are also easily accessible. In Athens, the station is centrally located in the city, and there is a connection to the metro and the city buses. In Thessaloniki, many bus lines stop at or near the train station, and it is not far from its center. So moving to the stations does not take much time. Specifically, according to the respondents, the average time for the students in Athens to reach the Railway Station is 30′–45′, which coincides with the dominant response (45.7%). For the students in Thessaloniki, the average and dominant answer is 15′–30′ (48.7%). Finally, the train passenger does not need to be sufficiently early since the seats are numbered, and the seat number is listed on the ticket, so he/she needs to be at the station only 5′–10′ earlier to find his seat and arrange his luggage. To summarize, the total door-to-door journey travel time by train from Athens to Thessaloniki and vice versa is estimated to be 5 to 7 h.

The train ticket price is fixed, so the traveler does not need to have booked it a long time before, only so early that there is availability. It is even possible for students to adjust the price depending on the class they choose to travel, and on time as the evening train is cheaper. Thus, a student can pay €27.80 (first class) or €22.80 (second class) on a regular route if studying in Athens and staying in Thessaloniki and vice versa (50% discount) and €41.50 (first class) and €34 (second class) if he/she is just a student (25% discount). The evening fare is €17.60 (first class) or €12.60 (second class) if studying in Athens and staying in Thessaloniki and vice versa (50% discount) and €26.30 (first class) and €18.80 (second class) if he/she is just a student (25% discount). Regarding the door-to-station traveling, the average cost to the Railway Station in Athens is €3–6, with the most frequent answer being €0–3 (78.3%). The average cost to the Railway Station in Thessaloniki is €3–6, with the most frequent answer being €0–3 (78.7%). With all this in mind, traveling by train can cost around €19–40 for a student studying in Athens and staying in Thessaloniki and vice versa and €25–54 for the other students.

Finally, the train is one of the safest and most comfortable ways to travel. In the matter of comfort, the luggage and the cramping of the space can be considered as disadvantages. However, there are other amenities, such as the canteen service in the wagon and the power plugs on the seats. From the results of the questionnaire, reaching the Railway Station in Athens needs one (1) (29.5%) or two (2) (61.2%) means of transportation. It seems that traveling to and from the Railway Station in Athens can cause fatigue, which also affects travelers' comfort experience.

Travel by Intercity Bus

The distance from the Intercity Bus Station in Athens to the Intercity Bus Station in Thessaloniki is 495 km. The journey takes about 6 h, depending on the departure and arrival stops and traffic conditions. Frequencies are around nine (9) to ten (10) daily, and occasionally more are added during high demand. Apart from the main Intercity Bus Stations, there are also secondary stops both in Athens and Thessaloniki, namely "Pedion tou Areos" and "Monastiriou," respectively. One to two times a day, there are also other stops included: at the Port of Piraeus in Athens and at the Chalkidiki's intercity bus station in Thessaloniki, which undoubtedly serves many travelers. The Intercity Bus Stations in Athens and Thessaloniki are relatively easily accessible. The Intercity Bus Stations in Athens is not so central to the city and is only connected by urban bus lines. The solution for this is given by the "Pedion tou Areos" bus stop, which is in the center of the city and serves people mainly from the Northern Suburbs of Athens. In Thessaloniki, many bus routes end up at the Intercity Bus Station. For the intercity buses to/from Athens, there is a stop at "Monastiriou" bus stop, which is also not far from the Railway Station of Thessaloniki and its center. So, moving to the Intercity Bus Stations does not always take much time. Specifically, according to the respondents, the average time that students need to reach the Intercity Bus Station in Athens is 30′–45′, which coincides with the main answer (31.8%), and for the Intercity Bus Station in Thessaloniki the average and dominant answer is 30′–45′ (38.7%). Finally, the intercity bus passenger does not need to be far enough ahead of time as the seats are numbered, and the seat number is listed on the ticket, so he/she needs to be at the station only 5′–10′ earlier to find his seat and arrange his luggage. To summarize, the total door-to-door journey time by bus from Athens to Thessaloniki and vice versa is estimated to be 7 to 7.5 h.

The intercity bus fare is fixed, so the traveler does not need to book it long before, only as early as necessary for availability. There are no classes or price changes depending on the time of departure, and the student discount is 50% for every student, whether or not they study in one of the two cities. Thus, a student can purchase a student ticket for this route for €19, while considering the 10% discount for online purchases is becomes €17.10. The questionnaire also shows the average cost of traveling from home to the Intercity Bus Station in Athens to be €3–6 with the most frequent answer being €0–3 (63.6%). The average cost of a trip to the Intercity Bus Station in Thessaloniki is €3–6 with the most frequent answer being €0–3 (74%). With all this in mind, traveling by an intercity bus can cost from €23 to €31 for students.

Finally, the intercity bus is a relatively safe way of travel, but it does not offer much comfort. The tightness of the space, as well as maintaining the same position for a long time, are disadvantages that one might think before choosing the intercity bus.

However, other amenities are offered, such as that the drivers are loading and unloading luggage, and that luggage is locked in the track of the bus without the need for supervision. Moreover, there are a few power plugs and free Wi-Fi available.

Furthermore, the questionnaire shows that the majority of the students who responded need two (2) (44.2%) or three (3) (27.9%) different means of transportation to get to the Intercity Bus Station in Athens. It follows that traveling to and from bus stations in Athens can cause fatigue, which also affects travelers' comfort experience.

Travel by Private Vehicle

The center of Athens is about 502 km by car from the center of Thessaloniki. The duration of the journey depends solely on the speed that each driver drives, ensuring comfort and safety. If, for example, a driver drives at an average speed of 80 km/hr, the total time would be 6 h and 15 min, while if he/she drives at an average speed of 120 km/hr, then the total time would be 4 h and 23 min, not counting toll delays, etc. It is important to note that these times are measured without taking into account the human factor, which means that a driver is likely to need a 15′–30′ break in the middle of the trip, considering that truck drivers have to take a 45-min break after driving for 4.5 consecutive hours.

The cost of traveling by private vehicle is calculated as follows: The toll for a regular car is a total of €31.25. In addition, supposedly that the vehicle consumes 7.5 lt/100 km, it consumes about 38 lt of gasoline (that costs €1.58/lt) throughout the trip. So, overall, the cost is €90.71 (€31.25 for tolls and €59.46 for gasoline). Of course, these costs are also volatile and depend on the consumption rate of each car and fuel. For Liquefied Petroleum Gas (LPG), calculating that 8.25 lt/100 km is consumed, thus totaling 41.39 lt of LPG, and a cost of €0.80/lt, then the total travel cost for a car with LPG fuel will be €64.36 (€31.25 for tolls and €33.11 for LPG). For diesel, the journey amounts to approximately the same amount of money as the average consumption is 4.5 lt/100 km, that is in total 22.58 lt of diesel that costs 1.4 €/lt, resulting in a total cost of €62.86 (€31.25 for tolls and €31.61 for diesel). Finally, it is important to note that traveling by car is the only way of travel that offers a significant reduction in cost if there is more than one passenger. So if someone is traveling to/from Thessaloniki with four (4) more passengers and diesel fuel, it can only cost €12.60 if sharing the total cost, while passengers can set their departure, arrival, and brake time, and tailor the trip to their individual needs and preferences.

In terms of safety, traveling by car provides the least, since the human factor plays the most significant role, but this does not mean that this form of travel is considered unsafe. Regarding comfort levels, this can again be adapted by the passengers themselves, the driver, and the transport vehicle to the advantage that the passenger enters only in one means of transportation door-to-door.

Finally, the results of the questionnaire survey showed that from the 321 students who answered, 18 (5.6%) preferred riding alone, and 72 (22.4%) preferred to ride with passengers (the answers included the other means of transportation too).

Questionnaire Results on the Athens – Thessaloniki Connection

In the question which way of travel do they choose by having in mind all of their criteria, the 321 students answered as follows: 100 (31.3%) airplane, 92 (28.7%) train,

72 (22.4%) private vehicle with passengers, 39 (12.1%) intercity bus, and 18 (5.6%) private car as a sole passenger aka driver.

Summing It Up
In Fig. 1, the importance of the criteria when choosing the way to travel is displayed. In Table 1, a summary of all the information mentioned above is presented.

Table 1. Summary of time, cost, number of means, and comfort for all routes.

	To ATH	ATH to SGK	From SKG	Total	Comfort	Safety
Airplane	Airport	Airplane	Airport	Door-to-door	High	High
Average Time	45′–60′	60′–90′	30′–45′	3.5–4.5 h	–	–
Average Cost	€6–12	€45–200	€6–12	€60–215	–	–
No. of means	1–2	1	1	3–4	–	–
Train	Station	Train	Station	Door-to-door	Medium	High
Average Time	30′–45′	250′–330′	15′–30′	5–7 h	–	–
Average Cost	€3–6	€12.6–41.5	€0–3	€19–54	–	–
No. of means	2	1	1	4	–	–
Bus	Station	Bus	Station	Door-to-door	Low	Medium
Average Time	30′–45′	360′	30′–45′	7–7.5 h	–	–
Average Cost	€3–6	€17.1–19	€3–6	€23–31	–	–
No. of means	2–3	1	1	4–5	–	–
Car	Door	Private car	Door	Door-to-door	Low	Low
Average Time	–	4.5–6 h	–	4.5–6 h	–	–
Average Cost	–	€13–90	–	€13–90	–	–
No. of means	–	1	–	1	–	–

4 Conclusions

The cost of the trip is the decisive factor that most influences the choice of students, followed by the time of the journey, and in the third place is comfort. The time needed from home to station and vice versa and the connection between the two are less important, as only a minority of respondents answered "much" or "too much." Despite the high cost of traveling by airplane, most students preferred it, which at first sight is strange because the cost factor was found to be the most important. This finding seems contradictory, but it is not: in fact, students travel mainly by air when they have obtained a cheap ticket, either by purchasing it much earlier or by an offer. Otherwise, they travel by train, bus, or passenger car, which are cheaper solutions.

The vast majority of respondents reside in points associated with public transportation departures. However, in Athens, a not-insignificant 16.8% does not have this capability, and the corresponding rate for Thessaloniki's students is 9.6%, which is not low if we consider that the city is smaller, and the distances are shorter.

Due to the central position of the Railway Stations, both in Athens and Thessaloniki, the cost and duration of traveling to and from them are very low, which undoubtedly contributes to the increased attractiveness of train. Travel time from or to the airport is very important in Athens (63% need more than 45′) but significantly shorter for Thessaloniki (only 31.3% need more than 45′). The majority of Athenians use two (2) or more means of public transportation to reach the Athens International Airport (56%), Railway Station (70%), and Intercity Bus Station (75%). In Thessaloniki, this question was not asked because there is only one means of public transportation.

What is very interesting is the dispersion of the answers about the cost of moving from home to the departure station for the two cities. In Athens, the answers about the transport costs for the Athens International Airport, the Railway Station, and the Intercity Bus Station are evenly reduced as the prices go up. In Thessaloniki, the Thessaloniki Airport dominates with €0–3 (61.3%). However, the next largest response rate is €12–20 (18%), which is very different from the first. This is also the case for the Intercity Bus Station and Railway Station in Thessaloniki, with the most frequent amount being €0–3 (74% and 78.7% respectively) compared to the next one, which is €6–12 (10.7% and 10.7% respectively). This shows that in Athens the farther away from the station the amount is gradually increasing, while in Thessaloniki, unless one is served by the city bus, which is the only form of urban public transport of the city, there is no other option but to give a lot more money, probably for a taxi.

For the Athenians, the Intercity Bus Station proves not to be adequately accessible, with almost 40% of the respondents saying it takes more than 45′ to get there from their home. Similar is true for Thessaloniki, where 30% of the respondents need more than 45′ to travel to the Intercity Bus Station. This is comparable to the 37% for Thessaloniki Airport. Obviously, there is a need for major improvements in the connections with the Intercity Bus Stations, especially in Athens.

References

1. Kanellaidis, G., Abacoumkin, C.: Perceptions of safety of intercity travel modes using factor analysis methods. Transp. Q. **37**(3), 411–418 (1983). https://hdl.handle.net/2027/mdp.39015021808616?urlappend=%3Bseq=421
2. Tsafarakis, S., Gkorezis, P., Nalmpantis, D., Genitsaris, E., Andronikidis, A., Altsitsiadis, E.: Investigating the preferences of individuals on public transport innovations using the Maximum Difference Scaling method. Eur. Transp. Res. Rev. **11**(1), 3 (2019). https://doi.org/10.1186/s12544-018-0340-6
3. Nalmpantis, D., Roukouni, A., Genitsaris, E., Stamelou, A., Naniopoulos, A.: Evaluation of innovative ideas for Public Transport proposed by citizens using Multi-Criteria Decision Analysis (MCDA). Eur. Transp. Res. Rev. **11**(1), 22 (2019). https://doi.org/10.1186/s12544-019-0356-6
4. Papadima, G., Genitsaris, E., Karagiotas, I., Naniopoulos, A., Nalmpantis, D.: Investigation of acceptance of driverless buses in the city of Trikala and optimization of the service using Conjoint Analysis. Util. Policy **62**, 100994 (2020). https://doi.org/10.1016/j.jup.2019.100994
5. Athens Public Transport Organization Homepage. http://www.oasa.gr/. Accessed 30 Mar 2020

System Performance and Assessment

A Measure Generator Tool for Sustainable Urban Mobility

Athanasios Karageorgos, Giannis Adamos(✉), and Eftihia Nathanail

Traffic, Transportation and Logistics Laboratory, University of Thessaly, Pedion Areos, 38334 Volos, Greece
atkarage@gmail.com, giadamos@civ.uth.gr, enath@uth.gr

Abstract. The present paper details the development of a designated software that facilitates the selection of the appropriate measures to achieve high level sustainable urban mobility. The main objective of this tool is to combine the capabilities of the city, the geographical location of the properties throughout the city and the expectations of stakeholders (i.e. budget, time horizon, hardness and objectives) to come up with a list of realistic measures and policies that will be ready to be implemented. A pilot version of the tool has been developed using Microsoft Excel formulas and Visual Basic scripts. A case study for the city of Volos, Greece was undertaken to assess the executability, robustness and efficiency of the framework in real-life conditions. The pilot trial revealed that the developed SUM Measure Generator Tool can successfully extract a variety of measures and solutions addressing stakeholders' needs and their vision for the city.

Keywords: Smart cities · Transport solutions · Infrastructure · Operation · Computerized platform

1 Introduction

Nowadays, cities are inhabited by 55% of the world's population; this figure is estimated to increase up to 70% by 2050. High density urbanization has a detrimental impact on citizens' quality of life and health, environment, i.e. noise and air pollution, social interactions and economy. Approximately 70% of the European citizens live in cities that serve their daily transport needs through outdated infrastructure and transport system. Urban mobility is vital and yet responsible for about 40% of the CO_2 emissions and up to 70% for the rest of the pollutants. European cities are overwhelmed by the ever-increasing levels of transport demand and the necessity of strategic strengthening for transport systems is particularly demanded (United Nations Department of Economic and Social Affairs, 2018).

The principles of sustainable cities were identified in the early 90's with the publication of Green Paper on Urban Environment, emphasizing the need to identify possible lines of action to improve urban areas. The White Paper on Transport Policy, published in 2001, declared that sustainable development of cities should account for economic, environmental and social aspects as well (European Commission, 2001). In 2006, a review of the White Paper suggested that it would be beneficial for the

E. G. Nathanail et al. (Eds.): CSUM2020, AISC 1278, pp. 475–484, 2021.
https://doi.org/10.1007/978-3-030-61075-3_47

scientific community to study and exchange knowledge with regards to best practice and legislation on public transport services. To this end, better legal frameworks for public transport operations should be provided. The Green Paper on Urban Transport was adopted on September 2007 and intended to initiate a discussion on the basic issues of urban mobility, focusing on the primary standard: "*to be effective, urban mobility policies must be based on an integrated approach*" (European Commission, 2007). The Green Paper of 2009 highlighted the need for creating cities that are more accessible and secure, greener and smarter, i.e. addressing new technologies and intelligent systems (European Commission, 2009a). The same year, the Action Plan of European Commission on urban mobility introduced 20 measures for public authorities to promote sustainable mobility, among which are the following: development of Sustainable Urban Mobility Plans (SUMPs), provision of high quality and affordable public transport, information provision, safe and easy access to all, promotion of lower, cleaner and zero emission vehicles, internalization of external costs, optimization of existing infrastructure and exploration of new fund sources and Intelligent Transport Systems (ITS) use (European Commission, 2009b). Later, in 2011, the White Paper "Roadmap to a Single European Transport", was published setting goals up to 2050 targeting the reduction of greenhouse emissions by 80–95% below 1990 levels, introducing new and more efficient technologies for vehicles and traffic management and optimizing the performance of multimodal logistic chains. The main initiatives pointed by the White Paper is to establish procedures from SUMPs establishment, developing road user charging and access restriction schemes and best practice guidelines to better monitor and manage urban freight flows (European Commission, 2011).

The European Commission recommends that European cities of all sizes should adopt the concept of SUMPs. This concept establishes guidelines for integrated urban transport planning for sustainable cities, emphasizing on the reduction of emissions, congestion and delays, meliorating the quality of life and making transport services affordable and accessible by everyone. SUMPs should contain measures and policies that improve the efficiency and resilience of urban transport for an effective and sustainable outcome. Many projects and initiatives were launched (ELTIS, CIVITAS, CH4LLENGE, ADVANCE) challenging European cities in planning and implementing SUMPs. One of the main guidelines in developing SUMPs is to encourage citizens and all relevant stakeholders, to get involved in transport planning, evaluating and testing evidence and addressing uncertainty.

In the context of this paper, the planning process of a SUMP is discussed with particular focus on determining a proper selection of measures towards sustainable urban transport. As addressed by the "Measure Selection Manual" of the European Platform of Sustainable Urban Mobility Plans, measure selection is the process of indicating the most appropriate and cost-effective policy measures to achieve the SUMP's vision and at the same encounter potential barriers and difficulties. The selection of the proper measures is a rather challenging process. First of all, there is large number of measures that may be available to cities, including infrastructure improvement, provision of new public transport solutions, effective management of the existing road network and enhancement of public acceptance. At the same time, it is expected that stakeholders, i.e. politicians and decision makers form their own vision of the city, however it is commonly proved that their ideas or solutions are not the most cost-effective. On the other hand, the most cost-effective interventions usually require more time to be implemented, or they are lacking funding or consensus. Grouping individual measures is an important component of the successful performance of a SUMP, which also should be more than a wish-list of measures (CH4LLENGE, 2019).

It is commonly known that the importance of decision making and planning towards sustainable mobility is essential and has significant long-term impacts on economy, environment and society. Towards this direction, the aim of this paper is to introduce the architecture and pilot application of a software facilitating the optimum selection of sustainable urban mobility measures. The tool, hereinafter "SUM Measure Generator Tool", addresses the morphology and existing infrastructure of a city and incorporates them into stakeholders' vision to provide an individualized selection of measures, sorted by order of implementation and level of adversity.

2 Tool Framework

In the following paragraphs, the analytical framework used to develop the designated software is introduced. In particular, this section details the methods used to assess the structure of the dataset of measures, explains the categorization of criteria used to validate the measures and clarifies the main functions of the tool.

2.1 Database Development

Specific platforms and initiatives have been investigated in an attempt to construct a global and up-to-date dataset of measures. The main sources consulted in this research are the U.S. Department of Transportation, specifically the Smart City challenge (U.S. Department of Transportation, 2019), the SuM4All initiative by the World Bank (Sustainable Mobility for All, 2019) and the CiViTAS platform funded by the EC Horizon 2020 program (CiViTAS, 2019). The output of this review was a database that constitutes of 120 unique measure records, covering all the related solutions that can be ready for implementation towards sustainable urban mobility.

2.2 Data Classification

Having finalized the dataset, each measure is then assigned at each of the following categories: layer classification, stakeholder, time horizon, budget, transport mode, city morphology, existing transport infrastructure, other city features, type of implementation and objectives. A brief description of each category follows:

- Layer Classification: All measures are categorized according to the level of orderliness in the existing infrastructure, policies and assets of the city. Therefore, *"Establishing Groundwork"* refers to measures that should be implemented first by a city with minimum or none efforts towards sustainability. The *"Revitalization of City"* contains the state in which the measures implemented are targeting core solutions to a more sustainable model. *"State of Art Improvements"* refers to all the measures that are delivered with advanced technologies, meaning that the city is at a state where an investment in high-end solutions is affordable. A list of indicative measures per group and layer of classification is presented in Tables 1, 2 and 3.

Table 1. List of indicative measures per layer of classification: Establishing Groundwork.

Group of measures	Indicative measures
Electrification of vehicles (EVs)	EVs education campaigns and promotional events
	Priority parking for EVs at parking spots in the city
ICT systems	Free Wi-Fi on public transport
Intelligent, sensor-based infrastructure	Real-time parking availability for parking operators
	Pedestrian detection at bus stops, density sensors
Municipality policy strategies	Improve local access to transport hubs including bus and train stations, ports and airports
	Procure contractors on a competitive basis, using packaging of batches of projects to attract multiple capable contractors
	Develop tools for measuring the accessibility of different locations and evaluate how various transport and land use changes will affect accessibility
Partnerships	Strategic business models and partnering opportunities
Safety and security	Redesign public spaces to increase road safety around schools
Smart land use	Pedestrianize city centre roads
Urban analytics	Optimizing public transit routes
Urban deliveries and logistics	Off-peak hour last mile delivery distribution
	Designated parking and staging areas for freight vehicles
User-focused mobility services and choices	Rehabilitation of historic/touristic landmarks areas
	Exclusive parking areas for tourist coaches in the city
	Establish public transport services during the night in the city
	Collaboration with taxi drivers to assess a potential shared service framework
	Smart payment card to support integrated ticketing

- Stakeholder: Refers to which stakeholder could take the initiative to implement one specific measure. Stakeholders are divided into: *citizen, public authority, public transport operator, academia, private sector* and *supply chain stakeholder*.
- Time Horizon: Refers to the average time needed for the complete implementation of a specific measure. So, there are *1–3 years*, *3–5 years* and *5–10 years*.
- Budget: Refers to a span of budget required for the measure to be implemented. Low is 0–100.000€, Medium is 100.000€–1.000.000€ and High is more than 1.000.000€.
- Transport Mode: All measures are characterized by the transport mode they are referring to: *private car, freight (freight vehicles in general), bike-scooter, bus, train, walk, ship, boat, taxi, airplane* or in some cases *none* of them.
- City Morphology: This criterion assesses whether a measure can be implemented in a city or not, accounting for morphological restrictions. For example, if a city is

Table 2. List of indicative measures per layer of classification: Revitalization of City.

Group of measures	Indicative measures
Electrification of vehicles (EVs)	Solar panels for EVs charging
	Bus fleets' electrification
ICT systems	Development of fiber optic network
Intelligent, sensor-based infrastructure	Real time parking data
	Environmental sensors/road weather information systems
Municipality policy strategies	Build complete multimodal networks
	Support vehicle-sharing community networks
	Invest in railway and port electrification
Safety and security	Neutralize accident black spots
Smart land use	Expand bicycle network
Sustainable terminal design	Intermodality planning
	Environmental design requirements for terminal structures
Urban analytics	Congestion and traffic analytics
	Integration of traffic management systems
	Integrated data exchange
Urban deliveries and logistics	Car sharing, car pooling
	Park and ride facilities
	Delivery freight lockers space zones

Table 3. List of indicative measures per layer of classification: State of Art Improvements.

Group of measures	Indicative measures
Automation	Vehicle-to-Vehicle platooning of two trucks
	Loop low speed electric autonomous vehicles (AV) shuttles in the centre of the city
	Low speed electric AV shuttles in industrial areas
	Low speed electric AV shuttles in first-last mile to terminals
	EV automated low-speed shuttle vehicles for first-last mile public transit access
Connected vehicles	Vehicle-to-infrastructure technologies
	Dedicated short-range communications
	Mobileye technology on buses
	Transit signal priority
Electrification of vehicles (EVs)	Electric bus rapid transit service
	Solar and wind farms for charging electric buses
Smart land use	Bus rapid transit lanes
Urban deliveries and logistics	Last mile distribution with drones
	Weight-in-motion technologies

highland, a bike network would be very challenging or even impossible to be developed. City morphology is divided in *coastal city, mountain city, lowland city, island city, inland city, lake city, river city.*

- Existing Transport Infrastructure: Measures are classified based on the requirements in existing transport infrastructure to make the implementation possible. A city could already provide *road* network, *port, rail* network, *airport, inland waterways, pedestrian walkways, bike network* or *parking areas.*
- Other City Features: Some measures are referring to specific city features such as a city is *tourist centre, global city (economic term), historic centre, religion centre,* if there's university *campus, industry* activity, if a city is *natura* protected *city,* if *advanced telecommunications* have been developed and lastly if a city is an *agricultural city.*
- Type of Implementation: All measures are distinguished between *soft, hard or soft-hard* type. In particular, intangible infrastructure is defined as *soft* implementation, physical structural development is *hard* and a combination of both if characterized as *soft-hard.*
- Objectives: The objectives are drawn from the "Methodology for the 19 WBCSD-SMP2.0 Indicators" report by the World Business Council. They are presented as a comprehensive set spanning in four dimensions: Global Environment, Quality of life in the city, Economic Success and Mobility System Performance. The research carried out within Sustainable Mobility Project 2.0 resulted in a set of 19 indicators (World Business Council, 2020). After some adjustments, the final set of objectives of this research is the following:
  - Improve accessibility for mobility impaired groups
  - Increase accessibility to transport modes
  - Reduce air pollution emissions
  - Decrease noise hindrance
  - Improve safety (combining safety and fatalities indicators)
  - Improve access to mobility services
  - Improve quality of public area
  - Improve urban functional diversity
  - Reduce commuting travel time
  - Increase economic opportunity
  - Net public finance efficiency
  - Mobility space usage efficiency
  - Decrease emissions of greenhouse gases
  - Decrease congestion and delays
  - Improve energy efficiency
  - Increase opportunity for active mobility
  - Improve intermodal integration & connectivity
  - Increase comfort and pleasure
  - Improve security
  - Increase public awareness

2.3 SUM Measure Generator Function

The main purpose of SUM Measure Generator is to provide a list of proposed measures with respect to the user-defined inputs. A logical process is responsible for matching the user-defined inputs to the database of measure records describes before. The output of this process is a factsheet containing the list of suggested measures, along with some vital information on each one of them. The proposed measures enlisted in the factsheet are also sorted by priority of implementation, in order to achieve the most effective results (Fig. 1).

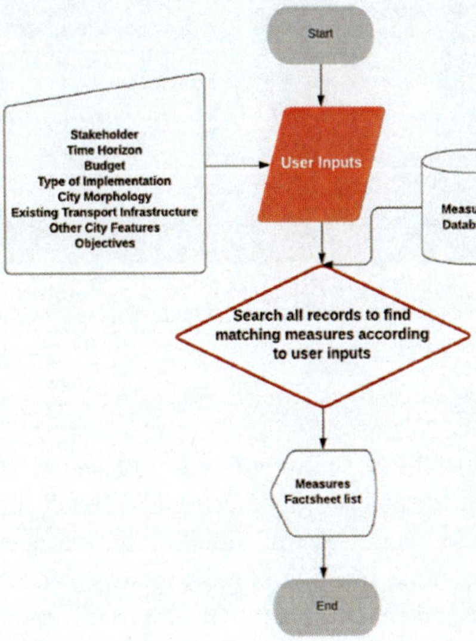

Fig. 1. Flowchart of measure generator tool.

3 SUM Measure Generator Development

A pilot version of the software is developed to visualize, analyze and modify the framework based on the results. Microsoft Excel functions and Visual Basic scripts were deployed to efficiently implement the framework.

3.1 User Interface

User-defined parameters are declared in a customized interface sheet. For each category, the user must select at least one of the criteria from the drop-down list. Given that at least one criterion has been selected for each category, the user can then press the "Generate Measures" button to launch the measure generator process and export the measures factsheet or press the "Clear Criteria" button to clear all fields and start again (Fig. 2).

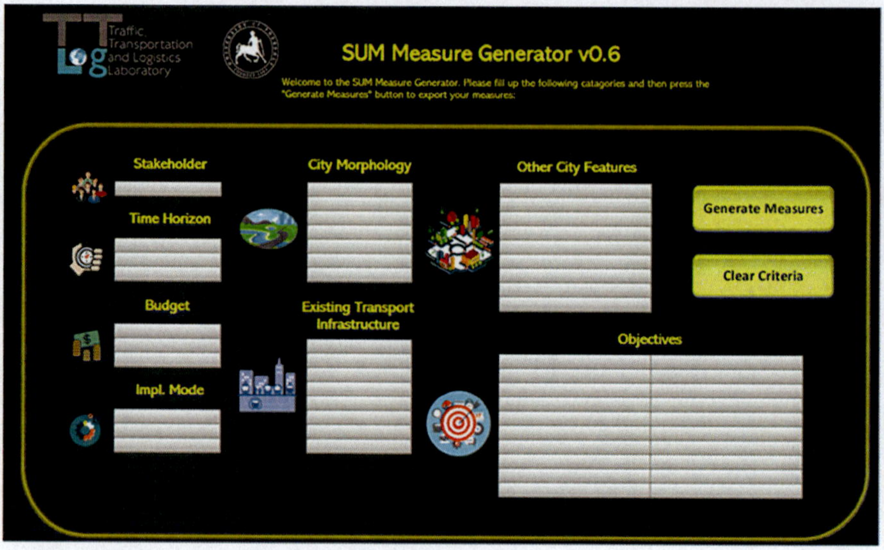

Fig. 2. User interface.

3.2 Code Design Analysis

In order to implement a matching search for each measure, from the database of measures, a combined formula was developed using MS Excel functions. This formula searches for the user-defined value of this field in the respective category of the database sheet. Running the same formula simultaneously for all of the categories of inputs, it can provide the factsheet with the "Groups" cell, the "Measures" cell, the "Layer Classification" cell, the "Transport Mode" cell, the "Time-horizon" cell, the "Budget" cell, the "Type of Implementation" cell and the "Objectives" cell. If the IF function doesn't find anything matching it returns "No" to each cell.

In order to run the formula for each row of the database, find and present only the found results, a macro is put in motion. This macro enables the generator tool to search all rows of the database, return the results to the factsheet, filter and delete the "No" rows, sort the final table by a layer classification order and wrap the text into the cells.

3.3 Measures Factsheet Interface

A custom factsheet containing the list of measures suggested by the SUM generator tool and a brief description of them has been designed. The main layout of this sheet is a table of contents that is filled with matching results from the macro execution. There is also a "Clear Measures Form" enabling the user to clear all output data and start again from the beginning with new inputs (Fig. 3).

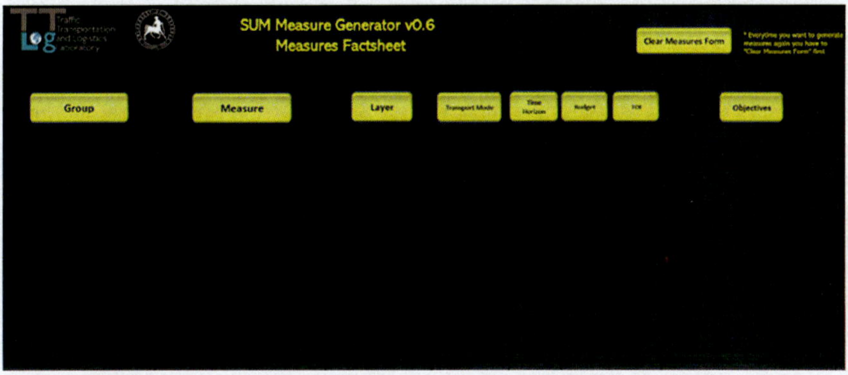

Fig. 3. Measures' factsheet interface.

4 Pilot Trial: City of Volos Case Study

In order to test the pilot model's efficiency, a trial run of SUM Measure Generator tool was executed. The user input value to the stakeholder field, is "Public Transport Operator", the Time-Horizon is "1–3" years, it is assumed that the stakeholder is focusing on "Low" and "Medium" Budget and "Soft" measures are chosen for the Type of Implementation. The city of Volos is "Coastal City", "Lowland City" and an "Inland City". The additional features of the city are that the city is "Tourist Centre", it has "Campus" and "Industry" and lastly it has "Advanced telecommunication" infrastructure. As for the objectives, the stakeholder is looking for "Access to mobility services", "Accessibility for impaired", "Air pollution emissions", "Comfort and pleasure", "Commute Travel Time", "Congestion and delays", "Energy efficiency", "Greenhouse emissions", "Intermodal integration & connectivity", "Mobility space usage", "Public Awareness", "Quality of public area" and "Urban functional diversity". The tool managed to extract all the valuable measure suggestions according to the inputs of the stakeholder user. Some good examples of this output are mentioned below:

- *Free Wi-fi on Public Transport*: This measure is an Information and Communication Technologies (ICT) Systems solution, it refers to "Establishing Groundwork" Layer of Classification, it can be implemented on "Bus" and "Train" transport modes and in "1–3" years, it has a "Medium" Budget value, it's a "Soft-Hard" measure due to intangible and physical infrastructure implementation and it can resolve "Access to mobility services", "Comfort and pleasure" and "Urban functional diversity".
- *Improve the quality and safety standards of public and private as well as formal and informal public transport operations, such as service frequency, reliability, cleanliness, and safe driving practices, and implement bus lanes and other bus priority measures*: This measure is a "Design and Deployment" solution, it refers to "Establishing Groundwork" Layer of Classification, it can be implemented on "Bus" and "Train" transport modes in "1–3" years, it has a "Medium" Budget

value, it's a "Soft-Hard" measure due to intangible and physical infrastructure implementation and it can address objectives "Access to mobility services", "Comfort and pleasure" and "Urban functional diversity", "Accessibility for impaired", "Affordability", "Mobility space usage", "Intermodal integration & connectivity" and "Commute Travel time".

- *Establish green procurement practices so that environmental criteria play a role in determining which firm gets a service contract*: This measure is an "Educate & Inform Strategies" solution, it refers to "Establishing Groundwork" Layer of Classification, it can be implemented on "ALL" transport modes and in "1–3" years, it has a "Low" Budget value, it's a "Soft" measure and it refers resolving "Public Awareness".

5 Conclusions

The pilot trial for the city of Volos validated that the SUM Measure Generator Tool can efficiently extract a variety of measures and solutions according to the stakeholders' needs and the specifications of the city. For further exploitation of the SUM Measure Generator Tool, the development can be accomplished by a web application form in order for users to get better access and experience. Additionally, each measure could be accompanied with a citation or a web link, allowing the user to investigate good practices or guidelines for implementing.

References

CH4LLENGE - Addressing the four key challenges of sustainable urban mobility planning. http://www.sump-challenges.eu/. Accessed 20 Oct 2019

European Commission: White paper – European transport policy for 2010: time to decide (2001)

European Commission: Green paper – towards a new culture for urban mobility (2007)

European Commission: Green paper TEN-T: a policy review (2009a)

European Commission: Action Plan on Urban Mobility (2009b)

European Commission: White paper – roadmap to a single European transport area "towards a competitive and resource" (2011)

United Nations Department of Economic and Social Affairs: World Urbanization Prospects - The 2018 Revision (2018)

Sustainable Mobility for All: Global Roadmap of Action – Toward Sustainable Mobility (2019)

World Business Council: SMP2.0 Sustainable Mobility Indicators – 2nd Edition (2020)

CiViTAS cleaner and better transport in cities thematic groups. https://civitas.eu/thematic-groups . Accessed 20 Oct 2019

U.S. Department of Transportation. https://www.transportation.gov/smartcity. Accessed 20 Oct 2019

Estimation of the Willingness to Pay for Road Safety Improvements and Its Correlation with Specific Demographic, Psychological, and Behavioral Factors

Evangelia Beli and Dimitrios Nalmpantis(✉)

School of Civil Engineering, Faculty of Engineering, Aristotle University of Thessaloniki, PO Box 452, 541 24 Thessaloniki, Greece
dnalba@civil.auth.gr

Abstract. This paper attempts to investigate the correlation of the willingness to pay for road safety improvements with several factors, such as gender, age, involvement in traffic accidents, etc. The willingness to pay shows the determination of people towards the goal of drastically reducing traffic accidents. In order to estimate the willingness to pay for road safety improvements, a contingent valuation questionnaire survey took place, in which 452 drivers of different age groups, economic statuses, and levels of education participated. The main aim of the survey was to investigate the correlation of the willingness to pay for the implementation of three scenarios of road safety improvements, in the form of traffic accident reduction rate, with several factors. The main statistically significant results are the following: a) in regard to gender, women would pay more for road safety improvements compared to men; b) in regard to mental state, people who avoid driving when they are affected by strong emotions, such as anger and distress, would pay more for road safety improvements; and c) in regard to age, as the age increases the self-reported driving ability increases as well, a finding that was not expected. Another result is that there are indications that people who have been involved in traffic accidents would pay more for road safety improvements compared to those that have not been involved and, moreover, the amount of money they would pay increases according to the severity of the accidents; nevertheless, this finding needs further research regarding its statistical significance.

Keywords: Contingent Valuation Method (CVM) · Traffic accidents · Gender · Age · Mental state · Experience

1 Introduction

Many factors may contribute to a traffic accident that sometimes may even be lethal. Road safety is a vital issue, and the absence of road safety measures would lead to an increase in road accidents. There are millions of accidents every year around the world, and, unfortunately, we face the need to quantify both the consequences of traffic accidents and the Willingness To Pay (WTP) of people to alleviate them with road safety improvements.

E. G. Nathanail et al. (Eds.): CSUM2020, AISC 1278, pp. 485–494, 2021.
https://doi.org/10.1007/978-3-030-61075-3_48

As traffic accidents are considered a leading cause of death worldwide, the European Union (EU) and the member states, as well as international organizations, take measures and create policies for road safety improvements.

The purpose of this paper is to estimate the WTP of people to improve road safety. The research that was carried out in this context attempted to assess the WTP of the participants and to find characteristics and factors that possibly affect it (e.g., age, level of education, parenting, previous experience of traffic accidents, etc.).

According to the World Health Organization (WHO) [1], 1.35 million people die every year globally, and almost 50 million are injured from traffic accidents. Traffic accidents are ranked among the 10 most frequent death causes. In 2000, traffic accidents were in 10th place, with 1.5 million deaths. In 2016, they had risen to 8th place, with 1.4 million deaths, of which 74% being men or boys [2]. Unfortunately, road accidents are the first cause of death among the ages of 5–29 [1]. If no immediate and effective action is taken, fatal traffic accidents are predicted to become the world's 5th leading cause of death, with 2.4 million deaths annually. The economic impact of road accidents is calculated at 1% to 3% of countries' Gross Domestic Product (GDP) (almost 500 billion dollars). Reducing traffic accident victims and fatalities will reduce hassle, "unlock" growth, and free resources for more productive use [3].

In 2016, 29% of deaths were due to traffic accidents, with low-income countries having the biggest mortality rate, with 29.4 deaths per 100,000 residents, while the global rate was 18.8 deaths per 100,000 residents. The rate for middle-income and high-income countries was lower. Especially in high-income countries, deaths from traffic accidents were not even ranked in the top 10 causes of death [2]. 93% of all global accident-related deaths take place in low- and middle-income countries, despite the fact that these countries possess 60% of the world's vehicles. In contrast to high-income countries, people that come from lower socioeconomic areas are more likely to be involved in road accidents [4]. In Greece, the respective rate in 2016 was 9.2 deaths per 100,000 residents [5].

Willingness to pay is an approach, which scientists started using extensively in the 1990s. In the case of road safety and according to the approach of this paper, the aim is to find the amount of money that people are willing to pay for road infrastructure improvements. Of course, exposing human life to danger is not something to be calculated exclusively in economic terms. Still, it is interesting to know how much people are willing to pay according to the probability of traffic accident reduction [6].

The method we used is the Contingent Valuation Method (CVM) method. According to Mitchell and Carson [7], CVM consists of questions, which try to elicit people's preferences for certain public goods through the detection of how willing they are to pay for specific improvements on them [7]. It is a simple and "flexible" valuation method that is used extensively in Cost-Benefit Analysis (CBA) studies and the evaluation of environmental impact [8].

In 2006, Hultkrantz, Lindberg, and Andersson evaluated human life, in terms of improved road safety in an urban area in Sweden, with the use of the CVM. That was achieved with the estimation of a private good's value (e.g., a device) and a public good's value (e.g., a road safety program). The research found that people were more willing to pay for a private good than for a public good [9].

This result was expected considering that, by definition, "public good" is a good or service that is both non-excludable and non-rivalrous, and this means that even a free-rider can use it without having to pay for it.

In 1988, the Department for Transport (DfT) of the United Kingdom (UK) changed the method that was used for the evaluation of fatal road accidents. From 1968, the evaluation was conducted based on the loss of productivity, the medical cost, and the estimation of human cost (e.g., pain, sadness, and hardship). Since 1988, the evaluation is conducted based on the WTP, namely the amount of money people are willing to spend for the reduction of the possibility of a fatal accident. According to this new method, direct cost (e.g., medical cost) was added to the amount of the evaluation of the WTP to calculate the total value of death prevention [10].

In 2016, Haddak, Lefèvre, and Havet researched the evaluation of non-fatal traffic accident injuries. In particular, they tried to estimate the WTP of French households to improve road safety and reduce the risk of non-fatal injuries resulting from traffic accidents. They found that the respondents' previous experience of traffic accidents seemed to play a significant role in strongly influencing the evaluation of non-fatal injuries [11].

In 2014, Ainy, Soori, Ganjali, and Baghfalaki tried to estimate the cost of road accidents with the WTP approach in Iran in 2013. The survey involved samples of people divided into four categories: a) passengers in vehicles, b) pedestrians, c) car drivers, and d) motorcyclists. The mean WTP was 2,612,050 Iranian Rials (IRR) ($\approx$\$62), and the statistical value of life was estimated, according to 20,408 fatalities, at approximately 402 trillion IRR ($\approx$\$9.5 billion). The cost of traffic accidents in Iran in 2013 constituted 6.64% of its Gross National Income (GNI), which was well above the world average [12].

2 Methods

A questionnaire survey was carried out, in which 452 people participated to gather the required data. The CVM questionnaire was a structured questionnaire developed by the authors. The response menu, depending on the questions, included multiple-choice questions, Likert scale questions, and slider questions.

The questionnaire was disseminated through social media (e.g., Facebook) and a students' e-mailing list of the Aristotle University of Thessaloniki (AUTh), asking the respondents to spread it even further. Thus, a modified snowball sampling technique was used for the dissemination of the CVM questionnaire.

The collected data was used to perform descriptive and inferential statistical analyses. For the descriptive statistical analysis, the "IBM SPSS Statistics 25" and "Microsoft Excel 2016" software programs were used to create graphs and tables and to calculate the percentages, means, and medians that are presented below. For the inferential statistical analysis, the same tools were used to correlate variables and evaluate their significance ($a = 0.05$) with bivariate analysis (Pearson's r was extracted and two-tailed t-test was carried out). Additionally, the Kruskal-Wallis test and Dunn's multiple comparison test were performed using the "GraphPad Prism 8" software package.

3 Results and Discussion

The first question was which the gender of the respondent was. The respondents answered as follows: 46.9% "men", 52.43% "women", 0.22% "other", and 0.55% "I prefer not to answer". Due to the very low percentages of the answers "other" and "I prefer not to answer", we did not consider these answers for the rest of our analysis. Regarding the respondents' age, most of them were between the age of 20 and 25 with the mean age being 33 years old.

The three main slider questions of the survey were about the amount of money that the respondents are willing to spend each year, for five (5) years, on three scenarios. The 1st scenario was about the implementation of a road safety program that would reduce the traffic accident rate by 30%, the 2nd scenario would reduce the traffic accident rate by 45%, and the 3rd scenario would reduce the traffic accident rate by 60%. The following Fig. 1, Fig. 2, and Fig. 3 show the relative frequency of every selected amount.

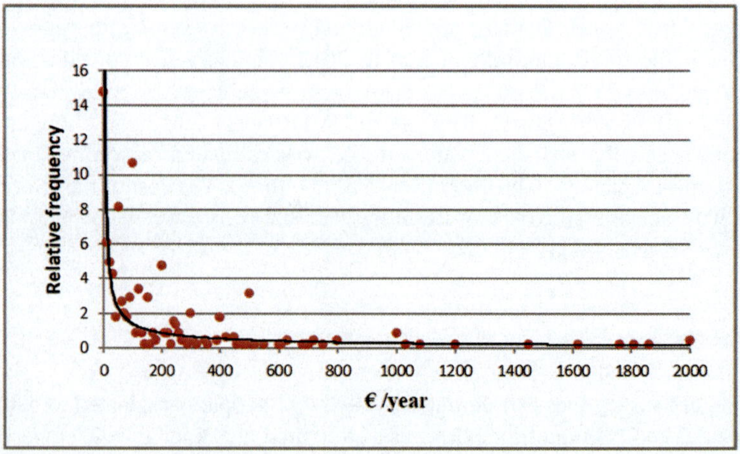

Fig. 1. The variance of the selected amounts of money (30% reduction in traffic accident rate).

In the 1st scenario, the average (mean) amount of money was €178.84 per year, the median was €100 per year, and the most selected answer was €0 per year (with a relative frequency of 14.80%). In the 2nd scenario, the average (mean) amount of money was €241.79 per year, the median was €120 per year, and the most selected answer was €0 per year (with a lower relative frequency of 11.10%). In the 3rd scenario, the average (mean) amount of money was €297.81 per year, the median was €170 per year, and the most selected answer was once again €0 per year (with a much lower relative frequency of 9.30%).

In order to prove that the means of the amounts of money "given" for the three scenarios are statistically significantly different, the Kruskal-Wallis test was carried out. Moreover, Dunn's multiple comparison test showed that each scenario is different from the others.

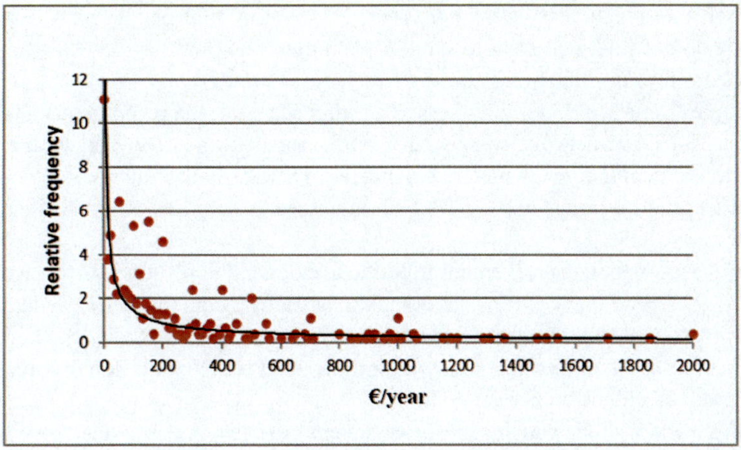

Fig. 2. The variance of the selected amounts of money (45% reduction in traffic accident rate).

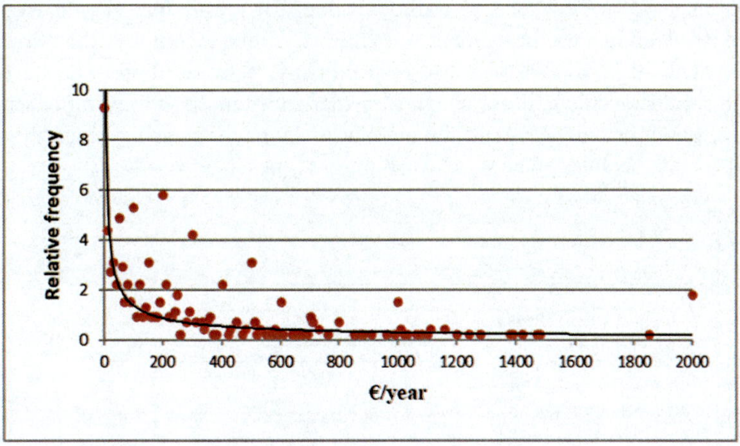

Fig. 3. The variance of the selected amounts of money (60% reduction in traffic accident rate).

Concerning the question about the region of origin, people from smaller areas (such as villages) tend to be willing to spend more money on road safety than people from larger areas (such as metropolises) (p-value = 0.02 and Pearson's r = −0.109).

Another question was about having an income or not. The percentage of people without an income was 40.27% and that of those with an income was 59.73%. These results were expected because many respondents were students who live from parental help and do not have their own income. From the analysis, we concluded that people without an income are likely to spend more money on the improvement of road safety than people with their own income.

Although this result may seem unexpected, we know that people who earn their own money do not find it so easy to spend it, and they feel that they have already paid their fair share through taxation.

At the question about having witnessed a traffic accident, the respondents answered as follows: 36.28% had never witnessed a traffic accident, 41.59% had witnessed at least one traffic accident with material damages, 18.58% had witnessed at least one traffic accident with injuries, and 3.54% had witnessed at least one traffic accident with casualties.

At the question about involvement in traffic accidents, the respondents answered as follows: 51.99% had never been involved in a traffic accident, 40.71% had been involved in at least one traffic accident with material damages, 7.08% had been involved in at least one traffic accident with injuries, and 0.22% had been involved in at least one traffic accident with casualties.

At the question about causing traffic accidents, the respondents answered as follows: 82.08% had never caused a traffic accident, 16.15% had caused at least one traffic accident with material damages, 1.55% had caused at least one traffic accident with injuries, and 0.22% had caused at least one traffic accident with casualties.

Considering the involvement in traffic accidents, it seems that people who have never been involved in a traffic accident are willing to spend less money than those who have been involved in at least one. More specifically, between those who have been involved in a traffic accident, those who have experienced more serious traffic accidents are likely to pay more money, with the amount of money increasing according to the seriousness of the accident. The downward trend is presented in Fig. 4.

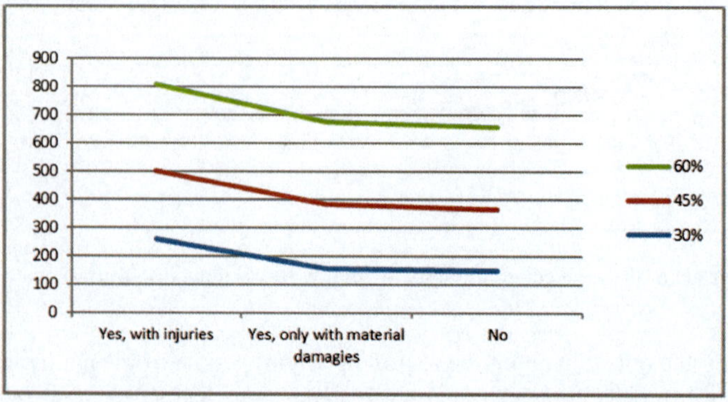

Fig. 4. The downward trend of traffic accident involvement impact per reduction rate scenario.

Additionally, there was a question about self-reported driving ability. 87.70% of the respondents stated that they think of themselves as above the average drivers, and the rest 12.30% as below the average drivers. More specifically, as the age increases, the self-reported driving ability increases too (p-value = $4.84 \cdot 10^{-10}$ and Pearson's $r = 0.287$). This result was unexpected and promising regarding the awareness of new drivers.

Another group of questions was about how some emotions (such as anger and distress) or situations (such us workplace confrontation, confrontation within the family, fatigue, and alcohol influence) would prevent the respondent from driving (Fig. 5). Alcohol influence is the situation under which most of the respondents would avoid driving (72.60%). The relation between avoiding driving due to anger and the amount of money willing to pay in the 3rd scenario (accident rate reduction by 60%) was found to be proportional and statistically very significant: as avoiding driving due to anger increases, the average (mean) amount of money willing to pay increases too (p-value = 0.003 and Pearson's r = 0.139). Regarding the 2nd scenario (accident rate reduction by 45%), the proportional increase was also statistically very significant (p-value = 0.006 and Pearson's r = 0.130). Similar statistically very significant results were also found in the 3rd scenario of avoiding driving due to distress (p-value = 0.008 and Pearson's r = 0.125).

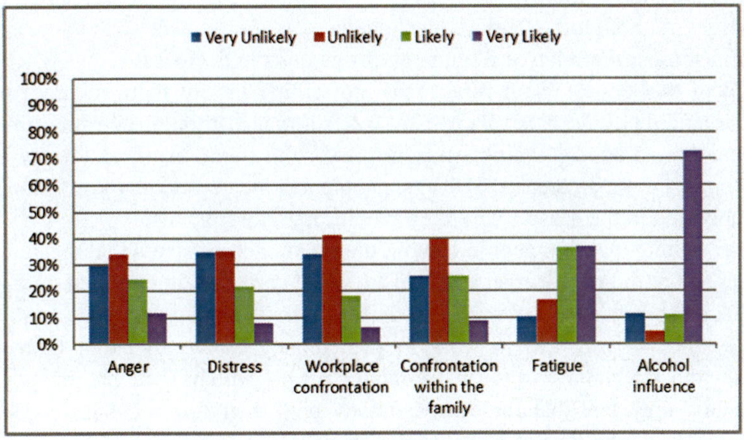

Fig. 5. Likelihood of avoiding driving per mental state.

When it comes to gender, in all three scenarios, women were willing to pay more money for road safety than men. In the 1st scenario, the average (mean) amount of money that women were willing to pay was €189.02 per year, while men were willing to pay €157.59 per year. In the 2nd and 3rd scenarios, the amounts of money willing to pay were €284.38 and €339.66 for women and €195.61 and €252.87 for men, respectively. The fact that women were willing to pay more money than men is statistically very significant in the 2nd scenario (p-value = 0.004 and Pearson's r = 0.134) and statistically significant in the 3rd scenario (p-value = 0.016 and Pearson's r = 0.113).

In the last question, we asked the respondents to state their preference for the implementation of three road safety programs. In this case, they had to rank them from 1 (most desirable) to 3 (least desirable). The 1st program would reduce the accident rate by 10%, and the required amount of money was 300 euros (one-time payment). The 2nd program would reduce the accident rate by 25%, and the required amount of money was 500 euros (one-time payment). The 3rd program would reduce the accident rate by 40%, and the required amount of money was 680 euros (one-time payment).

The top ranked choice was the 3rd program, as it was the first choice of the respondents. The 2nd program was the second choice of the respondents, and the 1st program was the third choice of the respondents. These are promising results!

4 Conclusions and Recommendations for Future Research

Many interesting conclusions can be drawn from our results. Starting with the three main slider questions, we calculated the average (mean) amount of money that people were willing to pay for a given accident rate reduction. For a 30% accident rate reduction, people were willing to pay €178.84 per year (for five years). We showed that as the accident rate reduction increases, the amount of money willing to pay increases too (€241.79 for the 2nd scenario and €297.81 for the 3rd scenario). It seems that people choose wisely. Although the "zero euros" choice was top-ranked, as the accident rate reduction increases, the relative frequency of this answer decreases.

We found that people from rural areas are willing to pay more money for road safety. This finding could be explained by the fact that in rural areas (due to the bad road's condition, limited policing, speeding, etc.) the traffic accident fatality rate is higher compared to urban areas [13]. Thus, people feel the need to be more "protected", and probably this is the reason why they would pay more money.

Another finding was that people without their own income stated that they are more likely to pay more money for road safety. That was a somehow unexpected finding, but the fact that most of these people were students and lived on parental help clears things up. People who do not have to work to gain money usually spend it more easily compared to those who have to work. Moreover, people who have an income pay taxes and, probably, they feel that they have already paid their fair share for road safety compared to those who do not have an income.

One of our most exciting results is the one where the selected amount of money was correlated to previous traffic accident experience. We found that people who had never experienced a traffic accident would pay less money than those who had been involved in one or more traffic accidents. Moreover, between those who had been involved in a traffic accident, those who experienced more serious accidents are more likely to pay more money. Probably, the reason for that is that those without previous experience of a traffic accident are more likely to believe that they will never be involved in one, as most respondents think they are better drivers than the rest.

In the analysis of the correlation of age and self-reported driving ability, we found that as the age increases, self-reported driving ability increases too. This finding is promising because it shows that some attitudes gradually change among the youth, and young people start to realize that they are not invincible behind the wheel, probably due to new approaches to driver training, as the use of simulators, etc. [14].

Another finding was about the correlation of avoiding driving per mental state and the selected amount of money willing to pay for three scenarios. First of all, alcohol influence was found to be the most deterrent factor, probably due to law enforcement and the fact that people can clearly understand that they are incapable of driving under the influence after many years of relevant road safety campaigns. People who stated that they are more likely to avoid driving when they are under the influence of strong emotions are willing to pay more. We can assume that people who know they are affected by strong emotions, such as anger and distress, feel more vulnerable to traffic accidents and, therefore, value road safety more. These people are more self-conscious and, therefore, more responsible for themselves and others.

Concerning the correlation between gender and money willing to pay for the scenarios, we found that women are willing to pay more money than men. In all three scenarios, both the average (mean) and the median of the money willing to pay were higher for women than men. That may be due to the fact that women tend to be more reserved, calm, and they think twice before an unsafe action, such as risky driving, and so they value road safety more than men [15].

About the last question of our survey, it seems that people prefer to give more money for the accident rate to be reduced more. The respondents chose wisely, as they selected the answer with the highest cost-benefit ratio. Considering the fact that Greece is in a straitened economic situation and the fact that most respondents were Greeks, it is really impressive that people chose the most expensive option, and that is an indicative sign of how people anticipate road safety improvements and consider them to be a crucial necessity.

Future research could include more factors (e.g., in the case of Greece the impact of the previous economic crisis on the WTP would be of high interest), retrieve more answers to have more statistically significant results, employ methods that reduce bias such as multicriteria analysis [e.g., 16, 17], conjoint analysis [e.g., 18–20], etc.

In conclusion, it is necessary to raise awareness on road safety issues through campaigns and, primarily, through education and training. Although Greece has made significant improvements in traffic accident reduction in recent years, it still has not reached other European countries' standards. Through the years, many road safety programs have been implemented in many European countries, and Greece follows with promising results. Nevertheless, road safety improvements should first start at the individual level and this is an investment worth paying for. As we found in our survey, people are more than willing to do so!

References

1. World Health Organization: Global status report on road safety 2018. World Health Organization, Geneva (2018). https://www.who.int/violence_injury_prevention/road_safety_status/2018/en/
2. World Health Organization Webpage: The top 10 causes of death, 24 May 2018. https://www.who.int/news-room/fact-sheets/detail/the-top-10-causes-of-death. Accessed 07 Feb 2020
3. World Health Organization: Global plan for the decade of action for road safety 2011–2020. World Health Organization, Geneva (2011). https://www.who.int/roadsafety/decade_of_action/plan/plan_english.pdf
4. World Health Organization Webpage: Road traffic injuries, 7 February 2020. https://www.who.int/news-room/fact-sheets/detail/road-traffic-injuries. Accessed 07 Feb 2020

5. World Health Organization Webpage: Global Health Observatory (GHO) data (2016). https://www.who.int/gho/road_safety/en/. Accessed 07 Feb 2020
6. Kourtis, E.M.: Modelling the economic impact of road accidents in Greece. Master's thesis, National Technical University of Athens, Athens (2018). https://www.nrso.ntua.gr/geyannis/wp-content/uploads/Kourtis-ad75.pdf
7. Mitchell, R.C., Carson, R.T.: Using Surveys to Value Public Goods: The Contingent Valuation Method. Resources for the Future, Washington D.C. (1989)
8. Venkatachalam, L.: The contingent valuation method: a review. Environ. Impact Assess. Rev. **24**(1), 89–124 (2004). https://doi.org/10.1016/s0195-9255(03)00138-0
9. Hultkrantz, L., Lindberg, G., Andersson, C.: The value of improved road safety. J. Risk Uncertain. **32**(2), 151–170 (2006). https://doi.org/10.1007/s11166-006-8291-z
10. O'Reilly, D., Hopkin, J., Loomes, G., Jones-Lee, M., Philips, P., McMahon, K., Ives, D., Soby, B., Ball, D., Kemp, R.: The value of road safety: UK research on the valuation of preventing non-fatal injuries. J. Transp. Econ. Policy **28**(1), 45–59 (1994). https://www.jstor.org/stable/20053023
11. Haddak, M.M., Lefèvre, M., Havet, N.: Willingness-to-pay for road safety improvement. Transp. Res. Part A Policy Pract. **87**, 1–10 (2016). https://doi.org/10.1016/j.tra.2016.01.010
12. Ainy, E., Soori, H., Ganjali, M., Le, H., Baghfalaki, T.: Estimating cost of road traffic injuries in Iran using willingness to pay (WTP) method. PLoS ONE **9**(12), e112721 (2014). https://doi.org/10.1371/journal.pone.0112721
13. National Highway Traffic Safety Administration: Traffic safety facts 2012 data: rural/urban comparison. National Highway Traffic Safety Administration, Washington D.C. (2014). https://crashstats.nhtsa.dot.gov/Api/Public/ViewPublication/812050
14. Nalmpantis, D., Naniopoulos, A., Bekiaris, E., Panou, M., Gregersen, N.P., Falkmer, T., Baten, G., Dols, J.F.: "TRAINER" project: pilot applications for the evaluation of new driver training technologies. In: Underwood, G. (ed.) Traffic and Transport Psychology: Theory and Application, pp. 141–156. Elsevier, Oxford (2005)
15. Beli, E., Nalmpantis, D.: Peer pressure on road safety: the case of the Aristotle University of Thessaloniki, Greece. In: Proceedings of the 8th Transport Research Arena (TRA 2020), Helsinki, Finland, 27–30 April 2020. https://doi.org/10.5281/zenodo.3708323. (Conference canceled)
16. Nalmpantis, D., Roukouni, A., Genitsaris, E., Stamelou, A., Naniopoulos, A.: Evaluation of innovative ideas for public transport proposed by citizens using multi-criteria decision analysis (MCDA). Eur. Transp. Res. Rev. **11**(1), 22 (2019). https://doi.org/10.1186/s12544-019-0356-6
17. Bhandari, S.B., Nalmpantis, D.: Application of various multiple criteria analysis methods for the evaluation of rural road projects. Open Transp. J. **12**, 57–76 (2018). https://doi.org/10.2174/1874447801812010057
18. Tsafarakis, S., Gkorezis, P., Nalmpantis, D., Genitsaris, E., Andronikidis, A., Altsitsiadis, E.: Investigating the preferences of individuals on public transport innovations using the maximum difference scaling method. Eur. Transp. Res. Rev. **11**(1), 3 (2019). https://doi.org/10.1186/s12544-018-0340-6
19. Tsoukanelis, A., Genitsaris, E., Nalmpantis, D., Naniopoulos, A.: Conjoint analysis for the optimization of a potential flexible transport service (FTS) in the region of Zagori, Greece. In: Nathanail, E.G., Karakikes, I.D. (eds.) CSUM 2018. Advances in Intelligent Systems and Computing, vol. 879, pp. 478–486. Springer, Cham (2019). https://doi.org/10.1007/978-3-030-02305-8_58
20. Papadima, G., Genitsaris, E., Karagiotas, I., Naniopoulos, A., Nalmpantis, D.: Investigation of acceptance of driverless buses in the city of Trikala and optimization of the service using conjoint analysis. Util. Policy **62**, 100994 (2020). https://doi.org/10.1016/j.jup.2019.100994

Evaluation of a Ride-Sharing Service in Thessaloniki: The Perspective of Both the Service Provider and the Users

Georgia Ayfantopoulou, Maria Natalia Konstantinidou, Neofytos Boufidis(✉), and Josep Maria Salanova Grau

Hellenic Institute of Transport- Centre for Research and Technology Hellas, Thessaloniki, Greece
{gea, mariakon, boufidisneo, jose}@certh.gr

Abstract. The concept of using shared transportation modes like bikes or cars, called as shared mobility, is a growing innovative trend in advanced cities. It aims at offering sustainable and effective transportation alternatives while improving air quality and congestion in urban environments. In the city of Thessaloniki, a shared mobility system was developed and tested for several months. This scheme is based on taxi/ride-sharing and offers shared trips from one suburb, Kalamaria, and one exurb, Thermi, to the city center and vice versa. The users are registered via a mobile app and are clustered into groups according to origin, destination and desired time of their trip. For each group, a trip is scheduled and broadcasted to the taxi association with all relevant details. The trip is assigned to a nearby vehicle, in order not to generate extra vehicle-kms. The service was monitored and assessed thoroughly by analyzing all the backend server data, a database with all requested and realized trips, taxi allocation details, notifications sent to users and drivers and delays with respect to the trip schedule. Further, floating car data are acquired from taxi OBUs and the users' smartphones enabling detailed analysis of trips. From the customer perspective, a service assessment regarding perceived quality, rate of satisfaction, problems or improvement suggestions was performed via questionnaires distributed to users.

Keywords: Ride-sharing · Data analytics · Service assessment

1 Introduction

In recent years the increased consumption of goods has raised concerns over ecological and societal impact. The need of people to be embedded in the local society and communal consumption created the concept of "collaborative consumption"/"sharing economy". This concept seems to provide an appealing alternative to consumers [1].

Sharing economy developments in the transport sector have important implications on the formulation of transport policies as they influence the future demand and are predicted to be one of the largest sectors in terms of revenues in many countries. Urban ride-sharing apps and parking sharing platforms are growing at over 35% per year in U.K. [2].

E. G. Nathanail et al. (Eds.): CSUM2020, AISC 1278, pp. 495–504, 2021.
https://doi.org/10.1007/978-3-030-61075-3_49

2 Literature Review

In this section the concept of sharing economy is supported by examples applied in the transport sector within Europe and around the world. Ride-sharing, car-pooling and car-sharing are the most famous types of passengers transport sharing applications. Ride-sharing typically refers to a service where an individual can order a ride through an online/mobile platform. The passenger travels sharing a private vehicle driven by its owner under a fee arrangement [3]. One of the most successful business models of ride-sharing companies is Uber.

Operating in more than 500 cities around the world, Uber service attracts a large number of travellers with the 16–34 age group to be the more frequent users [3]. Similar services appeared in the ride-sharing services such as the Lyft whose rides increases from 53 million in 2015 to 160 million in 2016 [3]. New companies try continually to enter this sector occupying a position in the market share (e.g. Sidecar in the U.S.A) [3]. Except of the start-ups, also well-established firms start to follow the "servitisation" trend offering transport as a service instead of selling vehicles. Rather than just selling products, Daimler and BMW, following on from Zipcar, turn beyond their traditional target groups by renting access to products [4].

Car-pooling is defined as every journey executed by a privately owned car with two or more people belonging to different households. The passengers share both the car use and the journey's expenses, but it is more an informal agreement than a typical commercial arrangement [3]. Since the number of cars used for each journey is decreased, the car-pooling scheme could have important implications on road traffic reduction. Conner-Simons [5] studying the effects of a taxi sharing scheme in New York based on the car-pooling concept predicted an about 75% reduction of taxi traffic. Other positive effects found in the literature of taxi fleets used in car-pooling services are the reduced emissions, the reduced fares and the more efficient routes [6].

Since the number of users is a critical factor of success for the car-pooling schemes, such services are typically implemented in densely populated urban areas where the number of potential users is larger. Brimont et al. [7] conclude that there is also a demographic distinction between the members of a car-pooling and a car-sharing system since the first ones are often blue-collar workers of lower income.

One of the world's largest car-pooling initiatives is BlaBlaCar founded in 2006. BlaBlaCar known as a long distance, peer-to-peer car-pooling service, currently employs over 600 people in 22 countries and counts about 35 million members. The company' model goes beyond the concept of car-pooling for short trips as the majority of its members are drivers who offer the available seats in their cars also for long trips. The BlaBlaCar's platform has registered users, who create a profile and can be assessed by the passengers as soon as the trip ends. Highly trusted members become "ambassadors".

There were examples of car-pooling services operated through bulletin boards and word of mouth before the Internet era, but the sharing models have been updated due to the responsiveness and the flexibility that information and communication technology (ICT) can provide. The matching of the demand and supply is completed through sharing platforms that operate as electronic marketplaces where a matching between riders and drivers is achieved in real time. The prediction of the demand and consequently the determination of travel patterns in transport sector are accomplished through location sharing services that allow users to check-in using GPS enabled devices [8]. Much of the data being generated as yet is not being exploited.

Opposed to car-pooling scheme, car-sharing does not require participants to travel together [9]. It involves the use of the vehicle on a short-term hire basis. The process of sharing can be operated by companies where vehicles are used by members or by individuals who hire ("share") their car for a fee at a minutely charge.

3 The Context and Challenge in Thessaloniki

Thessaloniki is the largest city in the northern part of Greece. The population in its greater area reaches the number of 1,006,730 citizens and its area coverage is 1,455.68 km2. The average density in Thessaloniki is estimated to 665.2 inhabitants per km2 and the total number of private vehicles including cars, heavy vehicles as well as motorcycles is higher than 777,544 [10]. Thessaloniki's critical role in the development of the social, financial, and commercial sectors in the national and the greater Balkan level can be explained by its geographical location and the existence of a developing transportation hub within the city's area.

3.1 General Mobility Context

Based on household phone surveys, 58% of the citizens in Thessaloniki have a driving license and 84% of the population owns at least one private car. Furthermore, each household is comprised of an average number of 3.03 persons and an average number of 1.75 driving license holders [11]. Figure 1(a) captures the car ownership percentages per number of owned cars. This results in 1.35 cars per household, which are used for the daily trips from/to work and leisure activities.

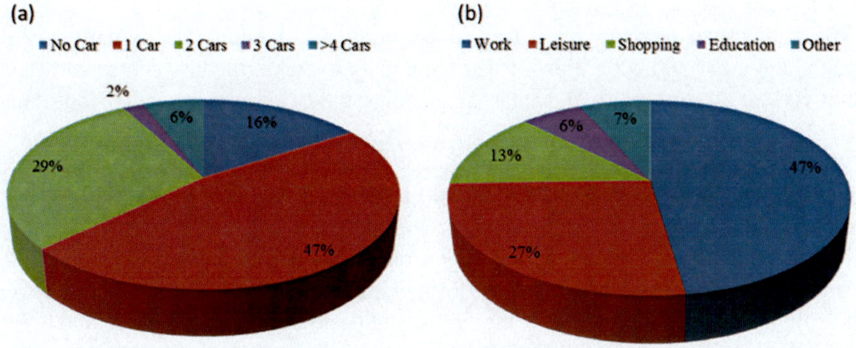

Fig. 1. Car ownership per number of owned cars in Thessaloniki (a) and trip purpose in Thessaloniki (b).

Each citizen of Thessaloniki executes an average number of 2.08 trips. 89.4% of the respondents answered that they usually execute one trip for work, education or leisure and one trip for returning home. As shown in Fig. 1(b), 47.6% of the trips are conducted for work and 26.8% for leisure. Among the other trip purposes, the results show 12.9%, 5.8% and 6.8% for shopping, education and other purposes respectively [11]. Most working places and leisure zones are located in the city centre.

Figure 2(a) presents the modal split analysis. 75% of trips is executed with private vehicles including private cars, motorcycles and taxis, while 23% is executed with public transport (PT) and 2% with non-motorized modes of transport (NMT) [11].

Based on the survey, the average vehicle occupancy is 1.44. As depicted in Fig. 2(b), 65% are single occupancy vehicles; while 35% of the vehicles travel with at least 2 passengers (28% and 6% of the vehicles travel with 2 and 3 passengers respectively, while 1% travels with more than 4 passengers). Regarding the distribution of vehicle type 77% are private vehicles, 5% are motorcycles, 2% are taxis and 16% are vans and trucks [11].

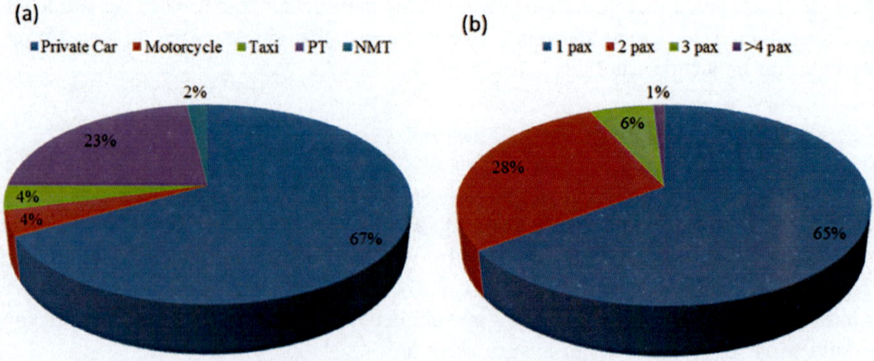

Fig. 2. Modal split (a) and vehicle occupancy (b) in Thessaloniki

The total travel demand for a typical weekday in all municipal departments is estimated around 1,300,000 vehicle trips. On a daily average, the city centre attracts a total of 10% of all trips, while the areas surrounding the city centre attract an additional 30% [11].

With regards to the temporal profile of the measured traffic volume, the peak traffic hours are observed between the 08:00–09:00 in the morning and 16:00–17:00 in the afternoon. The two peaks are quite concentrated (one hour of duration), and the congestion remains high between the two peaks. Concerning the morning peak hour traffic, the municipality of Thessaloniki is the origin area for 32% of all trips and more specifically the city centre (MD1) is the origin and destination area for 4.9% and 8.7% of all trips respectively [11].

3.2 Mobility Challenge in Thessaloniki

From the above facts it is clear that the municipality of Thessaloniki, and more concretely its city centre attracts a significant number of trips. In addition, there is a large amount of intra-commuting trips from people crossing the city centre in both directions. The longitudinal shape of Thessaloniki and its urban design ended up in four major axeσ, three of which cross the city centre.

These needs are mostly covered by private vehicles due to the strong car dependency of most Greeks and the low level of service of the Public Transport. As a result, the congestion levels of the city centre are high during most of the day.

The public authorities have tried to reduce congestion by improving the traffic lights system along one of the three arterials in 2012 with the installation of an adaptive traffic light system. In addition, the activities of the police are mostly concentrated in this axis aiming at reducing illegal parking, which reduces by 25% the capacity of this arterial almost during the whole day.

The concept of sharing economy seems to be a solution towards congestion decrease aiming at improving the use of resources, not only with regards to the existing road infrastructure, but also the individual infrastructure assets such as the vehicles themselves. At national level there are car sharing initiatives driven by the users themselves, while at local level, a few initiatives have already started. Since there is already the interest of commercial fleets to enter the sharing economy, the idea of shared-taxi is a clear opportunity to develop a new transport service in Thessaloniki.

The mobility eco-system in Thessaloniki is really mature; the Public Authorities collaborate actively with public and private stakeholders in implementing large-scale cooperative ITS pilots[1] in the city with the support of professional fleets; info-mobility services based on state-of-the-art technologies[2] which are largely used by the citizens of Thessaloniki have already been developed; a metro is under construction and will soon complement the public transport adding more competition, which will hopefully end up to an overall increased level of service. All these elements are basic for the preparation of a "Mobility as a Service" scheme in Thessaloniki, which is in line with the Sustainable Urban Mobility Plan of the city.

3.3 The Mobility Service in Thessaloniki

The developed shared mobility scheme is based on a taxi/ride-sharing service which started on May 2019 and aims at reducing the traffic congestion in the city center by reducing the commuting trips from 2 zones located at the eastern part of the city, one suburb and one exurb. A set of taxis of the TaxiWay taxi association collects users of

[1] COMPASS4D (http://www.compass4d.eu/), CO-GISTICS (http://cogistics.eu/), C-Mobile and SAFER-LC (started in May 2017).

[2] Mobithess (http://www.mobithess.gr/), easytrip (http://www.easytrip-project.eu/).

the Thessaloniki regions and transports them to the city center and back to their home aggregating as much as possible the trip origins and destinations aiming at concentrating trips in a few vehicles. The service is provided via a mobile application, developed by the Hellenic Institute of Transport (HIT). Users create their profile and indicate the days, times, and points of origin and destination of the desired route.

The computing system that hosts the mobility service clusters the users with matching trip characteristics (origin, destination and time). More precisely, the algorithm performs an automated, daily allocation of users in groups which are scheduled to travel together with the same taxi. This allocation occurs the day before the trip, at around 9 p.m. local time. The users who want to opt out and not be considered for next day's trips can update their preferences before the groups are scheduled. A scheduled trip may also be cancelled via the mobile application, so all scheduled trips are classified as either completed or cancelled.

4 Evaluation Results

4.1 Service Provider's Perspective

A database was developed to store non-sensitive and non-personal data generated by the users of the mobility service, in order to enable real time monitoring and also offline analysis and impact assessment. A web-based data analytics dashboard (https://galileoapi.imet.gr/dashboard) was developed for efficient monitoring of the system's operation and key performance indicators, for instance the number of active users and the total number of completed trips. The offline analysis investigated more complicated questions under the generic scope of assessing the service's impact, for instance analyzing user loyalty and trip quality indicators.

This analysis uses data captured from the first day of the pilot test on 16 May 2019, until 3 January 2020. Throughout this period 46 individuals were registered, however, one of them did not book any trip with the service, bringing the total number of users to 45. The registrations occurred over the pilot period, meaning that not all individuals were registered from day one.

We define a person-trip as a single one-way transfer of a person with the service, irrespective of the number of passengers. In total, 8.202 such trips were scheduled but 5.926 were completed, a percentage of about 72%. As depicted in Fig. 3, users are loyal to the service, as the vast majority of them realized more than 60 trips; 4 of them travelled at least 260 times. It is also found that the average user completes 7 person-trips per week, and the most frequent user averaged 10 person-trips per week.

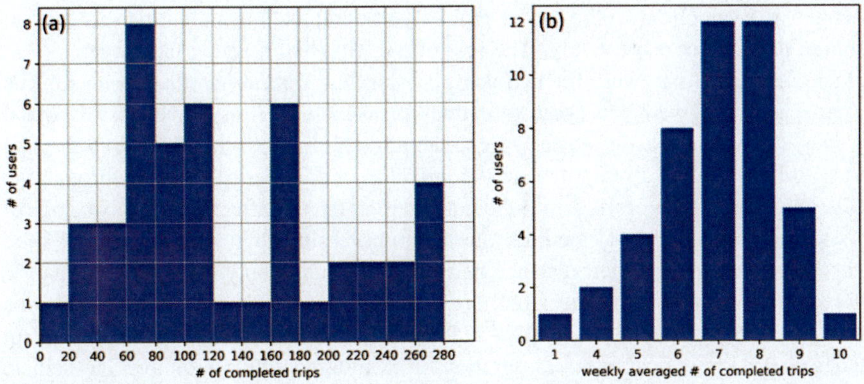

Fig. 3. Distribution of number of users with respect to completed trips (a) and with respect to the weekly averaged trip frequency (b).

Roughly the 28% of scheduled person-trips were cancelled. Figure 4 shows that the distribution of users with respect to cancelled trips is skewed right, which clearly indicates that a few users have cancelled several trips; six users are found to have cancelled between 100 and 160 each. Overall, the mean cancellation frequency amounts to 3 person-trips weekly.

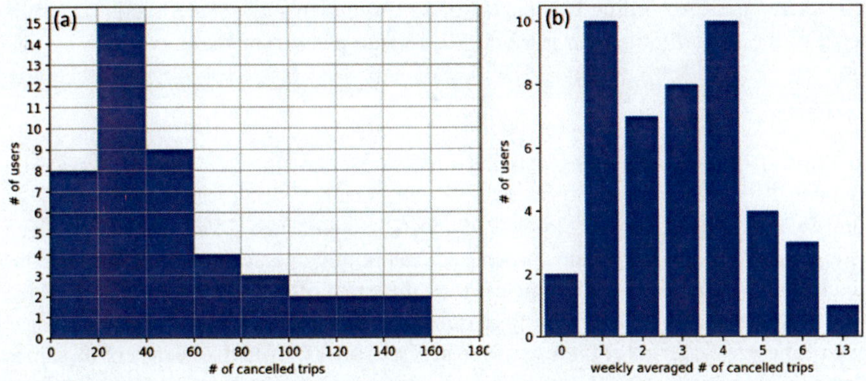

Fig. 4. Distribution of number of users with respect to cancelled trips (a) and with respect to the weekly averaged trip cancellation frequency (b).

The daily schedule of vehicle trips is produced during the allocation of passenger groups in vehicles. In case all group users cancel their scheduled trip, the vehicle trip is subsequently also cancelled. Throughout the study period, 4.457 vehicle-trips were scheduled, but approximately 80% of them were actually completed, since 873 were cancelled. Another negative effect of user-trips cancelation is the decrease of vehicle occupancy. The overall occupancy for all completed vehicle trips stands at 1.65

passengers per vehicle, with the 31.4% and 17% of completed vehicle trips serving two and three passengers respectively. The rest of the trips had a single passenger.

The completed trips were tracked using Global Navigation Satellite Systems. The raw geospatial data were processed to extract the vehicle trips as sequences of points. However, it was not feasible to fully generate all vehicle trips, mainly due to the fact that taxi drivers did not always declare the initiation or completion of a trip through the mobile app timely. Therefore, trip indicators were extracted after applying two logical filters and an outlier removal method. The filters necessitate the duration of trips to be at least three minutes and their distance between three and twenty km. The interquartile (IQR) statistical method to detect outliers was applied, which removes observations below $Q_1 - 1.5 \times IQR$ or above $Q_3 + 1.5 \times IQR$. The clean data reveal a mean trip distance of 9.5 km and a mean trip duration of 28 min, resulting in an average vehicle speed slightly above 20 km/h.

After the end of each trip, the users were requested to individually rate the service (five-star scale) and leave comments. It is found that a rating was issued in 53.8% of the total passenger trips, but, remarkably, only eleven comments were issued. Seven of those were very positive, stating that the service was either fine and as expected or even superb. The only negative comment reported a 15-min delay of the taxi. The rest of the comments were either neutral (with the text 'OK') or not relevant to the service rating. The average rating is 4.8 out of 5. Lastly, a short digital questionnaire was shared amongst the users after the completion of the pilot, in an attempt to compare travel time with the shared taxis service and privately owned vehicles. The users were requested to consider door to door travel time, including all delays for example parking or waiting for the shared taxi. According to the 87% of 23 respondents, the shared taxis service is faster. For the majority of them it saved 10–15 min per trip or more.

4.2 Users' Perspective

A final evaluation questionnaire was delivered to users of the taxi sharing service on December 2019. The sample consists of 32 participants (9 males and 23 females), of whom 28% belong in group "25–34", 47% in "35–44" and 25% in "45–54". The questionnaire doesn't investigate further socioeconomic characteristics. The questionnaire consists of questions about perceived quality, rate of satisfaction, problems faced and improvement suggestions. The final question aimed to investigate the willingness to pay of the respondents for the use of the service. It was formed as a matrix of values and response categories and the participants are asked to evaluate the various amounts (from 1 to 25€) that are willing to pay for the service in a four Likert scale ("Too cheap", "Cheap", "Expensive", "Too expensive").

The time saving in their daily trips is chosen by the majority of the users (75%) as the main reason for participating in the pilot taxi-sharing service while the cost saving, the congestion reduction and the ability to do other things during the trip are stated by 10%, 3% and 3% of the respondents respectively. Among the other reasons that are stated by 9% of the participants are the low quality of the public transport, safety and other personal reasons. Regarding the main disadvantages of the service users stated the cost (34%), the non-dynamic booking of the service (28%) and the limited availability of the taxi vehicles (16%). More than 90% of the users are satisfied or very satisfied with the mobile application, drivers' behavior, the time punctuality and the reliability of the service.

In terms of the most common problems that the users had to deal with, 44% stated some malfunctions of the mobile application, 22% stated non timely assignment of the route to a vehicle while 25% stated that they faced no problem during the service's use. All the users stated that the responsiveness of Galileo team to problems occurred during the pilot operation was very high and that they are very satisfied (81%) or satisfied (19%) with the overall experience of taxi sharing use. An improvement suggestion mentioned by some of the users is the appearance of the drivers' phone number in the mobile application in order the communication with the drivers to be facilitated in cases of unexpected events such as the time delays. Finally, 91% of the participants are interested in using the taxi sharing service after the end of the pilot period.

For the estimation of WtP in the final question a van Westendorp Price Sensitivity Model was developed [12]. The answers of each of the response category ("Too cheap", "Cheap", "Expensive" and "Too expensive") are graphed and the intersections points of the four curves reveal interesting findings for the respondents' willingness to pay. The intersection points of "expensive" and "too cheap" and "cheap" and "too expensive" curves are identified as Point of Marginal Cheapness (PMC) and Point of Marginal Expensiveness (PME) respectively and the range between these two points is the Range of Acceptable Prices (RAP). The range from 2.00 to 3.50€ is defined as the Range of Acceptable Prices for the use of the taxi-sharing service.

5 Conclusions

Thessaloniki is a typical example of a city with a relatively low level of public transport service and a strong dependence of citizens to their privately owned vehicles. The city centre attracts a very large number or trips on a daily basis, therefore a shared-taxis mobility scheme was designed as an alternative transport mode for citizen travelling there.

Several metrics converge to the fact that the attempt was successful. Over several months of testing, an increasing number of users participated in close to four thousand vehicle trips. Almost 85% of them completed 60 to 280 trips and the average frequency was seven trips per week. The overall user rating is 4.8 out of 5.

Survey analyses indicate that the level of service is sustainable in terms of both cost and time. The service was overall faster than a privately owned vehicle and the WtP results conform to a real-world share taxis implementation cost.

The research team will strive to support the shared vehicle scheme in the future, enrolling more passengers. The main future objective is to improve the vehicle occupancy in order to maximize the level of service per vehicle-km. This can be achieved in numerous ways, for instance with incentives for loyalty and for users who rarely cancel scheduled trips.

In addition, statistical methods, for instance Multiple Linear or Logistic Regression models, can be applied to model the survey responses and provide a better understanding of the contribution of each variable.

Acknowledgments. The research was inspired by the pilot implementation of a new taxi-sharing service developed under the European funded project Galileo 4 Mobility - Fostering the adoption of GALILEO for Mobility as a Service. The project has received funding from the European Global Navigation Satellite Systems Agency under grant agreement No 776381. The authors also express their gratitude to the participants who devoted time to fill out the questionnaires which provided the data that enabled the realization of this research. This study could not be realized without the effort and cooperation of TaxiWay taxi association and its drivers.

References

1. Hamari, J., Sjöklint, M., Ukkonen, A.: Sharing economy: why people participate in collaborative consumption. J. Assoc. Inf. Sci. Technol. **67**(9), 2047–2059 (2016)
2. PWC: Shared benefits (2016). https://www.pwc.co.uk/issues/megatrends/collisions/sharingeconomy/future-of-the-sharing-economy-in-europe-2016.html. Accessed 27 July 2020
3. Standing, C., Standing, S., Biermann, S.: The implications of the sharing economy for transport. Transp. Rev. **39**(2), 226–242 (2018)
4. Cusumano, M.A.: How traditional firms must compete in the sharing economy. Commun. ACM **58**(1), 32–34 (2014)
5. Conner-Simons, A.: Study: carpooling apps could reduce taxi traffic 75%, News (2016). https://www.csail.mit.edu/ridesharing_reduces_traffic_300_percent. Accessed 27 July 2020
6. Zhang, D., He, T., Zhang, F., Lu, M., Liu, Y., Lee, H., Son, S.H.: Carpooling service for large-scale taxicab networks. ACM Trans. Sens. Netw. **12**(3), 1–35 (2016)
7. Brimont, L., Demailly, D., Saujot, M., Sartor, O.: The new collaborative mobility actors: from promises to challenges for public authorities. In: Study. New Prosperity. 2nd ed., pp. 1–36. The Institute for Sustainable Development and International Relations (IDDRI), Paris (2016)
8. Chen, Z., Schintler, L.A.: Sensitivity of location-sharing services data: evidence from American travel pattern. Transportation **42**(4), 669–682 (2015)
9. Circella, G., et al.: What effects U.S. passenger travel? Current trends and future perspectives, pp. 1–76. California Department of Transportation and National Centre for Sustainable Travel, California (2016)
10. Census Data: (2012). www.statistics.gr. Accessed 27 July 2020
11. Mitsakis, E., Stamos, I., Salanova Grau, J.M., Chrysochoou, E., Iordanopoulos, P., Aifadopoulou, G.: Urban mobility indicators for Thessaloniki. J. Traffic Logist. Eng. **1**(2), 148–152 (2013)
12. VanWestendorp, P.: NSS-price sensitivity meter (PSM)-a new approach to study consumer perception of price. In: Proceedings of the ESOMAR Congress, Venice (1976)

Discussing the Role of Traffic Safety in Sustainable Urban Mobility Plans Using Spatial Analysis Techniques

Panagiotis Tzouras, Stefanos Tsigdinos(✉), Christos Karolemeas, and Efthimios Bakogiannis

National Technical University of Athens, 15780 Zografou, Greece
stef.tsigdinos@gmail.com

Abstract. In recent years, EU has placed great importance on the safety of road users (real and perceived). In this context today in Greece, around 180 municipalities are implementing SUMPs for first time and therefore a primary identification of the hazardous points is made through the mapping of traffic incidents. This article presents the results of the mapping of traffic accidents in five metropolitan municipalities of Attica (Athens, Piraeus, Marousi, Kifisia, Chalandri), and their analysis using GIS tools. Network kernel density analysis was performed to determine the spots where a high concentration of severe accidents appeared, as well as spatial autocorrelation using Moran index and Hot-Spot Analysis in terms of time, driver's age and type of vehicle involved. The results indicate the hazardous points in the study areas and their particular characteristics. Finally, it is noted that the aforementioned analysis can contribute to the design of feasible solutions in order to improve road safety and at the same time, create a safe and sustainable transport system in each of the study cities.

Keywords: Traffic safety · Spatial analysis · SUMPs

1 Introduction

Contemporary cities face several serious challenges questioning their sustainability and efficiency [1]. The car-oriented approach that influenced transport planning process in cities mainly after 1950s [2], resulted in urban systems with intense traffic congestion, urban sprawl, environmental pollution, inaccessible and unattractive public spaces, movement barriers (e.g. arterials passing through central areas) and especially road accidents [3–5]. Focusing on road accidents and crashes, we should strongly mention that in recent years European Union (EU) has placed great importance on road users' safety (real and perceived), as it has considerable influence both in the operation of transport systems and in citizens' quality of life [6]. Tellingly, according to the European Commission [7], more than 35.000 people were killed and more than 1.5 Mio. Were injured in the year 2009 in European countries. Focusing on Greece, in the year 2017, Hellenic Statistical Authority (ELSTAT) recorded approximately 750 deaths, 700 serious and 12.200 minor injuries in approximately 10.800 incidents [8].

E. G. Nathanail et al. (Eds.): CSUM2020, AISC 1278, pp. 505–514, 2021.
https://doi.org/10.1007/978-3-030-61075-3_50

Under these circumstances, a new perspective regarding cities and their future has emerged during recent years; this concept is no other than Sustainable Mobility [9]. Sustainable mobility describes a system that meets transport needs through promoting public transport, walking and cycling and, at the same time, enhances environmental integrity, social equity and economic efficiency. A new type of strategic plan, proposed by EU, that integrates all possible measures pointing towards an alternative mobility level, is the Sustainable Urban Mobility Plan (SUMP) [10]. It is worth mentioning that road safety is an integral part of sustainable mobility [11, 12] and SUMPs in particular. Today in Greece, around 180 municipalities are implementing SUMPs; by mapping the traffic incidents, the hazardous points are identified for the first time.

In this context this research outlines the results of the mapping procedure of traffic accidents in five metropolitan municipalities of Attica (Athens, Piraeus, Marousi, Kifisia and Chalandri) and their analysis using GIS tools. Network kernel density analysis was performed to determine the spots where a high concentration of severe accidents appeared, as well as spatial autocorrelation using Moran index in terms of time, driver's age and type of vehicle involved. The data of the research were derived from Hellenic Statistical Authority (ELSTAT) and concern the years 2012–2017.

2 Literature Review: Spatial Analysis of Traffic Accidents

A plethora of researchers have invested great efforts in studying the impacts of various risk factors [e.g. 13, 14] and road safety measures [15] and have developed or adopted several mathematical methodologies to approach crash prediction problems [e.g. 16] or road-road safety site prioritization problems [e.g. 17]. In the aforementioned context, in order to decrease traffic accidents and improve road safety levels, it is vital to understand how, where and when traffic accidents occurred, as these accidents are subject to both spatial and temporal variations [18]. A thorough understanding of spatial patterns can contribute significantly to tackle accidents effectively [19, 20]. Spatial analyses in road safety usually include the examination of crashes through taking into consideration their absolute or relative locations [5]. The way in which researchers select and define the spatial units of analysis influences directly the aim of the study, the interpretability of results and the data preparation as well [21].

Spatial analysis of point events, known as point pattern analysis (PPA), has been broadly examined by spatial scientists and a notable variety of methods have been developed for detecting high-risk sites (i.e. "hot spots"). The PPA methods can be classified into two broad categories [22, 23]. The first category includes methods which examine the first-order effects of a spatial process and the second category contains methods that examine the second-order effects of a spatial process [19]. Precisely, the first group focuses on the underlying properties of point events and measures the variation in the mean value of the process. It involves methods such as Kernel Density Estimation, Quadrat Count Analysis, etc. The second group mainly refers to the spatial interaction (dependency) structure of point events for spatial patterns, and includes methods such as Spatial Autocorrelation, Nearest Neighbor Statistics, G function, F function, K function, etc. In this research we focus on Kernel Density Estimation and Spatial Autocorrelation methods:

Kernel Density Estimation (KDE) is one of the most popular methods for analyzing the first order properties of points' distribution [22, 24]. It should be noted that this method is not a direct analytical method; on the contrary it constitutes an interpolation technique [25]. The KDE estimates the density within a range (bandwidth) of each observation to represent the value at the center of the window. Within the bandwidth, the KDE weighs nearby objects more than far ones based on a kernel function. The KDE generates a density of the events (discrete points) as a continuous field (e.g., raster). By using the density (or average attributes) of nearby objects to represent the property at the middle location, the KDE captures the very essence of location: it is not the place itself but rather its surroundings that make it special and explains its setting [26]. Delmelle et al. [27] mentioned that the KDE is an approach that leads to intuitive visualization and exploration of data.

The planar KDE has been used widely for traffic accidents "hot spots" analysis and detection. Some examples include: study of urban cyclists' traffic hazard intensity [28] wildlife-vehicle accident analysis [29, 30], pedestrian crash zones detection [31] highway accident "hot spot" analysis [32], etc. However, Kernels are projected over 2-D spaces, while road crashes occur in the road network which is a 1-D linear area [19]. Hence, in order to overcome this discrepancy, KDE has been expanded to network KDE approach [33] in which the network is now represented as fundamental units of equal network length (termed pixels). The studies utilized this method [e.g. 19, 34–36] concluded to results regarding each segment of the network, thus achieving greater accuracy.

In simple terms, spatial dependence essentially refers to events at a location (e.g. road crashes) being highly influenced by events at neighboring locations. It is typically measured via spatial autocorrelation metrics [5]. According to Griffith [37] spatial autocorrelation is the correlation between the values of a variable, which derives strictly from the proximity of these values in the geographic space. This approach is contradicting the general assumption of independent observation in the domain of classic statistics. Another similar definition proposed by Legendre and Legendre [38] mentions: "Spatial features have the tendency to be dependent. A phenomenon known as spatial autocorrelation can be defined as the property of random variables to take values over distance that are more similar or less similar than expected for randomly associated pairs of observations, due to geographic proximity".

Spatial autocorrelation can be used to describe and compare the spatial structure of a variable [39]. Specifically, this method contributes to [40]: a) test on model mis-specification, b) measure of the strength of the spatial effects on any variable, c) test on assumptions of spatial stationarity and spatial heterogeneity, d) identification of spatial clusters, e) understand the influence that the geometry of spatial units has on a variable, f) test on hypotheses about spatial relationships, g) identify the outliers, both spatial and non-spatial.

3 Methodology

This study adopts a quantitative approach in order to illustrate and analyze the spatial dimension of traffic accidents comprehensively. For this purpose, we used two different methods in order to obtain more efficient and representative results of the existing situation. The methods were Kernel Density Estimation (KDE) and Spatial Autocorrelation. It should be mentioned that this approach ensures a wide view of different aspects concerning the spatial dimension of accidents (e.g. clustering, interactions, etc.). After the application of the aforementioned methods, the paper presents these findings in the form of diagrams, and maps, thus evolving the understanding of the current usage patterns. The secondary data used were derived from Hellenic Statistical Authority (ELSTAT) and refer to traffic accidents concerning the years 2012–2017 in 5 municipalities in the metropolitan area of Athens. The database obtained included attributes regarding the location and the time of the accident, as well as the identity of street users involved. We should note that a pre-process was essential, in order to organize, and validate the accuracy of the data.

Firstly, the purpose of using KDE is to identify the hazardous locations (i.e. clusters) in the road network. It is essential to note that for the estimation of the planar Kernel density in each point of the grid two basic parameters are required, namely: kernel function (k-function) and bandwidth (i.e. search radius). The purpose of the k-function is to measure the so called "distance decay effect" [41]. In this study, the Bitweight function, which resembles better the Gaussian distribution function, is used in all the kernel density estimations. Regarding the estimation of kernel densities in the network space, we used the same general function. In our approach, the network kernel function is defined for two cases: (1) the location s does not coincide with a node and (2) the location coincides with a node [33, 42]. The selection of the right bandwidth is surely not an easy process. The range of the commonly used bandwidths starts at 50 m for urban areas and goes up to 500 m for highways [36]. In this study, we tested 6 different values, i.e. 50, 100, 200, 300, 400 and 500; by observing visualized output maps, the bandwidth of 200 m is selected and recommended for the identification of hazardous locations in the road network. Another important parameter for the network KDE is the maximum segment length (or pixel size). In the SANET tools manual, a rule of thumb is mentioned and recommended to the users; according to it, the pixel size has to be 10 times smaller than the selected bandwidth. In this study, this rule was respected; therefore, the maximum segment length is equal to 20 m.

Finally, another important issue that has puzzled researchers is related with the selection of the right density threshold(s) [35]. Initially, all the accidents and the road networks from all the municipalities were imported in QGiS and kernel densities were computed using v.kernel tool from GRASS GIS. At next, descriptive statistics, of kernel densities were estimated; by performing a X^2 test; it was observed that for a confidence interval of almost 95%, the density observations are gamma distributed. Using the cumulative probability function of Gamma distribution, we approached and finally determined the values of four thresholds. If the density at location s is higher than the first, second, third and fourth threshold, then this density value is also higher than the 85%, 90%, 95% and 99% of all density values, respectively. In the second stage, a dataset of densities was created for each municipality and for each road hierarchy level. Following the aforementioned approach, for each municipality and for each hierarchy level, four thresholds were determined.

Secondly in our analysis for spatial autocorrelation we use two different methods in order to produce better and more cohesive results. The first method calculates the index of Moran's I (Global and Local), which is a common statistical diagnostic tool that is defined as the division of the spatial variance with the total spatial variance of each variable [43, 44]. The main contribution of Moran's I is the evaluation whether a spatial pattern is clustered, random or dispersed. It works on both feature locations and features values simultaneously. The second method is the Hot-Spot Analysis which calculates the Getis-Ord Gi* statistics [40]. Specifically, this tool works by looking at each feature within the context of neighboring features. It identifies statistically significant hot spot as a location having high value and surrounded by high valued neighbors as well.

In both methods (applied in ArcGIS environment) we used the same parameters, thus ensuring the adoption of a common view between the results of each method. Precisely, for the conceptualization of spatial relationships, the inverse distance method was selected, as the closer two traffic accidents are in space, the more likely they are to interact with each other. Regarding the distance method; Manhattan distance method was chosen, since it is more appropriate for urban environment and when travel is restricted to a street network. Finally, the threshold distance was set to 3200m based on the size of the municipalities.

4 Results

The main results from the network kernel density estimation analysis could be concluded in the following: the highest mean and the maximum kernel density value were reported in the municipality of Athens. Furthermore, the median value is not equal to 0 as it happens in the other study municipalities; thus, there are not many segments in Athens with zero kernel density. The mean density value of Piraeus is approximately 4.5 times lower than Athens. The variance is also low compared to the municipality of Athens. Higher concentration of traffic accidents was reported in the municipality of Marousi compared to the other northern municipalities. In all study areas, the kernel densities in residential streets were lower than the ones in primary, secondary and collector roads. In addition, only in the residential streets, the median value is equal to zero; therefore, in the other road hierarchy levels, there are a few road segments, where no traffic accidents have been recorded in the examined period. Finally, no significant differences in the kernel densities can be observed between primary and secondary roads.

By applying the method described in the previous chapter, four thresholds per municipality were estimated (see Table 1). Road segments or spots with densities that are higher than the 99% of density values of each dataset were noticed using black color (black spots) (see Fig. 1-NW). As it can be seen, in order to characterize a spot as black in Athens, higher density values are required. In addition, it is observed that kernel densities are relatively high in the central areas of each municipality.

Table 1. Presentation of threshold values.

	k shape	θ scale	thre1 (85%)	thre2 (90%)	thre3 (95%)	thre4 (99%)
Athens	0.3	43.7	28.6	40.1	61.7	117.6
Piraeus	0.1	22.0	5.0	8.7	16.7	40.5
Kifisia	0.1	21.2	0.6	1.9	6.2	24.0
Marousi	0.1	22.6	2.5	5.2	12.0	34.2
Chalandri	0.1	17.5	1.7	3.6	8.6	25.5
Primary roads	0.4	49.2	37.6	51.3	76.7	141.4
Secondary roads	0.3	50.7	35.9	49.6	75.3	141.0
Collector roads	0.3	32.7	20.8	29.3	45.3	86.8
Residential streets	0.1	26.6	4.8	8.9	18.0	45.9
All municipalities	0.1	45.0	11.3	19.1	36.0	85.3

Specifically, high concentrations have been reported around the center of Athens, in primary roads in Piraeus, in the central part of Kifisia and in primary arterial roads in Chalandri. The spatial pattern is almost similar for all the study areas.

An alternative approach is to use different thresholds for each different road hierarchy level. In Fig. 1-NE, the black spots seem to spread out of the main arteries of the metropolitan area of Athens. In residential streets, the thresholds values are lower. Hence, a comparatively high concentration of traffic accidents in a spot is not an obligatory condition for calling it black. Problematic junctions mainly in the road network of Athens can be identified clearly now; again, many of the black spots are concentrated around central Athens and particularly in some secondary and collector roads. In addition, according to this analysis some junctions (primary roads intersecting) in the central part of Piraeus are quite problematic. In the road networks of the rest municipalities, a few "dangerous" junctions can be observed.

Regarding the findings of the Spatial Autocorrelation Analysis, we should highlight the following: The results of the "Global Moran's I" analysis show significant spatial autocorrelation for severity (Moran's $I = 0.0232$, $p < 0.01$), season (Moran's $I = 0.02986$, $p < 0.01$) and time (Moran's $I = 0.02298$, $p < 0.01$) of the accidents, revealing the existence of potential patterns in their spatial distribution. On the other hand, the patterns of the age of victim and the vehicle type involved do not appear to be significantly different than random. Furthermore, when it comes to the calculation of the "Local Moran's" Index, we should mention that in total, 14 high-high clusters were detected concerning the severity of accidents. They were observed on the primary road network of the metropolitan area, which includes roads that carry high traffic volume. Also, most of the high-low outliers were close to low-low clusters, while low-high outliers were mainly located near the high-high clusters (see Fig. 1-SW).

Finally, when using the Getis-Ord Gi* statistic for the severity of accidents, 141 hot spots (99% confidence) were detected. As it can be observed from Fig. 1-SE, most of the hot spots are located on the primary and secondary network, showing that high traffic volume and high speed (especially at night) may cause severe accidents.

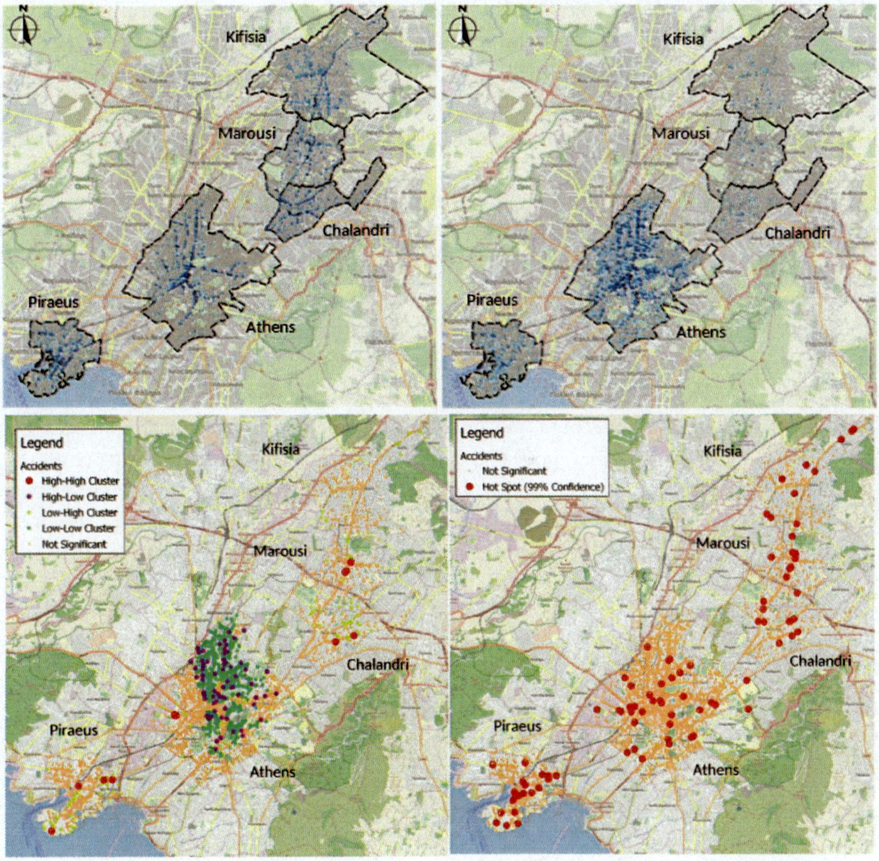

Fig. 1. Network KDE (NW-Municipality and NE-Hierarchy) and Spatial Autocorrelation (SW-Local Moran's I and SE- Hot Spot Analysis, both concerning severity).

5 Discussion and Conclusions

The current research deals with the spatial dimension of road accidents by using two different methods (KDE and spatial autocorrelation). These methods as applied in the five municipalities of Athens' Metropolitan Area indicate some valuable findings. The KDE method cannot produce absolute results, since its output depends highly enough on the initial parameters (i.e. bandwidth, distribution, etc.) and the selected thresholds. In this study, the bandwidth was chosen to be equal to 200 m and a methodology to define thresholds using Gamma distribution was presented. To identify hazardous spots in the road network of a municipality area efficiently, a planner should use different thresholds per road hierarchy level. By using "global" thresholds, a planner can make comparisons between similar municipalities or areas in terms of concentration of black spots in order to understand well enough the seriousness of traffic safety problems in the study area. Apart from KDE, the spatial autocorrelation methods unveiled that the spatial distribution of the accidents is correlated with severity, season, and time (e.g. peak hours). Hence, it would be substantial for SUMPs to formulate flexible measures adjusted to different seasons, and time periods, or to implement measures prioritized according to the severity of the accidents. As it seems the efficient analysis of existing road accidents provides valuable information about dangerous points or corridors.

In an attempt to bridge the gap between analysis and planning procedure of a SUMP, we suggest some example measures or interventions corresponding to the three dimensions of traffic safety (Risk, Severity and Exposure): The analysis revealed a noticeable spatial pattern where black spots are mainly concentrated in the central cores of the study area. Therefore, the creation of ring road zones will reduce through movements, thus decreasing the exposure in the city centers. Moreover, in arterial roads (mainly high risk linear clusters), planners should decide appropriate solutions including: a) construction of roundabouts in case of serious movement conflicts, b) widening of sidewalks or other pedestrian oriented interventions, c) construction of cycle tracks, when the coexistence with cars in cycle lanes seems dangerous and d) bus lanes. These interventions contribute significantly to reducing accidents' risk i.e. probability of traffic accident occurrence. Furthermore, planners should also pay attention to hazardous clusters by considering the enhancement of horizontal and vertical signing indicating lower speed limits. In central areas especially, where spatial analysis identified significant hotspots, they should propose the creation of traffic calming areas with a speed limit up to 20–30km/h. These measures will deal effectively with the severity of the accidents. Finally, municipalities which implement road safety measures, have to develop a traffic safety observatory in order to monitor and evaluate the progress of road accidents, and hence reconsider their further actions.

In general, it is explicit that road safety has undoubtedly a great role to play in SUMPs. Hence, the methods we applied can be fundamental tools, used both in analysis and monitoring process. Definitely, these tools add an extra dimension to traffic safety analysis and they are easy-to-use for transport and urban planners with basic GIS knowledge. For this reason, this research has notable value and could be considered as a small but at the same time important step towards a truly integration of road safety in SUMPs. At last, spatial analysis of road accidents and their connection with SUMPs cannot be fully argued in one single research. Another take could be the implementation of more spatial analysis techniques and afterwards the combination of the results in order to produce an integrated evaluation index. Moreover, further research can deal with the formulation of models aiming to predict the spatial distribution of accidents and factors affecting it (e.g. geographically weighted regression). Finally, new studies can use our findings, in order to propose planning solutions.

References

1. Jabareen, Y.: Sustainable urban forms: their typologies, models and concepts. J. Plan. Educ. Res. **26**(1), 38–52 (2006)
2. Marshall, S.: Streets & Patterns. Spon Press, Abbingdon (2005)

3. Nieuwenhuijsen, M., Khreis, H.: Car free cities: pathway to healthy urban living. Environ. Int. **94**, 251–262 (2016)
4. Soni, N., Soni, N.: Benefits of pedestrianization and warrants to pedestrianize an area. Land Use Policy **57**, 139–150 (2016)
5. Ziakopoulos, A., Yannis, G.: A review of spatial approaches in road safety. Accid. Anal. Prev. **135**, 105323 (2020)
6. World Health Organization: Preventing injuries in Europe: From international collaboration to local implementation (2010)
7. European Commission: Towards a European road safety area: policy orientations on road safety 2011-2020. EC, Brussels (2010)
8. RSO-NTUA: Greece 2007–2017: Basic road safety figures. RSO-NTUA (2018)
9. Banister, D.: The sustainable mobility paradigm. Transp. Policy **15**(2), 73–80 (2008)
10. Wefering, F., Rupprecht, S., Buhrmann, S., Bohler-Baedeker, S., Consult-Forschung, R.: Beratung GmbH: Guidelines. Developing and Implementing a Sustainable Urban Mobility Plan. European Commission (2014)
11. Cre, I.: Road Safety-A key issue for SUMPs. Civitas Insight (2016)
12. Castro-Nuño, M., Castillo-Manzano, I.J., Fageda, X.: The role of road safety in a sustainable urban mobility: an econometric assessment of the Spanish NUTS-3 case. Int. J. Sustain. Transp. **12**(3), 205–217 (2018)
13. Theofilatos, A., Yannis, G.: A review of the effect of traffic and weather characteristics on road safety. Accid. Anal. Prev. **109**, 36–46 (2014)
14. Papadimitriou, E., Filtness, A., Theofilatos, A., Ziakopoulos, A., Quigley, C., Yannis, G.: Review and ranking of crash risk factors related to the road infrastructure. Accid. Anal. Prev. **125**, 85–97 (2019)
15. Elvik, R., Vaa, T., Hoye, A., Sorensen, M.: The Handbook of Road Safety Measures. Emerald Group Publishing, Bingley (2009)
16. Lord, D., Mannering, F.: The statistical analysis of crash-frequency data: a review and assessment of methodological alternatives. Transp. Res. Part A Policy Pract. **44**(5), 291–305 (2010)
17. Lee, J., Abdel-Aty, M.: Macro-level analysis of bicycle safety: focusing on the characteristics of both crash location and residence. Int. J. Sustain. Transp. **12**(8), 553–556 (2018)
18. Loo, B.P., Anderson, T.K.: Spatial Analysis Methods of Road Traffic Collisions. CRC Press, Boca Raton (2015)
19. Xie, Z., Yan, J.: Kernel density estimation of traffic accidents in a network space. Comput. Environ. Urban Syst. **32**, 396–406 (2008)
20. Choudhary, J., Ohri, A., Kumar, B.: Spatial and statistical analysis of road accidents hot spots using GIS. In Proceedings of the 3rd Conference of Transportation Research Group of India (CTRG), Kolkata, India (2015)
21. Imprialou, M.I.M., Quddus, M., Pitfield, E.D., Dominique, L.: Re-visiting crash–speed relationships: a new perspective in crash modelling. Accid. Anal. Prev. **86**, 173–185 (2016)
22. Bailey, T.C., Gatrell, A.C.: Interactive Spatial Data Analysis. Longman, Essex (1995)
23. O' Sullivan, D., Unwin, D.J.: Geographic Information Analysis. John Wiley, Hoboken (2002)
24. Silverman, B.W.: Density Estimation for Statistics and Data Analysis. Chapman Hall, London (1986)
25. Anderson, T.K.: Comparison of spatial methods for measuring road accident "hot-spots": a case study of London. J. Maps **3**(1), 55–63 (2007)
26. Wang, F., Antipova, A., Porta, S.: Street centrality and land use intensity in Baton Rouge, Louisiana. J. Transp. Geogr. **19**, 285–293 (2011)

27. Delmelle, E.C., Thill, J.C., Ha, H.H.: Spatial epidemiologic analysis of relative collision risk factors among urban bicyclists and pedestrians. Transportation **39**(2), 433–448 (2012)
28. Delmelle, E.C., Thill, J.C.: Urban bicyclists: A spatial analysis of Adult and Youth Traffic Hazard Intensity. Transp. Res. Rec. **2074**(1), 31–39 (2008)
29. Krisp, J., Durot, S.: Segmentation of lines based on point densities-An optimization of wildlife warning sign placement in southern Finland. Accid. Anal. Prev. **39**(1), 38–46 (2007)
30. Mountrakis, G., Gunson, K.: Multi-scale spatiotemporal analyses of moose-vehicle collisions: a case study in northern Vermont. Int. J. Geogr. Inf. Sci. **23**(11), 1389–1412 (2009)
31. Pulugurtha, S., Krishnakumar, V.K., Nambisan, S.S.: New methods to identify and rank high pedestrian crash zones: an illustration. Accid. Anal. Prev. **39**(4), 800–811 (2007)
32. Erdogan, S., Yilmaz, L., Baybura, T., Gullu, M.: Geographical information systems aided traffic accident analysis system case study: city of Afyonkarahisar. Accid. Anal. Prev. **40**(1), 174–181 (2008)
33. Okabe, A., Satoh, T., Sugihara, K.: A kernel density estimation method for networks, its computational method and a GIS-based tool. Int. J. Geogr. Inf. Sci. **23**(1), 7–32 (2009)
34. Loo, B.P., Yao, S., Wu, J.: Spatial point analysis of road crashes in Shanghai: a GIS-based network kernel density method. In: Proceedings of the 19th International conference on geoinformatics (IEEE), Shanghai, China, pp. 1–6 (2011).
35. Mohaymany, A.S., Shahri, M., Mirbagheri, B.: GIS-based method for detecting high-crash risk road segments using network kernel density estimation. Geo-spatial Inf. Sci. **16**(2), 113–119 (2013)
36. Bil, M., Andrasik, R., Janoska, Z.: Identification of hazardous road locations of traffic accidents by means of kernel density estimation and cluster significance evaluation. Accid. Anal. Prev. **55**, 265–273 (2013)
37. Griffith, D.: Spatial Autocorrelation and Spatial Filtering: Gaining Understanding Through Theory and Scientific Visualization. Springer, Heidelberg (2003)
38. Legendre, P., Legendre, L.: Numerical Ecology. Elsevier, Amsterdam (1998)
39. Overmars, K., de Koning, G., Veldkamp, A.: Spatial Autocorrelation in multi-scale land use models. Ecol. Model. **164**(2–3), 257–270 (2003)
40. Getis, A.: Spatial autocorrelation. In Fischer, M., Getis, A. (eds.) Handbook of Applied Spatial Analysis: Software Tools, Methods and Applications, pp. 255–278. Springer, Heidelberg (2010)
41. Zhixiao, X., Yan, J.: Kernel density estimation of traffic accidents in a network space. Comput. Environ. Urban Syst. **35**(5), 396–406 (2008)
42. Harirforoush, H., Bellalite, L.: A new integrated GIS-based analysis to detect hotspots: a case study of the city of Sherbrooke. Accid. Anal. Prev. **130**, 62–74 (2019)
43. Rura, M., Griffith, D.: Spatial statistics in SAS. In: Getis, A., Fischer, M. (eds.) Handbook of Applied Spatial Analysis: Software Tools, Methods and Applications, pp. 43–52. Springer, Heidelberg (2010)
44. Bivand, R., Muller, W.G., Reder, M.: Power calculations for global and local Moran's I. Comput. Stat. Data Anal. **53**(8), 2859–2872 (2009)

The Impact of Megatrends on the Transition from Car-Ownership to Carsharing: A Delphi Method Approach

Dimitrios Papanaoum[1], Georgios Palantzas[1,2], Theodoros Chrysanidis[1], and Dimitrios Nalmpantis[1,2(✉)]

[1] School of Science and Technology, Hellenic Open University, Parodos Aristotelous 18, 263 35 Patras, Greece
[2] School of Civil Engineering, Faculty of Engineering, Aristotle University of Thessaloniki, PO Box 452, 541 24 Thessaloniki, Greece
dnalba@civil.auth.gr

Abstract. The purpose of this paper is to examine the new reality that is formed through the gradual transition from car-ownership to carsharing and carpooling. In this new reality, car use is gradually detached from its ownership, and there are challenges and prospects that have to be examined in order to become opportunities. Everyone will be able to use a car without having it bought, either as a co-passenger or as a user of a carsharing or a carpooling service. The sharing economy is offering more and more possibilities towards this direction, exploiting the widespread use of smartphones worldwide. Carsharing services like DriveNow of BMW and Car2Go of Daimler, ride-hailing services like Uber and Lyft, and carpooling applications seem to change the landscape of car mobility radically. What are the megatrends that seem to be formed in the future? What are the challenges and the prospects of this newly formed reality? Regarding these and other relevant questions, this paper tries to give some insights by exploring the effect of megatrends on the demand for carsharing and carpooling services, after a questionnaire survey amongst experts using the Delphi method. It was found that the most critical megatrends for the next decade will be: a) Innovative and affordable technologies, b) Automation, and c) Internet of Things; the most important trend that will affect the demand for carsharing will be "Automation," and the most important trend that will affect the demand for carpooling will be "Artificial Intelligence."

Keywords: Carpooling · Ride-hailing · Disruptive · Mobility · Futurology

1 Introduction

Car use seems to be gradually disconnecting from car-ownership. Anyone can use a car without having bought it, either as a co-passenger or as a user of a carsharing service. The sharing economy offers many opportunities in this direction, taking advantage of the widespread use of smartphones. Carpooling applications, services such as Uber and Lyft, carsharing services such as the Bavarian Motor Works' (BMW) DriveNow and Daimler's Car2Go seem to change the landscape of mobility radically.

E. G. Nathanail et al. (Eds.): CSUM2020, AISC 1278, pp. 515–524, 2021.
https://doi.org/10.1007/978-3-030-61075-3_51

1.1 Carsharing

The general rule of sharing economy in consumption is that it focuses on access to goods and services through sharing and not through ownership [1]. Carsharing agrees with the general idea of sharing economy, and it is a sustainable mode of transportation. Carsharing is essentially a scheme that gives vehicles' shared access to many users at different times to move occasionally.

Usually, carsharing trips have a short duration within the urban complex. Users are registered club members or corporate service users and qualify for carsharing. This service is provided 24 h a day, and the vehicle is booked, received (or reached in free-floating systems), and returned by the user himself (or parked wherever in free-float systems), without any time commitment. The user pays for the service available, including fuel costs and vehicle insurance [2].

The first founding of a carsharing service was in 1948 in Switzerland. Other countries followed, such as France with Procotip, the Netherlands with Witcar, England and Sweden with Bilpoolen. By 1980, carsharing had been extended to Denmark, Italy, and Ireland [2]. In the American continent, carsharing did not go much further in the 1980s, but today it is developing as elsewhere, e.g., in South America, Asia, Australia, and of course, Europe [2].

In the carsharing sector, there are big companies that are usually supported by large car manufacturers, such as Daimler's Car2Go and BMW's DriveNow. The charge is proportional to the time of use of the vehicle (time charge) and charged per minute, hour, or day. Charges include rental costs, fuel costs, insurance, and parking in authorized locations (or even on-road in free-floating systems). The route is one-way so that it can be used by subsequent users [2].

Carsharing is one of the pillars of smart transportation infrastructure, as it is expected to reduce traffic congestion, parking demand, and pollution in cities. Boldrini, Bruno, and Laarabi (2019) have studied the sociodemographic and urban activity indicators associated with variations in carsharing demand in 10 different European cities [3]. Meelen, Frenken, and Hobrink (2019) have conducted a geographical analysis of sustainable transitions in the Netherlands in order to find the weak spots for carsharing. Their study emphasizes equally on the spatial heterogeneity of niche, regime, and landscape [4].

1.2 Carpooling

The occasional carpooling also fits in with the general idea of sharing economy and co-operative consumption, and it is also a sustainable mode of transportation, as there is a reduction in moving vehicles and car trips.

Carpooling is a form of transportation sharing economy in the form of two to three or more passengers per motor vehicle. Passenger transport coincides with the main purpose of the driver, which is to arrive at the destination without aiming to gain profit. The benefits it gives to passengers are time and money savings, through the high availability of vehicles and the sharing of road toll costs and fuel [2, 5].

The first examples of carpooling appeared in Los Angeles, CA, and San Francisco, CA, in 1914 with minibuses (jitneys). The next milestone in the history of carpooling was in 1941–1945 during World War II with the promotion of the common use of vehicles again with the aim of saving energy. The third appearance of carpooling again coincided with the oil energy crisis in 1970 [2].

Carpooling can be categorized into occasional carpooling, preplanned carpooling, and real-time carpooling by matching passenger preferences with those of the driver in real-time [2]. The rapid spread use of smartphones gives huge prospects for real-time carpooling. Ride-hailing companies like Uber Technologies Inc. And Lyft Inc. Are known for providing vehicle drivers that collect prospective and unknown travelers using applications based on distance, direction, and travel time algorithms [6].

Moreover, carpooling, as well as carsharing, help to achieve the goal of a 2000-W society [7]. Scarinci et al. (2017) investigated the social characteristics and travel habits of the population living in a case study area. Then, they evaluated the modal shift induced by transportation policies such as carsharing, carpooling, and car-free district zones. The resulting energy consumption was evaluated and compared with the 2000-W society limit. They concluded that only a set of measures combining car use reduction (e.g., carsharing and carpooling), increase in walking and cycling, and reduction in the total travel distance could achieve this goal [7].

1.3 Previous Research, Scope, and Objectives

We searched for previous research on carsharing and carpooling on Scopus indexing service with the keywords "carsharing," "carpooling," in several forms, i.e., with or without a dash, separately or jointly, together with "Delphi" in titles, abstracts, and keywords. We found just one paper on carsharing with autonomous cars [8]. This paper [8] focuses on the acceptability of such a future service. On the other hand, our paper focuses on the effect of megatrends on the demand for conventional carsharing and carpooling services in the next decade in Europe. Thus, it seems that there is a gap in the literature, and our paper might have a positive contribution.

The scope of this paper is an attempt to decode the transition from car-ownership to carsharing and carpooling through a Delphi method approach, as it is appropriate for forecasting future trends and evolutions, exploring the effect of megatrends on the demand and expansion of carsharing and carpooling services.

2 Methodology

2.1 Megatrends

A prerequisite for examining how future megatrends will affect carsharing and carpooling is first to identify them. A similar task has been performed for public transport in the frame of the Horizon 2020 "Collective Innovation for Public Transport in European Cities" (CIPTEC) project [9, 10, 11, 12, 13], in which 10 megatrends were identified [14], to which we added five (5) more megatrends, which we considered to be important [15]. These megatrends are the following:

1. “Enhancement of urban governance” [ENH] as the power is decentralized from central authorities and is concentrated on the local authorities. More and more decisions are now being made locally [14].
2. “Globalization” [GLOB] as the physical and economic frontiers appear to be fading. The mobility of goods and people is increasing, and travel is becoming more comfortable and more affordable [14].
3. “Sharing economy, individual empowerment, and social innovation” [SHARE] is another observed megatrend. In the case of transportation, the shared use of transport vehicles is gaining ground over their ownership [14].
4. “Population aging” [AGE] is a decisive megatrend as life expectancy is growing and the birth rate declines. Urban centers are called upon to manage a larger proportion of older people [14].
5. “Transforming families and household sizes” [FAMSIZE] is a megatrend observed within the European population having no children at all or at older ages. Divorces are increasing, and family and household sizes are decreasing [14].
6. “Urbanization and urban sprawl” [URBAN] is a consequence of the growth of urban centers and urban sprawl, as more people are now living in urban areas. Some people are moving from central cities to less dense areas and are dependent on vehicles. That creates new demand for better and comprehensive transport services [14].
7. “Sustainable lifestyles” [LIFEST] is a megatrend resulting from the fear of consuming wasted energy, the amount of waste generated, and carbon footprint [14].
8. “Innovative and affordable technologies” [INNOTECH] is a megatrend that is generally observed, as Information Technology (IT) and Information and Communications Technology (ICT) are ubiquitous, and free use of wireless Internet is offered in public places. Users can organize their travel more effectively, and the travel experience becomes more efficient and enjoyable [14].
9. “Environmental degradation” [ENV] due to increasing demand for natural resources leads to their depletion. Ecosystems are disrupted or destroyed, the climate is changing, and the environment is getting worse. Vehicle fuels are becoming more expensive [14].
10. “Harmonization of legislation in EU” [EU LAW] is another megatrend observed in a competitive transport system, which increases mobility, removes borders in key areas, and enhances growth and employment [14].
11. “Artificial Intelligence” (AI) [INTEL] is a megatrend that can significantly increase the function of carsharing and carpooling. Deploying algorithms can make route and passenger selection easier and even merge routes [15, 16].
12. “Automation” [AUTO] is a megatrend that expects full functionality, continuous access, and supervision of production that can be assisted by AI. Connected and Autonomous Vehicles (CAV) are expected to play a significant role in carsharing [15, 16].
13. “Global warming” [WARM] is a megatrend that can play a significant role in carsharing and carpooling. The acceleration of the greenhouse effect can have significant consequences for the entire planet, and the use of shared services may reduce GreenHouse Gas (GHG) emissions that accelerate the problem [15].

14. "Terrorism" [TERR] acts as a deterrent in situations of sharing goods and services. Suspicion and fear are not allies of the sharing economy. It is, therefore, a significant trend that needs to be investigated [15].
15. "Internet of Things" (IoT) [IOT] is a megatrend where all physical and digital things can be controlled by smartphones and connected devices. The physical part remains natural as a form, but its digital image is controlled through the Internet. User connection and route planning are also configured online [15].

2.2 General Framework

The methodology followed was to develop a questionnaire to investigate the effect of megatrends with the convergence of the experts' opinions on carsharing and carpooling, using the Delphi method. The questionnaire consists of four sections:

1. The first section deals with the evolution of the recognized megatrends over the next decade in the European area.
2. The second section deals with how the demand for carsharing services will evolve over the next decade, under the impact of the recognized megatrends.
3. The third section deals with how the demand for carpooling services will evolve over the next decade, under the impact of the recognized megatrends.
4. The fourth section deals with the socio-economic profile of the respondents and some of their demographics.

The respondents had the following answer options in each question section: a) large increase, b) moderate increase, c) small increase, d) small reduction, e) moderate reduction, and f) large reduction, which were graded from + 6 to -1, respectively. The medians, averages (means), standard deviations, scores (i.e., mean value normalized to 100%), and rankings were calculated in both rounds.

Our approach was "open," i.e., there was no specific consensus threshold, as there were many objective limitations to the number of possible rounds, i.e., a strict timeframe in which the research should be completed being realized in the frame of a postgraduate diploma thesis, and the fact that in Round 2 we received less than half answers compared to those from Round 1.

2.3 1st Delphi Round

The questionnaire survey was conducted between May 28 and July 18, 2018. The questionnaire was developed using the Google Forms platform, and it was targeted at transport experts. It was sent mainly to European research experts involved in carsharing and carpooling research projects, to avoid transport engineers who had no expertise in carsharing and carpooling. A total of 152 online questionnaires were sent, and only 20 experts replied.

From the 20 respondents, regarding their affiliation they were: eight (8) transportation consultants, six (6) from academia, three (3) experts from the private transport sector, two (2) experts from the public transport sector, and one (1) was a researcher.

Regarding the geographical distribution they were: four (4) from Greece, three (3) from Germany, three (3) from the United Kingdom, three (3) from Italy, two (2) from the Netherlands, one (1) from Belgium, one (1) from Austria, one (1) from Spain, one (1) from New Zealand, and one (1) from the United States of America (USA). Regarding their gender, thirteen (13) were men and seven (7) were women; gender misrepresentation of women is a well-known issue in the transportation sector. Regarding their expertise on carsharing and carpooling, their self-assessment was as follows: six (6) very high expertise, eight (8) high expertise, five (5) some expertise, and one (1) no expertise (keep in mind that all of them were transportation experts).

2.4 2nd Delphi Round

Round 2 of the Delphi questionnaire continued with only 12 of the 20 experts.

From the 12 respondents, regarding their affiliation they were: four (4) from academia, three (3) transportation consultants, two (2) experts from the private transport sector, two (2) experts from the public transport sector, and one (1) was a researcher. Regarding the geographical distribution they were: two (2) from Greece, two (2) from the United Kingdom, two (2) from Italy, one (1) from Germany, one (1) from Belgium, one (1) from Austria, one (1) from Spain, one (1) from New Zealand, one (1) from the USA, and none (0) from the Netherlands. Regarding their gender, seven (7) were men, and five (5) were women. Regarding their expertise on carsharing and carpooling, their self-assessment was as follows: five (5) very high expertise, four (4) high expertise, two (2) some expertise, and one (1) no expertise.

In Round 2, we presented the results of Round 1 to the respondents in the form of mean values, standard deviations, medians, and pie charts showing the percentage of each answer from Round 1, aiming to achieve convergence of views.

3 Results and Discussion

The results of Round 2 are presented below in Table 1, Table 2, Table 3, Fig. 1, and Fig. 2. Scores are mean values normalized to 100%.

Table 1 shows that, according to the experts, the megatrends that appear to be increasing dramatically in Europe over the next 10 year are the following: a) Innovative and affordable technologies, b) Automation, c) Internet of Things, and d) Artificial Intelligence.

Table 2 shows that the demand for carsharing in the next decade will be most affected by Automation. Fig. 1 shows that the following megatrends: a) Automation, b) Innovative and affordable technologies, and c) Internet of Things are expected to increase significantly in Europe over the next decade and, at the same time, achieve a high impact on the demand for carsharing.

Table 3 shows that the demand for carpooling in the next decade will be most affected by Artificial Intelligence. Fig. 2 shows that the following megatrends: a) Automation and b) Artificial Intelligence are expected to increase significantly in Europe over the next decade, and at the same time, achieve a high impact on the demand for carpooling.

Table 1. Megatrends (1st Section, 2nd Round).

How will the recognized trends in Europe evolve over the next decade?	Median	Mean	Standard deviation	Score %	Rank
8. Innovative and affordable technologies	6	5.42	0.79	90.28	1
12. Automation	6	5.42	0.67	90.28	1
15. Internet of Things	6	5.42	0.67	90.28	1
11. Artificial Intelligence	5	5.17	0.72	86.11	4
6. Urbanization and urban sprawl	5	4.83	0.94	80.56	5
4. Population ageing	5	4.75	1.14	79.17	6
3. Sharing economy (…) innovation	5	4.75	0.87	79.17	6
2. Globalization	5	4.75	0.87	79.17	6
10. Harmonization of legislation in EU	4	4.58	0.79	76.39	9
13. Global warming	5	4.58	0.67	76.39	9
7. Sustainable lifestyles	4	4.50	0.90	75.00	11
1. Enhancement of urban governance	4	4.25	0.45	70.83	12
5. Transforming (…) household sizes	4	4.25	1.29	70.83	12
9. Environmental degradation	4	3.83	0.72	63.89	14
14. Terrorism	3	3.67	1.23	61.11	15

Table 2. The effect of megatrends on the demand for carsharing (2nd Section, 2nd Round).

How will the demand for carsharing services evolve over the next decade under the influence of recognized trends?	Median	Mean	Standard deviation	Score %	Rank
12. Automation	5	5.00	0.74	83.33	1
8. Innovative and affordable technologies	5	4.92	0.79	81.94	2
6. Urbanization and urban sprawl	5	4.83	0.94	80.56	3
15. Internet of Things	5	4.83	0.83	80.56	3
7. Sustainable lifestyles	5	4.75	0.97	79.17	5
3. Sharing economy (…) innovation	4	4.58	0.79	76.39	6
10. Harmonization of legislation in EU	4	4.42	0.79	73.61	7
11. Artificial Intelligence	5	4.42	0.90	73.61	7
2. Globalization	4	4.25	0.97	70.83	9
1. Enhancement of urban governance	4	4.25	0.62	70.83	9
4. Population ageing	4	4.17	1.11	69.44	11
13. Global warming	4	4.00	0.43	66.67	12
5. Transforming (…) household sizes	4	3.92	0.79	65.28	13
9. Environmental degradation	4	3.92	0.51	65.28	13
14. Terrorism	3	3.00	0.74	50.00	15

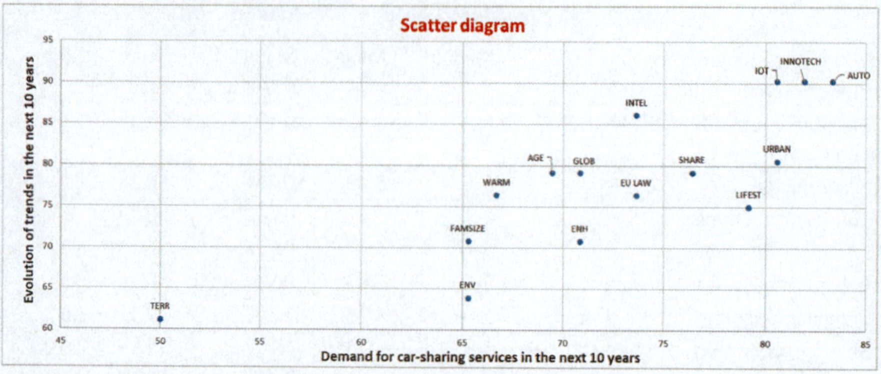

Fig. 1. The effect of megatrends on the demand for carsharing in the next decade.

Table 3. The effect of megatrends on the demand for carpooling (3rd Section, 2nd Round).

How will demand for carpooling services evolve over the next decade under the influence of recognized trends?	Median	Mean	Standard deviation	Score %	Rank
11. Artificial Intelligence	5	4.83	1.19	80.56	1
7. Sustainable lifestyles	5	4.75	0.87	79.17	2
3. Sharing economy (…) innovation	5	4.75	0.45	79.17	2
12. Automation	5	4.75	0.87	79.17	2
6. Urbanization and urban sprawl	5	4.67	0.65	77.78	5
8. Innovative and affordable technologies	5	4.58	1.08	76.39	6
15. Internet of Things	4	4.58	0.79	76.39	6
10. Harmonization of legislation in EU	4	4.50	0.67	75.00	8
13. Global warming	4	4.33	0.65	72.22	9
2. Globalization	4	4.17	0.58	69.44	10
1. Enhancement of urban governance	4	4.17	0.39	69.44	10
5. Transforming (…) household sizes	4	4.17	0.72	69.44	10
9. Environmental degradation	4	4.17	0.58	69.44	10
4. Population ageing	4	4.00	0.60	66.67	14
14. Terrorism	3	3.25	0.45	54.17	15

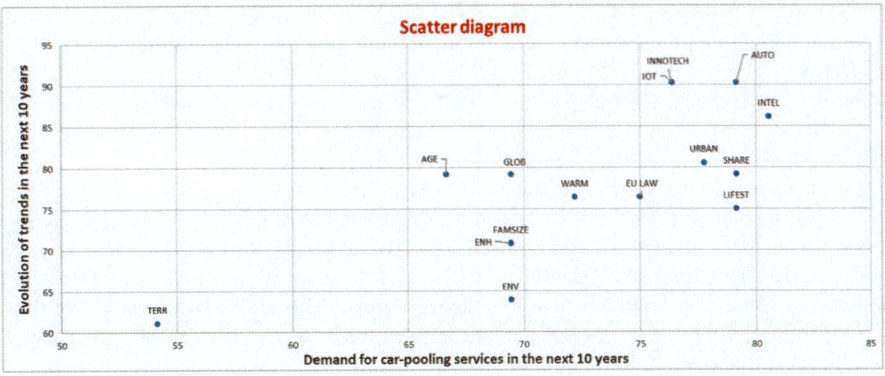

Fig. 2. The effect of megatrends on the demand for carpooling in the next decade.

4 Conclusions and Recommendations for Further Research

In general, it seems that the innovative aspects of new developments will play a key role in the demand for carsharing and carpooling. All of the above should be taken into account by public authorities and institutions, the academic community, consulting firms, the transport industry, and stakeholders in general, to catch up with the developments in carsharing and carpooling, by developing relevant infrastructure and know-how. The aim is to take advantage of the challenges and to create opportunities for the transition from car-ownership to carsharing and carpooling, for the rapid decline in the number of vehicles in use, the improvement of environmental indicators, and the promotion of the quality of life in urban centers over the next decade.

The present paper had many limitations, mainly concerning its limited implementation time, which might have had an impact on the sample size, especially regarding the responses of Round 2. In case there were more responses, we would be able to draw conclusions per category of respondents, after determining a specific consensus index threshold. The survey could be conducted at regular intervals, e.g., every five years or less, to examine the changes that have occurred with the advancement of technology, which is now increasing geometrically.

References

1. Botsman, R., Rogers, R.: What's Mine is Yours: the Rise of Collaborative Consumption. HarperCollins, New York (2010)
2. Avtzi, P., Liakopoulou, S., Mavromati Kakana, M.: Investigation of feasibility and characteristics of creating a carpooling system for university and technological institutions' students in Thessaloniki. Diploma thesis, Aristotle University of Thessaloniki, Thessaloniki (2015). https://www.tsrg.gr/wp-content/uploads/2015/12/Διπλωματική-Carpooling-Αβτζή-Π.-Λιακοπούλου-Σ.-Μαυρομάτη-Κ.pdf

3. Boldrini, C., Bruno, R., Laarabi, M.H.: Weak signals in the mobility landscape: car sharing in ten European cities. EPJ Data Sci. **8**, 7 (2019). https://doi.org/10.1140/epjds/s13688-019-0186-8
4. Meelen, T., Frenken, K., Hobrink, S.: Weak spots for car-sharing in The Netherlands? the geography of sociotechnical regimes and the adoption of niche innovations. Energy Res. Soc. Sci. **52**, 132–143 (2019). https://doi.org/10.1016/j.erss.2019.01.023
5. Liakopoulou, S., Mavromati Kakana, M., Avtji, P., Genitsaris, E., Naniopoulos, A.: Investigating the preferences of students towards the creation of a carpooling system serving the academic bodies of Thessaloniki city. Transp. Res. Procedia **24**, 425–432 (2017). https://doi.org/10.1016/j.trpro.2017.05.091
6. Lazo, A.: How the daily commute is going to change. The Wall Street Journal Website (2016), https://www.wsj.com/articles/how-the-daily-commute-is-going-to-change-1461550158, Accessed 30 Mar 2020
7. Scarinci, R., Rast, F., Bierlaire, M.: Needed reduction in mobility energy consumption to meet the goal of a 2000-watt society. Transp. Res. Part A Policy Pract. **101**, 133–148 (2017). https://doi.org/10.1016/j.tra.2017.05.010
8. Merfeld, K., Wilhelms, M.P., Henkel, S., Kreutzer, K.: Carsharing with shared autonomous vehicles: uncovering drivers, barriers and future developments – A four-stage Delphi study. Technol. Forecast. Soc. Chang. **144**, 66–81 (2019). https://doi.org/10.1016/j.techfore.2019.03.012
9. Collective Innovation for Public Transport in European Cities Homepage. https://ciptec.eu/, Accessed 30 Mar 2020
10. Tsafarakis, S., Gkorezis, P., Nalmpantis, D., Genitsaris, E., Andronikidis, A., Altsitsiadis, E.: Investigating the preferences of individuals on public transport innovations using the Maximum Difference Scaling method. Eur. Transp. Res. Rev. **11**(1), 3 (2019). https://doi.org/10.1186/s12544-018-0340-6
11. Nalmpantis, D., Roukouni, A., Genitsaris, E., Stamelou, A., Naniopoulos, A.: Evaluation of innovative ideas for Public Transport proposed by citizens using Multi-Criteria Decision Analysis (MCDA). Eur. Transp. Res. Rev. **11**(1), 22 (2019). https://doi.org/10.1186/s12544-019-0356-6
12. Stamelou, A., Genitsaris, E., Nalmpantis, D., Naniopoulos, A.: Investigating potential synergies among social entrepreneurship and public transport through experts' consultation in Greece. In: Nathanail, E., Karakikes, I. (eds.) Data Analytics: Paving the Way to Sustainable Urban Mobility. CSUM 2018. Advances in Intelligent Systems and Computing, vol. 879, pp. 496–503. Springer, Cham (2019). https://doi.org/10.1007/978-3-030-02305-8_60
13. Genitsaris, E., Stamelou, A., Nalmpantis, D., Naniopoulos, A.: A criteria-based evaluation framework for assessing public transport related concepts resulted from collective Intelligence approaches. In: Nathanail, E., Karakikes, I. (eds.) Data Analytics: Paving the Way to Sustainable Urban Mobility. CSUM 2018. Advances in Intelligent Systems and Computing, vol. 879, pp. 526–537. Springer, Cham (2019). https://doi.org/10.1007/978-3-030-02305-8_64
14. Angelidou, M., Genitsaris, E., El Aissati, H., Naniopoulos, A.: The effect of major market and societal trends on public transport in European cities. Transp. Res. Procedia **24**, 105–112 (2017). https://doi.org/10.1016/j.trpro.2017.05.075
15. Papanaoum, D.: From car ownership to carpooling and carsharing: trends' investigation using the Delphi method. Master's thesis, Hellenic Open University, Patras (2018). https://apothesis.eap.gr/handle/repo/40511
16. He, W., Hwang, K., Li, D.: Intelligent carpool routing for urban ridesharing by mining GPS trajectories. IEEE Trans. Intell. Transp. Syst. **15**(5), 2286–2296 (2014). https://doi.org/10.1109/TITS.2014.2315521

Urban Mobility Transition to Sustainability: A System Dynamics Approach

Vasiliki V. Georgatzi(✉) and Yeoryios Stamboulis

Department of Economics, University of Thessaly, Volos, Greece
{vageorgatzi,vageorgatzi}@uth.gr

Abstract. Urban Mobility (UM) faces transition, in an era of global transformation under the pressures of increasing urbanization, aging and saturated infrastructures, extreme pollution levels, and technological disruption. Until now, researchers have analyzed the transition of UM through a variety of approaches, focusing mainly in technological issues. Existing studies are mostly qualitative, while quantitative ones usually have partial focus, without considering a whole system being in transition. Policies, initiatives and strategies may result in converging, conflicting or even competing outcomes, as a result of the interaction of the different actors belonging to the subsystems of an examining system. We consider UM as a socio-technical regime in transition, prone to external and internal pressures and equipped with adaptive capacity for a more or less successful transition towards new modes of UM. We illustrate the interactions of the system using system dynamics methodology that helps us capture the dynamics of the multiple interactions taking place at different levels and interacting helices (government, academia, industry, civil society) during the transition process.

Keywords: System dynamics · Socio-technical transition · Urban mobility · New urban mobility modes · Innovation helices

1 Introduction

Transport is fundamental to our economy and society, as it enables economic growth and job creation. Furthermore, urban mobility (UM) is vital for economic activity and for the quality of life of citizens. As much as 80% of humanity's greenhouse gas (GHG) emissions is produced in urban areas, where more than one-half of the world's population now lives [1]. In this paper, we consider UM as a socio-technical system in transition. Socio-technical systems take several years to evolve, establish coordination, stability and path dependencies and become resistant to change [1–4]. We adopt the multi-level perspective (MLP) that assumes multiple socio-technical landscape dynamics, multiple regimes and multiple niche innovations [4]. Single shocks or external pressures that destabilize the regime take place at the landscape level. Landscape pressures are these that contribute to transitions either positively or negatively (e.g. economic recession and high oil prices, rising of house prices, climate change mitigation, etc.). Each new UM mode (e.g. car-sharing, cycling, walking, public transport (PT), MaaS, etc.) seeks to develop its regime, involving the development of

E. G. Nathanail et al. (Eds.): CSUM2020, AISC 1278, pp. 525–538, 2021.
https://doi.org/10.1007/978-3-030-61075-3_52

different niche innovations, complementary assets (e.g. AFVs, ICT, etc.) and institutional arrangements, as well as the development of and interaction between different aspects of the constituent subsystems. We adapt the model of the evolution of the UM socio-technical system in terms of a Quadruple Helix Innovation Model in order to explore the transition pathways that emerge through the dynamic interactions of the different helices within each UM mode/regime. Finally, in this work in progress, we use System Dynamics, in order to map (and subsequently simulate) the interactions that drive the transition from one regime to another.

The rest of the paper is structured as follows: a concise literature review is presented in Sect. 2, the methodology is presented in Sect. 3, the causal loop model of the interactions taking place within the UM landscape is presented and discussed in Sect. 4 and finally, we conclude, in Sect. 5.

2 The Socio-technical System of Urban Mobility

UM is a complex field [5, 6] involving changes in peoples' lifestyles and the availability of alternative travel modes. The way alternative modes of transport will unfold in the future is still not certain and depends on a variety of different elements, such as technological, social, mobility and regulatory trends and developments within the system or external [7]. Transitions to alternative modes of travel "will involve adjustments in the established order and changes in technology, consumer habits, markets, business models, infrastructure and cultural significance" [2]. These adjustments, according to Adeniyi, Cooper and Swan, "will include embedding strategies that can foster structural changes in technology, institutions and practices" across the sector. At the same time aligning policies are necessary to mobilize the transition to a low carbon economy [2, 4, 8–10], so as to overcome the inertia and the barriers that existing socio-technical systems, regimes and infrastructures pose due to lock-ins and path dependencies.

Several conceptual approaches have been used to analyze the transportation system's transition (multi-level perspective -MLP, transition management, innovation systems, strategic niche management, etc.), utilizing various methodologies (Agent-based modeling - ABM, System Dynamics, econometrics, statistic models, etc.). Whilst the MLP can in principle account for a broader range of social phenomena, studies that utilize the MLP tend to focus on the development and diffusion of new technologies [11]. Strategic niche management constitutes an ex-post analysis [12] of transition cases. Transition management is a rather complicated approach to apply and it is usually biased towards the incumbent regime actors [13]. The innovation systems (IS) approach explores how system elements affect the development, diffusion, and use of new innovations. Most of these previous studies focus on the diffusion of alternative vehicle technologies, as well as on the impact of these technologies on climate change and pollution, while a significant amount of them also focuses on the impact of policies both on the diffusion of alternative fuel vehicles and climate change mitigation regarding transportation.

We could categorize previous studies in three clusters: the ones that examine the environmental challenges that transition pathways face, those that focus on technological substitution (from internal combustion engine - ICE, to alternative fuel vehicles - AFVs), and finally those that deal with business model innovations in transport. The findings of the literature review that we conducted, shows that there is consensus amongst researchers regarding the sign of impact (positive/negative) of one factor on another on all cases researched.

Environmental Challenges: Studies in this category focus on the environmental/sustainability impact of transport, e.g. on CO_2 emissions or other pollutants. Some support that we have to move to other modes of transport. For example, Han and Hayashi [14] examined intercity passenger transport in China, and argue that moving to railway transport instead of private car will lead to a reduction of CO_2 emissions. How factors affect the transition to sustainability according to these studies is presented in Table 1.

Table 1. Factors that can mitigate the environmental impact of transport (Source: Authors' own elaboration based on literature review)

Category	Factor	Influence (+/ −)	References
Policy	Fuel taxes	+	[14]
Economic	High cost of past investment in private car	−	[14]
	Cost of purchasing AFVs	−	[30]
Behavioural	Lack of knowledge about the implications of AFVs	−	[14]
	Environmental consciousness	+	[30]
Demografic	Population	−	[32]
	Income	+	[32]
Availability	No of available taxis/ buses	+	[32]
	Bus lines mileage	+	[32]
	Rail traffic milleage	+	[32]

Technological Substitution: Most of the studies that examine transition in the transport sector seem to focus on technological substitution and the factors affecting the adoption of alternative technologies. Financial incentives such as subsidies, reductions in registration taxes and policies such as the ones for scrapping the old conventional vehicles and GHG & ZEV regulations, seem to be an efficient way to push the AFVs acquisition [15–23]. Also, marketing and word of mouth affect the willingness of consumers to choose AFVs [18, 23, 24]. Literature findings on how factors could affect the demand for AFVs are presented in Table 2.

Business Model Innovation (BMI): At the same time, other researchers examine the impact of the emergence of new business models along with the new modes of transport that become available. Sarasini and Linder [11] argue that mobility modes such as car-sharing, are an emerging mobility business model that is expected to reduce

Table 2. Factors influencing the adoption of AFVs (Source: Authors' own elaboration based on literature review)

Category	Factor	Influence (+/−)	References
Economic	Fuel/ energy prices	−	[24, 31, 33]
	Operating cost	−	[16, 23, 24, 34]
	Scale of ICE	−	[24, 34]
	Cost of AFVs	−	[33, 35]
Technical	Increasing average vehicle life span	+	[31]
	Increase driving range	+	[24, 33]
	No of available types of AFVs	+	[19]
	Research and development	+	[24]
Demographic	Income	+	[31]
	Increasing population	+	[31]
Complementary	Lack of recharging infrastructures	-	[23, 24, 31, 33, 35]
	Maintenance services	+	[24]
	ICT use		[34]
Policies	Public funding	+	[15–17, 19, 20, 22–24]
	ZEV & GHG target credits	+	[31]
Behavioural	Willingness of consumers to adopt AFVs (increasing)	+	[24]
	Safety (increasing)	+	[24]
	Personal experience	+	[24]
Environmental	Ecological impact	+	[24, 34]
	Environmental awareness of consumers	+	[33]

environmental impact by 30%. Also, they conclude that BMI may act as both a market and non-market strategy, providing as an example the field of mobility, where "non-market strategies are important when experimenting with new business models" [11]. A few years earlier, Geum, Lee and Park [25] argued that the use of car sharing may reduce energy consumption, while at the same time the increasing use of car-sharing may improve traffic conditions and therefore lead to higher use of car-sharing. The factors facilitating or hampering the transition to new modes of transport are summarized in Table 3.

To sum up, we have noticed that previous studies have focused mostly on the effects of specific transport segments such as urban transport [26, 27], transport modes [11, 28–29] or road transport [30], individual policy measures, e.g. support of public mobility services, taxes or scrappage schemes [14, 28] and particular technologies like biofuels or electric vehicles [15, 31]. Existing research does not provide a "helicopter view" of the interactions that take place within the system amongst the regimes of alternative modes of urban mobility (AMUMs) and the various helices. We believe that there is still need for a system-wide perspective of how a new socio-technical system emerges from the transition process, which is taking place through these interactions. We aim to contribute to this by developing a comprehensive tool for policy makers, showing how different subsystems interact and the effects that these interactions have in the transition process of UM.

Table 3. Factors that influence the adoption of alternative modes of transport service provision (Source: Authors' own elaboration based on literature review)

Category	Factor	Influence (+/ −)	References
Policy	Existence of proper legislation	+	[28, 36]
	Political will	+	[28, 37, 38]
	Taxes	−	[11, 28]
Economic	Funding	+	[28, 36]
	Fuel prices	−	[37]
Demographic	No of available fleet/quality of fleet	+	[29, 37]
	No of available parking lots	+	[29]
	Income	−	[29, 38]
	Population	+	[29, 38]
	Education	+	[38]
Technical	New technologies, ICT	+	[11, 27, 28, 38]
	Compatibility with existing infrastructures/ Infrastructure availability	+	[37, 38]
	Integrated fare / Payment solutions (debit/credit cards, etc.)	+	[27, 37]
Behavioural	Presence of shared vision for alternative modes	+	[28]
	People's attitude and habits	−	[28]
	User acceptance	+	[37, 38]
Entrepreneurial	BMI (business model innovation)	+	[11]
	Existing BM	−	[11]
Environmental	GHG emissions	+	[27, 38]
	Noise	+	[27, 38]
	Traffic congestion	+	[27, 38]

3 Methodology: System Dynamics Modelling of the UM Quadruple Helix

Regimes span across different fields, such as science, technology, policy, user practices, markets, cultural and symbolic meanings of technology and industry networks [39]. Within the multi-regime system of UM different regimes co-exist and interact to cover the different needs of the population, while three different interactions would be noticed amongst the different regimes, competition, symbiosis or integration interactions [4]. Competition interactions are those that will drive the incumbent system to a shift of its relative size, e.g. a "modal shift" from private ownership to car sharing. Regime symbiosis means that the different regimes interact, but are relative distinct from each other. Integration means that the regimes closely interact to form a new coherent system. A combination of various modes of mobility can be integrated into intermodal transport systems (e.g. MaaS).

In our work, we combine the n-tuple Helix Innovation Model approach (pioneered by Etzkowitz & Leydesdorff [41]), and more specifically the Quadruple Helix Innovation Model, with the socio-technical transition theory, using System Dynamics methodology, in order to delineate the interactions within and between the helices. We choose to use the System Dynamics methodology as it gives us the possibility to examine causal loops instead of just individual causal relations (such as the Bayesian Network where only directed acyclic graphs (DAG) are possible). For the needs of our analysis, we perceive transition as an interplay of actors "organized" in helices/subsystems, according to the N-tuple Helix Innovation Model by Leydesdorff [40]. The Quadruple Helix Innovation Model that we adopt, extends the interactions of the three main helices namely Academia, Industry and Government, which still seem to have a technology-push bias, extended by including the aspect of use/application of the social sphere by adding the helix of Society [42]. In other words, we perceive the socio-technical regime of UM to involve four interacting spheres (modeled as subsystems), 1) Industry (I), 2) Academia (A), 3) Government (G), and 4) Society (S) (Table 4).

Table 4. Quadruple Helix Model actor's actions (Source: Author's own elaboration based on [43])

Academia (Research institutes and universities)	Industry (Companies and entrepreneurs)	Government	Society (Users, consumers)
- Research and development actions - Knowledge development - Workforce development	- Experiment with, implement and provide demand for new technologies - Participate in applied R&D and demonstration in cooperation with academia	- Fund research and development and education - Legitimation by legislation - Create conducive policies and markets - Raise public awareness	- Demand for new technologies - Public movement and demonstrations - Testing and acceptance of new technologies - Legitimation of further policy regarding the increase of quality of life

Based on the Quadruple Helix Innovation Model as applied above for the case of UM, we use system dynamics tools to map the interactions within the system. The system dynamics simulation method, allows us to illustrate "the feedback processes, which, along with stock and flow structure, time delays, and non-linearities, determine the dynamics of the system" ([44, p.12]) and comprehend how nonlinear behavior emerges. System dynamics (SD) is a combination of (feedback) control theory and computer simulation which examines system behavior caused by endogenous structure [44], i.e. the dynamics of a system result from the interaction of its elements (factors, actors) represented in the model. Feedback and time delays render system behavior counterintuitive and, thus, poorly understood by business managers and policy makers who often make decisions where the outcome is the opposite of their expected aims [45]. The SD methodology combines qualitative and quantitative modelling. Shepherd [46] argues that qualitative models are best built with the input of the pertinent stakeholders and are generally represented by causal loop diagrams (CLDs). The development of a CLD is part of the model building process. It connects entities included in the system by causal relationships and as the diagram develops feedback loops become evident. These may be either positive (reinforcing) or negative (correcting or balancing) feedback loops. When a system contains both balancing and reinforcing loops then a dynamic equilibrium may be reached [46]. The role of qualitative analysis using the CLD, is important in the policymaking process as the authors state, especially in transport policy analysis where many key decision factors are uncertain or unknowable [16].

New modes of mobility compete to get established and a new "experimental/unknown" system emerges. As shown in Fig. 1, we move from State 1 to a new State 2 where the composition and structure of the socio-technical regime has changed [47]. Moving from the urban mobility system in time 1 (UMS_{t1}) to the urban mobility system in time 2 (UMSt2), new knowledge is created and used, while at the same time new technologies are available and capable to reform the urban mobility system, user preferences change, and policies and regulations are introduced to control this process of change. For example, users experience changes regarding the modality of mobility, while at the same time the value proposition of the mobility industry is changing, adding a novel feature, where users purchase services (e.g. car sharing, MaaS, etc.) rather than physical artefacts (e.g. cars, moto, etc.).

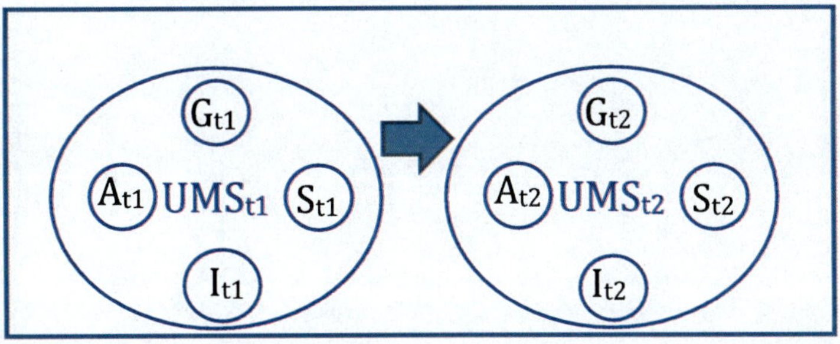

Fig. 1. Transition from an incumbent socio-technical system to a new one (Source: Authors' own elaboration)

Currently the incumbent regime in the socio-technical system of urban mobility is the privately-owned car with internal combustion engine (partially substituted by hybrid or electric) and as Sheller writes:

"Cars will not easily be given up just (!) because they are dangerous to health and life, environmentally destructive, based on unsustainable energy consumption, and damaging to public life and civic space. Too many people find them too comfortable, enjoyable, exciting, even enthralling. They are deeply embedded in ways of life, networks of friends and sociality, and moral commitments to family and care for others" [48] (p. 236).

Thus, it is difficult for people to refrain from the private car and many elements have to change in their current routine. In order to check what plays significant role and may facilitate or hinder the trajectory away from private car and towards alternative modes of mobility, we present a causal-loop-diagram (CLD) that captures the dynamics of change and the interactions taking place at different levels (i.e. technology, infrastructure investments, diffusion of use, etc.) and spheres (helices) and may lead to the establishment of alternative UM socio-technical system.

System dynamics frameworks is used to model, analyze and examine system behavior under different scenarios. It has the unique feature of distinguishing - physical and information - stocks and flows, and illustrating the feedback loops that form the structure of systems and determine emergent system behaviour. Thus, it provides a framework where non-linear behavior emerges from linear relations (expressed in linear mathematical functions). The transition to the new UM system, will be manifest by the changes in the levels of use of alternative modes of urban mobility (AMUMs). These changes will bear - hopefully positive - environmental impact [11]. It is crucial that AMUMs are expected to use more sustainable vehicle technologies such as in a manner that can further reduce ecological impacts.

4 The System Dynamics of Urban Mobility Transition

A stylized visualization of the generic feedback loops is provided in Fig. 2. In the CLD developed for the need of this work, we use factors from the four different helices, where the color of each factor represents the helix that it belongs to.

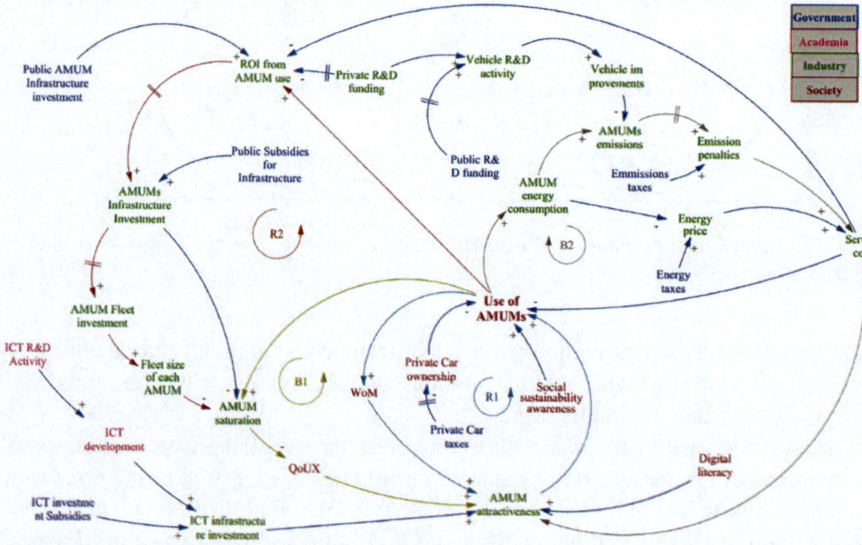

Fig. 2. CLD showing main feedback loops noticed in AMUMs diffusion. Source: own work using Vensim®.

We notice four main feedback loops: two balancing loops (B1, B2) and two reinforcing ones (R1, R2). In B1, we can see that the AMUMs attractiveness is one of the main factors that influence the AMUMs use. As the quality of user experience (QoUX) increases, the attractiveness of the AMUMs improves and the use of AMUMs is getting increased as well. However, as the use of AMUMs increases, the saturation for each mode is reached and is able to affect negatively the AMUMs attractiveness.

As the use of the AMUMs changes (B2), the energy consumption from the mobility through each mode is changing, and the emissions caused by each mode are also influenced both to the same direction. If emissions are increasing, then the relevant penalties are also rising, leading to increases of the cost of mobility. This increase of the service cost may influence the attractiveness of the AMUMs negatively and restrict their use.

The first reinforcing loop (R1) highlights the significance of "branding" and the role that word-of-mouth (WoM) may have in commuters' behavior, which belongs in the social sphere. The word-of-mouth may influence the attractiveness of the AMUMs either way; if the word-of-mouth is positive then the influence is going to be positive and the attractiveness of the mobility mode is going to increase. Changes in the attractiveness of each AMUM will in turn influence their level of use.

The second reinforcing loop (R2), highlights the industry helix, showing that the more we use each of the mobility modes, the more return on investment (ROI) mobility service providers will enjoy or the opposite. High ROI may lead to further investment on infrastructure, which -in turn- facilitates the investment on new fleet. As the fleet size increases the AMUMs saturation is decreasing leading to higher QoUX. The improvement of the QoUX leads to a higher AMUM's attractiveness and, consequently, to increase of the use of AMUMs.

Beyond the factors that constitute the four main loops of our CLD, other important factors that can influence the shift of the users towards other mobility modes are mapped. For example, the public and the private funding are two of the most important factors, both for the technology improvement that is necessary so as to improve the efficiency of the vehicle stock that is used, and for the development of the infrastructure that is essential so as to facilitate the function of the mobility servers/operators. The policies that governments decide to impose are also an essential factor influencing the way that people and industries tend to make decisions. Here, we chose to use mainly policy factors that may affect the cost of mobility services. Behavioural factors play a significant role as well, as they could influence the way that population choose the mode of their mobility. Sustainability awareness is one of them. Policies that can distract the incumbent regime, like car ownership taxes are important, as they cause destabilization. As most of the AMUMs are dependent on ICTs, the ICT R&D investment and activity are crucial for the development of complementary ICT technologies and infrastructure.

5 Conclusions and Future Research

While most of current research focuses on the micro-dynamics of particular emergent disruptive innovations, here we examine the transition of UM as a whole. It is not yet clear which mix of new mobility modes or 'Smart Mobility' innovations will be sufficient for the transition away from a socio-technical system dominated by the private car regime and the internal combustion engine. In this paper, we use the qualitative system dynamics tool, the causal loop diagram, to map the interactions within and amongst the four helices, conceived as agents/subsystems, involved in this transition process. The main contribution of our paper is that it does not focus on a specific technology or mode of mobility as most of the previous research do. We attempt to comprehend the transition of the system as a whole and the way that actors can influence the system's pathway to a new sustainable and multifarious system that aims to serve social needs and adopt to users' preferences.

Through the literature review, we understand that the transition process of the UM socio-technical system is a multifactor process. These factors can be either, economic, technical, demographic, behavioural, entrepreneurial, environmental, political, or regarding the complementary assets and the infrastructure needed. Each of these factors pertain one or more helices and change in one of these can cause changes in one or more of the four helices. Through the modelling approach, and the model built using factors corresponding to the four helices, we investigate the drivers that can cause the destabilization and the transformation of the UM system as we know it. We highlight how the behavior of different actors is influenced by different factors. Consequently, we are capable to determine the behavior and interactions at the level of AMUMs (regimes) and helices (as quasi-actors) that are involved in the development of a new UM system. In the new UM system, a new different mobility mode mix will emerge. This different mix and the modification on the level of use of each mode will show us the transition.

Understanding system behavior will facilitate future research, first to investigate to what extent alternative modes of UM may address the problems that citizens and cities face with mobility activity, and more specifically environmental issues; and second, to explore the challenges that different subsystems will face from potential alternative mobility mode adoption. Thus, we would be able to create a useful participatory policy-making tool for transition management (adoption of AMUMs, complementarities, synergies, etc.) and co-evolution of spheres (helices) that may lead to the establishment of more sustainable UM regimes and UM system by extension.

This work consists a research in progress, so our future steps include to build the SD simulation model using stocks and flows, with quantified mathematical equations and calibrate it with historical data and conduct behavioral and structural sensitivity analysis. For future research, we aim to examine more factors regarding the economic situation and demographic characteristics such as skills availability, digital literacy, etc. We believe that modelling and simulation will endow transition research with a more reflective quality, and will help us answer questions such as: 1) How and to what extent can all alternative UM modes co-exist?, 2) What synergies emerge or are needed so as for new UM modes to get established?, 3) Which changes can lead an AMUM to get established or not?, and 4) How the elements of each helix can influence the trajectory of each UM mode and of the landscape in general?

Acknowledgments. The authors would like to thank the three anonymous reviewers, for helpful critical feedback in an earlier version of this paper. This research work is co-financed by Greece and the European Union (European Social Fund—ESF) through the Operational Programme "Human Resources Development, Education and Lifelong Learning" in the context of the project "Strengthening Human Resources Research Potential via Doctorate Research" (MIS-5000432), implemented by the State Scholarships Foundation (IKY).

References

1. Feng, Y.Y., Chen, S.Q., Zhang, L.X.: System dynamics modeling for urban energy consumption and CO2 emissions: a case study of Beijing China. Ecol. Model. **252**, 44–52 (2013). https://doi.org/10.1016/J.ECOLMODEL.2012.09.008
2. Adeniyi, I., Swan, J.C.W.: Barriers to low carbon transitions and energy system innovations. In: 14th International Postgraduate Research Conference 2019: Contemporary and Future Directions in the Built Environment, Manchester, pp. 547–557 (2019)
3. Papachristos, G.: Transition inertia due to competition in supply chains with remanufacturing and recycling: a systems dynamics model. Environ. Innov. Societal Transitions **12**, 47–65 (2014). https://doi.org/10.1016/j.eist.2014.01.005
4. Geels, F.W.: Low-carbon transition via system reconfiguration? a socio-technical whole system analysis of passenger mobility in Great Britain (1990–2016). Energy Res. Soc. Sci. **46**, 86–102 (2018). https://doi.org/10.1016/j.erss.2018.07.008
5. Jittrapirom, P., Marchau, V., van der Heijden, R., Meurs, H.: Future implementation of mobility as a service (MaaS): Results of an international Delphi study. Travel Behaviour and Society **21**, 281–294 (2018). https://doi.org/10.1016/j.tbs.2018.12.004 (In Press)
6. Kamargianni, M., Li, W., Matyas, M., Schäfer, A.: A critical review of new mobility services for urban transport. Transp. Res. Procedia **14**, 3294–3303 (2016). https://doi.org/10.1016/j.trpro.2016.05.277
7. Polydoropoulou, A., Pagoni, I., Tsirimpa, A., Roumboutsos, A., Kamargianni, M., Tsouros, I.: Prototype business models for Mobility-as-a-Service. Transp. Res. Part A Policy Pract. **131**, 149–162 (2019). https://doi.org/10.1016/j.tra.2019.09.035
8. Geels, F.W.: The multi-level perspective on sustainability transitions: Responses to seven criticisms. Environ. Innov. Societal Transitions **1**(1), 24–40 (2011). https://doi.org/10.1016/j.eist.2011.02.002
9. Geels, F.W., Sovacool, B.K., Schwanen, T., Sorrell, S.: Sociotechnical transitions for deep decarbonization. Science **357**(6357), 1242–1244 (2017a). https://doi.org/10.1126/science.aao3760
10. Geels, F.W., Sovacool, B.K., Schwanen, T., Sorrell, S.: The socio-technical dynamics of low-carbon transitions. Joule **1**(3), 463–479 (2017b). https://doi.org/10.1016/j.joule.2017.09.018
11. Sarasini, S., Linder, M.: Integrating a business model perspective into transition theory: the example of new mobility services. Environ. Innov. Societal Transitions **27**, 16–31 (2018). https://doi.org/10.1016/j.eist.2017.09.004
12. Schot, J., Geels, F.W.: Strategic niche management and sustainable innovation journeys: theory, findings, research agenda, and policy. Technol. Anal. Strat. Manag. **20**(5), 537–554 (2008). https://doi.org/10.1080/09537320802292651

13. Lachman, D.A.: A survey and review of approaches to study transitions. Energy Policy **58**, 269–276 (2013). https://doi.org/10.1016/j.enpol.2013.03.013
14. Han, J., Hayashi, Y.: A system dynamics model of CO2 mitigation in China's intercity passenger transport. Transp. Res. Part D Transp. Environ. **13**, 298–305 (2008). https://doi.org/10.1016/j.trd.2008.03.005
15. Vergis, S., Turrentine, T.S., Fulton, L., Fulton, E.: Plug-in electric vehicles: a case study of seven markets (2014). https://escholarship.org/uc/item/5ps3z0f5
16. Stepp, M.D., Winebrake, J.J., Hawker, J.S., Skerlos, S.J.: Greenhouse gas mitigation policies and the transportation sector: The role of feedback effects on policy effectiveness. Energy Policy **37**(7), 2774–2787 (2009). https://doi.org/10.1016/j.enpol.2009.03.013
17. Shepherd, S., Bonsall, P., Harrison, G.: Factors affecting future demand for electric vehicles: a model based study. Transp. Policy **20**, 62–74 (2012). https://doi.org/10.1016/j.tranpol.2011.12.006
18. Zhang, T., Gensler, S., Garcia, R.: A study of the diffusion of alternative fuel vehicles: an agent-based modeling approach. J. Prod. Innov. Manag. **28**(2), 152–168 (2011). https://doi.org/10.1111/j.1540-5885.2011.00789.x
19. Janssen, A., Lienin, S.F., Gassmann, F., Wokaun, A.: Model aided policy development for the market penetration of natural gas vehicles in Switzerland. Transp. Res. Part A Policy Pract. **40**(4), 316–333 (2006). https://doi.org/10.1016/j.tra.2005.06.006
20. Ulli-Beer, S., Bosshardt, M., Dietrich, P., Wokaun, A.: What alternative drive-train technologies and policies are needed to meet a 50% CO 2 reduction target? The case of the EU-fleet (No. 2009–01–0319). SAE Technical Paper (2009). https://doi.org/10.4271/2009-01-0319
21. Armenia, S., Baldoni, F., Falsini, D., Taibi, E.: A system dynamics energy model for a sustainable transportation system. In Paper delivered at the ISDC Conference, Seoul, South Korea (2010)
22. Santos, G., Behrendt, H., Maconi, L., Shirvani, T., Teytelboym, A.: Part I: Externalities and economic policies in road transport. Res. Transp. Econ. **28**(1), 2–45 (2010). https://doi.org/10.1016/j.retrec.2009.11.002
23. Kwon, T.H.: (2012) Strategic niche management of alternative fuel vehicles: a system dynamics model of the policy effect. Technol. Forecast. Soc. Chang. **79**(9), 1672–1680 (2010). https://doi.org/10.1016/j.techfore.2012.05.015
24. Struben, J., Sterman, J.D.: Transition challenges for alternative fuel vehicle and transportation systems. Environ. Plan. B Plan. Des. **35**(6), 1070–1097 (2008). https://doi.org/10.1068/b33022t
25. Geum, Y., Lee, S., Park, Y.: Combining technology roadmap and system dynamics simulation to support scenario-planning: a case of car-sharing service. Comput. Ind. Eng. **71**, 37–49 (2014). https://doi.org/10.1016/j.cie.2014.02.007
26. Jifeng, W.A.N.G., Huapu, L.U., Hu, P.E.N.G.: System dynamics model of urban transportation system and its application. J. Transp. Syst. Eng. Inf. Technol. **8**(3), 83–89 (2008). https://doi.org/10.1016/S1570-6672(08)60027-6
27. Pangbourne, K., Mladenović, M.N., Stead, D., Milakis, D.: Questioning mobility as a service: unanticipated implications for society and governance. Transp. Res. Part A Policy Pract. **131**, 35–49 (2019). https://doi.org/10.1016/j.tra.2019.09.033
28. Karlsson, I.C.M., Mukhtar-Landgren, D., Smith, G., Koglin, T., Kronsell, A., Lund, E., Sarasini, S., Sochor, J.: Development and implementation of Mobility-as-a-Service – a qualitative study of barriers and enabling factors. Transp. Res. Part A Policy Pract. **131**, 283–295 (2019). https://doi.org/10.1016/j.tra.2019.09.028

29. Kim, K.: Can carsharing meet the mobility needs for the low-income neighborhoods? lessons from carsharing usage patterns in New York City. Transp. Res. Part A Policy Pract. **77**, 249–260 (2015). https://doi.org/10.1016/j.tra.2015.04.020
30. Barisa, A., Rosa, M.: A system dynamics model for CO2 emission mitigation policy design in road transport sector. Energy Procedia **147**, 419–427 (2018). https://doi.org/10.1016/j.egypro.2018.07.112
31. Walther, G., Wansart, J., Kieckhäfer, K., Schnieder, E., Spengler, T.S.: Impact assessment in the automotive industry: mandatory market introduction of alternative powertrain technologies. Syst. Dyn. Rev. **26**(3), 239–261 (2010). https://doi.org/10.1002/sdr.453
32. Lei, X., Zhang, J., Li, J.: A system dynamics model for urban low-carbon transport and simulation in the city of Shanghai, China. AISS Adv. Inf. Sci. Serv. Sci. **4**(1), 239–246 (2012). https://doi.org/10.4156/aiss.vol4.issue1.31
33. Kieckhäfer, K., Wachter, K., Spengler, T.S.: Analyzing manufacturers' impact on green products' market diffusion–the case of electric vehicles. J. Cleaner Prod. **162**, S11–S25 (2017). https://doi.org/10.1016/j.jclepro.2016.05.021
34. Köhler, J., Whitmarsh, L., Nykvist, B., Schilperoord, M., Bergman, N., Haxeltine, A.: A transitions model for sustainable mobility. Ecol. Econ. **68**(12), 2985–2995 (2009). https://doi.org/10.1016/j.ecolecon.2009.06.027
35. Köhler, J., Turnheim, B., Hodson, M.: Low carbon transitions pathways in mobility: applying the MLP in a combined case study and simulation bridging analysis of passenger transport in the Netherlands. Technol. Forecast. Soc. Chang. **151**, 119314 (2018). https://doi.org/10.1016/j.techfore.2018.06.003
36. Smith, G., Sochor, J., Karlsson, I.M.: Mobility as a Service: development scenarios and implications for public transport. Res. Transp. Econ. **69**, 592–599 (2018). https://doi.org/10.1016/j.retrec.2018.04.001
37. Moradi, A., Vagnoni, E.: A multi-level perspective analysis of urban mobility system dynamics: what are the future transition pathways? Technol. Forecast. Soc. Chang. **126**, 231–243 (2018). https://doi.org/10.1016/j.techfore.2017.09.002
38. Bernardino, J., Aggelakakis, A., Reichenbach, M., Vieira, J., Boile, M., Schippl, J., Krail, M.: Transport demand evolution in Europe–factors of change, scenarios and challenges. Eur. J. Fut. Res. **3**(1), 1–13 (2015). https://doi.org/10.1007/s40309-015-0072-y
39. Geels, F.W.: Processes and patterns in transitions and system innovations : refining the co-evolutionary multi-level perspective. Technol. Forecast. Soc. Chang. **72**, 681–696 (2005). https://doi.org/10.1016/j.techfore.2004.08.014
40. Etzkowitz, H., Leydesdorff, L.: The Triple Helix–University-industry-government relations: a laboratory for knowledge based economic development. EASST Rev. **14**(1), 14–19 (1995). https://ssrn.com/abstract=2480085
41. Leydesdorff, L.: The triple helix, quadruple helix,…, and an N-tuple of helices: explanatory models for analyzing the knowledge-based economy? J. Knowl. Econ. **3**(1), 25–35 (2012). https://doi.org/10.1007/s13132-011-0049-4
42. Carayannis, E.G., Barth, T.D., Campbell, D.F.J.: The Quintuple Helix innovation model: global warming as a challenge and driver for innovation. J. Innov. Entrepreneurship **1**, 1–12 (2012). https://doi.org/10.1186/2192-5372-1-2
43. de Coninck, H., Puig, D.: Assessing climate change mitigation technology interventions by international institutions. Clim. Change **131**(3), 417–433 (2015). https://doi.org/10.1007/s10584-015-1344-z
44. Sterman, J.D.: Business Dynamics: Systems Thinking and Modeling for a Complex World. Irwin McGraw-Hill, New York (2000)

45. Casey, T.R., Töyli, J.: Dynamics of two-sided platform success and failure: an analysis of public wireless local area access. Technovation **32**(12), 703–716 (2012). https://doi.org/10.1016/j.technovation.2012.08.003
46. Shepherd, S.P.: A review of system dynamics models applied in transportation. Transp. B Transp. Dyn. **2**(2), 83–105 (2014). https://doi.org/10.1080/21680566.2014.916236
47. Papachristos, G.: A system dynamics model of socio-technical regime transitions. Environ. Innov. Societal Transitions **1**(2), 202–233 (2011). https://doi.org/10.1016/j.eist.2011.10.001
48. Sheller, M.: Automotive emotions: feeling the car. Theory Cult. Soc. **21**, 221–242 (2004). https://doi.org/10.1177/0263276404046068

Children's Safe and Sustainable Independent Mobility

A Comparison of International Practices and the Situation in Greece

Garyfallia Katsavounidou(✉)

Aristotle University of Thessaloniki, 54124 Thessaloniki, Greece
gkatsavou@plandevel.auth.gr

Abstract. Children's independent mobility (CIM), especially regarding their everyday travel to and from school, is an essential component of an equilibrated childhood. It affects positively many aspects of their lives: physical health, social and cognitive skills, and overall sense of wellbeing. In the postwar period the percentage of children travelling independently to school declined dramatically in developed countries, especially due to traffic danger. An urban environment that does not provide for safe routes to school intensifies the use of cars for short rides, creates congestion in school zones, renders streets even more prone to traffic accidents and it is certainly unsustainable, from a climate-change point of view. Due to its multilayered importance, research on CIM, originally a subject of environmental psychology, has lately been expanding in many other fields, such as public health, transport design, urban design, and planning, thus providing valuable qualitative and quantitative data that have steered changes in policies and practices around the world. In Greece, children as pedestrians are killed or injured in road accidents in numbers that continue to be tragically high; at the same time, children are being chauffeured to and back from school by their parents on an everyday basis, in what we could call "ridiculous car trips". However, CIM has so far been quite an overlooked area of research. The paper presents findings from studies and selected examples of good practices that have promoted children's independent mobility around the world and compares them with existent situation in Greece.

Keywords: Children's independent mobility · Children's safety from traffic · Children's wellbeing · School travel · Good practices · Greece

1 Introduction

From the late 1990's onwards, research on children's everyday lives and on how the quality of the urban environment intersects with the wellbeing of children has much advanced as an interdisciplinary field of inquiry. While until then it was primarily examined through the lens of environmental psychology and sociology [1], it is

E. G. Nathanail et al. (Eds.): CSUM2020, AISC 1278, pp. 539–549, 2021.
https://doi.org/10.1007/978-3-030-61075-3_53

currently recognized as an important area of research also for urban design, urban planning and transport design [2]. Children's Independent Mobility (CIM) is an overall concept to describe, analyze and design for safe and sustainable ways for children to travel independently, that is, without being driven at the back seat of a car. In this paper, we will mostly discuss the spatial aspects that frame children's everyday mobility patterns, and particularly the role of urban design and urban policies in shaping – and changing – these patterns, through: a) a review of the key findings of studies on CIM around the world; b) a presentation of successful cases of positive change towards safety for children using the streets; and c) an attempt to compare these theoretical and practical approaches with the existing situation in Greece.

To our knowledge, no comprehensive quantitative or qualitative studies exist about how children travel to and from school in Greek cities and towns, how the design of urban streets affects their safety and shapes their mobility patterns, or how these patterns differ from neighborhood to neighborhood; nor has there been any evaluation of the many repercussions of the phenomenon related to public health, the economy and the environment. A field study contacted by the author [3] regarding the child-city relationship in middle childhood, based on ethnographic methodology,[1] revealed how negatively the "threat of the car" influences children's mobility. Through their field observations, the participants emphasized how unfriendly open public space is for them: absence of sidewalks, sidewalks blocked by cars, cars blocking crossings, motorcycles in pedestrian streets, lines of cars creating a visual barrier between the child walking on the sidewalk and the street (see Fig. 1). Unfortunately, such experiences constitute the norm in Greek cities and towns; like "canaries in a coalmine",[2] children's experiences showcase the level of wellness of a society. It seems that in the Greek context, the lack of research on CIM is strongly related to an overall disregard of children as users of urban space and citizens, in a broader sense.

[1] The study took place between 2009–2011 in the town of Veria in northern Greece (population 47,000), and involved eighty-one children participants, aged from 9 to 12 years old, living in five different districts of the town, who were asked to photograph the places of their everyday lives and then write their comments about them, creating a photo diary.

[2] For the pioneer 'geographer of children' William Bunge [4], children are like "canaries in a coalmine", meaning that they reflect the pressures of the urban environment with greater sensitivity due exactly to their vulnerability. Bunge was the first to introduce the concept of "geographical expedition" for doing research on the everyday spaces of children.

Fig. 1. Two photos taken by children participants in ethnographic study on the use of urban space during middle childhood. Captions given by the children who took them are "Cars flood the sidewalks" (l) and "This is a play space in my neighborhood, but it is so full of cars" (r).

2 The Shrinking of Children's Independent Mobility and Its Repercussions

Colin Ward, the British sociologist and author of the seminal book *The Child in the City* [5], includes a passage of a childhood memory of the author Alfred Parr, who grew up in Bergen, Norway, at the turn of the twentieth century. When his mother sent him on an errand to buy fish, little Alfred would walk from home to the train station for fine to ten minutes, buy a ticket, watch the coal-powered train pull in, board the train, go across the bridge separating the fishboat harbor from the sail harbor, then go through a tunnel, arrive at the terminal and get off, walk past the central park of the town (where he would stop to watch the band playing), stroll by the central shopping district, pass by the fire station and look inside through the open gates, pass by the "ancient" buildings of the Town Hall square, enter the fish market, explore, select the fish, haggle about the price, pay and then take the same route back home. The extraordinary thing, and what makes such a trip quite unimaginable to be done by a child on his/her own in our times, is that he was four years old.

This passage already shows us two very important aspects related to CIM: first, that childhood is a social construction [6], meaning that the child is capable of what the society, at a given time and space, believes he/she is capable of, and secondly, that children's everyday travel exceeds the utilitarian movement from point A to point B and in reality includes a complex set of experiences – sensory, social, emotional. Programmed as he/she is, in early years, to absorb as much new knowledge as his/her brain is capable of, the child transforms travel into an opportunity to get a rich array of valuable input – especially when this travel is done unaccompanied. It is worth noting that when adults remember their childhoods, usually the most intense memories are from places outdoors, while being on their own or with friends – that is, without adult supervision [7]. In a global study [8], responses on what is a "good" environment by children in eight different countries, living in neighborhoods with a variety of characteristics, coincided: affordability of places to play and safe and independent mobility were the main aspects of "city-friendliness".

Nonetheless, studies across cultures and countries have shown that over the past 50 years there is a significant decrease in freedom of children to travel on their own, compared to previous generations [9]. Historical research has shown that the generations of children living in the United States have experienced a significant shrinking in the range of independent mobility (how far away from their homes they, or their parents, consider safe to roam about) and an increase in the age thresholds they attain these ranges [10]. While in 1969 eighty-nine percent of children aged five to fourteen, living within a mile from school, walked or biked, in 2009, this percentage fell to a mere thirty-one percent [11]. High percentage of children being driven to school is a common pattern in most Western countries, across continents and cultures, from Australia [12] to Italy [13] and from Canada [14] to Hong Kong [15].

The case of Australia is typical of this shift. In a similar way to the United States, only with a time difference in the occurrence of change, the percentage of children who travel independently to school has decreased from sixty-one present in the 1990's to thirty-two percent in 2012 [16]. In a field study of children aged between nine and thirteen, living in Perth, Western Australia, researchers applied a mixture of methods (GPS tracking, activity diaries kept by the children, surveys, and photo/pictorial collages) in order to understand the connection between affordances and mobility patterns [12]. Despite a wide range of spaces in the vicinity, the shrinking of independent mobility inhibited children from using them. Instead, they were being chauffeured in places for play and recreation located much farther from their homes.

The obstacles and restrictions in children's mobility (admittedly, not only because of traffic danger, but also due to security concerns from the parents' part, especially in large cities) have a negative impact on their mental and emotional health, for example their social skills and behavior [17]. They also greatly affect the level of their physical activity, of which active travel to and from school is a big part. From a health perspective, walking is the simplest and easiest form of physical exercise, and the most readily incorporated in everyday activities [18]. Thus, an environment that provides good conditions for walking is a positive contributor to people's health, in terms of reducing the risk of obesity and other related health conditions.

3 Children's Safety from Traffic as a Social Imperative

According to Pooley [19], there are seven main factors that influence CIM. The first four are related to the child: age, gender, socioeconomic status and family structure; the other three are external influences: the prevailing transport technology, the urban structure, and transport networks available. Likewise, the matrix of parameters related to the acute decline in CIM is cross-sectional. Social perceptions about childhood have changed drastically through the twentieth century and parental patterns have shifted towards over-protection [20]; family relations, extended work schedules for adults and a continuous decrease of free time of children have also influenced how, where to, and for how long children travel in their neighborhood. But, if we take all these social parameters on one side, the other side – i.e. the physical characteristics of the urban environment and the prevailing means of transport – seems to weigh much more. In a reciprocal way, what the social sciences classify as "domestication" of childhood has its origins in how cities and towns in the postwar era became dramatically dangerous for children as pedestrians, due to the domination of the private car.

Indeed, many advanced Western European countries began to face the problem of child mortality due to traffic already in the 1960's. For example, from 1955 to 1971, Denmark, today a bulwark of sustainable mobility, had the highest percentage in children's deaths due to traffic accidents in Western Europe. Schools in Odense became the pioneer in changing this, by drawing maps of children's routes and mapping the dangers along these routes. Data were then used by the city to make the urban environment safer for children, through the designation of slow-speed areas, creation of traffic islands, and separate bike paths. Since then, there has been an impressive increase in the percentage of students walking and/or cycling to school, while decreasing the total number of crashes by eighty-two percent. The Odense example became the inspiration for the Safe Routes to School Program in the United States [11].

Similarly, back in 1972, in Dutch cities, nowadays considered among the most pedestrian- and bicycle-friendly in the world, pedestrian deaths by motor vehicles had reached record levels, with more than 3,300 people dead, 500 of whom were children. A year before, the six-year-old daughter of journalist Vic Langenhoff had died in such an accident, hit by a speeding car while walking on her way to school. After this tragic death, her father ran a series of articles on the national newspaper *De Tijd*, entitled "Stop de Kindermoord" – or "Stop the Child Murder". He emphasized that a society in not just until a child is free to run to meet his or her father without risking being hit by a car [21]. This polemical article became a national sensation and raised awareness in Dutch society that allowing cars to take over urban streets meant allowing children to be murdered. People's reaction took the form of active protest, picket marches, and activist demonstrations, demanding a safer and more children-friendly environment (see Fig. 2). In response to this citizens' movement, a national official committee was formed to examine the problem. A few years later, in 1976, Dutch legislation on street design integrated the principles and specific design requirements of the *woonerf*, which has by now become a standard type of street in many countries around the world, known as "home zone", "complete street" or "calmed-traffic street".

Fig. 2. "Stop the Child Murder" protests in the Netherlands in 1972. Source: https://lcc.org.uk//pages/holland-in-the-1970s

Another example of synergy between citizens' movements and public administration is that of the Safe Routes to School (SRTS) Program in the United States [11]. The SRTS program is based on the same principle with the Odense initiative: when it comes to CIM, it is the local school community who should act as the main agent and coordinator. The program started as a community initiative in the Bronx, NY, in 1997 and as grassroots efforts intensified, it was adopted by the Federal Government in 2005; as a recognition of the importance of CIM, the SRTS program has received a federal budget of 1.2 billion dollars over seven years. The program is locally based and includes the "4 E's": Education, Encouragement, Enforcement, Engineering:

1. Education: the philosophy of the program is that adults should be educated on the issue of CIM as much as students. Adults, as drivers, are responsible for abiding to road safety rules, but also, as the ones making decisions about their children's mobility choices, they need to be educated about the benefits of walking and cycling for their children's health. They also need to become aware of the negative impact driving has on the environment – at a local and at the global level.[3]
2. Encouragement: students and the local community become aware of the positive effects of walking and biking through student contests, school activities, and via organizing special days and events, such as the "Walk to School Day", an idea that started in Great Britain in 1994 and is now internationally celebrated.
3. Enforcement: sometimes it is difficult to enforce the obvious, that is, abiding to the speed limits and yielding to pedestrians. An innovative concept called Pedestrian Safety Enforcement (PSE) was first tried in Redmond, Washington in 1998. Based on people's complaints about violations by drivers, special PSE operations by the police department are held on specific sites. A plainclothes officer attempts to cross and officers document if motorists yields – if not, citations are issued. The program has significantly increased police officer awareness of the problem.
4. Engineering: last but not least, walking and biking to and from school largely depends on whether the urban design along the child's route is human-centered, with streets providing for ample space to walk, separate bike paths, and sufficient traffic-calming measures – such as speed bumps, raised crossings, traffic islands etc.

Nonetheless, trips on foot or by bike continue to be a risk for children in the US: the leading cause of death in the US between 2010–2014 for children and adolescents (aged five to nineteen) was unintentional injuries, of which sixty percent were from traffic-related accidents. In a study of travel behavior during school drop-off in one hundred Toronto elementary schools, it was found that risky behavior around schools, in the so-called "drop-off zones", was pervasive [14]. A much more successful strategy has been lately applied in Norway: the "heart zone" [22]. The measure was first implemented in 2016 in seventy-seven schools in Bergen (the same town where little Alfred Parr grew up), and, having positive results, it has been adopted in school areas in the capital, Oslo. The "heart zone" is a completely car free zone around a school,

[3] For example, data shows that car engines idling at the perimeter of schools is a major source of pollution (idling engines consume 3.5 L of gasoline an hour; 12% of urban smog is attributable to idling vehicles). A "No Idling" campaign in Toronto, Ontario, was organized as a social marketing effort, using posters, stickers, printed material, and involving volunteer parents and school staff.

where it is not permitted to drop off or pick up pupils by car. Dropping-off takes place outside of such a zone, in designated spaces. With this, and other strategies, Norway achieved its "Vision Zero" in 2019 – that is, zero pedestrian or bicyclist deaths.

4 The State of Children's Safety and Mobility in Greece

Despite a significant reduction in road accidents and fatalities between 2010 and 2017, Greece continues to have one of the worst scores in the European Union, with sixty-nine deaths per million inhabitants, while the EU median is forty-nine [23]. On the issue of pedestrians' safety, an analysis of the annual statistical reports on the number of victims in road accidents paints a very grim picture regarding pedestrian casualties: between 2010 and 2018, a total of 1389 people died while walking in Greek streets. On average, about nine percent of the casualties were children and young people (ages 0–24): in the period 2010–2018, 117 lost their lives, 238 were severely injured and 3833 suffered minor injuries (see Table 1).

Table 1. Pedestrians' casualties in traffic accidents in Greece.

	Ages 0–24			All ages		
Year	Deaths	Severe injuries	Minor injuries	Deaths	Severe injuries	Minor injuries
2010	9	37	503	179	265	2350
2011	21	37	493	223	210	2175
2012	18	40	447	170	208	2057
2013	13	24	417	151	159	1895
2014	9	24	403	125	141	1929
2015	9	30	394	128	139	1833
2016	15	16	407	149	111	1810
2017	12	16	376	118	108	1894
2018	11	14	393	146	101	1818
2010–2018	117	238	3833	1389	1442	17761

Source: Hellenic Statistical Authority (https://www.statistics.gr/), author.

As far as independent mobility goes, again due to lack of data, all we know, simply by watching any elementary school entrance of any Greek city or town in the morning, is that many students are driven to school by a parent/guardian. Given the fact that urban districts in Greece fall into the compact city typology [24] and that in most urban schools the radius of the school district is less than one and a half kilometer, this kind of trip can be called both unnecessary and unsustainable.

Children's safe and sustainable mobility doesn't seem to be a pressing issue in Greek public conscience, either. It arises as such only momentarily, for instance whenever there is a fatal road accident involving a child – who was trying to cross the street, or waiting at a bus stop, or running onto street to catch a ball. Even then, in the coverage of such tragic accidents, the media and public discussion platforms often put the blame on the "lack of education on road safety at schools". In pedagogical terms, Greek society teaches children to be fearful and always careful with cars instead of finding a solution to the harmful impacts of traffic. Similarly, school activities regarding CIM usually involve talks by police officers, informing children about road rules and signage etc. But "why is it the child who is burdened with the responsibility of road safety when they are not the ones creating it?" [2, p. 47].

In Greece there is a tendency to present traffic danger as a *problem of the children*, separate from the physical parameters of the urban environment (absence of substantial sidewalk, lack of pedestrian crossings etc.) and the tragic incompetence of drivers. Policy makers evade the real problem: that the built environment allows for harmful and risky behavior from the part of the adults. Instead of "educating" the child to adjust himself/herself to the patterns of motor traffic, examples around the world show that the best practices entail a modification of the environment where children grow up, through changes in traffic rules and a pedestrian-friendly street design [25].

A positive step towards this direction was made in Greek national legislation in 2013, with the publication of the Ministerial Decision entitled "Technical Guidelines for traffic interventions in urban environment for application in school zones and in areas with heavy traffic in the context of improving road safety" (TG) [26]. In its first part, it outlines the list of documentation data needed to be collected, on accidents (categorized by age of victims and by cause and circumstances of the accident), traffic (load, flows, pedestrian behavior, etc.), geometric/physical and environmental attributes of the streets around the school; and economic data. However, there is no obligatory analytical phase, since, according to the TG, local municipality "is knowledgeable of the local conditions and traffic problems, at any case." In the second part, there are detailed descriptions of the existing various types of traffic-calming measures, such as speed-stoppers, noise-making strips, reflectors ('cat's eyes'), anti-parking pollards and snake-shaped street design. Finally, there are examples of how the retrofitting of streets in the perimeter of schools should be done, using a mix of the mentioned measures, such as vertical signs, raised pedestrian crossings, railings in front of the school entrance, as well as the introduction of a "children's movement guide", mimicking the "blind people's guide", a mandatory element in sidewalks in Greece.

To this day, almost seven years after becoming mandatory, these guidelines have not been applied in their entirety in any school district, probably since no budget has been provided to municipalities to cover implementation costs, but also because they are not easily applicable in the constrained geometric characteristics of streets in compact and/or historic cores of Greek cities and towns. And, perhaps most importantly, the issuing of the Technical Guidelines has not been supported by any education, encouragement or enforcement on CIM, which should accompany changes in the engineering of the environment. Nonetheless, the Sustainable Urban Mobility Plans (SUMP) which are currently in process in many Municipalities in Greece present an excellent opportunity for including and promoting the concept of CIM.

5 Conclusion

The implications of the shrinking of CIM worldwide cannot be overstated. In terms of physical health, the decline in everyday physical activity is associated with obesity and diabetes in children. From a psychological point of view, travelling around the city by one's self, being aware of the environment, encountering many different other people, is crucial for a child's emotional growth and sense of independence. Cognitively, the experience of walking on one's own is invaluable in gathering information about "how the world works"; especially for the growing mind that absorbs new knowledge like a sponge, daily travel offers a wealth of new experiences. And lastly, from a pedagogical point of view, active travel at a young age, in the form of walking, cycling and using public transport, shapes lifelong sustainable mobility patterns.

Although CIM is of great importance for children's wellbeing, there is hardly any research on the subject in the Greek context. Existent legislation and practices in Greece are rudimentary and haphazard, and interventions are sporadic and without an overall strategy. There is a great discrepancy between how CIM is approached in Greece and in countries where practices have been successful (see Table 2).

Table 2. Differences between successful practices on CIM and existent practices in Greece.

Successful practices	Existent situation in Greece
The problem of CIM came to the forefront by bottom-up movements and media coverage; change becomes a social imperative	No existent bottom-up movements; haphazard and misleading media coverage; no coordinated collective effort towards change
Generous public funding of CIM as a field of research and practice	No public funding for research and/or physical implementation of CIM
Data collection on CIM (including maps, routes, drawn in the scale of each school) as important and necessary step for action	Data available is only of statistical nature and about accidents; no data, quantitative and qualitative, on children's mobility patterns
Strategic approach towards CIM, based on data, using participatory methodology and defining specific goals	No overall strategy; Legislation on "how-to" technical specifications without preliminary analysis or people's involvement
Holistic approach of intervention: raising community awareness, physical improvements, education, enforcement	Interventions on the physical environment only, in a "public works" mentality
Regarding education: informing all actors involved, and raising awareness on adults (parents, school staff, police officers)	Regarding education: police officers informing children of road safety rules to abide to

Best practices around the world show us that safety for pedestrians – including children – is improved through education, encouragement, good street design, lowering of speed limits, ample sidewalks, and, ultimately, reducing the number of cars in urban streets. These practices are based on concrete scientific data and on quantitative and qualitative studies. They are most successful when social awareness is raised through education: when the society is well-informed about the causes of the problem and the possible solutions. There must be a systematic effort in the Greek academic world to bring the issue of children's safe and sustainable independent mobility to the forefront, through interdisciplinary research, based on synergies between the social sciences on one hand and urban and transport design on the other.

References

1. Spencer, C., Woolley, H.: Children and the city: a summary of recent environmental psychology research. Child Care Health Dev. **26**(3), 181–198 (2000)
2. Waygood, E.O.D., et al.: Transport and child well-being: an integrative review. Travel Behav. Soc. **9**, 32–49 (2017)
3. Katsavounidou, G.: Child, city, play: A polyphonic biography. Doctoral Dissertation. University of Thessaly (2012)
4. Bunge, W.: Fitzgerald: Geography of a Revolution. First published in 1971. University of Georgia Press, Athens (2011)
5. Ward, C.: The Child in the City. Pantheon Books, New York (1978)
6. Ariès, P.: Centuries of Childhood: A social history of family life. Random House, New York & Toronto (1962)
7. Sebba, R.: The landscapes of childhood: the reflection of childhood's environment in adult memories and in children's attuites. Environ. Behav. **23**, 395–422 (1991)
8. Chawla, L. (ed.): Growing up in an Urbanising World. Earthscan, London (2000)
9. Hillman, M., et al.: One False Move: A Study of Children's Independent Mobility. Policy Studies Institute, London (1990)
10. Gaster, S.: Historical changes in children's access to US cities: a critical review. Child. Environ. **9**(2), 34–55 (1992)
11. Center for Health Training and the National Highway Traffic Safety: Safe Routs to School: Practice and promise, US Department of Transportation (2004)
12. Babb, C., et al.: Children's active travel, local activity spaces and wellbeing: a case study in Perth. WA. Travel Behav. Soc. **9**, 81–94 (2017)
13. Prezza, et al.: The influence of psychological and urban factors on children's independent mobility and relationship to peer frequentation. J. Commun. Appl. Soc. Psychol. **11**(6), 435–450 (2001)
14. Rothman, L., et al.: The school environment and student car drop-off at elementary schools. Travel Behav. Soc. **9**, 50–57 (2017)
15. Leung, K.Y.K., Loo, B.P.Y.: Association of children's mobility and wellbeing: a case study in Hong Kong. Travel Behav. Soc. **9**, 95–114 (2017)
16. Schoeppe, S., et al.: Australian children's independent mobility levels: secondary analyses of crosssectional data between 1991 and 2012. Child. Geogr. **14**(4), 408–421 (2016)
17. Hüttenmoser, M.: Children and their living surroundings: empirical investigations into the significance of living surroundings for the everyday life and development of children. Child. Environ. **12**(4), 403–413 (1995)
18. Hwang, J., et al.: Social inequalities in child pedestrian traffic injuries: differences in neighborhood built environments near schools in Austin, TX, USA. J. Transp. Health **6**, 40–49 (2017)
19. Pooley, C., et al.: A Mobile Century?. Ashgate Publishing, Hampshire (2005)

20. Holloway, S., Valentine, G. (eds): Children's Geographies: New Social Studies of Childhood. Routledge, London & New York (2000)
21. Feddes, F., de Lange, M.: Bike city Amsterdam: how Amsterdam became the cycling capital of the world. Utigeverij Bas Lubberhuizen, Amsterdam (2019)
22. Tinga, R.: Oslo experiments with car free 'heart zones' around schools. ELTIS Homepage (2018). https://www.eltis.org/discover/news/oslo-experiments-car-free-heart-zones-around-schools, Accessed 03 June 2020
23. European Commission: Road Safety 2017: How is your country doing? http://ec.europa.eu/roadsafety/, Accessed 09 Mar 2020
24. Papagiannakis, A., et al.: Transit-oriented development in the southern European city of Thessaloniki introducing urban railway: typology and implementation issues. Eur. Plan. Stud. (2020). https://doi.org/10.1080/09654313.2020.1724267
25. Davis, A., Jones, L.J.: Children in the urban environment: an issue for the new public health agenda. Health Place **2**(2), 107–113 (1996)
26. Ministerial Decision, Greek Ministry of Infrastructure and Transport: Technical Guidelines for traffic interventions in urban environment for application in school zones and in areas with heavy traffic in the context of improving road safety (ΔΜΕΟ/Ο/3050 ΦΕΚ 2302/Β΄/16-08-2013) (2013)

Smart Cities

Megatrends that Affect Sustainable Mobility Planning and Their Implications on Sports Tourism: The Case of the Authentic Marathon, Athens

Eleni Anoyrkati[1(✉)], Thanos Kenanidis[2], and Kostas Alexandris[2]

[1] Coventry University Enterprises Ltd., Puma Way, Coventry CV1 2TT, UK
e.anoyrkati@coventry.ac.uk

[2] Aristotle University of Thessaloniki, Thermi, 570 01 Thessaloniki, Greece
{kenanidi,kalexand}@phed.auth.gr

Abstract. The ongoing trends and future challenges faced by the increased mobility demand lead to the development of sustainability measures that would support greener, accessible and safer transport means.

According to the World Tourism Organisation, Sports Tourism represents the fastest growing sector in global tourism contributing to a great extend to an increased need for sustainable mobility. Many destinations and host locations already possess the appropriate infrastructure for accessible and seamless mobility to enhance the tourists' experience and capitalize on this global opportunity.

This paper aims to analyse the importance of sustainable mobility implementation on the development of sports' tourism sector integrating as a case study the Athens Marathon: The Authentic. Using the Delphi technique based on transport experts' opinions, the Megatrends affecting sustainable mobility are identified and prioritized. The prioritized Megatrends were assessed by the Athens Marathon key stakeholders using standardized open-ended exploratory interviews to investigate to what extend the sustainable mobility Megatrends relate and affect the transport planning for the successful implementation of the Marathon. Policy recommendations were put forward to pave the way to the application of a sustainable mobility system that will also correspond to the needs of this event.

Keywords: Sustainable mobility · Trends · Sports events

1 Introduction

The European Commission's objective is to achieve a 'Smart, green and integrated transport', therefore the transport thematic priority is treated as a holistic research area that incorporates societal needs and not only technological priorities. The main focus is to develop a resource efficient transport system that respects the environment, offers enhanced mobility options with less congestion, better safety and greater security. However, all this needs to be accomplished 'At a time of public budget constraints, major demographic changes and increasing global competition' (EC 2010, p. 2)

E. G. Nathanail et al. (Eds.): CSUM2020, AISC 1278, pp. 553–561, 2021.
https://doi.org/10.1007/978-3-030-61075-3_54

Policy decisions should be based on the best, in the sense of accuracy, available educated projections of where the future is going. Over the last years, the foresight science has made huge steps in actually developing and enhancing a number of methodologies that can be used. (Popper et al. 2012). According to Vejlgaard (2008), Megatrends are cultural, economic, political and technological changes that have not yet happened and their effects or implications are reflected on the whole or almost entire society'.

Previous research has shown that the organization of sports events tend to impose several challenges to the host cities with regards to transportation. The most common is the increase the volume of tourists over a specific period that results on increased traffic volumes (Humphreys et al. 2017), while especially in the case of major and mega-events they do have a huge influence on urban and transport planning through the requirements they impose on cities (Kassens 2009). Manelli (2010) suggests that apart from the additional traffic flows, the major events also create extra pollution so one of the challenges is to also promote a more sustainable mobility that can also assure normal transport operations for the citizens too during the event

Although the literature so far has examined the impact of major events on the transportation system of the host cities, there is no previous research that actually suggests a plan that can be implemented from a policy perspective to support the transition from a transport burden event to the achievement of sustainable mobility.

The paper provides policy directions based on Megatrends that impact the sustainable mobility planning of Athens Marathon. Sustainable mobility issues are a topic of strategic importance for everyday life, with major impact in the life of human beings and its quality, especially when short-term events impose certain challenges and burdens in the transport system.

2 Research Objectives

The main research questions addressed in this paper are:

- What are the current Megatrends affecting the passenger sustainable mobility system?
- What are the most important (ranking) Megatrends affecting the passenger sustainable mobility system?
- Which of these Megatrends need to be considered for sustainable mobility planning for Athens authentic Marathon?
- What are the policy directions that need to be followed in order to achieve sustainable mobility during the organization of Athens Marathon?

3 Literature Review

3.1 Megatrends Affecting Sustainable Mobility

Demographic changes are leading to increased demand for mobility while energy consumption and emissions need to be reduced providing at the same time accessible to all mobility solutions (EC 2017). Structural demographic variables such the decline of fertility rate, increase of life expectancy, population ageing, growth of single households and emigration have detrimental impact on mobility patterns (ESPON 2010).

Ageing population is becoming a global challenge with a prediction of persons aged 60 or over that will reach two billion by 2050 (UNFPA 2012). Shrestha et al. (2017) reports that there is an increased mobility by older people who actually travel mainly for healthcare purposes.

May and Marsden (2010) support that the continuous urbanisation which is expected to grow even further by 2050, does not come with an equivalent grow and development of the transport infrastructure. If transport infrastructure was developed with the same pace as the urban expansion then all transport modes (motorized and non-motorized) could join forces in offering co-modal solutions that support urbanisation and improve the quality of life (Kamga 2015). This is particularly witnessed in megacities where more than half of the inhabitants need to commute to city centers combining the use of more than one transport modes (Mather et al. 2011).

The advances on Information and Communication Technologies have had a tremendous impact on transportation as the digital transformation led to the adoption of the Internet of Things (IoT), Big Data, artificial intelligence and cloud systems in the mobility operating systems both on the technical side but also on the development of new business and operating models (Anoyrkati et al. 2016). The mobility concepts such as Mobility as a Service (MaaS), E-Mobility, Traffic Management as a Service (TMaaS), Platform as a Service (PaaS), and Infrastructure as a Service (IaaS) are expected to dominate the future and replace the traditional modes of transport. According to a research conducted by the EC funded project OPTIMISM (2013) though these concepts can deliver solutions that can tackle social and economic challenges, there is an increased concern about issues surrounding data privacy, fair competition and passengers' protection.

3.2 Sustainable Mobility Planning Considerations for Sports Events

According to Parker (2019), sports' tourism is fastest developing sector of the travel industry. The impacts of sports tourism and in particular of the organisation of major events, tend to occur to the overall urban environment. These events create a number of extra trips which – in some cases- are incompatible with the capacity of the transport systems (Rubert 2010) Another particularity of these events is that they are temporarily and therefore any transport planning decisions need to be based on fixed end date solutions.

The literature suggests that the planning and operation of sports events needs to take into consideration and try to minimise possible negative impacts such traffic: these should be directed towards sustainable mobility and development. Authors recommend that proper transport strategies should be built to tackle the mobility issues surrounding these events (Getz 1997; Dwyer et al. 2000; Gnoth and Anwar 2000; Breen et al. 2001; Bowdin et al. 2001; Jones 2001; Wood 2006).

Humphreys et al. (2017) examined the Major League Baseball (MLB) games impacts and they suggest that the traffic increases dramatically during the events and the fact that these are usually organised in large metropolitan cities causes a greater affect in the inhabitants too. Manelli (2010) looked upon Olympics and Football International Tournaments and concluded that public transportation is the main means that it is being used by the tourists during the events. Menezes (2014) investigated the World Cup event and suggested that the coordination has key role in transport planning of sport events. This refers to a constant dialogue between the organizers of the event, government, planners and users of transport. Similarly, Kassens (2010), who focused his analysis on Olympic Games, supports the same path of communication between the different actors but also introduces the idea of investigating and aligning Megatrends of government policy, economic power, technology and social with both transportation and sports events. Sustainable mobility and public transportation should be at the heart of transport planning decisions has been also suggested by Malhado et al. (2013). More specifically, public transport should be available because the perceptions of 'travel time' and 'convenience' influence tourists' attitudes and might be important targets for promoting behavioral change towards sustainability

4 Methodology

Secondary Research for Developing of the List of Megatrends: The trends were categorized according to the 2005 World Summit on Social Development where sustainable development was the result of three forces: economy, society and the enlivenment. The literature review was conducted on the basis of the term 'Megatrends', reviewing both grey and academic literature.

The literature review derived a long list of Megatrends connected to sustainable mobility. These were further categorized into clusters and provided input in the Delphi consultation. A total of fifty-two trends were identified.

Experts' Survey on Megatrends: During the conduct of this research, a number of experts participated using the Delphi method to deliver their views on the Megatrends that support the application of sustainable mobility. Delphi has been reported as a method that has been frequently used for both qualitative and quantitative foresight research (Rowe and Wright 1999). The experts who participated in the Delphi were from policy, academia and industry.

The steps involved in this research were the ones introduced by Cyphert and Gant (1971), Brooks (1979), Ludwig (1997) and Custer et al. (1999). These included three iterations:

1. First Online Questionnaire: A questionnaire was used to ask experts to identify the main drivers for sustainable mobility. A literature review was conducted prior to that in order to derive the long list of trends and Megatrends.

2. Online consultations: During this phase, the experts were contacted and asked to read and review the results of the first phase.
3. Second Expert Online Questionnaire: The last phase included the final ranking of the Megatrends in terms of their importance in achieving sustainable mobility.

A total of 59 experts representing 17 European countries participated in the Delphi survey.

Survey about the Athens Authentic Marathon: The second study focused on the sustainable mobility Megatrends that should be considered on mobility planning for Athens Marathon. Therefore, to obtain an understanding on what extend these Megatrends are still relevant in transport policy planning in the case of extraordinary short-lived events such as the Athens Authentic Marathon; Ten interviews were conducted with key stakeholders from the organization committees of the Marathon using standardized open-ended exploratory semi-structured interviews. This type of interviews has offered benefits both to respondents (giving then a greater sense of engagement in the interaction) and to researchers (giving us more richly textured data on the Megatrends connection to Marathon) (Singer et al. 2017).

5 Results: Ranking of Megatrends Affecting Sustainable Mobility

The most important Megatrends identified, in order of importance, are: the development of large metropolitan areas, ageing society, urbanization, charges (eg congestion), inadequate infrastructure investments, pricing, financial recession, taxation (on fuels), sustainable development unemployment rate, renewable energy options.

During the third iteration of the Delphi method, the ranking revealed that the most predominant trend was the large metropolitan cities followed by urbanisation. The order of the ranking has been: urbanisation, ageing society, financial recession, sustainable development, unemployment rate, Inadequate infrastructure investments, Taxation of fuels, Pricing, Charges (e.g. for congestion), Renewable energy options, International Trade.

6 Impact of Sustainable Mobility Megatrends on the Athens Authentic Marathon

The AMA (Authentic Athens Marathon) is a big part of sports tourism in Greece according to the number of participants (over 50,000 runners) and their attendants. Since 1983 the race has been held every year, reaching 2019 (37th marathon, 5-day program) with over 60,000 runners, 20,300 of them were marathon runners and 12,000 of them were tourists from 120 countries (AMA 2019). The AMA, since 2005, has been focusing heavily on supporting sustainable development by introducing measures on recycling and creating offsets on the carbon dioxide emissions created by airplane travel.

Ten key stakeholders of the AMA were interviewed in an effort to identify key measures that can be introduced to support sustainable mobility during the AMA implementation. The sample consisted of the chairs of the organising committee and serial Marathoners.

The stakeholders suggested that large metropolitan cities, urbanisation, financial recession and sustainable development are the main challenges that they are relevant with the transportation issues around Marathon. The urban sprawl combined with the density of population in Athens is one of the main challenges as the congestion and pollution is already evident even without the existence of the AMA. Concerns about environmental protection and sustainable development have been increased. The AMA is already trying to offset the pollution caused by the increased tourists this period. Lastly, the financial recession has two sides: tourists are willing to spend less, therefore, public transportation would be a preferred option. And on the other side, infrastructure investments, from a local and national authorities' side, are also limited.

According to the AMA stakeholders, the main obstacle in the smooth transportation of runners and tourists is their transfer to the starting point of the Marathon where there is no public transportation offered. Unlikely, to the other Marathons, AMA has a different starting and ending point. This practically does not create huge traffic problems from the operation of the Marathon within the city but only traffic related to the increased number of tourists. The only traffic that has been observed is during the 'rush hours' which is practically the time when the athletes and their companions travel to the Marathon area as well as the use the public transport to the same direction by the inhabitants. The use of private cars to the starting point is not possible, because there is no parking space on the spot and the return is difficult due to the traffic. Regarding the route of the marathon, there is no organization on the side road so that the transition of coaches, relatives and tourists is facilitated, in various passages of the route. So, the main problems are around the transportation of tourists and runners to and from the Marathon route and not much around the traffic in Athens.

The stakeholders suggested that the creation of cycling lanes and hiking trails in the city of Athens will support the decongestion but also a cycling route along the route of the Marathon for coaches and tourists will be useful too. 'On-demand' transport was chosen as the most preferred option amongst the stakeholders that can include for example electric bikes or public transportation. Social media and aps informing tourists about routes, congestion and transportation available, was also suggested.

7 Conclusions and Policy Recommendations

The starting point of this research was the importance of Megatrends in the application of sustainable mobility.

This paper examined the connection between the Megatrends that affect the application of sustainable mobility on the mobility planning of short-lived events such as the Athens Marathon. Using a Delphi exercise the key Megatrends were identified and categorized in order of importance. Thereafter, key stakeholders from the organization committee of the Marathon were interviewed in order to identify the relevance of the sustainable mobility Megatrends on the mobility planning of the Marathon.

Some key sustainable mobility Megatrends are the development of large metropolitan cities, urbanisation, ageing society, financial recession, sustainable development. While the key Megatrends related to the mobility planning of the Marathon almost concise with the ones on sustainable mobility (urbanisation, development of large metropolitan cities, financial recession and sustainable development).

Therefore, policies need to be introduced and further enhanced for the trends that appear to be most influential in achieving sustainable mobility in AMA. The policies should include measures, such as the development sustainable development directed to the growth of smart cities by introducing a better urban governance. This refers to a pressure to provide enhanced public transport services to citizens for the period of the AMA with a well-integrated public transport involving rail, metro, tram and bus combined with on-demand solutions. An information campaign taking place in various places in the city such as bus stations should also be introduced, informing the citizens that it will be preferable to avoid travelling certain hours when the Marathon participants are travelling to the starting point.

With regards to addressing sustainable development trends, policies need to reinforce sustainable decision-making and to establish it as mainstream thinking. A social movement needs to evolve in culture and effective ways for communicating sustainable values need to be developed (e.g. through marketing, advertising, influencing, incentivising, etc.). Strategies for sustainable development are not stand-alone measures as they need to be applied in conjunction with other policy measures while collaboration between relevant stakeholders (authorities, transport operators, service providers, etc.) should be encouraged. Policy tools such as incentives, compulsory measures or mechanisms encouraging voluntary involvement are needed to overcome the lack of co-operation between main stakeholders. A clear and appropriate definition of the roles played by each stakeholder, including passengers, road users, and the organizational committee of the AMA is also necessary. It is, therefore, recommended that further effort be put into the development of appropriate business models for the successful implementation of sustainable development during the AMA and beyond.

As discussed above, the conversion of Athens into a smart city can be a measure that tackles problems arising from the development of large metropolitan cities and the increased urbanization. Smart cities are characterised by digital networking which can tackle problems such as traffic with making flows more efficient. However, this requires implementation of advances in IT such as a more comprehensive coverage with the next generation of mobile communication networks (5G). Policies that support the development of the next generation of mobile communication can enhance multimodal transport information, management and payment systems. Within the context of a total digital transportation of the mobility sector, as mentioned above, there are concerns about Big Data and security. To avoid abuse of power and data by service providers, a clear and cohesive legislation should be put in force to protect personal rights and achieve data transparency.

References

European Commission: Europe 2020 Flagship Initiative Innovation Union. SEC (2010) 1161 edn. European Commission, Brussels (2010)

Popper, R., Teichler, T.: Practical Guide to Mapping Forward-Looking Activities (FLA) Practices, Players and Outcomes, European Foresight Platform, Deliverable 2.3 (2012)

Vejlgaard, H.: Anatomy of a Trend. McGraw-Hill, New York (2008)

Humphreys, B.R., Hyunwoong, P.: Professional Sporting Events and Traffic: Evidence from US Cities. Working Paper No. 17-05] West Virginia University (2017)

Kassens, E.: Transportation Planning for Mega Events: a model of urban change. PhD thesis, Massachusetts Institute of Technology, USA (2009)

Manelli, S.: Mobility management and public transport for Mega Events. In: Seventh Venice International Conference, Italy (2010)

European Commission: An Agenda for a Socially Fair Transition towards Clean, Competitive and Connected Mobility for all, COM (2017) 283 final edn. European Commission, Brussels (2017)

European Commission: Future Orientations for Cities: Final Report. ESPON ERDF, Luxembourg (2010)

UNFPA: Ageing in the Twenty-First Century: A Celebration and a Challenge. United Nations Population Fund, New York (2012)

Shrestha, B.P., Millonig, A., Hounsell, N.B., McDonald, M.: Review of public transport needs of older people in European context. J. Pop. Ageing **10**(4), 343–361 (2017)

May, A., Marsden, S.: Urban Transport and Mobility. International Transport Forum, Leipzig (2010)

Kamga, C.: Emerging Travel Trends, High-Speed Rail, and the Public Reinvention of U.S. Transportation. Transp. Policy **37**, 111–120 (2015)

Mather, M., Pollard, K., and Jacobsen, L. A.: Reports on America: First Results from the 2010 Census. Population Reference Bureau, Washington, DC (2011)

Anoyrkati, E., Maraš, V., Bugarinovic, M., Paladini, S.: Paving the way to sustainable mobility. In: International Conference on Traffic and Transport Engineering, pp. 777–784. Belgrade University, Serbia (2016)

OPTIMISM project: Impact of best ICT practices on mobility patterns, Deliverable 4.2 of the project OPTIMISM (2013). http://www.optimismtransport.eu/blog/category/publications, Accessed Feb 2020

Parker, R.: The Remarkable Growth Of Sport Tourism. https://humankinetics.me/2019/04/04/the-remarkable-growth-of-sport-tourism/ublished, Accessed 15 Mar 2020

Rubert M, Da Siliva, L.: Strategies for transport during sports mega-events and their degree of importance. In: XVI PANAM, Lisbon, Portugal (2010)

Getz, D.: Event Management and Event Tourism. Cognizant Communications, New York (1997)

Dwyer, L., Mellor, R., Mistilis, N., Mules, T.: A framework for assessing tangible and intangible impacts of events and conventions. Event Manag. **6**, 175–189 (2000)

Gnoth, G., Anwar, S.A.: New Zealand bets on event tourism. Cornell Hotel Restaurant Adm. Q. **41**, 72–83 (2000)

Breen, H.A., Bull, A., Walo, M.: A comparison of survey methods to estimate visitor expenditure at a local event. Tour. Manag. **22**, 473–479 (2001)

Bowdin, G.I., Mcdonnell, J.A., O'toole, W.: Events Management. Butterworth-Heinemann, Oxford (2001)

Jones, C.: Mega-events and host-region impacts: determining the true worth of the 1999 world cup. Int. J. Tour. Res. **3**, 241–251 (2001)

Wood, E.H.: Measuring the social impacts of local authority events: a pilot study for a civic pride scale. Int. J. Nonprofit Sector **11**, 165–179 (2006)

de Menezes, T.R., de Souza, J.F.: Transportation and urban mobility in mega-events: the case of Recife. In: XVIII Congreso Panamericano de Ingeniería de Tránsito, Transporte y Logística. Procedia - Social and Behavioral Sciences, vol. 162, pp. 218–227 (2014)

Mendes Malhado, A.C., Rothfuss, R.: Transporting 2014 FIFA world cup to sustainability: exploring residents' and tourists' attitudes and behaviors. J. Policy Res. Tour. **5**(3), 252–269 (2013)

Rowe, G., Wright, G.: The delphi technique as a forecasting tool: issues and analysis. Int. J. Forecast. **15**(4), 353–375 (1999)

Cyphert, F.R., Gant, W.L.: The Delphi technique: a case study. Phi Delta Kappan **52**, 272–273 (1971)

Brooks, K.W.: Delphi technique: expanding applications. North Central Assoc. Q. **3**(54), 377–385 (1979)

Ludwig, B.: Predicting the future: have you considered using the Delphi methodology? J. Extension **35**(5), 1–4 (1997)

Custer, R.L., Scarcella, J.A., Stewart, B.R.: The modified Delphi technique: a rotational modification. J. Vocat. Tech. Educ. **15**(2), 1–10 (1999)

Eleanor Singer, E., Couper, M.: Some methodological uses of responses to open questions and other verbatim comments in quantitative surveys. Surv. Res. Center Univ. Michigan **11**(2), 115–134 (2017)

Co-creation Techniques and Tools for Sustainable and Inclusive Planning at Neighbourhood Level. Experience from Four European Research and Innovation Projects

Angelidou Margarita[1(✉)], Fróes Isabel[2], Karachaliou Eleni[1], and Wippoo Meia[3]

[1] Q-Plan International, Thessaloniki, Greece
mangelidou@qplan-intl.gr
[2] Copenhagen Business School, Frederiksberg, Denmark
[3] Waag Technology and Society, Amsterdam, The Netherlands

Abstract. This paper presents outcomes from the cluster workshop of the Horizon2020 SUNRISE, MUV, Cities4People and Metamorphosis projects, and specifically the workshop session for the use of co-creation tools and techniques across the four Horizon2020 projects. The workshop was organised by CIVITAS SATELLITE in cooperation with the Innovation and Networks Executive Agency of the European Commission on the 19th of November 2019 at the premises of INEA in Brussels. Several different participation techniques & tools are being used by the projects, including co-creation workshops, serious games, interviews, simple questionnaires at the place of the intervention, field trips, school scans and thematic walks/scans, cultural probes, focus groups, etc.. In general lines, the projects have opted for practical, hands-on tools rather than online and complicated tools. A series of key learnings about the practical use of those tools emerged; co-creation activities should try to incorporate diverse neighbourhoods and settings; it is preferable that only one challenge is addressed at a time; evidence is an excellent prompt for discussion; discussion topics should be important to the participants, rather than the organisers; silent observation helps achieve a deeper understanding; and methods and tools that involve the physical presence of the users at the place of intervention are most effective.

Keywords: Cities · Sustainable and inclusive urban planning · Transport planning · Co-creation · Communities

1 Introduction

During the past decade, planning practices, as well as Research and Innovation (R&I) practices, have been undergoing two fundamental changes. The first one regards the increasing engagement of users as co-creators and co-implementers of the desired planning solutions. To this end, the arsenal of participation tools and techniques that

E. G. Nathanail et al. (Eds.): CSUM2020, AISC 1278, pp. 562–572, 2021.
https://doi.org/10.1007/978-3-030-61075-3_55

exist within practitioners' and researchers' reach has expanded without precedence. There is a tool for each stage of the co-creation process, addressing from the identification of needs, the setting of priorities, the ideation of the concept, up until the co-design and co-evaluation of solutions. Contemporary thinking posits that it is important to use a broad variety and different means for citizens and stakeholders to participate in the co-creation process, as this will ensure a more inclusive representation of the citizens' ideas, needs and concerns and agency from citizens that feel that their opinion 'counts'. Moreover, it will ensure a higher degree of innovation, as more ideas are brought to the table. The whole process is expected to lead to more sustainable results which are endorsed by larger parts of the population.

The second change that has been taking place regards the increasing digitization and the penetration of digitised tools in all planning and R&I practices. This digitization addresses different stages of the process and serves different functions, ranging from data collection, data processing, data-driven decision making. While digitization offers undeniable benefits, such as cost reduction and analytical accuracy, it does not come without risks. For example, the risk of exclusion of important or marginal population groups and the risk of missing the finer aspects of the participants' needs, aspirations and ideas are not to be overseen.

These topics are some of the key points discussed at the cluster workshop of the Horizon2020 SUNRISE (GA 723365), MUV (GA 723521), Cities4People (GA 723194) and Metamorphosis (GA 723375) projects, organised by CIVITAS SATELLITE in cooperation with the Innovation and Networks Executive Agency (INEA) of the European Commission (EC), which took place on the 19th of November 2019 at the premises of INEA in Brussels. In this paper, we present the main findings of the workshop session specifically dedicated to the use of co-creation tools and techniques across the four Horizon2020 projects, with representatives from all the projects, as well as representatives from CIVITAS SATELLITE, INEA and other policy makers. This session was one of the four workshop sessions that took place at this meeting.

To this end, we first present the theoretical underpinnings of using participation techniques and tools for urban and mobility planning at neighbourhood level. We then describe the data collection setting and process. We advance to the presentation of the analytical findings of the workshop. The last section offers the major conclusions from the performed exercise.

2 Co-creation Techniques and Tools for Planning at Neighbourhood Level.

2.1 Co-creation Techniques and Tools

In this section we offer a definition of co-creation and cover some of the historical aspects behind co-creation and design thinking approaches. The historical aspects are followed by a presentation of the tools and how they were applied in the four neighbourhood projects.

We suggest the definition of co-creation as the creation, development and deployment of ideas and solutions emerging from a collaborative process among a group of key project theme's stakeholders (citizens, city representatives, designers, companies, makers, etc.). This collaborative process ensures that the ideas and solutions are implemented in the context they emerge and selected based on an informed and democratic process. In order to achieve these results, projects need to bring together groups of people from different areas, expertise and perspectives to work towards understanding current issues that may exist, as well as ideating to create solutions to tackle these issues, or even to innovate and create scenarios to be tried out and prototyped to test their validity and impact.

First, it is valuable to remind ourselves that co-creation tools and techniques are not new. They have evolved from psychology and sociology fields, making their way into consultancies and creative industries, as a need to reach, communicate, work with and understand people. In its early days it became widely known as 'participatory design', emerging from within Scandinavian contexts in the late 60s. It involved understanding how factory workers used machines, and adapting the machines to fit their use accordingly, and it soon became recognised as an important step when designing factory machinery [1]. This approach was well acknowledged and revived within design industries in the nineties, more specifically with the exponential growth of digital products and services, which require to be used by different target groups at various contexts. As part of this technological evolution, design practice shifted its approach from designing for people to designing with people; in other words, developing research and design tools and methods to include people in their design process. These tools were merged into one cluster, defining the co-creation tools and methods and consequently, the design thinking process. Despite its spread and history (the design thinking approach has been applied for over three decades by various companies, agencies and consultants such as IDEO, live-work, Engine, etc.), design thinking and co-creation methods gained a momentum in the first two decades of the 21st century. The tools, which have been applied widely and successfully in the private sector, have now started to catch up within the public sector, coming to play a key part in calls for proposal within EU tenders and current EU projects [2, 3].

Though tools are essential aids in the facilitation of co-creation, to ensure a full implementation of co-creation, a project-team also needs to have full understanding of the entire process, as well as the mindset and skills to improvise and adapt to uncertainties, specific situations and people. In this paper we focus primarily on tools, but we are very much aware of the need for specific skills to have a successful implementation of co-creation.

The four projects share the co-creation approach, which was the red thread going across the four neighbourhood projects, both surprising and challenging cities to develop unique solutions to urban planning and development, however each with their own specific focus. In short, the co-creation tools can be divided by specific purpose (research, team-building, ideation, assessment, validation) and they are described in the following.

1. Research: The research tools cover different ways to learn about citizens and stakeholders' needs, perceptions, habits, preferences, etc. These tools can be deployed at different times of the projects, however, the projects should always

commence with them, as the starting point for understanding current problems, needs and opportunities.

2. Team-building: These tools are deployed to bring the groups together, learning about each other and creating a common language to be used during the project process. More importantly, these tools are responsible for creating a sense of trust, equality and reliability, leading to a positive collaborative group.
3. Ideation & development: These tools are primarily applied towards creating other perspectives and options, which might not lay obvious. They engage participants into playing with perspectives through various media (drawing, talking, role-playing, etc.) to develop ideas and scenarios that fit into solving or addressing issues uncovered through the research tools.
4. Assessment: The assessment tools help participants gain a perspective on their ideas, creating an understanding about their feasibility, impact and positioning in regards to other ideas and the project budget and timeline.
5. Validation: Validation tools are used towards grasping how the ideas and solutions work as planned and how they might set a new standard for the community; they can be used before, during and after carrying out the deployment of the chosen solutions.

These tools, when combined, create a holistic approach throughout the projects and facilitate their development and implementation. In the next section, we elaborate how some of these tools were applied in the planning process of the neighbourhood projects.

2.2 Using Co-creation for Planning at Neighbourhood Level

The versatility of the co-creation tools has been well demonstrated through the distinct focus of each of the projects. Despite the common neighbourhood, public space and mobility thread, each project had its own lens and therefore, required a tailored plan to carry out not only the activities, but their processes as a whole.

Due to the array of stakeholders and projects' goals, the organisation of dates and events need to be well-scheduled and planned to achieve their highest impact. For example, in the case of the Cities-4-People project, development tools, such as road-maps were combined with blueprinting, bringing an even more detailed structure of the whole process, including which tasks need to be executed before, during and after each of the projects' interventions. Collaborating using these tools was a way to bring out the knowledge emerging from within the participants' own roles to define and organise the project interventions. City officials and transport authorities knew of regulations, citizens and associations' representatives knew of other related community or school events and had specific wishes. By working together in describing the steps and requirements for each of the project phase, the various stakeholders have had a platform to discuss and coordinate their actions and these tools were facilitators towards a better communication and planning process.

A key aspect of these processes is the tacit knowledge created and residing among the members of the city representatives and their local community who have taken part in the projects. If not well documented, this knowledge can be easily lost by a change of employees or community participants. Therefore, to make sure this knowledge is secured, shared and replicated, some of the co-creation tools, such as roadmaps and blueprints, can help achieve exactly that; make the knowledge and project experience tangible towards future developments using related processes, so it can be shared with other employees and communities.

Collaborating with citizens and their associations throughout the ideation, planning and execution of the project events, cities gain a wider reach, which in return make these projects more visible while also giving citizens an aspect of agency and ownership, key to a higher project impact. Overall, the breadth of co-creation tools have proven to not only to facilitate projects involving a wide range of groups at neighbourhood level, but also to engage and inform city officials and citizens of the various opportunities lying dormant through usual top-down approaches. Furthermore, by applying such tools, cities have the opportunity to also learn from their citizens, who might be able to contribute with their own set of skills, which would be otherwise unknown.

3 Methodology

The cluster workshop of the Horizon2020 SUNRISE (GA 723365), MUV (GA 723521), Cities4People (GA 723194) and Metamorphosis (GA 723375) projects, titled "Co-creating Urban Mobility in Neighbourhoods" was organised by CIVITAS SATELLITE in cooperation with the Innovation and Networks Executive Agency (INEA) of the European Commission (EC), and took place on the 19th of November 2020 at the offices of INEA in Brussels, Belgium. A series of presentations and interactive sessions during the entire day of the workshop took place, addressing policy objectives related to sustainable planning of mobility and use of land with an eye for innovation at neighbourhood level; the core ideas behind the projects and their approaches to co-creation, impact assessment and scale-up; a World café workshop comprising four dedicated tables running sessions that addressed different topics; and a final roundtable presenting and discussing the workshop outcomes. A total number of 27 participants attended and contributed to the activities of the cluster workshop, representing policy makers as well as researchers, industrial partners, and citizen associations participating in the projects.

The World Café is a creative technique that creates an environment of open conversation and action by facilitating dialogue and knowledge and ideas sharing. The typical World Café format includes a first stage where the participants discuss among them and a second stage where results are summarised and discussed. The first stage takes place at so-called "café tables", each of which is overseen by a moderator and is dedicated to the discussion of a specific topic. Participants are called to change tables at regular intervals, thus passing through as many tables as possible and discussing as many different topics as possible. Table moderators, on the other hand, stay at the same table, take notes and summarise the previous conversations to the newcoming table participants. In this way, ideas are cross-fertilised as the first stage of the World Café progresses. During the second stage all the main ideas are collated and presented by the moderators to the participants in a plenary session. The final outcomes are presented to the participants, a cross fertilisation across topics take place, and new possibilities are explored [4].

In this paper we present the results of the workshop session about "Participation techniques and tools to engage citizens and co-create at neighbourhood level", moderated by Margarita Angelidou from Q-Plan (Cities4People) and Salvatore Di Dio from Push (MUV). In total, 25 stakeholders -quadruple helix representatives from all the projects, as well as CIVITAS SATELLITE, INEA and other policy makers- participated in the session across four tables and respective sessions.

The participants were first introduced to the concepts that this session aimed to address. They set out by discussing that it is important to provide a broad variety and different means for citizens and stakeholders to participate in the co-creation process, because this will ensure:

- a more inclusive representation of the citizens' ideas, needs and concerns
- an agency from citizens that feel that their opinion 'counts'
- a higher degree of innovation, as more ideas are brought to the table
- more sustainable results, suggested and endorsed by the majority of the population

Yet, the use of co-creation techniques and tools also creates challenges:

- how to identify the technique/tool that best serves the purpose of the co-creation exercise?
- how to ensure access of all population groups?
- how to get a representative mix of participants?
- how to approach sensitive/marginalized population groups?
- how to balance the use of online/offline tools?
- how to sustain interest, avoid participant fatigue and cultivate communities of interest/practice?

In sequence, the participants were presented with a questionnaire and asked by the moderators to reflect upon the following questions:

- ***Q1. Which techniques and tools have you used in your projects and how?*** Name the techniques and tools (Fig. 1). Explain the stage in which they were used (for example co-creation, co-implementation, co-evaluation), the target groups, for how long they were used (hours? days? months?) and how many participants joined. Was participation open or upon invitation? What kinds of financial and human resources were required (e.g. for preparation, implementation and follow-up, logistics, equipment, techniques)?
- ***Q2. What challenges did you encounter in using those tools?*** What would you change if you had the chance to start over? Selecting the most appropriate? Obtaining permissions? Setting up? Appropriate duration? Attracting input? Analysing collected data?
- ***Q3. Ultimately, which are the success factors for the effective utilization of techniques & tools for co-creation at neighbourhood level?***

A lively discussion followed, lighting up free interaction among the participants, who exchanged ideas and experiences. Following the World café methodology, the moderators collected, grouped and synthesised those ideas into organised concepts which they presented in the plenary session that followed.

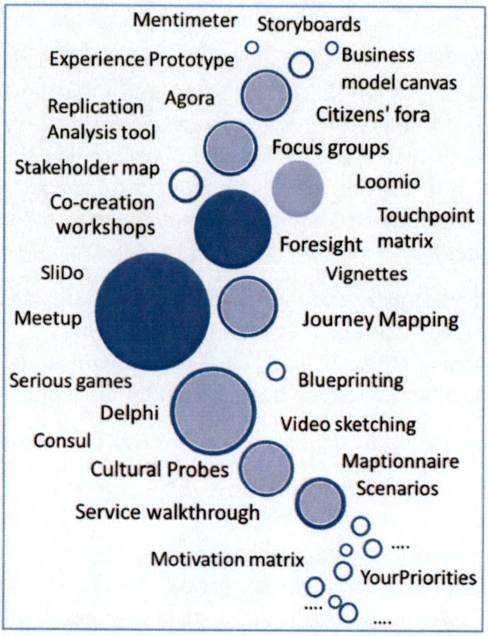

Fig. 1. Examples of techniques and tools presented to the workshop participants (authors' development).

4 Findings

4.1 Techniques and Tools Used

Several different participation techniques and tools are being used by the projects, including co-creation workshops (broad range of specific methods and formats), serious games, interviews, simple questionnaires at the place of the intervention, field trips, school scans and thematic walks/scans/signalling (especially for minors and challenged users), public stands, open fora, cultural probes, focus groups, Delphi and YourPriorities [5].

In general lines, projects have opted for practical, hand-on tools rather than online and/or complicated tools. Some participants expressed the belief that online methods and tools are less effective than face-to-face ones (Fig. 2). Moreover, online tools are seen as especially time consuming; they need to be active for a considerable amount of time in order to attract adequate input from participants. They serve, though, in cases where physical accessibility of participants is restricted. In any case, offline and online methods have to be in right balance.

The downside of one-on-one interactions and the use of face-to-face methods and tools, on the other hand, is cost. They require time and human resources, hence they are expensive. Some projects encountered challenges in getting enough personnel to moderate the workshops and administer the interviews on a voluntary basis.

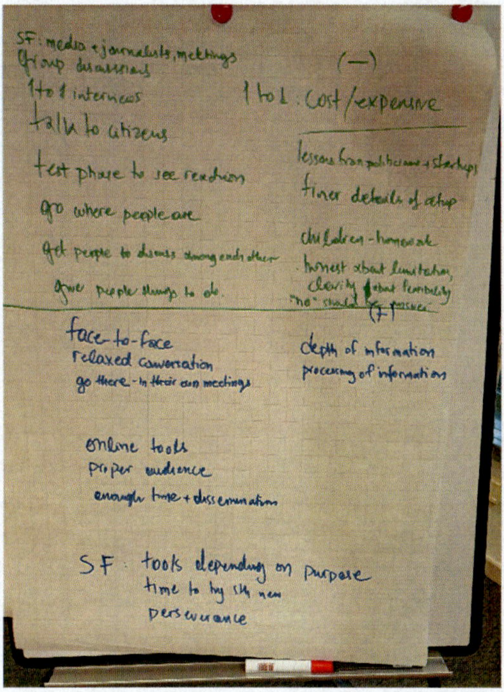

Fig. 2. Collection of ideas during the workshop session (authors' records).

4.2 Deciding Which Participation Techniques and Tools to Use to Engage Citizens and Co-create at Neighbourhood Level

Participation techniques and tools should be appropriately selected and combined to serve the purpose of the co-creation stage and scope, the involved participants, as well as the desired outcomes. One step further, the appropriate mix of such tools should be selected, in a way that allows equal access of all user groups to the co-creation process. This includes the mix of online/offline techniques, since many participants lack access to online tools. Diversity should also be sought in terms of physical settings.

Moreover, the use of user-centered approaches that include immersion in the natural environment and social events of the users is very important. Most project teams prefer to go themselves to the natural environments of the participants and incorporate co-creation in the meetings of the communities. Immersion allows for a substantial observation of attitudes, perceptions and challenges. Thus, face-to-face relationships, one-on-one interviews and relaxed conversations establish relationships of trust and cooperation.

In co-creation, time is deemed either an asset or a liability. The participants should not be pressured, but instead given enough time to think about their input. It is important to respect the learning curve of the participants and the social skills that they are developing during the co-creation process. In using the tools, it is important to start early, securing enough time that will allow for trials. A test phase to see reactions is helpful; if something does not work, there is flexibility to try something new.

Regarding dissemination, enough time should be foreseen to disseminate the events, implement the activities reaching an advanced degree of information depth, process the information and then draw proper conclusions. In any case, it helps to bring in the media and keep them engaged. Positive experiences were reported by bringing in journalists and organising open meetings and group discussions where media representatives are invited. Professional and advanced marketing techniques to attract and sustain participant interest are also welcome.

4.3 Implementing the Exercise

The finer details of the setup are of increasing importance, since they offer the means to make the participants feel valued. Such details include the place, time and welcoming of the participants.

The implementation process can start from evidence. This evidence will provide prompts and triggers for the participants to engage more quickly and substantially in the co-creation process. There exist instances of evidence that easily trigger reactions, such as, for example, the coverage of green spaces in the city. Every co-creation instance should ideally address one challenge at a time and hands-on topics that are easy to comprehend. Only this way will the participants have the ability to reflect, talk, and immerse deep into the essence of the topic. This will help achieve a certain depth of information, which is required to address a specific challenge. Moreover, the selection of an important and possibly sensitive topic for the community will attract interest and participation.

During the co-creation process, it is important to get participants to discuss amongst themselves. This way the conversation carries on and the input is both richer and cross-fertilised. Besides conversation, silent observation can prove especially effective during the prototyping activities, because it allows the moderators to listen to the discussion of the participants and understand which issues matter most to them. It is also important to see participants not as mere input providers, but to actually engage them in doing or producing something. They can also have some 'homework' to take away – a practice that was successfully adopted by one of the projects.

Visual tools are also very helpful. The use of maps, pins, pictures and pictograms (especially for specific and vulnerable population groups) are particularly helpful. It is also advised to use methods and tools that involve the physical presence of the users at the place of intervention. It is helpful to visit the place of the intervention with the user. It was suggested to walk along with them around the physical place of the intervention. This will help them identify and communicate points on the spot.

Last, but not least, it is very important to document the names of the participants. It often happens that returning participants do not remember the views they have expressed in previous co-creation activities. Documentation should also be kept for reasons of transparency and as a way to protect the administrators of the co-creation exercise. All related measures should be designed in advance.

4.4 After the Exercise

Critically, tools are not what make co-creation successful: it is the people that know how to implement them, in a logical way, following a process. The co-creation exercise should thus have a follow-up, especially if it has involved face-to-face participation techniques and tools, such as field trips, focus groups, workshops, which encourage the development of personal relationships. The follow up would ideally include the results of the activity.

In general, it is advised to make the outputs publicly available in the form of open knowledge. This will sustain the interest of the participants and keep the public conversation ongoing. It is also an active expression of the intention for transparency and openness.

4.5 Ultimately, Why Co-create for Planning at Neighbourhood Level?

Compared to other solutions, co-creation can work better in the domain of planning at neighbourhood level because it has an anthropological fundament. It addresses the human aspect of place ownership. People take individual decisions, but space belongs to everyone and that's why co-creation can work better. It is not the car of the building, but the user that is affected by the decisions being made in this domain. Hence personal beliefs, emotions and convenience interfere.

More particularly, the right to move and use public space is a very emotional issue. Each and every one of us should have the right to be access public spaces and to move freely therein. Sustainability at neighbourhood level has practical, daily life implications for all of us. It touches everyone and has social implications. It defines our schedule and lifestyle. The users have a high stake, they have something to lose and to sacrifice, which is their usage and occupation of space in the way they want and in sequence the ability to plan their schedule according to their convenience.

In parallel, sustainable use of space at neighbourhood level brings together many stakeholders and hence provides a fruitful field for cooperation. It is a zone where anyone can participate.

How we use public space and how we choose to move is also a form of a 'status symbol'. Depending on how quickly, conveniently, costly, cleanly and freely we move in space, we perceive our social status and position ourselves in society. For example, people get very emotional about their "right" to use the car; the ownership of a private vehicle and the ability to park it exactly out of our homes is seen by many as a symbol of power and freedom.

Finally, it is something that many people believe they have no significant power over, that they cannot change it. They feel far from this decision-making process. Hence when they are given the opportunity to express their views, they really want to take advantage of it.

5 Conclusions

All the examined projects include an ample variety of co-creation techniques and tools, fittingly selected to serve the objective of each planning stage. Many of these tools can be found in the so-called 'Co-creation Navigator' [6] provided by WAAG society, and aiming to serve as a reference guide to co-creation tools. This guide was developed by WAAG in the context of the two CIVITAS projects C4P and MUV and is currently being populated with some of the methods and tools developed in SUNRISE.

All in all, it seems that co-creation in the planning domain in Europe is increasingly gaining ground and becoming a well-established norm, especially in innovation-related projects. Although this may seem normal today, it is worth remembering that 10 or 20 year before the situation was much different, with citizens and other stakeholders usually being only consulted or not involved at all in European countries that did not have such a tradition.

Yet it emerges that project managers retain trust in the more direct, immersive and interactive co-creation techniques and tools, and especially the ones that require direct engagement with the stakeholders and immersion in their physical environment. Although an abundance of digital tools to support the process is available, they are less preferred compared to the more direct ones. It is thought that digital tools risk missing the finer details and aspects of user needs, perceptions and requirements. Hence a safe conclusion can be made about the preference for simple and hands-on tools.

That said, co-creation offers techniques and tools that serve better the objectives of sustainable urban planning and mobility at neighbourhood level, due to the user-centered and anthropological nature of both. Mobility and the use of public space are both a 'personal' and a 'social' matter; they represent major conflict areas, where personal needs, preferences and beliefs are called to adjust to the sake of broader well-being. Moreover, they shape the identities of people and the identities of places.

References

1. Sanders, E.B.N., Stappers, P.J.: Co-creation and the new landscapes of design. CoDesign **4**(1), 5–18 (2008)
2. Cassia, F., Magno, F.: Public services co-production: exploring the role of citizen orientation. Int. J. Qual. Serv. Sci. **1**(3), 334–343 (2009)
3. Nevens, F., Frantzeskaki, N., Gorissen, L., Loorbach, D.: Urban transition labs: co-creating transformative action for sustainable cities. J. Cleaner Prod. **50**, 111–122 (2013)
4. Sloccum, N.: Participatory Methods Toolkit: A practitioner's manual. viWTA, King Baudouin Foundation & UNU/CRIS, Belgian Advertising (2003)
5. YourPriorities tool. https://www.yrpri.org, Accessed 01 Mar 2020
6. WAAG Society: Co-creation Navigator. https://ccn.waag.org/, Accessed 01 Mar 2020

Modelling Urban Mobility During the Recession: The Case of Athens

Christos Kokkalis and Ioanna Spyropoulou(✉)

National Technical University of Athens, 9 Iroon Polytechniou Street, 15780 Zografou-Athens, Greece
chris.kokkalis.21@gmail.com, iospyrop@central.ntua.gr

Abstract. The Greek economy suffered a recession in the recent years, referred to as the Greek economic crisis, and as a result mobility was affected. The aim of this paper is to investigate the impact of the economic crisis on urban mobility, represented through traffic conditions. As a reference case the city of Athens is employed; a city that relies heavily on private motorized transport modes. Towards this objective, the variation of traffic conditions, employing the quantity of traffic flow is explored. The investigated time period was between 2006 and 2018, including a period before and during the economic crisis, as well as a period that demonstrates slight economic growth. The analysis considers aggregate traffic flow, as well as flow sub-populations serving different trip purposes. Three trip purposes are explored: work, entertainment and holidays. For the needs of this study, specific arterials were selected which serve a large proportion of the commuting and entertainment trips performed in the city of Athens, and traffic data were collected from loop detectors at dedicated locations on these arterials. The collected data required thorough processing as it contained noise, and general linear models were designed to capture the effect of various parameters, including year, trip purpose, time of day and so on, on traffic conditions. The results from this analysis provided insights about the variation of traffic conditions, considering the explored parameters.

Keywords: Traffic flow · Traffic conditions · Economic crisis · Recession · Trip purpose · Traffic pattern

1 Introduction

Greece experienced an economic recession which officially started in 2008. This followed the general recession experienced by several western countries between 2008 and 2010, but demonstrated a substantially higher duration compared to that of the majority of other countries. As a result, the everyday life and habits of the population were greatly affected, resulting in the modification of their mobility characteristics. This is expected to be demonstrated by a reduction in the number of trips performed, as well as a possible shift from private cars to less expensive transport modes. Thus, several elements related to the operational performance of transport systems are expected to have been affected.

E. G. Nathanail et al. (Eds.): CSUM2020, AISC 1278, pp. 573–583, 2021.
https://doi.org/10.1007/978-3-030-61075-3_56

Rešetar et al. [1] noted that as a result of the global crisis which also affected Croatia during the period of 2008–2013, the number of registered vehicles remained unchanged in contrast to the previous and following years when an increasing trend was observed. Furthermore, the reduced number of new vehicle sales resulted in an increase of the vehicle fleet age during the respective period. Another noted difference was the relative increase of diesel cars compared to petrol cars, justified by the lower fuel prices and the reduced fuel consumption that diesel cars offer. This is anticipated to have affected environmental conditions, a result also noted in a study of Spanish motorways [2]. In particular, Sobrino and Monzon [3] identified the different factors determining Greenhouse Gas (GHG) emissions. These were traffic activity and fuel economy which were determined by socioeconomic factors such as employment, population and Gross Domestic Product (GDP). Traffic activity involved the total distance traveled and motorization, while fuel economy in turn determined energy efficiency and fuel consumption mix. Fleet composition, was a contributing factor which was determined by motorization and energy efficiency. Results indicated that during a four year period (2007–2010), a reduction in GHG emissions in the transport sector was observed following a 17 year increase. This was partly attributed to the reduction in travelled distances and the number of trips, which were associated with the reduction of employment rate and income per capita. Traffic noise, comprise yet another element describing environmental conditions. Analysis over a 9 year period (2005–2013) with the employment of traffic, speed and noise measurements collected at dedicated locations of a peripheral motorway in Athens indicated that noise levels were reduced as a result of the reduction of traffic flows [4]. Several studies have focused on the research of the impact of the financial crisis on road safety. Noland and Zhou [5] explored the factors affecting the number of fatalities in US states during a 31 year period (1984–2014). Amongst the contributory factors household median income and inequality of income distributions were identified. Wegman et al. [6] noted that the observed increased rate of fatalities reduction over the years in Sweden and Great Britain is associated to the reduction of GDP and increase in unemployment rates. Mpogas et al. [7] employed a before (2005–2009) and after (2010–2014) study and observed a reduction in accident fatalities during the second period which was also associated with the reduction in GDP and the increase of unemployment rates. This reduction was noted to be substantially higher in non-urban areas.

The effect of the economic crisis on other transport modes has also been explored. Dobruszkes and Van Hamme [8] noted that the changes in the air passenger traffic volume are in line with the global economic recession, and affected countries in the developed world more than emerging countries. Azadian [9] designed a regression model to describe cargo traffic in Florida. Model results indicated a reduction in years 2008 and 2009, an initial recovery in 2010 and a growth trend in the following years (2012–2016). On the other hand, maritime cargo transport volumes in three Greek major ports were not found to be affected by the explored macroeconomic indicators representing the economical situation in Greece [10].

Although an immediate effect of the recession on transport involves changes in traffic conditions, limited research explores the change of relevant indicators. The aim of this research is to explore the impact of the Greek economic crisis on traffic conditions in urban areas, and determine whether this effect is dependent on trip purpose.

Section 2 presents data relevant to the Greek economic crisis from national statistics and other traffic related studies, while data collection is presented in Sect. 3. Section 4 presents the adopted methodology and the analysis results, while in Sect. 5 the main findings of the study are discussed.

2 Greece in Numbers During the Recession

2.1 Macroeconomic Data

This section presents macroeconomic to depict the economical situation during the study period. Figure 1 presents GDP and GDP per capita based on data collected by the Hellenic Statistical Authority.

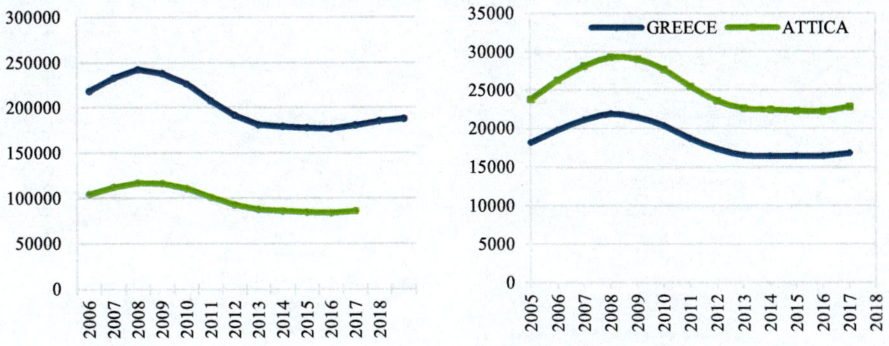

(a) GDP (10^6 €) (b) GDP per capita (€)

Fig. 1. GDP and GDP per capita.

Data shows a similar trend between Greece and Attica Prefecture with GDPs presenting an annual increase between 2005 and 2008. After 2008, the trend is decreasing up to year 2013. The period between 2013 and 2016 does not present substantial changes, and a small increase is observed in year 2017. Attica is associated with a higher GDP per capita compared to the whole of Greece, as it comprises the financial and political centre of Greece. Figure 2 presents unemployment and employment rates, as these are anticipated to affect the number of performed trips both as a result of reduced household income but also following a reduction of work related trips.

Results present similar trends to GDP indicators. Unemployment increases in the period of 2008–2013, following which a decreasing trend is observed. Employment (represented by the number of employees) begins to decrease in 2008 for the Greek population and in 2009 for the population of Attica, and a slight increasing trend commences in 2013, which becomes even more evident after 2017. Last, the number of registered vehicles, as well as, fuel consumption are presented in Fig. 3.

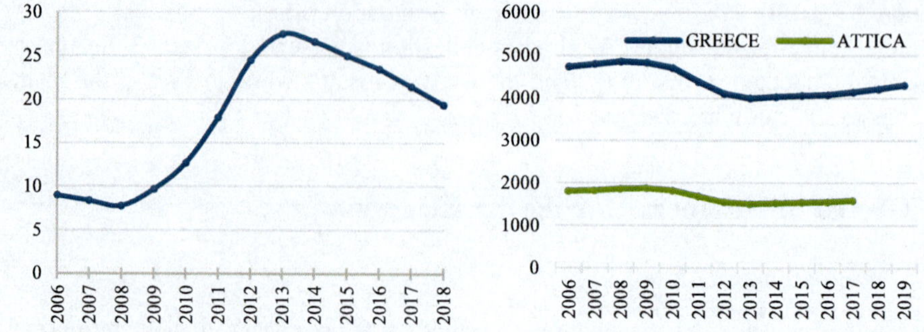

(a) Unemployment rate (%) (b) Number of employees (10^3)

Fig. 2. Unemployment and employment indicators.

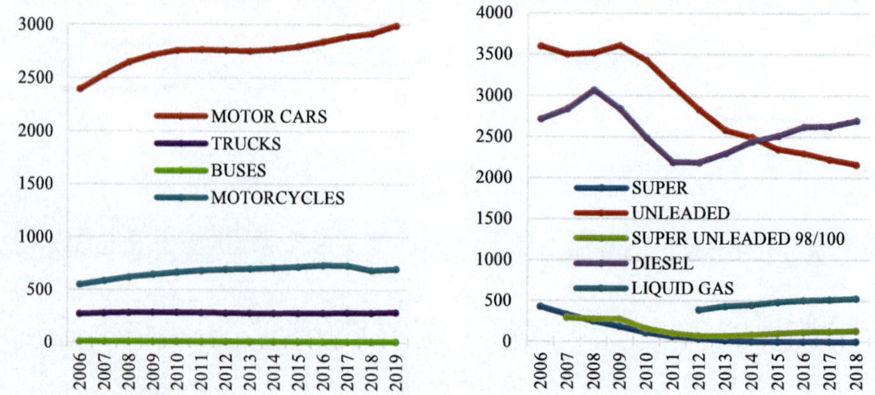

(a) Vehicle fleet (10^3) (b) Fuel consumption (10^9 tons)

Fig. 3. Vehicle fleet in Attica and fuel consumption in Greece.

The effects of the economic crisis started to become evident considering vehicle fleet in 2008, with a reduction of the increase rate for passenger cars, while in 2012 and 2013 vehicle fleet slightly decreased. A smooth increase followed from 2014 onwards. Motorcycles, being a cheaper transport mode considering both purchase and use cost, presented an increase throughout the crisis years. Fuel consumption, involving all motorised modes (considering transport of people and goods), presented a reduction between 2009 and 2012, and a gradual increase from 2013. This presents a general picture of motorised mobility in Greece, and it can be assumed that people's mobility in Athens using motorised modes follows more or less a similar pattern.

2.2 Traffic Data

Sermpis et al. [11] estimated the effect of the economic crisis on a number of traffic indicators for the period 2009–2012 considering several major arterials of the city of Athens. Results indicated a reduction of mean daily traffic flow in all explored arterials which ranged between 6% and 16%. The differences in the estimated values are a result of inherit arterial characteristics. Mean speed during peak hours was found to be higher in all explored arterials, and ranged between 2% and 20%. Last, a reduction of travel times considering specific routes was also observed, and was about 20% on average during the day and about 60% at peak hour periods. Prevedouros and Halkias [12] developed time-series regression models to forecast short term and long term traffic demand on a peripheral motorway in Athens, considering GDP, fuel price, vehicle sales and Athens Stock Exchange index, with satisfactory results. Papagiannakis et al. [13] conducted a revealed preference survey in Thessaloniki, the second largest city of Greece, to determine the effect of the crisis on trip characteristics. Trip purpose, trip frequency and modal split were explored. The financial crisis had a greater effect on entertainment or shopping trips compared to work trips. Furthermore, a substantial reduction was observed to trips performed by private cars (12% as a driver and 3% as a passenger) and reduction in the use of taxis (2%) was also observed, while cheaper transport modes and presented an increase in the modal split: public transport (12%), motorcycles (2%), bicycle (1%) and walking (3%).

3 Data Collection

3.1 Field Survey Design

Athens metropolitan area, i.e. the study area, consists of central Athens and its suburbs (north, south, east and west). The majority of businesses are located in or close to central areas, whilst the suburbs mainly comprise residential areas [14]. Its road network mainly follows a radial system, with several arterials connecting the city centre with the suburbs, and some smaller ones connecting the aforementioned ones. Most of the arterials are signalised. Athens is also served by a peripheral toll-road. Public transport including 3 metro lines, a light rail service, a tram service and a dense bus route network serve trips in the explored area.

For the needs of the study traffic flow data at dedicated locations during the period of 2006–2018 (excluding the period between July 2015 and October 2016, during which the traffic management centre had ceased its operation) was obtained. The data in the form of traffic flow, mean speed and occupancy measurements (from inductive-loops and video detection loops) was provided from the Traffic Management Centre of the Attica Prefecture. The selection of the spatio-temporal characteristics of the selected data was made considering the characteristics of the explored traffic network and the different explored trip purposes i.e. work, entertainment and holidays. For each trip purpose representative arterials, the relevant days of week and times of day serving the specific trip purpose were determined. For example, for work trips data was collected for weekdays that were working days (excluding national days, holidays and strike days) and for time periods between 07.00–10.00 and 14.00–20.00, and involved measurement sites on links mainly connecting the suburbs to the city centre. In total, data from 38 locations was collected (22 for work, 12 for entertainment and 12 for holiday trips). The following figure presents the data collection sites.

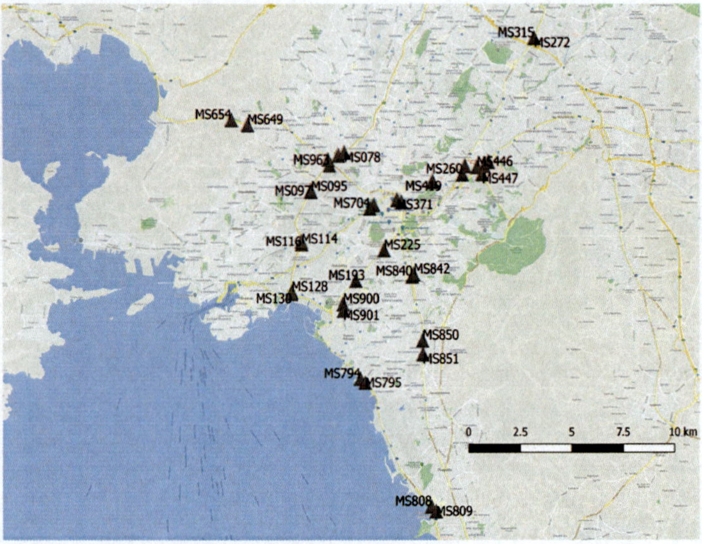

Fig. 4. Data collection sites.

The total number of selected days was 18.038 for work trips, 5.504 for entertainment trips and 1.311 for holiday trips. Data was thoroughly processed to exclude measurements errors and mitigate data noise, as much as possible.

3.2 Traffic Patterns

The existence of specific traffic flow patterns was explored through the production of several diagrams. Figure 4 presents mean annual traffic flow values for the three explored trip purposes, for indicative and representative considering each trip purpose measurement sites (MS) (Fig. 5).

Data indicated differences between the different measurement sites resulting from inherit site characteristics. Work related trips (the presented data involved arterials linking the city centre with the suburbs) exhibited rather similar patterns, while entertainment trips (the presented data involved links close to the city centre, connecting nightlife areas) and trips related to holidays (the presented data involved arterials linking with major motorways that connect Athens with South and North Greece) did not provide a clear pattern. In general, a reduction trend was observed after 2008 or 2009, and a slight increase trend towards the end of the investigated period.

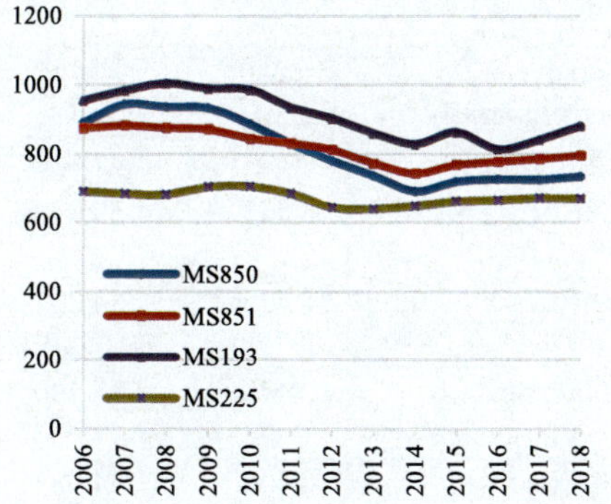

(a) Mean traffic flow (veh/hour) for work trips

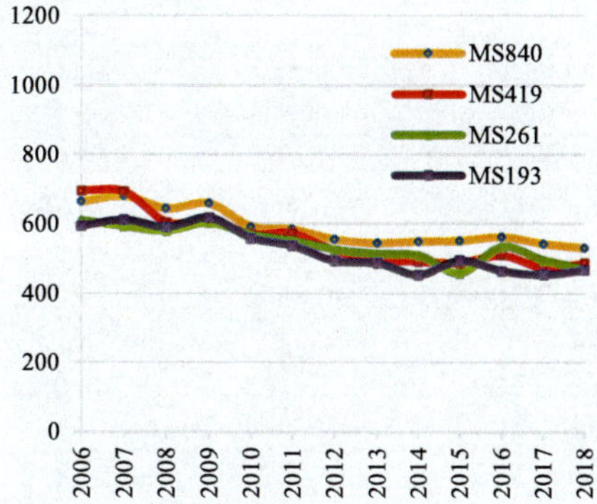

(b) Mean traffic flow (veh/hour) for entertainment trips

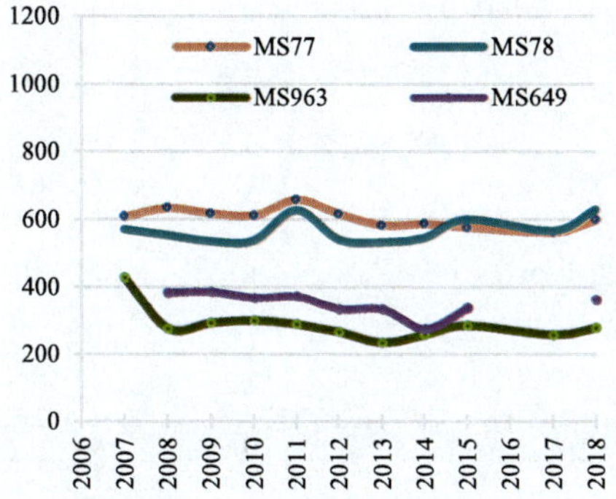

(c) Mean traffic flow (veh/hour) for Easter holidays

Fig. 5. Traffic flow trends throughout the explored periods for different trip purposes.

4 Modelling Traffic Flow

General linear models were designed to model hourly traffic flow in the explored area, as they are widely used to represent count data. GLM assumes that the dependent variable Y_i is expressed as a linear function of the independent variables x_i, corresponding coefficients β_i and an error term ε_i through the following equation.

$$Y_i=\beta_0+\beta_1 * x_1+\beta_2 * x_2+\cdots+\beta_n * x_n+\varepsilon_i \text{ , and } \mu_i \; [\![=E(Y]\!]_i) \qquad (1)$$

The dependent variable is the $\log_e$ value of traffic flow, while the independent variables include temporal parameters: time of day, day of week, month of year, spatial parameters: proximity to the city centre and road characteristics (number of lanes and operation of traffic signals) and trip purpose.

Table1. Traffic flow model.

Variable	Estimate value	t-value	Variable (interactions)	Estimate value	t-value
Intercept	5.436	562.957	2006:work	0.473	42.836
Time [1.00–2.00]	−0.395	−78.701	2007:work	0.557	59.978
Time [2.00–3.00]	−0.863	−171.443	2008:work	0.445	84.237
Time [3.00–4.00]	−1.214	−240.066	2009:work	0.448	48.590
Time [4.00–5.00]	−1.202	−237.832	2010:work	0.406	44.054
Time [5.00–6.00]	−0.564	−112.465	2011:work	0.370	40.210
Time [6.00–7.00]	0.260	51.921	2012:work	0.307	33.381
Time [7.00–8.00]	0.789	157.743	2013:work	0.273	29.657
Time [8.00–9.00]	0.914	182.809	2014:work	0.248	26.962
Time [9.00–10.00]	0.890	177.993	2016:work	0.281	22.119
Time [10.00–11.00]	0.867	173.422	2017:work	0.225	24.181
Time [11.00–12.00]	0.870	174.048	2018:work	0.256	27.389
Time [12.00–13.00]	0.879	176.036	2006:entertainm	0.511	32.111
Time [13.00–14.00]	0.904	180.795	2007:entertainm	0.491	48.743
Time [14.00–15.00]	0.937	187.526	2008:entertainm	0.487	48.798
Time [15.00–16.00]	0.948	189.642	2009:entertainm	0.470	46.523
Time [16.00–17.00]	0.934	186.756	2010:entertainm	0.444	44.116
Time [17.00–18.00]	0.952	190.440	2011:entertainm	0.384	38.271
Time [18.00–19.00]	0.933	186.597	2012:entertainm	0.320	31.575
Time [19.00–20.00]	0.896	179.268	2013:entertainm	0.336	33.678
Time [20.00–21.00]	0.840	168.113	2014:entertainm	0.315	31.417

(*continued*)

Table1. *(continued)*

Variable	Estimate value	t-value	Variable (interactions)	Estimate value	t-value
Time [21.00–22.00]	0.766	153.140	2016:entertainm	0.277	10.444
Time [22.00–23.00]	0.492	98.269	2017:entertainm	0.075	6.065
Time [23.00–24.00]	0.238	47.551	2018:entertainm	0.102	8.098
Day [Tuesday]	−0.016	−6.910	2007:holidays	0.232	19.317
Day [Wednesday]	0.008	3.440	2008:holidays	0.182	16.607
Day [Thursday]	0.037	15.730	2009:holidays	0.220	20.189
Day [Friday]	0.132	22.245	2010:holidays	0.151	14.155
Day [Saturday]	0.132	21.175	2011:holidays	0.048	4.327
Day [Sunday]	0.032	4.997	2012:holidays	−0.016	−1.460
Month [March]	−0.009	−3.320	2013:holidays	−0.024	−2.095
Month [April]	−0.006	−2.229	2014:holidays	−0.038	−3.539
Month [May]	0.025	−.158	2017:holidays	−0.023	−2.032
Month [June]	0.081	29.967			
Month [July]	0.075	27.835			
Month [August]	−0.147	−32.739			
Month [September]	0.002	7.635			
Month [December]	−0.039	−14.369			
Area [non-central]	−0.258	−137.062			
Traffic Signals [no]	0.290	139.361			
Number of lanes [3]	−0.023	−13.356			
R^2 = *0.758*					

Model design was performed using the R programming language, and Table 1 presents the results of the GLM model considering all collected data. The model presents a good fit, and results indicate the effect of various parameters on traffic conditions. It should be noted though, that the data involved days and times serving specific trip purposes, and thus results do not present the general pattern of traffic flow in the city of Athens. For example, Saturday data is used only for entertainment trips, thus data involves late evening hours only. At the same time Sunday data is used for holiday trips, thus data involves mainly late afternoon hours were holiday return trips are performed. The value of this model lies in the interaction effects that represent the annual variation of traffic flow for each trip purpose.

5 Discussion

This study presents urban mobility considering traffic flows for a time period before, during and after the Greek recession. According to the macroeconomic indicators Greece experienced a 5 year period (2008–2013) of financial recession, a period during which economy demonstrated a stable pattern (2014–2016), while slight economic growth emerged again in 2017. At the same time, the financial situation affected relevant mobility indicators. Passenger car purchase rate was reduced during the recession, with the greatest effect being evident towards its end, while a smooth increase appeared in the following years. Fuel consumption, a more direct mobility indicator, presented a one year lag (reduction commenced in 2009) and a one year head start (increase commences in 2012).

The aim of the study was to identify the impact of the crisis on urban mobility as represented through traffic conditions in the city of Athens, considering different trip purposes – namely, work, entertainment and holidays. Results indicated that traffic flow presents a roughly similar trend to the general trend depicted from the macroeconomic indicators. In particular, traffic flow was reduced in 2008 compared to 2006 and 2007, and the reduction continued at least until 2013. Differences in traffic flow patterns were observed between the different measurement sites, which is a result of inherit site characteristics, and between the different trip purposes.

The linear model results, presented traffic flow distribution considering spatio-temporal characteristics. It must be noted that the collected data involves sites and specific time periods where trips serving each of the explored purposes is anticipated to be performed. The interaction terms (year)*(trip purpose) is an indicator of the annual variation of traffic flow considering the different trip purposes. The reduction of work related trips commences in 2008, and an increasing trend appears in 2016 and 2018. Traffic flow considering entertainment trips commences to decrease in 2007, with its lowest value being in 2017, while in 2018 an increase is observed. Last, considering holiday related trips, traffic flow presents a continuous reduction, which is lower in 2017 compared to 2014. Furthermore, in total holiday trips are ones mostly affected by the crisis, as they presented the highest reduction rates. Work related trips presented the highest reduction rate in 2008, and a smooth reduction pattern in the following years. Entertainment trips presented a smoother reduction rate with the highest values being exhibited in 2011 and 2012.

Study results indicate that the economic crisis resulted in a reduction of traffic flow during the recession but also during the stability period, and that this effect varies between the different trip purposes. Future research includes employing time-series analysis to capture both annual and seasonal variations, considering the design of distinct models between the different trip purposes.

References

1. Rešetar, M., Pejić, G., Lulić, Z.: Changes and trends in the Croatian road vehicle fleet – need for change of policy measures. Transp. Policy **71**, 92–105 (2018)
2. Pérez-López, P. Gazol, C.M., Oliver-Solà, J., Hueli, S., Teresa Moreira, M., Feijoo, G.: Greenhouse gas emissions from Spanish motorway transport: key aspects and mitigation solutions. Energy Policy **60**, 705–713 (2009)
3. Sobrino, N., Monzon, A.: The impact of the economic crisis and policy actions on GHG emissions from road transport in Spain. Energy Policy **74**, 486–498 (2014)
4. Vogiatzis, K., Kopelias, P.: Benefits and limitations toward a sustainable road environment during the years of economic recession. Int. J. Sustain. Dev. Plan. **10**(5), 701–712 (2015)
5. Noland, R.B., Zhou, Y.: Has the great recession and its aftermath reduced traffic fatalities? Accid. Anal. Prev. **98**, 130–138 (2017)
6. Wegman, F., Allsop, R., Antoniou, C., Bergel-Hayat, R., Elvik, R., Lassarre, S., Lloyd, F., Wijnen, W.: How did the economic recession (2008–2010) influence traffic fatalities inOECD-countries? Accid. Anal. Prev. **102**, 51–59 (2017)
7. Mpogas, K., Kopelias, P., Mitropoulos, L., Kepaptsoglou, K.: Road safety in urban areas in Greece during economy downturn. A before-after comparison. Transp. Res. Procedia **24**, 228–234 (2017)
8. Dobruszkes, F., Van Hamme, G.: The impact of the current economic crisis on the geography of air traffic volumes: an empirical analysis. J. Transp. Geogr. **19**, 1387–1398 (2011)
9. Azadian, F.: Assessing the effect of urban socioeconomic factors and the financial crisis of 2008 on domestic air cargo traffic in Florida. Transportation **47**, 223–241 (2020)
10. Konstantakis, K.N., Papageorgiou, T., Christopoulos, A.G., Dokas, I.G., Michaelides, P.G.: Business cycles in Greek maritime transport: an econometric exploration (1998–2015). Oper. Res. Int. J. **19**, 1059–1079 (2019)
11. Sermpis, D., Chaziris, A., Vorvolakos, U.: Traffic data variations in major arterials of Athens. In: Conference of Transportation in Economic Recession Periods, Hellenic Institute of Transportation Engineers, Athens, Greece (2013). (in Greek)
12. Prevedouros, P.D., Halkias, B.M.: Expressway traffic demand forecasts in the volatile economic environment of Greece. Transp. Res. Procedia **15**, 607–619 (2016)
13. Papagiannakis, A., Baraklianos, G., Spyridonidou, A.: Impacts of the economic crisis on modal choice and trip frequency of urban transport system. In: Proceedings 7th International Congress on Transportation Research, Athens, 5–6 November 2015 (2015)
14. Sermpis, D., Babis, C., Theofilis, I.: The impact of a transit strike on the traffic patterns in the Athens road network. In: Proceedings of the 11th World Conference on Transport Research Society, Berkeley, CA, United States (2007)

A User Acceptance Survey of Pay-How-You-Drive Urban Pricing Schemes

Kyriaki Christovasili, Eleni Mantouka(✉), and Eleni Vlahogianni

National Technical University of Athens, 5 Iroon Polytechniou Street, Zografou Campus, 157 73 Athens, Greece
elmant@central.ntua.gr

Abstract. Urban congestion pricing is widely regarded as an effective way of traffic management in order to relief metropolitan centers from heavy traffic, reduce emissions, air pollution and as a means to promote public transport usage. Until recently, pricing strategies were formed mostly by taking into account peak hours and exemption provisions (e.g., for taxis or clean-fuel vehicles). However, recent advancements in smartphone sensing have enabled the monitoring of driving behavior, a practice which can act as an additional parameter for the determination of urban tolls. The aim of this paper is to investigate drivers' perception towards PHYD urban pricing schemes and identify factors that may affect their acceptance in the city of Athens using a questionnaire survey. Our preliminary results have shown that men tend to be more aggressive and exceed speed limits than female drivers do. The findings reveal that women are more likely to accept PHYD urban pricing schemes. In addition, commuters are also more likely to accept such a measure contrary to those travelling for other purposes. Finally, eco-minded drivers are more willing to accept PHYD pricing systems. Such results can be really useful for both researchers and decision and policy makers who are willing to design user friendly and easy acceptable pricing systems.

Keywords: Pay-How-You-Drive · Road pricing · Stated preference survey · Clustering

1 Introduction

The urban transportation system, in an environment with increased population density and economic activity, is a key element of economic growth and equality. Also, due to its impact on traffic congestion and greenhouse gas (GHG) emissions, it constitutes a major issue for urban areas with high dependency on car usage. In order to deal with such problems, urban pricing has been suggested as an improvement solution, which in many cases has faced a hostile environment due to a lack of confidence in what will eventually be implemented [1]. Urban toll schemes are rising again as a prominent measure for managing traffic, reducing emissions and funds for new construction, and promoting public transportation. Urban tolling is not a new concept. The most well-known schemes are found in London and Stockholm, where their impact has been

E. G. Nathanail et al. (Eds.): CSUM2020, AISC 1278, pp. 584–594, 2021.
https://doi.org/10.1007/978-3-030-61075-3_57

reflected to reduced volume by 25% and has improved air quality [2]. A flat rate pricing scheme is well known and easy to apply for both policy makers and users. Nowadays, people want different pricing options, especially if they are willing to change some of their driving habits [3].

Most existing pricing schemes can be divided into: Congestion-based, Distance - Based and Pay-As-You-Drive insurance (PAYD). The Congestion-based, is a tax applied to specific areas, the Distance-based, implies a higher price per km travelled and the PAYD, is based on the drivers' mileage. The more you drive, the more you pay, and the less you drive, the more you save. This system is able to achieve goals such as being fair, affordable, improving road safety, reducing various traffic problems and the impact of GHG emissions [4].

A current effective demand management can lead to an efficient future mobility management. Reduced congestion, less emissions and the quality of urban life, benefits all. The technologies of telematics contribute to the introduction of new pricing schemes which serve the needs of authorities worldwide.

Traditionally, pricing schemes are applied in order to promote more sustainable travel modes (such as public transport) and prevent reckless use of private vehicles [5, 6]. Nevertheless, some studies have shown that this measures do not have the anticipated impact on the car use [7] and therefore, in recent years, usage-based pricing schemes arise, which borrow the main aspects of the well-known PHYD and PAYD insurance schemes. Since we cannot prevent drivers from using their own car, we can at least motivate them to drive properly and in a more sustainable way.

The following Fig. 1, illustrates the existing pricing schemes and the innovative concept of PHYD, which encourages drivers to improve their driving behavior in order to pay lower premiums.

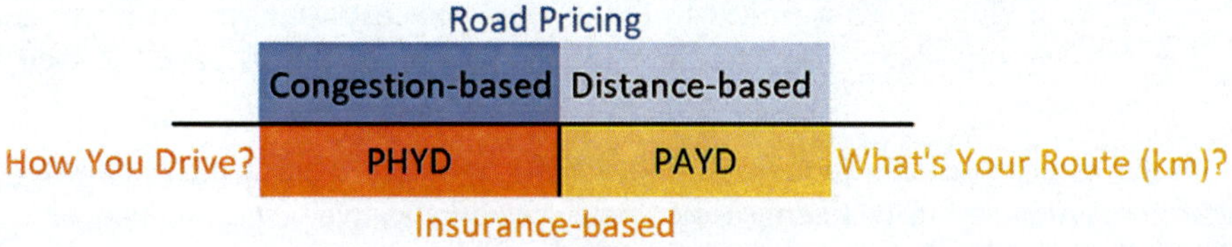

Fig. 1. Pricing schemes.

Taking advantage of this new pricing schemes, urban congestion toll systems can be redefined and become more user-friendly. Previous studies [7], has shown that the success of such services depends on a number of factors ranging from people's characteristics, attitudes, and opinions, but also from the public's perception of the scheme and corresponding benefits, e.g. to reduce car usage, improve air quality and so on. To this end, the aim of this paper is to investigate drivers' perception towards PHYD urban pricing schemes and identify factors that may affect their acceptance. Thus a more user friendly and acceptable pricing system may be designed.

2 Methodology

The acceptance investigation of a new concept, such as the application of PHYD pricing schemes, requires the collection of stated preferences data which are able to capture the perceptions and opinions of the crowd before the measure is actually applied. Therefore, a questionnaire was designed at first and then an online survey was carried out. Once the data was collected, clustering methodology was applied in order to identify groups of potential users of the system which share common characteristics.

2.1 Questionnaire Design and Survey Execution

In order to capture the views of drivers regarding PHYD pricing schemes as well as their perception on the benefits of pricing schemes in general, a questionnaire survey was designed. The stated preferences survey was conducted in from 7/11/2019 until 20/02/2020 and the total number of questionnaires collected is 390.

The questionnaire consists of 5 parts and includes 5 scenarios and 26 questions. In the first part, the respondents were asked which are their preferred means of transport (e.g. car, bicycle or public transport). In the second part, the mobility profile of each driver was captured (e.g., trip purpose, average travel distance, average travel cost, perceived driving behavior).

- Trip purpose: In transport demand analysis, movements are usually characterized by the purpose they serve, leisure, personal affairs, work, etc. [8].
- Average travel distance: this characteristic emerges by the drivers' exposition to traffic accident risk.
- Average travel cost: in the present research the travel cost is defined as the cost of public transport ticket and the fuel cost of private vehicles.
- Perceived driving behavior: refers to the drivers' choices in real time which are linked to the probability of getting involved in a traffic accident, based on their driving way.

In the third part of the questionnaire, questions regarding drivers' perceptions on traffic congestion and its countermeasures were asked. Indicatively, each individual is asked to indicate, between a set sum of answers provided, the impacts of congestion, which are the most appropriate countermeasures, and the benefits of urban road pricing systems. Afterwards the respondent is asked to point out if he would accept the measure of urban tolls, the level of importance, and the benefits that come with such a measure. Finally, the acceptance of an electronic application which can collect personal data with regards to the driving behavior is being investigated, and the extent to whether or not each driver would take advantage of the tips provided in order to improve his driving behavior, drive more eco-friendly and save money. The fourth part, includes scenarios with different cost and travel time values in order to identify whether the drivers will change their mobility patterns or stay firmly at the same behavior. Finally, demographic characteristics were collected as well.

2.2 User Profiling Through Clustering

As mentioned before, we attempted to identify groups of users with common characteristics, who are more likely to accept and get engaged with the proposed system. In order to do so, K-modes clustering technique was applied.

Clustering is a well-known task of dividing a set of observations into a number of groups so that the observations within the same group are similar. The most widely used clustering technique is K-means clustering, where a cluster can be thought as a group of data points whose interpoint distances are small compared with the distances of points outside of the cluster. For each data point X_n, a corresponding set of binary indicator variables $r_{nk} \in \{0.1\}$ are introduced, where $k = 1, \ldots.., K$ describing which of K clusters the data point X_n is assigned to, so that if a data point is assigned to cluster k then $r_{nk} = 1$, and $r_{nj} = 0$ for $j \neq k$. Then, an objective function is defined, given by:

$$J = \sum\nolimits_{n=1}^{N} \sum\nolimits_{k=1}^{K} r_{nk} \|x_n - \mu_k\|^2 \tag{1}$$

which represents the sum of squares of the distances of each data point to its assigned vector μ_k, where μ_k represents the center of the kth cluster. The goal is to find values for $\{r_{nk}\}$ and the $\{\mu_k\}$ so as to minimize J [9]. In the case where all data used are categorical, an alternative to k-means technique is the k-modes clustering. The k-modes algorithm tries to minimize the sum of within-cluster distance from the mode of that cluster, summed over all clusters. In the case of k-modes the distance used in the Hamming distance which is mostly used in information theory.

3 Implementation and Findings

3.1 Analysis of Responses

This section provides an overview of the data collected from the questionnaire with the aim to point out the characteristics of the sample. In addition, some of the most interesting findings with regards to users' driving behavior, environmental concerns and their willingness to accept a new road pricing scheme are being discussed. The socio-demographic characteristics of the sample are shown in the following Table 1:

Table 1. Socio-demographics of the sample.

Variable	Example	Percentage of sample
Gender	Male	56.15%
	Female	43.85%
Age	Young (18–34)	59.23%
	Middle-aged (35–54)	33.85%
	Elderly (>55)	6.92%
Income	Low (<10.000)	35.90%
	Medium (10.000–25.000)	53.59%
	High (>25.000)	10.51%

A sample of 390 people is used, which is well distributed among males and females. For the rest of the analysis only respondents who hold a driver's license and use their car at least once or twice a week are taken into consideration (Fig. 2).

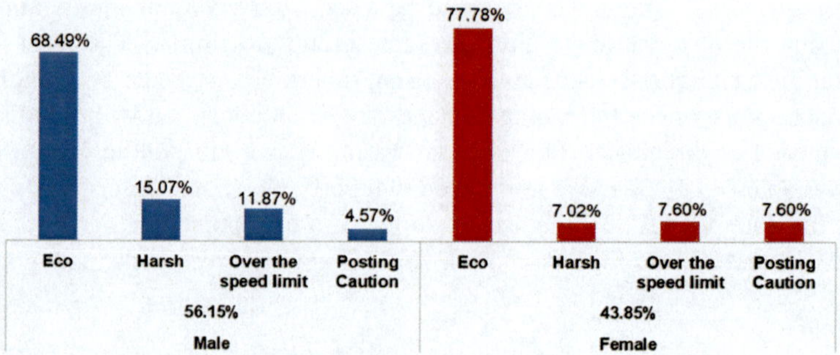

Fig. 2. Male and female drivers' perceived behavior.

In the figure above, it can be clearly seen that there is a significant difference in driving behavior between the two sexes. 15% of the male drivers stated that they drive in an aggressive manner while the responding percentage for females is 7%. Men also stated that they exceed the speed limits more often than women do. Men's driving experience affects their driving skills and reactions by creating a sense of personal safety and confidence as stated in [10].

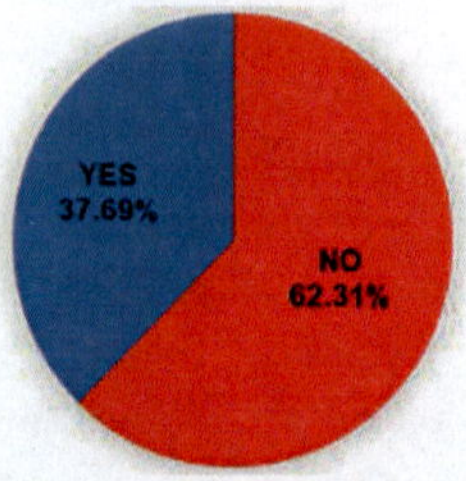

Fig. 3. Urban tolls acceptance.

As it can be seen in Fig. 3, most of the respondents (62.3%) are not willing to accept the urban tolls as a measure of regulating traffic congestion. Studies from several cities, show that the acceptance tends to increase after the implementation [11].

More specifically and according to Fig. 4, drivers with lower income are less likely to accept the measure of urban tolls, while drivers with higher income do not have a clear tendency to accept or reject the measure. Road pricing is expected to receive more support from higher income groups as time is of the essence [12].

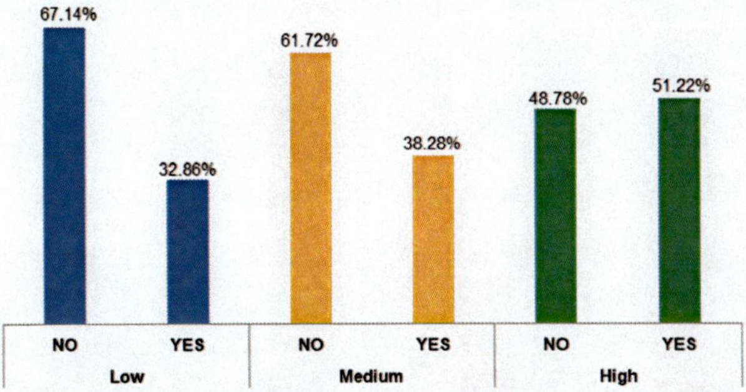

Fig. 4. Relation between urban tolls acceptance and income.

Furthermore, according to main economic theories and in contrast with previous studies [12], the influence of occupation on the acceptability of road pricing seems to be remarkable. Unemployed people and government employees do not accept urban tolls as it can be seen in Fig. 5. This is an expected result, according [5], as the income is an influencing factor for the acceptance in the city of Athens. On the other hand, students and self-employed people, seem to be more progressive and are more likely to accept the measure.

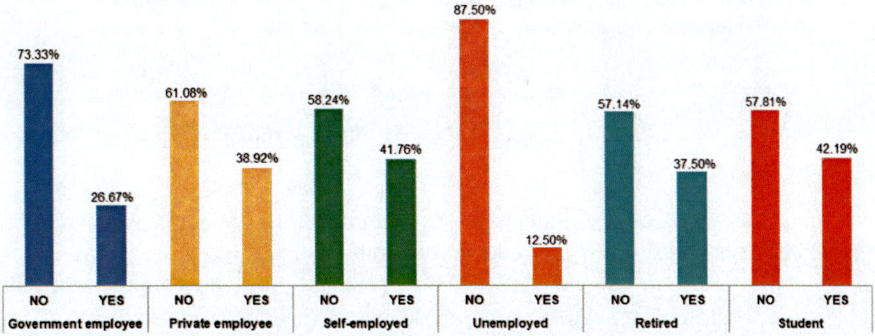

Fig. 5. Relation between urban tolls acceptance and occupation.

[12], in their study mention that urban road pricing is not considered to be an efficient measure of improving ones' driving behavior. Although, with regards to Fig. 6, drivers who perceive that they drive in an eco-friendly manner are more likely to accept the proposed system. This result indicates that people who are eco-minded are more likely to accept measures that lead to reduced air pollution and improved environmental conditions of cities (Fig. 7).

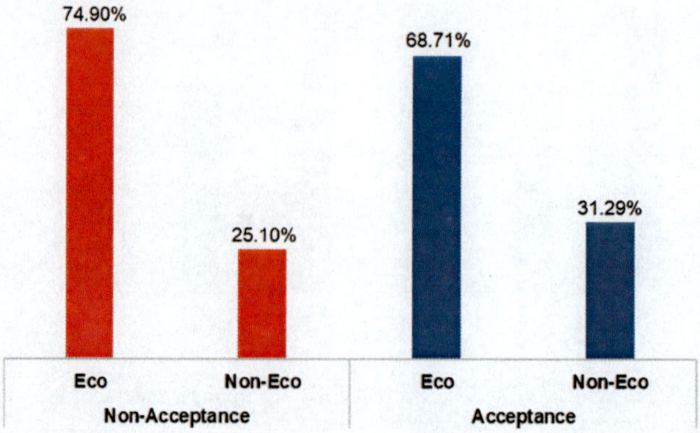

Fig. 6. Relation between urban tolls acceptance and driving behavior.

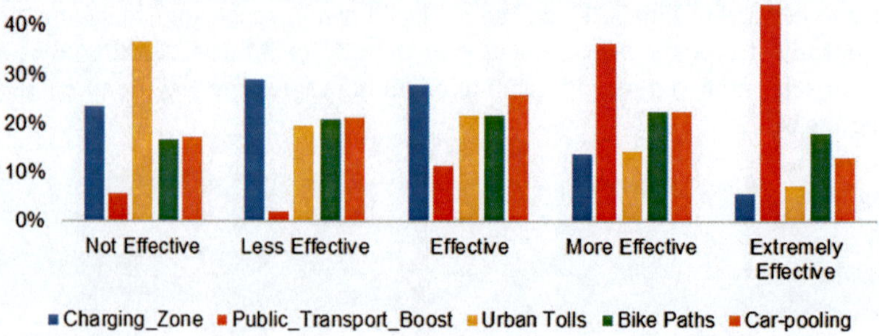

Fig. 7. Traffic congestion reduction measures stated efficiency.

All in all, and as displayed in the figure above, urban tolls are not considered to be the most effective measure for the reduction of traffic congestion while the improvement of public transport can be. According to [11], the most popular and efficient measure in order to reduce queues is the improvement of the public transport system with a percentage of those in favour reaching up to 80%.

3.2 Analysis of Responses

In order to identify groups of drivers in the sample who share the same characteristics k-modes clustering was performed. The number of clusters as emerged from the Calinski-Harabasz criterion was set to $k = 3$. The variables used for the analysis are gender, trip purpose and time travelling. Results indicated that there are three main groups of drivers in the dataset.

- Male commuters who drive to work during peak hours constitute cluster 1.
- Female commuters who drive to work during peak hours constitute cluster 2.
- And finally, female commuters who drive for personal purpose in non-peak hours constitute cluster 3 (Fig. 8).

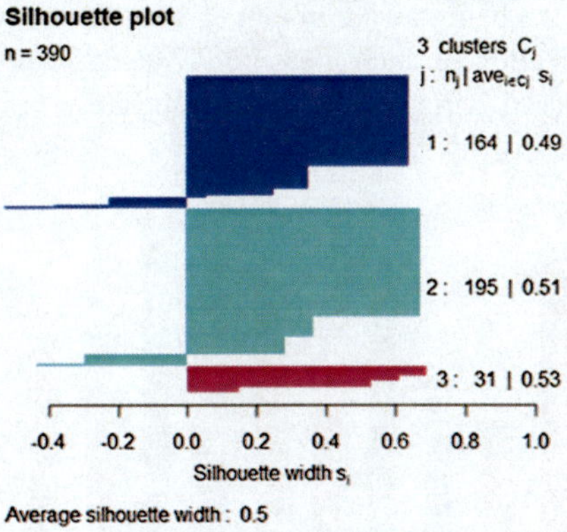

Fig. 8. Silhouette plot.

The average silhouette value is 0.5, which indicates well separated clusters (Fig. 9).

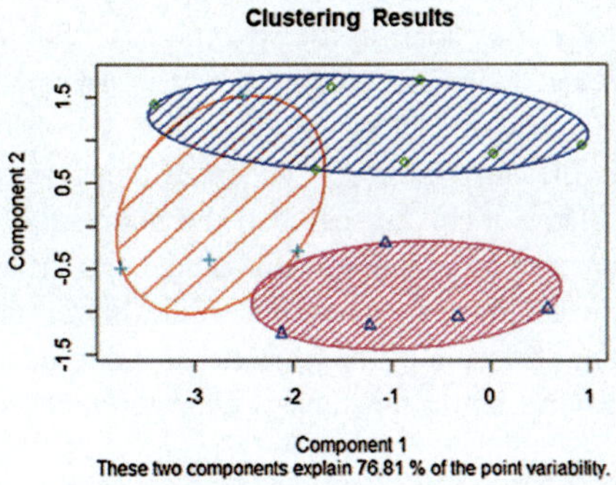

Fig. 9. Clustering results plot in 2d.

Subsequently, the acceptance rates of each of the identified group of drivers is investigated. Finally, 4 scenarios are created:

- Scenario 1: Eco-Driving and Urban Tolls Acceptance
- Scenario 2: Eco-Driving and Non-Urban Tolls Acceptance
- Scenario 3: Non-Eco-Driving and Non-Urban Tolls Acceptance
- Scenario 4: Non-Eco-Driving and Urban Tolls Acceptance

Figure 10 presents the correlation of having knowledge on the concept eco driving and urban tolls acceptance. Results taken from this figure are coming along with those seen earlier in Fig. 3, as in both cases especially high ratios of non-acceptance are noticed. However, when taking a closer look at clusters 1 and 2, for men and women respectively, where both are being aware of eco driving, it can be noticed that men are less objected in the application of such measure. Whereas, when not taking into consideration the factor of eco-driving, women are more likely to accept.

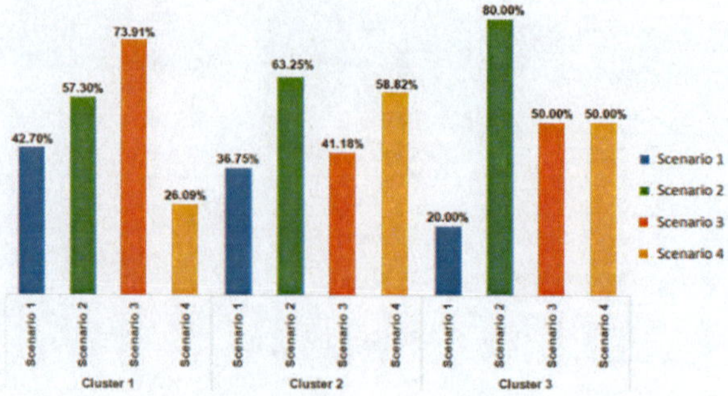

Fig. 10. Relation between clusters and scenarios

In detail, drivers of the first cluster, whether they have knowledge of the eco driving or not, are leaning towards to the non-acceptance of the tolls.

In cluster 2, it is worth noting that, in the specific case where drivers are aware of eco driving, they have almost twice as much higher ratio towards non-acceptance from those who accept such a measure. However, for those who do not know what eco driving is, the urban tolls measure is mostly accepted at a rate of 58.8%.

Finally, in the third cluster, it can be seen that the majority of the drivers are aware of eco driving and at the same time they do not accept the urban tolls. Such result is expected as their travel takes place in non-peak traffic hours and thus some of them may have the possibility to choose a different route or mode of transport. Drivers who do not know what eco driving is, do not have a clear tendency to accept or reject the measure.

Having in mind the analysis conducted above, the result that can be extracted is that the awareness of eco driving cannot act as a factor that drives or either determines the acceptance of urban tolls. It is safe to say that the acceptance of such a measure is mostly connected with factors like as: gender, time and scope of travelling.

Figure 11, reveals that in cluster 1, men drivers do care more about eco driving advices. In cluster 2 women do prefer advices about saving money. In general, in all three clusters people care about eco driving and the improvement of their driving behavior. Privacy issues, such as the abuse of personal data, are viewed as another obstacle to the acceptability of the measures. However, people will allow an application to use the sensors of their mobile phone in order to collect information for their driving behavior. Men tend to trust more the use of such an application.

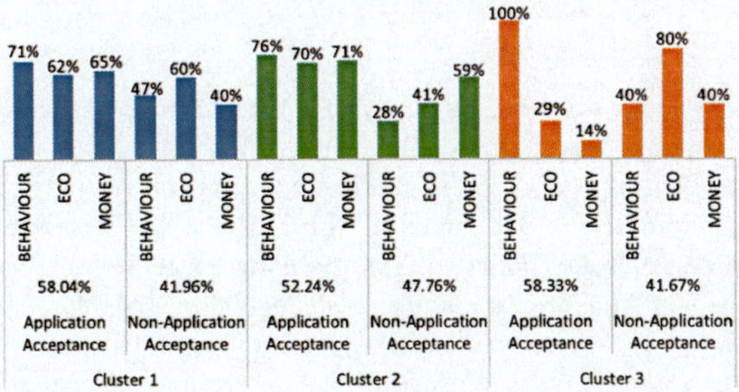

Fig. 11. Relation between clusters – application and application advice.

3.3 Discussion and Conclusions

Studies present that pricing strategies are formed mostly taking into consideration peak hours and exemption provisions. The aim of this paper is to investigate drivers' perception towards PHYD pricing schemes and identify critical characteristics that can lead to their acceptance. Firstly, a questionnaire is designed and distributed to a random but representative sample of 390 participants in Attica region, Greece. This questionnaire, includes both revealed and stated preference questions regarding the acceptance of urban tolls, the effects of congestion and the use of a smartphone application in order to record and collect information about driving behavior.

The conducted survey presents a main finding: this being that people will not easily accept urban tolls. This particular measure is considered not to be an effective way for the reduction of traffic congestion as the improvement of public transport can be.

As far as the two genders are concerned, male are more aggressive drivers and exceed speed limits than female do. Women are more likely to accept PHYD urban pricing schemes. Another important finding is that the income can dictate the acceptance of urban tolls. Drivers with lower incomes or even unemployed do not accept such a measure. Drivers with higher income are equally divided between acceptance and non-acceptance. In this paper, a user profiling took place. From each profile emerged that higher percentages of men tend to trust more the use of an application which collects driving behavior data and do care about eco driving advices. Women do prefer advices about saving money. Finally, the awareness of eco driving cannot act as a factor that drives or either determines the acceptance.

To the best of our knowledge, the present study is one of the very first times that a discrete choice experiment towards the acceptance of urban tolls based on driving behavior is carried out. That is the main contribution of the study.

The present research keeps taking place with the aim of increasing the sample and drawing safer conclusions. The questionnaire can generate various correlations that lead to conclusions about drivers' behavior and characteristics that may influence the acceptance of the urban tolls measure.

References

1. Viegas, J.: Making urban road pricing acceptable and effective: searching for quality and equity in urban mobility. Transp. Policy **8**(4), 289–294 (2001)
2. Croci, E., Ravazzi Douvan, A.: Urban road pricing: a comparative study on the experiences of London, Stockholm and Milan, Bocconi IEFE Centre for Research on Energy and Environmental Economics and Policy, p. 381. Working Paper Series ISSN (2016)
3. Litman, T.: Pay as you drive pricing and insurance regulatory objectives. J. Insur. Regul. **23** (3), 36–50 (2005)
4. Litman, T.: Pay-as-you-drive vehicle insurance premiums into used-based charges. Victoria Transport Policy Institute, pp. 36–53 (2004)
5. Rentziou, A., Milioti, S., Gkritza, K., Karlaftis, M.: Urban road pricing: modeling public acceptance. J. Urban Plan. Dev. **137**(1), 56–64 (2011)
6. Schaller, B.: New York City's congestion pricing experience and implications for road pricing acceptance in the United States. Transp. Policy **17**(4), 266–273 (2010)
7. Liu, Z., Shiwakoti, N., Bie, Y.: Measuring the public acceptance of urban congestion-pricing: a survey in Melbourne (Australia). Collab. Urban Transp. **33**(4), 902–912 (2018)
8. Nair, H.S., Bhat, C.: Modeling trip duration for mobile source emissions forecasting. J. Transp. Sci. Technol. **6**, 17–32 (2003)
9. Bishop, C.: Pattern Recognition and Machine Learning, pp. 78–90. Springer, New York (2006)
10. Tsoxos, G., Kehagia, F., Satrazemis, A.: Greek drivers' driving behaviour and the impact on road safety. Patra, s.n., pp. 1–12 (2005)
11. Eliasson, J.: Road Pricing in Urban Areas. Stockholm: s.n (2002)
12. Schade, J., Schlag, B.: Acceptability of Urban Transport Pricing. Goverment Institute for Economic Research, Helsinki (2000)

Evaluating Pedestrian Environments: Evidence from Small Cities in Greece

Georgios Barmpas[1(✉)], Georgios Georgiadis[2], Anastasia Nikolaidou[2], Rafail Katkadigkas[1], and Dimitrios Tsakiris[1]

[1] Lever Development Consultants S.A., 26th October Str., 43, 54627 Thessaloniki, Greece
giorgos.barbas@lever.gr

[2] Transport Engineering Laboratory, Department of Civil Engineering, Faculty of Engineering, Aristotle University of Thessaloniki, 54124 Thessaloniki, Greece

Abstract. Sustainable urban mobility is now a well-established concept in European cities which, during the last decade, they continuously put significant effort into elaborating sustainable urban mobility plans and upgrading their systems in favor of environmentally friendly transport modes. Walking is being gradually established as a primary pillar of urban mobility by being promoted as an attractive transport solution for both door-to-door and last mile trips. In order to increase walking modal shares, cities should provide a high-quality pedestrian mobility environment both in terms of related services and infrastructures. This study presents several results from a structured walkability audit process for evaluating pedestrian mobility conditions in urban areas. The study examines the case of small cities in Greece which so far have been characterized by a more car-oriented transport development and experience comparatively higher difficulties into securing appropriate funding for sustainable transport investments. In this study's context, the proposed walkability audit has been performed in three (3) small Greek cities which represent different urban development models, i.e. a city with a physical barrier (Trikala), a historical small city with particular town layout (Mesologgi) and a city with high inclinations (Kavala). For each city case, key pedestrian quality and quantity indicators, such as sidewalks operating width per reference network length, quality of sidewalk pavement and pedestrians' perceived safety and security, are measured and discussed. Comparative analysis of these indicators demonstrates critical strengths and weaknesses in each city and assists into prioritizing appropriate pedestrian mobility improvement actions.

Keywords: Walking · Walkability · Audit · Sustainable mobility · Small cities · Trikala · Kavala · Messologi

1 Introduction

Various modes of transport are suitable for different purposes, but all are key parts of an urban transport system. Private cars have an advantage for long to medium distance trips, while public transport is more capable on transporting large numbers of people to

E. G. Nathanail et al. (Eds.): CSUM2020, AISC 1278, pp. 595–605, 2021.
https://doi.org/10.1007/978-3-030-61075-3_58

activity centers. Walking, as a mode of transport, is ideal for travelling short distances, and thus a large number of urban journeys are performed on foot, and even those using other transport modes have to also begin and finish their journeys with walking. About 95% of short distances under 100 m are covered on foot [1].

Conventional transportation planning practices tend to favor motor vehicle transport improvements over mobility management solutions to transportation problems. These approaches treat walking as a minor transport mode and find only modest benefits from supporting walking activity. In addition, walking is a mean of experiencing and interacting with the local environment and wider society in a way not possible when using other modes of transport, particularly motorized transport. The advantages of walking, such as low cost and availability to all people, make it the most preferred and common form of physical activity [2].

In this context, during the last years great emphasis is given to the study of transport systems that improve pedestrians' quality of mobility. Urban planners try to develop pedestrian-friendly cities with areas more attractive to walkers; to increase citizen's physical activity and support healthy lifestyles. Vulnerable street users are becoming the center of urban mobility and the evaluation of pedestrian level of service the main objective of recent research concerning urban transport [3].

Walking is a critical component of an urban mobility system, and improved walkability and increased walking can provide significant benefits to society. Walkability is defined as a measure of how friendly an area is to walk. Streets with high walkability can achieve health, environmental, and economic benefits for their users. Factors influencing walkability include the existence or absence and quality of footpaths, sidewalks or other pedestrian rights-of-way, traffic and road conditions, land use patterns, building accessibility and safety, among others. In general, walkability is an useful concept in sustainable urban design [4].

The aftermath of European Commission's White Paper of Transport (2011) have introduced a new transport planning philosophy among European cities, based on the principles of sustainable development. A Sustainable Urban Mobility Plan (SUMP) is a long-term strategic plan designed to satisfy the mobility needs of people and businesses in cities and their suburbs for a better quality of life. During the past decade a significant number of European cities have elaborated their SUMPs, setting goals and designing measures in favor of sustainable means of transport, such as pubic transport, bicycle, pedestrians, pooling and sharing schemes etc.

Based on central government monetary incentives, many Greek cities begun their SUMP's elaboration since 2016. These plans are considered the first steps in strategic transport planning for most of these cities. A major part of these efforts focuses on the analysis and improvement of cities' pedestrian facilities. In this process, various kinds of data, such as sidewalks' width, pavements' level of maintenance, sidewalk accessibility facilities (ramps) and pedestrian volumes are being collected and examined.

In this context, the present paper presents a structured walkability audit process as a tool for the monitoring, evaluation and improvement of the walking conditions in an urban area. The paper examines the case of three (3) small Greek cities with different characteristics but with common feature the application of policies which so far are promoting the use of car at the expense of sustainable modes of transport such as walking. Analysis of the walkability audit results indicates the appropriate interventions for the improvement of walking conditions in each case study.

The remainder of the paper is as follows. Next section reviews the concept of walkability and the corresponding audits which have been developed so far. Section 3 explains the proposed walkability audit for small Greek cities and the areas where it was implemented. The audit results are presented and discussed in Sect. 4. Section 5 summarizes the main conclusions.

2 Walkability Audits

Improved walkability improves accessibility, provides consumer and public cost savings, increases street livability, improves public health and supports strategic economic development, land use and equity objectives. A variety of methods can be used to evaluate these impacts. When considering walkability as a broader concept, it includes quantitative objective characteristics (infrastructure, land use etc.) as well as qualitative features such as subjective sense of safety, aesthetic sensibility etc. These indicators are often evaluated in the form of an index which is critical in terms of data reduction (the values of the individual features usually cannot be reconstructed from the final values of an index). There is weighting of individual characteristics (which is partly at the discretion of the researcher) and it often lacks spatial or socio-demographic differentiation [5].

There is an extensive bibliography related to the development of walkability audits [6, 7, 8], where a variety of parameters that influence walking is considered. The use of reliable audits to assess walkability is a relatively new technique but provides a good way to operationalize walkability concepts and may enable planners to identify specific sites for walkability improvements. The needs of pedestrians can be elicited through the use of tools to assess the environment or the "walkability" of a particular route using some kind of scoring or checklist approach. These approaches usually use best practice for determining what pedestrian factors to include and the tools are often used by transport authorities and consultants.

Pedestrian Environment Review System (PERS) permits users to evaluate a range of factors about a route or crossing [10]. The aggregation of these results enables the operator to audit walking routes and identify specific features which should be improved. The Level of Service (LoS) methodology for pedestrians, used by Gallin [11] is another approach which provides an "overall measure of walking conditions on a route, path or facility" examining design parameters (e.g. sidewalk width), location parameters (e.g. connectivity) and user parameters (e.g. pedestrian volume). A weighting system is then applied recognizing certain factors that are more important to pedestrians. The LoS approach has also been used to assess pedestrian trip quality based on more qualitative factors such as enclosure, articulated buildings and proximity to traffic [12].

Other methodologies for assessing the walkability of the pedestrian environment focus on checklists or observations that are completed by the pedestrians themselves. Walkable America [13] produced a "walkability" checklist, which asks residents to assess their local community. Living Streets [14] produced a community street audit which focuses on identifying what communities want to be improved. More recently mobile methods are increasingly being used. Definitions of such methods are varied and evolving rapidly from a range of disciplinary perspectives and providing potential for a trans-disciplinary approach.

The integration of walkability assessments is potentially a valuable complement in urban planning processes through considering important quantitative and qualitative aspects of walking in cities. Recent literature shows a variety of approaches involving discrepancies in the definition of walkability, the factors which contribute to it, and methods of assessing them. Especially, in the framework of these audits it is attempted to correlate the walking behavior with parameters referring to the built and natural environment such as physical characteristics of roads, pedestrian infrastructure, traffic, etc. by measuring the influence of these parameters on walking through the use of statistical tools and Geographical Information Systems (GIS). A great majority of these studies assess the walkability of road segments in defined urban areas identifying as walkability the quality of sidewalks, while other studies examine the walkability in special areas such as surrounding schools [15].

Regarding the Greek cities, several walkability audits were developed the past decade as a result of extensive research [16] or as tools in order to assess the existing situation of a case study network [17, 18].

3 Material and Methods

3.1 Assessment Methodology of Pedestrian Facilities

In order to assess the existing pedestrian facilities in a structured and comparable manner, we accordingly developed a walkability audit with the aim to objectively assess the physical environment as a whole and collect credible information about the pedestrian environment in particular. This proposed audit considers several generally approved walkability audit methodologies taking into account previous relevant research regarding the parameters that affect pedestrians' environment both in sidewalks and footpaths, with a series of modifications for a proper adaptation in Greek cities' conditions.

The proposed walkability audit is summarized in Table 1. The first column of Table 1 lists the eleven (11) parameters which are examined in the audit process. These parameters examine the infrastructure's geometry (width, available height, etc.), as well as its operational characteristics, such as accessibility, perceived security and comfort. The second column describes the required details for their measurement. The parameters are matched to their values after detailed observation of sidewalk's characteristics, in addition with in situ measurements of specific dimensions (width, obstacles coverage, available height, etc.) with common equipment.

Table 1. Walkability audit parameters and measurement details.

#	Examined parameter	Details
1	Total sidewalk width	Sidewalk width is the main parameter affecting pedestrian's comfort and safety. The Greek legislation sets a minimum 1.50 m sidewalk width
2	Sidewalk's available height	Point by point or linear height obstacles may reduce pedestrians' comfort and safety and are appropriately observed and measured
3	Sidewalk pavement's level of maintenance	A general assessment of the sidewalk's pavement quality. The main sub-parameters considered are: (a) number or extend of any damaged tiles, (b) traction, (c) unified (or not) structure/materials along the sidewalk, (d) drainage and (e) Intense altitude difference along the sidewalk
4	Pedestrian crossing quality	Existence of the required number of crossings, assessment of its proper placement on the section as well as its level of maintenance
5	Sidewalk accessibility – pedestrian ramps	Existence and functionality (quality) of the ramps
6	Guide for visually impaired people	Existence and design quality of a guide for visually impaired people along the sidewalk
7	Green space	Existence of trees and plants along the sidewalk and how it affects pedestrians' comfort
8	Sidewalk's equipment	Quality and quantity of several pieces of sidewalk equipment that improve pedestrian's comfort, i.e. benches, bins, shading installations, separation from traffic when appropriate etc.
9	Pedestrian obstacles	Possible permanent (street's furniture, trees, signs, etc.) or temporary (merchandise, tables, illegal parking, etc.) objects on the sidewalk which reduce the effective width and/or affect pedestrians speed and comfort
10	Estimated pedestrian volume	Pedestrian volumes can be counted in detail by in situ measurements. However, for the needs of this audit, the volumes can be also estimated according to the street characteristics
11	Estimated pedestrian's perceived security	Pedestrian's perception of a sidewalks' security. The sub parameters examined are: (a) Street's lighting and (b) Street's commercial character

3.2 Data Collection and Process

Three small Greek cities with different urban fabric characteristics were selected as case studies for the implementation of the proposed walkability audit, based on the availability of the required data:

- Trikala is a small city in central Greece with less than 61,000 residents. The city is built onto Thessaly's plain and its geomorphology, which includes almost zero inclinations, is ideal for cycling and walking. According to the recent SUMP of Trikala, citizens are using 32,8% car, 28,9% motorcycle, 2% public transport, 10,3% bicycle and 26% walking in their daily trips [16]. During the past decade, local authorities' efforts focused on upgrading Trikala into a modern sustainable city by introducing "Smart City" applications.

- Kavala is a small city in northern Greece, built along the seafront. Its population is 54,000 residents according to the latest census (2011). The city's main characteristics, which drastically affect its transport system, are high inclinations in combination with narrow streets, observed in the most districts of the city. This situation affects road safety, parking as well as pedestrian facilities. Kavala's modal split is 54,0% car, 4,1% taxi, 16,3% motorcycle, 7,0% public transport (bus) and 17,5% walking, according to the city's SUMP [17].
- Messologi is a small city in western Greece with less than 13,000 residents. It is a medieval city, built on a peninsula in the middle of a sea-lake. The crucial characteristic about Messologi's urban area has to do with the historical city's (city's center) narrow roads (5–6 m width), which characterize its transport system. The small size of Messologi in addition with zero inclinations, act in favor of bicycle and walking. According to a recent online survey during Messologi's SUMP elaboration [18], the city's modal split is estimated at 57,9% car, 2,7% motorcycle, 19,5% bicycle, 3,8% public transport (bus) and 16,1% walking.

In all the above cities we performed the walkability audit through in situ measurements on a weekday morning, for the sidewalks of the central area and/or main commercial streets in each city case. The identification of the segments that are representative of the main sidewalks and walkways for each case study was conducted using street maps of the selected cities and a series of on-site observations. Thus, the study areas cover the main pedestrian infrastructure for each city center. The length of the pedestrian network examined in each city is presented in Table 2. The observations were collected in audit forms for each sidewalk section on a street. The breakdown of a street's sidewalks was defined according to the city blocks.

The post data collection analysis included the creation of a digital database with the results of the audit for each sidewalk section. The next step was a geospatial visualization of the database on a GIS environment. The shape files including network dimensions (were provided by the Hellenic Statistic Agency). Finally, basic validation and calibration procedures were conducted for a more accurate analysis.

4 Results and Discussion

The assessment of the existing walking facilities in the three cities returned both quantitative and qualitative results. In this section, to save space, selected results of each city case are examined through a comparative analysis. For each city case, a set of indicators regarding their existing pedestrian facilities was selected for comparison. Indicators were calculated based on general characteristics (e.g. population, modal split) for each city as well as on the results of the walkability audit survey.

Table 2 reports the results of these indicators. The first column contains the indicators' names, while the second column explains the variables considered for their measurement. The rest of the columns include the indicators' values for Trikala, Kavala and Messologi respectively.

Table 2. Comparative analysis on pedestrian facility indicators.

Indicator	Details	Trikala	Kavala	Messologi
Population	City's residents	61,000	54,000	13,000
Modal split-Walking	Percentage of city's daily trips	26%	17,5%	16,1%
Pedestrian streets	Percentage of pedestrian streets in the city's road network	1.5%	2%	2.6%
Central area sidewalks examined	Total length of the central area's sidewalks examined (m)	9,485	14,830	8,012
Pavement width	Percentage of sidewalks with operational width equal or more than 1.5m	63%	69%	38%
Pavement quality	Percentage of sidewalks with medium to high maintenance level	61%	65%	23%
Pedestrian crossings	Percentage of sidewalk crossing with proper placement and level of maintenance	50%	52%	23%
Accessibility - Ramps	Percentage of sidewalks with properly designed ramps	20%	47%	10%
Accessibility – Guide for visually impaired	Percentage of sidewalks with properly designed path for visually impaired people	0%	30%	0%
Obstacles	Percentage of sidewalks free of obstacles that affect pedestrians' comfort	59%	52%	38%
Safety & security	Percentage of sidewalks with high perceived safety & security	74%	88%	82%

Table 2 shows that more than 30% of the sidewalks in a central area of the three cities have insufficient width (Fig. 1), when examined according to the national guidelines (minimum operational width = 1.5 m) [19]. This percentage may be increased if the minimum width threshold also considers the respective pedestrian volumes. Affected by an intensity of land uses, the central sidewalks often have temporary or permanent obstacles, such as lighting columns, signs, litter bins, advertisement signs and other street furniture as well as café/restaurants' furniture consisting of tables and chairs which reduce the available operational width below the appropriate minimum standard, indicating the need for specific measures. This situation highly reduces pedestrians' comfort and safety, when combined with insufficient sidewalk width in all cities examined. An evaluation of the above elements' necessity and maintenance condition should take place in order to remove those that are considered as unnecessary and might constitute a threat to the pedestrians' safety due to their poor condition. Furthermore, a stricter enforcement policy concerning the public space occupation by the cafés and the other relevant stores should be applied.

A significant part of central areas' sidewalks has a low maintenance level, especially in Mesologgi. Local authorities could not invest in sidewalk regeneration during the past decade (financial crisis), resulting in old and degraded pavements which have a negative impact in comfort and safety.

Vulnerable users of the streets were not a priority on previous street designs. As a result, the examined cities are characterized by insufficient accessibility regarding pedestrian infrastructure (insufficiently designed ramps and very low percentage of roads with guide for visually impaired). The absence of ramps and guides for the mobility and vision impaired pedestrians respectively are considered as the most significant weaknesses, highlighting thus the emergence for small-scaled limited budget interventions in order to make the footpaths accessible to all. Specifically, ramps should be installed, in accordance with the official national specifications, in almost all the entrances of the footpaths as they provide greater accessibility while they are favored by all types of pedestrians. Moreover, the guides for the vision impaired pedestrians should be constructed along all the footpaths.

However, there are additional findings regarding each city's pedestrian facilities that cannot be adequately presented through simple indicators. Specifically:

- Kavala is a city with high inclinations at most of its street sections with North-South orientation. Thus, inevitably, a significant portion of its pedestrian network in the perimeter of the city's center is developed in the form of steps (about 10% of Kavala's street network is steps [17]). Even though steps can be considered an exclusive infrastructure for pedestrians, their low to zero accessibility and comfort reduce their usage.
- Trikala has a major physical barrier running through the city. River 'Litheus' divides the city's central area in two parts, which are connected with five bridges in total (3 of them are accessible only by walking and cycling). Even though such barriers may cause uneven development of land uses between the two banks, in Trikala this is not a case. This can be pointed out by considering pedestrian infrastructure as well. Pedestrian infrastructure is almost equally divided between the north (54%) and south (46%) part of the city's central area. Both pedestrian areas are living streets with services, commercial uses and entertainment.
- Messologi has a city center with extremely narrow street network. As a legacy from its past, the old city of Messologi has a dense urban core, resulting in various weaknesses such as lack of parking spots, insufficient to zero sidewalk width, orientation difficulties etc. Part of the road network in the southern city center has been modified into a woonerf zone with traffic calming measures, while the center of the old city has about 1,300 m of small pedestrian streets forming a grid around the central square.

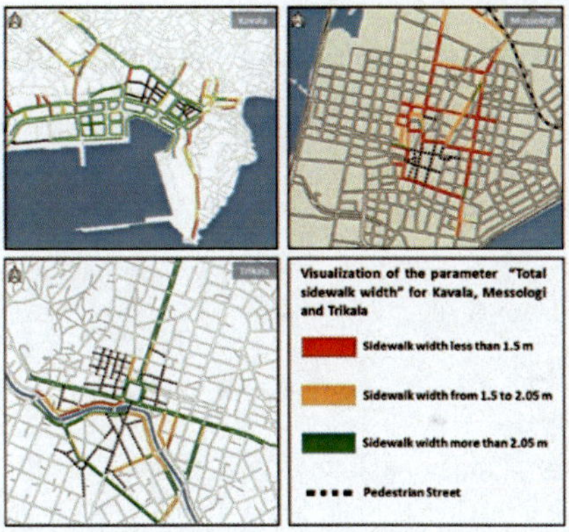

Fig. 1. Visualization of the parameter "Total sidewalk width" for Kavala, Messologi & Trikala.

5 Conclusion

The improvement of pedestrian convenience, mobility and safetyis an crucial step for the promotion of sustainable mobility in urban areas. One of the reasons people are reluctant to walk is the lack of appropriate and attractive walking corridors in cities. Although the main goal of municipalities in most European countries, including Greece, is to encourage walking, the quality of paths and pedestrian networks is not sufficient to do so. Sidewalks and pavements in cities tend not to be appropriate in terms of their design, the facilities they serve or their visual attributes. Therefore, it is essential to provide municipalities with suitable analytical tools, to help them evaluate the appropriateness and effectiveness of walking paths within their jurisdictions. Towards this direction, walkability audits constitute a useful tool.

In the context of the current research, a walkability audit was developed based on previous relevant research findings and taking into consideration the specific features of Greek cities. This paper presents the implementation of the proposed walkability assessment methodology to three small Greek cities with different characteristics. More cities will follow in future projects. Based on the results of the survey, aspects that need special attention include the low level of service for people with disabilities, the poor quality of walking environment, the insufficient width available for pedestrian movement, the quite frequent presence of obstacles as well as the perceived level of safety and security.

The proposed walkability audit is an easy to use tool but yet comprehensive in order to provide consistent and meaningful results. The outputs of its application may act as a guide for urban planners and decision makers who wish to develop and implement a successful walking and walking path strategy. It can be also useful to city planners and designers for increasing the awareness about sidewalk conditions among local policy makers, prioritizing investments and upgrading the sidewalk facilities. Walkability assessment of pedestrian infrastructure would justify devoting more funding to walking facilities and programs, shifting road space from traffic and parking lanes to sidewalks and paths, policies to create more walkable land use patterns, and greater efforts to manage motor vehicle traffic to improve walking safety and comfort.

References

1. United Nations: Shanghai Manual – A guide for sustainable development in the 21th century, Bureau international des expositions, Shanghai World Exposition Executive Committee (2010)
2. NZ Transport Agency: Pedestrian planning and design guide. New Zealand Transport Agency, Wellington (2009)
3. Christopoulou, P., Pitsiava-Latinopoulou, M.: Development of a model for the estimation of pedestrian level of service in Greek urban areas. Procedia Soc. Behav. Sci. **48**, 1691–1701 (2012)
4. Litman, T.A.: Economic value of walkability. Transp. Res. Rec. **1828**(1), 3–11 (2003)
5. O'Connor, D., Borscheid, M., Reid, O., Lochlainn, M.: Northside partnership "Healthy Community" pilot area walkability audit. In: Proceedings of the Irish Transport Research Network (2015)
6. Boarnet, M.G., Day, K., Alfonzo, M., Forsyth, A., Oakes, M.: The Irvine-Minnesota inventory to measure built environments development. Am. J. Prev. Med. **30**(2), 144–152 (2006)
7. Lee, S., Talen, E.: Measuring walkability: a note on auditing methods. J. Urban Des. **19**(3), 368–388 (2014)
8. Millington, C., Thompson, C.W., Rowe, D., Aspinall, P., Fitzsimons, C., Nelson, N., Mutrie, N.: Development of the Scottish walkability assessment tool (SWAT). Health Place **15**, 474–481 (2009)
9. Pikora, T., Bull, F., Jamrozik, K., Knuiman, M., Giles-Corti, B., Donovan, R.: Developing a reliable audit instrument to measure the physical environment for physical activity. Am. J. Prev. Med. **23**, 187–194 (2002)
10. Streetaudit - PERS - TRL Software. https://trlsoftware.com/products/road-safety/street-auditing/streetaudit-pers/. Accessed 05 Mar 2020
11. Gallin, N.: Quantifying pedestrian friendliness – guidelines for assessing pedestrian level of service. Road Transp. Res. **10**(1), 47–55 (2001)
12. Jaskiewicz, F.: Pedestrian level of service based on trip quality. In: Transportation Research Circular E-C019, TRB, National Research Council, Washington, D.C. (2000)
13. America Walks: Steps to a Walkable Community A Guide for Citizens, Planners, and Engineers, Schwartz Engineering PLLC (2012)
14. Charity, L.S.: DIY Community Street Audit Pack. Living Streets (2005)
15. Sdoukopoulos, A., Verani, E., Nikolaidou, A., Tsakalidis, A., Gavanas, N., Pitsiava-Latinopoulou, M., Mikiki, F., Mademli, E., Pallas, C.: Development and implementation of walkability audits in Greek medium-sized cities: the case of the Serres' city center. Transp. Res. Procedia **24**, 337–344 (2017)
16. Christopoulou, P., Pitsiava-Latinopoulou, M.: Development of a model for the estimation of pedestrian level of service in Greek urban areas. Transport Research Arena – Europe (2012)
17. Galanis, A., Eliou, N.: Development and implementation of an audit tool for the pedestrian built environment. In: Transport Research Arena – Europe (2012)
18. Tsakalidis, A., Sdoukopoulos, A., Gavanas, N.: Assessing the walking conditions in pedestrian networks: the case of the city center in Larissa, Greece. In: 4th International

Conference on Environmental Management, Engineering, Planning and Economics (CEMEPE) (2013)
19. Municipality of Trikala: Sustainable Urban Mobility Plan of Trikala 2020–2030. Trikala (2019). (in Greek)
20. Municipality of Kavala: Sustainable Urban Mobility Plan of Kavala. Kavala (2020). (in Greek)
21. Municipality of I.P. Messologiou: Sustainable Urban Mobility Plan of I.P. Messologiou. Messologi (2020). (in Greek)
22. Hellenic Ministry of Environment, Physical Planning and Public Works: General Guidelines – Designing for everyone (in Greek)

A Comparative Gap Analysis for Electromobility and Alternative Fuels

Foteini Orfanou(✉), Panagiotis Papantoniou, Eleni Vlahogianni, and George Yannis

National Technical University of Athens, 5 Iroon Polytechneiou, 15773 Athens, Greece
{forfanou, ppapant, elenivl, geyannis}@central.ntua.gr

Abstract. Electromobility and alternative fuels are considered as key solutions towards the reduction of emissions and energy savings in order to achieve more sustainable and environmentally friendly transportation systems. European countries are trying to achieve significant progress in promoting these new technologies in order to ensure better life and air quality for their residents. However, the number of electric vehicles is still low and not all European cities have made worth mentioning progress towards this direction. The objective of this research is to identify the gaps hindering the promotion of electromobility and alternative fuels in 9 European regions from 8 European countries: Italy, Slovenia, Greece, Finland, Norway, Latvia, Belgium and Romania. The analysis is focusing on three thematic areas: Business, Governance and Research and Innovation Strategies for Smart Specialization (RIS3) and each area is divided in various aspects such as charging Infrastructure, e-Vehicle fleet, incentives, technology, campaigns, legislation, enforcement, education, research and Innovation. The analysis is based on estimating the progress achieved in each aspect using a scale from 1 to 10 (10 indicates the highest level of performance and level 1 the lowest). Results indicate that there are regions with significant progress in the field of electromobility and others that still should elaborate a lot to increase their performance in most of the analyzed aspects. The findings assessment will help policy makers to prioritize their effort on the areas that have the highest gap until the desired level in electromobility is achieved and identify other regions with similar electromobility issues.

Keywords: Gap analysis · Electromobility · Alternative fuels · European Regions

1 Introduction

Electric vehicles and other alternative fuel vehicles are promising alternatives to reduce CO_2 emissions providing better air quality for the city residents and the road users as well as increasing energy efficiency. Additionally, electric vehicles are more quiet than conventional cars and even silent in low speeds improving life quality and reducing noise levels. This advantage of electric vehicles has also psychological benefits for the drivers, passengers and other road users as the noise reduction or absence brings about less anxiety, less frustration and better and potentially less aggressive driving behavior.

E. G. Nathanail et al. (Eds.): CSUM2020, AISC 1278, pp. 606–615, 2021.
https://doi.org/10.1007/978-3-030-61075-3_59

Production, sale and use of electric cars has already started in the European Region and their penetration rate is being slowly increased during the last years [1]. More specifically, the biggest increase since 2008 occurred in 2017, when the sales of the battery electric vehicles (BEV) increased by almost 50% compared to 2016 while for plug in hybrid electric vehicles the corresponding increase was 35% for the same time period (Europe Environment Agency, 2018). Concerning registrations of these two types of electric vehicles in the European region, they consist 0.7% of the total number cars. Despite their advantages, the number of electric vehicles in Europe is still limited and various measures are taken by the EU members for increasing the fleet of the eco-friendly cars [2]. Different factors cause this low penetration rate such as applied policies, lack of financial and non- financial incentives, high price of electric vehicles compared to the conventional ones, lack of charging infrastructure, low public awareness [3, 4].

The increase of the number of electric vehicles is a major target of the European Union within the framework of more environmentally friendly, cleaner and greener transportation systems. Forecasts made by the International Energy Agency support that the share of EVs in the European region can be up to 23% in 2030 concerning all road transport vehicles except of two and three wheelers. Additionally, the White Paper on Transport sets the aim of reducing CO_2 emissions in transportation by 60% by 2050 [5]. Achieving this target means formulation of a concrete methodology in local, regional or national level for effectively developing actions favoring the promotion of the new technologies. This methodology includes the analysis of the current situation in terms of electromobility and alternative fuels diffusion, a GAP and SWOT analysis as well as formulation of recommendations for promoting the sustainable technologies.

The objective of this research is presenting the results of the GAP analysis conducted in 9 European Regions concerning electromobility and alternative fuels. Gaps will be detected in order to identify the progress achieved as well as reveal and clarify the fields need to be further enforced. Three thematic areas have been selected and consist the basis for the GAP analysis: business, governance and Research and Innovation Strategies for Smart Specialization (RIS3). For each thematic area individual specific indicators are defined and evaluated for each region using a scale from 1 to 10. Finally, the results of the analysis for each region will be compared in order to reveal similarities along with differences among them. The paper is organized as follows: Sect. 2 includes the description of the thematic areas and their indicators, Sect. 3 the results of the GAP analysis while in Sect. 4 conclusions are summarized and presented.

2 Methodology

2.1 Gap Analysis

Gap Analysis is conducted in order to reveal the current situation and the deficiencies prevailing in a specific sector or field. It gives an overview about the progress that has been achieved and whether the goals defined have been fulfilled. Through the gap analysis the deviation between the current and the target status is defined and clarified enabling the formulation and set of the next steps and actions that should be planned and implemented in towards the goal satisfaction. In the present study, nine European

Regions are analyzed and for each one a gap analysis was conducted where the three thematic areas (business, governance and RIS3) were evaluated. The nine European Regions are: Province of Brescia (Italy), Calabria region (Italy), Regional Development Agency of Gorenjska (Slovenia), Region of Attica (Greece), Flemish Government Department Environment (Belgium), Regional Council of Kainuu (Finland), Rogaland County Council (Norway), Bucharest-Ilfov Regional Development Agency (Romania) and Zemgale Planning Region (Latvia). Each of the three thematic fields was subdivided in various aspects and each one was evaluated on a scale from level 1 to 10. Level 10 indicates the highest level of performance or implementation and level 1 the lowest. For the evaluation of these indicators a group of stakeholders was formulated in each project partner region. The stakeholders are experts with deep knowledge in the field of electromobility and alternative fuels working in the industry, in the infrastructure and (public) service, regional public authorities, in business association and in the academia/research. This particular synthesis of the stakeholder group allows the more objective evaluation of the individual indicators.

2.2 Thematic Areas

The three thematic areas are defined below (Tables 1, 2 and 3):

- Business: Market take-up of sustainable mobility from alternative fuels
- Governance: Needs, requirements and policies to enhance sustainable mobility from e-mobility and alternative fuels
- Research and Innovation Strategies for Smart Specialization (RIS3): e-mobility in relation to RIS3 Smart Specialization Strategy documents

Thematic Area 1: Business

Electric vehicles can offer significant economic and environmental advantages as they are expected to have reduced maintenance and operational costs as well as zero CO2 emissions and reduced to zero level of noise in very low speeds. Business is one of the main areas that will play important role in the diffusion of electromobility and alternative fuels.

Table 1. Aspects of the thematic area Business.

Business aspect	Description
Charging infrastructure	Available charging stations already installed in the Region
e-Vehicle fleet	Number of electric vehicles in the Region
AF vehicles fleet	Number of vehicles using Alternative fuels in the Region
Technology	Existence of the available technology to support the use of electric and Alternative fuels vehicles
Incentives	Given to the industry/companies in order to motivate and encourage them to be involved in these aspects
Campaigns	Campaigns organized by the industry and the private sector for raising public awareness about electromobility and Alternative fuels

Incentives should be given to industry and enterprises in order to be involved in this field and contribute in promoting electric vehicle, enhance their characteristics and performance and increase user acceptance and willingness to use them. Market take up means how many people will shift from diesel and conventional vehicles to electric and alternative fuel ones and therefore the target is to achieve a continuous higher percentage. Additionally, business could also significant contribute in the necessary infrastructure, required for the easier and more comfortable use of electric vehicles. Joint investment programs should be proposed in order to establish an efficient and publicly accessible charging network.

Thematic Area 2: Governance

It is significant that the people ruling a municipality, a region or a country are interested in transforming the prevailing transportation system into a more sustainable and environmental friendly one aiming at improving air and life quality of the residents as well as creating healthier conditions during the fulfilment of their transportation needs. Policies to expand the use of electric mobility and formulation and adoption of the appropriate legislation framework favoring and encouraging the use and purchase of electric vehicles is the first step for increasing users' willingness to use vehicles being equipped with this new technology. Public authorities and the state should take care of the establishment of a privately and publicly accessible charging infrastructure which is considered as a significant measure for adopting the use of EVs [6]. The charging network enables electric vehicle users to move within and outside of the city easily and more comfortable without anxiety and frustration about where to charge their cars if necessary.

Table 2. Aspects of the thematic area Governance.

Governance aspect	Description
Legislation	The existing legal framework and policies established favoring the use and purchase of electric vehicles
Enforcement	Enforcement of regulatory changes, laws, directives and standards for the promotion of electromobility and increase of the share of electric vehicles and alternative fuels in the fleet
Education	Inform students about the advantage of electromobility and alternative fuels
Incentives	Incentives given to the private and public sector by the state and the public authorities
Campaigns	Campaigns organized by public authorities, regions and municipalities for raising public awareness about electromobility and alternative fuels

Moreover, the creation of a fast charging station network within the country will also encourage tourists to visit the country with their electric vehicles and should also be a priority for the state. These actions along with the efficient improvements and upgrades of the infrastructure requires the strong cooperation between governance and business and support of the entrepreneurial decision of integrated e-mobility productions. Funding is also an important issue concerning actions about promoting electromobility and alternative fuels like research and innovation activities as it will be further described in the next section (Thematic Area 3: RIS3). Finally, communication strategies including information and education campaigns organized by public authorities, regions and municipalities brings about raise in public awareness about electromobility and alternative fuels as potential users may have an incomplete or distorted view about these technologies [3, 7, 8].

Thematic Area 3: Research and Innovation Strategies for Smart Specialization (RIS3)

Research and Innovation Strategies for Smart Specialization – RIS3 are integrated, place – based economic transformation agendas, taking into consideration the special and individual characteristics of each region. The mission of the formulation of these strategies is summarizing in the following 5 axes [9]:

- They focus policy support and investments on key national/regional priorities, challenges and needs for knowledge-based development.
- They build on each country/region's strengths, competitive advantages and potential for excellence.
- They support technological as well as practice-based innovation and aim to stimulate private sector investment.
- They get stakeholders fully involved and encourage innovation
- They are evidence-based and include sound monitoring and evaluation systems

Table 3. Aspects of the thematic area RIS3.

RIS3 aspect	Description
Research	Intensiveness of research activities on technologies related to electromobility and alternative fuels concepts
Innovation	Level of innovation and experimentation in projects related to electromobility and alternative fuels
Synergies	Level of involvement of various stakeholders in projects related to electromobility and alternative fuels
Strategic plan	Formulation of a regional strategic plan concerning the diffusion of electromobility and alternative fuels
Monitoring	Existence of a sound monitoring and evaluation system in projects related to electromobility and alternative fuels

Research and Innovation Strategies involves research activities on various technologies concerning every sector (transportation, agriculture, economy, etc.) and the participation of research departments and institutions. The RIS3 concept indicates flexible and dynamic innovation strategies aiming to regional firms and productive system and tries to avoid fragmentation efforts in the field of innovation support. It refers to all three priorities of Europe2020 – sustainable, smart and inclusive growth – and guides priority-setting in national and regional innovation strategies, as well as cross-border cooperation where appropriate. The strategies are formulated based on each country's/region's strengths, competitive advantages and potential for excellence and the main focus areas include policy support and stimulation of private sector investments and technological support.

3 Results

3.1 Business

Charging infrastructure is an aspect where 4 out of 9 regions have achieved significant progress (Table 4). The well-developed and sufficient charging infrastructure in Flanders, Bucharest – Iliov, Rogaland and Gorenjska is a very important step towards the increase of the number of electric and alternative vehicles. On the other hand, the regions of Calabria, Kainuu, Brescia and Zemgale need to elaborate more on the installation of charging stations and the creation of an appropriate network. Especially, for Gorenjska, it is controversial that despite the available charging infrastructure and facilities in the region, the number of electric vehicles is extremely low. Region of Attica evaluated the charging infrastructure with only 2 out of 10 indicating that charging facilities and have to be increased and installed in various locations within the region to enable the use of an EV.

Table 4. Comparative gap analysis for the thematic area Business.

European Region	CI*	EV** Fleet	AFV*** Fleet	Technology	Incentives	Campaigns
Brescia	5	4	2	8	5	2
Calabria	4	3	1	5	4	4
Gorenjska	9	1	1	4	6	2
Attica	2	2	3	4	2	3
Flanders	8	5	5	5	7	4
Kainuu	4	3	1	5	3	2
Rogaland	8	8	8	7	9	6
Bucharest-Ilfov	8	5	5	4	8	4
Zemgale	5	2	2	4	2	3

** CI: Charging Infrastructure, ** EV: Electric Vehicles, *** AFV: Alternative Fuels Vehicles*

Despite the fact that some of the regions provide the necessary infrastructure the number of electric vehicles is still low in all of them except of the Rogaland where the EV fleet was evaluated with 8. It is interesting to investigate why the number of EVs is not higher in the regions where enough charging stations have been installed favoring the use of EVs. Exactly the same picture for the number of alternative fuels vehicle fleet, with the Rogaland Council achieving the highest score. Province of Brescia has achieved significant progress in developing the required technology to support the use of electric vehicles along with the Rogaland while for all the others regions this aspect got an average score. Concerning incentives, it is worth mentioning that in Flanders, Bucharest – Iliov and Rogaland, lot of incentives are given to the industry and the public sector in order to motivate and encourage their people to get involved in the electromobility concept and improve the technology they use. Attica, Kainuu and Zemgale are the regions with the lowest score (2 and 3) in this aspect.

Finally, the highest deficiency for all regions is identified in the aspect of the informative campaigns. Campaigns for promoting the new technologies for a more sustainable and environmental friendly transportation system are not a strong asset for any of the analyzed regions, a fact that indicates that industry should be more active in this field as it consists a valuable tool for getting potential users familiar with electromobility and alternative fuels. The lack of campaigns can explain the low penetration rates of electric and alternative fuels vehicles in some regions as people do not support or are not aware of the technology and they didn't have the opportunity of getting familiar with these new technologies by using or driving test vehicles.

3.2 Governance

In the Governance sector, the Rogaland has significantly high scores in almost all aspects while Flanders and Gorenjska have achieved significant progress in terms of education – incentives- campaigns and legislation – enforcement – incentives respectively (Table 5). The appropriate legislation framework is a very important tool for encouraging the purchase and the shift towards more sustainable authorities. Additionally, enforcement of regulatory changes, laws, directives and standards for the promotion of electromobility and increase of the share of electric vehicles and alternative fuels in the fleet can also contribute towards more sustainable transportation system.

Similarly to the business sector, incentives given to the public sector and campaigns organized by regional or local authorities are not a priority for most of the regions. It is important to understand the importance of informing public about the existence of alternative technologies as well as their advantages and therefore raise public awareness. Low evaluation scores are also observed in the aspect of education for most of the regions with Brescia and Attica having the lowest ones.

Table 5. Comparative gap analysis for the thematic area Governance.

European Region	Legislation	Enforcement	Education	Incentives	Campaigns
Brescia	6	7	2	3	3
Calabria	5	5	5	4	5
Gorenjska	8	8	5	8	5
Attica	3	2	2	2	3
Flanders	5	4	8	8	8
Kainuu	7	5	7	4	2
Rogaland	8	8	6	10	7
Bucharest-Ilfov	6	6	5	6	4
Zemgale	2	3	6	2	5

3.3 RIS3

The third thematic area is mainly focusing in research and technology advances as well as the strong cooperation of the various involved stakeholders (synergies). Table 6 shows that most of the regions achieve significant progress in almost all aspects apart from Attica as the overall score is around 2 out of ten. More specifically, the province of Brescia has made remarkable steps in the aspect of research activities and the intense participation of research institutions and universities in the development of technologies for making the use of electric vehicles and alternative fuels more convenient and beneficial for the users. Additionally, innovation and experimentation is also encouraged and highly progressing as it as was assessed with 8 out of 10. Almost the same scores are observed for Flanders, Kainuu and Gorenjska in the aspects of research and innovation. It is interesting to mention that the scores for the region of Kainuu were very low in the previous thematic areas. The level of involvement of various stakeholders is very low in the Province of Brescia (Synergies were evaluated with 4/10) and its increase should be a priority for the region as one of the main goals of RIS3 is the involvement and strong cooperation among different sectors and is one of the key factors for the efficient promotion and development of electric vehicles. The same results concerning synergies between various stakeholders are also presented in Table 6 for the regions of Calabria, Attica and Zemgale. For Calabria and Attica, the aspect of research and innovation are also not a strong asset and a lot of effort should be put for improving this indicator. Finally, Bucharest – Iliov region has an average evaluation score for all aspects of RIS3 as the scores are fluctuating around 5. One of the basic and easy steps for further progress is the promotion of synergies among different sector and a broad network of experts and their strong cooperation is considered a necessity not only for promoting sustainable technologies but also in other various fields.

Table 6. Comparative Gap Analysis for the thematic area RIS3.

European Region	Research	Strategic Plan	Synergies	Innovation
Brescia	8	6	4	8
Calabria	5	5	5	5
Gorenjska	7	9	6	8
Attica	3	2	2	2
Flanders	8	7	7	7
Kainuu	8	9	7	8
Rogaland*				
Bucharest-Ilfov	6	6	7	6
Zemgale	4	7	3	3

* *Rogaland does not have yet Research and Innovation Strategies for Smart Specializations*

4 Discussion and Conclusions

Electromobility and alternative fuels are considered the key solutions for dealing with the continuously increasing energy consumption and emissions. The effective promotion of these sustainable technologies highly depends on identifying and detecting the aspects where a region lacks of knowledge and ideas so that targeted actions and measures are taken, designed and implemented. The Gap Analysis conducted in 9 European regions, within the framework of this study, revealed these deficiencies in terms of three thematic areas (business, governance and RIS3) and their indicators.

The results showed that most of the analyzed regions need to elaborate more in most and important indicators such as the formulation of appropriate and favorable legislation framework, the enforcement of laws and directives, as well as the development of sufficient infrastructure by installing charging stations in key locations. The increase of incentives, such as financial incentives are considered to increase the electric vehicle market take up [10]. Two of the aspects that all regions received low scores are the organization of campaigns and the strong synergies. For promoting the environmentally friendly technologies, it is necessary to inform public about their characteristics and advantages as well as ensure their safety. Furthermore, the cooperation between the involved stakeholders is an important step. From the 9 European regions analyzed in the present study, Rogaland is the most advanced one, achieving high scores in all aspects indicating the significant progress achieved and the good level of electromobility promotion. Flanders, Gorenjska and Bucharest are achieving high scores in the majority of the aspects while Brescia is trying to achieve a good effort. Attica, Zemgale, Kainuu and Calabria are the regions with the lowest scores in almost all aspects indicating that industry and government as well as researcher should increase their involvement in the electromobility field and take actions and measures towards this direction.

The results of the present research can be used as knowledge, ideas and policy exchange among the regions while cooperation between them can also be activated. Besides, they will be used as the basis for the development of recommendations favoring the promotion of electromobility in the regions in each thematic area and afterwards in the formulation of a regional action plan including concrete actions that will be adopted and implemented by the regions but it can also be an inspiration for any other local, regional or national authority supporting electromobility and alternative fuels.

Acknowledgment. The research is based within the framework of e-MOPOLI project (Electro MObility as driver to support POLicy Instruments for sustainable mobility) funded by the European Union within the framework of the Interreg Europe Programme.

References

1. European Environment Agency Homepage. https://www.eea.europa.eu/
2. Cansino, J., Sánchez-Braza, A., Sanz, M.: Policy instruments to promote electro-mobility in the EU28: a comprehensive review. Sustainability **10**(7), 2507 (2018)

3. Coffman, M., Bernstein, P., Wee, S.: Electric vehicles revisited: a review of factors that affect adoption. Transp. Rev. **37**, 79–93 (2017)
4. Sierzchula, W., Bakker, S., Maat, K., Van Wee, B.: The influence of financial incentives and other socio-economic factors on electric vehicle adoption. Energy Policy **68**, 183–194 (2014)
5. European Commission. Roadmap to a Single European Transport Area—Towards a Competitive and Resource Efficient Transport System. White Paper (2011)
6. Mersky, A.C., Sprei, F., Samaras, C., Qian, Z.S.: Effectiveness of incentives on electric vehicle adoption in Norway. Transp. Res. Part D Transp. Environ. **46**, 56–68 (2016)
7. Zhang, Y., Yu, Y., Zou, B.: Analyzing public awareness and acceptance of alternative fuel vehicles in China: the case of EV. Energy Policy **39**(11), 7015–7024 (2011)
8. Thiel, C., Alemanno, A., Scarcella, G., Zubaryeva, A., Pasaoglu, G.: Attitude of European car drivers towards electric vehicles: a survey. JRC report (2012)
9. European Union Homepage. https://ec.europa.eu/
10. Bühne, J.A., Gruschwitz, D., Hölscher, J., Klötzke, M., Kugler, U., Schimeczek, C.: How to promote electromobility for European car drivers? Obstacles to overcome for a broad market penetration. Eur. Transp. Res. Rev. **7**(3), 30 (2015)

Defining and Prioritizing Indicators to Assess the Sustainability of Mobility Systems in Emerging Cities

Juan Camilo Medina[1(✉)], Jorge Pinho de Sousa[1], and Edgar Jimenez Perez[2]

[1] INESC TEC, Faculty of Engineering, University of Porto, Rua Dr. Roberto Frias, 4200-465 Porto, Portugal
up201610839@fe.up.pt, jorge.p.sousa@inesctec.pt
[2] Centre for Air Transport Management, Cranfield University, College Road, Wharley End, Bedfordshire MK43 0AL, UK
E.Jimenez@cranfield.ac.uk

Abstract. Mobility and urban planning are among the main problems that cities face in order to achieve sustainable development. The significant economic and social changes in emerging cities, as well as the urgency of environment protection, make sustainable urban mobility planning a very critical issue. Although sustainability is difficult to be measured directly, it can be assessed through a system of parameters that attempt to reflect its multiple aspects. The definition of adequate indicators constitutes an essential part of this process, even if the literature is already abundant in available metrics. This research proposes a set of indicators, grouped in different dimensions, as a tool for the assessment of sustainability in mobility systems, for the specific context of emerging cities. These dimensions and indicators are based on a comprehensive literature review on sustainable urban mobility indicators, subsequently subjected to validation with experts through surveys, semi-structured interviews and statistical analysis, in order to identify the indicators significance and their priority within the selected dimensions.

Keywords: Sustainable urban mobility · Measurement indicators · Mobility planning · Multicriteria Decision Aid (MCDA)

1 Introduction

There is a generalized and increasing interest in the concepts of sustainability and sustainable transportation. Sustainability refers to a balance of economic, social and environmental goals [1], and reflects the essential human desire to improve and protect our world. The concept emphasizes the integrated nature of human activities and the need for coordinated decisions between different jurisdictions, sectors and groups [2].

In such context, transportation systems have a significant impact on the different dimensions of sustainability [3]. Moreover, transport is widely considered as a sector with significant positive and negative externalities affecting society, the environment

E. G. Nathanail et al. (Eds.): CSUM2020, AISC 1278, pp. 616–625, 2021.
https://doi.org/10.1007/978-3-030-61075-3_60

and the economy, and in the particular case of metropolitan regions, it is directly connected to the sustainable development of cities [4].

Therefore, addressing urban mobility is a priority for government agencies, due to the increasing travel times and the lack of accessibility, generated by larger rates of population growth and motorization [5]. Until now, public policies in emerging cities have promoted the use of cars, thus increasing congestion with a number of negative consequences, including an inefficient economy, productivity losses, pollution and traffic-related fatalities and injuries [6].

Hence, the increasing interest in making transportation environmentally, economically and socially sustainable, has created the need for the design of sustainable transportation indicators to be used for planning purposes [1]. In line with these concerns [3] concluded that "*traditional transportation performance metrics tend to focus on vehicle mobility and congestion and, in practice, fail to assess the degree to which transportation planning leads to sustainable outcomes. Lacking appropriate metrics, transportation managers and policy-makers seldom have sufficient information to make decisions explicitly taking into account the sustainability concerns*".

On the other hand, urban growth patterns have been changing and therefore attention has shifted to other types of cities, namely the so-called emerging cities, i.e. cities with less than two million inhabitants but with high growth rates, and where the major transformations of urban space are occurring [7]. This new pattern of settlement is creating great opportunities, but also generating new, huge challenges. Hence, emerging cities must achieve a sustainable development, by promoting new growth strategies in addition to improving their resilience and to adapt to climate change [7].

Nevertheless, currently, no standardized set of sustainable transportation indicators is available [1]. There is, therefore, an urgent need for decision-makers to have measurement support tools, such as sets of indicators, organized in groups, with consistent definitions, accessible data and technically measurable, and adequate for evaluating the impact of different policies in the sustainability of mobility systems.

In this context, and given the high levels of involved uncertainty, along with a variety of relevant practical aspects, a successful sustainability assessment will strongly depend on a careful and sound selection of indicators [1]. Consequently, this research aims at providing guidance on which measuring tools can improve the sustainable policy design, beyond just proposing a sounder list of indicators for measuring sustainability of mobility systems in emerging cities.

In line with these concerns, and based on an extensive literature review, we have selected and defined 42 sustainable transport indicators, grouped in 5 dimensions (environment and human health; economic and social; operational; fiscal and governance; and mobility system effectiveness and land use).

These indicators are composed by qualitative and quantitative sets of measurable factors, allowing cities to evaluate their performance and assess progress over time. They also help to guide policy design, and planning and management activities, across multiple sectors and stakeholders.

The proposed selection of sustainability indicators for transportation will hopefully provide an useful framework for measuring of the sustainability performance of transportation systems, and for the development and improvement of strategies to eliminate their negative impacts.

In this research, two Multicriteria Decision Aid (MCDA) methods were used to validate the selection of indicators and to prioritize those indicators and dimensions, through double iteration surveys and semi-structured interviews with experts from academia/research, consultancy and government organizations. Those methods were the well-known Analytic Hierarchy Process (AHP) and a fuzzy variant of AHP (FAHP).

In this paper, Sect. 2 presents the framework developed to assess the sustainability performance of mobility systems in emerging cities. Section 3 describes the proposed sustainability measurement indicators and their final weighting, according to a set of prioritization techniques. Finally, Sect. 4 presents a set of conclusions and opportunities for further research.

2 A Framework to Assess the Sustainability of Mobility Systems in Emerging Cities

A promising approach to design mobility systems is planning through indicators. Indicators are capable of generating information to support decision-making, and enable tracking and monitoring of goals, benefits, effectiveness and efficiency of the proposed actions [8].

According to [9] "*an indicator is a variable, based on some measurements, representing as accurately as possible a phenomenon of interest*". Indicators are selected and defined to measure progress towards an objective [10]. The OECD defines sustainable transportation indicators as statistical measures that give an indication of the sustainability of social, environmental and economic development [9]. Yet, indicators must be carefully defined in order to avoid ambiguous conclusions [11].

In this research, we have designed a framework to be used as a tool for the assessment and planning of mobility services, in terms of sustainability. This framework is based on a comprehensive literature review on sustainable urban mobility, and on redefining and consolidating all relevant elements, taking into account the advantages and weaknesses of currently available solutions (see Fig. 1).

The fundamental concepts of sustainability are often illustrated through the three spheres of sustainability, emphasizing the integrated nature of environmental, social, and economic sustainability [12]. Nonetheless, sustainability in the context of transportation does not simply refer to the transportation system itself; it is also about understanding the broader impacts of that system [3]. The proposed framework encompasses a five-dimensional approach, with indicators that cover the traditional three pillars of sustainability, along with some complementary aspects.

Environment and Human Health. Over the last decades, we have all become increasingly aware of the environment's limited ability to cope with the unrestrained development of humanity. [13] concluded that "*air and water pollution, as well as climate change, are having a significant effect on human health and quality of life, in some of the world's largest developing economies*". Along these lines, this dimension aims at measuring the impact of activities related to urban transport on the environment and the population.

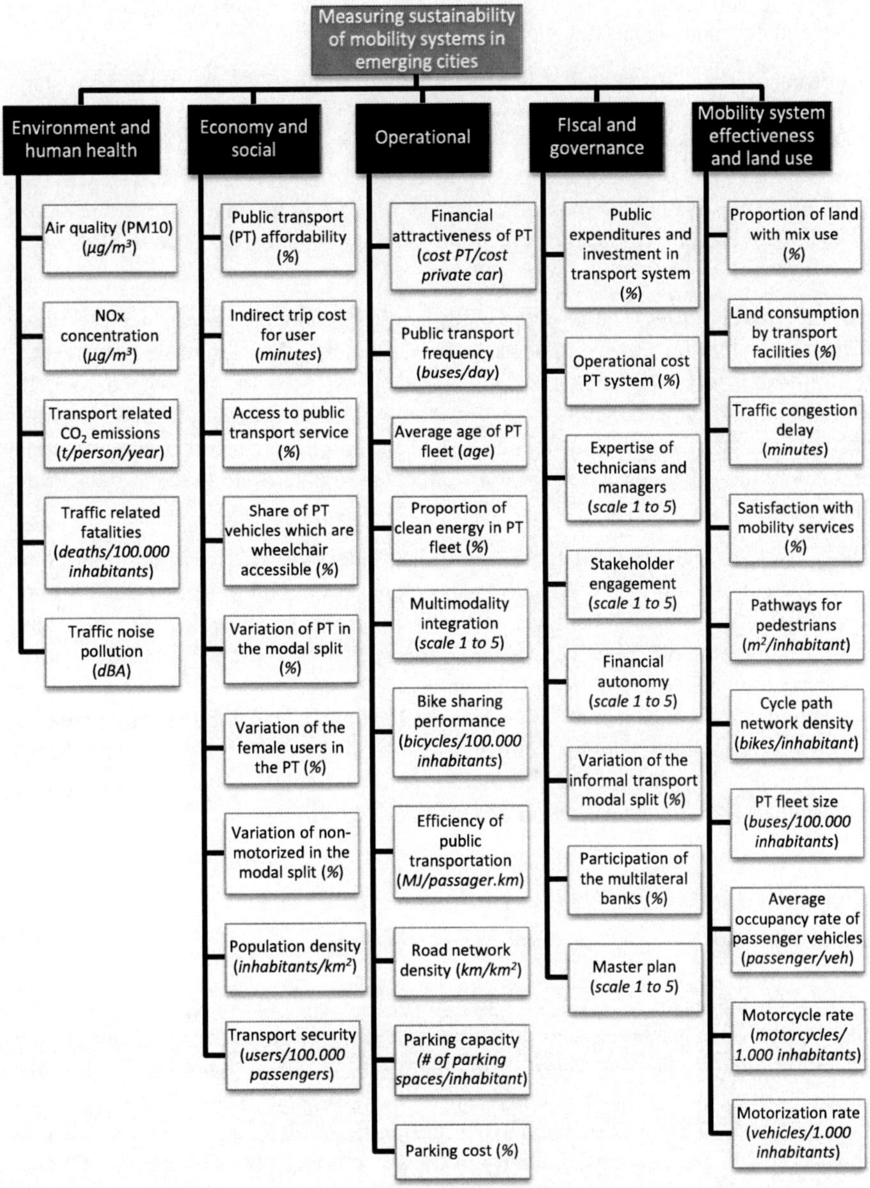

Fig. 1. Proposed framework, with dimensions (level 1, in black) and indicators (level 2, in white), for assessing sustainability of mobility systems in emerging cities (level 0, in gray).

Economy and Social. [3] stated that "*to be sustainable, a transportation system must be reasonably affordable for every individual, in terms of the monetary and time costs associated with using the transportation system, and in terms of social sustainability, as a way to promote social equity*". Hence, this dimension analyzes potential economic vulnerabilities associated to the mobility system, and the ability of this system to promote equality and social inclusion.

Operational. Urban sustainability indicators follow in general more traditional standards. Therefore, according to [8] "beyond dealing with isolated social, economic and environmental aspects, new indicators must tackle plans involving features such as integration, long-term planning and wide-spread ranges of actors". Hence, a system that operates efficiently is one that maximizes access, while minimizing waste and resource usage. This dimension provides technical attributes to evaluate the operation of urban mobility systems.

Fiscal and Governance. The growing importance of municipal governance and fiscal performance factors for citizens [14] has motived the consolidation of this dimension, with the integration of topics not previously taken into account, and with the redefinition of existing subjects. Thus, this dimension aims at assessing the way public spending is managed, in transport-related initiatives, and the degree of independence in planning activities of those entities that are expected to contribute to the sustainability of the system.

Mobility Systems Effectiveness and Land Use. Effectiveness is an important component of the definition of sustainability; therefore, any mobility system that is considered "sustainable" must also have the capacity to be effective [3]. For that, this dimension integrates different metrics that support the transport policies formulation processes in reducing externalities of the mobility system.

In summary, we expect that this conceptualization of the different sustainability dimensions, consolidated in our framework, can help public (regional and local) authorities and policy makers to better understand sustainability issues, and to design more sustainable and inclusive mobility services.

3 Prioritizing Measurement Indicators to Assess Sustainability

For validation purposes, the proposed framework was circulated, in three iterations, among experts from academia, and consultancy and government organizations. In the first iteration, *preliminary phase*, we have organized five 2-hour recorded face-to-face meetings, in English, with sustainability experts from Copenhagen (Denmark), Malmo and Lund (Sweden), between March 2019 and April 2019, with the purpose of discussing on how to measure sustainability in transportation systems of emerging cities. In the second iteration, considered as an *exploratory phase*, the goal was to open a formal discussion about the definition of each element of the proposed structure – the information gathered via an online survey (with 105 respondents) was analysed using descriptive statistics, to redefine and assess the proposed structure.

Then, a *confirmatory phase* used two MCDA methods, used to prioritize the elements in the framework. Both, an AHP-based questionnaire and semi-structured interviews were responded by 19 experts in sustainability topics. Pairwise comparison evaluations between the dimensions and indicators were used to compute weights at both levels.

As stated by [15] "*the AHP approach is based on an additive weighting process, in which several relevant criteria and alternatives are explored through their importance. AHP has been extensively applied by academics and professionals in many fields and problems, as it is the case of transport engineering*".

The top level of the hierarchy is associated with the global goal considered in the problematic situation under analysis, and each level denotes the factors contributing to the associated upper levels. Meanwhile, the bottom level contains the considered alternatives or actions. AHP allows the comparison of factors, with the importance of individual factors being relative to their effects on the problem solution, and the priority list of the considered alternatives to be researched [16].

However, AHP has been criticized for its inability to deal with imprecision and subjectivity, in the pairwise comparison process [17]. To overcome these problems, several works have tried to integrate *fuzzy theory* with AHP. The Fuzzy AHP (FAHP) approach can deal with the vagueness of linguistic judgments by establishing an imprecision prioritization [18]. Therefore, FAHP sets provide more efficient and realistic results, because they mimic human reasoning more closely, overcoming limitations of bounded rationality [19].

Consequently, the second step in our *confirmatory analysis* involved evaluating the proposed framework, applying the Fuzzy AHP method for level 0 (measuring sustainability in mobility systems) and for level 1 (dimensions). As a result, weights were determined for each level, thus leading to a suitable evaluation framework to measure sustainability of mobility systems in emerging cities, as described in Tables 1, 2, 3, 4, 5 and 6.

Table 1. Final weighting of the dimensions, in the proposed structure.

Dimensions	Priority (%)	Rank
Environment and human health	47.85	1
Economy and social	23.81	2
Mobility system effectiveness and land use	14.14	3
Operational	8.35	4
Fiscal and governance	5.86	5

The obtained results clearly show the importance of assessing sustainability in terms of environmental and human health, as well as the concern for traffic related fatalities. Moreover, even though the fiscal and governance dimension obtained the lowest priority, this dimension has indicators that are quite relevant, due to a permanent shortage of investment resources and the resulting need for adequate fiscal management.

Overall, translating sustainable transport and sustainability values into a set of indicators (grouped in several dimensions) allows decision makers to identify and focus on the main challenges that mobility systems face to achieve sustainability. With a

Table 2. Final weighting of the indicators for the environment and human health dimension

Indicator	Priority (%)	Rank
Traffic related fatalities (*deaths/100,000 inhabitants*)	32.29	1
Air quality (PM10) (*μg/m³*)	31.65	2
Transport related CO_2 emissions (*t/person/year*)	20.59	3
NOx concentration (*μg/m³*)	10.79	4
Traffic noise pollution (*dBA*)	4.68	5

Table 3. Final weighting of the indicators for the economy and social dimension

Indicator	Priority (%)	Rank
Access to public transport service (%)	21.24	1
Public transport (PT) affordability (%)	17.92	2
Variation non-motorized in modal split (%)	10.98	3
Variation PT in the modal split (%)	10.48	4
Transport security (*users/1,000 passengers*)	9.29	5
Indirect trip cost for user (*minutes*)	8.36	6
Share PT which are wheelchair accessible (%)	8.13	7
Population density (*inhabitants/km²*)	7.81	8
Variation of the female users in the PT (%)	5.80	9

Table 4. Final weighting of the indicators for the mobility system effectiveness and land use dimension

Indicator	Priority (%)	Rank
Pathways for pedestrians (*m²/inhabitant*)	21.66	1
Satisfaction with mobility services (%)	20.20	2
Cycle path network density (*bikes/inhabitant*)	17.79	3
Proportion of land with mix use (%)	9.55	4
PT fleet size (*buses/100,000 inhabitants*)	8.30	5
Average occupancy rate of passenger vehicles (*passengers/vehicle*)	6.57	6
Traffic congestion delay (*minutes*)	5.50	7
Land consumption by transport facilities (%)	4.91	8
Motorization rate (*vehicles/1,000 inhabitants*)	2.97	9
Motorcycle rate (*motorcycles/1,000 inhabitants*)	2.55	10

more detailed and comprehensive assessment of the entire system and its components, structuring actions and strategies to improve sustainability become easier to design and implement.

In short, the proposed framework serves as a guidance of how sustainability can be measured in transportation systems, being easily adaptable to different city contexts.

Table 5. Final weighting of the indicators for the operational dimension

Indicator	Priority (%)	Rank
Efficiency public transportation (*megajoule/passager.km*)	16.68	1
Multimodality integration (*scale 1 to 5*)	16.68	2
Public transport frequency (*buses/day*)	15.88	3
Attractiveness of public transport (*cost public transport/cost private car*)	13.03	4
Proportion of clean energy in PT fleet (%)	11.61	5
Bike sharing performance (*bicycles/1,000 inhabitants*)	10.24	6
Average age of PT fleet (*age*)	7.25	7
Road network density (*km/km*^{2})	4.03	8
Parking cost (%)	2.48	9
Parking capacity (*number of parking spaces/inhabitant*)	2.14	10

Table 6. Final weighting of the indicators for the fiscal and governance dimension

Indicator	Priority (%)	Rank
Public expenditures and investment in transport system (%)	21.89	1
Master plan (*scale 1 to 5*)	20.86	2
Operational cost PT system (%)	17.77	3
Expertise technicians and managers (*scale 1 to 5*)	13.48	4
Financial autonomy (*scale 1 to 5*)	9.84	5
Stakeholders engagement (*scale 1 to 5*)	7.34	6
Participation of the multilateral banks (%)	5.33	7
Variation of the informal transport modal split (%)	3.50	8

The frame work can be useful for transportation managers in policy making processes, contributing to the generation of bottom-up (rather than top-down) policies that come from lower-level government organizations. Therefore, in cities with lower institutional capacity, this prioritization can be used to determine which interventions would have the greatest impact. And this is valid even for quite different geo-spatial contexts, as it is the case of Europe, where top-down policies can be designed taking into account the specific features of each transportation system.

4 Findings and Conclusions

Issues related to sustainability and sustainable transport are of critical importance for mobility planners and policy makers. Consequently, these actors need to have tools for measuring sustainability by evaluating the different components of the urban mobility systems that can affect their performance. Based on the extensive literature on the subject, this paper recommends a hierarchical framework, refined with expert-based validation using MCDA techniques.

The framework presented in this paper will simplify the task of evaluating sustainable transportation and has the potential to be adapted to more specific contexts of different emerging cities, presenting a wide list of variables, validated by experts and analyzed with sound approach.

Beyond this sustainability measurement structure for urban mobility systems of emerging cities, the development of a clear methodology to evaluate the relationships between the variables that compose the framework, and their response to implemented actions by the decision makers, stands out as an interesting opportunity for future research.

Acknowledgements. The first author was funded by project NORTE-08-5369-FSE-000038, supported by the Northern Portugal Regional Operational Program (NORTE 2020), under the PORTUGAL 2020 Partnership Agreement, through the European Social Fund (ESF).

References

1. Ibadi, M.K.: Developing a standardized sustainable transportation indicators set. J. Univ. Babylon **25**(5), 1779–1786 (2017)
2. Litman, T.: Well measured: developing indicators for comprehensive and sustainable transport planning (2005)
3. Zheng, J., Garrick, N.W., Atkinson-Palombo, C., McCahill, C., Marshall, W.: Guidelines on developing performance metrics for evaluating transportation sustainability. Res. Transp. Bus. Manag. **7**, 4–13 (2013). https://doi.org/10.1016/j.rtbm.2013.02.001
4. Sdoukopoulos, A., Pitsiava-Latinopoulou, M., Basbas, S., Papaioannou, P.: Measuring progress towards transport sustainability through indicators: analysis and metrics of the main indicator initiatives. Transp. Res. Part D Transp. Environ. **67**, 316–333 (2019). https://doi.org/10.1016/j.trd.2018.11.020
5. Kunz, M., Kujawa, H., Neckel, A., Fuga, T.: Urban mobility: method of analysis of the adequacy of public policies in cities based on the principles proposed by the UN. J. Civ. Eng. Archit. **12** (2018). https://doi.org/10.17265/1934-7359/2018.05.002.
6. Burinskienė, M., Gaučė, K., Damidavičius, J.: Successful Sustainable Mobility Measures Selection (2017). https://doi.org/10.3846/enviro.2017.102
7. Terraza, H., Rubio Blanco, D., Vera, F.: De ciudades emergentes a ciudades sostenibles (2016)
8. Lima, J.P., da Silva Lima, R., da Silva, A.N.R.: Evaluation and selection of alternatives for the promotion of sustainable urban mobility. Procedia Soc. Behav. Sci. **162**, 408–418 (2014). https://doi.org/10.1016/j.sbspro.2014.12.222.
9. Joumard, R., Gudmundsson, H.: Indicators of environmental sustainability in transport: an interdisciplinary approach to methods. European Commission (2010)
10. Litman, T.: Sustainable transportation indicators: a recommended program to define a standard set of indicators for sustainable transportation planning (2008)
11. Tam, E.K.: Challenges in using environmental indicators for measuring sustainability practices. J. Environ. Eng. Sci. **1**(6), 417–425 (2002)
12. Nichols, J., Garrick, N., Atkinson-Palombo, C.: Framework for Developing Indicators of Sustainability for Transportation Planning No. 09-2829 (2009)

13. Kan, H., Chen, R., Tong, S.: Ambient air pollution, climate change, and population health in China. Environ. Int. **42**, 10–19 (2012)
14. Banco Interamericano de Desarrollo: Guía Metodológica Programa de Ciudades Emergentes y Sostenibles: Tercera edición (2016)
15. Ghorbanzadeh, O., Moslem, S., Blaschke, T., Duleba, S.: Sustainable urban transport planning considering different stakeholder groups by an interval-AHP decision support model. Sustainability **11**(1), 9 (2019)
16. Saaty, R.W.: Decision Making in Complex Environment: The Analytic Hierarchy Process (AHP) for Decision Making and the Analytic Network Process (ANP) for Decision Making with Dependence and Feedback. Super Decisions, Pittsburgh (2003)
17. Kahraman, C.: Fuzzy Multi-criteria Decision Making: Theory and Applications with Recent Developments, vol. 16. Springer, New York (2008)
18. Aikhuele, D.O., Souleman, F.S.: Application of fuzzy AHP for ranking critical success factors for the successful implementation of lean production technique. Aust. J. Basic Appl. Sci. **8**, 399–407 (2014)
19. Moslem, S., Ghorbanzadeh, O., Blaschke, T., Duleba, S.: Analysing stakeholder consensus for a sustainable transport development decision by the fuzzy AHP and interval AHP. Sustainability **11**, 3271 (2019). https://doi.org/10.3390/su11123271

Mobility as a Service: Implications for Spatial and Social Cohesion

Monika Themou[1(✉)], Foteini Mikiki[2], and Maria Markou[3]

[1] National Technical University of Athens,
Ministry of Energy and Environment, Athens, Greece
monika.themou@gmail.com
[2] Hellenic International University, Municipality of Serres, Serres, Greece
[3] National Technical University of Athens, Athens, Greece

Abstract. The article discusses spatial and social cohesion in relation to what MaaS promises. It starts from general principles based on Aristotle in which citizens and the polis aim at the common good -*ευδαιμονία*- which is attainable along with the highest virtue, justice, guiding both individual and collective actions. Starting from spatial elements of collective urban life as the road and the square it is argued that for social cohesion common aims are necessary. Regarding digital information, which is crucial today and concerns MaaS directly, the important issue is whether the dissemination of information is guided or not by market criteria. It is concluded that MaaS can function properly only if it supplements mass transportation media and only thus can contribute to social and spatial cohesion.

Keywords: MaaS · Spatial and social cohesion · Spatial democracy · Justice · Common good

1 Mobility and Communality

It is certain that all cities today suffocate from high volume of vehicles, but it is also certain that many factors which have been important in determining urban mobility have significantly changed, while new factors are expected to play an important role in the future. Discourses on urban mobility are proliferating, Maas, customer centered planning, data platforms, investment projects in urban mobility, door-to-door journey planning, travel experience offers, walk and bike, reduction of CO_2 emissions, etc. In the present paper we will attempt to examine to what extend MaaS, which promises also social and spatial cohesion, can indeed contribute in this direction [6, 3, 18, 15].

To have enduring social bonds the existence of common aims is necessary. This invariably entails a political framework allowing participation in decision making related to public goods. Mobility can be considered a public good because it corresponds to the ability of the citizen to use the city. S/he becomes in this way an active member in the multiple networks of the community using transport networks.

To attain a common aim a series of actions is required. Are, however, such procedures available to the dweller of contemporary European cities? While there is lip service to participation in the commons in fact there is a long road to really attain

E. G. Nathanail et al. (Eds.): CSUM2020, AISC 1278, pp. 626–632, 2021.
https://doi.org/10.1007/978-3-030-61075-3_61

participatory planning capable of expressing the common good. The evaluation of all aims and actions in terms of quantitative economic variables flattens social values. Politics is powerless and chaotic precisely because it does not incorporate future aims corresponding to accepted values. The lack of such aims makes politics susceptible to transient local, national and/or global interests aiming at profits. We would need a human-centered politics and policies which would refer to humanity as a whole.

2 Justice and the City

For Aristotle man is a 'political animal' and the aim of an organized community is the security of life and the attainment of self-sufficiency and 'happiness' (*ευδαιμονία*). Once life has been secured, happiness can only be attained within the community. For Aristotle justice is the highest virtue and encompasses all other virtues. It forms social morals and when justice is secured, *ευδαιμονία* can be also secured. *Paideia*, education, aims at knowledge but also the attainment of the virtue of justice.

Justice is realized also through the active participation of the citizen. The participation of the citizen in all activities of the city -*πόλις*- is the defining element of the citizen as such. Participation, therefore, is a fundamental element of the political community -of the polis. Participation is also the main element of democracy as a polity. All these values -justice, participation, the pursuit of the common good- are interconnected and if one is not fulfilled they all are disturbed [1]. *Ευδαιμονία* depends on this interdependence, with the role of justice predominant. For justice to be attained, aims aspiring to a 'common good' are necessary. Aristotle adds that laws are necessary to secure such aims - aspiring to the common good. Thus the question emerges to what extend laws today really aim at the common good.

Today, the notion of the 'common good' is often found in policies only nominally, in full contrast with the aims they actually promote. Particular interests are often the real motive, under the supposed pursuit of the common good. Evaluating everything through economic variables distanced thought from what is important and of quality leads to global destruction. Poverty and marginalization as also the mindless depletion of natural resources in the name of economic dominance of the few, represent the threat to humanity. Thus emerges as an unavoidable need a transformation of thought. In this transformation the important is the way knowledge is organized because it provides the rules of the constitution of society. We need to recreate citizenship at the local scale and to create citizenship at the global scale, after we connect the different homelands with Earth as the overall homeland and after we understand the vital relation of humanity to nature [10].

3 Streets and Squares

Returning to the city, we can argue that the connection between spatial cohesion and citizen's participation and is necessary and is evident in the history of the cities. Urban space expresses, through its form, existing socioeconomic and cultural conditions. The production of space, even if it is made through urban planners' models, when it actually materializes is transformed and tends to represent social reality. In particular, public spaces within the city are pivotal in sustaining social and spatial cohesion, especially public spaces as the street and the widening of the street which created the square. The square as a widening of the street *-plateia/agora-* emerges when democracy emerges in ancient Athens. Both the square and the street –odos-, as spaces of meeting, interaction, interchange, play an important role as elements of spatial cohesion. Participation, a defining element of democracy, is the outcome of communality; the street and the wide street/square are the principal expressions of this communality [16].

The streets and the squares acquired their full significance in the six thousand years from the settlement of Catalhoyuk to classical Athens. Catalhoyuk, which dates to 6500–6000 years BC, has the amazing characteristic of buildings fully in contact one to another, no roads and entrance to the houses through the rooftops with ladders. The circulation is through the roofs of the houses [4]. Probably reasons of fortification imposed this form, since in its lower level there are no openings, just walls. From that settlement to the city of Athens at c. 450 BC in which the widening of the road of Panathinaia forms the agora, where commerce is intermingled with religious rites and functions of the polity –democracy, many other cities exhibit public spaces where collective life is manifested [5].

It can be argued that streets and squares constitute an urban factor and are deployed in three-dimensional space in a dialectic relationship to buildings. Streets, widenings of streets, small squares and squares create between the buildings a network of continuous movement and social interconnection. The point to be stressed is that when open space is commercialized, public spaces lose their importance, lose their social function for a city with citizens. This occurs because the activities which take place have the inherent dynamic to modify through material and often non-material relations spatial cohesion.

Planned space and its form express all sociopolitical and cultural elements and makes evident which elements need to be reexamined to attain spatial cohesion. It is important to mention that the domain ruled by Ancient Sparta had a network of roads possibly better than that of Ancient Athens in terms of physical characteristics of the roads. Roads with sufficient width are necessary primarily for the movement of the military and for the movement of carriages, and such are the roads in the domain of Sparta [11]. However, the combination of trajectories and buildings in Athens is unique and reflects its polity –democracy.

Streets in classical Athens acquire width according to their importance in the 'polis' and have a specific dialectic to the built space which reflects the social and cultural integration of movement within the city. The best example is the *peripatos* beneath the Acropolis, which is integrated within the Dionysian theater in a dialectical way expressing its historical evolution. The *peripatos* is integrated within the theatre stressing thus the specific importance of the street corresponding to the movement of citizens through time. In the organization of space in the southern slope of the Acropolis one can clearly distinguish the right of the Athenian citizen to participate in all aspects of urban/public life which he himself was co-formulating [5].

Which is, however, the principal element defining the dynamic of the street and the square? For the street is movement, the mapping of the city and the reading through action of its everyday life and all that it entails. For the square is remaining on, action and the sense of belonging to the community.

4 Urban Mobility and MaaS

People travel from antiquity to our days for the same reasons. Commerce, war, worship, education, cooperation, tourism. The distance to be travelled determines whether overnighting outside the place of permanent residence will be required or not. Using this criterion we can define the difference between travel and journey. Journeys are usually for everyday activities and for relatively small distances while travel requires spending nights outside home and requires more time.

But how is movement transfigured today? A significant difference in todays' societies is that connecting with others does not necessarily requires physical movement. Technology has enabled the creation of networks which transfer information, text and images, and allow interaction in real time without physical presence [11].

Thus, there are three factors determining the form of public spaces today and especially roads. Their importance for collective activities, their role as transportation systems and the way they are/can be transformed to information.

The coexistence of many persons in the same street does not necessarily imply that communal activities and communal bonds are present. Crowed commercial streets or streets where many people are simply co-present do not constitute a place for democracy, do not obliterate poverty and do not rectify inequalities and social divisions. For any project that has to do with urban mobility if we want it to contribute to democracy, common aims are necessary, aims converging on the common good for the polis and the citizens.

To reclaim streets from the domination of the private car and to open them up to pedestrians and bikes, to reconstruct them with bioclimatic materials are definitely steps for a 'better' image of the city and small steps towards a better quality of life but not contribute to social and spatial cohesion [14]. MaaS as part of reorganizing transportation with criteria the particularity of individual journeys transposes the same logic as that of using private cars to other type of vehicles [9]; the variety of urban life that is present in a journey, disappears. To attain social integration policies/actions should be geared to collective rather than individual aims. The complexity and the importance of life in motion should not be lost in the search of functioning tools. Lack of resources and outright poverty will not allow many sectors of the population to use MaaS services and thus inequality and marginalization do remain. There is also a possibility that MaaS, oriented towards profit, will undermine the amelioration of public transport.

MaaS promises every journey to be 'an experience', for whom however? For the necessary urban journeys rarely someone drives a car for the pleasure the experience of driving allows. A private car is used when mass transport means do not exist, are not sufficient/adequate or cannot provide a comfortable journey. Those who drive for the pleasure of the experience of driving will choose another type of vehicle. Beyond limiting the use of private cars and redirecting mobility to walking and the use of the bike, to indeed limit pollutant emissions what is required is to limit the use of vehicles and to strengthen the use of public transport. It is not enough to use 'clean energy'. It is also important to limit the overall energy required.

5 MaaS and Information Networks

We could focus now on how the road is presented trough informational networks. The overflowing of information is the characteristic element of the beginning of the new millennium. Technology, especially in transportation, is promising processes that will make travelling easier, multimodal and nonpolluting. In this respect MaaS indeed is promising to provide a multitude of ways for going from point A to point B. MOD (Mobility on Demand) on the other hand provides services for the duration the user requires travel services. Flexibility is the main advantage in these novel approaches.

The provision of information through a digital platform is a significant new development but the crucial question is who and following which criteria manages the platform. If information and the way it is connected to other spatial activities within the city is managed through market criteria the existence of the platform itself does not contribute to some form of overcoming economic class divisions [17]. Streets and journeys proposed may be determined by market decisions creating exclusions for journeys, streets and even whole areas which for some reason will not be proposed. For example, interstates within cities, as in the case of the Los Angeles metropolitan area, are preferred not only because they usually allow for a faster journey but also because poorer areas are avoided.

Information management remains a crucial problem. The UITP (International Association of Public Transport) considers that information from the network of mass transport media should not be made available to private companies [2]. It would be better MaaS to function as a service connected to public transport because if not, it would take the form of a product available through the market with all the negative consequences it would entail. Thus, MaaS would function in a way complementing public transport and not as antagonistic to it. Of course, having access to the use of a vehicle from the part of individuals who cannot own private cars because of the financial burden it imposes, addresses social inequality and contributes to social cohesion to the extent that ownership or not of a private car can be considered inequality [19].

In the case of large, usually global, information providers such as Google it can be stated that 'whilst an increasing proportion of the urban subaltern may have access to digital communications devices, their ability to challenge or disagree with this process (producing space through digital information) is increasingly limited [7]. Therefore, the power of Google's algorithms and databases -like many others- is leaving its own particular mark of spatial-informational injustice on history' [13].

The choice and the content of the routes proposed by digital platforms cannot be left to market criteria [10]. Every street is a spatial factor which includes a variety of information corresponding to the public culture. The actual network of roads cannot be replaced by a network of informational representation of urban space according to unknown criteria and aims. Every street is a spatial factor of urban life incorporating a multitude of meanings corresponding to collective life. Digital platforms appear as being constructed allowing equal access for all to the multitude of urban information, in fact however, they aim at the maximization of users and profits and create gaps regarding the information provided [13]. The 'right to the city' Lefebvre has proclaimed [8] has also to be present in the informational representation of the city. A right for a just informational representation has, in our days, to be a central element of the overall right to the city and is necessary to combat the commercialization present the dominant platforms. Justice is the highest virtue for the polis in Aristotle. Justice in information access and in controlling the sources of information is paramount.

6 Mobility and Social/Spatial Cohesion

The principles of designing transportation systems are successful when they recognize the necessary elements and correlate them to the aim of achieving mobility without exclusions. More generally, a just socio-spatial distribution of mobility opportunities can be attained only through continually expanding, and strengthened networks of public transport with accessible cost. The right to mobility should not be limited by the medium of transport, geographic/special characteristics, economic inequalities and personal particularities. To that aim is necessary to have a strong, extensive and multimodal framework of public transport. The private car industry, wanting to claim a significant part of urban mobility, proposes specially equipped vehicles incorporating advanced information systems ready for MaaS. However, if we simply replace simple private cars with these new ones while the overall journeys with cars remain the same or are increased then the cities will not reduce the number of circulating cars to provide much needed social space. In addition, the commercialization of urban mobility instead of ensuring access to all, will increase social divisions and inequalities.

Current developments characterized by generalized commodification of urban activities destroyed the existing solidarity and help to the others. The market promotes only maximization of material interests and obliterates the -free- provision of friendship and solidarity [10]. Social cohesion is based on five variables, spatial, economic, environmental, cultural and especially political [12]. The present-day economic crisis and the crisis of the coronavirus pandemic shows how only a satisfactory public transport system can serve society in a situation of crisis [2]. Only a network of public transport can and should have as a guiding principle spatial democracy and service to all. As with every aspect of community life in times of crisis, in small towns, cities, metropoles, individuals are affected to an extensive degree when there are no services that support the inhabitants as a whole.

Maybe, therefore, the necessary transformation of thought to which we referred in the beginning of the article - to recreate citizenship at the local scale and to create citizenship at the global scale, after we connect the different homelands with Earth as the overall homeland and after we understand the vital relation of humanity to nature- can come about as a result of unpredictable conditions such as those facing all today with the corona virus pandemic.

Transport networks are a necessary condition for spatial and social cohesion but not a sufficient one. If spatial and social cohesion-the two are in fact indissociably linked- is the goal, it should be posed as such across all policy levels, from the particular to the strategic, from the local to the regional. Moreover, spatial and social cohesion cannot be limited to a neighborhood: it can only refer to the city as a whole, the suburbs, the region, indeed the national territory.

The planning for the ideal urban mobility is part of aiming for a better society. What is required is to recognize existing problems and potentialities, as well as activating elements that would advance happiness (*ευδαιμονία*) and justice.

References

1. Aristotle: Nicomachean Ethics. Zitros, Athens (2006). (in Greek)
2. Arthur D Little future lab/UITP: The Future of Mobility 3.0: Reinventing mobility in the era of disruption and creativity (2018)
3. Becker, H., et al.: Assessing the welfare impacts of shared mobility and mobility as a service (MaaS). Transp. Res. Part A **131**, 228–243 (2020)
4. Efthumiou, M.: World History: Man Against Nature, Athens (2003)
5. Ficuciello, L.: La Strada di Athene, Scuola Archeologica Italiana di Atene, Pandemos, Athens (2008). (in Italian)
6. Jittrapirom, P., et al.: Mobility as a Service: a critical review of definitions, assesments of schemes and key challenges. Urban Plan. **2**(2), 13–25 (2017)
7. Keivani, R.: A review of the main challenges to urban sustainability. Int. J. Urban Sustain. Dev. **1**(1–2), 5–16 (2010)
8. Lefebvre, H.: The Right to the City, Koukkida, Athens (2006 [1968]). (in Greek)
9. Litman, T.: Community Cohesion as a Transport Planning Objective. Victoria Transport Policy Institute, Victoria (2018)
10. Morin, E., Nair, S.: A Politics of Civilization. Livanis, Athens (1998). (in Greek)
11. Pikoulas, G.: Travelling in ancient Greece. In: Vigopoulou, I. (ed.) Travel: From the Ancients to the Moderns, EIE, Athens (2003). (in Greek)
12. Sassen, S.: The city: its return as a lens for social theory. City Cult. Soc. **1**, 3–11 (2010)
13. Shaw, J., Graham, M.: An informational right to the city? Code, content, control and the urbanization of information. Antipode **49**(4), 907–927 (2017)
14. Stigendal, M.: Cities and Social Cohesion. Malmo University Publications, Malmo (2010)
15. The Institution of Engineering and Technology: Could mobility as a service solve our transport problems?, London (no date)
16. Themou, M.: The importance of the typology of the square. M.A thesis, South Bank University, London (1993)
17. UCL Energy Institute/Department of Transport: Feasibility Study for "Mobility as a Service" concept in London, London (2015)
18. Wong, Y.Z., et al.: Emerging transport technologies and the modal efficiency network: a case for mobility as a service (MaaS). Institute of Transport and Logistics Studies, The University of Sydney, Sydney (2018)
19. Ziccardi, A.: Poverty and urban inequality: the case of Mexico City metropolitan region. ISSJ, UNESCO, Oxford (2016)

Social Networks and Traveller Behavior

Toward Active Transport as a Utilitarian and Recreational Form of Sustainable Urban Mobility

Parsa Arbab[1(✉)], Javier Martinez[2], Sherif Amer[2], and Karin Pfeffer[2]

[1] School of Urban Planning, College of Fine Arts, University of Tehran, Tehran, Iran
parsaarbab@ut.ac.ir
[2] PGM Department, ITC Faculty, University of Twente, Enschede, The Netherlands
{j.a.martinez,s.amer,k.pfeffer}@utwente.nl

Abstract. The ongoing growth of motorized transport modes poses serious challenges to urban environments such as air and noise pollution, adverse health impacts, and improper allocation of space in the city. Over time, the paradigms of transport planning have shifted gradually toward active transport as a utilitarian and recreational form of sustainable urban mobility with desirable ecological, economical, and social properties. Active travel modes, concerning the physical activity of locomotion, generally include walking and bicycling, which have environmental benefits such as a decrease of carbon and reduction of traffic congestions. Besides, active transport, as a form of an active lifestyle, is one of the most usual physical activities which have numerous benefits regarding health and wellbeing, including reducing risk factors of chronic diseases, such as overweight and obesity, and increasing happiness, contentment, and engagement. Therefore, active transport has been increasingly considered in transportation and urban planning studies as an alternative for motorized transport and a sedentary lifestyle. Studies have shown that the built environment can facilitate active transport by walking or cycling to destinations and accordingly contributes to residents' physical activity. Hence, this study aims to explore and explain the association between the built environment and active transport by urban residents through a literature review. Insights such as there are important natural/physical and macro/micro environmental characteristics that encourage the tendency for active transport as well as socioeconomic and sociodemographic attributes differently affect this indicative relationship, are subsumed in a conceptual framework that can guide future empirical studies.

Keywords: Active transport · Physical activity · Built environment · Sustainable urban mobility · City

E. G. Nathanail et al. (Eds.): CSUM2020, AISC 1278, pp. 635–644, 2021.
https://doi.org/10.1007/978-3-030-61075-3_62

1 Introduction: Active Transport as an Active Lifestyle

Sustainable urban mobility as a critical issue faced by cities globally, needs restructuring toward non-motorized modes and physical activity with multiple beneficial and positive effects on the function of the city and human health (Skayannis et al. 2017, 2018). Active transport modes such as walking or cycling pose one of the most promising solutions to cope with increasing challenges arising from motorized transportation (Hackl et al. 2019). Therefore, cities aim at shifting transportation solutions toward more environment- and people-focused means of mobility. This transition - reducing motorized transport and promoting active transport - also has public health benefits in the short and long term (Nieuwenhuijsen and Khreis 2016).

Active transport, active mobility, active commuting, or active travel as a form of an active lifestyle is one of the most usual physical activities concerning their associated increases in human body energy expenditure. As shown by recent findings of the World Health Organization (WHO), physical inactivity is a significant risk factor affecting global mortality, and hence an active lifestyle is essential (An et al. 2019; Vich et al. 2019). Therefore, the Global Action Plan on Physical Activity has been adopted by WHO in 2018 to ensure that all people have access to safe and enabling environments and to diverse opportunities to be physically active in their daily lives, as a means of improving individual and community health and contributing to the social, cultural and economic development of all nations (WHO 2018).

Accordingly, "World Health Organization advocates strategies that target entire populations, including the design of environments to promote physical activity for transportation and recreation as part of everyday life, or active living" (Day 2018, p. 303). These strategies have identified active transport as an alternative for minimizing sedentary lifestyles, reducing risks of consequent chronic diseases such as obesity and diabetes, and improving health standards by increasing physical activity levels (Gao et al. 2018; Vich et al. 2019).

Meanwhile, improvements in the built environment characteristics are also essential for health-oriented urban planning and management, as built environments have long-lasting, positive or negative effects on the health outcomes and effects for entire populations (Day 2018; Sallis et al. 2018). Poor quality of sidewalks, limited access to recreational facilities such as parks, and lack of available nearby destinations are possible causes of inactivity and decreasing physical activity/active transport. Next, the built environment characteristics vary in supporting physical activities, including active transport. Bornioli et al. (2019) state, for instance, that "a strategy to promote active mobility in the built environment can be constructed around safety, comfort, and moderate sensory stimulation by targeting the micro elements that prevent them" (p. 200). Therefore, active transport, which relates to physical activity, health, and chronic disease prevention, has been considered increasingly important in transportation and urban planning studies.

Many studies show evidence of the significant relationship between the neighborhood-scale built environmental characteristics, such as density, land-use, and connectivity, and travel behavior centered on active transportation modes. The pervasive effect of density, land-use, type of urban fabric, and slope of terrain on urban mobility, especially concerning automobile travel reduction and non-auto trip encouraging, are some of the findings that are being assessed by these studies. Safe walking and cycling accessibility concerning well-connected streets, walking paths, sidewalks, bike infrastructures and bike lanes, traffic safety, and less motorized traffic are other neighborhood or street features that have been considered (An et al. 2019; Ferrer and Ruiz 2018; Gao et al. 2018; Helbich 2017; Lindelow et al. 2017).

In short, there is a crucial opportunity for urban planners to plan and design cities and neighborhoods which are conducive to physical activity, especially by active transport modes. Nevertheless, this goal will not be realized without a proper understanding of the association between the built environment characteristics and physical activity patterns, especially active transport. The quality and nature of diverse types of neighborhood and also street scale-built environment may diversely affect active lifestyle among residents (An et al. 2019). Thus, findings should be analyzed in more depth, both theoretically and practically, to gain deeper insight into the relationship between the built environment characteristics and the active transport by residents in urban neighborhoods (Gao et al. 2018). Supplementing research with neighborhood and street scale-built environment measures could acquire more reliable and valid understanding and filling the knowledge gap concerning the significant relationship between environmental characteristics and active transport. Therefore, facing the challenge of increasing physical inactivity and subsequently overweight and obesity, yet it is necessary to explain the role and impact of the built environment, aimed at increasing physical activity and active transport and reversing that growing threatening tendency in terms of motorized transport and passive lifestyle (Lin 2018).

So, while there are already multiple studies on the relationship between the built environment characteristics and active transport, we lack a comprehensive overview of the findings. To address this gap, this paper reviews the recent literature on how the built environment characteristics affect active transport (walking, cycling) by urban residents. Based on the review, it develops a conceptual framework that can guide future empirical studies on this significant association.

2 Approach and Methodology

Since we have been interested in recent scientific articles, we limited our search to sources published in 2015 or later. They were identified through six databases, including PubMed, Sage, ScienceDirect, Springer, Taylor & Francis, and Wiley. Search keywords for articles' titles and also abstracts were "active commuting," "active mobility," or "active transport." Moreover, we used simultaneous combinations of "built environment," "neighborhood environment," "neighborhood qualities," or "physical environment" and "cycling," "physical activity," or "walking." After careful reading of title, abstract, and keywords, we considered 19 articles, listed in the reference section, relevant for analyzing the relationship between the built environment and active transport as a type of physical activity. The articles were analyzed by exploring the main factors and characteristics. Insights were subsumed in a conceptual framework to explain the significant association between the built environment and active transport.

The literature review serves to develop a conceptual framework. It provides a starting point for all who are interested in a particular field (Pare and Kitsiou 2016), the discussed relationship between the built environment and active transport. More specifically, it prepares a well-structured overview in a specific area, active transport as a utilitarian and recreational form of sustainable urban mobility concerning the built environment characteristics. On this base, it also outlines the implications and interpretations of the findings to present the evidence on a meta-level and discover areas in which more research is required, essential for developing theoretical and conceptual frameworks (Snyder 2019; Van Wee and Banister 2016).

3 Active Transport and the Built Environment

Recent international policy frameworks on environment and health attach more importance to promoting walking and cycling as usual forms of physical activity (Kahlmeier et al. 2020). Advancing active transport is the main aim of many European cities regarding improving environmental qualities at the local level and subsequently increasing the health and wellbeing of their residents (Ferreira et al. 2016). In the duality of active and passive modes, walking and cycling have been combined into a category of active modes that refer to physical behaviors that are affected differently by different correlates of urban form. While some Anglophone research has been published, European studies still are mainly underrepresented concerning urban form correlates and urban design variables (Helbich 2017). So, as mentioned earlier, defining research on contextualized ways in which specific features of the built environment affect the physical activity/active transport at higher levels, is also considerable. They can identify what we learn, uncover gaps in knowledge, and direct future research (Day 2018).

Studies have shown that walkable and cyclable neighborhoods encourage active transport (walking or cycling) to destinations and accordingly can contribute to residents' physical activity (Sallis et al. 2018). Lindelow et al. (2017), for instance, state that "the travel behavior of residents in a neighborhood can partly be explained by the fact that residents have selected to live in a neighborhood that they perceive lives up to their preferences of, for instance, walkability. Consequently, neighborhoods with a large share of walking could be understood as consisting of residents that have chosen to live where they perceive walking to be feasible, pleasant, etc., in addition to the built environment itself encouraging walking" (pp. 520–521). So, on the one hand, people are looking for a place to encourage their physical activity/active transport. On the other, location also can motivate people to do physical activity/active transport.

Active transport counters the emerging sedentary lifestyle issue. Some practical findings have shown that people who do active transport in forms of walking or cycling are likely to be achieving physical activity levels significantly more than the minimum recommended levels. Meanwhile, active transport is positively associated with some socioeconomic variables such as age, income, and bicycle ownership, and some built environment characteristics such as land-use and urban density. So, the various effects of individual factors and neighborhoods' spatial physical, social and environmental qualities should be taken into consideration (Ferreira et al. 2016; Fishman et al. 2015).

On this base, the focus on active transport as a type of physical activity for all age groups, which allows them to change the inactive and sedentary lifestyles through their favorite intensity, requires significant classifications concerning purposes, mechanisms of the impact, and finally factors and characteristics. Regarding the purposes, two modes of active transport, including walking and cycling, can be further divided into two categories including transport or travel-related, aims at reaching a destination, and leisure or recreational-related, addresses achieving entertainment (Gao et al. 2018; Wang et al. 2016).

Regarding the mechanisms of the impact, some physical built environment characteristics act as motivators or incentives, and some act as barriers or obstacles to physical activity/active transport. The motivators and barriers that affect each other are as following (Wang et al. 2016):

1. Motivators or incentives: opportunities including availability and suitability of facilities and shortening the distance, safe accessibility such as improving personal security, and improving transport safety, and physical settings in terms of increasing comfort level and provision of supporting facilities;
2. Barriers or obstacles: opportunities barriers such as limited foot and cycling paths and lack of land for recreation, accessibility barriers including longer travel distance, poor access to the facilities, and no interesting destinations, safety barriers in terms of unsafe foot or cycling paths, traffic safety, and security of exercise place, and physical setting barriers such as lack of pleasant routes, discomfort, and no supporting facilities.

They could be classified as facilitators or motivators and barriers or deterrents to the active transport. For example, lack of car parking space at the destination, pleasant walking routes, short distances to destinations, pedestrian streets, and driving restriction zones or stress due to traffic congestion are the primary facilitators or motivators to the decision to walking as the primary form of active transport. On the other hand, insecurity due to crime, absence of people, inadequate lighting at night, or walking along a conflictive area, presence of traffic lights, walking along large avenues, lack of sidewalks, and steep streets are the main barriers or deterrents (Ferrer and Ruiz 2018).

According to the influential factors and characteristics, first, it is necessary to identify and distinguish between personal and social factors as well as natural and built environmental characteristics that have a significant role regarding active transport (Wang et al. 2016). The proportion of green space such as parks and agricultural and natural areas, water spaces, daily maximum air temperature, daily precipitation sum, and daily average wind speed are the critical natural ones. Building density, land-use diversity, street density, number of bus stops, and distance to high-quality public transport nodes are important built environmental variables (Gao et al. 2018).

Also, socioeconomic status disparities are essential as effect modifiers of the relationship between the built environment characteristics and health-related outcomes (Lee et al. 2015; Sallis et al. 2018). The socioeconomic levels or individual factors around the city and place of residence are expected to be related to physical activity as well as active transport. They can be considered as a percentage of foreign residents, unemployed, part-time workers, university graduates, homes occupied by their owners, households with a parking space, and median income (Feuillet et al. 2015). Gender, age, household structure, education, ethnicity, and car ownership are other major individual variables concerning the significant relationship between the built environment and active transport (Gao et al. 2018).

Generally, the built environment factors which are associated with transportation-based physical activities are as follows, that their adequacy or appropriate condition can act as a stimulus and their lack or inappropriate status can act as an obstacle to active transport (Feuillet et al. 2015):

1. Land-use and facilities including the percentage of area covered by individual housing, collective housing, vegetation cover, as well as proximity facilities density;
2. Level of walkability and bikeability such as walk and bike paths conditions and bike-sharing facilities;
3. Public transport availability in terms of the distance to the nearest subway, bus, or train station from each home.

Many research studies provide evidence for the relationship between neighborhood design and active transport and indicate that the impacts happen at the neighborhood level (Ferrer and Ruiz 2018). Accordingly, Lee et al. (2015) state that "there is growing evidence that neighborhood environment, such as green space, parks, and pedestrian environment, is associated with physical activity and various health outcomes, especially obesity-related diseases. However, among the possible factors contributing to physical activity and obesity-related diseases, little is known about the urban neighborhood environment, such as slopes or street patterns, and trigger factors that encourage residents to walk" (p. 1205). So, all the factors should be evaluated in two mutual hypotheses regarding the mentioned barriers or deterrents and facilitators or motivators. The first hypothesis is that obesogenic environments[1] in terms of low-walkable and automobile-oriented neighborhoods with few facilities for physical activity/active transport may direct residents to be inactive based on spending more time in their cars or doing more sedentary recreation, including television viewing and computer gaming. Regarding increasing physical activity as a critical health-based strategy, the alternative hypothesis states that the physical activity/active transport as possible mechanism/mechanisms for achieving health outcomes is influenced by the neighborhood environment (Lee et al. 2015; Sallis et al. 2018).

On the other hand, Ferrer and Ruiz (2018, p. 111) argue that "in addition to meso-scale (or neighborhood scale) built environmental factors such as residential density, land-use mix or street connectivity, special attention should be given to micro-scale (or street level) built environment characteristics, such as the presence of trees, the width of the sidewalks, and the quality of the streets, as the roles of micro-scale elements are not well understood due to limited data availability." They have analyzed factors of the built environment affecting the decision to walking in the form of a short trip as less than 30–45 min walking distance. On this base, the main characteristics of the built

[1] Obesogenic environments describe specific aspects of living environments which facilitate overeating relative to need and partaking in sedentary activities. They are characterized as involving a great preponderance of motorized transport and sedentary occupations and encouraging the consumption of high-fat and energy-dense foods (Poortinga et al. 2011; Ulijaszek 2018).

environment influencing walking as active transport are safety from crime (street lighting, people's presence, cleanliness), traffic safety (traffic volume and speed and times of crossing waiting), walking facilities (sidewalk width, obstacles), aesthetics (green elements, buildings, noise), convenience, and other perceptions (car parking availability, hills, and pedestrian volume, open and wide spaces, and length) (Ferrer and Ruiz 2018).

In another study, Zandieh et al. (2016) indicated that outdoor walking level is the most common type of health-beneficial physical activity associated with the built environment in a residential neighborhood. However, most previous analyses have considered macro built environment characteristics as inclusive design and structure, including residential density, mixed land-use, and route connectivity. Accordingly, it is also necessary to focus on micro built environment characteristics, which can be modified easier than macro ones. They include safety (well lighting, people's presence, and crime rate), pedestrian infrastructure (traffic condition, sidewalk condition, and amenities) and aesthetics (trees, attractive sights, and buildings) in the neighborhood. Moreover, spatial inequalities in perceived built environment characteristics may affect disparities regarding neighborhood support for walking (Zandieh et al. 2016).

Furthermore, regarding the widely accepted influence of environmental factors on usual physical activities, including active transport modes, the conceptual framework should be contextualized to local conditions. Besides, local targeting of health-based policies might be more effective in promoting physical activity/active transport. It seems that the association of the individual or sociodemographic attributes, as well as the built and natural environmental characteristics, with active transport, as a type of physical activity, is context-specific. So, it is necessary to analyze the area-specific significant association between the built environmental and socioeconomic factors and active transport in forms of walking and cycling in different locations with diverse specifications (Feuillet et al. 2015; Gao et al. 2018).

4 Discussion: Toward a Conceptual Framework

The crucial role of the everyday living environment in personal and public health has become more and more important in recent decades due to the predominance of mechanized life and, accordingly, sedentary lifestyles. The built environment can motivate the physical activity of individuals as an inevitable necessity or demotivate them as a serious challenge. Active transport as a type of usual physical activity is one of the critical areas in this vision due to its multiple roles and functions in addition to promoting health through physical activity. Therefore, focusing on active transport is a vital opportunity for researchers in various fields, including urban planners and designers, who also need to analyze the mechanism of its effectiveness and improvement as a great response. Such studies will also provide a reasonable basis for the formulation and application of future efficient policies.

So, it is necessary to explore and explain the significant associations between the built environment characteristics and active transport. To this end, the first step in our research is to develop a conceptual framework for addressing and clarifying the various aspects of this relationship. Based on the literature review, we obtained the following insights. First, active transport as the most usual physical activity can be regarded as walking, cycling, and other forms based on a human-powered street vehicle. Second, active transport can have different purposes as utilitarian transport or travel for reaching a destination and attractive leisure or recreation for achieving entertainment. Meanwhile, if we focus on the association between active transport and substantial characteristics, categorizing factors into two categories of motivators, incentives, and facilitators, on the one hand, and barriers, obstacles, and deterrents, on the other hand, can well direct us to achieve the practical results.

Furthermore, as Fig. 1 shows, focusing on the factors and characteristics requires some critical points to consider. On the one hand, socioeconomic and sociodemographic attributes such as gender, age, household structure, education, income, and car ownership differently affect the relationship between the built environment and active transport. On the other hand, there are significant environmental characteristics that encourage or inhibit the tendency for active transport. These characteristics can be analyzed in two distinct ways, with the possibility of integration. First, we can scrutinize the physical environmental aspects such as residential density, route connectivity, safety, and sidewalk condition. Secondly, we can focus on the natural ones, including slope, air temperature, and presence of trees.

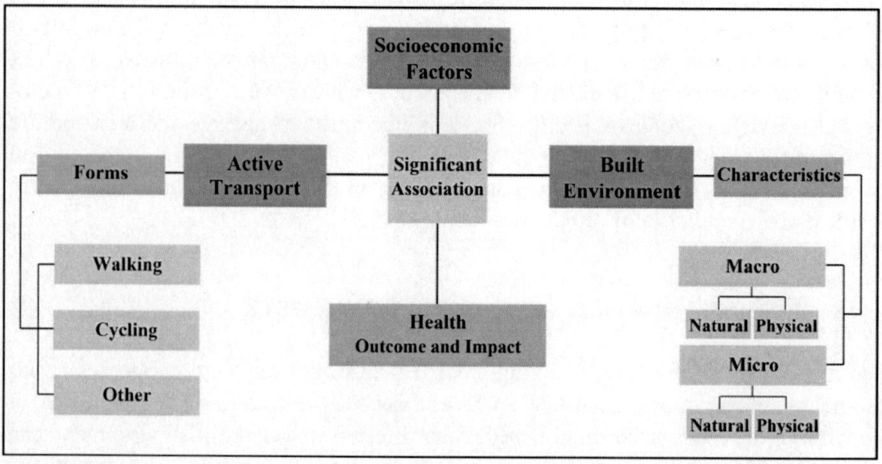

Fig. 1. Conceptual framework on the significant association between the built environment and active transport.

Meanwhile, we can peruse all of them at different scales; from the perspective of macro and meso characteristics which act at the city or neighborhood scale such as residential density, route connectivity, slope, and air temperature, while the micro characteristics have a vital role at the place or street level including safety, sidewalk condition, and presence of trees. The proposed conceptual framework aims to improve the understanding of the association between the built environment characteristics of active transport in a context-specific manner, which clarifies disparities in the evidence.

So, based on this conceptual framework, performing empirical studies in different and diverse contexts, concerning distinct urbanism and various lifestyle, completes this path. Efficient and innovative solutions, rather than general and perhaps ineffective, will be the consequence of this orientation.

References

An, R., Shen, J., Yang, Q., Yang, Y.: Impact of built environment on physical activity and obesity among children and adolescents in China: a narrative systematic review. J. Sport Health Sci. **8**(2), 153–169 (2019)

Bornioli, A., Parkhurst, G., Morgan, P.L.: Affective experiences of built environments and the promotion of urban walking. Transp. Res. Part A Policy Pract. **123**, 200–215 (2019)

Day, K.: Physical environment correlates of physical activity in developing countries: a review. J. Phys. Act. Health **15**(4), 303–314 (2018)

Ferreira, I.A., Johansson, M., Sternudd, C., Fornara, F.: Transport walking in urban neighbourhoods—impact of perceived neighbourhood qualities and emotional relationship. Landscape Urban Plan. **150**, 60–69 (2016)

Ferrer, S., Ruiz, T.: The impact of the built environment on the decision to walk for short trips: evidence from two Spanish cities. Transp. Policy **67**, 111–120 (2018)

Feuillet, T., Charreire, H., Menai, M., Salze, P., Simon, C., Dugas, J., Hercberg, S., et al.: Spatial heterogeneity of the relationships between environmental characteristics and active commuting: towards a locally varying social ecological model. Int. J. Health Geograph. **14**(1), 1–14 (2015)

Fishman, E., Bocker, L., Helbich, M.: Adult active transport in the Netherlands: an analysis of its contribution to physical activity requirements. PLoS ONE **10**(4), 1–14 (2015)

Gao, J., Kamphuis, C.B.M., Dijst, M.: Helbich, M: The role of the natural and built environment in cycling duration in the Netherlands. Int. J. Behav. Nutr. Phys. Act. **15**(1), 1–16 (2018)

Hackl, R., Raffler, C., Friesenecker, M., Kramar, H., Kalasek, R., Soteropoulos, A., Wolf-Eberl, S., Posch, P., Tomschy, R.: Promoting active mobility: evidence-based decision-making using statistical models. J. Transp. Geogr. **80**, 1–13 (2019)

Helbich, M.: Children's school commuting in the Netherlands: does it matter how urban form is incorporated in mode choice models? Int. J. Sustain. Transp. **11**(7), 507–517 (2017)

Kahlmeier, S., Racioppi, F., Götschi, T., Castro, A., Cavill, N.: Chapter Fifteen - The who health economic assessment tool for walking and cycling: how to quantify impacts of active mobility. In: Nieuwenhuijsen, M.J., Khreis, H. (eds.) Advances in Transportation and Health, pp. 329–342. Elsevier, Amsterdam (2020)

Lee, H., Kang, H.M., Ko, Y.J., Kim, H.S., Kim, Y.J., Bae, W.K., Park, S., Cho, B.: Influence of urban neighbourhood environment on physical activity and obesity-related diseases. Public Health **129**(9), 1204–1210 (2015)

Lin, L.: Leisure-time physical activity, objective urban neighborhood built environment, and overweight and obesity of Chinese school-age children. J. Transp. Health **10**, 322–333 (2018)

Lindelow, D., Svensson, A., Brundell-Freij, K., Hiselius, L.W.: Satisfaction or compensation? The interaction between walking preferences and neighbourhood design. Transp. Res. Part D Transp. Environ. **50**, 520–532 (2017)

Nieuwenhuijsen, M.J., Khreis, H.: Car free cities: pathway to healthy urban living. Environ. Int. **94**, 251–262 (2016)

Pare, G., Kitsiou, S.: Methods for literature reviews. In: Lau, F., Kuziemsky, C. (eds.) Handbook of eHealth Evaluation: An Evidence-Based Approach, pp. 157–179. University of Victoria, Victoria (2016)

Poortinga, W., Gebel, K., Bauman, A., Moudon, A.V.: Neighborhood environment, physical activity and obesity. In: Nriagu, J.O. (ed.) Encyclopedia of Environmental Health, pp. 44–53. Elsevier, Burlington (2011)

Sallis, J.F., Conway, T.L., Cain, K.L., Carlson, J.A., Frank, L.D., Kerr, J., Glanz, K., Chapman, J.E., Saelens, B.E.: Neighborhood built environment and socioeconomic status in relation to physical activity, sedentary behavior, and weight status of adolescents. Prev. Med. **110**, 47–54 (2018)

Skayannis, P., Goudas, M., Rodakinias, P.: Sustainable mobility and physical activity: a meaningful marriage. Transp. Res. Procedia **24**, 81–88 (2017)

Skayannis, P., Goudas, M., Crone, D., Cavill, N., Kahlmeier, S., Mitsiadi, V.: Health related benefits of non-motorised transport: an application of the health economic assessment tool of the world health organisation to the case of Trikala, Greece. In: Nathanail, E., Karakikes, I. (eds.) Data Analytics: Paving the Way to Sustainable Urban Mobility, CSUM 2018. Advances in Intelligent Systems and Computing, pp. 779–796. Springer, Cham (2018)

Snyder, H.: Literature review as a research methodology: an overview and guidelines. J. Bus. Res. **104**, 333–339 (2019)

Ulijaszek, S.: Physical activity and the human body in the (increasingly smart) built environment. Obes. Rev. **19**(S1), 84–93 (2018)

Van Wee, B., Banister, D.: How to write a literature review paper? Transp. Rev. **36**(2), 278–288 (2016)

Vich, G., Marquet, O., Miralles-Guasch, C.: Green streetscape and walking: exploring active mobility patterns in dense and compact cities. J. Transp. Health **12**, 50–59 (2019)

Wang, Y., Chau, C.K., Ng, W.Y., Leung, T.M.: A review on the effects of physical built environment attributes on enhancing walking and cycling activity levels within residential neighborhoods. Cities **50**, 1–15 (2016)

World Health Organization (WHO): Global action plan on physical activity 2018–2030: more active people for a healthier world. World Health Organization, Geneva (2018)

Zandieh, R., Martinez, J., Flacke, J., Jones, P., van Maarseveen, M.: Older adults' outdoor walking: inequalities in neighbourhood safety, pedestrian infrastructure and aesthetics. Int. J. Environ. Res. Public Health **13**(12), 1–24 (2016)

How Public Transport Could Benefit from Social Media? Evidence from European Agencies

Georgios Georgiadis, Anastasia Nikolaidou(✉), Ioannis Politis, and Panagiotis Papaioannou

Transport Engineering Laboratory, Department of Civil Engineering, Faculty of Engineering, Aristotle University of Thessaloniki, 54124 Thessaloniki, Greece
nikolaid@civil.auth.gr

Abstract. Currently, there has been a research focus on exploiting the information that can be extracted from social media platforms for scientific purposes. Social media data is dynamic, user-generated, and includes a combination of spatial, temporal, and textual information. Mining and analyzing transport information from social media is a relatively new field and it can be used during transportation planning and management, as well as for supporting the achievement of specific transport policy goals. In addition, social media provide public transport agencies with an unparalleled opportunity to connect with their customers. This paper explores the use of social media among public transport agencies in Europe. In this context, a relevant literature review was combined with original findings obtained from a survey on selected public transport agencies. Relevant information was derived from online sources, including blog posts, websites, online journals and publications. The survey results provided an integrated overview of the use of social media among the selected agencies. Public transport operators use social media mainly for: (a) real-time updates and information, (b) information to customers regarding services, fares, and services disruptions, (c) engaging citizens by handling complaints and inquiries, (d) employee recognition and recruitment of staff and (e) video entertainment and contests. Twitter was most commonly used for short communications and service updates, Facebook for short announcements and service updates, but also for community building and branding. Organizations employ YouTube and Instagram to build community support, whereas LinkedIn was used for networking and recruitment.

Keywords: Social media · Public transport · Big data · Real-time information · Social media policy

1 Introduction

Social media refers to websites and applications that are designed to encourage users to share their experiences, opinions and sometimes their location quickly, efficiently, and in real-time [1]. Social media began as mostly customers phenomena, but currently are

E. G. Nathanail et al. (Eds.): CSUM2020, AISC 1278, pp. 645–653, 2021.
https://doi.org/10.1007/978-3-030-61075-3_63

widely used by organizations (commercial, non-profit, and public) for both internal and external communication.

The existing uses of social media in the transport sector can be categorized based on the purpose of their use into the following categories: (a) event detection; (b) understanding mobility and activity patterns; (c) measuring customer's satisfaction; (d) information and communication by transport authorities and public agencies; and (e) conducting transport surveys [2]. Social media platforms are seeing increasing adoption, as they provide a cost-effective, reliable, and timely information exchange channel with riders [3].

In addition, research shows that there is less systematic understanding of specifically who uses what social media platform and for what kind of purposes. This paper aims to fill this gap by first identifying and synthesizing the relevant research and practice literature on the usage of social media. Then, by utilizing a survey on selected public transport agencies in Europe, an analysis is made on the actual social media usage by these organisations. Through this survey, a better understanding of the types of communications and messages through social media emerges. This paper focuses particularly on the frequency and intensity of information exchanges and interactions between agencies and their riders as well as potential customers (e.g., real-time alerts, service disruptions, system updates, etc.).

2 Literature Review

Social media are a comparatively new arena for public agencies that seek to influence the way people think about public services in a crucial domain such as transport provision. Social media have helped organizations built direct and transparent relationships with users, and evidence showed that they help improve transport services and operations [4]. In general, agencies use the following types of social media platforms: social and business networking, blogging, micro-blogging, media and document sharing sites [5].

The Urban Transportation Monitor (UTM) surveyed transportation organizations about their use of social media. When asked which social media platforms they use, about half of the respondents named Facebook and Twitter. Twitter was most commonly used for brief communications and service updates. Facebook has been used for announcements and service updates, but also for meeting notices, community-building, and commercialization. Moreover, YouTube videos covered a wide range of topics, including how-to-ride information, project updates, agency promotions and news. Organizations promoted more in-depth discussion through blogging sites, while LinkedIn was used for networking and recruiting purposes [6].

In a recent study conducted in USA, regarding the use of social media amongst public agencies, it was found that many public transport agencies have a high presence in different social media sites, indicating a significant change in how government agencies interact with their customers, compared to the traditional means of communication used in the past (newspapers, leaflets, promotion through TV channels [7]. Facebook interactions focus on showing how organizations are engaging with members of the community, and Twitter is more rider-focused with consistent real-time updates about services and disruptions. Other social media platforms may serve specific purposes, such as Instagram, Flickr or Pinterest for communication with the use of photos or short videos, and LinkedIn for professional networking purposes [8].

Recent studies have found that engaging in interactive communications through social media could improve overall the image of an agency [9]. Interactive and immediate messages that respond to questions, concerns, and comments were categorized as of more positive sentiment, regardless of actual quality of services, although agencies responded to a limited number of tweets per day.

In 2012, a report from the Transit Cooperative Research Program (TCRP), identified five types of social media use for public transport, including [10]:

- Real-time updates for sharing real-time service information and alerts to riders.
- Sharing of public information about services and fares.
- Citizen engagement and involvement where the interactive ability of social media is used in order to connect with passengers.
- Employee recognition for identifying current and recruiting new employees.
- Entertainment of passengers through songs, videos, and contests.

In general, social media are considered lately as a worthwhile investment for public transportation agencies and organizations. However, a survey on selected USA transportation providers found that although 94% of them use some forms of social media, only one third of them use these social networking tools based on a strategy or a social media implementation plan [5].

Despite the widespread use of social media in the field of public transport, the goals, strategies, objectives and effectiveness of the use of social media still seem vague and inconsistent. Social media presence may represent a significant investment, in terms of money and time, of staff resources in maintaining sites, responding to users' feedback and comments and keeping updated the official social media sites of each organization [10].

3 Social Media Use in European Public Transport Sector

Based on the literature review presented in the previous section and in the context of the present study a survey was conducted to investigate the intensity and purpose of social media use in a representative sample of European public transport agencies.

3.1 Data Collection and Process

The sample of public transport agencies in our survey was selected in a two (2) stage process. First, in order to ensure the representativeness of our sample, we explored relevant statistics provided by the Mobility in Cities Database (MCD) of the International Association of Public Transport (Union Internationale des Transports Publics (UITP)) [11]. This database contains, inter alia, key public transport figures from 35 European metropolitan areas for 2012. Figure 1 shows that we selected an initial sample of ten (10) European metropolitan areas that stand for relatively high, medium and low public transport supply figures within MCD.

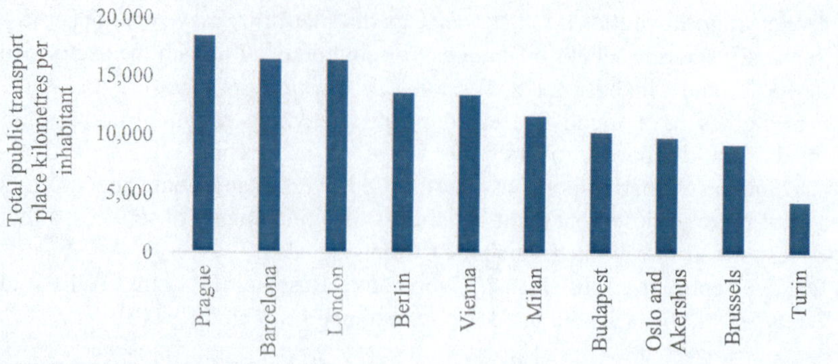

Fig. 1. Public transport supply figures for the selected European metropolitan areas [11].

Secondly, we identified all public transport operators and authorities in each metropolitan area and we followed a data-driven approach to select those agencies which had a greater social media activity compared to their other public transport counterparts in the same metropolitan area. Table 1 shows the final sample of the ten (10) public transport agencies we examined. Population figures are derived from MCD [11].

Specific information regarding the use of social media by the selected transport agencies was gathered. Five (5) social media platforms were considered which, based on the literature review, are commonly used by transport organisations. These platforms are:

- Social and professional networking sites, i.e. Facebook, Instagram and LinkedIn.
- Micro-blogging sites, i.e. Twitter
- Media sharing sites, i.e. YouTube.

For the selected transport organisations, all posts publicly available via their official social media accounts were accessed between January 2019 and December 2019. Customer acceptance and popularity of social media accounts for each agency was determined by registration date and number of followers (Table 2). Total number of posts and average number of posts per month were calculated so as to highlight the intensity of social media activity in each agency (Table 3). All public posts were further examined to be classified according to their content in six (6) categories, i.e.: service info (static and real time), agency news, other news, contests and promotions as well as job listings.

Table 1. Sample of European public transport agencies examined.

Name of public transport agency (holder of social media accounts)	Type of agency	Metropolitan area	Population	Public transport modes operated or supervised
Prague Public Transport (MHD)	Operator	City of Prague	1,246,000	Bus, tram, metro
Transports Metropolitans de Barcelona (TmB)	Operator	Barcelona Metropolitan Area	3,220,000	Bus, metro
Transport for London (TfL)	Authority	Greater London Area	8,310,000	Bus, tram, suburban rail, ferry
Verkehrsverbund Berlin-Brandenburg (VBB)	Authority	State of Berlin	3,375,000	Bus, tram, metro, suburban rail
Wiener Linien (WL)	Operator	City of Vienna	1,741,000	Bus, tram, metro
Azienda Trasporti Milanesi (ATM)	Operator	Milan Metropolitan area	2,123,000	Bus, tram, metro, suburban rail
Budapest Transport Privately Held Corporation (BKV)	Authority	Municipality of Budapest	1,727,000	Bus, tram, metro, suburban rail, ferry
Ruter	Authority	Oslo and Akershus County	1,169,000	Bus, tram, metro, suburban rail, ferry
STIB (Société des Transports intercommunaux bruxellois)	Operator	Brussels Capital region	1,154,000	Bus, tram, metro, suburban rail
Gruppo Torinese Trasporti (GTT)	Operator	Turin Metropolitan Area	1,515,000	Bus, tram, metro, suburban rail

3.2 Results and Discussion

Tables 2 and 3 show that while all public transport agencies use social media platforms, the corresponding popularity and activity rates may differ substantially. Overall, TfL and TmB are the most engaged organisations to social media activities. Both of them have profiles in all five (5) social media networks while TfL has the highest number of followers and TmB has the highest number of posts per month. Contrary to these, BKV has the lowest number of followers and posts per month with presence in only three (3) social media platforms.

More specifically, Table 2 indicates that, in general, the oldest accounts are associated with Facebook and Twitter and the most recent ones with LinkedIn and Instagram. Despite their comparatively longer tradition, Facebook and Twitter are not necessarily used by all agencies. TmB, TfL, Ruter and STIB were the first to launch social media activities in 2009. A remarkable number of followers (over 2 millions) appears for the Twitter account of TfL. Overall, either Facebook or Twitter have the highest popularity rates within the five (5) platforms considered in each agency.

The lowest figures for the number of followers belong to YouTube but this is not always the case since there were agencies where YouTube had a considerable popularity rate compared to the other social media sites (e.g. TmB, TfL and WL). Despite being the youngest one, Instagram is gradually emerging as a key social media platform with increasing popularity for almost all agencies which were examined. Instagram's dynamics are further confirmed in Table 3, since: (a) the total number of Instagram's posts surpasses YouTube ones in all agencies, though YouTube accounts have oldest registration dates and (b) the average number of Instagram posts per month is higher than LinkedIn and YouTube corresponding ones and sometimes even greater when compared to Facebook ones (e.g. TfL). Table 3 also highlights Twitter as the platform that concentrates the greatest activity in terms of updates and posts.

Table 2. Findings from the online survey of official social media accounts- Popularity.

Public transport agency (holder of social media accounts)	Registration date					Number of followers				
	Facebook	Twitter	LinkedIn	Instagram	YouTube	Facebook	Twitter	LinkedIn	Instagram	YouTube
MHD	2011	2014	–	2015	2015	29,871	12,700	–	6,202	940
TmB	2009	2010	2019	2017	2010	79,712	211,700	13,322	13,500	45,300
TfL	2012	2009	2016	2013	2007	423,926	2,400,000	164,439	128,000	47,400
VBB	2013	2011	–	2017	2016	4,634	12,300	–	518	6
WL	2010	2010	2018	2013	2007	99,320	29,600	1,963	28,700	51,400
ATM	–	2012	2019	2017	2011	–	283,600	32,995	29,500	1,380
BKV	2013	–	–	2018	2014	15,745	–	–	3,281	1,190
Ruter	2010	2009	2015	2014	2010	60,030	29,500	5,433	5,524	490
STIB	2010	2009	2016	2018	2010	88,683	130,400	17,222	1,476	2,730
GTT	2010	2010	–	2019	2016	35,132	16,400	4,592	1,584	86

Figure 2 demonstrates the diversity of information which is published by all ten (10) public transport agencies and its distribution against the five (5) social networks which we explored. Apart from Twitter, all other social media are mostly devoted to sharing agencies' news. This category of information refers to agencies' public transport investments, critical modifications of services, activities' reports as well messages on sustainability issues regarding the contribution of public transport to the enhancement of quality of life, congestion reduction, improvement of safety and environmental conditions etc. Facebook is the only platform which accommodates all six (6) categories of information considered. Facebook is mostly used to deliver information on various agency news, other issues concerning local communities as well as contests and promotions for branding. Twitter is mostly employed for brief communications, service updates in real-time and for providing information about unexpected events such as service delays, disruptions or traffic accidents affecting particular lines or stations. This fact probably explains the relatively large size of Twitter posts and followers in Tables 2 and 3. Instagram is mainly used for sharing photos and short videos on agencies' news. YouTube videos cover topics which usually deal with information about agencies' initiatives and project updates. LinkedIn is mainly used for posting detailed information regarding agencies' news and projects as well as for networking and recruiting purposes. Figure 2 also indicates that, in practice, public transport agencies do not rely on social media platforms to upload static service information, such as timetables, network maps, hours of service etc. Finally, in all social media platforms, agencies frequently take the opportunity to communicate with and to respond to comments from their customers.

Table 3. Findings from the online survey of official social media accounts- Activity.

Public transport agency (holder of social media accounts)	Total number of posts [all years]					Average Number of posts per month [2019]				
	Facebook[a]	Twitter	LinkedIn[a]	Instagram	YouTube	Facebook	Twitter	LinkedIn	Instagram	YouTube
MHD		8,632		533	70	42	143	–	23	1
TmB		35,800		1,687	914	109	284	22	64	6
TfL		102,100		2,210	579	55	106	25	88	9
VBB		8,403		297	1	5	55	–	14	0
WL		33,900		748	247	34	30	18	17	8
ATM		105,600		353	40	–	255	7	9	1
BKV		–		263	44	37	–	–	6	6
Ruter		51,400		280	111	3	11	13	2	2
STIB		83,000		349	391	29	600	14	21	11
GTT		5,201		36	7	16	183	–	3	0

[a]Facebook and LinkedIn do not provide historical data on the total number of posts

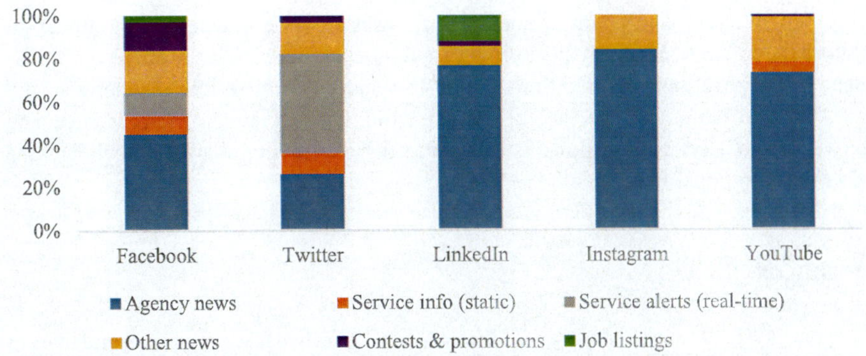

Fig. 2. Distribution of posts per content and social media platform for all agencies

4 Conclusion

Public transport agencies have invested in integrating social media into their businesses and communities and consider social media to be an important channel to interact and respond to public transport riders as well as other citizens. However, literature review findings imply that the use of social media by public transport agencies is not as widespread as in the case of other commercial organizations, and that the potential applications are limited into mainly uploading information, promoting, and trying to engage in dialog with the public.

In this paper we investigated the utilization of social media in European public transport sector by examining the corresponding official accounts of a representative sample of ten (10) agencies. Empirical findings indicate that social media strategies may greatly vary, in terms of platforms adopted and frequencies of posts, within the public transport agencies considered. The most active agencies were found in metropolitan areas with relatively higher public transport supply figures but, overall, both popularity and activity rates of social media accounts are not clearly related to the size of public transport operations.

Social media are mainly used to keep public informed on agencies' activities and importance of using public transport. However, the purpose, content and frequency of posts depend to a large extent on the type of the social media platform. Twitter is mostly used for frequent and short announcements whereas Facebook mostly provides service information and updates about agency news. YouTube and Instagram are mainly used for sharing videos and photos whilst LinkedIn is a rather professional networking site. Furthermore, Twitter is frequently used as a real-time information channel, LinkedIn assists into staff recruitment while Instagram was found to be a relatively new, promising and attractive social media platform in European public transport industry. It was also made evident that public transport agencies prefer their official sites rather than their social media accounts for uploading timetables, routes, maps and other static info for their services.

Further in-depth assessment of social media usage by the selected public transport operators could be achieved by performing topic-specific interviews as next step of the present survey, highlighting additional details on successful practices, challenges, and lessons learned. In such case, useful insight would be gained on the existence and effectiveness of any tailored policies regarding the use of the available social media platforms by the operators.

References

1. Kaplan, A.M., Haenlein, M.: Users of the world, unite! the challenges and opportunities of Social Media. Bus. Horiz. **53**, 59–68 (2010). https://doi.org/10.1016/j.bushor.2009.09.003
2. Nikolaidou, A., Papaioannou, P.: Utilizing social media in transport planning and public transit quality: survey of literature. J. Transp. Eng. Part A Syst. **144**, 04018007 (2018). https://doi.org/10.1061/jtepbs.0000128

3. Cottrill, C., Gault, P., Yeboah, G., Nelson, J.D., Anable, J., Budd, T.: Tweeting transit: an examination of social media strategies for transport information management during a large event. Transp. Res. Part C Emerg. Technol. **77**, 421–432 (2017). https://doi.org/10.1016/j.trc.2017.02.008
4. Collins, C., Hasan, S., Ukkusuri, S.V.: A novel transit rider satisfaction metric: Rider sentiments measured from online social media data. J. Public Transp. **16**, 21–45 (2013). https://doi.org/10.5038/2375-0901.16.2.2
5. Liu, J.H., Shi, W., Elrahman, O.A. (Sam., Ban, X. (Jeff), Reilly, J.M.: Understanding social media program usage in public transit agencies. Int. J. Transp. Sci. Technol. **5**, 83–92 (2016). https://doi.org/10.1016/j.ijtst.2016.09.005
6. Liu, J.H., Ban, X., Elrahman, O.A.: Measuring the impacts of social media on advancing public transit, p. 60 (2017)
7. Bregman, S., Watkins, K.E.: Best Practices for Transportation Agency Use of Social Media. CRC Press, Boca Raton (2013). https://doi.org/10.1201/b15768
8. Mainka, A., Hartmann, S., Stock, W.G., Peters, I.: Government and social media: A case study of 31 informational world cities. In: Proceedings of the Annual Hawaii International Conference System Science, pp. 1715–1724 (2014). https://doi.org/10.1109/HICSS.2014.219
9. Schweitzer, L.: Planning and social media: a case study of public transit and stigma on Twitter. J. Am. Plan. Assoc. **80**, 218–238 (2014). https://doi.org/10.1080/01944363.2014.980439
10. Bregman, S.: TCRP Synthesis 99: Uses of Social Media in Public Transportation (2012)
11. UITP: Mobility in cities database. https://www.uitp.org/MCD. Accessed 21 July 2020

Improving Mobility Services through Customer Participation

Sérgio-Pedro Duarte(✉), Marta Campos Ferreira,
Jorge Pinho de Sousa, Jorge Freire de Sousa, and Teresa Galvão

INESC TEC, Faculty of Engineering, University of Porto,
Rua Dr. Roberto Frias, 4200-265 Porto, Portugal
{s.duarte,mferreira,jfsousa,tgalvao}@fe.up.pt,
jorge.p.sousa@inesctec.pt

Abstract. In their quest for sustainability, cities design and deploy smart mobility solutions aiming to improve the efficiency and management of transportation systems and to provide better services to citizens. Those solutions are often based on Information and Communication Technologies (ICT) and on digital services, but their maintenance and management are a greater challenge than their implementation. Problems can be difficult to identify since they can be exogenous or endogenous to the service provider. Usually, in their effort to maintain good service levels, companies implement complex and expensive information systems that use sensors to monitor infrastructure and hardware but ignore other sources of valuable information. In a digitalized world, customers easily report problems that are a cause of lower quality of service and worse user experience. However, for several reasons, service providers do not always pay due attention to these complaints. As communication channels are already open, we claim that customer participation through these reports can be used to significantly enhance the delivery and quality of mobility services. In this work we propose a methodology that takes advantage of customers' participation in the maintenance and management of smart city solutions. With this methodology, we aim to redesign the process of customer interaction with service providers in order to improve the overall efficiency and the service experience. Our research is based on a case study from a public transport service in the metropolitan area of Porto, in Portugal.

Keywords: Urban mobility · Service design · Collaboration · Public transport · Sustainability · Digitalization

1 Introduction

For several years, cities have been implementing smart solutions to improve many of their services. The case of transportation is especially relevant due to its important role in the daily life of citizens and to its significant environmental impacts. In fact, although other sectors are steadily decreasing emission levels, that is not the case of the transport sector [1]. That is surely one of the reasons why cities want to attract more passengers to public transport, and why they need, therefore, to rethink their services.

E. G. Nathanail et al. (Eds.): CSUM2020, AISC 1278, pp. 654–663, 2021.
https://doi.org/10.1007/978-3-030-61075-3_64

The link between sustainability and technology is visible in the evolution of the concepts of *smart city* and *sustainable city*. Currently, some authors adopt the term *smart sustainable city* since it reflects both technological and sustainable perspectives [2]. Some of these technological solutions are naturally linked to the current high levels of digitalization, and terms such as *digital city* and *information city* were quite common before the universal adoption of the term *smart city* [3].

In the new technology-enabled service context, there are plenty of channels, social media, and smart services, through which customers and service providers interact. According to the service-dominant (S-D) logic, value is co-created with customers as value-in-use. Irrespective of the degree of intangibility involved in a given offer, "the customer is always a co-creator of value" [4]. Customers assume, therefore, an active, connected and informed role on the value creation process. This paradigm is opposed to the traditional goods-dominant (G-D) logic, that considers customers as passive consumers, and where goods are created by firms and distributed to consumers [5].

Therefore, the success of digital solutions adoption strongly depends on the correct (re)design of processes and on the implementation of a user-centred mind set. Users and other stakeholders are engaged and encouraged to share and use their experiences as a way to generate more effective and meaningful solutions [6].

In this context, [7] identify eight subtopics, under the "leveraging service design" research priority, that need further research. One of those eight topics consists of "involving customers through participatory design and codesign to enhance service experience". In line with these concerns, this paper aims at exploring more active ways to engage public transport customers in the service design process. Customers have valuable knowledge and opinions, and are, in general, willing to contribute to the decision-making processes. At the same time, they wish to share their experience and to participate in the service delivery process. However, the potential of these contributions is often overlooked, and methodologies that enable effective customer participation are missing. In the end, if their contribution is not recognized or encouraged, customers quickly lose motivation and give up participating.

This paper proposes a methodology that leverages customers' engagement to improve the transport service through their active participation in the service design processes. The methodology proposed resorts to Business Process Reengineering (BPR) assuring the efficiency of internal processes [8], while the S-D logic sets the ground for co-creation processes. This mixed approach was first proposed for the design of an IS and is being used in the development of a IS framework in the context of urban mobility services [9]. As an illustrative example for the application of the proposed methodology, we use data from the urban transport operators of the city of Porto, Portugal. This data (obtained in 2019) consists of customers' reports on the use of a mobile ticketing application (Anda), received by the customer support service through various channels, such as the mobile ticketing application, telephone, social media or email.

The remainder of the paper is organized as follows. The next section presents the proposed methodology. Section 3 provides an illustrative example, and Sect. 4 presents the main conclusions of the work and some opportunities for future research.

2 A Methodology for Improving Urban Mobility Services

The design of a technology-based service is always an opportunity to redesign organizational processes, taking advantage of digitalization. In fact, Enterprise Architecture Frameworks usually include BPR and the development of information systems (IS) that support new business processes. Service Design (SD) research has also been adopting Business Process Management (BPM) methods. While SD focus on improving customer experience, BPR focus on organizing business processes to improve service levels. The methodology presented in this paper uses a BPR approach and a SD mindset. In other words, we redesign internal business processes so that we can improve customer experience.

As in every service, mobility services must have their business processes aligned with a given strategy, and an IS that supports those processes. If a service has a customer-centric strategy, that should be reflected in their organizational processes. To promote value co-creation, customers should be allowed to participate in the design of the mobility service and in the associated decision-making processes [10]. So, if customers are willing to share their experience, by reporting positive and negative aspects of the service, that information should be considered on maintenance and management plans. These plans are developed to manage priorities in the way problems and suggestions are handled. They include the correction of operational problems such as software problems, and infrastructure malfunctions, but also strategic and tactical issues, such as scheduling and accessibility to vehicles (ramps, time of opening/closing doors, etc.).

The contact between customers and the service provider usually occurs through multiple channels centralized in a customer support department. Redesigning processes related to this interaction may benefit both customers and service providers. Here, we briefly present the methodology used to improve the mobility service, after the implementation of changes in the organizational processes. As referred, the improvement of an existing mobility service can benefit from data generated by existing helpdesk channels. Therefore, it is necessary to redesign processes and create mechanisms to incorporate that data efficiently.

The methodology developed has a waterfall part and an iterative part (see Fig. 1). Instead of reviewing all the processes, we only select the ones that are expected to benefit from the available data. Thus, before redesigning processes, we collect and analyze existing data. Then, the processes are redesigned and validated by the actors, before being implemented.

Since there is a record of reports sent by customers (e.g., complaints, suggestions, and requests), it is possible to identify what processes need to be redesigned. Reports originate support cases and address different issues that will trigger different actions in the back-office, and impact in different ways the maintenance and management plans.

By analyzing the existing cases, we can also create categories according to the content of each report and its urgency. While some complaints may be urgent and need immediate attention (e.g., infrastructure malfunctions), others may need approval (e.g., reimbursing the price of a ticket); and suggestions will also require analysis and

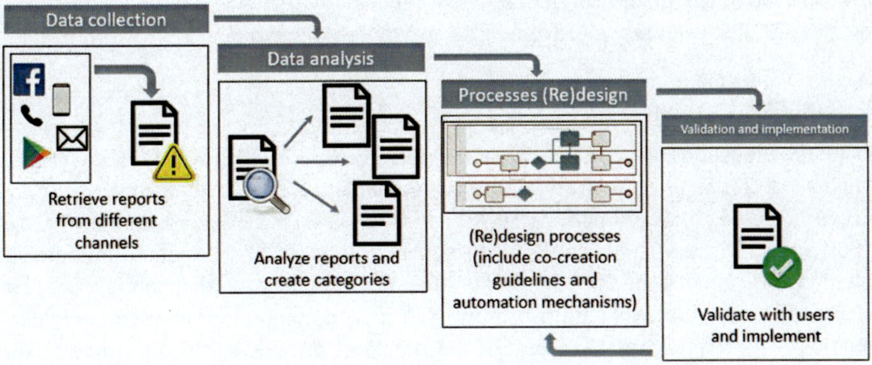

Fig. 1. Methodology for improving urban mobility services

eventual approval. These categories are used during the redesign of the processes and, afterwards, the same categories will help to classify new incoming reports.

Identified processes must be redesigned to assure that the input from customers reaches decision-makers. We have developed co-creation guidelines that shall be taken into account during process redesign. Besides, mechanisms should be considered here to automate some tasks, using the available technological capabilities. For instance, text-mining can help to automatically assign a category and a priority to a case, thus leaving more time for the customer support to interact with customers. Part of the process cannot be automated because it depends on the activities of the backend. This includes giving feedback to the customers, showing in this way their input is essential. This will motivate them to continue to send reports, thus feeding the maintenance plan.

Co-creation is also present in the redesign of the processes, as the actors are the ones that validate them. Process redesign and validation are the iterative part of the methodology. When all processes are approved, they are supposed to be implemented by adapting the IS accordingly. The implementation should be managed closely. It must include tests, and a period of adaptation that involves performing both (as-is and to-be) processes. During that period, the categories attributed by the IS must be verified and checked for correction. This can be performed by randomly selecting some cases. This evaluation procedure needs to be negotiated with the decision-maker before implementation. This methodology can be applied to other, new categories, identified later.

3 Illustrative Example

The proposed methodology can be materialized in several practical examples. As a proof of concept, this section explores a case study that uses data from the urban transport operators from Porto, Portugal. This data consists of reports sent by customers (the passengers) concerning the use of a mobile ticketing application called Anda.

The next subsections present the context and, then, describe the application of the different phases of the methodology. Throughout the text we explain how information can be incorporated in the service delivery process of transport operators and may contribute to a continuous improvement of the mobility service.

3.1 Context

The public transport network of the Metropolitan Area of Porto (AMP) covers an extension of 1,575 km^2, with 1.75 million inhabitants. The system consists of 12 operators: 157 bus lines (73 public (STCP) and 84 private), 82 light rail stations (Metro do Porto) and 25 train stations (CP – Portuguese Railways). The Andante Intermodal System (SIA), created in 2003, promotes the intermodality of the public transport system. All 12 operators that are part of the SIA have common fare rules and ticketing technology. Passengers may use any of the transport providers with the same ticket regardless of the number of transport operators involved in a trip. The electronic fare system in AMP is open (ungated) and uses an entry-only Automated Fare Collection (AFC) system based on a distance structure. The fare media consists of contactless travel cards (Andante), and fares are defined according to a zone structure. To perform a journey, customers must load the travel card with the correct travel ticket for their trip.

Recently, in June 2018, a new fare media was introduced, which consists of a mobile application called Anda. It is based on a check-in/be-out approach, where the customer intentionally checks-in at a vehicle/station, by tapping the mobile phone on a Near Field Communication (NFC) reader. During the journey the mobile application is continuously communicating with the Bluetooth Low Energy (BLE) beacons installed inside the buses or at light rail and train stations, allowing for a micro-location of the customer [11]. When the mobile phone loses the signal of the beacons, it automatically closes the journey. A fare optimization algorithm then calculates the fare to be paid by the customer, which is charged at the end of the month.

The main advantage of the Anda application, when compared to the more traditional Andante travel card, is the fact that customers do not need to have previous knowledge about routes or fares [12]. Moreover, at the end of the month, customers are charged the minimum possible total value, considering the fare rules in force. However, like any new digital solution, the system still has some bugs and weaknesses, and, from the customers' point-of-view, adapting to a new ticketing media takes time.

Since its full deployment, customers using the Anda application interact daily with the customer support service, through several channels, such as telephone, email, Facebook, Google play and physical stores. This interaction can have several purposes, such as asking questions, reporting errors, or making suggestions for improvement. The information from the various channels is collected in a single platform, to be further processed and analyzed. Depending on the content of the report, it can be assigned to the customer support, to a maintenance team, or to the management team. In our case study, maintenance teams refer to the partners of the Anda system that include, among others, the mobile application developers, technical support for the NCF readers and BLE, and the transport operators. The management team is responsible for analyzing and taking decisions about more complex cases.

3.2 Data Collection and Data Analysis

As an illustrative example, we have used data from the reports of Anda customers from 2019. During this year, a total of 8,624 cases were recorded on the customer support platform. Among other information, the cases include the motive of the report, the channel used, and the type (request, inquiry, or complaint). From all reports, 1,006 had the type and channel as "*undefined*" and were therefore ignored, leaving a total of 7,618 to be analyzed (see Tables 1 and 2). Clearly, the Anda app has become the main channel for reporting.

Table 1. Type of report

Type of report	
Complaint	4,919
Inquiry	1,886
Request	813
Total	7,618

Table 2. Channel of report submission

Channel	
App	3,119
E-mail	569
Phone	1,873
Facebook	58
Google	4
AMT	3
Technical support CMS	20
Flyer	12
Consultation	1,690
Total	7,618

In what concerns the motive of the report, the most common situation is related to the start of the journey ("*check-in*") with 2,123 cases, with "*redo enroll*" appearing in the third place, with 1,092 cases (see Table 3). By analyzing the motives mentioned in the reports, we have identified three main categories: i) urgent errors; ii) usage and usability problems; and iii) suggestions and not critical or immediate errors.

Table 3. Type of reported cases for ANDA_Redo enroll and ANDA_Check-in.

Motive	Type	Number of reports	Total
ANDA_Redo enroll	Complaint	674	
ANDA_Redo enroll	Request	257	
ANDA_Redo enroll	Inquiry	161	1,092
ANDA_Check-in	Complaint	2,034	
ANDA_Check-in	Request	10	
ANDA_Check-in	Inquiry	79	2,123

Urgent errors refer to problems that hinder the operations. They may happen if a passenger is not able to start a journey when tapping the smartphone, due to a malfunction of the infrastructure. These problems need to be solved as soon as possible, and do not require any approval from management. They should, therefore, be assigned to one of the maintenance teams.

If the journey does not start due to problems with the mobile application, it is necessary to contact the customer, to check if it is possible to solve the problem without the intervention of the maintenance team. This contact is done by the customer support. If the problem requires technical support, the case should be assigned to the maintenance team of the mobile application. Since this type of problems prevents the passenger to travel, these cases should have high priority in the maintenance plan. The situations that can be solved by the passenger with the help of the customer support are included in the category of *usage and usability problems*.

Finally, *suggestions* represent potential improvements to the service, in the passenger's perspective. However, suggestions need to be analyzed in order to evaluate their feasibility and implications, and can naturally be the basis for further improvements to the service.

These are the categories that have been identified until now. If more categories are identified in newer cases, other processes should be redesigned.

3.3 Processes Redesign

There are two main weaknesses in the current process (see Fig. 2): the fact that the customer support must analyze all the cases, and the fact that no feedback to the customer is assured after the problem has been handled. The lack of feedback makes the customer lose motivation and probably stop reporting further issues. Moreover, if the customer support is involved in all cases, the delay in responses will possibly be higher.

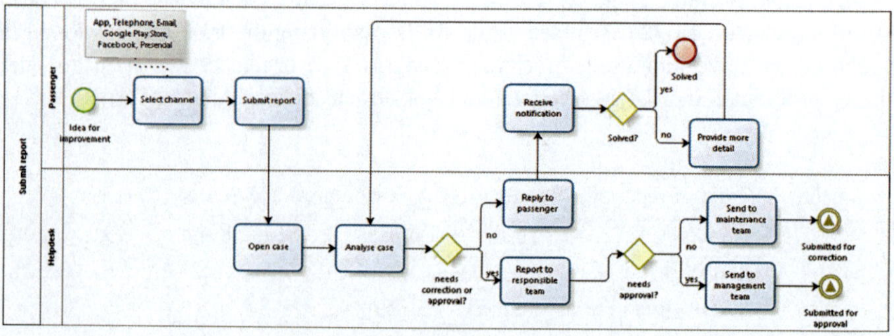

Fig. 2. Submit report process (AS-IS).

To reduce these delays, we have redesigned the process, so that the category is automatically attributed to the report, by the system, and not manually (see Fig. 3). It is important to simplify this part of the process because two of the most common reported issues (*Check-in* and *Redo enroll*) may hinder the beginning of the trip since, in these cases, the customer cannot use the mobile application, and may have to buy a regular physical ticket. If this happens too often, the customer is likely to stop using the app. Thus, this automatic classification procedure allows an immediate allocation of the case to the team responsible for handling it.

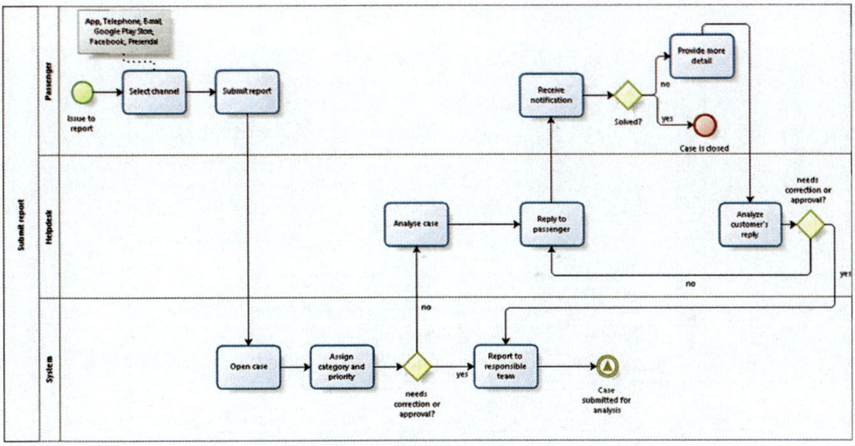

Fig. 3. Submit report process (TO-BE).

Another advantage is that this automation reduces the workload of the customer support, thus providing more time for cases that require interaction with the customer (usage and usability problems). This small change in the process may have significant impact in the service, because it improves the communication with the customer and improves the information sent to the maintenance and management teams. This automation may also help in assigning each case to the right team, according to the category of the case, depending on if it needs approval or not (see Fig. 4).

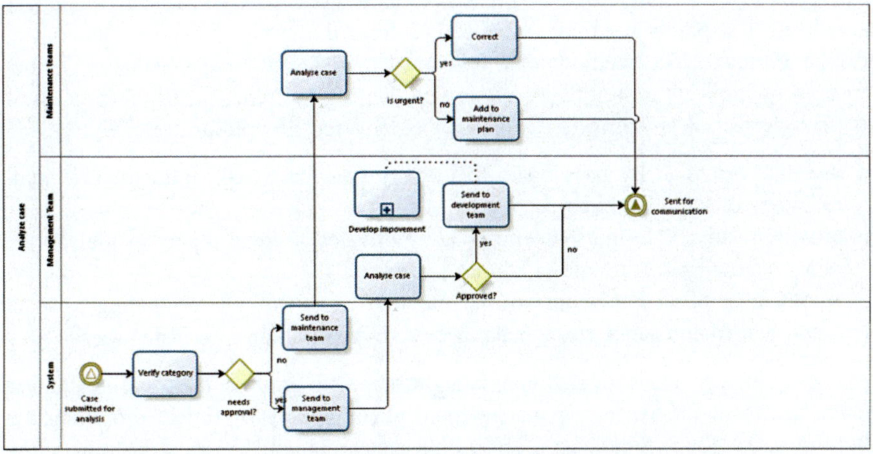

Fig. 4. Analyze case process (TO-BE).

As proposed in the methodology, these processes must be validated by the relevant actors to assess their viability. For instance, in our situation, we have already identified that the result of the case should be sent by the customer support and not automatically, as shown in Fig. 5.

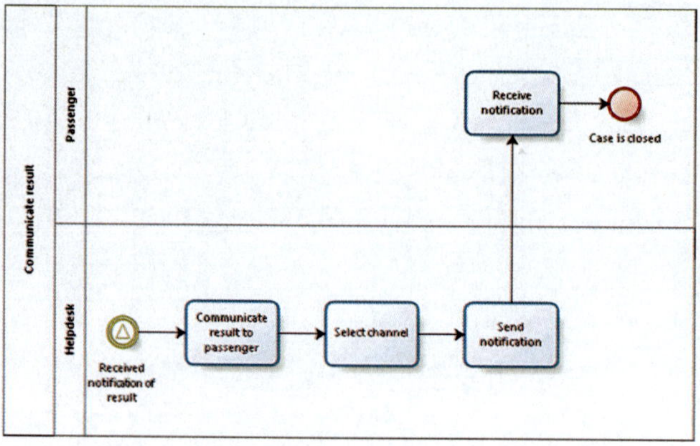

Fig. 5. Communicate result process (TO-BE).

4 Conclusions and Future Work

The success of innovative digital solutions to improve mobility services strongly depends on the level of acceptance from the customers. This acceptance will be higher if the service is designed according to their expectations. Co-creation methods help to engage the customers in improving service experience and efficiency.

This paper proposes a methodology that leverages customer engagement in improving mobility services and promotes participation in the service design processes and outcomes. The methodology, that resorts to BPR and is based on an S-D logic, was materialized in a practical example using data from the urban transport operators of Porto, in Portugal. The methodology was used to redesign processes for customer support, thus improving the quality of service. During the redesign of the processes, three main aspects were considered to foster co-creation of solutions, namely:

i. use data provided by passengers to identify problems (e.g., infrastructure malfunctions or software bugs);
ii. provide feedback so that passengers are motivated to keep reporting their experience (showing them that their input is essential);
iii. automate helpdesk processes so that the staff can focus on responding to open cases, and maintenance teams have access to information in a faster way.

Redesigning processes for customer support helped to improve the way information is integrated in maintenance and management plans, mainly by timely providing the right information to the right actors. The whole system will only work if customers (the passengers) go on reporting problems, followed by fast reactions and responses.

Additionally, the collected data also showed that some problems concerning the usage of the mobile application could also be solved with an almost real-time help of the customer support. This suggests that the task "submit report" can be redesigned to include a small questionnaire in order to provide the most complete information about the case. If the problem can be solved by the customer alone, the system will redirect to already existing FAQ, or integrate a *chatbot* feature in the application. These requirements would obviously require further developments of the current solution.

In the future, this methodology can be applied to more detailed service processes, including the interaction of passengers with the different channels. For that purpose, Service Design methods and models, such as the Service Experience Blueprint, can be beneficial, as they map the service experience with multi-channel services.

Acknowledgments. Sérgio-Pedro Duarte is funded by Fundação para a Ciência e Tecnologia (FCT), Portugal, through grant PD/BD/128064/2016. Jorge Freire de Sousa, Jorge Pinho de Sousa and Teresa Galvão are partially funded by the ERDF - European Regional Development Fund through the Operational Programme for Competitiveness and Internationalisation - COMPETE 2020 Programme and through FCT within project PTDC/ECI-TRA/32053/2017 and 7 POCI-01-0145-FEDER-032053.

References

1. Union, European: Stat. Pocketbook **31**(4), 2017 (2017)
2. Ahvenniemi, H., Huovila, A., Pinto-Seppä, I., Airaksinen, M.: What are the differences between sustainable and smart cities? Cities **60**, 234–245 (2017)
3. De Jong, M., Joss, S., Schraven, D., Zhan, C., Weijnen, M.: Sustainable-smart-resilient-low carbon-eco-knowledge cities; Making sense of a multitude of concepts promoting sustainable urbanization". J. Clean. Prod. **109**, 25–38 (2015)
4. Vargo, S.L., Lusch, R.F.: Service-dominant logic: continuing the evolution. J. Acad. Mark. Sci. **36**(1), 1–10 (2008)
5. Vargo, S.L., Lusch, R.F.: From repeat patronage to value co-creation in service ecosystems: a transcending conceptualization of relationship. J. Bus. Mark. Manag., 1–11 (2010)
6. Holmlid, S.: Designing for Resourcefulness in Service: Some Assumptions and Consequences, pp. 151–172. Lapl. Univ. Press, January 2012
7. Ostrom, A.L., Parasuraman, A., Bowen, D.E., Patrício, L., Voss, C.A.: Service research priorities in a rapidly changing context. J. Serv. Res. **18**(2), 127–159 (2015)
8. Davenport, T.H.: Process Innovation: Reengineering Work through Information Technology (1993)
9. Duarte, S.P., De Sousa, J.F., De Sousa, J.P.: EmC-ICDSST 2019 on Decision Support Systems : Main Developments & Future Trends A Conceptual Framework for an Integrated Information System to enhance Urban Mobility (2019)
10. Ferreira, M.C., Dias, T.G., Cunha, J.F.: An in-depth study of mobile ticketing services in urban passenger transport: state of the art and future perspectives. In: Smart Systems Design, Applications, and Challenges, IGI Global, pp. 145–165 (2020)
11. Ferreira, M.C., Dias, T.G., Cunha, J.F.: Is Bluetooth low energy feasible for mobile ticketing in urban passenger transport? Transp. Res. Interdiscip. Perspect. **5**, 100–120 (2020)
12. Ferreira, M.C., Dias, T.G., Cunha, J.F.: Codesign of a Mobile Ticketing Service Solution Based on BLE. J. Traffic Logist. Eng. **7**(1), 10–17 (2019)

Engaging Residents of Thessaloniki on Sustainable Mobility Through a Citizens' Panel: Considerations and Implications from a Methodological and Practical Perspective

Vasiliki Amprasi(✉), Evangelos Genitsaris, Aristotelis Naniopoulos, and Dimitrios Nalmpantis

School of Civil Engineering, Faculty of Engineering, Aristotle University of Thessaloniki, PO Box 452, 541 24 Thessaloniki, Greece
vasiampr@civil.auth.gr

Abstract. The original core concept of sustainable mobility is based on shifting travelers from private and motorized vehicles to other soft and more eco-friendly modes of mobility. The fulfillment of this aim indicates the importance of the human factor in terms of knowledge, awareness, perceptions, attitudes, intentions, and behavior. Citizens' Panels, comprised by a small group of people representative of different socio-demographic backgrounds, can be used as a managerial and planning "tool" by public authorities and planners on a medium or long term period and for a variety of reasons, such as: to challenge and test their infrastructure and services proposals and plans, collect people's values, ideas, and opinions, co-shape future mobility visions, increase public's engagement, form and train volunteering "ambassadors" of policies, etc. Since 2018, a local Citizens' Panel on Sustainable Mobility, the case study of this paper, was formed in Thessaloniki in the frame of the PE4Trans/Interreg Europe project, accounting for three physical meetings and some social media actions so far. Following a brief review of the literature on the participatory planning processes, the general scope, and the process steps (recruitment, group characteristics, etc.) adopted for the establishment of this Panel are analyzed. Next, the special aim of each Citizens' Panel meeting and the overall rationale of them are explained, accompanied by a description of the techniques and methods applied during the conversation and debating activities. Finally, the outcomes of each Citizens' Panel are presented being critically discussed in relation to the initial targets set and the methodology adopted.

Keywords: Public engagement · Participatory planning · Co-creation

1 Introduction

During the last years, there is a tendency in governance that citizens should be integrated into decision-making processes regarding policies that affect their quality of life significantly.

E. G. Nathanail et al. (Eds.): CSUM2020, AISC 1278, pp. 664–673, 2021.
https://doi.org/10.1007/978-3-030-61075-3_65

The importance of adopting more inclusive participatory planning processes has been recognized by policymakers and planners [1]. These processes comprise specific methods employed to incorporate stakeholders, including citizens' concerns, needs, and values in the final outcome, and ensure their greater contribution to the decisions taken [2]. Adequate interaction and actual consideration of everyone's ideas are needed towards the development of positive solutions and the prevention of manipulation phenomena. The aim is to enable active participation by all the members of the group involved and achieve consensus. A typical example is the co-creation process which is a user-centered approach based on the "collective" creativity of two or more people aiming to jointly produce innovative ideas that add value to a good or service [3].

Engaging stakeholders and the public may be challenging. Still, it leads to numerous advantages: effective decision-making, different perspectives, cost-effectiveness and/or a more equitable distribution of resources, shared ownership, greater long-term success, and relationships of trust [1, 4]. On the contrary, there are difficulties due to conflicting interests, specified technical knowledge required for complex transport decisions, legitimacy of the engagement process, and heterogeneity of participants [4].

The challenge of engaging not only stakeholders but also citizens in policy dialogue in the field of sustainable transport planning was shared by a group of institutions with common regional needs. As a result, the "Public Engagement for Sustainable Public Transport" (PE4Trans) project was launched. PE4Trans project is co-funded by the European Regional Development Fund (ERDF) in the frame of the territorial cooperation Interreg Europe Programme 2014–2020. The most important thing about this research project is that citizens are invited to contribute with ideas for sustainable public transport to help experts and policymakers to develop public transport solutions/policies, which are sustainable and contribute to the reduction of the carbon footprint. The bottom-up approach it adopts is implemented through the local Citizens' Panels that are formed in the partners' territories and their planned meetings. In order to achieve and maintain a high level of citizens' engagement throughout the process, the same group of people/participants are assigned to attend all the foreseen meetings.

The aim of this paper is to present the case of the Citizens' Panel on Sustainable Mobility, which has been formed since 2018 in Thessaloniki, Greece, as an interesting example of citizens' engagement to a co-creation participatory process on transport. The demonstration of the case study in Sect. 2 is structured as follows: Firstly, there is a brief description of the main characteristics of the study area in order to outline the demographic and mobility context. The scope and the preparatory actions that were carried out for the successful establishment of the Panel are also stated. Next, the actual implementation details are explained, including the aim of each meeting, the brief description of the methods used along with some general comments, followed by the results of each meeting. In Sect. 3, the conclusions of the paper are critically presented.

2 Case Study: Citizens' Panel on Sustainable Mobility

2.1 Study Area

The Metropolitan Area of Thessaloniki is located in a privileged position in Northern Greece. It includes 11 Municipalities, which cover an area of 1,455 km^2 with a population of 1,012,297 inhabitants [5]. Within its boundaries, there is the second-largest urban area of Greece, Thessaloniki, being the capital of the Region of Central Macedonia. This area hosts about 100,000 students and attracts tourists and visitors throughout the year, mainly on weekends.

According to data regarding the modal split in Thessaloniki, the majority of trips are realized by car (67%). 23% of trips are made with public transport modes, 4% with motorcycles, 4% with taxis, and 2% with non-motorized active transport modes, such as walking or cycling [6]. It is worth noting that, in the frame of a survey conducted, citizens of Thessaloniki stated they were reluctant to change their mobility habits and reduce car usage because of the lack of alternatives [7]. The only public transport mode available to address the mobility demand in the Metropolitan Area of Thessaloniki is buses. During summer, there is a boat service added in the Thermaikos Gulf, but mainly for recreational purposes. In the near future, three new public transport modes are planned to be integrated into the currently inefficient public transport system: an underground metro network, an urban boat service operating throughout the year and not only during the summer months, and a suburban railway at the western part of the city.

2.2 Process for the Establishment of the Citizens' Panel (*Preparation Phase*)

The main goal of the Citizens' Panel is to elaborate specific solutions and actions towards the improvement of public transport policies, by including and engaging citizens in policy-making processes, such as the design and implementation of sustainable transport strategies. The concept of the case study is to involve the same group of people, the so-called "Members of the Panel," to all the stages of the participatory process, that is, to all the physical meetings/workshops foreseen. The success of such a panel depends strongly on the heterogeneity and inclusiveness achieved during the identification and selection of the mixture of participants. Hence, the first and most important step towards the formation of the local Citizens' Panel group on Sustainable Mobility in Thessaloniki was the appropriate recruitment of the participants.

Considering the above, emphasis was given to the invitation process, which began much earlier than the realization of the first meeting and targeted to reach as many citizens as possible. Therefore, many different communication channels were used: appearances on media, contact with chambers and unions, creation of a social media page dedicated to this purpose, press releases reposted on websites, etc. The large number of over 60 applicants allowed the recruitment of a very diverse and heterogeneous group of people of all ages, incomes, education levels, sexes, vulnerable or not users, with or without disabilities, etc. In particular, among the 28 non-expert participants, there were two people with disability, mothers with strollers, students, retired, people from various municipalities of the Metropolitan Area of Thessaloniki, etc.

The 28 applicants who were selected as "Members of the Panel" received, along with the email of acceptance, the "inspiration material", which included information about the workflow of the meetings as well as thoughts/pictures/definitions/interpretations of a sustainable future on transport. It was considered as a "warm-up material," which should be provided to the participants prior to the workshop in order for them to be prepared and informed about the workshop's theme and their expected contribution [8]. It is worth mentioning at this point that a special logo was created for the Panel (see Fig. 2a) in order to nudge its Members and engage them more with their "mission" by developing the feeling of belonging in a team and being part of a formal process. This logo was designed to be used in any local Panel-related activities that would take place in the frame of the PE4Trans project. Moreover, 10 citizens were appointed as alternates.

After the establishment of the Citizens' Panel, the preparation of its physical meetings followed. They were realized in the form of co-creation workshops using some of the existing relevant methods and techniques found in the literature [9, 10, 11]. In general, the methods used at each workshop should be carefully selected considering its goal and objectives, the background of the participants, the theme, the time available, and any other specific requirement or resource. The methods used for the meetings of the local Panel on Sustainable Mobility in Thessaloniki are described below.

2.3 Analysis of the Citizens' Panel's Meetings (*Implementation Phase*)

Since 2018, that the Citizens' Panel on Sustainable Mobility was established in Thessaloniki, three physical meetings took place, another one is planned to be organized in May 2020, and the last one in spring 2021. The process followed for their implementation, their goal, and the connection of their sequence is shown in Fig. 1.

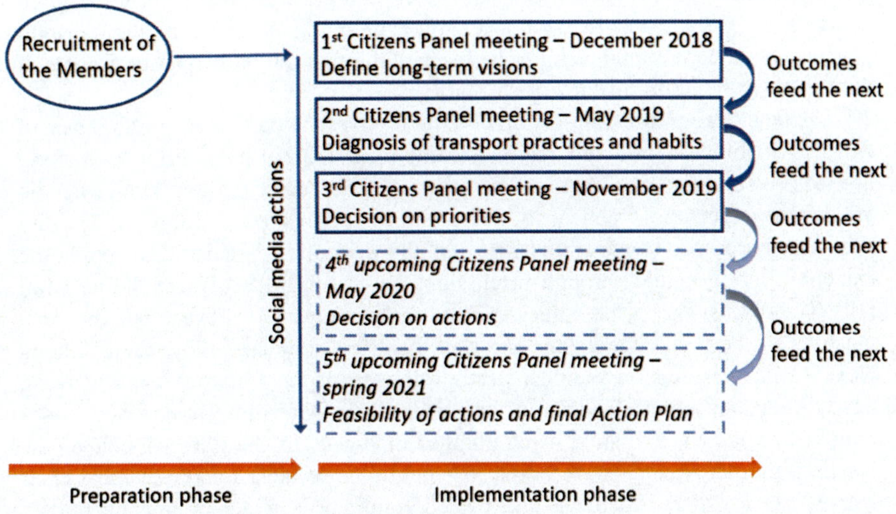

Fig. 1. The process followed for the accomplishment of the Citizens' Panel "mission" and aims.

The meetings were (and are going to be) held in the form of workshops applying co-creative methods in order to adopt a bottom-up planning approach and stimulate everyone's active participation. The participants were divided into **working groups** during the workshops. Highly experienced facilitators and table moderators were appointed to manage the diversity of the participants' profiles, guide the working groups, and collect all input. Other implementation details of each one of the workshops are analyzed below and indicative photos are depicted in Fig. 2.

The social media page created especially for the Citizens' Panel was fed with relevant content for the engagement of the participants between the workshops.

- **1st Citizens' Panel physical meeting (CP1):** As seen in Fig. 1, the aim of the first meeting was to define the long-term **visions** for a sustainable future. After a short presentation for the familiarization of the participants with the concepts of visions and Sustainable Mobility, they were asked to imagine and describe in the form of a vision the main characteristics of the ideal local situation related to mobility, but reflecting all aspects as quality of life, environment, security, etc. They were encouraged not to limit their imagination and present any idea, feasible or not. Inspiring pictures to foster "out of the box thinking" were distributed. Even so, it was found that it might be a bit difficult for citizens to envision the future freely and without restrictions from the present. The method used for the core co-creation session was **brainstorming** in the working tables. This technique is usually used for the stimulation of creative thinking and the generation of many ideas in a short period of time [11]. The results of each table group were presented in plenary. The **Dot-voting** method was implemented for the prioritization of the visions derived: all visions were listed on a board, and participants voted for their favorites by placing colored stickers-"votes" on them. It is considered one of the easiest and fastest methods for the evaluation of ideas [11]. Then, each one of the top 5 visions was assigned to a working table to be further elaborated. These top 5 visions, presented in the next subsection of this paper, comprise the first step in the road-mapping process towards the Action Plan for the improvement of public transport policies, which is the final result to be reached through the participatory process of the Citizens' Panel on Sustainable Mobility.
- **2nd Citizens' Panel meeting (CP2):** CP2 focused on the participatory **diagnosis** of citizens' mobility habits and regional component issues/needs. Prior to making decisions on priority axes or proposing actions, it is necessary to gain knowledge on the occurred transport practices as well as on the motivations behind the mobility behaviors observed. Towards this direction, the workshop started with mapping the citizens' driving values through discussions based on the **Schwartz Value Map** [12]. The discussion on the values enabled the participants to investigate the deep meaning of them and understand the way their decisions are influenced by them. Next, **brainstorming** on regional needs followed in the frame of each working table. After presenting all ideas in the plenary, the **Dot-voting** took place, and a joint list of the most important **challenges** emerged. In the next session of the meeting, the participants were asked to redefine those most important local challenges, as, in order to achieve unexpected solutions, a wider or unconventional problem definition is needed. This session was realized in three rounds using the

World Café method [13, 14]. Each table was assigned with a different topic, the participants changed tables at every round in order to be involved with all the topics, and the "host" of each table remained at the same table throughout the process in order to help the moving "guests" to be familiarized to the discussed concepts. A different problem-solving technique was implemented per round. During the first round, the participants shared thoughts on the challenge that was assigned to their team using the 1st technique. For the second and third round, they moved clockwise to the next table, according to the World Café method, in order to enrich the topics initially assigned to the other teams, while the 2nd and 3rd selected techniques were used respectively for the analysis.

- **3rd Citizens' Panel meeting (CP3):** This workshop was focused on setting **priorities** as well as taking some initial decisions on the transport behaviors of the **target groups**, which require a change in order to bring optimal impact. Taking the participatory planning one step further, key stakeholders of the city's transport sector were also involved and worked jointly with the Citizens' Panel. A presentation was delivered as a reminder of the work done so far by the Citizens' Panel on Sustainable Mobility. In order to outline the target groups of urban travelers to focus on, the participants worked in plenary and clustered all the ideas that emerged. As a result, six (6) "personas" were co-created, each one with specific characteristics. They were further analyzed using the **Empathy Map (EM)** technique [16]: the EM template (see Fig. 2c) was gradually filled in three rounds by including the persona's profile, what he/she hears, feels and thinks, sees, says and does, what are his/her pains and gains. This technique aims to understand the target group's professional and personal characteristics, needs, motivations, and influential factors. Following the **World Café** method, the participants changed tables clockwise in each of the three rounds, while the table moderators always stayed at the same table as the "host" of the next moving working group. During the last round, they were asked to summarize the conclusions of the empathy map analysis in one sentence in the form presented in the next subsection.
- **4th Citizens' Panel meeting (CP4):** The objective of the workshop is to design and to agree on **actions** addressing the target groups of CP3 while considering the visions of CP1. Its implementation is planned to take place in May 2020. However, it has already been designed. Firstly, a presentation on theoretical psychological background regarding how to achieve a change of citizens' behaviors and habits will be delivered to equip the participants with fundamental principles on the subject. Then, the **World Café** method will be implemented like in the previous meetings: participants will work in groups changing table clockwise at every round to identify barriers and drivers and develop a portfolio of actions. This way, everyone will contribute to the ideas for actions dedicated to each target group. Table moderators will ensure whether the actions are in line with the barriers and drivers of the target group.
- **5th Citizens' Panel meeting (CP5):** The objective of the 5th and final meeting is to validate the proposed actions of CP4 in terms of **feasibility**. In addition, it aims to **fine-tune** the outcome (Action Plan) and citizens' contribution to the participatory process of the Citizens' Panel on Sustainable Mobility of this paper.

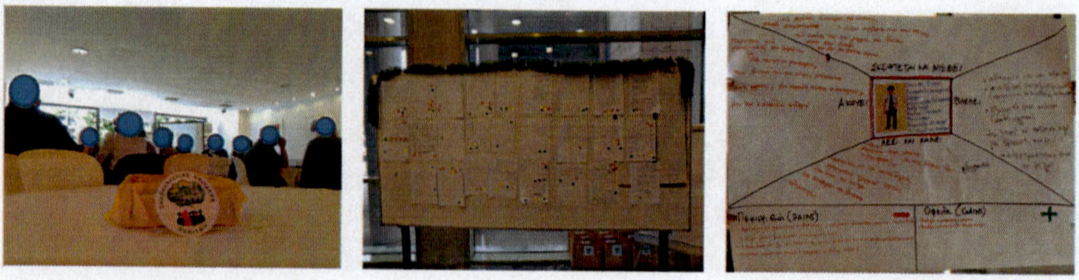

a b c

Fig. 2. a) Citizens' Panel special logo in front of the working groups, b) Example of dot voting session, and c) Example of the empathy map method.

2.4 Results of the Citizens' Panel's Meetings

After the implementation of each Citizens' Panel's meeting, the results were elaborated by the project team, and they were further analyzed to achieve the continuity of the process and the desired built-on effect. Thus, according to the objective set and the expected form of outcome of each workshop, the following findings were derived:

- **1st Citizens' Panel physical meeting (CP1):** The top 5 voted **visions** were:
  —*RE-THESS*: technology-oriented vision focusing on the use of smart solutions on transport towards a REsilient, REnnovative, REvolutionary Thessaloniki.
  —*A sustainable city with the inclusion and equal participation of all citizens:* the main idea was accessibility; inclusion of all citizens' needs to the transport policies and strategies along with the establishment of participatory processes.
  —*A car-free city*: car-free historical center of Thessaloniki with green spaces.
  —*THESS-ALAS:* exploitation of the area of Thermaikos bay with the development of sea transport.
  —*A city that starts with a "C," I choose to stay in my city!!:* City in a two-way relationship with its inhabitants: "we take care of our city so that our city takes care of us" motto.
- **2nd Citizens' Panel meeting (CP2):** The most important **values** that were recognized to motivate citizens and influence their automatized or rationalized decision-making behaviors were freedom, respect to everyone's needs in terms of equality and accessibility of people with disability, protection of the environment, security and safety connected to personal responsibility, health, and education. The main **challenges** considered to be of highest priority were the following:
  - Appropriate *education* and *awareness-raising* regarding every aspect of life and mobility, such as proper behavior, change of mentality/mindset of road users, mobility habits, environmental protection, etc.
  - *Restructuring and advancing the public transport modes* with emphasis on innovation and Intelligent Transportation Systems (ITS) [17, 18]. Moreover, general re-organization was discussed extensively regarding cleanliness, frequency, accessibility, reliability, new transport modes, multimodality, etc.

- *Infrastructure* of poor quality that is insufficient and discourages pedestrians and cyclists from using it. Consequently, it ends up operating against the desirable shift towards environment-friendly means of transport.
- Lack of proper *organization, holistic,* and integrated strategic planning.

Emphasis, also, was given to the need for supporting academic research with the view to empower and establish a sustainable transport system in the city.

- **3rd Citizens' Panel meeting (CP3):** By the end of CP3, the whole analysis conducted through the empathy maps for each **target group** was summarized in one statement in the form: "*We want to change [current behavior] of [target group] so that they [positive behavior].*" The main **statements** produced towards the achievement of behavior change for more sustainable mobility habits were:
  - We want to change the daily car use of *parents with little kids* (persona Urania) so that they increase the use of public transport or bicycles due to environmental and health reasons aiming to set a good example for their children.
  - We want more awareness-raising campaigns for drivers and citizens regarding mobility for *people with difficulties (elderly) and disability* (persona Danae).
  - We want to change *teenagers'* (persona Leonidas) influencers. We want to multiply "good" behaviors among teenagers in order to influence their friends' behavior so that they move al together, taking advantage of vehicle sharing.
  - We want to change the behavior of *people who choose the car for trips for personal purposes* (persona Alexander) to influence them to commit themselves to use public transport with a specific X frequency per month or week.
  - We want to change the daily transport routine of *people living in areas outside of the public transport system network* (persona Persephone) so that they can coordinate their trips with their neighbors and carpool to work.
- **4th Citizens' Panel meeting (CP4):** The expected outcome is a pool of **actions** focusing on the aforementioned target groups, keeping in mind the visions of CP1, the challenges of CP2, and the statements of CP3 as general guidelines. These aim to improve the current policies by contributing to the task of policymakers in order to enhance sustainable public transport and change citizens' transport behaviors.
- **5th Citizens' Panel meeting (CP5):** Based on the financial and institutional restrictions, not all actions designed during CP4 will be feasible to be incorporated into the Action Plan. A careful selection will take place. As a result, the outcome expected to be derived from CP5 is to fine-tune the actions selected by the pool of CP4 and to agree on the final form of the **Action Plan** in terms of feasibility.

3 Discussion and Conclusions

The workflow adopted a holistic approach of public engagement in policy design processes starting from defining the long-term direction *(visions – the WHY),* recognizing the current situation *(motivating values and local challenges – WHERE we stand),* going on to identifying and analyzing the involved stakeholders *(target groups – WHO)* until deciding on the measures should be taken *(actions – HOW).*

The co-creation meetings were successfully organized and implemented so far. The venues of such meetings should be carefully selected to meet the functional criteria considering the number of the participants, the accessibility, as well as the aesthetic to enhance their creative way of thinking (e.g., CP2 took place in an art gallery).

The invitation process for the recruitment should be planned in detail beforehand in order to ensure the broad range of participants and should be launched much earlier than the 1st meeting. The diversity of profiles of those selected as "Members" in our Panel enabled the identification of various perspectives on a series of topics and encouraged the debate, which is essential for the consensus process to follow. However, keeping them engaged is the most challenging task, and much effort and time should be dedicated to it. A strategic plan was set up during the preparation phase by our project team to provide incentives for participation for each meeting, such as farewell gifts through a lottery and engaging/dissemination material with the special CP logo. Even so, their commitment to attending all the meetings was not ensured. Furthermore, posts regarding the outcome of the meetings, photos, and relevant news on sustainable mobility are uploaded on the Citizens' Panel social media page to keep the interest alive, but the interaction and dialogue triggered are limited. Unavoidably people who are not familiar with the Internet and social media are excluded by these digital initiatives.

Regarding the people's opinion on the process, the results of the evaluation questionnaires distributed at the end of the co-creation workshops demonstrate the high satisfaction rate of participants and their willingness to be involved in other relevant initiatives, too. It is true that the interest shown by the citizens exceeded by far the initial expectations in a country like Greece with no strong participatory culture.

The overall conclusion through Citizens' Panel on Sustainable Mobility is that the participatory concept can be wisely exploited in the fields of transport as long as interaction and actual consideration are ensured.

Acknowledgments. This paper is based on the findings of the research activities of the project "Public Engagement for Sustainable Public Transport" (PE4Trans), which is co-funded by the European Union through the European Regional Development Fund (ERDF) and national funds, under the territorial cooperation Interreg Europe Programme 2014–2020 [19].

References

1. Leyden, K.M., Slevin, A., Grey, T., Hynes, M., Frisbaek, F., Silke, R.: Public and stakeholder engagement and the built environment: a review. Curr. Environ. Health Rep. **4** (3), 267–277 (2017). https://doi.org/10.1007/s40572-017-0159-7
2. Cascetta, E., Pagliara, F.: Public engagement for planning and designing transportation systems. Procedia Soc. Behav. Sci. **87**, 103–116 (2013). https://doi.org/10.1016/j.sbspro.2013.10.597
3. Naniopoulos, A., Genitsaris, E., Stamelou, A., Kostopoulos, I., Aissati el., H., Schmitz, W.: Co-creating innovation: concepts and ideas for Public Transport resulted through participatory processes applied in four different urban areas of Europe. In: Proceedings of the 7th Transport Research Arena (TRA 2018), Vienna, Austria, 16–19 April 2018

4. Quick, K.: Public participation in transportation planning. In: Garrett, M. (ed.) Encyclopedia of Transportation: Social Science and Policy, pp. 1132–1137. Sage Publications, Thousand Oaks, CA (2014)
5. Hellenic Statistical Authority Webpage: 2011 Population-Housing Census. https://www.statistics.gr/en/2011-census-pop-hous. Accessed 20 Mar 2020
6. Mitsakis, E., Stamos, I., Salanova Grau, J., Chrysochoou, E., Iordanopoulos, P., Aifadopoulou, G.: Urban mobility indicators for Thessaloniki. J. Traffic Logistics Eng. **1** (2), 148–152 (2013). https://doi.org/10.12720/jtle.1.2.148-152
7. Office of Urban Resilience of the Metropolitan Development Agency of Thessaloniki: Preliminary resilience assessment of Thessaloniki. Office of Urban Resilience of the Metropolitan Development Agency of Thessaloniki, Thessaloniki, Greece (2016). https://drive.google.com/file/d/0B_zbMuz0nwfxMDE2Z2w1aWNxRnc/view
8. Bertini, P., Plumpley, E.: Co-creation: designing with the user, for the user, 2 December 2014. UX Booth Webpage. https://www.uxbooth.com/articles/co-creation-designing-with-the-user-for-the-user/. Accessed 20 Mar 2020
9. DesignKit Webpage: Co-creation session. https://www.designkit.org/methods/33. Accessed 30 Mar 2020
10. Stratigea, A.: Theoria kai methodoi symmetochikou schediasmou [Theory and methods of participatory planning]. Hellenic Academic Libraries Link, Athens, Greece (2015). https://hdl.handle.net/11419/5428
11. Genitsaris, E., Stamelou, A., Nalmpantis, D., Naniopoulos, A.: Applying the co-creation process for improving Public Transport in the city of Thessaloniki. In: Proceedings of the 8th International Congress on Transportation Research (ICTR 2017), Thessaloniki, Greece, 28–29 September 2017
12. Schwartz, S.H.: An overview of the schwartz theory of basic values. Online Readings Psychol. Culture **2**(1), 1–20 (2012). https://doi.org/10.9707/2307-0919.1116
13. The World Café Webpage: World Cafe method. https://www.theworldcafe.com/key-concepts-resources/world-cafe-method/. Accessed 31 Mar 2020
14. Genitsaris, E., Roukouni, A., Stamelou, A., Nalmpantis, D., Naniopoulos, A.: Co-creating innovative concepts to address crucial trends and challenges that public transport faces in Thessaloniki. In: Ketikidis P., Solomon, A. (eds.) University-Industry Links: Coproducing Knowledge, Innovation & Growth: Proceeding of the 10th International Conference for Entrepreneurship, Innovation and Regional Development (ICEIRD 2017), pp. 159–166. ICEIRD, Thessaloniki, Greece (2017). https://iceird.eu/2017/conference-proceedings
15. Mind Tools Webpage: Problem solving: solving complex business problems. https://www.mindtools.com/pages/main/newMN_TMC.htm. Accessed 31 Mar 2020
16. Ferreira, B., Silva, W., Oliveira, E., Conte, T.: Designing personas with empathy map. In: Proceedings of the 27th International Conference on Software Engineering and Knowledge Engineering (SEKE 2015), Pittsburgh, PA, 6–8 July 2015. https://doi.org/10.18293/SEKE2015-152
17. Tsafarakis, S., Gkorezis, P., Nalmpantis, D., Genitsaris, E., Andronikidis, A., Altsitsiadis, E.: Investigating the preferences of individuals on public transport innovations using the Maximum Difference Scaling method. Europ. Transp. Res. Rev. **11**(1), 3 (2019). https://doi.org/10.1186/s12544-018-0340-6
18. Nalmpantis, D., Roukouni, A., Genitsaris, E., Stamelou, A., Naniopoulos, A.: Evaluation of innovative ideas for Public Transport proposed by citizens using Multi-Criteria Decision Analysis (MCDA). Europ. Transp. Res. Rev. **11**(1), 22 (2019). https://doi.org/10.1186/s12544-019-0356-6
19. Public Engagement for Sustainable Public Transport Homepage. https://www.interregeurope.eu/pe4trans/. Accessed 31 Mar 2020

Investigating the Travel Information-Seeking Behavior for Daily Trips in a Greek Medium Sized City

Maria Karatsoli(✉) and Eftihia Nathanail

Department of Civil Engineering, University of Thessaly,
Pedion Areos, 38334 Volos, Greece
makarats@uth.gr

Abstract. The deployment of Information and Communication Technologies (ICTs) have created new options for travel information seeking. Dissemination of travel information to travelers could have immediate benefits such as reduction of travelers' uncertainty, congestion alleviation, reduction of travel time and cost. However, daily trips in a medium sized city are often based on habitual patterns and not on received travel information. By addressing the factors that urge the users to access travel information in a medium sized city, an information provision strategy could be developed, to serve the needs of the commuters and improve the daily traffic situation. The main objective of this paper is to investigate the factors that affect the travel information seeking in the urban environment of Volos city, in Greece. A digital questionnaire was formulated to investigate the seeking of travel information for daily trips and its impact on commuters' mobility and travel choices. The final sample size comprised 108 users. The analysis of the data was done through descriptive and inferential statistics. Results showed that online sources of travel information are more preferable by the respondents. Moreover, information about traffic congestion or rerouting is often used by non-sustainable mode users, while sustainable mode users are more interested in information about public transport itineraries.

Keywords: Urban mobility · Journey planning · Travel information

1 Introduction

Unfamiliar destinations, unknown traffic conditions and unreliable public transport timetables could lead to time-consuming and unsafe daily trips. Travel information contributes to a better perception and analysis of a travel situation. This information could change travel behavior and lead to the enhancement of convenience, safety and efficiency of travel. An optimized travel plan could offer a faster, safer and more environmentally friendly trip. Travel information systems encompass a wide range of technologies aiming at the provision of up to date travel information to travelers. Navigation applications that inform about route alternatives, real-time information about public transport and information-based strategies on social networks encourage commuters to make right travel choices. Although commuters make their own trip

E. G. Nathanail et al. (Eds.): CSUM2020, AISC 1278, pp. 674–683, 2021.
https://doi.org/10.1007/978-3-030-61075-3_66

choices, these decisions affect the whole network and consequently other commuters at a given time.

The potential of travel information to affect travelers' behavior led to an increased interest in its role on travelers' choices [1, 2]. The impact of traffic information on travel choices is well stated also in literature [3, 4]. The study of Asselin et al. [5] suggests that the use of new technologies could have immediate results and can assist with trips that necessitate familiarity with roads [6]. However, when travel information is available travelers have to decide whether to follow it or not. The reliability of information is a key feature on individuals' decision making. A reliable information system is important, since inconsistent information may have reverse results and reduce the probability of complying with it [7]. In this study, it was found that travelers are more likely to follow pre-trip information compared to en-route information.

Regarding the factors that affect travel decisions, previous studies found that socio-economic characteristics [8], travel- related characteristics, [9, 10] and commuters' attitudes [3, 11] play an important role on travelers' final decision. The study of [12], proved that commuters with flexible work schedule are more likely to change their daily activities based on available information. Godwin et al. [13] showed that pre-travel information seeking behaviors of public transport users are affected by trip frequency, sociodemographic characteristics, trip characteristics and information source. In their study, it was also mentioned that the understanding of smartphones or other devices usage for information seeking, is important for the development of information provision strategies that serve the needs of public transport passengers.

In this study, an attempt is made to investigate the factors that cause the travel information seeking in a medium sized city. To achieve that a questionnaire survey was distributed in the Polytechnic School of University of Thessaly. Aim of the research is to identify to what extend travel information is used in an urban environment and which are the reasons for that. The frequency that users make changes in trips and plans when travel information is available is also examined. By addressing the factors of travel information seeking, information providers could be able to understand better what commuters search for and when. The findings could lead to an improved service delivery [14].

2 Methodology

2.1 Survey Design and Data Collection

An online survey was carried out to investigate the factors that affect the travel information seeking in the urban environment of a medium-sized Greek city. The questionnaire was distributed in addition to a GPS- travel diary survey, in order to collect socio-demographic and trip information and to investigate the impact of travel information. The survey was built on Survey Monkey, owing to its integrated and simple environment that the software offers. It was disseminated to students and employees of the Polytechnic School of University of Thessaly, in Volos, Greece. The questionnaire was written in both Greek and English, since the survey was disseminated also to Erasmus students. The questionnaire collection period was approximately one month, from the beginning of December 2019 till beginning of January 2020, while the promotion of the survey was carried out during transport-related lectures and in the public spaces of the Polytechnic school.

The questionnaire consisted of three parts. The first part referred to the daily trips of the participants, in which data regarding the trip distance and duration from and to University, the flexibility at arrival time and the familiarity with traveling in Volos city, were collected. The next part aimed to investigate the intention and the factors that affect the travel information seeking. In parallel, the impact of travel information on commuters' mobility and travel choices was investigated. The last part recorded the socio-economic characteristics of the respondents, by collecting personal information such as gender, age, education level, employment status, etc.

2.2 Data Analysis

The analysis of the data was done through descriptive and inferential statistics. In the first case, sample characteristics, such as age, gender and occupation were addressed by estimating the frequency distribution per characteristic and the median values. In the second case, the statistical analysis of the responses was realized using non-parametric tests. The Mann-Whitney two-sample U-test was performed to assess differences between the samples in characteristics measured on the 5-point scale. Participants responded by rating items ranging from 1 (lower scores) to 5 (higher scores) with 3 as a midpoint. This scale was used as it increases the variance in the measurements and allows a greater differentiation in the results [15]. The familiarity with traveling in Volos city, the importance of different factors in travel route choice and the impact of information on daily trip decisions were measured on the 5-point scale (Not at all 1, Slightly 2, Moderately 3, Very 4, Extremely 5). The participants responded on a 5-point scale (Never 1, Seldom 2, Sometimes 3, Often 4, Always 5) if they seek any travel information when traveling in Volos city. The rating of the main daily commute as regards characteristics such as shortest route, environmentally friendly route, economical route, safety, comfort etc. was measured on a 5-point Likert scale (Very dissatisfied 1, Dissatisfied 2, Slightly satisfied 3, Satisfied 4, Very satisfied 5).

3 Results

3.1 Sample and Daily Trip Characteristics

The current survey was distributed in addition to a GPS- travel diary survey, in order to collect socio-demographic and trip information and to investigate the factors and impact of travel information seeking. Specifically, it was decided to recruit only people from the University community as they would be easier to be reached and more committed to participate. Hence, all respondents were students and employees of the Polytechnic School of University of Thessaly, in Volos, Greece. The final sample size comprised 108 users, who fully completed the questionnaire. 52% of them are men and the rest 48% are women. 84% of the respondents belong to the age group of 18–23, 13% of them to 24–40 and 3% to 41–65. In addition, 83% of the participants are students, 8% are students with a part-time job, 4% are students with a full-time job, 3% belong to University staff and the rest 2% are researchers. The high percentages in the age group of 18-24 years old and in the occupation group of students was expected as the questionnaire survey was carried out at the university campus of the Polytechnic school. The place of origin of 23% of participants is Volos and the rest 77% are from other cities. 95% of the respondents live in Volos. 30% of them live in Volos for less than 4 years, 42% for 4 to 6 years and 28% for more than 7 years.

Table 1 summarizes the characteristics of respondent's daily trips. The familiarity with traveling in Volos city, the flexibility in arrival time at workplace/university and the ownership of driver's license are also included. Most of the participants use car (32%) or walking (31%) for their daily trips. 27% of the participants stated that are extremely familiar with traveling in Volos city and 33% that have not driver's license.

3.2 Preferable Source and Type of Travel Information

Conventional data sources such as radio, Advanced Traveler Information Systems (ATIS) such as Variable Message Signs (VMS) and in-car navigation, ICTs, such as smartphones and mobile devices with Internet connection, compose a rich information landscape that travelers can benefit from. Travelers often consult multiple sources to get enough information and assistance for their trips [16].

The respondents were asked about the preferable information source for receiving information when travel to three destinations (residence, activity location and workplace/university). The respondents could choose more than one of five sources (navigation applications i.e. Google maps, radio, social media, VMS, in-vehicle GPS navigator) or none if they do not prefer to use any information. 47% of them stated that they prefer to use information from navigation applications when travel to their residence and 39% responded that they don't prefer to use any information on their way back home. 58% stated that they prefer to use navigation applications when travel to an activity location or to workplace/university. Social media was a preferable source for participants that travel to an activity location with percentage 28%. 12% and 27% of the participants stated that they don't prefer to get any information on their way to an activity location or workplace/ university, respectively. Radio, VMS and in vehicle GPS navigators were the least preferred sources for all destinations, with percentages less than 1%. The results are in line with findings in literature, in which travelers prefer more straightforward sources of information [7]. In the context of travelers' information-seeking behaviors, the shift from conventional travel information sources to online sources is well researched in the study of Kambele et al. [17] and Kim et al. [18].

The participants were asked to select one or more types of information that would like to receive before and/or during their daily trips. 67% stated that would like to receive information about traffic conditions, 65% about travel times, 56% about traffic incidents, 52% about alternative routes, 41% about economic ways of travel and 31% about environmentally friendly routes.

Table 1. Trip characteristics.

Variables	Level	%
Transport mode	Car (Diesel,Gasoline/electric, hybrid or new technology)	32/1
	Motorcycle	7
	Bicycle	12
	Electric bicycle	1
	Public Transportation	16
	On foot	31
Trip distance estimation from residence to workplace/university	<= 2 km	53
	>2 km	47
Trip duration estimation from residence to workplace/university	<= 10 min	38
	10 < min <=15	37
	>15 min	25
Trip distance from workplace/university to residence	<= 2 km/> 2 km	50
	>2 km	50
Trip duration estimation from workplace/university to residence	<= 10	40
	10 < min <=15	31
	>15 min	29
Familiarity with traveling in the city of Volos	Not at all	1
	Slightly	6
	Moderately	32
	Very	34
	Extremely	27
Is your arrival at workplace/university flexible?	Not at all	4
	5–15 min	65
	16–30 min	9
	31–60 min	3
	I work with a flexible schedule	19
Driver's license	No	33
	Yes, owner less than 1 year	8
	Yes, owner from 1 year to 5 years	44
	Yes, owner more than 5 years	15

3.3 Importance and Rating of Trip Characteristics

The Importance Performance Analysis (IPA) was used to determine the importance and the quality of participants' daily trips. The respondents were asked to rate the importance and the performance of their commuting characteristics. Ten attributes of a daily route were examined. The participants rated their daily routes in terms of comfort, safety, transport mode, travel time, potential delays, traffic congestion and in terms of shortest, environmentally friendly, economical and scenic route. The importance was measured on a 5-point scale (Not at all 1, Slightly 2, Moderately 3, Very 4, Extremely

5) as well as the performance (Very dissatisfied 1, Dissatisfied 2, Slightly satisfied 3, Satisfied 4, Very satisfied 5).

The IPA graph consists of four quadrants and each of them suggests a different approach. The Quadrant I includes attributes with the greatest weakness that require actions in order to improve the quality and the performance of the attributes. Attributes that fall in Quadrant II have great quality which should be preserved. The Quadrant III includes attributes where low priority is given by the respondents. Attributes positioned in the Quadrant IV are less important for the respondents but of high performance [19].

The following graph depicts the ten characteristics that were examined in this study. The environmentally friendly route attribute falls in the Quadrant III, showing that the respondents give less priority in such a route, rating also its quality as low. Respondents' trips perform low quality in terms of traffic congestion, potential delays and scenic features. Despite the fact that participants stated that these characteristics are moderately important for them, a slight change could show that these attributes require actions for improvement. Travel time, safety and shortest route attributes play an important role for the respondents, so actions should be kept in order to preserve their high performance. Conclusions about the transport mode, comfort and economical route attributes cannot be safely drawn in this study (Fig. 1).

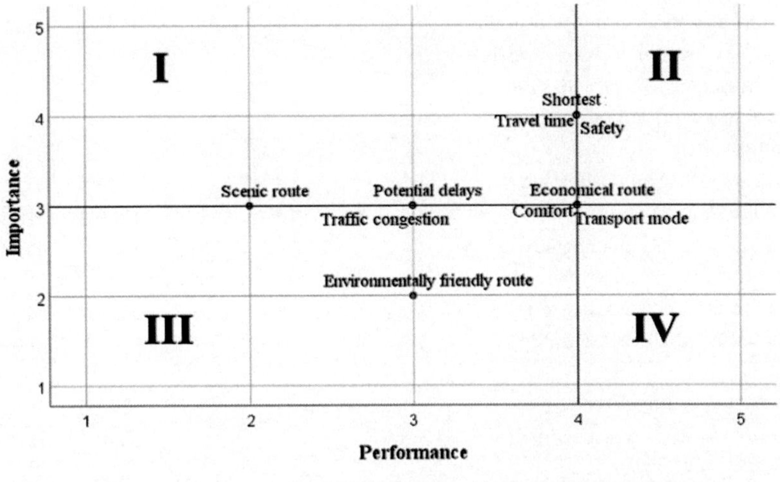

Fig. 1. Importance of commuting characteristics and rating of current situation.

3.4 Travel Information Seeking Parameters' Rating

Parameters related to the frequency of travel information seeking between different origin and destination points (four parameters) and based on different factors (eleven parameters) were further examined. The rating was measured on a 5-point scale (Never 1, Seldom 2, Sometimes 3, Often 4, Always 5). The sample was separated in non-sustainable and sustainable mode users. The separation was done in order to check to what extent the parameters related to travel information seeking affect the sustainable and non- sustainable mode users. The category of sustainable mode users includes the participants that use sustainable modes for their daily commute. In this survey, public transport, cycling, riding scooter and walking were considered as sustainable modes of transport. The category of non- sustainable mode users includes participants that use non- sustainable mode of transport for their mobility (car, motorcycle or taxi). 60% of the participants use a sustainable mode of transport for their daily trips and the rest use a non-sustainable way of transport. The exact percentages for each mode of transport are included in Table 2.

Table 2. Parameters' rating between non-sustainable and sustainable mode users.

Parameters	Groups			
	Non-Sustainable	Sustainable	N-S vs. S	
	Median	Median	U	p-value
Do you get travel any information from:				
-your residence to your workplace/university	1	1	1046.5	**0.008***
- your workplace/university to your residence	1	1	1089.0	**0.008***
- your residence to an activity location	2	3	1048.0	**0.022***
- your activity location to your residence	2	2	1185.5	0.158
When traveling in the city of Volos, to what extend would you seek information				
- to reach an unknown destination	4	4	1315.0	0.590
- to find the shortest route	2	3	1286.5	0.469
- to find the most environmentally friendly route	2	2	1329.5	0.649
- to find the most economical route	3	2	1331.5	0.680
- to stay informed about the traffic congestion	3	2	1063.5	**0.029***
- to estimate your travel time	3	3	1264.0	0.381
- to identify the ideal transport mode according to your needs	2	2	1038.5	**0.019***
- to be informed on public transport itineraries	2	3	892.5	**0.001***
- to decide upon your departure time	3	3	1209.5	0.223
- for dynamic re-routing	3	2	1074.0	**0.035***
- when reliable information is provided	3	3	1149.0	0.108

M: average rating, SD: Standard Deviation, *statistically significant (p-value < 0.05)

Table 2 presents an overview of the median values of the fifteen variables and the test results of the comparisons between non-sustainable and sustainable mode users.

Results are described through the U and p-value, indicating the strength of the respective evidence. According to Table 1, seven out of the fifteen hypotheses tested were found significant. Results showed that both non-sustainable and sustainable mode users never seek any travel information from their residence to their workplace/university and vice versa. However, this is not the case when it comes to traveling from residence to an activity location. Sustainable mode users get travel information more often compared to non-sustainable mode users. Non-sustainable mode users would seek more often for information about traffic congestion and dynamic re-routing when traveling in Volos compared to non- sustainable mode users. This can be explained due to the fact that non- sustainable modes are mainly affected by congestion. Sustainable mode users would search information about public transport itineraries more often compared to non-sustainable mode users. The fact that public transport users belong to this category can explain this frequency between the two groups. Finally, both groups would seldom seek information to identify the ideal transport mode according to their needs when traveling in Volos city.

3.5 Impact of Travel Information on Commuters' Decisions and Travel Choices

Concerning the possible influence of travel information on commuters' final decisions, the respondents were asked to rate on a 5-point scale (Not at all 1, Slightly 2, Moderately 3, Very 4, Extremely 5), to what extend their choices may be affected by travel information. Figure 2 shows the number of participants per scale point for six different aspects of changes on travel choices. Based on the figure, the availability of travel information, can affect users more often in choosing a particular route. To a smaller extend, the travel information can make users to cancel their plans or change their activities. This may be related to the fact that in both cases radical changes in the initial plans are required.

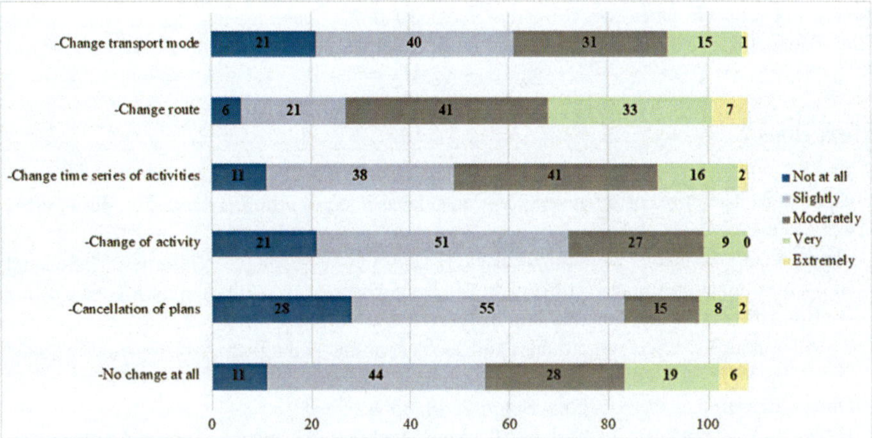

Fig. 2. Impact of travel information on participants' travel choices.

4 Conclusions

The study focuses on identifying the factors that affect travelers' information seeking in a medium sized city and the impact of available information on travelers' daily trip schedule. Findings of the questionnaire survey indicate that users mostly prefer online sources of travel information such as navigation applications or social networks compared to conventional sources. Most of the respondents stated that they are interested in information about traffic conditions and travel times. A shortest route, safety and travel time were rated as very important by the respondents. However, an environmentally friendly route is slightly important for them. Results showed that both non-sustainable and sustainable mode users never seek travel information from their residence to their workplace/ university and vice versa. Sustainable mode users get information from residence to an activity location more often compared to non-sustainable mode users. Regarding the impact of information on participants' choices, a route change or a change of time series of activities is more possible compared to cancellation of plans or change of activity. Information providers should take into account commuters' and trip characteristics in order to develop a reliable information system. Factors that lead to travel information seeking reveal the type and the way how information should be provided in order the commuters to follow it.

4.1 Limitations

This study entails limitations. The survey was carried out at a University campus, thus the majority of the respondents are students and belong to the young age group. This may bias the results while it is believed that they do not allow sufficient investigation of the behavioral differences among respondents of different occupations and ages.

Acknowledgements. «This research is co-financed by Greece and the European Union (European Social Fund- ESF) through the Operational Programme «Human Resources Development, Education and Lifelong Learning» in the context of the project "Strengthening Human Resources Research Potential via Doctorate Research" (MIS-5000432), implemented by the State Scholarships Foundation (IKY)».

References

1. Avineri, E.: The effect of reference point on stochastic equilibrium. Transp. Sci. **40**(4), 409–420 (2006)
2. Chorus, C.G., Arentze, T.A., Timmermans, H.J.P.: A comparison of regret-minimization and utility maximization in the context of travel mode-choices. In: 87th Annual Meeting of the Transportation Research Board, Washington, DC (2008)
3. Choocharukul, K.: Effects of attitudes and socioeconomic and travel characteristics on stated route diversion: structural equation modeling approach of road users in Bangkok. Transp. Res. Rec. J. Transp. Res. Board **2048**, 35–42 (2008)
4. Tseng, Y.-Y., Knockaert, J., Verhoef, T.: A revealed-preference study of behavioral impacts of real-time traffic information. Transp. Res. Part C **30**, 196–209 (2013)

5. Asselin-Miller, N., Biedka, M., Gibson, G., Kirsch, F., Hill, N., White, B., et al.: Study on the deployment of C-ITS in Europe: final report framework contract on impact assessment and evaluation studies in the field of transport study on the deployment of cooperative intelligent transport systems (C-ITS) in Europe (2008)
6. Chowdhury, S., Giacaman, N.: En-route planning of multi-destination public-transport trips using smartphones. J. Public Transp. **18**(4), 31–45 (2015)
7. Ramos, G., Frejinger, M. E., Daamen, W., Hoogendoorn, S.P.: A revealed preference study on route choices in a congested network with real-time information. In: Proceedings of the 13th International Conference on Travel Behaviour Research – Toronto, pp. 15–20 (2012)
8. Chorus, G.C., Molin, J.E.E., Van Wee, B.: Use and effects of advanced traveller information services: a review of the literature. Transport Rev. **26**(2), 127–149 (2006)
9. Tam, M.L., Lam, W.H.K.: Modeling the market penetration of personal public transport information system in Hong Kong. J. Intell. Transp. Syst. **9**(2), 81–89 (2005)
10. Al-Deek, H., Chilakamarri Venkata, S.R.C., Flick, J.: Impact of dynamic message signs on satisfaction and diversion of toll road travelers. Transp. Res. Rec. **2099**(1), 1–13 (2009)
11. Farag, S., Lyons, G.: What affects pre-trip public transport information use? empirical results of a qualitative study. In: Proceedings of the Transportation Research Board Annual Meeting, Washington DC (2008)
12. van Bladel, K., Bellemans, T., Janssens, D., Wets, G.: Activity travel planning and rescheduling behavior: empirical analysis of influencing factors. Transp. Res. Rec. **2134**(1), 135–142 (2009)
13. Yeboah, G., Cottrill, D.C., Nelson, D.J., Corsar, D., Markovic, M., Edwards, P.: Understanding factors influencing public transport passengers' pre-travel information seeking behavior. In Public Transport **11**(1) (2019)
14. Nelson, J.D.: Perfect information and ITS. In: Ison, S., Cowie, J. (eds.) The Routledge Handbook of Transport Economics, pp. 62–73. Routledge, Abingdon (2018)
15. Krosnick, J.A., Presser, S.: Question and questionnaire design. In: Wright, J.D., Marsden, P. V. (eds.) Handbook of Survey Research. Elsevier, San Diego (2009)
16. Tang, Q., Hu, X.: Triggering behavior changes with information and incentives: an active traffic and demand management-oriented review. In: Advances in Transport Policy and Planning, Chap. 8, vol. 3, pp. 209–250 (2019)
17. Kambele, Z., Li, G., Zhou, Z.: Travelers' information-seeking behaviors. J. Travel Tourism Mark. **32**(1–2), 141–152 (2015)
18. Kim, H., Xiang, Z., Fesenmaier, D.R.: Use of the internet for trip planning: a generational analysis. J. Travel Tourism Mark. **32**(3), 276–289 (2015)
19. Martilla, J.A., James, J.C.: Importance–performance analysis. J. Mark. **41**(1), 77–79 (1977)

The Rise of Run-Commuting as a Form of Transportation: Research on the Characteristics and Spatial Needs of These Trips

Apostolos Anagnostopoulos(✉)

Highway Laboratory, Department of Civil Engineering, Aristotle University of Thessaloniki, 54124 Thessaloniki, Greece
aposanag@civil.auth.gr

Abstract. Running has been totally integrated into the daily life of many people around the world. Every year amateur runners participate in running competitions while they are training in urban public spaces. In this framework, a new transportation practice involves running trips for commuting purposes. Several studies highlight walking and running as activities of major importance for a healthy life and a livable city. Run-commuting is a new trend that uses running as a form of transportation. However, the spatial needs of run-commuting activity and the infrastructural necessities are fully unknown yet. The main objective of this study is to understand and critically evaluate running as a mode of transport. The paper commences with the role of running in active transportation. The benefits, risks, and characteristics of run-commuting trips are presented according to the literature review. In the second part of the paper, real trajectories of run-commuting trips are analyzed. Findings indicate the behavioral patterns and safety issues of these trips. The paper closes with a discussion of the conclusions, including recommendations for the design of running-friendly cities.

Keywords: Run-Commuters · Healthy city · Public space · Fitness applications

1 Introduction

According to the World Health Organization (WHO) [1] a healthy and active city identifies and promotes the value of active living and physical activity. During the last years, environmental sustainability and people's health are getting great awareness. The weaknesses of rapid urbanization, especially on greenhouse gas emissions and the urban quality of life are getting attention more than ever. Transport is a kind of specification area that contributes significantly to global warming. Upon this, policy makers promote sustainable traffic modes, such as cycling electric scooters and walking, as a mode of transportation due to the benefits regarding traffic emissions and congestion.

E. G. Nathanail et al. (Eds.): CSUM2020, AISC 1278, pp. 684–693, 2021.
https://doi.org/10.1007/978-3-030-61075-3_67

The proportion of people exercising in informal ways like running, walking and cycling rapidly grew in the last decade. Running has been totally integrated into the daily life of many people around the world. Every year amateur runners participate in running competitions while they are training in urban public spaces. In this framework, a new transportation practice involves running trips for commuting purposes.

There is growing public interest in run-commuting. A lot of research has been conducted regarding walking and cycling commuting-activities. However, the spatial needs of runners and especially of run-commuting activity and the infrastructural necessities are fully unknown yet.

Data from smart applications used by the citizens of cities facilitate transportation planners and policymakers to create a new notion of cities. Therefore, cities can adapt to the behavior of their residents while their citizens are involved in the design process even if they do not know it. This is the general concept of Smart Cities. Fitness applications can be a useful tool regarding this process.

The main objective of this paper is to understand and critically evaluate running as a mode of transport. The method followed includes the following steps: i. Literature review about the role of running in active transportation. In this context, the benefits, risks, and characteristics of run-commuting trips are presented. ii. Case study analysis by means of spatial analysis of run-commuting trips. More particularly, an approach is proposed regarding the use of available fitness data to assist urban planners and transportation engineers to identify the spatial distribution of run-commuting trips in cities. iii. Discussion and development of recommendations for the design of running-friendly cities, based on the conclusions of the previous parts.

2 Running as a Form of Transportation

2.1 The Rapid Rise of Running Activity

Running activity has been widely known for years. Its beginning is related to competitive events over various distances. However, nowadays it represents one of the most popular leisure activities worldwide [2].

Back to 1896, running as an athletic event was part of the first Olympic Games in Athens. In this framework, running as a sport activity was only accepted among professional athletes. Running in public was considered as a waste of time. This situation was reversed on 1960. Recreational running and jogging became more popular and athletes started to train on public spaces. Health-related benefits of this physical exercise were more known to the public. Running activities were supported by politicians as well. The number of marathons' participants increased dramatically. However, the recession of 1990s can be characterized as a phase of stagnation of sports generally and thus of recreational running as well.

Since then, the situation has been reversed. The number of running events and participants has extremely been extended during the last decade. According to [3] running activity is characterized by two periods of extremely popularity. The first period rep-resents a product of the cultural and fitness revolution while the second one is characterized by social, health-oriented, and economic aspects as well [3].

During the second period and especially in the last decade, a new aspect of running activity has started which is related to commuting trips. According to Simon Cook [4] several moments of the mobility of everyday life can be characterized as forms of running-transport for many years now. People are in a hurry to cross the roads and catch buses for their commuting trips. These emergency-runners differ significantly from run-commuters. However, they are engaged in this form of transportation.

2.2 The Role of Running in Active Transportation

Definition. Active transport general refers to any form of human-powered transportation, such as walking, jogging, running, cycling, skating, etc. [5]. Run-commuting is a new transportation practice that has been identified the last years on developed countries [6]. It can be defined as the daily activity of going from home to work and vice versa by running or jogging.

Nowadays, more and more people have started to use the activity of running for commuting purposes beyond recreational purposes. Thus, this activity can be characterized as a mean of transport which looks like walking activity. However, even if the main difference between these activities is the speed, there are several constraints such as the physical fitness of commuters and the distance between work and home place.

Benefits. The benefits of active transport, especially of run-commuting, as they have been cited in the literature, include that:

Health. According to [7], healthy people are more inclined to travel actively. Several studies indicate the benefits of active commuting [8]. As the run-commuting is an active activity, these benefits can be associated with running as well. Active commuters are likely to be fitter and less likely to be over-weight [8]. The potential health benefits of active transportation on urban areas are highlighted in many studies [9–11]. Most of them indicate the significant reduction of mortality risk and chronic illnesses such as Type 2 diabetes, depression, and Alzheimer's disease. The shift in activity can easily be sustained on a daily routine for long term-periods against structured exercise programs, providing similar health benefits [12].

Economy. Except for the multiple health benefits for human nature, there are many benefits of economic interest. According to a research conducted recently in London [13], it is stated that if citizens of London walked or cycled for twenty minutes a day, except for the health benefits, this would save from the National Health Service of England about 2 billion Euros in treatment costs. Moreover, this would contribute to wider benefits for the economy and society.

Running events of a large scale became a business enterprise the last decades and their marketing perspectives are promising. The willingness to pay for people that travel to participate in running events in Europe, is high for those with high income [14]. Proper run-friendly urban infrastructures can attract more sport tourists and give prominence to cities through sight-running (city running tours).

Aesthetics. The character of a city is mainly determined according to its aesthetic appeal. Run-friendly streetscape and proper street furniture design can impress citizens and improve the aesthetic quality of public spaces. Upon this, remarkable practices across the world are highlighted: The architects Sandra Janser and Elisabeth Koller proposed and installed on 2010 the project *Ready. Steady. Go.* within the Jakomini area of Graz in Austria. The idea consists of painted streets in the form of a running track, which among other benefits, increase the attractiveness of the streets. Moreover, the installed running tracks on Margaret Island (Budapest) and on Jumeirah Beach (Dubai) contribute to high aesthetic quality of cities' public spaces.

Environmental. Transport emissions are responsible in a significant level for the environmental issues that many public and private administrations face. Air pollutants that are at high levels, especially in the city centers due to congested road streets, are harmful to human health. It is claimed that road transport emissions are responsible for the 20% of Europe's total CO_2 emissions [5]. A shift away from cars in respect of active transport, contribute to microclimate improvement and attractive environments. Thus, running can be a promising option for active commuting. Except for the improvements on air quality, the reduction of the noise pollution of motorized traffic, contributes to directly health benefits.

Risks. On the other hand, the risks that run-commuters face, include the following:

Safety. Running is an activity full of safety risks. According to [15], the lack of light on streets results on social, traffic and physical safety issues. Commuters cannot feel safe and be easily visible from other road users. Upon this, Collinson and Hockey [16] identified other road users as potential causes of injury. The visibility of the surface of the path on which commuters run is reduced as well due to poor lighting. According to this, a significant percentage of runners do not run on the dark [17]. The importance of the running surface due to injury hazards is highlighted in many studies as well [18–20]. Finally, a recent study indicates that running frequency is affected by infrequent impediments, such as threats by other people [21]. Run-commuters are vulnerable road users as they interact with traffic of high speed and mass. Thus, they suffer the most severe consequences with other road users and cannot protect themselves.

Health. Several studies indicate the negative impacts on health of run-commuters. Specifically, a recent study [22] used carbon monoxide as a traffic pollution indicator to compare the exposures associated with seven common transport modes (walking, running, cycling, bus, train, car, and motorized scooter). The results suggest that even if air pollution exposures are the highest for motorized modes, the active mode commuters and especially runners, experienced the highest doses.

Moreover, the impacts of the coronavirus (COVID-19) outbreak, may require a revision of the way people travel. During 2020 countries over the world declared a social distance of 1.5 m between people to reduce the spread of the virus. However, this distance is sufficient under various circumstances [23]. A recent study [24] considered the potential aerodynamic effects by runners' movements in the absence of wind. It concluded that runners avoid the substantial droplet exposure either by running in a staggered arrangement or by keeping 10 m distance in case of running behind other runners.

Level of Service. The built environment such as mixed land uses, and the presence of facilities lead to higher levels of physical activities such as walking according to several studies [25]. However, the attractiveness for running is reduced as many interactions with other road users are caused. According to [18], walking, running and cycling activities share the same public spaces. The coexistence of these three distinct road user groups may significantly reduce the level of comfort for runners in running. Moreover, the same study defined the cycle of public urban accessibility. According to this, the three aforementioned activities define three concentric circles representing the distances of the accessibility on urban public spaces. Pedestrians are active mainly in the city center, while cyclist's paths extend to places outside of the urban fabric. Runners belong to both categories.

A recent approach [26] of modeling and simulating runners in urban public places indicate the need for the adoption of strategy regarding the development of running-paths on urban spaces.

3 Identification of Run-Commuting Trips Characteristics

A review on the behavior of recreational runners is necessary for the identification of run-commuters' behavior. A limited number of studies deals with the behavior and the preferences of the people that choose to run in urban public spaces. According to [27], the routes people choose to run, are strictly positively associated with the presence and quality of green and open spaces. Runners avoid running on slippery roads and paths were people walk with dogs [16]. Moreover, the choice of routes is compatible with specific patterns. Runners tend to choose the same routes for their activities. They also avoid routes that are associated with presence of dogs, reduced street lighting, bad weather conditions, existence of busy traffic roads and shared paths with pedestrians and bicycles [18]. According to a recent survey [17], in Europe, 36% of people who belong at the age of 15 to 65 years old are active runners. Amateur runners are training on a daily basis in urban public.

The run-commuting activity is a new rapidly growing transportation practice. According to available fitness application data of run-commuting trips [28], the total number of running-commutes over the world has been increased by 65% in 2018 (21 million). The median covered distance for the total of these trips was 6.6 km.

Results of an international survey [29] highlight the required equipment of run-commuters on their trips. They usually carry with them a backpack, work clothes, their lunch, and a fitness/GPS watch. The 68% of the surveyors replied that the place where they work does not have showers, while the 61% served by a nearby facility (gym).

J. Gehl [30] identifies three types of outdoor activities. Run-commuting merge the characteristics of necessary and optional activities. A well-designed public space allows and enhances a multitude of activities to take place comfortably. Figure 1 represents graphically the connection between outdoor quality and outdoor activities [30]. An increase in outdoor quality boosts the optional and necessary activities.

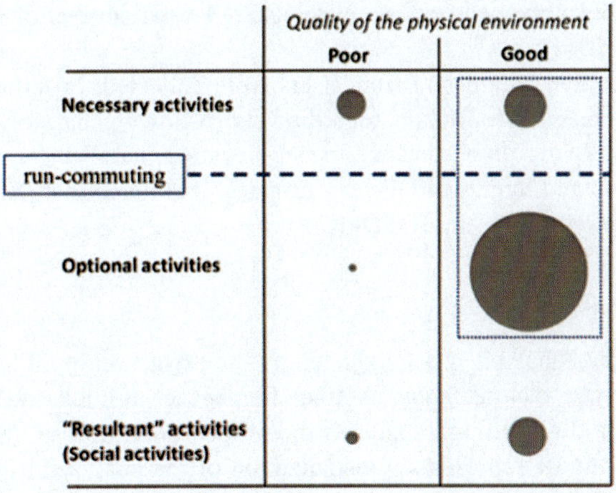

Fig. 1. Connection between outdoor quality and outdoor activities.

4 Analysis of Run-Commuting Trips

To better identify the characteristics of run-commuting trips, real trajectories of run-commutes were analyzed. A geospatial big data analysis was conducted extracting GPS track data from physical activity tracking apps that runners use in their commuting trips.

4.1 Case Study City

London was selected as the case study city. According to [28], London topped the world's list of cities with the most active run commuters on 2017. Cervero [31] indicate London as the epicenter of the run-commuting movement as well. This city is characterized by congested roads, crowded subways, legions of marathons and workplaces that provide remarkable facilities such as showers and gyms. Thus, the significant growth of public interest in this mode of transportation is assumed to be acceptable.

Run-commuting in London picked up during the 2012 Olympic, due to saturated transport networks [31]. Upon this, according to the projection of the average traffic speed on central London, it is anticipated that till the end of 2020, run-commuters will be as fast as vehicles' speeds during inter-peak periods, due to traffic calming measures and traffic congestion [28, 32].

4.2 Data Collection

Fitness data collection and its usage in urban and transport planning are highlighted in many studies [26]. For the purposes of the study, available data from physical activities that fitness apps [28, 33] provide were collected. Run-commuting trips were identified using the word-key "commute" on the search engine of the fitness applications. As the physical activities of most users are private, limited access on the total database is available. The final sample was randomly selected. A total number of 180 routes of 55 users was extracted.

The basic available parameters from users were collected. Specifically, the developed database included information regarding the following data of run-commuting trips: date, time, moving time, distance, pace, elevation, calories, runners' comments and finally, latitude and longitude GPS coordinates. It is noted that personal data were excluded from the database due to GDPR.

4.3 Data Analysis

A geospatial big data analysis was conducted by the extraction of GPS track data from the physical activity tracking apps. A total number of million location data from running routes was digitized and analyzed in an open GIS software [34].

The following figure represents a visualization of the analyzed trips in the city of London (a). An example of typical run-commuting trips from a random run-commuter, during different days, from point A (home) to point B (work) is graphically represented as well (b).

According to the heatmap of Fig. 2, it can be observed that the resultant spatial distribution of run-commuting trips is quite distinct. Most origin or destination points are in the central business district (CBD) of the city. Commuters prefer green spaces (parks) and run parallel to the river. Moreover, it is observed that a typical trip of a run-commuter differs during the days.

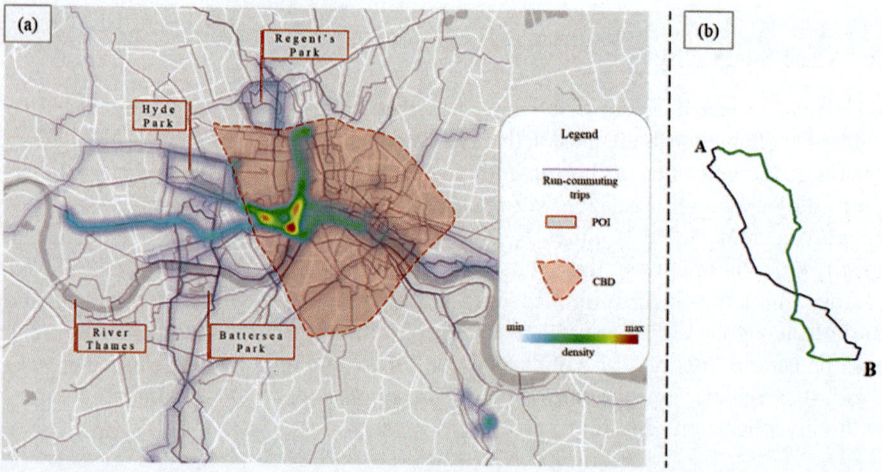

Fig. 2. Heatmap of run-commuting trips in London (a) and typical trips of a run-commuter (b).

The street typology, the distance and the patterns of the trips were analyzed. Routes were further analyzed in respect of safety issues. The number of intersections that run-commuters crossed during their routes was calculated, as a safety index for the evaluation of traffic exposure.

4.4 Findings

The main findings of the analysis are presented in the following Table 1.

Table 1. The main findings of the analysis.

Characteristics of run-commuting trips	
Average distance (km)	7.2
Average number of potential vehicle-runner conflict points	23
Route choice pattern	different routes during the days
Preferable street typology	paths on green spaces

According to the results, the average distance of run-commuting trips is 7.2 km. Run-commuters use green spaces such as parks in their trips, while they avoid congested road-networks. Upon this, an extensive analysis of the origin-destination points of each trip indicated that run-commuters follow different routes than the proposed shortest routes for vehicles. Street typology, aesthetics, continuity, and safety are some acceptable reasons why they choose alternative routes. Finally, their routes differ over the days. It is noted that a distinct difference of run-commuting trips against leisure running trips is the shape of the route. Recreational runners usually finish their activity at the starting point of the route (loop) [18]. Most of the comments that runners reported, highlight their positive experience. Specifically, many of them reported that:

"I was going to get the bus and I'm glad I didn't", "Running is hard: Beatles White Album in the ears", "Best way to beat the protest traffic"

5 Discussion

The run-commuting activity is a new transportation practice which is still largely under research. This paper has made a comprehensive effort to provide a clearer view of how this practice can transform a city.

According to the literature review, run-commuting trips are rapidly growing providing several benefits on health, economy, aesthetics, and the environment. However, there are many risks as well which need further research for the proper integration of run-commuting trips in city's infrastructures. Run-commuters are vulnerable road users as they interact with traffic of high speed and mass. The analysis of real run-commuting trajectories highlights that runners are exposed to traffic and danger situations many times during their trips. They avoid shared spaces, as they cannot easily interact with other vulnerable users.

An approach to use available fitness data to assist urban planners and transportation engineers to identify the spatial distribution of run-commuting trips in cities is proposed. The results of the spatial analysis indicate that run-commuters avoid traffic congested routes and prefer green spaces to make their trips, using alternative routes during the days. However, they are exposed to traffic many times per route. As their speed is larger than pedestrians, a new approach should be considered for the evaluation of their safety.

As run-commuting continues to become known, comprehensive strategies are required. Cities' streets need to be flexible to meet the changing demands of new transport modes. Effective policies and design principles will increase the safety, acceptability, and attractiveness of run-commuting. Proposals include making streets safer and healthier for running by installing crossings, more greenery and adopting traffic calming measures. Properly enforced reductions in speed limits or zones can reduce potential vehicle-runner conflicts. Running paths can be integrated on urban public spaces with proper and safe for running activities surfaces, providing accessibility and continuity. Finally, proper public facilities such as showers, water taps and rest areas such as pocket parks should be considered to the design principles. The design of runnable cities can be a promising future city-living strategy.

References

1. Edwars, P., Tsouros, A.: A healthy city is an active city: a physical activity planning guide. World Health Organization (2008)
2. Yair, G.: The commitment to long-distance running and levels of activity: personal or structural? J. Leisure Res. **22**, 231–237 (1990)
3. Scheerder, J., Breedveld, K., Borgers, J.: Running across Europe. The Rise and Size of one of the Largest Sports Markets. MacMillan (2015)
4. Cook, S.: Running as Transport: a geographical provocation. Master thesis, Royal Hollowey, University of London, UK (2014)
5. Hunkin, S., Krell, K.: Promoting active modes of transport: a policy brief from the policy learning platform on low-carbon economy. In: INTERREG Europe (2019)
6. Cook, S.: Run-commuting. In: Minicars, Maglevs, Mopeds: Modern Modes of Transportation Around the World. ed./Selima Sultana; Joe Weber. Santa Barbara, p. 255 (2016)
7. Woodcock, J., Edwards, P., Tonne., C, Armstrong, B.G., Ashiru., O, Banister., D.: Public health benefits of strategies to reduce greenhouse-gas emissions: urban land transport. Lancet **374**(9705), 1930–43 (2009)
8. Gordon-Larsen, P., Nelson, M.C., Beam, K.: Associations among active transportation, physical activity, and weight status in young adults. Obes. Res. **13**(5), 868–875 (2005)
9. Andersen, L.B., Schnohr, P., Schroll, M., Hein, H.O.: All-cause mortality associated with physical activity during leisure time, work, sports, and cycling to work. Arch. Intern. Med. **160**(11), 1621 (2000)
10. Morris, J.N., Clayton, D.G., Everitt, M.G., Semmence, A.M., Burgess, E.H.: Exercise in leisure time: coronary attack and death rates. Br. Heart J. **63**(6), 325–334 (1990)
11. Hamer, M., Chida, Y.: Active commuting and cardiovascular risk: a meta-analytic review. Prev. Med. **46**(1), 9–13 (2008)
12. Dunn, A.L., Marcus, B.H., Kampert, J.B., Garcia, M.E., Kohl, H.W., Blair, S.N.: Comparison of lifestyle and structured interventions to increase physical activity and cardiorespiratory fitness. JAMA **281**(4), 327 (1999)
13. Transport for London (TfL): Healthy Streets for London: Prioritizing walking, cycling and public transport to create a healthy city, London (2017)
14. Wicker, P., Hallman, K.: A multi-level framework for investigating the engagement of sport volunteers. Europ. Sport Manage. Q. **13**(1), 110–139 (2013)

15. Reiling, M., Dolders, T.: Running Amsterdam. Designing a runner friendly city. Master thesis, Landscape Architecture and Spatial Planning, Wageningen University and Research Centre (2015)
16. Collinson, A., Hockey, J.: From a certain point of view: sensory phenomenological envisionings of running, space and place. J. Cont. Ethnograph **44**(1), 63–83 (2015)
17. Syvonate: Reasons to run survey (2008). http://www.everythingaboutrunning.asics.eu. Accessed 03 April 2020
18. Anagnostopoulos, A., Anagnostopoulou, S.: Research on the integration of running paths in urban public spaces by using big data: Case study of Larissa city. In: Proceedings of the 2nd Pan-Hellenic Conference for Public Space, Thessaloniki, Greece (2019)
19. Collinson, A.: Running the routes together: corunning and knowledge in action. J. Contemp. Ethnography **37**, 38–61 (2008)
20. Collinson, A., Hockey, J.: Working out' identity: distance runners and the management of disrupted identity. Leisure Stud. **26**(4), 381–398 (2007)
21. Ettema, D.: Runnable cities: how does the running environment influence perceived attractiveness, restorativeness, and running frequency? Environ. Behav. **48**(9), 1127–1147 (2016)
22. Peyroux, E., Bussen, L., Costello, S., Dirks, K.: Traffic pollution while commuting - Does commute mode matter? (2015)
23. Setti, L., Passarini, F., Gennaro, D., Barbieri, P., Perrone, M., Palmisani, J., Gilio, A., Piscitelli, P., Miani, A.: Airbone transmission route of COVID-19: why 2 meters/6 feet of inter-personal distance could not be enough. Int. J. Environ. Res. Public Health **17**, 2932 (2020)
24. Blocken, B., Malizia, F., van Druenen, T., Marchal, T.: Towards Aerodynamically Equivalent OVID19 1.5 m Social Distancing for Walking and Running (2020). Preprint at http://www.urbanphysics.net/Social%%2020Distancing%20v20_White_Paper.pdf
25. Brian, E., Susan, L.: Built Environment correlates of walking: a review. Med. Sci. Sports Exerc. **40**(7), S550–S566 (2008)
26. Anagnostopoulos, A.: Runnability: Modeling runners behavior at urban public spaces. In: Proceedings of the 47th European Transport Conference (ETC), Dublin, Ireland (2019)
27. Karusisi, N., Bean, K., Oppert, J.N., Pannier, B., Chaix, B.: Multiple dimensions of residential environments, neighborhood experiences, and jogging behavior in the RECORD Study. Prev. Med. **55**, 50–55 (2012)
28. Strava. http://strava.com. Accessed 03 Apr 2020
29. TRC: International Survey of Run Commuters. Atlanta, GA, USA (2014)
30. Gehl, J.: Life Between Buildings: Using Public Space. Van Nostrand Reinhold, New York (1987)
31. Cervero, R.: Running to Work. ACCESS Magazine **1**(48), 24–30 (2016)
32. Transport for London (TfL): Travel in London: Report 10 (2017)
33. Mapmyrun. https://www.mapmyrun.com. Accessed 03 Apr 2020
34. QGIS Geographic Information System. http://qgis.org. Accessed 03 Apr 2020

Deployment of a Mobile Public Transport Information Application and the Operators' Perspective

Eleni Antoniou[1], Eirini Kastrouni[2], Ismini Stroumpou[1], Alexandros Papacharalampous[1], and Alexandros Deloukas[2](✉)

[1] AETHON Engineering Consultants, Athens, Greece
[2] ATTIKO METRO S.A, Athens, Greece
adeloukas@ametro.gr

Abstract. This paper discusses an innovative application for seamless transportation and an ecosystem of models and algorithms for Public Transport user choice simulation and data analytics, all developed within the Shift2Rail My-TRAC project. Its core is the understanding of the way travelers make choices for their trips. It is assumed that travel choices depend not only on the transportation system's capabilities, but also on the user's perception of the transportation service, as well as on their emotional state with respect to travelling. Furthermore, My-TRAC facilitates engagement of multiple stakeholders by seamlessly integrating services and creating connections between rail/bus operators and travelers. The My-TRAC project has developed the following components:

1. My-TRAC app: an advanced travel companion application;
2. Operators' Portal (OP): interface for the Transport Service Providers (TSPs). The OP is a gateway for TSPs to connect and communicate with My-TRAC
3. My-TRAC platform: the orchestrator of the aforementioned components

This paper focuses, first, on the OP description, the data needs of operators arising from the My-TRAC app and the possible uses of My-TRAC backend database for Public Transport planning operations. Secondly, this paper evaluates the findings from the Athens' pilot, in comparison to the other project pilots conducted in Lisbon, Barcelona, and the Netherlands. A TSP subscribed to the OP portal can request anonymized and aggregated behavioral data stored in My-TRAC, and can provide data regarding real-time changes. The OP utilizes static and real-time GTFS information to achieve its functionalities, in accordance to the data requirements of Google.

Keywords: Mobile public transportation information application · Public transport operators · My-TRAC

E. G. Nathanail et al. (Eds.): CSUM2020, AISC 1278, pp. 694–703, 2021.
https://doi.org/10.1007/978-3-030-61075-3_68

1 Introduction

The main goal of My-TRAC is to develop an innovative transportation services platform for the rail sector, designed for public and private travelers and operators, in order to provide an improved passenger experience. The My-TRAC platform is designed to provide interfaces for both travelers and operators: the former receive improved trip planning information customized to their needs and preferences, while the latter benefit from receiving access to anonymized and aggregated behavioral data, to improve demand forecasting models. The different components of My-TRAC are presented in Fig. 1.

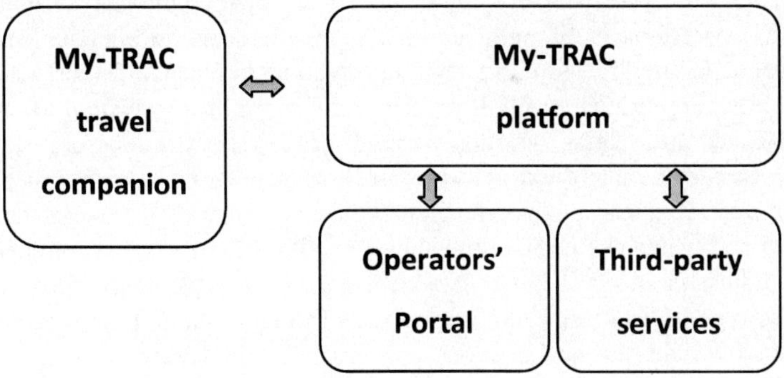

Fig. 1. My-TRAC components.

The My-TRAC platform collects and joins information from various sources: (i) PT operators for schedules and actual information (i.e., delays, disruptions), (ii) MaaS providers (i.e., car-sharing, bike-sharing, taxi services), (iii) service- and trip-related datasets (i.e., parking availability, crowd density at stations, security), and correlate this information with users' preferences and emotional state. The My-TRAC platform is designed to provide personalized information to the user, in a bid to improve the quality of the provided service and therefore, the travelers' level of satisfaction. The complexity of the aforementioned operations is sizeable, encompassing the generation and integration of various models and algorithms.

This paper describes in detail only one of the main components of My-TRAC, namely the Operators' Portal. The OP displays the anonymized and aggregated data collected via the My-TRAC app, and provides operators with the option of using interactive components in order to communicate with their passengers (e.g. notification about an upcoming delay, either scheduled or not).

The main objectives for the development of the portal are as follows:

- provide the operators with a connection to the My-TRAC environment;
- provide real-time and historic behavioral analytics data of travelers to Public Transport (PT) operators. This data can be used to assist operators with their planning operations and optimize their services;

- allow operators to upload useful real-time data [1] following the GTFS standard about alerts, announcements or disruptions, in order to help travelers with their everyday trips (e.g. suggesting alternative modes and/or routes), while also increasing reliability of the My-TRAC app.

2 Background

Recently, there has been massive growth in availability of transportation Internet and Communication Technologies (e.g. smartphone apps etc.) [2]. However, it is evident that, compared to scientific developments, practical big data applications are relatively limited, or applied at a relatively lower pace. The direct outcome of this is that transportation system users do not benefit from the current use of big data [3]. For PT users, arrival/departure prediction time and information on planned detours/disruptions and schedule during special events is a very useful piece of information [4].

Travel-related information is not only useful for travelers, but also for planners and operators, who may be interested in metrics such as travel time, transfer time, number of passengers etc. Past research has shown that there is a great need to understand travel behavior variability and evolution of travel patterns of PT users in order for operators to temporally adjust their service to satisfy demand during peak hours. Smartcard and other similar type of data can often be very useful to understand these variations, and provide useful insights to the operators. For instance, this type of analysis was performed in Rotterdam, the Netherlands, using data fusion of smartcard & Global System for Mobile (GSM) communications data [5]. Similar system-level transit information has been collected in various other cities: in Seoul, South Korea, using data from the Automated Fare Collection (AFC) system [6]; in Montreal, Canada, using similar data for a span of 51 weeks [7];and in Gatineau, Canada, clustering of multi-week data extracted from 35.4 million Smart card transactions [8].

3 Operators' Portal

The Operators' Portal aims to be the gateway for PT operators to communicate with the My-TRAC app and vice versa. It will constitute a tool for the operators to have access and download useful travelers' behavioral data, in order to improve strategic planning and dynamic operations, and to provide updates in the form of real-time GTFS data, informing users about alerts or changes.

By using the portal, operators will have access to real-time and historic data concerning behavioral analytics of passengers, and will provide data that are targeted to passengers aiming to give information on delays and changes on the schedule and network. There is no app or tool that handles real-time GTFS, making the Operators' Portal a unique data entry and analytics app. The added value of the portal, compared to other web apps, is that the operators can exchange and receive useful data, in a user-friendly way, regardless of their IT background. The portal allows TSPs to upload real-time GTFS data, as opposed to others, that handle only static GTFS. The connection of

My-TRAC app and portal simplifies the communication between TSPs and passengers, since any incident or urgent event is communicated to them, via the app. The portal also provides access to anonymized and aggregated data (adhering to GDPR rules) from real users of My-TRAC app (passengers), and makes them available to TSPs in an easy-to-read form that can be used in strategic operations and planning. Currently, operators have to conduct expensive and time-consuming surveys and data analyses in order to retrieve similar data. Behavioral data from the portal will help operators provide passengers with travel options tailored to their preferences. With the addition of Operators' Portal, My-TRAC is not just a travelling app, since it involves TSPs into the process, and connects them with the Public Transport passengers.

The two main functionalities of the portal are described in detail in the following sections. Fig. 2 presents the sitemap (blue boxes) and the functionalities (orange boxes) of each page.

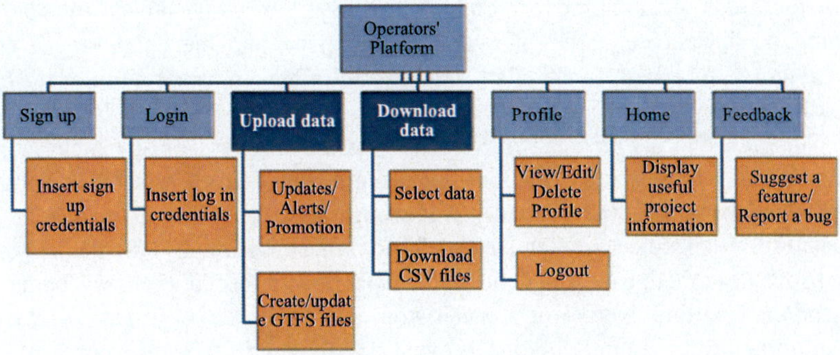

Fig. 2. Operators' Portal sitemap.

3.1 Functionality #1: Data Upload

This functionality responds to one of the objectives of the project, which is to upload useful real-time GTFS data about alerts, announcements or disruptions, to help travelers with their everyday trips, while also increasing the reliability of the My-TRAC app. The operators enter the My-TRAC portal in order to upload useful data about the operations of the PT, such as changes on the timetable of a trip, or an urgent message that a specific train stopped working. They can choose among "Trip updates", "Service alerts" and "Promotional messages" (Fig. 3), depending on the type of information to be uploaded. The first two categories are based on real-time GTFS [1], while the last one consists a new functionality that allows operators to promote important messages to My-TRAC users through the app.

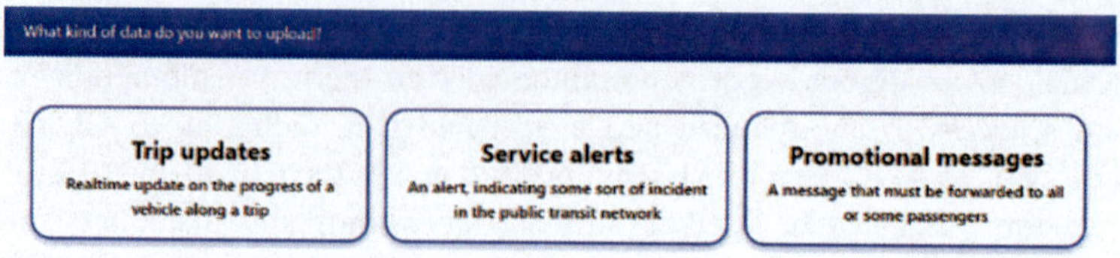

Fig. 3. Upload data page.

The system categorizes the uploaded data (station-, route-, delay-related, etc.) in the corresponding tables, and stores them for future use. My-TRAC has collected the GTFS files for Belgium, Greece (Athens), the Netherlands, Catalonia (Barcelona) and Portugal (Lisbon), allowing for the presentation of a list with all the available IDs and routes that an operator would need to modify or update, depending on the country. Some of the fields are required or conditionally required, while others are optional. Depending on the choice, different input forms appear, and the operator can select which category suits best to the data that needs to be uploaded (Table 1). More specifically:

Trip updates (Fig. 4) are real-time updates on the progress of a vehicle along a trip. They usually represent fluctuations or future changes in the timetables, or even cancelled or added trips to the schedule. There are 4 sections: i) Trip Descriptor, where the updated trip is specified, ii) Vehicle Descriptor where (optionally) the vehicle of the specific trip is described, iii) Stop Time Update that constitutes the real-time update for arrival and/or departure events for a given stop on a trip, and iv) Event Time where timing information for a single predicted event is provided in the corresponding input fields, based on the GTFS Real time Reference.

Service alerts (Fig. 4) indicate some sort of incident in the PT network. Typically, they are urgent messages (and not permanent changes) that should be communicated to the users through the My-TRAC app, either as direct messages ("popups") or used to update the recommendations provided by the My-TRAC models. Service alerts affect only some travelers travelling on a specific route, line or vehicle. Thus, dissemination of the service alert should involve only affected individuals. This assists on enhancing the precision of My-TRAC app, leading to better services for the travelers. Service alert form consists of i) Alert description where details of the alert are provided, ii) Time range that specifies that the alert will be displayed within the given time range, iii) Entity selector that indicates exactly which parts of the network this alert affects and iv) Trip Descriptor where the affected trip is specified. Below, the operator can also choose the cause, the effect and the relation between the trip and the static schedule.

Promotional messages (Fig. 4) on the other hand, are short messages, containing advertisements or announcements that the operators wish to promote to travelers. The operators can also specify the time range and the frequency of the messages' appearance, the group age that these messages respond to, and the mode choice. This way, the messages will be forwarded only to specific travelers, improving the user experience of My-TRAC users. This functionality is crucial for the OP business model.

When the operators submit the completed fields, the GTFS files are automatically created/updated. Information about updates and future events are communicated to My-TRAC app users and/or other operators that need to stay informed about the services of the given PT operator.

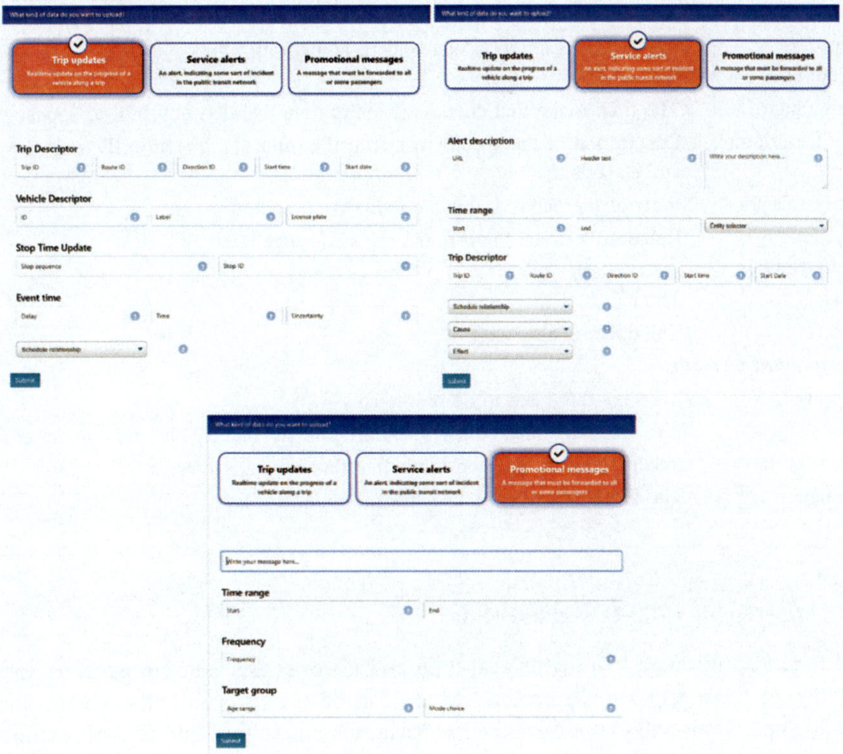

Fig. 4. Trip updates form (top left); Service alerts form (top right); Promotional messages form (bottom).

Table 1. Upload data definition.

DATA input fields	Definition
Trip updates	
Trip Descriptor	Description of the specific trip (trip ID, route ID, direction ID, start time and start date)
Vehicle Descriptor	(optional) Description of the vehicle of the specific trip (ID, label)
Stop Time Update	Real-time update for arrival and/or departure events for a given stop on a trip (stop sequence, stop ID)
Event time	Specifications of the time of the event/update (delay, time, uncertainty)

(continued)

Table 1. (*continued*)

DATA input fields	Definition
Schedule relationship	Relation between the trip and the static schedule
Service alerts	
Alert description	(optional) Description of the alert, along with a link
Time range	The time range that covers the entire time the alert is useful to appear
Trip Descriptor	Description of the specific trip (trip ID, route ID, direction ID, start time and start date)
Entity selector	Parts of the network this alert affects
Schedule relationship	Relation between the trip and the static schedule
Cause	The cause of the alert
Effect	The effect of this alert
Promotional messages	
Message	Message that need to be promoted
Time range	The time range that covers the entire time the message is useful to be seen
Target group	Specification of age and mode that this message refers to
Frequency	The frequency of time that this message appears

3.2 Functionality #2: Download Data

This functionality aims to provide real-time and historic data concerning behavioral analytics of travelers to PT operators, which can be used to assist them with their planning operations and to optimize their services. Obtaining this kind of data is crucial for the improvement of the PT services, since it assists with their strategic planning operations and continuous improvement of their services. They are usually hard to obtain, since the collection and analysis is an expensive and time-consuming procedure. My-TRAC app clusters travelers according to their behavioral patterns, and summarizes information of the most frequently repeated activities and actions, specifying their preferences.

The operators can retrieve real-time and historic data concerning behavioral analytics of travelers and travel patterns (e.g. how many students by gender used a specific metro station, during a day, for the past month, thus enabling the operator to provide more effective station ads). The operators can download the data required in a CSV file, for further usage. They have the possibility to download the data they need by selecting specific route, mode and dates. They can also select aggregation of the available data based on gender, day of the week, card type and occupation (Figs. 5, 6). The aggregated data is an additional source of information for the operators, for improving and updating their PT operations. The timetables, routes and general services of operators will be always up to date, offering the best experience for the travelers and users of My-TRAC app, since the operators will have a better overview of travelers' behavior. The retrieved data may also include crowdedness information, allowing the operators to allocate PT capacity more efficiently (Table 2).

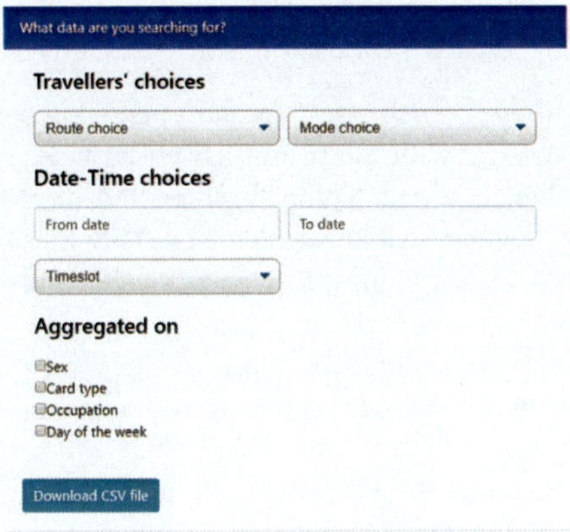

Fig. 5. Download data page.

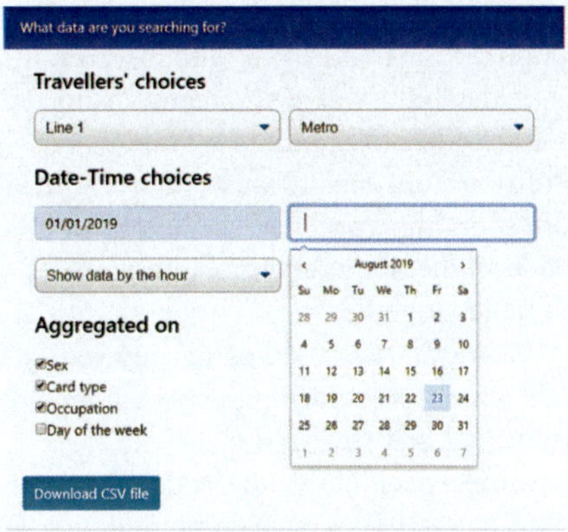

Fig. 6. Download data page (alternative screen).

Table 2. Download data definition.

DATA input fields	Definition
Route choice	(Optional) The specific route of the travelers
Mode choice	(Optional) The specific PT mode of the travelers
Date choice	Date range of the required data
Timeslot	Data presented are divided by 1, 2 or 6 h
Aggregated on	Aggregation of travelers (can be more than one option)

3.3 Technical Background

The project management method that was followed for the development of the portal was the Agile methodology, a process that encourages iterative planning and continuous adaptation to changing requirements throughout the process, by measuring and evaluating the status of a project [9].

The development of the code was continuously integrated into a shared repository, allowing the quick and easy detection of errors, leading to a rapid delivery of a high-quality project. Furthermore, pair programming was another agile technique that was implemented throughout the development, ensuring the quality and accuracy of the code through cross-examinations and thorough review.

PHP was the main programming language used for the development of the web portal, in combination with HTML for the structure and the appearance of the web pages and CSS for pages' styling like colors, fonts and layouts. Alongside with HTML and CSS, Javascript was also used for interactivity with the web pages, as the three technologies combined, constitute the core of web browsers and applications. In addition, other libraries and frameworks were integrated to improve performance such as Bootstrap and by extension StartBootstrap, which provide design templates for buttons, forms and even whole web portals.

A database was also created using *phpMyAdmin,* which is a free and open source tool intended to handle the administration of MySQL with the use of a web browser. With this tool and the use of MySQL, the database along with the tables and their corresponding rows were created, modified or deleted. All the tables of GTFS files reside in the above-mentioned database, with the corresponding foreign keys that link the tables together.

BitBucket was used as the repository where all the different versions of the source code were stored during the portal development. In this way, every change of code can be tracked and reversed.

More specifically, the source code that constitutes the Operators' Portal is composed of 24 PHP scripts combined with HTML, 6 Javascript scripts and 3 CSS scripts, in total 4092 lines of written code.

4 Conclusions, Feedback, and Next Steps

The portal aims to be the gateway for Public Transport operators to communicate with My-TRAC and vice versa. With its two main services, 'upload' and 'download' data, the portal is expected to serve as an integral tool for operators, that is going to be useful for the smooth and undisturbed operation of Public Transport. When the operators provide their data to the portal, they contribute to the improvement of My-TRAC's app and consequently their customers' experience, while when they extract aggregated user data, they have the opportunity to update and improve their service.

The OP was completed in September 2019, and its main functionalities were presented to Public Transport operators of Greece, Norway, Spain and Portugal during the My-TRAC consortium meeting (October 2019; Delft, the Netherlands). During this workshop, both a live demo and a hands-on exercise took place. Following the presentation, a discussion between the operators and the consortium took place, where the operators provided feedback, rating the portal and its services using a questionnaire. The results of the discussion and the questionnaire combined, showed that TSPs found the portal quite interesting and easy to-use (rating of 3.5 out of 5), as they would integrate it to their services, considering a small submission fee. They also stated their interest to collect characteristics of app users, such as gender or age, as well as using the portal to notify users about service disruptions (100% of TSPs), upcoming events (50% of TSPs), or new service promotion (50% of TSPs). According to the TSPs, the most important feature of the portal was the service alert notification, as well as the option to download useful aggregated data quickly. Furthermore, they expressed the need to be able to upload GTFS data from a file source, and not just manually. Finally, they would highly recommend the portal to their colleagues (100% of TSPs). There were also some productive comments about the technical aspects of the portal, and the appearance (UI/UX design) of the portal.

Similar comments were received during the Athens intra-operators' focus group which took place in May 2020 with the Athens Public Transport Organization (O.A.S. A. S.A.), with some additional comments on the usefulness of crowdedness information within particular stations, in light of the latest COVID-19 developments. The aforementioned feedback helped with the improvement of the OP operations, and the quantification of its added value.

The Operators' Portal is going to be constantly updated and improved. More specifically, GTFS input fields will be based on real GTFS files from Belgium, Greece (Athens), the Netherlands, Catalonia (Barcelona) and Portugal (Lisbon);GTFS from other countries will be added, so that My-TRAC's Operators' portal will become a necessary tool for operators all over Europe; GTFS files will be produced automatically, in the *Upload data* page, so that they can be stored for future use, in My-TRAC app; and data retrieved from My-TRAC pilots in said areas, are going to be processed and lead to re-evaluation of the functionalities and operations of the portal.

References

1. Google LLC, Google Transit APIs - GTFS Realtime Overview, https://developers.google.com/transit/gtfs-realtime/. Accessed 26 Aug 2019
2. Gössling, S.: ICT and transport behavior: a conceptual review. International J. Sustain. Transp. **12**(3), 153–164 (2018)
3. Yap, M., Munizaga, M.: Workshop 8 report: Big data in the digital age and how it can benefit public transport users. Res. Transp. Econ. **69**, 615–620 (2018)
4. National Academies of Sciences: Engineering, and Medicine: Use and Deployment of Mobile Device Technology for Real-Time Transit Information. The National Academies Press, Washington, DC (2011)
5. de Regt, K., Cats, O., Van Oort, N., Van Lint, H.: Investigating potential transit ridership by fusing smartcard and global system for mobile communications data. Transp. Res. Rec. **2652**(1), 50–58 (2017)
6. Wu, L., Kang, J.E., Chung, Y., Nikolaev, A.: Monitoring multimodal travel environment using automated fare collection data: data processing and reliability analysis. J. Big Data Anal. Transp. **1**(2–3), 123–146 (2019)
7. Deschaintres, E., Morency, C., Trépanier, M.: Analyzing transit user behavior with 51 weeks of smart card data. Transp. Res. Rec. **2673**(6), 33–45 (2019)
8. Viallard, A., Trépanier, M., Morency, C.: Assessing the evolution of transit user behavior from smart card data. Transp. Res. Rec. **2673**(4), 184–194 (2019)
9. Team Linchpin. https://linchpinseo.com/the-agile-method/. Accessed 02 Sept 2019

Traffic Emissions and Environmental Impacts

Emissions Estimation for Obsolescing Bus Fleets: Problems and Advances

Maria Vittoria Corazza[1(✉)], Paulo Cantillano Lizana[2], Daniela Vasari[2], Enrico Petracci[2], and Marco Pascucci[3]

[1] Sapienza University of Rome, DICEA, Via Eudossiana 18, 00184 Rome, Italy
mariavittoria.corazza@uniroma1.it
[2] Pluservice Srl, Strada Statale Adriatica Sud 228, 60019 Ancona, Senigallia, Italy
{d.vasari,e.petracci}@pluservice.net
[3] Pascucci Consulting, Viale Pinturicchio 93, 00196 Rome, Italy
mp@ingmarcopascucci.it

Abstract. When it comes to simulate traffic emissions, problems arise: which is the best model to use, which data are needed and whether they are all available, whether the process is time-consuming and/or calls for expert, etc. It is not surprising, then, that public transport operators might be hesitant in estimating the emission package of the bus fleets they manage. Moreover, the underestimation of the local fleet potential in producing pollution (in case of small fleets) on the one hand, and the reluctance to estimate it when the fleet is composed by a good amount of old-generation vehicles on the other, are further elements of uncertainty. However, transit operators' obligation to manage and maintain clean bus fleets is becoming more and more imperative and in Italy recent regulations subsidize transit companies to purchase buses with cleaner performance. The assessment of how much pollutant is the fleet to renovate becomes, then, crucial. To this end, an emission-estimation model for bus fleets is developed and described in the paper, with the research goal to facilitate the emissions calculation among transit operators. The procedure is embedded in an already successfully-implemented software for the management and maintenance of bus fleets. The model moves from well-known assessment methods (namely COPERT and IVE) but addresses issues like long-protracted mileage and age of vehicles which make it especially useful when "old" fleets emissions are to be assessed. A case study, in a middle-size city in Italy, is also described to highlight the model development.

Keywords: Emissions · Buses · Obsolescence

1 Introduction

Bus companies are frequently concerned about environmental aspects, but improving bus operations from this point of view is a challenging issue. Budget constraints impose restrictions on the purchase of more environmentally friendly vehicles, also because vehicles with alternative fuel systems are more expensive than traditional ones, even if this difference is mitigated by lower maintenance costs for the former [1]; electrification requires a series of technological choices and infrastructure investments [2] which must

E. G. Nathanail et al. (Eds.): CSUM2020, AISC 1278, pp. 707–717, 2021.
https://doi.org/10.1007/978-3-030-61075-3_69

give economic returns over short time horizons [1, 3], and at the same time doubts about the viability of a full replacement or conversion of conventional fleets by/into electric ones can be raised; subsidies or funding to purchase cleaner vehicles are limited or one-off; and eventually the operators' understandable reluctance to have more different technologies to manage. At the same time, according to a UITP survey, around 57% of the fleets worldwide is composed by pre-Euro V vehicles, with an average age of 6.9 years, although few countries (Russia, Brazil and China) affect the average age significantly [4]. The snapshot is, then, that of too many old and polluting buses circulating in urban areas. Italy is no exception: more than 70% of urban buses are diesel-fueled, 51% pre-Euro V, with an average age of 11 years [5].

The research question is, then, whether it is possible to operate mixed EURO-engines fleet, with a not negligible amount of old vehicles, in a more environmentally-conscious way, by controlling their emissions. This question originates from a specific circumstance, in Italy: the recent National Sustainable Mobility Strategic Plan (NSMSP) subsidizes the transition towards newer and cleaner fleet within an 8-year horizon, by granting 2.5 billion Euros per year to purchase more eco-friendly vehicles (i.e. non-diesel). The NSMSP creates a unique opportunity, but since the non-diesel vehicles are more expensive than the conventional ones, it is forecast that the amount of buses that can be purchased will be lower if compared to the amount of diesel buses that could be purchased with the same subsidy. The downside is evident: with a smaller amount of new vehicles in, the aging process will progress with an average of 17.5 years to be reached in 2033 [5]. The research goal is then to help bus operators to assess the pollution levels emitted by the fleets they manage and support them in the decision whether to renew or simply control their fleet performance by improving maintenance process, within the overall goal to contribute to decrease emissions.

Moving from all of the above, this paper describes the experience of Olbia, a small-medium sized Italian town, with the development of an integrated emission model based on the local city bus fleet, with the aim of identifying a simplified approach based on the combination of different methodologies. The model, described next, refers to mixed fleets with different EURO engines. Results are described, to provide indications on further research activities.

2 Emissions from Bus Fleets: Recurring Procedures

2.1 Assessment via Models

There are different models for estimating traffic emissions, and they are all well known, as well as for Portable Emission Measurement Systems (PEMS). As regards the former, the scientific literature relating to public transport vehicles mostly refers to the first decade of the 2000s, while recent developments refer mainly to scenarios in which the presence of more recent vehicles is dominant [6, 7]. As far as PEMS are concerned, even if, from a practical point of view, they show advantages compared to laboratory tests, their accuracy can be questioned [8]. However, a proper assessment of traffic emissions is a *condicio sine qua non* for the development of sustainable mobility policies, and also a constant driver in this process. Coherently, the advantages of clearer fleets are acknowledged also by many transit operators in their willingness to innovate [9] and switch towards non-conventionally-fueled modes [3, 10]. But all of the above is also to be associated with other problems faced by transit operators. Often they are not able to directly simulate the emissions of the fleets they manage because data process could be demanding and results not easy to interpret. More issues to put up with are the operators' underestimation of the local fleet potential in producing pollution (especially in case of small fleets) and/or their reluctance to estimate it when the fleet is composed by a not negligible amount of old-generation vehicles. There may be implicit difficulties in choosing the most appropriate emission model, and the data collected so far have allowed the development of different methodologies, with different degrees of accuracy, based on the types of inputs associated with the emission profiles under different operating conditions [11, 12]. Simulations software abound and emission inventories have progressively become more comprehensive. Yet, the choice is difficult as the model are many. Each is based on specific emission records, hardly interchangeable. This means that the inventories must be as much comprehensive as possible, including the vastest range of parameters to render simulated operational conditions in the most possible reliable way. Hence, in pursuing the advantage to create a unique "reference" at European level, the COPERT model was developed (coordinated by the European Environmental Agency – EEA), and has progressively improved until its 2019 version.

2.2 Dealing with the Mileage Degradation Within COPERT

COPERT is a model developed on the basis of an extended usage, belonging to the "group" of models based on average speeds, which estimates emissions in hot and cold start conditions, plus the "non-exhaust" component for many categories of vehicles. COPERT is actually a standard method adopted in all member states of the European Union for emissions calculation.

COPERT basic inputs, along with environmental and emission factors, also include vehicles engines and fuel specifications. Emission factors, in turn, describe the performance of a vehicle depending on various parameters (speed, acceleration, deceleration, weather conditions and grade in different urban driving cycles). From all these inputs it is possible to calculate the pollutants package emitted by the vehicle in hand.

Among all such inputs, an important one is the age of the vehicle, as it is assumed that newer vehicles should pollute less than vehicles with protracted mileage. Accordingly, COPERT includes the "Mileage degradation", an option that describes how annual miles driven as vehicles get older lead to more pollution: for example, over 80.000 km there is an increase by 60% of CO and NOx and an increase by 30% on NMVOCs in vehicles with technology EURO I and EURO II [13]. However, this option is not available for all the types of vehicles, and do not apply for diesel buses. Taking a EURO III bus with an average yearly mileage of 50,000 km as an example, the amount of CO_2 emitted over a period of 8 years simulated by COPERT is constant. It is rather intuitive that this is far from reality, as already observed elsewhere in literature [14]. Therefore, if the pollutant emission packages for buses with protracted mileage are to be assessed, then COPERT needs to be complemented with additional simulations.

Ignoring the mileage aspect can produce an underestimation of emissions packages. In order to solve this aspect, it is possible to carry out mixed evaluations based on the comparison between COPERT and other models. Among them, the IVE (International Vehicle Emission) model appears to be suitable. IVE can be classified inside the Modal Models group. It was designed in the United States for developing countries, but nowadays a large amounts of applications worldwide are available [15–17]. IVE methodology is fully described in [17–19], and for the issue in hand, suffice it to say that IVE includes more realistic mileage range, also applicable to diesel buses, according to three classes: below 79,000 km, between 80,000 km and 161,000 km and over 161,000 km. In turn, IVE has other downsides, if compared to COPERT: the above three mileage classes can be too vast when modeling small operations; EURO VI is not contemplated and so is the effect of lubricants; data collection can be difficult without GPS.

3 A Simplified Emission Model for Mixed Bus Fleets

What discussed above highlights an aspect: there is not a model that is able to give an answer to any possible scenario. And considering that from the practical point of view public transport operators are not able to use different models so as to handle any specific case pertaining to the fleets that they operate, then the goal is the identification of a simplified procedure in order give them a straightforward tool. To this aim, a simplified but reliable emission calculation methodology has been developed, starting from the COPERT one and integrating it with the IVE similar procedure for buses with protracted mileage, thanks to the IVE three mileage classes, previously mentioned. The opportunity was provided by the requirement to improve a software designed to plan bus fleets regular maintenance and control budgets [20] which is successfully used by many Italian transit operators. Upgrading the software with the emission calculator meets a two-pronged requirement: on the one hand, it facilitates the assessment of the emission packages for transit operators who manage the most diverse types of bus fleets (small and big; with mixed technologies, i.e. diesel and CNG, and EURO-compliant engines; new and old vehicles with different ranges of mileage); on the other, when evidencing high emission levels, it enables to run more effective maintenance activities and even to renew the most polluting vehicles, in line with the NSMSP mandate. At the same time, to be user-friendly, the function with the emission calculation should be the least demanding, with easy-to-collect inputs and few data entries.

Thus, the methodology to simplify emissions calculation was developed to meet two main aspects: i) to be scientifically-sound but, at the same time, reliable, in order to be embedded in the software already available, and be easily used by the operators; ii) to be implemented in the majority of the cases (mixed bus fleets operating in urban environments), in Italy and elsewhere. The methodology was based on the following steps:

- Creation of the specific emission inventory database for EURO IV and V engines based on the IVE one

- Complement it with the COPERT emission inventory to include in the emission calculation also buses with EURO VI-compliant engines
- Development of a bilinear regression model to develop a simplified calculation process (via a specifically designed spreadsheet), from which specific algorithms for the software can be further originated.

Results were tested and calibrated on a real case study, i.e. the provincial town of Olbia (around 60,000 inhabitants) in the Sardinia island, further described. This urban area is particularly relevant for the case in hand as operations are affected by seasonal peaks, different weather conditions occur, the local fleet is composed by a mix of different Euro-engine vehicles, and mileage is consistent to the three classes above-mentioned in Sect. 2. The simplified model is specifically targeted for the estimation of CO_2, NO_x, CO, PM, SO_x and NH_3, to comply with the national and supranational mandatory assessment of the level of air quality in urban areas. To be noted that a preliminary COPERT-based calculation for this software for a similar pollutant package was already developed within a project funded by the European Commission and successfully tested on a very small amount of buses and restricted operations, as described in [20]. These results were also considered while preparing tests and validation.

3.1 Data Process and Validation

COPERT is the EEA emission calculator, and the necessity of evaluating discrepancies between this model and IVE led to validate IVE under parameters that are common for EURO IV and EURO VI standard performance. In order to do that, four types of pollutants (CO_2, NO_x, CO and NH_3) were considered, excluding PM10 and SO_x not estimated in COPERT and simulated only by IVE last version. Likewise, since IVE does not include EURO VI performance, data processed are from EURO IV and V items, common to both models. For the validation, input data are summarized in Table 1 and refer to EURO IV and V diesel buses (15 – 18t), with fuel metals contents in different percentage. From these data, COPERT estimates the pollutant amount emitted in t/year. With these outputs available, IVE can be run to calculate emissions according to the model's the three classes of mileage. Running IVE requires additional parameters entries: average hourly speed, amount of engine ignitions, Vehicle Specific Power (VSP), Engine Stress (ES) and "bins". VSP is a parameter which correlates speed, accelerations, road grade and altitude to consider aerodynamic drag and rolling resistance [21, 22], whereas ES describes the operational condition of engines. After collecting data to "feed" VSP and ES via GPS, it is possible to get "bins", i.e. 60 emissions variations, since a correction factor is associated with each bin or VSP condition. Data to feed VSP and bins are available in literature, and those found in [23–25] enabled to find an appropriate set for this case. Once all of the above is set, it is possible to run simulations for the three mileage categories and get the IVE emission estimations.

Variations were expected, and among these it is worth noticing that: i) for CO_2, both COPERT and IVE provide a similar output when the vehicle mileage is over 80.000 km; ii) CO outputs are markedly different; iii) for NO_x, both show similar

Table 1. Input validation data for EURO IV and V buses.

Main validation data	Value	Unit
Operational input parameters		
Mean activity (as annual average)	60,000	km
Life Cumulative Activity (as sum of annual average)	300,000	km
Share - Urban Peak	15	%
Share - Urban Off Peak	85	%
Speed - Urban Peak	15	km/h
Speed - Urban Off Peak	25	km/h
Fuel input parameters		
Energy Content	42.6950	MJ/kg
H:C ratio	1.86	
O:C Ratio	0	
Density	840	Kg/m^3

results for EURO IV standards. Nonetheless, for the case of EURO V a factor f to regularize data is required ($f = 2$); the same applies for CO ($f = 5$) and NH_3 ($f = 0.15$ and 0.55 respectively for EURO IV and V).

Simulations resulted from both COPERT and IVE are reported in Table 2.

Table 2. Comparison between COPERT and IVE models.

Pollutant emitted [t/year]	EURO IV		EURO V	
	Mileage Class < 79 K km			
	IVE	COPERT	IVE	COPERT
CO_2	21.850	59.094	21.850	53.394
NO_x	0.3587	0.3777	0.2044	0.4224
CO	0.0128	0.0776	0.0128	0.1348
NH_3	0.0012	0.0002	0.00126	0.00066
	Mileage Class 80–160 K km			
Pollutant emitted [t/year]	IVE	COPERT	IVE	COPERT
CO_2	54.192	59.094	54.1926	53.3945
NO_x	0.3636	0.3777	0.2072	0.4224
CO	0.0134	0.0776	0.01349	0.1348
NH_3	0.0012	0.0002	0.00126	0.00066
	Mileage Class > 161 K km			
Pollutant emitted [t/year]	IVE	COPERT	IVE	COPERT
CO_2	53.823	59.094	53.823	53.3945
NO_x	0.3710	0.3777	0.2115	0.4224
CO	0.0144	0.0776	0.0144	0.1348
NH_3	0.0012	0.0002	0.00126	0.00066

3.2 The Application to the Case Study

The integrated simulation model was tested towards the real operation conditions. The data collection and the testing activities came from a bus fleet of 12 types of diesel vehicles, operating different routes, in Olbia. The vehicles were, in average, more than 8 years old, with a total mileage ranging between 420,000 and 510,000 km. Having vehicles with such high mileage enabled also the study of conditions far above the 161,000 km threshold, thus contemplating the common situation of buses with very high ranges. Along with the operational features presented in Table 3 for each vehicle, the simulation contemplated the use of auxiliaries for the total duration of the daily operations. These and other vehicles data enabled to run the integrated model 108 times per bus line, resulting into total of 1,404 simulations. A complete database of the emission package assessed for the whole analyzed fleet was then obtained. Consequently, a series of scenarios was built, so as to associate the emitted pollutants with the real operational conditions. The scenarios included different sets of options concerning: i) driving cycles (short urban, long urban, and extra urban); ii) speed (12.5, 25, 30 km/h respectively); iii) hourly ignitions (2, 1, 0.5 events respectively); iv) weather (air temperature: 20, 25, 30°C); average road slope (grade: 0, 1, 2, 3%). This enabled the creation of 468 matrices, one for each vehicle under the three mileage classes and by each EURO standard analyzed and for each of the six pollutants considered.

Table 3. Operational Features of the Tested Fleet (vehicles selected as example).

Veh. ID	Max grade (m)	Operation time (hh: mm)	Ignitions (event/day)	Avg. speed peak hour (km/h)	Avg. speed off-peak (km/h)	Peak time duration (hh:mm)
1051	10	17:20	38	25	30	3:15
1052	65	12:10	12	31	38	3:23
1053	50	11:15	15	20	26	2:31

The identified scenarios were also designed to be included in the maintenance software. As mentioned above, the initial goal was to give to the software users the possibility of evaluating emissions by just introducing a minimal amount of information based on the options described above. Once these values are set, the model locates a unique matrix among 156 matrices available with different parameter dataset for each of the six pollutants mentioned. Each row of the matrix is associated to the three mileage classes. Therefore, according to the options entered by the user, it is possible to calculate the daily emissions of a vehicle under the parameters set. Customers can update the obtained value by planning changes in operations, so to have the final amount of pollutant emitted yearly by the selected fleet.

4 Main Findings

Some specific trends were highlighted under the resulting set of simulations carried out for the case study. As regards carbon dioxide relevant results can be associated with the mileage variation. Even if emissions of CO_2 for both EURO IV and V engines are increasing with the total mileage as expected, after 80,000 km the trend can be considered practically stable for all the driving cycles. If a sensitivity analysis is carried out to observe how temperature and grade affect this pollutant emission, an increase around 4–5% is reached whenever the temperature rises every 5 °C. More impacting is the slope variation: for each grade (in %), emissions of CO_2 increase by 5 to 6% in the long urban and extra urban cycles. Summarizing all these results, it is possible to highlight for CO_2 a very simple trend (Fig. 1).

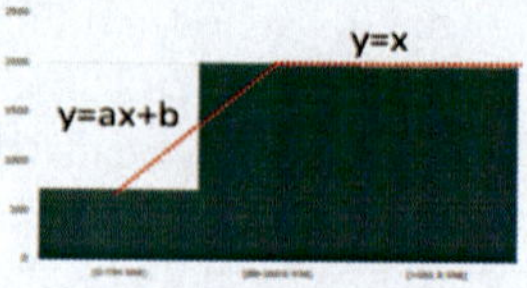

Fig. 1. Carbon Dioxide emissions trends.

For what concerns NO_x emissions, the trend is different (Fig. 2), with an increase of 1.4% when passing from the lowest mileage range to the higher, and up to 2% when reaching the third mileage class. When temperature is considered, at 20 °C, no relevant effects can be assessed. But when temperature reaches 25 °C, for EURO IV engines emissions are slightly reduced; this becomes more evident for the EURO V ones, where a 7% less is observed, if compared to the 20°C situation. At 30 °C, the decrease is 5% less for EURO IV and 6% less for EURO V in comparison with the 20°C status.

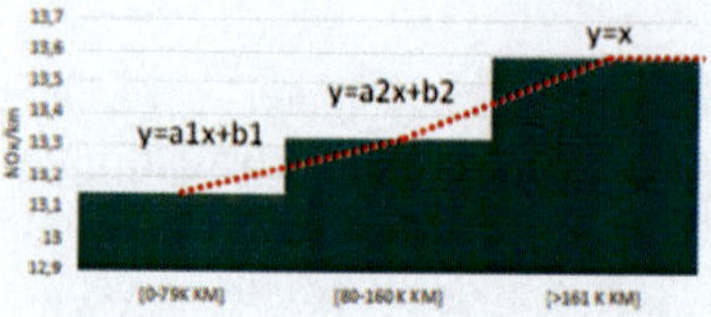

Fig. 2. Nitrogen Oxides emissions trends.

The additional observation that the total emissions for long urban cycle are smaller than the short urban and extra urban cycles, can be explained by the relationship between the emission and speed, already noted in literature for passenger cars [26] and here confirmed by the integrated model for the buses. CO emission trends are similar to the NO_x's, with a 4.5% increase for both EURO standards, when passing from the first mileage class to the second, and a 6% increase, from the second to the third mileage class. However, beyond 161,000 km it is safe to assume that there is no additional increase. Of note is the increase of emissions by 3% when passing from 25 °C to 30 °C and similarly a per-kilometer increment by 1.5% and 3.3% for every grade respectively for short urban and extra urban cycles. Particulate Matters (PM) emissions trends are consistent with what observed in literature for passenger cars: to be noted that, after 160,000 km of mileage, the growth is exponential. Sulphur Oxides emissions have a different behavior (Fig. 3), decreasing with mileage and this is due to the composition of SO_x which is 98% SO_2, easily degradable after 80,000 km [27]. Ammonia (NH_3) appears in the cold-start emissions stage and is particularly harmful as a PM2.5 "precursor". No variation due to mileage can be detected as if the age of the vehicle would not affect the NH_3 emitted, whereas it decreases while the temperature raises.

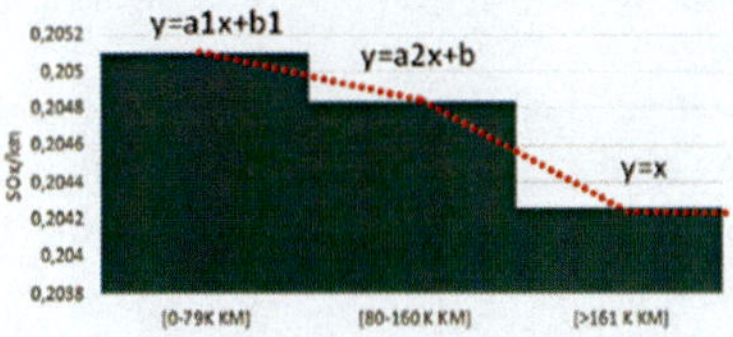

Fig. 3. Sulphur Oxides emissions trends.

5 Conclusions

Providing an easy tool to calculate emissions for buses and analysing emission trends according to mileage, EURO performance, and driving environment can really help transit operators to understand the level of sustainability of the fleets they manage, and optimize operations when aware that temperatures, slopes or engine performance can impact emissions, with a positive impact on the communities served. The integrated tool proved, thus far, to be reliable, user-friendly and widely implementable to a vast range of buses, thanks to the development its specific emission inventory. A further development of the model based on Artificial Neural Networks is currently under study.

References

1. Stempien, J.P., Chan, S.H.: Comparative study of fuel cell, battery and hybrid buses for renewable energy constrained areas. J. Power Sources **340**, 347–355 (2017)
2. Meishner, F., Sauer, D.: Technical and economic comparison of different electric bus concepts based on actual demonstrations in European cities. IET Electrical Systems in Transportation, p. 14, January 2020
3. Bousse, Y., et al.: Electrification of public transport in Europe: vision and practice from the ELIPTIC project. IEEE Xplore (2018). https://doi.org/10.1109/EEEIC.2018.8494518
4. UITP, "Global bus survey" leaflet, May 2019
5. ASSTRA, "Investire nel TPL: Scenari e fabbisogni" https://www.trasporti-italia.com/citta/convegno-nazionale-asstra-investire-nel-tpl-scenari-e-fabbisogni/37236

6. Nanaki, E.A., et al.: Environmental assessment of 9 European public bus transportation systems. Sustain. Cities Soc. **28**, 42–52 (2017)
7. Jiménez, F., Román, A.: Urban bus fleet-to-route assignment for pollutant emissions minimization. Transp. Res. Part E: Logistics Transp. Rev. **85**, 120–131 (2016)
8. Giechaskiel, B., et al.: Framework for the assessment of PEMS (Portable Emissions Measurement Systems) uncertainty. Environ. Res. **166**, 251–260 (2018)
9. Musso, A., Corazza, M.V.: Visioning the bus system of the future: stakeholders' perspective. Transp. Res. Rec. **2533**, 109–117 (2015)
10. Corazza, M.V., et al.: From EBSF to EBSF_2: a compelling agenda for the bus of the future. A decade of research for more attractive and sustainable buses. In: EEEIC 2016 - International Conference on Environment and Electrical Engineering, art. no. 7555479IEEE Xplore, pp. 621–626 (2016)
11. Vicente, et al.: Road vehicle emission factors development: a review. Atmos. Environ. **70**, 84–97 (2013)
12. Borge García, R., Rafael, et al: Development of road traffic emission inventories for urban air quality modeling in Madrid (Spain). In: 21st International Emission Inventory Conference "Air Quality Challenges: Tackling the Changing Face of Emissions", 13–16 April, 2015, San Diego (EEUU), pp. 1–36 (2015)
13. Boulter, P.G.: Emission factors 2009: Report 6 - deterioration factors and other modelling assumption on road vehicles. TRL, London (2009)
14. NAEI - National Atmospheric Emission Inventory, "Method for applying emission degradation correction factors for the Copert 4 Nox for light duty petrol vehicles", Department for Business, Energy and Industrial Strategy London (2012)
15. Nagpure, A.S., Gurjar, B.R.: Development and evaluation of Vehicular Air Pollution Inventory model. Atmos. Environ. **59**, 160–169 (2012)
16. Hongzhao, D.O., et al.: A research on the vehicle emission factors of real world driving cycle in Hangzhou city based on IVE model. Automot. Eng. **33**(12), 1034–1038 (2011)
17. Guo, H., et al.: Evaluation of the International Vehicle Emission (IVE) model with on-road remote sensing measurements. J. Environ. Sci. **19**(7), 818–826 (2007)
18. Lents, J., Davis, N.: IVE model user's guide, model and data files. Technical report, US Environmental Protection Agency (2009). http://www.issrc.org
19. Davis, N., et al.: Development and application of an international vehicle emissions model. Transp. Res. Rec. **1939**, 156–165 (2005)
20. Corazza, M.V., et al.: Testing innovative predictive management system for bus fleets: outcomes from the Ravenna case study. IET Intel. Transport Syst. **12**, 286–293 (2018)
21. Jiménez-Palacios, J.L.: Understanding and quantifying motor vehicle emissions with vehicle specific power and TILDAS remote sensing. Thesis (Ph.D.), Massachusetts Institute of Technology, Dept. of Mechanical Engineering (1999)
22. Zhai, H., et al.: Vehicle-specific power approach to speed- and facility-specific emissions estimates for diesel transit buses. Environ. Sci. Technol. **42**(21), 7985–7991 (2008)
23. Liao, R., et al.: Analysis of emission effects related to drivers' compliance rates for cooperative vehicle-infrastructure system at signalized intersections. Int. J. Environ. Res. Public Health **15**(1), 122 (2018)
24. Onchang, R., et al.: Changes of air pollution and climate forcing emissions due to fuel switching to gasohol in motorcycle fleet in an urban area of Thailand. EnvironmentAsia **10**(2), 94–104 (2017)
25. Lai, J.: Development of city-specific driving cycles for transit buses based on VSP distributions: case of Beijing. J. Transp. Engineering **139**(7), 749–757 (2013)

26. Lozhkina, O.V., Lozhkin, V.N.: Estimation of nitrogen oxides emissions from petrol and diesel passenger cars by means of onboard monitoring. Transp. Res. Part D: Transport Environ. **47**(1), 251–264 (2016)
27. Cooper, B.: Sulphate emissions from automobile exhaust. Platin. Met. Rev. **20**(2), 38–45 (1976)

Evaluation of the Aesthetic Impact of Urban Mass Transportation Systems

Christos Pyrgidis(✉), Antonios Lagarias, Ioannis Garefallakis, Ioannis Spithakis, and Michele Barbagli

Civil Engineering Department, Aristotle University of Thessaloniki, Thessaloniki, Greece
pyrgidis@civil.auth.gr

Abstract. Visual nuisance is one of the environmental consequences caused by the operation of an urban mass transportation system and especially that of a surface system, such as a tramway and a BRT (Bus Rapid Transit), or that of an elevated system, such as a monorail. It constitutes a design and construction parameter of the system and it entails the system's aesthetics as a whole. It is of interest to the system's users but mainly to all the residents of the urban area that the system will be running through. By studying the current approaches for evaluating visual nuisance as is conducted in large transport projects today, a lack of an objective evaluation method can be observed. This paper presents a relatively objective and numerical method for evaluating the visual nuisance caused by each of these three transportation systems. The methodology proposed is based on assigning points to each of the different structural elements of these systems and then evaluating the entire system based on its overall score. The findings of this research may be applied during the bidding process for or at the design stage of a new system, at the evaluation of an existing system, or finally for the evaluation of corrective interventions aimed at upgrading an existing system.

Keywords: Visual nuisance · Tramway projects · Tramway aesthetics · Monorail projects · Bus rapid transit projects

1 Introduction

Visual nuisance constitutes a design and construction parameter of a transportation system and it entails the system's aesthetics as a whole, since that determines the degradation or improvement of the landscape in which the system is to be integrated in [1–3].

The evaluation of new urban mass transportation projects and the comparison of different possible solutions within a proposed system, require in an increasing amount the evaluation of the resulting visual nuisance [4, 5].

By studying the current approaches for evaluating visual nuisance, as such evaluations are conducted in large transport projects today, two major shortcomings have been identified [1–3]:

E. G. Nathanail et al. (Eds.): CSUM2020, AISC 1278, pp. 718–727, 2021.
https://doi.org/10.1007/978-3-030-61075-3_70

- The lack of an objective and quantitative method for evaluating visual nuisance. On the contrary, there exist a number of guidelines and best practices from various institutes, which cannot however constitute but a generalized and theoretical basis.
- The lack of an objective supervisory/regulatory authority that evaluates the visual nuisance of projects

The topic of visual impact of urban guided mass transit, and of urban mass transportation in general, have been subject to vast studies. Providing an exhaustive overview of results and lessons learned would go far beyond the limits of the scope of this paper. A good generic overview of the issue of visual impact of public transport is presented in [6]. In recent literature [1, 7] a relatively objective and numerical methodology for evaluating the visual nuisance caused by a tramway as a whole has been proposed. A pilot application of this methodology was undertaken for the tramway of Athens which is in operation. In reference [2] the aforementioned methodology, after appropriate modifications, was proposed for evaluating the visual nuisance caused by a monorail. This modified methodology was applied to the case of the Marconi Express monorail in Bologna, Italy, which is expected to start its operation within the end of 2020.

In this paper, the aforementioned methodology is further adapted in order to be applicable to a third urban mass transportation system, namely that of Bus Rapid Transit (BRT). An overview of the architecture of this innovative methodology is given, while for each individual step of the methodology a comparative analysis of the required for each transport system (tram, monorail, BRT) data, is attempted.

The findings of this research may be applied during the bidding process for or at the design stage of a new tramway/monorail/BRT system, at the evaluation of an existing system, or finally for the evaluation of corrective interventions aimed at upgrading an existing system [1–3]. They are mainly of interest to the assessors of urban railway systems, as the proposed methodology can replace their qualitative decision process with a quantitative approach for the evaluation of visual nuisance. They are also of interest to designers as it provides them with a list of best practices for the reduction of visual nuisance. The proposed methodology is also useful in order to provide decision makers with simple and solid quantitative data. In any case, it constitutes a scientific tool to support the applicability verification of environmental impacts of any specific urban guided mass transport system [3].

The proposed method finds limitations in its maturity level, not being adopted yet in a critical mass of studies in order to collect return of experience about robustness of the algorithm.

2 Proposed Methodology for Evaluating Visual Nuisance

The methodology proposed is already developed in a great extent in references [1, 2] and [3]. It is based on assigning points to each of the structural elements of these systems and then evaluating the entire system based on its overall score [7]. More specifically the following steps were considered:

1. The structural elements of a tramway, of a monorail and of a BRT system that contribute to visual nuisance were identified and recorded (Table 1).

Table 1. Structural elements of a tramway, of a monorail and of a BRT system that contribute to visual nuisance- Weighting factor per element (compilation for [1] and [2]).

Structural element	Tramway	Monorail	BRT
Exterior image of the rolling stock	$w_i = 3.0$	$w_i = 2.7$	$w_i = 2.8$
Interior image of the rolling stock	$w_i = 1.3$	$w_i = 1.9$	$w_i = 1.8$
Stops/Stations	$w_i = 2.7$	$w_i = 2.8$	$w_i = 2.8$
Electrification system	$w_i = 2.8$	n.a	n.a
Superstructure covering materials	$w_i = 2.7$	n.a	$w_i = 2.0$
Corridor separation techniques	$w_i = 2.3$	n.a	wi = 1.9
Signaling equipment	$w_i = 1.4$	n.a	$w_i = 1.2$
Depot(s)*	$w_i = 1.4$	n.a	$w_i = 2.0$
Pillars	n.a	$w_i = 3.1$	n.a
Guidance beams	n.a	$w_i = 2.7$	n.a
Emergency escape ways	n.a	$w_i = 2.3$	n.a
Sum	**17.6**	**15.5**	**14.5**

*In monorails depots, even though they may provoke visual nuisance, were excluded from the structural elements to be assessed. This choice was based on the arguments that are situated in a single point along the route and are usually located outside the core urban area of a city or in areas of lower aesthetic significance (e.g. industrial zones).

2. The different available aesthetic solutions for each of the structural elements and for each of the three examined systems were identified and recorded (Tables 2, 3 and 4 :1st column).
3. These aesthetic solutions were then categorized qualitatively into five aesthetic categories Aesth: O, A, B, C, D (Tables 2, 3 and 4 : 2nd column). In order to rank the aesthetic solutions in the five aesthetic categories the following criteria were taken into account:

   - The concealment of the structural element from the line of sight of observers
   - The limitation in number or in size of the different parts of the structural element
   - The providence during the design of the structural element for the reduction of its visual nuisance, meaning a design that considers the aesthetics of the element as a priority.

4. A score of Visual Nuisance Points VNPs was attributed to each of these aesthetic categories (0 to 4) (Table 5).
   A weighting factor w_i concerning the contribution to visual nuisance of each of the structural elements to the entire tramway/monorail/BRT system was identified (Table 1) [2, 7, 8]. In order to ascertain the impact of each structural element on the visual nuisance of the system as a whole, a series of interviews were conducted.

Table 2. Available aesthetic solutions per structural element of a tramway and their association with a specific aesthetic category (adapted for [5] - further elaboration).

Structural element – Aesthetic solution	Aesthetic category
Exterior image of the rolling stock	
• Modern vehicles which are designed exclusively for the system they are made for, while taking into account the existing character of the urban area that the system will be located in	A
• Modern vehicles with an innovative design that does not take into account the existing character of the urban area	B
• Conventional vehicles without any distinctiveness in their design	C
• Use of trams that are longer than 65 m	Degrade by one category
Interior image of the rolling stock	
• Innovative design, with large open spaces and no advertisements	A
• Large windows and adequate open spaces at eye level, limited use of advertisements	B
• Small windows or obstruction of the passengers' sight with many elements at eye level or extensive use of advertisements.	C
Stops	
• Mainly small and discreet stops. Limited size of structural elements and mainly use of glass or thin metallic parts	A
• Stops with a distinctive design that are integrated in the urban area in which they are constructed. The design takes into account the reduction of visual nuisance through the use of transparent or thin parts	A
• Stops with a distinctive architectural design, with the use however of large structural elements that hide part of the sky or the urban area	B
• Conventional stops with large structural elements. No effort to reduce visual nuisance	C
• Placement of advertisements on the surfaces of the stop (over 50%)	D
Electrification system	
• No use of catenary wires and electrification poles. Ground level electrification (free catenary system) [9]	O
• Effort to limit the amount of catenary wires per track	B
• Effort to limit the amount of electrification poles, use of existing buildings to support catenary wires	B
• No effort to reduce the amount of catenary wires or electrification poles.	C
Tramway superstructure covering materials	
• Use of cover elements (turf or colored stones) that have a visual continuity with the surrounding landscape, meaning that the cover materials appear to be a continuation of the surrounding ground [10]	A
• Use of cover elements or colored stones without however taking into account the visual continuity with the surrounding landscape	B
• Tramway corridors with no covering materials	C
Tramway corridor separation techniques	
• Separation (from the other means of transport) with the use of small structural elements that are designed to improve the area's aesthetics (vegetation or well-designed elements)	A
• Separation with the use of small structural elements, poles or fences	B
• Separation with the use of large structural elements, poles or fences or other solid non-transparent elements that are over 1m in height	C
• Placement of advertisements on the structural elements used for separation	D
Signaling equipment	
• Effort to limit the use of signaling equipment or use of a distinctive design of poles and signs	B
• Use of conventional signaling poles	C
Depot(s)	
• The depot is placed outside of the urban area	O
• The depot is placed within the urban area but there is consideration for its aesthetics	B
• The depot is placed within the urban area but there is no consideration for its aesthetics	C

Table 3. Available aesthetic solutions per structural element of a monorail system and their association with a specific aesthetic category (adapted for [6] - further elaboration).

Structural element – Aesthetic solution	Aesthetic category
Exterior image of the rolling stock	
• Modern vehicles which are designed exclusively for the system they are made for, while taking into account the existing character of the urban area that the system will be located in	A
• Modern vehicles with an innovative design that does not take into account the existing character of the urban area	B
• Conventional vehicles without any distinctiveness in their design or any association with the characteristics of the urban landscape.	C
• Use of long train sets (> 70 m)	Degrade by one category
Interior image of the rolling stock	
• Innovative design, with large open spaces and no advertisements	O
• Large windows and adequate open spaces at eye level, limited use of advertisements	A
• Small windows or obstruction of the passengers' sight with many elements at eye level or extensive use of advertisements.	B
Stations	
• Integration of stations within existing buildings without significant alterations to their face.	O
• Stations with use of transparent or thin parts in combination with a concurrent limitation to their size to one minimally satisfying their functional requirements as well as a placement near ground level so as to avoid hiding part of the sky.	A
• Stations with use of transparent or thin parts or with a limitation to their size to one minimally satisfying their functional requirements or with a placement near ground level so as to avoid hiding part of the sky.	B
• Conventional stops with large structural elements. No effort to reduce visual nuisance	C
Pillars [6]	
• Limited size of pillars with the use of thinner cross-sections.	B
• Adoption of one or even a combination of the following ways (covering the pillars with some form of vegetation or with reflective panels)	B
• No consideration for reducing the visual nuisance caused by pillars.	C
• Construction of pillars along higher altitude areas that are visible throughout the city (in case that this solution is not chosen for panoramic view reasons).	Degrade by one level
Choice of route [11]	
• If a route offering scenery of a particular value to the passengers is chosen (Choice of a route offering scenery of a particular value to the passengers)	Reduction of 0.5 VNP points from the final value of VN factor
Escape ways	
• No escape way present	O
• Escape way existing, but concealed within the guidance beam overall gauge	A
• Escape way existing and not concealed within the guidance beam overall gauge	B
Guidance beams	
• Use of beams with reduced width and height (lower than 0.7m and 1.4m respectively).	B
• Conventional monorail beam	C

Table 4. Available aesthetic solutions per structural element of a BRT system and their association with a specific aesthetic category.

Structural element – Aesthetic solution	Aesthetic category
Exterior image of the rolling stock	
• Modern buses which are designed exclusively for the urban area (city) they are made for, while taking into account the existing character of the urban area (city)	A
• Modern buses with a new and innovative design that does not take into account the existing character of the urban area (city)	B
• Conventional buses without any distinctiveness in their design or in relation to the existing character of the city	C
• Use of buses that are longer than 18.5 m	Degrade by one category
Interior image of the rolling stock	
• Innovative design, with large open spaces and no advertisements	A
• Large windows and adequate open spaces at eye level, limited use of advertisements	B
• Small windows or obstruction of the passengers' sight with many elements at eye level or extensive use of advertisements.	C
Stops	
• Mainly small and discreet stops. Limited size of structural elements and mainly use of glass or thin metallic parts	A
• Stops with a distinctive design that are integrated in the urban area in which they are constructed. The design takes into account the reduction of visual nuisance through the use of transparent or thin parts	A
• Stops with a distinctive architectural design, with the use however of large structural elements that hide part of the sky or the urban area	B
• Conventional stops with large structural elements. No effort to reduce visual nuisance	C
• Placement of advertisements on the surfaces of the stop (over 50%)	D
BRT superstructure covering materials [9]	
• Use of cover elements (colored bus lanes)	A
• BRT corridors with no covering materials (Conventional pavement)	C
BRT corridor separation techniques	
• Separation with the use of small structural elements that are designed to improve the area's aesthetics (vegetation or well-designed elements)	A
• Separation with the use of small structural elements, poles or fences	B
• Separation with the use of large structural elements (poles or fences or other solid non-transparent elements)	C
• Placement of advertisements on the structural elements used for separation	D
Signaling equipment	
• Effort to limit the use of signaling equipment or placement of multipurpose poles or use of a distinctive design of poles and signs	B
• Use of conventional signaling poles	C
Depot(s)	
• The depot is placed outside of the urban area	O
• The depot is placed within the urban area but there is consideration for its aesthetics	B
• The depot is placed within the urban area but there is no consideration for its aesthetics	C

Table 5. Visual Nuisance Points per aesthetic category (compilation for [1–3]).

Aesthetic category	**VNP**
Solution Category O Concealment of the structural element from any observers	0
Solution Category A No option for concealment – Effort to limit the size of the element that is observable and simultaneous significant effort for an aesthetically pleasing design of the element	1
Solution Category B No option for concealment – Either an effort to limit the size of the element that is observable or significant effort for an aesthetically pleasing design of the element	2
Solution Category C No option for concealment – No effort to limit the size of the element that is observable and no effort for an aesthetically pleasing design of the element	3
Solution Category D Increase of visual nuisance through the use of larger structural elements or use of additional elements that are not functionally necessary and obstruct the sight of an observer (advertisements, large non-transparent elements)	4

Table 6. Evaluation of the total Visual Nuisance caused by a tramway, a monorail and a BRT system (compilation for [1, 2] and further elaboration).

Total VN value Tramway	Total VN value Monorail	Total VN value BRT	System evaluation
0.84–1.68	0.42–1.31	0.94–1.79	Visual Nuisance Qualitative Category I The tramway/monorail/BRT has reduced to a large extent the visual nuisance it causes. It has taken this parameter into consideration at the design level and has chosen effective solutions in partially or totally concealing the structural elements from observers. At the same time a priority was given to its tasteful design. It has a low negative impact on the image of the urban area while at the same time it includes visually pleasant elements
1.68–2.52	1.31–2.20	1.79–2.64	Visual Nuisance Qualitative Category II The tramway/monorail/BRT has partially reduced the visual nuisance it causes. It has limited the size and intrusiveness of some elements and has improved their aesthetics. It has a medium negative impact on the image of the urban area. There might be the need for individual corrective actions in some of the areas in which it operates
2.52–3.36	2.20–3.10	2.64–3.51	Visual Nuisance Qualitative Category III The tramway/monorail/BRT system has taken few or no measures in reducing the visual nuisance it causes. Its structural elements limit the line of sight of observers to a large extent, while their design is neutral or unpleasant. The railway system has a high negative impact on the image of the urban area and is in need of corrective actions to limit the visual nuisance it causes

The interviewees (12 for the tramway, 16 for the monorail and 6 for the BRT) were all engineers of various disciplines, including architects, urban design engineers and traffic engineers. No ordinary users or observers were interviewed, the reasons for this choice are:

- A questionnaire to normal users or observers would have needed to be approached in a consistent way from a statistical point view. The pool of interviewed persons should have been identified in size and composition in order to be representative of public opinion of the specific project. This would have meant to define a specimen of the population of a specific location big enough to be statistically relevant, and representative in terms of age, gender, social and economic status, economic conditions, user profile of the system etc. of all potential users and observers. This would be far out of the scope of this paper.
- A questionnaire to normal users would have a validity only when applied to users of a specific project. Responses will be biased by local factors, and this would not be useful for the generic approach of this paper.

The interviews resulted in the weighting factors shown in Table 1 by using the following approach:

- The interviewees were presented with the suggested methodology
- Each of the structural elements was given a grade from 3 (least impact) to 10 (most impact) by the interviewees based on the impact that particular element has on the visual nuisance caused by the system.
- The mean grade for each structural element was calculated. The mean grade was then divided by a reference value (the minimum grade of 3) and rounded to the nearest decimal point.

5. A formula for estimating the overall level of Visual Nuisance VN that is caused by the operation of a tramway/monorail/BRT system was proposed (Eq. 1) [7]

$$VN = \frac{\sum_i (w_i \times VNP_i)}{\sum_i w_i} \tag{1}$$

Where:

- VNPi: is the Visual Nuisance Points of every structural element i of the tramway/monorail/BRT. They are dependent on the qualitative category of the aesthetic solution chosen to reduce the visual nuisance caused by structural element i.
- w_i: is the weighting factor that defines the level of influence every structural element i has on the visual nuisance caused by the system as a whole. It is dependent on the size of the structural element, its construction site, its final location and the influence it exerts on the perception of observers.

6. The overall VN score that is derived from the application of this formula, may be used to evaluate a tramway/monorail/BRT system as a whole using three qualitative categories (I, II or III) as shown in Table 6

Taking into account for the 3 reviewed types of urban transport systems:

- The weighting factors w_i per structural element, as those are presented in Table 1
- The different available aesthetic solutions for each of the structural elements in combination with the aesthetic category in which they belong (O, A, B, C, D), as those are presented in Tables 2, 3 and 4
- The special circumstances, such as choice of route or length of vehicles, under which aesthetic categories are downgraded or upgraded, as those are presented in Table 2, 3 and 4
- Equation 1

The following maximum and minimum achievable Visual Nuisance Points per system may be derived:

- Tramway: Max 3.36 Min 0.84
- Monorail: Max 3.10 Min 0.42
- BRT: Max 3.51 Min 0.94

Based on these limit values the overall Visual Nuisance score may be used to evaluate a tramway/monorail/BRT system as a whole using three qualitative categories (I, II or III) as shown in Table 6.

7. Table 6

The proposed methodology was based on three basic assumptions [1–3]:

- Regardless of any subjective visual pleasure caused at the sight of some structural elements, it is assumed that every structural element of the three examined systems, except the ones that are concealed, causes visual nuisance.
- The structural elements of the systems are considered to be independent to one another. This means that any changes in the visual aspects of one structural element do not affect any other structural element, but only the system as a whole.
- The existing landscape is considered as "ideal". This means that the former state of the landscape is not taken into account. This methodology investigates the best practices for the systems themselves and does not go into any comparison with the former state of the landscape.

3 Conclusions

In this paper a methodology is proposed for evaluating the visual nuisance caused by three urban mass transportation systems: A tramway, a (straddled) monorail and a BRT.

The findings of this research may be applied during the bidding process for or at the design stage of these systems, at the evaluation of an existing system, or finally for the evaluation of corrective interventions aimed at upgrading an existing system. They are mainly of interest to the assessors of urban railway systems, as the proposed methodology can replace their qualitative decision process with a quantitative approach for the evaluation of visual nuisance. They are also of interest to designers as it provides them with a list of best practices for the reduction of visual nuisance. In any case, it constitutes a scientific tool to support the applicability verification of environmental impacts of any specific project [2, 12].

It may be concluded that of all the structural elements of a tramway system, the exterior image of the rolling stock, the electrification system, the stops and the tramway superstructure covering materials play the most vital role in how intrusive a tramway system is to the aesthetics of an observer (both a user and an outside observer).

Concerning the monorail, the pillars play the most vital role in how intrusive a monorail system is. This type of elements may susceptible to improvements regarding their visual design and intrusiveness. Additionally, in case where the monorail serves transportation of passengers through areas that are of particular interest in terms of view, the innovation / futuristic factor of the monorail can, contrary to the norm, be considered as a suitable and pleasant presence.

Finally, in the case of a BRT, the exterior image of the rolling stock and the stops are the structural elements that produce the highest optical nuisance.

References

1. Pyrgidis, C., Lagarias, A., Dolianitis, A.: The aesthetic integration of a tramway system in the urban landscape – evaluation of the visual nuisance. In: 4th Conference on Sustainable Urban Mobility – CSUM2018, Skiathos Island, Greece, 24–25 May 2018
2. Pyrgidis, C., Barbagli, M.: Evaluation of the aesthetic impact of urban monorail systems. In: 11th International Monorail Association Annual Conference (Monorailex 2019), Chiba, Japan, 24–26 November 2019
3. Pyrgidis, C.: Applicability verification - a supporting tool for undertaking feasibility studies on urban railway systems. In: 9th International Congress ICTR 2019, Athens, 24–25 October 2019
4. LI/IEMA: Guidelines for Landscape and Visual Impact Assessment. Spon, London (2002)
5. EPD: EIAO Guidance Note No. 8/2010 – Preparation of Landscape and Visual Impact Assessment under the Environmental Impact Assessment Ordinance, Hong Kong (2010)
6. Musso, A., Piccioni, C.: Interazioni con il territorio e l'ambiente, § B3.3.3 – Impatto visivo. In: Corona, G., Festa, D.C. (eds.) Trasporto Pubblico Locale: Risorse, Pianificazione, Esercizio. Edited by , 1st edition, EGAF edizioni, Forlì (2015). ISBN 978-88-8482-631-2
7. Lagarias, A.: The aesthetic integration of railway systems in urban space: a methodology for evaluating visual nuisance. Project of final thesis, Aristotle University of Thessaloniki, Greece (2015)
8. Anastasiou, I., Nikolos, A.: Technical and operational applicability verification of a BRT system. Project of final thesis, Aristotle University of Thessaloniki, Greece (2019)
9. Novales, M.: Overhead wires free light rail systems. In: 90th TRB Annual Meeting, Washington (2011)
10. Zantopoulos, C., Pyrgidis, C., Sapounas, D.: Design, construction and cost evaluation of tramway superstructure. In: 8th International Conference on Transport Research, Thessaloniki (2017)
11. Garefallakis, J., Spithakis, J.: Technical and operational applicability verification of a monorail system. Project of final thesis, Aristotle University of Thessaloniki, Greece (2019)
12. Pyrgidis, C.: Railway Transportation Systems: Design, Construction and Operation, 1st edn. CRC Press, Boca Raton (2016)

A Methodological Approach for Estimating Urban Green Space: The Case of Thessaloniki, Greece

Alexandros Sdoukopoulos(✉)

Transport Engineering Laboratory, Faculty of Engineering, Department of Civil Engineering, Aristotle University of Thessaloniki, 54124 Thessaloniki, Greece
asdoukop@civil.auth.gr

Abstract. Sustainable Urban Mobility Plans (SUMPs) are considered as one of the most valuable policy tools for adapting the Sustainable Development Goals to the urban level and alter the way people move and live towards more sustainable patterns. SUMPs are strategic plans that focus on the performance of the transport system, but contrarily to traditional transport studies, they also thoroughly examine the interactions between transport and its interdependent systems, such as land-uses. However, the transition from theory to practice points out that a considerable number of SUMPs give little attention to significant land-use related issues, such as the available green space, mainly due to the lack of available data and/or easy-to-use calculation tools. In this context, the current paper presents a methodological approach for estimating Urban Green Space (UGS). As the proposed approach aims at illustrating in an actual manner how "green" a city is, it moves beyond the areas that are characterised in urban development plans as green. Thus, it takes into consideration many different types of green spaces, e.g. public and private, street, institutional, residential, parks. The developed methodological approach is a semi-automated GIS-based process that utilises the robust method of the Normalised Difference Vegetation Index (NDVI) and uses remote sensing imagery as well as cadastral data. The proposed methodological approach was implemented in the seven Municipalities of the urban agglomeration of Thessaloniki, and the results highlight significant differences in the available green space per capita between the examined areas.

Keywords: Urban Green Space · NDVI · Thessaloniki

1 Introduction

After three decades of considerable work done both by countries and the United Nations (UN), the sustainability agenda, which was first introduced by the Earth Summit in Rio de Janeiro in 1992 [1] remains more topical than ever. In the light of global challenges such as climate change, environmental degradation, and social inequality, the UN adopted in 2015 the new "2030 Agenda for Sustainable Development", which encompasses 17 Sustainable Development Goals (SDGs) [2]. SDGs are the outcome of a long-term,

E. G. Nathanail et al. (Eds.): CSUM2020, AISC 1278, pp. 728–738, 2021.
https://doi.org/10.1007/978-3-030-61075-3_71

evolutionary process and constitute a comprehensive blueprint for altering the way people live and move towards more sustainable patterns [2].

Given the great role cities are undertaking in terms of population (55.3% of the world population lives in cities), social equity (urbanisation often produces diverse forms of exclusion and deprivation), economic development (80% of the global GDP is accounted by cities) and environmental performance (cities face multiple broad environmental challenges and generate negative externalities), the adaption of SDGs to the urban level constitutes an issue of primary importance, yet a complex and challenging task [3–4].

The development of Sustainable Urban Mobility Plans (SUMPs) is a critical step towards bringing cities in line with the SDGs, namely making cities "inclusive, safe, resilient and sustainable" [2]. SUMPs are strategic plans that aim at satisfying mobility needs for an enhanced quality of life [5]. They focus on the performance of the transport system, but contrarily to the traditional transport studies, SUMPs also thoroughly examine the interactions between transport and its interdependent systems, such as land-uses. Moreover, SUMPs are not an exclusive domain of transport planners, as other experts such as urban planners, environmental engineers and marketing managers are also actively involved in the planning process [5, 6].

Despite the above, the transition from theory to practice points out that a considerable number of SUMPs experience difficulties in fully embracing the sustainable transport planning principles, mainly as the combined result of very tight time schedules, limited budget and resources, lack of available data as well as easy-to-use calculation tools. In this context, SUMPs often disregard the two-way relationship between transport and land-use system [7] and, thus, give little attention to significant land-use related issues, such as the available Urban Green Space (UGS).

Although the benefits of UGS are widely recognised and accepted, its calculation, mapping and assessment remain a challenge [8], especially within the usually tight framework of SUMPs. Hence, SUMPs are often called upon to face dilemmas such as "should this land become a park or a parking lot?", by relying only on traffic data and not on the complete picture.

In this framework, the scope of the current research is to develop a simple, easy-to-implement and reliable methodological approach to illustrate and estimate UGS, and subsequently apply it to the seven Municipalities that comprise the urban agglomeration of Thessaloniki, Greece. The described methodology utilises the concrete method of the Normalised Difference Vegetation Index (NDVI) and constitutes a semi-automated process that involves the use of an ad hoc, user-friendly GIS toolbox. It should be pointed out here that this toolbox aims at becoming a valuable tool both in the context of SUMPs, and other environmental researches.

2 Urban Green Space: Definition and Benefits

Although Urban Green Space is at first glance a term easily understandable, it is quite a challenge to be defined in a clear and unequivocal manner, due to its great variation and complexity [9]. Moreover, according to the scientific field and the context of each research, UGS is usually classified into several categories based on its type, form, functionality and ownership [10]. The review carried out by Taylor and Hochul [9] highlighted that six different types of definition could be identified in the literature, whilst a great share of relevant publications do not specify green space explicitly [9]. Within the scope of the current paper, Urban Green Space is defined as "any vegetation found in the urban environment, including urban forests, parks, open spaces, parklets and small pocket parks, street trees, private residential gardens and green roofs, except for sporting fields as well as informal green in vacant land" [11, 12]. This wide-ranging consideration of UGS types, which goes far beyond the green areas described in the urban development plans, reflects the significant role and impact of each green element in an urban green network, especially in compact cities like the Southern European ones [13–16].

As far as the benefits of Urban Green Space are concerned, UGS is a vital component of every city, as it provides crucial ecosystem services and delivers multiple benefits to the community [17, 18]. To begin with, UGS upgrades the quality of the environment of a city considerably. Green space provides natural, positive soundscapes, while at the same time, it contributes to the reduction of traffic noise [19, 20]. Moreover, UGS is considered as an effective air pollution mitigating measure, as vegetation absorbs specific air pollutants both from the atmosphere and the indoor environment [21, 22]. Furthermore, UGS helps to reduce urban heat island effects and moderate temperature [23]. According to several authors, the availability of Urban Green Space is highly correlated with great benefits to public health, including reduced mortality rates, improved mental health and wellbeing [24–28]. Also, UGS is often used for leisure and recreational activities, enhancing thus everyday life [29]. Lastly, green space is a critical element in developing a city's branding strategy and cultural identity [30].

3 Methodological Approach

The methodological approach followed in this research falls within the broad category of the remote sensing techniques. Its main goal is to enable even an unexperienced GIS user to calculate accurately through NDVI how green an urban area is.

NDVI is a robust vegetation index delivering reliable results, widely applied in a number of various studies such as monitoring of green space, status of growing crops, etc. [31]. It is based on the reflectance properties of vegetation in red and near-infrared wavelengths and is calculated as follows:

$$NDVI = (NIR - Red)/(NIR + Red) \tag{1}$$

Where *Red* and *NIR* stand for the spectral reflectance measurements acquired in the red (visible) and near-infrared regions, respectively [32]. Very low values of NDVI (<0.1) correspond to soil, sand and rock; moderate values (0.2–0.5) indicate low vegetation like shrubs, lawn and grassland; while high values (>0.6) represent dense vegetation such as temperate and tropical trees [31].

However, estimating UGS and extracting vegetation through NDVI in a GIS environment is a demanding task consisting of several intermediate steps. To this end, an ad hoc GIS toolbox was built in ESRI ArcGIS Model Builder, consisting of 3 successive tools, with a view to automate the vast majority of the necessary steps and minimize the required user actions as well as input data.

More specifically, as a starting point, the first tool, namely "NDVI and Initial Selection" should be executed, and the Red and NIR bands of the multispectral remote sensing imagery should be imported. This tool has as a predefined vegetation threshold the value 0.2 as multiple studies converge on this [32]; however, users could easily modify it and set a new value. The workflow of the tool mentioned above is visualised in Fig. 1.

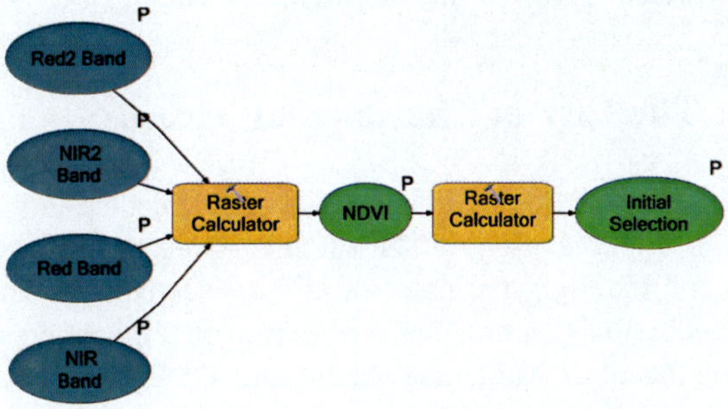

Fig. 1. The geoprocessing workflow of the "NDVI and Initial Selection" tool.

As a second step, the "Remove Shadows" tool should be run. The shadow effect caused by buildings is quite intense, especially in urban areas, and has a significant impact on the results of NDVI. Hence, based on a visual/sampling check to the output of the previous step, specific NDVI values corresponding to shadow pixels have to be inserted in order to be removed. The workflow of the second successive tool is illustrated in Fig. 2.

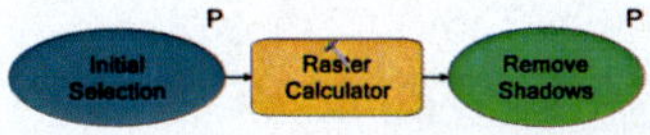

Fig. 2. The workflow of the "Remove Shadows" tool.

At the final stage, one of the two alternative tools described below should be selected, based on data availability. More precisely, the first tool, being more basic, involves the minimum input data, whilst the second one, as more advanced, requires a more comprehensive database. Thus, the "Calculation of UGS" tool requires as input, a polygon feature class defining the study area, while on the other hand, the "Optimised Calculation of UGS" tool requires additionally three polygon feature classes representing: a) buildings, b) land plots, and c) all allowed land-uses, but open spaces and areas described as green in urban plans. The extra feature classes involved in the last tool are used to leave out from consideration the green area in sporting fields and vacant lots. It should be pointed out here that both tools deliver a polygon feature class which illustrates the UGS of the examined area as well as a Microsoft Excel file with the total area of UGS given in square meters. These two outputs could greatly facilitate both further spatial analyses and calculation of relevant indicators, such as the "Green area per capita". The geoprocessing workflows of the two tools are presented in Fig. 3.

The "Calculation of UGS" tool is appropriate for making a rough estimation of UGS in an area where available data are inadequate, while it is also proposed for calculating UGS with sufficient accuracy in urban areas with a limited number of vacant lots. On the contrary, "Optimised Calculation of UGS" tool approaches more thoroughly the assessment of UGS, as it excludes from the final output specific categories of green such as the informal green in vacant lots, being thus entirely consistent with the definition of UGS given in this research.

4 Case Study: The City of Thessaloniki, Greece

The methodological approach described above was implemented in the Municipalities comprising the urban agglomeration of Thessaloniki, Greece, with a view to estimate the available UGS and highlight the potential differences between the various areas. Thessaloniki is the second-largest urban area of Greece in terms of area (approximately 60 km^2), population (nearly 1 Mio. Inhabitants) and GDP (over 15,000 Mio. Euro) [33, 34]. At the same time, it constitutes the capital and administrative centre of the region of Central Macedonia and a major commercial gateway for the Balkans.

The urban agglomeration of Thessaloniki consists of seven Municipalities, namely Ampelokipi-Menemeni, Kalamaria, Kordelio-Euosmos, Neapoli-Sykies, Pavlos Melas, Pilea-Chortiatis, and Thessaloniki. It should be noted here that the municipal boundaries of Pilea-Chortiatis extend far beyond the functional boundaries of the urban agglomeration. Hence, only the municipal units of Pilea and Panorama were considered in this case. In this framework, industrial areas or areas not being part of the urban core, such as the Southwest part of the Municipality of Ampelokipi-Menemeni and the Nothern area of the Municipality of Pavlos Melas namely Philothei, respectively, were also excluded from the analysis.

In the present research, a relatively recent (2016) high-quality four-band imagery, with a resolution of 0.5 m was used to estimate UGS in the study area. Moreover, the latest population census (2011) carried out by the Hellenic Statistical Authority was utilised for calculating the indicator "Green area per capita". As far as the other input data are concerned, the feature classes illustrating the municipal boundaries as well as the buildings were again derived from the Hellenic Statistical Authority, while the polygon feature classes representing the land plots were obtained from the Hellenic Cadastre. Lastly, the land-uses feature classes were created from scratch based on the relevant maps which are included in the respective urban plans published in the official Greek Government Gazette.

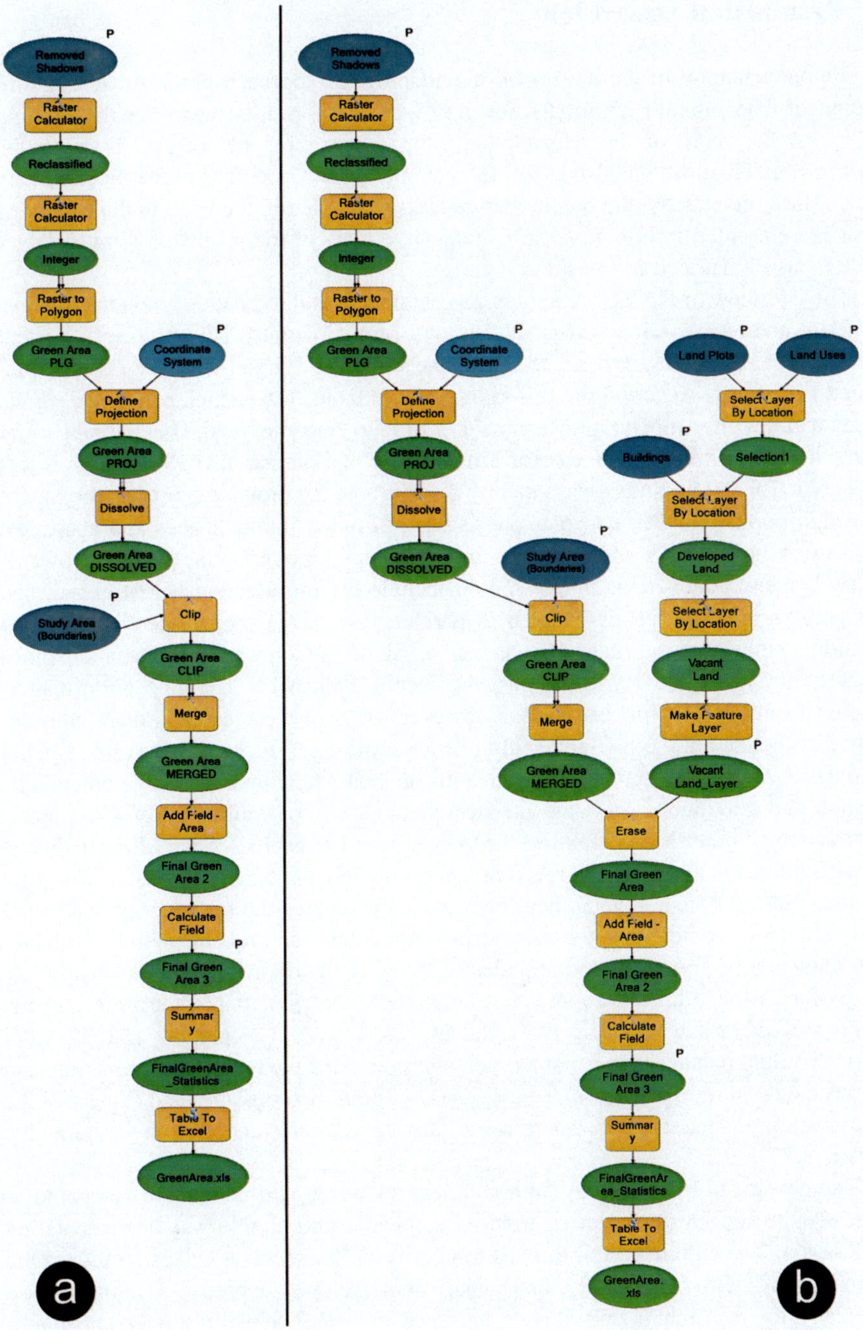

Fig. 3. The workflows of the "Calculation of UGS" and "Optimised Calculation of UGS" tools (3a and 3b respectively).

5 Results and Discussion

The implementation of the developed methodological approach in the urban agglomeration of Thessaloniki highlights not only interesting results regarding the existing UGS and the score of the relevant indicator "Green area per capita" between the various Municipalities but also useful facts about the pros and cons of the methodology itself. More specifically, the output feature classes containing the UGS in the examined areas are presented in Fig. 4, while an indicative view of the results on a much larger scale is also illustrated in the same Figure.

The overview of the UGS mapping gives a first, visual indication about the unequal distribution of green area across the various Municipalities. This finding is further specified and quantified by the scores of the indicator "Green area per capita", presented in Table 1. According to this table, the available "Green area per capita" in the Municipalities of Ampelokipi-Menemeni, Kordelio-Euosmos and Thessaloniki is far below ten sq. m, while in the case of Municipality of Kalamaria, UGS is just over this threshold. The above findings are mainly the result of the prevailing settlement pattern (i.e. high population density areas consisting of limited public spaces and apartment blocks with no gardens), and underline the emergency for altering the priorities towards a new "green" policy. This policy should include the introduction of both small and large-scale green interventions, such as parklets and new parks, respectively, and it should be embedded in urban plans as well as SUMPs. Concerning the Municipalities of Neapoli-Sykies and Pavlos Melas, the results indicate a moderate performance, ranging from 13 to 19 m^2 per capita. However, UGS in these cases is disproportionately distributed among the municipal units, as some areas such as Pefka and Polichni account for the largest share of green area inside their Municipalities, while others, like Neapoli and Stavroupoli, are characterised by much lower availability of green areas. Contrary to all other Municipalities, the "Green area per capita" in the Municipality of Pilea-Chortiatis is notable high, reaching approximately the 50 m^2. The latter is mostly the result of a different development plan, which promotes the low-density residential uses with detached houses and extensive private gardens, and it is indicative of a better environmental performance and an enhanced quality of life in general.

As far as the followed approach is concerned, the pilot implementation and the analysis of the results indicate the following facts. The "Calculation of UGS" tool strikes the right balance between the necessary input data, the required process time and the accuracy of the results, in study areas with limited vacant land, such as the Municipality of Thessaloniki, where the difference in the results of the two alternative tools is 8.7%.

The developed methodology, although less advanced and innovative compared to other remote sensing techniques, provides a reliable and simple way for calculating UGS easily. The automation of the great majority of the necessary steps facilitates the estimation of UGS significantly, and makes the developed approach a valuable tool, especially for researchers and experts involved in comprehensive studies and plans, such as SUMPs, where multiple variables and parameters have to be considered. On the other hand, the approach described above requires increased computing power, while it

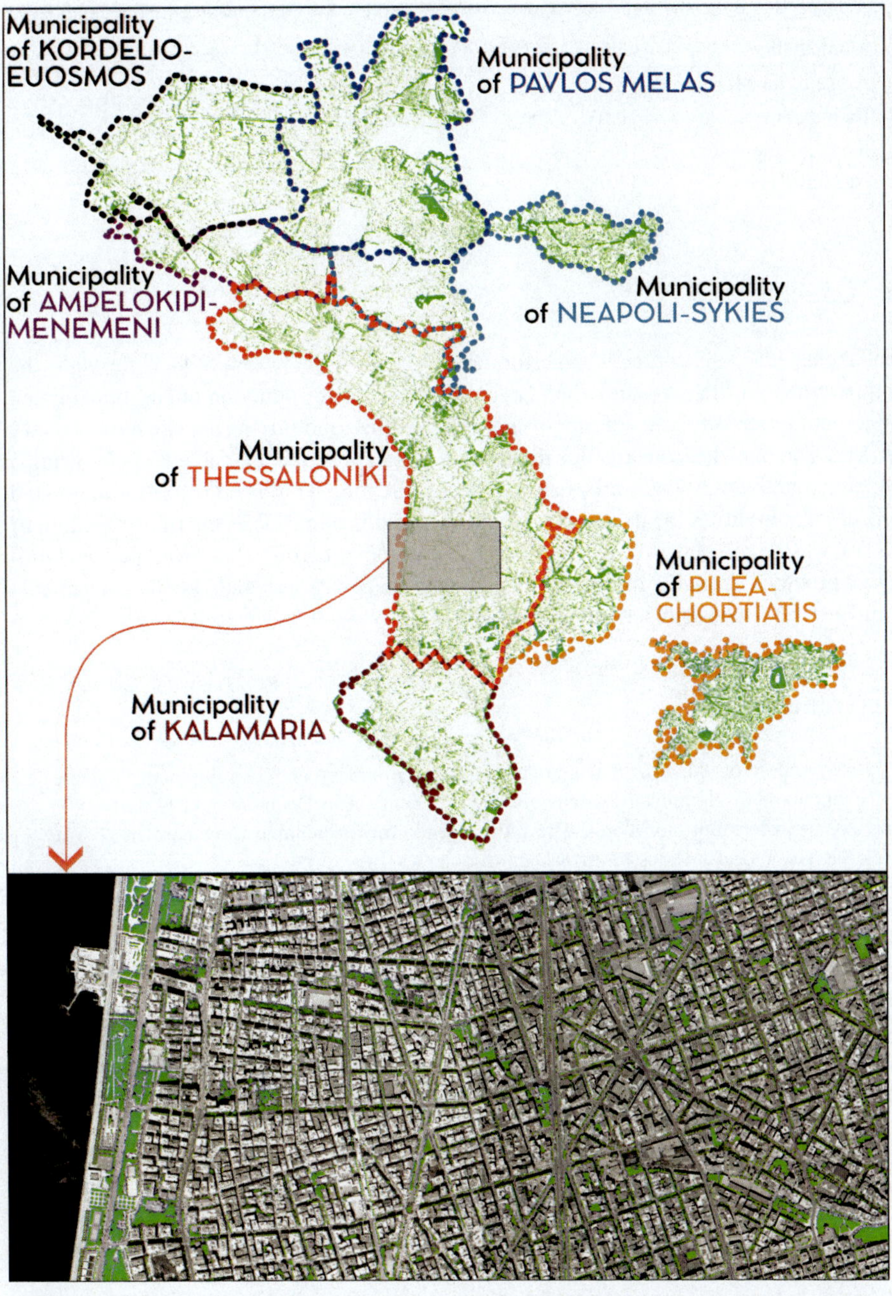

Fig. 4. UGS in the seven Municipalities constituting the urban agglomeration of Thessaloniki.

is also time-consuming. Moreover, as NDVI is sensitive to seasonal variability, the selection of imagery date plays a key role in delivering solid results.

Table 1. Scores of the indicator "Green area per capita" for the considered Municipalities.

Municipality	Green area (sq. m)/cap	Municipality	Green area (sq. m)/cap
Ampelokipi-Menemeni	6.08	Pilea-Chortiatis	49.53
Kalamaria	10.03	Pavlos Melas	18.87
Kordelio-Euosmos	8.19	Thessaloniki	7.15
Neapoli-Sykies	13.08		

6 Conclusions

Enhancing the environmental performance of an urban area either through the improvement of the existing Urban Green Space or the introduction of additional green areas, comprise an issue of major importance that should be considered in every SUMP. The first step towards this direction is to calculate the available UGS by using a concrete methodology as well as an easy-to-handle tool. To this end, a semi-automated process that involves the use of an ad hoc GIS toolbox and utilises the robust method of NDVI was developed in the context of the current research. The developed methodology provides a reliable means of calculating UGS easily and could become a valuable tool for relevant researchers and experts.

References

1. UN: Report of the World Commission on Environment and Development, Agenda 21: Programme of Action for Sustainable Development – Rio Declaration (1992)
2. UN: Transforming our World: The 2030 Agenda for Sustainable Development (2015)
3. UN: The World's Cities in 2018-Data Booklet (2018)
4. UN-Habitat: Urbanization & Development: Emerging Futures-World Cities Report (2016)
5. Rupprecht Consult: Guidelines for Developing and Implementing a Sustainable Urban Mobility Plan, Second Edition (2019)
6. Perra, V.M., Sdoukopoulos, A., Pitsiava-Latinopoulou, M.: Evaluation of sustainable urban mobility in the city of Thessaloniki. Transp. Res. Procedia **24C**, 329–336 (2017)
7. Chang, J.S.: Models of the relationship between transport and land-use: a review. Transp. Rev. **26**(3), 325–350 (2006). https://doi.org/10.1080/01441640500468432
8. Swanwick, C., Dunnett, N., Woolley, H.: Nature, role and value of green space in towns and cities: an overview. Built Environ. **29**, 94–106 (2003)
9. Taylor, L., Hochuli, D.F.: Defining greenspace: multiple uses across multiple disciplines. Landscape Urban Plan. **158**, 25–38 (2017)
10. Vargas-Hernández, J.G., Pallagst, K., Zdunek-Wielgołaska, J.: Urban green spaces as a component of an ecosystem. In: Dhiman, S., Marques, J. (eds.) Handbook of Engaged Sustainability. Springer, Cham (2018)
11. Kabisch, N., Haase, D.: Green spaces of European cities revisited for 1990–2006. Landscape Urban Plan. **110**, 113–122 (2013)
12. Haaland, C., van Den Bosch, C.K.: Challenges and strategies for urban green-space planning in cities undergoing densification: a review. Urban Forest. Urban Greening **14**(4), 760–771 (2015)

13. Peschardt, K.K., Schipperijn, J., Stigsdotter, U.K.: Use of small public urban green spaces (SPUGS). Urban Forest. Urban Greening **11**, 235–244 (2012)
14. WHO: Urban Green Space Interventions & Health-A review of impacts & effectiveness (2017)
15. Tzoulas, K., Korpela, K., Venn, S., Yli-Pelkonen, V., Kaźmierczak, A., Niemela, J., James, P.: Promoting ecosystem and human health in urban areas using green infrastructure: a literature review. Landscape Urban Plan. **81**(3), 167–178 (2007)
16. Kabisch, N., Strohbach, M., Haase, D., Kronenberg, J.: Urban green space availability in European cities. Ecol. Ind. **70**, 586–596 (2016)
17. Norton, B.A., Coutts, A.M., Livesley, S.J., Harris, R.J., Hunter, A.M., Williams, N.S.: Planning for cooler cities: a framework to prioritise green infrastructure to mitigate high temperatures in urban landscapes. Landscape Urban Plan. **134**, 127–138 (2015)
18. Wolch, J.R., Byrne, J., Newell, J.P.: Urban green space, public health, and environmental justice: The challenge of making cities 'just green enough.' Landscape Urban Plann. **125**, 234–244 (2014)
19. van Renterghem, T., Forssén, J., Attenborough, K., Jean, P., Defrance, J., Hornikx, M., Kang, J.: Using natural means to reduce surface transport noise during propagation outdoors. Appl. Acoust. **92**, 86–101 (2015)
20. Groenewegen, P., van den Berg, A., de Vries, S., Verheij, R.: Vitamin G: Effects of green space on health, well-being, and social safety. BMC Public Health **6**(1), 149 (2006)
21. Mueller, W., Steinle, S., Pärkkä, J., Parmes, E., Liedes, H., Kuijpers, E., Pronk, A., Sarigiannis, D., Karakitsios, S., Chapizanis, D., Maggos, T.: Urban greenspace and the indoor environment: pathways to health via indoor particulate matter, noise, and road noise annoyance. Environ. Res. **180**, 108850 (2020)
22. Nowak, D.J., Crane, D.E., Stevens, J.C.: Air pollution removal by urban trees and shrubs in the United States. Urban Forest. Urban Greening **4**, 115–123 (2006)
23. Nowak, D.J., McHale, P.J., Ibarra, M., Crane, D.E., Stevens, J.C., Luley, C.J.: Modeling the effects of urban vegetation on air pollution. In Gryning, S., Chaumerliac, N. (eds.) Air pollution modeling and its application XII, 399–407. Plenum Press, New York (1998)
24. Hartig, T., Mitchell, R., de Vries, S., Frumkin, H.: Nature and health. Annu. Rev. Public Health **35**, 207–228 (2014)
25. Gascon, M., Triguero-Mas, M., Martínez, D., Dadvand, P., Rojas-Rueda, D., Plasència, A.: Residential green spaces and mortality: a systematic review. Environ. Int. **86**, 60–67 (2016)
26. Triguero-Mas, M., Dadvand, P., Cirach, M., Martínez, D., Medina, A., Mompart, A., Basagaña, X., Gražulevičienė, R., Nieuwenhuijsen, M.J.: Natural outdoor environments and mental and physical health: relationships and mechanisms. Environ. Int. **77**, 35–41 (2015)
27. Orban, E., Sutcliffe, R., Dragano, N., Jöckel, K.H., Moebus, S.: Residential surrounding greenness, self-rated health and interrelations with aspects of neighborhood environment and social relations. J. Urban Health **94**(2), 158–169 (2017)
28. Klompmaker, J.O., Janssen, N.A., Bloemsma, L.D., Gehring, U., Wijga, A.H., van den Brink, C., Lebret, E., Brunekreef, B., Hoek, G.: Residential surrounding green, air pollution, traffic noise and self-perceived general health. Environ. Res. **179**, 108751 (2019)
29. Kabisch, N., Qureshi, S., Haase, D.: Human-environment interactions in urban green spaces-a systematic review of contemporary issues and prospects for future research. Environ. Impact Assess. Rev. **50**, 25–34 (2015). https://doi.org/10.1016/j.eiar.2014.08.007
30. Sandström, U.G., Angelstam, P., Khakee, A.: Urban comprehensive planning-identifying barriers for the maintenance of functional habitat networks. Landscape Urban Plann. **75**, 43–57 (2006). https://doi.org/10.1016/j.landurbplan.2004.11.016
31. Bhandari, A.K., Kumar, A., Singh, G.K.: Feature extraction using normalized difference vegetation index: a case study of Jabalpur city. Procedia Technol. **6**, 612–621 (2012)

32. Wong, M.M.F., Fung, J.C.H., Yeung, P.P.S.: High-resolution calculation of the urban vegetation fraction in the Pearl River Delta from the Sentinel-2 NDVI for urban climate model parameterization. Geosci. Lett. **6**(1), 2 (2019)
33. Hellenic Statistical Authority Homepage. https://www.statistics.gr/. Accessed 03 Oct 2020
34. IMF Homepage. https://www.imf.org/external/index.htm. Accessed 03 Oct 2020

Evaluating Urban Mobility Sustainability Through a Set of Indicators: The Case of the City of Lamia, Greece

Maria Polyzou[1], Georgios Palantzas[1,2](✉), and Dimitrios Nalmpantis[1,2]

[1] School of Science and Technology, Hellenic Open University, Parodos Aristotelous 18, 263 35 Patras, Greece
palatzas@civil.auth.gr
[2] Faculty of Engineering, School of Civil Engineering, Aristotle University of Thessaloniki, PO Box 452, 541 24 Thessaloniki, Greece

Abstract. Sustainable mobility is the ability to meet society's need to move freely, communicate, gain access, trade, and establish relationships without sacrificing other essential human or ecological values, today or in the future. Sustainable mobility includes both the transportation of goods and people. Sustainability, when applied to urban mobility, can be measured by various sets of indicators. This paper presents the application of the indicators produced by the Sustainable Mobility Project 2.0 (SMP 2.0) of the World Business Council for Sustainable Development (WBCSD) in the city of Lamia, Greece. Accessibility for mobility-impaired groups, affordability of public transport for the poorest, fatalities' reduction, access to mobility services, mobility space usage, comfort, and pleasure are some of the 16 indicators calculated and estimated in the frame of this work. The indicators were calculated by a) conducting a relevant public opinion survey on a sample of 380 citizens, b) performing spatial data analysis, and c) studying existing transport studies and data. The set of indicators is used to evaluate the current situation of the mobility system in the city of Lamia, Greece, and to identify priority measures, practices, and policies for their improvement. By using them, the Municipality and other local authorities can identify where the strengths and weaknesses lie in the urban mobility system and assess potential actions. If repeated over time, this application will reveal the improvement or the worsening of the local urban mobility system, as well as the measures with the most significant impact on the system as a whole.

Keywords: Urban sustainable mobility · Transportation · Quality of life

1 Introduction

The need for efficient urban mobility has received increasing attention worldwide in terms of planning, organizing, applying, and investing in improved and sustainable mobility, delivering as a consequence to the cities, enhanced productivity, attractiveness, and overall quality of life [1].

E. G. Nathanail et al. (Eds.): CSUM2020, AISC 1278, pp. 739–750, 2021.
https://doi.org/10.1007/978-3-030-61075-3_72

The sustainability of urban mobility has various dimensions and definitions [2–6]. Indicatively, according to the World Business Council for Sustainable Development (WBCSD) [7] "sustainable mobility is the ability to meet society's need to move freely, gain access, communicate, trade and establish relationships without sacrificing other essential human or ecological values, today or in the future." A more simple definition has been provided by the European Conference of Ministers of Transport (ECMT) [8]: "a sustainable transport system is one that is accessible, safe, environmentally-friendly, and affordable."

Particularly for the evaluation of transport systems and urban mobility, a broad range of indices and performance indicators have been developed, proposed, and stand out as necessary tools and means for monitoring mobility conditions and assisting decision-makers in policy implementation. Such systems of indicators can be found in the relevant literature [9–12], while there are also reviews of the relevant literature [13–15]. Thereby, the several approaches followed on the necessary type and number of performance indicators for assessing urban mobility highlights the fact that creating a single framework/strategy for sustainable urban mobility indicators is not possible. Each city has its own built and natural environment, and transport challenges may vary according to geographical, environmental, economic, political, and other factors.

Such a set of 19 indicators has been developed through the research carried out within the Sustainable Mobility Project 2.0 (SMP 2.0) during 2013–2016, a project launched under the auspices of the WBCSD. In this paper, an attempt has been made to evaluate the sustainability of the mobility conditions in the small-medium sized Hellenic city of Lamia, Greece, with the use of this comprehensive set of indicators.

2 Overview of the Study Area

The study area of this research is Lamia, a typical medium-sized city in central Greece. Lamia's administrative area, as a municipal unit, extends over an area of 413 km^2. At the same time, according to the results of the most recent (2011) census carried out by the Hellenic Statistical Authority (HSA) [16], its population reaches approximately 75,000 inhabitants.

Today, Lamia is the capital of the regional unit of Phthiotis and the Region of Central Greece, being also one of the most significant economic and administrative centers of the latter. According to the latest data (2017) from the HSA [16], the Gross Domestic Product (GPD) per capita in Phthiotis regional unit is €14,113 (16% lower than Greece's average).

The latest traffic studies of the area have been conducted in 1990 and 2000, while the local Sustainable Urban Management Plan (SUMP), in accordance with the European Union's (EU) 2013 Urban Mobility Package [17], is under development. The city lies on the outskirts of a hill and a castle, where the relief is steep with long slopes while the broader area of the center grows on gentle slopes with a relative flatness that allowed the creation of roads with mild geometric features.

The available infrastructure for pedestrians and people with reduced mobility is generally poor, in terms of adequate sidewalks and open public green spaces, while there is no cycling network. The shape of the road network is radial, while the center concentrates the most important administrative and commercial functions of the city. The poor geometric features of the streets in the downtown area, the concentration of the economic activity in the city center, in combination with the dominant role of private car in daily mobility and the lack of alternative bypass routes and parking spaces, lead to traffic congestion, long delays, and noticeable burden of air quality during peak periods [18]. Currently, the municipal area is served by a fleet of 26 buses, standard and minibuses, which operate in 24 routes [19].

3 Methodology

For the estimation and the adjustment of the sustainable mobility indicators' set of the SMP 2.0 project for the city of Lamia, Greece, and the evaluation of the current mobility conditions, a methodological approach was implemented, according to the following four phases: a) understanding of the selected set of indicators, b) data gathering, c) data analysis and calculation of the indicators, and d) conducting a SWOT analysis and proposing priority solutions for assessing the city's current position and prospects for improvement of mobility sustainability.

3.1 Understanding of the Selected Set of Indicators

The WBCSD has previously developed the sustainable mobility set of indicators applied in the city of Lamia, Greece, through the SMP 2.0 project towards supporting cities in developing fact-based and integrated SUMPs.

Based on the sustainable mobility approach described by Banister (2008) [3], the indicators' set was structured around four dimensions of sustainable mobility. The first three are according to the wide-spread Triple Bottom Line (TBL) concept, which focuses sustainable development on the harmony between Global environment (G), Quality of life in the city (Q), and Economic success (E), while the performance of the Mobility system (M) itself was added as a fourth dimension [20].

The SMP 2.0 project has resulted, among others, in a set of 19 indicators, which combine various criteria, such as Green-House Gas (GHG) emissions, noise hindrance, affordability of public transport, and accessibility for deficiency groups [21]. In some cases, the indicators' have an "impact on two, three, or even four sustainable mobility dimensions. For example, congestion increases air pollution (G), provokes a waste of time for the passenger (Q), and has high associated costs (E)" [22].

The methodological application of the indicators includes all modes of road and rail transport for passengers and goods, excluding air transport and shipping, as, in most cities, the sustainability impact of these modes is beyond the scope of urban governance [20]. The methodology and the calculation method of the indicators are analytically described in a relevant WBCSD report [22].

During the SMP 2.0 project, these indicators have been calculated in Bangkok, Thailand; Campinas, Brazil; Chengdu, China; Lisbon, Portugal; Indore, India; and Hamburg, Germany. For the two latter cities, relevant reports have been published online [23, 24], and a sustainable Mobility Planning tool was subsequently developed, designed to increase the availability and user-friendliness of the sustainable mobility indicators, the mobility solutions, and the toolbox of the SMP 2.0 project [25].

The European Commission endorsed this set of indicators in April 2016, encouraging cities to use them for measuring and improving their mobility footprint [26] and its General Directorate for Mobility and Transport (DG MOVE) funded the Sustainable Urban Mobility Indicators (SUMI) project with the aim to help European cities to use the SMP 2.0 indicators [27].

3.2 Data Gathering

Four methodologies have been used for data gathering. Input data for each parameter calculation were originally based on either field measurements or population surveys and interviews. Some data were stored in existing databases, and other data needed some geographical analysis through Geographic Information System (GIS) software. Subsequently, the collected data were organized in spreadsheets and a GIS database.

- *Using existing databases*: The following indicators: a) Affordability of public transport for the poorest quartile, b) Fatalities, c) Economic opportunity, and d) Net Public Finance, were based on local, regional, and national databases, studies, reports, laws, and regulations such as the two traffic studies of 1990 and 2000 (TS), the Master's thesis of Niafa and Arvanitis (MS) [18], official data of the HSA and the local Traffic Police (TP), legislative and regulatory provisions on salaries and bus tickets' fare (LR), and data of the local Public Transport Operator (PTO) [19].
- *Field measurements*: The Congestion and delays' indicator was calculated through a number of measurements in specific city's points, to represent typical areas where solutions should be targeted. Measurements were made on 10 major roads for both car and public transport modes, during morning and evening peak-hours, and free-flow conditions.
- *Surveying and interviews*: A questionnaire survey, via a questionnaire based on both online and on-field interviews, was conducted for the following indicators: a) Accessibility for mobility-impaired groups, b) Quality of public area, c) Commuting travel time, d) Intermodal integration, e) Comfort and pleasure, f) Security, and g) Economic opportunity. A sample of 380 answers was collected, which is statistically significant given the population of the city ($\sim$75,000 inhabitants) for 50% response distribution, 5% margin of error, and 95% confidence level. The structure of the questionnaire and the set of the questions were based and adapted by the one suggested by the SMP 2.0 project [21].
- *GIS analysis*: A GIS software was used to calculate three parameters based on spatial data: a) Access to mobility services, b) Urban functional diversity, and c) Opportunity for active mobility. Specifically, digital and analog data sets (maps) were converted into a suitable digital format for GIS use.

Due to the lack of available and reliable data on specific traffic characteristics, such as total vehicle-kilometers per capita, annual passenger and ton-km, distance driven by transport modes annually, as well the limited available time for the execution of this research, three indicators were not calculated. Thus, in total 16 indicators were calculated. These indicators are a) Emissions of GHGs (well-to-wheels GHG emissions by all city passenger and freight transport modes, expressed in t-CO_2eq per capita), b) Air-polluting emissions (air-polluting emissions of all passengers and freight city transport modes, expressed in kg NO_x eq./cap per year), and c) Energy efficiency (total energy consumed for city transport, expressed in MJ per km).

3.3 Data Analysis and Calculation of the Indicators

The calculation of each of the indicator parameters was done according to the methodology described in the relevant report [21]. The definition of the applied indicators and their representative parameters/indices to be calculated are presented in Table 1.

The main assumptions made during the calculation of the values of the urban mobility indicators are the following: a) the Gross Domestic Product (GDP) parameter (GDP of the city per year) included in the calculation of the Net Public Finance indicator refers to the regional unit of Phthiotis, as GDP data was not available for the city, and b) concerning the calculation of the Noise hindrance indicator, 50 measuring points were selected and located in various living environments (center, ring road, schools, hospitals, etc.). The measurements were executed during daytime and peak-hours, without, however, correcting the possible disturbances of other sources of noise that might have been disturbing the measurements.

3.4 Conducting a SWOT Analysis and Identification of Priority Solutions

The final phase of the applied methodology is based on the estimated values of the indicators and consists of a) the conduct of a Strengths, Weaknesses, Opportunities, and Threats (SWOT) analysis, and b) the setting of a preliminary list of potential solutions to manage the sustainable mobility challenges in Lamia.

Table 1. Overview of the applied set of mobility indicators.

ID	Indicator	Definition	Calculated parameter/index
1	Affordability of public transport for the poorest people	Share of public transport cost for fulfilling basic activities of the household budget for the poorest quartile of the population	Affordability Index of public transport [% of household income]
2	Accessibility for mobility impaired groups	Accessibility for deficiency groups to transport and transport services	Averaged score of accessibility of city transport [%]
3	Noise hindrance	Percentage of population hindered by city transport noise	Noise hindrance index [% of population]
4	Fatalities	Fatalities by road and rail transport accidents in the city	Fatality rate [# per 100.000 population per year]
5	Access to mobility services	Percentage of population living within walking distance of public transport or shared mobility system	Appropriate access Index [% of population]
6	Quality of public area	Reported social usage of streets and squares and subjective appreciation of the public area quality	Averaged score of public area quality appreciation and sociability [%]

(*continued*)

Table 1. (*continued*)

ID	Indicator	Definition	Calculated parameter/index
7	Urban functional diversity	Percentage of presence of out of 10 spatial functions related to daily activities except for work in grids of 1 km x 1 km	Functional diversity score [%]
8	Commuting travel time	Duration of commute to and from work or an educational institution	Average commuting time per person [minutes/day]
9	Economic opportunity	Degree of accessibility to the job market and education system	Averaged score of restriction of economic opportunity [%]
10	Net Public Finance	Net government and other public authorities' revenues from transport-related taxes and charges minus operational and other costs per GDP	Net Public Finance indicator of the city transport [%]
11	Mobility space usage	Proportion of land use, taken by all city transport modes, including direct and indirect uses	Land use for mobility applications [m^2/capita]
12	Congestion and delays	Delays in road traffic and in public transport during peak-hours compared to free-flow travel	Congestion and delay index (percentage delay during peak hours) [% of delay]
13	Opportunity for active mobility	Length of roads/streets with sidewalks, bike lanes, and pedestrian zones related to the total length of the city network	Share of road length adapted for active mobility [%]
14	Intermodal integration	Availability of intermodal connections and quality of the interchange facilities	Average satisfaction score quality of interchanges [%]
15	Comfort and pleasure	Physical and mental comfort of citizens while using the urban transports/services	Averaged score of comfort and pleasure of city transport [%]
16	Security	Risk of crime in urban transport	Averaged satisfaction score of security of city mobility [%]

4 Results and Discussion

The calculated values of the 16 mobility indicators are presented in Table 2, where also the unit and the scale span of the values according to the SMP 2.0 project's methodology, and the relevant score from 0 (worst) to 10 (best), are mentioned. A first comparison of the indicators' analysis shows that six (6) out of the 16 indicators (ID 2, 6, 12, 13, 14, and 16) are returning relatively low sustainable mobility scores (<5.0), and addressing these indicators could significantly improve the sustainable mobility footprint of the city.

Table 2. Results from the calculation of the selected set of mobility indicators.

ID	Indicator	Values	Scale span & Units	Score
1	Affordability of public transport for the poorest people	17.0%	0: A.I. > 35% of household income 10: A.I. < 3.5% of household income	5.7
2	Accessibility for mobility impaired groups	42.6%	0: 0 [% satisfaction] 10: 100 [% satisfaction]	4.3
3	Noise hindrance	13.7%	0: $\geq$ 70 [% of population] 10: 0 [% of population]	8.0
4	Fatalities	3.84	0: 35 [fatalities/100.000 capita] 10: 0 [fatalities/100.000 capita]	8.9
5	Access to mobility services	95.0%	0: 0 [% population] 10: 100 [% population]	9.5
6	Quality of public area	34.9%	0: 0 [% usage & satisfaction] 10: 100 [% usage & satisfaction]	3.5
7	Urban functional diversity	60.9%	0: average score 0 [%] 10: average score 100 [%]	6.1
8	Commuting travel time	26.2	10: $\leq$ 10 [minutes per day] 0: $\geq$ 90 [minutes per day]	8.7
9	Economic opportunity	80.1%	0: 100 [%] 10: 0 [%]	8.0
10	Net Public Finance	0.12%	0: $\leq$ (-2,5) [% of GDP] 10: $\geq$ 0 [% of GDP]	10
11	Mobility space usage	52.7	0: $\geq$ 125 (m^2/capita) 10: $\leq$ 25 (m^2/capita)	7.2
12	Congestion and delays	2.95%	0: $\geq$ 3.0 [% delay] 10: $\leq$ 1.25 [% delay] (relation peak hour/normal travel time	0.3
13	Opportunity for active mobility	65.1%	0: 0 [% road length] 10: $\geq$ 200 [%]	3.3
14	Intermodal integration	33.4%	0: 0 [% satisfaction] 10: 100 [% satisfaction]	3.3
15	Comfort and pleasure	61.3%	0: 0 [% satisfaction] 10: 100 [% satisfaction]	6.1
16	Security	53.0%	0: 0 [% satisfaction] 10: 100 [% satisfaction]	5.3

"Intermodal integration" received a poor score. As there is minimal provision for car or bicycle parking, and no metro or tram, this makes it very difficult for citizens to have a convenient choice of transport mode. Taking also into consideration that the provisions for pedestrians are also limited, the intermodal connectivity of city transport should be ranked by the city's authorities as a priority issue. Furthermore, many respondents were not satisfied with the frequency of connecting urban public transport and the integration of schedules of trip' information and ticketing.

The very high score of "Access to mobility services" demonstrates the strong dynamic and prospects for the public transport system to enhance further its share and quality of services, improving in parallel the score of most of the indicators.

The mean score of "Affordability of public transport for the poorest people" should encourage the PTO to assess possible initiatives for making public transport more available to people with lower incomes.

The maximum score of "Net Public Finance" reflects both the high affordability to sustain mobility related expenditures and the possibilities for relevant investments.

The score for "Urban functional diversity" was not good, as there is not adequate coverage of commercial and civil functions within walking distance, showing that more improvements of Sports & Recreation and Park & Green functions are needed.

There is a low score also for "Opportunity for Active Mobility" due to relevant poor infrastructure. The addition of sufficient dedicated bicycle lanes and sidewalks, and a sound provision of pedestrianized zones, could practically improve the score.

The almost zero score of "Congestion and delays" shows the extremely low level of service of the road network during the peak-hours and the need for solutions.

Although "Mobility space usage" has a good score, the city would benefit from a more balanced land use planning, as there are few parking places assigned with respect to the city area.

"Comfort and pleasure" has an acceptable score with quite prospects for improvement. The respondents were dissatisfied with public transport in general due to the lack of real-time information and the availability of space for baggage. Quantity and location of parking spaces is the main reason to respondents' dissatisfaction when driving in the city. Regarding cycling, respondents are mainly dissatisfied with the lack of adequate and safe parking areas. Considering walking in the city, respondents are dissatisfied with many aspects such as the existence, quality, and width of sidewalks.

"Quality of public area" has a low score as people do not like the lack of public spaces for physical activities and the poor greenery, safety, and child-friendliness.

Concerning "Security," the responses showed that people feel significantly less safe at night for all modes of mobility than during the daytime. High noise levels were measured in the city center and outside of some schools during peak hours.

The responses of "Accessibility for mobility impaired groups" gave back an unacceptable score. Pregnant women were dissatisfied with the availability of relevant parking spaces and safe access on foot to bus stops. Older travelers and physically impaired people were mainly unsatisfied with the number of seating places in buses. Almost half of the visually impaired people were dissatisfied with guidance and warning systems. Moreover, a SWOT analysis was conducted (Table 3).

Table 3. SWOT analysis regarding the sustainable mobility challenges in Lamia.

Strengths	Weaknesses
• The majority of the population lives within walking distance from the public transport system • The adequate mix of spatial functions creates proximity of mutually interrelated activities • Presence of a ring road • Low commuting time to/from work or education • There are only minor difficulties in accessing the job market and education system	• Narrow double direction roads in the city center • Extensive use of private vehicles and trucks • Poor quality of pedestrian infrastructure • Insufficient cycling infrastructure • Low quality of public areas • Poor arrangement of the parking system • Poor connection between different transport modes and points of interest
Opportunities	**Threats**
• Arrangement of the parking system • Further extension of the minibus lines • Arrangement of improved infrastructure for pedestrians, cyclists, and accessibility for all • Real-time information on-board and at bus stops • Enhancement of greenery/physical activity areas	• The descending trend of GDP per capita due to the economic crisis • Absence of a SUMP or a recent traffic study • Fuzzy willingness to change travel habits

Despite the fact that Lamia faces heavy traffic congestion during peak-hours, lack of parking spaces, and inadequate pedestrian infrastructure, the SWOT analysis indicates that the city comprises some significant strengths and opportunities for meeting priority indicators (e.g., those having a score under 5.0) managing its mobility challenges. These strengths ground mainly on its urban, spatial, and transportation characteristics. These characteristics should be accordingly assessed and exploited by the local authorities in terms of identifying and applying potential solutions for improving significantly both the use of soft modes, i.e., walking and cycling, and public transport.

Planning and application of suitable solutions are advisable to follow an identification and understanding of similar issues and relevant good and bad practices in other cities and urban areas. Based on the above, a list of priority solutions was filtered and categorized into five (5) major categories, as follows: a) Improve accessibility of the city center and its connectivity with district functions (e.g., enhance availability and quality of pedestrian infrastructure, dedicated bicycle lanes, parking fee enforcement, alternative routes such as the ring road [see 28]); b) Improve the quality of public area (e.g., increase and ameliorate green open spaces [see 29]); c) Improve public transport (e.g., passenger-friendly bus stops, reorganize city bus routes, full accessibility [see 30, 31]); d) Manage congestion (e.g., parking areas, pedestrian zones, park and ride, enforcement by traffic police, environmental awareness); and e) Technology solutions (e.g., real-time displays/apps for arrival time at bus stops [see 30–32]).

5 Conclusions

In the framework of this research, the sustainability of the mobility conditions of a Hellenic medium-sized city was estimated using a set of urban mobility indicators proposed by the SMP 2.0 project and a relevant SWOT analysis. The values of the indicators highlighted specific issues that need substantial improvement, while the SWOT analysis showed some prospects for enhancing mobility and meeting current challenges. The identified priority solutions can stimulate in an integrated way a shift towards cleaner and more sustainable mobility practices in the city, such as walking, cycling, and public transport, supported in parallel by the use of new technologies.

References

1. Cats, O.: The robustness value of public transport development plans. J. Transp. Geogr. **51**, 236–246 (2016). https://doi.org/10.1016/j.jtrangeo.2016.01.011
2. European Commission: Special Eurobarometer 406: Attitudes of Europeans towards urban mobility. European Commission, Brussels (2013). https://ec.europa.eu/commfrontoffice/publicopinion/archives/ebs/ebs_406_en.pdf
3. Banister, D.: The sustainable mobility paradigm. Transp. Policy **15**(2), 73–80 (2008). https://doi.org/10.1016/j.tranpol.2007.10.005
4. Zheng, J., Garrick, N.W., Atkinson-Palombo, C., McCahill, C., Marshall, W.: Guidelines on developing performance metrics for evaluating transportation sustainability. Res. Transp. Bus. Manage. **7**, 4–13 (2013). https://doi.org/10.1016/j.rtbm.2013.02.001
5. Black, J.A., Paez, A., Suthanaya, P.A.: Sustainable urban transportation: performance indicators and some analytical approaches. J. Urban Plan. Develop. **128**(4), 184–209 (2002). https://doi.org/10.1061/(ASCE)0733-9488(2002)128:4(184)
6. Organisation for Economic Co-operation and Development: Policy instruments for achieving environmentally sustainable transport. OECD Publishing, Paris (2002). https://doi.org/10.1787/9789264176249-en
7. World Business Council for Sustainable Development: Mobility 2001: World mobility at the end of the twentieth century and its sustainability. World Business Council for Sustainable Development, Geneva. https://www.wbcsd.org/contentwbc/download/2633/33225
8. European Conference of Ministers of Transport: Assessment and decision making for sustainable transport. OECD Publications, Paris (2004). https://www.itf-oecd.org/sites/default/files/docs/04assessment.pdf
9. Lee, D.: Transit cost and performance measurement. Transp. Rev. **9**(2), 147–170 (1989). https://doi.org/10.1080/01441648908716717
10. Lautso, K., Toivanen, S.: SPARTACUS system for analyzing urban sustainability. Transp. Res. Rec. J. Transp. Res. Board **1670**(1), 35–46 (1999). https://doi.org/10.3141/1670-06
11. World Business Council for Sustainable Development: Mobility 2030: Meeting the challenges to sustainability. World Business Council for Sustainable Development, Geneva (2004). https://www.oecd.org/sd-roundtable/papersandpublications/39360485.pdf
12. Zito, P., Salvo, G.: Toward an urban transport sustainability index: an European comparison. Eur. Transp. Res. Rev. **3**(4), 179–195 (2011). https://doi.org/10.1007/s12544-011-0059-0

13. Costa, P.B., Morais Neto, G.C., Bertolde, A.I.: Urban mobility indexes: a brief review of the literature. Transp. Res. Procedia **25**, 3645–3655 (2017). https://doi.org/10.1016/j.trpro.2017.05.330
14. Braga, I.P.C., Dantas, H.F.B., Leal, M.R.D., de Almeida, M.R., dos Santos, E.M.: Urban mobility performance indicators: a bibliometric analysis. Gestão & Produção **26**(3), e3828 (2019). https://doi.org/10.1590/0104-530x3828-19
15. Litman, T.: Developing indicators for comprehensive and sustainable transport planning. Transp. Res. Rec. J. Transp. Res. Board **2017**(1), 10–15 (2007). https://doi.org/10.3141/2017-02
16. Hellenic Statistical Authority Homepage. https://www.statistics.gr. Accessed 16 Mar 2020
17. European Commission Webpage. https://ec.europa.eu/transport/themes/clean-transport-urban-transport/urban-mobility/urban-mobility-package_en. Accessed 16 Mar 2020
18. Niafa, G., Arvanitis, G.: Update of measures for facing the transport problem of Lamia (Master's thesis). Aristotle University of Thessaloniki, Thessaloniki (2007). (in Greek)
19. Public Transport Operator of Lamia Webpage. http://www.astikoktellamias.gr/page/grammes-kai-dromologia. Accessed 16 Mar 2020
20. Gillis, D., Semanjski, I., Lauwers, D.: How to monitor sustainable mobility in cities? Literature review in the frame of creating a set of sustainable mobility indicators. Sustainability **8**(1), 29 (2016). https://doi.org/10.3390/su8010029
21. World Business Council for Sustainable Development: Integrated sustainable mobility in cities, a practical guide (SMP 2.0 project Final Report). World Business Council for Sustainable Development, Geneva (2016). https://www.wbcsd.org/contentwbc/download/1073/13798
22. World Business Council for Sustainable Development: Methodology and indicator calculation method for sustainable urban mobility (SMP 2.0 project Report). World Business Council for Sustainable Development, Geneva (2015). https://www.wbcsd.org/contentwbc/download/3006/38273
23. World Business Council for Sustainable Development: Project report for the city of Indore, India (SMP2.0 project Report). World Business Council for Sustainable Development, Geneva (2016). http://docs.wbcsd.org/2016/01/Indore_SMP_Report.pdf
24. World Business Council for Sustainable Development: Abschlussbericht der City Task Force Hamburg. World Business Council for Sustainable Development, Geneva (2016). http://docs.wbcsd.org/2016/03/Sustainable_Mobility_Project_Hamburg.pdf. (in German)
25. World Business Council for Sustainable Development Homepage. http://www.wbcsdsmp.org. Accessed 16 Mar 2020
26. European Commission: Study on urban mobility – Assessing and improving the accessibility of urban areas: Final report and policy proposals. Publications Office of the European Union, Luxemburg (2017). https://doi.org/10.2832/443101
27. Rupprecht Consult – Forschung & Beratung Webpage. https://www.rupprecht-consult.eu/project/sumi.html. Accessed 16 Mar 2020
28. Gkavra, R., Nalmpantis, D., Genitsaris, E., Naniopoulos, A.: The walkability of Thessaloniki: citizens' perceptions. In: Nathanail, E.G., Karakikes, I.D. (eds.) CSUM 2018. AISC, vol. 879, pp. 191–198. Springer, Cham (2019). https://doi.org/10.1007/978-3-030-02305-8_23
29. Nalmpantis, D., Lampou, S.-C., Naniopoulos, A.: The concept of woonerf zone applied in university campuses: the case of the campus of the Aristotle University of Thessaloniki. Transp. Res. Procedia **24**, 450–458 (2017). https://doi.org/10.1016/j.trpro.2017.05.071
30. Tsafarakis, S., Gkorezis, P., Nalmpantis, D., Genitsaris, E., Andronikidis, A., Altsitsiadis, E.: Investigating the preferences of individuals on public transport innovations using the maximum difference scaling method. Eur. Transp. Res. Rev. **11**(1), 1–12 (2019). https://doi.org/10.1186/s12544-018-0340-6

31. Nalmpantis, D., Roukouni, A., Genitsaris, E., Stamelou, A., Naniopoulos, A.: Evaluation of innovative ideas for public transport proposed by citizens using multi-criteria decision analysis (MCDA). Eur. Transp. Res. Rev. **11**(1), 1–16 (2019). https://doi.org/10.1186/s12544-019-0356-6
32. Papadima, G., Genitsaris, E., Karagiotas, I., Naniopoulos, A., Nalmpantis, D.: Investigation of acceptance of driverless buses in the city of Trikala and optimization of the service using conjoint analysis. Utilities Policy **62**, 100994 (2020). https://doi.org/10.1016/j.jup.2019.100994

Impact Assessment of Climate Change on Coastal Transport Systems in the Greater Thessaloniki Area

Apostolos Papagiannakis(✉) and Konstantinos Ntafos

School of Spatial Planning and Development, Faculty of Engineering, Aristotle University of Thessaloniki, 54124 Thessaloniki, Greece
apa@plandevel.auth.gr

Abstract. Long-term planning and operation of transport systems should more than ever consider climate change and extreme weather conditions. There are multiple risks such as coastal and urban floods, sea level rise, very high and extremely low temperatures, drought and wind. The aim of this article is to assess the impacts of sea level rise so as to identify the transport infrastructure vulnerabilities in the Greater Thessaloniki area, in a coastal zone of 60 km long and 2 km wide from the seashore, which is located within five different municipalities. Based on cartographic data obtained from Climate Central's Surging Seas Risk Zone Map, land use data from Corine Land Cover and population data from Hellenic Statistical Authority, two sea level rise scenarios of 0.5 and 1 m were simulated using GIS. According to the most likely scenario introduced by Intergovernmental Panel on Climate Change of the United Nations, that is 0.5 m rise, until 2100 1.87% of the total length of the coastal road network will be covered by the sea, while in the worst-case scenario of the 1 m, the percentage will be 3.07%. At the same time, the interruption of the road access to the airport in both scenarios, the vulnerability of some parts of the port, and their potential inability to operate are highlighted. The research findings indicate the need to plan and construct resilient transport systems as well as to coordinate and implement specific climate change adaptation measures for transport infrastructures in Thessaloniki coastline.

Keywords: Climate change impact assessment · Sea level rise · Transport infrastructure vulnerability · Transport resilience · Thessaloniki coastline

1 Introduction

Current and projected impacts of climate change pose serious challenges to transport systems. There is high confidence that climate change will threaten transport systems both with extreme weather and severe changes. There are a lot of risks, such as coastal and urban floods, rising sea levels, very high and extremely low temperatures, drought and wind [1]. Long-term planning must ensure sustainability, adequate operation and reliability of all transport networks and infrastructures. However, existing transport systems have been designed and operated based on past climatic conditions. At the same

E. G. Nathanail et al. (Eds.): CSUM2020, AISC 1278, pp. 751–759, 2021.
https://doi.org/10.1007/978-3-030-61075-3_73

time, they are characterized by complex and non-linear relationships and interdependencies between their various internal and external components, which increase their complexity as well as the requirement for the maintenance of a dynamic, balance and adequate function [2]. The vulnerability of a transport system is defined as the sensitivity to events that could lead to a significant reduction in the level of service and accessibility [3]. Resilient transport systems have the ability to anticipate threats, absorb impacts, recover and adapt after a persistent environmental disturbance or even an economic, environmental or social crisis with a minimum cost [4]. As a result, they can maintain an operational level, which is critical in supporting all human activities [5].

Adaptation and mitigation are complementary resilience strategies for reducing and managing climate change risks in transport systems [6]. Mitigation measures are considered to be very important for the reduction of both greenhouse gas emissions and the environmental impact of the transport sector. At the same time, adaptation measures are necessary to address the impacts of climate change. The two levels of action differ in the spatial scale to which they are implemented, as mitigation refers to the macroscopic or international scale, while the adaptation concerns the regional and local scale [7]. An adaptation strategy for terrestrial transport systems should take into consideration the sensitivity of the infrastructure to disruptions or degradation that could significantly reduce their operation and efficiency, the impacts of this disruption/degradation and the actions that should be taken to restore or maintain the system and achieve the desired state [8]. When developing adaptation strategies, it is important to understand that their fundamental objective does not only concern networks and infrastructures, but also the population as a whole [9].

Rising sea levels, rising forces and high ripples during severe weather events can erode the coastline resulting in the collapse and destruction of coastal transport infrastructure. By incorporating elements of flexibility and agility into the design and construction could help transportation systems achieve resilience goals. Flexibility relates to expected or anticipated changes, while agility to changes that are unexpected or unpredictable. For example, flexibility in the context of sea level rise refers to the system's inherent abilities to react and dampen impacts if coastal flood intensity is within an expected range. Agility refers to the system's inherent abilities to respond to the effects of an unpredictable and unprecedented rise in water levels (e.g. unexpected and rapid melting of ice or the occurrence of a tidal wave) [2]. According to the European Environment Agency, there are three methods that can encourage a fundamental transition in the transport system towards resilience [10]:

- Increasing awareness and knowledge.
- Promoting innovation to achieve a transition through long-term visions.
- Integrated spatial approach to transport systems.

In this context, the present article investigates the impact of sea level rise on the transport systems of the Greater Thessaloniki coastal zone. The second section presents the study area and the applied research methodology. The third section examines the simulation results of two alternative sea level rise scenarios (+0.5 and +1 m). The fourth section discusses resilience and adaptation measures for transport infrastructure, and finally, the fifth section contains the conclusions of the research.

2 Study Area and Methodology

According to [11], approximately 33% of the Greek population reside in coastal areas, located 1–2 km from the coast. About 80% of industrial activities, 90% of tourism and leisure activities, 35% of agriculture (usually of high productivity), fishing and aquaculture, as well as a significant part of infrastructure (ports, airports, etc.) are also located in coastal zones. Based on these data, the study area was delineated within a 60 km long and 2 km wide zone along the coastline. This zone comprises a significant part of the Greater Thessaloniki area, namely the municipalities of Thessaloniki and Kalamaria (characterized by a dense and continuous urban fabric), as well as the municipalities of Pylaia, Thermi and Thermaikos (located in the peri-urban and suburban area), with a total population of 590.275 inhabitants [12]. Within this zone there is a significant concentration of population, multiple land uses and productive activities. At the same time, major arterial roads as well as all important regional and national transport infrastructures, such as Thessaloniki's New Railway Station, Thessaloniki's Port, Macedonia Intercity Bus Station, Thessaloniki International Airport "Makedonia" and city bus terminals, are located there.

The objective of the methodology is twofold: to estimate the extent and to map the portion of the study area that will potentially be covered by water in the case of two alternative sea level rise scenarios. The Intergovernmental Panel on Climate Change (IPCC) predicts that sea level will rise by 30 to 60 cm by 2100, even if the climate commitments made under the Paris Agreement are met [13]. However, if emissions continue to rise, following current trends, then the mean sea level is projected to rise by 84 cm until 2100 and up to 5.4 meters until 2300. The two simulated scenarios, which are based on the projected global mean sea level rise by IPCC [13] and the European Environment Agency [14] for the region of Southeast Europe and the Mediterranean, are as follows:

Scenario 1 (medium to low impacts): Projected sea level rise of 0.5 m for the period 2081–2100, with 1986–2005 as a base period, assuming an average global temperature rise by 1.8°C, according to IPCC's intermediate scenario of the Representative Concentration Pathways (RCP4.5) [6] and the World Climate Research Program's fifth phase of the Coupled Model Intercomparison Project (CMIP5) [15].

Scenario 2 (extreme impacts): Projected sea level rise of 1 m for the period 2081–2100, with the same base period (1986–2005), according to the worst-case scenario RCP8.5 of the IPCC [6] for future global average temperature rise by 3.7°C.

Initially, the web application of Climate Central [16] was used, and more specifically the Risk Zone Map mapping tool that enables the simulation of sea level rise scenarios based on satellite images with a resolution of 90 meters that derive from National Aeronautics and Space Administration (NASA). Existing land uses were obtained from Corine land cover system 2018 [17]. Moreover, digital cartographical material and demographic data from Hellenic Statistical Authority [12] were used for mapping urban blocks and population. Finally, the road network and the layout of transport infrastructure derived from the openstreetmap application. The abovementioned spatial data and information were added to ArcGIS [19] to calculate the area of the coastline (in square meters) and the total length of the road network within the study area that are expected to be inundated, in relation to the current situation. Moreover, urban blocks as well as the respective population affected in the two selected scenarios were also identified.

3 Impact Assessment of Sea Level Rise

The results from the simulation of the two scenarios as well as the impact factor of sea level rise for each municipality within the study area are presented in Table 1 and Table 2. The impact factor corresponds to the coverage percentage of the road network's area and length, within the 2 km zone from the coastline. The impact factor of the total study area in case of Scenario 1 is 12.30%, while the one of Scenario 2 is 13.34%. In terms of road network coverage, the impact factor is 1.87% and 3.07% respectively.

Table 1. Impact of sea level rise on coastline area.

Municipality	Surface within study area (ha)	Scenario 1 Surface area under sea (ha)	Scenario 2 Road length under sea (ha)	Scenario 1 Impact factor (%)	Scenario 2 Impact factor (%)
Thessaloniki	1650.87	19.73	50.13	1.20	3.04
Kalamaria	646.07	5.34	9.82	0.83	1.52
Pylaia	614.46	22.26	37.26	3.62	6.06
Thermi	1403.98	629.96	652.06	44.87	46.44
Thermaikos	6923.96	705.32	750.15	10.19	10.83
Total study area	11239.35	1382.61	1499.41	12.30	13.34

In both scenarios, the suburban municipality of Thermi has the highest impact while the municipalities of Thessaloniki and Kalamaria have the least impact. In the municipality of Thessaloniki the effects of sea level rising are concentrated in particular areas of the continuous urban fabric that are located, as expected, on the coastal front. The affected road sections are important to the connectivity and operation of the main road network, such as Nikis Avenue, a minor arterial road running through the coastal front in city center, and Maria Callas, a coastal collector road serving densely populated areas. It is emphasized that based on Scenario 1, parts of the port's infrastructure and facilities at the piers are expected to be inundated, disrupting its operation. In Scenario 2 the phenomenon will be intensified as part of the national road that functions as the west gateway to the city is also affected (see Fig. 1). In the suburban municipality of Pylea, the affected areas are mainly coastal uncultivated land and industrial or commercial land uses. However, the most significant negative impacts are detected in the municipality of Thermi and particularly in the area where Thessaloniki International Airport is located along with some patches of arable land. In particular, the airport's facilities and runways as well as the main access road to it are fully inundated in both scenarios. In the suburban municipality of Thermaikos, areas of the continuous urban fabric, which are located in the settlements of Peraia, Neoi Epivates and Agia Triada, will also be inundated. Coastal agricultural land and wetlands under environmental protection (in the settlement of Epanomi) will be flooded as well.

Table 2. Impact of sea level rise on coastline road network.

Municipality	Road length study area (km)	Scenario 1 Road length under sea (km)	Scenario 2 Road length under sea (km)	Scenario 1 Impact factor (%)	Scenario 2 Impact factor (%)
Thessaloniki	495.26	1.20	3.31	0.24	0.67
Kalamaria	190.97	0.20	1.03	0.10	0.54
Pylaia	114.59	3.39	6.94	2.96	6.06
Thermi	181.17	8.40	12.95	4.64	7.15
Thermaikos	419.26	13.08	18.81	3.12	4.49
Total study area	1401.25	26.27	43.04	1.87	3.07

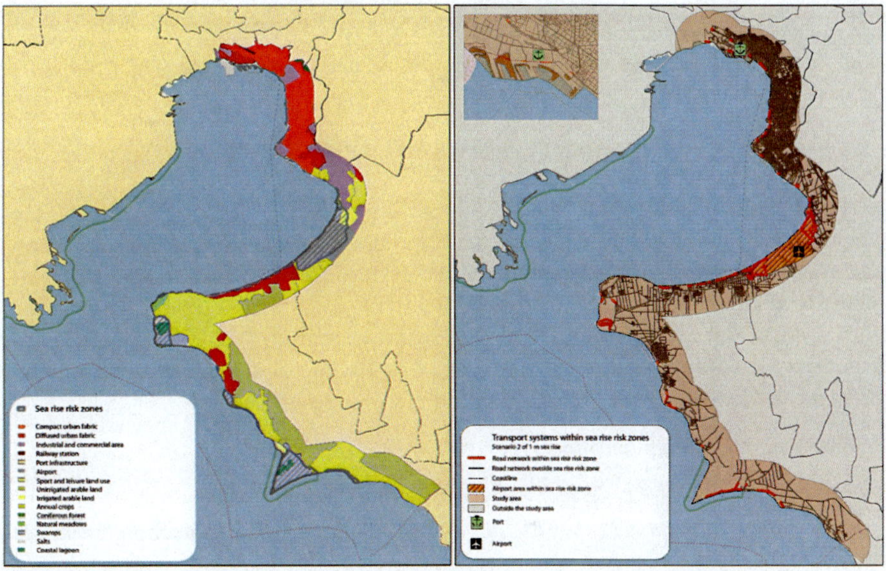

Fig. 1. Sea level rise impacts on land uses (left) and on transport infrastructures (right). Simulation of Scenario 1 of 0.5 m rise.

Regarding the affected population, negative impacts of Scenario 2 affect all the municipalities. These impacts are further exacerbated along the continuous urban fabric, and particularly in the eastern suburban settlements. In Scenario 1, specific coastal urban blocks in the city of Thessaloniki with a population of 723 people are flooded, while in the municipality of Thermaikos 1045 people are affected. In Scenario 2, the affected population ranges between 1443 and 1698 people respectively (see Fig. 2).

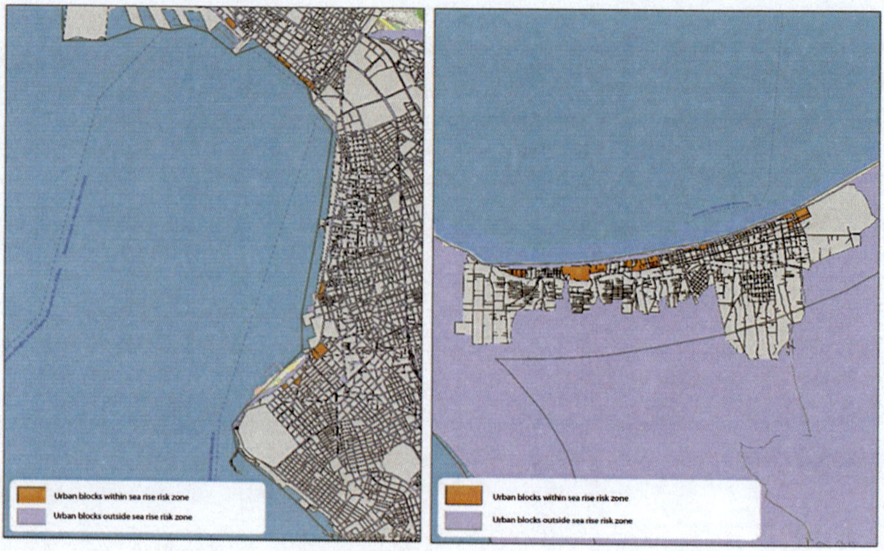

Fig. 2. Sea level rise impacts on residential areas. Simulation of Scenario 2 of 1 m rise.

4 Adaptation Measures for Transport Infrastructures

Based on the simulation results, the followings actions are considerate necessary: the development of an integrated resilience strategy into the metropolitan transport planning, as well as of specific adaptation measures for Thessaloniki's transport infrastructure to address the expected spatial effects of rising sea level. Following the proposed adaptation strategies and roadmaps by international [8, 20] and national [21–23] bodies and research initiatives, the following technical measures and adaptation actions for the road network, port and airport are proposed:

- *Construction of dams and flood defense schemes or elevation projects of coastal road networks in areas with high vulnerability.*

Appropriate interventions require the development of flood protection measures based on high-resolution spatial data to accurately assess the degree of vulnerability of the affected road segments and to prioritize their criticality in relation to the overall functioning of the metropolitan road network.

- *Planning of emergency routes or diversions in case of closure of road segments.*

The selection of road routes passing through areas that are not vulnerable is necessary both to ensure access by emergency vehicles and to enable safe evacuation of the population.

- *Construction of dams-embankments to protect infrastructure at coastal airports from floods.*

Based on the scenarios, the area occupied by the airport is going to be significantly affected by sea-level rise. In line with international practice [24] implemented at international airports (Changi in Singapore, Schiphol in Amsterdam), it is necessary to carry out flood protection studies involving the design and construction of natural embankments as well as the construction of extensive drainage networks and water pumping systems.

- *Preparation of a vulnerability study for the port and examination of the need for relocation, redesign and reinforcement of seawalls for the protection of maritime transport infrastructures.*

Thessaloniki's port appears to be facing similar risks to the airport regarding Coastal Flooding Impacts. The Thessaloniki Port Authority, following the example of other ports of international or regional scale, should investigate the need for dam construction or see wall reinforcement. For example, the Port of Rotterdam Authority, the City of Rotterdam and the Ministry of Infrastructure and Water Management set a common goal of maintaining a sustainable, durable and safe port area [25]. As part of a co-operation program, a comprehensive adaptation strategy has been formulated focusing on raising the awareness of stakeholders on climate change issues, sharing knowledge and developing studies and projects for the protection of port facilities.

5 Conclusions

Over the last decades, the observed changes in climate has led to fluctuations in weather conditions as well as in the frequency and intensity of extreme weather events, which are causing increasing infrastructure damage, longer delays for passengers, and disruptions along with additional safety risks and higher operating costs [14]. Direct effects of climate change are related to disasters on infrastructure, anthropogenic and natural environment, while indirect ones take place after the occurrence of the phenomena and are related to disruption in economic and human activities.

Rising sea level as a result of climate change pose a significant threat to urban and suburban areas with an extensive coastline where basic urban and economic functions are concentrated, such as the Greater Thessaloniki area. Quantitative analysis and impact assessment at national and local level is a prerequisite for assessing the risk and vulnerability of transport systems. According to international bodies and organizations such as the United Nations, the IPCC and organized state initiatives, there is an urgent need to tackle climate change in the transport sector by developing resilience strategies that could be further specified in mitigation and adaptation measures for existing infrastructures. Based on [26], a resilient city has the capacity to absorb disturbances and reorganize its social, economic and technical systems while undergoing climate changes, so as to preserve essentially the same function, structure, identity, and feedbacks. The key resilience objectives for transport systems are to reduce vulnerability, susceptibility and fragility as well as to ensure robustness, accessibility and connectivity.

The present study examined two scenarios for sea level rise of up to 0.5 and 1 m to identify the degree of vulnerability and susceptibility of the coastal area and the transport systems in Thessaloniki. The simulation findings for both scenarios revealed that road segments of major urban and interurban highways, airport's access routes and facilities as well as certain parts of the port will be inundated. Densely populated residential areas on the coastal front will also be affected. It should be noted that the quantifications of impact factor represent an initial numerical approach that its degree of accuracy is related to the resolution accuracy (90 meters) of the satellite images used to simulate the scenarios. Further investigation is required in the context of flood protection studies with high-resolution spatial data. In any case, the findings indicate the coastal areas with the highest vulnerability and stress the need for specific adaptation measures, such as elevation of infrastructure, traffic diversion planning and flood defense schemes. In conclusion, the financial costs associated with the maintenance, prevention, protection and construction of new resilient and durable infrastructures are expected to be particularly high, due to expensive materials and high technical requirements. However, adaptation investments are estimated to be less costly than the costs needed to repair affected infrastructure in the case of no-measures be taken [27].

The introduction of the concept of resilience to the process of planning and constructing transport networks, as well as to the implementation of specific adaptation measures for infrastructures, should be a matter of priority for the near future. Public information and awareness as well as cooperation of the implicated authorities are a prerequisite for integrated and effective preparation in relation to the rising sea level and the emergence of extreme flood events.

References

1. Koetse, M.J., Rietveld, P.: The impact of climate change and weather on transport: an overview of empirical findings. Transp. Res. Part D Transp. Environ. **14**(3), 205–221 (2009)
2. Markolf, S.A., Hoehne, C., Fraser, A., Chester, M.V., Underwood, B.S.: Transportation resilience to climate change and extreme weather events – beyond risk and robustness. Transp. Policy **74**, 174–186 (2019)
3. Wan, C., Yang, Z., Zhang, D., Yan, X., Fan, S.: Resilience in transportation systems: a systematic review and future directions. Transp. Rev. **38**(4), 479–498 (2018)
4. Marchese, D., Reynolds, E., Bates, M., Morgan, H., Clark, S., Linkov, I.: Resilience and sustainability: similarities and differences in environmental management applications. Sci. Total Environ. **613–614**, 1275–1283 (2017)
5. Tamvakis, P., Xenidis, Y.: Resilience in transportation systems. Procedia Soc. Behav. Sci. **48**, 3441–3450 (2012)
6. Intergovernmental Panel on Climate Change (IPCC): Climate Change 2014, Synthesis Report. IPCC, Geneva (2014)

7. Tol, R.S.J.: Adaptation and mitigation: trade-offs in substance and methods. Environ. Sci. Policy **8**(6), 572–578 (2005)
8. United Nations Economic Commission for Europe (UNECE): Climate change impacts and adaptation for international transport networks. Expert group report, UN, Geneva (2013)
9. Taylor, M., Philp, M.: Adapting to climate change - implications for transport infrastructure, transport systems and travel behaviour. Road Transp. Res. **19**(4), 69–82 (2010)
10. European Environment Agency (EEA): Adaptation of transport to climate change in Europe. Publications Office of the European Union, Luxembourg (2014)
11. Kontogianni, A., Tourkolias, C., Damigos, D., Skourtos, M.: Assessing sea level rise costs and adaptation benefits under uncertainty in Greece. Environ. Sci. Policy **37**, 61–78 (2014)
12. EL.STAT.: Demographic characteristics Census (2011). http://www.statistics.gr/el/statistics/-/publication/SAM03/. Accessed 17 Febrauary 2019
13. Intergovernmental Panel on Climate Change (IPCC): Climate Change 2014, Impacts, adaptation, and vulnerability, Part B:Regional aspects, Contribution of working group II to the fifth assessment report of the IPCC. Cambridge University Press, New York (2014)
14. European Environment Agency (EEA): Climate change, impacts and vulnerability in Europe 2016, An indicator-based report, Publications Office of the European Union, Luxembourg (2017)
15. World Climate Research Programme. https://www.wcrp-climate.org/wgcm-cmip/wgcm-cmip5. Assessed 12 March 2019
16. Climate Central. https://www.climatecentral.org/. Accessed 25 May 2019
17. Copernicus Europe's Eyes on Earth: CORINE Land Cover (2018). https://land.copernicus.eu/pan-european/corine-land-cover/clc2018?tab=mapview. Accessed 19 Febrauary 2019
18. OpenStreetMap. https://www.openstreetmap.org/#map=9/41.0845/22.3572. Accessed 20–19 June 2019
19. Esri: ArcGis-The mapping and analytics platform. https://www.esri.com/en-us/arcgis/about-arcgis/overview. Accessed 22 May 2019
20. National Academies of Sciences: Engineering, and Medicine: Transportation Resilience. Adaptation to Climate Change. The National Academies Press, Washington DC (2016)
21. Ministry of Environment and Energy (MEE): National climate change adaptation strategy. Report of the MEE (in Greek), Athens (2016)
22. Giannopoulos, G., Gagatsi, E., Mitsakis, E., Salanova, J.M.: Risks and impacts of climate change on the transport sector. Environmental. Financial and social impacts of climate change in Greece, pp. 260–268. Report of the Climate Change Impact Assessment Committee, Athens (2011)
23. Stamos, I., Mitsakis, E., Salanova, J.M.: Roadmaps for adaptation measures of transportation to climate change. Transp. Res. Rec. J. Transp. Res. Board **2532**, 1–12 (2015)
24. Airports Council International (ACI): Airports' resilience and adaptation to a changing climate. ACI policy brief (2018)
25. Rotterdam climate initiative: Rotterdam climate change adaptation strategy. City of Rotterdam (2013)
26. Walker, B., Holling, C.S., Carpenter, S.R., Kinzig, A.: Resilience, adaptability and transformability in social–ecological systems. Ecol. Soc. **9**(2), 5 (2004)
27. Schweikert, A., Chinowsky, P., Espinet, X., Tarbert, M.: Climate change and infrastructure impacts: comparing the impact on roads in ten countries through 2100. Procedia Eng. **78**, 306–316 (2014)

How Ready Are Greek Consumers to Use Electric Vehicles?

Vasileios Lioutas(✉), Giannis Adamos, and Eftihia Nathanail

Traffic, Transportation and Logistics Laboratory, University of Thessaly, Pedion Areos, 38334 Volos, Greece
lioutasvasileios@gmail.com, giadamos@civ.uth.gr, enath@uth.gr

Abstract. Sustainable urban mobility is evolving rapidly, attempting to facilitate mobility and goods' transportation in an economic and environmentally friendly direction. Among the actions taking place worldwide to provide such solutions is the promotion of electromobility. In this paper, a systematic literature review is carried out focusing on the impacts of adopting Electric Vehicles (EVs) at a national and local (city) level and the incentives required to promote their adoption by future consumers. Some countries, such as Norway, have successfully established measures to foster the use of EVs and achieved remarkably high penetration rates. On the other hand, there are still countries, which face crucial issues regarding the adoption of EVs. One of them is Greece, where the low availability of charging infrastructure and the lack of incentives have created significant barriers towards the promotion of electromobility. To this end, a Pan-Hellenic survey was implemented addressing the beliefs, preferences and intentions of people to use electric vehicles in their daily life. The questionnaire was answered by 400 respondents and from the data collected, critical conclusions were revealed, as for the incentives needed to promote electromobility, the optimal location of charging infrastructure and the pricing of the provided services at a national and at a local (city) level.

Keywords: Electromobility · Sustainability · User acceptance · Optimal charging location · Questionnaire survey

1 Introduction

Electrification accelerates the efforts confronted with climate change and the use of renewable energies, reduces the dependence to fossil fuels and deploys the progress in electric vehicles technologies and infrastructure. In addition, electrification assists in addressing the regulations in terms of CO_2 and other pollutant emissions, set by the European Union (EU) and international organizations. Towards this direction, old and new automotive industries invest in Electric Vehicles (EVs) and try to deploy the significant progress in electromobility and energy efficiency. Furthermore, public incentives to support the market take-up of electric vehicles and the availability of charging infrastructure are strongly encouraged (ERTRACK et al., 2017).

E. G. Nathanail et al. (Eds.): CSUM2020, AISC 1278, pp. 760–769, 2021.
https://doi.org/10.1007/978-3-030-61075-3_74

Road transport decarbonization and emissions' elimination are the key factors for expediting EVs. In order to achieve the goal set by the European Union (EU) for 2030 to reduce greenhouse gas emissions by 40% (Herold et al., 2019), the penetration rates of EVs should be enlarged. Electrification will only be efficient if the electric energy used, comes from low carbon/neutral sources such as solar, wind, hydro and geothermal, otherwise nor the quality of the air or emissions will be improved. In many cases, when electricity comes from high carbon sources, the environmental impacts may be worse than using conventional vehicles, working on an Internal Combustion Engine (ICE) motor. By 2016, the EU's electricity generation was 295,8 g CO_2/kWh and 29% of the total amount produced, came from renewable sources (EEA, 2018).

The aim of this paper is to capture consumers' preferences and willingness to purchase and consequently use electric vehicles in their daily mobility. To this end, a Pan-Hellenic questionnaire survey was organized by the Traffic, Transportation and Logistics Laboratory of University of Thessaly. The survey was structured according to the principles of Theory of Planned Behavior (Ajzen, 1991), including measurement variables like behavioral beliefs and behavioral intentions, which can sufficiently predict persons' behavior towards a desirable direction (Nathanail & Adamos, 2013). Consumers are defined in this study as potential future buyers of an EV and desirable behavior would be to switch from ICE vehicles to EVs.

The rest of the paper is structured as follows: Sect. 2 includes background information on electromobility, incentives promoting the purchase of EVs and the situation in Greece. The methodological approach and data analysis are given Sect. 3, followed by results in Sect. 4. Section 5 summarizes the main findings of this research.

2 Background

2.1 Electromobility

Electromobility refers to the use of EVs, which are defined as vehicles that use one or more electric motors for propulsion. This vehicle category includes Battery EVs (BEV), Plug-in EVs (PEVs), Plug-in Hybrid EVs (PHEV) and Fuel Cell EVs (FCEV). BEVs refer to vehicles that use only chemical energy stored in a rechargeable battery pack to power the electric motor. PHEVs combine the all-electric powertrain of a BEV with an ICE. EVs and PHEVs can be summarized as PEVs as they both can be recharged from an external source of electricity (Spöttle et al., 2018).

Through electric mobility, vehicles are expected to eliminate local emissions, improving urban air quality and noise, still it is important to note that benefits resulting from the reduction of noise are restricted to 30 km/h (ERTRACK et al., 2017). However, recently, batteries' life cycle and their environmental footprint are being examined in literature. All researchers conclude that the key aspect for effective use is charging, which should be accomplished with renewable sources. Most charging stations are expected to be located around buildings inside the city grid and as a result energy consumption of the electric grid at a local level will be greatly considered in the near future, when EV's sales will be increased.

It was estimated that by 2020, there would globally be between 9 to 20 million EVs and this number will be around 40–70 million by 2025 (Fernández, 2019). Considering that there were 950 million ICE vehicles in 2015, it is obvious that even under the most optimistic scenario, EVs can reach up to 10% of vehicles globally (Fernández, 2019). The promotion of electromobility requires the provision of several incentives and/or interventions. First of all, charging infrastructure should be efficient and easily accessible, including the construction of an adequate number of charging points. Since 2010, when EVs were firstly introduced to the market, there has been a stable increase in charging stations, counting 102.861 in Europe in 2018 (Spöttle et al., 2018).

In addition to the above, charging time should be reduced, i.e. by installing fast charging infrastructure. Still, the relevant high costs and the negative impact on batteries' lifetime, remain significant barriers to the widespread use at houses and workplaces. Incentives promoting the purchase of EVs should be established, such as lower taxation, free parking and priviledged road access. The consumers' driving range anxiety is also an issue, but the majority of them believe that a driving range between 400–600 km would abolish these thoughts (Fernández, 2019). PEVs have become increasingly popular, reaching market shares of 39,2% in Norway, 2,2% in the Netherlands, 1,9% in the UK, 1,8% in France and 1,6% in Germany. It is worth mentioning that almost 4 out of 10 newly registered passenger cars in Norway in 2017 were PEVs (Spöttle et al., 2018).

2.2 Incentives Promoting the Purchase of EVs

Governmental support for electric mobility has been growing the last 10 years, but the number of EVs still remains low in many countries. As of 2017, several EU Member States had none or a few incentives in place, however this situation is expected to change, since the EU Member States are obliged to revise their legislation (Spöttle et al., 2018). These incentives take place at global, national and local level and contain consumer rewards, local motivations, charging infrastructure and complementary policies.

As far as consumers are concerned, governments can fund part of the EV's total cost and unshackle them of taxes, such as registration or road taxes. These economic factors differ among countries and are affected by citizens purchasing power. It is expected that in countries with a higher purchasing power, the number of EVs will be higher compared to countries with a lower purchasing power (Rietmann & Lieven, 2019). Households with a medium or low income do not seek to buy an EV, due its high actual cost (Neves et al., 2019). On the other hand, low operating costs of EVs and the taxation's benefits may counterbalance this situation. Therefore, it is obvious that economic factors affect more low and medium societal levels and are probably a matter of indifference for the high ones. Other social factors, such as education level, occupation and age can affect the purchase of an EV. Literature suggests, that people with a higher education level can adopt an EV easier (Neves et al., 2019).

Local incentives refer to traffic regulations and advantages for the drivers of PEVs, i.e. in Norway drivers can access highways, high occupancy vehicle lanes, toll bridges, ferries and roads free of charge. In addition, in Germany and Norway, PEVs have access to areas with free or reduced parking fees and can circulate through bus lanes (Rietmann & Lieven, 2019).

Charging infrastructure plays a crucial role regarding EVs' market share, since its availability is associated with PEV adoption. While high availability of charging infrastructure does not automatically lead to high adoption rates, low availability is an obstacle to the adoption of PEVs (Spöttle et al., 2018). For this reason, type, location and quantity of charging infrastructure should be defined. Globally, there are two main types of charging modes, normal and fast. Normal charging, which is also known as slow charging, can charge a PEV in 6–8 h (EEA, 2016) and is mainly located in houses, workplaces and public spaces. The cost of a single charging point rises up to 2.000€ (Spöttle et al., 2018). With fast chargers, a PEV can charge the 80% of its battery in an average of 20 min and the total battery package can be charged in 1–3 h (European Environmental Agency, 2016). Fast chargers are mainly located in public spaces due to their high cost, which rises up to 60.000€ (Spöttle et al., 2018). Where PEV drivers may charge is strongly related to their trip's destination. If home charging is available, there is a strong preference for this type of charging. In addition, drivers traveling for business purposes would charge at home and at work, since these destinations form up about the 50% of their trips. The required quantity of public charging stations can be calculated by the ratio of PEVs per charging point. In the EU, the ideal ratio of PEVs per charging point will lie between 10–16 (Spöttle et al., 2018).

Complementary policies include mainly initiatives to support electromobility and climate goals. Several organizations, such as "E-mobility NSR" have been established to promote electromobility solutions in the Netherlands, Belgium, Norway and other countries (Rietmann and Lieven, 2019). In addition to the above, activities that inform and educate citizens for the electrification of road transport have been initiated.

2.3 The Situation in Greece

Greece is one of EU Member States, which have significantly low growth in terms of electromobility. As of 2017, the country's PEV fleet consisted of 341 vehicles and 38 public accessible charging points. Moreover, the PEV market share of new passenger cars was only 0,2% in 2017 (Spöttle et al., 2018).

Greece needs to make progress in this sector, in order to be able to follow the growth of the other European countries. In 2019, the "National Plan for Energy and Climate" was released, pointing out that Greek automobile fleets are among the oldest in Europe and the adoption of PEVs is extremely low. To this end, it is expected that one billion euros will pe provided to develop measures to renew the fleet with cleaner vehicles (National Plan for Energy and Environment, 2019).

3 Method and Data Analysis

In order to capture consumers' perception regarding the adoption of EVs, a Pan-Hellenic survey was organized and data were collected through an online questionnaire. Based on the findings of literature review and the research work conducted in Lioutas (2020), the questionnaire enabled the assessment of sever groups of indicators, namely: beliefs regarding the progress made in the automotive sector, possibility to buy an EV, beliefs regarding the purchase of an EV, incentives regarding the purchase of an EV, charging of EVs, public authorities' initiatives regarding the charging of EVs and renewal of public authorities' fleets with EVs.

In total, 40 indicators were assessed by respondents on a 5-point Likert scale, with 1 being the lowest possible score and 5 being the highest. Furthermore, supplementary data were collected, regarding trip characteristics such as trip purpose, origin-destination, etc. and demographics like gender, age, education level, employment status and net-income per month. The survey was carried out between 21 December 2019 and 21 January 2020 via SurveyMonkey (www.surveymonkey.com) and was promoted through Social Media, University of Thessaly's platform, electronic media and newspapers. The responses recorded by SurveyMonkey were then exported to a database and were analyzed. In total, 436 responses were received, from which the valid ones were 400.

Data were analyzed through descriptive and inferential statistics. In the first case, a number of the sample characteristics, such as gender and age were addressed by estimating the frequency distribution per characteristic, using Microsoft Excel Software. In the second case, the statistical analysis of the responses was completed using non-parametric tests through IBM SPSS Statistics Software. A confidence level of 95% and confidence interval of 5% were assumed.

4 Results

4.1 Sample Characteristics and Mobily Patterns

The final sample size consists of 43,5% women and 56,5% men. More than half of the participants are up to 25 years old (51%), 20,5% of them are up to 40 years old and the remaining of them are older than 41 years old (28,5%). In addition, most of the respondents (65,8%) are highly educated, 33,8% of them have received a secondary level of education and the remaining 0,4% are primarily educated. As far as employment status is concerned, it was observed that 48,8% of the participants are employed, 44,8% students, 1,8% unemployed and the remaining 4,6% stated a different status. 41,5% of the respondents have a monthly net-income greater than 1.500€, 26,5% 1.001–1.500€, 21,5% less than 1.000€ and the remaining 10,5% preferred not to say.

Focusing on trip purpose at a local (city) level, it was noticed that most of the participants (48,5%) travel for work, 29% for education, 14,8% for leisure and the remaining 7,7% travel for shopping. Outside the city grid, most of the participants (52,3%) travel for holidays, 29,3% travel to visit family and the remaining 18,4% travel for work. As long as the frequency of traveling more than 100 km per month is concerned, the majority (37%) of respondents make less than one trips. Annually, 47% of respondents travel between 10–20 thousand km, 35,5% travels less than 10 thousand km and the remaining 17,5% travels more than 20 thousand km.

4.2 Consumers' Perceptions Towards Electromobility

Most respondents (88,5%) state that they are interested in the new technological progress in the automotive sector and 95,3% of them believe that EVs are a significant technological development. The vast majority (84,5%) would be interested to use an EV and 61,8% of them would abolish the usage of a conventional vehicle, if they owned an EV. These proportions address those participants who indicated cumulatively that they agree and totally agree with the above statements. When asked about the possibility to buy an EV, 53,5% of the sample responded positively and 47% of them would consume 10.000–20.000€ for such a purchase. Three quarters (76,5%) of the respondents stated that the installation of charging stations in their city would motivate them to buy an EV.

In addition, 91,3% of the respondents believe that EVs are friendlier to environment and 61% of them believe that operational cost is less, compared to conventional vehicles. On the other hand, sample seems not to be fully aware of aspects like safety, performance, maintenance cost and satisfaction of their travel needs resulting from the use of an EV, since when asked about these factors 40–59% of them answered neither positively nor negatively. Focusing on potential factors that would motivate consumers to purchase an EV, it seems that the main concern would be the availability of charging infrastructure (94,8%) and the strongest incentives would be low operating costs (92%), state subsidy (91,3%) and road taxes reduction (90,5%). Similarly, these percentages characterize those respondents who cumulatively agree and totally agree with the specific statements.

Regarding charging of EVs, 44,3% of respondents would pay up to 4€ to complete a full normal (slow) charge and 31,3% of them would pay between 4,1–8€ to complete a full fast charge (Fig. 1). Depending on the charging mode (1÷4), the full charge of an EV provides approximately autonomy of 100 km (EEA, 2016). Also, 47,8% of participants would charge their EV 2–3 times per week. Ideally, the absolute majority (91,8%) would perform a normal (slow) charge during night, while for 40,3% of them, the ideal time for a fast charge would be at noon (Fig. 2). In addition, 33,5% of responders would be worried sufficiently not to have their EV's battery fully charged, while for 40,3% of them, the most acceptance percentage before charging their EV's battery, would be 20%.

As for the ideal location of charging infrastructure at a local (city) level, respondents were asked to select up to three locations/places that best fit their needs or expectations. The most popular responses were public parking areas (45,5%), gas stations (43%), houses (39,3%), strategically positioned on-street sties (28%), workplaces (24,8%), shopping centers (23,8%) and terminals (22%). The total list of the respondents' preferences is listed in Fig. 3.

Focusing on possible initiatives of public authorities regarding charging of EVs, the majority of respondents stated that Municipalities should install normal (71%) and fast (91,5%) charging stations at public spaces. Respondents also believe that Municipalities should grant land from their own properties (76,8%) and cooperate with private companies (81%) for the installation of charging stations. In addition, almost all survey participants (91,8%) stated that public authorities should organize actions for the awareness-raising of citizens towards electromobility. Lastly, citizens believe that the

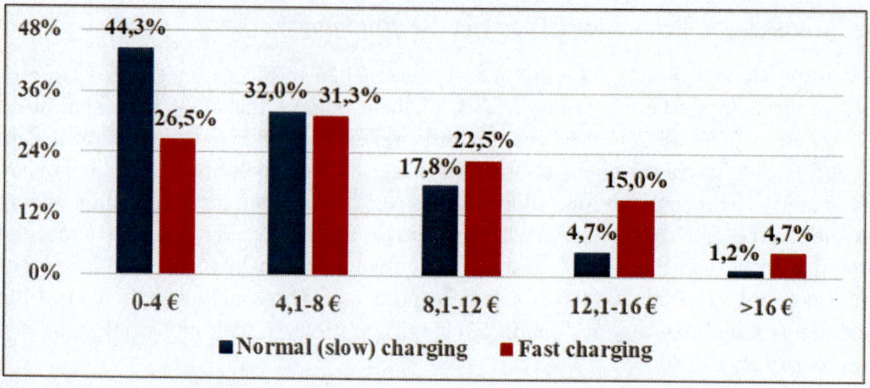

Fig. 1. Acceptable pricing for full charge per EV.

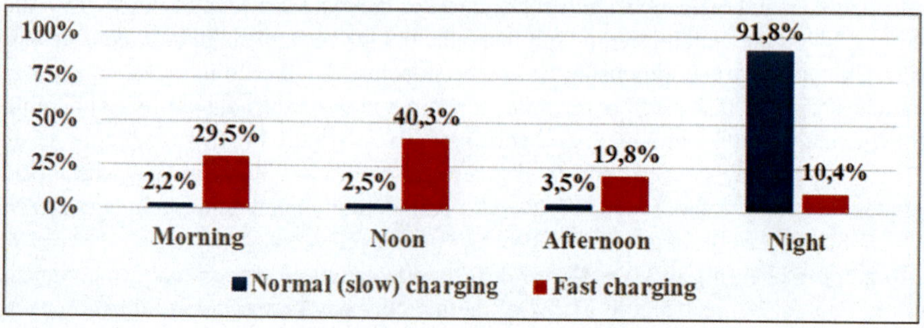

Fig. 2. Preferable charging timing.

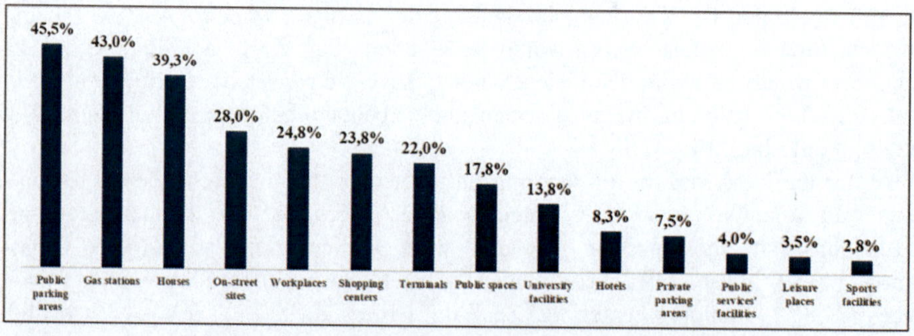

Fig. 3. Ideal locations for charging infrastructure.

upgrade of Municipalities' vehicle fleets with EVS would increase their green reputation, i.e. they would become more contemporary (87,3%), environmentally friendlier (92,8%), more attractive to tourists (62,8%) and they could work as a good case for other Municipalities (93,5%).

4.3 Consumers' Willingness to Purchase an EV

In order to investigate those parameters that seem to affect consumers' willingness to buy an EV, two-way ANOVA test was performed, and sample was grouped according to several characteristics such as gender, age, income, annual travel distance, vehicle age, etc. Indicatively, results showed that there were significant differences in the average willingness of consumers to buy an EV between combinations of income categories and vehicle (p-value < 5%), but not in the average willingness within income categories (p-value > 5%) or vehicle age (p-value > 5%). It was revealed that respondents of high income (> 1.500€) regardless the vehicle age, are quite willing to buy an EV compared to the rest potential consumers (Fig. 4).

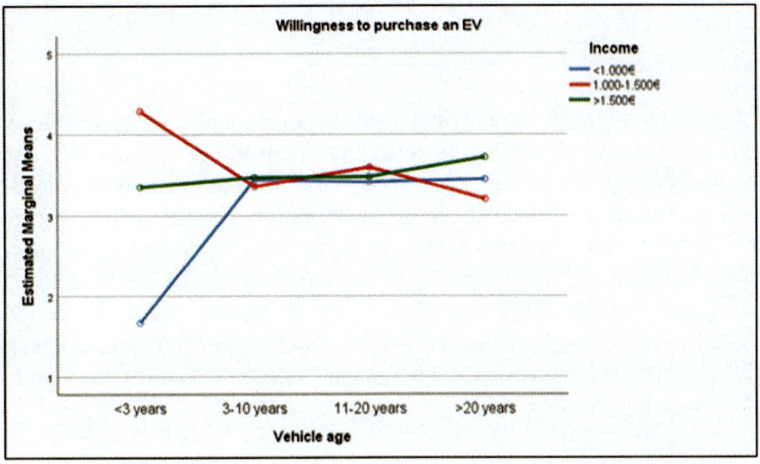

Fig. 4. Estimated willingness of consumers to purchase an EV.

The interrelationships between consumers' willingness to buy an EV with appropriate factors were also tested. In this case, constructs were built by combining measured indicators, using alpha test (Cronbach, 1951), where Cronbach a > 0,6. The first construct is technology acceptance, including variables such as "interest on new technological advances in automobility", "EVs are a very important technological advancement" and "interest on using an EV" (a = 0,742). Behavioral beliefs resulted from a combination of seven variables (a = 0,801), i.e. "EVs are best fitting my mobility needs", "EVs are environmentally friendlier", "EVs are safer", "EVs have lower operating costs", etc. Consumers' behavioral intentions towards buying an EV were formulated by six variables (a = 0,684), such as "state subsidy", "road taxes reduction", "GHG emissions reduction", "possession of an advanced technology vehicles", etc. Lastly, motivation refers specifically to the construction and operation of additional charging stations in the city and the degree that this intervention would affect consumers' decision to purchase an EV. It was observed that the overall willingness of consumers is more and statistically significantly related to "technology acceptance"

(β = ,467, p-value < 1%) and "behavioral beliefs" (β = ,409, p-value < 1%) and less related to "motivation" (β = ,274, p-value < 1%) (Table 1).

Table 1. Bivariate correlations of the individual factors and their relationship with the variable characterizing consumers' willingness to purchase an EV.

Factor	1	2	3	4	5
1. Willingness	–	–	–	–	–
2. Technology acceptance	,467**	–	–	–	–
3. Behavioral beliefs	,409**	,533**	–	–	–
4. Behavioral intentions	,356**	,493**	,476**	–	–
5. Motivation	,274**	,336**	,380**	,332**	–

***p-value < 1%, *p-value < 5%*

Based on the findings of the previous analysis a prediction model was developed, associating the overall willingness of consumers (dependent variable) with specific constructs (independent variables). It is noted that several alternative combinations of variables were tested, and that explaining better future consumers' willingness, i.e. significant contribution of variables in the prediction thus, higher value of adjusted R^2, is depicted in Table 2. The regression is significant (F(4, 399) = 38,321, p-value < 5%) and explains 27% of variance. Based on the values of the indicator Beta (Table 2), "technology acceptance" is the strongest predictor of willingness (p-value < 5%). Behavioral intentions and behavioral beliefs also contribute significantly to the total variances (p-value < 5%).

Table 2. Regression analyses of willingness of consumers to purchase an EV.

Factor	B	Std. Error	Beta	T	P-value
Technology acceptance	,518	,099	,293	5,248	<5%
Behavioral beliefs	,301	,099	,166	3,053	<5%
Behavioral intentions	,307	,114	,143	2,700	<5%
Constant	-1,343	,417		-3,221	<5%

Adjusted R^2= 27%, F(4, 399) = 38,321, p-value < 5%

5 Conclusions

The penetration rate of EVs in Greece is very low despite the fact that more of 50% of this survey's respondents stated that they would possibly purchase an EV. This situation highlights that the effectiveness of the current incentives is limited and that strong motivations should be established, i.e. state subsidy and road tax benefits. Charging infrastructure availability seems to be of high concern of consumers. As resulted from this research, citizens would prefer the construction of fast chargers rather than normal (slow) ones by the local authorities. This opinion moves alongside literature, since the installation of fast charging points demands high resources and thus they cannot be widely used at private places as in (ERTRACK et al., 2017, Spöttle et al., 2018).

It is also important to raise awareness of people towards electromobility, since they are mainly aware of the environmental benefits, but not of other characteristics of EVs, such as comfort and safety or operating and maintenance costs. For example, most respondents of this survey would spend up to 20.000€ to buy an EV when the starting prices are at approximately 25.000€.

Lastly, citizens believe that an integrated plan of developing and operating charging stations at strategic areas would improve the city's overall image and attractiveness, i.e. for tourists.

References

Ajzen, I.: The theory of planed behavior. Organ. Behav. Hum. Decis. Process. **50**, 179–211 (1991)

Cronbach, L.J.: Coefficient alpha and the internal structure of tests. Psychometrical **16**(3), 297–334 (1951)

ERTRAC, EPoSS & ETIP SNET: European Roadmap Electrification of Road Transport. 3rd Edition, Version 10 (2017)

EEA (European Environmental Agency) (2016). https://www.eea.europa.eu/about-us. Assessed 15 January 2020

EEA (European Environmental Agency): Overview of electricity production and use in Europe (2018)

Fernández, R.Á.: Method for assessing the environmental benefit of road transport electrification and its influence on greenhouse gas inventories. J. Cleaner Prod. **218**, 476–485 (2019)

Herold, A., Cook, V., Baron, Y., Cames, M., Gores, S., Graichen J., Kasten, P., Mehlhart, G., Siemons, A., Urrutia, C., Wolff, F.: EU Environment and Climate Change Policies, European Parliament, Policy Department for Economic, Scientific and Quality of Life Policies, Brussels (2019)

Lioutas, V.: A feasibility study on installing charging stations for electric vehicles in a medium sized city. Thesis, University of Thessaly, Department of Civil Engineering (2020)

Nathanail, E., Adamos, G.: Road safety communication campaigns: research designs and behavioral modeling. Transp. Res. Part F Traffic Psychol. Behav. **18**, 107–122 (2013)

National Plan for Energy and Climate. http://www.opengov.gr/minenv/. Assessed 15 January 2020

Neves, S.A., Marques, A.C., Fuinhas, J.A.: Technological progress and other factors behind the adoption of electric vehicles: empirical evidence for EU countries. Res. Transp. Econ. **74**, 28–39 (2019)

Rietman, N., Lieven, T.: How policy measures succeeded to promote electric mobility – worldwide review and outlook. J. Cleaner Prod. **206**, 66–75 (2019)

Spöttle, M., Jörling, K., Schimmel, M., Staats, M., Grizzel, L., Jerram, L., Drier, W., Gartner, J.: Research for TRAN Committee – Charging Infrastructure for Electric Road Vehicles. European Parliament, Policy Department for Structural and Cohesion Policies, Brussels (2018)

Traffic Calming Measures as a Tool to Revitalise the Urban Environment: The Case of Serres, Greece

Alexandros Sdoukopoulos[1](✉), Eleni Verani[1], Anastasia Nikolaidou[1], Ioannis Politis[1], and Foteini Mikiki[2]

[1] Transport Engineering Laboratory, Department of Civil Engineering, Faculty of Engineering, Aristotle University of Thessaloniki, 54124 Thessaloniki, Greece
asdoukop@civil.auth.gr
[2] Municipality of Serres, Serres, Greece

Abstract. Urban areas have a primary role in the world today. Almost half of the worldwide population lives in cities, while more than 80% of the global GDP is generated in these areas. On the other hand, cities account for 80% of the energy production and over 67% of the Greenhouse Gas Emissions. In this framework, city centres are generally considered as the most critical areas in urban and transport planning, as they attract a significant share of trips, and in tandem with the complex traffic problems and challenges such as congestion, they face serious environmental issues. In an attempt to mitigate the problems mentioned above, planners promote the sustainable mobility measures and principles, including the green modes of transport and the implementation of traffic calming interventions. In this context, the current paper presents a "before and after" study focusing on the traffic calming interventions that were implemented in the centre of the city of Serres, Greece during the last years. To this end, a traffic simulation model (VISUM), as well as a Road Emissions Calculating Model (COPERT), were developed, while an analysis of traffic accidents was also conducted. Moreover, the acceptance of the interventions among the citizens was also assessed through a questionnaire survey. The results pointed out that the interventions in the centre of Serres have met with great success as the quality of the environment and life, in general, was improved significantly.

Keywords: Sustainable urban mobility · Traffic calming measures · Urban regeneration

1 Introduction

The rapid urbanisation taking place during the last decades both in developed and developing countries has a profound impact on cities and the role they play. Approximately 55% of the world population currently lives in cities, while as far as Europe is concerned, this share reaches even the 75% [1, 2]. Cities are the engine of the economy, and at the same time, they comprise the enablers of innovation and creativity, as places of increased connectivity [3, 4]. On the other hand, cities often perform

E. G. Nathanail et al. (Eds.): CSUM2020, AISC 1278, pp. 770–779, 2021.
https://doi.org/10.1007/978-3-030-61075-3_75

poorly in terms of environmental quality, as they consume over the two-thirds of the world's energy and they also account for the vast majority of air pollutants and Greenhouse Gases emissions [5]. The abovementioned highlight the need for cities to become the drivers in improving air quality and contribute to preventing climate change in general.

In this context, the transition from the conventional design approach to the sustainable transport planning "era" could be considered as a key asset. Sustainable mobility has at its core the promotion of alternative, green modes of transport and aims at the creation of a "friendly" and an accessible by everyone, safe and secure urban environment of high aesthetics, with improved micro-climate, low emissions of air pollutants and low levels of traffic noise [6].

Towards this direction, traffic calming measures could play a vital role. According to Lockwood (1997), traffic calming measures are defined as "the combination of physical measures that reduce the negative effects of motor vehicle use, alter driver behaviour and improve conditions for non-motorised street users" [7]. More specifically, the implementation of traffic calming measures mainly aims at reducing traffic flows, making specific areas unattractive for motorised through-traffic as well as lowering travel speeds, in an attempt to encourage safer, more responsible driving [7–9]. In addition, traffic calming strategy also sets as goals the creation of an enhanced street environment, the promotion of walking and cycling by increasing actual and perceived safety level for non-motorised users, and the reduction of need for police enforcement [7–9].

Based on their main objective, traffic calming measures fall into two broad categories: a) those interventions designed primarily to address high speeds, and b) the restrictive measures planned to control traffic flows and volume [10, 11]. The various types of traffic calming measures could be classified into five main groups namely: a) *vertical deflections* (e.g. road humps, speed cushions, speed tables, plateau, rumble devices, raised intersections, raised crosswalks), b) *horizontal deflections* (e.g. pinch points, chicanes, lateral shifts), c) *carriageway constrictions* (traffic islands, pedestrian refuge), d) *other traffic management and traffic calming measures* (e.g. road closure, roundabouts, changed road surface, entrance treatment, shared space) and lastly, e) *speed limits* [7–10, 12]. As far as the applicability of these interventions is concerned, traffic calming measures could be installed either on a specific road network segment (black-spot approach) or the entire network (area-wide approach) [13].

Regarding the effectiveness of traffic calming schemes, several studies point out that traffic calming measures that result in lower speeds could also lead to fewer severe or fatal accidents [9, 14–16]. According to Taylor et al. (2000), a 1 mph reduction in average travel speed decreases by 5% the number of traffic accidents, while other studies found out that a reduction of 14.4 km/h could lead to reduced traffic accidents by 60–70%, improving thus road safety significantly [17, 18].

On the other hand, the impacts of traffic calming measures on air quality are ambiguous, as multiple parameters affect the results [17]. Litman (1999) highlighted that traffic calming measures that reduce travel speed could contribute to reducing air pollution, while on the contrary, those involving an increased number of stops may result in increased pollutant emissions [19]. Hence, a number of studies concluded that traffic calming measures benefit air quality, while others found that they could increase fuel consumption and emissions [17]. Concerning traffic noise, traffic calming measures could be regarded as a mitigating factor in the vast majority of cases. However, certain types of measures, along with specific conditions, such as a high share of HGVs, could have an adverse effect on noise levels [20].

In line with the main sustainable mobility principles and practices, the Municipality of Serres, Greece implemented during the last years several relevant interventions, such as the pedestrianisation of the central area of the city, the creation of a bicycle network, the establishment of numerous roundabouts as well as the installation of other traffic calming measures in a broad area [21, 22].

In this context, the current research presents a comprehensive assessment of the traffic calming interventions implemented in the city of Serres, in terms of traffic flows, travel speed, road safety, air pollution as well as the perceived level of satisfaction. To this end, traffic and emissions simulation models were developed utilising on-field measurements; spatial and statistical analyses of traffic accidents data were conducted; while a descriptive analysis of a questionnaire survey results was also carried out.

2 Description of the Study Area

Serres is considered as a typical medium-sized Meditteranean city located in Northern Greece. It constitutes the second largest city of the Region of Central Macedonia in terms of population and economic growth, while it also acts as a pole of attraction for the wider area as a significant cultural, commercial and industrial centre. Based on the latest census (2011) carried out by the Hellenic Statistical Authority, its population reaches a total of 52,287 inhabitants, whilst according to Eurostat, the GDP of the Regional Unit of Serres exceeds the 1,700 million Euro, corresponding to a value of approximately 10,000 Euro per capita [23, 24].

The prevailing land-use in the city of Serres is residential, while other various uses such as recreational, commercial, retail stores, banks, offices, etc. are also identified mainly in the Central Business District (CBD) [25]. CBD attracts over 50% of the daily trips, thus being under intense environmental pressures. Moreover, the significant number of trips having as origin or/and destination the city centre challenges the ability of the local transport system to provide an acceptable level of service mainly during peak periods. Even though the average trip length in the city of Serres is approximately 1.5 km, the vast majority of transport users rely heavily on private cars for their mobility needs, while the modal shares of the other modes of transport are much lower. Buses comprise the only means of Public Transport (PT) in the city and despite the adequate spatial coverage, play along with the non-motorised modes a minor role [21, 22].

Considering the negative externalities generated by the extensive use of private cars, namely traffic noise, congestion, road accidents, air pollution, environmental degradation, etc., the local authority of Serres adopted the sustainable mobility agenda. In this framework, during the last years, a bicycle network was established, and a series of traffic calming interventions was implemented in an attempt to discourage the use of private cars and mitigate their impact as well as promote walking and cycling as genuine alternatives to driving.

More specifically, the nine different interventions, which were implemented in the city of Serres and considered in this research, are the following:

- Pinch points, curb extensions and raised crosswalks in Tsalopoulou Str.;
- Road closure/pedestrianisation of Em. Andronikou Str.;
- Road closure/pedestrianisation of K. Karamanli Str.;
- Shared space/woonerf in N. Nikolaou, Dimogerontwn, and Eforon Str.
- Curb extensions and changed road surface in Ermou Str.;
- Raised crosswalks along Magnisias Str., near the campus of the International Hellenic University;
- Raised crosswalks along Kyprou Str., near the Metropolitan Cathedral;
- Roundabout in the intersection of Makedonomachon and Dorylaiou Str.; and
- Roundabout in the intersection of 8is Maiou and Komninon Str.

Lastly, it should be noted that the first five measures were all implemented in the city centre, while the others were widely scattered in different locations.

3 Methodological Approach

In the framework of the current research, a methodological approach was developed to evaluate the traffic calming measures implemented in the city of Serres during the last few years. The methodological approach comprises a comprehensive "before and after" study focusing on the following parameters/fields: a) traffic flows, b) travel speed, c) traffic accidents, d) Greenhouse Gases (GHG), pollutant emissions and fuel consumption, and e) level of acceptance, as perceived by the locals.

To this end, four different actions were deployed, namely the development of a macroscopic traffic simulation model; the implementation of a road emissions calculation model; the conduction of spatial and statistical analyses on traffic accidents data; and the descriptive statistical analysis of the questionnaire survey results. More specifically, a traffic simulation model was developed in the environment of PTV Visum. Visum is a comprehensive software package for strategic transport planning, traffic analyses and forecasts on urban, regional and national level [26]. The developed model consists of 960 links and 306 nodes covering the whole area of the city of Serres and utilises the on-field traffic volume, travel speed and traffic composition measurements, which were conducted in two time periods, referring to the prior (2014) and the post-interventions period (2019).

The measurements described above, along with the outputs of the developed traffic model (e.g. traffic flows and travel speed), were used as input data to estimate GHG and pollutant emissions through the COPERT (COmputer Programme to calculate Emissions from Road Transport) model. COPERT is the EU standard vehicle emissions calculator, being widely used in relevant researches and studies [27]. COPERT obtains the emission factors as a function of the average travel speed, while the basis for the model's calculations is a comprehensive vehicle database [27]. In the framework of the current research, the original vehicle database was adjusted to the characteristics of the local vehicle fleet of the city of Serres.

As far as traffic accidents are concerned, disaggregated accidents data were collected from the local Traffic Police Department for the respective time period (2014–2019). These data were initially processed and analysed in spreadsheets, while they were subsequently digitised and stored in an ESRI ArcGIS geodatabase in order to perform various spatial analyses.

Regarding the last action, a questionnaire survey was carried out, aiming at measuring at a five-point Likert scale the perceived level of acceptance of the examined traffic calming measures. The survey was conducted in October 2019 through the "Surveymonkey" online platform (Computer-Assisted Self Interviewing Method), and a sample of 65 valid questionnaires was collected [28]. The data were analysed in spreadsheets using descriptive statistics, and more specifically, mode and frequencies were applied as a measure of central tendency and variability, respectively.

It should be pointed out here that the potential impacts of the examined traffic calming measures on traffic flows, speed, accidents and pollutant emissions were evaluated in the following spatial levels: a) traffic calmed area, b) adjacent area, and c) wider reference area.

The workflow diagram of the followed methodological approach, illustrating the basic input data and actions, is given in Fig. 1.

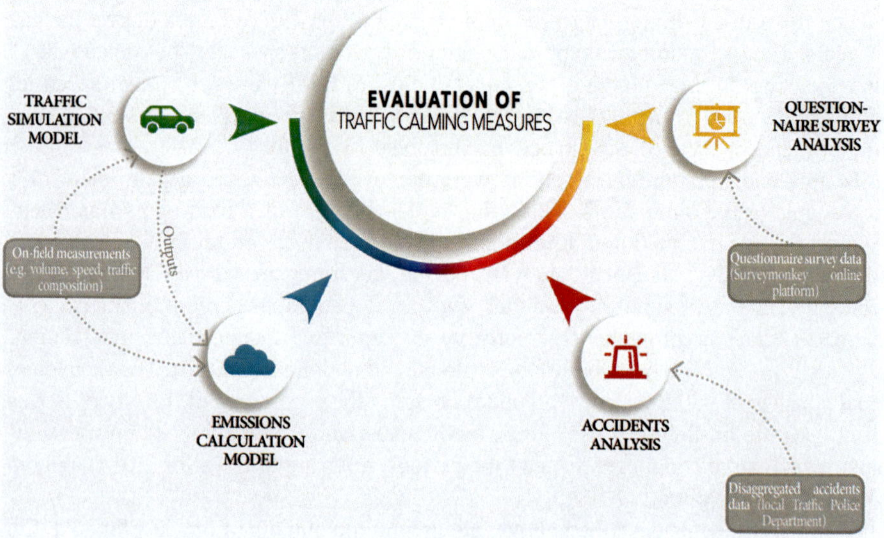

Fig. 1. The workflow diagram of the followed methodological approach.

4 Results and Discussion

The methodological approach described above was used to highlight the impacts of the traffic calming measures implemented in the city of Serres and evaluate their role as a tool for revitalising the urban environment. It should be noted that the presentation of the results does not refer to each of the installed traffic calming measures, yet it follows an area-based approach, as the impacts of the adjacent interventions could not be considered separately. Hence, the implementation of traffic calming measures is assessed in five different cases/areas, namely: a) the city centre (corresponding to *interventions 1–5*), b) Magnisias Str. (*intervention 6*), c) Kyprou Str. (*intervention 7*), d) the intersection of Makedonomachon and Dorylaiou Str. (*intervention 8*), and e) the intersection of 8is Maiou and Komninon Str. (*intervention 9*).

As far as the first case is concerned, the installed traffic calming measures are found to have a positive impact mainly on road accidents and travel speed, and subsequently on emissions and traffic flows.

More specifically, traffic flows are significantly reduced in the traffic calmed area (45.2%), while on the contrary, a small increase of approximately 7% is identified in the adjacent road axes. Overall, however, a 19.6% decrease is highlighted in the wider reference area. Similarly, fuel consumption, GHG and air pollutant emissions were decreased along the traffic calmed axes by shares ranging from 15% to 29%, while on the other hand, an increase on average by 10% is underlined in the adjacent area. Thus, the environmental performance in terms of pollutant emissions was barely improved in the broader reference area.

Travel speeds are considerably decreased both in the traffic calmed zone as well as the neighbouring streets, as a result of the traffic calming interventions and the slightly increased traffic volumes in the adjacent area, respectively. Concerning road safety, traffic calming measures resulted in a great reduction in the number of accidents, ranging from 57.9% in the traffic calmed area to 25.7% in the adjacent axes.

Regarding the perception on traffic calming measures, the vast majority (73.8%) of the participants in the survey are quite and very satisfied with the impact of the interventions on the examined area, while at the same time, they consider them as quite important steps towards increasing non-motorised trips and regenerating the urban environment.

As far as the second and third case are concerned, the installation of raised crosswalks in Magnisias and Kyprou Str. led to lower average travel speeds by approximately 6% and 3% respectively. Although the speed reductions are moderate, a remarkably improved traffic safety level is pointed out in both cases, corresponding to 100% and 88.9% fewer accidents, respectively. On the contrary, it should be underlined that traffic accidents are significantly increased in the adjacent area, thus indicating the emergence for remedial treatment in the neighbouring road axes to prevent further accidents “migration”.

The conversion of the four-legged intersection of Makedonomachon and Dorylaiou Str. into a roundabout (case 4), although explicitly contributed to the improvement of the local traffic conditions, also led to a 100% increase (from 2 to 4 in absolute terms) of traffic accidents. The latter is contrary to the widely proven and accepted findings that roundabouts are the safer type of intersection, and could be considered either as a matter of drivers’ difficulty in familiarising with changes in the road environment or a result of chance factors [29, 30].

In this context, although the reduction of accidents number (from one to none) in the recently constructed roundabout in 8is Maiou and Komninon Str. (case 5) is considered as positive, no clear and concrete conclusions could be drawn, as the inconsiderable number of accidents could have been easily affected by pure chance elements.

As far as the perceived level of safety is concerned, the 42.2% of the participants in the questionnaire survey state that roundabouts contribute highly to the improvement of traffic safety, nevertheless, they express their concerns regarding the vulnerable road users. In general, however, the respondents are quite satisfied with the introduction of the two roundabouts, while the great majority of them (57.2%) regard roundabouts as a valuable measure to enhance the urban environment.

The results are summarised in the infographic below (Fig. 2).

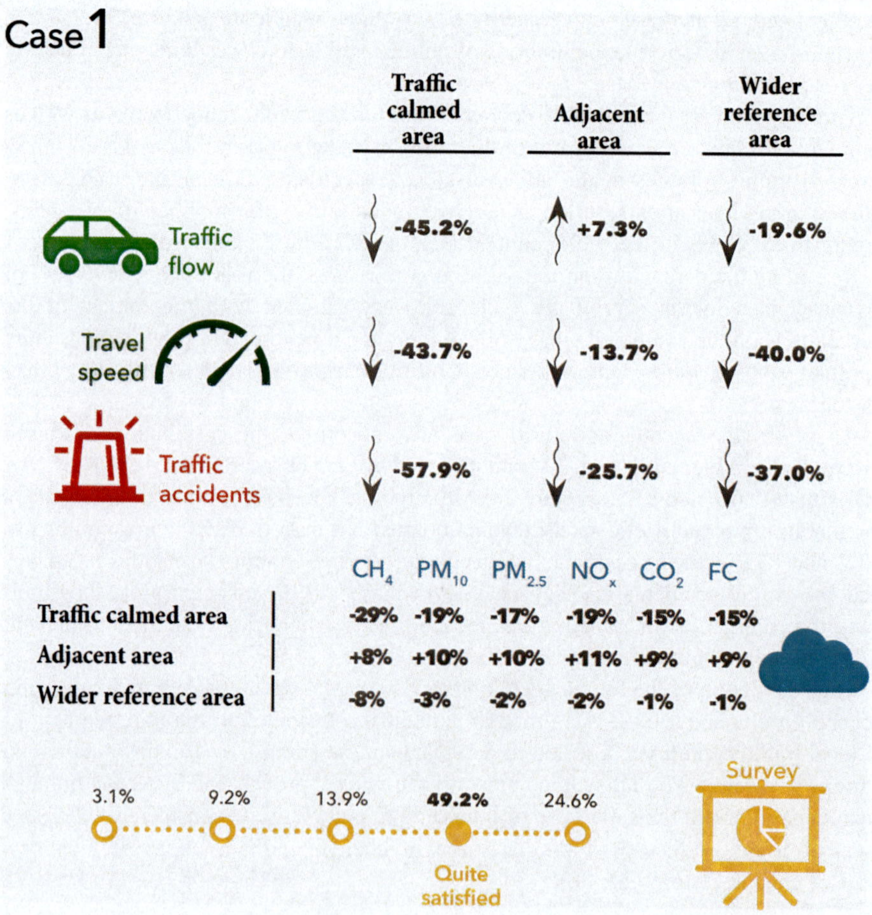

Fig. 2. Main results of the evaluation of the considered traffic calming measures.

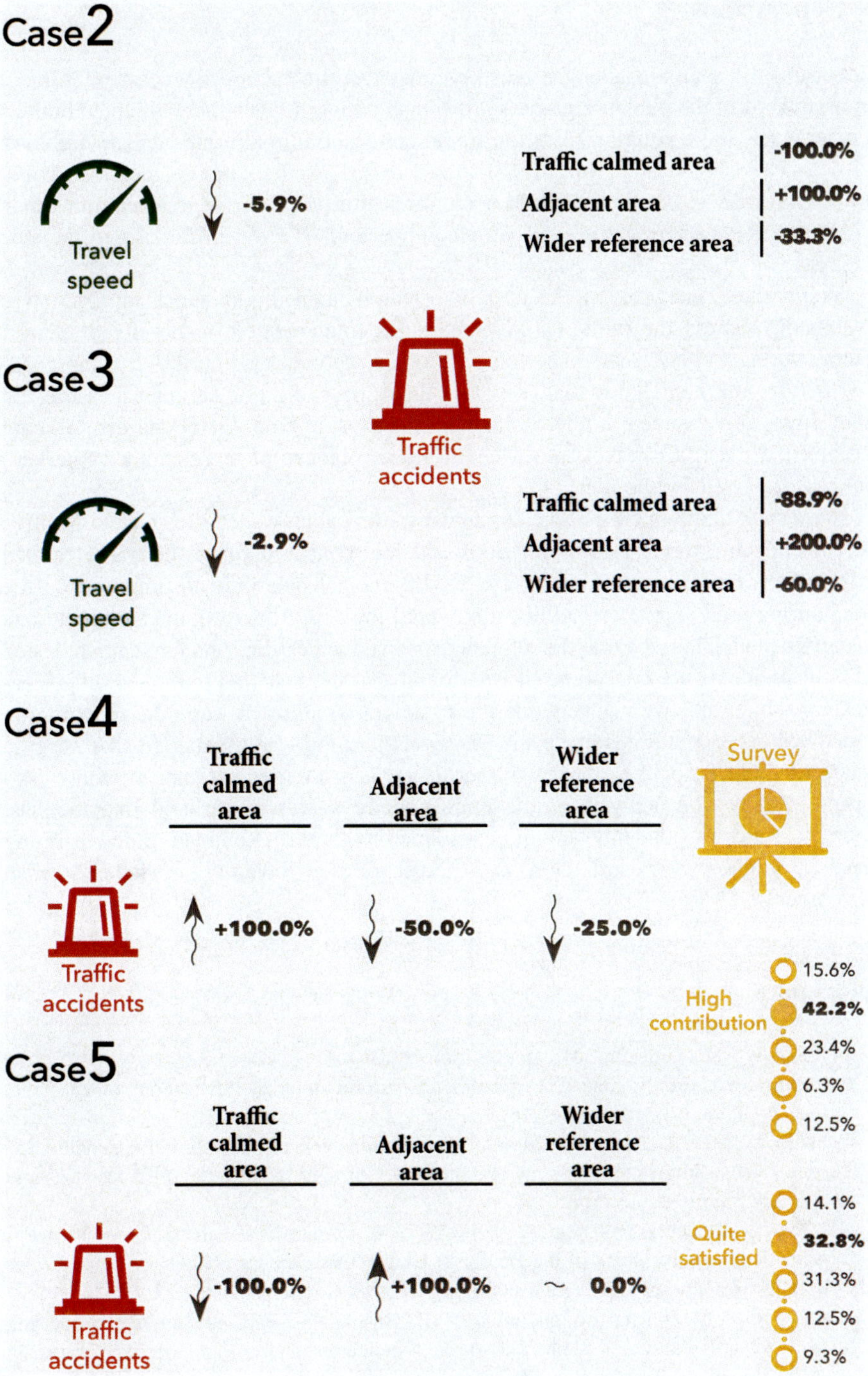

Fig. 2. (*continued*)

5 Conclusions

Addressing the major challenges that several cities are facing nowadays requires a paradigm shift in the planning process. Sustainable mobility is widely accepted both as an emergence and a solution to tackle traffic congestion, road accidents, air pollution and degradation of the environment. In this context, traffic calming measures could play a vital role as they constitute a useful and effective tool for regeneration urban space as well as altering the way people move and live towards more sustainable patterns.

Within the framework of the current research, a methodological approach was developed to assess the traffic calming measures implemented in the city of Serres, Greece, during the last years. The methodological approach comprises a comprehensive "before and after" study that considers the impact of traffic calming measures on traffic flows, travel speed, traffic accidents, fuel consumption, GHG and air pollutant emissions, while it also takes into account the acceptance of the examined interventions, as perceived by the locals.

The results highlighted that the examined traffic calming measures have a positive effect in reducing traffic flows, emissions and fuel consumption in the traffic calmed areas; however, considerable traffic and environmental pressures are imposed on the neighbouring road axes. Despite this, the overall impact of the examined interventions is clearly beneficial, given that the reference area is a wider zone. As far as travel speed and road accidents are concerned, it was found that the installed traffic calming measures contribute to lowering network speed as well as in many cases to an explicitly defined enhanced traffic safety level. However, it should be mentioned that specific attention should be paid to the neighbouring area in order to prevent accidents "migration". Lastly, the analysis of the questionnaire survey underlined that the vast majority of the respondents are quite satisfied with the examined traffic calming measures, while they regard them as a necessary step towards a revitalised urban environment.

References

1. UN: The World's Cities in 2018-Data Booklet (2018)
2. Eurostat: Urban Europe Statistics on cities, towns and suburbs. Publications office of the European Union, Luxembourg (2016)
3. Concilio, G., Li, C., Rausell, P., Tosoni, I.: Cities as enablers of innovation. In: Concilio, G., Tosoni, I. (eds.) Innovation Capacity and the City. Springer Briefs in Applied Sciences and Technology. Springer, Cham (2019)
4. EU, Directorate General for Regional Policy: Cities of tomorrow-Challenges, visions, ways forward. Publications Office of the European Union, Luxembourg (2011)
5. C40 Cities Homepage. https://www.c40.org. Accessed 04 Mar 2020
6. Sdoukopoulos, A., Pitsiava-Latinopoulou, M., Basbas, S., Papaioannou, P.: Measuring progress towards transport sustainability through indicators: analysis and metrics of the main indicator initiatives. Transp. Res. Part D **67**, 316–333 (2019)
7. Lockwood, I.M.: ITE traffic calming definition. ITE J. **67**, 22–24 (1997)

8. Granà, A., Giuffrè, T., Guerrieri, M.: Exploring effects of area-wide traffic calming measures on urban road sustainable safety. J. Sustain. Dev. **3**(4), 38–49 (2010)
9. Berthod, C., Leclerc, C.: Traffic calming in québec: speed humps and speed cushions. J. Civ. Eng. Archit. **7**(4), 456–465 (2013)
10. Barbosa, H.M.: Impacts of traffic calming measures on speeds on urban roads. Doctoral Dissertation, University of Leeds, Leeds (1995)
11. Harvey, T.: A Review of Current Traffic Calming Techniques. University of Leeds, Leeds (1992)
12. Gonzalo-Orden, H., Rojo, M., Pérez-Acebo, H., Linares, A.: Traffic calming measures and their effect on the variation of speed. Transp. Res. Procedia **18**, 349–356 (2016)
13. Elvik, R., Vaa, T.: The Handbook of Road Safety Measures. Elsevier Science, Oxford (2004)
14. Ghafghazi, G., Hatzopoulou, M.: Simulating the environmental effects of isolated and area-wide traffic calming schemes using traffic simulation and microscopic emission modeling. Transportation **41**, 633–649 (2014)
15. Ewing, R.: Traffic Calming: State of the Practice. Institute of Transportation Engineers, Washington (1999)
16. Grundy, C., Steinbach, R., Edwards, P., Green, J., Armstrong, B., Wilkinson, P.: Effect of 20 mph traffic speed zones on road injuries in London, 1986-2006: controlled interrupted time series analysis. BMJ **339**, b4469 (2009)
17. Ahn, K., Rakha, H.: A field evaluation case study of the environmental and energy impacts of traffic calming. Transp. Res. Part D **14**, 411–424 (2009)
18. Taylor, M., Lynam, D., Baruya, A.: The Effects of Drivers Speed on the Frequency of Road Accidents. TRL Report 421. Transport Research Laboratory, Crowthorne (2000)
19. Litman, T.: Traffic Calming Benefits, Costs, and Equity Impacts. Victoria Transport Policy Institute, Victoria (1999)
20. Vaitkus, A., Čygas, D., Jasiuniene, V., Jateikiene, L., Andriejauskas, T., Skrodenis, D., Ratkeviciute, K.: Traffic calming measures: an evaluation of the effect on driving speed. Promet-Traffic Transp. **29**(3), 275–285 (2017)
21. Consultants, A.K.K.T.: Urban Mobility Study. Municipality of Serres, Serres (2015)
22. Consortis: Sustainable Urban Mobility Plan. Municipality of Serres, Serres (2020)
23. Hellenic Statistical Authority Homepage. https://www.statistics.gr/. Accessed 09 March 2020
24. Eurostat Homepage. http://ec.europa.eu/eurostat/. Accessed 17 March 2020
25. Sdoukopoulos, A., Verani, E., Nikolaidou, A., Tsakalidis, A., Gavanas, N., Pitsiava-Latinopoulou, M., Mikiki, F., Mademli, E., Pallas, C.: Development and implementation of walkability audits in Greek medium-sized cities: the case of the Serres' city centre. Transp. Res. Procedia **24C**, 337–344 (2017)
26. PTV Group Homepage. https://www.ptvgroup.com/en/solutions/products/ptv-visum/areas-of-application/. Accessed 04 March 2020
27. Gkatzoflias, D., Kouridis, C., Ntziachristos, L., Samaras, Z.: COPERT 4: Computer Programme to Calculate Emissions from Road Transport-User Manual. Emisia SA, Aristotle University Thessaloniki, European Environment Agency, Thessaloniki (2007)
28. Surveymonkey Homepage. https://www.surveymonkey.com/. Accessed 04 Mar 2020
29. Daniels, S., Brijs, T., Nuyts, E., Wets, G.: Explaining variation in safety performance of roundabouts. Accid. Anal. Prev. **42**, 393–402 (2010)
30. Distefano, N., Leonardi, S., Pulvirenti, G.: Driving errors at roundabouts: a CHAID analysis. In: 32nd ICTCT Conference Proceedings, Warsaw (2019)

Development of an on-Spot Bio-Waste Screening Methodology with Vehicle Selection Using Multi-criteria Decision Analysis (MCDA): Implementation in the Municipality of Chalkis, Greece

Konstantinos Gkoulias[1], Georgios Palantzas[1,2], and Dimitrios Nalmpantis[1,2](✉)

[1] School of Science and Technology, Hellenic Open University, Parodos Aristotelous 18, 263 35 Patras, Greece

[2] School of Civil Engineering, Faculty of Engineering, Aristotle University of Thessaloniki, PO Box 452, 541 24 Thessaloniki, Greece

dnalba@civil.auth.gr

Abstract. Municipal Solid Waste (MSW) creates many problems, which arise both from proper and unconstrained disposal. The purpose of this paper is to examine the new scenery that develops in the area of MSW collection while applying the on-spot screening method for bio-waste. Bio-waste could be defined as the food waste generated by households, restaurants, food sector services, the food retail sector, and food processing companies. It also includes biodegradable waste from parks and gardens. For this purpose, a case study is used, that of the Municipality of Chalkis, Greece. There is a short description of the current situation regarding the management of MSW in the Municipality of Chalkis. Next, a specific area of Chalkis, i.e., Nea Artaki, is located, and the bio-waste collection is examined as a separate current. In this frame, there is a description and evaluation of alternative systems of on-spot bio-waste screening at the source, followed by the definition of any equipment needed, such as road garbage bins, in-house garbage bins, biodegradable bags, etc. in a systematic way. Next, using a Multi-Criteria Decision Analysis (MCDA) method, i.e., the Analytic Hierarchy Process (AHP), the appropriate transfer vehicle was chosen while taking into account the route and the deposition location, i.e., a vehicle with two separate garbage collector compartments of MSW and not a conventional garbage truck. The experience gained was standardized in a specific methodology formulated to be implemented by Greek municipalities in harmonization with the national waste management plan for the separate collection of bio-waste.

Keywords: Global Positioning System (GPS) · Municipal Solid Waste (MSW) · Analytic Hierarchy Process (AHP) · Organic waste · Chalkis

E. G. Nathanail et al. (Eds.): CSUM2020, AISC 1278, pp. 780–789, 2021.
https://doi.org/10.1007/978-3-030-61075-3_76

1 Introduction

The increasing quantities of Municipal Solid Waste (MSW) production in Greek cities, but also all over the world, the environmental management of MSW and the prospect of profit, have led to the separation of waste streams. One of these streams is that of organic MSW, or else bio-waste. After the industrial revolution and the evolution of medical science, the world's population has increased and continues to grow, as shown in Fig. 1.

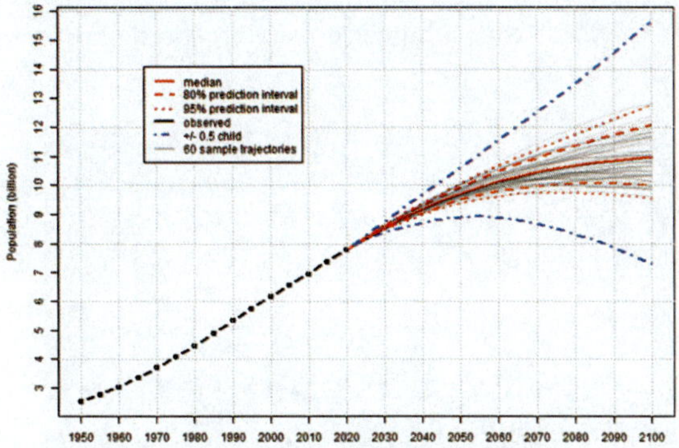

Fig. 1. World population prospects in 2019 [1].

The scale of the waste brought by the industrial revolution and the population growth, especially in urban centers, is of particular concern. The MSW production creates several problems, which arise both from their management and their uncontrolled disposal. The creation of biogas from organic residues, smoke from self-igniting waste with the dioxins it contains, the contamination of subsoil and aquifers by leachates, the move of bad odors and particles by the wind are some of them [2]. To address all these problems and to enable societies to reach, as soon as possible, zero-waste production, new measures are taken in the collection, transport, and management of solid waste.

The purpose of this paper is to examine the new landscape that is developing in the field of MSW collection by applying waste sorting at the source, i.e., on-spot, for one of the waste streams created, i.e., that of bio-waste. The experience gained from this work will be standardized in a specific methodology, formulated for implementation in Greek cities, in line with the Greek National Waste Management Plan (NWMP) for separate bio-waste collection.

2 Literature Review

Trabandt (2007) [3], after mentioning the discovery and the evolution of the Global Positioning System (GPS) in transport, suggests it could be used by garbage trucks: garbage bins are first placed at specific points, then the garbage trucks pass through a predesigned route, and at every garbage bin it marks its location electronically.

Thus, once the registration has been completed and the MSW volume has been calculated, the route in question is archived for future use. With the activation of GPS, when it is time to collect the waste from the specific route, as shown in Fig. 2, the waste collection is facilitated, as there are no delays, with almost fixed collection times and fewer distances traveled, with a higher collection speed and fewer stops [3].

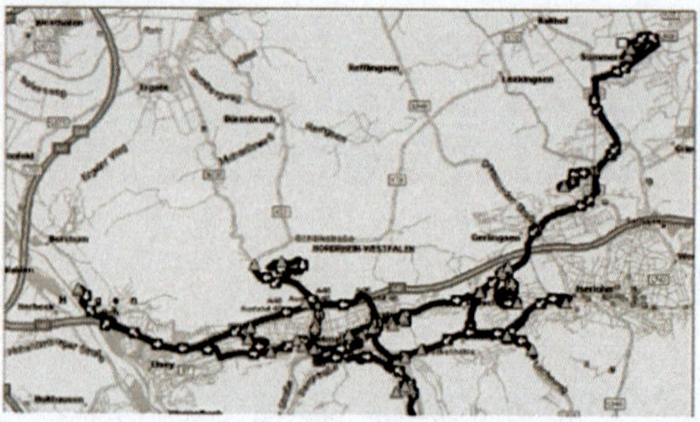

Fig. 2. Indicative optimized garbage collection route with the use of GPS [3].

The expected results of using GPS in the MSW collection are the following:

1. better route design data,
2. real-time knowledge of whether the vehicles are active or not,
3. greater efficiency,
4. better control,
5. reduction of costs, etc.

According to Gallenkemper, Becker, and Dornbusch (2007) [4], sorting in general and sorting waste collection have been desired for a long time with the ultimate goal of reusing materials. This is also depicted in the diagram of Fig. 3 [4].

The reason for MSW sorting is the reuse of materials that they contain. These materials will replace several of the currently used raw materials. This way, there will be direct economic benefits, but also indirectly ones, such as tax relieves. That is why current European legislation promotes the reuse of the materials that make up MSW. In Fig. 3, the diagram shows the German waste collection plan divided into three parts [4], as follows:

1. The first part deals with residues collected in two different streams, one for residues and one for bulky objects [4].
2. The second part deals with the collection of bio-waste bins, which includes green waste [4].

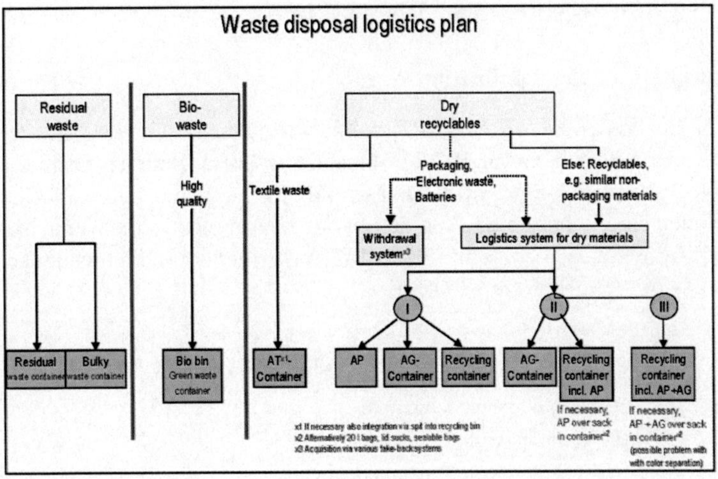

Fig. 3. Plan of future waste collection in different streams in Germany. (adapted from [4])

3. The third part deals with the collection of solid waste with three separate collection streams, one for textiles, one for electrical waste and electronic equipment, and one for waste, which can be recovered [4].

The parameters affecting the quality and quantity of waste in a city are the following: a) the structure of the city, b) the structure of the district, c) existing industries and craft industries, d) the social structure, e) the culture of the population, f) the level of education of the inhabitants, g) the size of the garbage bins, h) the MSW collection rate, i) the fees, j) the number of MSW streams for sorting at the source, and k) information and awareness of the public [4]. Similar parameters also hold for ports [5].

3 Methodology

We examined the case of the collection and transfer to the composting plant of the organic MSW for the Nea Artaki area of the Municipality of Chalkis.

We conducted a Multi-Criteria Decision Analysis (MCDA) using the Analytic Hierarchy Process (AHP), the final aim of which was to choose the most appropriate of the two types of garbage trucks proposed, which were the following: a) the one with two separate compartments (one for the organic and one for the remaining MSW), and b) the standard garbage waste truck. The most important criteria were chosen from an economic, social, and environmental point of view.

Taking into consideration all the above, a relevant implementation methodology was formulated, evaluated, and is proposed for use by Greek municipalities.

4 Application, Results, and Discussion

4.1 Description of the Application Area

The City of Chalkis is one of the Municipalities of mainland Greece, with an area of 412.38 km^2 and a population of 102,223 inhabitants (2011 National Census [6]).

In Table 1, the breakdown of the households in the Municipality of Chalkis per number of members is presented. The number of households in the Municipality as a whole is 36,373, with an average size of 2.81 members per household while the number of households of Nea Artaki is 3,384 [6].

Table 1. Households in the municipality of Chalkis in 2011 [6].

Households with 1 member	7,521
Households with 2 members	9,957
Households with 3 members	7,965
Households with 4 members	7,521
Households with>=5 members	3,409
Total electrified houses (households)	36,373
Average number of members per household	2.81

4.2 MSW Quantities Calculation Formula

When designing an on-spot sorting at the source system, it is recommended that an estimate should be calculated of the expected amounts of bio-waste that will be collected and transferred to the bio-waste treatment plant. In practice, this translates that for N number of households in an area, the expected amount of Π *(kg/week)* of bio-waste can be calculated by Eq. (1) [7], as follows:

$$\Pi = N \cdot P \times K \tag{1}$$

where:

- Π *(kg/week)*: the quantities of bio-waste expected to be collected weekly,
- N: the total number of households in the area served,
- P *(%)*: the percentage of households estimated to be active in the system one year after the implementation of the sorting at the source system, and
- K *(kg/week)*: a specific factor of collected bio-waste (from measurements) [7].

It should be taken into consideration that in order to assess the volume of the bio-waste collected, the specific weight is taken equal to 500 kg/m^3 [7].

4.3 Selection of a Sorting at the Source System

For the on-spot sorting at the source system, there are the following options: a) the door-to-door collection system, and b) the central garbage bin collection system.

The door-to-door collection system is mainly suitable for single houses or houses with a garden or a yard so that the garbage bin can be placed outside of the house. It is also not excluded that this method may be applied to apartment blocks if there is a shared space, e.g., pilotis, available [7].

The collection system of central garbage bins is based on a dense network of garbage bins, relatively close to each household. The inhabitants transfer the presorted waste to the bins, and it is collected from there. The difference with the door-to-door system is that these bins are common and serve the area in which they are placed [7].

4.4 Selection of Vehicles

Vehicles should be suitable for the collection of bio-waste, its transport, and unloading it at the treatment plant or a transshipment station.

Vehicles should be selected as follows:

- They should have sufficient capacity to collect as large quantities as possible on one route.
- Their capacity should not exceed the available quantity to be collected to achieve operational cost savings.
- They should be sealed and preferably closed, without excluding open vehicles.
- They should have a garbage bin lifting mechanism.
- They should be suitable for manual loading of containers, especially when door-to-door systems are in place.
- They should be able to disable the compression mechanism, which is not desirable when there are liquids.
- They should carry the logo of the on-spot sorting at the source program.
- Depending on the building features of the area served, they should be highly adaptive so that they can also be used on narrow roads.

Therefore, one proposal is to procure vehicles that have two compartments (Fig. 4), e.g., its right-hand part for the mixed waste and left-hand part for the bio-waste, so that with one predesigned route, both streams are collected.

Another proposal is the use of a vehicle, which loads the waste from the side with a mechanical arm, which is automatically driven when the vehicle is placed next to the garbage bin, but this is not recommended for Greece due to narrow roads.

4.5 Use of Global Positioning System (GPS)

Once the city's network has been digitally recorded, the collection vehicles should be equipped with a GPS to have a clear understanding of the location and routes they operate in real-time. Several functions are activated by the installation and activation of these tools, such as a) calculation of the length and time of the route of each vehicle, b) highly adaptive routes that can be changed at any given time if the conditions require it, c) optimization of routes when conditions require it, d) real-time location of the vehicles in the city, e) garbage bin location on each road, f) content of bins in kg or as percentages, and g) saving time and money (less fuel, damage, etc.) [8].

Fig. 4. Two-compartment garbage truck (COS.ECO) in Munich 2018.

4.6 Use of Geographic Information System (GIS)

For the use of GPS, a prerequisite is that digital Geographic Information System (GIS) backgrounds have been created at many different levels [9]. This digitization should include: a) the map of the area of application, b) the road network (width of roads, one-way streets, etc.), c) the points of interest (depots, schools, etc.), d) the garbage bin positions, e) the type and color of the garbage bins, f) the population density by region, g) the capacity of the collection vehicles, h) the estimated collection times, i) the number and time of stops of the vehicles, j) the number and time of staff breaks, etc. [9].

4.7 Estimation of Collection Time

The estimation of the time of collection is a necessary process in order to be able to estimate the required number of collection routes. The net collection time of collection, *X (min)*, can be estimated by Eq. (2), as follows [7]:

$$X = \Upsilon \times M \tag{2}$$

where:

- *X (min)*: is the net collection time,
- Υ: is the total number of external central bins, and
- *M (min/bin)*: is a specific net collection time factor per central bin [7].

For the door-to-door system, the collection time is given in relation to the type of garbage bins to be collected. Each garbage bin, other than those for which the lifting mechanism is used, corresponds approximately to one household. Therefore, the net collection time *X (min)*, can be estimated by Eq. (3) [7], as follows:

$$X = \Upsilon \cdot EM\Sigma \cdot T + Z \cdot M \tag{3}$$

where:

- *X (min)*: is the net collection time;
- *Υ*: the total number of ~10 lt kitchen containers to be distributed;
- *ΕΜΣ (%)*: the Estimated Participation Rate (%), i.e., the proportion of households expected to be actively involved in the system one year after the start of the on-spot sorting at the source system, where 45% can be used;
- *T (min/bin)*: a specific net collection time factor per bin (excluding two-wheeled bins), where the number of bins is equal to the participating households;
- *Z*: the number of two- or four-wheeled bins to be installed; and
- *M (min/bin)*: a specific net collection time factor per central bin [7].

4.8 Multi-criteria Decision Analysis for the Selection of Collection Vehicle

In order to choose the most appropriate collection vehicle, we used the Analytic Hierarchy Process (AHP) with the following criteria: a) time, b) procurement and operational cost, c) collection personnel and accident prevention, d) betterment of the quality of life, e) efficiency and credibility, and f) environmental protection. The AHP questionnaire was filled by nine (9) persons specialized in the collection and management of the MSW of the Municipality of Chalkis. We created the AHP hierarchical model using the SuperDecisions software of the Creative Decisions Foundation [10] (Fig. 5), with two alternatives: a) the garbage truck with a separate compartment for the bio-waste, and b) a common garbage truck.

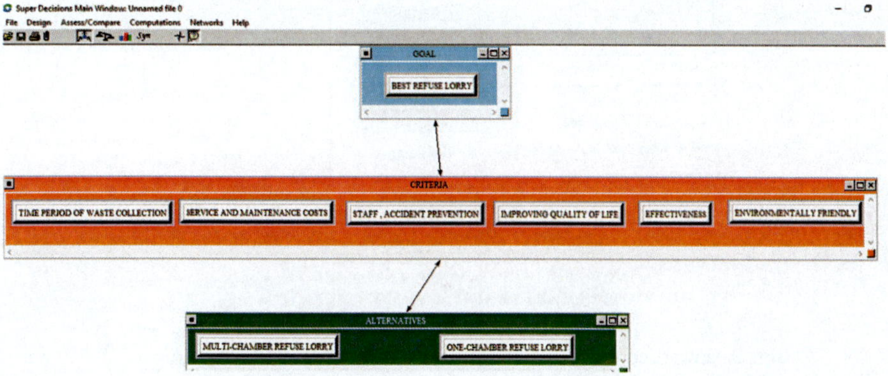

Fig. 5. The AHP hierarchical model.

Following an already known methodology [11, 12], we found that in our case, the garbage truck with a separate compartment for the bio-waste was a slightly better option compared to a common garbage truck (50.1% > 49.9%).

4.9 Proposed Methodology for Municipalities

The collection of bio-waste carries several changes to the collection procedure such as a) the choice of the garbage bins to be used that depends on the type of housing, i.e., houses or apartment buildings; b) the optimal route; c) the washing of the bins almost after each discharge; d) the choice of the best vehicle; etc.

According to each case, AHP may give the best solution for the collection vehicle. To sum it up, from the examination described above, a methodology was derived proposed for application by Greek municipalities to plan the bio-waste collection procedure, and it is presented in Fig. 6.

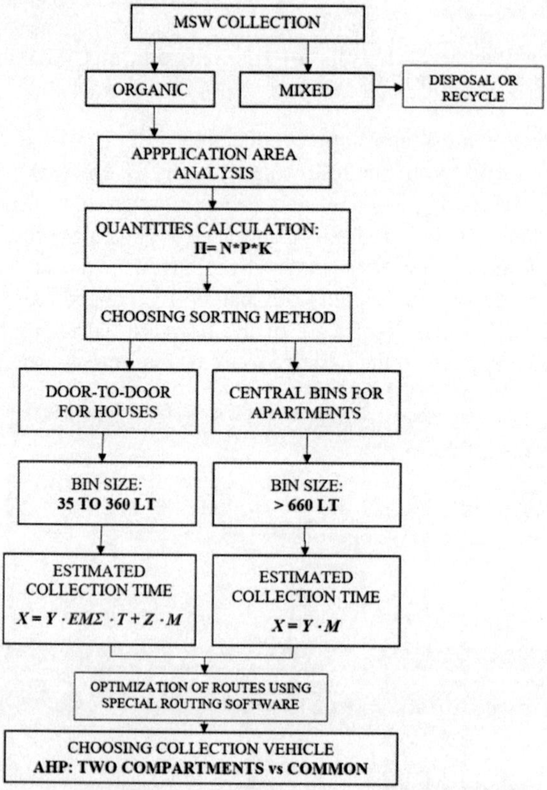

Fig. 6. Proposed on-spot bio-waste collection planning methodology.

5 Conclusions

The main conclusions of this paper are the following:

- brown bins, for bio-waste, should be placed next to the existing ones to avoid changes in drop off points for inhabitants;
- household composting bins should be provided to reduce discards of organic matter in standard mixed waste collection bins;

- biodegradable bags should be provided to households throughout the year to avoid odors; and
- special two-compartment garbage trucks should be considered, one for collecting bio-waste and one for the other MSW, as shown by the performed MCDA analysis.

References

1. United Nations Webpage: World population prospects 2019. https://population.un.org/wpp/Graphs/Probabilistic/POP/TOT/900. Accessed 30 Mar 2020
2. Kouimtzis, T., Fytianos, K., Samara-Konstantinou, K.: Chimeia perivallontos [Environmental chemistry]. University Studio Press, Thessaloniki (1998)
3. Trabandt, R.: GPS-Einsatz im bereich der entsorgungswirtschaft [GPS use in the field of waste management]. In: Urban, A., Halm, G., Weber, M. (eds.) Optimierung der Abfall-Logistik: Kassler Abfall-Logistik-Tape, pp. 121–132. Kassel University Press, Kassel (2005). http://www.uni-kassel.de/upress/online/frei/978-3-89958-145-4.volltext.frei.pdf
4. Gallenkemper, B., Becker, G., Dornbusch, H.-J.: Entwicklung der entsorgung [Development of disposal]. In: Urban, A., Halm, G., Morgan, R. (eds.) Weiterentwicklung der Abfallwirtschaft: Abfallwirtschaft ohne Duale Systeme?, pp. 9–16. Kassel University Press, Kassel (2007). https://www.uni-kassel.de/upress/online/frei/978-3-89958-300-7.volltext.frei.pdf
5. Palantzas, G., Naniopoulos, A., Nalmpantis, D., Theodossiou, P.: The "chain" management of ship-generated waste and cargo residues in the port of Thessaloniki. J. Mar. Environ. Eng. **8**(2), 161–169 (2005). http://www.oldcitypublishing.com/journals/jmee-home/jmee-issue-contents/jmee-volume-8-number-2-2005/jmee-8-2-p-161-169/
6. Hellenic Statistical Authority Webpage: 2011 Population-Housing Census. https://www.statistics.gr/en/2011-census-pop-hous. Accessed 30 Mar 2020
7. Athens-Biowaste: Odigos gia tin efarmogi, elencho kai axiologisi programmaton dialogis stin pigi kai kompostopoiisis vioapovliton [Guide for the implementation, control and evaluation of sorting programs at the source and composting of biowaste], Athens-Biowaste, Athens (2014). http://www.biowaste.gr/site/wp-content/uploads/2014/08/odigos_FINAL.pdf
8. Geocapture Webpage: Touren immer fest im Blick [Tours always in view]. https://www.geocapture.de/branchen/muellabfuhr. Accessed 30 Mar 2020
9. Konstantinidis, P.: Protaseis anadiorganosis tou systimatos aporrimmaton tou Dimou Nikaias me tin chrisi programmatos GIS-routing. Efarmogi senarion enos dyo kai trion revmaton astikon aporrimmaton [Recommendation proposals for the reorganization of the waste system of the Municipality of Nikaia using a GIS-routing program. Application of scenarios of one, two, and three streams of municipal waste]. Master's thesis, Harokopio University, Athens (2007). http://estia.hua.gr/browse/2430
10. SuperDecisions Homepage. http://www.superdecisions.com/. Accessed 30 Mar 2020
11. Nalmpantis, D., Roukouni, A., Genitsaris, E., Stamelou, A., Naniopoulos, A.: Evaluation of innovative ideas for Public Transport proposed by citizens using Multi-Criteria Decision Analysis (MCDA). Eur. Transp. Res. Rev. **11**(1), 22 (2019). https://doi.org/10.1186/s12544-019-0356-6
12. Vasileiadis, I., Nalmpantis, D.: Development of a methodology, using multi-criteria decision analysis (MCDA), to choose between full pedestrianization and traffic calming area (Woonerf Zone Type). In: Nathanail E., Karakikes I. (eds.) Data Analytics: Paving the Way to Sustainable Urban Mobility. CSUM 2018. Advances in Intelligent Systems and Computing, vol. 879, pp. 315–322. Springer, Cham (2019). https://doi.org/10.1007/978-3-030-02305-8_38

Smart Urban Logistics Systems

Validating Urban Freight Deliveries Through Traffic Microsimulation: An Experimental Study

Ioannis Karakikes(✉), Eftihia Nathanail, and Maria Karatsoli

Department of Civil Engineering, University of Thessaly,
Pedion Areos, 38334 Volos, Greece
iokaraki@uth.com

Abstract. Traffic microsimulation tools have been extensively used in evaluating urban freight transport systems. Aim of this paper is to develop a city wide urban traffic model that integrates the daily freight trips of a private courier company on the basis of a typical day. To achieve this, the traffic model is calibrated and validated based on travel time data from Volos' taxis GPS devices. The validated traffic model is then enriched by freight vehicles' routes in form of public transport lines so as to integrate the rest vehicular traffic with the considered freight traffic. Freight data were given by "ELTA Courier" company which provides door-to-door deliveries in the city of Volos, Greece. The produced model simulated the freight trips of six freight vehicles, four motorbikes and two minivans, performing 199 delivery stops in the examined typical day, during 09:00–20:00. The development of the urban freight traffic model is demonstrated through the case study of Volos, providing useful insights.

Keywords: City logistics · Last mile delivery · Modeling · Urban freight transport

1 Introduction and Background

Promoting and sustaining an effective Urban Freight Transport (UFT) system is much needed in cities today. Effective UFT systems ensure reliable, fast and convenient deliveries at the lowest cost for the recipient, and efficient and profitable operation for the carriers and shippers [1]. A sustainable UFT system should be capable of continuously balancing congestion, nuisance and externalities that occur to the city and can be achieved by an appropriate mix of different solutions [2].

Due to the complex nature of UFT systems, impacts from their deployment are difficult to be estimated unless by using simulation tools such as traffic microsimulation [3]. Traffic microsimulation models, with their dynamic nature, can generally provide more output than the conventional static demand-supply models [4]. Microscopic approaches are data intensive and hard to calibrate but are able to model in detail ITS and dynamic fleet management. In addition, the advancement of hardware and software has increased the computational performance of microscopic simulators

E. G. Nathanail et al. (Eds.): CSUM2020, AISC 1278, pp. 793–802, 2021.
https://doi.org/10.1007/978-3-030-61075-3_77

making them available for relatively larger studies. Modelling city logistics is based on the systems approach which requires high quality of traffic and freight data for identifying problems, as well as evaluating the results.

Although, it is more practical to experiment computer simulation models rather than with real networks, there is a plethora of studies indicating that further investigation of the emulation of the freight flows in urban environments is of particular interest. Barceló et al. [4] developed a Decision Support System (DSS) to assist analysts in the design and evaluation of real time fleet management in urban areas, based on an ad hoc implementation of the referenced Taniguchi's methodological framework. They used AIMSUN microscopic traffic simulator to emulate traffic dynamics and ITS applications. Teo et al. [5] presents a multi-agent systems (MAS) model to evaluate city logistics solutions for an urban road network in an e-commerce delivery system environment. Their future research intention is to improve the model by generating dynamic demand and incorporating dynamic travel time using micro-simulation software like VISSIM. Qureshi et al. [6] presented a micro simulation-based evaluation of the soft time windows variant of the Vehicle Routing Problem (VRP) with Vissim. Walker and Manson [7] showed in their study the development of a micro simulation traffic model, concluding that more telematics does not necessarily lead to more efficient urban logistics, since topography of urban street layout is an important contingency variable. Hosoya et al. [8] developed a micro simulation model of the metropolitan area of Tokyo, Japan in order to evaluate four logistics policies in Vehicle-Kilometer-Traveled (VKT) NOx and costs terms, considering also individual firm's behaviour and their characteristics.

Aim of this study is to demonstrate a step-by-step guide for developing an urban freight traffic simulation model in PTV Vissim. The model is meant to be used as a testbed based on which a courier company of Volos city will be able to evaluate its performance. This will help the company as well as the city authorities to better understand the current situation and based on the insights provided to investigate for further adoption of policy solutions towards system's sustainability.

2 Available Datasets and Methodology

Delivery data were requested and given by the private courier company "ELTA Courier" which provides door-to-door deliveries in the city of Volos. As the dataset indicated that deliveries were performed between 09:00 and 19:27 traffic data were respectively searched for the same time of day 09:00–20:00. Analytically, morning and noon peak hour traffic volumes and travel times were given by the Traffic, Transportation and Logistics Laboratory (TTLog) of the University of Thessaly based on a travel survey conducted in 2011, 2012 and 2013 aiming at developing an origin-destination matrix of travel demand in the city of Volos. Evening peak hour traffic volumes were taken from a diploma thesis recorded traffic volumes in major intersections in an effort to store city's traffic data in a structured way by using GIS [9]. Traffic signal programs were given by Volos Traffic Planning Department, while public transportation routes, stops and timetables were downloaded by the official website of Volos' public transportation (Astiko Ktel Volou). Figure 1 shows the methodological steps.

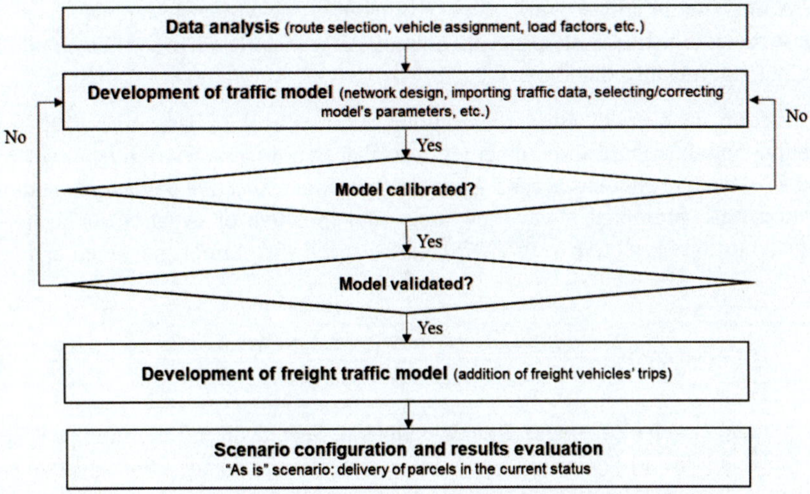

Fig. 1. Flowchart of methodological approach.

3 Calibrating and Validating the Model

Perfect matching between field data and model's results is nearly impossible and analyst's calibration efforts should be terminated when certain thresholds are reached. Those thresholds can differentiate based on the desired level of fidelity and resources [10]. A prevalent example of calibration thresholds has been formulated by the U.S. Department Of Transportation (DOT) following an update to the 2004 version of Wisconsin DOT (WisDOT) [11].

3.1 Calibration

Calibration was based on two performance measures; travel times and traffic volumes. 13 vehicle travel time sections and data collection points named onwards as "detectors" were coded at 13 characteristic points of the network to record model's travel times and volumes. The selection of those points was made based on detailed actual traffic data availability during the morning (09:00-10:00) and noon (14:00-15:00) peak hours, allowing comparisons with the values obtained from the simulation.

As a first step, average travel times were obtained from the model within the morning and noon peak hours in order to check whether the model needs calibration. As results can highly vary, a minimum number of simulation runs is necessary to achieve statistically confident results [12]. The deviations between the actual travel times and model's travel times were calculated as the average of the five absolute deviations for each detectors' section during morning and noon peak hours. The results of the initial five runs resulted in deviations beyond the 15% threshold, indicating a need for calibration. As of lack of more detailed data, calibration was performed based on the following actions:

i) detecting abnormalities owing to mistyping errors or wrong design.
ii) reasonable adjustment of speed distributions at a local level, when the simulation output significantly diverges.

As regards the second action, the criterion that was used to determine whether the simulation output significantly diverges was the sign of the respective travel time deviation in the five simulation runs. Specifically, if the respective deviations in all five simulation runs, presented always the same sign (positive or negative), adjustments were made to the respective section. Otherwise speed distribution remained as it was [13].

$$Deviation = \frac{ActualTravelTimes - VissimTravelTimes}{ActualTravelTimes}\% \tag{1}$$

For example, vehicles' travel time deviations at Sect. 2 during morning peak hour were +15%, +20%, +13%, +26% and +11% (average +17%) for the respective simulation runs. The aforementioned deviations indicate that vehicles are moving slower than what actual data indicated and therefore, their exported travel time measurements were higher than the respective actual travel times measurements. All vehicle types' speed distribution in this example was increased by 10 km/h, i.e. new speed distribution for cars 40 km/h, trucks 30 km/h and motorbikes 40 km/h. The speed distribution of sections which presented same sign was adjusted maximum by ±10 km/h, as anything higher than that would represent no real speed distributions.

Linear regression was performed with independent variable the actual travel times and dependent variable the calibrated and non-calibrated morning and noon travel times, respectively, exported from the model. Figure 2 depicts the results of the non-calibrated ($R^2_{morning} = 0.9132$, $R^2_{noon} = 0.9543$) and calibrated model ($R^2_{morning} = 0.9745$, $R^2_{noon} = 0.9738$).

It has to be noted that there were sections that still exceeded the 15% threshold during morning and noon peak hours. The exceedance in these sections cannot be attributed solely to insufficient calibration. The current coded road network in Vissim has undergo structural and operational changes compared to the road network of 2011–2013 when the traffic data were collected. Some of the major changes in this period of years are:

i) replacement of four major signalized intersections with roundabouts (2017–2019).
ii) new schedule and itineraries for the public transport buses (last update 09.2019).
iii) introduction of bus lane along Dimitriados and Iasonos streets, dedicated for the traffic of public transportation and taxis (introduced in 2015).
iv) reprogramming signalized intersections, one-way traffic rearrangements, pedestrianizations, etc. (within 2013–2019).

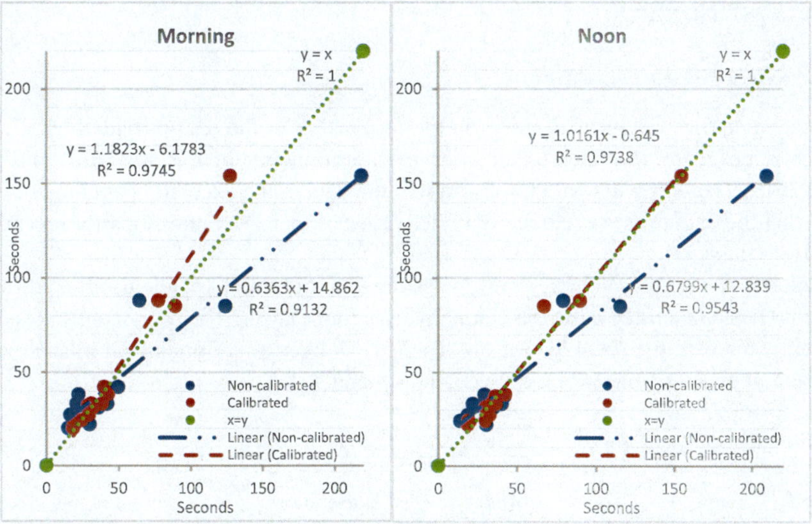

Fig. 2. Comparing non-calibrated and calibrated travel times in the morning (left) and noon (right) peak hours.

3.2 Validation

Validation process tests the ability of the model to represent the real system [14]. This is ensured by checking whether model's results fall within the accepted thresholds when being compared to an updated validation dataset. The dataset which was used to validate the model was travel time data from Volos' taxis GPS devices within the period 01.02.2019–31.05.2019. The selection of this period was made for two reasons. First, due to the limited number of taxis carrying activated GPS devices on 09.04.2019 (35 out of the 200 vehicles). Secondly, the nominal GPS tracking interval of the Taxi GPS system is equal to 30 secs which produces a low number of measurements, unable to produce reliable results for the validation of such an extended network. The selection of the period was based on the hypothesis that all these days are typical to 09.04.2019, thus only weekday data between 09:00 and 20:00 were considered in the validation process.

The average taxi travel times per section and interval were compared to the average travel times of the five simulation runs as exported by Vissim to reveal model's validation level. Based on the average deviations an overall 15% and 29% deviation is achieved during the peak and the off peak hours respectively, which is considered satisfactory considering the aforementioned changes to the road network, see Calibration subchapter.

4 "As is" Scenario

4.1 Delivery Data Description

An excel data file was formed based on the processing of the data provided by ELTA courier referring to the parcel deliveries of the company in the area of Volos on 09.04.2019. The following paragraphs present the data of the excel file per column. It is noted that the data relates exclusively to deliveries as there were no stops for receiving parcels.

Column A entitled "Vehicle" designates the type of freight vehicle used to perform the deliveries. On a daily basis the company uses four Dayang Daytona motorbikes and two minivans with the specifications of Table 1. Deliveries are organized in such way that each vehicle serves different areas and zip codes.

Table 1. ELTA courier's freight vehicles specs in Volos.

Brand	Model	Horsepower	Engine type	Note
Daytona Dy - (Motorbikes 1–4)	125 EFI CBS	8.5PS (6.3 kW; 8hp)	Gasoline	2015
Fiat Doblo - (Minivan 1)	1.3 Multijet	90PS (66 kW; 89hp)	Diesel	2009
Renault Kangoo - (Minivan 2)	1.9 L F-series	65PS (48 kW; 65hp)	Diesel	1997

Column B "Service no." shows the order number of each delivery in the daily delivery sheet, per freight vehicle. During the examined day, "Motorbike 1" was assigned 54 deliveries, "Motorbike 2" 68 deliveries, "Motorbike 3" 51 deliveries, "Motorbike 4" 47 deliveries, "Minivan 1" 82 deliveries and "Minivan 2" 48 deliveries. Column C "Address" designates the actual addresses of the stops made by the freight vehicles, while column D "Zip code" their respective zip codes. Column E "Weight" shows parcel's weight in kg which is to be delivered per stop. Column F "Payment on receipt" shows the amount of money in Euros that are to be collected by the courier from the receiver following the "payment on receipt" form of payment which is supported by the company. Moreover, each parcel's departure time from the station and delivery time to the recipient (columns H–L) was retrieved by typing in company's tracking system the unique voucher number (column G).

The current Total Load (cTL) carried by each vehicle during the delivery process was calculated based on the weights of each delivery. This was achieved by summing up the weights of deliveries departing at the same time. For every successful delivery the cTL was decreased by the weight of the delivered parcel, while in cases of failed deliveries the cTL remained the same and the parcel returned to the station for rescheduling its delivery. Note that the drivers attempt only one delivery in the day after a failed attempt of delivery, unless otherwise agreed between the recipient and the company. Column M "current Total Load" designates the cTL for the freight vehicles and shows the initial delivery load of the vehicle after departing the station.

4.2 Configuring the "as is" Scenario in PTV Vissim

Two new vehicle types (610: ELTA Motorbike and 620: ELTA Van) were created in Vissim to represent the freight vehicles. The first vehicle type represents the four Motorbikes. This type was assigned Vissim's default distribution attributes, i.e. accelerations/decelerations, for motorbikes. The second vehicle type represents the two Minivans, to which default distribution attributes for cars were assigned. Freight vehicle routes were coded in model as "Public Transport Lines". Public Transport Stops were used to represent delivery stops. As the modeled road network in Vissim is limited as compared to the real network, projections of the real addresses of the delivery stops were made to the closest point of the modeled road network (see Fig. 3).

Fig. 3. Delivery stops in their actual position (left, Google maps) and delivery stops as modeled (right, PTV Vissim).

This design, however, presents limitations, as it assumes that parking spaces are available in all delivery addresses. In the city of Volos due to low enforcement and urban design, there are many cases in which freight Motorbikes park shortly wherever there is adequate space e.g. in front of garage entrances, on the sidewalk, between parked vehicles, etc. This has clearly negative effect on e.g. walkability or urban aesthetics, but slightly affects the rest vehicular traffic. Furthermore, minivans are primarily used for remote deliveries or deliveries of heavy weight or bulk parcels. While remote deliveries do not deal with parking problems, heavy or bulk deliveries are made by blocking for several seconds a lane of the road which affects majorly the vehicular traffic. Deliveries can be also made by Minivans' parking on the pavement or pedestrian crossings, in front of garage entrances or reserved parking spaces, but modeling such events needs detailed study.

Freight vehicles exact dwell time to complete a delivery was unknown, thus a linear time distribution of 120 simulation seconds (2 min) to 300 simulation seconds (5 min) was assigned to each Public Transport Stop based on ELTA Courier estimations. The dwell time is considered as the time period in-between turning off/on the vehicle's engine.

4.3 Results Evaluation

The modeled road network of the "as is" scenario is a sub area of the real study area. This, in combination with the design of delivery stops mentioned above, the small number of modeled collector roads and the calibration and validation limitations may cause deviations between real and model's freight vehicles' distances and delivery durations.

To keep track of this deviation, three subscenarios were created namely "Real", "Real of Modeled Stops" and "Vissim". The differences between the subscenarios can be found in Table 2.

Table 2. Subscenarios of the "as is" scenario.

Characteristic	Real	Real of modeled stops	Vissim
Analysis based on ..			
Traffic simulation (PTV Vissim)			✓
Google maps	✓	✓	
Delivery stops considered			
Delivery stops in the area of modeled road network (199/343 of the study area)		✓	✓
All delivery stops in the study area (343/343)	✓		
Distance covered by freight vehicles ..			
for servicing 199/343 stops within the modeled road network (measured in Vissim).			✓
for servicing 199/343 stops within the modeled road network (measured in Google maps).		✓	
for servicing 343/343 stops in the study area (measured in Google maps).	✓		
Travel times ..			
for servicing 199/343 stops within the modeled road network (measured in Vissim).			✓
for servicing 199/343 stops within the modeled road network (measured in Google maps).		✓	
for servicing 343/343 stops in the study area (measured in Google maps).	✓		

The real distances covered by the actual freight vehicles arose by summing up the distance between two consecutive delivery stops in Google maps. The calculation of the real delivery durations has been performed by subtracting the departure time of each freight vehicle from its arrival time to the ELTA Courier station. The calculation of delivery durations in Vissim was performed by averaging the durations in the five simulation runs. The number of stops, distances covered by freight vehicles and delivery durations for all freight vehicles per subscenario can be seen in detail in Table 3.

Table 3. Subscenarios of the "as is" scenario.

Vehicles	Real	Real of modeled stops [1]	Vissim	Vissim deviation from [1] (%)
Distance covered by freight vehicles (km)				
M1	46.573	38.160	16.611	56%
M2	36.325	36.325	17.822	51%
M3	46.811	19.451	9.614	51%
M4	56.79	5.140	4.755	7%
V1	157.013	32.020	22.608	29%
V2	65.86	33.130	25.478	23%
Total	409.372	164.226	96.888	36%
Delivery duration by freight vehicles (hh:mm), (sec)				
M1	5:56, 21360	5:05, 18300	3:22, 12109	34%
M2	9:36, 34560	9:36, 34560	5:16, 18987	45%
M3	5:54, 21240	3:40, 13200	2:26, 8742	33%
M4	6:37, 23820	0:38, 2280	00:56, 3369	47%
V1	8:55, 32100	1:50, 6600	1:58, 7100	7%
V2	8:11, 29460	6:23, 22980	2:58, 10679	53%
Total	45:09, 162540	27:12, 97920	16:56, 60986	36%

5 Concluding Remarks

This paper developed and evaluated an urban freight traffic microsimulation model. Although extended microsimulation models require high resources for their development, it is proved that if they are properly configured, they can output evaluation results with high precision.

In this study, the comparison of "Real of Modeled Stops" subscenario and "Vissim" subscenario indicates that there is a 36% in average deviation as regards the distance covered and delivery duration. It is believed that this deviation derives from the simplifications made, i.e. limited number of collector roads within the modeled road network, projections of delivery stops, lack of detailed traffic data. A denser modeled road network would increase model's fidelity by demanding less projections of delivery stops, thus fewer simplifications.

Concerning further research on this model, UFT solutions should be configured and have their performance evaluated compared to the "as is scenario". This ex-ante evaluation will reveal potential positive or negative impacts the future solution may bring to the UFT system of Volos city.

Acknowledgements. This research is co-financed by Greece and the European Union (European Social Fund- ESF) through the Operational Programme « Human Resources Development, Education and Lifelong Learning » in the context of the project "Strengthening Human Resources Research Potential via Doctorate Research" (MIS-5000432), implemented by the State Scholarships Foundation (IKY).

This research has been based on data provided by ELTA Courier company (https://www.elta-courier.gr).

References

1. Rai, H.B., Lier, T.V., Meers, D., Macharis, C.: Improving urban freight transport sustainability: policy assessment framework and case study. Res. Transp. Econ. (2017). https://doi.org/10.1016/j.retrec.2017.08.005
2. ENCLOSE: Deliverable D5.2: A Framework for the definition and implementation of Sustainable Urban Logistics Plans in historic small-/mid-size towns (2015)
3. Taniguchi, E., Thompson, R.G., Yamada, T.: Emerging techniques for enhancing the practical application of city logistics models. Procedia-Soc. Behav. Sci. **39**, 3–18 (2012). ISSN 1877-0428
4. Barceló, J., Grzybowska, H., Pardo, S.: Vehicle routing and scheduling models, simulation and city logistics. In: Zeimpekis, V., Tarantilis, C.D., Giaglis, G.M., Minis, I. (eds.) Dynamic Fleet Management. Operations Research/Computer Science Interfaces Series, vol. 38. Springer, Boston (2007)
5. Teo, J.S.E., Taniguchi, E., Qureshi, A.G.: Evaluating city logistics measure in e-commerce with multiagent systems. Procedia–Soc. Behav. Sci. **39**, 349–359 (2012). ISSN 1877-0428
6. Qureshi, G., Taniguchi, E., Yamada, T.: A microsimulation based analysis of exact solution of dynamic vehicle routing with soft time windows. In: 7th International Conference on City Logistics, 7–9 June 2015, Mallorca, Spain (2015)
7. Walker, G., Manson, A.: Telematics, urban freight logistics and low carbon road networks. J. Transp. Geogr. **37**(2014), 74–81 (2014)
8. Hosoya, R., Sano, K., Ieda, H., Kato, H., Fukuda, A.: Evaluation of logistics policies in the Tokyo metropolitan area using a microsimulation model for urban goods movement. J. East. Asia Soc. Transp. Stud. **5**, 3097–3110 (2003)
9. Tsitsogiannis, L.: Storing traffic volumes of Volos road network using GIS. (Master's Thesis). University of Thessaly - Department of Civil Engineering, Volos (2017)
10. Park, B., Schneeberger, J.: Microscopic simulation model calibration and validation: case study of Vissim simulation model for a coordinated actuated signal system. Transp. Res. Rec. J. Transp. Res. Board **1856**, 185–192 (2003)
11. U.S. Department of Transportation: Traffic Analysis Toolbox Volume III: Guidelines for Applying Traffic Microsimulation Modeling Software 2019 Update to the 2004 Version, Washington, DC 20590, Federal Highway Administration (2019)
12. Tian, Z.Z., Urbanik, T., Engelbrecht, R., Balke, K.: Variations in capacity and delay estimates from microscopic traffic simulation models. Transp. Res. Rec. **1802**(1), 23–31 (2002)
13. Karakikes, I., Spangler, M., Margreiter, M.: Motorway simulation using bluetooth data. Transp. Telecommun. J. **17**(3), 242–251 (2016)
14. So, J., et al.: Building, Calibrating and Validating a Large-Scale High-Fidelity Microscopic Traffic simulation Model – A Manual Approach. Washington DC, 95th Annual Meeting of the Transportation Research Board (2016)

Technological Development in Small Intermodal Terminals: A Solution for a More Balanced Freight Transport?

Giulia Sommacal[1] and Federico Cavallaro[2](✉)

[1] Eurac Research, Institute for Regional Development, Viale Druso 1, 39100 Bolzano, Italy
giulia.sommacal@eurac.edu

[2] Interuniversity Department of Regional and Urban Studies and Planning, Politecnico di Torino, Viale Pier Andrea Mattioli 39, 10125 Turin, Italy
federico.cavallaro@polito.it

Abstract. Despite numerous efforts to balance the modal split of freight mobility, transalpine transport is mostly performed via road, causing negative consequences on the urban centres located along main infrastructural corridors. Combined Transport (CT) could contribute to a reduction of such pressure, but this form of transport is not considered competitive enough from operators. This paper assesses the role of technological development to improve the performances of CT. The analysis is focused on the transalpine corridor with the highest traffic volumes (Brenner) and considers both the linear and punctual infrastructure (respectively, the railway line and the intermodal terminal located near the city of Trento). After a description of the existing and planned infrastructural equipment and an analysis of the ITSs already adopted in other EU contexts, the suitability of the latter for Brenner is performed, distinguishing between a strategic and a tactical level. This analysis has relevant implications for urban mobility, since the infrastructure to access/leave the city suffers from huge congestion problems, also due to the coexistence of passenger and freight transport along the same routes. In this framework, the reduction of the road freight component can have positive effects on the overall circulation of vehicles.

Keywords: Combined Transport · Intermodal terminal · Brenner corridor · ITSs

1 Introduction

Overall traffic volumes continue to raise along the main transalpine corridors. This increase is mostly due to the heavy good vehicles (HGVs): with an average of 24,000 per day in 2018, they increased by 4% compared to the previous year [1]. The highest values have been registered along the Brenner corridor (Munich-Verona), the central part of the Scandinavian Mediterranean corridor, with an average of 7,200 HGVs per day (almost 1/3 of the entire Alpine arc) and an increase by +7.3% per year. Road freight transport imposes negative impacts on society and environment. In the EU-28,

E. G. Nathanail et al. (Eds.): CSUM2020, AISC 1278, pp. 803–813, 2021.
https://doi.org/10.1007/978-3-030-61075-3_78

the road component counts for about 83% of all transport externalities, generating a damage quantified at 4.8% of the continental Gross Domestic Product [2]. In particular, transport is responsible for a quarter of EU greenhouse gas (GHG) emissions. According to the European Green Deal objectives (2019), the achievement of a climate neutrality by 2050 would require a 90% reduction in transport emissions. Referring to the local air pollution, in 2018 annual average values of NO_2 exceeded the EU limit value of 40 μg/m^3 for several monitoring stations of the Brenner corridor. This led to economic sanctions from EU [1]. This data justifies the attempts promoted by the EU to shift a substantial part of freight transport from road to less polluting systems (such as rail and inland waterways). The Alpine regions, characterized by higher negative externalities in comparison to flat terrain, face a growing need to implement coordinated policies and measures for modal shift. Solutions adopted so far have not been sufficient to change the trend of the last years and integrative measures are required [3]. Combined Transport (CT) can contribute to increase the modal shift, but it is facing resistance from operators [4].

According to *EC (2017)* [5], intelligent transport systems (henceforth, ITSs) may contribute to increase the attractiveness of CT. This paper aims to identify the bottlenecks to the development of a strong and competitive CT along the Brenner corridor and proposes some tailored ITS solutions that may contribute to increase its attractiveness. The focus is on the Italian Region Trentino/Alto Adige, composed by the two autonomous Provinces of Bolzano/Bozen and Trento. This region, which is the central part of the corridor, is served by the intermodal terminal of Trento-Roncafort, of which the two provinces hold a share. The paper is organized as follows: Sects. 2 and 3 describe the intermodal terminal of Trento-Roncafort and the Brenner corridor in terms of infrastructural equipment and transport volumes. Section 4 outlines the solutions under development in this area and presents tailored ITSs that have been adopted in other contexts and that may be adopted also along the Brenner line. Finally, results are discussed in Conclusions, also in terms of urban mobility and in view of the main changes deriving from the introduction of the new high capacity (HC) railway line.

2 Intermodal Terminal of Trento-Roncafort

According to *Woxenius (1997)* [6], the Brenner corridor is characterized by six intermodal centres: two international terminals (Interporto Quadrante Europa-IQE-in Verona, IT and Riem in Munich, DE) and four small-scale terminals: Trento-Roncafort (IT), Brennersee (AT), Hall in Tirol (AT) and Wörgl (AT). Each of them presents different characteristics in terms of dimension, infrastructural equipment and volumes handled, which are coherent with the traffic generated and attracted by their catchment area. This section focuses on the Italian part of the Brenner corridor and describes the main locational, infrastructural and technical features of the intermodal terminal of Trento-Roncafort. Considering the Italian stretch of the Brenner corridor, this is the only alternative to the main terminal in Verona, which is the main node for EU rail freight traffic, but it is close to saturation. In this perspective, the terminal in Trento can increase the competitiveness of rail freight transport along the Brenner.

Trento-Roncafort is located near the motorway A22 exit of Trento Nord, in a strategic position: it is close to SR47 Valsugana (a road of primary importance that connects Trento with Padua and gives access to an area with several industries), to Provincial Road 235 and runs near the Brenner railway line. The terminal is managed by Interbrennero S.p.A. [7], a company with a majority of public shareholdings (>85%). The terminal has a surface of about 100 ha (95% of which is built) and provides different services: conventional transport, Unaccompanied CT (UCT) and Accompanied CT (ACT)- the only terminal on the Italian side of Brenner with such service. The infrastructural equipment is constituted of 150,000 m^2 for handling goods, 9 railway tracks (3 $\times$ 750 m, 4 $\times$ 650 m, 2 $\times$ 400 m), of which 7 are electrified and 1 track is for special/hazardous materials. Goods-containers, swap bodies and semi-trailers- are handled by 3 mobile tyre cranes, 3 forklifts, 1 road tractor, 1 van and 1 quargo. In the current layout, up to 20 pairs of trains can be worked per day. Furthermore, the terminal guarantees several main services, including those of security, customs, container maintenance, container repair, container cleaning, dangerous goods, reefer and trucking. Other services are aimed at HGVs arrivals and departures, floodlit parking area with 24 h video surveillance and with infrared barrier beams, VAT depot/warehouse, warehouses for shipping companies, couriers and for goods warehousing, multi-brand service centre, hotel, bar and restaurant, bank and insurance service, administration and convention centre.

Figure 1 analyses the transport demand in terms of number of trains and tons moved between 2000 and 2018 [7]. Until 2008 (except for 2004), the flow constantly increased, passing from 1,807 trains and 875,782 tons to 5,491 trains and 2.7 million tons. In 2009–2010, a stabilization is visible, followed by a dramatic increase in 2011 (6,715 trains, 3.7 million tons), the last year of adoption of the sectorial driving ban in Tirol. After that date, a constant decline in number of trains and tons movements has been registered. With 1,270 trains and 625,441 tons, in 2018 it reached the lowest values since the beginning of the century.

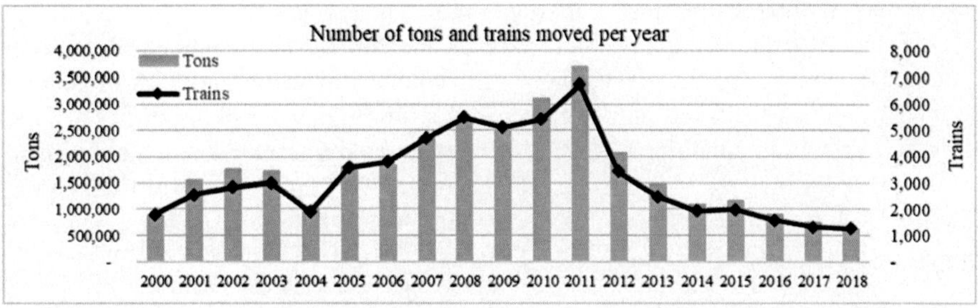

Fig. 1. Number of tons and trains moved per year in Trento-Roncafort [7].

The number of trains and tons in the last three years (Table 1) can be distinguished according to the type of services provided (ACT, UCT, conventional). The total amount of trains decreased by 16.9% between 2016 and 2017; by 3.4% between 2017 and 2018. The trains sum up to a total of 1,270 in 2018 with a share of 48% by ACT,

13% by UCT and 39% by conventional transport. In 2018, the main decrease has been registered for ACT (−23%), while an increase is visible for UCT (from 17 trains in 2017 to 161 trains in 2018). The decrease of ACT is due to the closure of the connection between Trento and Regensburg (DE) and to the contextual renewal works along part of the Brenner railway line. This limited the capacity of the line, with a reduction of the service between Trento and Wörgl (AT), the main (and from 2018, the only) connection for ACT along Brenner. When referring to the ITUs and tons handled, ACT is confirmed as the main service (51% of ITUs and 58% of tons transported per year in 2018), followed by conventional transport (39% of ITUs and 40% of tons registered per year) and by UCT (10% of ITUs and 2% of tons transported per year). In the last three years the load factor has decreased from 68.7% in 2016 and in 2017, to 65.4% in 2018.

Table 1. Transport volumes in Trento-Roncafort [7].

	Year	ACT	UCT	Conventional	Total	Variation 2017/2016	Variation 2018/2017
Trains	2016	1,121	–	462	**1,583**	−16.9%	−3.4%
	2017	796	17	502	**1,315**		
	2018	613	161	496	**1,270**		
Loading units (ITUs)	2016	18,243	13	7,719	**25,975**	−15.1%	−7.7%
	2017	13,278	400	8,370	**22,048**		
	2018	10,321	1,979	8,061	**20,361**		
Tons	2016	653,483	230	241.925	**895,638**	−16.6%	−16.3%
	2017	476,352	10,377	260,084	**746,813**		
	2018	362,139	11,658	251,644	**625,441**		

Regarding the main origins and destinations, the service Trento-Wörgl (together with Brenner-Wörgl), is the most important ACT. The service is provided on a daily basis by the Austrian company Rail Cargo Operator–Austria GmbH/ROLA. The maximum dimensions of the HGVs are 4 m height, 2.6 m width and 18.8 m length. The total weight of the truck may not exceed 40t, but higher weights (up to 44t) are allowed under certain conditions. Regarding conventional transport, the main terminal arrivals and departures are Monfalcone (IT), Livorno (IT), Pöls (AT) and Villach (AT), while for UCT the service is based on spot trains, with different origins/destinations.

3 The Brenner Corridor

The TEN-T policy, based on the *Regulation (EU) No 1315/2013,* aims to improve a European network of infrastructure. For rail, all terminals need to be efficiently linked to the linear infrastructure. This section complements the previous one, by describing the characteristics of the transport network connected with Trento-Roncafort.

3.1 The Brenner Railway Line

The Brenner railway line is the only freight line that crosses the Region Trentino/Alto Adige and that is linked with Trento-Roncafort. The total amount of tons transported at Brenner between 2013 and 2018 registered a constant increase, reaching 52.9 Mt, but the incidence of goods transported by rail (14.1 Mt, 27%) is low if compared with road transport (38.8 Mt, 73%). In 2018, an ambitious policy declaration [8] signed by the Tyrol-South Tyrol-Trentino Euregion has indicated two important milestones: balancing the transport modes by 2027 and reversing the current trend by 2035. Actual values indicate an opposite trend: in 2018 transalpine freight trains crossing the Brenner corridor were 19,142, −2.1% compared to 2017 [1]. Considering the Italian stretch, the main origin/destination is Brenner-IQE and vice versa, with more than 68% of all trains. Brenner-Trento (and vice versa) covers 4% of all freight trains movements.

In order to promote the development of CT at the regional level, the two autonomous provinces of Trento and Bolzano/Bozen have started to provide financial aids to cover part of the costs related to the use of the ACT and UCT services. Furthermore, the construction of the new HC railway line, which includes the Brenner Base Tunnel (BBT) between Innsbruck (AT) and Fortezza (IT), should be operative by 2028 and would guarantee a complete and faster connection to the Region thanks to the Southern access route between Verona and Fortezza.

3.2 The Brenner Motorway A22

A22 is the main road access to Trento-Roncafort, which is located close to the motorway exit of *Trento nord*. The analysis of freight volumes entering/exiting the toll booths of the city (i.e. *Trento nord*, *Trento sud* and *Trento centro*, with this last open for exit until mid-May 2011) can be helpful to understand the variation in freight demand and the absolute number of HGVs that may be potentially interested by an improvement of the terminal performances [9]. Figure 2 shows that the toll booth of *Trento nord* presents by far the highest values, more than double than *Trento sud*: this is coherent with the locational distribution of productive industrial and craft sectors, which are mostly concentrated in the northern part of the city. The dramatic increase registered in the last years (mostly for exits), together with the reduction of rail flows (see previous section), is leading to circulation problems along A22, with all related negative externalities [3]. Indirectly, this also causes relevant problems to the urban network of Trento, and particularly to the urban bypass, which is used both from HGVs to reach the main productive areas and by commuters that daily reach the city centre from the close valleys.

In this framework, the two autonomous provinces of Trento and Bolzano are working to revise the concession contract of the motorway, by including a more rigorous quantification of external costs in the tariff scheme and by introducing a specific environmental toll based on the Eurovignette Directive. The aim of this measure is to harmonize tolls with other transalpine corridors, thus reducing the detoured traffic, which is contributing to the high traffic volumes [10]. Revenues from the environmental toll (estimated at M€250) have to be used to upgrade the intermodal nodes of Trento and Isola della Scala (nearby Verona), as well as the Valdaro river port (Mantua).

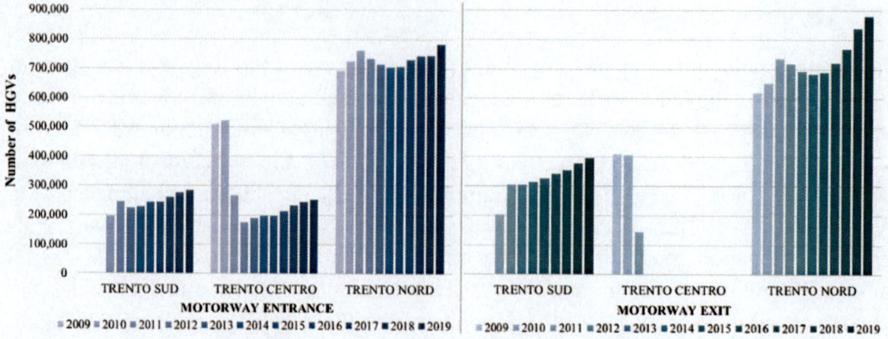

Fig. 2. HGVs at the A22 tool booths of Trento [9].

4 ICT Solutions for CT Along the Brenner

The decrease of volumes in the intermodal terminal of Trento-Roncafort (Sect. 2) reveals a loss of attractiveness for goods transported by rail. In addition to the traditional measures adopted to discourage road freight transport and to support CT operations described above, other innovative measures should be considered. ICT solutions can be a valid possibility: their role in freight transport as key enabler is well recognized at the EU and national level. The *Directive 2010/40/EU* aims to encourage the development of innovative transport technologies to create ITSs. However, a systematic study for the Brenner corridor is still lacking. To cover this gap, this section describes the measures that have already been adopted or planned (paragraph 4.1) both in the intermodal terminal of Trento-Roncafort and along the Brenner corridor. This is the basis for understanding which integrative solutions, already adopted in other territorial contexts, can be useful also for our case study (paragraph 4.2). For this analysis, a distinction between strategic and tactical solutions is provided [11].

4.1 Innovative Technologies, Digitalization and Future Developments

At strategical level, *Interbrennero S.p.A* and the Brenner HC railway line are already developing some interesting innovative solutions. As partner of the National Logistic Platform UIRNet, *Interbrennero S.p.A* promotes the interconnection between carriers, logistics companies, institutions, ports and logistics centres. UIRNet contributes to the development of smart logistics nodes through the modulation of the flow of goods on the net and of logistics services in the centres of modal shift [12]. Currently, however, this participation is only partial, since the terminal adopts different web-platforms to manage the various phases of CT, including the booking of the ACT service, the collection of goods, tracking & tracing, information on irregularities or about parking facilities. Second, in 2017 *Interbrennero S.p.A* signed a memorandum of understanding with the Italian railway infrastructure manager for an expansion of 5 ha of the terminal area. The new area will be equipped with 3 electrified railway tracks of 750 m for ACT. These tracks will be equipped with signals, which will allow the trains to link to the Brenner railway line without additional shunting, reducing handling times and costs. With this new scenario, up to 36 pairs of trains can potentially be worked per day. Summed to the current 20 pairs of trains, the capacity of the terminal can dramatically increase. In parallel to this initiative, the deployment of the ERTMS/ETCS standard on the Italian railway section of Brenner eases the interoperability between countries.

At a tactical level, the terminal users can take advantage of the light panels distributed along the A22 and at the toll booth of Trento Nord, which inform drivers about road, traffic and weather conditions (e.g. by indicating parking availability at rest areas). Furthermore, in the last years two technological initiatives on the relation Verona-Nuremberg-Rostock have been implemented along the Brenner corridor [13]: the *Wagons-sharing concept* and the *appliance of the First in-First-out principle on high frequent routes*. The first one, developed in a part of IQE, has introduced a new management scheme based on the optimization of trains' and wagons' arrivals and departures, according to a scheme largely adopted in passenger mobility. The second one, tested by TX Logistik in six different routes passing via Brenner, increases the performance of trains especially on frequently used transport routes.

4.2 New Potential ITSs

Such initiatives give evidence of recent efforts for the improvement of CT thanks to the technological development. An analysis of integrative ITSs already adopted in other contexts and suitable also for Brenner can be a further support for the enhancement of the service (Table 2). Measures are divided into 4 main categories (column 1), according to [14]. Then, their strategic/tactical nature is highlighted (column 2), together with a brief description and a concrete example (column 3) for each of them. Measures are then reported to the Brenner context, showing which of them have been already implemented or could be implemented (columns 4 and 5), based on a technical and financial analysis made in the EU-funded project SMARTLOGI [15].

Among the tactical ones, whose implementation could be an integration of the strategic measures presented above (e.g. m1.1), the usage of an autarchic GPS tracker in daily operations (m1.2), similar to the one installed on the route Herne (DE)–Verona (IT), permits the improvement of the tracking & tracing system with real time information. It may optimize the process on tracks but also in the terminal thanks to the automatic wagon order assignment via GPS (m4.10). This could be helpful if merged with an interaction platform such as *UIRNet* (m4.7), allowing all actors involved in the CT process to be informed about real circulation of vehicles. The electronic system for freight transport (m4.3) is another potential measure: it permits the connection of administrative, tax, customs and logistic procedures into a single technological platform, by integrating information on different types of transport, expediting the processes of the supply chain management. It leads to improve a communication flow along the transport chain, developing harmonized standards for data exchange and communication flows. Furthermore, these solutions can be integrated with other packages, which support terminal operations: the terminal management system (m4.8) and the fleet management system (m4.4). The first one, which depends on a software solution expediting internal order processes, strengthens loading activities. The second one optimizes the terminal organization, including vehicle acquisition and its disposal.

A second category of measures increases CT efficiency thanks to the adoption of innovative operational and technical solutions. An energy efficient train driving system (m2.2) or an automatic train route optimizer (m2.3) may contribute to optimize energy efficiency and to increase the train path capacity and the timetable stability, especially if adopted for critical points on the track network. The adoption of adequate country software packages for locomotives (m2.4) is ideal for transalpine CT: indeed, freight trains have to cross several countries (at Brenner: DE, AT, IT) and the inclusion of such software packages avoids the change of locomotives at the borders. Considering a broader temporal horizon, the driverless train operation (m2.1), already tested on the Betuweroute (NL-DE) by Alstom, may be also interesting for a highly congested line such as Brenner, since it allows an optimization of the circulation.

Referring to the terminal, the adoption of an automated freight train inspection (m4.1) would guarantee high-precision results and permits a faster wagon identification, reducing related costs and operational times related to this activity. Implementation costs are relatively low and could benefit from other ongoing experiences in Italian terminals (e.g., the port of Trieste). Another solution is related to the use of intelligent railway points (m4.5): frequently used points on the terminal tracks are fitted with sensors that transmit data to a central IT system in real-time, providing information about the condition of the main intersections, also allowing the identification of maintenance work or repairs at an early stage. Furthermore, the implementation of robotic and automated solutions (e.g. automated guided vehicles or flexible transhipment technologies for CT) in intermodal terminals (m4.9) would offer new solutions for loading/unloading, increasing safety and security, contributing also to the environmental sustainability. However, the limited amount of goods handled in Roncafort poses some doubts about their effectiveness of such measure.

Finally, the interaction with other infrastructure has to be mentioned. A closer synergy between Roncafort, A22 and the (sub)urban network of Trento with an exchange of information on suitable routes and trip indicators (e.g. distance, time, costs, parking) would make trips more sustainable (m.3.1). The development of a real time information for ACT service on the light panels distributed along the A22-motorway can reduce waiting and travel times for truck drivers. This system may interact with the existing online sources, which can inform truck drivers on parking facilities in real-time (m4.6).

Table 2. ITSs relevant for CT and their applicability at Brenner [15].

A) Railway				
Category and related measures (strategical – S; tactical - T)		Brief description+example	Adoption at Brenner	Potential implementation
1_Traffic management and control				
1.1_European Rail Traffic Management System (ERTMS)	S	A Europe-wide standardized automatic signalling, control and train protection system (ERTMS deployment along the *ScanMed* RFC being implemented by EU)	Under development	–
1.2_Autarchic GPS tracker	T	A self-powered and maintenance-free GPS tracker mounted in railway wagons (new GPS tracker tested along the Brenner corridor by TX Logistik)	Tested	Yes
2_Driver behaviour change				
2.1_Driverless train operation	S	A train driven automatically (tested on the Betuweroute by Alstom).	No	Yes
2.2_Energy efficient train driving system	S	A recommendation to the train driver for the train speed to be kept and the current kinetic energy (RCS-ADL solution developed in the Swiss railway network).	No	Yes
2.3_The automatic train route optimizer	S	The control programme optimises train management at problematic points on the track network (RCS-HOT solution developed in the Swiss railway network).	No	Yes
2.4_Software package for locomotives	S	Adequate country software packages for locomotives avoiding unnecessary changing drivers or locomotives (solutions tested along the Brenner by TX Logistik)	Tested	Yes
B) Intermodal terminal				
Category and related measures (strategical – S; tactical - T)		Brief description+concrete example	Adoption at Trento-Roncafort	Potential implementation
3_Navigation and traveller information				
3.1_Multimodal tour planning system	S	It provides the information used by operators or decision makers to initiate measures or improvements to the transport system (adopted in Duisport-DE)	No	Yes
4_Logistics and fleet management				
4.1_Automated freight train inspection	T	An automated inspection/identification of freight trains (TrainINSPECT system developed by KleinTech)	No	Yes
4.2_BlockChain	S	Optimization of business transactions and trading relationships at different level (TradeLens developed by Maersk Line and IBM)	No	No
4.3_Electronic system for freight transport	S	Paperless, electronic flow of information for a simple and harmonised procedure (Sinfomar platform activated by the Port of Trieste-IT)	No	Yes
4.4_Fleet Management System (FMS)	T	All operations from vehicle acquisition to disposal to satellite positioning and data communication to back office applications (FMS developed by the EMIXIS Company)	No	Yes
4.5_Intelligent railway point	T	Frequently used points on railway in the terminal fitted with sensors that transmit data to a central IT system in real-time (activated in the Port of Hamburg - DE)	No	Yes
4.6_Intelligent truck parking	T	Application to solve a high number of trucks for limited parking slots (IRU-TRANSPark platform)	Yes	Yes
4.7_Supply chain management system	S	Specialised modules devoted to demand forecasting, production, transportation, delivery and distribution (UIRNet platform in IT)	Yes	–
4.8_Terminal management system	T	A software solution expediting internal order processes integrates customer-side processes (e.g. bookings) and numerous additional functions (e.g. driver authorisations), (WOLF Platform)	Partially	Yes
4.9_Terminal Automated solutions	S	Automated solutions (e.g. transhipment technologies or autonomous vehicles) in the intermodal terminal (Port of Rotterdam - NL)	No	No
4.10_Terminal: software solution to support the wagons sharing concept	T	Software solution to optimize trains and wagons' arrivals and departures (tested in IQE Verona-IT)	No	Yes

5 Conclusions

The forthcoming opening of BBT is expected to increase rail capacity and reduce travel times, slopes, energy and related transport externalities. The use of longer and heavier trains (up to 750 m and 1,600 tons) will lower unitary costs of rail transport and make it more competitive. In this modified scenario, intermodal terminals have not to become the new bottlenecks of the system. Indeed, they need to guarantee efficient transhipment solutions, thanks to an adequate infrastructural equipment and a standardization of transport units and technologies. The trend registered in the past years in Trento-Roncafort (Sect. 2) reveals a loss of attractiveness that has to be inverted before the opening of the new HC railway line. In order to improve its infrastructural equipment, new tracks for ACT service and an extension of tracks to host 750 m trains have been planned. Such measures, even if necessary to give the minimum infrastructural equipment, risk to be insufficient. In this paper, we have discussed the opportunity to integrate them with different ICT solutions. A list of potential alternatives has been provided, based on a best practice analysis carried out in the framework of the EU-funded project SMARTLOGI. Even if their nature is different-some of them have a more strategical approach, whereas others are more tactic- their adoption may contribute to make the intermodal terminal more integrated with the new railway line. A more detailed economic and technical assessment could integrate the first inputs presented in this paper and reveal more rigorously which solutions would produce the best impacts. Ultimately, this approach could also reduce the number of HGVs circulating along the main (sub)-urban roads, a phenomenon that is crucial for the city of Trento: being part of a mountain region with a limited room for infrastructure, this creates several conflicts with commuters that daily travel to/from the city. Thus, this is not merely a freight-related issue; it is rather part of the entire mobility layout of the region.

Acknowledgement. The research leading to these results received funding from the Interreg V-A Italia-Austria Programme 2014–2020, under Project ITAT3008 “SMARTLOGI- Logistica transfrontaliera sostenibile e intelligente”.

References

1. iMONITRAF!, Annual Report 2019: Strengthening the partnership for the Alpine transport strategy. Homepage. http://www.imonitraf.org. Accessed 25 Mar 2020
2. European Commission: Handbook on the external costs of transport, Luxemburg (2019)
3. Nocera, S., Cavallaro, F., Irranca Galati, O.: Options for reducing external costs from freight transport along the Brenner corridor. Eur. Transp. Res. Rev. **10**, 53 (2018). https://doi.org/10.1186/s12544-018-0323-7
4. Cavallaro, F., Sommacal, G., Božičnik, S., Klemenčič, M.: Combined transport in the Alps: reasons behind a difficult acceptance and possible solutions. Res. Transp. Bus. Manage. (2020). https://doi.org/10.1016/j.rtbm.2020.100461
5. European Council: Council conclusions on the digitalisation of transport, Brussels (2017)
6. Woxenius, J.: Terminals–a barrier for intermodality? In: Nordic Transport Research's conference on Intermodal Freight Transport, Ebeltoft (1997)

7. Interbrennero Spa Homepage. http://www.interbrennero.it. Accessed 25 Mar 2020
8. Europaregion/Euregio: Resolution 01/2018 of the European Region Tyrol-South Tyrol-Trentino. Strategy for a modal shift of transport from road to rail on the Brenner axis (2018)
9. A22-Autostrada del Brennero Spa. https://www.autobrennero.it. Accessed 25 Mar 2020
10. Köll, H., Lange, S., Ruffini, F.V.: Detours of trans-alpine goods transport by road. Revue de Géographie Alpine. J. Alp. Res. [En ligne] **95**(1), 65–76 (2007)
11. De Luca, M.: Manuale di pianificazione dei trasporti, Franco Angeli Edizioni (2000)
12. UIRNET Homepage. https://www.pln.cloud/uirnet/en. Accessed 25 Mar 2020
13. AlpInnoCT project: Toolbox of Action (Handbook) (2019)
14. Amitran project, CO2 Assessment Methodology for ICT in Transport. D3.1: Methodology for classification of ITS (2014)
15. SMARTLOGI project: D.3.3.1-Best practices on incentives schemes and ICT tools for enhancing multimodal transport and customs fast corridors (2018)

Business Model Development Based on Sharing Systems and Data Exchange for Sustainable City Logistics

Michail Koutras(✉), Giannis Adamos, and Eftihia Nathanail

Traffic, Transportation and Logistics Laboratory, University of Thessaly, Pedion Areos, 38334 Volos, Greece
mixaliskoutras@gmail.com, giadamos@civ.uth.gr, enath@uth.gr

Abstract. Transport is adapting to the rapid development of technology, adopting new trends and focusing on people and environment. At the same time, the needs of citizens' mobility and goods' transportation are complex, requiring the involvement of different stakeholders under the appropriate partnerships. To this end, the development of innovative integrated business models, including the principles of circular economy, is vital. The aim of this paper is to propose a new business model framework, which facilitates the incorporation of emerging trends in city logistics, i.e. sharing systems and data exchange and the development of synergies amongst private and public actors. Towards this direction, the main findings of a literature review are presented, including new and emerging trends in transport, as well as the content and applications of sustainable and circular business models. These findings feed the development of the proposed business model for sustainable city logistics.

Keywords: Sustainability · Circular economy · Goods' distribution · Freight transport · Smart cities

1 Introduction

Freight transport constitutes a small part of the overall transport activities in city centers. However, the respective impacts on society, economy and environment are crucial, requiring immediate action through effective solutions and interventions. Traffic congestion, road accidents, air and noise pollution are among the most significant social and environmental impacts met in a city (EEA 2016).

Key actions to 2050 recognized by the European Union are the establishment of better operating models, the development of integrated mobility systems, data and infrastructure sharing, the development of public-private services and respective regulatory and procurement strategies (TRIMIS 2020). It is clear that there is a strong need for inclusive and sustainable cities that facilitate safe, efficient, accessible to all and pollution-free mobility of people and transhipment of goods. In order to achieve this aim and at the same time encounter data protection concerns, like commercial and personal data protection and ethical matters, skillful collaboration is required among all

E. G. Nathanail et al. (Eds.): CSUM2020, AISC 1278, pp. 814–823, 2021.
https://doi.org/10.1007/978-3-030-61075-3_79

stakeholders of the mobility ecosystem to develop innovative and integrated business models that address oriented retailing, goods' distribution patterns and shared logistics.

The aim of this paper is to propose an integrated framework, which facilitates the transition of traditional business models to a new approach, incorporating emerging trends in city logistics, like sharing systems and data exchange.

The rest of the paper is structured as follows: Sect. 2 includes background information on new and emerging trends in freight transport. A state-of-the-art review of business models is given in Sect. 3, followed by the development of the proposed circular business model in Sect. 4. Section 5summarizes the main findings of this research.

2 Background

2.1 New and Emerging Trends in Freight Transport

Currently 54% of the world population lives in urban centers and it is estimated that this proportion will reach 60% until 2030 (Oliveira et al. 2018) creating greater needs for transportation. Simultaneously, technology blooms and is used on a daily basis to improve the living standards and reduce the environmental impacts. Smart cities use Information and Communication Technologies (ICT), improve services and offer a lot of benefits to citizens and environment. Smart cities focus also on green initiatives, related to sustainability and environmentally friendly applications of ICT (Walvarens and Ballon 2013).

Marketing targets towards a better relationship between business and customer. The last years, the promotion of goods and services adopts the mechanisms of marketing, in order to create and capture value, using data to predict customers' choices and products' characteristics (France and Ghose 2018).

Businesses are performing a transition to the digital era with the broad use of technology. The conversion of technology from analog to digital is defined as digitalization. In this case, technology is used to improve productivity and speed up production by saving costs and reducing errors (Parida et al. 2019). However, it requires expertise, otherwise it acts like a limiting factor. Data protection is another element that concerns clients and the highest protection measures should be followed.

Internet of things facilitates the interconnection of different processes, allowing the appropriate determination, collection and communication of data (Metallo et al. 2018). It simplifies automation tasks; however, it has to confront data protection, which is a major issue for the consumers. E-logistics refers to the virtual planning of the transportation of goods and materials (Dębkowska 2017). Cloud architecture overcomes business cost and increases the speed of the transfers between various actors in logistics networks.

Lastly, crowdsourcing with the use of crowdsensing systems, i.e. smartphones combines the feedback from a "crowd" of persons to form a "bigger picture" of an interesting or emerging phenomenon (Heiskala et al. 2016).

2.2 Sharing Systems

New trends and technologies are already being used by private companies in addition to the traditional means of transport (McCoy 2018). The advances in technology allowed new opportunities for sharing systems, such as car or bike-sharing. Parking sharing, public transport, taxis and leasing can also be considered as trends in sharing systems. These types of transport seem to be capable to face current and future needs and improve mobility of people, movement of goods and eventually quality of life in cities. Behavior towards ownership is a crucial factor of the success of such schemes (Berger 2014). The challenge is that possession of a vehicle becomes redundant and people prefer to rent or lease a mode to cover their daily needs. In the case of leasing, users seem to maintain the vehicle in a better condition (Tunn et al. 2019). In fact, more than 50% of car owners in industrialized countries would share their vehicles with others (Berger 2014).

The above changes in behavior are part of the transition to a sharing economy; thus, an economy that promotes the collective consumption and the better use of resources and promises sustainable transport, effective supply chain and environmental advantages for cities.

On the other hand, the applications of sharing systems in freight transport are less frequent and the lack of data is one of the most important barriers. Freight transport seems in general less investigated by literature and sharing economy is indicated as a long-range concept. Still, there are some exceptions. For example, competition among ports has switched to competition at the level of shipping logistics (Van der Horst and De Langen 2008), therefore transport concerning the hinterland area of a port is crucial. An innovative mechanism is proposed in Caballini et al. (2016), where trips are shared among carriers ensuring positive environmental impacts in terms of congestion limitation and profit increase. Similarly, He et al. (2013), proposed a strategy for the improvement of the dynamics of container terminals with the adoption of truck sharing systems amongst different container terminals. As a result, sharing systems can tackle many efficiency challenges, such as the empty container truck trips problem, which burdens traffic inside and around the port (Theofanis 2007).

Outside the port area, such cases of sharing systems or vehicles are rarely met. Kim et al. (2016) explored the case of a Korean mixed feed company. The strategy proposed sharing trucks among the different factories and proved to be very useful in high-demand conditions. It was revealed that transport costs can be reduced by over 10%. Another example is e-cargo bikes, tested in Basil. This theoretical concept proves that safety training and infrastructure development are important issues, including also the aspect of decision making and the role of policy makers (Hess and Schubert 2019). Another promising service is truck platooning thus, a block of vehicles moving together in a short distance with between one another. They are considered promising in road safety aspects and as effective means to reduce fuel consumption (Van De Hoef 2015).

2.3 Data Exchange

The application of Information and Communication Technologies (ICT) in the transportation sector has led to the development of the concept of Intelligent Transportation Systems (ITS). However, the complexity of transportation systems is increasing and issues concerning traffic congestion grow. At the same time, the great amount of data generated is being underutilized (Ayfantopoulou et al. 2019). The European Union (EU) initiated several actions, promoting data sharing among companies. Since 2014, with the initiative «Towards a thriving data-driven economy», EU refers to the great value of Big Data for crucial sectors, such as health and transportation. In 2016, EU addressed the crucial issue of data protection with General Data Protection Regulation–GDPR.

Ayfantopoulou et al. (2019) proposed that standardization of data, as well as data formats, is essential; thus, unexplored data will emerge, innovative transportation services will be created and interoperability of ITS will be reassured. These datasets, concerning road safety, multimodal transportation and real time traffic data, could be crucial for traffic management and supply chain design. As an example, travel time reliability could be estimated, ensuring on-time deliveries and directly affect the growth of e-commerce and same-day deliveries (Yuksel et al. 2019). Using the available technology, data can be exploited by companies and solutions in modern transportation challenges can be found, i.e. blockchain (DHL 2020).

3 Business Models: State-of-the-Art Review

3.1 Overview

The complexity of modern companies and the challenges that they deal with, make it difficult for them to remain competitive and profitable. Recent research focuses on better understanding of the company resources and activities (Goyal et al. 2017). To this end, the application of appropriate business models is necessary.

The entry of the business model concept is set during the e-commerce growth and the creation of new forms of transactions in the 1990's (Timmers 2018). This approach was initiated to communicate complicated business ideas to investors within a short time frame (Zott et al. 2011). Nowadays, it works as a tool for systematic analysis and planning of the company (Doleski 2015), following the new trends, the growing economies and the global market needs. In general, business model is used to describe the overall structure and activities, which are required to achieve a strategic outcome (Goyal et al. 2017). Osterwalder and Pigneur (2010) argued that business model is like a blueprint for a strategy to be implemented through organization structures, processes and systems. Geissdoerfer et al. (2018) define business model as a simplified representation of the value proposition, value creation and delivery, and value capture elements and the interactions between these elements within an organizational unit. The basis of many business models is the business model Canvas by Osterwalder and Pigneur (2010), which consists of nine blocks, describing nine different elements: key partners, key activities, key resources, value propositions, customer relationships, channels, customer segments, cost structure and revenue streams.

3.2 Sustainable Business Models

Elements are closely related to challenges that the company has to tackle and they alternate as the company's targets alternate. The way that these objectives are achieved play also a very important role. The above are parts of business model innovation. Its success is determined by the ability of the organizations to adapt the business models to the environmental changes in terms of customer, technology, competition and macro-economic trends (Voelpel et al. 2006). An important target is sustainability and how it can be achieved in modern societies. Therefore, sustainable business models emerge, by incorporating principles that enhance sustainability or by integrating sustainability into their value proposition, value creation and delivery activities. Sustainable business models also incorporate multi-stakeholder management and hold a long-term perspective (Geissdoerfer et al. 2018). The presence of ICT is also strong. A sub-category of sustainable business models are green business models. The target is to develop green logistics, namely logistics which is environmentally and socially friendly as well as economically functional (Rodrigues et al. 2001). Green business models are also related with the reduced use of resources or the use of sustainable resources.

3.3 Circular Business Models

The latest trend in economics is sharing economy and in particular circular economy. An economic system promotes the sustainable development and minimizes resource input and waste, emissions and energy leakage out of the system (Geissdoerfer et al. 2018). Hence, circular business models abide by the principles of sustainable business models and incorporate closing, narrowing and slowing resource loops. In few words, circular business models comprise recycling measures (closing), efficiency improvements (narrowing) and use phase extensions (slowing) (Geissdoerfer et al. 2018). Maintaining the multi-stakeholder management, three basic principles are added: reducing, reusing and recycling (3Rs Concept) (Heyes et al. 2018). Another component is behavioral change, in which the use of services is dominating. Due to the repair, refurbishment and remanufacturing, the longevity of the product is improved (Nußholz 2018). Designing a closed resource loop is a difficult task and each part needs extreme attention. As an example, reverse logistics could cause great damage, while products return to the manufacturer (Kalverkamp and Young 2019). Therefore, the envisioning of each cycle is proposed, with the right tools and the different elements such as the value proposition, need re-definition after each product cycle (Nußholz 2018).

4 Business Model Development for Sustainable City Logistics

The development of the proposed business model follows the principles of sharing economy, like closing the resource loops. The concept of 3Rs is adopted alongside the re-assessment after each cycle. The basis of the model are the nine blocks/components of the business model canvas, proposed by Osterwalder and Pigneur (2010). Additionally, internalization of externalities (NOVELOG 2017) is included, since the value for customers' perspective is considered important and it is not adequately covered by the rest blocks. Human and planet are also added under value proposition. Lastly, evaluation and redesign are indicated as essential for a circular business model. The proposed business model is depicted in Fig. 1.

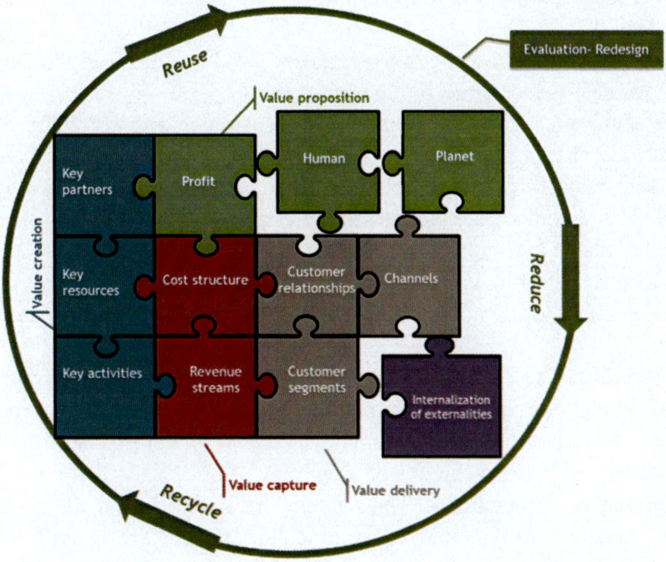

Fig. 1. Proposed business model for sustainable city logistics.

A brief description of the blocks of the proposed business model follows, while the detailed canvas of the model is depicted in Fig. 2:

- **Key partners**. This block includes companies involved in city logistics processes, such as vehicle renting or leasing companies, transport companies, fuel companies and insurance companies. These firms could share their data and/or vehicle fleets towards achieving sustainability in their operation.
- **Key activities**. The elements in this block describe the main activities that a company should accomplish in order to achieve the expected outcomes, by creating value for the customers. The use of new information and communication technologies, digitalization and Internet of Things are strongly recommended. This block also contains the two additional vital activities, vehicle sharing and data exchange.
- **Value proposition**. The elements in this block refer to the value that is created by the products or the offered services. The value is directly related with the customer needs. It is important to note that the value described in this block is the value that the customer is willing to pay for.

- **Key resources**. This block contains the main resources that the company processes or uses to create the value for its customers, including ICT equipment, human resources and infrastructure. Part of these resources can be rented, such as precision instruments.
- **Customer relationships**. Current or future relationships with customers are the content of this block. Dedicated personal assistance or dedicated user profiles are recommended. In many cases, companies offer a website or an application, making their services easily accessible by their customers.
- **Customer segments**. This block concerns the customers for which the company is creating value. The main objective is to satisfy a wide spectrum of the market, offering dedicated services, as for example renting many different types of vehicles.
- **Channels**. The specific block addresses the channels through which the customer segments are reached. The promotion of the company is also an important element.
- **Cost structure**. This block concerns the highest costs in the business model. It answers also the question, which resources and which activities are the most expensive.
- **Revenue streams**. These streams refer to the value that customers are willing to pay or they are already paying for. Renting, leasing or memberships are included. In circular economy, new revenue streams are created by refurbished or remanufactured products, minimizing the use of resources.
- **Internalization of externalities**. The content of this block is to describe how the existing or potential traffic and environmental impacts are caused by the key activities or using the key resources. Reducing traffic congestion, heavy vehicle traffic or air pollution emissions can be quantified, describing the value from the customers' perspective.
- **Evaluation and redesign**. This block addresses the elements, which help evaluating the results of the existing business model in order to define the key activities and resources that will be incorporated in the revised business model, considering the next cycle of products/energy. Revenue streams and externalities are the expected results.

The proposed business model can be applied by any company involved in city logistics, i.e. courier companies or catering shops. The added value of this approach is that facilitates efficiently the transition to the main principles of circular economy, addressing better use of resources and promotion of sharing schemes and environmentally friendly means of urban deliveries, i.e. electromobility and cargo bikes.

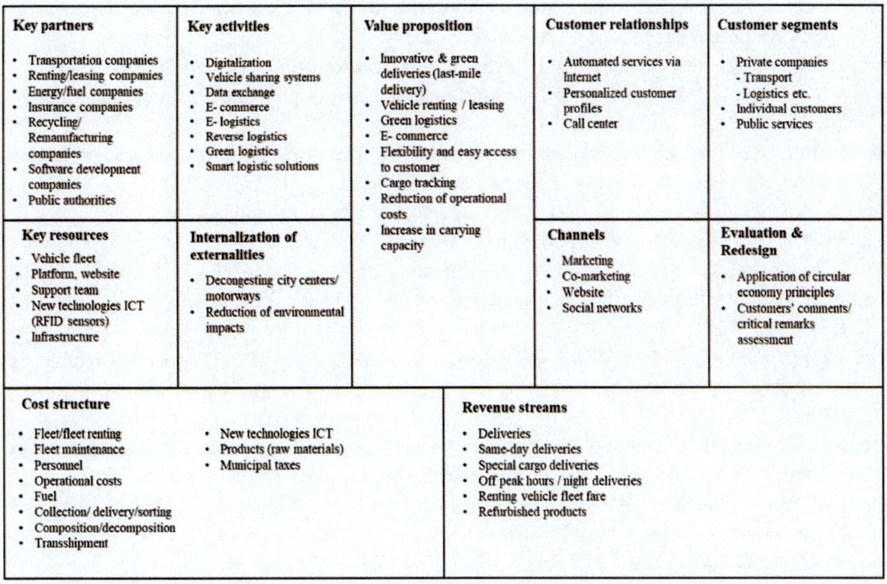

Fig. 2. Detailed canvas of the proposed business model for sustainable city logistics.

5 Conclusions

Freight transport burdens the quality of life in cities, resulting in intense impacts on environment, economy and society. The development of technology enables to create a win-win situation for companies and environment, providing green urban freight transport solutions and interventions. The success of this complex plan sets the application of a circular business model as a requirement to achieve coherent synergies amongst the involved stakeholders and facilitate the implementation of complex procedures. Incorporating two promising activities, i.e. sharing systems and data exchange, a human-centric circular business model was developed, enabling the determination of the required conditions for sustainable freight transportation in cities.

References

Ayfantopoulou G., Dolianitis A., Mylonas C., Stamelou A., Mitsakis E.: Transport Data Exchange Platforms: A European perspective. TRB Annual Meeting (2019)

Berger R.: Shared Mobility: How new businesses are rewriting the rules of the private transportation game. Roland Berger Strategy Consultants GMBH (2014)

Caballini, C., Sacone, S., Saeednia, M.: Cooperation among truck carriers in seaport containerized transportation. Transp. Res. Part E **93**, 38–56 (2016)

Dębkowska K.: E- logistics as an elements of the business model maturity in enterprises of the TFL sector. In: 7th International Conference on Engineering, Project, and Production Management. Procedia Engineering 182, 143–148 (2017) (2016)

DHL. https://www.logistics.dhl. Assessed 20 Jan 2020

Doleski, O.D.: Integrated Business Model: Applying the St. Gallen Management Concept to Business Models (2015)

European Environment Agency (EEA), https://www.eea.europa.eu. Assessed 20 Jan 2020

France, S., Ghose, S.: Marketing analytics: Methods, practice, implementation, and links to other fields. Expert Syst. Appl. **119**(2019), 456–475 (2018)

Geissdoerfer, M., Morioka, S., Monteiro de Carvalho, M., Evans, S.: Business models and supply chains for the circular economy. J. Clean. Prod. **190**(2018), 712- 721 (2018)

Goyal S., Kapoor A., Esposito M., Sergi B.: Understanding business model – literature review of concept and trends. Int. J. Competitiveness **1**(2) (2017)

He, J., Zhang, W., Huang, Y., Yan, W.: A simulation optimization method for internal trucks sharing assignment among multiple container terminals. Adv. Eng. Inform. **27**(4), 598–614 (2013)

Heiskala, M., Jokinen, J., Tinnilä, M.: Crowdsensing- based transportation services – an analysis from business model and sustainability viewpoints. Res. Transp. Bus. Manag. **18**, 38–48 (2016)

Hess, A., Schubert, I.: Functional perceptions, barriers, and demographics concerning e-cargo bike sharing in Switzerland. Transp. Res. Part D **71**, 153–168 (2019)

Heyes, G., Sharmina, M., Mendoza, J., Gallego-Schmid, A., Azapagic, A.: Developing and implementing circular economy business models in service-oriented technology companies. J. Clean. Prod. **177**, 621–632 (2018)

Kalverkamp, M., Young, S.: In support of open-loop supply chains: Expanding the scope of environmental sustainability in reverse supply chains. J. Clean. Prod. **214**, 573–582 (2019)

Kim, H., Kim, J., Yang, W., Lee, K., Kwon, O.: Effects of truck- shared delivery: a case of a korean mixed feed company. J. Int. Logist.Trade **14**(2), 165–173 (2016)

McCoy, K., Andrew, A., Glynn, R., Lyons, W.: Integrating Shared Mobility into Miltimodal Transportation Planning: Improving Regional Performance to Meet Public Goals. U.S, Department of Transportation (2018)

Metallo, C., Agrifoglio, R., Schiavone, F., Mueller, J.: Technol. Forecast. Soc. Chang. **136**, 298–306 (2018)

NOVELOG Deliverable D7.3. Business models for urban freight transport solutions (2017)

Nußholz, J.: A circular business model mapping tool for creating value from prolonged product lifetime and closed material loops. J. Clean. Prod. **197**, 185–194 (2018)

Oliveira, L., Barraza, B., Bertocini, B., Isler, C., Pires, D., Madalon, E., Lima, J., Viera, J., Meira, L., Bracarense, L., Bandeira, L., Oliveira, R., Ferreira, S.: An overview of problems and solutions for urban freight transport in Brazilian cities. Sustainability **10**, 1233 (2018)

Osterwalder, A., Pigneur, Y.: Business Model Generation: A Handbook for Visionaries, Game Changers, and Challengers. Wiley , Hoboken (2010)

Parida, V., Sjödin, D., Reim, W.: Reviewing literature on digitalization, business model innovation, and sustainable industry: past achievements and future promises. Sustainability **11**, 391 (2019)

Rodrigues, J.P., Slack, B., Comtois, C.: Green logistics (the paradoxes of). In: The Handbook of Logistics and Supply-Chain Management, p. 2 (2001)

Theofanis, S., Boile, M., Janakiraman, S., Naniopoulos, A.: Reducing unproductive empty container movements around marine container terminals: the role of a virtual container yard. In: IAME Conference, Athens, Greece, pp. 67–78 (2007)

Timmers, P.: Business models for electronic markets. Electron. Market. **8**, 3–8 (2018)

TRIMIS. https://trimis.ec.europa.eu/. Assessed 20 Jan 2020

Tunn, V.S.C., Bocken, N.M.P., van den Hende, E.A., Schoomans, J.P.L.: Business models for sustainable consumption in the circular economy: an expert study. J. Clean. Prod. **212**, 324–333 (2019)

Van der Horst M., De Langen P.: Coordination in hinterland transport chains: a major challenge for the seaport community. Marit. Econ. Logist. **10**, 108–129 (2008)
Voelpel, S.C., Leibold, M., Eckhoff, R.: A: The tyranny of the Balanced Scorecard in the innovation economy. J. Intellect. Cap. **7**, 43–60 (2006)
Walravens, N., Ballon, P.: Platform business models for smart cities: from control and value to governance and public value. IEEE Commun. Mag. **51**, 72–79 (2013)
Yuksel E., Bertini R., Ozkul S., Menon N., Unnikrishnan A., Garg R.: Leveraging a permanent freight data collection infrastructure and archive system: visualizing urban truck temporal and spatial dynamics. TRB Annual Meeting (2019)
Zott, C., Amit, R., Massa, L.: The business model: recent developments and future research. J. Manag. **37**(4), 1019–1042 (2011)

Validating the Simulated Impacts of Urban Freight Transport

Ioannis Karakikes(✉) and Eftihia Nathanail

Department of Civil Engineering, University of Thessaly, Pedion Areos, 38334 Volos, Greece
iokaraki@uth.gr

Abstract. This paper provides a comprehensive guide for assessing the impacts of an urban freight transport system through traffic simulation software. Based on a calibrated and validated city wide urban traffic model that integrates the freight trips of six freight vehicles in the examined day, aim of the paper is to establish an impact assessment framework that evaluates model's environmental and transport impacts.

The analysis of the freight trips showed that they account for 22009.2 g of CO2 emissions and 84.17 h of total network delays on a daily basis. Other impact assessment indicators' values such as average vehicle speed, first attempt successful deliveries, freight vehicles' number of roundtrips, load factor, utilization factor, emissions, traffic throughput, etc. were also computed.

Keywords: City logistics · Last mile delivery · Indicators · Model · Assessment

1 Introduction and Background

As cities population growth is increasing due to rapid urbanization, there is need to recognize the existence of mobility issues and understand their mobility situation in relation with indicators such as environmental compatibility, economic efficiency and quality of life. The basis for action was set in 1992 at the United Nations Conference on Environment & Development (Agenda 21 – Chapter 40, Action 40.6) in which it is mentioned that "Countries at the national level and international governmental and non-governmental organizations at the international level should develop the concept of indicators of sustainable development in order to identify such indicators" [1]. Indicators are the basis of benchmarking, a technique for performance management that allows measurement, comparison, identification of best practices and improvement [2]. Benchmarking not only reveals the present performance status of a system against others, but also provides information on the dimensions and scale of potential improvement and the best practices for implementing changes that effect improvement [3].

The usage of indicators, however, is associated with many issues that need to be addressed. In most of the cases evaluation can become a lengthy process, requiring a great amount of resources to draw relevant data. The absence of organized databases in conjunction with low frequent and unreliable data, constitute barriers for the

E. G. Nathanail et al. (Eds.): CSUM2020, AISC 1278, pp. 824–834, 2021.
https://doi.org/10.1007/978-3-030-61075-3_80

deployment of a comprehensive system of indicators able to monitor the progress of urban sustainability in an area [4].

Aim of this study is to provide a comprehensive guide for assessing the impacts of Urban Freight Transport (UFT) systems through a set of appropriate indicators. The assessment is based on data stemming from a calibrated and validated city wide urban traffic model that integrates the freight trips of six freight vehicles, four motorbikes and two minivans, performing 199 delivery stops in the examined typical day, during 09:00–20:00. The impact assessment of the produced urban freight traffic model is expected to help authorities to understand better their current UFT system and to enable comparisons when future solutions are to be ex-ante evaluated.

2 Previous Work and Methodology

The methodological steps to develop and assess the impacts of the urban freight trips are described in the following paragraphs. Previous work has given emphasis on model's development process.

Data Procurement and Analysis. Two different data sets were used in the traffic simulation model; (a) traffic data and (b) data of goods' deliveries. Data were provided on the basis of a typical day, during which traffic volumes and number of deliveries in the area of Volos are representative. Delivery data were requested and given by the private courier company "ELTA Courier" which provides door-to-door deliveries in the city of Volos. As the dataset indicated that deliveries were performed between 09:00 and 19:27 traffic data were respectively searched for the same time of day, i.e. 09:00–20:00.

Development of Traffic Simulation Model. The traffic model was developed using PTV Vissim microscopic traffic simulation software. Regional, arterial and characteristic collector roads that run through the study area, were selected to form model's road network, as modeling all city's road network would demand a very high amount of data and resources.

Model Calibration and Validation. Model's calibration and validation was based on the WisDOT criteria [5]. Model's results were compared with real-world data in five simulation rounds with different random seeds. To validate the model, the generated average travel times of the time sections were compared to real travel time data from taxi GPS devices.

Scenario Configuration. The "as is" scenario simulates the actual situation and concerns daily freight trips performed by ELTA courier vehicles. Parcels, in this scenario, are delivered door-to-door to the end-users' addresses by company's freight vehicles. Two new vehicle types (610: ELTA Motorbike and 620: ELTA Van) were created in Vissim to represent the four Motorbikes and two Minivans. Freight vehicle routes were coded in model as "Public Transport Lines". Public Transport Stops were used to represent delivery stops.

It is noted that, the modeled road network of the "as is" scenario does not extend to the boundaries of the study area. In addition, a high share of collector roads is not represented in the model as that would demand a very high amount of data and resources. Thus, projections of the addresses of 191 delivery stops were made to the closest point of the modeled road. 152 delivery stops were not considered in the model as their distance was higher than 300 m from any point of the network. As this assumption, can create high deviations between the real results and model's results, two additional subscenarios, namely "Real" and "Real of Modeled Stops" were formulated. The differences between the subscenarios can be found in Table 1.

Table 1. Subscenarios of the "as is" scenario.

Characteristic	Real	Real of modeled stops	Vissim
Analysis based on..			
Traffic Simulation (PTV Vissim)			✓
Google maps	✓	✓	
Delivery Stops considered			
Delivery stops in the area of modeled road network (199/343 of the study area)		✓	✓
All delivery stops in the study area (343/343)	✓		
Distance covered by freight vehicles			
For servicing 199/343 stops within the modeled road network (measured in Vissim).			✓
For servicing 199/343 stops within the modeled road network (measured in Google maps).		✓	
For servicing 343/343 stops in the study area (measured in Google maps).	✓		
Travel times			
For servicing 199/343 stops within the modeled road network (measured in Vissim.			✓
For servicing 199/343 stops within the modeled road network (measured in Google maps).		✓	
For servicing 343/343 stops in the study area (measured in Google maps).	✓		

Results Evaluation. The results of the "as is" scenario were presented and the benefits at a city level were estimated.

3 Impact Assessment of the "as is" Scenario

The evaluation of the "as is" scenario was performed based on a number of indicators through which environmental, and transport and mobility impacts were assessed. The selection of the indicators was based on the list of [6] which brought together all impact assessment indicators detected in the assessment frameworks presented by STRAIGHTSOL, BESTUFS, SUGAR, GRASS & C-LIEGE [7–9] (see Table 2).

Table 2. Evaluation indicators.

Criterion	Indicator	Explanation	Unit	Scenario value
Environment				
Emissions	CO	Total emissions produced based on fuel, technology and vehicles type over a given period	[g]	109.5*
	CO2		[g]	22009.2*
	NOx		[g]	74.8*
	PM		[g]	4.8*
	VOC		[g]	93.4*
Transport and Mobility				
Transport system	Total delays in traffic	Total delays due to the traffic of the freight vehicles over a given period	sim. sec.	303,038*
	Average vehicle speed	Average of total distance divided by the total travel time for all freight vehicles over a given period	km/h	29.31
UFT Fleet	Number of roundtrips	Number of roundtrips per vehicle over a given period	number	2.5
	First attempt successful deliveries	Degree that deliveries take place at a first delivery attempt over a given time period	%	92.5*
	No. of commercial light vehicles	Number of vehicles with a gross vehicle weight of no more than 3.5 tonnes	number	2
	No. of delivery mopeds	Number of delivery mopeds	number	4
	Traffic throughput	Total veh-km covered by freight vehicles a given time period	Veh-km	96.888*
	Load factor	Average load factor of the fleet over a given time period	%	24.9
	Vehicle utilization factor	Hours that vehicles are in service, e.g. deliveries, pick-ups, transporting, weighting, loading/unloading over 24 h	%	14*

*Scenario Value of "Vissim" subscenario.

3.1 Environmental Indicators

There is a number of widely adopted environmental impact assessment methodologies aiming at assessing the exhaust emissions of any pollutant. For example, the EMEP/EEA Emission Inventory Guidebook by the European Environment Agency (EEA) provides such a methodology. The Guidebook suggests three levels of methodologies based on input data availability; Tier 1, 2 and 3 [10]. The European Chemical Industry Council (Cefic) and the European Chemical Transport Association (ECTA) published in 2011, a book with guidelines for measuring and managing transport CO2 emissions from freight transport operations [11]. The objectives of these guidelines are i) the development of a common, simple, and precise method for the calculation of CO2 emissions from freight transport operations and ii) the assessment and promotion of industry best practices to reduce transport emissions. The Intergovernmental Panel on Climate Change (IPCC) refined in May 2019 the 2006 IPCC Guidelines for National Greenhouse Gas Inventories aiming among other at estimating the actual annual emissions [12]. Further to the methodologies there are common used emission calculation models, incorporated or not with simulation software, i.e. COPERT, EMFAC, MOBILE, MOVES, ARTEMIS, Handbuch (HBEFA), VERSIT+, EMV, etc. A detailed summary for each emission model can be found in Kousoulidou et al. [13].

Due to the size of the network it was decided to calculate the emissions of each freight vehicle via COPERT Street level [14]. COPERT considers that the freight vehicles are driving on a link with length equal to their covered distance and average speed which is computed as the total delivery duration divided by the covered distance. In that way, the computational effort is minimum and comparisons among freight vehicles are enabled (see Table 3).

Table 3. Comparing "Real", "Real of Modeled Stops" and "Vissim" CO, CO2 and NOx emissions for the "as is" scenario.

Input data									
	Distance covered by freight vehicles (km)			Delivery duration (h)			Average speeds (km/h)		
Sub scenario	Real	Real of Modeled Stops	Vissim	Real	Real of Modeled Stops	Vissim	Real	Real of Modeled Stops	Vissim
M1	46.573	38.16	16.611	5.93	5.08	3.36	7.8	7.5	4.9
M2	36.325	36.325	17.822	9.60	9.60	5.27	3.8	3.8	3.4
M3	46.811	19.451	9.614	5.90	3.67	2.43	7.9	5.3	4.0
M4	56.79	5.14	4.755	6.62	0.63	0.94	8.6	8.1	5.1
V1	157.013	32.02	22.608	8.92	1.83	1.97	17.6	17.5	11.5
V2	65.86	33.13	25.478	8.18	6.38	2.97	8.0	5.2	8.6
Sum	409.37	164.23	**96.89**	45.15	27.20	**16.9**	53.81	47.37	37.41
Output Data									
	CO (g)			CO2 (g)			NOX (g)		
Sub scenario	Real	Real of Modeled Stops	Vissim	Real	Real of Modeled Stops	Vissim	Real	Real of Modeled Stops	Vissim
M1	83.8	68.7	29.9	2952.7	2419.3	1053.1	7.9	6.5	2.8
M2	65.4	65.4	32.1	2303.0	2303.0	1129.9	6.2	6.2	3.0
M3	84.3	35.0	17.3	2967.8	1233.2	609.5	8.0	3.3	1.6
M4	102.2	9.3	8.6	3600.5	325.9	301.5	9.7	0.9	0.8
V1	59.0	12.0	10.2	52476.2	10701.6	8893.1	204.0	41.6	31.2
V2	29.7	14.9	11.5	25906.7	13032.0	10022.0	91.0	45.8	35.2
Sum	424.4	205.3	**109.5**	90206.9	30015.0	**22009.2**	326.7	104.2	**74.8**

3.2 Transport and Mobility Indicators

Total Delays in Traffic. To calculate the total delays due to the traffic of the freight vehicles the Vehicle Network Performance option was enabled in the "Evaluation" menu of PTV Vissim software for the period 1800 sim. sec. (corresponds to 09:00) to 41400 sim. sec. (corresponds to 20:00). Vehicle Network Performance was measured in two versions of the same Vissim model. The first version regards all the vehicular traffic (cars, motorbikes, trucks and buses) while the second all the vehicular traffic considering also freight vehicles' sub routes. To perform the evaluation, five simulation runs were performed for each version with the same random seeds so as to allow comparisons of the versions. Total delays due to freight vehicles arose by subtracting the total delays of the version without the sub routes from the version with the sub routes of freight vehicles (Table 4). It is noted, that the delay of a vehicle in a time step is the part of the time step in which the actual speed is less than the desired speed.

Table 4. Contribution of freight vehicles in total traffic.

	Average total delays (h)	Average total distances (km)	Average total travel times (h)
With the sub routes	1509.9 h	129,407.99	4444.6 h
Without the sub routes	1425.7 h	129,315.68	4426.9 h
(With) – (Without)	84.2 h	92.31	28.9 h

Note: "Average" signifies the average results of the five simulation runs.

Average Vehicle Speed. Speed is one of the factors that affect urban accessibility, component of transport system accessibility. In most of the cases higher speeds correspond to a higher number of performed deliveries, however they can be associated with higher fuel consumptions and safety issues [15]. In this study, average vehicle speed is calculated as an option of the Vehicle Network Performance menu option in Vissim and is measured by dividing the Total Distance by the Total Travel Time. In the "as is" scenario, the average speed for all vehicles is equal to 29.31 km/h.

Number of Roundtrips. The number of roundtrips indicate how many trips a freight vehicle has realized in order to deliver its daily volume. The lower the number of roundtrips performed, the more efficient the delivery process is, as a higher number of roundtrips adds more vehicle-kilometers from/to the station [15]. Table 5 shows the number of roundtrips per freight vehicle.

Table 5. Number of roundtrips per freight vehicle for the "as is" scenario.

Freight vehicle	Number of roundtrips	Departure time from the Station in sim. sec. (hh:mm)			
Motorbike 1	3	1: 3900 (09:35)	2: 24960 (15:26)		3: 33000 (17:40)
Motorbike 2	2	1: 4320 (09:42)		2: 22380 (14:43)	
Motorbike 3	4	1: 1800 (09:00)	2: 10560 (11:26)	3:15360 (12:46)	4: 24900 (15:25)
Motorbike 4	2	1: 3000 (09:20)		2: 15480 (12:48)	
Vehicle 1	1	1: 3480 (09:28)			
Vehicle 2	3	1: 2400 (09:10)	2: 15600 (12:50)		3: 26400 (15:50)
Average	2.5 roundtrips				

First Attempt Successful Deliveries. This indicator describes the degree to which deliveries take place at a first delivery attempt. If a delivery is not successful at the first attempt, then the driver leaves a notification coupon at the delivery address stating the exact hour of the attempted delivery and at the same time the recipient receives in his mobile phone a message/notification that an unsuccessful delivery was attempted. The parcel to be delivered in this case stays on the Station till the recipient pick it up from the Station or is delivered at a following delivery trip the same day or next, if agreed with the carrier agent at an extra cost. ELTA courier neither does offer any options of rearranging the delivery date for free nor deliver in local pick up points. Table 6 shows the percentage of first attempt successful deliveries per freight vehicle.

Table 6. First attempt successful deliveries percent per freight vehicle for the "as is" scenario.

Sub scenario/Freight vehicle	Real (%)	Real of Modeled Stops/Vissim (%)
Motorbike 1	54/54 (100%)	43/43 (100%)
Motorbike 2	59/68 (86.8%)	59/68 (86.8%)
Motorbike 3	45/51 (88.2%)	18/20 (90%)
Motorbike 4	31/40 (77.5%)	11/11 (100%)
Vehicle 1	82/82 (100%)	25/25 (100%)
Vehicle 2	41/48 (85.4%)	28/32 (87.5%)
Average	312/343 (91.0%)	184/199 (92.5%)

Zooming in the results it can be concluded that at least 7.5% of the delivery trips in Vissim are unnecessary. Considering an average delivery distance and travel time, this corresponds to a total daily loss of 36.8 km and 4.06 spend hours. These could be avoided if the recipient was able to choose the arrival window or delivery method.

Traffic Throughput. Traffic throughput expresses the total vehicle-kilometers (veh-km) covered by the freight vehicles [16]. Veh-km is a measure of traffic flow which is determined by the multiplying the number of vehicles on a given road or traffic network by the average length of their trips in kilometers (Table 7).

Table 7. Number of veh-km for the "as is" scenario.

Subscenario/Indicator	Real	Real of Modeled Stops	Vissim
Traffic throughput (veh-km)	409.372	164.226	96.888

Load Factor. In an attempt to determine the efficiency of the vehicles' trip, the load factor was estimated [17]. It is widely recognized that an increased load factor usually represents efficient transport of goods as the vehicle-kilometers needed to transport the same load are reduced [18]. However, there are criticisms in literature that a narrow decision making based on simple ratios can affect the economic and environmental sustainability of the trips [19]. Load factor can be measured in various dimensions i.e. weight, deck area, height, empty running or volume, depending on the accuracy requirement and the purpose. Focusing on the weight dimension, as weight-based data are the only available in the ELTA courier's system, different definitions and metric systems have been used.

In this study's dataset, all stops correspond to deliveries as there are no pick-ups. Thus, the average load accounts for the empty runs. For example, a full motorbike load of 50 kg leaving the station to reach a recipient 2 km away and returns empty will have a load factor of 50% (0.05 tonnes * 2 km = 0.1 t-km over 4 veh-km divided by 0.05 tonnes) [18]. ELTA Courier station was the starting point for all vehicle trips, as this is the only station of the company in Volos. Trip distances between the stops were retrieved manually by Google maps. Considering that for every successful delivery the load is decreased by the weight of the delivered parcel and assuming that the Maximum Load for motorbikes is equal to 0.055 tons (55 kg) and for Minivans 0.180 tons (180 kg), load factors including empty running trips are as follows (Table 8):

Table 8. Average load factors per roundtrip per freight vehicle for the "as is" scenario.

Freight vehicle	Load factor	Average load factor per roundtrip - (First trip total load in *kg*)			
Motorbike 1	18%	1: 44% (47.76)	2: 9% (11.6)		3: 1% (1)
Motorbike 2	33%	1: 47% (47.51)		2: 19% (26.62)	
Motorbike 3	15%	1: 18% (17.14)	2: 9% (10.41)	3: 21% (21.55)	4: 12% (12)
Motorbike 4	30.5%	1: 53% (52.02)		2: 8% (13.78)	
Vehicle 1	28%	1: 28% (136.65)			
Vehicle 2	24.7%	1: 59% (176.3)	2: 9% (36.24)		3: 6% (35.5)
Average	24.9%				

Vehicle Utilization Factor. One of the weaknesses of urban freight distribution when a shipper operates own vehicles is the low vehicle utilization factor which is associated with high costs [20]. The vehicle utilization factor is defined as the average number of hours a vehicle is in service per day and is computed by subtracting the idle time from the 24 h divided by 24 h. A vehicle is in service during deliveries, pick-ups, transporting and loading/unloading. Idle time is the average time per day that a vehicle is not utilized. In this study, the total delivery duration per freight vehicle has been computed. Considering that the loading time is a variable dependent on the number of deliveries to be performed in the next roundtrip and unloading time depends on the number of unsuccessful deliveries in the previous roundtrip, loading/unloading time can be computed as follows:

$$\sum_{k=0}^{K}\sum_{j=1}^{J} d_{kj}a + u_{kj-1}a \tag{1}$$

where,

k denotes the freight vehicle, *j* denotes vehicle's k roundtrip no. to be performed (see Table 5)
d is the number of deliveries, *u* is the number of unsuccessful deliveries, *a* is an empirical constant which equals to 1 min for parcels less than 5 kg and 1.5 min for heavier. The value of the constant was estimated by ELTA Courier.

The idle time derives by subtracting the total delivery duration and loading/unloading time per freight vehicle from the 24 h of a day. Freight vehicle's fleet utilization factor is equal to the average utilization factor of the six freight vehicles (Table 9). Vehicle utilization factors are calculated for all three sub scenarios.

Table 9. Comparing "Real", "Real of Modeled Stops" and "Vissim" vehicle utilization factors for the "as is" scenario.

Sub scenarios	Real	Real of Modeled Stops	Vissim
M1	28%	24%	17%
M2	45%	45%	27%
M3	28%	17%	12%
M4	30%	3%	5%
V1	43%	9%	10%
V2	37%	29%	15%
Average	35%	21%	14%

4 Concluding Remarks

Simulation models' purpose is to replicate adequately a working system in order to understand it better. Although extended microsimulation models require high resources for their development, once they are properly configured, they can output a high variety of evaluation results.

Based on a calibrated and validated city wide urban traffic model, this paper assesses the impacts of freight trips in an urban environment. Interpretation of the impacts at their current status can be achieved only when being compared to standards i.e. emission thresholds. In future, when these values are "before-after" compared, with "after" being the scenario of modeling a potential UFT solution, they will denote the improvement or deterioration of the after scenario as compared to the before. ELTA Courier freight vehicles are not the only freight vehicles circulating the urban road network. Knowing the percentage of ELTA Courier freight flows represented at a city level, can be a basis for calculating the UFT impacts of the whole UFT system.

Certain limitations when calculating specific indicators are mentioned in the main body. Concerning further research on this model, various UFT solutions will be configured and have their performance assessed compared to the "as is" scenario.

Acknowledgements. This research is co-financed by Greece and the European Union (European Social Fund- ESF) through the Operational Programme «Human Resources Development, Education and Lifelong Learning» in the context of the project "Strengthening Human Resources Research Potential via Doctorate Research" (MIS-5000432), implemented by the State Scholarships Foundation (IKY).

This research has been based on data provided by ELTA Courier company (https://www.elta-courier.gr).

References

1. United Nations: Sustainable development, Rio de Janeiro: United Nations (1992)
2. Geerlings, H., Klementschitz, R., Mulley, C.: Development of a methodology for benchmarking public transportation organizations: a practical tool based on an industry sound methodology. J. Clean. Prod. **4**, 113–123 (2006)
3. Henning, T., Muruvan, S., Feng, W., Dunn, R.: The development of a benchmarking tool for monitoring progress towards sustainable transportation in New Zealand. Transp. Policy **18** (2), 480–488 (2011)
4. May, A.D.: Encouraging good practice in the development of sustainable urban mobility plans. Case Stud. Transp. Policy **1**(3), 3–11 (2015)
5. WisDOT: Microsimulation Guidelines - Model Calibration (2002)
6. Nathanail, E., Mitropoulos, L., Karakikes, I., Adamos, G.: Sustainability framework for assessing urban freight transportation measures. Logist. Sustain. Transp. **9**(2), 16–36 (2018)
7. Balm, S., Browne, M., Leonardi, J., Quak, H.: Developing an evaluation framework for innovative urban and interurban freight transport solutions. Procedia Soc. Behav. Sci. **125** (2014), 386–397 (2014)
8. BESTUFS: Deliverable 5.2. Quantification of Urban Freight Transport Effects II (2008)
9. C-LIEGE: Final publishable report: Clean Last mile transport and logistics management for smart and efficient local Governments in Europe (2014)
10. EEA: EMEP/EEA air pollutant emission inventory guidebook 2019. Publications Office of the European Union, Luxembourg (2019)
11. ECTA/Cefic: Guidelines for Measuring and Managing CO2 Emission from Freight Transport Operations, Paris (2011)
12. IPCC: Refinement to the 2006 IPCC Guidelines for National Greenhouse Gas Inventories, Kyoto, Japan (2019)

13. Kousoulidou, M., Fontaras, G., Lonza, L., Dilara, P.: Overview of emission and traffic models and evaluation of vehicle simulation tools. Institute for Energy and Transport, Brussels (2013)
14. Emisia: COPERT Street Level (2018). https://www.emisia.com/utilities/copert-street-level/. Accessed 12 Dec 2019
15. Russell, G.T., Taniguchi, E.: City logistics and freight transport. In: Button, K.J., Hensher, D.A.,. Brewer, A.N. (eds.) Handbook of Logistics and Supply-Chain Management (Handbooks in Transport), vol. 2, pp. 393 – 405 (2017)
16. FHWA: Traffic Data Computation Method. Department of Transportation, Washington, DC: U.S. (2018)
17. Santén, V., Rogerson, S.: Shippers' transport efficiency: an approach for measuring load factor. Logist. Res. **11**(3), 1–15 (2018)
18. Teo, J.S., Taniguchi, E., Qureshi, A.: Evaluation of load factor control and urban freight road pricing joint schemes with multi-agent systems learning models. Procedia Soc. Behav. Sci. **125**, 62–74 (2014)
19. Arvidsson, N.: The milk run revisited: a load factor paradox with economic and environmental implications for urban freight transport. Transp. Res. Part A Policy Pract. **51**, 56–62 (2013)
20. Nathanail, E., Gogas, M., Papoutsis, K.: Best practices, implications and future needs. Deliverable D2.1 STRAIGHTSOL – STRAtegies, Thessaloniki: EUROPEAN COMMISSION (2012)

The Rise of the On-Demand Warehousing: Is the Greek Market Ready for This Change?

Elpida Xenou[1], Leonidas Parodos[1(✉)], George Tsoukos[2], Georgia Ayfantopoulou[1], and Zisis Maleas[1]

[1] Hellenic Institute of Transport- Centre for Research and Technology Hellas, Thessaloniki, Greece
{elpixenou, lparodos, gea, zisismaleas}@certh.gr
[2] New Technologies and Innovation Department, TREDIT S.A, Thessaloniki, Greece
g.tsoukos@tredit.gr

Abstract. The necessity for creating flexible and agile supply chains and adapt on the continuously changing and quite competitive e-commerce logistics market, has made warehousing a strategically critical component of the supply chains. In response to that, new logistics concepts such as on-demand logistics and shared warehousing popped up recently, which with the help of the e-marketplaces offer innovative logistics solutions through a single access point and support the logistics actors to: easily optimize the capacity of their warehouse, avoid the fixed costs of owning a warehouse; implement short-term supply chain planning and finally come closer to the end-customer. Although several on-demand warehousing practices can be seen worldwide, the Greek economic crisis delayed until recently the emergence of such trends so, little is known on how the Greek market would respond to such changes. In response to that, this paper introduces briefly the first on-demand warehousing platform to be operating in Greece and closes the knowledge gap on the Greek market's response following a three-step methodological process: conducting a dedicated questionnaire survey circulated to different types of businesses, arranging logistics expert groups and carrying out extensive desktop research. The paper shows that the Greek stakeholders are quite familiar to shared warehousing solutions and they are open to digital changes such as using an on-demand warehousing platform. However, they are less willing to apply such innovations to their daily business. Therefore, specific platform requirements for increasing the attractiveness of such tools to the Greek market are also highlighted.

Keywords: On-demand logistics · Shared warehousing · Virtual Freight Center · Flexible warehousing · E-Marketplace

1 Introduction

The rapid rise of the e-commerce during the recent years in addition with the highly competitive and constantly changing business environment, have brought significant changes in the way that companies respond to their customers (Kovacs and Kot 2016).

E. G. Nathanail et al. (Eds.): CSUM2020, AISC 1278, pp. 835–844, 2021.
https://doi.org/10.1007/978-3-030-61075-3_81

Starting from the USA and spreading worldwide, the need for flexible and agile supply chains, has become a necessity for long-term sustainability (Christopher 2000). Moreover, the excess increase of the e-business and the time sensitive customer requirements, over the last few years, have created the need for implementing "on demand logistics concepts" as well as developing cooperative logistics schemes (shared logistics) (Zhong et al. 2019; Kersten et al. 2018). In terms of warehousing, the growth of the transport sector together with the rapid development of the e-business and the direction towards next day (Amazon), or even within next hours deliveries (Amazon Prime), forced the companies to search for alternative solutions to the traditional warehousing in order to maintain a competitive advantage (DHL 2017). In response to that, one of the latest trends in supply chain management and logistics worldwide, with several start-ups to be noticed popping up in US and in Europe, seems to be the operation of e-marketplaces that provide on demand warehousing services.

In Greece though, the economic crisis and the recession of the past decade delayed the emergence of these trends up until recently. Although over the last few years, the e-commerce is significantly rising (ELTRUN 2017) and the people seem to be more familiar with technological tools for services exchange, little is known about how the market responds to on demand warehousing concepts while no similar e-marketplaces can be found. This paper aims to close this knowledge gap by introducing the development and operation of the first Virtual Freight Center in Greece which aims to facilitate the development of cooperative and on demand warehousing models by the Greek market and gaining clear insight about: 1) the willingness of the Greek market in adopting shared warehousing concepts, 2) the readiness of the Greek stakeholders in using such innovative tools in their daily operations and finally 3) which are the most preferable functionalities of such an e-marketplace and by which criteria the Greek logisticians would choose this tool for their daily operation.

This paper is structured in two main chapters. The first chapter presents initially an extended literature review on the new emerging concepts of shared economy and on-demand logistics, continuing to the introduction of a brand new concept, called "on demand warehousing" and the operation of a Virtual Freight Center in Greece and finalizing to the main research questions that this paper comes to answer. The next session presents the methodological process of the research and the final chapter presents the main findings and results.

2 Conceptual Background and Terminology

2.1 The Benefits of the on Demand and Sharing Economy in the Logistics Market

The term "on demand economy", also known in literature as "gig economy" and "sharing economy", was first introduced by technological start-up companies that penetrated the market the past decade. The concept behind this new trend was the provision of innovative solutions for connecting the demand for a service or product with the supply through a single access point and thus satisfying instantly the customers' demand (Dervojeda et al. 2013, ECORL 2015). The main differentiation from the "sharing economy" concept lies in the fact that in the sharing economy, the technology enables the cooperation among the companies and individuals by allowing them to share and benefit from the excess capacity of their assets (vehicles, warehouses, houses, parking spaces) (Melo et al. 2019). On this basis, innovative sharing economy start-ups advocate that economic efficiency and growth can be realized through appropriate accelerator programs that steer start-ups towards the adoption of sharing economy concepts that enable the use of underutilized and redundant resources in order to further scale-up (Grinevich et al. 2019; Rockstart 2019). Therefore, the "on-demand economy" attributes involve the responsive fulfillment of demand and of its variances, within predefined time-windows, while the "shared economy" attributes, the sharing of assets, capacity and resources. Finally, the "gig economy" attribute involves the temporal character of the economy, characterized by short-term & independent contracting.

So far, in the logistics sector, the "on demand logistics" services were mostly related to the "instant deliveries" services as one of the main outcomes of the rapid growth of the e-commerce economy. Particularly in EU, the on-demand deliveries are noticed to be one of the fastest growing on demand logistics sectors with an increase of the customer's participation in 2017 up to 110% and total spending to \$8.2 billion (Rockbridge Associates 2018). Instant deliveries were defined by Dablanc et al. (2017) as the provision of a "two hours window" delivery services, implemented by either a private actor or an independent contractor, following the connection of the demand and the supply through a digital platform (Dablanc et al. 2017). As on-demand deliveries are required to be responsive and fast, some orders are dispatched at very low utilization levels. This leads though to higher trip frequencies, more vehicles, traffic, noise, emissions and eventually costs. To counterbalance these externalities, new opportunities arise for shared logistics operations (Perboli et al. 2016). Among the advantages of sharing economy in logistics and supply chain management are the enabling of the companies to share cost-intensive assets such as warehouses, vehicles and information flows (Ocicka and Wieteska 2017) Logistics companies have started the past years to sharing assets mainly warehouses and transport capacities in order to increase their efficiency, better utilize their excess capacities and quickly respond to the need for on-demand deliveries due to the e-commerce development. For example, the innovative vehicle-sharing platforms that exist worldwide, allow the interested companies to share information about trucks, routes and load factors. The World Economic forum estimated that 15% of the trucking market would be using such kind of platforms and 20% of the warehouse market would move to shared agreements by 2025 (World Economic Forum 2016).

During the last decade, all companies that own or use warehouse facilities, either retailers, manufacturers or 3rd party logistics providers, face difficulties to determine warehousing strategies. Some organizations operate in very demanding environment with extremely seasonal fluctuations in the demand. For that reason, shared warehousing is getting industry attention over the last few years. However, finding and securing warehouse space for periods of excess demand can be a time-consuming, lengthy, and often frustrating process. These were also the main drivers that pushed towards the introduction of the "on demand warehousing" concept. The following session aims to provide clear insight on the main attributes of on demand warehousing and the existing experiences that can be found in the global market.

2.2 The Concept of on Demand Warehousing, Existing Experiences in Europe and in Greece

Warehouse on demand is filling the need for flexible, short-term warehouse space across a fast-paced capacity-crunched supply chain. The on-demand warehouse model helps to transform a traditionally fixed cost into a variable cost that enables greater flexibility in areas of the supply chain through cooperation and sharing capacities. A shared, on demand warehouse can be a flexible solution with short-term contracts (1–3 years) or even daily/weekly/monthly according customers' requirements. The main idea behind shared on demand warehousing comes from the need for connecting in real-time, logistics operators looking for storage space (demand) with existing storage facilities that are partially utilized by other companies (supply). According to a Flexe (2017) report on Warehouse Capacity and Trends, companies that use on-demand warehouse space to meet a single peak seasonal increase in inventory were found to have improved the utilization of the warehouse by approximately 100%. In addition, these companies reported that they were successful in reducing seasonal inventory and warehouse costs by 50%.

In response to the rising need for on-demand warehousing, companies worldwide have developed e-marketplaces which offer services beyond just technology platforms that match demand and supply. The e-marketplace in warehousing connects any company with additional warehouse space, to a company that needs it for a specific amount of time. The using company can utilize its excess capacity for short-term or long-term projects, which results in saving more money and having greater flexibility and convenience (Gartner 2019). The use of on-demand warehousing e-marketplaces can be compared to services like Uber and Airbnb.

In US a company called Flexe finds spare warehouse space for e-commerce merchants all over the US and is fast becoming the Airbnb of warehousing. In five years, Flexe has built a marketplace of spare storage space in 550 warehouses representing close to 2.3 million square metres without purchasing any assets. In addition, several e-market places developed worldwide (Stockbooking, Flowspace, Stowga, Stord, Ware2Go, a.o.) to deal with dynamic distribution and offer greater proximity to the customers. Some of the on-demand warehouse providers report that there are companies which now placing all their warehousing needs through on-demand platforms. It remains to be seen whether the traditional warehouse providers such as K+N, DHL and XPO will enter the highly competitive on-demand arena as the financial benefits for their customers as well as for their own assets are prevalent (Gartner 2019). The target of all such e-marketplaces is to facilitate the storage needs or the logistics services of their customers, to gain agility and generate overall savings.

Despite the development of such e-marketplaces for on-demand warehousing worldwide, Greek stakeholders have paid little attention so far. It is well-known that there are existing platforms in freight transportation (Timocom, Freights) in Greece, which constantly gain market share however, there are no such solutions for on-demand warehousing so far. In response to that, a research project under the name Warehouse Match & Optimize (WareM&O), co-financed by the European Union and Greek national funds through the Operational Program Competitiveness, Entrepreneurship and Innovation, develops a Virtual Freight Center for connecting in real-time, users that are in search for warehouse space (demand) with others that have excess or underutilized warehouse space (supply) through a single access point, a digital platform. The VFC aims at providing to a wide range of businesses the required visibility and information on the available logistics facilities and services in various regions of the country together with clever tools to make effective use of these areas and adapting a flexible and dynamic supply chain structure according to their needs and business goals.

For the successful implementation of this tool and the effective penetration after the project's completion in the Greek market, it was necessary to be in frequent communication with the Greek stakeholders and potential users of the platform in order to identify the potential interest for the VFC; the maturity level of the actors to the daily use of matchmaking platforms and finally to capture the main requirements and basic functionalities of the tool for making it attractive to the Greek entrepreneurs. The outcomes of this process constitute the main subject of this paper. In the following chapters the methodological process and the main findings of this research are presented.

3 Methodology

Our research approach was based on three different pillars as described below:

1. The implementation of expert group meeting with stakeholders that operate in the supply chain. The expert group was held in Thessaloniki and gathered at least fifteen field experts from different business sectors.
2. A dedicated questionnaire survey among different company types. The survey comprised of two parts. The first part collected data about warehousing in terms of size, needs, ownership status and load factor. The second part of the survey asked about their opinions regarding the willingness to adopt on-demand warehousing solutions and about the functionalities that such an e-marketplace should have in order to attract them as customers. The dedicated questionnaire shared to the potential users in three different ways. First, it was shared during the expert-group to the participants for completion. Then, it was shared electronically through a google-doc form to a large database of potential interested companies. Finally, the questionnaire distributed to the participants of a dedicated logistics conference, SOLUTIONS VI, in Thessaloniki.
3. A comprehensive mapping and desktop analysis of the Greek market in terms of warehousing. Data analysis for both questionnaires and expert group was made in Python programming language.

4 Findings and Implications

In Greece, the logistics infrastructure is concentrated in three key, discrete and interconnected centers in Attica, Viotia and Thessaloniki. However, a lack of high standard facilities for the provision of high level logistics services is noticed. More specifically, there are 243 warehouse and storage companies, deploying 4232 facilities in Central Macedonia and Attica region (Eurostat 2016). In addition, the total number of companies recording in transportation and storage were 62878, having more than 65623 facilities in the same regions. It can be seen from the statistics above, that there are plenty potential customers for an on-demand warehousing e-marketplace at theoretical level. Given the fact that no previous research has been done for the Greek market, our paper aims to reveal the maturity of the Greek market to adopt such tools for the daily business.

The main findings from the expert group have shown that one fundamental attribute for the VFC's attractiveness is the provision of increased visibility in terms of warehousing availability along the whole Greece. The logistics and warehousing experts share the opinion that there is need to enhance the visibility along their supply chains in conjunction with the development of collaborative models in order to enable Greek businesses to achieve the necessary flexibility and respond rapidly to the constantly-changing challenges of the national and global economy. In addition, the experts admitted that even though there are notable potential advantages from the use of an on-demand warehousing platform, the Greek market is not yet familiar with such innovative tools, thus resulting in difficulties in adopting such solutions for their daily business.

In order to enhance and verify the preliminary results derived from the expert group, we conducted a questionnaire which was shared to all business sectors relevant with logistics and supply chain industry.

4.1 Sample's Main Characteristics

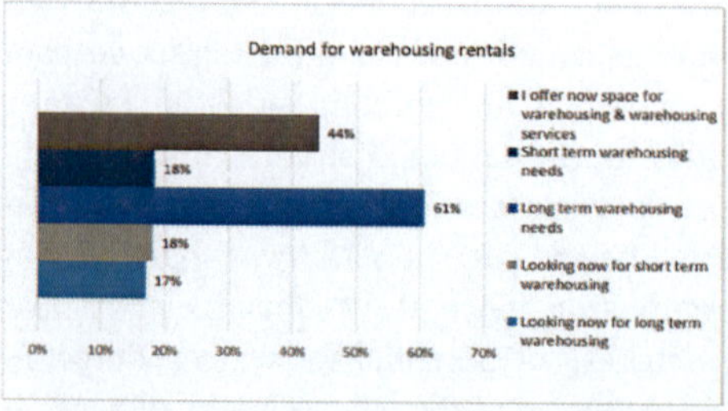

Fig. 1. Demand for warehousing rentals.

The initial sample consisted of 90 Greek logisticians which were occupied in 69 dominant Greek companies, supply chain operators. Based on the above sample, the main findings in this survey; the dominant business sector that was participated is industrial and manufacturing (58%), while 3PL companies (30%) follow and the remaining 12% are retailers, transport operators and supply chain business related companies. In order to understand in-depth the potential demand for the VFC's services, the respondents were asked to record the number, the size and load factor of warehouses that they operate. The main findings show that only 6% of them do not operate at that moment a warehouse and is in search for the appropriate facility. The remaining 96% is operating either their owned (36%) or rental warehouse facilities (28%) or both (31%). In total, 140 warehousing facilities were documented, where approximately 65% of them were big facilities with more than 2500 m^2 capacity (Fig. 1).

The next and final part for understanding in depth the participants' needs and identify the potential gap that this service comes to fill in, is to investigate the current demand for warehousing facilities. Hence, the following figure highlights that almost the 35% of the respondents is at the moment looking for a new facility where half of them search for short term warehousing solutions. In terms of supply, there is a 44% that offers space or/and warehousing services, which shows that besides the 3PL companies, also the remaining business sectors are willing to offer space for shared warehousing.

4.2 Main Findings About the Interest for Shared Warehousing Solutions and the VFC

The main philosophy of the Virtual Freight Center lies in the development of cooperative warehousing solutions for maximizing the load factor and the corresponding revenues of the supply chain companies and the provision of on-demand and flexible warehousing rental services for facilitating the short-term warehousing and satisfying appropriately the customer's demand. In line with this aim, the potential users of the VFC should be 1) interested in using this platform and 2) willing to cooperate and either have excess warehousing capacities for short-term or long-term or they are in search for warehouse space in order to satisfy their demand. Interestingly, both the survey and the expert group showed that regarding the first point, mentioned above, almost the 80% of the respondents are in favor for the operation of such a tool, while a 13% is uncertain (Fig. 2).

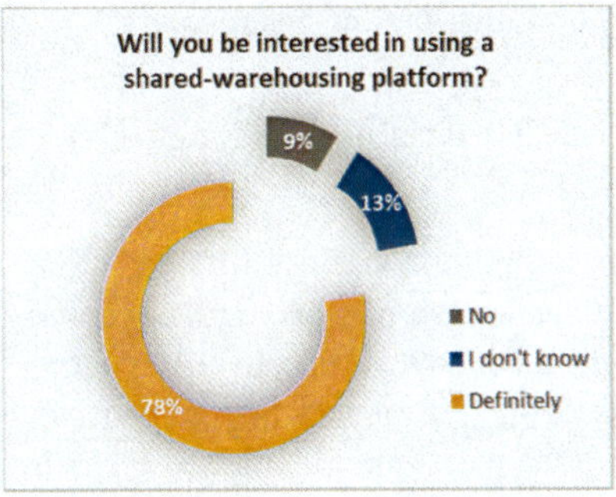

Fig. 2. Interest in using shared warehousing platform.

Now in terms of willingness to cooperate, the main findings observed, reveal a lack of consistency in the respondent's opinion comparing to the strong interest expressed above for the VFC. More specifically, the following figures show that only the approximately 22% and 25% of the respondents are willing to either share part of their

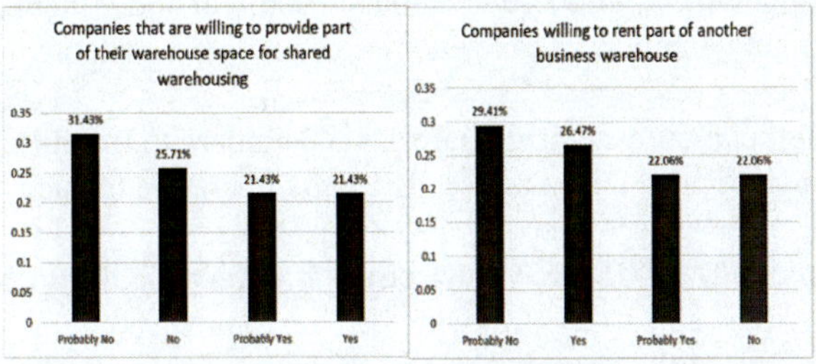

Fig. 3. Willingness of the Greek market to implement shared warchousing solutions.

available warehousing facilities for cooperative warehousing or rent part of the warehousing facilities of another company, excluding the 3PL companies. However, a corresponding percentage of 22% respectively is skeptical but positive towards this concept (Fig. 3).

For clarifying the reasons for this lack of readiness in adopting shared warehousing concepts, a deeper analysis on the potential correlations with the sample's characteristics was crucial and showed that vast majority of non-positive answers are highly connected with the load factor of the companies' warehousing facilities. More specifically, the participants who have higher load factors (up to 86%) are more reluctant to adopt shared warehousing solutions, mainly because they don't see any opportunities for further profit, which constitutes a quite logical explanation. On the other hand, the corresponding load factor rate for those who are willing to explore the cooperative is at 79%. Further analysis on the non-positive respondents in the frame of the expert group revealed that also the lack of clear understanding on the main benefits and potential impact of this VFC on a company's logistics operations, is also one of the reasons. In any case, the outcomes are quite encouraging for the Greek market and show a market that although it is not mature enough at the moment, is open to changes for improving the effectiveness of its warehousing and distribution methods.

4.3 Which Are the Main User Requirements and Criterial for Making the VFC Attractive to the Greek Market?

Having identified the positive reaction of Greek market in the existence of a VFC, the next goal was to capture the parameters that will make such a platform attractive and easy to use. Taking that into account, the participants that showed a positive or neutral interest in the operation of the VFC (the correspondent 91% of the total responses), were asked to choose from a list of user requirements the most important based on their current warehousing needs. The 47% of the respondents pointed out that a dynamic and constantly growing database of warehousing repositories would help them to find quickly and easy the available spaces In addition, the stakeholders would like to have a tool that provides quick and easy finding of the right warehouse space in the appropriate cost (43%). Furthermore, 37% of the respondents would prefer to have a tool that can quickly and easily match the demand and supply of warehouse availabilities or warehouse services. Moreover, 35% of the sample would like to use the platform also as an observatory of prices of services and warehouse locations which will allow them to map and record the current market condition and have a more complete view about the offers.

Finally, in terms of the most crucial criteria for the Greek stakeholders to use this tool, three were the main criteria identified: 1) Ease of use: The daily operation of such a tool should be very as simple as possible, even for users that have no previous warehousing or logistics experience; 2) WMS/ERP connectivity: A very significant criterion is to ensure that the tool will be able to connect with the intercompany systems such as WMS/ERP, and finally 3) Cost of the service: The cost of the service should be affordable for the companies in order to choose this tool as a solution to their warehousing problems.

5 Conclusion and Future Areas of Research

The present paper, introduces to the Greek stakeholders the concept of shared economy on-demand warehousing, and tries to fill the existing knowledge gap on the following: What is the level of maturity of the Greek market in adopting shared and on demand warehousing solutions? Are the Greek stakeholders interested in using a Virtual Freight Center in their daily operations and finally which are the most crucial and preferable functionalities of an on-demand e-marketplace in order to make it attractive to the Greek market.

To accomplish this goal, the close communication with the Greek logisticians through expert groups was necessary as well as the implementation of an extensive data collection through both a dedicated questionnaire handing out to the potential users of the VFC and a desktop research.

The main results of this study can be summarized as follows: Greek stakeholders seem to be very interested in using a shared-warehousing platform, but are less willing to implement such solutions to their daily business. Although the majority of the respondents is not yet determined to use such a platform, almost 45% is an optimistic and positive percentage to begin with showing that Greek market is open to digital changes. The warehouse load factor seems to be inseparable with the stakeholders' willingness to adopt shared warehousing. In addition, issues related to data management, increased supply chain visibility, specific platform requirements and selection criteria were highlighted by the respondents to develop an attractive and easy to use platform.

The results presented in this paper are part of the main research outputs of the project WareM&O which is co-financed by the European Union and Greek national funds through the Operational Program Competitiveness, Entrepreneurship and Innovation.

Since on demand warehousing is a relatively new concept, there are several opportunities for future areas of research. Some indicative future research questions that could be addressed are: Which are the implications of on-demand warehousing solutions to the urban environment? How can the city respond to minimize the potential negative externalities of the exploitation of all possible real estate, including small areas located in the city center, for on-demand logistics operations?

References

Christopher, M.: The agile supply chain. Ind. Mark. Manage. **29**, 37–44 (2000)

DHL Trend Research: Sharing Economy Logistics (2017)

Dablanc, L., Morganti, E., Arvidsson, N., Woxenius, J., Browne, M., Saidi, N.: The rise of on-demand 'Instant Deliveries' in European cities. Supply Chain Forum Int. J. **18**(4), 203–217 (2017)

Dervojeda, K., Verzijl, D., Nagtegaal, F., Lengton, M., Rouwmaat, E., Netherlands, P.W.C., Monfardini, E., Frideres, L., Luxembourg, P.W.C.: Business Innovation Observatory - The Sharing Economy Accessibility Based Business Models for Peer-to-Peer Markets. European Commission (2013)

ECORL: Comparative Study on Sharing Economy in EU and ECORL Consortium Countries, pp. 1–47 (2015)

ELTRUN: Annual survey in eCommerce and On-line Marketing. In: 13th eBusiness Forum, ELTRUN (2017)

FLEXE: Warehouse Capacity Economics and Trends report. White paper (2017)

Gonzalez, D., Aimi, G., Johnson, J., Rogerson, C.: Gartner's "Supply Chain Brief: How Shippers Can Benefit from the Fast-Growing On-Demand Warehouse Marketplace Sector" (2019)

Grinevich, V., Huber, F., Karataş-Özkan, M., Yavuz, Ç.: Green entrepreneurship in the sharing economy: utilising multiplicity of institutional logics. Small Bus. Econ. **52**(4), 859–876 (2017)

Kersten, W., Blecker, T., Ringle, C.M.: The road to a digitalized supply chain management: smart and digital solutions for supply chain management. In: Proceedings of the Hamburg International Conference of Logistics (HICL), epubli GmbH, Berlin (2018)

Kovács, G., Kot, S.: New logistics and production trends as the effect of global economy changes. Pol. J. Manage. Stud. **14**, 115–126 (2016)

Melo, S., Macedo, J., Baptista, P.: Capacity-sharing in logistics solutions: a new pathway towards sustainability. Transp. Policy **73**, 143–151 (2019)

Ocicka, B., Wieteska, G.: Sharing economy in logistics and supply chain management. LogForum **13**(2), 183–193 (2017)

Perboli, G., Rosano, M., Gobbato, L.: Decision support system for collaborative freight transportation management: a tool for mixing traditional and green logistics. In: 6th International Conference on Information Systems, Logistics and Supply Chain, ILS Conference, Bordeaux, France, 1–4 June 2016

Rockbridge Associates.: The On-Demand Economy Continues to Grow – Rockbridge (2018). https://rockresearch.com/on-demand-economy-continues-to-grow-2/

Rockstart.: Web and Mobile Program. https://www.rockstart.com/emerging-tech/web-mobile/. Accessed 21 Feb 2020

World Economic Forum: World Economic Forum White Paper Digital Transformation of Industries: In collaboration with Accenture. Logistics Industry, pp. 1–31 (2016)

Zhong, Y., Guo, F., Wang, Z., Tang, H.: Coordination Analysis of Revenue Sharing in E-Commerce Logistics Service Supply Chain with Cooperative Distribution. SAGE Open (2019)

Planning the Future: Innovative Technology for City Logistics

Afroditi Anagnostopoulou[1(✉)], Evangelos Spyrou[2], Aggelos Aggelakakis[1], and Maria Boile[1]

[1] Centre for Research and Technology Hellas (CERTH), Hellenic Institute of Transport (HIT), Thessaloniki, Greece
a.anagnostopoulou@certh.gr

[2] School of Electrical and Computer Engineering, Aristotle University of Thessaloniki, Thessaloniki, Greece

Abstract. This paper describes the needs to develop proper infrastructure and improved solutions for reducing emissions of last mile distribution and determining vehicle routing optimization with emphasis in city logistics 4.0 that consider urbanization, volatility as well as quality of life and climate change. To this end, the development of an online application is proposed to automate the vehicle routing optimization of electric light goods vehicles (LGVs) in an attempt to enhance the design and management of city logistics.

Keywords: V2X communication · Electric vehicle routing problem (EVRP) · Realistic case study

1 City Logistics

The economic vitality of the cities is increasingly dependent on the effectiveness of city logistics. Goods transport and delivery play a key role in the economic activity which is affected by regional, national and global business networks. Efforts to reduce traffic and carbon emissions are being made by regional authorities and national governments (COP21 [1] and COP23 [2]). Companies also seek to reduce operational costs and the related production of Green House Gas (GHG) emissions adopting "on demand" economic activities since the bloom of e-commerce [3]. Innovative tools and advanced technological solutions utilized in urban transportation logistics to capture the dynamic aspects of the current environment. These tools and solutions are able to provide real-time data for trucks and vehicles as well as accurate information about roadway conditions, traffic information etc. [4, 5]. Special equipment, sensors and infrastructure provide traffic data and historical information about traffic volumes. A large amount of data is thus continuously collected and may complement the historic traffic data of a certain region. The value of big data is to support traffic monitoring and management, enhance the system's reliability, support fleet operations and improve environmental performance.

Economic transformation and societal change render policy-oriented expertise in analysis of trends and developments to learn from successful practices and contribute to

E. G. Nathanail et al. (Eds.): CSUM2020, AISC 1278, pp. 845–852, 2021.
https://doi.org/10.1007/978-3-030-61075-3_82

the existing know-how. As such, aspects of infrastructure, regulatory and technology should be incorporated to urban freight transport management approaches [6]. As such, innovative technological advances (i.e. Internet of Things - IoT, Information and Communication Technology - ICT and Intelligent Transport Systems - ITS) should be considered in modeling city logistics including also operational restrictions [7–9].

Their effectiveness should ensure the commitment of both public and private stakeholders that support the implementation of modern measures capable in achieving sustainable development regarding freight transport activities. For this reason, proper key performance indicators (KPIs), both quantitative and qualitative, should be proposed for the performance measurement of sustainable urban freight transport to provide guidelines suitable for the stakeholders of both private and public sector in order to assess the implementation of a set of measures and policies [10]. Recent trends in advanced technologies foster innovative solutions in city logistics such as crowd shipping [11] that aim to reduce the total number of vehicles for distribution, parcel lockers [12] that aim to reduce costs of last mile delivery, and green vehicles [13] that aim to reduce GHG emissions of deliveries in the city. However, their practical implementation requires laboratory experiments in testing them, in order to provide estimations and quantitative results that could support their real-life applications.

The aim of this paper is to examine the use of electric LGVs in city distribution and as such, an online application is developed based on advanced technologies to automate the vehicle routing optimization process. The remainder of this paper is organized as follows: Sect. 2 describes how vehicle routing optimization process is developed based on the innovative online solution. Then, Sect. 3 presents the results of the laboratory experiments conducted and in Sect. 4 conclusions are drawn.

2 Online Application of Vehicle Routing Process

The focus of this paper is given on a modification of the Electric Vehicle Routing Problem [14], which denotes that all delivery demands (shipment of products to customers) are initially located at a central depot and vehicles exchange information with Vehicle to Everything (V2X) communication. The aforementioned problem is an NP-hard since it is a natural generalization of the GVRP [15]. Therefore, substantial computational effort is required for determining optimum or near optimum solutions even for medium size instances. Given a homogeneous fleet of capacitated vehicles, the goal is to design a set of vehicle routes, in order to satisfy the delivery requirements of a set of geographically scattered customers. Furthermore, the design aims to ensure that the remaining battery of each vehicle is enough, in order to return to the depot either directly or via a charge station.

Each customer i has a known delivery demand d_i and it must be serviced within a predefined time window $[a_i, b_i]$ that models the earliest and the latest times during the day when service can take place. As such, vehicles must remain at the customer locations during the service, while there is a waiting time wt_i if a vehicle arrives before the customer's earliest time window. Each customer must be visited only once by exactly one vehicle, while the load of a vehicle must not exceed vehicle's maximum capacity Q at any time along its route. Vehicles could visit charging stations more than once and should remain parked there for a predefined period of time h_i for charging. Finally, the fleet of vehicles has a maximum level of battery charge M.

Algorithm 1 Multi-restart Solution Framework

Require: μ // User-defined parameters
Ensure: Best encountered feasible solution $s*$

while *Predefined CPU time consumption* **do**
  $s_i \leftarrow$ Create Initial Solution // Greedy Randomized Construction
  $s_{i'} \leftarrow$ Guided Local Search // Intensification Local Search
  while *improvement is observed* **do**
    $s_r \leftarrow$ Ruin
    $s_{rr} \leftarrow$ Recreate
    $s_{rr'} \leftarrow$ Guided Local Search
  end while
  $s* \leftarrow s_{rr'}$ // Store Best Solution
end while

In an effort to automate the vehicle routing optimization process, an Open Source Routing Machine engine is used to present the derived routing plan derived by a multi-restart solution framework (Algorithm 1). The proposed framework consists of a greedy randomized construction algorithm for building initial solutions, a Guided Local Search algorithm (GLS) [16] tuned for intensification search and several operators that are used to destruct and reconstruct solutions following a ruin-and-recreate principle. The suggested construction heuristic utilizes the generic sequential insertion-based construction scheme of Solomon [17], while it is enhanced with a probabilistic mechanism similar to greedy randomized adaptive search procedures [18]. Finally, the suggested ruin-and-recreate mechanism is used to restart and diversify the local search process towards promising regions of the solution space.

A simple multilevel architecture is developed offering different services for companies' operators and end-users that involves the portal in which customers could provide their data and the main server in which the optimization algorithm is implemented. The corresponding flowchart (Fig. 1) presents the process for optimizing the vehicle routing of electric LGVs and the main interface of portal is presented in Fig. 2. In more detail, the developed online application follows a sequence of steps as they are presented in Fig. 3. Python scripts are used to parse data step by step and a leaflet.js is connected with an Open Source Routing Machine engine to provide at step 3 a map with the final sequence of customers that each vehicle should serve.

3 Online Application of Vehicle Routing Process

An empirical study is executed and a real data set is provided by a logistics company based in Thessaloniki Region. Six charging stations and 25 customers each with specific operational constraints (i.e. demand and time windows) are considered and the main objective is to minimize the total number of vehicles required and the total journey time of the routing plan. The total available energy of a 'fully charged' electric

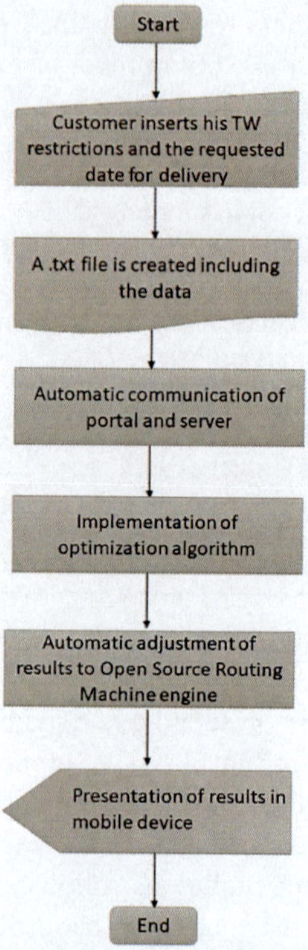

Fig. 1. Flowchart

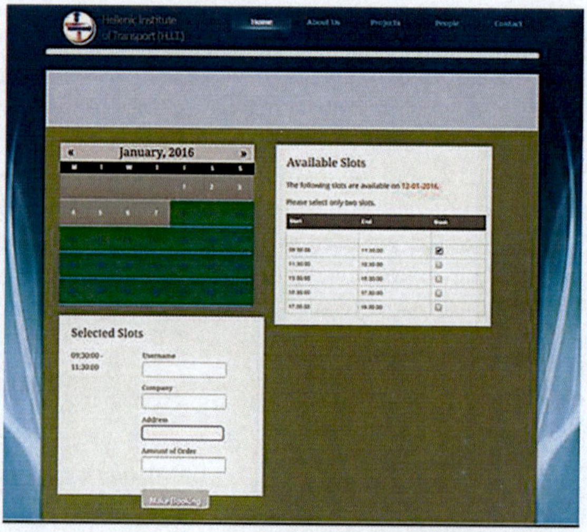

Fig. 2. Main interface of portal

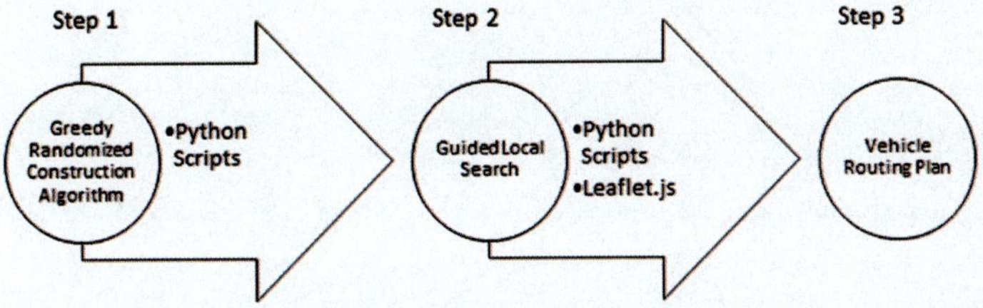

Fig. 3. Steps of implementation

vehicle is 24,000 Wh and consumption rate is defined at 0.2 Wh/m. In addition, the "fast" vehicle charging provides 80% of total energy and the average vehicle speed in the wider urban area is estimated at 52 km/h. The vehicle fleet is assumed to be homogeneous and the maximum capacity of each vehicle is set to 150 units of products. Finally, opening hours are set from 8:00 am to 18:00 pm.

The online application is tested and the final routing plan is depicted in Fig. 4. The experimental results prove the effective implementation of electric vehicles in the urban environment. More specifically, 2 itineraries are required and information on energy consumption and remaining energy for each vehicle are provided in Fig. 5. The vehicle of itinerary 2 needs to visit a fast charging station in order to serve all the assigned customers to it. We observe that the remaining energy in both vehicles is enough for additional routes and vehicles could be used either for servicing more customers within the day or for the next day deliveries.

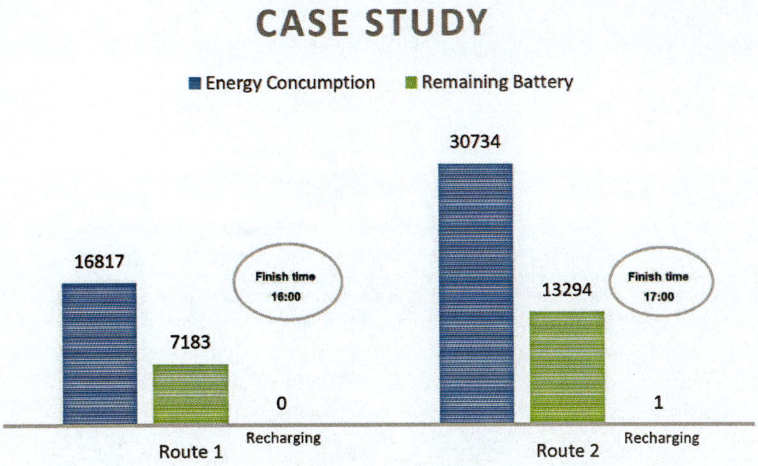

Fig. 4. Routing plan

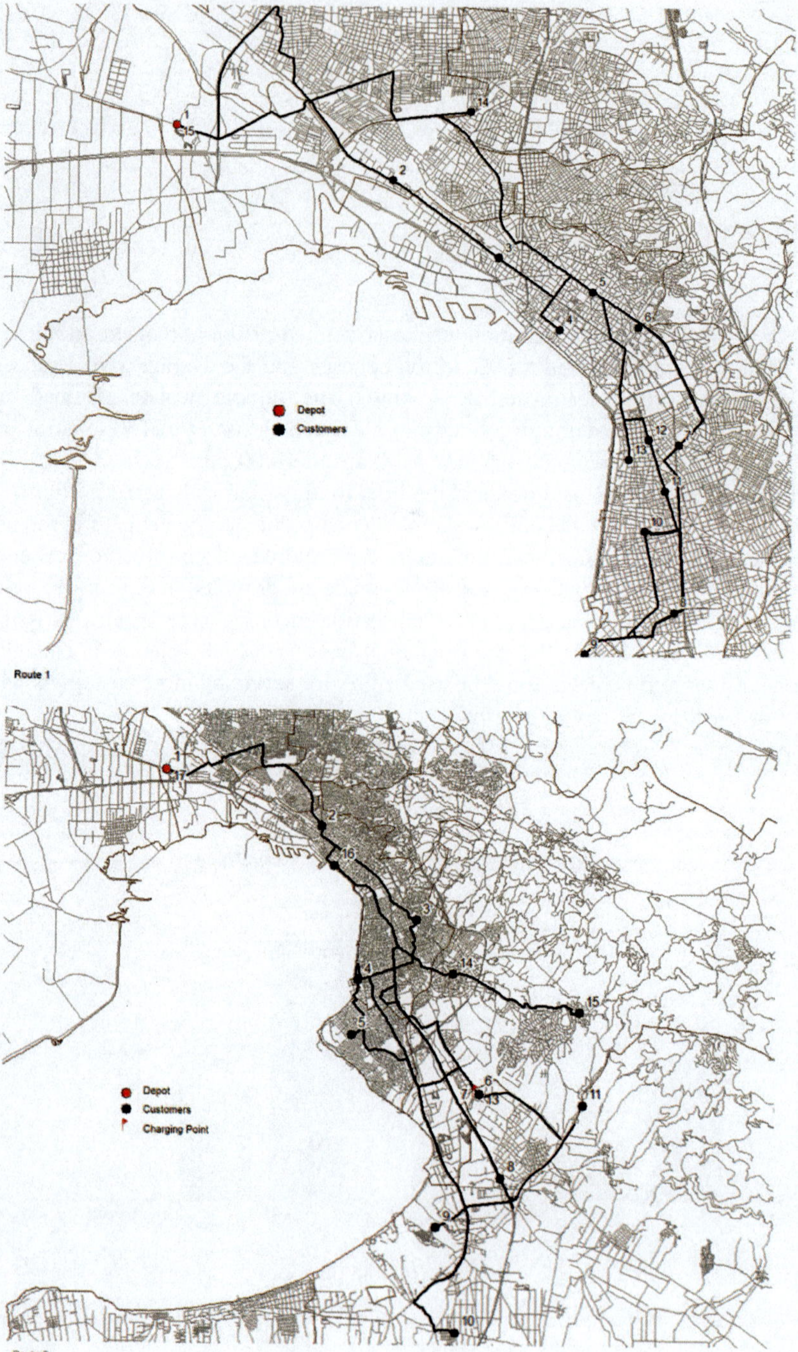

Fig. 5. Final routes

4 Conclusions

The dynamic environment of urban transportation logistics renders the development of more sophisticated solutions that allow real-time communication among company, customers and vehicles. This paper presents how an online application could automate distribution processes and communication between company and customers as well as company and vehicles. This study focuses on using electric LGVs based on realistic data for achieving cost efficiency in terms of fleet size and travel time. A worth-pursuing research direction is the development of extra services that offer dynamic navigation for urban freight distribution considering real-time traffic congestion in the road network and customized customer requests based on their operational needs.

Acknowledgements. This research has been co-financed by the European Regional Development Fund of the European Union and Greek national funds through the Operational Program Competitiveness, Entrepreneurship and Innovation, under the call RESEARCH - CREATE - INNOVATE (project code: T1EDK - 04012).

References

1. COP21. http://www.cop21paris.org/. Accessed 25 Feb 2020
2. COP23. https://cop23.com.fj/. Accessed 25 Feb 2020
3. EU ETS Handbook. https://bit.ly/39CirCR. Accessed 25 Feb 2020
4. Taniguchi, E., Thompson, R.G., Yamada, T.: New opportunities and challenges for city logistics. Transp. Res. Procedia **12**(12), 1 (2016)
5. Letnik, T., Marksel, M., Luppino, G., Bardi, A., Božičnik, S.: Review of policies and measures for sustainable and energy efficient urban transport. Energy **163**, 245–257 (2018)
6. PIARC Report, Public Sector Governance of Urban Freight Transport. https://bit.ly/2JEbExY. Accessed 25 Feb 2020
7. Liu, S., Zhang, Y., Liu, Y., Wang, L., Wang, X.V.: An 'Internet of Things' enabled dynamic optimization method for smart vehicles and logistics tasks. J. Clean. Prod. **215**, 806–820 (2019)
8. Kitjacharoenchai, P., Min, B.C., Lee, S.: Two echelon vehicle routing problem with drones in last mile delivery. Int. J. Prod. Econ. **225**, 107598 (2019)
9. Mendes, R.S., Wanner, E.F., Martins, F.V.: Optimization of the vehicle routing problem with demand responsive transport using the NSGA-II algorithm. In: 2016 IEEE 19th International Conference on Intelligent Transportation Systems (ITSC), pp. 2657–2662. IEEE (2016)
10. Nathanail, E., Mitropoulos, L., Karakikes, I., Adamos, G.: Sustainability framework for assessing urban freight transportation measures. Logist. Sustain. Transp. **9**(2), 16–36 (2018)
11. Punel, A., Ermagun, A., Stathopoulos, A.: Studying determinants of crowd-shipping use. Travel Behav. Soc. **12**, 30–40 (2018)
12. Iwan, S., Kijewska, K., Lemke, J.: Analysis of parcel lockers' efficiency as the last mile delivery solution-the results of the research in Poland. Transp. Res. Procedia **12**, 644–655 (2016)

13. Oliveira, C.M.D., Albergaria De Mello Bandeira, R., Vasconcelos Goes, G., Schmitz Gonçalves, D. N., D'Agosto, M.D.A.: Sustainable vehicles-based alternatives in last mile distribution of urban freight transport: a systematic literature review. Sustainability **9**(8), 1324 (2017)
14. Anagnostopoulou, A., Boile, M., Theofanis, S., Sdoukopoulos, E., Margaritis, D.: Electric vehicle routing problem with industry constraints: trends and insights for future research. Transp. Res. Procedia **3**, 452–459 (2014)
15. Erdoğan, S., Miller-Hooks, E.: A green vehicle routing problem. Transp. Res. Part E Logist. Transp. Rev. **48**(1), 100–114 (2012)
16. Voudouris, C., Tsang, E.: Guided local search. Eur. J. Oper. Res. **113**, 80–119 (1998)
17. Solomon, M.M.: Algorithms for the vehicle routing and scheduling problems with time window constraints. Oper. Res. **35**, 254–265 (1987)
18. Feo, T., Resende, M.G.C.: Greedy randomized adaptive search procedures. J. Global Optim. **6**, 109–154 (1995)

Location Planning of Small Consolidation Centers in the City of Volos

Konstantinos Mpogas, Eftihia Nathanail(✉), and Ioannis Karakikes

University of Thessaly, Pedion Areos, 38334 Volos, Greece
kmpogas@gmail.com, enath@uth.gr

Abstract. As urbanization increases by the years the needs of the society, in terms of quantity of urban freight deliveries, become more demanding. Urban freight transport and goods distribution become the main challenges that a city has to address, owing to the impacts that such operations create to traffic, environment and economy. Decision makers and transport operators have to seek into implementing smart solutions towards alleviating traffic congestion, accidents and environmental deterioration, maintaining at the same time a good level of service. The establishment of small Urban Consolidation Centers (UCC) in strategically selected locations, along with the use of flexible and environmentally friendly vehicles for "last mile distribution" can contribute positively towards the city's sustainable mobility. The paper studies the establishment of small UCCs in Volos city and assesses their performance, implementing facility location models and multi-criteria analysis and feasible location models for location/allocation problems. Results show that the operation of two small UCCs saves great costs per mile of freight distribution, while the operation of more than two is not profitable.

Keywords: Urban consolidation centers · Facility location models · P-median · Multi-criteria analysis · WSM · Optimization

1 Introduction

United Nations in their report [1] highlight challenges that societies are facing as urbanization growth increases. While in 1950 just three out of ten people lived in urban centers globally, in 2014 the percentage was 54% and it is predicted that by 2050 it will approach 66%. This trend was explained by Dociu and Dunarintu [2] who stated that urban centers offer opportunities for economic and social development, participation in modern productive activities as well they are key points for research and innovation. As urbanization increases, the quantity of good deliveries in cities and service level become more and more demanding, resulting consequently in continuous increasing traffic demand. The environmental problems caused by freight traffic in big urban centers is a global major issue, as commercial vehicles produce a substantial amount of emitted pollutants [3]. Taking into account that the majority of urban centers are highly dense and a lot of freight deliveries are made during peak hours, traffic conditions in these areas rapidly deteriorate [4].

E. G. Nathanail et al. (Eds.): CSUM2020, AISC 1278, pp. 853–867, 2021.
https://doi.org/10.1007/978-3-030-61075-3_83

Considering the EU predictions [5] that freight transportation will grow by 80% in 2050, it is of vital importance for stakeholders to find and implement sustainable solutions in the field of city logistics. To achieve that Allen et al. [6] noted that urban planners, transport companies and supply chain stakeholders, have to develop and implement policies on:

- Commercial vehicles accessibility in urban centers
- Last mile delivery
- Consolidation and distribution of goods

Business organizations responded to this challenge by consolidating freight. There are many types of freight consolidation. According to a report by Transport for London (TfL) [7], there are two approaches for consolidation solutions solutions in U.K. and internationally; behavioral and physical (Table 1).

Table 1. Consolidation solutions

Behavioral solutions	Physical solutions
Procurement led solutions (Collective & collaborative procurement, Increasing the size of orders, Bunching orders, Nominated carriers)	Urban consolidation centers
Upstream supply chain	Micro-consolidation centers
Click & collect at store	Locker boxes/locker banks
	Pick up drop off (PUDO) parcel shop

The most popular consolidation solution is the use of Urban Consolidation Centers (UCC) [8–10]. Although there are a lot of advantages generated by the operation of UCCs, location is the critical factor for their success, as they have to be attractive and financially efficient [11]. Minimization of costs (land, fleet, taxes), accessibility increase and assurance of timely and reliable services are factors associated to the location planning, and thus shape the sustainability of a UCC.

2 Literature Review

2.1 Urban Consolidation Centers

According to Browne et al. [11] a UCC is a logistics facility which is situated close to the area that it serves, for example, an urban center, an entire town, or a specific site. Long haul Heavy Good Vehicles (HGVs) of various modes arrive at the UCC to unload their cargo. Freights are then sorted, bundled and consolidated into smaller vehicles for distribution. For the last mile delivery, more and more UCCs use small electric and other eco-friendly types of vehicles in order to improve the environmental conditions of the cities. The introduction of such types of freight fleets leads to a decline in greenhouse gas emissions and local air pollution [12], however, it is clear that any such action is more effective, if it is part of a municipality's overall mobility strategy.

A succeed type of UCC, is the concept of Binnenstadservice (BSS) which started operating in April 2009 in Nijmegen. To be attractive it focused on servicing small retail shops without fee for the "basic service" while it offered other paid special services, like storage (small retailers often have not enough space for their products), home deliveries (for large goods), value added logistics including retour logistics (clean waste) and possible e-tailing services [13]. BSS uses electric bicycles and natural gas trucks to deliver the freight at the retail shops with impressive social, economic and environmental impacts, i.e. the number of trucks and the vehicle kilometers decreased, while further reduction is anticipated if more retailers decide to participate in the scheme.

2.2 Urban Micro-Consolidation Centers

A recent trend in the field of urban freight consolidation is to bundling commodities near to delivery points. This strategy is implemented usually with the use of light commercial vehicles - especially electric – or cargo bicycles, and scopes to facilitate the freight distribution in urban centers. As a consequence, the urban area is benefited as the reduction of HGVs has a positive impact on the traffic conditions and the environmental indicators. Due to the limited capacity of the electric freight vehicles and the small number of available sites in the urban centers, UCCs are small-sized while they serve small and dense urban centers. Browne et al. [14] name these types of UCCs as "micro-consolidation centers". An analysis of Janjevic and Ndiaye [15] have separated micro-consolidation initiatives into six different typologies according to the type of logistical platforms and the logistical scheme used for their operation:

- vehicle reception points
- goods reception points
- urban logistics boxes
- urban micro-consolidation centers in combination with suburban centers
- transshipment points in combination with suburban centers
- mobile depots.

An innovative example of urban micro-consolidation centers is the Citylog Project "BentoBox" in Berlin and Turin [16]. The BentoBox gives also a sender the possibility to send a parcel, so it can be used for both the collection and for the delivery of parcels [17]. Particularly in Berlin, its new collecting and distributing system for express courier services offers a high efficient and eco-friendly option to store and distribute different kind of cargo. Collections are possible every time of the week and the box is outdoor and indoor useable like shopping centers and other buildings. The pilot project pointed out a declined car traffic, which minimizes the carbon emissions significantly.

In 2009 a major supplier of stationery and other office supplies to businesses decided to try a new urban distribution concept for the center of London. Instead of using only a UCC which is located in the suburban and HGVs for the urban freight deliveries they chose to locate a micro-consolidation center for further consolidation of goods [18]. Its location was nearby the city center which is the delivery area and was used as a transshipment point for the freights of HGVs. Instead of performing the distribution with HGVs in the center, electric vans and tricycles were used. The analysis of the before – after situation indicated that, although the distance travelled per parcel rose substantially in the distribution area, due to the far smaller load limits of electric vehicles as compared to diesel vans, the total kilometers and the CO_2 emissions per parcel delivered, fell by 20% and 55% respectively.

Another succeed urban micro-consolidation concept is the "La Petite Reine" which started its operation in Paris in 2001 to provide delivery services for carriers and retailers [19]. It used clean delivery vehicles, such as electrically-assisted cargo tricycles and electric vans, adapted to dense urban centers. The tricycles could go anywhere including areas that small vans and other light commercial vehicle could not. These vehicles were zero emission, silent, and agile. Another similar concept in Paris is the Chronopost Concorde [20]. Chronopost Concorde is an innovative organization of parcel deliveries in the 7th and 8th boroughs of Paris using eco-friendly vehicles as well as an Urban Logistics Space (ULS). This facility is located in the Concorde underground parking lot, below the Place de la Concorde and it is connected with a hub outside of Paris. Instead of performing deliveries with big trucks, consolidation takes place into the center of Paris and the last mile deliveries are made by using a fleet of electric vehicles.

2.3 UCC Location Planning

There are many different types of methods for location planning. The most common are mutli-criteria analyses and facility location models. According to Roy [21] when considering a distinct set of alternatives described by certain criteria, there are four different types of analyses that can be performed to provide significant support to decision makers:

- Identify the best alternative or choose a limited set of the best alternatives
- Classification of alternatives
- Classification/grouping of alternative solutions in predetermined expatriate groups
- Identify the main distinctive features, alternatives and perform the analysis based on these characteristics.

Facility location models have been used in different disciplines [22]. [23–26] proposes a classification scheme for three categories of location models; i) continuous location models, ii) network location models and iii) discrete location models. Continuous location models deal with the geometric representation of reality. They consider the problem of location, in the context of finding location sites within a continuous space, without specific points. Network location models face the problem of location, provided that the points of demand and the facilities that supply them, are located only on a predefined network or on a given graph with nodes and links. In the context of this structure, the given data are, demand and routes. Although there is a view that demand can only exist at the nodes of the network, there are models of this form that accept that there is demand anywhere at the links [27]. Discrete location models are those in which facilities and requirements are in distinct positions. These problems therefore determine the location of the facility and distribute the demand points to one or more facilities. So, they are often called location/allocation problems. In the theory of these models according to [28] there are the four main problems:

- P-median problem
- P-center problem
- Uncapacitated facility location problem, and
- Quadratic assignment problem.

There is a number of studies about location planning of UCCs in literature. Awasthi et al. [29] present a multi-criteria decision making framework for urban distribution centers' location planning under a fuzzy environment. As a first step, the criteria for evaluating possible locations for distribution centers are identified. Then, candidate locations are tested against those criteria, while ratings by decision makers for the criteria and the potential locations are provided. For the selection of the most appropriate locations Fuzzy TOPSIS is used to determine aggregate scores for all potential locations. As a final step, sensitivity analysis is conducted to evaluate the influence of criteria weights on final decision.

By definition an urban micro-consolidation center ought to be cost-effective and environmentally friendly. Sopha et al. [30] performed a structured analysis which considers the trade-offs between the total cost and the total CO_2 emission. Their study proposed a spatial analysis for the selection of potential UDC locations, considering infrastructure, proximity to market, land availability, land cost, land slope, and disaster risk. The proposed places were examined through multi-objective mixed-integer linear programming to find the optimal solution, in terms of number and locations of the UDC, in accordance to the total costs and CO_2 emissions.

A research by Ji et al. [31] proposed a local optimization model for urban micro-distribution centers which are connected with other primary consolidation centers. Several Secondary Distribution Centers (SDCs) which mainly do express deliveries, are distributed in locations of the service area to cover all demand points. Their capacity to serve daily volumes is limited while each SDC has a specific model for trucks. Truck loads and mileage are limited too. The suggested model normalizes the volume and weight of consolidation units, taking into account the real road network conditions while it predicts current and future demands. This a mixed integer programming model - with important practical significance - which tries to reach the lowest overall cost.

In discrete facility location problems, the available options of the potential sites where new facilities are to be located are limited. A well-known example is the p-median problem which is that of locating p facilities to minimize the demand weighted average distance between demand nodes and the nearest of the selected facilities. Drożdziel et al. [32] used p-median to find the optimal location for a local urban distribution center and then used "VRP Spreadsheet Solver" software, to optimize both the fleet number and the vehicle routing problem. As a next step, they run the same software for the current distribution center, concluding that the proposed location of the p-median model becomes more suitable for operation of a local distribution center, as the number of fleets and the total vehicle-kilometers are less than the existing.

3 Methodology

The scope of this study is achieved through the following steps. The first step was the identification of the criteria to determine the candidate sites for locating one or more urban micro-consolidation centers. The determined criteria were: i) Proximity to the city center in order to be possible the quicker service for the majority of businesses, ii) Land availability for ensuring the smooth operation of the consolidation process, iii) Accessibility for securing comfortable mobility conditions. The second step was the determination of twelve candidate sites taking into account the above criteria and especially the land availability. Because of the lack of a lot of available sites, was not run a research analysis but a direct choice of them by the authors. Also, it was conducted the identification of those businesses which will be serviced by the urban micro-consolidation centers and their incorporation into a geographical information system, with their precise location coordinates. The third step was the evaluation of the candidate sites through a multi-criteria analysis, running the Weighted Sum Model (WSM). The criteria which were used were: i) Land cost, ii) Proximity to the market, iii) Proximity to the suppliers, iv) Capacity, v) Accessibility, and Environmentally sensitive area. The next step is the calculation of the criteria weights, implementing Analytical Hierarchy Process (AHP) [33]. Pairwise comparison was conducted by a group of experts, who attributed relevant importance of one criterion as compared to another. Consistency of the scoring process was verified through the calculation of the Consistency Ratio (CR), which is the ration of the Consistency Index (CI) and the Random Consistency Index (RI), through the formula:

$$CR = CI/RI \tag{1}$$

For CR values smaller or equal to 10%, the attributed weights are considered consistent. Finally, after the implementation of multi-criteria analysis and the classification of the results, the top three candidates were used as an input to SITATION software to optimize the number and the location of the centers under a location-allocation process, via P-median model. For running the location-allocation problem were essential two sites at least, but it was noticed that the second and third ranked sites of WSM, were very close for both normalization types were used in order to run the model, so were chosen three sites.

4 The Case of Volos

4.1 Purpose of Study

Purpose of this study is to design an integrated consolidation system which is based on a main UCC located in the wider urban area, connected to micro-consolidation points operating in the city. The implementation of the system is being tested in the city of Volos. A thorough description of urban consolidation centers is conducted, emphasizing in the cooperation between them in the structure of hob-spoke system. Methods implemented in literature are presented. Further, the integrated system design is detailed and various assessment methodologies are implemented. Finally, conclusions about the alternative designs performance are drawn and a comparison among the various methodologies is discussed. In this direction, we assume that a UCC is located in the industrial zone of Volos. The freight, which will be delivered to businesses in the central business district, will be consolidated at the UCC, forwarded to the urban micro-consolidation center closest to the businesses, and finally consigned to the final destination. The goods are assumed to be distributed in the urban area by eco friendly vehicles. The problem to be solved concerns a location-allocation problem, which defines the number, location, and demand points or the businesses within the jurisdiction of the micro-consolidation centers. Multi-criteria analysis was used to assess the performance of the system.

4.2 Study Area

Volos is a city located in central Greece with a population of 125.248 residents in 2011 [34]. The local economy primarily depends on commerce, services and tourism. There are two industrial areas at the west and northwest city entrance just a few kilometers away from the urban center. These areas include many steel manufactures, metal recycling, steel wire ropes production, and lime manufactures, while 3 km eastern of the city there is a large cement production industry with cargo ship terminal. In the city center both a passenger and cargo port is located, which is the third largest in Greece. Additionally, the city of Volos is served by the "Nea Anchialos" National Airport, which started its operation in 2010 and is located 26 km from Volos. Finally, the city is connected with the railway network of trainOSE.

4.3 Data

An extended analysis of the official data of Magnesia chamber [35] indicated that 1,962 retailers operate in Volos, providing food, drinks, medicines, cosmetics and food/drink service businesses (Table 2). These businesses are to be served by the micro-consolidation centers (Fig. 1).

Table 2. Number of business in Volos city per type.

Books/Printing	Groceries	Clothes/ Footwear/ Fabric	Electronic s/m size	Cosmetics/ Perfumes/ Pharmacies	Domestic use	Watches jewelry eyewear	Food/drinks
55	540	339	60	35	70	79	794

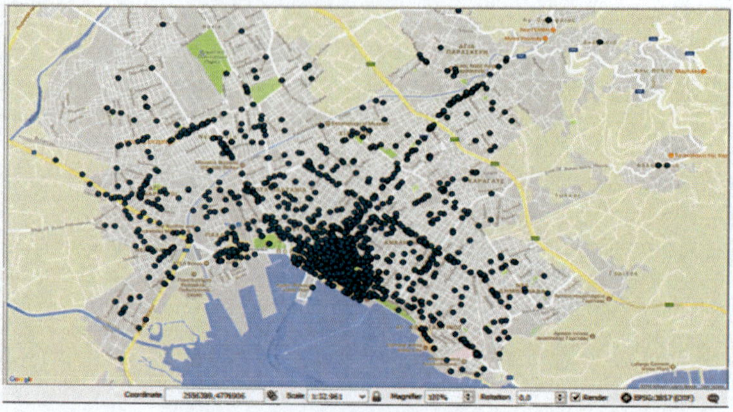

Fig. 1. Illustration of Volos' businesses in QGIS environment.

Another data set that was taken into account was the daily (or weekly) freight flows that businesses necessitate. This data set was collected through a survey focusing on freight distribution in Volos [36].

Businesses were grouped based on their commercial type. For every group, average weekly flows were estimated, along with the type of vehicles used to deliver them. Given that there are several types of light commercial vehicles (LCVs) and taking into account that as LCV in European Union is considered a van with a gross vehicle weight of no more than 3.5 metric tons (tonnes), an average type was determined with a representative load capacity. It has been calculated that a 3.5t van has a 1t payload [37], and taking into account that in car market there are LCVs with max payload of almost 500 kg, it was assumed for this study's average type vehicle, a payload of 600 kg. Based on Cruijssen's [38] estimate that the average loading factor of freight minivan vehicles is 56%, the assumption for this study was that delivery vehicles deliver about 300 kg per trip. Through the multiplication of the average weekly flows from the survey by the average payload of each delivery was calculated the weekly flow of each business in Volos, so by dividing this outcome by 7, the final daily flow of each business was estimated to 1070 tones. Due to the limitation of the overall number of business which can used as an input in SITATION Software, a zoning system was formulated based on the transportation axes that accommodate traffic in the central business district area. Therefore, instead of businesses was used a limited number of zone areas as an input for representing the demand points of the p-median model. The final step was to calculate the and the total tonnage to be delivered per zone and their centroids' coordinates in order to be specified as the demand points (Fig. 2). It is noted that the zones where formulated based on the road network design, so the number of businesses in every zone are different.

The six criteria which were considered in the multicriteria evaluation methods, were weighted from a group of four transportation experts all being stakeholders of the city of Volos. Based on the pairwise comparison, Table 3 was formulated.

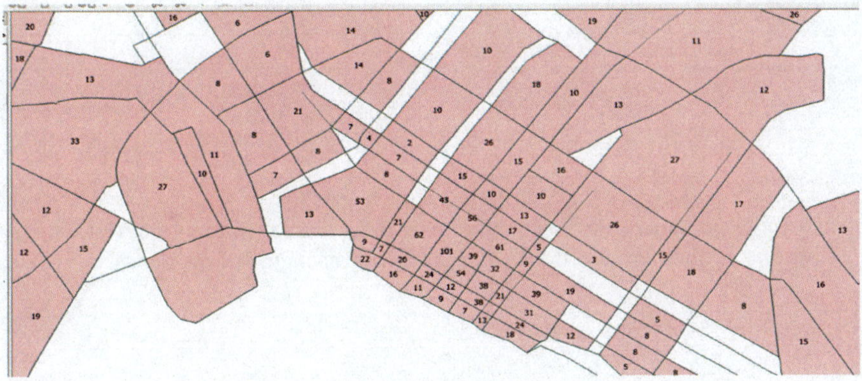

Fig. 2. Zones.

Table 3. Criteria comparison matrix.

Criteria	Land cost	Proximity to the market	Proximity to suppliers	Capacity	Accessibility	Env/ly sensitive area
Land cost	1,00	0,50	4,00	3,00	1,00	3,00
Proximity to the market	2,00	1,00	5,00	4,00	2,00	4,00
Proximity to suppliers	0,25	0,20	1,00	0,33	0,20	0,50
Capacity	0,33	0,25	3,00	1,00	0,33	2,00
Accessibility	1,00	0,50	5,00	3,00	1,00	3,00
Env/ly sensitive area	0,33	0,25	2,00	0,50	0,33	1,00
Sum	4,92	2,70	20,00	11,83	4,87	13,50

The values of criterion "Land cost" for each alternative location were calculated as the product of the zone price, the structure factor and the facade coefficient. The "Proximity to the market" values resulted from the QGIS heatmap. The heatmap estimated how close a candidate location is to every business of Volos via the "Kernel Density Estimator" (KDE) method [39]. The "Proximity to suppliers" values were estimated based on the distance between the UCC and each micro-consolidation center. The "Capacity" of each alternative was determined by the available area of each location, while the "Accessibility" resulted quantitively from the expert team, based on the urban mobility conditions. Finally, for the criterion "Environmentally sensitive area" each location was rated qualitatively through a 1–5 scale with 1 corresponding to a Green park and 5 to an Appropriate location, as depicted in Table 4 [40].

Table 4. Quality scale for "Environmentally sensitive area" criterion

1	2	3	4	5
Green park	square	Close (<50 m)	Far (>50 m)	Appropriate locations (ports, parking buildings, rail stations)

The twelve candidate sites were introduced in QGIS Software in order to estimate their coordinates and they are shown in Fig. 3 with background the zones.

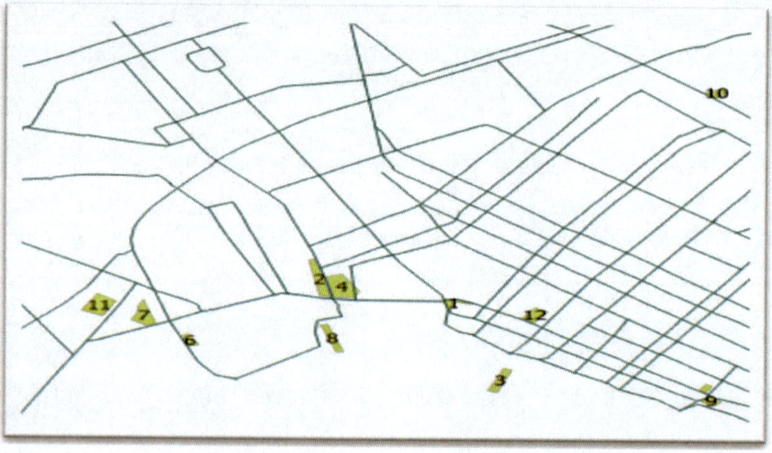

Fig. 3. Twelve candidate sites.

The first location that is proposed is the parking lot located at the point where are connected the streets Iasonos, Dimitriados, Lambraki and November 2nd. The second one is a part of the privately owned space of OSE that has an entrance on both Lambraki and Papadiamantis streets. The third location proposed is a place of the passenger terminal at the port of Volos. The fourth option for the installation of the micro-consolidation center is part of Riga Fereou Square. The fifth possible choice is the location where a multi-storey car park is already under construction at the junction of Dimitriados and Filellinon streets. The sixth alternative proposed is the plot behind the KTEL of Volos which is enclosed by the local streets Skra, Zachou and Almyrou. The seventh possible place is located on Sekeri Street opposite the building of the Department of Urban and Spatial Planning, and Regional Development. The eighth alternative analyzed is the commercial port of Volos while the ninth is the park of Agios Konstantinos and specifically the area where recreational areas (cafes) operate. The tenth and eleventh possible areas are a part of Chiliadou Park and a plot of land which is located on Athens Avenue correspondingly. The last alternative is Panepistimiou Square, which is located on Dimitriados Street.

5 Results

5.1 Multicriteria Evaluation of Micro-Consolidation Centers Location

Table 5 presents the calculated weights of the six criteria. It is noted that their CR was 3%, which testifies the consistency in experts' scoring.

Table 5. Criteria weights.

Criteria	1	2	3	4	5	6	Weights
1	0,20	0,19	0,20	0,25	0,21	0,22	21,16%
2	0,41	0,37	0,25	0,34	0,41	0,30	34,54%
3	0,05	0,07	0,05	0,03	0,04	0,04	4,69%
4	0,07	0,09	0,15	0,08	0,07	0,15	10,19%
5	0,20	0,19	0,25	0,25	0,21	0,22	22,00%
6	0,07	0,09	0,10	0,04	0,07	0,07	7,42%

Table 6 includes the values of the criteria for each of the 12 alternatives.

Table 6. Criteria values for the micro-consolidation centers.

Location	Criteria					
	Quantitive values				Qualitive values	
	Land cost	Proximity to the market	Proximity to the suppliers	Capacity	Environmentally sensitive area	Accessibility
	Cost (€/m^2)	Index	Minutes	Sq. meters		
1	3499,2	976	7	600	4	2
2	1296	738	5	2100	3	3
3	4185	1015	8	2750	4	4
4	3061,8	738	5	3600	1	3
5	3499,2	958	8	720	3	2
6	1296	303	5	540	4	4
7	777,6	303	3	2100	4	3
8	1200	738	5	7800	5	4
9	4957,2	958	8	900	1	2
10	1846,8	479	11	900	1	2
11	1296	212	5	7800	4	4
12	3499,2	1130	7	1440	2	4

Furthermore, values of Table 6 were normalized based on Eqs. 2 and 3 and weighted based on the weights of Table 5. The weighted normalized values are presented in Table 7.

$$x_{i,m,j} = \frac{X_{i,m,j}}{\max\limits_{j} X_{i,m,j}}, \quad x_{i,m,j} = \frac{\min\limits_{j} X_{i,m,j}}{X_{i,m,j}}, \tag{2}$$

$$x_{i,m,j} = 1 - \frac{\max\limits_{j} X_{i,m,j} - X_{i,m,j}}{\max\limits_{j} X_{i,m,j} - \min\limits_{j} X_{i,m,j}}, \quad x_{i,m,j} = 1 - \frac{X_{i,m,j} - \min\limits_{j} X_{i,m,j}}{\max\limits_{j} X_{i,m,j} - \min\limits_{j} X_{i,m,j}}, \tag{3}$$

Where $x_{i,m,j}$ represents the normalized value of each alternative for every criterion and $X_{i,m,j}$ represents the quantitive or qualitive value of each alternative for every criterion.

Table 7. WSM results.

Location	Max/Min normalization		Min Max/Max Min normalization	
	WSM	Rank	WSM	Rank
1	0,54	7	0,44	9
2	0,62	4	0,59	5
3	0,68	2	0,67	3
4	0,53	8	0,48	7
5	0,52	10	0,41	10
6	0,53	9	0,53	6
7	0,60	5	0,48	8
8	0,79	1	0,82	1
9	0,48	11	0,30	11
10	0,38	12	0,26	12
11	0,60	6	0,60	4
12	0,68	3	0,69	2

5.2 Location – Allocation

In order to optimize the location/allocation problem of establishing urban micro-consolidation centers in Volos, the top three rated centers were used as service points. These are alternatives 3, 8 and 12. Other input data in the model were the businesses grouped 105 clusters, comprising the demand, represented in the mathematical network as the respective areas' centroids and the demand in tones for each one of them. Seven scenaria were formulated assuming operation of one, two and three micro-consolidation centers. In each scenario, the centers' location and areas of jurisdiction were determined, through the solution of P-median problem. Demand is completely satisfied in all scenaria. Table 8 shows that the best solutions for servicing the businesses are scenario 4 and 7, with sites 3,8,12 and 8, 12, respectively. As the number of established micro-consolidation centers is higher in scenario 4 than in scenario 7, the associated cost is higher in the first scenario, as compared to the latter. Therefore, scenario 7 is preferred. The assignment of businesses to a center of the optimal solution is presented in Fig. 4.

Table 8. Scenarios solved via P-median model.

Solutions	Alternatives	Average distance	Cost per mile	Total covered demand (tones)
Scenario 1	3	0.477	510.2	1070
Scenario 2	8	0.744	795.9	1070
Scenario 3	12	0.432	462.8	1070
Scenario 4	3–8–12	0.361	386.5	1070
Scenario 5	3–8	0.387	414.8	1070
Scenario 6	3–12	0.432	462.8	1070
Scenario 7	8–12	0.361	386.5	1070

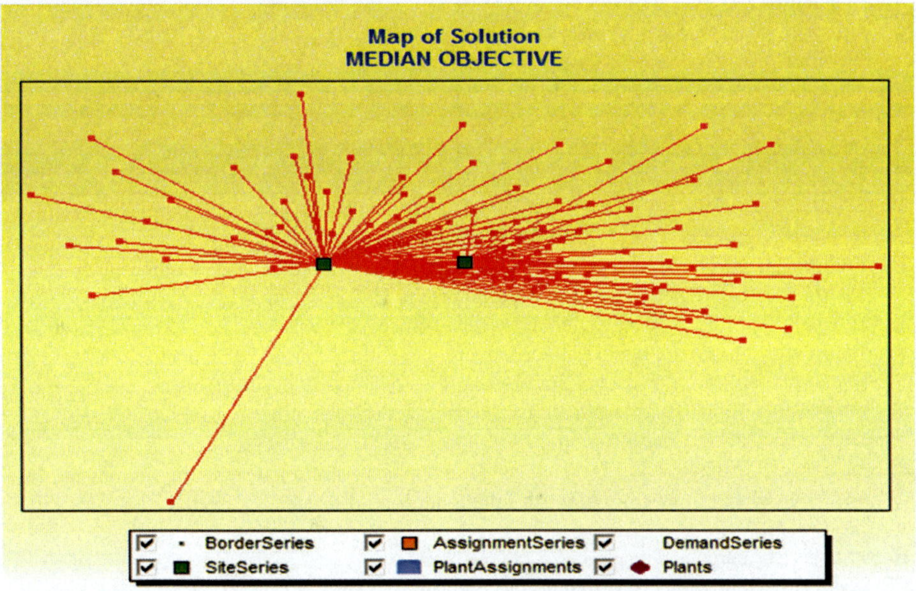

Fig. 4. Map of optimum scenario.

6 Conclusions

The use of p-median model revealed both the optimal location(s) and operating number of urban micro-consolidation centers. A useful conclusion that can be drawn is that a higher number of urban micro-consolidation centers, is not necessarily the optimal choice as total costs and transport costs can significantly be increased. This can be easily seen in this study, through scenarios 4 and 7. Both scenarios cover 100% of the demand, have the same costs per mile and average covered distance, however scenario 7 assumes three urban micro-consolidation centers whereas scenario 4 assumes two, thus the first scenario is preferred over the second.

References

1. United Nations, Department of Economic and Social Affairs, Population Division: World Urbanization Prospects: The 2014 Revision, Highlights (ST/ESA/SER.A/352) (2014)
2. Dociu, M., Dunarintu, A.: The Socio-Economic Impact of Urbanization. Int. J. Acad. Res. Account. Finan. Manag. Sci. **2**(1), 47–52 (2012)
3. Sonnabend P.: Logistics Innovation for the Urban Domain, BESTUFS-II Final Conference, Athens 11–13 June (2008)
4. Patier, D., David, B., Chalon, R., Deslandres, V.: A new concept for urban logistics delivery area booking. Procedia Soc. Behav. Sci. **125**, 99–110 (2014)
5. EU Publications, Connecting Europe's citizens and businesses (2014)
6. Allen, J., Thorne, G., Browne, M.: BESTUFS good practice guide on urban freight transport. Manual, BESTUFS (2007)
7. https://tfl.gov.uk/corporate/publications-and-reports/freight
8. Chwesiuk, K., Kijewska, K., Iwan, S.: Urban consolidation centres for medium-size touristic cities in the Westpomeranian Region of Poland. Procedia Soc. Behav. Sci. **2**, 6264–6273 (2010)
9. Van Duin, J.H.R., Muñuzuri, J.: Urban Distribution Centers. In: Taniguchi, E., Thompson, R.G. (eds.) City Logistics: Mapping the Future, pp. 115–134 (2015)
10. Gogas, M.A., Nathanail, E.: Evaluation of urban consolidation centers: a methodological framework. Procedia Eng. **178**, 461–471 (2017)
11. Browne, M., Sweet, M., Woodburn, A., Allen, J.: Urban freight consolidation centres: Final report. Technical report, Transport Studies Group, University of Westminster (2005)
12. Foltyński, M.: Electric fleets in urban logistics. Procedia Soc. Behav. Sci. **151**, 48–59 (2014)
13. van Rooijen, T., Quak, H.: Local impacts of a new urban consolidation centre – the case of Binnenstadservice. nl. Procedia Soc. Behav. Sci. **2**, 5967–5979 (2010)
14. Browne, M., Allen, J., Leonardi, J.: Evaluating the use of an urban consolidation centre and electric vehicles in central London. IATSS Res. **35**(1), 1–6 (2011)
15. Janjevic, M., Ndiaye, A.B.: Development and application of a transferability framework for micro-consolidation schemes in urban freight transport. Procedia Soc. Behav. Sci. **125**, 284–296 (2014)
16. https://cordis.europa.eu/result/rcn/58755_en.html
17. Quak, H.J.: Improving urban freight transport sustainability by carriers – Best practices from The Netherlands and the EU project CityLog. Procedia Soc. Behav. Sci. **39**, 158–171 (2012)
18. Leonardi, J., Browne, M., Allen, J.: Before-after assessment of a logistics trial with clean urban freight vehicles: a case study in London. Procedia Soc. Behav. Sci. **39**, 146–157 (2012)
19. http://www.bestfact.net/wp-content/uploads/2016/01/CL1_091_QuickInfo_La-Petite-Reine-16Dec2015.pdf
20. TURBLOG. Transferability of urban logistics concepts and practices from a world wide perspective - Deliverable 3.1 - Urban logistics practices – Paris Case study (2011)
21. Roy, B.: Méthodologie Multicritère d'Aide à la Décision. Economica, Paris (1985)

22. Aikens, C.H.: Facility location models for distribution planning. Eur. J. Oper. Res. **22**, 263–279 (1985)
23. Lee, S.M., Green, G.I., Kim, C.: A multiple criteria model for the location–allocation problem. Comput. Oper. Res. **8**, 1–8 (1981)
24. Puerto, J., Fernandez, F.R.: Multicriteria mini-sum facility location problems. J. Multicriteria Decis. Anal. **8**, 268–280 (1999)
25. Ross, G.T., Soland, R.M.: A multicriteria approach to location of public facilities. Eur. J. Oper. Res. **4**, 307–321 (1980)
26. Hamacher, H.W., Nickel, S.: Classification of location models. Location Sci. **6**, 229–242 (1998)
27. Daskin, M.S.: Network and Discrete Location: Models, Algorithms and Applications. Wiley, Hoboken (2013)
28. Mirchandani, P.B., Francis, R.: Discrete Location Theory. Wiley-Interscience Series in Discrete Mathematics and Optimization. John Wiley & Sons, Inc., New York (1990)
29. Awasthi, A., Chauhan, S.S., Goyal, S.K.: A multi-criteria decision making approach for location planning for urban distribution centers under uncertainty. Math. Comput. Model. **53**, 98–109 (2011)
30. Sopha, B.M., Asih, A.M.S., Pradana, F.D., Gunawan, H.E., Karuniawati, Y.: Urban distribution center location: combination of spatial analysis and multi-objective mixed-integer linear programming. Int. J. Eng. Bus. Manag. **8**, 1–10 (2016)
31. Ji, Y., Yang, H., Zhang, Y., Zhong, W.: Location optimization model of regional express distribution center. Procedia Soc. Behav. Sci. **96**, 1008–1013 (2013)
32. Droździel, P., Wińska, M., Madleňák, R., Szumski, P.: Optimization of the post logistics network and location of the local distribution center in selected area of the Lublin province. Procedia Eng. **192**, 130–135 (2017)
33. Saaty, T.L.: Decision making with the analytic hierarchy process. Int. J. Serv. Sci. **1**(1), 83–98 (2008)
34. Hellenic Statistical Authority (ELSTAT), on line reports at http://www.statistics.gr
35. Chamber of Magnesia. http://www.cci-magnesia.gr/en/home
36. Hatzistamatis, N., Nathanail, E., Karakikes, I.: Can night deliveries actually work in a medium size Greek city? In: 9th International Congress on Transportation Research, Athens, Greece (2019)
37. Browne, M., Rizet, C., Allen, J.: A comparative assessment of the light goods vehicle fleet and the scope to reduce its CO2 emissions in the UK and France. Procedia Soc. Behav. Sci. **125**, 334–344 (2014)
38. Cruijssen F.: Collaboration concepts for co modality (CO^3) (2012). http://www.co3-project.eu/
39. Fleming, C.H., Calabrese, J.M.: A new kernel density estimator for accurate home-range and species-range area estimation. Methods Ecol. Evol. **8**, 571–579 (2017). https://doi.org/10.1111/2041-210X.12673
40. Nathanail, E.: Developing an integrated logistics terminal network in the CADSES. Trans. Stud. Rev. **14**(1), 125–146 (2007)

Human Factors

Sustainable Travel Behavior and Perspectives on the Daily Commute – A Questionnaire Survey in a Typical Mid-Sized Greek City

George Botzoris[1](✉), Athanasios Galanis[2], Panagiotis Lemonakis[3], and Maria Giannopoulou[1]

[1] Department of Civil Engineering, Democritus University of Thrace, Kimmeria Campus, 67100 Xanthi, Greece
{gbotzori,mgian}@civil.duth.gr

[2] Department of Civil Engineering, International Hellenic University, Terma Magnesias, 62124 Serres, Greece
atgalanis@ihu.gr

[3] Department of Civil Engineering, University of Thessaly, Pedion Areos, 38334 Volos, Greece
plemonak@uth.gr

Abstract. This study examines the relationship between commuters' travel behavior and sustainable transport modes infrastructure in the city of Volos (a mid-sized Greek city) during the years of economic crisis (2008–2017). The scope of this survey was to study how the travel behavior has changed during the examined years and if there is a preference to sustainable transportation. The study was based on a questionnaire survey conducted on a random sample of 293 individuals through personal interviews in the year 2017. The questionnaire was formed of 32 questions divided into four parts. This study supports that citizens have changed their travel behavior in the city during the examined years, favoring the use of sustainable transport modes. The recovery of Greek economy from economic crisis should not be an excuse to increase use of private vehicles but an opportunity to change commuters travel behavior in urban areas and promote sustainable transport modes with relative societal, environmental and economic benefits.

Keywords: Sustainability · Transportation · Planning · Demand · Urban · Survey

1 Introduction

Economic crisis in Greece has affected the daily habits of citizens. Due to that fact, citizens attempt to reduce the transport cost for their daily utilitarian and recreational trips both in urban and suburban areas. Commuters attempt to increase the use of sustainable transport modes (walk, bicycle, public transport) or increase teleworking because they are willing to reduce the necessary cost for transport. Residents in urban areas which offer well-organized and high quality public transport drive half as many

E. G. Nathanail et al. (Eds.): CSUM2020, AISC 1278, pp. 871–881, 2021.
https://doi.org/10.1007/978-3-030-61075-3_84

annual kilometers; own half as many private cars, use public transport, walk and bike much more than residents in car-dependent communities [1, 2].

Well-organized and high quality public transportation systems may affect travel activity in ways that provide health benefits, including reduced traffic accidents and pollution emissions, and increased affordability which can reduce financial stress to households with lower income [3]. The sustainability of transport systems is a very important activity to be addressed and a growing number of initiatives around the world focus on the definition and measurement of sustainability in transportation planning and infrastructure provision [4].

This study examines the relationship between commuters' travel behavior and sustainable transport modes infrastructure in the city of Volos (mid-sized Greek city) during the years of economic crisis in Greece (2008–2017) presenting results about their travel behavior and perspectives on the daily commute. Relative studies have examined this relationship in selected Greek cities [5–7]. Furthermore, this study presents an evaluation of the roundabouts in the city. Improvement of road infrastructure and construction of roundabouts can present many benefits for the commuters.

2 Literature Review

2.1 Sustainable Transportation: Walking

Walking is a sustainable transport mode available to everybody without cost. Citizens will prefer to increase walking in urban areas for short distance trips if the built environment for pedestrians is safer and more convenient. Cities that favor walking provide more benefits for their citizens such as a higher level of personal safety and road safety, selection of multiple transport modes and better accessibility to destinations.

The benefits of walking promotion for urban and regional level are numerous. Pedestrians do not pollute the air, make transport noise or occupy large area of urban space. Preference to walk in urban areas and pedestrians' behavior depends on many factors [8]. Shay et al. proposed two groups of factors that can influence walking: motivation and ability [9]. The "motivation" factors relate to social or personal characteristics. The motivation factors can be operational only with the presence of the "ability" factors in order to increase walking among citizens.

There are also major factors that affect a commuter in order to decide to reach his destination travelling on foot such as travel time and distance [10]. Pedestrians walking speed is low; affecting the travel distance which is usually limited to 1–2 km. Factors such as the value of time and the personal image are usually the critical ones for a citizen's choice to walk. Especially, professionals with higher income are not willing to spare time selecting to walk or use public transportation. Furthermore, one additional factor for citizens to increase walking is personal safety [11]. Women usually avoid walking during night time and select other transport modes or even choose not to travel at all. Finally, parents are reluctant to encourage their children to walk because they consider that face road and personal safety issues [12].

2.2 Sustainable Transportation: Bicycling

Bicyclists are vulnerable road users because they face higher risk for their road safety when they travel in the streets. Bicycle is a sustainable transport mode and commuters will prefer to increase bicycling if the built environment is safer and more convenient, mainly for urban trips of short or medium distance. The road infrastructure for bicycling differs among cities of a country or among districts of an urban agglomeration. Many differences are recognized related with cultural, geographical and economic factors. The built environment differs between the neighborhoods, the central business district and other commercial-industrial areas of a city. It is necessary for the commuters to bike with safety and convenience in the entire urban area and travel to all destinations. Municipality officials and stakeholder can cooperate in order to improve the bikeability level of their city.

Researchers have focused on route quality issues in order to promote bicycling, such as traffic conditions, bicycle and vehicle lane design, signalization, slope, curb and surface conditions, safety, lighting, weather, environmental factors and accessibility to land uses [13, 14].

Many methodologies and instruments are developed in order to identify the bikeability level of the built environment. The measurement of the level of service or similar indices, and the assessment of bicyclists' safety level based on route-related variables is a type of instrument. Furthermore, other methodologies examined the bicycling level of service on road segments in order to provide an index of the accommodation level of urban and rural roads to bicycle travel [15]. The reliability and validity of these instruments is not fully examined [16]. Small number of studies examined environmental characteristics related with bikeability, like comfort, satisfaction, and level of stress [17]. Some of the main barriers for bicyclists are the insufficient or unsafe bicycling infrastructure, motorists' traffic flow and speed, shortage of bicycling amenities, and undesirable conditions of land use [18–22].

3 Questionnaire Survey

3.1 Methodology and Data Collection

The survey took place in the city of Volos, a mid-sized coastal port city, sixth most populous of Greece, situated midway on the Greek mainland. Volos is a city with tradition in bicycle use for daily urban trips. Its flat terrain and the existence of the port present many benefits for the commuters, since they can travel across the seaside, from one part of the city to the other, without mixing with motorized transport modes.

The survey was based on a questionnaire survey conducted during the year 2017 on a sample of 293 individuals. For the conduction of this survey we selected a sample of participants randomly and as representatively as possible, based on their demographic data. In this survey, the participants answered anonymously and this encouraged them to answer more freely in questions regarding their personal income. Furthermore, this survey was conducted with personal interviews and that helped the participants to receive clarifications if necessary.

The required sample size (n) was calculated based on the equation [23, 24]:

$$n \geq N \cdot \left[1 + \frac{N-1}{p \cdot (1-p)} \cdot \left(\frac{d}{z_{a/2}} \right)^2 \right]^{-1} \quad (1)$$

where:

n: the number of the sample of a survey.

N: the number of overall population.

d: the acceptable margin of error (or confidence interval), it represents how close the answers of the sample reflect the views of overall population. The smaller the margin of error is, the most confidence one should have that the results of the survey reflect the entire population (in our research ± 5%).

$z_{\alpha/2}$: parameter, expressed as a percentage, which represents how confident we can be that the sample of a survey represents the overall population within the acceptable margin of error (in our research $z_{\alpha/2}$ = 1.645 for a confidence level of 90%).

p: the proportion of individuals (with a specific characteristic) falling into the group for which we are interested within the overall population. If no previous experience exists, then a value p = 50% is considered as the worst case.

Based on Eq. (1) and due to the fact that the population of the greater area of the city of Volos is around 125,000 inhabitants, the minimum number of questionnaires to be completed was calculated equal to 271 (n = 271).

The questionnaire was formed of 32 questions divided into four parts. In the first part (questions 1–8), the participants stated their demographic data, possession of private vehicles and bicycles, and the level of their personal income. In the second part (questions 9–21), the participants stated the usage level of the examined transport modes, road safety and personal safety issues in the city, between the years 2008 and 2017. In the third part (questions 22–29), the participants answered questions about the presence of pedestrian and bicycle infrastructure in the city, the usage level, the necessity for improvement and expansion of the network, and their willingness to financially support relative projects. In the fourth part (questions 30–32), the participants evaluated the roundabouts in the city.

After the data collection was completed, followed the data analysis and export of results and conclusions. The data were analyzed with the use of the IBM SPSS® Statistics and the analysis was based on the descriptive statistics, derived by the sample, in order to describe the basic features of the participants concerning their travel behavior and their opinion about the sustainable transport infrastructure in the city of Volos.

3.2 Results

The results of the survey are presented in the following Tables 1, 2, 3, 4, 5, 6, 7, 8, 9 and 10. More specific, in Table 1 the demographic data of the survey are presented, where as in Table 2 is presented the participants' level of income comparing the years 2008 and 2017. In Table 3 is presented the possession of private vehicles and bicycles comparing the years 2008 and 2017, and in Table 4 the use of transport modes comparing the years 2008 and 2017. In Table 5 is presented the reduction of private vehicle use and willingness to change with another transport mode comparing the years 2008 and 2017, and in Table 6 the reasons to increase walking and cycling comparing the examined years. In Table 7 the perceived level of pedestrians and bicyclists' road and personal safety comparing the examined years is presented. Finally, in Tables 8, 9 and 10 are presented the evaluation of the bikeway network, the pedestrian infrastructure, and the roundabouts' functionality in the city of Volos.

Table 1. Demographic data (questions 1 to 4).

Question 1: Age the year 2017 (%)		Question 3: Profession (%)		Question 4: Education (%)	
				Primary school	3.8%
≤29	21.5%	Public servant	21.5%	Secondary school	19.5%
30-39	29.4%	Private employee	35.2%	High school	36.5%
40-49	26.3%	Self-occupant	19.5%	University	40.2%
50-59	12.6%	Unemployed	13.3%	Question 2: Sex (%)	
60-69	10.2%	Retired	10.5%	Male	53.2%
				Female	46.8%

Table 2. Level of personal income comparing the years 2008 and 2017 (questions 5a and 5b).

Question 5a: Personal income in the year 2008 (before taxes)				
<5,000€	5,001€–10,000€	10,001€–15,000€	15,001€–20,000€	>20,000€
35.4%	18.9%	18.7%	15.9%	11.1%
Question 5b: Personal income in the year 2017 (before taxes)				
<5,000€	5,001€–10,000€	10,001€–15,000€	15,001€–20,000€	>20,000€
49.7%	22.6%	16.2%	8.4%	3.1%

Table 3. Possession of private vehicles and bicycles comparing the years 2008 and 2017 (questions 6 to 8).

Question 6: Possession of private vehicle				
Year 2008:		90.1%	Year 2017:	83.3%
Question 7: Possession of private vehicles per family				
Number	0	1	2	>2
Year 2008:	8.5%	52.6%	29.7%	9.2%
Year 2017:	13.3%	57.7%	23.9%	5.1%
Question 8: Possession of bicycle				
Year 2008:		50.9%	Year 2017:	65.9%

Table 4. Use of transport modes comparing the years 2008 and 2017 (questions 9 to 13).

Question 9a: Use of private vehicles for utilitarian trips

Year	Almost never	Rarely	Sometimes	Often	Almost always
2008	22.84%	13.58%	20.37%	21.60%	21.60%
2017	18.52%	15.43%	20.99%	22.22%	22.84%

Question 9b: Use of private vehicles for recreational trips

Year	Almost never	Rarely	Sometimes	Often	Almost always
2008	20.37%	16.05%	20.99%	27.78%	14.81%
2017	19.14%	21.60%	30.86%	22.22%	6.17%

Question 10a: Use of public transportation for utilitarian trips

Year	Almost never	Rarely	Sometimes	Often	Almost always
2008	31.48%	27.16%	17.90%	12.35%	11.11%
2017	26.54%	22.22%	20.37%	13.58%	17.28%

Question 10b: Use of public transportation for recreational trips

Year	Almost never	Rarely	Sometimes	Often	Almost always
2008	29.63%	22.22%	21.60%	16.05%	10.49%
2017	20.37%	22.22%	23.46%	20.99%	12.96%

Question 11a: Use of taxi for utilitarian trips

Year	Almost never	Rarely	Sometimes	Often	Almost always
2008	33.95%	16.05%	18.52%	22.84%	8.64%
2017	37.04%	19.75%	18.52%	14.81%	9.88%

Question 11b: Use of taxi for recreational trips

Year	Almost never	Rarely	Sometimes	Often	Almost always
2008	35.80%	18.52%	16.67%	17.28%	11.73%
2017	35.19%	19.75%	19.14%	17.90%	8.02%

Question 12a: Bicycling for utilitarian trips

Year	Almost never	Rarely	Sometimes	Often	Almost always
2008	38.89%	12.96%	15.43%	21.60%	11.11%
2017	30.86%	14.81%	22.22%	22.84%	9.26%

Question 12b: Bicycling for recreational trips

Year	Almost never	Rarely	Sometimes	Often	Almost always
2008	32.72%	17.90%	16.67%	21.60%	11.11%
2017	30.86%	18.52%	20.37%	16.05%	14.20%

Question 13a: Walking for utilitarian trips

Year	Almost never	Rarely	Sometimes	Often	Almost always
2008	12.96%	20.37%	22.22%	18.52%	25.93%
2017	14.20%	15.43%	24.07%	21.60%	24.69%

Question 13b: Walking for recreational trips

Year	Almost never	Rarely	Sometimes	Often	Almost always
2008	12.96%	17.28%	21.60%	16.05%	32.10%
2017	11.73%	12.96%	23.46%	16.05%	35.80%

Table 5. Reduction of private vehicle use and willingness to change with another transport mode comparing the years 2008 and 2017 (questions 14 and 15).

Question 14a: Personal use of private vehicle use in the city between 2008 and 2017					
Utilitarian trips			Recreational trips		
Reduction	Same use	No opinion	Reduction	Same use	No opinion
42.3%	45.4%	12.3%	52.2%	34.5%	13.3%
Question 14b: Opinion concerning the general reduction of private vehicle use in the city					
Utilitarian trips			Recreational trips		
Reduction	Same use	No opinion	Reduction	Same use	No opinion
44.4%	28.3%	27.3%	46.1%	27.6%	26.3%
Question 15a: Willingness to change private vehicle with another transport mode (trips < 2 km)					
Utilitarian trips			Recreational trips		
Walk	Bicycle	Bus	Walk	Bicycle	Bus
32.1%	36.5%	31.4%	35.8%	37.2%	27.0%
Question 15b: Willingness to change private vehicle with another transport mode (trips > 2 km)					
Utilitarian trips			Recreational trips		
Walk	Bicycle	Bus	Walk	Bicycle	Bus
23.2%	35.5%	41.3%	31.1%	34.1%	34.8%

Table 6. Reasons to increase walking and cycling comparing the years 2008 and 2017 (questions 16 and 17).

Question 16: Reasons to increase walking during the years 2008–2017				
Infrastructure improvement	Road safety	Personal safety	Income reduction	Health benefits
1.4%	1.7%	0.7%	63.1%	33.1%
Question 17: Reasons to increase bicycling during the years 2008–2017				
Infrastructure improvement	Road safety	Personal safety	Income reduction	Health benefits
3.1%	2.4%	0.7%	59.7%	34.1%

Table 7. Perceived level of pedestrians and bicyclists' road and personal safety comparing the years 2008 and 2017 (questions 18 to 21).

Question 18: Pedestrian personal safety				
Year 2008	Day	87.4%	Night	70.3%
Year 2017		62.1%		33.8%
Question 19: Bicyclist personal safety				
Year 2008	Day	96.6%	Night	81.6%
Year 2017		76.5%		58.4%
Question 20: Pedestrian road safety				
Year 2008	Day	53.6%	Night	43.3%
Year 2017		51.2%		41.3%
Question 21: Bicyclist road safety				
Year 2008	Day	67.9%	Night	51.2%
Year 2017		64.2%		26.3%

Table 8. Evaluation of the bikeway network in the city of Volos (questions 22 to 26).

Question 22: The presence of bikeway network in the city is considered as					
Positive		Negative		No opinion	
81.6%		5.8%		12.6%	
Question 23: Use of bikeway network for trips in the city					
Almost never	Rarely	Sometimes		Often	Very often
34.8%	24.2%	18.8%		14.7%	7.5%
Question 24: Necessity to improve the bikeway network in the city					
None	Slightly	Moderate		Much	Very Much
6.1%	10.6%	16.4%		42.3%	24.6%
Question 25: Necessity to expand the bikeway network in the city					
None	Slightly	Moderate		Much	Very Much
2.4%	13.3%	26.3%		38.2%	19.8%
Question 26: Annual funding of bikeway network projects in the city					
0 €	1€–10€	11€–20€	21€–30€	31€–50€	>50€
6.1%	31.1%	20.8%	18.8%	14.7%	8.5%

Table 9. Evaluation of the pedestrian infrastructure in the city of Volos (questions 27 to 29).

Question 27: Necessity to improve the pedestrian infrastructure in the city					
None	Slightly	Moderate		Much	Very much
1.0%	3.8%	18.4%		42.7%	34.1%
Question 28: Priority area to improve the pedestrian infrastructure in the city					
Central district		Main streets		Local streets	
64.8%		25.6%		9.6%	
Question 29: Annual funding of pedestrian infrastructure projects in the city					
0 €	1€–10€	11€–20€	21€–30€	31€–50€	>50€
6.1%	24.6%	24.9%	23.2%	14.7%	6.5%

Table 10. Evaluation of the roundabouts in the city of Volos (questions 30 to 32).

Question 30: Agreement with roundabouts construction in the city				
Agree: 57.3%		Disagree: 13.3%		No opinion: 29.4%
Question 31: Evaluation of roundabouts positive impact in road safety				
None	Slightly	Moderate	Much	Very much
13.7%	21.8%	28.3%	22.2%	14.0%
Question 32: Evaluation of roundabouts positive impact in transport convenience				
None	Slightly	Moderate	Much	Very much
5.1%	14.3%	36.2%	27.0%	17.4%

4 Conclusions

A thorough quantification and causal correlation between the economic crisis (expressed as per capita GDP) and the variables that measures the citizens' transportation habits requires the estimation of income elasticities for the various transport modes, accordingly quantitative data before and during economic crisis (i.e. traffic volume at selected road cross-sections, annual number of passengers of various transport modes, etc.). In most mid-sized Greek cities such data are unavailable; consequently, the questionnaire survey was the only possible method to monitor the above parameters.

The present paper examines the change of commuters' behavior for utilitarian and recreational urban trips towards sustainable transport modes in the city of Volos, by comparing the year 2008 (before the economic crisis) and the year 2017 (economic crisis continued, but it was in recession). The main conclusions are the following:

- Personal income of respondents was severely reduced during the examined years (Table 2).

- Respondents have reduced their private vehicle use and they have stated a general reduction of private vehicles use during the examined years (Tables 4 and 5).
- Respondents have stated that they are willing to use sustainable transport modes for their daily urban trips. Furthermore, a change of travel behavior in favor of the sustainable transport modes is noticed during the examined years (Tables 4 and 5).
- Personal income reduction and health benefits are the main factors that affect the increase of walking and bicycling in the city (Table 6).
- Road and personal safety issues affect commuters to walk or bike mainly during nighttime in the city (Table 7).
- Respondents favor the expansion and improvement of bicycle and pedestrian infrastructure in the city, and they are willing to fund sustainable transport projects in the city (Tables 8 and 9).
- Respondents agree with the construction of roundabouts, but they moderately evaluate the positive impact in road safety and travel convenience (Table 10).

This study supports that there are numerous factors that affect commuters travel behavior. The change of citizens' income and the existence of properly designed and well-maintained sustainable transport infrastructure can alter commuters' travel behavior in urban areas in favor of sustainable transportation.

References

1. Arrington, G.B., Sloop, K.I.: New transit cooperative research program research confirms transit-oriented developments produce fewer auto trips. ITE J. **79**(6), 26–29 (2010)
2. Litman, T.: Understanding transport demands and elasticities: How prices and other factors affect travel behavior. Victoria Transport Policy Institute (2019)
3. Litman, T.: Evaluating public transportation health benefits. Victoria Transport Policy Institute (2018)
4. Mihyeon Jeon, C., Amekudzi, A.: Addressing sustainability in transportation systems: definitions, indicators, and metrics. J. Infrastruct. Syst. **11**(1), 31–50 (2005)
5. Galanis, A., Botzoris, G., Siapos, A., Eliou, N., Profillidis, V.: Economic crisis and promotion of sustainable transportation: a case survey in the city of Volos. Transp. Res. Procedia **24**, 241–249 (2017)
6. Botzoris, G.N., Galanis, A.T., Giannopoulou, M., Varlami, E.: Sustainable transportation planning and traffic noise in urban environment: case survey in a typical mid-sized Greek city. In: 4th International Conference 'Energy in Transportation', Athens, Greece (2019)
7. Botzoris, G.N.: Economic crisis and its impact on sustainable urban transport. Int. J. Econ. Finan. Issues **10**(1), 33–41 (2020)
8. Gkavra, R., Nalmpantis, D., Genitsaris, E., Naniopoulos, A.: The walkability of thessaloniki: citizens perceptions. In: 4th Conference on Sustainable Urban Mobility-CSUM 2018, pp 191–198. Springer, Cham (2018)
9. Shay, E., Spoon, S.C., Khattak, A.J.: Walkable environments and walking activity. Report to the Southern Transportation Center, University of Tennessee (2003)
10. Mackett, R.L.: Policies to attract drivers out of their cars for short trips. Transp. Policy **8**, 295–306 (2001)
11. Easton, H., Smith, F.: Reducing crime on public transport. Nacro-Crime and Social Policy Sector (2003)

12. Bradshaw, R., Jones, P.: The family and the school run: What would make a real difference? University of Westminster (2000)
13. Landis, B., Vattikuti, V., Ottenberg, R., McLeon, D., Guttenplan, M.: Modeling the roadside walking environment: Pedestrian level of service. Transp. Res. Rec. **1773**, 81–88 (2001)
14. Moritz, W.: Adult bicyclists in the United States: characteristics and riding experience in 1996. Transp. Res. Rec. **1636**, 1–8 (1998)
15. Jensen, S.U.: Bicycle tracks and lanes: a before-after study. In: Proceedings of Transportation Research Board 87th Annual Meeting, Washington, D.C. (2008)
16. Moudon, A.V., Lee, C.: Walking and bicycling: an evaluation of environmental audit instruments. Am. J. Health Prom. **18**, 21–37 (2003)
17. Landis, B.W., Vattikuti, V.R., Brannick, M.T.: Real-time human perceptions: towards a bicycle level of service. Transp. Res. Rec. **1578**, 119–126 (1997)
18. Goldsmith, S.A.: Reasons why bicycling and walking are not being used more extensively as travel modes. U.S. Department of Transportation, Washington D.C. (1993)
19. Litman, T.: Pedestrian and bicycle planning: a guide to best practices. Victoria Transport Policy Institute (2000)
20. Galanis, A., Eliou, N.: Development and implementation of an audit tool for urban bikeway road network. In: Transport Research Arena (TRA) 5th Conference: Transport Solutions from Research to Deployment, Paris, France (2014)
21. Galanis, A., Papanikolaou, A., Eliou, N.: Bikeability audit in urban road environment: case study in the city of Volos, Greece. Int. J. Oper. Res. Inf. Syst. **5**, 21–39 (2014)
22. Stamatiadis, N., Cafiso, S., Pappalardo, G.: A comparison of bicyclist attitudes in two urban areas in USA and Italy. In: 4th Conference on Sustainable Urban Mobility-CSUM 2018, pp 272–279. Springer, Cham (2018)
23. Kotoula, K.M., Sialdas, A., Botzoris, G., Chaniotakis, E., Grau, J.M.S.: Exploring the effects of university campus decentralization to students' mode choice. Periodica Polytechnica Transp. Eng. **46**(4), 207–214 (2018)
24. Profillidis, V., Botzoris, G.: Modeling of Transport Demand: Analyzing, Calculating, and Forecasting Transport Demand, 1st edn. Elsevier, Oxford (2018)

A First Look at E-Scooter Users

Alexandra Raptopoulou[1], Socrates Basbas[1], Nikiforos Stamatiadis[2], and Andreas Nikiforiadis[1](✉)

[1] School of Rural and Surveying Engineering, Aristotle University of Thessaloniki, 54124 Thessaloniki, Greece
anikiforiadis@topo.auth.gr

[2] Department of Civil Engineering, University of Kentucky, Lexington, KY, USA

Abstract. Micromobility, defined as short-distance transport, is recently experiencing a global growth. The primary goal of this new mobility trend is to provide environmentally friendly options for trips that cannot be accomplished with public transport and reduce automobile dependence. A popular transport mode is electric scooters (e-scooters) and they have been making inroads in several countries and cities on a global scale. However, there is little guidance regarding their operations. The purpose of this research is to evaluate the users' attitudes and behavior of e-scooters in Thessaloniki, Greece. In-person and electronic surveys were completed targeting persons who have used an e-scooter more than once. The findings indicate that users most frequently ride on bicycle lanes and on roads without bike lane, while they prefer to use them for recreational purposes and not for commuting. Concerning the issues that would prevent them from using an e-scooter, greater importance is assigned to the bad behavior of car drivers, while bicycle lanes is considered the most efficient way for improving the use of e-scooters. Finally, slight differences were found on the behavior of users based on their age and gender, but people of different age and gender share similar attitudes.

Keywords: Micromobility · E-Scooters · Urban mobility

1 Introduction

Urbanization is on the rise and it is projected that by 2050 an additional 2.5 billion people will be living in cities [1]. Public transport cannot address current mobility needs and the greater reliance on the personal automobile to satisfy mobility needs has led to increased road congestion [2]. In addition, the goal of reducing pollutant emissions and, by extension, the greenhouse effect, has led to the development of a new form of mobility, coined as micromobility [3]. These new modes of transportation are attempting to address and provide innovative solutions to problems such as congestion, reducing the cost of travel while saving energy, and achieving faster travel to nearby destinations [4]. The introduction of micromobility can provide coverage to mobility needs of residents in areas where there is no public transport, or the service frequency is too low [2]. Bruce [5] estimated that micromobility services are capable of replacing more than 4 trillion miles worldwide, and their use has grown exponentially in recent

E. G. Nathanail et al. (Eds.): CSUM2020, AISC 1278, pp. 882–891, 2021.
https://doi.org/10.1007/978-3-030-61075-3_85

years. A National Association of City Transportation Officials (NACTO) survey showed that in 2018 there were 84 million trips completed using a micromobility mode [6].

One such micromobility mode is the electric scooters (e-scooters), which first appeared in 2017 in California [3]. E-scooters have gained wide global acceptance generating great enthusiasm and a positive outlook for addressing mobility needs [7]. For this reason, they have appeared in many countries around the world and are in high demand [8]. However, their wide use raises several concerns. Many users drive e-scooters on the sidewalks due to lack of infrastructure, such as bicycle paths or lanes, because they feel safer on them [2]. Another issue is their parking locations, which is often on sidewalks and at spots that block pedestrian movement especially those with mobility disabilities [9]. These issues have resulted in negative public reactions, such as destroying them or throwing them into the sea as an indication of their displeasure with them [10]. It is worth noting that there is also an increase in the number of crashes involving e-scooters and some of them have been fatal. This situation has led some cities to create guidelines, regulations and legal frameworks to address their operation setting permissible speed limits, permitted locations for their operation, and appropriate parking areas in order to establish a suitable and safe environment for their operation [11].

This study examined the attitudes of e-scooter users in the city of Thessaloniki, Greece in order to understand their behavior and issues associated for their use. The findings of this work will address an important aspect of the continued use of e-scooters in Greece and provide a basis for developing possible policies for addressing their efficient and safe operation.

2 Literature Review

Micromobility is related to the completion of a short movement at low speed that provides an easy and quick access to travel whenever necessary. The desired results of using a micromobility mode to travel are to reduce air pollution and traffic congestion [12]. In general, the term micromobility refers to the various movements carried out by bicycle (conventional or electric) and e-scooters. These are characterized as “first and last mile” journeys and allow individuals to choose from many different routes depending on their needs, something that public transport cannot easily accomplish [13].

The most recent mode of micromobility is the e-scooter, which first appeared in 2003, but due to the dominance of bike-sharing systems they were not very successful. There are many companies that have invested more than $ 5.7 billion over the past 4 years in shared e-scooters [14]. The typical operation requires the installation of an app in the user’s mobile phone that is used to locate available e-scooters on a map and then using the app unlock the e-scooter to use it and then lock it again once the trip is completed. Costs vary among systems and typically there is a cost associated with activating the e-scooter and a charge per distance. Once the trip is completed, users can simply leave the e-scooter wherever they wish [15].

The increasing use of e-scooters globally has forced governments to respond quickly to the consequences of their expanded use [2]. For this purpose, new laws and regulations were introduced for their smooth integration into traffic, since it was deemed necessary to inform and educate users on the proper use of e-scooters.

The training of e-scooter users is considered in many countries as a means for improving safety through reduction of crashes, creating a harmonious coexistence with other modes of transport and pedestrians, and identifying proper use of the devices. Educational booklets and pamphlets describing the laws, rules and safety precautions that must be observed are typical means of informing users. In addition, transportation agencies have also organized events where they answer users' questions and complaints, developed web platforms to provide feedback on questions, placing signage that prohibits the entry of e-scooters on specific areas, and distributing free helmets [9]. Additional efforts include use of posters on public transport and parking areas that provide good conduct practices and safety rules [16].

Many countries have led to the modification and adoption of new regulations and legal frameworks to incorporate e-scooters into existing traffic. These typically include the minimum age of the user, the permitted speed limit, the allowable areas for driving and parking, and areas where they are prohibited. Any infringement is punishable by a fine. A typical example is France, which established in September 2019 that e-scooter riders should be over 8 years old and can drive at top speeds of 25 km/h. In particular, Paris, France has prohibited driving on the sidewalks and any infringement is punishable with a €135 fine while parking in the middle of sidewalks that prevents pedestrian flow as well as blocking crosswalks and entry doors carries a €35 fine [17]. In another example, the city of Lyon, France has requested that all e-scooters operating in the city to have a GPS built in that automatically reduces the speed to 8 km/h when the e-scooter is in a pedestrian zone [18]. In Germany, the speed of e-scooters cannot exceed 20 km/h and the minimum age for the user is 14 years. They are allowed to operate on the roads and bicycle lanes but are not allowed on sidewalks. In addition, by June 2019, all e-scooters are required to have a self-adhesive sticker which constitutes annual insurance; helmet use currently is not mandatory [19]. A notable example is that of the United Kingdom where unlike other countries driving e-scooters is only allowed in a private area and banned on sidewalks under Act 1835 [20]. Regarding Greece, a legislative framework for e-scooters will be adopted in the coming months to incorporate them into the Traffic Regulations. According to the Ministry of Infrastructure, the most likely scenario is that e-scooters traveling at speeds of 25 km/h are considered to be motorized while those that are driven at speeds of 6 km/h are considered pedestrians. There will also be restrictions and limitations of use on certain areas [21].

The literature review shows that there is a somewhat uncontrolled operation of e-scooters and cities and national governments are racing to establish a set of guidelines that would define their efficient and safe operation. This study will provide a meaningful contribution to understanding the attitudes of the users in the Greek environment and aid in the development of a set of rules that could address their behavior and be responsive to their needs.

3 Study Area and Data Collection

E-scooters appeared recently in Greece and in particular in Thessaloniki, creating a new mobility trend. Thessaloniki has a population of 1,110,551 residents and is the second largest city in Greece [22]. It is a coastal city with a major commercial port, a popular tourist destination, and attracts a large number of students because of the existence of many Universities. It also includes a 12 km bike lane network [23] and the only public transport in the city is public buses.

A research study was undertaken to understand the increased demand for e-scooters as transport mode and the attitudes and behaviors of users. This research was based on the collection of questionnaires that were administered both in person and electronically targeting users of e-scooters. The respondents were persons who used an e-scooter more than once. The questionnaire was organized in a manner that identified the frequency of use, the duration and purpose of trips, typical parking areas after the use, factors that would affect the frequency of use, and socioeconomic characteristics of the users.

The in-person survey was used during the July-October 2019 period resulting in 160 responses. This approach was initially considered as more appropriate in order to allow for clarification of the questions posed to the respondents. The team also realized that the survey questions did not require any additional explanation and it was determined that the use of an electronic questionnaire would allow for a faster data collection and reach a much larger audience. Additional 111 questionnaires were collected. Thus, the total number of questionnaires completed is 271.

4 Survey Results

4.1 Descriptive Statistics

Most of the respondents are males (68.6%) and most are fairly young (73.4% in the 18–27 age class). Most respondents belong in middle-income class (48% with incomes in the range 7,000–24,000 €). The results from the web survey indicated that most respondents were employed (40.5%) and students (36%). Most could be also considered as having a high education level, since university students and bachelors degree holders comprised most of the respondents (35.1% and 33.3% respectively).

According to the users, the most common way to find an e-scooter is by walking (91.1%). The majority of users responded that they use the e-scooter in combination with other means of transport less than once a week (55.7%). In the event that an e-scooter is not available, the most common travel alternative is walking (43.9%). The two most common areas where users ride their e-scooters are the bicycle lane (42.4%) and on a road without bike lane (34.3%). As for their parking, 70.8% responded that they park them at the edge of the sidewalk. E-scooters are not used as a basic means of travel and for leisure trips it is mainly used for 10 min (33.6%). This survey found that the trip purpose is mainly for entertainment and shopping, and that users rent them less than once a month (36.9%). Furthermore, they never use e-scooters for education purposes (67.5%) and work purposes (66.4%).

In a ranking scale, where 1 corresponds to the less important and 6 corresponds to the most important factor for avoiding using an e-scooter, the users responded that bad driving behavior (mean = 4.51), weather conditions (mean = 4.37) and traffic congestion (mean = 4.19) are the most important issues that would prevent them from using an e-scooter. Less important are inadequate connection with other means of transport (mean = 2.46) and high levels of air pollution (mean = 1.97). In a scale 0 (not at all important) to 3 (very important), the most important intervention to improve e-scooter usage is to build bicycle lanes (mean = 2.84) while improvement of road conditions seems to be very important (mean = 2.52). The existence of road lighting (mean = 2.17) and road markings for e-scooters (mean = 2.00) are considered to be important as well. Less important are parking lots (mean = 1.69) and connection with other means of transport (mean = 1.36).

4.2 Comparative Analysis by Gender and Age

Table 1 presents the more common place for riding an e-scooter by gender. It can be seen that females are more likely than males to ride on roads without bike lanes or on bike lanes or on the sidewalk, while males are much more likely to ride on pedestrians-cyclists shared spaces. Table 2 also presents the most common place for riding an e-scooter, but by the age class of the users (two age classes were considered appropriate for the comparative analysis, in order to have approximately equal distribution in the different classes). The percentages show that older riders prefer roads without bike lanes or separated bike lanes, while younger riders prefer sidewalks or shared spaces.

Table 1. More common place for riding e-scooter by gender (percentages)

Location	Male	Female
On a road without bike lane	33.3	36.5
On a bike lane	40.9	45.9
On the sidewalk (without bikes)	3.8	7.1
On a common pedestrian-bicycle area	22.0	10.6

Table 2. More common place for riding e-scooter by age (percentages)

	Age	
Location	<=36	>36
On a road without bike lane	33.3	42.9
On a bike lane	42.1	46.4
On the sidewalk (without bikes)	5.3	–
On a common pedestrian-bicycler area	19.3	10.7

Figure 1 and Figure 2 present the frequency (0: Never, 1: Less than once a month, 2: 1–5 times a month, 3: 1 day a week, 4: 2–4 times a week, 5: More than 4 days a week and 6: Daily) of using an e-scooter per trip purpose by gender and age respectively. It can be seen that the percentages are approximately the same for the two genders regarding education and work trip purposes, but males tend to use a little bit more frequently the e-scooters for recreational purposes. The comparative analysis for the different age groups indicates that younger people use much more frequently the e-scooters for all trip purposes.

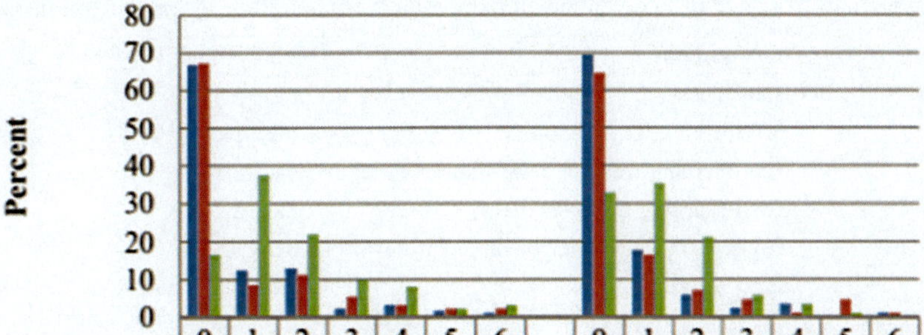

	Male							Female						
	0	1	2	3	4	5	6	0	1	2	3	4	5	6
■ Education	67	12	13	2,2	3,2	1,6	1,1	69	18	5,9	2,4	3,5	0	1,2
■ Work	67	8,6	11	5,4	3,2	2,2	2,2	65	17	7,1	4,7	1,2	4,7	1,2
■ Recreation	17	38	22	10	8,1	2,2	3,2	33	35	21	5,9	3,5	1,2	0

Fig. 1. Frequency of using an e-scooter per trip purpose and by gender (percentages).

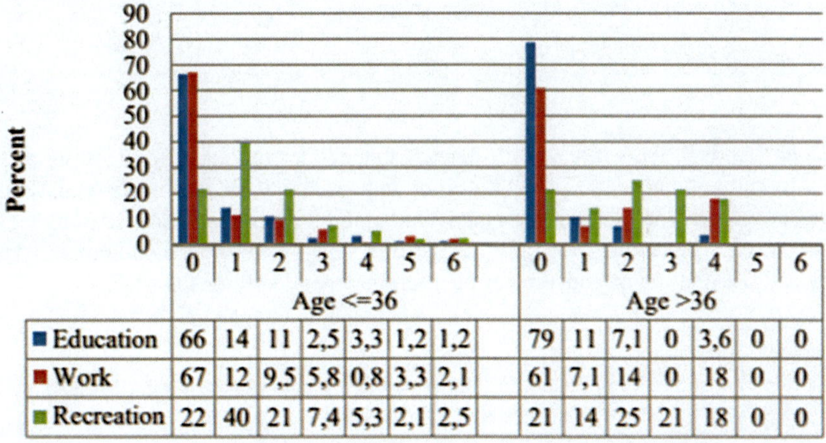

	Age <=36							Age >36						
	0	1	2	3	4	5	6	0	1	2	3	4	5	6
■ Education	66	14	11	2,5	3,3	1,2	1,2	79	11	7,1	0	3,6	0	0
■ Work	67	12	9,5	5,8	0,8	3,3	2,1	61	7,1	14	0	18	0	0
■ Recreation	22	40	21	7,4	5,3	2,1	2,5	21	14	25	21	18	0	0

Fig. 2. Frequency of using an e-scooter per trip purpose and by age (percentages).

Table 3 shows the importance that the respondents assign to the different interventions for improving the use of e-scooters in the urban environment. Males and females seem to prioritize the interventions in the same way. However, females give higher importance to all the interventions comparing to males. Also, according to Table 4, a similar prioritization of interventions can be identified for the different age groups.

Table 3. Importance of interventions for improving e-scooter use by gender (means).

Interventions	Male	Female
Bike lanes	2.82	2.87
Parking lots for e-scooters	1.62	1.86
Improvement of road conditions	2.44	2.71
Road marking for e-scooters	1.89	2.25
Connection with other means of transport	1.31	1.48
Road lighting	2.09	2.33

Table 4. Importance of interventions for improving e-scooter use by age (means).

	Age	
Interventions	<=36	>36
Bike lanes	2.83	2.93
Parking lots for e-scooters	1.72	1.43
Improvement of road conditions	2.50	2.68
Road marking for e-scooters	2.04	1.64
Connection with other means of transport	1.40	1.07
Road lighting	2.19	2.00

Table 5 and Table 6 indicate the factors that hinder the users from riding an e-scooter more frequently. As in the case of the importance of the interventions, it seems that the attitudes of males and females are similar and they assign greater importance to the bad driving behavior, the traffic congestion and the weather conditions. These factors are also the major concern for all users independently of their age.

Table 5. Importance of factors for hindering the e-scooter use by gender (means).

Factors	Male	Female
Traffic congestion	4.11	4.36
Bad driving behavior	4.46	4.61
Weather conditions	4.46	4.16
High level of air pollution	1.97	1.98
Desirable destinations are far away	3.51	3.33
Bad connection with other means of transport	2.42	2.55

Table 6. Importance of factors for hindering the e-scooter use by age (means).

	Age	
Factors	<=36	>36
Traffic congestion	4.24	3.75
Bad driving behavior	4.50	4.61
Weather conditions	4.40	4.14
High level of air pollution	1.90	2.61
Desirable destinations are far away	3.47	3.29
Bad connection with other means of transport	2.45	2.61

5 Discussion

This research was completed in Thessaloniki, Greece aiming to identify usage and issues of e-scooters based on responses from users. The data collected using both in person and web questionnaires.

The results showed that males use e-scooters slightly more frequently than females in the city of Thessaloniki and this is in agreement with prior research like the 2019 San Francisco Municipal Transportation [24]. The main trip purpose for the users was entertainment or shopping, which is different than the 2018 findings from Portland, Oregon where the majority was for trips with education as purpose [9]. This difference could be attributed to the fact that most university campuses in Thessaloniki are in the city center and access to them is easier on foot, since most students reside in close proximity to the various campuses. Of interest here is the fact that the lack of infrastructure forces users to utilize roadways and sidewalks instead of bicycle lanes. It should be noted that e-scooters are typically used for short trips lasting between 5 to 15 min, which is in agreement with the San Francisco study [24]. The short-term usage could be attributed to the shorter distances in Thessaloniki and the layout of the city. Also, users responded that they mainly park an e-scooter at the edge of the sidewalk. The most important intervention for increasing and improving e-scooter usage is considered the development of bicycle infrastructure that also underscores their lack. On the other hand, the least important intervention is the connection with other means of transport. The most important factors that would prevent them from using an e-scooter are bad driving behavior and weather conditions, while the least important is the high levels of air pollution.

It is generally observed that users' opinions do not differ much based on gender and age. It is worth mentioning that females prefer more to ride on bike lanes in relation to males. Furthermore females use e-scooters less frequently for recreational purposes. The comparative analysis showed that younger people use much more frequently the e-scooters for all trip purposes. It seems that the factors that deter the users from riding an e-scooter more frequently are similar. Users consider that the most important factors are bad driving behavior, traffic congestion and weather conditions.

It is apparent that such efforts should be continued with the collection of additional questionnaires to allow for a more robust statistical analysis and conclusion as well as obtaining answers from a more diverse sample. Moreover, the findings here support the need for improved bicycle infrastructure to allow for a safer and more efficient circulation of e-scooters. The data also supports the need for user training and education as well as the development of guidelines of how e-scooters could be incorporated in the existing traffic network.

References

1. CB Insights. https://www.cbinsights.com/research/report/micromobility-revolution/. Accessed 3 Feb 2020
2. Deloitte. Insights. https://www2.deloitte.com/us/en/insights/focus/future-of-mobility/micro-mobility-is-the-future-of-urban-transportation.html. Accessed 8 Feb 2020
3. Polis: Macro managing Micro mobility. Discussion Paper. Polis, Brussels (2019)
4. Gubman, J., Jung, A., Kiel, T., Strehmann, J.: Shared E-Scooters: Paving the Road Ahead Policy Recommendations for Local Government. Agora Verkehrswende (2019)
5. Medium. https://medium.com/micromobility/the-bull-case-for-micromobility-15c43ebf7381. Accessed 10 Feb 2020
6. NACTO: Shared micromobility in the U.S.: 2018. National Association of City Transportation Officials (2019)
7. Populus: The micro-mobility revolution: The introduction and adoption of electric scooters in the United States. San Francisco (2018)
8. Lee, M., Chow, J., Yoon, G., Yueshuai He, B.: Forecasting e-scooter competition with direct and access trips by mode and distance in New York City. arXiv: General Economics (2019)
9. Portland Bureau of Transportation: 2018 E-Scooter Findings Report. PBOT, Portland (2018)
10. The New York Times. https://www.nytimes.com/2019/06/17/opinion/electric-scooters-nashville.html. Accessed 8 Feb 2020
11. BBC. https://www.bbc.com/news/world-europe-49248614. Accessed Mar 2 2020
12. Shaheen, S., Cohen, A.: Shared Micromobility Policy Toolkit: Docked and Dockless Bike and Scooter Sharing. Transportation Sustainability Research Center, UC Berkeley (2019)
13. DuPuis, N., Griess, J., Klein, C.: Micromobility in Cities: A History and Policy Overview. National League of Cities (2019)
14. Forbes. https://www.forbes.com/sites/carltonreid/2019/03/18/bicycling-take-a-hike-the-micromobility-revolution-will-be-motorized/#5c09777c135d. Accessed 10 Feb 2020
15. Fong, J., McDermott, P., Lucchi, M.: Micro-Mobility, E-Scooters and Implications for Higher Education. UPCEA Center for Research and Strategy (2019)
16. City of Santa Monica. https://www.santamonica.gov/press/2018/08/22/santa-monica-launches-public-education-campaign-on-e-scooter-safe-rules-of-the-road. Accessed 3 Feb 2020
17. European Consumer Centre France. https://www.europe-consommateurs.eu/en/consumer-topics/on-the-road/electric-scooters-in-france/?fbclid=IwAR0Xfuo9xpqHym-dAKRnqbATLoB7HbUMzZrFtCWPSldiswEe2VflERrEX-c. Accessed 12 Feb 2020
18. Eltis. https://www.eltis.org/discover/news/e-scooter-regulations-germany-and-france. Accessed 11 Feb 2020
19. European Transport Safety Council. https://etsc.eu/germany-and-france-to-regulate-e-scooters/. Accessed 11 Feb 2020
20. U.K Department of Transport. https://www.gov.uk/government/publications/powered-transporters/information-sheet-guidance-on-powered-transporters. Accessed 12 Feb 2020
21. Kathimerini. https://www.kathimerini.gr/1059770/article/epikairothta/ellada/hlektrika-patinia-triwn-taxythtwn-ston-neo-kok. Accessed 15 Feb 2020

22. Elstat. https://www.statistics.gr/el/statistics/-/publication/SAM03/. Accessed 5 Feb 2020
23. Nikiforiadis, A., Basbas, S.: Can pedestrians and cyclists share the same space? The case of a city with low cycling levels and experience. Sustain. Cities Soc. **46**, 101453 (2019)
24. San Francisco Municipal Transportation Agency: Powered Scooter Share Mid-Pilot Evaluation. SFMTA, San Francisco (2019)

Driving Behaviour and Road Safety at Signalised Intersections in Sicily and Thessaloniki

Socrates Basbas[1], Tiziana Campisi[2](✉), Giovanni Tesoriere[2], Antonino Canale[2], and Panagiotis Vaitsis[1]

[1] School of Rural and Surveying Engineering, Aristotle University of Thessaloniki, 54124 Thessaloniki, Greece
pvaitsis@topo.auth.gr

[2] Faculty of Engineering and Architecture, University of Enna Kore, Cittadella Universitaria, 94100 Enna, Italy
tiziana.campisi@unikore.it

Abstract. Road safety plays an important role in cities as far as sustainability and resilience are concerned . A large number of studies focus on the evaluation of the safety level of road users and especially vulnerable road users. In the framework of this paper an analysis was carried out concerning one of the most important problems at signalized intersections, which are the RLR (Red Light Running) and the YLR (Yellow Light Running) violations. This problem has to do with the behaviour and attitude of drivers as far as their compliance with Highway Code regulations is concerned. The existing situation in relation to RLR/YLR violations was investigated in the city of Thessaloniki, Greece and in Sicily, Italy. Field counts took place in each area followed by data analysis. Two sets of counts (Sicily and Thessaloniki) were used in the framework of this paper in order to investigate the above-mentioned phenomenon. Finally, a comparison between the results concerning both areas was made together with an attempt to investigate through inferential statistics the extent to which both the traffic volume and the category of vehicles affect RLR/YLR. The results can be used as a basis in order to calibrate behavioural models.

Keywords: Road safety · Drivers · Traffic lights · Red Light Running (RLR) · Yellow Light Running (YLR) · Driving behaviour

1 Introduction

A large number of road accidents take place in the area of signalised intersections [1]. One of the major safety issues at signalised intersections concerns the Red Light Running (RLR) and the Yellow Light Running (YLR). This phenomenon has to do with the crossing of an intersection when the traffic light is red or yellow respectively. These violations are responsible for a significant number of road accidents which result in a large number of deaths, heavy and light injuries as well as property damages. According to NHTSA's Fatality Analysis Reporting System, RLR is the cause for 752 injuries per 100.000 people per year [2]. This type of violation is increasingly higher

E. G. Nathanail et al. (Eds.): CSUM2020, AISC 1278, pp. 892–900, 2021.
https://doi.org/10.1007/978-3-030-61075-3_86

compared to the other types of road accidents which lead to fatal crashes [3, 4]. Finally, a higher number of fatalities due to this phenomenon, even in a shorter period of time in relation to the USA, take place in China [5].

The yellow phase of traffic lights plays a significant role in the safety level of the intersections. Over 50% of the crashes which take place are due to the violation of this phase [6]. The decision of the drivers to stop at or cross the intersections is affected by various factors such as how distant the crossing is, the attention the driver pays to the traffic lights, the speed of the vehicle etc. Some reasons which lead to the crossing of the intersections by the drivers include the distraction of their attention [7, 8], alcohol consumption [7], speed and aggressive driving [6]. Also, the inappropriate design of the time of the traffic lights' cycle or the implementation of an inappropriate cycle in the traffic lights are two more reasons which increase the RLR violations [6]. Some measures which can reduce these violations include message warning signs, signal timing adjustment and autonomous vehicle technologies [9, 10]. The prediction of potential RLR/YLR violations is a primary target before the implementation of measures against this phenomenon. This prediction cannot be easily made due to the fact that it concerns many environmental and behavioral factors [11].

However, the violations of RLR/YLR can be predicted by using Probit models [12] or Logit models [13, 14]. In order to use these types of models, a large amount of data is necessary in order to identify the key factors which are related to the infrastructure and the drivers' behaviour. Other methodologies used to estimate this phenomenon include machine learning techniques [15]. Several studies concerning these types of infringements are using different methodologies [16–18].

The present paper aims to examine the current situation in relation to RLR/YLR violations which take place in Thessaloniki, Greece and in Sicily, Italy. Field counts took place in each area followed by data analysis. Finally, a comparison between the results concerning both areas was made together with an attempt to investigate the extent to which traffic volume and category of vehicles affect RLR/YLR.

2 Materials and Methods

Observational surveys were carried out in the city of Thessaloniki, Greece and in Sicily, Italy. A total of 16 intersections (9 in Thessaloniki and 7 in Sicily) were used in order to perform the analysis and obtain the main findings. One major difference is the equipment that has been used. In Sicily, audiovisual equipment was used for the recording of violations while in Thessaloniki field counts were made with the use of observers at each intersection. Thessaloniki's data were collected in the year 2018 while data in Sicily were collected in the year 2019. Random sampling is used in both surveys.

The data analysis was carried out by using the SPSS v23.0 software. Descriptive and inferential statistics were used to analyze the collected data. Firstly, the key factors and features of RLR/YLR violations were identified using descriptive statistics for both areas. Moreover, inferential statistics were used in order to identify the variables (key features) which affect this phenomenon. Results are presented separately for each area and then a comparison is made between the two in order to observe whether traffic volume and category of vehicles play a significant role in the creation of these types of violations.

3 Description of the Undertaken Research

3.1 Thessaloniki

The first survey was conducted in March 2018 in Thessaloniki which is the second largest city in Greece with 1,006,730 inhabitants in its greater area [19]. Data collection was made in three corridors which are located in the eastern part of the city [20]. Namely, these three corridors are 25th Martiou Rd. (Corridor 1), Petrou Syndika Rd. (Corridor 2) and Markou Mpotsari Rd. (Corridor 3) and they are all characterized by high traffic volumes on a daily basis. Each one of the three corridors has one lane per direction used by the traffic. Counts were made in three intersections along each corridor and they include the number of RLR/YLR violations, the category of the vehicles involved in these violations and the traffic volume.

The categories of the vehicles include private cars, trucks, buses and motorcycles/bicycles. The survey was carried out during the Spring of 2018 between 9:00 a.m. and 11:00 a.m., which is actually the peak period. Violations of RLR/YLR in each intersection are only recorded for vehicles that move along the corridors.

3.2 Sicily

In Sicily, 6 intersections were monitored which are located in the central-southern area of the island respectively. The first 3 are in the town of Enna (Area 1), the two following are in the city of Caltanissetta (Area 2) and the last one is in the city of Agrigento (Area 3). They are characterized as intense traffic corridors that allow access to the cities. The first survey was conducted in the city of Enna between January and February 2018 [21]. The city of Enna is inhabited by approximately 30 thousand people. Enna has a very intense commuter flow of about 10 thousand university students [22]. The data was collected in three corridors located in the southernmost part of the city. The three corridors in particular are the Viale Diaz located in the upper part of the city, Monte Cantina, Scifitello and all of them are characterized by high volumes of traffic on a daily basis. Each one of the three corridors has one lane per direction used by traffic. The counts were made in three crossings and include monitoring the same parameters as in the case of Thessaloniki. The data of the next two intersections located in Caltanissetta (titled via Leone XXIII and via Rosso di San Secondo) and Agrigento (via Ugo la Malfa) located in the southernmost area of Sicily were also observed. These investigations were conducted in the first months of 2019. The data were collected from 06:00 to 22:00 with the peak period being between 8:00 and 9:00 in the morning and between 17:00 and 18:00 in the afternoon/evening. Each intersection was monitored for 21 days.

4 Survey Results

4.1 Thessaloniki's Results

The results which derived from the descriptive and inferential statistics are presented in this sub-section. Figure 1 presents the relation between the violation of RLR and YLR for each corridor. Corridor 1 is placed in the left part of the Figure while the other two follow. The YLR violations are considerably higher in the case of corridor 2.

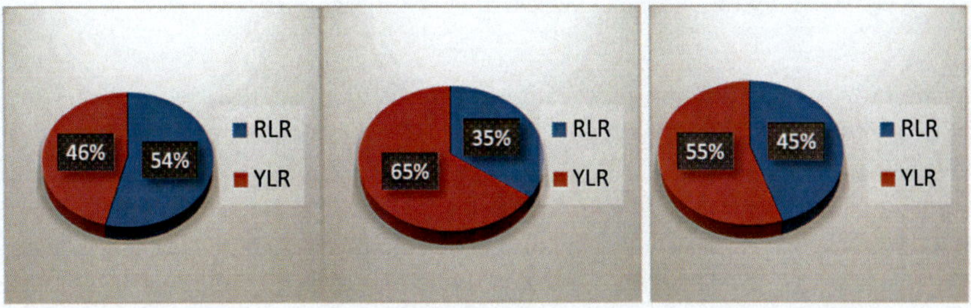

Fig. 1. RLR/YLR violations.

In Fig. 2 the categories of vehicles which are involved in the YLR violations are presented. The results are nearly similar for the three corridors, something which was highly expected due to their proximity. The YLR violations mainly refer to private cars.

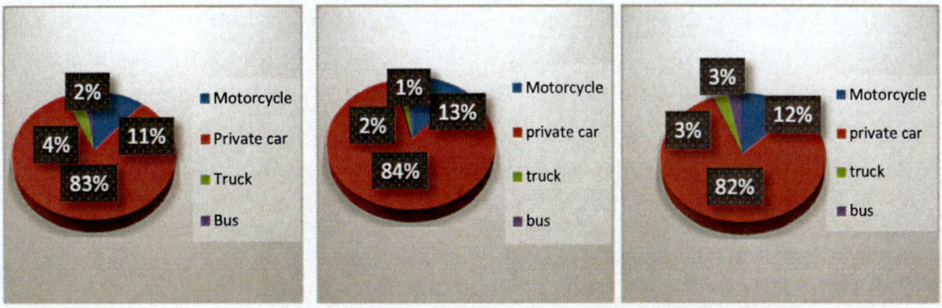

Fig. 2. YLR per vehicle category in each corridor.

As far as the RLR violations are concerned, it is worthwhile to see the high percentage of motorcycles in all the corridors as it can be observed in Fig. 3.

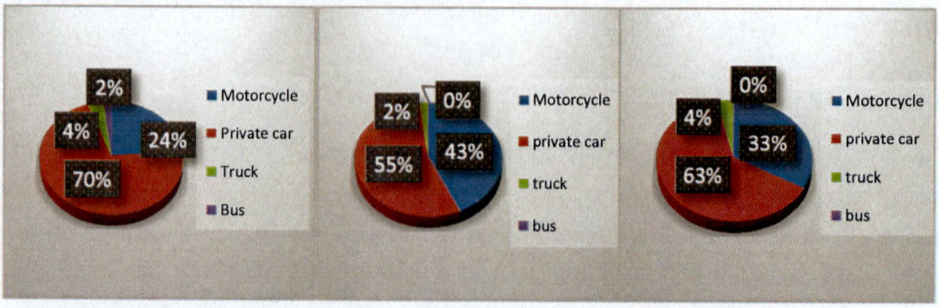

Fig. 3. RLR per vehicle category in each corridor.

In the framework of this research, inferential statistical analysis was carried out in order to identify the factors which affect RLR/YLR. Statistical analysis shows that the variable "vehicle category" and "RLR violations" have a strong correlation. The same phenomenon can be observed in the YLR violations too. The statistical test which was used in order to perform this kind of analysis was the one-way ANOVA test. The next correlation refers to the variables "traffic volume" and "RLR violations". The inferential statistic test which was used was Pearson r. The results indicate that there is a correlation between them as it has been expected. The same result derives from the Pearson r test for the "traffic flow" and the "YLR violations".

4.2 Sicily's Results

The results concerning Sicily's data are presented in this section. Descriptive and inferential statistics were used in order to analyse the current situation concerning the violations of RLR/YLR in three areas of Sicily and also to identify the key factors which affect this phenomenon. In each of the following Figures, the results are presented from left to right for the areas of Enna, Caltanissetta and Agrigento respectively. Figure 4 presents RLR in relation to YLR in the three areas in Sicily. As it can be observed, the YRL violations are slightly higher that the RLR in Enna while in Caltanissetta the RLR has its highest value. Finally, in Agrigento RLR and YLR are equal.

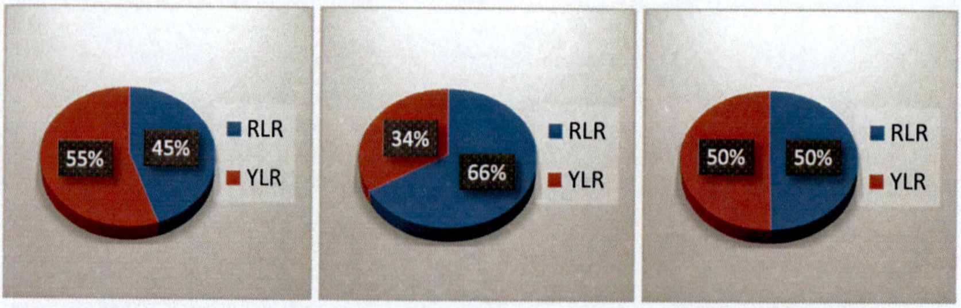

Fig. 4. RLR/YLR violations a) Enna b) Caltanissetta 2 c) Agrigento.

The categories of vehicles which were involved in the YLR violations are presented in Fig. 5. The majority of these violations are mostly made by private cars in each one of the three areas. It is interesting to see that motorcycles and trucks represent a significant percentage of these violations in Enna in relation to the other two areas. This fact can be justified by the increased presence of motorcycles and trucks in Enna in relation to the other two areas.

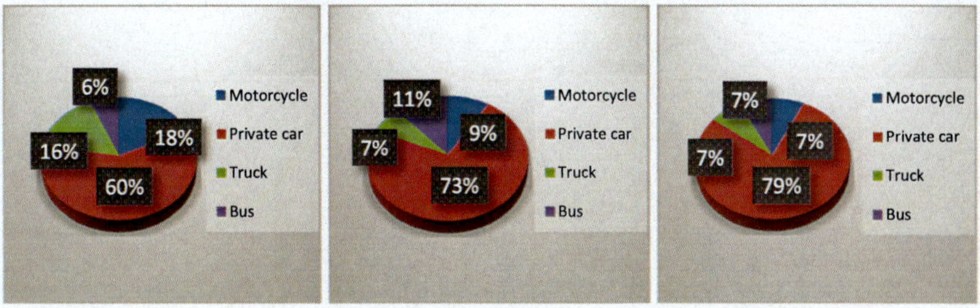

Fig. 5. YLR per vehicle category in each area.

As far the RLR violations are concerned, private cars have the highest percentage (Fig. 6). Both Figs. 5 and 6 have similar values.

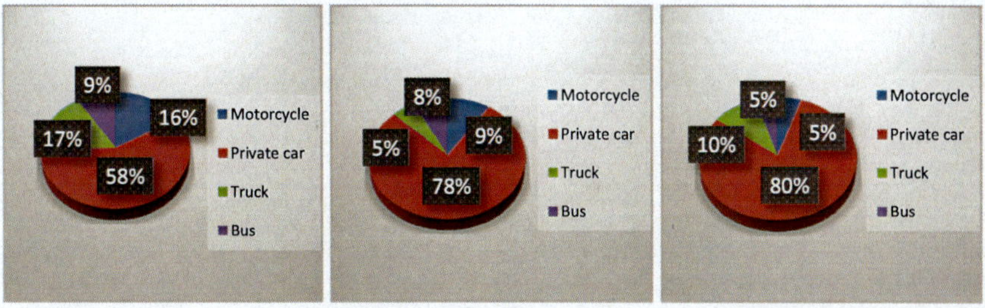

Fig. 6. RLR per vehicle category type in each area.

Inferential statistical analysis was carried out, as mentioned above, in order to identify the features that affect RLR/YLR. The first correlation refers to the variables "vehicle category" and "RLR violations". The one-way ANOVA statistical test shows a strong correlation between them. The more violations are made, the more private cars are involved.

The second correlation concerns the "YLR" and the "vehicle category". The one-way ANOVA statistical test was used. The same results as with the RLR violations are applied in this case. The other correlations refer to "traffic volume" and the "RLR and YLR violations". For this correlation the Pearson r statistical test was used. The results indicate that there is a strong correlation between them. The results of the inferential statistical analysis are presented in Table 1.

Table 1. Results of the inferential statistical analysis.

One-way ANOVA test between RLR violations and vehicle category (Thessaloniki)

	Sum of Squares	df	Mean Square	F	Sig.
Between Groups	864,917	3	288,306	4,907	,032
Within Groups	470,000	8	58,750		
Total	1334,917	11			

One-way ANOVA test between YLR violations and vehicle category (Thessaloniki)

	Sum of Squares	df	Mean Square	F	Sig.
Between Groups	1772,917	3	590,972	47,278	,000
Within Groups	100,000	8	12,500		
Total	1872,917	11			

Pearson r between RLR and traffic volume (Thessaloniki)

		RLR	Traffic volume
RLR	Pearson Correlation	1	,816**
	Sig. (2-tailed)		,001
	N	12	12
traffic volume	Pearson Correlation	,816**	1
	Sig. (2-tailed)	,001	
	N	12	12

Pearson r between YLR and traffic volume (Thessaloniki)

	Traffic volume	YLR
Pearson Correlation	1	,983**
Sig. (2-tailed)		,000
N	12	12
Pearson Correlation	,983**	1
Sig. (2-tailed)	,000	
N	12	12

**. Correlation is significant at the 0.01 level (2-tailed).

One-way ANOVA test between RLR violations and vehicle category (Sicily)

	Sum of Squares	df	Mean Square	F	Sig.
Between Groups	95752,000	3	31917,333	5,27	,027
Within Groups	48372,667	8	6046,583		
Total	144124,667	11			

One-way ANOVA test between YLR violations and vehicle category (Sicily)

	Sum of Squares	df	Mean Square	F	Sig
Between Groups	63508,667	3	21169,556	5,340	,026
Within Groups	31712,000	8	3964,000		
Total	95220,667	11			

Pearson r between RLR and traffic volume (Sicily)

		RLR	traffic volume
RLR	Pearson Correlation	1	,962**
	Sig. (2-tailed)		,000
	N	12	12
traffic volume	Pearson Correlation	,962**	1
	Sig. (2-tailed)	,000	
	N	12	12

Pearson r between YLR and traffic volume (Sicily)

	RLR	traffic volume
Pearson correlation	1	,894**
Sig. (2-tailed)		,000
N	12	12
Pearson Correlation	,894**	1
Sig. (2-tailed)	,000	
N	12	12

**. Correlation is significant at the 0.01 level (2-tailed).

5 Discussion

This paper presents a comparison of RLR and YLR violations at signalized intersections in Thessaloniki, Greece and Sicily, Italy. Three main corridors were used in this research in Thessaloniki while three areas were used in Sicily.

The results obtained from Thessaloniki's data indicate that there are less RLR violations in relation to YLR violations in two out of three corridors. The private car has the highest involvement (more than 80%). As far as the RLR violations are concerned, it should be noted that the percentages of the motorcycles which are involved in these violations are significantly higher than those in the case of YLR. From the inferential statistics it can be concluded that the increase of traffic volume leads to an increase of the RLR/YLR violations.

As far as Sicily's results are concerned, there is a different result in each area related to RLR/YLR violations. The majority of RLR violations involve private cars (at a percentage of over 70%) with an exception for the city of Enna where the respective percentage is lower. It should be noted that in Enna the RLR violations of trucks and motorcycles have a very high percentage in relation to the other two areas. Moreover, similar results can be obtained for the YLR violations in the three areas of Sicily. According to the results of the inferential statistics, it seems that there is a strong correlation between the "category of vehicles" and "RLR/YLR violations". Also, there is a correlation between the "traffic volume "and the "RLR/YLR violations", similar to Thessaloniki's results.

By performing a comparison between the results of the two surveys, it is noted that most of the RLR and YLR violations are made by drivers of private cars. The RLR violations in relation to YLR one differ from corridor to corridor or area to area. Another fact is that traffic volume plays a significant role to the propensity of this phenomenon.

The data limitations of both surveys prevent any conclusions of general applicability to be made. However, it is expected that these findings can help researchers in their effort to better understand the RLR/YLR violations in the urban environment. Future research will include the development of a statistical model (Probit or Logit) to predict the RLR and YLR violations in both Sicily and Thessaloniki.

Acknowledgements. We would like to express our sincere thanks and appreciation to the students (academic year 2017–2018 of the Interdisciplinary Postgraduate Program "*Design, Organization and Management of Transport Systems*" which was organized by the School of Civil Engineering and the School of Rural & Surveying Engineering, Aristotle University of Thessaloniki, in collaboration with the Hellenic Institute of Transport, for their assistance with the collection of data in the city of Thessaloniki.

References

1. Huang, H., Wang, D., Zheng, L., Li, X.: Evaluating time-reminder strategies before amber: common signal, green flashing and green countdown. Accid. Anal. Prev. **71**, 248–260 (2014)
2. NHTSA. NHTSA's Fatality Analysis Reporting System (FARS) Reports. National Highway Traffic Safety Administration, Washington DC, USA (2012)
3. Retting, R.A., Williams, A.F., Preusser, D.F., Weinstein, H.B.: Classifying urbancrashes for countermeasure development. Accid. Anal. Prev. **27**(3), 283–294 (1995)

4. Retting, R.A., Williams, A.F.: Characteristics of red light violators: results of afield investigation. J. Saf. Res. **27**(1), 9–15 (1996)
5. Wang, X., Yu, R.: A field investigation of red-light-running in Shanghai, China. Transp. Res. Part F **37**, 144–153 (2016)
6. Yang, Z., Tian, X., Wang, W., Zhou, X., Liang, H.: Research on driver behavior in yellow interval at signalized intersections. Math. Prob. Eng. **2014**, 8 (2014)
7. Zhang, Y., Yan, X., Li, X., Wu, J., Dixit, V.V.: Red-light-running crashes' classification, comparison, and risk analysis based on general estimates system (GES) crash database. Int. J. Environ. Res. Public Health **15**(6), 1290 (2018)
8. Li, P., Li, Y., Guo, X.: A Red-light running prevention system based on artificial neural network and vehicle trajectory data. Comput. Intell. Neurosci. **2014**, 1–11 (2014)
9. Bonneson, J., Middleton, D., Zimmerman, K., Charara, H., Abbas, M.: Intelligent detection-control system for rural signalized intersections. No. FHWA/TX-03/4022-2 (2002)
10. Ragland, D.R., Zabyshny, A.A.: Intersection decision support project: taxonomy of crossing-path crashes at intersections using GES 2000 data. In: Safe Transportation Research & Education Center (2003)
11. Baratian-Ghorghi, F., Zhou, H., Zech, W.C.: Red-light running traffic violations: a novel time-based method for determining a fine structure. Transp. Res. Part A **93**, 55–65 (2016)
12. Sheffi, Y., Mahmassani, H.: A model of driver behavior at high speedsignalized intersections. Transp. Sci. **15**(1), 50–61 (1981)
13. Papaioannou, P.: Driver behavior, dilemma zone and safety effects at urban signalized intersections in Greece. Accid. Anal. Prev. **39**(1), 147–158 (2007)
14. Newton, C., Mussa, R.N., Sadalla, E.K., Burns, E.K., Matthias, J.: Evaluation ofan alternative traffic light change anticipation system. Accid. Anal. Prev. **29**(2), 201–209 (1997)
15. Jahangiri, A., Rakha, H., Dingus, A.T.: Predicting red-light running violations at signalized intersections using machine learning techniques. Accid. Anal. Prev. **96**, 316–328 (2015)
16. Porter, E.B., England, J.K.: Predicting red-light running behavior: a traffic safety study in three urban settings. J. Saf. Res. **31**(1), 1–8 (2000)
17. Ren, Y., Wang, Y., Wu, X., Yu, G., Ding, C.: Influential factors of red-light running at signalized intersection and prediction using a rare events logistics regression model. Accid. Anal. Prev. **95**, 266–273 (2016)
18. Papaioannou, P., Basbas, S., Kokkalis, A., Vasiliadou, J.: Traffic violations and risk perception in Greece. In: Proceedings of the 4th International Conference "Global Safety", ZVD, DVR – Deutscher Verkehrssicherheitsrat e.V., PRI – La Prevention Routiere Internationale, Bled, Slovenia, pp. 157–163 (2000)
19. ELSTAT (Hellenic Statistical Authority). Population & Building Census. ELSTAT, Athens (2011)
20. Amprasi, V., Chrysis, K., Garyfalou, M.I., Karagouni, K., Kotsiou, P., Makrydaki, M., Mohamed, A., Demogianni, S., Papaioannou, G., Papakonstanti, E., Stoupa, E., Takatzoglou, K., Vaitsis, P., Vlahakis, D.: Assignments for Spring Semester, Course: Design and Management of Traffic and Parking Systems, Interdisciplinary Postgraduate Program "Design, Organization and Management of Transport Systems". School of Civil Engineering, School of Rural & Surveying Engineering, Aristotle University of Thessaloniki, in collaboration with the Hellenic Institute of Transport (2018)
21. Campisi, T., Tesoriere, G., Canale, A., Basbas, S., Vaitsis, P., Nikiforiadis, A., Nikolaidis, M.: Comparison of Red-Light Running (RLR) and Yellow-Light Running (YLR) traffic violations in the cities of Enna and Thessaloniki. Transp. Res. Procedia **45**, 947–954 (2020)
22. Istat, D.S.: Popolazione residente—Censimento 2011—Superficie Delle Abitazioni Occupate da Persone Residenti. Istat. Italian National Institute of Statistics. http://daticensimentopopolazione.istat.it/, Accessed 9 Nov 2016

Recording and Evaluation of Motorcyclists' Distraction of Attention in Urban Areas

Panagiotis Lemonakis[1(✉)], Myrofora Koroni[2], Eleni Misokefalou[1], and Nikolaos Eliou[1]

[1] Department of Civil Engineering, University of Thessaly, Pedion Areos, 38334 Volos, Greece
{plemonak, emisokef, neliou}@uth.gr

[2] Department of Environmental Engineering, International Hellenic University, 14th km Thessaloniki - N. Moudania National Road, 57001 Thermi, Greece
myrtokoroni@gmail.com

Abstract. Over the last years, distracted driving constitutes a considerably increasing road safety problem with disastrous results and it possesses a leading position among the accidents causes. The present study deals with motorcycle users' distraction due to out of the vehicle factors. Considering exterior factors as the most significant, we can group them in four categories: built roadway, situational entities, the natural environment, and the built environment. All these contribute to the setup of a very dangerous environment by increasing driver's distraction and inattention. This research is based on a medium scale experimental procedure which took place in urban areas in Greece. The distraction of the motorcyclist' attention is evaluated via a continuous recording of his gaze which acts as the main indicator regarding driver's performance. The main objective of this paper is to find out and evaluate the main factors of distraction of motorcyclists' attention in urban road segments and intersections. The results of this type of research procedures are very useful as a tool to encourage the adaptation of more precise regulations with regard to the road infrastructure, the placement of roadside elements, etc.

Keywords: Driving distraction · Motorcyclists · Naturalistic research · Road safety

1 Introduction

The visual distraction of driver's attention during the implementation of the driving is a reaction to visual stimuli, activated and developed in a different way in each person, depending on a variety of factors. It is detected in all drivers with floating rate and frequency of appearance. In each case, the results of this kind of distraction of attention are critical for the driving task, the driver's safety as well as the rest of road users.

During the past decade several studies has implied that distracted driver is one of the leading causes of fatal and serious road accidents. At the same time, scientific research has focused on drivers' distraction putting under research mostly car drivers. Road safety audit in the urban road environment during the last years includes apart

E. G. Nathanail et al. (Eds.): CSUM2020, AISC 1278, pp. 901–911, 2021.
https://doi.org/10.1007/978-3-030-61075-3_87

from the vehicles' drivers, the motorcycle ones [1]. At this direction, driver distraction causes as well as details of the process have to be studied on motorcyclists as well in order to be able to improve road environment and to create safe road networks.

The present paper examines the comprehension of distraction concerning a vulnerable category of road users, the distraction motorcycle drivers. The distraction of attention is evaluated via a continuous gaze recording which acts as the main indicator. The main objective of the research is to find out and evaluate the main factors of distraction of motorcyclists' attention in urban road network. Considering out-of-the-vehicle factors as highly significant, the study deals with distraction caused by those factors, including the ones which are not related with the driving task, resulting in impairments to driver attention and decision-making ability and hence the driver's overall performance. In this direction the paper presents the side effects of numerous road elements, such as information signs, roadside advertising, etc. to driver attention.

The study is based on a medium-scale experimental procedure which took place in a medium sized Greek city and it is part of a larger research project conducted by the University of Thessaly, Department of Civil Engineering. The experimental naturalistic data were collected by a total number of 11 motorcyclists with the use of Gaze Intelligence gaze tracker equipment, which is capable of making continuous data recording of driver's gaze direction. The gaze analysis was based on the records of the number of glances at every selected element of the route, as well as the total time that a specific point (whereas point in the paper implies a selected road element) captured driver's gaze. The analysis offers conclusions on different aspects - driver, road element and the road in total - comparing and categorizing the results appropriately.

1.1 Theoretical Background

In the first International Conference on Distracted Driving the scientific community agreed on a definition for distracted driving: "*Distraction involves a diversion of attention from driving because the driver is temporarily focusing on an object, person, task, or event not related to driving, which reduces the driver's awareness, decision-making, and/or performance, leading to an increased risk of corrective actions, near-crashes, or crashes*" [2].

The investigation of the main causes of distraction highlights two basic categories of distractor generators: those coming from the interior of the vehicle (internal factors) and those from the external environment (external factors) which can be grouped in four major subcategories: built roadway, situational entities, natural environment, and built environment [3].

Psychological tools such as Neisser theories, Gestalt theory etc. have been used from scientists in order to explain the phenomenon of driver distraction [4]. The most holistic approach is based on a multilevel procedure which is activated each time the attention is distracted based on many factors [5]. This multilevel procedure takes into consideration the several forms that distraction has - visual, cognitive, biomechanical and auditory [6].

Especially for advertisement it should be noted that the primarily reason of the existence of advertisement signs is to capture driver's eye in order to transmit a message. As a result, driver's attention is distracted from the driving task for a significant period of time that may not be accepted in terms of traffic safety. This disturbed attention, which is more intense in case of changeable messages event [7], diminishes driver's ability to respond and avoid a crash caused by an unexpected event. Several design guidelines regarding advertising placement have been published in many countries while in some of them roadside advertising is not permitted at all. However, financial interest on advertising causes a continuous pressure resulted an increased number of advertisement signs along the roads [8].

1.2 Frequency of Driver Distraction

The need for this research arises not only by the lack of relevant approaches but also from international and domestic statistics that rank driver distraction very high on the list of causes of accidents. The importance of the issue emerges from data which shows distraction from a secondary task as a cause of serious accidents as well as crashes. Both accident statistics and experimental studies conclude that distracted driving possesses a leading position among accident causes.

The most characteristic research in the field was carried out by the Virginia Tech Transportation Institute (VTTI) for NHTSA, the "100- Car Naturalistic Driving Study" [9]. During the 100-Car Naturalistic Driving Study, driver involvement in secondary tasks contributed to over 22% of all crashes and near-crashes recorded during the study period which is confirmed by accident data from United States [10]. In Greece, the official, most recent statistics published by the Greek Traffic Police for the first semester of 2019 give to driver distraction high ranking among all other accident causes as they show that 6% of the total fatal accidents were due to driver's distraction while other causes or types may have a strong relationship with the distraction issue. At this point it should be noted that a significant percentage of the accidents is reported as an accident of an unknown cause (18.5%) or under investigation (52,7%) [11].

In an attempt to study the phenomenon in depth certain studies focus on the external factors and the significant impact in driving task. A review of various accident databases resulted that external distractors are responsible for 10% of all accidents [12]. Particularly for advertising, many studies have identified them as a significant cause of driver distraction and traffic accidents [12–15]. This is confirmed also by a naturalistic research conducted by the University of Thessaly which not only proves the distraction caused by advertising but also perform a quantitative analysis of the mono-factor effects of various characteristics to show their effects on distraction time [4].

Motorcyclists' fatal accidents statistics highlight the importance of focusing on the specific road users. In Greece 37.9% of those accidents involve motorcyclists while a percentage of 36.9% of all accidents take place in urban areas [11].

2 Methodology

2.1 Selection of the Appropriate Method

The most suitable research methodology has been detected via a comparative study between the allocated methods [15]. The most popular among the available methods are included in four basic categories:

- Studies based on elements of accidents
- Experimental studies (simulator studies or studies in a test tracks)
- Observational-naturalistic studies (observation of determined point or use of special equipment vehicles)
- Questionnaires studies

Furthermore, there are some kinds of methods that are not included in any of the previous categories like Peripheral Detection Task and Visual Occlusion. The assessment of each method in the basis of its application in distraction studies depends on the driver category, prevailing conditions and available means is very important.

The method considered the most appropriate, for this kind of studies, belongs in the observational-naturalistic category.

2.2 Recruitment

The recruitment of the participants who carried out the experiment was based upon two parameters; the level of their driving experience and the variability of their demographic criteria. Due to the fact that male riders are over-represented in motorcycle accidents and they constitute the typical rider gender, all of the experiment subjects were male whose residence was adjacent to the tested road section. They used the urban network of the city where the measurements took place in a daily basis for both commuting and recreational purposes. In order to prevent any accidents during the execution of the experiments the participants were chosen to have increased riding experience and be in perfect health condition. Their age varied between 40 and 55 years old whilst their riding experience was minimum 5 years since they obtained a legal riding license. It is worth noting that the latter was a prerequisite from the insurance company of the motorcycle.

Before the commencement of the measurements the participants drove the motorcycle that was used during the experiment for such time until they felt familiar with its handling. That time ranged for few minutes for the experienced riders to few days for the inexperienced ones. That stage of the procedure was very crucial to safely complete the measurements since the size and weight of the motorcycle required adequate riding skills.

Aiming at a large data set of eye movements, efforts were put to ensure the maximum possible number of test riders. The authors addressed to motorcycle clubs seeking for volunteers and ended up with 11 riders who met the requirements described above. Although the initial target was to recruit 20 riders to carry out the measurements this number proved to be rather ambitious.

2.3 Recording Equipment

The recording of riders' visual behavior performed using the eye tracking glasses of Gaze Intelligence. Two cameras are installed at the inner side of the glasses, recording the behaviour of the eyes and one camera in the front captures the external environment. The main component accompanied by two pair of Infra-Red Filtering spare lenses of different shading. The purpose of these lenses was to filter the intense sunlight and reduce the effect of infrared waves. Either of the two set of lenses were installed depending on the weather conditions. However due to the fact that the external light sources and particular the sunlight intensity was not constant during the field measurements some of the data are lacking temporarily consistent accuracy.

With the help of an assistant and after putting the proper lens and a nose rest on the main tracking component the volunteer rider wore a flip up helmet in open face mode allowing thus a convenient adaptation of the recording equipment beneath the helmet. Afterwards, through a USB portal the eye tracking device was connected to a notebook. The next steps were to run the software that supports the hardware and calibrate the system based on three points mode at a distance of 3 m ahead of the rider. The calibration procedure performed for each individual independently. Finally, the assistant pressed the button to start the recording of the measurement, closed the lit of the notebook and put it in a rucksack hanging at the back of the rider who then ignited the motorcycle's engine (a Suzuki Vstrom DL 650) and started his route. The file of the measurement was a video file format (i.e. *.avi) which integrated the outcome of the three cameras described above. The successive directions of eye movements are depicted with a colored circle (Fig. 1).

Fig. 1. Snapshot of output video file.

2.4 Experimental Environment

The environment where the field measurements conducted was a mid-sized Greek city, the city of Arta. A road section of approximately 4 km passing through the city center was chosen as it meets the requirements of the experiment, gathering a great number of fixed and potential situational sources of riding distraction. The elements of the road that were analyzed belong in 5 categories: 1. Road signs, 2: Advertising related signs, 3. Combination of road signs with non-related to driving causes of distraction, 4. Signs with changeable message and 5: Other entities such as a gas station, a kiosk etc.

The measurements performed during working hours when the city center was crowded enough and more specifically between 27/12/2019–30/12/2019 from 10:00 to 14:00. The weather conditions were ideal for the needs of the experiment with prevailing moderate temperatures and winds whereas the cloudy ambient light conditions pro-vided smooth lighting conditions. However as stated before, short intervals of intense sunlight interrupted the cloudy conditions, posing a threat to the validity and accuracy of the recorded data. This undesirable situation was mitigated with the use of the proper pair of lenses.

3 Results

3.1 Overview

It is well known that scientific research is based on observing phenomena in order to collect and analyze the information needed so as to make useful and generalizable conclusions, to the maximum extent possible. Consequently, it was necessary to use descriptive analysis in order to present the basic characteristics, to identify trends and to form a cause-and-effect relationship. A total of 805 records have been produced as the results of measurements of 11 drivers, after data cleaning. A 4 level analysis of the data was performed aiming to offer conclusions regarding the issue under research. At this point it should be noted that distraction time intervals are classified into four groups: 0–0.7 s, 0.71–1.6 s, 1.61–2.0 s and anything more than 2 s [4]. The present paper examines the diversion of attention from driving due to specific visual stimuli for unsafe time intervals greater than 0.7 s which is the threshold for safe reaction times during the execution of the driving task.

3.2 Analysis of Significant Distractions of Attention Per Driver

The frequency of distraction per driver, for time intervals > 0.7 s, >1.6 s and > 2 s is presented in Fig. 2, while Fig. 3 presents the average time intervals for each of the previous mentioned cases.

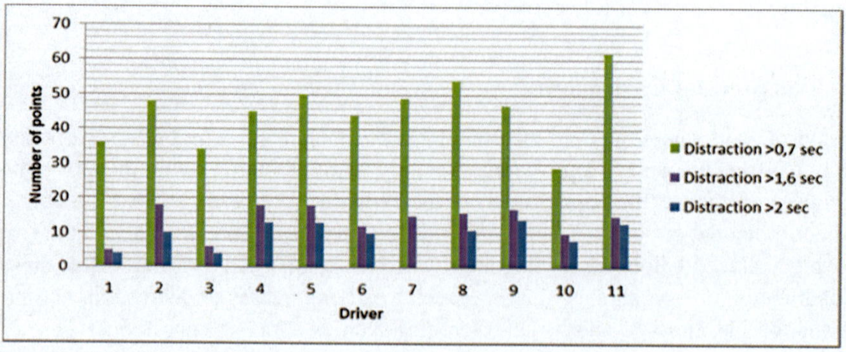

Fig. 2. Frequency of significant distractions per driver.

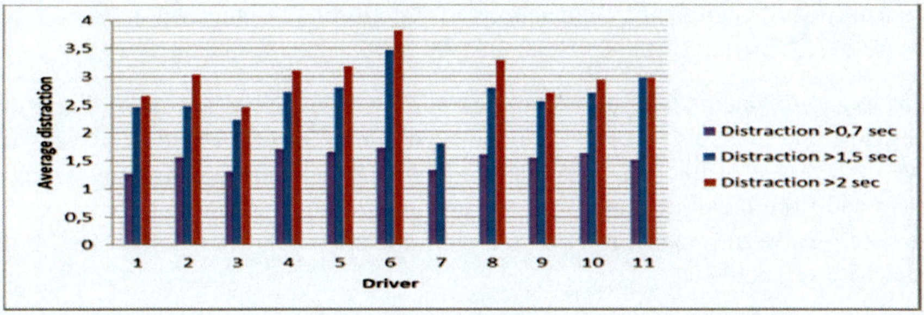

Fig. 3. Average time of significant distractions per driver.

From the two diagrams it is obvious that all drivers are distracted for non-safe time intervals during the implementation of the driving task while the number of distractions may differ among drivers. Furthermore, the fact that the minimum average distraction time among the participants is 1.2 s shows that road elements attract the driver's attention not only simultaneously but for a period of time that may permit an accident to take place.

3.3 Analysis of Significant Distractions of Attention Per Road Element Category

Figure 4 presents the average time intervals of distraction per road element category, for time intervals >0.7, >1.6 s and >2 s. The diagram includes the following categories: 1. Road signs, 2: Advertising related signs, 3. Combination of road signs with non-related to driving causes of distraction, 4. Signs with changeable message and 5: Other entities such as a gas station, a kiosk etc. From the analysis it is obvious that advertising possesses the leading position among distraction causes under study, both in terms of frequency of appearance and in average time intervals of distraction. Furthermore, signs with changeable messages also concentrate a great number of glances as well as the maximum average times of distraction.

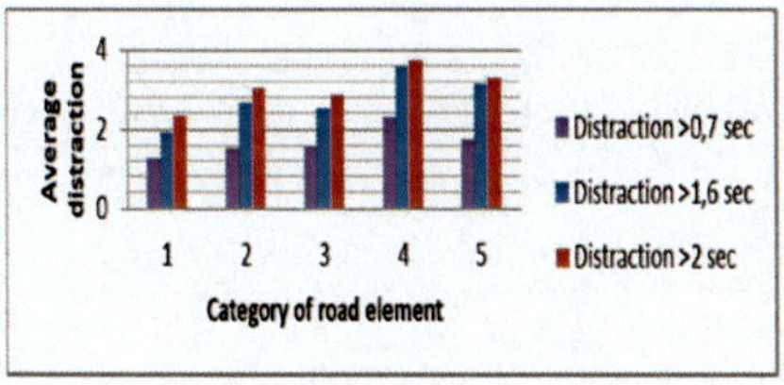

Fig. 4. Average time of significant distractions per road element category.

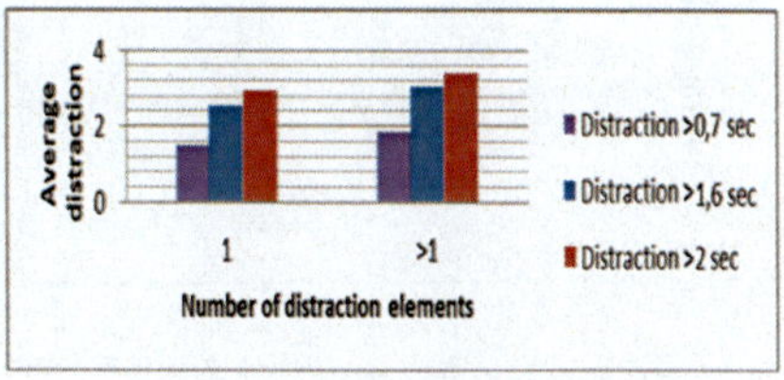

Fig. 5. Average time of significant distractions per number of elements.

3.4 Analysis of Significant Distractions of Attention Per Number of the Elements

The average time intervals of distraction per number of elements acting as distraction causes, for time intervals > 0.7, >1.6 s and >2 s is presented at Fig. 5. This analysis shows the average distraction from the one hand for distractions caused by a single element and from the other hand from distractions caused by multiple elements at the same time. The results appear to be increased in the latter case for a percentage of 26, 20 and 15% respectively.

3.5 Analysis of Significant Non Related to Driving Distractions

Both frequency and average times of distractions that are not caused from elements related to the task of drivers indicate a great excess of the safe - in terms of road safety - threshold (Table 1).

Table 1. Frequency and average time of non-related to driving distractions.

	Distraction >0.7 s	Distraction >1.6 s	Distraction >2 s
Frequency	429	130	95
Average time	1.58	2.74	3.12

Furthermore, as it is shown at Fig. 6, average time for each driver regarding the same distraction causes confirms the primary suspicious regarding the dangerous role of advertising and raises concerns about the initial intend of some driver's to get more distracted from non-driving related causes than others.

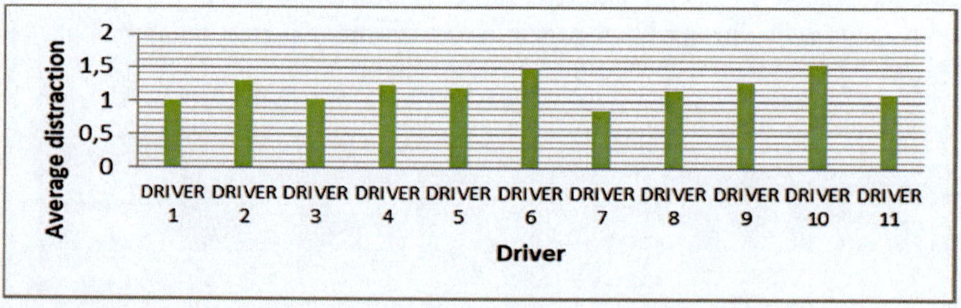

Fig. 6. Average time of significant non-related to driving distractions per driver.

4 Conclusions

The main objective of this paper is the detection of driver distraction due to road environment-based factors, especially those that are not related to the main task of driving. The focus of the present research on motorcyclists is due to their participation in road accidents and the lack of attention that had been given to these vulnerable road users from the scientific community.

The methods commonly used to study driver distraction are not all feasible or effective to the same extent. For this research a naturalistic method was selected. The chosen method serves the goal of the research and allows the continuous data recording with its main advantage being the fact that driving is as close to the real thing as possible. Thus, the results are characterized by a high degree of reliability and validity. In order to conclude regarding specific potential factors of driver distraction, a quantitative analysis was performed.

Among the restrictions of the research an emphasis has to be given on the lack of demographic data as it is the main reason for not performing an inductive statistical approach. Additionally, the small possibility of the researcher to create different driving scenarios as well as to control the environmental conditions is among the disadvantages of the method. Finally, the sensitivity to changes (e.g. lightness conditions) could be mentioned as disadvantage of the eye tracker.

The analysis of 805 cases shows that distractions for time intervals greater than 0.7 s are detected at all drivers during the driving task. While the number of distractions may differ among drivers, the minimum average distraction time among the participants, which is 1.2 s, shows that road elements attract the driver's attention not only simultaneously but for a period of time that may permit an accident to take place. Furthermore, both elements that are related to the driving task as well those that are not related to the driving task attract drivers attention, with the second group to lead the way.

Advertising related elements, especially with changeable messages concentrate a great number of glances as well as the maximum average times of distraction. In Greece, the relevant legislation regarding road advertising is not applied properly. The current legislation has to adapt research results in order to form a comprehensive policy to reduce visual pollution near the roads, aiming to the improvement of road safety by eliminating driver's visual distraction of attention [1]. In order to achieve this goal the collaboration with experts such as psychologists and doctors is necessary.

Discussions about the proximity of the objects along the road have to be made regarding the presence of multiple elements at the same time. Also, it is critical to specify the relationship between distraction and accident risk for each road under study.

Certain cases were detected with average distraction times significantly higher than average. Cases like these should be further analyzed. Additionally, elements that attract the attention for time intervals that are crucial for road safety must be analyzed in terms of usefulness and necessity as well as in terms of proximity.

The results of this type of research procedures are very useful as a tool to encourage the adaptation of more precise regulations with regard to the road infrastructure, the placement of roadside elements, etc. This research will be extended, in the future, not only to enrich the current data base in order to perform a more advanced statistical analysis but also to cover suburban road environments as they constitute a very risky environment due to high speeds.

Funding. This research study was conducted under the post doc scholarship supported by University of Thessaly and funded by Stavros Niarchos Foundation.

References

1. Misokefalou, E., Eliou, N., Galanis, A.: Driving distraction of attention audit of motorcycle and bicycle users in urban areas. Case study in the city of Volos, Greece. Paper presented at the 5th International Congress on Transportation Research, Volos, Greece (2010)
2. Hedlund, J., Simpson, H.M., Mayhew, D.R.: International Conference on Distracted Driving-Summary of Proceedings and Recommendations. Paper presented at the International Conference on Distracted Driving, Toronto (2005)
3. Horberry, T., Edquist, J.: Distractions outside the vehicle. In: Στο: Driver Distraction: Theory, Effects and Mitigation, pp. 215–228. CRC Press, Florida (2008)
4. Misokefalou, E.: Investigation into and modeling of factors influencing drivers' attention and impacts on road safety. Ph.D. thesis, University of Thessaly, Volos, Greece (2014)
5. Lee, J.D., Regan, M.A., Young, K.L.: What drives distraction? Distraction as a breakdown of multilevel control. In: Στο: Drver Distraction. Theory, Effects, Theory, and Mitigation, pp. 41–56. CRS Press, Taylor and Francis Group, Boca Raton, London, New York (2009)
6. Ranney, T.A., Garrott, W.R., Goodman, M.J.: National Highway Traffic Safety Administration driver distraction research: past, present and future. Paper presented at the 17th International Technical Conference on Enhanced Safety of Vehicles, Amsterdam (2001)
7. Oviedo-Trespalacios, O., Truelove, V., Watson, B., Hilton, J.A.: The impact of road advertising on driver behavior and implications for road safety: a critical systematic review. Transp. Res. Part A **122**, 85–98 (2019)
8. Herrstedt, L., Greibe, P., Andersson, P.: Roadside advertising affects driver attention and road safety. In: 3rd International Conference on Driver Distraction and Inattention, Gothenburg, Sweden, 4–6 September 2013 (2013)
9. Klauer, S.G., Dingus, T.A., Neale, V.L., Sudweeks, J.D., Ramsey, D.J.: The impact of driver inattention on near-crash/crash risk: an analysis using the 100-car naturalistic driving study data. Report No. DOT HS 810 594, National Highway Traffic Safety Administration, Washington, D.C (2006)
10. NHTSA's National Center for Statistics and Analysis: An examination of driver distraction as recorded in NHTSA databases, Research note. Report No. DOT HS 811 216, National Highway Traffic Safety Administration, Washington, D.C. (2009)
11. Greek Police. https://www.astynomia.gr. Accessed 26 Mar 2020
12. Wallace, B.: Driver distraction by advertising: genuine risk or urban myth? Municipal Eng. **156**, 185–190 (2003)

13. Stutts, J.C., Reinfurt, D.W., Staplin, L., Rodgman, E.A.: The Role of Driver Distraction in Traffic Crashes. AAA Foundation for Traffic Safety, Washington, D.C. (2001)
14. Young, K., Mahfoud, J.M., Stanton, N.A., Salmon, P.M., Jenkins, D.P., Walker, G.H.: Conflicts of interest: the implications of roadside advertising for driver attention while driving: skill and awareness during inspection of the scene. Transp. Res. Part F **12**, 381–388 (2009)
15. Misokefalou, E., Eliou, N.: Comparative analysis of drivers' distraction assessment methods. Paper presented at the 22nd ICTCT Workshop, Towards and Beyond the 2010 Road Safety Targets-Identifying the Stubborn Issues and their Solutions, Leeds, UK (2009)

Urban School Travel – Understanding the Critical Factors Affecting Parent's Choices

Kornilia Maria Kotoula[1](✉), George Botzoris[2], Georgia Ayfantopoulou[1], and Vassilios Profillidis[2]

[1] Hellenic Institute of Transport, 6th km Charilaou-Thermi Road, 57001 Thessaloniki, Greece
{nilia, gea}@certh.gr

[2] Department of Civil Engineering, Democritus University of Thrace, Kimmeria Campus, 67100 Xanthi, Greece
{gbotzori, vprofill}@civil.duth.gr

Abstract. Travelling to and from school forms mobility habits and travel behavior aspects of students from a very young age, also adopted in later life. Parents are the key players of the whole mode choice process as in most cases they are the ones to decide how and by which transport mode their children will complete their everyday school trips. Understanding parents' perceptions on different travel modes and studying the motives behind the mode choice decision in school trips, is a rather essential issue as it may provide useful information to policy-makers, transport and spatial planners on how to overcome possible barriers and difficulties in order to satisfactory cover all students' future mobility needs. The paper provides an extensive literature review regarding a wide range of factors found to influence students' travel, following a statistical exploratory factor analysis of a questionnaire survey took place in Thessaloniki, Greece. The initial analysis of the sample identifies key themes while it also develops a comprehensive picture of caregivers' experiences about travel mode choice to school in a typical Greek urban environment. Some interesting findings verify that socio-economic and household demographic factors, built-environment variables, and parents' attitudes regarding their daily trips and mobility habits, are important factors affecting the school mode choice procedure.

Keywords: School transportation · Mode choice · Travel behavior

1 Introduction

Travelling to school and back, forms mobility habits and travel behavior aspects of students from a very young age, also adopted in their later life. School students travelling by private vehicle are more likely to present a negative attitude towards other transport modes as adults, while the adoption of policies and the implementation of specific infrastructures that enhance the active transportation (walking and bicycling) may lead to a behavioral change which maintains throughout the students' adulthood [1].

E. G. Nathanail et al. (Eds.): CSUM2020, AISC 1278, pp. 912–922, 2021.
https://doi.org/10.1007/978-3-030-61075-3_88

In the mode choice process, parents are the key players as in most cases they are the ones to decide how and by which transport mode their children will travel to school, while most of times they even prefer to differentiate their own mobility habits in order to meet their children mobility needs [2]. On the contrary, university students are the ones to decide on their own as they form a social group, essentially autonomous in terms of mode choice decision making [3].

The above-mentioned issues combined to the fact that the use of private vehicles for school trips has increased in recent decades against alternative transport modes, pose an imperative need to understand the parents' perceptions regarding the use of different travel modes. A further analysis of the motives behind the parents' mode choice decision on school trips, may lead to solutions' and strategies' proposals which may reverse this trend in the near future, while it will also provide useful information to policy-makers, transport and spatial planners on how to overcome possible barriers and difficulties in order to satisfactory cover all students' future mobility needs.

The objective of this study is to investigate the critical factors that affect parents' choices regarding their children school trips and was initially motivated by the observation that there is a striking lack of research on this topic in Greece. The paper is structured as described below: firstly, an extensive literature review is presented regarding the relevant research; following a description of the study subjects as well as the data sources is provided; following basic empirical results are presented and discussed; finally, conclusions are provided and future directions are summarized.

2 Literature Review

Students' age and gender appear to directly affect the mode choice in school transportation as they are considered key factors in most relevant studies. Parents of students aged 12–18 years express significantly less anxiety regarding walking and bicycling to school than for younger age groups [4], while escorting students to school is reduced proportionally to age [5]. As many studies conclude, older students are most likely to walk to school than the younger ones [6–8], while, it is also more common for older students to not travel by car but by alternative transport modes [9, 10]. However, there are studies not correlating students' age with mode choice [11, 12].

In addition to the age, a difference is observed regarding the transport mode used in school trips between girls and boys. According to several studies [10, 13, 14], girls are less likely to make use of alternative transport modes, while other argue that there is no correlation between gender and the use of alternative transport modes [15, 16].

Parents' professional status has also been found as another factor influencing students' mobility patterns. A working mother in the family environment has a catalytic effect on the increase of students driven to school [17], while fathers' working hours seem not to be related to the mode choice decision process [9, 17].

Two additional factors found to affect parents' mode choice decision is the family income and the car ownership index. These two factors seem to be inextricably linked, as the number of vehicles in a household depends directly on family income. The car ownership index also determines the number of trips that can potentially be executed during a typical weekday. As many studies conclude, students of high family income are less likely to travel alternatively to and from their school unit [18, 19]. Similarly, research studies [20, 21] argue that an increase in the number of vehicles available in a household increases possibilities for students be driven to school.

The distance of students' residence from school is extensively examined and proven to be a determining factor in the mode choice decision process. Increasing distance reduces the likelihood for students to travel by alternative transport modes [18, 22], and long distances are associated with less independent school mobility [23–25].

The built environment and the parents' perception regarding the safety levels of students' neighborhood, also plays a critical role in the selection process. High traffic loads, high speeds, a congested road network of wide axes and the need of crossing overloaded intersections, seem to negatively affect parents in choosing alternative transport modes for their children's school trips [4, 7, 19]. The reconstruction of sidewalks and the installation of traffic control devices seem to enhance alternative mobility in areas where such interventions take place [26], while the existence of organized bicycle paths enhances the choice of bicycle as a school transport mode [4].

Safety is another crucial factor in the school mode choice decision process. Traffic conditions and criminality seem to be the two predominant factors that greatly influence parents' decision. Fears and worries about the provided safety levels in school transportation have been reported in numerous studies [27, 28], even though fatal pedestrian and cyclist accidents during the school hours show a sharp decline from 1987 to 2009 [29]. On the other hand, parents' concerns about criminality are based more on social standards that have been developed rather than real risks. Kidnapping cases account for only 2% of violent crimes among young people and 4% of all kidnappings in school districts [30]. However, parents are often worried about the worst, even in cases they know the chances of adduction are extremely low [31]. Extra interpersonal risks such as bullying, gang activity and homelessness are reasons parents often report. Parent' fears and concerns largely explain the fact that girls are less likely to walk or bike to school and to be accompanied by older adults [13, 32].

3 Research

3.1 Case Study

For the research needs and based on the literature review findings, a questionnaire survey was designed and conducted in order factors affecting parents' mode choice decision on school trips in the city of Thessaloniki to be defined. The survey addressed parents of both public primary and high school students. The authors decided not to include private schools as in this case the transport system is organized differently with most of parents to prefer the school bus provided by the private school unit as they consider it safer than any other school transport mode [33]. Regarding the study area, Thessaloniki is the second-largest city in Greece with over one million inhabitants in metropolitan area, while the area where the current research focuses numbers approximately 930,000 inhabitants (around 100,000 primary and high school students).

The survey took place from May to June 2019 and from September to November 2019 and the minimum sample size n (512 students) was defined using the Eq. (1) for N = 100,000 students, p = 50%, d = ±5%% and $z_{\alpha/2}$ = 1.96 for confidence level 95% [34]:

$$n \geq N \cdot \left[1 + \frac{N-1}{p \cdot (1-p)} \cdot \left(\frac{d}{z_{a/2}}\right)^2\right]^{-1} \tag{1}$$

The respondents completed the questionnaires by participating in personalized interviews or by replying to invitations sent to them via their e-mail address. The questionnaire is structured in 3 parts. In the first one questions regarding the parents' socio-demographic characteristics and students' personal characteristics are included, while in the second part issues related to the current school travel procedure are investigated. Three main topics are examined and analyzed in the last part, related to:

- parents' perception regarding specific parameters appear to motivate them in the mode choice decision process. For that, eighteen factors are presented, and parents are called to rate them using a typical 5-level Likert scale in order the level of significance for each factor to be defined,
- parents' perception regarding the built environment safety where students travel. For that, thirteen statements are presented and parents are called to declare the level of agreement or disagreement for each one by using a 5-level scale,
- parents' travel habits regarding the use of motorized and non-motorized transport modes. In this last part of the questionnaire, fifteen relative statements are provided in order possible links between the parents' attitudes towards specific transport modes and the final mode choice decision to be defined.

4 Descriptive Analysis

Regarding the first part's analysis, the 2/3 of the sample are women. Most of respondents belong in the age group of 40–49 years old (66.2%), are full time employees (85.4%), holding a bachelor's degree (35.2%). Most parents possess a driver license (78.5%) and the car ownership index is estimated 1.55 vehicles/household. Regarding students, 54.1% are girls. Concerning the distance from the school unit it seems that 60.7% of students live in a distance less than 1 km and 13.5% live in a distance more than 3 km from the school unit, a fact that probably prohibits them to walk to school.

Differences are found in the mode choice between the morning and the afternoon shift, while interest present the differentiation of parents' responses regarding the current transport mode used for school trips and the one they would prefer if circumstances were different (Table 1). Significant differences are noticed in the '*Walking with friends*', '*School bus*' and '*By parent's car*' options. The '*waking with friends*' option increases by 7% in the afternoon shift, probably due to the fact that parents feel more comfortable and safe during this hour of the day, permitting therefore their children travel more independently. Frequencies of responses indicate a large difference between the current transport mode choice and the preferred one with the most significant to be noticed regarding the option *"I drive him/her to school"*. While 42.4% of parents choose private vehicle for the completion of the school trip in the morning, only 11.1% would prefer to use it if circumstances were different. The 31.3% of difference corresponds to the 1/3 of the sample and it seems to be allocated in the preferable option of walking with friends (13.3% increase) and use of school bus paid by the school unit (12.6% increase).

Table 1. Differences between the current and the preferable transport mode during the morning and afternoon shift.

Transport mode options	Morning shift			Afternoon shift		
	Current transport mode	Preferable transport mode	Difference	Current transport mode	Preferable transport mode	Difference
Walking, alone	10.0%	11.1%	1.1%	11.0%	11.1%	0.1%
Walking, with friends	15.6%	28.9%	13.3%	22.6%	30.0%	7.4%
Walking, with me	16.0%	18.2%	2.2%	18.4%	17.5%	−0.9%
Walking, accompanied by adults	5.8%	5.3%	−0.5%	7.4%	5.4%	−2.0%
Urban bus	2.0%	3.6%	1.6%	3.4%	3.2%	−0.2%
School bus	7.2%	19.8%	12.6%	16.4%	20.7%	4.3%
By parent's car	42.4%	11.1%	−31.3%	18.6%	10.1%	−8.5%
By a friend's car	1.0%	0.0%	−1.0%	1.8%	0.0%	−1.8%
Bicycle	0.0%	2.0%	2.0%	0.0%	2.0%	2.0%
Taxi	0.0%	0.0%	0.0%	0.4%	0.0%	−0.4%

Regarding parents' perception on specific parameters found to motivate them in the mode choice decision process, in Fig. 1 the cumulative frequencies of each parameter on both sides of the scale are depicted. Safety, student's age and distance from the residence to the school unit are the most important factors concentrating high percentages of significance with travel time, parents' working hours and students' safety to follow, while gender and income are the less significant factors.

Concerning the cumulative frequencies of elements depicted in Fig. 2, these are related to the level of significance parents attribute to variables linked directly with the built environment. The majority of parents (79%) agree that *"The existing infrastructure cannot protect a cyclist"*. A high percentage of parents (56.5%) agree with the statement *"Residences in the neighborhood are in good condition"* and 46.2% agree or strongly agree that *"There are no vandalism trails in the neighborhood"*. Parents seem to keep a rather neutral position regarding the statement *"The route from residence to school is safe"*, while their perception about the sidewalks safety is rather negative.

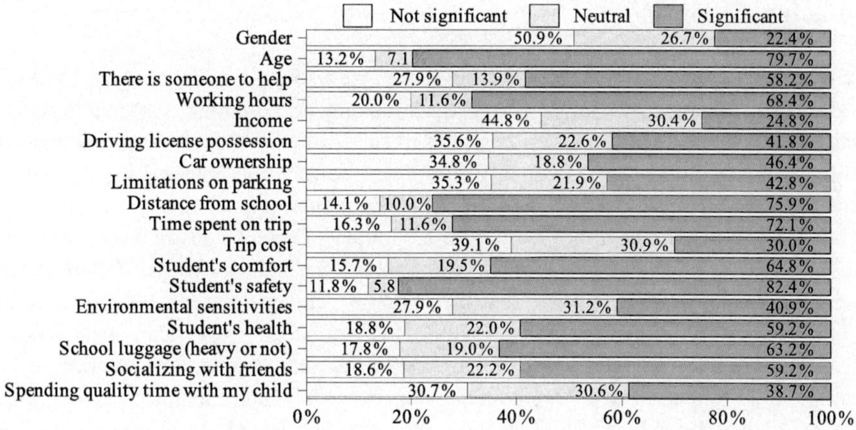

Fig. 1. Cumulative frequencies regarding parents' motives in the mode choice process.

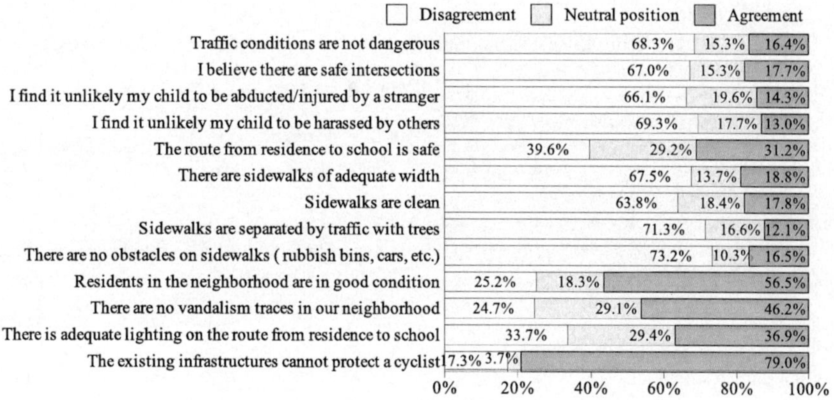

Fig. 2. Cumulative frequencies regarding parents' perceptions on built environment.

Finally, Table 2 presents the level of parents' agreement and disagreement in fifteen statements related to motorized and non-motorized transport modes.

Parents in their large majority are not satisfied with time consistency and comfort of the provided urban transport services as almost 56.6% of the sample believes that the urban bus is not a reliable transport mode and almost 50% avoids using it for trips within the city. Regarding the use of private vehicle, 80.4% of parents agree that traffic congestion do bothers them and 91.3% agree that *"Driving to school contributes to traffic congestion"*. While 57.6% agree that *"Students driven by car to school are more possible to be dependent on it"*. As for parents' perception related to the bicycle use, 91.4% would like their children to use alternative transport modes for their school trips while 88.3% believe that *"Walking or cycling to school increases students' physical activity"*.

5 Factor Analysis

Following an Exploratory Factor Analysis (EFA) was adopted to investigate whether the observed variables (49 items of the questionnaire) could form factors. Initially, variables were tested regarding their correlations (use of Pearson coefficient) and results revealed a large number of statistically significant correlations making it possible to group the variables into factors. For the EFA, principal axis factoring method was deployed using the direct oblimin rotation technique (Table 3). In total 41 variables were included on the factors (weights over 0.05). The Kaiser-Meyer-Olkin index concerning the sampling adequacy was high (0.864), and the Bartlett test concerning the homogeneity of variances across factors was statistically significant ($p < 0.001$) indicating that factor analysis could be applied to the data sample. The rotation technique identified 9 factors with eigenvalues greater than 1 which include variables with factor loadings greater than 0.5. The scree plot also confirmed the existence of nine factors. These nine factors account for 61% of the total variation of data. The factors were labeled based on the conceptual framework of variables.

Table 2. Parents' attitudes towards motorized and non-motorized transport modes.

Statement	Totally disagree	Disagree	Neither agree nor disagree	Agree	Totally agree	Mean	s.d
Travelling to school on foot/bike is a good way my child to become familiar with the neighborhood	0.6%	7.2%	11.7%	63.9%	16.6%	3.89	0.78
I would like my child to travel to school on foot or by bike under the appropriate circumstances	1.4%	2.1%	5.1%	51.2%	40.2%	4.27	0.77
Travelling to school on foot or by bike is a way to increase my child's physical activity	0.4%	0.6%	10.7%	50.6%	37.7%	4.25	0.70
Driving my child to school may lead to car use addiction	2.2%	16.1%	24.1%	41.2%	16.4%	3.54	1.02
Driving to school contributes to traffic congestion	1.2%	2.0%	5.5%	57.2%	34.1%	4.21	0.73
Driving is more comfortable than walking/cycling	13.9%	26.8%	30.7%	22.3%	6.3%	2.81	1.13
I like driving within the city	16.6%	20.9%	33.2%	21.7%	7.6%	2.83	1.17
Owning a car makes my life more comfortable	3.7%	6.4%	17.0%	55.9%	17.0%	3.76	0.94
I use my car even for short distances	14.5%	27.5%	24.2%	26.4%	7.4%	2.85	1.18
Car ownership is a prestige symbol	28.6%	30.5%	26.4%	9.2%	5.3%	2.32	1.14
Traffic congestion doesn't bother me	38.6%	41.8%	6.3%	5.9%	7.4%	2.02	1.16
I like to use the urban bus for travelling within the city	18.2%	31.3%	25.0%	19.1%	6.4%	2.64	1.17
The urban bus is a very reliable transport mode	28.1%	28.5%	24.0%	17.4%	2.0%	2.37	1.12
I am satisfied with the comfort of the urban bus	38.8%	35.4%	16.0%	5.9%	3.9%	2.01	1.07
I am satisfied with the time consistency of the urban bus services	41.7%	33.0%	14.1%	5.9%	5.3%	2.00	1.13

Table 3. Exploratory factor analysis results.

Variables	Factor: description, (eigenvalue/% of variance explained)
1. Safety	Parents' motivation to select a transport mode for the school trip (9.555/22.420%)
2. Travel time	
3. Working hours	
4. There is someone to help me in my child's transportation	
5. Age	
6. Convenience	
7. Distance	
8. Socialization	Parents' motivation to select transport mode regarding the child's physical and mental health (4.522/10.150%)
9. Health	
10. Luggage weight	
11. Environmental sensitivities	
12. Quality time	
Variables	*Factor: description, (eigenvalue/% of variance explained)*
13. Satisfied with the comfort of urban bus services	Parents' attitude towards the use of urban bus (3.172/6.849%)
14. Urban bus is a reliable transport mode	
15. Satisfied with time reliability of urban bus services	
16. I like travelling by urban bus within the city	
17. Transport mode, School – Residence)	Mode choice (2.444/5.017%)
18. Preferable transport mode, School – Residence	
19. Transport mode Residence – School	
20. I like driving within the city	Parents' attitude towards the use of private vehicle (2.223/4.422%)
21. I use my car for all trips within the city	
22. Driving is more comfortable than walking/bicycling	
23. Owning a car makes my life comfortable	
24. I would prefer my child to walk/bike to school under different circumstances	Parents' attitude towards walking and bicycling (1.962/3.821%)
25. Walking/bicycling to school is a way my child be familiar with environment	
26. Walking or cycling to school increases students' physical activity	
27. Car ownership	Parents' motivation for private car use (1.621/3.197%)
28. Driving license possession	
29. There are no parking limitations outside my home or the school unit	
30. There are no trails of vandalism in the neighborhood	Neighborhood safety (1.495/2.629%)
31. Residences of neighborhood are in good conditions	
32. The neighborhood the student travels is safe	

(*continued*)

Table 3. (*continued*)

Variables	Factor: description, (eigenvalue/% of variance explained)
33. Sidewalks have sufficient width	Sidewalks and route safety (1.193/2.024%)
34. Sidewalks are clean	
35. Sidewalks are separated from traffic with trees	
36. There are no obstacles on the sidewalks	
37. The crossings are safe	
38. Traffic conditions are not dangerous for the student	
39. It is unlikely my child to be harassed by others	
40. It is unlikely that my child will be injured or abducted by a stranger	
41. Adequate traffic lights	

6 Conclusions

According to the survey's results, predominant transport mode for school trips is the private vehicle. However, a mode shift could be achieved if conditions allowed, posing walking first in the selection list and the use of hired school bus second. Safety, age, distance, travel time and working hours are the most significant factors, while travel costs and income seem not to have any essential effect. Regarding the built environment, the lack of appropriate infrastructure for walking and cycling seems to mostly affect negatively parents to allow students either walk or bike to school. In addition, the inconsistency of the current urban transport system as well as the low provided comfort levels are found the main factors affecting their negative attitude towards the use of urban bus, while under different circumstances the majority prefer their children to walk or bike to school, as this increases their physical activity. The EFA defined nine clustered factors indicating as a next step a Confirmatory Factor Analysis to be deployed for the creation of a measurement model which will investigate the variables' validity and reveal relevant covariances between some of the factors and some of the errors of observed variables. As a final step, the development and analysis of a Structural Equation Model (SEM) is proposed for defining direct and indirect influences and interactions between the factors and the mode choice decision.

References

1. Carlin, J.B., Stevenson, M.R., Roberts, I.: Walking to school and traffic exposure in Australian children. Aust. N. Z. J. Public Health **21**(3), 286–292 (1997)
2. Ehteshamrad, Sh.: Analysis of joint mode choice behavior of parents for work and children for school trips. Ph.D. dissertation, Tarbiat Modares University (2017)

3. Kotoula, K.M., Sialdas, A., Botzoris, G., Chaniotakis, E., Grau, J.M.S.: Exploring the effects of university campus decentralization to students' mode choice. Periodica Polytechnica Transp. Eng. **46**(4), 207–214 (2018)
4. Kerr, J., Rosenberg, D., Sallis, J., Saelens, B., Frank, L., Conway, T.: Active commuting to school: associations with environment and parental concerns. Med. Sci. Sports Exerc. **38**(4), 787–794 (2006)
5. Sheiner, J.: School trips in Germany: gendered escorting practices. Transp. Res. Part A Policy Pract. **94**, 76–92 (2016)
6. Noland, R., Park, H., Von Hagen, L.A., Chatman, D.: A mode choice analysis of school trips in New Jersey. In: 89th Annual Meeting of the Transp. Research Board, Washington (2010)
7. Seraj, S., Sidharthan, R., Bhat, C., Pendyala, R., Goulias, R.: Parental attitudes towards children walking and bicycling to school: a multivariate ordered response analysis. In: 91st Annual Meeting of the Transportation Research Board, Washington D.C. (2012)
8. Yang, Y., Markowitz, E.: Integrating parental attitudes in research on children's active school commuting: evidence from a community school travel survey. J. Transp. Res. Board **2318**(1), 116–127 (2012)
9. McDonald, N.: Children's travel: patterns and influences. University of California, Berkeley, United States (2005)
10. Zwerts, E., Wets, G.: Children's travel behavior a world of difference. In: 85th Annual Meeting of the Transportation Research Board, Washington, D.C. (2006)
11. Merom, D., Tudor-Locke, C., Baumana, A., Risselb, C.: Active commuting to school among NSW primary school children in implications for public health. Health Place **12**(4), 678–687 (2006)
12. Dellinger, A.M., Staunton, C.E.: Barriers to children walking and bicycling to school - United States, 1999. Morb. Mortal. Wkly. Rep. **51**(32), 701–704 (2002)
13. O'Brien, M., Jones, D., Sloan, D., Rustin, M.: Children's independent spatial mobility in the urban public realm. Childhood **7**(3), 257–277 (2000)
14. Evenson, K., Huston, S., McMillen, B., Bors, P., Ward, D.: Statewide prevalence and correlates of walking and bicycling to school. Arch. Pediatr. Adolesc. Med. **157**, 887–892 (2003)
15. Su, J.G., Jerrett, M., McConnell, R., Berhane, K., Dunton, G., Shankardass, K., Wolch, J.: Factors influencing whether children walk to school. Health Place **22**, 153–161 (2013)
16. Carlin, J.B., Stevenson, M.R., Roberts, I.: Walking to school and traffic exposure in Australian children. Aust. N. Z. J. Public Health **21**, 286–292 (1997)
17. DiGuiseppi, C., Roberts, I., Li, L.: Determinants of car travel on daily journeys to school: cross sectional survey of primary school children. Br. Med. J. **316**, 1426–1428 (1998)
18. Ermagun, A., Samimi, A.: Active transportation mode choice behavior across genders in school trips. In: 91st Annual Meeting of the Transportation Research Board, Washington, D.C. (2012)
19. Mitra, R., Buliung, R.N.: Intra-household travel interactions, the built environment and school travel mode choice: an exploration using spatial models. In: 91st Annual Meeting, Transportation Research Board, Washington D.C. (2012)
20. Copperman, R.B., Bhat, C.R.: An analysis of the determinants of children's weekend physical activity participation. Transportation **34**(1), 67–87 (2007)
21. Mackett, R.L.: Letting children be free to walk. In: 12th Annual Conference on Walking and Liveable Cities (WALK 21), Vancouver, Canada (2011)
22. Cervero, R., Duncan, M.: Walking, bicycling, and urban landscapes: evidence from the San Francisco bay area. Am. J. Public Health **93**(9), 1478–1483 (2003)
23. Fyhri, A., Hjorthol, R.: Children's independent mobility to school, friends and leisure activities. J. Transp. Geogr. **17**(5), 377–384 (2009)

24. Carver, A., Timperio, A., Crawford, D.: Parental chauffeurs: what drives their transport choice? J. Transp. Geogr. **26**, 72–77 (2013)
25. Curtis, C., Babb, C., Olaru, D.: Built environment and children's travel to school. Transp. Policy **42**, 21–33 (2015)
26. Boarnet, M.G., Anderson, C.L., Day, K., McMillan, T.E., Alfonzo, M.: Evaluation of the California safe routes to school legislation: urban form change and children's active transportation to school. Am. J. Prev. Med. **28**(22), 134–140 (2005)
27. Martin, S., Carlson, S.: Barriers to children walking to or from school - United States, 2004. Morb. Mortal. Wkly. Rep. **54**(38), 949–952 (2005)
28. McDonald, N.C., Aalborg, A.E.: Why parents drive children to school: implications for safe routes to school programs. J. Am. Plan. Assoc. **75**(3), 331–342 (2009)
29. National Highway Traffic Safety Administration: School Bus Driver In-Service Safety Series (2011)
30. Finklehor, D., Ormrod, R.: Kidnapping of Juveniles: Patterns from NIBRS. Juvenile Justice Bulletin. U.S. Department of Justice: Office of Justice Programs: Office of Juvenile Justice and Delinquency Prevention (2000)
31. Ahlport, K.N., Linnan, L., Vaughn, A., Evenson, K.R., Ward, D.S.: Barriers to and facilitators of walking and bicycling to school: formative results from the non-motorized travel study. Health Educ. Behav. **35**(2), 221–244 (2008)
32. Valentine, G.: Oh yes I can. Oh no you cannot: children and parents' under-standing of kids' competence to negotiate public space safely. Antipode **29**(1), 65–89 (1997)
33. Sakellariou, A., Kotoula, K.M., Morfoulaki, M., Mintsis, G.: Identification of quality indexes in school bus transportation system. Transp. Res. Procedia **24**, 212–219 (2017)
34. Profillidis, V., Botzoris, G.: Modeling of Transport Demand: Analyzing, Calculating, and Forecasting Transport Demand, 1st edn. Elsevier, Oxford (2018)

Comparison of Driver's Behavior in Greece and Palestine (West Bank)

Jameel Al-Karablieh and Fotini Kehagia(✉)

Civil Engineering Department, Aristotle University of Thessaloniki, 54124 Thessaloniki, Greece
fkehagia@civil.auth.gr

Abstract. The Driver Behavior Questionnaire is one of the most widely used instruments for measuring self-reported driving behaviors. The aim of this research is to conduct a comparative analysis of driving behavior based on the results of the AUTh Driver Questionnaire (ADQ) in a representative sample of drivers in Greece and Palestine (West Bank). ADQ was created in order to find a new comprehensive form which includes several aspects relating the attitudes of drivers and evaluating their driving performance. Data on a range of attitudes towards road safety were collected using an online survey. A total of 830 responses were obtained for each country. The Questionnaire consists of two main divisions; the first section includes information regarding socio-demographic information of drivers in terms of gender, age category, level of education, etc. and the second section comprised of 50 questions relating characteristics of drivers' behavior. The data analysis included of two main parts: descriptive analysis and factor analysis, which comprises of exploratory factor analysis (EFA) and confirmatory factor analysis (CFA). Through EFA; the criteria used to determine the number of factors were the Kaiser criterion of eigenvalues over 1.0, the Cattell Scree plot, parallel analysis, and the interpretability of factors. The analysis resulted eleven and eight factors had eigenvalues over 1.0 in Greece and Palestine samples respectively. The results of the study highlight the differences in the drivers' behavior indicating differences in national culture and the commitment of drivers to traffic regulations.

Keywords: Driver's behavior · Drivers' questionnaire · Drivers' attitudes · Violation · Factor analysis

1 Introduction

Every year the lives of approximately 1.35 million people are cut short as a result of a road traffic crash (WHO 2018). Road traffic crashes continue to present a serious public health challenge (Rowe et al. 2015). Analyses of traffic accidents indicate that human factors are a sole or a contributory factor in road traffic accidents (Lewin 1982; Bener et al. 2008). It has been estimated that the human factor is responsible for 85–90% of road accidents (Lewin 1982; Sarbescu 2013). Thus, many studies have focused on the relationship between the human factor and driving behavior (Dahlen and White 2006; Sarbescu 2013). Driver behavior is a very complex matter that is influenced by one's knowledge, abilities, and skills on the one hand and personality traits such as volition,

E. G. Nathanail et al. (Eds.): CSUM2020, AISC 1278, pp. 923–934, 2021.
https://doi.org/10.1007/978-3-030-61075-3_89

values, and motives on the other. Jashua and Garber (1992) has detected that the most common accident type resulted from drivers' faults. Drivers often engage in behaviors that pose a risk to both themselves and to other road users. While many of these unsafe actions are active, conscious rule violations, others are the result of errors due to inexperience, momentary mistakes or inattention. Intentional or not, both rule violations and deficiencies in memory, judgment, or situational awareness can and do contribute to traffic collisions (Stanton and Salmon 2009; Cordazzo et al. 2014). Based on Reason's extensive work on the human contribution to disaster across a wide range of situations, the Driver Behavior Questionnaire (DBQ), it was designed as a self-report measure of the behaviors that may increase risk of crash involvement (Reason et al. 1990; Rowe et al. 2015). DBQ is the most widely used and applied form in many countries on variety categories of drivers such as United Kingdom, USA, China, Australia, Sweden, Greece, Netherlands, Spain, France, New Zealand, Turkey, Qatar and United Arab Emirates. However, the factorial structures of the DBQ as well as the number of items vary between different drivers' cultures and nations (Martinussen et al. 2013). Since its development, the DBQ has been applied in various researches regarding risky and/or aggressive driving, and road accidents (Parker et al. 1995; Iversen and Rundmo 2002; Machin and Sankey 2008).

Many studies have found support for the original three-factor structure of DBQ which is consisting of lapses, errors and violations. But there are other studies which applied DBQ, getting another aspects which are relating to drivers behaviors; where they found that aggressive violations, ordinary violations and lapses were applicable, although not firmly stable across countries (Martinussen et al. 2013). But, the DBQ model was created at a time when some of modern technologies were not used or widespread such as using of maps navigation GPS, mobile phones and others. The future versions of the DBQ must contain the items regarding mobile phone to examine the influence of using it on the behavior of drivers while driving (Maslac et al. 2017). Thus, there is necessity to create a new comprehensive and modern model, which contains of the attitude of drivers and evaluation their behaviors during driving of vehicles by diversity of questions. For that it, a new model issued and named AUTH driver questionnaire (ADQ) for evaluation of drivers' behaviors. Whereas, ADQ adopted mostly on pattern which was used on DBQ. ADQ is new form and created to evaluate of drivers' behaviors via 50 different questions which are relating to drivers of vehicles, and it was applied on drivers as the first time in Greece and Palestine.

2 Methodology

2.1 Preparation of Questionnaire

The questionnaire form prepared to distribute online and included various aspects of driver behavior. It consists two main divisions; the first section includes of socio demographic information of drivers in terms of gender, age category, level of education, marital status, suffering of disability, wearing of glasses and doing of traffic accidents with or without injury. The second section comprises of 50 questions which are evaluation of drivers' behaviors and their commitments to satisfy of traffic safety.

ADQ available in languages of Greek and Arabic which used in Greece and Palestine respectively. Furthermore, it is available in English language to be used for any country.

2.2 Distribution of Questionnaire

The questionnaire form was distributed online randomly by sending the questionnaire link via email and social media such as Facebook, Whatsapp and Viber, where it was sent and shared to different categories of drivers in Greece and Palestine. In each country, a total of 830 of drivers' participants who completed filled of questionnaire.

3 Data Collection

Data of filled questionnaires included the answers of 830 respondents which collected online; where, the first part was for driver information and included eight miscellaneous questions; four questions for personal information and another four questions answering by dichotomous scale yes or no. About the answers of the second part of questions was used Likert scale of five points from always to never. Whereas, data of respondents' answers obtained online and were prepared as appropriately forms to analyze by statistical methods as required.

4 Analysis of Data

The stage of data analysis of ADQ included of three main parts: descriptive analysis, exploratory factor analysis (EFA) and confirmatory factor analysis (CFA), which applied for collected data of 830 respondents of drivers in Greece and Palestine. The analysis of data was performed by the statistical software package IBM SPSS for the descriptive analysis and EFA. In addition, M-Plus software was used for CFA.

4.1 Descriptive Analysis

The 830 participants in Greece and Palestine were answered the items of the questionnaire. The descriptive analysis of the results of both in Greece and Palestine was presented in Table 1.

The items of questions were answered by five points of Likert scale that comprised of always, very often, sometimes, rarely and never which represented by numbers of 1, 2, 3, 4 and 5 respectively. There are differences in the values of mean and standard deviation of drivers' answers to the questions. This gives an indication that there are variety in attitude and behavior of drivers between Greece and Palestine.

Table 1. Results of the analysis is concerning the mean and standard deviation of answers of respondents.

Item No.	Item	Greece		Palestine	
		Mean	Std. Dev.	Mean	Std. Dev.
Q1	I comply the instructions and regulations of traffic	1.53	0.811	1.48	0.711
Q2	I haven't committed a traffic violation of law	1.92	0.875	1.72	0.763
Q3	I give the priority and open the way for emergency vehicles such as police, ambulance, civil defense, etc.	1.16	0.693	1.37	0.689
Q4	Watching the traffic and warning signs on the road increase my attention while driving	1.61	0.838	1.69	0.737
Q5	knowing that there is traffic control including intelligent surveillance improve my behavior and my commitment to traffic instruction while driving	1.90	0.874	1.61	0.79
Q6	I think that increasing the traffic awareness of drivers through audio and visual means will reduce the violation of traffic laws, traffic accidents and achieves the required traffic safety	1.85	0.893	1.88	0.752
Q7	I park the vehicle in suitable places which permitted by law	1.81	0.858	1.7	0.673
Q8	I use seat belt while driving	1.31	0.704	1.8	0.801
Q9	I make sure that all passengers fastening the seat belts before beginning of trip	1.58	0.835	2.21	0.952
Q10	I prefer driving a vehicle on the road during the daytime rather than driving at nighttime	2.79	0.948	2.18	0.863
Q11	I prefer driving the vehicle during the daytime because the vision is better for other vehicles and surrounding environment	2.74	1.146	2.06	0.926
Q12	I prefer driving the vehicle in the lighted roads at evening time than the roads that are not lit, considering that they are furnished with lines, marks and reflectors required	2.52	1.029	2.09	0.824
Q13	I prefer driving the vehicle on the road during good weather conditions with clear visibility comparison to other weather conditions	1.88	0.999	2.02	0.98
Q14	The bad weather conditions and the resulting blurred vision or risk of slippage, for example, affect my performance positively when I drive a vehicle on the road, such as increasing attention, reducing speed, and not violating traffic laws	1.97	0.978	2.05	0.98

(*continued*)

Table 1. (*continued*)

Item No.	Item	Greece		Palestine	
		Mean	Std. Dev.	Mean	Std. Dev.
Q15	The bad weather conditions reduce my overall use of the vehicle on the road	3.16	1.018	3.25	0.868
Q16	The temperature rising through summer season is increasing my tiredness, fatigue and reduce attention while driving a vehicle on the road	3.09	0.969	2.99	0.761
Q17	The system of cooling and heating effects on me positively while driving the vehicle	2.44	1.04	2.36	0.9
Q18	I maintain a clear vision during driving by taking care of the cleanliness of the glass	1.63	0.896	1.64	0.778
Q19	I don't drive the vehicle on the road while I am under the influence of alcohol or narcotic drugs that were prevented to take during driving	1.41	0.792	1.36	0.784
Q20	I drive the vehicle on the road while feeling tired, drowsy or sick	4.09	0.86	3.63	0.949
Q21	I drive the vehicle on the road while feeling angry or challenging others	3.88	0.91	3.43	0.904
Q22	I don't take regular breaks while driving long distances	3.91	0.956	3.4	0.895
Q23	I take stimulants that help you wake up while driving when feeling tired especially when driving for long distances	4.33	0.917	3.85	1.078
Q24	The traffic jam is affecting on my behavior during driving, such as feeling angry or tired, etc.	3.00	0.967	2.73	0.766
Q25	I talk by cell phone while driving a vehicle on the road	3.76	0.865	2.55	0.917
Q26	I use the Internet via a Smartphone, such as texting and reading messages, social networking applications, browsing and watching the screen while driving a vehicle on the road	4.08	0.907	3.57	0.92
Q27	I use alternatives of phone device while driving such as a Bluetooth headset or speaker in the vehicle	3.61	1.238	2.47	0.897
Q28	I think that using smart applications to identify road and locations improves your performance and behavior while driving on the road	2.29	1.097	2.45	0.843

(*continued*)

Table 1. (*continued*)

Item No.	Item	Greece		Palestine	
		Mean	Std. Dev.	Mean	Std. Dev.
Q29	I reduce the speed of the vehicle to appropriate speed when approaching specific areas such as intersections, curves and slopes, etc. on the roads	1.48	0.778	1.35	0.679
Q30	I decrease the speed of vehicle and stop when approaching the pedestrian crossing areas	1.39	0.784	1.46	0.76
Q31	I commit to the appropriate speed required while driving on the roads whether or not there are signs for speed limit	1.83	0.913	1.67	0.767
Q32	I cross the intersection of a light signal in the case of yellow light while driving the vehicle	3.99	0.982	3.95	0.888
Q33	The delay in getting the desired place or urgency for any reason motivate you to increase speed or violate some traffic laws to reduce the time to reach	4.24	0.848	3.68	0.945
Q34	I reduce vehicle speed when approaching bumps on roads	1.80	0.926	1.62	0.764
Q35	I inspect daily my vehicle, such as water, oil engine and tires, etc., before driving the vehicle	2.54	0.91	2.21	0.786
Q36	I don't drive the vehicle which has a mechanical failure that may cause the vehicle to stop or not to be controlled	1.53	0.83	1.63	0.831
Q37	I check regularly and repair the vehicle in a timely manner without delay it	1.91	0.875	1.79	0.742
Q38	I eat and drink while driving a vehicle on the road	3.92	0.917	3.28	0.843
Q39	I talk and engage with passengers while driving a vehicle on the road	2.88	0.887	2.75	0.726
Q40	I smoke while driving a vehicle on the road	4.19	1.032	3.72	1.318
Q41	The existence of announcements of all kinds besides the roads, is affecting on my attention and distract my concentration while driving the vehicle	3.48	0.967	3.36	0.961
Q42	I observe the screen that shows the speed of the vehicle, engine information, fault codes, etc.	1.42	0.723	2.36	0.834
Q43	I drive on the off-road, not paved or unsafe to reach the required location	4.35	0.877	3.64	0.83

(*continued*)

Table 1. (*continued*)

Item No.	Item	Greece		Palestine	
		Mean	Std. Dev.	Mean	Std. Dev.
Q44	I specify the route that you will take and estimate the expected time of trip to get the required location before start driving of vehicle	2.19	0.925	2.45	0.82
Q45	I make sure that the first-aid kit is available in my vehicle	2.04	0.984	2.29	1.061
Q46	I take care of the presence of emergency supplies in the vehicle such as fire extinguisher, equipment for changing the wheels and the searchlight for lighting in the dark, etc.	1.60	0.87	1.86	0.966
Q47	I prefer the system of participation with other vehicles or using public transport more than using of the private car for economic saving	3.36	0.968	3.76	1.083
Q48	I prefer driving the vehicle which has less fuel consumption such as Hybrid Cars that are less impactful for environmental pollution	2.50	1.08	2.52	1.09
Q49	I check and estimate the amount of fuel available in the vehicle and its adequacy to be used before starting the journey and arriving required location	1.69	0.908	1.81	0.884
Q50	The low income and economic crises affect and reduce of using my vehicle in mobility	2.93	1.053	3.34	0.976

4.2 Factor Analysis

Exploratory Factor Analysis is a statistical approach for determining the correlation among the variables in a dataset. This type of analysis provides a factor structure (a grouping of variables based on strong correlations). In general, an EFA prepares the variables to be used for creation structural equation modeling (Haig 2010; Yong and Pearce 2013). EFA is a widely utilized and broadly applied statistical technique in the sciences. EFA with varimax rotation as the method used for data analysis, and of the researchers who report their criteria for deciding the number of factors to be retained for rotation, a majority use the Kaiser criterion. Furthermore, study design, data properties, and the questions to be answered all have a bearing on which procedures will yield the maximum benefit (Costello and Osborne 2005). Several methods have been proposed for determining the number of factors. Most of studies dropped the factors whose eigenvalues are less than one since these provide less information than is provided by a single variable and according to the scree graph, which is probably the most popular method for determining the number of factors, but it is subjective, causing different people to analyze the same data with different results. Varimax rotation is the most popular orthogonal rotation technique (Costello and Osborne 2005). EFA with Varimax rotation was applied in the study as a first step to determine the main variables. EFA is a technique within factor analysis and identifying the relationships between measured variables. This branch of analysis is implemented regarding driving performance and driver behaviors in order to investigate which observed variables are most highly correlated with the common factors and how many common factors are needed to give an adequate description of the data (Papantoniou 2015). EFA was used for ADQ for Greece and Palestine.

The case study of Greece; through applied of EFA and estimation the stability of data sample for factor analysis by KMO and Bartlett's test. It was found that there is high correlation between the variables depending on the resulted values of Kaiser-Meyer-Olkin = 0.938 and sig. of Bartlett's Test = 0.000. There were 11 factors generated within eigenvalue greater one and these factors explained 63.513% of total variance.

According to the results; all values of factors loading are acceptable where they are more than 0.3 which ranges between (0.348–0.861); thus each construct satisfies the convergent validity. Moreover, it includes the eigenvalues which are more than 1 which ranged (1.034–14.816); and it comprises of Cronbach's Alpha which were all acceptable and greater than 0.5 where they ranged from (0.505–0.924); it is indication all factors are reliable and there is a good internal consistency among items of the questionnaire. Considering to the results of EFA; it indicates the ADQ which applied in Greece, is successful and acceptable. The factors were named based on the set of questions which contained and numbered from 1 to 11 respectively as following; 1. Positive behavior, 2. Weather conditions and daytime, 3. Commitment of traffic rules, 4. Smart devices, 5. Errors, 6. Ordinary violation, 7. Essential supplies, 8. Economic status, 9. Distraction, 10. Attention and 11. Feeling. An evaluation of the inter-correlation between the factors that resulted by analysis for Greece is shown in Table 2.

The case study of Palestine; through applied of EFA and estimation the stability of data sample for factor analysis by KMO and Bartlett's test. It was found that there is high correlation between the variables depending on the resulted values of Kaiser-Meyer-Olkin = 0.947 and sig. of Bartlett's Test = 0.000. There were 8 factors generated within eigenvalue greater one and these factors explained 63.445% of total variance. Depending on the results; all values of factors loading are acceptable where they are more than 0.3 which ranges between (0.354–0.810); thus each construct satisfies the convergent validity. Furthermore, it comprises of the eigenvalues which are more than 1 which ranged (1.093–16.059); and it comprises of Cronbach's Alpha which were all acceptable and greater than 0.5 where they ranged from (0.607–0.919); it is indication all factors are reliable and there is a good internal consistency among items of the questionnaire. In addition, according to the results of EFA; it indicates the ADQ which applied in Palestine, also is successful and acceptable. The factors were named and numbered from 1 to 8 respectively as following; 1. Commitment of Traffic Rules, 2. Weather Conditions and Attitude, 3. Essential Supplies, 4. Distraction, 5. Economic Status, 6. Ordinary Violation, 7. Feeling and 8. Errors.. An evaluation of the inter-correlation between the factors that resulted by analysis for Palestine is shown in Table 3.

A confirmatory factor analysis is carried out in order to examine the fit of the model established in the EFA (Martinussen et al. 2013). It is used to test whether measures of a construct are consistent with a researcher's understanding of the nature of that construct (or factor). Furthermore, the objective of CFA is to test whether the data fit a hypothesized measurement model. CFA is a quantitative data analysis method that belongs to the family of structural equation modeling (SEM) techniques. Through CFA will determine the significance of the analysis. The Comparative Fit Index (CFI) is equal to the discrepancy function adjusted for sample size. CFI ranges from 0 to 1 with a larger value indicating better model fit, Tucker-Lewis Index (TLI) ranges from 0 to 1 and; Root Mean Square Error of Approximation (RMSEA) is related to residual in the model and RMSEA values range from 0 to 1 with a smaller RMSEA value indicating better model fit. In addition to other values of Standardized Root Mean Square Residual (SRMR), X^2/df and significant value (p). If model fit is acceptable, the parameter estimates are examined. The ratio of each parameter estimate to its standard error is distributed as a z statistic and is significant at the 0.05 level. Unstandardized parameter estimates retain scaling information of variables and can only be interpreted with reference to the scales of the variables. Standardized parameter estimates are transformations of unstandardized estimates that remove scaling and can be used for informal comparisons of parameters throughout the model. Standardized estimates correspond to effect-size estimates (Prudon 2015).

The case study of Greece, CFA was performed in order to test the fit of ADQ model which comprised of 11 factors. The results of structure of 11 factors showed that the factors inter-correlate to explain of drivers' behaviors. No items loaded on more than one factor. The goodness of fit indices suggests satisfactory for model including of the whole sample which included 11 factors that had 50 items. Furthermore, the resulted values of CFA were given as indication of correlation between the factors, and the required the values of fit indexes as following:

CFI = 0.844, TLI = 0.829, RMSEA = 0.053, SRMR = 0.083, X^2/df = 18083.582/1225 and $p = 0.000$.

The case study of Palestine, CFA was performed in order to test the fit of ADQ model which comprised of 8 factors. The results of the structure of 8 factors showed that the factors inter-correlate to explain of drivers' behaviors. No items loaded on more than one factor. The goodness of fit indices suggests satisfactory for model including of the whole sample which included 8 factors that had 50 items. Furthermore, the resulted values of CFA were given as indication of correlation between the factors, and the required the values of fit indexes as following:

CFI = 0.743, TLI = 0.725, RMSEA = 0.070, SRMR = 0.102, X2/df = 19181.269/1225 and $p = 0.000$.

As shown in the results of questionnaire analysis that was applied in both Greece and Palestine, that it fulfilled the main requirements for its applicability to various categories of drivers. In addition, the difference in results, especially the number of factors that resulted in each country, gives an indication that there is a diversity in the behavior of drivers according to countries difference.

5 Inter-correlations Between the Factors

This part of the study includes an evaluation of the inter-correlation between the factors that resulted by analysis for Greece and Palestine as showing in Table 2 and Table 3.

Table 2. Inter-correlation between factors of ADQ in Greece.

Factor	1	2	3	4	5	6	7	8	9	10	11
1	1										
2	.508**	1									
3	.781**	.485**	1								
4	−.235-**	−.292-**	−.201-**	1							
5	−.530-**	−.378-**	−.510-**	.454**	1						
6	−.581-**	−.481-**	−.521-**	.328**	.484**	1					
7	.539**	.254**	.472**	−.116-**	−.324-**	−.441-**	1				
8	.128**	.072*	.142**	.054	−.101-**	−.177-**	.285**	1			
9	−.383-**	−.364-**	−.326-**	.393**	.433**	.475**	−.228-**	−.101-**	1		
10	.464**	.246**	.416**	−.095-**	−.316-**	−.312-**	.434**	.152**	−.232-**	1	
11	.242**	.280**	.208**	.006	−.091-**	−.146-**	.126**	.067	−.090-**	.115**	1

** Correlation is significant at $p < 0.01$
* Correlation is significant at $p < 0.05$

Based on the values which shown in Table 2 for Greece, there are correlations between twelfth factors within different levels. Moreover, there are significant between most of the factors. Whereas, most of the correlations were at significant level $p < 0.01$ and the others at $p < 0.05$ exception between factors of 4 and 8; in addition, between 8 and 11.

Table 3. Inter-correlation between factors of ADQ in Palestine.

Factor	1	2	3	4	5	6	7	8
1	1							
2	.600**	1						
3	.566**	.457**	1					
4	−.119-**	.077*	.118**	1				
5	−.173-**	−.040	−.165-**	.022	1			
6	−.331-**	.010	−.184-**	.318**	.200**	1		
7	.339**	.426**	.222**	.037	.159**	−.100-**	1	
8	.575**	.429**	.409**	−.190-**	−.244-**	−.392-**	.190**	1

** Correlation is significant at $p < .01$
* Correlation is significant at $p < .05$

Furthermore, according to the values which shown in Table 3 for Palestine, there are correlations between eighth factors within different levels. Also, there are significant between most of the factors. Whereas, most of the correlations were at significant level $p < 0.01$ and the others at $p < 0.05$ exception between factors of 2 and 5, 2 and 6, 4 and 5; in addition, between 4 and 7.

The results of inter-correlation between the factors in Greece and Palestine, give an indication of there are consistency among the factors.

6 Conclusions

The study included a new model using to assess drivers' behaviors in various aspects through variety of questions. It was applied on different categories of drivers of two different countries in terms of continent, language, culture, regulations, etc. ADQ was filled by various categories of drivers in terms of gender, age, education level, etc., in Greece and Palestine. There were variation in the mean values of drivers' answers of ADQ, which gave an indication of different behavior of drivers between Greece and Palestine. EFA resulted different number of factors, whereas the analysis resulted 11 and 8 factors for Greece and Palestine respectively. The different results of factors numbers and distribution of questions, gave an indication that there are differences of drivers' attitudes and behaviors between Greece and Palestine. ADQ is reliable in Greece and Palestine; where all the values of factors which resulted by EFA were acceptable; the loading of questions were higher than 0.30 and the Cronbach's alpha of factors were higher than 0.50 in both countries. CFA were mostly acceptable of both countries depending on the values which resulted by the analysis. Moreover, It was found that there are correlations between the factors as well as between the factors and the items of questionnaire. Among the most important differences between the two countries are the culture, applicable laws and traffic regulations, road infrastructure, traffic awareness for drivers, economic situation, etc. These variances were a basis for comparing the behaviors of the various categories of drivers in Greece and Palestine and obtaining results that give an indication of the impact of different countries on road safety by evaluating drivers' performance and behaviors while driving vehicles.

References

Bener, A., Ozkan, T., Lajunen, T.: The driver behaviour questionnaire in Arab Gulf countries: Qatar and United Arab Emirates. Accid. Anal. Prev. **40**(4), 1411–1417 (2008)

Cordazzo, S., Scialfa, C., Bubric, K., Ross, R.: The driver behavior questionnaire: a North American analysis. J. Saf. Res. **50**(2014), 99–107 (2014)

Costello, A., Osborne, J.: Best practices in exploratory factor analysis: four recommendations for getting the most from your analysis. Pract. Assess. Res. Eval. **10**(7), 1–9 (2005)

Dahlen, E., White, R.: The big five factors, sensation seeking, and driving anger in the prediction of unsafe driving. Pers. Individ. Differ. **41**(5), 903–915 (2006)

Haig, B.: Abductive Research Methods; Exploratory Factor Analysis. International Encyclopedia of Education (Third Edition) (2010)

Iversen, H., Rundmo, T.: Personality, risky driving and accident involvement among Norwegian drivers. Pers. Individ. Differ. **33**(8), 1251–1263 (2002)

Jashua, C., Garber, J.: A Causal Analysis of Large Vehicle Accidents Through Fault- Tree Analysis. Risk Anal. **12**(2), 173–187 (1992)

Lewin, I.: Driver training: a perceptual-motor skill approach. Ergonomics **25**(10), 917–924 (1982)

Machin, A., Sankey, K.: Relationships between young drivers' personality characteristics, risk perceptions, and driving behavior. Accid. Anal. Prev. **40**(2), 541–547 (2008)

Martinussen, L., Blomqvist, L., Moller, M., Ozkan, T., Lajunen, T.: Age, gender, mileage and the DBQ: the validity of the Driver Behavior Questionnaire in different driver groups. Accid. Anal. Prev. **52**(2013), 228–236 (2013)

Maslac, M., Antic, B., Pesic, D., Milutinovic, N.: Behaviors of professional drivers: validation of the DBQ for drivers who transport dangerous goods in Serbia. Transp. Res. Part F **50**(2017), 80–88 (2017)

Papantoniou, P.: Risk factors, driver behavior and accident probability. The case of distracted driving. Doctoral thesis. Department of Transportation Planning and Engineering, School of Civil Engineering, National Technical University of Athens (2015)

Parker, D., Reason, T., Manstead, R., Stradling, S.: Driving errors, driving violations and accident involvement. Ergonomics **38**(5), 1036–1048 (1995)

Prudon, P.: Confirmatory factor analysis as a tool in research using questionnaires: a critique. Compr. Psychol. **4**(10), 1–19 (2015)

Reason, T., Manstead, A., Stradling, S., Baxter, J., Campbell, K.: Errors and violations on the roads – a real distinction. Ergonomics **33**(10–11), 1315–1332 (1990)

Rowe, R., Roman, G., McKenna, F., Barker, E., Poulter, D.: Measuring errors and violations on the road: a bifactor modeling approach to the Driver Behavior Questionnaire. Accid. Anal. Prev. **74**(2015), 118–125 (2015)

Sarbescu, P.: Psychometric properties of the Manchester Driver Behavior Questionnaire in Romania: validation of a cross-cultural version. Int. J. Traffic Transp. Psychol. **1**(1), 20–27 (2013)

Stanton, N., Salmon, P.: Human error taxonomies applied to driving: a generic driver error taxonomy and its implications for intelligent transport systems. Saf. Sci. **47**(2009), 227–237 (2009)

Yong, A., Pearce, S.: A beginner's guide to factor analysis: focusing on exploratory factor analysis. Tutorials Quant. Methods Psychol. **9**(2), 79–94 (2013)

Investigating the Correlation of Mobile Phone Use with Trip Characteristics Recorded Through Smartphone Sensors

Panagiotis Papantoniou[1(✉)], Armira Kontaxi[1], George Yannis[1], and Petros Fortsakis[2]

[1] National Technical University of Athens, 5 Heroon Polytechniou Street, 15773 Athens, Greece
{ppapant, geyannis}@central.ntua.gr, akontaxi@mail.ntua.gr

[2] OSeven Telematics Limited, 27B Chaimanta Street, 15234 Athens, Greece
pfortsakis@oseven.io

Abstract. The Internet of Things (IoT) constantly offers new opportunities and features to monitor and analyze driver behaviour through wide use of smartphones, effective data collection and Big Data analysis. In that framework, the aim of this paper is to investigate the impact of mobile phone use on driving behaviour and road safety through the investigation of driving analytics collected by smartphone sensors. For this purpose, a 100-driver naturalistic experiment was carried out and an innovative data collection scheme using a smartphone application was exploited in order to record the respective driving performance data. Statistical analyses were carried out using linear mixed binary logistic regression models in order to investigate the correlation of mobile phone use with trip characteristics, such as speed, duration, harsh events, etc. Exposure metrics found to be significantly associated with the probability of mobile phone use are total trip distance and driving on workdays and during rush hours. Additionally, the increase of the average speed is found to reduce the probability of mobile phone use while driving.

Keywords: Mobile phone · Driving behaviour · Naturalistic experiment · Smartphone sensors · Mixed binary logistic regression

1 Background and Objectives

Road safety is a complicated scientific field of transport research of high importance, as road accidents constitute a major problem in a socioeconomic level. There is a significant number of risk factors identified in literature, which affect accident probability. The most important risk factors recognized in literature (WHO 2018; Elvik 2004) are human factors (speeding, distracted driving, driving under the influence of alcohol and other psychoactive substances etc.), unsafe road infrastructure, unsafe vehicles and inadequate law enforcement of traffic laws. Among them, human factors are likely to be the most crucial cause of road traffic fatalities and injuries every year and therefore the importance of studying how these factors can affect traffic risk is high (Salmon et al. 2011).

E. G. Nathanail et al. (Eds.): CSUM2020, AISC 1278, pp. 935–944, 2021.
https://doi.org/10.1007/978-3-030-61075-3_90

Regarding distracted driving, mobile phone use and complex conversation appear to be the most critical risk factors (Papantoniou et al. 2015). Given that mobile usage is an inevitable part of everyday driving process and is expected to increase over the years (Stutts et al. 2001), its impact on driving behaviour in traffic and road safety is particularly important and should be further investigated. Literature so far has showed that when the driver is using the phone while driving, this is reflected to the behaviour expressed in terms of loss of control, response to incidents, or accident occurrence (Bellinger et al. 2009; Fitch et al. 2015; Dingus et al. 2016).

Although the importance of driver monitoring is established in the transportation field, researchers have been struggling with the difficulty of collecting accurate real-time driving data by adopting low-cost collection and processing methods. In that environment, the high penetration rate of smartphones and social networks nowadays provide new opportunities and features to monitor and analyze driver behaviour. The use of smartphones allows for drivers to be recorded under normal driving conditions and without any influence from external parameters, resulting at being considered as one of the most appropriate methods for the assessment of driving behaviour (Ziakopoulos et al. 2020).

Many studies have shown interesting results using data collected through smartphone sensors under naturalistic driving conditions. Researchers aim at examining the effect of various driving behaviour indicators on driver performance and cash risk (Tselentis et al. 2018; Papadimitriou et al. 2019; Paefgen et al. 2012) or at identifying aggressive and dangerous driving profiles through a clustering approach (Mantouka et al. 2018; Ma et al. 2017; Meseguer et al. 2013).

In light of the aforementioned, the aim of the current research is the utilization of high-resolution smartphone data for the investigation of the impact of detailed trip characteristics on the use of mobile phone while driving. To that end, a naturalistic driving experiment was carried out in order to model the probability of the mobile phone use while driving in all type of roads, and separately in each road type (urban, rural, highway).

2 Data Collection

In order to achieve the research objective, an innovative smartphone application developed by OSeven Telematics (www.oseven.io) was exploited aiming to record driver behaviour using the hardware sensors of the smartphone device. One hundred and sixteen drivers participated in the related experiment during a 2-months timeframe and a large database of several thousand trips is created. The solid integration platform for collecting, transferring raw data and recognizing the driving behaviour metrics via ML algorithms is also developed by OSeven Telematics. Data collected from the application has been utilised in earlier research papers which also feature additional details regarding the application (Tselentis et al. 2018; Papadimitriou et al. 2019). The standard procedure that is followed every time a new trip is recorded by the application is clearly presented in Fig. 1.

Fig. 1. The OSeven data flow system.

For every trip a driver completes, a large amount of data is recorded, transmitted through WiFi or cellular network and valuable critical information such as metrics, features, highlights and driving score is produced in order to evaluate driving profile and performance. The exposure indicators that are available include indicatively duration (seconds), total distance (mileage), type(s) of the road network used, given by GPS position and integration with map providers e.g. Google, OSM, (highway, rural or urban environment) time of the day driving (rush hours, risky hours) and weather conditions. Moreover, the driving indicators which can reliably quantify the risk associated with a specific driving behaviour are the following: speeding (distance and time of driving over the speed limit and the exceedance of the speed limit), driver distraction (caused by smartphone use during driving), number and severity of harsh events number and severity of harsh events (braking and acceleration), harsh cornerings, driving aggressiveness (e.g. braking, acceleration).

Finally, it must be noted that since privacy and security consist two critical principles in the field of telematics, the OSeven platform has very clear privacy policy statements for the end users covering the type of data collected, the reason data is collected for, the time that data is stored and the procedures for data security based on encryption standards for data in transit and at rest. All this is done using state-of-the-art technologies and procedures in compliance with standing Greek and European personal data protection legislation (GDPR).

3 Theoretical Background

In the present analysis, the mobile phone use while driving is examined and correlated with trip characteristics recorded through smartphone sensors. The examined variable can be available either as the percentage of time using the mobile phone per trip driving duration or as a binary regarding the entire trip in a form of yes/no aspect. Herein, the binary variable is selected for modelling the use of mobile phone.

Since, the dependent parameter consists of a binary variable, binary logistic regression is selected as the appropriate analysis method. A binary logistic regression estimates the probability that a characteristic is present given the values of explanatory variables; $\pi = \Pr(y = 1|X = x)$. The mathematical formula provides the odds of this event occurring, when correlated with various independent variables. The odds are expressed by the logit link function as follows:

$$logit(\pi_i) = logit\frac{\pi_i}{1-\pi_i} = \beta_0 + \Sigma\beta_\iota x_\iota \quad (1)$$

The related outcome is specified as:

$$y_i = \beta_0 + \Sigma\beta_\imath x_\imath + e_{0i} \tag{2}$$

However, one may also consider that in the present dataset there are repeated measurements (trips) over the same units (drivers). Therefore, in order to capture personal driver traits, such as personality and experience, which affects their driving style, random effects are introduced to Binary logistic regression Models in order to extend them as Mixed Binary logistic regression Models. In other words, Mixed Binary logistic regression Models are formulated by using the fixed effects Binary logistic regression Models as basis and then adding random effects to the equation. The formulation of the linear mixed effects model, assuming a random intercept reflecting the repeated measurements (i) over drivers (j), is as follows:

$$y_{ij} = \beta_{0j} + \Sigma\beta_\imath x_{\imath j} + e_{0ij} \tag{3}$$

$$\beta_{0j} = \beta_0 + u_{0i} \tag{4}$$

4 Results

4.1 Descriptive Statistics

Overall, during the 2-months timeframe of the experiment 11.987 trips from a sample of 100 car drivers have been recorded. However, for the present analysis a crucial criterion was set; all drivers chosen to be included in the analysis were required to have driven at least for 40 trips. This number approximately equals the typical monthly number of working trips for a driver assuming that each driver drives 2 trips a day for 5 working days a week. This number is reasonable to filter out drivers for which there are not enough observations, and it is also the 'industrial' criterion set by OSeven to start providing driver evaluation. As a result, from the 100 car drivers, 82 were ultimately selected creating a large dataset of 11.398 trips. While this might seem like a considerable loss of driver diversity, only 589 trips in total were removed from the initial dataset (resulting in a 4.91% reduction).

Before presenting the models development, explanatory descriptive analysis of the data is implemented, allowing for an overview of the examined risk factor; the use of the mobile phone while driving. In this context, the use of the mobile phone in this subsection is presented as a share of trip time during which mobile phone was used (% of trip duration), taking into consideration that it constitutes a common risk metric that reveals unsafe driving behavior patterns among the drivers (Lipovac et al. 2017). A first interesting observation is the different frequency of mobile use with respect to the type of road network. Specifically, it is found that the share of mobile use is more frequent on urban and rural roads in comparison with highways, which seems logical, taking into consideration the driving pattern in highways; longer distances covered and high speeds developed (Fig. 2).

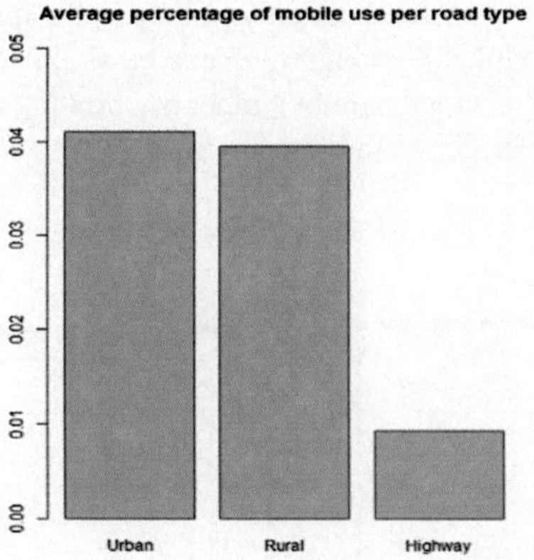

Fig. 2. Average percentage of mobile phone use per road type.

Furthermore, Fig. 3 illustrates the average percentage of mobile use of the sample collected on a driver basis (left) and on a trip basis (right). It is evident that the majority of drivers show an average of 2% mobile phone use during their trips, which is reflected in the histogram developed for the number of trips, as well. However, it should be noted that there are a few trips where the share of mobile phone use reaches a maximum value equal to 60% of the total trip duration, indicating that this may mostly concerns short trips; the trip duration of the collected sample varies from 61 to 12,480 s with a standard deviation of 918 s.

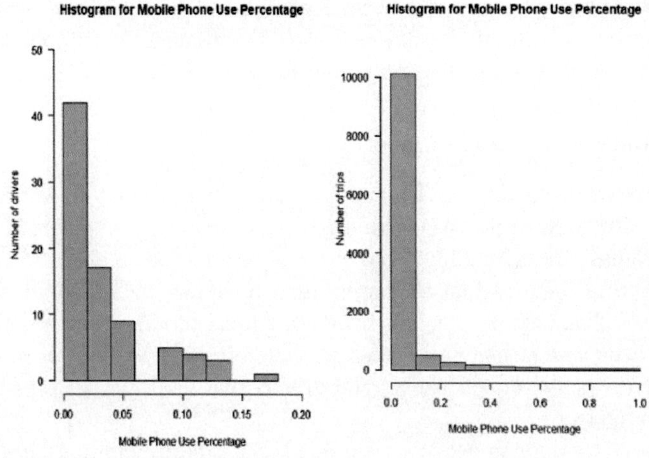

Fig. 3. Histogram of the percentage of time using the mobile phone per trip driving duration per driver (left) and trip characteristics (right).

Another interesting finding concerns the correlation of mobile phone use and harsh events recorded during a trip duration. Harsh acceleration, harsh braking and harsh cornering events can be considered highly significant indicators that reveal an aggressive and unsafe driving behaviour (Vaiana et al. 2014). Figure 4 presents a scatter plot illustrating the average number of harsh brakings (left) and harsh accelerations (right) per trip duration, against the share of mobile phone use during the trip on a driver basis.

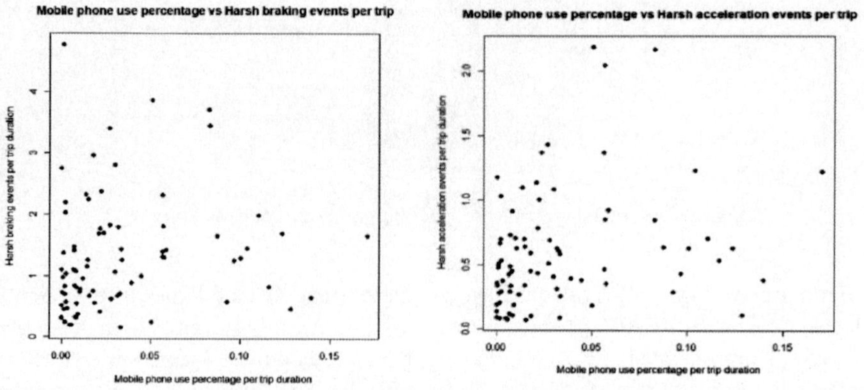

Fig. 4. Percentage of time using the mobile phone per trip driving duration in relation with the average harsh braking (left) and harsh acceleration (right) events per trip duration.

Overall, the scatter plots reveal a nonlinear relationship between the two driving behaviour indicators for both event types, resulting in a similar pattern despite the difference in the average number of harsh brakings and harsh accelerations, respectively. More specifically, harsh events are slightly less frequent when mobile phone is used either for a small share of the trip, or for the largest share of the trip.

4.2 Mixed Binary Regression Models

In order to model the expected use of mobile phone while driving, models in a Mixed Binary Regression framework were calibrated, as previously explained. Since the OSeven application allows for a high resolution, big-data oriented collection scheme, it was attempted to include random effects in order to capture the unique driving behaviour traits for each driver. This entails having a critical minimum sample of trips for each driver to achieve a meaningful outcome. Therefore, a screening was made among participant drivers, as described above, and drivers that had over 40 trips each were selected for the analysis.

Mixed Binary Regression Models were fitted in R-studio (with the lme4 package) via maximum likelihood. A number of models were tested with different configurations in the collected parameters in both fixed effects and random effects. Specifically, the fixed effects regard the explanatory variables, while the random effect regards the model intercept. A model for all trips was developed (overall model) as well as models for each different road type (urban, rural and highway).

The selected variables were chosen after taking into account the following: high statistical significance, low correlation among them, and final rational interpretation of their impact on the dependent variable. Due to the last criteria, authors did not include harsh events in the model, as harsh events would be the result of a distracted driving and not the opposite. Table 1 provides a description of the parameters that were selected for models.

Table 1. Description of the parameters used in the models.

Independent variable	Description
Trip distance	Total trip distance (km)
Workday	Workday (0 = yes, 1 = no)
Morning rush	Morning rush hour 06:00–10:00 (0 = yes, 1 = no)
Afternoon rush	Afternoon rush hour 16:00–20:00 (0 = yes, 1 = no)
Average speed	Average driving speed per trip (km/h)

The final models are presented in Table 2. Modelling results regarding the use of mobile phone while driving (yes/no) reveal interesting findings; the parameters of trip distance, workday and afternoon rush have all been determined as statistically significant and positively correlated with the use of mobile phone. In the same context, average speed and morning rush are statistically significant and negatively correlated with the use of mobile phone.

Table 2. Mixed binary logistic models

Parameter	Overall model		Urban road		Rural road		Highways	
	B	P-value	B	P-Value	B	P-value	B	P-value
Intercept	−1.613	<0.001	−2.313	<0.001	−2.752	<0.001	−6.457	<0.001
Trip distance	0.051	<0.001	0.182	<0.001	0.095	<0.001	0.025	<0.001
Workday	0.174	0.003	0.176	0.005	0.174	0.008	–	–
Morning rush	−0.354	<0.001	−0.385	<0.001	−0.44	<0.001	−0.704	<0.001
Afternoon rush	–	–	0.121	0.046	0.127	0.045	–	–
Average speed	−0.01	<0.001	−0.007	0.017	0.008	<0.001	0.037	<0.001
Random effect (variance of random intercept)	1.475	<0.001	1.515	<0.001	1.472	<0.001	1.763	<0.001
Number of obs	11398		11398		11398		11398	
Number of drivers	82		82		82		82	
AIC	10888.3		9517.6		9002		1637.7	

More specifically, the aforementioned results could be further interpreted, calculating the relative risk ratio of every variable and thus measuring the increase in the probability of mobile phone use while driving. The exposure metric of trip distance seem to increase the odds of mobile phone use during the trip; the effect is found to be higher in urban areas, less in rural and the least in highways, as the respective odds ratios exp (B) are 1.20, 1.10, and 1.02. Driving on workdays compared to driving on weekends also increases the odds of mobile phone use, for all the developed models apart from the highway model, where the variable is not statistically significant. This impact may be interpreted by the fact that the busy daily schedule make drivers use more often their mobile phone than on weekends, when the purpose of the trip is not work-related.

In the same context, rush hours have also a significant impact on mobile phone use. Particularly, it is found that driving during morning rush hours (06:00–10:00) compared to the rest of the day decreases the odds of using the mobile phone during the trip. This can be explained by the fact that very early in the morning there are less mobile work-related calls. The impact of the morning rush hours appears to be higher in highways than in the other types of road, as well as the overall model. On the contrary, driving during afternoon rush hours (16:00–20:00) increases the odds of using the mobile phone while driving. The effect is significant only in model of urban and rural roads.

Finally, regarding the driving behaviour indicator of average speed, some conflicting results are revealed. Average speed per trip was found to be negatively associated with the odds of mobile phone use on all road types and in urban areas (odds ratios exp(B = −0.010) and exp(B = −0.007) equal to 0.99 respectively) as drivers adapt compensatory behaviour, confirming existing studies. However, when driving in rural areas and highways, it seems that the higher the average speed the higher the odds of mobile phone use. This is probably explained by the fact that, while in urban area the driver is aware of unexpected incidents that occur often and reduces the speed while talking on the mobile phone, outside urban areas, where the events occur less often, the compensatory behaviour of the drivers weakens and the speed is higher.

5 Conclusions and Future Research

This present research aimed to investigate the impact of mobile phone use on driving behaviour and road safety through the investigation of driving analytics collected by smartphone sensors. In order to achieve that objective, a naturalistic driving experiment was carried out in order to examine distracted driving as expressed by the use of mobile phone while driving. Results substantiate correlations of mobile phone use with specific exposure and driving behaviour indicators, namely the average driving speed, the trip distance as well as driving on workdays and rush hours.

Furthermore, the present study does not only investigate the mobile phone use, but also allows for mobile detection while driving by means of mobile sensors. This is very essential, taking into consideration the high penetration rate of smartphone use in our everyday life. Future research should go one step further and focus on real-time detection of mobile phone use providing new possibilities to driver feedback, aiming at the reduction of mobile phone use while driving and therefore improving the overall driving behaviour.

Future research should also focus on the improvement of the accuracy of the models, by testing the developed models to different samples and traffic conditions. Furthermore, with respect to mobile phone use (yes/no use during the trip, share of trip duration, frequency of phone use over the trip etc.), analyses per gender, age, history of accidents, self–assessment, driving experience and more demographic characteristics could be undertaken in order to capture any particular trends found in the categories of these parameters.

Acknowledgements. This research is co-financed by Greece and European Union (Social European Fund) through the Operational Program “Human Resource Development, Education and Lifelong Learning”, within the action “Strengthen Postdoctoral Researchers - Second Phase” (MIS 5033021), implemented by the State Scholarships Foundation (IKY).

The authors would like to thank OSeven Telematics, London, UK for providing all necessary data exploited to accomplish this study.

References

Bellinger, D.B., Budde, B.M., Machida, M., Richardson, G.B., Berg, W.P.: The effect of cellular telephone conversation and music listening on response time in braking. Transp. Res. Part F **12**(6), 441–451 (2009)

Dingus, T.A., Guo, F., Lee, S., Antin, J.F., Perez, M., Buchanan-King, M., Hankey, J.: Driver crash risk factors and prevalence evaluation using naturalistic driving data. Proc. Natl. Acad. Sci. **113**(10), 2636–2641 (2016)

Elvik, R., Christensen, P., Amundsen, A.: Speed and road accidents. An evaluation of the Power Model. TØI report, 740 (2004)

Fitch, G.M., Bartholomew, P.R., Hanowski, R.J., Perez, M.A.: Drivers' visual behaviour when using handheld and hands-free cell phones. J. Saf. Res. **54**, 105-e29 (2015)

Lipovac, K., Đerić, M., Tešić, M., Andrić, Z., Marić, B.: Mobile phone use while driving-literary review. Transp. Res. Part F Traffic Psychol. Behav. **47**, 132–142 (2017)

Ma, C., Dai, X., Zhu, J., Liu, N., Sun, H., Liu, M.: Driving sense: dangerous driving behavior identification based on smartphone autocalibration. Mobile Inf. Syst. **2017**, 1–15 (2017)

Mantouka, E.G., Barmpounakis, E.N., Vlahogianni, E.I.: Mobile sensing and machine learning for identifying driving safety profiles (No. 18-01416) (2018)

Meseguer, J.E., Calafate, C.T., Cano, J.C., Manzoni, P.: Driving styles: a smartphone application to assess driver behavior. In: 2013 IEEE Symposium on Computers and Communications (ISCC), pp. 000535–000540. IEEE (2013)

Paefgen, J., Kehr, F., Zhai, Y., Michahelles, F.: Driving behavior analysis with smartphones: insights from a controlled field study. In: Proceedings of the 11th International Conference on Mobile and Ubiquitous Multimedia, pp. 1–8 (2012)

Papadimitriou, E., Argyropoulou, A., Tselentis, D.I., Yannis, G.: Analysis of driver behaviour through smartphone data: the case of mobile phone use while driving. Saf. Sci. **119**, 91–97 (2019)

Papantoniou, P., Papadimitriou, E., Yannis, G.: Assessment of driving simulator studies on driver distraction. Adv. Transp. Stud. **35**, 129–144 (2015)

Salmon, P., Young, K., Lenné, M., Williamson, A., Tomasevic, N.: The Nature of Errors made by Drivers. Austroads Publication No. AP–R378/11. Austroads Ltd., Australia (2011)

Stutts, J.C., Reinfurt D.W., Staplin L., Rodgman E.A.: The role of driver distraction in traffic crashes. Report Prepared for AAA Foundation for Traffic Safety (2001)

Tselentis, D.I., Vlahogianni, E.I., Yannis, G.: Comparative evaluation of driving efficiency using smartphone data. Methodology **32**, 33 (2018)

Vaiana, R., Iuele, T., Astarita, V., Caruso, M.V., Tassitani, A., Zaffino, C., Giofrè, V.P.: Driving behavior and traffic safety: an acceleration-based safety evaluation procedure for smartphones. Modern Appl. Sci. **8**(1), 88 (2014)

World Health Organization. Global Status Report on Road Safety (2018)

Ziakopoulos, A., Tselentis, D., Kontaxi, A., Yannis, G.: A critical overview of driver recording tools. J. Saf. Res. **72**, 203–212 (2020)

Attitudes and Preferences of University Student Bicyclists: The Tale of Two Greek Cities

Nikiforos Stamatiadis[1], Andreas Nikiforiadis[2](✉), Socrates Basbas[2], Pantelis Kopelias[3], Elpida Karantagli[2], Anastasia Sitra[3], and Nikos Mantas[3]

[1] Department of Civil Engineering, University of Kentucky, Lexington, KY, USA
nick.stamatiadis@uky.edu

[2] School of Rural and Surveying Engineering, Aristotle University of Thessaloniki, AUTh Campus, 54124 Thessaloniki, Greece
anikiforiadis@topo.auth.gr

[3] Department of Civil Engineering, University of Thessaly, Pedion Areos, 38334 Volos, Greece

Abstract. Recently, the share of bicycles among transportation modes has been rising. This is also the case in university campuses that have seen an increase of bicycling as a transport mode. Transportation authorities globally encourage bicycling as a way for pollution reduction and alleviation of traffic congestion. However, lack of bicycle facilities could significantly hinder increase in bicycling levels. This study used in-person surveys to evaluate the attitudes and desires of bicyclists in two university campuses in Greece: Thessaloniki and Volos. The goal of the survey was to identify barriers to bicycling in both cities and pinpoint facility aspects that could be improved. Descriptive statistics and hypothesis testing were utilized in the analysis of the survey data to determine the level of agreement among participant responses and document their attitudes and preferences. Lack of bicycling facilities was the main impediment that respondents in both cities identified: a finding that agrees with prior research. In addition, the aggressive behavior of car drivers is considered an important obstacle in both cities. The survey findings support to the notion that bicyclists everywhere have similar attitudes about what the types of improvements required for increasing bicycling and enhancing their experiences. In addition, local conditions and practices have an impact on the relevance of specific issues.

Keywords: Bicycling · University campus · Surveys

1 Introduction

Recently, the share of bicycles among transportation modes has been rising. Trends in the European Union (EU) support this, where the number of bicycles per 1,000 inhabitants ranges from 52 in the Czech Republic to 1,000 in the Netherlands [1]. Data shows that in the Netherlands 26% use bicycles for their transportation needs and in

E. G. Nathanail et al. (Eds.): CSUM2020, AISC 1278, pp. 945–953, 2021.
https://doi.org/10.1007/978-3-030-61075-3_91

Denmark, 19% do so [2]. Among EU countries there is gender parity with respect to attitudes about bicycling: the same percentages are noted between genders when opting to use a bicycle to fulfill transportation needs [3]. The only difference among EU countries is the frequency of use where in some areas it is used daily (such as the Netherlands and the Scandinavian region), while elsewhere it is used less frequently [1].

Bicycling has been encouraged as an alternative transportation option globally due to potential health benefits, decline of emissions, and relieving of congestion. It is therefore essential that transportation agencies need to improve and install appropriate infrastructure to support the current bicycling levels.

Infrastructure that is appropriate for bicycling has been proven to be an encouragement for bicycling [4]. Such infrastructure would provide a safe environment for bicycling. Transportation authorities need to be cognizant of issues that bicyclist perceive and develop plans to address them accordingly in order to deliver the required types of bicycle facilities and increase the number and frequency of bicycle trips. Surveys are a common tool used to identify bicyclists' mindsets and detect necessary enhancements.

This study analyzes surveys of university student bicyclists in Thessaloniki and Volos, Greece to determine their opinions and issues as they relate to bicycling. The surveys mainly designed to identify barriers to bicycling in both cities and determine the types of interventions bicyclists prefer aiming to expand usage of bicycling. The comparison of the results between the two surveys could reveal whether bicyclists have similar apprehensions and likings across the country and identify how these could be influenced by living in a city with (Volos) or without (Thessaloniki) adequate bicycle facilities. Comprehension of these aspects will allow for developing a more targeted plan to address more appropriately bicycling needs.

2 Background

The share of bicycling as a principal transportation mode has seen a continued increase in the past decades. In the 1990–2010 period, this increase has been more prominent in large urban areas, at least in the USA and Canada, with a rate twice as large as that observed nationally [4]. This growth has been ascribed to facility improvements and promotion of bicycling-related programs. Common actions include the installation and extension of bicycle facilities, completion of traffic calming measures, increased accommodation of bicycles in transit, and creation of bicycle sharing programs. A notable increase in bicycling levels was noted in a study of 14 cities that had recently updated and/or expanded their bicycle infrastructure [5]. The importance and effects of public policy on increasing bicycling was noted in the study conclusions. The study also noted that a wide-ranging, all-inclusive package of interventions is essential in increasing bicycling.

Common route selection elements for bicyclists include trip distance, left-turn frequency, intersection traffic volumes, and grades [6]. It can then be posed that route selection is preference-based and bicyclists tend to avoid perceived high-hazard locations (e.g., intersections and left turns). The extent to which bicyclists are aware of hazardous situations or familiar with alternative routes is not determined. It is also understood that trip length is more important for commuter bicyclists than other route selection elements. In general, bicyclists favor separate facilities or bicycle lanes [7–9]. Commuter bicyclists favor bicycle lanes over other facility types because they are along the vehicular network and typically offer more direct routes [10]. Past research has supported the positive relation between bicycling and the presence of bicycle facilities [11, 12].

A variety of survey types have identified bicyclists' attitudes and needs, and have been useful for identifying policy or infrastructure interventions to enhance bicycling. Along with large-scale mail and phone questionnaire-driven surveys, a variety of other tools have been leveraged to collect quantitative and qualitative data on bicycling and walking behavior and preferences. Included among these tools are trip diaries, in which participants record the details of bicycling trips over a defined period, and intercept surveys, where participants are asked questions at a fixed location. The National Highway Traffic Safety Administration (NHTSA) periodic National Survey of Bicyclist and Pedestrian Attitudes and Behaviors is one example of a survey that evaluates the attitudes and perceptions of survey participants towards bicycling and walking [13]. Large-scale surveys, such as this one, allow for a comparison of trends over time and have verified the increase in bicycling as a primary mode of transportation between 2002 and 2012. The Pedestrian and Bicycle Survey (PABS) [14] is a mail survey that has been utilized in San José, California, to document travel behavior within the community. Policymakers and planners have used its results to shape planning actions of the community's transportation agencies.

The National Documentation Project (NDP) is a survey that records use and demand of facilities for bicyclists and pedestrians annually [15]. Typical tools that NDP employs to approximate pedestrian and bicyclist needs include and the collected data offers transportation authorities data that is difficult to collect for their analysis and evaluation of infrastructure improvements. Similar surveys have been also conducted in Europe to comprehend the concerns of urban travelers and identify policies to be initiated to achieve the goals set forth in Urban Mobility plan [16]. The survey identified transport behaviors, issues in cities, and concepts on enhancing travel. Most respondents (50%) denoted they never use a bicycle instead of a car and only 12% stated they choose bicycles on a daily basis [3].

A number of countries and cities in Europe conduct periodic surveys at the national and local level to assess the needs for bicyclists. For example, in Germany, surveys are conducted intermittently to collect data on bicyclists' satisfaction with cycling conditions and cycling policy [17]. In the Netherlands, a system called Cycle Balance (Fietsbalans) uses a suite of methodologies to evaluate the quality of bicycling infrastructure and services and inform policy and planning [18]—surveys measure bicyclist satisfaction with current infrastructure, bicycles are instrumented to assess route quality, and a database review gauges facility usage. Locally, the city of Copenhagen has collected statistics and information on bicycling since the 1990s and has used these data to identify needed infrastructure and policy changes.

Preference surveys are a common tool for comprehending bicycling levels but there are no comparative studies that systematically compare the perceptions and mindsets of bicyclists across areas. This gap is addressed here, and in doing so much needed understandings of differences among bicycle travel and usage between cities is identified.

3 Methodology

Researchers developed a questionnaire survey to collect responses about bicyclist attitudes, opinions about bicycle facilities, and preferences for a smartphone app that would assist bicyclists. The survey targeted student audiences in two university cities in Greece: Thessaloniki and Volos. Volos has a much denser bicycle network as it compares to that of Thessaloniki, since Thessaloniki has only 2 km more bicycle lanes (12 km in Thessaloniki and 10 km in Volos) but a much larger area. The survey was conducted in the two cities to contrast the respondents' responses and establish the potential impact of infrastructure extent on their views.

The questionnaire questions were developed after reviewing previous surveys given in other parts of the world to address similar research goals. The survey was kept brief to maximize the number of participants. There were four major units that identified demographics of the respondents, bicycle usage and barriers to bicycling, and a list of items that would help increase bicycle use as a transport mode. The ability to include additional responses was provided as well. A 4-point Likert scale was used without ties to require a decision.

Researchers statistically analyzed survey results to estimate the level of agreement in the rankings of each group of participants and compare answers from the two cities. The Mann-Whitney U test was selected, since it was considered appropriate given the type of available data [19]. The Mann-Whitney U test is the non-parametric equivalent of the t-test for independent samples. Thus, it makes no assumptions about the distribution of the variables and allows for comparisons of samples of unequal sizes. In general, statistical significance at the 95% confidence level was assumed to be acceptable ($p\text{-value} < 0.05$). For this study, the null hypothesis (H_o) was that both groups provided equal ratings.

The survey was administered in person throughout both campuses. The total number of respondents was 577 with 300 in Thessaloniki and 277 in Volos.

4 Survey Results

4.1 Demographics

The questions on participant demographics showed similar and approximately equal levels of gender participation in the two cities. Also, the age distribution was similar, since both surveys targeted university students. In both cases approximately 88% of the respondents belonged in the 18–24 age group, with the remaining 12% belonging in the 25–34 age group.

4.2 Frequency of Use

A question about weekly trip length revealed differences between the bicyclists in the two cities. In Thessaloniki, 68% of the respondents indicated that they do not bicycle at all, while only 10.3% bicycle on average more than 10 km within the week. In Volos, the corresponding responses were 83.4% and 9.4%. However, the respondents in Volos were much more willing to use bicycle if the city environment was safer for bicycling. More specifically, 48% of the respondents in Volos stated that they would use bicycle as their primary mode of transport daily, 40.1% would also like to use it daily for accessing the university and 19.9% would like to use it daily for shopping and leisure, if the city environment was bicycle-friendly. The corresponding numbers for the respondents in Thessaloniki were much lower: 18.6%, 23.5%, and 6.8%. These large differences could be partially attributed to the fact that distances in Thessaloniki are typically much longer.

4.3 Impediments to Bicycling

The following group of questions sought to shed light on impediments to bicycling. There were two questions, where the first focused on issues regarding the infrastructure and the second focused on issues regarding the roadway environment, which could prevent students from bicycling.

The most important reason for avoiding cycling, in both cities, was the absence of appropriate infrastructure (lack of bicycle lanes) (Fig. 1). In the case of Thessaloniki, 67.3% of the respondents highlighted the lack of bicycle lanes as the most important reason for not using bicycle, while the corresponding percentage for the case of Volos is 49.8. Taking also into account the presence of narrow bicycle lanes as infrastructure deficiencies preventing bicycle use, the aforementioned values climb to 72% for Thessaloniki and 52.7% for Volos. The second most cited impediment, as it can be understood by the mean ranks, was the pavement condition for both cities. The ranking of the remaining roadway reasons differs between the two cities, since in Thessaloniki the narrow travel lanes were the third highest ranked justification respondents gave for not bicycling, while in Volos the lack of spots for bicycle parking was the third most important impediment with 13%. Overall, the top two choices found to be the same for both cities, however, the rank of the other impediments is different.

Mann-Whitney U-test results showed that there is no statistically significant difference between the respondents from Thessaloniki and Volos for the rankings related to street lighting ($p = 0.052$) and presence of many intersections ($p = 0.192$), since the null hypothesis cannot be rejected. However, it was rejected for the no bicycle parking ($p = 0.000$), no bicycle lanes ($p = 0.000$), narrow travel lanes ($p = 0.001$) and poor pavement ($p = 0.050$). This indicates the importance that each of these impediments have within the specific population and reflect overall similar perceptions.

Through the survey it was also examined which factors of the roadway environment affect the bicycling tendencies of the university students. It can be understood that aggressive driving is a key obstacle to bicycling for the respondents in both cities (Fig. 2). The percent of respondents that highlighted this as the most significant reason, is 36.3 in the city of Thessaloniki and 31.8 in the city of Volos. High traffic volumes

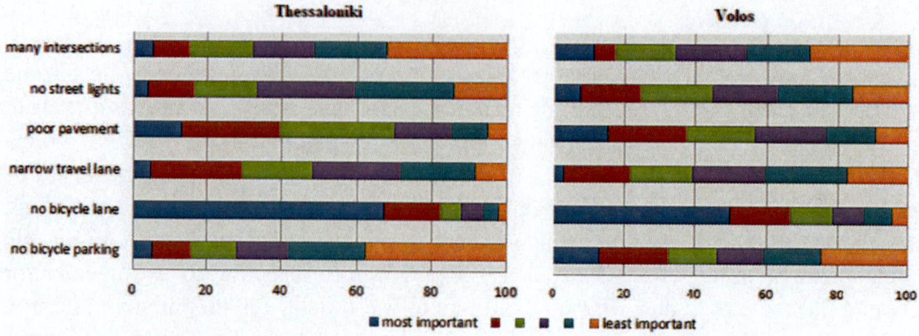

Fig. 1. Roadway reasons for not bicycling (percentages).

were the second most frequently cited reason in Thessaloniki (19.7%). In Volos, the second most frequently cited roadway environmental impediment for bicycling was weather (29.2%). Modal connectivity and air pollution found to be in both cities the least significant reason for avoiding bicycling.

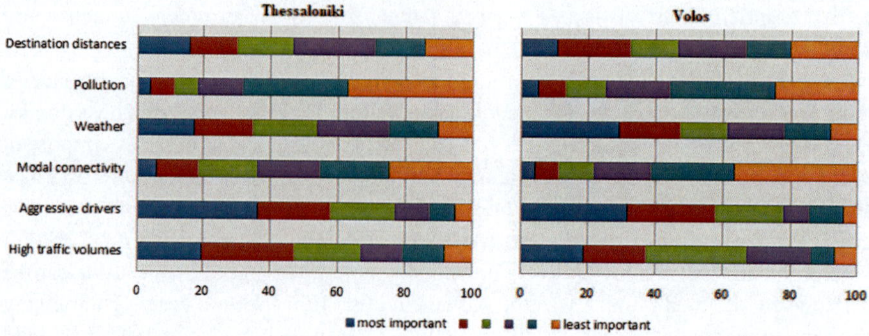

Fig. 2. Roadway environmental reasons for not bicycling (percentages).

Based on the Mann-Whitney U-tests, a statistically significant difference in the rankings given by respondents in Thessaloniki and Volos for weather, pollution, and modal connectivity (i.e., $p < 0.05$), can be identified. On the other hand, no statistically significant difference was found for high traffic volumes ($p = 0.537$), aggressive drivers ($p = 0.557$) and destination distances ($p = 0.378$).

Figure 3 presents the results of the final question, which asked the respondents to rate potential interventions for eliminating impediments to bicycling. University students both in Thessaloniki and Volos assign greater rates to the development of new bicycle lanes. For 86.3% of the respondents in Thessaloniki and 76.2% of the respondents in Volos the development of new bicycle infrastructure is the most important intervention. The second most important intervention according to the university students in both cities is the improvement of the pavement conditions. Based on the abovementioned findings, it is indicated that university students share the view that developing or improving infrastructure will significantly encourage bicycle usage.

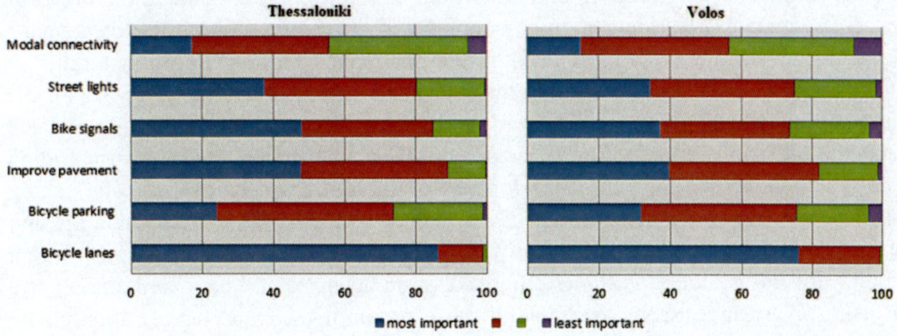

Fig. 3. Importance of interventions to increase use of bicycling (percentages).

The Mann-Whitney U-tests indicated that it is not possible to reject the null hypothesis of equal rank between the participants from the two cities for bicycle parking ($p = 0.178$), street lights ($p = 0.191$) and modal connectivity ($p = 0.657$). The null hypothesis was rejected (i.e., $p < 0.05$) for bicycle lanes, improve pavement and bicycle signals. Survey results confirmed that respondents prioritize infrastructure improvement and that targeted infrastructure improvements have the potential to increase bicycling.

5 Discussion

The present paper attempts to examine the attitudes and preferences of university students bicyclists is two cities in Greece, i.e. Thessaloniki and Volos. More specifically, survey respondents evaluated the factors that prevent them from bicycling and they also assessed the potential of various interventions in overcoming barriers to safe bicycling trips.

University students from Volos stated greater willingness for bicycling on a daily basis than those from Thessaloniki, possibly due to the fact that the trips in Volos are shorter. However, currently students from both cities are avoiding using a bicycle for their daily trips: a fact that could be attributed to the lack of adequate infrastructure.

The respondents in Thessaloniki and Volos highlighted to a large extent the fact that the lack of appropriate bicycling infrastructure is a key obstacle to bicycle usage; enhancing the results of previous research [20]. The assignment of rates in the potential interventions also strengthened the conclusion regarding the importance of developing bicycling infrastructure for the formation of more bicycle-friendly cities.

Regarding suggestions for future research, a more detailed investigation of how demographic attributes affect bicyclists' attitudes and perceptions could be very useful. The fact that the analysis described in this paper focuses on university students, who are in a large extent a homogenous population, does not allow a more thorough examination of demographic influences. Additional surveys should be administered to collect representative data for the entire population of each city. Multi-round surveys, applying Delphi technique, can be also essential for identifying the impact of the local context [21]. Also, for future research it would be interesting to examine the needs of university students regarding mobile apps, as well as the possible utility of a smartphone app whose tools could be used to help map and plan bicycling routes [20].

Overall, the survey results prove that bicyclists share similar views regarding the interventions which are needed for promoting bicycling, regardless the city and the context. Authorities attempting to increase the share of bicycles among transportation modes in their jurisdictions must develop new or improve existing infrastructure, taking into account methodologies, which assist planners to assess the infrastructures and design them properly [22]. However, in the existing economic conditions, where funding for infrastructure is limited, transportation agencies and local authorities have to justify every investment, particularly those promoting bicycle usage. Smartphone apps that could provide needed information to bicyclists can be used to collect data that agencies can analyze to identify potential interventions.

Acknowledgements. This work was completed as part of a Fulbright Fellowship grant.

References

1. European Conference of Ministers of Transport: Safety in road traffic for vulnerable users. Organisation for Economic Co-operation and Development OECD, Paris (2000)
2. Directorate General for Internal Policies: The Promotion Of Cycling. European Parliament, Brussels (2010)
3. Directorate General for Mobility and Transport: Attitudes of Europeans towards urban mobility. Eurobarometer 406 (2013)
4. Pucher, J., Buelher, R., Seinee, M.: Bicycling renaissance in North America? an update and re-appraisal of cycling trends and policies. Transp. Res. Part A **45**, 451–475 (2011)
5. Pucher, J., Dill, J., Handy, S.: Infrastructure, programs, and policies to increase bicycling: an international review. Prev. Med. **50**, 106–125 (2010)
6. Broach, J., Dill, J., Gliebe, J.: Where do cyclists ride? a route choice model developed with revealed preference GPS data. Transp. Res. Part A **46**, 1730–1740 (2012)
7. Kang, L., Fricker, J.: Bicyclist commuters' choice of on-street versus off-street route segments. Transportation **40**, 887–902 (2013)
8. Buehler, R., Pucher, J.: Cycling to work in 90 large American cities: new evidence on the role of bicycle paths and lanes. Transportation **39**, 409–432 (2012)
9. Nikiforiadis, A., Basbas, S.: Can pedestrians and cyclists share the same space? the case of a city with low cycling levels and experience. Sustain. Cities Soc. **46**, 101453 (2019). https://doi.org/10.1016/j.scs.2019.101453
10. Aultman-Hall, L., Hall, F., Baetz, B.: Analysis of bicycle commuter routes using geographic information systems: implications for bicycle planning. Transp. Res. Rec. **1578**, 102–110 (1998)
11. Dill, J., Carr, T.: Bicycle commuting and facilities in major US cities: if you build them, commuters will use them-another look. Transp. Res. Rec. **1828**, 116–123 (2003)
12. Parkin, J., Wardman, M., Page, M.: Estimation of the determinants of bicycle mode share for the journey to work using census data. Transportation **35**, 93–109 (2008)

13. Schroeder, P., Wilbur, M.: 2012 National Survey of Bicyclist and Pedestrian Attitudes and Behavior. Volume 1: Summary Report. National Highway Traffic Safety Administration, Washington, DC (2013)
14. Forsyth, A., Krizek, K.J., Agrawal, A.W.: Measuring walking and cycling using the PABS (Pedestrian and Bicycling Survey) approach: a low-cost survey method for local communities. Mineta Transportation Institute, San Jose (2010)
15. Gilmore Research Group: Public Attitude Survey of Bicycle and Pedestrian Planning. Wilbur Smith Associates Bellevue, Washington, DC (2007)
16. National Bicycle and Pedestrian Documentation Project. http://bikepeddocumentation.org/index.php, Accessed 15 Feb 2020
17. Thiemann-Linden, J., Mettenberger, T.: Cycling Quality Management and Evaluation in Europe. Federal Ministry of Transport, Building and Urban Development, Berlin (2012)
18. Check, J., Schutt, R.K.: Survey research. In: Check, J., Schutt., R.K. (eds.) Research Methods in Education, pp. 159–185. Sage Publications, Thousand Oaks (2012)
19. Siegel, S., Castellan, N.J.: Nonparametric Statistics for the Behavioral Sciences, 2nd edn. McGraw-Hill, New York (1988)
20. Stamatiadis, N., Pappalardo, G., Cafiso, S.: Use of technology to improve bicycle mobility in smart cities. In: 5th IEEE International Conference on Models and Technologies for Intelligent Transportation Systems, Napoli (2017). https://doi.org/10.1109/mtits.2017.8005636
21. Cafiso, S., Di Graziano, A., Pappalardo, G.: Using the Delphi method to evaluate opinions of public transport managers on bus safety. Saf. Sci. **57**, 254–263 (2013). https://doi.org/10.1016/j.ssci.2013.03.001
22. Nikiforiadis, A., Basbas, S., Garyfalou, M.I.: A methodology for the assessment of pedestrians-cyclists shared space level of service. J. Clean. Prod. **254**, 120172 (2020)

Infrastructure Resilience

Urban Transport Plans in Chile: The Inclusion of Sustainable Modes of Transportation in Public Infrastructure Projects

Marco Mendieta Ávila(✉) and Juan José Pons

Universidad de Navarra, 31009 Pamplona, Spain
mmendieta.1@alumni.unav.es

Abstract. The urban transport plans in Chile have had a successful development in the last decades, mainly due to economic growth and to better practices in public policies, but, above all, to the central administration's need for planning and organizing the state's traffic. This article aims to demonstrate how this country adopted an urban mobility culture to include more sustainable modes of transportations in its public infrastructure. Firstly, this study exposes the situation of urban transportation before the assumption of new policies by the central government, in which the main issue lied on poor management and a lack of organization. Followingly, an analysis of the results of projects developed after such reforms, is presented. Overall, deriving from these results, this study provides evidence for the following: an increase in the inclusion of sustainable modes of transportation in Chile's infrastructure projects, and a slight decrease in the relative importance of private modes of transportation, compared to the past.

Keywords: Plans · Mobility · Sustainability · Policies · Modes

1 Introduction

In some Latin American cities like São Paulo, Mexico city, Santiago de Chile, Bogotá, Quito, Medellín and others, the approach to urban transport management is mainly related to traffic engineering [7], which is characterized by giving priority to private motorized transport. Civil engineering is seen as the main area of propulsion and project planning for urban changes and as the recipient of large investments related to the infrastructure of transportation. Greater interdiscipline is, therefore, lacking in such projects. This is especially revealed when their development, and their results, fail to contemplate many of the users' basic needs, particularly when these projects directly impact the general population [13]. The participation of society and the opinion of the citizens for whom these programs are conceived therefore become necessary in order to achieve the objectives and the proper functioning of mobility in order to avoid management and efficiency failures, as was the case of the Transantiago transport system [13].

It is increasingly evident that sustainability is an element that can no longer be ignored, given that the axes on which it is based (social, environmental and economic

E. G. Nathanail et al. (Eds.): CSUM2020, AISC 1278, pp. 957–969, 2021.
https://doi.org/10.1007/978-3-030-61075-3_92

axes) affect many sectors, especially in large cities where there is an exponential growth of the population that derives into a lack of organization, pollution and social inequality [2].

In Chile, the lack of organization, the absence of the state and the poor management of mobility have driven its authorities to create institutions in charge of setting the guidelines to correct these issues. Nonetheless, many mistakes were made in their commencements, characterized by a lack of efficiency in the institutional management, failure to define responsibilities and the scarcity of monitoring and control over the assumed commitments [13]. The intention of this study is, therefore, to show that, despite all the aforementioned, sustainable modes of mobility were gradually included in the projects. Note that this will be done without diving into the details of efficiency in their bidding, construction or operation stages.

1.1 Urban Growth and Its Effects on Mobility

According to the 2017 Census of the National Statistics Institute [5], about 88% of the approximately 17 million inhabitants of Chile, lives in urban areas.

Although there is currently a population growth rate of only 1%, the lowest in its history; 62.4% of the population is concentrated in the central regions of Metropolitana, Biobío and Valparaíso, which makes up for a high demand for mobility, particularly in Metropolitana

[9] which contains 41% of all Chileans; that is, about 7.4 million people [5].

Urban growth in Chile is reflected in the construction of new residential areas, expanding the distance between homes and traditional urban centers, which often forces habitants to travel continuously increasing distances [16]. Effectually, there is a higher growth of new homes in peri-urban areas, attributed to lower costs, than in the city center, on which there is great dependence in terms of labor and services [2].

Consequently, in the country, as well as in the entire region, there is an increasing motor- ization rate, which is among the highest in Latin America. According to the United Nations, through ECLAC (Economic Commission for Latin America and the Caribbean), until 2010, Chile had almost 120 cars for every thousand habitants, surpassed only by Mexico, Brazil and Uruguay [3]. Only a few years later, in 2017, a motorization rate of approximately 198 vehicles per thousand is observed [6].

In Greater Santiago, alone, there was an increase from 1.1 million cars in 2010 to more than 1.4 in 2015. This represented an increase in the fleet of automobiles of approximately 27% in 5 years [8]. Furthermore, it is expected that by 2025 the number of cars will double, reaching 2.7 million, with a motorization rate close to 360 vehicles per 1,000 habitants [10]. The predominant use of automobiles has unequivocally proven to entail a negative effect on the environment, given that the pollution of cities is directly related to private motorized mobility.

According to a study by the University of Chile, one of the main sources of air pollution in the cities of Chile is the use of vehicles, and, consequently, the control of these pollutants is contrasted with the growth rates of motorization [4]. In this same study, the authors expose the benefits of compact cities and the use of public transport, as measures to recycle and to mitigate air pollutants.

At the same time, the growth of investment in private motorized transportation infrastructures and the non-investment in sustainable mobility may contradict the country's environmental goals. It is worthy to note that Chile has presented, in the Paris climate agreement in 2015, the goal of reducing the intensity of carbon emissions 30% below those of 2007 [14].

1.2 Towards a Sustainable Urban Mobility

According to data from SECTRA, projects that contemplate the use of sustainable modes in urban mobility have been carried out for decades. This study extracts those that correspond to 1998, as the starting year of the analysis, from their website.

One of the main projects for the reconversion of mobility, directed to improving the management of urban transportation in the capital, has been Transantiago, which allowed for substantial changes in the formalization of the service and the unification of itineraries in the first decade after of 2000. Although this urban transport service has had both success stories as well as setbacks, it has allowed organizing the mobility of Santiago in a relatively short period.

The Transantiago entered operation in 2007, leaving behind the complex management of the itinerary network. Until 2005, Santiago was served by 8,000 buses throughout small and informal entrepreneurs. That was a chaotic service producing traffic congestions and several accidents [1].

Santiago's transportation plans have combined Transantiago with the metro system. The new lines under construction and expansion are serving the city in coordination with urban buses, in addition to the connection that exists with pedestrian and bicycle mobility.

According to what was observed in its documents, the Ministry of Transport and Telecom- munications (MTT) has taken into account intersectoral participation while drafting the National Transport Policy, published in 2013, although without the rigor of the law. Before the aforementioned national policy, the "Santiago 2025 Transportation Master Plan" was launched in 2012. In it, the main drivers for rethinking people mobility and the logistics freight stand out.

According to this document, the Chilean capital is undergoing abrupt changes in mobility due to the increase in motorization, a fact that creates pressure for an increase of investment in roads and in required space [9].

The plan, however, leans towards providing a more prominent role to sustainable modes of transportations. The reasons for this are: - The space that is necessary to increase road infrastructure is nonexistent, and citizens oppose the expropriations that would be required for such purpose. - There is a state budget limitation for this type of investments. - Sustainability, contemplated in the government's transportation policy, will be promoted through the use of mass public transport [9].

2 Description of the Problem and State of the Art

It is well known that, in Latin America, sustainable urban mobility is a pending issue within infrastructure projects.

Research that specifically discusses the sustainable modes of transportation used in Chile's public infrastructure projects has not been found. Moreover, a review of official websites of the Chilean government, and the municipalities of its cities and regions, shows no comprehensive plans for sustainable mobility other than chapters or sections of projects which only consider them partially.

The collection of such information was based on the SECTRA digital library and on information exchanged through e-mail with SECTRA, which has evidenced the absence of the work towards sustainable urban mobility plans (SUMP) that can be found in other countries.

With this in mind, a decision has been taken to review their public infrastructure projects, some of which have proven to contain elements of SUM. In addition, certain measures taken by SECTRA have been observed, such as transitioning toward more sustain- ability in their planning along with some degrees of participation of all parties involved in the projects.

The details of these tasks will be specified in the methodology section.

The main objective of this study will, therefore, focus on exposing the inclusion of sustainable mobility modes in Chile's public infrastructure projects, over a period of 18 years (from 1998 to 2015). The results will be presented in three axes: thematic, temporal and spatial.

3 Sources and Methodology

This study has been carried out through the analysis of the extensive documentation pro- vided by the Secretariat of Transportation Planning (SECTRA), under the Ministry of Transport and Telecommunications (MTT). This institution defines itself as a technical organization for transport planning [11].

The information gathered from these infrastructure projects is only processed quantitatively, in order to demonstrate whether there is a change in sustainable culture, as mentioned above, at least with regards to the inclusion of sustainable modes of transportation. From the data available on this website, three large groups of projects have been selected: The Urban Transportation Plans (UTP), the Urban Bicycle National Plans (UBNP) and the Traffic Management Plans (TMP). In total there are 448 projects affecting 64 cities in the country.

3.1 Urban Transport Plans (UTP)

UTP's (also referred to as master plans) are plans that seek to improve mobility in large and medium-sized cities through investments that can be very large. These plans are carried out in accordance with the development project that the city or region requires, and with the purpose of "esponding to the current and future mobility needs of the city's inhabitants" [12]. They mainly consist in road infrastructure projects, with a given priority, in many cases, to private transportation. However, occasionally, they also contemplate the provision of facilities for public transportation or bicycle mobility, and the technologies for the organization of transit. Of the 448 projects, 372 correspond to this group of plans, representing 83% of the total (Fig. 1).

3.2 National Plan for Urban Bicycle Projects (UBNP)

In 2012, the Ministry of Transport and Telecommunications created the "Unit for the Pro- motion of Bicycle Use" as a result of years of work in promoting this mode of transportation, and following the need of centralizing the management of projects aimed at their use and the infrastructure that is directly related [17]. In the section "Promoting Sustainable Trans- port" [17] of its website, SECTRA organizes and summarizes, by regions and cities, the information on the country's bicycle lanes. These projects fall outside of the UTPs bicycle infrastructure, and are referred to, by the secretariat, as the "bike lanes of road projects." Of the 448 projects studied, 43 cor- respond to the UBPN, which is equivalent to almost 10% of the total analyzed.

3.3 Traffic Management Plans (TMP)

These plans aim to assess and improve the mobility situation in cities by giving priority to pedestrians, users of public transportation and cyclists, while dimensioning private transportation and its relationship with other modes of mobility. The TMPs are gen- erally developed in cities that are small or that do not have needs that require high investment infrastructure projects, such as is the case for UTPs. Of the 448 projects analyzed, 33 correspond to this group, representing 7%.

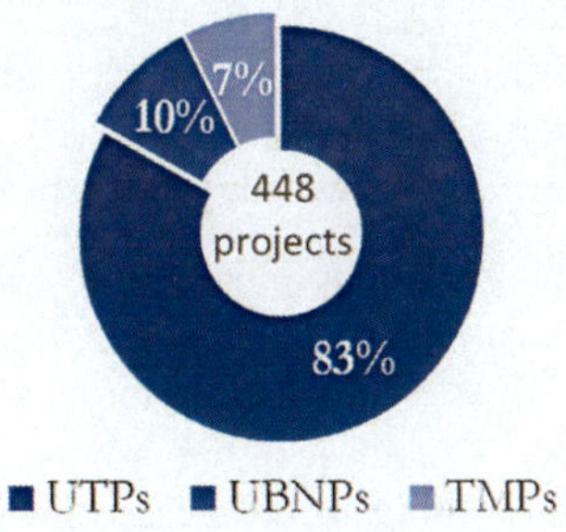

Fig. 1. Distribution of analyzed projects according to the type of plan.

3.4 Data Extracted from Projects

UTPs have sections on the SECTRA website, where the main details of the projects to be executed are specified. These are presented by region.

When accessing each region, a window that presents all these projects with the descrip- tions of their tasks, among other information, is displayed. Each description has been used to review the details and purpose of the infrastructure projects. Similar steps have been taken in the revision of the UBNPs and the TMPs.

For the general analysis of this study, the following parameters of each project have been considered (Fig. 2): region, city, year of completion-publication, project and details. The last field, that of details, is the one that consists in the assessment of mobility modes in each project.

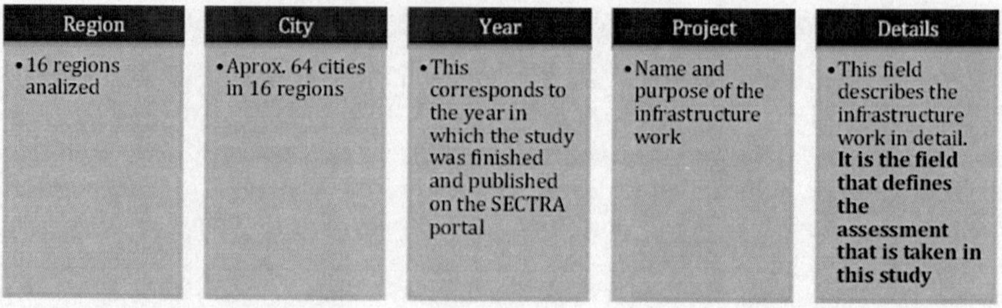

Fig. 2. Parameters analyzed for the infrastructure projects.

The "details" field describes each project even though. In some projects, the consideration of sustainable modes of transportation is not demonstrated. Nevertheless, they do evidence the preference for at least one mode, which is generally for private transportation. Another important detail is that each project may possibly favor one or more modes of transportation, whether sustainable or not, since the assumption that each project promotes development in only one way should not be made (there is not a one-to-one correspondence). The 448 projects available on the SECTRA website have been reviewed, in order to have knowledge of their content and to be able to classify each one of them according to the mode of transportation that they are trying to promote. These are classified into four groups:

- Projects to create or improve infrastructure linked to private motor vehicles.
- Projects to create infrastructure for cyclists.
- Projects aimed at creating or improving public transportation modes.
- Pedestrian mobility development projects.

The realized study covers years 1998 to 2015, since it is during this period that the information available on the SECTRA website is most complete, particularly in reference to the needs of this study. A database of the projects has been created, where the projects have been classified according to the degree in which they meet the aforementioned criteria.

4 Partial Results

The following is the quantitative analysis carried out on the transportation plans according to the three main axes:

- Thematic: review the projects by mode of transportation to examine which one was favored.
- Temporal: observe the inclusion of sustainable modes of transportations over time, identifying the period or periods in which substantial variations have occurred.
- Spatial: territorially analyze changes of mobility in Chilean regions, in order to explore the priority given by different transportation plans to sustainability, and to identify trends across time.

4.1 Thematic Analysis

After classifying the SECTRA information, the analysis is summarized in the following table:

According to Table 1, the plans that favor a greater number of modes of transportation are the UTPs, not only because the total number of projects exceeds that of other types of plans, but also because, within UTPs, most projects include measures referring to more than one mode of transportation.

Table 1. Favored modes according to the analyzed plan.

Type of plan	Private v.	Bicycles	Collective t.	Pedestrians
UTP	328	50	91	38
UBNP	0	43	0	0
TMP	26	27	7	18
Total	354	120	98	56

Altogether, 354 projects favor the private vehicle, representing 79% and followed by bicycles with 120, which are equivalent to 27%. Then, collective transport, with 98 projects, corresponds to 22% and, finally, those that include pedestrians have 56 projects, making it the mode of least impact with close to 13%.

As for private transportation, this is facilitated in 328 projects in the UTPs (Table 1) which constitutes the main evidence for the high priority given to this mode, along with the 26 projects in the TMPs.

As discussed above, UBNPs favor only bicycles. Throughout the study period, 43 projects of this type have been identified.

In the TMP, the least favored mode of transportation is collective transport, with only 7 projects that have included it. Nevertheless, this group of plans also includes the promotion of cycling (27 projects) and pedestrian mobility (18 projects).

It is necessary to highlight the prioritized modes of the analyzed plans, since this helps to perceive the weight given to them in these projects. Figure 3 shows how the private transportation mode surpasses the other three.

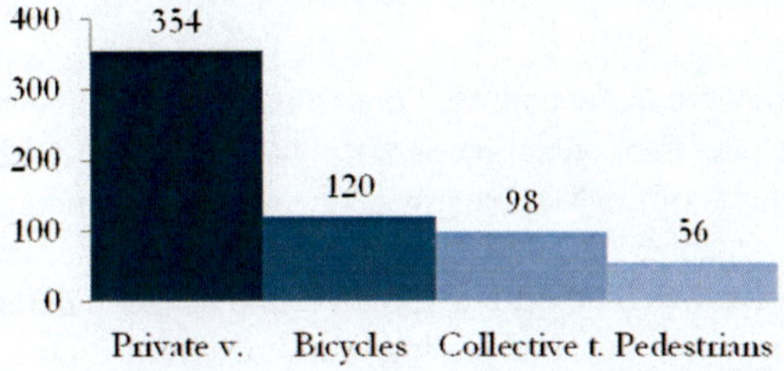

Fig. 3. Number of projects favored according to modal distribution

4.2 Temporal Analysis

Considering the evolution of mobility modes over time, which is one of the main objectives of the study, one can observe that the inclusion of these has changed quantitatively over the period of 1998–2015 (Fig. 4). The motivation for analyzing this axis was to detect changes in the rate of inclusion that could provide relevant information.

The period of analysis, is therefore, divided into two: the first from 1998 to 2006 and the second from 2007 to 2015. The reason for dividing the time sample is in order to contrast, between both periods, the priorities that have been given to the modes of transportation when projecting infrastructure works.

Between 1998 and 2006 the private mode of transportation was clearly considered as a priority. This continues to be seen in the following period, despite the increasing presence of sustainable projects. Nevertheless, a fact that became relevant in the second period is the inclusion of bus mobility, specifically in 2002. Both the increase in bus mobility and the increase in car facilities observed this during year coincide with the project of urban transport and road network improvements in Gran Concepción, in the Biobío region.

In the second period of analysis, one can note much more sustainable changes than in the first. The effective inclusion of these modes of transportation is observed, despite that the private vehicle mode remains the most privileged. A notable fact is the high number of inclusion of projects that promote mobility on bicycles, which increased from just 19 in the first period to more than 100 in the second (Fig. 4).

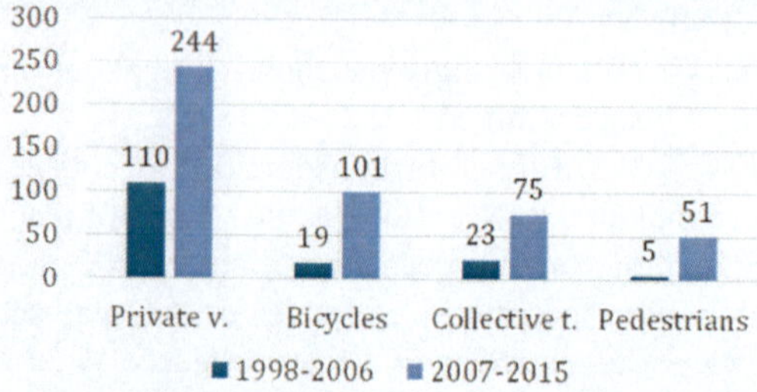

Fig. 4. Evolution of the number of projects, over time, according to the mode of transportation they favor.

In year 2012, there were many contributions to the increase of sustainable modes of transportation. These are due, in large part, to the "Santiago 2025 Transportation Master Plan", a document of the MTT that was released in December of that year. This plan was intended to organize the Chilean capital and its metropolitan area in the long term. The main incentive was that, despite the investments made in the past, the current high congestion and the forecast of a doubling of motorization for the period from 2012 to 2025 (from 1.3 to 2.7 million cars), created the need for a management guide and for strategic investment in infrastructure which required joint and medium term planning [9].

Of the 137 projects belonging to 2012, the year the plan was launched, 78 were planned for the Metropolitan region, which corresponded to almost 57% of what was projected for the entire country that year.

2014 was the year with the largest number of projects that facilitated the use of bicycles, mainly due to plans concerning national bicycle routes. This study has taken this year as the year of inclusion of bicycle mobility in these projects because, while the SECTRA website shows them as projected for the period 2010–2014, the exact year for each project is not detailed. Therefore, the last year of that period was selected for their inclusion in the analysis, despite the fact that their execution is likely to take much longer. Subsequently, of the 92 projects registered in 2014, 56 correspond to cycling infrastructure, that is, 61% of all projects developed that year.

In Chile, many sustainable mobility initiatives have been promoted, mainly the use of bicycles in Santiago and in other important regions of the country. Proof of this is the 7% increase in the use of this mode of transportation in the last decade [12]. Nonetheless, in many areas, it is lower-income users who make use of these means of mobility for economic reasons or due to poor connectivity with activity centers, which merely reflects social inequality and a lack of a comprehensive mobility policy, even in the most sustainable ways [15].

The inclusion of the collective transportation mode in Chilean projects is less frequent than that of cycling mode. From 1998 to 2015 there are a total 98 projects with express facilities for collective transport, of which only 23 are found before 2006. From the remaining 75 (Fig. 4), that were developed from 2007 to 2015, 21 correspond to the "Santiago 2025 Transport Master Plan", approved in 2012. Among other actions, this plan contemplates the extension of existing metro lines and the construction of new ones, the creation of more kilometers of suburban trains, and improvements and construction of bus lines, including the installation of a 5 km cable car system [9].

Finally, the projects that have included pedestrians are 56 in the entire study: 5 until 2006 and 51 from 2007 onwards.

4.3 Spatial Analysis

In the spatial analysis, the events cited above are observed, distributed among the country's regions. It can be seen how Chile is making progress in including sustainable modes in its transportation projects, despite the prevalent emphasis on private transportation, starting from the central regions and extending to the north and south.

Figure 5 shows the evolution over time of the three sustainable modes of transportation as well as the private transportation mode. For the purposes of spatial analysis, the final mapping covers the same periods of the previous analysis, from 1998 to 2006 and from 2007 to 2015.

The government's progress in considering the bicycle mode and particularly pedestrians in its projects throughout the country is highlighted.

Until 2006, only 5 of the 16 regions analyzed considered the development of cycling mobility in their projects: Maule, Biobío, Araucanía, Los Ríos and Los Lagos.

This does not deny the possibility that other regions might have considered this before the analyzed period. However, this falls outside the scope of this analysis. From 2007 and until the end of the analysis, the projects that contemplate bicycles are seen practically throughout the entire Chilean geography, as opposed to the previous period during which these were mostly concentrated in the center of the country. Until 2015, the highest numbers are found in the Maule and Araucanía regions.

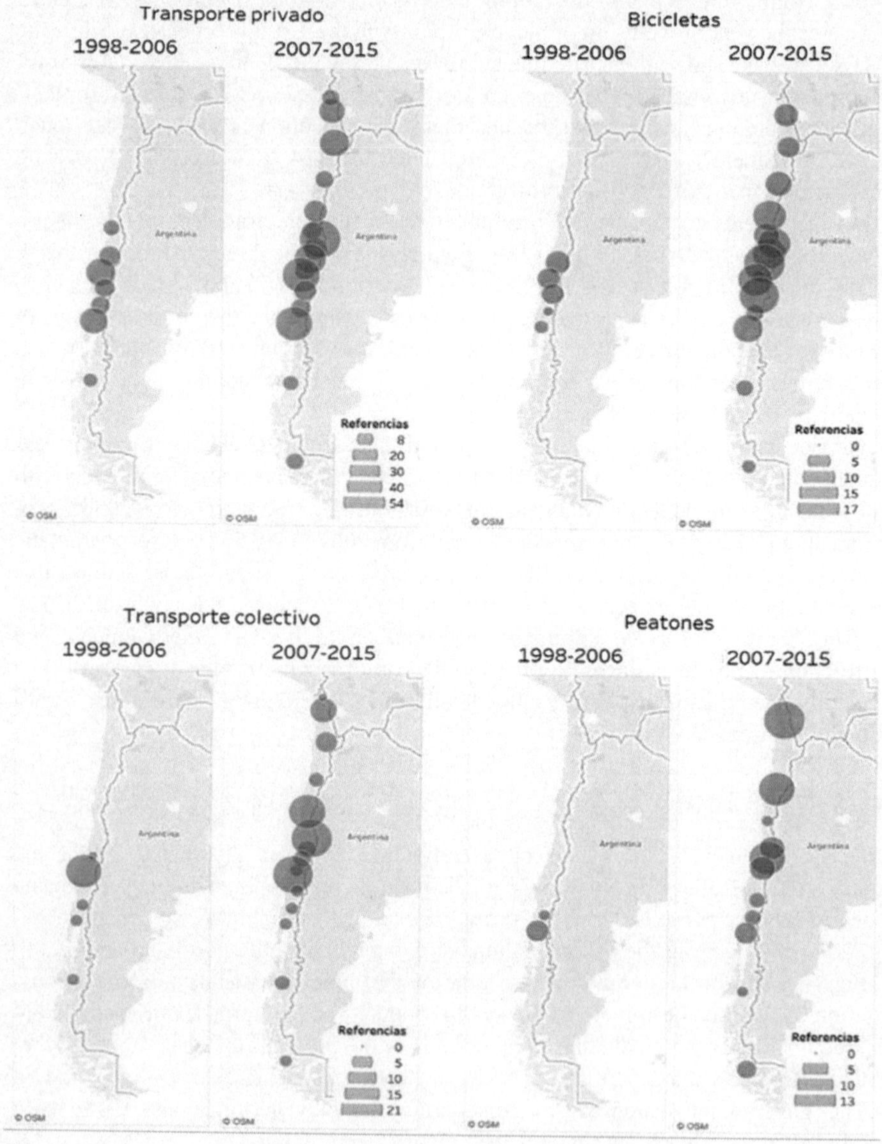

Fig. 5. Evolution of the number of transportation projects by urban area, type of modality and period.

The pedestrian mode is the one that stands out the most in its evolution from one period to the next, since until 2006 there were barely 23 projects that included this mode, but up to 2015 there is an increase up to 26 throughout the country, with a geographical distribution quite accentuated in the northern-central region. The Tarapacá, Atacama and O'Higgins regions have included this mode of transportation in their projects the most.

In collective transport, more projects are seen in the central regions of Metropolitana, Coquimbo and Biobío.

On the other hand, it has been verified that private transportation also followed the same path of dispersion, from the center to the extremes, which, despite the already high number of projects observed before 2006, increased even more until 2015. The lead in this mode of transportation is held by the Metropolitan, Biobío and Los Lagos regions.

Although this is not a sustainable mode of transportation, it is often decisive in ensuring that most of the settlements that are furthest from the capital, and therefore geographically isolated, can benefit from the road and highway networks that allow access to the opportunities that are concentrated in larger cities.

5 Discussion, Conclusions and Final Results

The study of the urban mobility planning trajectory in Chile over the last two decades has revealed a clear growth in the number and importance of the projects developed in the country. These are increasingly diverse and now extend to further regions and cities. From 1998 to 2006, 93% of the projects included in this study had prioritized private transportation, whereas that percentage fell to 74% in the 2007–2015 period. From this data, it appears that, even though the number of projects that aim to improve motorized transportation mobility has grown greatly, their relative importance has decreased significantly. Nevertheless, it remains the main endeavor of urban mobility planning.

In parallel, alternative means of transportation (public transport, bicycles and pedestrians) have grown even more strikingly, as shown in Table 2. This table represents the main result of this paper.

Table 2. The periods of cultural change in the mobility of Chile. Main result of this study.

Period	Private v.	Bycicles	Collective t.	Pedestrians
1998–2006	93%	16%	19%	4%
2007–2015	74%	31%	23%	15%

For example, according to SECTRA, bicycle mobility has been favored by the project to construct 632 km of bicycle routes and paths (474 km have already been approved and 158 km are still under study). With this, a network of bicycle facilities, and all kinds of "cycles", has been established. This was initially promoted by the cyclists themselves for their displacements and has been later taken over by the municipalities and governments of many regions, by including them in the projects of transportation across the country.

Chile follows the urban mobility pattern of the capitals of Latin America, where the automobile is the main protagonist and in whose cities the priority in investment in infrastructure is mainly aimed at facilitating the movement of private transportation which the maximum possible fluidity [7].

Also, the existence of citizen participation has not been observed in Chilean transportation plans since the beginning of the projects' planning. Of course, an item of "implementation of participation mechanisms" is introduced [12] in the Methodology section of the Urban Transport Plans but without specifying the degree of participation of citizens.

This aspect of participation is key, since it is a requirement for citizens and stakeholders to take charge of the sustainable urban mobility plan (or other mobility plans used by the country) and of the policies that it promotes. In this way, the probability of public acceptance and support increases, and the risks in decision-making and in the implementation of the plan are minimized [14].

It would be interesting - creating a new strand of research for these issues - to analyze the degree of success of projects according to the level of citizen participation.

References

1. Gomez-Lobo, A.: Transantiago: una reforma en panne. Trabajos de investigacíon en políticas públicas – Universidad de Chile **4**, 1–14 (2007). http://repositorio.uchile.cl/handle/2250/144835
2. Salazar-Burrows, A., Ugarte, C., Osses, P.: Exclusíon social asociada al transporte y su relacíon con la distribucíon de la densidad de poblacíon en la provincia de Melipilla, Regíon Metropolitana de Santiago de Chile. Revista de Geografía Norte Grande (59), 145–164 (2014)
3. de Miguel, C.: Desafíos en la medición del consumo y produccíon sostenible. In: Presentado en 8o Seminario regional de los indicadores de los Objetivos de Desarrollo del Milenio de CEPAL, pp. 1–25 (2014). https://www.cepal.org/MDG/noticias/seminarios/9/53699/1CEPALCdeMiguel.pdf
4. Romero, H., Irarrázaval, F., Opazo, D., Salgado, M., Smith, P.: Climas urbanos y contaminacíon atmosférica en Santiago de Chile. Revista EURE **36**(109), 35–62 (2010). http://www.eure.cl/index.php/eure/article/view/101/509
5. Instituto Nacional de Estadísticas. "Estadísticas Vitales – Informe Anual 2016". Instituto Nacional de Estadísticas (2018). https://www.ine.cl/
6. Instituto Nacional de Estadísticas. "Estadísticas Económicas – Transporte y Comunicaciones". Instituto Nacional de Estadísticas (2019). https://www.ine.cl/estadisticas/economicas/transporte-y-comunicaciones?categoria=Anuarios
7. Quintero-Gonzalez, J.R.: Del concepto de ingeniería de tránsito al de movilidad urbana sostenible. Ambiente y Desarrollo **21**(40), 57–72 (2017). https://doi.org/10.11144/javeriana.ayd21-40.citm
8. Ministerio de Transporte y Telecomunicaciones. "Informacíon de Transporte Urbano en ciudades de Chile (2010–2015)". Ministerio de Transporte y Telecomunicaciones - Coordinacíon de Planificacíon y Desarrollo (2016). https://www.mtt.gob.cl/

9. Ministerio de Transporte y Telecomunicaciones - Secretaría de Planificacíon de Transporte. “Plan Maestro de Transporte Santiago 2025”. MTT-SECTRA (2012). http://www.sectra.gob.cl/biblioteca/detalle1.asp?mfn=3067
10. Ministerio de Transporte y Telecomunicaciones - Secretaría de Planificacíon de Transporte. “Informe Ejecutivo. Encuesta Origen Destino de Viajes” (2012). http://www.sectra.gob.cl/biblioteca/detalle1.asp?mfn=3253
11. Ministerio de Transporte y Telecomunicaciones - Secretaría de Planificacíon de Transporte. “Política Nacional de Transportes”. Ministerio de Transporte y Telecomunicaciones (2013). http://www.mtt.gob.cl/wp-content/uploads/2013/05/documento-politica.pdf
12. Ministerio de Transporte y Telecomunicaciones - Secretaría de Planificacíon de Trans- porte. “Planes de transporte urbano” (2019). http://www.sectra.gob.cl/planestransporteurbano/definicion.htm
13. Figueroa, O., Orellana, A.: Transantiago: gobernabilidad e institucionalidad. Revista EURE **33**(100), 165–171 (2007)
14. Organizacíon para la Cooperacíon y el Desarrollo Económicos. “Brechas y estándares de gobernanza de la infraestructura pública en Chile: Análisis de Gobernanza de Infraestructura” (2017). Editions OCDE. http://dx.doi.org/10.1787/9789264286948-es
15. Salas, M.R.: La bicicleta como modo de transporte que visibiliza el acceso desigual a la movilidad cotidiana: el caso de Santiago, Chile. Revista de Urbanismo **39**, 1–26 (2018). https://doi.org/10.5354/0717-5051.2018.49157
16. Ureta-Icaza, S.: Manejando por Santiago. Explorando el uso de automóviles por parte de habitantes de bajos ingresos desde una óptica de movilidad sustentable. Revista EURE **35** (105), 71–93 (2009)
17. SECTRA. Secretaría de Planificacíon de Transporte, 2019. “Fomento de Transporte Sustentable” (2019). http://www.sectra.gob.cl/transportesustentable/transportesustentable.htm

Smart Infrastructure for Shared Mobility

Christos Gioldasis[1](✉) and Zoi Christoforou[2]

[1] COSYS-GRETTIA, Univ Gustave Eiffel, IFSTTAR, 77447 Marne-la-Vallée, France
christos.gioldasis@ifsttar.fr
[2] Department of Civil Engineering, University of Patras, Panepistimioupoli Patron, 265 04 Patras, Greece
zoi.christoforou@upatras.gr

Abstract. Micromobility has emerged as a green mode of transport, which offers an innovative way to handle persistent traffic planning problems such as the last-mile connections with public transit. The integration of micromobility in urban infrastructure is a key challenge faced by public authorities, policy and urban planners. In this paper we present the results of a face-to-face street travel survey among e-scooter users in Paris about infrastructure-related issues. The great majority of the respondents are not owners of e-scooters, but they use the free-floating electric scooters (FFES), which is an increasingly popular shared mobility scheme. The responses have been analyzed statistically and cross-tabulations have been produced. They demonstrate that the users perceive riding on the cycle lane, the sidewalk (excluding cycle lane) and the sidewalk on a cycle lane as safe. On the contrary, they perceive as less safe driving the e-scooters on the cycle lane, the shared lane for bus/taxi/bicycle and the mixed traffic lane. Nonetheless, it is found that the users drive faster at the types of road infrastructure, which they consider less safe.

Keywords: Micromobility · Safety · Infrastructure · E-Scooters · Survey

1 Introduction

Micromobility is an emerging transportation mode often defined according to vehicle characteristics such as weight [1] or electric batteries [2]. In the past few years, micromobility is enjoying a major surge in popularity thanks to the development of companies providing free-floating e-scooters in many cities all over the world [3].

Its rapid development led to concerns about the proper integration of micromobility in the urban networks along with the traditional modes of transport [4]. Another growing concern is the level of safety, which micromobility provides to its users, as well as to the other users of the road infrastructure [5]. Nonetheless, the available studies in this field remain limited. A survey among e-scooter users was launched in Paris to address the need for a better understanding of this emerging mode of transport. The respondents were asked questions about the use of e-scooters and their behavior when driving an e-scooter. In this paper, we examine the perceived safety of e-scooter riders while considering different urban infrastructures, i.e. sidewalk, pavement, bike lanes.. The remainder of this paper is structured as follows: Sect. 2 reviews existing

E. G. Nathanail et al. (Eds.): CSUM2020, AISC 1278, pp. 970–979, 2021.
https://doi.org/10.1007/978-3-030-61075-3_93

studies on e-scooter safety and the impact of infrastructure on safety; Sect. 3 presents the site of the survey, the methodology used and the characteristics of the sample; Sect. 4 presents the findings on perceived safety levels per infrastructure; Sect. 5 presents the conclusions of our work and suggestions for future work.

2 Background

2.1 Micromobility and Safety

Evidence shows that almost half of the free-floating e-scooter riders were engaged in illegal and risky driving such as rider not wearing a helmet, riding on the road, or doubling a passenger [6]. Risk-taking driving behavior may have consequences on e-scooter riders or passengers, pedestrians, and bicyclists. To assess that impact, clinical studies were conducted based on hospital reports in Israel [7] and in California [8]. The Israeli study demonstrated that most of the injured in crashes involving e-scooters were the e-scooter riders themselves, with pedestrians being the second most affected group. The importance of the age factor was also ascertained in this study. The Californian study found that e-scooter related injuries vary in severity and highlighted the low use of helmet among the injured. The authors also studied the age of the users. It seems that many users are below the minimum age requirement set by the operators, i.e. 18 years old.

These studies focus on human factors (e.g. age, gender) and their impact on safety as well as with the types of injuries observed in accidents involving e-scooters. The role of infrastructure, however, has not been explored so far, even though it is a well-known risk factor for other vulnerable road users (pedestrians and cyclists). In Sect. 2.2 we present studies which investigate the impact of the infrastructure on safety of soft mobility modes of transport.

2.2 Road Infrastructure and Safety of Soft Modes

Road infrastructure characteristics related to safety mainly refer to the number of lanes, the number of legs of a given intersection, the presence of right turns and the number of entering lanes, [8]. In [9], the authors study the safety of bicyclists and report that speed reducing measures such as raised bicycle crossings (e.g. on a speed hump) and the deflection of cycle track between 2 and 5 m away from the main carriageway lead to reduced crash risk [9]. Another positive factor seems to be the time since the first implementation [10]. The number of observed conflicts decreases significantly after the first year as users get familiarized with the new infrastructure. The existence of median (e.g. tall grass, trees), the physical separation between roadway and bikeway, and illumination are associated with lower risk of bicyclists being hit by motor vehicles [11]. Injury severities tend to be more acute at intersection locations where a combination of horizontal and vertical curves exists [12]. Similar findings are not reported for pedestrians [13].

3 Survey

The factors leading to accidents involving e-scooters are a focal point of the existing studies on micromobility. However, to the best of the authors' knowledge, the role of infrastructure has not been taken into consideration in any of these studies. Given the fact that the role of infrastructure is proved to be critical in the case of all other road modes,, we designed and undertook a survey to investigate e-scooter riders' perception of infrastructure-related risks. The survey was launched in Paris in 2019 because e-scooters enjoy high popularity in the city. The method of face-to-face interviews was chosen to increase the sample reliability and representativity. In addition to the infrastructure-related questions, the questionnaire included questions on socio-demographic characteristics and general mobility patterns. We use descriptive statistics and cross-tabulations to analyze and present the responses.

3.1 Site of the Survey

The survey was launched in May 2019 in Paris. The metropolitan area of Paris has a landmass of 12000 km^2 and is inhabited by about 11.1 million citizens. Seven sites were preselected for the face-to-face interviews covering different district types. The sampling method consisted in the second person arriving at a preselected point once the previous interview was completed. The preselected points include two unmixed residential areas in the inner suburbs (Issy-les-Moulineaux), Paris CBD (La Defense), two leisure-time sites (the banks of the Seine and the quays of St Martin canal), two major central mass transit hubs (Châtelet and République), and the area of Station F, which is regarded as the biggest startup campus in Paris. At Issy-les-Moulineaux and at La Defense the face-to-face interviews took place only on week-days, while at all other points the face-to-face interviews took place at week-days and at week-ends.

3.2 Socio-Demographic Characteristics of the Sample

In total, 459 e-scooter users were interviewed. 68.4% of the respondents were female. The great majority of the respondents were young. 0.4% of the respondents were younger than 12 year-old, 8.8% were aged between 12 and 17 years old, 44% were between 18 and 24, 27.6% were aged between 25 and 29, 13.6% were aged between 30 and 34, 2.8% between 35 and 39 years old, 2.8% between 40 and 44 years old, 3.6% between 45 and 49 years old, and 2.8% were aged above 50 years old. Only 7% of the respondents owned their own e-scooter. Regarding their last trip, 87.2% of the respondents declared that their last trip was in the centre of Paris, not in the suburbs.

Concerning their education, the majority of the respondents have higher education diplomas. 11% of them declared that they have none or before high school diploma, 11% declared that they have professional degree under high, 25% declared that they have general high school diploma, 21% declared that they have college credit, 30% declared that they have master's degree and above, and 3% did not answer. Asked about their household status, 45% responded that they live alone, 22% responded that they are a childless couple, 17% responded that they are a couple with child(ren), 3%

responded that they live alone with one or more children, and 12% responded that they live with roommate(s).

4 Results

The respondents of the face-to-face interviews were asked to assess travelling six types of road infrastructure: the cycle lane (CL), the sidewalk (excluding cycle lane) (S), the sidewalk on a cycle lane (SC), the road on a cycle lane (R), the shared lane for bus/taxi/bicycle (SL) and the mixed traffic lane (ML).

In particular, the interviewees were asked to express their perception of safety offered by six types of road infrastructure by rating them from 1 to 5, where 1 stands for the least safe, 3 for average safe and 5 for the most safe. The responses reveal the perception of safety, which does not necessarily coincide with the actual level of safety provided by each type of infrastructure. Figure 1 presents the responses of the interviewees. We notice that the cycle lane and the sidewalk are considered to be the safest, whereas the shared lane for bus/taxi/bicycle and the shared mixed traffic lane are considered to be the least safe. The response of the users may be explained by the existence of heavier vehicles, which are perceived as a potential risk but may also obstruct visibility. Frequent stops and larger dead angle can also explain this finding. We note here that the responses illustrated in Fig. 1 concern e-scooter riders having used an e-scooter on all 6 types of infrastructure.

Moreover, the interviewees were asked to declare the average speed of their last e-scooter trip per infrastructure type by rating them from 1 to 4. 1 stands for speed below 4 km/h, 2 stands for speed between 4 and 10 km/h, 3 stands for speed between 11 and 20 km/h and 4 stands for speed over 20 km/h. Figure 2 presents the responses of the interviewees. We notice that declared speed is higher on shared lane for bus/taxi/bicycle and on mixed traffic lane, even though they are considered as less safe. On the contrary, lower declared speeds are observed on the infrastructure types that are perceived as safer.

Furthermore, the interviewees were asked to estimate the time spent on each infrastructure as a fraction of their total trip duration. The scale used went from 1 to 4 with 1 standing for less than 10%, 2 for 10–50%, 3 for 51–90%, and 4 for over 90%.

Figure 3 presents the relevant results. Responses indicate that the riders combine several road types over a single trip and tend not to ride exclusively on a specific infrastructure type. This finding comes to underline the discontinuity of safe cycle paths within dense urban contexts which consists a significant risk factor. Switching from one road infrastructure to another increases risk because of the merging maneuvers, and physical obstacles.

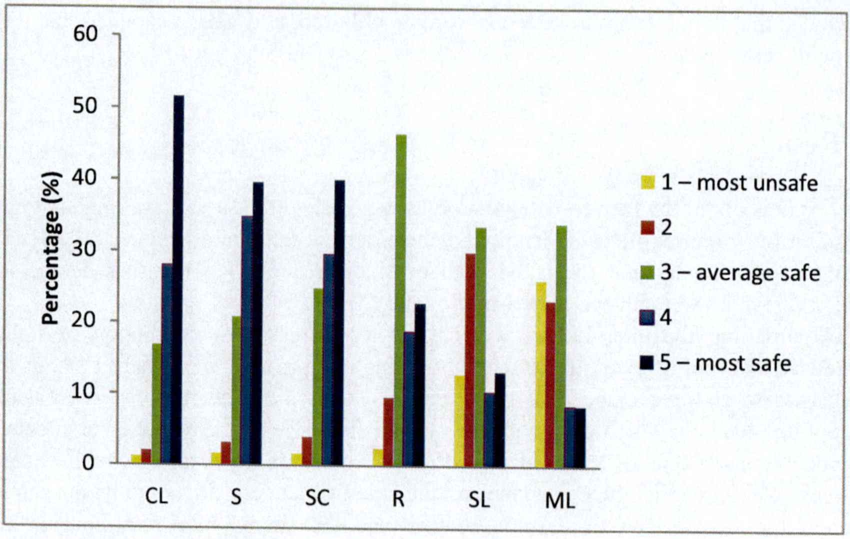

Fig. 1. Safety perception of riding on various types of road infrastructure.

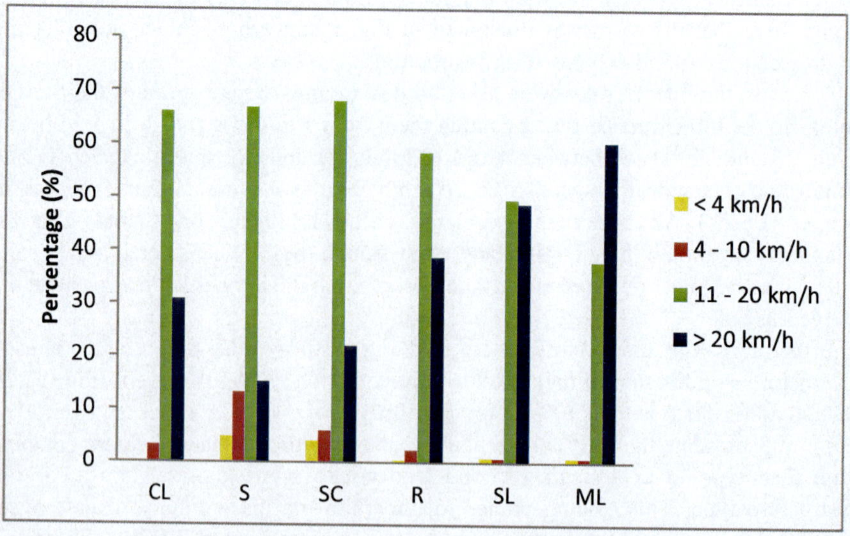

Fig. 2. Declared e-scooter speed while riding on various types of road infrastructure.

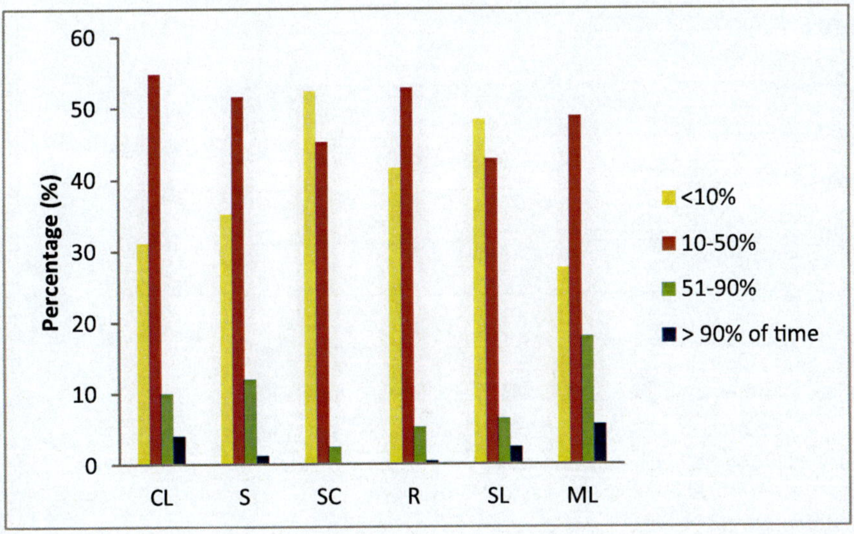

Fig. 3. Time spent on various types of road infrastructure as a percentage of the total duration of the trip.

Furthermore, we crossed safety-related findings to trip characteristics, namely trip duration and declared speed. In Fig. 4, 5, 6 and 7, we chose to present the results for the two 'extreme cases': the types of infrastructure that users classed as least and most safe (mixed traffic and cycle lane respectively). The cross-tabulations are given in absolute numbers of respondents and refer only to the interviewees who declared to have used all types of infrastructure (N = 252). Regarding the time spent, riders seem to pass more time on cycle lanes, if available, irrespectively to the perceived level of safety (Fig. 4). Mixed traffic lanes are avoided mainly by those who ranked it as highly unsafe (Fig. 5). Cycle lanes make rides confident and cycle faster (Fig. 6) while mixed traffic lanes encourage speeding independently to perceived safety level (Fig. 7). We note here that keeping up with the average general traffic speed is not necessarily a risk factor. Overall results are consistent and confirm (i) the need to expand existing cycle lanes and (ii) the necessity to further observe and explore specific behavioral issues, as speeding, that cannot be easily understood and interpreted.

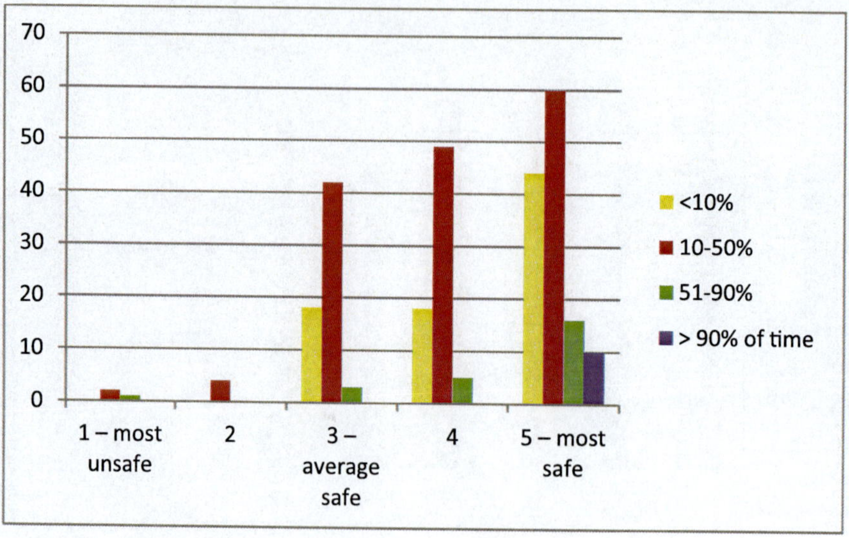

Fig. 4. Time spent and perceived safety level – cycle lanes.

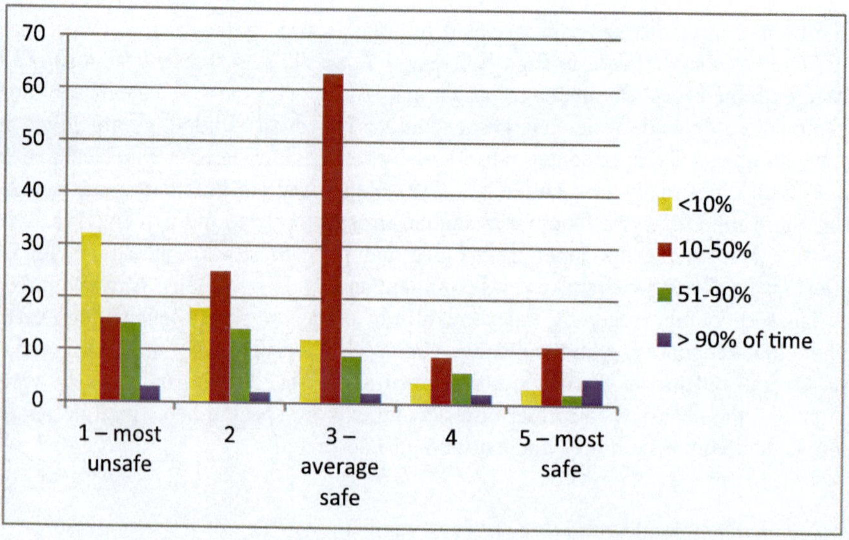

Fig. 5. Time spent and perceived safety level – mixed traffic lane.

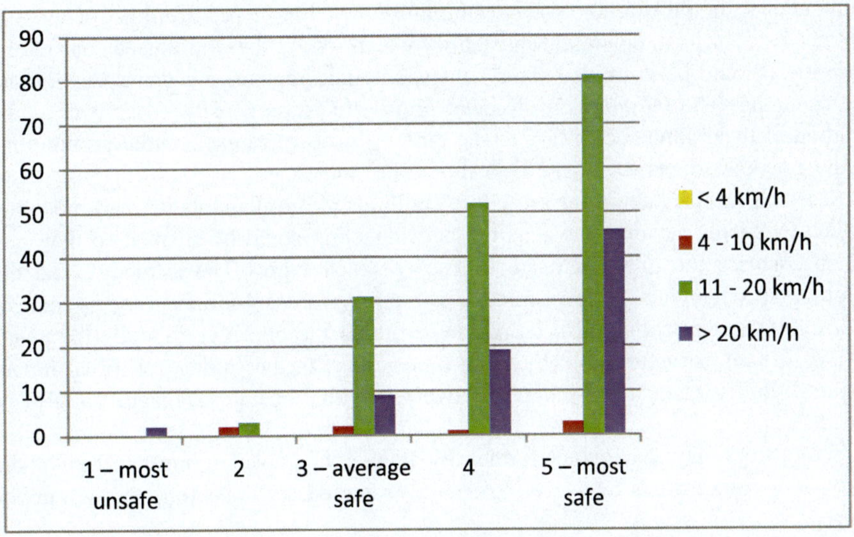

Fig. 6. Declared speed and perceived safety level cycle lane.

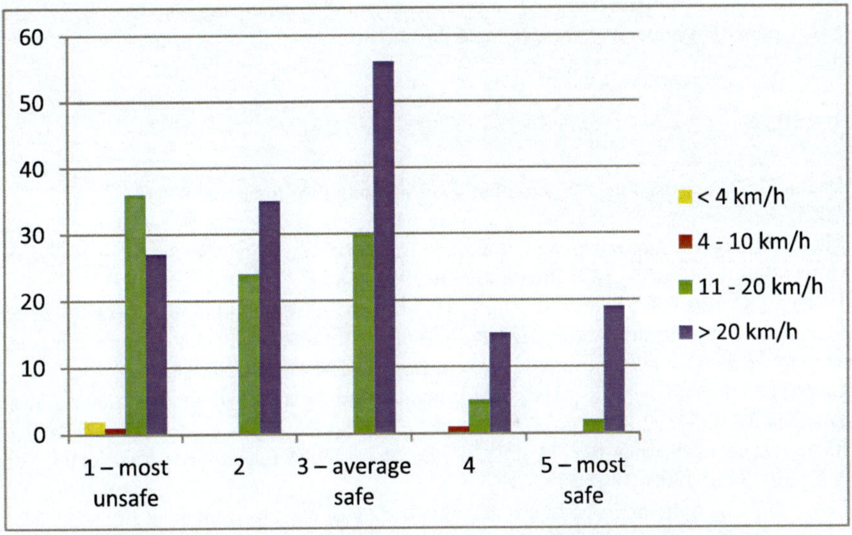

Fig. 7. Declared speed and perceived safety level – mixed traffic lane.

5 Conclusions – Future Work

This paper focused on the pereception of safety and use of various types of infrastructure by e-scooter riders. The findings revealed that the use of a type of infrastructure might not be affected by the perception of safety. Risky behaviours such as the switching types of infrastructure during a trip have also been revealed. Consequently, future studies should examine ways to minimize switching types of infrastructure and to smooth merging. Potential legislation which could be introduced by public authorities should be examined. This legislation could regulate e-scoter speeding and lane changing. The introduction of speed limits should be considered as well as the variation of these limits according to the type of infrastructure. Potential prohibitions for the e-scooter riders to change lane should be studied.

Research findings about infrastructure could also inform simulation and modeling studies. E-scooter movement simulation and modeling could be adapted so that they will incorporate the different characteristics of each type of infrastructure and the merging, lane changing and infrastructure switching process. Moreover, the upgrade of urban infrastructure should also be researched, with the objective to study the continuity of the infrastructure used by the e-scoter rider. In the studies for urban infrastructure, the creation of e-scooter networks, with separate signage, should be considered.

Future work may also include further research on the impact of infrastructure on the safety of e-scooter riders by analyzing data for reported accidents (e.g. through police reports).

Acknowledgements. This research is part of the ORNISIM project (https://ornisim-ifsttar.fr/) that is supported by La Foundation Maif (https://www.fondation-maif.fr/). The authors would like to thank IESEG and University of Patras students and PhD candidates who were involved in the preparation, translation, and realization of the survey.

References

1. Dediu, H.: The three eras of micromobility. Micromobility Industries (2019)
2. Maiti, A., Vinayaga-Sureshkanth, N., Jadliwala, M., Wijewickrama, R.: Impact of Urban Micromobility Technology on Pedestrian and Rider Safety: A Field Study Using Pedestrian Crowd-Sensing (2019). arXiv preprint arXiv:1908.05846
3. Heineke, K., Kloss, B., Scurtu, D., Weig, F.: Micromobility's 15,000-mile checkup (2019). McKinsey & Company Automative & Assembly: https://www.mckinsey.com/industries/automotive-andassembly/our-insights/micromobilitys-15000-mile-checkup
4. Hardt, C., Bogenberger, K.: Usage of e-scooters in urban environments. Transp. Res. Procedia **37**, 155–162 (2019)
5. Sikka, N., et al.: Sharing the sidewalk: a case of E-scooter related pedestrian injury. Am. J. Emerg. Med. **37**(9), 1807-e5 (2019)
6. Haworth, N.L., Schramm, A.: Illegal and risky riding of electric scooters in Brisbane. Med. J. Aust. **211**(9), 412–413 (2019)
7. Siman-Tov, M., Radomislensky, I., Israel Trauma Group, Peleg, K.: The casualties from electric bike and motorized scooter road accidents. Traffic Inj. Prev. **18**(3), 318–323 (2017)
8. Gomes, S.V.: The influence of the infrastructure characteristics in urban road accidents occurrence. Accid. Anal. Prev. **60**, 289–297 (2013)
9. Schepers, J.P., Kroeze, P.A., Sweers, W., Wüst, J.C.: Road factors and bicycle–motor vehicle crashes at unsignalized priority intersections. Accid. Anal. Prev. **43**(3), 853–861 (2011)

10. Phillips, R.O., Bjørnskau, T., Hagman, R., Sagberg, F.: Reduction in car–bicycle conflict at a road–cycle path intersection: evidence of road user adaptation? Transp. Res. Part F Traffic Psychol. Behav. **14**(2), 87–95 (2011)
11. Yan, X., Ma, M., Huang, H., Abdel-Aty, M., Chaozhong, W.: Motor vehicle–bicycle crashes in Beijing: Irregular maneuvers, crash patterns, and injury severity. Accid. Anal. Prev. **43**(5), 1751–1758 (2011)
12. Moore, D.N., Schneider IV, W.H., Savolainen, P.T., Farzaneh, M.: Mixed logit analysis of bicyclist injury severity resulting from motor vehicle crashes at intersection and non-intersection locations. Accid. Anal. Prev. **43**(3), 621–630 (2011)
13. Pulugurtha, S.S., Sambhara, V.R.: Pedestrian crash estimation models for signalized intersections. Accid. Anal. Prev. **43**(1), 439–446 (2011)

Assessing the Compliance of Existing Cycling Route Infrastructure Against National Guidelines in Greece

Georgios Georgiadis[1(✉)], Efthimios Bakogiannis[2], Aristomenis Kopsacheilis[1], Georgios Barmpas[3], and Ioannis Politis[1]

[1] Transport Engineering Laboratory, Department of Civil Engineering, Faculty of Engineering, Aristotle University of Thessaloniki, 54124 Thessaloniki, Greece
ggeorgiadis@civil.auth.gr
[2] Department of Geography and Regional Planning, School of Rural and Surveying Engineering, National Technical University of Athens, Heroon Polytechniou Street, 9, Zographou Campus, 15780 Athens, Greece
[3] Lever Development Consultants S.A, 26th October Street, 43, 54627 Thessaloniki, Greece

Abstract. The eve of the 2020s finds European cities in a process of shifting citizens' mobility preferences towards more sustainable transport means. Cycling is an emerging alternative to motorised modes, particularly due to its lower operating cost and its contribution to healthier lifestyles and containment of climate deregulation from transport sector. Towards establishing cycling as a commuting mode and decreasing private car use, city authorities are increasingly investing in delivering integrated cycling route networks, pro-cycling policies and related services. However, poor and inadequate network design may act as a deterrent to cycling, due to safety, accessibility or quality concerns. To address this, the Greek Ministry of Transport in 2016 issued a set of technical guidelines for cycling network infrastructures in order to standardize their design process. In response to these new guidelines, this paper proposes a methodology for the assessment of existing cycling infrastructure through the utilization of a structured checklist. This checklist considers critical components of bicycle networks, such as their signage, width, inclinations, lighting, intersections, types of separation and interface with other modes and examines them against the minimum requirements described in national standards. The methodology was piloted in the existing cycling networks of Athens and Trikala, Greece. Results indicated certain design flaws and the attributes that did not comply with national standards. The proposed methodology could assist planning authorities into monitoring the quality of cycling infrastructure in Greek cities and indicating the aspects that do not meet the national set thresholds along with highlighting the appropriate interventions.

Keywords: Cycling · Evaluation methodology · Sustainable Urban Mobility · Cycling route network · Infrastructure resilience

E. G. Nathanail et al. (Eds.): CSUM2020, AISC 1278, pp. 980–990, 2021.
https://doi.org/10.1007/978-3-030-61075-3_94

1 Introduction

The progressive degradation of environmental conditions has made government officials worldwide to prioritize their environmental agendas. While the parameters influencing environmental conditions vary, the negative contribution of car use is evident. In numbers, transport is accountable for 24.7% of the total greenhouse emissions worldwide, while road transport is the major contributor with 71.7% [1]. In response to the observed problems, the European Commission (EC) is aiming to limit its road transport emissions below 20%, in relation to 2008 levels, by 2030 [2] and to make EU a climate-neutral continent by 2050 [3]. Sustainable Urban Mobility Plans (SUMPs) have been introduced EU-wide, as the cornerstone of EC's policy towards sustainability in its transport systems. Most European countries have already taken action towards improving the sustainability of their transport systems, limiting the domination of the private car in urban networks.

Cycling is among the most preferable alternative travel options to the private car, since it combines minimal emissions with relatively high speed and promotes a healthier lifestyle for users. Many countries along with Greece, have started to develop cycling infrastructure, in order to incentivize car users to shift their preferable transport mode into cycling. New cycling infrastructure is considered as a key player in increasing cycle use [4] along with the implementation of car-restricting policies. Additionally, the upgrade of existing infrastructure has also a positive effect on the take-up of cycling by users, since it contributes to increased safety and comfort levels [5]. However, the introduction of any new cycling infrastructure is raising the question of its suitability to accommodate cycling demand while at the same time ensuring high quality levels. In response to this, many countries in EU and worldwide tried to standardise cycleway design by introducing detailed technical guidelines [6, 7]. Acknowledging the progress made EU-wide, the Greek government has issued a set of national guidelines that aim at high quality infrastructure through the establishment of cycleway design standards.

This paper considers the Greek guidelines and aims to propose and apply a structured evaluation checklist that can be utilised by transport practitioners as a cycling infrastructure quality appraisal tool. In this study, the proposed tool is applied to the existing cycleways in Athens and Trikala, Greece. Evaluation results indicate the relative strengths and weaknesses of the two cycle networks and highlight the importance of the recommended evaluation methodology into successful decision-making and prioritisation of cycling improvement measures.

2 Background

Recognizing the need to adjust to the new EU mobility paradigm, in 2016 the Greek government organized a funding scheme to assist the development of local SUMPs by allocating €9 million to 150 Greek municipalities. The variety of cycling benefits in urban environments made pro-cycling measures preferable against other mobility measures included in those SUMPs. Therefore, many Greek municipalities have started introducing new or upgrading existing cycling infrastructure to improve the bicycle's utility against other private transport means thus increasing cycling mode share, which currently is substantially lower in relation to other EU countries [8].

In order to establish a common framework in cycleway infrastructure design and thus achieve the necessary safety and comfort levels, the Greek Ministry of Transport issued in 2016 a set of national guidelines that aim to assist authorities in the process of introducing new or upgrading and evaluating existing cycleway infrastructure. These guidelines provide thresholds that cover the most basic aspects of a cycling infrastructure. Specifically, they include numeric thresholds that refer to cycleway's various operational characteristics, such as width, safety distances, height clearance etc. The guidelines also provide details about the integration of the cycling infrastructure to the street environment, through detailed drawings that give practitioners the opportunity to examine specific details such as the cycleway's segregation to motorised traffic, its interface with pedestrians and public transport, the proper design on intersections and many others. Regarding the signage of cycleways, detailed pictures of the required signs are included in a special section of the guidelines [9]. The proposed tool considers all the above characteristics equally on the infrastructure evaluation process.

3 Evaluation Methodology for Cycling Infrastructure

In order to examine the conformity of the design of existing and future cycling infrastructure against the newly introduced national guidelines, we developed a structured checklist, as a tool that can be used for in situ examination procedures and for designing new networks. This checklist examines a plethora of elements that cover different operational characteristics of cycling infrastructure and compares them against their corresponding national standards' thresholds. It is summarized in Table 1.

Table 1. Checklist for the evaluation of existing cycling infrastructure.

Element	Feature	Location	Value	Guideline threshold
Width	Design width	–	*in metres*	*in metres* (Greek design guidelines: Sects. 3.3, 3.4, 3.5) [9]
	Effective width	Per section **or** kilometre distance from start point	*in metres*	
Directness	Number of intersections where cyclists must give way	Total length of cycleway	Number of intersections	*minimum*
Coherence	Number of cycleway breaks	Total length of cycleway	Number of breaks per kilometre due to implementation gaps or obstacles	*minimum*

(*continued*)

Table 1. (*continued*)

Element	Feature	Location	Value	Guideline threshold
Integration to road network infrastructure	Cycleway direction	Per section **or** total length of cycleway	One-way; Contra-flow one-way; Two-way	Not defined
	Cycleway type	Per section **or** total length of cycleway	Cycle lane; Advisory cycle lane; Contra-flow cycle lane; Cycle track; Cycle street	Not defined
	Segregation structure	Per section **or** total length of cycleway	Width, height (if any)	width, height (Greek design guidelines: Sects. 3.3, 3.4, 3.6) [9]
Parking	Number of parking spots	Per section **or** kilometre distance from start point	Number of parking spots, Number of accommodated bicycles	Not defined
	Distance between parking spots (m)	Per section **or** total length of cycleway	*in metres*	25–50 m
	Type of parking spots	Per section **or** kilometre distance from start point	Short-term; Long-term	Not defined
	Parking spot design	Per section **or** kilometre distance from start point	Dimensions & style	According to guideline drawings (Greek design guidelines: Sect. 8) [9]
Cyclists' traffic space	Height clearance	Per section **or** kilometre distance from start point	*in metres*	>2.5 m
	Side safety distances	Per section **or** kilometre distance from start point	*in metres*	*in metres* (Table 3-3, Greek design guidelines, Sect. 3.2.2) [9]

(*continued*)

Table 1. (*continued*)

Element	Feature	Location	Value	Guideline threshold
Inclination	Inclination	Per section **or** total length of cycleway	%	maximum 5% or <3% if continuous
Visibility	Visibility	Per section **or** kilometre distance from start point	No visibility; Limited visibility for lateral moving vehicles; Limited visibility for parallel moving vehicles; Sufficient visibility	Sufficient visibility at total cycleway length
Lighting	Lighting	Per section **or** kilometre distance from start point	Sufficient lighting; Limited lighting; Total absence of lighting	Class S/A
Intersections	Intersection design	Intersection name **or** kilometre distance from start point	Compatible; Incompatible design	According to drawings (Greek design guidelines: Sect. 6) [9]
	Cyclists' priority	Intersection name **or** kilometre distance from start point	Yes; No	Yes
Surface	Material	Per section **or** kilometre distance from start point	Material (concrete, asphalt, pavement slabs, etc.)	According to guidelines (Greek design guidelines: Sect. 10) [9]
	Maintenance level	Per section **or** kilometre distance from start point	high; medium; low	*high*
Signage (Horizontal)	Solid lines	Per section **or** kilometre distance from start point	Yes; No (for each sign)	According to design guidelines drawings (Greek design guidelines: Sect. 9.2) [9]
	Bicycle symbol			
	Give-way symbol			
	Intersection sign			
	Crossing lines			

(*continued*)

Table 1. (*continued*)

Element	Feature	Location	Value	Guideline threshold
Signage (Vertical)	Directional signs	Per section **or** kilometre distance from start point	Yes; No (for each sign)	Not defined
	Intersection approaching signs			Sign codes: K-15, K-16, K-17
	Crossing approaching signs			Sign codes: Π-21α, Π-21β, Π-21γ
	Bicycle road sign			Sign codes: P-54, P-65, P-66
	Bicycle parking sign			Sign codes: Π-31δ
	Bicycle lane sign			Sign codes: Π-122, Π-122α, Π-123
Interface with Public Transport	Number of stops/stations	Per section **or** kilometre distance from start point	Number of stops/stations per kilometre	Not defined
	Cycleway design at bus stops/stations	Per stop	Compatible; Incompatible design	according to guidelines drawings [9]
Land uses	Commercial - Recreational - Mixed land uses	Per section **or** kilometre distance from start point	type of land use	Not defined
	Green space	Per section **or** kilometre distance from start point	Yes; No	Not defined
Motorised Traffic	Vehicle speed limit	per section **or** total length of cycleway	<30 km/h; <50 km/h; >50 km/h	Dependent on cycleway type (Greek design guidelines: Sects. 3.3, 3.4, 3.5) [9]

The first column of Table 1, lists the main categories of the characteristics that are mentioned in the national guidelines and need to be examined, for a cycleway infrastructure examination to be complete. The elements cover many different aspects of a cycleway, such as its integration in the streets' environment, the main dimensions of cyclists' traffic space, the interface with other transport means, the signage and lighting adequacy etc. The second column specifies the main features that can be either qualitatively or quantitatively measured for each of the first column's elements. The third column of Table 1 refers to the locations where these features should be either measured or observed. By noting-down observations in specific locations along any examined cycleway, the proposed checklist is made functional for on-site appraisal, since it provides the opportunity to be completed in situ and highlight the scale of infrastructures' inconsistencies compared to the standards set by the national guidelines. Observed values are filled into the fourth column of Table 1. Since, certain features cannot be expressed using numeric values, in some cases a qualitative scale is used instead (e.g. for visibility and maintenance level). In the last column of Table 1, the values of the fourth column are compared against the values defined in the issued national guidelines. For certain features, the guidelines do not propose a specific value or a range of values. In such cases, the threshold which was adopted reflects the corresponding advice for their design. For instance, in order to achieve a direct cycleway, the number of intersections where a cyclist must give way must be kept to the minimum.

4 Pilot Application

Pilot applications of the proposed evaluation tool were conducted for the cycle networks of Athens in 2017 and Trikala in 2019. Both of these networks were constructed before the 2016 national corresponding guidelines. Municipality of Athens has a population of more than 660,000 residents. The city's geomorphology is characterized by mild to medium inclinations which generally favor cycling. According to the most recent survey, citizens were using 53% car, 37% public transport, 2% bicycle and 8% walking in their daily trips [10]. Until now, only a small portion (5 km of the 27 km long) of the Athens metropolitan bicycle network has been implemented. In this pilot application we considered the 2.2 km part of the cycle network which lies within Athens municipality. On the other hand, Trikala is a small city of Greece with less than 61,000 residents. The city's geomorphology includes almost zero inclinations and is ideal for cycling and walking. According to the recent SUMP of Trikala, citizens are using 32.8% car, 28.9% motorcycle, 2% public transport, 10.3% bicycle and 26% walking in their daily trips [11]. Local authorities so far have not managed to compose the existing cycle lanes and tracks into an integrated network. In total, Trikala has a 6 km long cycle network. For this pilot application we only considered the part of network, which forms a ring surrounding the central part of the city. This central ring of cycle lanes has a total length of 1.7 km and hosts the highest cycle traffic volumes compared to the other city's infrastructure segments.

Table 2 summarizes the evaluation results for the examined cycle routes. In Athens, the effective width of the cycle track complies with the minimum guidelines' standard, but the safety distances are rather small. The number of cycleway breaks is low since the cycle route almost exclusively coincides with one single street. Parking facilities are generally satisfactory but no long-term parking spots are provided and their design does not follow guidelines' advices. Maintenance level was high. The cycle track provides convenient access to a metro station and passes through a mostly residential area. In Trikala, the cycle lane's actual width is not adequate for two-way movement. Cycleway breaks are comparatively more due to the shape of the examined segment and its intersections with many central streets. Parking spots are not adjacent to the cycle lane. Segregation structures do not generally comply with safety standards. Maintenance level was medium. Since the infrastructure is in the city's center it can appropriately provide access to important places of interest and public transport services.

In both cycleways, the installed signage, as well as the design and priority status of all intersections with motorized transport, do not follow guidelines and thus constitute the most critical safety concerns. Height clearance, vehicle speed limit, inclination, visibility and lighting are acceptable in both areas.

Table 2. Evaluation results in the cities of Athens and Trikala.

Category	Characteristic	Athens	Trikala
		Observation	Observation
Width	Design width	2.0 m	2.0 m
	Effective width	2.0 m	1.5 m
Directness	Number of intersections where cyclists must give way	100%	100%
Coherence	Number of cycleway breaks per km	1.36	4.24
Integration to road network infrastructure	Cycleway direction	Two – way	Two – way
	Cycleway type	Cycle track	Cycle lane
	Segregation structure	Not observed	Insufficient dimensions
Parking	Number of parking spots	13 spots, 78 bicycles	Available spots – not adjacent to infrastructure
	Distance between parking spots (m)	50–300	
	Type of parking spots	Short-term	Short-term spread across the city
	Parking spot design	inappropriate design	inappropriate design
Cyclists' traffic space	Height clearance	> 2.5 m	> 2.5 m
	Side safety distances	0.1 m	0.2 – 0.3 m

(continued)

Table 2. (*continued*)

Category	Characteristic	Athens	Trikala
		Observation	Observation
Inclination	Inclination	< 3%	< 3%
Visibility	Visibility	Sufficient visibility	Sufficient visibility
Lighting	Lighting	Sufficient lighting	Sufficient lighting
Intersections	Intersection design	Incompatible design	Incompatible design
Surface	Material	asphalt	asphalt
	Maintenance level	high	medium
Signage (Horizontal)	Solid lines Bicycle symbol Give-way symbol Intersection sign Crossing lines	Insufficient signage Only bicycle symbol signs observed per 10 m	Insufficient signage Only solid lines & crossing lines observed in most spots
Signage (Vertical)	Directional signs Intersection approaching signs Crossing approaching signs Bicycle road sign Bicycle parking sign Bicycle lane sign	Insufficient signage Infrastructure & intersection approaching signs observed at the required spots	Insufficient signage Infrastructure & crossing approaching signs observed in most spots
Interface with Public Transport	Number of stops/stations per km	0.91	3.03
	Cycleway design at bus stops/stations	Not examined – P.T. stop not adjacent to cycle infrastructure	Not examined – P.T. stop not adjacent to cycle infrastructure
Land uses	Commercial - Recreational - Mixed land uses	Residential land use Sporadic commercial & leisure land use	Mixed land use Commercial, leisure & city services
	Green space	Low public space available Low green space available	Plenty public space available Plenty green space available
Motorised Traffic	Vehicle speed limit	<50 km/h (acceptable)	<50 km/h (acceptable)

5 Conclusion

Safe, comfortable and convenient cycling infrastructure significantly encourages bicycle use. In Greek cities, cycling modal shares are very low.

Existing cycle routes were not developed under common standards and they are generally fragmented and poorly maintained. The Greek technical guidelines of 2016 contain specific standards and drawings for the construction and upgrading of new and existing cycling infrastructures. The assessment tool proposed in this paper, is based on a structured checklist of 16 elements, associated to the cycleways' most important design and operational features. The assessment methodology provides guidance to transport practitioners and local authorities on how to quantify these features, their measurements' locations and their threshold values according to the national guidelines. The proposed tool can be used for monitoring the progress of upgrading the existing cycling infrastructures in all Greek cities. Tool's usefulness was highlighted by its application to the Athens and Trikala cycling networks. Both infrastructures are not adequate, but cycling share is higher in Trikala apparently due to the relatively limited city size and urban sprawl. The obtained results assist into indicating the necessary interventions in each network, which, inter alia, include the redesign of intersections with motorized traffic, the installation of proper signage and the increase of side safety distances. Comparing the evaluation results to other national guidelines [7], the observed cycle lane widths are in most of the cases acceptable. Given the insufficient evaluation results on additional cycleways' design features, future research could focus on the examination of more characteristics against other EU-countries' guidelines. Such process could highlight differences in the approach adopted by EU countries and subsequently assist Greek authorities in revising their guidelines to better comply with EU-wide cycling standards and with Greek cities' urban fabric characteristics.

References

1. European Environment Agency: The first and last mile—the key to sustainable urban transport, Transport and environment report (2019)
2. European Commission: White Paper on Transport. Brussels (2011)
3. European Commission: The European Green Deal. Brussels (2019)
4. Prins, R.G., Panter, J., Heinena, E., Griffin, S.J., Ogilvie, D.B.: Causal pathways linking environmental change with health behavior change: natural experimental study of new transport infrastructure and cycling to work. Prev. Med. **87**, 175–182 (2016)
5. Pucher, J., Buehler, R.: Safer cycling through improved infrastructure. Am. J. Public Health **106**(12), 2089–2091 (2016)
6. European Commission. https://ec.europa.eu/transport/themes/urban/cycling/guidance-cycling-projects-eu/cycling-infrastructure-quality-design-principles/existing-guidance-and-standards_en, Accessed 14 Mar 2020
7. Yan, X., Wang, T., Ye, X., Chen, J., Yang, Z., Bai, H.: Recommended widths for separated bicycle lanes considering abreast riding and overtaking. Sustainability **10**, 3127 (2018)
8. European Union: Special Eurobarometer 406: Attitudes of Europeans towards urban mobility. Brussels (2013)

9. Hellenic Ministry of Infrastructure, Transport and Networks: Technical guidelines for cycling infrastructure, Government Gazette of the Hellenic Republic Issue Nr. 1053 (2016)
10. EPOMM. http://www.epomm.eu/tems/cities.phtml, Accessed 15 Mar 2020
11. Municipality of Trikala: Sustainable Urban Mobility Plan of Trikala 2020–2030. Trikala (2019). (in Greek)

New Challenges for Combined Urban Planning and Traffic Planning in Greek Cities. The Case Study of Karditsa

Vasilios Eleftheriou[1(✉)], Efthimios Bakogiannis[1], Avgi Vasi[1], Charalampos Kyriakidis[1], and Ioannis Chatziioannou[2]

[1] National Technical University of Athens (NTUA), 9, Ir. Polytechneiou Str, 10682 Athens, Greece
bileleutherioy@gmail.com
[2] Instituto de Ingeniería (UNAM), Circuito Escolar, Ciudad Universitaria, 04510 Mexico City, Mexico

Abstract. Over the last two decades, the center of gravity of urban-planning is thrown in sustainability and the various strategies that can be properly applied to transform modern cities into sustainable urban cores. The approach to sustainability is that the dominant capitalist mode of development exacerbates social and economic disparities within and among countries, leading to the destruction of significant natural resources. In that context, Greek cities should adjust their shot-term policies to a long-term strategy that contribute not only in improving their morphology and functionality but also their natural environment. This goal can be implemented by conducting a series of new strategic plans, like Local Spatial Plans (L.S.Ps) and Sustainable Urban Mobility Plans (S.U.M.Ps.), that promote a combined and holistic planning approach which is something new in Greek urban-planning system and that is the reason why questions are raised about how can policy makers manage the transition from a car-centric city to a sustainable city, in terms of legislative framework. This paper attempts to enlighten this issue by focusing on a case study of Karditsa. Karditsa is a medium-sized city where large-scale cycling interventions have successfully been implemented fifteen years ago, changing, since then, completely the direction of urban planning in Greece. Thus, it serves as an ideal example to present how the new holistic planning approach will be applied. Proposals are made in order for the city to face emerged challenges and provide solutions on sustainable mobility by promoting mixed-use streets, as the next step toward the existing planning projects.

Keywords: Karditsa · Local Spatial Plans · Sustainability · Long term strategies

1 Karditsa Today. Traffic and Urban Profile

1.1 Introduction

The city of Karditsa, despite repeated reports of its traffic peculiarities, is not an exception, but also a victim of 'unplanned' development. The common technique of

E. G. Nathanail et al. (Eds.): CSUM2020, AISC 1278, pp. 991–1000, 2021.
https://doi.org/10.1007/978-3-030-61075-3_95

development is oriented towards easy profit by continually integrating new areas around the city and building without taking into account the continuous improvement of its architectural image but the maximum "utilization" of the land. Its plotting plan at the time of its implementation could easily accommodate the on-foot trips and the territorial displacements of few carriages that corresponded to the city's detached houses. Nowadays, the same plotting plan is called in vain to provide a solution to the car-based needs of the apartment buildings. All of this was encouraged in 1994 via the revision of Karditsa's City Plan by increasing the width of roads in a direction favoring, according to world standards, driving by car. The purpose of this article is not to create yet another traffic study, but firstly to integrate all the existing ones and transform the traffic features and advantages of the city, so that Karditsa to serve as an example for other Greek Cities concerning traffic perspective and daily life of its inhabitants. Secondly, Karditsa to perfectly comply with modern traffic patterns and through that to highlight the necessity of comprehensive Urban Planning and Traffic Planning.

1.2 City's Urban Profile

The population of the main residential tissue of Karditsa is around 40,000. Its main feature is the availability of many free spaces that are gradually being handed over to the Municipality for use, but without, for the time being, a clear plan for their use. The most typical example of unclear logic of land use is the site of the former Loumaki military camp, along with the completion of the city's peripheral ring, while its perimeter is entirely occupied by agricultural land. However, it should be noted that regarding traffic, Karditsa has invested with positive results in separate traffic practices and particularly in utilizing the city's bicycle tradition. A turning point for the city's traffic reality is the 1970s, when the car began to make its presence in the city much more aggressive, gradually displacing the traditional means of transportation, such as bicycle and carriage, leading essentially to the complete disappearance of the latter. On the contrary, the bike, due to its many advantages and amenities as a medium (ease of acquisition, no need for other equipment, etc.) and the favorable geomorphology of the city have never deviated from the basic traffic habits of the city's people, but, decreased characteristically. After all, the aforementioned revision of the City Plan in 1994, as well as the traffic study of the 1980s, were fully subservient to this reality and evolution, giving uncontrolled space to the car and not taking into consideration alternative modes of transportation. Nevertheless, the bike, never in the history of the city, crossed the fine line that made it from a basic means of transport to a means of recreation or a means by which its users spent their leisure time and was always the first choice for the people's territorial displacement. The bike is considered to be identical to the city's tradition and physiognomy, and is still the key for upgrading its urban environment. It is one of the most modern transport planning tools that seeks to address environmental, economic, social cohesion and quality of life issues. Nevertheless, despite the fact that the Municipality of Karditsa has for many years been a model city in terms of the degree of utilization, integration and promotion of bicycles as a key means of transportation (see Panhellenic Conference 17–19/9/2004), has a characteristic inconsistency in promoting alternative sustainable means of transport such as walking and public transport, as well as improving the parameters that will lead to increased and safer bicycle use. The goal of making Karditsa a less car-friendly city requires the existence of very specific, long-term and multi-level policies that firstly will stop shielding themselves behind any positive conclusions in relation to other Greek cities. Secondly, will perceive and attempt to solve, even "collisionally", any problems that have arisen, in the overall city profile, from the uncontrolled car usage. These are policies which, in combination with other, larger-scale and bulk projects, such as the completion of the regional ring, which is now entering its final phase and will significantly reduce the flows in the city, along with the overall improvement of the railway network that connects the city with the rest of Greece will contribute to the pursued goal of improving the overall image of the city.

1.3 City's Traffic and Road Profile

Karditsa has a well-organized Heppodamean system that gives its characteristic rectangular grid, with well-constructed blocks and its corresponding road network. Mainly due to its flat geomorphology, which was also reflected in the policies implemented in the city for at least the last 20 years, gradually canceling out the particularities that shaped its structure. Evidence of this is the few shells which, especially in the city center, are still reminiscent of its past, with the corresponding traffic footprint of this parameter. An element of this finding is the fact that housing has been essentially displaced from the city center, which is occupied by a number of other, nuisance or not, land uses (commerce, leisure, services), this is a fact that not only is capable of generating a large number of movement, but also is able to essentially shape both of its different regions, its densely populated center and its sprawling region. Finally, both its center and its periphery are covered by a large number and area of free spaces which is also a characteristic of the city and largely reflects the identity of the city, and the distribution of land uses over the years and because of financial crisis is also evident. Finally, both its center and its periphery are covered by a large area of free space, which is also a feature and largely reflects the city′s identity. Furthermore, the distribution of land use over the years and due to its economic crisis is also evident.

The definition of land uses in a city or part of it, is one of the most crucial elements of urban planning practice and an important part of any project of any scale (from planning to neighborhood level). Land uses constitute the functional structure of a city, constitute city centers, organize social life and social spaces, influence the aesthetics of the structured space, determine the qualities of the public space, shape the conditions of land use, significantly affect values land, determine the safety of residential tissue, shape the network of urban mobility and environmental conditions in space [1].

The study of the General Development Planning (G.D.P.) carried out in 2013 (Office Partnership: "Balsamo Mitsi Valassa") organizes the city, its main residential, pre-Kapodistrian, web in 6 districts and 13 City Areas, with the corresponding and required provision for the basic required land uses and functions that form districts and neighborhoods. The boundaries of the above tissue are mostly identified with the existing peripheral artery, while in the southern part, where the unfinished artery is scheduled to be completed in the near future, the boundaries are identical to the existing railway network [2]. The two main road axes with the largest number of flows are Karaiskaki/Papandreou and Taliadourou-Ypsilantou. In particular, the first axis constituted a natural barrier of penetration of the center to the west side of the city, in the residential area, an element that is much more strongly observed in the north and south directions, and together with Trikalon Street they create a series of linear centers, with concentration of a large number of local and over-local use. The distribution of land uses in the manner described above follows, the institutionalized path, with few exceptions. The housing land use is located and distributed around the city's urban center, and defines to a large extent the city's traffic profile. Finally, in the center of the city, besides professional land uses, can be appreciated a large number of important free spaces, including the Pausilipos Grove, which was foreseen by the first city plan in the 19th century as an opportunity for a regional boulevard and public parks on the outskirts of the city.

2 Planning Till Today

2.1 The "General Development Plan" (G.D.P.)

The Study of G.D.P. of the enlarged Municipality of Karditsa was prepared in 2012–2015 in 2 phases (A&B) and aimed at defining the areas of productive and developmental uses. Its objective was to upgrade the study area and improve the living conditions of its users. At first Phase (A), which was a reflection of the current situation, was used to record planning and spatial data, along with population trends, infrastructures and land uses within the area of study, while in the second phase (divided in B1 and B2 sub-phases), can be appreciated the procession of data (B1) and the submission of the final proposal (B2). During this second phase, the proposal included, among other things, the de-congestion of the city's center, the stimulation of new linear centers and the consolidation of neighborhood areas with the gradual removal of bulky uses and the encouragement of pedestrian movements. The proposals for the change of the institutionalized G.D.P from the above-mentioned G.D.P are indirectly seeking the corresponding redistribution of travel needs that will arise between these new land uses and their distribution, in the fields of Education, Health, Administration, Culture-Sport and leisure, as well as directly submitting proposals for specific, traffic-based interventions. The previously mentioned lines of action form part of a unified proposal that seeks to reduce the car flows in Karditsa's main arteries, while at the same time stimulates the usage of sustainable transportation modes (cycling, walking and public transport).

2.2 The Research Programme of N.T.U.A.

The title of the research program, which was drawn up between 2004–2007, was "Investigation of methods of applying combined urban and traffic policies to create conditions of sustainable mobility". Its main objectives were to create mild traffic conditions in the city center by shielding it from the threat of the car while stimulating sustainable mobility, with construction interventions and away from police-type methods. The main way and goal of pursuing this goal was proposed to be the solution of the three rings, that is, in addition to the existing one (the peripheral-external ring), two others (interiors) to be created via the demarcation of the existing roads and avenues [3].

It is a fact that Karditsa has implemented many and important interventions in the past years that made it known throughout Greece, while under certain conditions this promotion exceeded the country's limits. Among these interventions can be found projects and initiatives whose construction sought to modify their respective mild traffic conditions and promote scale and environmentally friendly modes of transport, such as walking and cycling. In this context, the construction of an integrated bicycle network was a nationwide prototype that has been a guide for many cities in Greece regarding sustainable urban mobility [4].

What will strongly threaten Karditsa's character and quality of life in public spaces in the coming years is the car and its usage. The main traffic and urban "enemy" of the city today is the large number of flows, and in particular, the unnecessary part of them. Despite the short length of most trips, its predominant mentality that makes self-evident the use of the car, is the main difficulty and at the same time the most damaging pattern to the urban environment [5]. In any case, it should not be forgotten that, the final goal of planning and particularly the new generation of urban planning tools (L.S.P's), is to reduce the running costs of the city especially the one generated by traffic.

It is a given to modern (contemporary) cities that urban planning is driven by its traffic needs [6]. Public transport and the ability to enjoy walking and biking are also key principles for future development, along with public transport usage, a key disadvantage of which, (at least) in Karditsa's case, is the separation of the city's functions and the low construction density which objectively generates a large number of displacements. A critical starting point for a sustainable mobility strategy for a city and especially for the city of Karditsa is the definition of a perimeter around the center, within which restrictive parking and traffic policies will be implemented. The center of the city deserves to be given to pedestrians and cyclists and other environmentally friendly means of transportation for the protection of city's microclimate, which is a strong element of the city's identity and the basis for strengthening social interactions among the city's inhabitants.

The vision of Karditsa for less car usage started some years before, when the municipal authority did not hesitate to "clash" with those car users who were unable to grasp the wrong attitude of uncontrolled car use (i.e. for journeys with an average length not exceeding 0.5 km). Karditsa had realized that the road environment is not only the background but it also includes the activities happening to it, such as vehicle transit and the presence of people. The overwhelming, if any, enthusiasm for recognizing this venture, as well as the existence of different perceptions between people belonging to different election cycles (see closing municipal parking lots) while underestimating the necessity of continuing that planning reality, created such interventions in Karditsa, which in the medium to long term, have led to a renewed saturation of the city center by car.

2.3 The Urban Mobility Study for Low Cost Projects in Karditsa. Karditsa just Before S.U.M.P.

The development of Karditsa over the years, did not differ in the slightest from the model of development of other Thessalian (and not only) cities and so the city from the dirt roads and the ground-floor houses, was transformed into a city of high-rise buildings, large cross-sections of roads and saturation of cars and motor vehicles in general, with its corresponding imprint on its architectural and traffic profile and in the daily lives of its inhabitants. During the 1980s, when the first sidewalks on the city's main streets began, reactions were widespread, even though at that time the private car dependence was still at an early stage, as well as the consequences of such a development. This early refurbishment of the fragmented, till then, public space leads to the “lazy” re-appearance of the bicycle on the streets. Several years later, the construction of the first network of bicycle lanes launched this transportation mode, integrating it into the daily lives of the overwhelming majority of city dwellers.

3 Suggestions. Proposal for Pilot Implementation of a New “Ring” Type in the City Center

3.1 Traffic and Development. Karditsa Besides It’s S.U.M.P

It is therefore a common assumption that a diffferent traffic practice is necessary for the area that is attempting to be protected within the concept of sustainability. Of course, it should not be left unmentioned the current economic, productive system and the consumer standards it creates [7].

Table. 1. Car ownership index. Number of cars per 1000 inhabitants [8].

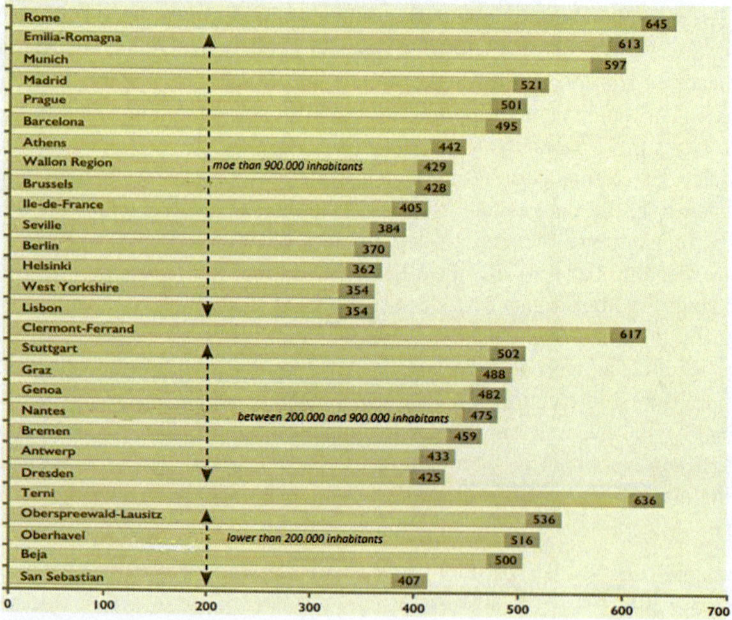

The dominant capitalist mode of production and development exacerbates social and economic disparities within and between countries, especially between developed and developing countries, leading to the destruction of some of the world's natural resources. Nowadays, the large private car ownership index, in the largest European, and not only, cities, has played an important role in the corresponding increase in travel which is made using this particular mean of transport. The upgrade of public transport necessarily leads to restrictions on some freedom of movement of the car leading to a reduction in the degree of its use, in contrast to the policies implemented in recent years which have promoted the car to the detriment of all other means.

As far as the study area is concerned, as a typical example of a medium-sized Greek city, in terms of traffic parameters and shortly before the development of the corresponding S.U.M.P and L.S.P the main comparative advantage for Karditsa is the fact that, by utilizing similar research and design tools of the past, it has identified this area which should at this stage be protected. In this context, short and medium-term interventions can have short-term, particularly significant long-term effects.

3.2 Proposal for Small-Sized Town Planning and Traffic Interventions

It has therefore been established, based on the existing hierarchy of the city's road network, that the major road axes which are essentially the recipients of the largest volume of city traffic are specific and already identified. Among these, the role of Ypsilantou Street, throughout its length, is pivotal, mainly because of the size of the through-flows that are realized using it. Both the existing traffic studies and practices, as well as the literature referring to Karditsa's traffic profile, place particular importance on this section, inter alia, on the extent to which it has been utilized for cross-section of the city's central area. The existing studies, including the one described above, which is essentially a small S.U.M.P, because of the financial data of its preparation period that did not allow for the financing of an integrated project of this type, today, (even economic) conditions are much more mature, also the future studies must take into account the conditions created by the unobstructed and extensive use of this road. The example of the construction of the main part of the Karditsa bicycle network on the North side of the road and its entire length should be a guide. The decrease of the used (from the cars) width on the road and the reduction of parking spaces that would have caused such a development, during the period of application, was the cause of intense confrontations between the car users and the Municipal Authority of that period. Nowadays, much lower cost and construction requirements could greatly counter traffic flows from that point, without excluding car access and all other means of access anywhere on the road, along it's entire length.

The construction (or at least placement) of fixed elements - curbs in a curved arrangement at the intersection of Ypsilantou St. And Iroon Polytechneiou St., in such a way that the moving vehicles on Ypsilantou St. at that point will be pushed to the Iroon Polytechneiou St. Section, in the Southern part of it and the migrants in the Northern part of Iroon Polytechneiou St. Will be pushed to the Eastern part of Ypsilantou St. Respectively, as described in the map below, it is estimated that will reduce by more than 50% the traffic flows that make use of this particular road for city transit, while the overall traffic flows on that particular road will be reduced to the same degree.

Alternatively, it is proposed to utilize the existing, current section of the road by the TAXIs, which already have a parking lot on Iroon Polytechniou St., North of Ypsilantou St. And Public Transport, utilizing exactly the same infrastructure. This ensures uninterrupted flow of the entire length of Ypsilantou St. For the utilization of adjacent, on-road, land uses. Its use is discouraged for city-wide transit, and the "recalcitrant" traffic congestion will be pushed to the adjacent existing road network without charging, at any point, the neighboring residential areas, the south limit of this approach will be Lappa Road., which is the corresponding proposed, from earlier studies, southern boundary of the intermediate ring. Additional interventions should be made at the intersection of Taliadourou St. And Karaiskaki St. Streets in such a way as to encourage the push of cars that will cross the city center, on, off-center ring, roads, also like Lappa St.

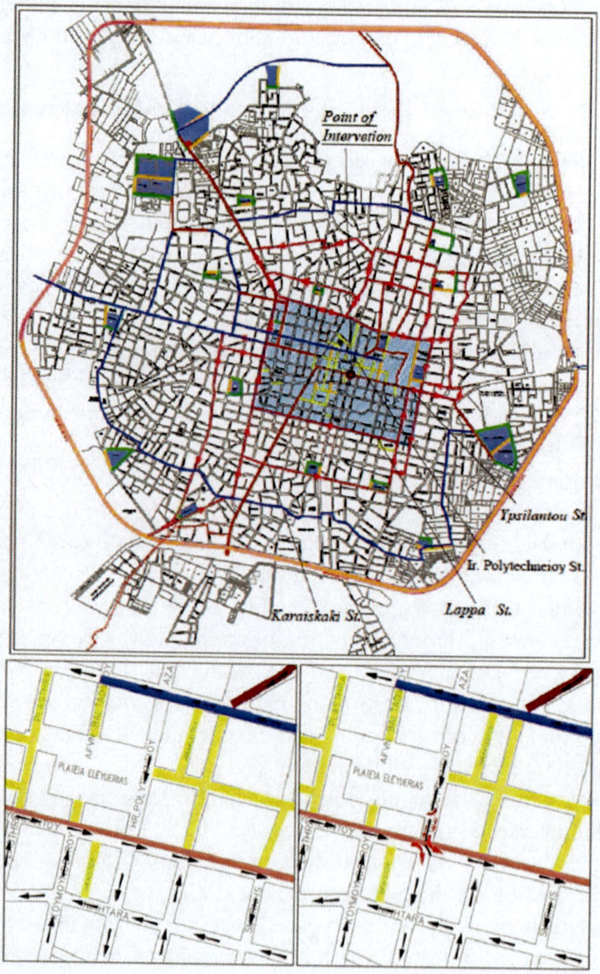

Fig. 1. Above: location of the proposed intervention in the city, Bellow: Left: Existing traffic, Right: The proposed intervention.

3.3 Pilot Implementation of a New "Ring" Type in the City Center

Finally, is proposed the pilot use of a new type of "ring" with the aim of limiting unnecessary movements and entrance of vehicles towards the city's central area, as defined in the aforementioned study carried out for sustainable mobility in the center of Karditsa. This ring measure could limit not only the traffic number of vehicles in question, such as the well-known example of Athens, but a series of other filters such as the number of occupants in each vehicle. Specifically, it is proposed to be established days of the week in which it is encouraged (and not required) to enter the city center of vehicles with more than one occupant, depending on the degree of acceptance and effectiveness, with the possibility of increasing either the days of the week or the number of occupants, or even the optional character of the program. This practice, at least on a pilot basis, will initially be encouraged and not enforced by police-type measures and fines, however, if not respected, a set of policy incentives can be defined in order the approach to be more accurately determined. Then follows a policy that was similarly followed by today's Municipal Authority and is oriented towards the reduction of Cleaning Fees for businesses and entrepreneurs who have proven, select and promote bicycle use to and from work, if the above mentioned measure is implemented in addition to the same 'reward' an extra measure could be offered in cooperation with relevant operators. Indicatively, is suggested a kind of discounts on public transport fees/tickets (Urban - Intercity Bus, Train) or even tolls discounts on specific routes, especially towards destinations that count with insufficient or even non-existent public transport coverage.

Obviously, due to the privatized status of almost all public transport, as well as large road axes, this is objectively much more complex, nevertheless, since the costs will be quite small at first, it can be accepted by both the authorities and the Municipal Authority itself, to the availability of which will be the maintained database with the bonus point of the migrants, as well as the issuance of the corresponding type of voucher, which will certify this facility-return. Commuters- beneficiaries will also have the option of not making use of this facility and the corresponding amount, if provided for reimbursement in the annual budget of the Municipality, may be reimbursed, always respecting the conditions and relevant legislation, for some public purpose (e.g. stray care, infrastructure improvement) and these commuters - in a formal way, to be rewarded in particular for demonstrating their sustainable commuting habits. This measure requires, in addition to scientific acceptance a close cooperation with the Municipality of Karditsa or its respective implementation municipality. Of course, in the event of its implementation, additional unnecessary disruptions will arise from users who will enter exclusively to make use of its 'benefits', this should of course be done in conjunction with the implementation of all other, constructive and legislative measures which have been proposed in all previous, corresponding, applications but also to future ones that aspire to give a new form to the traffic and the living profile of Karditsa's center and whole city, such as (pending start-up). S.U.M.P. and also L.S.P.

4 Conclusion

Nowadays, out of the cities, the first thing that missing is space. The reason is not the lack of urban-metropolitan space existence but the overwhelming occupation by the uncontrolled and unnecessary use of the car and the environmental footprint of this abuse [9]. After all, a city with fewer cars does not mean less commuting nor productivity constraints, but it may mean exactly the opposite, only in more humane conditions. It means a city with a greater absolute number of trips, with much happier and healthier population due to active transport, than means much less space occupation and neutralization of barrier effects. This will enable a larger part of the population to move and especially vulnerable groups (people with disabilities, children, the elderly, pregnants etc.) to be part of the city's social tissue more frequently [10]. That is, many of those who are now on the fringes of the city. After all, they are the ones who keep a city alive, because they are the ones who come into contact with space. The aim and need for the multitude of Greek cities is to achieve an integrated strategic planning that will combine urban and traffic planning and policies [11]. A series of traffic interventions which will lead the city as close as possible to its intended, each time, Urban Purpose starting with the improvement of the living conditions of its inhabitants. The main starting point of all this, especially at the case of the city of Karditsa may be the existing goodwill of city share bicycles system and bicycles lanes.

References

1. Serraos, K.: Organization of uses and activities during the space design process, Scientific meeting. N.T.U.A, Athens, Greece (2017)
2. Valassa, V.: The General Urban Plan (G.U.P.) of Karditsa, Karditsa (2015).
3. Vassi A., Bakogiannis E., Christodoulopoulou G., Siti M., Kyriakidis Ch.: The Urban Mobility Study for low cost projects in Karditsa, Athens (2016).
4. Bakogiannis, E., Kyriakidis, C., Siti, M., Eleftheriou, V.: Four stories for sustainable mobility in Greece. Transp. Res. Procedia **24**, 345–353 (2017)
5. Bakogiannis, E.: Instructions for S.U.M.P.s, Grigori Publications (2016).
6. Vlastos, Th., Milakis, D.: Urban planning VS transportation. From divergence to convergence. ORSA and SRSE/NTUA, Athens (2006)
7. Sarigiannis, I.: Introduction to the history and theory of the city. Symmetria Publications, Athens (1987)
8. Vlastos, Th.: A different public transport in a different city. European Commission, ECOS Ouverture Programme, Development company of the Municipality of Athens (2003)
9. Vlastos, Th..: Elegy of the Urban Solid, Article on "Environment and Law" Law Library, 3, 435–448 (2012).
10. Siolas, A., Vassi, A., Vlastos, Th., Kyriakidis, C., Bakogiannis, E., Siti, M.: Methods, tools and urban planning applications. Kallipos Publications, Athens (2015)
11. European Commission (EC): Cities of tomorrow, Challenges, visions, ways forward. COM (2011). Final edn. Brussels (2011).

Evaluating Fastest Path Procedures on Roundabouts by Extracting Vehicle Trajectories from Unmanned Aerial Vehicles

Apostolos Anagnostopoulos(✉) and Fotini Kehagia

Highway Laboratory, Department of Civil Engineering,
Aristotle University of Thessaloniki, 54124 Thessaloniki, Greece
{aposanag, fkehagia}@civil.auth.gr

Abstract. Unmanned aerial vehicles are one of the most emerging technologies being used recently in transportation field. The collection and analysis of high-resolution data such as vehicles' trajectories are challenging concerning the analysis of driving behavior for use on optimal road design. At present, the expected speeds of vehicles on roundabouts are calculated based on fastest path analysis techniques that several guidelines provide. This paper presents the results of experimental research be conducted as a first step in developing a model for the prediction of vehicle paths on roundabouts. The research aims to define the basic path elements of vehicles maneuvers at multilane roundabouts on free-flow conditions. An analytical guiding framework regarding the use of UAVs for data acquisition in traffic analysis is presented. The proposed methodology follows a simple and low-cost way of extracting detailed naturalistic vehicle trajectories data through UAVs. Video image processing techniques and geographic information system methods are used and provide a useful tool for obtaining spatiotemporal phenomena at two multi-lane roundabouts. The level of accuracy of the method is evaluated by minimizing the errors due to camera instability and distortion effects. The results of the experiment highlight the usefulness of the proposed UAV and video processing technique for further driving behavior analysis. The analysis that was conducted indicates differences between observed vehicles trajectories and existing fastest paths procedures.

Keywords: Roundabout · Vehicle path · Vehicle trajectories · Traffic safety

1 Introduction

Roundabouts are an increasingly appealing alternative form of at-grade intersections because of their multiple advantages regarding traffic safety, operational efficiency, atmospheric emissions and aesthetics [1]. Due to this fact, in the last decades roundabouts have gained increased political acceptance worldwide and are now a viable alternative for intersection design.

Their implementation on road network is a solution in cases where problems regarding traffic safety or capacity are occurred. Before-and-after studies of roundabout conversions indicate that modern roundabouts are safer than previous intersection

E. G. Nathanail et al. (Eds.): CSUM2020, AISC 1278, pp. 1001–1011, 2021.
https://doi.org/10.1007/978-3-030-61075-3_96

treatments [2, 3]. The most important reason that nowadays roundabouts have been a widespread design solution, is their effectiveness to road safety by the reduction of the number of road fatalities [1, 4].

The design of the geometry of a roundabout involves choosing between trade-offs of safety and capacity. The appropriate selection of the geometric parameters for the roundabouts design can enhance the operational performance and traffic safety. A well-designed roundabout operates most safely when its geometry forces vehicles to enter and circulate at slow speeds, achieves smooth speed by requiring vehicles to negotiating the roundabout along a curved path. Design solutions with low deflection tend to have low performance levels [1]. Drivers invade the adjacent lanes on multilane roundabouts with high speeds, searching for direct and comfortable paths. This can lead to several traffic safety issues.

According to [5], the fastest path modeling is a critical component of safe roundabout design. As the construction of roundabouts is quite new in Greece, there is a limited research regarding the driving behavior on roundabouts. Greek guidelines adopt the American guidelines procedures regarding the construction of fastest paths for speed control on roundabouts.

This paper presents the results of experimental research be conducted as a first step in developing a model for the prediction of vehicle paths on roundabouts. The research aims to define the basic path elements of vehicles maneuvers at multilane roundabouts on free-flow conditions. A method regarding the acquisition of real vehicles' trajectories with the use of unmanned aerial vehicles (UAVs) is proposed as well. Specifically, the questions addressed in this paper are:

- How UAVs can be used to extract detailed naturalistic vehicle trajectories?
- Do the real vehicle paths on free flow conditions at multilane roundabouts conform to those that international guidelines indicate?
- Can vehicle trajectories patterns be identified according to the drivers' movements on multilane roundabouts?

2 Literature Review

The assessment of the vehicle path curvature is one of the most critical parameters being used on the design process of a roundabout. According to several studies, a geometric design that allows vehicles to approach and negotiate a roundabout in high speeds can result in high possibilities of angle and rear-end crashes [6–8]. Thus, a roundabout operates most safely when its geometry forces vehicles to reduce their speeds along the curved path. The relative vehicle speeds on conflict points are decreased, contributing to safer maneuvers and interactions between the road users.

The vehicle path deflection is the main parameter used by standards and guidelines in order to control speeds through the roundabout. The negotiation speeds of vehicles movements are determined according to the radii of the fastest path that is allowed by the geometry. This is the smoothest, flattest path of least travel time made by a vehicle traversing the roundabout in free flow traffic conditions. According to [9] the fastest path can be defined as a representation of the most used trajectory by drivers under free flow conditions, when minimizing their driving discomfort.

However, there is not a specific rule regarding the methodological approach for this roundabout performance check. Several national European, US and Australian standards and guidelines were examined and analyzed for the purposes of this paper. Most of them have particular limitations and differ on the definition, construction procedure and radii estimation of the relative fastest paths. As a result, transportation engineers and agencies over the world use different guidelines and methods for the design process of roundabouts in the lack of a straightforward procedure.

2.1 International Vehicle Path Deflection Methods on Roundabouts.

The identified approaches that are followed by the international roundabout design guidelines in order to ensure vehicle path deflection are:

- Approach A: Measurement of path radii according to the construction of vehicle fastest paths through the roundabout. This procedure is followed by American, Greek, French, UK, Spanish, Sweden and Australian guidelines [1, 10–15].
- Approach B: Calculation of vehicle fastest path radii according to measured geometry features of the roundabout. This procedure is followed by Dutch, Belgian, Slovenian, Croatian, and Serbian guidelines [16–20]. Equation 1 is used for the calculation of path radius.

$$R = [(0.25 * L)^2 + (0.5 * (U + 2))^2]/(U + 2) \quad (1)$$

  L and U represent specific distances on roundabouts for all cases except Croatian guidelines [19]. It is noted as well that this method is proposed for single lane and turbo roundabouts. Belgian guidelines present a similar method regarding multilane roundabouts.
- Approach C: Checking the achieved deflection according to measured geometry features of the roundabout. This procedure is followed by German, Italian and Swiss guidelines [21–23].

It is noted that fastest path design procedures that belong in the first two categories, are recommended by the majority of national guidelines. However, differences are occurred regarding the category of the roundabout, the considered critical movements of vehicles on the roundabout and the minimum clearances to particular geometric features of the roundabout.

2.2 Fastest Path Construction Techniques and Procedures

An objective and reliable fastest path is comprised of a series of consecutive reverse spiral curves that are tangent to each other [1, 4]. The short length of tangent reflects the time it takes for a driver to turn the steering wheel. There are two main drawing techniques: (a) the freehand and (b) the computer-aided technique. In the first case, the engineer draws by hand a natural representation of the way a driver negotiates the roundabout on scaled drawing of the intersection. The minimum radii are measured using curve templates either by digitalize the sketch in CAD software. This method reflects better the anticipated driving behavior. However, requires high skills and expertise while it is difficult to use on the design process and thus is discouraged. On the other hand, the computer-aided technique is based on a drafting software (CAD). Both methods require an iterative design procedure for the selection of proper roundabout design geometric parameters. There are two main techniques that are adopted for creating fastest paths in CAD:

- Cubic splines technique or Wisconsin method [24]. This method is proposed by the US guidelines [1] and is adopted in UK guidelines as well.
- Tangent reverse curves (arcs) technique or ACHD method [25]. It is mentioned that these paths are not intended to trace the actual fastest paths graphically, as spirals are replaced by arcs and tangents.

The vehicle path is formed in such a way that the appropriate offsets to particular geometric elements of the roundabout (such as raised curbs, roadway centerlines and painted edge lines) are considered. These offsets differ in the aforementioned guidelines which results in different formed fastest paths.

3 Proposed Method to Study Vehicle Trajectories

UAVs are recently being used in transportation field to monitor and analyze the traffic flow. The advantages that this method offers regarding traffic data acquisition are great of importance. As the camera is in the air, the drivers' attitude is not distracted by the equipment [26]. According to this, extracted data represent naturalistic driving behavior which is significant for the study of vehicle trajectories. Moreover, UAVs require less experience and training in order to be controlled while they can handle a wide field of view, covering large areas quickly. These assets result in a saving-time, low-cost and non-infrastructure-based technique for acquiring individual vehicle trajectories, compared to other methods. However, there are many factors as well that influence the performance of this process. Among them, weather conditions (e.g. rain), technical issues (e.g. low battery duration) and regulatory issues (e.g. no-fly zones) are the most critical to be mentioned.

According to the existing literature, many studies that have been conducted recently are dealing with traffic data acquirement through unmanned aerial vehicles. The methodology that is used in these studies can be divided into three categories regarding the video processing technique that is used: a) the manual process, b) the semi-automatic process and g) the automatic process technique. The first two methods are more accurate and offer many flexibility advantages, however, they are time-consuming. On the other hand, the third method promises quicker results by using detection techniques and tracking algorithms with minimum manpower involved. Nevertheless, requires high expertise and knowledge of computer vision techniques, while its accuracy on high spatial data such as vehicle trajectories sometimes requires manual effort for the correction of missed trajectories.

The proposed technique that was used for the purposes of this study is based on UAV data collection followed by a manual process. The following steps were followed for the conducted surveys of data collection: (a) UAV video recording, (b) Video processing and (c) Coordinates extraction of vehicles trajectories. Each step is further divided into several sub-steps. The whole process is briefly described below.

3.1 Survey Material and Equipment

Two types of equipment were used in order to acquire reliable vehicle trajectories data: (a) a quadcopter UAV and (b) a RTK GNSS receiver.

The selected UAV is able to capture videos up to C4K analysis with a frame rate of 60 fps and high-resolution images (5472 × 3078). It's about a low-cost UAV and requires less experience in order to be controlled. Its camera is attached to a gimbal, which stabilize shots. Moreover, its weight (1368 gr) and size give contribute to better maneuverability, while low altitudes flights achieve high spatial resolution. The RTK GNSS receiver that was used provides reliable and high-accuracy data collection. The position of selected GCPs can be determined in centimeter-level accuracy in real world conditions. Specifically, the accuracy of this equipment for examined of survey is 8 mm + 1 ppm (horizontal) and 15 mm + 1 ppm (vertical).

3.2 UAV Data Collection Procedure

Site Selection. During the first step, specific criteria were adopted regarding the selection of proper roundabouts in order to be analyzed. Location characteristics and design elements of the intersections were considered. Two multilane roundabouts located near the urban limits of the city of Larissa were selected for the analysis. Both have been constructed recently in Greece (2019) and rely on different geometric features.

Preparation of the Surveys. Firstly, ten homogeneously distributed characteristic natural ground control points (GCPs) (such as corners of manholes and tactile pavements, intersections of white pavement markings etc.), were identified in the areas of interest. As the study area is defined by the roundabout layout, GCPs were identified in roundabouts' approaches and inside the central island. The collection of high-accuracy GNSS coordinates of the GCPs (horizontal position and altitude coordinate) was conducted using the GNSS RTK receiver on the field. The selected geographic reference was the "GGRS87/Greek Grid".
GCPs of well-known coordinates increase significantly the absolute accuracy of the analysis. It is required to be visible in all acquired frames. Their use intends to correct errors due to UAV tilt (wind forces affect) and camera lens distortion issues. The use of natural GCPs has the advantage of the flexibility to conduct surveys over different periods of time. The restraint of UAVs low battery and the requirement for visible GCPs for each survey means that permanent natural ground control points are a good solution. Figure 1 shows the natural GCPs on a case study roundabout and the georeferencing process that is described below.

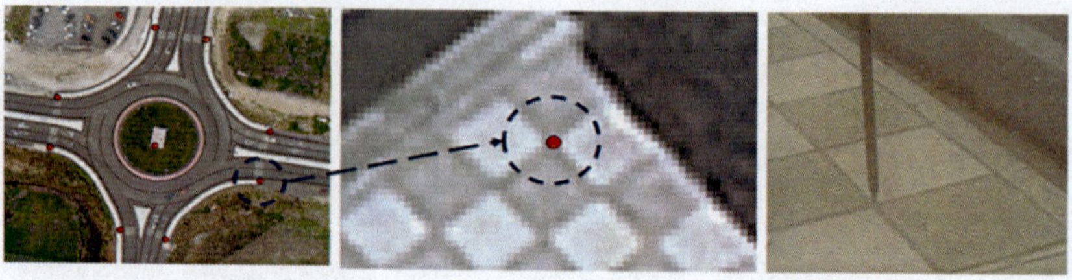

Fig. 1. Georeferencing of frames with natural GCPs.

UAV Survey Execution. Roundabouts' performance was recorded with the use of the UAV during summer and autumn of 2019. Two types of data were collected in this experiment: (a) data relies on the geometric elements of the roundabout and (b) data regarding the kinematic characteristics of the vehicles. High-resolution (4096 × 2160) images and videos were acquired by using the UAV in specific flight altitude and vertical point of view. Low flight altitudes (roundabout layout view) and high-resolution frames optimize the required time regarding the video processing and increase the accuracy. Moreover, recordings of a nadir point of view are required in order to minimize camera errors. Finally, the location of the UAV was as stable as possible in order to minimize the errors in the stabilizing process.

Field measurements were selected to be conducted during of-peak periods, in order to ensure free flow speed conditions. Real vehicles trajectories in unobstructed traffic conditions were collected. Weather conditions were stable.

3.3 Data Processing

Vehicle trajectories were extracted from UAV videos following specific steps. These steps aimed to: (a) reduce any bias and increase the accuracy and (b) digitize and study vehicle motion paths.

Firstly, no significant frames from videos (take-off and landing) were removed. A stabilization procedure was followed [27]. Proper filters made videos smoother while camera shakiness was eliminated. Specific events were identified (vehicle through movements on free flow speed conditions) and video frames were extracted per second for each examined video. Lens distortion of each frame was corrected [28]. The acquired images were georeferenced in high accuracy using an open-source GIS software [29]. The total accuracy of this process evaluated through the average of the RMSE between the real location and the final georeferenced location. The average of the RMSE was between the values of 0.28 and 0.6 for each frame. For the selected analysis of the video, this corresponds to maximum error of 2.8 cm.

Afterward, a digitization process was conducted regarding the geometric elements of the roundabouts and the vehicle trajectories. Specifically, the center of the front bumper of each vehicle was identified and the coordinates were extracted. Mathematical interpolation based on the known coordinates was used in CAD software using spline curves so the discrete data to be transformed into continuous functions. Twenty vehicle paths per each examined straight movement were analyzed.

4 Analysis of Fastest Paths on Roundabouts

Theoretical fastest paths of the multilane roundabouts were drawn by using a CAD tool according to the proposed method of [24]. Constructed fastest paths according to American guidelines [1] that adopted by Greeks [10] are presented in Fig. 2.

Geometric features and road markings from the extracted frames were digitized. The properly offsets regarding the selected approach were determined and finally the theoretical fastest paths of the two multilane roundabouts were drawn.

The entry path radius (R_1), the circulating path radius (R_2) and the exit path radius (R_3) were calculated for the examined through movements by measuring the minimum radii of fitted arcs on proper portions of the constructed fastest paths. In case of the entry path radius (R_1), the length of the fitted arc per each through movement was at least 20 m [1].

4.1 Real Trajectories Compared to Theoretical Vehicle Paths

The density of the extracted trajectories of vehicles on though movements at free flow speed conditions is presented in the following figure. The first two paths of the figure belong to roundabout A, while the third one belongs to roundabout B. The constructed theoretical fastest paths are presented as well.

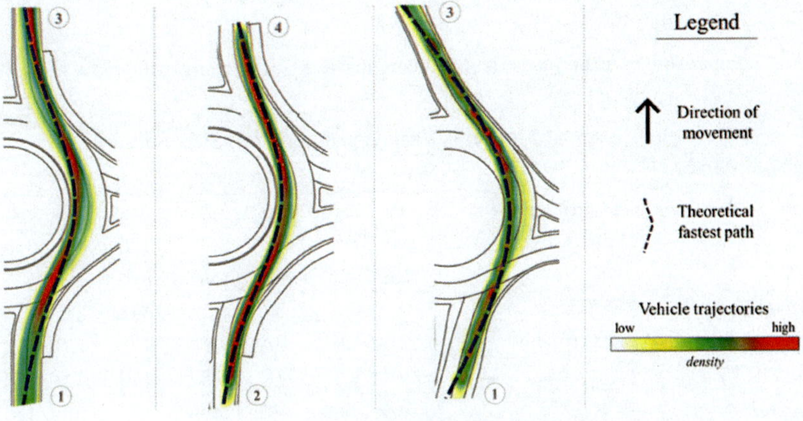

Fig. 2. Density of extracted vehicle trajectories.

According to Fig. 2 it can be noted that vehicles tend to follow the curve of theoretical paths. Specifically, in the second case (2–4), it seems that drivers follow close enough the theoretical path. However, in all cases there are many deviations of real vehicle trajectories to the theoretical fastest paths. Moreover, it is observed that the vehicles trajectories cover a considerable part of the roundabout area, which is related to safety issues. Upon this, the following figure confirm the differences that are observed regarding the lateral positions of vehicles at the entrance, in the middle and at the exit of their path. Four parts were defined per each section for the two lanes.

According to this, different driving behavior patterns are observed. Improper and unexpected paths are created and that probably can cause unexpected behavior and several conflicts and therefore potential danger (Fig. 3).

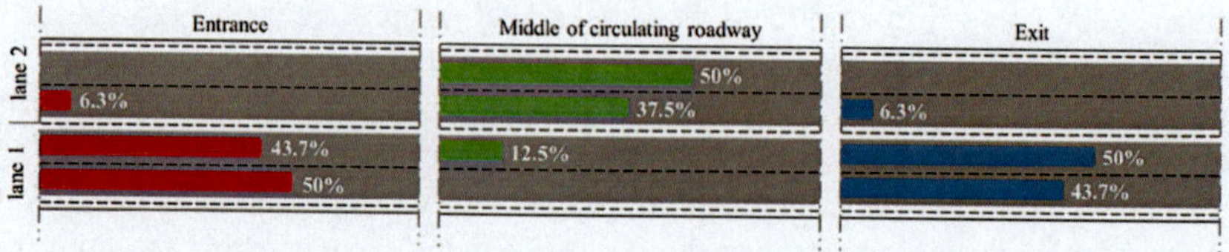

Fig. 3. Illustration of lateral positions of vehicles.

The radii of paths were measured for each examined vehicle path at the entrance, in the middle and at the exit of the roundabout. Specifically, tangent arcs were adopted to spline curves in such a way that they do not deviate from actual paths. The distances of vehicle paths from the raised curb at the entrance, at the exit and from the central island at the middle were measured as well. The exact radii values of real vehicle paths compared to the radii of theoretical paths are presented in the following table. American, UK and Belgian guidelines were used for the comparison as it can be considered that they mostly represent the different procedures used by the international standards for the construction of fastest paths regarding the multilane roundabouts.

Table 1. Comparison of experimental and theoretical paths regarding path radii and safety distances.

Analyzed roundabout	Through movement	Examined method	Vehicle path radii (m)/Safety distance (m)					
			Entry		Circulating		Exit	
A	1–3	Experiment	51.1	3.26	26.0	3.0	61.0	2.58
		American	119.0	1.5	38.2	1.5	120.2	1.5
		UK	≤ 100	min. 1	–	min. 1	–	min. 1
		Belgian	37.0	1	37.0	1	37.0	1
A	2–4	Experiment	62.3	1.91	33.0	2.54	77.3	2.01
		American	106.1	1.5	38.3	1.5	110.3	1.5
		UK	≤ 100	min. 1	–	min. 1	–	min. 1
		Belgian	37.8	1	37.8	1	37.8	1
B	1–3	Experiment	184.6	2.89	35.3	2.74	267.5	2.21
		American	226.4	1.5	49.3	1.5	233.7	1.5
		UK	≤ 100	min. 1	–	min. 1	–	min. 1
		Belgian	35.3	1	35.3	1	35.3	1

According to Table 1, it can be concluded that the theoretical radii values are nearest to experimental radii values in the middle of the roundabout. The calculation of vehicle fastest path radii according to measured geometry features of the roundabout [17] gives better results, which means that this method can predict better the radii in the middle of the roundabout and therefore the expected vehicle speeds. It is mentioned that this method is limited to specific values of angles (approximate near 180°) between the legs of through movements. However, the fastest path of roundabout B has 226°.

At the entrance and the exit of the roundabout significant differences are observed, especially regarding the measured path radii according to American guidelines. Moreover, the safety distances of the vehicles from the raised curbs are not close to the values that guidelines propose. Specifically, the minimum average value that is observed is 1.91 m which is 0.5 m away from the proposed value of 1.5 m.

5 Conclusions

The vehicle path deflection is the main parameter used by standards and guidelines in order to control speeds through the roundabout. The fastest path modeling is the main process adopted by the majority of the international guidelines regarding the control of vehicle speeds. However, a straightforward procedure regarding this roundabout performance check on the design process does not exist. Limitations and differences on the construction procedure and radii estimation of fastest paths are observed. The results of experiments conducted on two multilane roundabouts in Greece, indicate differences between observed vehicles trajectories and existing fastest paths procedures.

According to this, it is important to note that the already existed models and the respective techniques depicted in international guidelines, may not be feasible or even applicable in the case of Greece. The fact that different driving behavior patterns are observed, can cause potential danger due to unexpected behaviors and several conflicts.

By definition, the fastest paths on multilane roundabout cut across all lanes on the circulating carriageway. This means that drivers change lanes twice. Several conflicts are created as drivers follow improper vehicle paths through the roundabout. The ability for drivers to pass through the roundabout in different ways, creates many traffic safety issues. This is an acceptable reason why two-lane roundabouts are reconstructed to turbo roundabouts in many European countries. More case study roundabouts should be considered for further analysis and more observed trajectories of vehicles should be analyzed in order to develop a reliable vehicle path model for multilane roundabouts.

Finally, the proposed method for extracting detailed naturalistic vehicle trajectories data through UAVs provide a useful tool for obtaining spatiotemporal phenomena at intersections. Further driving behavior analysis is required. The acquisition of vehicles' kinematic characteristics such as speed, acceleration, etc. is necessary in order to optimize the roundabout design procedure. Automatic process techniques should also be tested.

References

1. TRB: NCHRP Report 672. Roundabouts: An Informational Guide, 2nd edn. TRB, Washington, D.C. (2010)
2. Schoon, C.C., van Minnen, J.: The safety of roundabouts in the Netherlands. Traffic Eng. Control **35**(3), 142–148 (1994)
3. Silva, A.B., Vasconelos, L., Santos, S.: Moving from conventional roundabouts to turbo-roundabouts. Procedia Soc. Behav. Sci. **111**, 137–146 (2014)
4. TRB: NCHRP Report 572. Roundabouts in the United States. TRB, Washington (2007)
5. FHWA: An Informational Guide. FHWA-RD-00-067, Portland, Oregon, USA (2000)
6. Arndt, O.K., Troutbeck, R.: Relationship between unsignalised intersection geometry and accident rates. In: Proceedings of the 3rd TRB International Symposium on Highway Geometric Design, Chicago, IL (2005)
7. Montella, A.: Identifying crash contributory factors at urban roundabouts and using association rules to explore their relationships to different crash types. Accid. Anal. Prev. **43**, 1451–1463 (2011)
8. Turner, S., Wood, G., Roozenberg, A.: Accident prediction model for roundabouts. In: Proceedings of the 22nd ARRB Conference, Canberra, Australia (2006)
9. Silva, A.B., Seco, A.: Trajectory deflection influence on the performance of roundabouts. In: Proceedings of the European Transport Conference (ETC), Strasbourg (2005)
10. Greek Ministry of Infrastructures and Transports: Road Design Guidelines Manual (OMOE): Roundabouts – K3, 2nd edn., Athens, Greece (2012)
11. SETRA: The Design of Interurban Intersections on Major Roads: At-grade Intersections. Bagneux, France (1998)
12. Kennedy, J.: International comparison of roundabout design guidelines. TRL (2007)
13. Ministerio de Fomento [Ministry of Public Works and Transport]: Recomendaciones sobre glorietas [Guidelines for Roundabouts], Spain (1999)
14. Trafikverket [Swedish Transport Administration]: Publikation 2015:087. VGU (Vägar och gators utformning) [Guidelines for the design of roads and streets], Sweden (2015)
15. Austroads: Guide to Road Design, Part 4B Roundabouts. Sydney, NSW, Australia (2011)
16. CROW: Eenheid in Rotondes [Standards for Roundabouts], publicatie 126 (1998)
17. Lauwers, D.: Verkeerstechnisch ontwerp van rotondes. In: Vademecum Rotondes, Ministerie van de Vlaamse Gemeenschap, Brussel (1997)
18. Ministrstvo za promet: TSC: Krožna križišča. Ljubljana, Slovenia (2011). (in Slovenian)
19. Građevinski fakultet Sveučilišta u Rijeci: Smjernice za projektiranje kružnih raskrižja sa spiralnim tokom kružnog kolnika na državnim cestama, Rijeka (2014). (in Croatian)
20. Putevi Srbije (PS): Priručnik za projektovanje puteva u Republici Srbiji: Funkcionalni elementi i površine puteva. Kružne raskrnice. Beograd (2012). (in Serbian)
21. FGSV: Bemessungsfahrzeuge und Schleppkurven zur Überprüfung der Befahrbarkeit von Verkehrsflächen, Koln, Germany (2001). (in German)
22. Italian Ministry of Infrastructures and Transports: Guidelines for the Design of Road Intersections, Rome, Italy (2006)
23. VSS: Carrefours, carrefours giratoires, Zurich, Swiss (1999)
24. Wisconsin DOT: FDM 11-26 Attachment 5.1 Roundabout Critical Design Parameters Document (2011)
25. ACHD Roundabout Design Guide, Appendix A: ACHD Fastest Path Procedure, Ada County Highway District, pp. A1–A26 (2011)

26. Anagnostopoulos, A., Kehagia, F.: CAVs and roundabouts: research on traffic impacts and design elements. In: Proceedings of the 47th European Transport Conference (ETC), Dublin, Ireland (2019)
27. Deshaker. https://www.guthspot.se/video/deshaker.htm. Accessed 03 Apr 2020
28. Adobe Lightroom. https://lightroom.adobe.com/. Accessed 03 Apr 2020
29. QGIS Geographic Information System. https://qgis.org. Accessed 03 Apr 2020

Re-thinking Transport Infrastructure Investments: The Case of Addis Ababa, Ethiopia

Yohanan Ermias Bekele(✉)

Addis Ababa, Ethiopia

Abstract. By examining the potential effects that the Addis Ababa Light Rail Transit (AALRT) project may have on city-dwellers, this paper aims to assess possible approaches to inclusive transport planning in rapidly growing cities of the Global South.

Keywords: Public transport · Public transport sector · Transport infrastructure project · Light rail · Social exclusion · Transport planning · Public transport system · Social urbanism project · Infrastructure project · Urban transport · City authority · Social inclusion · Transport system · Africa · Mass transit system · Transport planning process · Ethiopian railway corporation · Addis Ababa · Light rail transit project · African city · Mass transit · Social urbanism

1 Introduction

Across Sub-Saharan Africa, foreign investments in transport networks, telecommunication and energy production have made significant contributions to national development. Some critics however, have labeled the booming interest and involvement in infrastructure development in Sub-Saharan Africa as a form of neo-colonialism; where large-scale infrastructure investments are being made exchange for access to valuable natural resources and land. Casual observation of public perceptions also indicates that these investments are often short-term and provide limited human development (E.g. education, training, long-term employment).

According to African Development Bank estimates (2010), “cities will be home to 65% of the continent’s population by 2060, up from 40 percent in 2010, putting pressure on transport networks that are already under huge strain” [16]. The movement of most goods and people in many of these African cities is currently carried out through an eclectic range of motorized and non-motorized transport modes including walking, animal haulage and other formalized modes. The overwhelming majority of public transport throughout many sub-Saharan African cities however is the informal ‘minibus’. And even while these modes of transport continue to be the lifeblood of many African cities, transporting men, women, children, and the elderly at affordable prices, “the taxi network is of course not without its failings, particularly in regards to passenger safety, environmental sustainability, and meeting commuter demand in off-peak periods” [26].

E. G. Nathanail et al. (Eds.): CSUM2020, AISC 1278, pp. 1012–1021, 2021.
https://doi.org/10.1007/978-3-030-61075-3_97

It is now becoming clearer than ever that organized citywide mass transit systems will be required, if rapidly growing African cities are to cope with the influx of inhabitants and consequent gridlock. Cities including "Nairobi, Kampala and Luanda are among the capitals where light rail systems, bus rapid transit, trams or water taxis in the planning. But, as their policy-makers have found, there is no easy solution for affordable, efficient public transport" (ibid). As such, this paper aims to ask why newly introduced transport projects which promise to curb the enormous demand on the public transport system often fail to address issues of social exclusion, inequality, deployment and maintenance. In this analysis, I will explore what governance and business models, social networks and stakeholder behaviors can contribute to the creation of long-term employment, private sector support and community participation in the maintenance and overall resilience of current and future infrastructure projects.

While many African countries, are trying to work towards middle income statuses, their transport planning policies and investments often have limited focus on addressing issues of social exclusion. To illustrate the potential to address some of these issues, this paper will reflect on a light rail infrastructure project currently under way at a rapid pace in Addis Ababa, Ethiopia. This project is the first of its kind in Ethiopia and could potentially serve as an example of high quality pro-poor public transport in the global south. This case was also chosen to illustrate the current move towards planning organized mass transit systems in many cities on the continent.

While this paper focuses on a single transport infrastructure project in Addis Ababa, Ethiopia, the implications of such projects could be applied to other cities on the continent where similar spatial transformations are occurring. By examining the potential effects that the Addis Ababa Light Rail Transit (AALRT) may have on the diverse range of citizens in Addis Ababa, one might be able to find that some sections of the population will have adequate access to the newly built transit service while others will continue to remain on the margins.

This paper will also examine these socio-spatial effects of the AALRT project by expanding on several existing theoretical concepts in the field, primarily the 'right to the city'; an approach to urban planning with strong links to the paradigms of accessibility, social urbanism and social exclusion. Through a contextualized understanding of these concepts, one may be able to see how city authorities can "move away from the traditional systems-based approach to transport provision, towards a more people-focused and needs-based social policy perspective" [11].

2 A Hybrid Approach: Bridging the Right to the City with Accessibility, Social Inclusion and Social Urbanism

Projects like the AALRT and other major projects around the city are currently bringing about significant changes in the city's expansion and sprawl pattern as well. In rapidly formalizing cities like Addis Ababa, "new transport infrastructure generally energizes real estate markets and stimulates new activities and patterns of land use." [3]. These changes primarily benefit those citizens living in the formal city, where planning procedures are often most pertinent. This leaves out a large section of the population that often has difficulty accessing these real estate markets and other opportunities. It is therefore imperative that planning procedures and interventions take place to ensure the participation of and benefit to the city's most marginalized populations.

With a clearer understanding of accessibility, social inclusion and social urbanism, it is now possible to frame the right to the city in the context of the Addis Ababa case. Transport infrastructure projects like the AALRT can potentially benefit from adopting a right to the city approach in their operations and maintenance plans. By exploring ways in which they can bring about improved accessibility and develop related projects targeting the city's poorest inhabitants, the Ethiopian Railway Corporation (ERC) and Addis Ababa City Administration (AACA) could be in a position to have increased ridership and community support from local populations.

There is potential here for the AALRT project to be the center for a series of urban projects that involve "educational and cultural facilities, environmental upgrading and new public spaces" [3]. Such 'social urbanism' projects can utilize the participation of communities located around each station and vary in scale and effect. By adjusting it's mandate beyond simply providing a public transport service and adopting principles from the paradigm of social urbanism, the ERC and AALRT project can contribute to building "community self-esteem and sense of belonging", but also of 'leverage projects' that "lead to a profound social transformation" [3]. We must not forget however that there are "many obstacles to achieving these high ends [in] a city which is increasingly socially unequal and spatially segregated, with high levels of poverty, a growing social housing crisis, structural unemployment, alarming levels of informality and underemployment, and restricted opportunities for young people" (ibid, p. 53). Catalan urban planner Oriol Bohigas once spoke of "the creation of sensations of social inclusion and the right to enjoy the city; the exercise of – if not full – at least a decent level of citizenship" [4]. That is the challenge ahead of the AALRT project and AACA.

Social exclusion vis a vis the right to the city, recognizes "the need for the participation of people in societal governance processes" [8], in addition to its focus on the distributional aspects of urban governance. This participation can in part be achieved through the implementation of a series of social urbanism projects (as will be highlighted in an example later in this paper) that catch the public's attention and encourage their active involvement. The right to the city approach should further discussions on "enabling people to undertake a range of activities, and thus whether the transport system is currently allowing equal access to such activities for everyone" (ibid, p. 268). We must not forget however, that in cities that are "increasingly socially unequal and spatially segregated, with high levels of poverty, a growing social housing crisis, structural unemployment, alarming levels of informality and underemployment, and restricted opportunities for young people" [3], achieving this high ideal, will take generations of careful planning, monitoring and conscious urban governance practices.

By adopting a hybrid approach that; recognizes that some inhabitants will have better access to a new transport system while others will remain excluded; and prioritizes the needs of the city's poorest citizens through a series of quality urban interventions, city authorities could set an example for pro-poor infrastructure and urban governance in the region. This interdependent 3-pronged approach (See Appendix, Fig. 1) will require the ERC and it's partners to a) foster community participation in the maintenance and operation of the AALRT; b) develop diverse financing mechanisms to fund urban projects adjacent to the AALRT; and c) ensure equal access to the AALRT by the city's most often marginalized citizens (I.e. Handicapped, elderly, labourers etc.).

Based on elements from the aforementioned theoretical concepts, this hybrid approach "should also promote a discussion about allowing all users equal engagement and involvement in decisions about how the transport system serves their needs, how it meets their view of being included [8].

In the case of Addis Ababa for example, this hybrid approach can potentially be realized through a series of ERC-community meetings to discuss neighborhood specific accessibility needs and transport services (E.g. Short haul transport links etc.). Such initiatives could spark discussions and serve as a level playing ground for active community participation in decisions about how the transport system serves the needs of specific localities. This approach can also encourage partnerships by recommending a series of activities that can be initiated by the ERC (E.g. educational entertainment sessions, recreational vacations, movie presentations, children's theatricals, and tours of the system– which are open to the community) [14].

This hybrid approach will also require the ERC and other stakeholders to consider extending the public transport service to accommodate specific demographic that is often overlooked in the largely technocratic field of transport planning. Making considerations for the Addis Ababa's poorest, least educated and disadvantaged (often new rural emigrants) citizens will come at a cost and likely require careful planning, but ensuring their 'right to the city' through this transport mode will contribute to equitable city planning practices in Ethiopia and the rest of the continent. That being said, this hybrid approach being proposed is being developed with a set of underlying assumptions.

3 The Addis Ababa Light Rail Transit Project: Analysis and Observations

Today, Ethiopia, and Addis Ababa specifically plays a significant geo-political role in the region, serving both as the political capital for the continent and as a port and resource corridor from coastal Djibouti to Juba, South Sudan. The greater Addis Ababa region has a population of roughly 4 million and is subdivided into 10 Sub cities and 116 formally independent administrative units called Woredas (See Appendix, Fig. 2). Consequently, the demand for public transport has spiked and existing minibus and bus systems are having trouble meeting demand. In response to this the federal government partnered with the French rail industry and conducted a series of feasibility studies for Bus Rapid Transit and Light Rail development.

In conjunction with the city's move towards modernity, the Ethiopian Railway Corporation (ERC) has partnered with the China Railway Eryuan Engineering Group (CREEC), and is currently about to complete the first phase of implementing a Light Rail Tram system2. Running across the city in four directions, this project will serve as the capital's mass transit backbone, transforming mobility in a city where "nearly 90% of the population travels on foot, or by squeezing in to buses and taxis" (BBC 2013). "The three-section 70% low-floor vehicles will have a maximum speed of 70 km/h" [2] running on three lines: Defense Forces Hospital - Ayat Village (17.3 km), Meskel Square - Kality (16.2 km) and Lideta - Menilik Square (3.9 km) (ERC) (See Appendix, Fig. 3).

Thus far this large-scale citywide project has created 3000 jobs in the construction of the ground infrastructure and an additional 150 high skill employment in assembling the tramcars. 800 CREC employees, who are also responsible for training subcontracted local workers that will implement the second and final phase of the project, are overlooking this phase of the project. Contractors have also been conducting training for local staff, and the "Institute of Technology has opened at Addis Ababa University specializing in engineering…sending promising undergraduates to Russia, India and China to continue their education" [10]. With urban unemployment as high as 18% (CSA, 2012), major infrastructure projects like the AALRT have brought with them hopes of large-scale employment in the localities that they exist.

The first observation present in this analysis addresses the need to ensure patrons of the AALRT consist of those both in the peri-urban interface of the city as well as those living in more deprived areas of the inner city. Well-organized business interests centered around the core of the city often tend to dominate those massive coalitions found in the rest of the city. "Large groups in the population are generally weaker and more incoherent than small groups" [7], making it important for communities to mobilize in smaller, well managed groups that can articulate their priorities and negotiate with city authorities on mutual arrangements for joint ticketing/transfers, routes, pricing etc.

There is also potential for the AALRT project to be the center of a series of urban projects that involve "educational and cultural facilities, environmental upgrading and new public spaces" [4] Such 'social urbanism' projects can utilize the participation of communities located around each station and vary in scale and effect. For example, the ERC can partner with the Ethiopian Institute of Architecture, Building and Construction (EIABC) to design and construct public spaces that are unique to the communities surrounding each station. This could serve a double function as community event spaces and potential tourist attractions, with the prospect of generating revenue for the ERC and community groups.

It is also worth noting that well-organized business interests centered around the core of the city can also often tend to dominate those massive coalitions found in the rest of the city. "Large groups in the population are generally weaker and more incoherent than small groups" [7], making it important for communities to mobilize in smaller, well managed groups that can articulate their priorities and negotiate with city authorities on mutual arrangements for joint ticketing/transfers, routes, pricing etc. Such community mobilization could be especially effective in more isolated informal pockets of the city. Strong community ties at the neighborhood association level, can be leveraged in order for ERC to provide a better-quality service, and communities be better included in the transport planning process.

Furthermore, projects like the AALRT are also "complex to implement because their integration is always difficult, investments are heavy and the whole process involves so many different actors with sometimes divergent interests, including operators, infrastructure managers, residents, users, communities and institutions, etc." [25]. With a fleet of approximately 15,000, the informal minibus taxi network is an extensive "demand-driven public transport system that operates without routes maps, timetables or central co-ordination…and in Addis comprises thousands of independent taxi operators serving a multitude of …unofficial bus routes" [26] (See Appendix, Fig. 3). Given the limited capacity and role that authorities currently have had in the maintenance and management of the public transport sector, "it remains to be seen whether these two vastly different modes of public transport can operate side by side – and how such a transformation is going to be effectively managed and coordinated" (ibid). As such the introduction of a structured mass transit system will very likely require a regulatory body like the Ministry of Transport to coordinate the flow of ridership in the city and manage interactions between the existing informal taxi network and the perspective mass transit mode.

Such a management and coordination body, separate from the ERC could consist of taxi owners association members, the AALRT maintenance contractor, business associations and other community groups to ensure that there are minimal conflicts in the management of the city's public transport sector. Another entity that will also need to be considered in this regard is the Anbessa Bus Service, which is the city's only existing formal public bus service. Anbessa has a current fleet of 500 + high capacity buses and "operations are financially supported by the City administration that pays a subsidy for each passenger carried" [9]. The city has recently purchased additional domestically assembled articulated buses to support it's aged fleet and will likely be directly competing with the AALRT for passengers (especially those in lower income brackets).

Given the already diverse and complex networks of the existing public transport sector in the city, and the significant modal share of the formal and informal bus networks (See Appendix, Fig. 4). Authorities comprising such a coordinating body may need to explore the benefits and costs of such a network and "develop appropriate incentive mechanisms to influence the behavior of its operators. This may require some experimentation to identify the most effective mechanisms, which is an appropriate response when dealing a complex-adaptive system" [26]. Putting together such a joint forum will indeed be a challenge but has been identified as by officials to be essential to the longevity of this mass transit project.

4 Conclusion

By proposing 'the right to the city' as a guiding principle for the implementation of a public transport system like the AALRT, this paper has examined why the socio-economic impacts of transport infrastructure projects in the Global South often overlooked in light of increased technical innovation and progress; and proposes a hybrid planning paradigm that identifies and prioritizes the needs of the most underprivileged users. This analysis has resulted in a context-applicable approach in order "to engage with the cumulative effects of informalised economies, settlements, social-cultural systems and political cultures" [17] that are present in many African cities.

Improving equity amongst the citizens of any city is no easy task, and can take generations to achieve. However, by taking the necessary steps to ensure that all citizens' right to the city is respected and that access to services is not determined by socio-economic status, actions in the public transport sector can indeed have a profound effect on bridging the gap between the urban poor and rich. As such, "all measures described above are conducive to an improved transportation system, which in turn will enhance the city's equity, harmony if taking people as the center of the new transport paradigm" [6]. By putting people at the core of transport planning processes, city authorities can avert potential costs in maintenance and operations. As users are made aware of appropriate civic behavior, a collective consciousness will develop therefore making public transport not just a means of mobility, but rather a comfortable, convenient and pleasurable urban transport experience.

Major stakeholders in the public transport sector like ERC and the AACA now have the opportunity to show "that urban planning and transport need to go together so the growth of cities will be better controlled" [16]. If the city's rail transport planners act soon enough to ensure that infrastructure projects like the AALRT are inclusive and participatory, they can set an example, and more importantly a standard for other African cities to follow, avoiding inter alia, "the problems of pollution and congestion found in mega-cities in Latin America and Asia" (ibid).

The completion of the AALRT project "will mean a dramatic shift towards a centralized model of public transport for the city" [26], and with the committed leadership that already exists in the ERC and the potential social and economic impact of the AALRT project, Addis Ababa's public transport sector has the opportunity to educate, foster positive lifestyles and civic behavior and become "a model for state-owned social and commercial enterprises, both in administrative management and social responsibility" [14] in Sub-Saharan Africa.

While this analysis has aimed to provide a comprehensive exploration of possible approaches to inclusive transport planning in rapidly growing cities of the Global South, it is worth acknowledging that information gaps do exist in this analysis, and the author is fully aware that these gaps will have consequences for the line of argument presented and the conclusions reached in this paper. However, given the theoretical focus of the dissertation, and the rigorous amount of desk research and secondary sources reviewed, this document serves as an exploration of the potential social impacts that large transport infrastructure projects can have on some of the most underprivileged communities in urban Ethiopia.

Moreover, this paper has tried to show that the major transport infrastructure projects like the AALRT do indeed have an important role to play in "determining social outcomes for different sectors of modern society, in terms of both the absence of adequate transport services and the impact of the transport system on individuals and communities" [12]. As such, this analysis intends to serve as a catalyst for ensuring that such infrastructure projects simultaneously make efforts towards addressing some of the often-neglected issues of social exclusion and accessibility, in transport planning processes of cities in the Global South.

Appendix

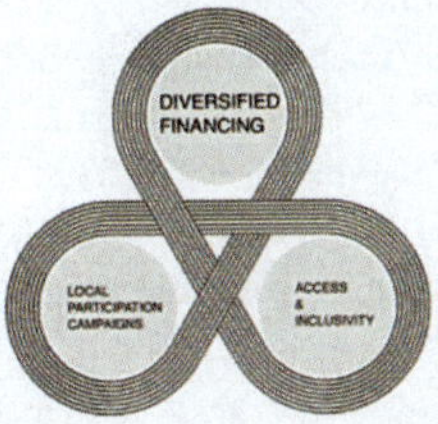

Fig. 1. Hybrid Approach to Inclusive Planning.

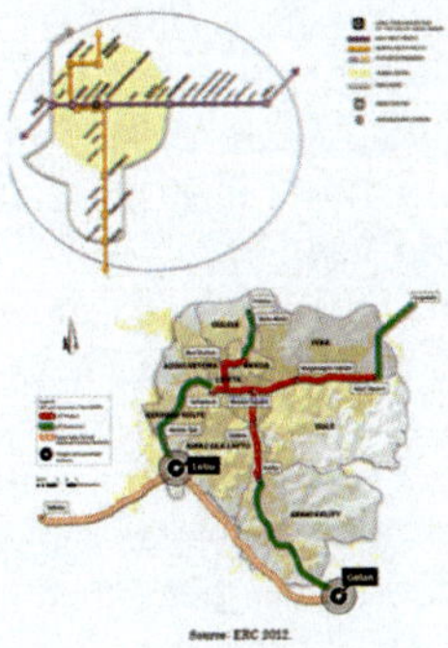

Fig. 2. AALRT Planned Routes, Stations and Phased extensions by Woreda "Administrative Units".

Fig. 3. Addis Ababa Existing and planned Minibus Routes.

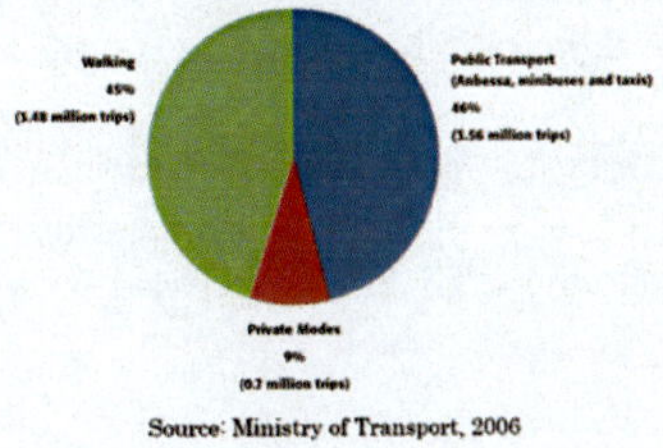

Fig. 4. Addis Ababa Modal Share for Public Transport.

References

1. Addis Ababa City Planning Project Office (AACPPO): Transportation challenges in a booming city: coordination of the mass transit network and urban development in Addis Ababa. Background to CODATU XV 22–26 October 2012 The role of urban mobility in (re) shaping cities. Agence France de Development (2012)
2. Barrow, K.: CNR to supply trams for Addis Ababa LRT. Int. Railway J., 17 March 2014 (2013). http://www.railjournal.com/index.php/africa/cnr-to-supply-trams-for-addis-ababa-lrt.html. Accessed 07 Aug 2014
3. Brand, P., Davila, J.: Mobility innovation at the urban margins: Medellín's Metro cables. City J. (2011). http://www.bartlett.ucl.ac.uk/dpu/metrocables/dissemination. Accessed 08 July 2014
4. Brand, P., Davila, J.: Metro cables and 'social urbanism': two complementary strategies. In: Dávila, J. (ed.) Urban Mobility and Poverty, pp. 46–54. Lessons from Medellín and Soacha, Colombia, DPU-UCL and Universidad Nacional de Colombia (2013)
5. Ethiopian Railway Corporation (ERC): Addis Ababa Light Rail Transit Project. Presentation by Mr. Yehualaeshet Jemere (MSc. CEng) Chief Officer, Construction and Project Execution Department, ERC (2012). Accessed July 2012
6. GIZ Challenges of urban transport in developing countries. GIZ Sourcebook on Sustainable Urban Transport for Policymakers in Developing Cities. GIZ - Sustainable Urban Transport Programme (SUTP), October 2010. www.sutp.org. Accessed 05 Apr 2014
7. Harvey, D.: Social Justice and the City. Blackwell Publishers, pp. 96–118 (1988). Harvey, D.: The right to the city. New Left Rev. **53**(September/October) (2008)
8. Hodgson, F.C., Turner, J.: Participation not consumption: the need for new participatory practices to address transport and social exclusion. Transp. Policy **10**(4), 265–272 (2003)
9. IBIS Transport Consultants. Study of urban public transport conditions in Addis Ababa, Ethiopia, March 2005. Public Private Infrastructure Advisory Facility (2005). www.ppiaf.org/sites/ppiaf.org/files/…/UrbanBusToolkit/…/Addis.doc. Accessed 06 July 2014
10. Jobson, E.: Transport: Riding the rails in Ethiopia and Kenya' The Africa Report. 25 March 2014 (2014). http://www.theafricareport.com/East-Horn-Africa/transport-riding-the-rails-in-ethiopia-and-kenya.html. Accessed 08 Aug 2014
11. Lucas, K.: The role of transport in the social exclusion of low-income populations in South Africa: a scoping study. In: Proceeding of Transportation Research Board Annual Meeting, 10–15 January, Washington DC (2010)
12. Lucas, K., Tyler, S., Christodoulou, G.: The Value of New Transport in Deprived Areas: Who benefits, how and why? Joseph Rowntree Foundation (2008). http://www.jrf.org.uk/sites/files/jrf/2228-transport-regeneration-deprivation.pdf. Accessed 09 Aug 2014
13. Ministry of Works and Urban Development (MWUD): Urban Transport Planning Manual. Mathewwos Consult: Architectural, Engineering, Urban Planning, Urban Development and Urban Management Consultants (2006)
14. De Medellin, M.: Cultura Metro (2014). https://www.metrodemedellin.gov.co/index.php?option=com_content&view=article&id=75&Itemid=113&lang=en. Accessed 07 Aug 2014
15. Mulligan, C.: The case for light rail, London, Jun, 0, 09:40 – Waterfront Conference Company (2003)
16. Obiero, M., Mbanga J., Redvers, L.: Transport: Can African Trains Take the Strain? The Africa Report, 26 July 2013 (2013). http://www.theafricareport.com/East-Horn-Africa/transport-can-african-trains-take-the-strain.html. Accessed 05 June 2014

17. Pieterse, E.: Rethinking African Urbanism from the Slum. LSE Cities Article Series, November 2011 (2011). http://lsecities.net/media/objects/articles/rethinking-african-urbanism-from-the-slum/en-gb/. Accessed 08 Aug 2014
18. Preston, J., Rajé, F.: Accessibility, mobility and transport-related social exclusion. J. Transp. Geogr. **15**(3), 151–160 (2006)
19. Railway Gazette: Addis Abeba light rail civil works to be completed this year, 4 October 2013 (2013). http://www.railwaygazette.com/news/news/africa/single-view/view/addis-abeba-light-rail-civil-works-to-be-completed-this-year.html. Accessed 09 Aug 2014
20. Serebrisky, T., Gomez-Lobo, A., Estupiñán, N., Munoz-Raskin, R.: Affordability and subsidies in public urban transport: what do we mean, what can be done? Transp. Rev. **29**(6), 715–739 (2009)
21. Tolon, U.W. Comparison of Urban Upgrading Projects on Development Cooperation in Ethiopia, November 2008. World Bank. UPC Commons (2008). https://upcommons.upc.edu/pfc/bitstream/2099.1/6374/8/07.pdf. Accessed 15 July 2014
22. Tyler, N. (ed.): Accessibility and the Bus System: From Concepts to Practice. Thomas Telford, London (2002)
23. Ujorha, T.: Lagos: China and the Light Rail Project. Trademark Southern Africa, 17 January 2012 (2012). http://www.trademarksa.org/news/lagos-china-and-light-rail-project. Accessed 06 June 2014
24. UN HABITAT: Addis Ababa Urban Profile. Participatory Slum Upgrading Programme in the African, Caribbean and Pacific Countries. UNON, Publishing Services Section, Nairobi, ISO 14001:2004-certified (2008)
25. World Bank, Sub-Sahara African Transport Policy Programme (SSATP) (2014). www.ssatp.org. Accessed 17 Aug 2014
26. Young, A,: Cities in motion: Addis Ababa already has a great network minibus taxis. The Guardian Online, 19 February 2014 (2013). http://www.theguardian.com/cities/2014/feb/19/cities-in-motion-addis-ababa-minibus-taxi. Accessed 08 June 2014

Addressing Street Network Accessibility Inequities for Wheelchair Users in Fifteen European City Centers

Alexandros Bartzokas-Tsiompras(✉), Yannis Paraskevopoulos, Aglaia Sfakaki, and Yorgos N. Photis

Faculty of Rural and Surveying Engineering, National Technical University of Athens, 9 Iroon Polytechneiou Street, 157 80, Zographos, Greece
abartzok@mail.ntua.gr

Abstract. Although thirteen percent of the population living in European cities are affected by some kind of disability it is not evident whether pedestrian streetscapes efficiently support their mobility needs or more interventions are required in order to facilitate and ensure their equal access to city life and services. In this paper, we focus to people using wheelchairs and compare, in terms of accessibility, centrality measures (i.e., angular choice) for fifteen European downtown areas and for two types of networks. Namely, the city-center-wide and the wheelchair users' street network and according to the following criteria: presence of sidewalks, pre-and-post crossing curb ramps, and efficient sidewalk width. To this end, we construct composite inequity indicators between the above types of spatial networks, based on bivariate spatial autocorrelation clustering of centrality analysis results and the share of wheelchair users' street network per city. The findings of this work suggest that there is unequal spatial distribution between accessible and inclusive pedestrian streetscapes and thus decreased levels of centrality for people with the above disabilities are highlighted in eleven out of the fifteen studied European city centers networks.

Keywords: Wheelchair users · Space syntax · Urban mobility · Urban accessibility

1 Introduction

According to the W.H.O [1] about 15% of the global population has some form of disability, and 70 million people need a wheelchair. In comparison, in Europe, 12.7% of (EU-27) people has a basic activity difficulty (e.g., seeing, walking etc.) and dwells in cities [2]. There are ongoing pressures in cities (e.g. rapid urbanization) that exacerbate the impacts of unfriendly and low accessibility built environments on disadvantaged people's lives [3, 4]. Evidence shows that people with disabilities living in inaccessible cities have lower social participation and activity [6], decreased levels of quality of life [7], and increased levels of poverty and social exclusion [8]. Additionally, disabled people often opt to live in suburbia rather than downtown areas and present different mobility patterns in comparison to able-bodied counterparts [3]. However, policy institutions, such as the United Nations and the European

E. G. Nathanail et al. (Eds.): CSUM2020, AISC 1278, pp. 1022–1031, 2021.
https://doi.org/10.1007/978-3-030-61075-3_98

Commission, recommend the promotion of inclusive accessibility planning frameworks and design practice. Thus, cities can embrace and invest in universal access to the public realm and streets [9] and in turn, everyone will be free to walk, cycle, use a baby stroller or a wheelchair. In this paper, we focus on disabled pearsons' ability to participate to city life and thus compare their spatial patterns of accessibility levels to those of pedestrians in the contexts of eleven European city centers.

2 Background

Even though numerous researchers are comparing place-based urban access around Europe [11] they do not consider different population groups with mobility limitations such as wheelchair users [12]. Moreover, a significant part of the research, although qualitative [3], is mainly focused on the built environment attributes that affect people with disabilities, while little is known about wheelchair users' mobility patterns. For example, Kirchner et al. [4] demonstrated that more than 80% of manual wheelchair users in New York City frequently mentioned as environmental barriers problems with sidewalk pavements, issues with puddles, problems with construction and curb cuts. A UK study [12] surveyed the accessibility to the city center of Swansea, UK, and explored the attitudes of wheelchair users. They found that most wheelchair users travel by car (76%), whereas only 9% use the pavements on their way to downtown. Furthermore, more than 60% rated people on sidewalks, access to shops, and the lack of curb cuts as major or prohibitive city center obstacles. Additionally, one out of three wheelchair users, considered peripheral streets of the central area, as problematic.

On the other hand, there are some model-based analytical approaches concerning wheelchair users' accessibility. Matthews et al. [13] after extensive research on wheelchair users' perceptions, developed a GIS-based application called MAGUS (Modeling Access with GIS in Urban Systems). This GIS solution helped wheelchair users to navigate in the city as well as to create a decision support system for planners to identify hostile and low accessibility urban environments. Similarly, Sobek & Miller [14] released a web-GIS platform called U-Access, and they found that the shortest optimal route in a case study at the University of Utah campus was significantly longer for wheelchair users than non-disabled individuals. A study [15] from Cuenca in Ecuador used a mobile application to audit a sample of streetscapes and found that the number of highly inaccessible streets for wheelchair users is between four to five times higher than these of non-disabled people. Finally, researchers used spatial network analysis methods to evaluate spatial connectivity and accessibility indicators (e.g., Pedshed ratio, accessible network length) for Lisbon's city center [16]. The results emphasized either on wheelchair users or to those without disabilities, and the findings demonstrated that a non-disabled person always has equal or better than a wheelchair user's accessibility, a result that they called it "disability-induced accessibility disparity."

In essence, although we know that disabled people deal with hostile pedestrian streetscapes [16], we haven't yet explored this in a European context. To this end, the main aim of this study is to compare spatial accessibility levels and pedestrian infrastructure for wheelchair users in fifteen European downtown areas. We employ classical accessibility and geographical analysis methods, such as space syntax and spatial autocorrelation, to compare and associate centrality levels between city-center-wide street networks and street networks designed to accommodate the needs of wheelchair users. More specifically, we adopt the angular choice measure, as a proxy of access to city life and central places, and capitalize on spatial clustering analysis results of these values to create composite indicators of street-level accessibility inequities in fifteen European central districts from twelve countries. This is the first research that sheds light on the accessibility of disabled people and practically answers the question whether wheelchair users can equally access and use downtown areas for their daily activities by focusing on more than one European city. To this end, findings as well as the proposed methodological approach of this work contribute to the emerging transport and environmental justice literature and can assist policymakers to get a better understanding of the inclusiveness of street-level accessibility planning around Europe.

3 Methodology and Methods

This work utilizes data from the National Technical University of Athens Walkability Database [17], containing a plethora of street level walkability related indicators, measured by a virtual microscale audit tool for pedestrian streetscapes. In order to analyze centrality inequities in urban networks for both wheelchair users and able-bodied persons we adopt a four-stage methodological framework mainly based on the space syn-tax network and local spatial autocorrelation analysis methods.

3.1 Stage 1: Definition of Urban Street Networks and Spatial Data

Our approach requires the construction and comparison of two forms of urban street network. The elemental network of open spaces between buildings (except from the formal public spaces of a city such as parks, squares etc.) as the entire network of street segments (EN) and the wheelchair users' street network (WN). Following relevant literature [4, 5, 20] the WN contains street segments that fulfill specific criteria that adapt to the equitable use in sidewalk design concept [20]. Namely, these are the presence of sidewalk (Safety), the presence of curb ramps at the pre-and-post crossing (Usability), and efficient sidewalk width (Attractiveness), meaning that three individuals can comfortably walk simultaneously and in parallel to each other from the beginning until the end of the sidewalk.

3.2 Stage 2: Space Syntax Analysis – Spatial Network Centrality

Space syntax is rooted in graph theory and the idea of urban morphology [10]. It measures the geometry and topology of the urban grid, and in turn, quantifies the human activity in urban space [10]. Hence, capitalizing on typical space syntax techniques, we calculate for each type of urban street network, i.e., EN, and WN, the centrality indicator of angular choice, since it is the most reliable proxy for the through-movement potentials of the network [5]. Classic accessibility measures are based on street networks and road-center-line maps. On the other hand, space syntax and angular choice, builds on graph analysis and converts street segments to nodes as well as street junctions to edges (for example see Stavroulaki et al. [21]). It counts the number of times each street segment falls on the 'shortest path' (in terms of angular deviation) between all pairs of segments within a selected distance and thus symbolizes the level of network accessibility. Technically, we apply it to each type of network per city, using a 400-m radius of local centrality. The calculations were carried out using QGIS 2.16.3 software and the Space Syntax Toolkit plugin (v. 0.2.1).

3.3 Stage 3: Spatial Analysis – Spatial Autocorrelation

The third stage refers to the spatial analysis and spatial autocorrelation of the space syntax measures and results in order to create inequity indicators describing the level of spatial accessibility per each type of network and city. We utilize Kernel Density Estimation (KDE) [18], with a 30 m bandwidth/search radius and a 10 m cell size, to create a common analytical framework for the two different networks' centralities. Grid-based centrality values per each network are then used as weights in the Standard Deviational Ellipses (SDE) [18] method. SDE is a geostatistical indicator of centrography and measures the size of dispersion of events around their mean center. Thus, identical spatial patterns of centralities per network are identified in cities where the two ellipses overlap (SDE Ratio is equal to 1). However, SDEs do not provide a solid basis for analytical spatial inequality analysis. Hence, we opted to perform Bivariate Local Spatial Autocorrelation (Bivariate LISA) using the Global and Local Moran's I [18] on GeoDa 1.12.1.161 software. This method quantifies and maps the variables of statistically significant spatial clusters and/or outliers and it is widely used in transportation inequality research [17]. To this end, Bivariate LISA helps us to understand the interconnection of centrality values between networks.

3.4 Stage 4: Indicators Analyzing Spatial Networks Inequities

Based on spatial autocorrelation results and using the share of areas per spatial cluster/outlier per city, we formulate ratios of high to low cluster groups and a composite indicator that assesses the level of accessibility inequality between the two different urban street networks. The first step includes the calculation of the '*wheel street network (WN) centrality ratio (WNC)*' and the 'entire *street network (EN) centrality ratio (ENC)*.' '*WNC*' describes the extent of high WN centrality clusters (H_{WN}) compared to the low WN centrality clusters (L_{WN}). In the same manner, '*ENC*' designates the extent of high EN (H_{EN}) centrality clusters compared to the low EN centrality clusters (L_{EN}). These two quantify the proportion of spatially concentrated highly "*central*" segments per each type of network per city. '*WNC*' and 'ENC' are defined as follows:

$$WNC = \frac{H_{EN}H_{WN} + L_{EN}H_{WN}}{L_{EN}L_{WN} + H_{EN}L_{WN}} \tag{1}$$

$$ENC \frac{H_{EN}H_{WN} + H_{EN}L_{WN}}{L_{EN}L_{WN} + L_{EN}H_{WN}} \tag{2}$$

Finally, we compute the '*Inequity in Centrality of WN Index*' (IWNC) and the '*Adjusted Inequity in Centrality of WN Index*' (AIWNC). '*IWNC*' (Eq. 3) refers to the equilibrium between the centralities' of the two networks. In other words, IWNC reflects the degree to which the network-driven central or non-central places of the downtown area (EN) are also central or non-central for the WN. However, AIWNC (Eq. 4) is a more comprehensive measure since it is weighted by the share of the WN length in the study area (SWNL). AIWNC represents more accurately the size of inequity that wheelchair users are usually facing with street-level accessibility.

$$IWNC = \frac{ENC}{WNC} \tag{3}$$

$$AIWNC = IWNC \times SWNL \tag{4}$$

With the adjusted metric, we both encourage wheelchair-accessible centrality and promote the idea of inclusive street planning in the city.

4 Results

The highest share of WN length (SWNL) is found in Barcelona's Example district (Fig. 3), where 97% of the EN has been effectively engineered with curb ramps, sidewalks and efficient pavement width, followed by Copenhagen (68%), Valencia (66%), Rotterdam (64%), and Gothenburg (61%). In any other city center, the WN length share ranged from 59% (Budapest) to 23% (Plovdiv). Also, Rome (35%), Bucharest (25%) and Athens (24%) belong to the most limited networks' access group. However, half of the downtown areas have roughly more than 49.9% (median) of their streets in the WN due to the high proportions of segments that do not meet the criterion of efficient sidewalk width (more details about these indicators are depicted online in bar charts: https://doi.org/10.6084/m9.figshare.12448625).

Concerning EN angular choice we observed that in all cases, the highest values were located in areas with high pedestrian and vehicular traffic. However, the two patterns of centrality per each network are distributed in a highly different way, as can be seen by the standard deviation ellipses (SDE) in Fig. 3. Only Barcelona shows almost identical spatial patterns, since the spatial distribution of centrality values between the compared networks overlap (SDE ratio = .99, see Fig. 2). Besides, Copenhagen, Rotterdam, and Antwerp presented somewhat similar spatial patterns. In contrast, Athens illustrated by far the smallest SDE ratio (.45), which translates into that the city has a significantly unorganized, scattered, and limited WN.

After checking for global spatial autocorrelation (Global Moran's I) [18] and statistically significant clustering of centrality values per each type of network per city, we proceeded with the bivariate spatial autocorrelation analysis (Local Moran's I) [18]. The results of this process unveiled a more detailed picture of the policy efforts towards accessibility for people with mobility impairments (see Fig. 1). All investigated European central districts demonstrated that the highest share of their area is characterized as a Low-Low (L-L) cluster (37.8% > L-L < 54.44%, see Fig. 1), which means that they are non-central places for both types of networks. This result was expected, as the spatial distribution of high angular choice values is often clustered around a limited number of central urban axes. Nevertheless, Barcelona has the largest cluster of High-High (H-H) centrality values, which includes 30.7% of the total downtown area, followed by

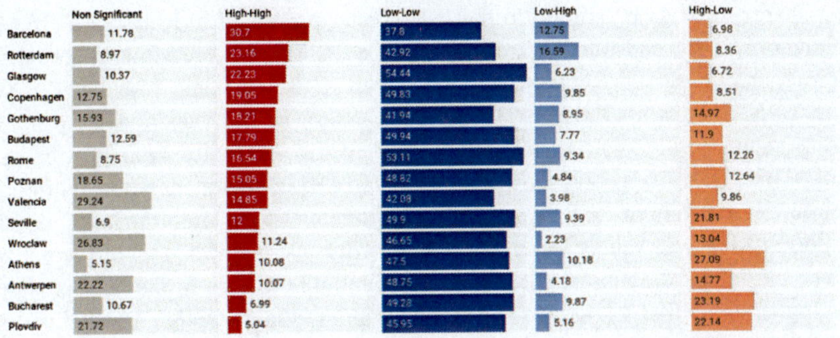

Fig. 1. Share of Bivariate (EN vs. WN) Local Moran's I clusters of centrality values per city

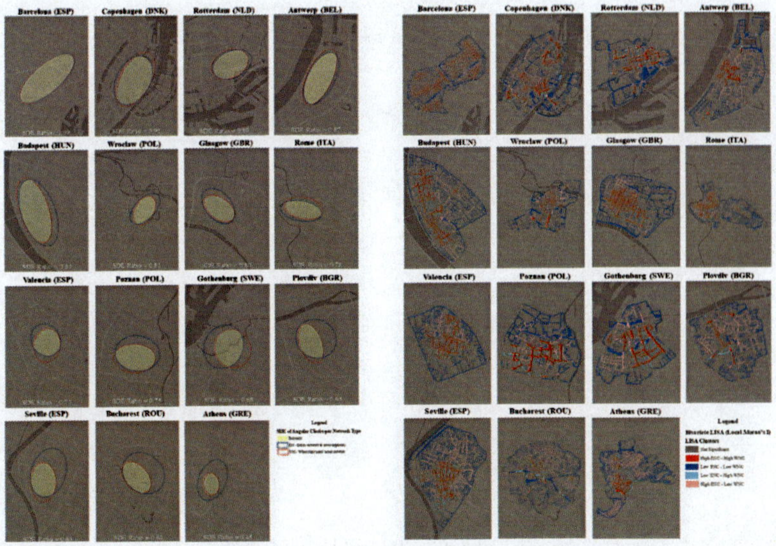

Fig. 2. Maps of Bivariate Local Moran's I results & SDE of centralities in EN & WN per city

Rotterdam and Glasgow, with 23.16% and 22.23%, respectively. On the other hand, Bucharest and Plovdiv presented the smallest area of H-H clusters that cover less than 7% of their area. However, the H-L clusters are recognized as highly problematic places since they present increased centrality of EN, but at the same time, in WN, centrality is significantly low. Notably, these areas need immediate policy attention in order to ensure access [9] for all. In this regard, four cities, i.e. Athens, Bucharest, Plovdiv, and Seville, demonstrated large shares of H-L clustering areas and with more than 20%. On the contrary, five cities, i.e., Glasgow, Barcelona, Rotterdam, Copenhagen, and Valencia, highlighted H-L clustering performance in less than 10% of their area. The L-H cluster, pertains to areas that, although they are not central for EN, they are highly central for the engineered WN. However, areas located in the L-H clusters are probably related to mainly residential communities of each downtown district. The highest shares of areas belonging to the L-H cluster presented in Rotterdam (16.6%), Barcelona (12.75%), and interestingly in Athens (10.18%), whereas the lowest shares found in Poznan (4.84%), Antwerp (4.18%), Valencia (3.98%), and Wroclaw (2.23%).

The WNC indicator (see Fig. 3) helped us assess if the clustering area of high centrality values in the WN is analogous to that of low centrality values of the same network. Thus, Barcelona (.97) showed a particularly equal treatment in the distribution of high to low clusters of its WN, as well as Rotterdam (.78). Unfortunately, all other cities presented WNC values lower than .5. Plovdiv was identified as the worst performance city since the clustering area of low centrality values in the WN is significantly higher than that of its high centrality values. On the other hand, the ENC evaluated if the clustering area of high centrality in the EN is also analogous to clusters of low centrality in the same network. Again Barcelona scored the highest ENC with .75, followed by Gothenburg and Athens with .65 and .64, respectively. High values of ENC in these three cities denote that their current street networks have well-engineered geometry with high levels of centrality in a broad area of the city center.

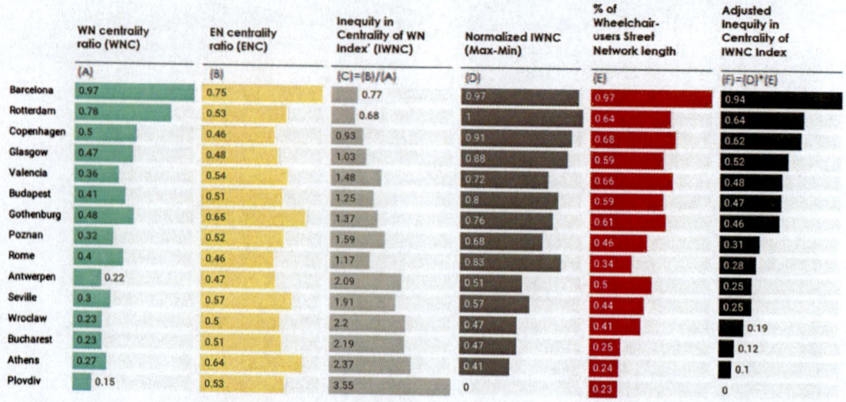

Fig. 3. % of WN length (red) & inequality indicators in centralities of WN& EN per city

Figure 3 summarizes the results of all centrality ratios and the computed inequality indicators. Therefore, three cities, namely Rotterdam, Barcelona, and Copenhagen, show the prevail of the WNC over the ENC (IWNC < 1). This finding means that the engineered WN is always better than the EN in terms of centrality, although WN is smaller than EN. Furthermore, Glasgow is the only city that presents an equal balance of centrality ratios (IWNC $\simeq$1). All other cities always have worse WNC than ENC

(IWNC > 1). Interestingly, this finding means that in eleven out of fifteen European cities, the EN offers significantly better levels of access to non-disabled persons, but when it comes to people with disabilities, the centrality values are unfairly lower. For instance, Plovdiv, holding the highest level of inequality, has a 3.55 times more underdeveloped and less central WN compared to the able-bodied persons EN.

Lastly, AIWNC takes into account the normalized IWNC weighted by SWNL, demonstrating that Barcelona, Rotterdam, and Copenhagen are the leading cities in providing equally central and extensive networks for wheelchair users, too. Meanwhile, Plovdiv, Athens, Bucharest, Wroclaw, and Seville are the worst cases for wheelchair users since either their WN is highly underdeveloped, or the clustering of high centrality values of their WN is still smaller than the EN can offer to people with disabilities.

5 Discussion and Conclusions

This work is the first geographical approach that points out the chasm between inclusive and accessible street network planning around Europe. Our findings made clear that a significant amount of policy efforts is still required to make sidewalks barrier-free, safe, and attractive for all. Therefore, we conclude that many European urban areas are far from realizing the ambitious goals of the New Urban Agenda that promise access to city life for all on an equal basis. To our knowledge, there is no other study comparing the centrality levels of street networks in accordance to the mobility conditions of street users of 15 European downtown areas. In particular, individual case studies have been investigating pedestrian networks in specific European areas, such as Lisbon's downtown, and demonstrating the striking size of accessibility disparity between a non-disabled and a wheelchair user in the city [16]. However, European-wide comparisons between microscale characteristics of pedestrian streetscapes of downtown areas have been recently made by Bartzokas-Tsiompras et al. [17] in eight capital cities and indicated the high level of inequity in walkability spatial distribution.

In this research, we also constructed a new set of quantitative indicators for European street networks designed for disabled people. We found that almost in all city centers, except for Barcelona, more than 32% of each street network is not sufficiently adaptable to the needs of wheelchair users. Interestingly, in some city centers, for example, Plovdiv and Athens, more than 75% of their street segments are not being planned to support wheelchair users. Based on space syntax and spatial autocorrelation methods, we explored spatial disparities in the distribution of centrality values (i.e., angular choice measure) between the EN and the WN. We found that a large number of cities provide networks with significantly different levels of centrality to people with different mobility conditions (i.e., wheelchair users). For example, the spatial clustering of WN centrality in Plovdiv is 3.5 times more unequal compared to that of an enabled-bodied person street network, while in Glasgow, centralities are almost equal between the two types of networks. Additionally, except for Rotterdam, Barcelona, and Copenhagen, in all other areas, the WN creates smaller clusters of high centrality values compared to that of the EN per city. Taking into consideration the results of the adjusted inequity ratio (AIWNC), which includes not only spatial patterns of network centrality but also the share of WN length, we concluded that Barcelona is by far the most accessible city for wheelchairs. Additionally, it shows the best spatial distribution of network centrality in terms of wheelchair users streets. In contrast, eleven out of fifteen European downtowns, show highly unequal street-level accessibility for wheelchair users and demonstrate more under-developed or problematic pedestrian infrastructure. For instance, Plovdiv, Athens, Bucharest, Wroclaw, and Seville were identified as the worst-performing cities for wheelchair access and environmental barriers in street networks having multiple impacts on the lives and the social inclusion of wheelchair users. The above-mentioned results are in accordance with the social model of disability [19] in which society is disabling physically impaired people. Consequently, European policymakers should invest in developing more intuitive standards that promote universal access [9] in street design and ensure equitable use in sidewalks [20].

There are certain hypotheses and simplifications adopted in our approach. Firstly, by focusing on wheelchair users we excluded all other forms of disability from our concept. We argue however, that it can be easily applied to the majority of them. Secondly, although the equitable use in sidewalks framework [20] includes more than 30 sidewalk design factors, we created WNs based on a limited set of variables. This was due to the limited data sources we had available for the studied cities, as well as the labor-and-time-intensive process required to collect them. Hopefully, in the future, proposed crowdsourcing solutions producing built environment data from wheelchair sensors might be a critical advantage for expanding our approach in several other cities with more precise datasets [22]. And finally, angular choice values near the boundary of each case study area unavoidably suffer from the edge-effect [5].

Although we applied a robust methodological framework this could be significantly improved by adopting more detailed accessibility modelling approaches, such as agent-based or utility-based models, and thus further explore spatial access of disabled people to public transport, health, education, and leisure activities. These insights will allow policymakers to get a better understanding of the disabled urban environments which undoubtedly still exist in modern cities.

Acknowledgments. This research is co-financed by Greece and the European Union through the Programme « Human Resources Development, Education and Lifelong Learning » in the context of the project "Strengthening Human Resources Research Potential via Doctorate Research" (MIS-5000432), implemented by the State Scholarships Foundation (IKY).

References

1. World Health Organization: Better health for people with disabilities: infographic. https://www.who.int/disabilities/infographic/en/. Accessed 23 Mar 2020
2. EUROSTAT: Population by type of disability, sex, age & degree of urbanization (2011). https://data.europa.eu/euodp/en/data/dataset/6qNy29oFl0WA7dbPo7TFg
3. Taylor, Z., Józefowicz, I.: Intra-urban daily mobility of disabled people for recreational and leisure purposes. J. Transp. Geogr. **24**, 155–172 (2012)

4. Kirchner, C.E., Gerber, E.G., Smith, B.C.: Designed to deter: community barriers to physical activity for people with visual or motor impairments. Am. J. Prev. Med. **34**, 349–352 (2008)
5. Vaughan, L., Geddes, I.: Urban form and deprivation: a contemporary proxy for Charles Booth's analysis of poverty. Radical Stat. **99**, 46–73 (2009)
6. Clarke, P.J., Ailshire, J.A., Nieuwenhuijsen, E.R., de Kleijn-de Vrankrijker, M.W.: Participation among adults with disability: the role of the urban environment. Soc. Sci. Med. **72**(10), 1674–1684 (2011)
7. Williams, G., Willmott, C.: Higher levels of mobility are associated with greater societal participation and better quality-of-life. Brain Inj. **26**(9), 1065–1071 (2012)
8. Pinilla-Roncancio, M.: The reality of disability: multidimensional poverty of people with disability and their families in Latin America. Disabil. Health J. **11**(3), 398–404 (2018)
9. Manley, S.: Creating an accessible public realm. In: Preiser, W.F.E., Smith, K.H. (eds.) Universal Design Handbook, vol. 2, pp. 17/5–17/12. McGraw-Hill, New York (2010)
10. Hillier, B., Burdett, R., Peponis, J., Penn, A.: Creating life: or, does architecture determine anything? Archit. Behav. **3**(3), 233–250 (1987)
11. Bartzokas-Tsiompras, A., Photis, Y.N.: Measuring rapid transit accessibility and equity in migrant communities across 17 European cities. Int. J. Transp. Dev. Integr. **3**(3), 245–258 (2019)
12. Bromley, R.D.F., Matthews, D.L., Thomas, C.J.: City centre accessibility for wheelchair users: the consumer perspective and the planning implications. Cities **24**(3), 229–241 (2007)
13. Matthews, H., Beale, L., Picton, P., Briggs, D.: Modelling Access with GIS in Urban Systems (MAGUS): capturing the experiences of wheelchair users. Area **35**(1), 34–45 (2003)
14. Sobek, A.D., Miller, H.J.: U-access: a web-based system for routing pedestrians of differing abilities. J. Geogr. Syst. **8**(3), 269–287 (2006)
15. Orellana, D., Bustos, M.E., Marín-Palacios, M., Cabrera-Jara, N., Hermida, M.A.: Walk'n'roll: mapping street-level accessibility for different mobility conditions in Cuenca, Ecuador. J. Transp. Health (100821) (2020)
16. Vale, D.S., Ascensão, F., Raposo, N., Figueiredo, A.P.: Comparing access for all: disability-induced accessibility disparity in Lisbon. J. Geogr. Syst. **19**, 43–64 (2016)
17. Bartzokas-Tsiompras, A., Tampouraki, E.M., Photis, Y.N.: Is walkability equally distributed among downtowners? Evaluating the pedestrian streetscapes of eight European capitals using a micro-scale audit approach. Int. J. Transp. Dev. Integr. **4**(1), 75–92 (2020)
18. O'Sullivan, D., Unwin, D.J.: Geographic Information Analysis, 2nd edn. John Wiley, Hoboken (2010)
19. Union of the Physically Impaired Against Segregation: Fundamental Principles of Disability. Union of the Physically Impaired Against Segregation, London (1976)
20. Aghaabbasi, M., Moeinaddini, M., Asadi-Shekari, Z., Shah, M.Z.: The equitable use concept in sidewalk design. Cities **88**, 181–190 (2019)
21. Stavroulaki, G., Marcus, L., Berghauser Pont, M., Nilsson, L.: Representations of street networks in space syntax: towards flexible maps and multiple graphs. In: Proceedings of the 11th Space Syntax Symposium, Lisbon, Portugal, pp. 174.1–174.16 (2017)
22. Mascetti, S., Civitarese, G., El Malak, O., Bettini, C.: SmartWheels: detecting urban features for wheelchair users' navigation. Pervasive Mob. Comput. **62**, article no. 101115 (2020)

Check for updates

Investigation of Vehicle Swept Path in Roundabouts

Andromachi Gkoutzini(✉), Panagiotis Lemonakis, George Kaliabetsos, and Nikolaos Eliou

Department of Civil Engineering, University of Thessaly, Pedion Areos, 38334 Volos, Greece
{andgkout, plemonak, gekalia, neliou}@uth.gr

Abstract. Due to their evident benefits roundabouts are widespread among the congested urban road networks. The primary aims of the roundabouts are to improve safety, traffic and environmental conditions, which derive from their proper design. Vehicle swept path analysis is considered a crucial parameter in the design of a roundabout. The accurate prediction of the vehicle's movement ensures a safe route and identifies the conflict areas, at the entry, exit and along the circulatory roadway especially in a 2-lane roundabout where accidents are over-represented. The purpose of this paper is to investigate vehicle's turning path in 2-lane roundabouts. The large trucks are of particular interest, because they demand more space in their motion. Consequently, the dimensions of the roundabouts are determined by the turning paths of the heavy vehicles and buses. Therefore, in order to perform swept path analysis, the turning paths of different type of large vehicles in roundabouts with different diameter have to be computed. Moreover, the overlap of the turning paths of two vehicles traveling simultaneously in a two- lane roundabout will be investigated. In the recent years, roundabouts have also become increasingly popular in Greece. In fact, in the city of Volos four roundabouts have been constructed in the last two years. The two of them have been chosen to be investigated in terms of geometry and safety based on the preceding analysis.

Keywords: Turning path · Roundabouts · Swept path

1 Introduction

Intersection related accidents are over-represented, despite the fact they constitute only a small portion on the road networks. This finding triggered the researchers to further investigate the root causes of such accidents and seek for alternatives that are capable to effectively compromise safety and junction capacity. The outcome was the introduction of roundabouts which provide safer interactions of the road users compared to the traditional priority junctions [1]. Fundamental element of the roundabout is the reduced speed of the incoming traffic and the smaller number of the conflict points [2]. However poor design or inconsistent construction might mitigated the benefits derived from the use of roundabouts and on several occasions the accident rates are increased after the conversion of a traditional intersection to a roundabout [3]. Besides the required

E. G. Nathanail et al. (Eds.): CSUM2020, AISC 1278, pp. 1032–1041, 2021.
https://doi.org/10.1007/978-3-030-61075-3_99

driving workload sustained by truck drivers negotiating a roundabout is greater compared to signalized intersections resulting in overturn events of large vehicles. As discussed by Pecchini et al. in their study, decisive factor in order to avoid accidents adjacent to roundabouts is the proper three-dimensional geometric layout of the roundabout [4].

Swept path and lateral displacement investigation especially on semi and full trailer configurations is crucial for the safety of the vehicles traveling in low and highway speeds. Economic and energy factors led the manufactures to construct vehicles with larger capacity initiating the use of multiple trailers attached to the tractor unit posing a threat to the stability of the truck – trailers system though [5]. On the other hand, poor design elements can result in hazardous situations for both the articulated vehicles and the rest of the road users. Therefore, an in-depth analysis of various vehicles' tracking performance is essential in order to ensure the safe operation of various design elements such as roundabouts, parking lots etc.

As implied earlier, key element of a robust swept path analysis is the introduction of the design vehicle which corresponds to the largest vehicle expected to use the roundabout for normal use. However, in extreme cases oversize and/or overweight vehicles might require to also pass through the roundabout which accommodation would lead to costly and probably dangerous for the regular road users' solutions. Wider traffic lanes and little deflection angles favor higher speeds on the one hand but on the other hand vehicles larger than the design one would negotiate the roundabout unhindered. On the contrary abrupt changes to road geometry impel the drivers to reduce speed but hinder the oversize and/or overweight truck drivers to safely pass through the roundabout. The compromise solution with the use of truck apron does not always ensures higher safety levels as pointed out before and thereby, alternative techniques must be sought e.g. opposite direction travel and fully traversable center island [6].

A comprehensive design of a roundabout is primary dependent on two criteria; the design or oversize/overweight vehicles must be able to safely pass the circulatory roadway without encroaching adjacent lanes or any outer edges of the intersection and the faster vehicles must be discouraged from traveling at inappropriate speeds. The latter has a profound effect not only in roundabout but also in road design generally [7]. However, these two criteria interact with each other with opposing results and hence the roundabout designer's goal is twofold; to provide adequate clearance for the design vehicles but also to prevent drivers from traveling at high speeds. These fundamental principles are further divided to specific checks that must be performed in order to evaluate design consistency including checks to the fastest path, path overlap, truck turning movements, stopping sight distances and intersection sight distances. When the requirements are not met then proper changes should be considered e.g. adjustments of the size of the inscribed circle diameter, adjustments of entry radius etc. and this iterative procedure shall be repeated from the start [8].

The purpose of this paper twofold. Firstly, it introduces a methodology to draw the fastest paths of large vehicles at roundabouts and calculate the speeds they can travel with. As mentioned before, heavy vehicles negotiating roundabouts in high speeds may result in serious injuries and hence the knowledge of the path and speed for each type of them individually are of significant importance. Secondly within the framework of the paper specific checks will be conducted at two recently designed and constructed roundabouts in the city of Volos. The checks are based on the methodology developed to draw the fastest path of various types of heavy vehicles and the international literature.

2 Methodology

2.1 Choosing a Design Vehicle

The design vehicle is selected according to the position of the intersection (i.e. urban or intercity, freight/industrial area), to the structure of the traffic flow (the percentage of heavy vehicles) and to the traffic volume. It is important to determine the category and type of vehicles that frequently pass through the designed or evaluated roundabout and examine which of them has the greatest swept path width regarding critical turning movements [9]. A design vehicle is not intended to represent an average or standard vehicle in its category, but the maximum physical dimensions and maximum surface area that the vehicle requires during its maneuvering, compared with other vehicles in its class [10]. In this study, a tractor-semitrailer with total length 16.50 m, trailer length 13.60 m and width 2.55 m, a three-axle bus with total length 15.00 m and width 2.50 m and a passenger car with total length 4.74 m and width 2.00 m were used as design vehicles. The dimension of the tractor semitrailer and the bus are according to the European Committee Directive 2002/7/EC.

2.2 Fastest Path Construction

The fastest path is the smoothest, flattest path that a vehicle can take through the entry, round the central island and through the exit of the roundabout under free flow conditions. In order to control the geometry and the safety, and to limit the speed of the vehicles through the roundabouts, the fastest path must be drawn for all approaches and all movements [11], as shown in Fig. 1(b). In particular, a fastest path is comprised of a series of consecutive arcs that are tangent to each other or have a common tangent section equal to the length of the vehicle (specifically in the case of heavy vehicles) and passes through prescribed distances away from the edges of the roundabouts elements [12]. According to AASHTO [13] (and the Greek standard/guidelines [14]), for the construction of the fastest path a vehicle with 2 m width is used and the minimum safety distance from the roadway centerline or concrete curb and painted edge line is 0.5 m. Consequently, the centerline of the vehicle path should be drawn by taking into account the following restrictions: 1 m away from edge line of the splitter island, 1.5 m away from a right curb on entrance and exit and 1.5 m away from the curb of the central island, as shown in Fig. 1.

The left turn maneuver represent generally the slowest fastest path of all possible whereas the fastest path could be the through movement or the right- turn depending the geometry of roundabout [10, 15]. In order to design the fastest paths in a Double-Lane Roundabout, the line markings are ignored, meaning that the vehicle is able to encroach on the adjacent travel lane.

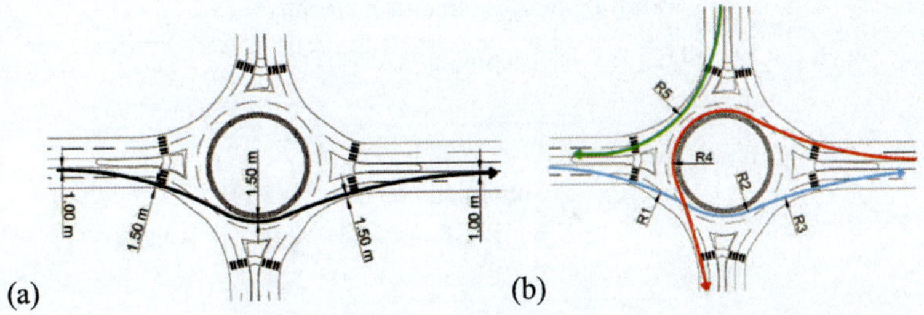

(a) (b)

Fig. 1. Construction of fastest path in Double - Lane Roundabout

2.3 Design Vehicle Swept Path Analysis

The way to evaluate the geometric characteristics of the roundabouts is to draw the vehicles' turning paths, namely the path of the left front overhang and the path of the right rear overhang. In this paper, swept path analysis was conducted with the use of Autodesk Vehicle Tracking software, due to the fact that its simulation results are the most accurate [16]. Swept path analysis ensures that conditions for unobstructed vehicle movements on roundabout entry and exit are achieved.

The safety in Double-lane roundabouts is mainly checked in the case of vchicles moving side by side. An extremely case is, when two buses are required to track side-by-side through a striped roundabout. This may cause partial overlap of the swept paths of the vehicles entering the roundabout from the same approach, but in practice it is not a real problem because larger vehicles such as tractor-semitrailers usually require the whole width of the circular carriageway, hindering any traffic by their side [17], and hence their swept paths are analyzed alone.

2.4 Fastest Path Critical Radii and Design Speed

As mentioned in Sect. 2.2 the fastest path consists of a series of arcs. The evaluation of a roundabout in safety terms requires the check of the radii of these arcs in order to ensure proper entry, circulating and exit speeds, sufficient entry path deflection and avoidance of trucks overturn. Because of that, the value of five critical radii (R_1, R_2, R_3, R_4, R_5) illustrated in Fig. 1(b) must be checked for each leg [12–15].

Based on the value of the radii and the maximum speed that a vehicle is able to travel along the drawn arcs, it must be ensured that the speed difference for any consecutive arcs do not differ by more than 20 km/h according to the Greek guidelines [14]. Table 1 presents the equations in force to calculate the travel speed of the vehicle based on the radius of the arcs. More specifically according to the Dutch and the Croatian guidelines the equation to calculate vehicle's speed is the same for all critical radii, as shown in Eq. (1) [12, 18] whereas AASHTO takes into account the side friction factor f and superelevation e as well as shown in Eq. (2). Greek guidelines and NCHRP [19] adopted AASHTO approach in a simplified form by assuming a fixed value of the superelevation and side friction effect (Eqs. (3) and (4)).

Table 1. Speed - Radius equations

According Dutch guideline [12, 18]		According AASHTO	
Radii	Equations	Radii	Equations
R_1, R_2, R_3, R_4, R_5	$V_i = 7.4 \cdot \sqrt{R_i}$ (1)	$V_i = \sqrt{127 \cdot R_i \cdot (f \pm e)}$ (2)	
		Simplified equation [14, 19]	
		R_1, R_3, R_5 and e = +0.02	$V_i = 8.7602 \cdot R_i^{0.3861}$ (3)
		R_2, R_4 and e = −0.02	$V_i = 8.6164 \cdot R_i^{0.3673}$ (4)

2.5 Procedure to Design the Paths of the Design Vehicles

The purpose of this case study is two-fold. Firstly, to determine a step by step procedure for the construction of the fastest and the turning path process for vehicles others than the one specified in the guidelines and secondly to conduct a safety consistency check. Initially, as described in Sect. 2.1, the appropriate design vehicles are selected depending on the characteristics of the case study area. The fastest path for a vehicle of 2 m width which corresponds to a passenger car is then drawn according to the guidelines and the five critical radii (R_1, R_2, R_3, R_4, R_5) are calculated for each approach, as shown in Fig. 2.

Fastest paths are drawn by hand. The freehand technique can provide a natural representation of the way a driver negotiates the roundabout, with smooth transitions connecting curves and tangents [11], but it requires significant engineering experience and skill [12]. The analytical design process is described below, in conjunction with Fig. 2:

- Step 1: A line parallel to the left approach edge at an offset 1 m and a curved line parallel to the edge of right curb at an offset 1.5 m are drawn.
- Step 2: A circle with radius R_2 at an offset 1.5 m from the curb of the central island are drawn.
- Step 3: For the construction of the entry or exit paths, a circular curve is drawn tangentially to the line and curves of Steps 1 and 2. The fastest path is planning by joining the arcs of circular curves of Steps 2 and 3, which are formed between the common tangent nodes.

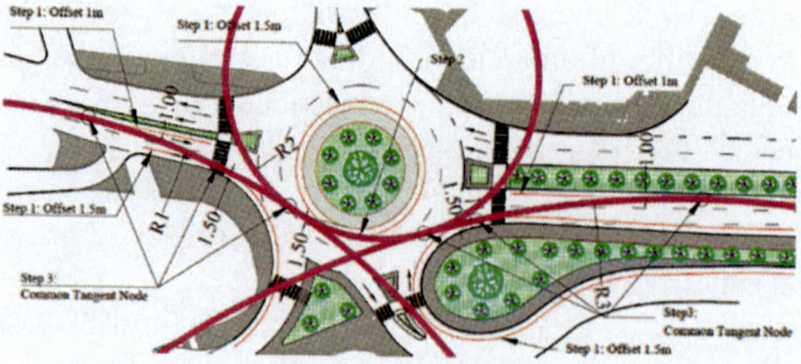

Fig. 2. Construction of Fastest Path for Double-lane Roundabout

It is worth mentioning that since the construction of the fastest path does not take into account the existence of line markings, the process is the same for both the single and the double lane roundabout. This condition is applicable when designing the fastest paths of other design vehicles which derive from the modification of the passenger car's fastest path. However, this procedure is subject to the following limitations: any point of the innermost and outermost path of the design vehicle should keep lateral clearance of 0.50 m from the edge lines of the legs and 1.00 m from both the central island and the inscribed circle of the roundabout [20]. Passenger buses should be accommodated within the circulatory roadway without tracking over the truck apron, which could jostle bus occupants [19]. Having estimated the paths of the design vehicles, the speed is calculated according to the Eqs. (1), (3) and (4).

At the end, the swept path analysis of two design vehicles traveling round the double lane roundabout side by side is performed in order to identify eventual overlaps. For the construction of these turning paths it is assumed that the drivers stay in the correct lane. It should also be noted that the fastest path methodology does not represent expected vehicle speeds, which can vary substantially based on the vehicles construction, the driving abilities and the tolerance for gravitational forces [12].

3 Case Study

3.1 Roundabout Area

In this case study, two Double- Lane Roundabouts in the city of Volos in Greece were used to conduct fastest paths and paths analysis of design vehicles in order to cope with the general problems of path planning. The one roundabout case concerns the area near the city and intercity bus terminal and the other concerns the area near the Volos port citizen customs, as shown in Fig. 3. Both of them constructed in 2018 in order to solve traffic and safety problems of the area. The geometric properties of the two roundabouts are shown in the following Table 2. Both of them serve 5 approach legs whereas their apron width is 3 m.

Table 2. Roundabouts Geometry Properties

Roundabout	Bus terminal	Port customs
Radius Central Island	9.50 m	10.00 m
Inscribed Circle Diameter	45.00 m	46.00 m
Circulatory Roadway Width	10.00 m	10.0 m

3.2 Results and Analysis

Swept Path Analysis

The swept path analysis that was performed for the design vehicles described in Sect. 2.1, showed that on the one hand a car traveling at the left lane of the circulatory roadway and a truck or a bus traveling next to it at the right lane is feasible but on the other hand the extreme occasion of two trucks (tractor-semitrailer) or two buses circulating side-by-side is not attainable. Nevertheless, the simultaneous use of both traffic lanes from two large vehicles side by side, is not recommended even if none of them encroaches on the adjacent lane for safety reasons [21] although under certain considerations this practice is acceptable [19].

Fastest Paths & Speed Results

The construction of the fastest paths of the two roundabouts according to Sect. 2.5 showed that the exact fastest path is a matter of personal judgement. The paths drawn in Fig. 3 corresponds to the passenger car, the intercity bus and the tractor semi-trailer respectively.

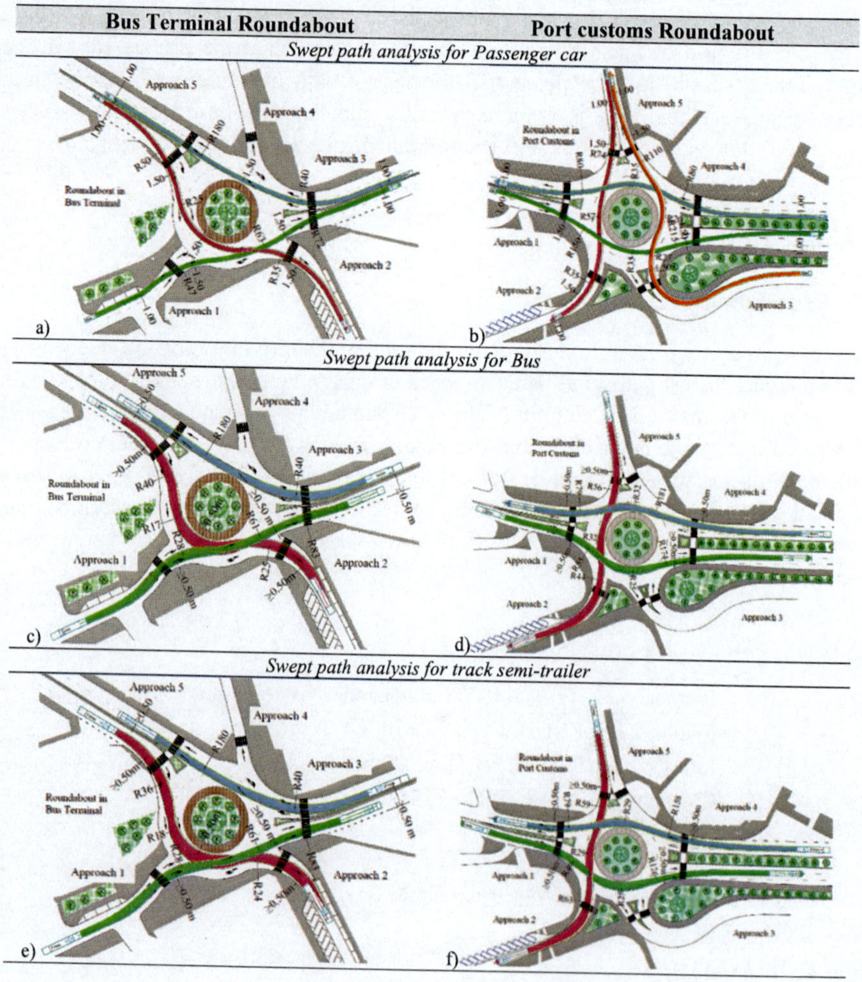

Fig. 3. Design of fastest paths

The speeds of the design vehicles calculated according to the drawn fastest path for the design vehicles (Fig. 3) i.e. when the vehicles pass through the roundabout in free flow conditions without interfering or conflict with other vehicles. In order to conduct speed tests, the smallest radius *R* of each fastest path is determined and along with the Eqs. (1–4) a critical speed is calculated. This speed has to conform with the overall design of the roundabout otherwise speed reduction measures should be considered [12]. The results are encompassed in Table 3.

Table 3. Comparison of the calculated speeds at the roundabouts

	R (m)	V (km/h) *(eq. 3/4)*	V (km/h) *(eq. 1)*	R (m)	V (km/h) *(eq. 3/4)*	V (km/h) *(eq. 1)*	R (m)	V (km/h) *(eq. 3/4)*	V (km/h) *(eq. 1)*
Direction	1–3			3–5			5–2		
Bus Terminal Roundabout									
Passenger Car	47	38.74	50.73	40	36.40	46.80	25	28.11	37.00
Bus	28	31.71	39.16	40	36.40	46.80	17	24.39	30.51
Truck Semi-trailer	28	31.71	39.16	40	36.40	46.80	18	24.91	31.40
Port Customs Roundabout									
Passenger Car	35	31.80	43.78	35	31.80	43.78	57	38.04	55.87
Bus	25	28.11	37.00	32	30.77	41.86	32	30.77	41.86
Truck Semi - trailer	25	28.11	37.00	29	29.68	39.85	29	29.68	39.85

In the Approach 1 of Bus Terminal Roundabout, the design technique of shifting the approach alignment further toward to the left of the roundabout center was used in order to enhance the entry deflection technique. This type of entry encourages balanced lane use, but at the same time it contributes to the overturn while it restricts the offtracking of large vehicles. Hence, bus's and tractor semi-trailer's fastest path of Approach 1 is significantly thicker compared to the passenger car's (green paths) and hence R_1 in the former case is less than in the latter case (28 m and 47 m respectively).

Furthermore, the type of the entrance in the Approach 3 of Port Customs Roundabout is inappropriate for the maneuvering of large vehicles because of the existence of the Splitter Island. Consequently, the geometry of the approach legs has a significant impact in the fastest paths and swept path analysis of design vehicles.

The large angle between Approach 3 and Approach 5 in Bus Terminal roundabout shown in Figs. 3(a), (c), (d) enables the drivers to travel faster (cyan path). Therefore, in this specific occasion the fastest path is the one for the right-turn movement instead of the through movement.

According to Table 3 in Port Customs Roundabout, high speeds are likely to be observed due to the large radius of the approach curve and speed discrepancies between consecutive arcs is not always less than 20 km/h. In addition, the estimation of the speed using Eq. (1) is 20–32% higher than the estimation of the speed with the Eq. (3) and (4).

4 Conclusions

Based on the literature review and the preceding analysis, the design of fastest path is based on the judgment of the engineer and on the geometry of the roundabout. The geometry of any approach is a factor that influences the entry speed, the fastest path and the swept path analysis of each design vehicles, as results from the Approach 1 of Bus Terminal Roundabout and the Approach 3 of Port Customs Roundabout.

Swept path analysis in roundabouts of this case study shows that, two trucks or two buses could not track side by side. Furthermore, the swept paths of three-axle bus with length of 15.00 m have almost the same (and sometimes greater) offtracking with the semi-trailer. Therefore, it requires larger space while turning and hence, its use as a design vehicle especially in urban roundabouts where bus traffic is expected, is highly recommended. Lastly, the estimation of speed according Dutch and Croatian guideline equations gives 20–32% greater results than according Greek and US guideline equations.

The implementation of certain measures and techniques might address any limitations to the movement of the large vehicles at the roundabouts due to geometric design inconsistencies e.g. Opposite Direction Travel, Fully Traversable Center Island. Future research oriented to such measures can confirm whether they are feasible or not.

References

1. Campbell, D., Jurisich, I., Dunn, R.: Improved Multi-lane Roundabout Designs for Cyclists. Land Transport New Zealand, Wellington (2006)
2. Owen, A.: Speed control at roundabouts – use of maximum entry path radii. In: 23rd ARRB Conference – Research Partnering with Practitioners, Adelaide Australia (2008)
3. Mussone, L., Matteucci, M., Bassani, M., Rizzi, D.: Traffic Analysis in Roundabout Intersections by Image Processing, Milano (2011)
4. Pecchini, D., Roncella, R., Forlani, G., Giuliani, F.: Measuring driving workload of heavy vehicles at roundabouts. Transp. Res. Part F **45**(2017), 27–42 (2016)
5. Fancher, P.S., Segel, L., Winkler, C.B., Ervin, R.D.: Tracking and stability of multi-unit truck combinations. The University of Michigan Transportation Research Institute (1984)
6. Godavarthy, R.P., Russell, E., Landman, D.: Using vehicle simulations to understand strategies for accommodating oversize, overweight vehicles at roundabouts. Transp. Res. Part A (2016). http://dx.doi.org/10.1016/j.tra.2016.03.002
7. Federal Highway Administration: Roundabouts: An international guide. FHWA, Virginia (2000)
8. Tennessee Department of Transportation: TDOT Roadway Design Guidelines, Chapter 4 - Guidelines for Design of Roundabouts (2015)
9. Džambas, T., Ahac, S., Dragčevic, V.: Design of turbo roundabouts based on the rules of vehicle movement geometry. J. Transp. Eng. (2016). https://doi.org/10.1061/(ASCE)TE.1943-5436
10. Harwood, D.W., Torbic, D.J., Richard, K.R., Glauz, W.D., Elefteriadou, L.: NCHRP-REPORT 505: Review of Truck Characteristics as Factors in Roadway Design. Transportation Research Board, Washington, DC (2003)

11. Montella, A., Turner, S., Chiaradonna, S., Aldridge, D.: Proposals for improvement of the Italian roundabout geometric design standard. In: SIIV - 5th International Congress - Sustainability of Road Infrastructures (2012)
12. Ahac, S., Džambas, T., Dragčević, V.: Review of fastest path procedures for singlelane. In: 4th International Conference on Road and Rail Infrastructure, Croatia (2016)
13. AASHTO: A Policy on Geometric Design of Highways and Streets. American Association of State Highway and Transportation Officials, Washington D. C. (2011)
14. Ministry of Infrastructure, Transport and Networks (MITN): Road Work Studies Instructions, Volume 10, Part2: Roundabouts (RWSI), Athens (2011)
15. Montella, A., Chiaradonna, S., Turner, S., Aldridge, D.: Proposals for improvement of the Italian roundabout geometric design standard. In: SIIV - 5th International Congress - Sustainability of Road Infrastructures, pp. 189–202 (2012)
16. Kaliabetsos, G., Lemonakis, P., Gkoutzini, A., Eliou, N.: Parametric investigation of vehicle minimum turning path. In: 9th International Congress on Transportation Research, Athens (2019)
17. Brilon, W.: Roundabouts: A State of the Art in Germany, Seattle (2014)
18. Surdonja, S., Dragcevic, V., Deluka-Tibljas, A.: Analyses of maximum-speed path definition at single-lane roundabouts. J. Traffic Transp. Eng. (Engl. Ed.) **5**(2), 83–95 (2018)
19. Rodegerdts, L., Bansen, J., Tiesler, C., Knudsen, J., Myers, E., Johnson, M., Moule, M., Persaud, B., Lyon, C., Hallmark, S., Isebrands, H., Crown, R.B., Guichet, B., O'Brien, A.: NCHRP Report 672: Roundabouts: An Informational Guide, 2nd edn. Transportation Research Board, Washington, D.C (2010)
20. Bezina, S., Stanceric, I., Ahac, C.: Design vehicles and roundabout safety- review of Croatian design guidelines. In: Transport Infrastructure and Systems, pp. 237–244. Taylor & Francis Group, London (2017)
21. Waddell, E., Gingrich, M.A., Lenters, M.: Trucks in roundabouts: pitfalls in design and operations. ITE J. **79**(2), 40–45 (2009)

Digitalization and Data Sharing

Covid-19 Transport Analytics: Analysis of Rome Mobility During Coronavirus Pandemic Era

Stefano Brinchi[1], Stefano Carrese[4], Ernesto Cipriani[4], Chiara Colombaroni[2], Umberto Crisalli[3](✉), Gaetano Fusco[2], Andrea Gemma[4], Natalia Isaenko[2], Livia Mannini[4], and Marco Petrelli[4]

[1] Rome Mobility Agency (Roma Servizi per la Mobilità), Via L. Gaurico 9, 00143 Rome, Italy

[2] DICEA, Sapienza University of Rome, Via Eudossiana 18, 00184 Rome, Italy

[3] Department of Enterprise Engineering, Tor Vergata University of Rome, Via del Politecnico 1, 00133 Rome, Italy
crisalli@ing.uniroma2.it

[4] Department of Engineering, Roma Tre University, Via Vito Volterra 62, 00146 Rome, Italy

Abstract. Dealing with emergencies is never an easy task. Coping with the Covid-19 emergency was and still is an incredibly tough and unprecedent challenge for mankind, not only from a healthy point of view, but also from a socio-economic point of view, where a key role will be played by a reliable and sustainable transport of people and goods. Professional and researchers from all disciplines have been required to adapt their models, methods and tools to support decision makers in coronavirus emergency days. This paper reports the experience of the authors for the analysis of the Rome mobility system to estimate the impacts of the progressive lockdown imposed by the government. It focuses on data available for monitoring, which have been used to carry out transport analytics both for private and public transport. Such analytics have been used for planning the reboot of the transport system to be ready for the post Covid-19 era.

Keywords: Covid-19 · Transport analytics · Transport modelling · Big data

1 Introduction

The dramatic Covid-19 epidemic induces the extraordinary reduction of people mobility as the unique way to prevent infection and disease of many other people as well as to avoid an exponential spread of the virus, when diagnosis and care solutions are not available yet.

The complete shutdown of non-essential activities, personal isolation and social distancing are extreme "lockdown" measures that impact considerably on social and economic patterns of involved communities with catastrophic effects similar to wars.

E. G. Nathanail et al. (Eds.): CSUM2020, AISC 1278, pp. 1045–1055, 2021.
https://doi.org/10.1007/978-3-030-61075-3_100

In the lockdown phase, monitoring people movements is fundamental to prevent the spread of the virus; however, it is equally important to analyze the mobility patterns that have been deeply modified in the spatial and temporal distribution, in the choice of the transport mode as well as in the route choice behavior.

From a system of unstructured activities because of the lockdown, researchers are asked to recompose the 'puzzle' of the activities, by including transport mobility ones, and to recalibrate their models and tools to support the design and management of transport systems to be ready for the post-Covid era that, at first, will certainly include the respect of the limits imposed by healthy requirements and technical constraints of the transport system.

This paper reports some of the results obtained by a working group, which includes the authors, aiming at supporting the Rome Mobility Agency [1] to:

- analyze the Rome mobility system and estimate the impacts of the progressive lockdown imposed by the government;
- plan the reboot of the mobility system by calibrating and applying transport simulation models to design, for example, public transport services able to ensure passengers social distancing required in the first post-Covid months.

In particular, this paper presents the analysis of the Rome mobility system, as the usage of this analysis to plan the reboot of the transport system is still in progress.

At this time, the major data providers are supplying reports that illustrate data analytics on people's mobility in most Countries, regions and even towns in the world [2–5] but they do not present the higher resolution level required to analyze the mobility within cities. For this reason, this paper presents an analysis of mobility data at the traffic zoning level (more than 1300 traffic zones) adopted in the modelling system of Rome Mobility Agency, aiming at exploring lockdown trip mobility patterns among traffic zones of the metropolitan area of Rome. Such a time-spatial representation enables the application of decision support systems to predict the impacts of future actions on the post-Covid mobility.

The paper is structured as follows: Sect. 2 describes the study area and its transport data monitoring system; Sect. 3 presents the transport analytics for both road and public networks, focusing on the considered lockdown levels; Sect. 4 summarizes conclusions and future perspectives of this research.

2 The Metropolitan Area of Rome

2.1 The Study Area

The urban area of the Municipality of Rome extends over about 1,285 km^2 and is characterized by about 2.9 million inhabitants; they represent approximately half population of the whole Lazio region (5.6 million), concentrated in just 8% of the regional surface. Similar population densities can be observed in London and Paris.

Adding the metropolitan area to the city (Municipality of Rome plus its province), the population grows up to more 4.6 million of inhabitants in about 5,352 km^2. During the morning rush hour of a typical workday, such population generates a total of 743 thousand trips of which, more than 50% (about 550,000) are within the city. It is worth-noting that while the use of private vehicles is higher in the most peripheral areas (72% of the overall demand), the opposite happens for public transport, which presents a sporadic use in the outermost areas (25–26% of modal choices), which increases in the most central area (about the 32%).

2.2 Transport Networks and Services

Over the about 8,000 km of the existing roads, the main road network extends to about 400 km where 50% of the traffic passes and the 30% of accidents occur; the largest part of such a network (300 km) is within the GRA, a circular freeway approximately 68 km long which separates the city from its suburbs.

As for the transit system, a radial metro network extends for a total of about 60 km, currently with two interchanges in the city centre (Termini rail station and San Giovanni metro station). Additional seven rail lines connect the surrounding suburban areas to the city centre: five of these rail lines create a half circle inside the GRA known as the "rail ring". The urban mass system is integrated with a bus network providing an extensive and widespread coverage of the service, and it is mainly composed by about 200 low-medium frequency lines.

2.3 The Transport Data Monitoring System

The Control Mobility Centre of the Rome Mobility Agency performs the tasks of monitoring, managing and controlling urban traffic through its various subsystems. Such tasks are mainly oriented to the preparation of databases to feed both data analysis and traffic simulation models as well as to provide reliable mobility information services.

The Agency uses various types of sensors both traditional and innovative, which can be summarized as follows: electromagnetic loops; gates of limited traffic zones; Floating Car Data (FCD), i.e. GPS positions and speeds of cars and heavy vehicles; Bluetooth and WiFi data; turnstiles data; AVL (Automatic Vehicle Location) data of public transport vehicles; traffic video cameras; integrated accident recording system; crowdsourced traffic reports.

Such technologies feed the OpenData webservices available on the Agency website [1], where also data related to the public transport are treated.

The most relevant data reported above have been used to estimate the impacts of Covid-19 on the mobility of Metropolitan Area of Rome.

The results of the analysis of such data produced the Transport Analytics presented in the following Sect. 3 and weekly updated in [6].

3 Transport Analytics

3.1 Rome Lockdown at a Glance

The progressive lockdown imposed by the government authorities (national, regional and local) is briefly described in Table 1. The reader should note that this table reports only information impacting on the Rome mobility system, which the following analysis focused on.

Table 1. Covid-19 lockdown timeline in Rome.

Level	Date	Shutdown description
L0	23th February 2020	Strict quarantine (red zones) imposed in towns of northern regions. Travelling to the north regions is not recommended
L1	4th March 2020	schools are closed while universities open for research purposes only. Social distancing in crowded places is recommended
L2	11th March 2020	Bars, restaurants and public food courts are closed. Refraining from unnecessary travel is recommended. Social distancing is mandatory. Working remotely is highly suggested, except for unavoidable services. Limited Traffic Zones are disabled (all traffic allowed in the city center)
L3	13th March 2020	All public parks and gardens are closed. No outdoor sport activities allowed. Residents can walk their dogs only nearby home. Stay home is imposed except for unavoidable needs
L4	22th March 2020	All "non-essential" shops and activities are closed. Food and drink retailers, banks, chemists, petrol stations, newsagents, as well as other particular categories defined "essential" to support coronavirus emergency are exempted from the ban. Residents are still allowed to shop for food, to go to work (if their job belongs to "essential" categories and it is impossible to do remotely) or to move for unavoidable needs (e.g. support to elderly who are not self-sufficient)

The imposed Italian-style prohibitions on everyone leaving their homes can be seen in terms of the percentage variation of the mobility players (workers and students) at the different lockdown milestones (L0, L1, L2, L3, L4) w.r.t. the pre-Covid scenario, which influence both the trip generation and the trip distribution according to the imposed limitations (see Fig. 1). The pre-Covid scenario, below assumed as the reference (R) one, is set to the average mobility levels and features measured in the workdays of the first two weeks of February 2020. Then, the impact of the different lockdown milestones can be summarized as follows: in L0 all intercity trips (mainly involving national railway stations and Rome airports) are importantly downsized; in L1 study trips miss; in L2 retail and recreational trips switched off; in L3 the mobility for sports and leisure is forbidden; in L4 only unavoidable trips are allowed.

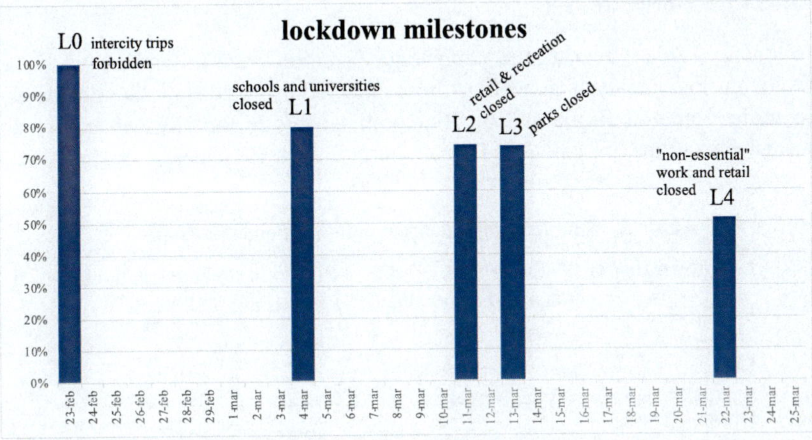

Fig. 1. Covid-19 lockdown milestones and reduction of mobility players.

The effects of such restrictions in people activities can also be seen by analyzing at a glance the people presence in specific POIs (Point of Interest) captured by smartphone data and published by Google [2] as pictured in Fig. 2. Such figure illustrates mobility trends for the categories: retail & recreation (i.e. restaurants, cafes, shopping centers, theme parks, museums, libraries, and movie theaters), grocery & pharmacy (i.e. grocery markets, food warehouses, farmers markets, specialty food shops, drug stores, and pharmacies); parks (i.e. national parks, public beaches, marinas, dog parks, plazas, and public gardens); transit stations (i.e. public transport hubs such as subway, bus, and train stations); workplaces and residential. Figure 2 outlines the percentage reductions of activities according to the different lockdown milestones as well as the increase percentage of people staying at home due to the imposed prohibitions.

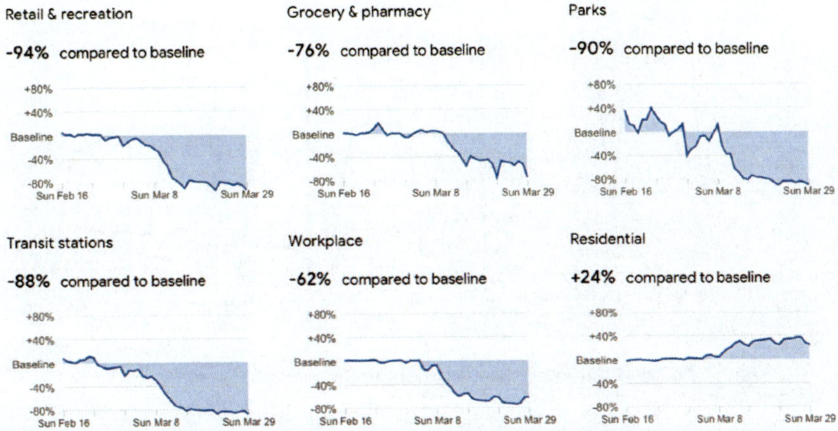

Fig. 2. Example of lockdown effects on personal activities at regional level. (Lazio region, Google [2]).

3.2 Road Network Analysis

The following Table 2 reports the percentage variations of cars and heavy vehicles resulting from FCD analysis as well as those detected by fixed stations at gates of the limited traffic zones of Rome and by Bluetooth sensors in the city centre for pedestrians. Such variations are calculated w.r.t. the reference R (see Sect. 3.1).

Table 2. Summary of lockdown mobility trends in Rome.

Lockdown (level)	FCD (cars)	FCD (heavy vehicles)	Bluetooth (pedestrians)	Fixed stations (all vehicles)
L0	−4%	−4%	−11%	−1%
L1	−16%	−1%	−28%	−1%
L2	−68%	−19%	−78%	−58%
L3	−84%	−56%	−87%	−84%
L4	−80%	−38%	−93%	−76%

By focusing on FCD also pictured in Fig. 3, we can observe that differences are nearly negligible when quarantine has been imposed in northern towns of Italy (L0). The same can still be detected for heavy traffic in L1 milestone; differently, light traffic is affected by the shutdown of educational institutions, as expected. Both categories are characterized by substantial reductions with the adoption of social distancing and remote working measures (L2), although heavy traffic reduces less because it represents the backbone of freight and logistic supply. L3 (parks closure) large reductions are not insightful for vehicles, since this milestone was imposed to reduce recreational gatherings during weekends, so it is mainly oriented to pedestrians.

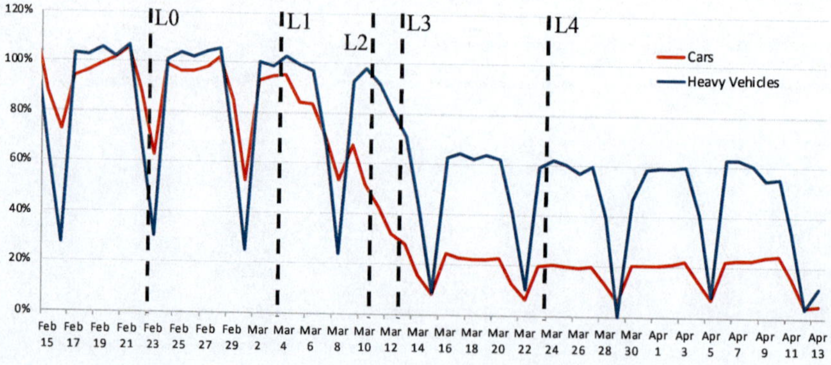

Fig. 3. Private cars and Heavy Vehicles traffic evolution (FCD data).

The adoption of L4 restrictive measures, banning any non-essential activity, implies biggest decreases (with the exception of L3) with respect to the reference for both traffic categories.

The sprawl of the FCD sample in the study area is described in Fig. 4, which shows the variation in terms of trip generations and attractions per traffic zone of the study area. The comparison between L4 (right side) and L0 (left side) allows to point out both variations and spatial distribution of the traffic generated and attracted by still-open "essential" activities.

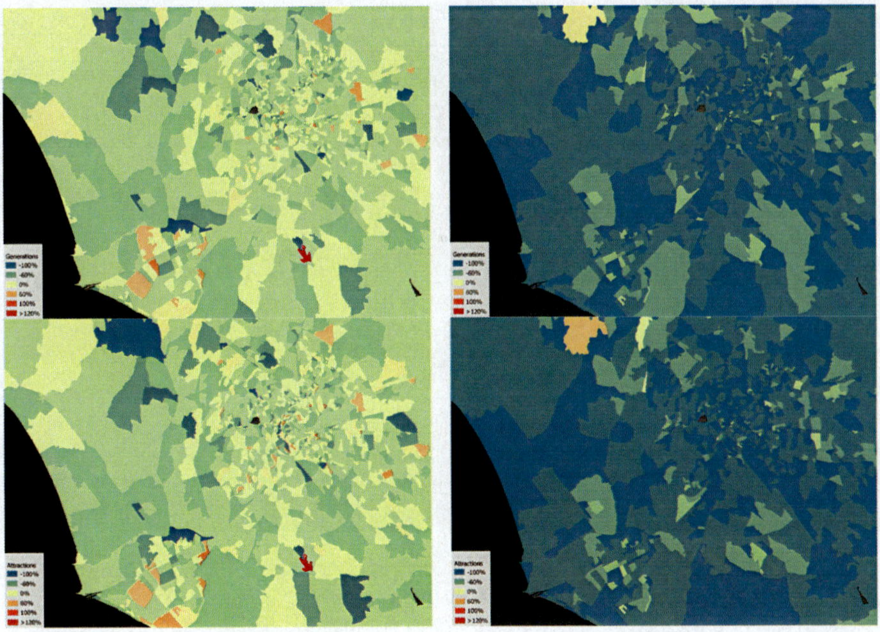

(L0) (L4)

Fig. 4. Effects of lockdown on trip generations and attractions (FCD data).

As FCD represents a sample of the whole traffic, absolute values of traffic flows can be detected by fixed stations located at strategic points of the road network. Then, Fig. 5 shows the percentage reduction of traffic volumes per vehicle type (Motorcycles, Cars, Light and Heavy Vehicles) detected by video cameras located around the city center, whose total reduction (all vehicles) is reported in Table 2. The reader should note that the absence of traffic variation up to L1 and the difference (about 10%) w.r.t. FCD can be explained by the positions of the fixed stations which mainly monitor the city centre, while FCD also includes the suburbs. Moreover, Fig. 5 shows the decrease for different vehicle classes at different lockdown milestones. In particular, Cars and Light Vehicles exhibit a very similar trend, with no significant decline until L1 and progressive reductions as 31% on L2, 36% on L3, up to 62% on average in the workdays after L4. Motorcycle traffic shows a similar trend until L3 and a wider reduction of about 18% after L4. A very different trend characterizes Heavy Vehicles, which are not affected by the restrictions addressed to people but are indirectly due to the reduced activities of shops and offices. Indeed, heavy traffic reduces only as 73% on L2, 62% on L3, and about 50% on L4 and later on.

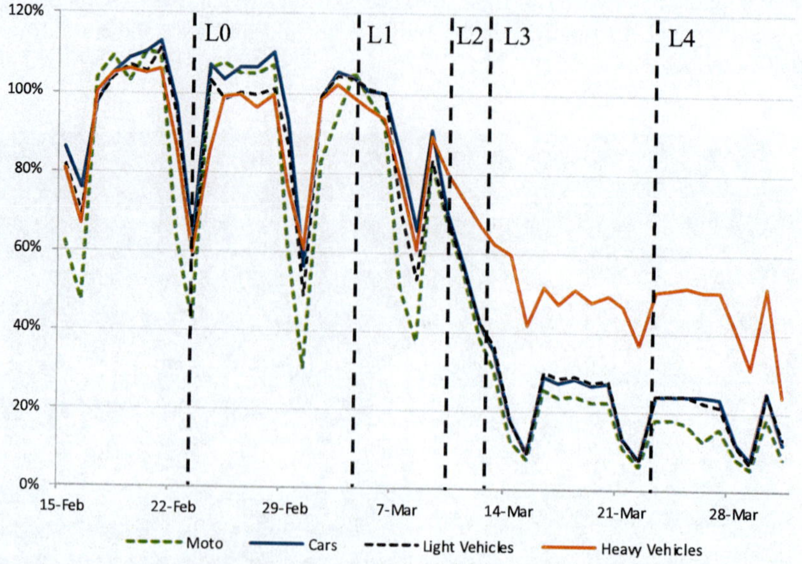

Fig. 5. Evolution of vehicle traffic detected by fixed stations.

By deepening on pedestrian traffic, Fig. 6 describes results of the Bluetooth data analysis, which shows a similar trend of variation as in the previous analyses. Specifically, reductions are noteworthy even in L0 milestone (11%) since it occurs in a workday. Then, increasing reductions are detected both in L1 (28%) and in L2 (78%) milestones. L3 ban affects specifically pedestrians as shown by the large reduction of the 87%, which is extremely wider if compared with the value of the same day of the previous week (falling in L1) when the reduction was only the 10%. The adoption of the most restrictive measures of L4 reflects in the maximum reduction of pedestrian movements (93%).

3.3 Public Transport Analysis

Passenger trends on the three metro lines (A, B, and C) and their extensions (B1 and Lido), again compared to the average values observed in the working days of the first two weeks of February, is reported in Fig. 7. The lines that cross the central area of the town and connect opposite semi-peripheral areas (A, B, and B1) show very similar trends and have the highest reductions of passengers with respect to the reference R: 22% on L2; 10% on L3, and 5% on average in the workdays after L4. Less wide reductions are observed on the C and Lido lines that connect the City center with the

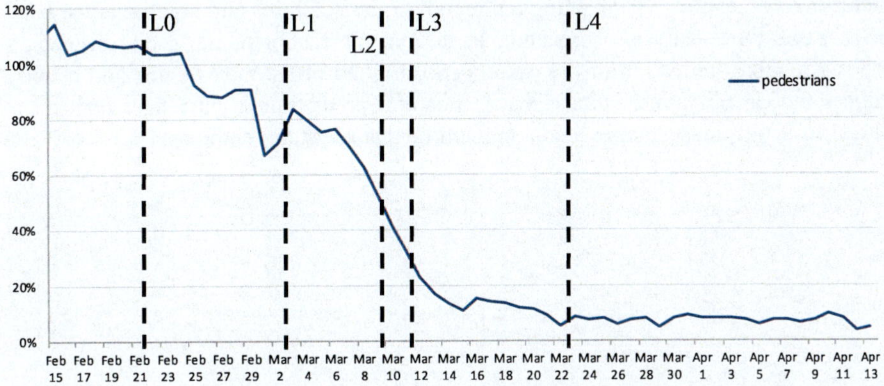

Fig. 6. Pedestrian traffic evolution (Bluetooth data).

most external areas of the metropolitan area: 37%, 21%, and 12% on C line and 27%, 14%, and 7% on Lido line, respectively on L2, L3, L4 and later on.

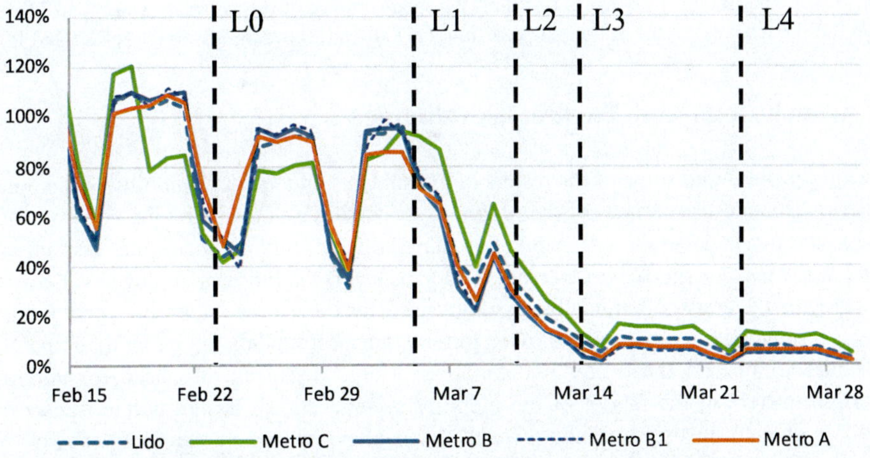

Fig. 7. Evolution of the number of daily passengers in the metro lines in Rome.

Such different trends for the two kinds of metro lines highlight different impacts of lockdown measures on different town areas. Such differences reflect also on the time distribution of trips originated from different zones. A particular analysis is detailed in Fig. 8, which compares the number of passengers entered the stations of A line in the morning and in the afternoon on March 10 (when L1 was imposed) and March 13 (when L3 was imposed). While the reduction of passengers in the morning peak hour is uniformly distributed in the two days at all stations, a lower correspondence holds for afternoon trips. Also, a stronger difference is observed for Spagna station, which is located in a luxury shop area in the historic city center. This remark provides interesting suggestions for planning the reboot of the activities and highlights the need for a spatial-time analysis of demand. In fact, in the reboot phase, social distancing imposes a strong reduction of the passengers admitted in the metro trains and heavily reduces the capacity of the line. Hence, a careful prediction of passenger demand is crucial to avoid unacceptable loads onboard or queues at the entrance.

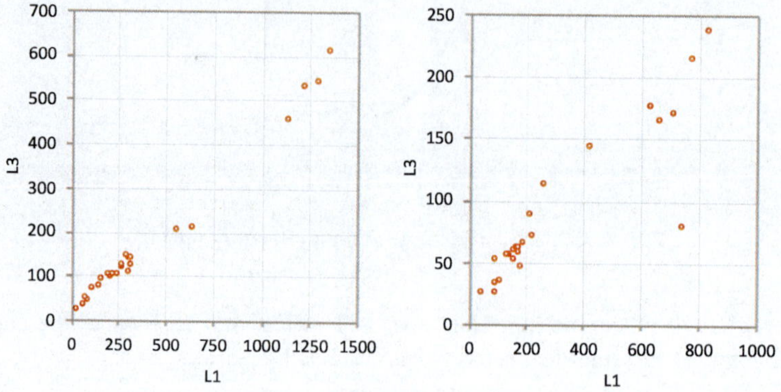

Fig. 8. Comparison of passengers boarding the metro line A in the morning peak hour 7:00–8:00 (on the left) and in the afternoon peak hour (17:00–18:00) during a workday of L1 and L3

4 Conclusions and Further Developments

This paper presented some of the results of the analysis on the Rome mobility system in order to estimate the impacts of the progressive lockdown imposed by the government due to the Covid-19 epidemic. Aiming at analyzing the problem at a higher resolution level w.r.t. usual large-scale (e.g. region or city level) analytics, a zoning level compliant with the transport modelling system used by the Rome Mobility Agency has been introduced. Such analyses allowed to feed transport models to predict the impacts of future actions on the post-Covid mobility. They have been also used to support decision makers in the reboot of the mobility system, e.g. to design public transport services able to ensure passengers social distancing required in the first post-Covid months. The application of such models is in progress; it represents the main future development of this research.

References

1. Rome Mobility Agency (Roma Servizi per la Mobilità), homepage. https://www.romamobilita.it. Accessed 18 Apr 2020
2. Apple, Mobility Trends Reports. https://www.apple.com/covid19/mobility. Accessed 17 Apr 2020
3. Google, COVID-19 Community Mobility Reports. https://www.google.com/covid19/mobility. Accessed 17 Apr 2020

4. Moovit, Impact of Coronavirus (COVID-19) on Public Transit usage. https://moovitapp.com/insights/en/Moovit_Insights_Public_Transit_Index-countries. Accessed 17 Apr 2020
5. TomTom, COVID-19 Update. https://www.tomtom.com/covid-19. Accessed 17 Apr 2020
6. Rome Mobility Agency (Roma Servizi per la Mobilità), Covid 19: impatto sulla mobilità. https://romamobilita.it/it/covid-19-impatto-sulla-mobilita. Accessed 18 Apr 2020

Digitalization in Freight Transport Services: Balkan Area

Attila Akac[1(✉)], Afroditi Anagnostopoulou[1], and Dimitrios Nalmpantis[2]

[1] Hellenic Institute of Transport, Centre for Research and Technology Hellas, 1, G. Kasimati street, 18531 Piraeus, Greece
{akac.attila,a.anagnostopoulou}@certh.gr

[2] School of Civil Engineering, Faculty of Engineering, Aristotle University of Thessaloniki, PO Box 452, 541 24 Thessaloniki, Greece
dnalba@civil.auth.gr

Abstract. New information technologies and international digitalization trends have led to the shifting of the norm in recent freight transport systems. Electronic marketplaces and digital freight transport platforms have emerged since the early 2000s' as a result of the advances in information technology, plus Business to Business (B2B) and Business to Customer (B2C) hubs have developed new tools to facilitate product and information exchange and support the negotiation, contracting, and settlement processes. In particular, freight logistics platforms have been largely investigated and operated by numerous members of the global freight industry, especially in North America, Australia, and Europe, focusing on load and truck booking/auctioning, load matching algorithms, and freight planning services. This paper proposes a categorization and analysis scheme which (i) presents the latest state of practice in freight transport platforms at a global level and (ii) focuses on their offered services in an attempt to categorize these platforms based on the applied services. In addition, the freight exchange platforms adopted in the Balkan region are analyzed following the proposed scheme. A mapping process is achieved, and critical issues are identified in developing a transnational freight logistics platform for the Balkan area.

Keywords: Transport digitalization · Freight logistics services · State of practice in freight transport platforms · Groupage

1 Introduction

Freight transport is a major global market estimated to reach 17.45 billion USD by 2023 [1], which could be increased by modern technology, e.g., the Internet of Things (IoT) and digitalization solutions. Electronic marketplaces and digital freight transport platforms have emerged since the early 2000s as a result of the advances in information

E. G. Nathanail et al. (Eds.): CSUM2020, AISC 1278, pp. 1056–1065, 2021.
https://doi.org/10.1007/978-3-030-61075-3_101

technology, plus Business to Business (B2B) and Business to Customer (B2C) hubs have developed new tools to facilitate product and information exchange and support the negotiation, contracting, and settlement processes. These tools could focus on optimizing cost efficiency, enhancing a sustainable mobility environment in EU, and delivering Cooperative Intelligent Freight Transport Systems (C-ITS) [2].

Specific emphasis on this applied research paper has been given to the road freight transport sector that involves two main stakeholder groups (user profiles): (i) the industries interested to transfer cargo or utilities (referred as clients or auctioneers) and, (ii) the providers of carriage services towards the clients/auctioneers (referred as carriers). Customized digital tools have been developed based on the needs of both categories of users.

In particular, freight logistics platforms have been largely investigated and operated by numerous members of the global freight industry, especially in North America, Australia, and Europe, focusing on load and truck booking/auctioning, load matching algorithms, and freight planning services. Although, in this applied research study an emphasis has been given to specific countries of the Balkan region (Albania, Bulgaria, and Greece) due to their high share of road freight transport activity (more than 60% of national modal split for freight transport), their recent national strategic key challenges and potential cooperative freight transport activities [2, 3].

Specifically, in Albania and Bulgaria a targeted national strategy has been set to address two significant issues: (i) enhancing private-public partnerships in transport sectors to increase multinational activities in the freight logistics sector and, (ii) modernize current C-ITS national infrastructure by promoting the establishment of new tools to reduce logistics costs in both national and transnational levels [2, 3]. These issues affect their low achievement in the Logistics Performance Index (LPI) score (2.6 for Albania and 3.0 for Bulgaria with a maximum of 5.0) compared to the average EU level. Greece has different key challenge areas focusing on rail freight transport infrastructure, and this is the main reason for Greece's underachieving performance in the EU logistics sector (3.2 LPI score compared to EU average level of 3.6 LPI score) [2, 4, 5]. Thus, there is a high necessity to upgrade existing tracking/tracing tools and customs policy with new digital applications. As such, the aforementioned countries need innovative mitigation measures in order to assure a competitive freight transport market compared to the standards of average EU level freight transport markets. Taking this into consideration, these countries are further investigated in terms of their potential increase in their freight transport activities.

This paper proposes a categorization and analysis scheme which: (i) presents the latest state of practice in freight transport platforms at a global level and (ii) focuses on their offered services in an attempt to categorize these platforms based on the applied services. In addition, the freight exchange platforms adopted in the Balkan region (specifically in Greece, Bulgaria, and Albania) are analyzed following the proposed scheme. As such, a mapping process is achieved, and critical issues are identified to be considered in developing a transnational freight logistics platform for the Balkan area.

The remainder of the paper is organized as follows: Sect. 2 presents a summary of the existing freight exchange platforms for both categories of users at global and EU level. These platforms are categorized according to the type of services they offer (i.e., freight management tools, information repositories or booking/auctioning platforms) and they are described in Sect. 3. Following, Sect. 4 presents the digital freight transport services in the Balkan area, and in Sect. 5, the respective results are summarized. Finally, in Sect. 6 conclusions are drawn and recommendations for future research are provided.

2 Freight Exchange Platforms at Global and EU Level

Freight transport industries have started to digitalize their services between the early 2000s with the rational initialization of web-based software applications in executional activities of key industry players. Especially in North America (USA & Canada), supply chain companies became early adopters of online freight exchange initiatives making this a common procedure. Nowadays, freight exchange platforms give access to user profiles through multiple platforms (desktop, smartphones, and tablets) with special emphasis on the development of mobile smartphone applications. More specifically, a global state-of-practice market research analysis has been made during this applied research study in order to identify the strong global markets in freight transportation and cluster these applications in a group of proposed categories according to their offered services. This analysis gives an overview of the freight exchange platforms allocation on key markets (continent and country location) along with the category of the services provided. At this point, an important disclaimer should be referred: all the platforms recorded in the following analysis have an English display interface of their services. Thus, there is a high chance that relevant freight exchange platforms offered in common languages, such as Spanish, Chinese, French, or others, may not be included in this analysis.

Overall, the majority of the freight exchange platforms examined are from the EU region and the Northern American countries (USA & Canada). As can be seen in Fig. 1, in total **65 freight exchange platforms** that are offered in English have been identified. The Asian platforms offered in English were based in India and offered transnational freight transport services. Focusing on the European area, it can be observed that the key players on the freight transport digital industry are United Kingdom (UK) and Germany (see Fig. 2), which offer their services in a pan-European level as their location is vital for the current Trans-European Transport (TEN-T) network.

In the same context, all freight exchange platforms recorded in this study have been grouped into a set of categories. The proposed categories have been created according to the type of provided services. These groups will be presented in the following section.

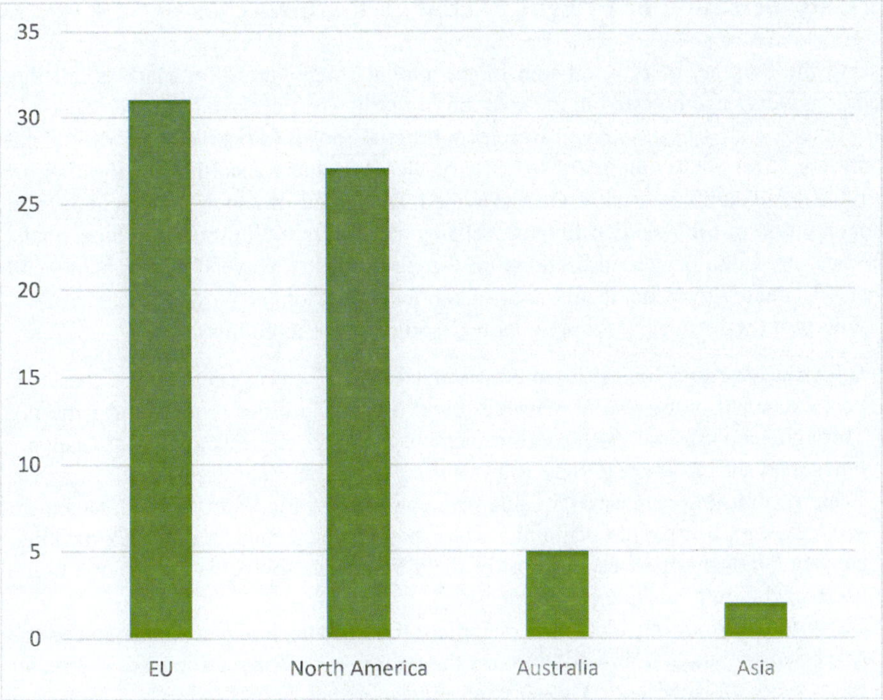

Fig. 1. Freight exchange platforms distribution across continents.

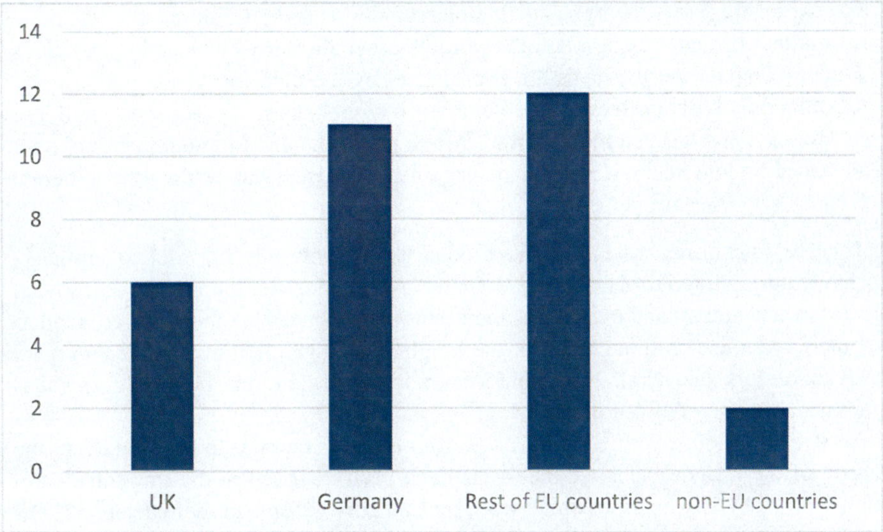

Fig. 2. Freight exchange platforms distribution across the EU region.

3 Categorization of Freight Exchange Platforms

During the continuous development of the global freight transport markets, all key market players have invested in the creation of unique services and digital products. In this context, a significant amount of relevant digital tools has been created with slight alterations. Thus, it is important to classify these products according to the type of individual and combinatorial services provided. During this research study, a high-level categorization of the recorded freight exchange platforms (65 in total) has been made. The outcomes could offer indications on the key industry players' requirements and potential clients' goals on an EU and Balkan level later on.

The proposed categories with a short description are as follows:

1. **Informative repositories (forums/channels):** The platform operates as a message/post exchange online portal between the two users' profiles (clients and carriers). Through such a portal, both user profiles publish their requests, questions, and any other relevant information they want to share with the other users.
2. **Freight planning services:** The platform contains a variety of freight management services, e.g., truck route planning, cargo space management, cargo/truck tracking, etc. The platform operates as a host of the different platforms that are offered to the users in different packages and prices.
3. **Cargo/truck booking platform:** The platform operates in the same context as the first category but, additionally, it offers the option of booking a truck (for clients) or cargo types (for carriers). The booking process is handled via the platform owner and the respective partners, via the platform itself or on an external source.
4. **Auction platform:** The platform operates in the same way as in category 3 but with an important difference: auctions replace booking reservations between the users. The platform follows the same auctioning regulations and procedures as in the other sectors that use auctioning platforms. In this case, all transactions and reservations are confidential, and the deals are coordinated by the platform owner.
5. **Advanced freight services:** This category is a combination of categories 2 and 3 or categories 2 and 4. In short, freight exchange platforms that fit into this group, offer advanced freight reservation and management services that could offer different privileges to their users/members.

The aforementioned categories have been used as clusters in order to groupage effectively the acquired data (see Fig. 3). In the last decade, smartphones became a vital asset for every human, and their use is increasing rapidly against other devices, such as computers or tablets. Although one of the freight exchange platforms' users group are truck/van drivers (carriers), the evolution of these freight transport platforms has limited smartphone applications (see Fig. 4).

As it can be observed, only ¼ of platforms have mobile apps. Moreover, the majority of the freight exchange platforms have been assigned in the first three categories, with an evenly distribution between the category groups. In more detail, 32% of the current web-based platforms offer just information for the user groups and 38% of the platforms offer either freight management services or booking reservation products.

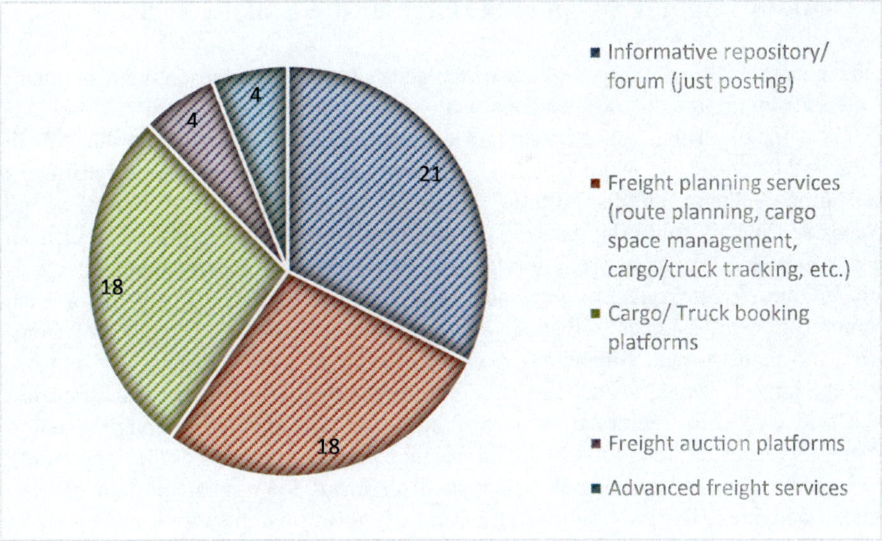

Fig. 3. Results of freight exchange platforms categorization.

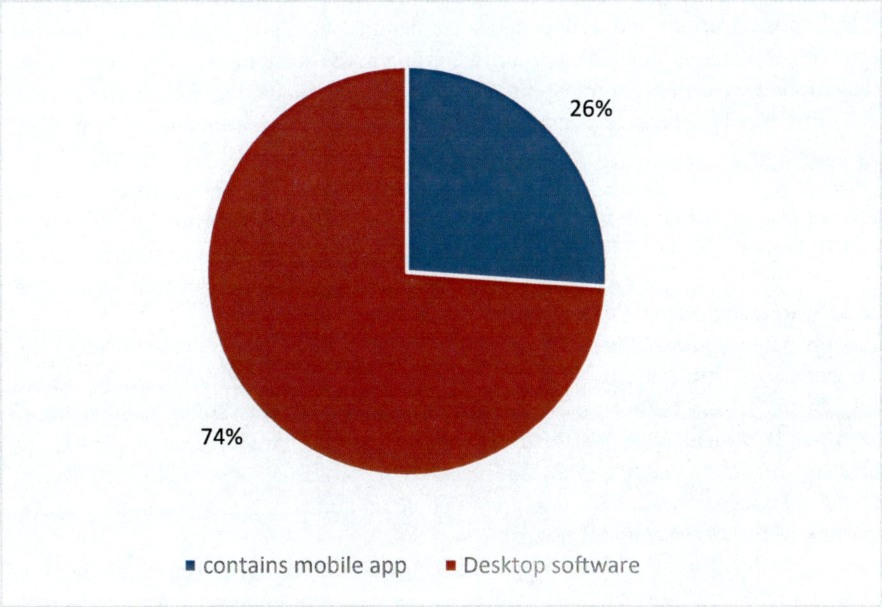

Fig. 4. Freight exchange platforms that include mobile apps.

4 Digital Freight Transport Services in the Balkan Area

In this chapter, following an overview of the pertinent literature presenting a summary of the existing freight market systems at the global level presented in the previous section, a focus on the respective freight exchange platforms adopted in the Balkan region will be presented. More specifically, the investigated freight exchange platforms cover three Balkan countries: Greece, Bulgaria, and Albania. These countries were investigated in this study due to the availability of industry stakeholders' input from these countries. The input from forwarders (carriers) and transport companies was gathered during interview sessions held by their representative chambers of each country (i.e., Professionals' Chamber of Thessaloniki, Bulgarian Chamber of Commerce and Industry, and Albanian Chamber of Commerce and Industry).

Regarding forwarders and transport companies in each country, the transport industry mainly uses traditional ways for freight transport services, although recently, the majority of users started to use digital freight transport platforms or they are willing to become early adopters of newly launched platforms for a small portion of their logistics activities. The most significant example in this research study was the case of Albania.

Currently, carriers and clients in Albania prefer two leading online platforms (i.e., "TRANSMOGUL" and "URSUSforwarding") mostly for their online information/communication tool and use their booking options for international cargo shipping. The majority of transactions and arrangements are handled with conventional methods (via corporate networking, phone deals, etc.) according to the stakeholders. On the contrary, Bulgarian and Greek freight transport user groups have pursued an increased usage of trustworthy pan-European and national freight transport service initiatives with slight differentiations.

Bulgarian freight industry players have exploited major pan-European freight exchange platforms that are based outside of Bulgaria. These platforms (i.e., "Speditor.NET," "Trans.eu," and "TIMOCOM") offer different freight transportation services along with cargo/truck booking services, which are used for both national and international cargo shipments.

In the same context, Greek freight transport industry stakeholders use platforms with worldwide applications (e.g., TIMOCOM) and platforms produced by Greek entrepreneurs (e.g., "Nestcargo," "Freights," etc.). As the forwarders and clients in Greece can be considered as early adopters (based on the interview outcomes) like the respective Albanian partners, they are eager to approach newly established freight exchange services provided by young entrepreneurs and adjust their operational tasks according to the current digital trends.

In depth, the aforementioned freight exchange platforms have been highlighted by the interviewed key industry stakeholders as the most commonly used tools in their respective countries. Specifically, these platforms are briefly described below:

- *"TRANSMOGUL" platform:* The TRANSMOGUL platform [6] is a new initiative that has been recently initialized (within 2019) in Albania to provide simple and efficient communication between forwarders and transport companies. It aims to ensure full capacity usage of transport providers and enhance the efficiency of payment procedures. It is offered both as an online platform and as a mobile application for easy access.

- *"URSUSforwarding" platform:* The URSUSforwarding platform [7] is also used in the Albanian area mainly for international services and by large freight forwarder companies. It acts as a mediator company for transport providers and has access to more than 150 cargo transport units. URSUSforwarding offers economical and custom-tailored transport solutions that enhance flexibility and high quality of service for shippers.
- *"Speditor.NET" platform:* The Speditor.NET platform [8] has been developed in Bulgaria and is a separate operation platform. It provides a large number of services both at the state level, national level, and throughout the EU. Speditor.NET is the most popular platform amongst Bulgarian freight forwarders and transport operators [8].
- *"Trans.eu" road transport platform:* The Trans.eu platform [9] is a transport exchange platform with modern tools facilitating transport management. Over 6,000 shippers offer cargo, 25,000 carriers create community, and 9,000 freight forwarders actively use the platform. The platform is relatively new, but it quickly gains users among the Bulgarian transport and forwarding companies.
- *"TIMOCOM" platform:* The TIMOCOM platform [10] was developed by Jens Thierman in Germany, and it has high activity in Bulgaria and Greece. As a freight forwarder, he was searching for a pan-European procurement platform for transport jobs, but he could not find what he was looking for; so, he decided to build it himself. The result was the TIMOCOM Soft- und Hardware GmbH. The company began developing its own freight and vehicle exchange. TIMOCOM GmbH is a mid-sized Information Technology (IT) and data specialist, with more than 500 employees from over 30 countries, and representative offices in Poland, the Czech Republic, and Hungary. As the provider of Europe's first Smart Logistics System, they are expanding the logistics industry using smart, safe, and simple solutions to meet the daily challenges faced by their customers, who represent a neutral network of over 43,000 verified companies.
- *"Nestcargo" platform:* The Nestcargo platform [11] was founded in September 2013 by Truckbird Private Company as a cargo marketplace that brings in touch shippers with carriers. Truckbird PC's vision is to create a pan-European cargo marketplace that reduces "empty routes" in road, sea, and air transport and also contributes towards the elimination of transport's carbon footprint. Therefore, it developed a business tool (i.e., Nestcargo platform) that creates value for the international transport industry, while aiming to create new jobs. With Nestcargo, companies can find trustworthy transport shipping companies to ship their goods. Whether it is road, sea, or air cargo transport, Nestcargo's global network of carriers may help every company to find the right carrier at the right price to transfer their goods.

- *"Freights" platform:* The Freights exchange platform [12] was established by Freights SA (a Thessaloniki based startup focusing on freight loads and trucks' exchange) and is engaged in electronic management for transportation and inter-transport services, offering immediate and innovative solutions in all kind of business needs. Freights is developed to be an innovative freight exchange platform that combines freight forwarders with transporters to accomplish reliable freight and truck exchanges.

5 Results

In this applied research study, an extensive review on the recent freight exchange platforms has been made in a global level, in the EU area and, more specifically, in three countries of the Balkan region (Albania, Bulgaria, and Greece). Overall, sixty-five **(65) freight transport platforms were identified** at the global level and **seven (7)** of them **are further examined in the participating countries**. As can be observed, Albanian and Greek freight industry players have a preference for national initiatives rather than the respective global ones. Additionally, only Albanian freight partners focused their digital freight exchange actions at the national level rather than the other two examined countries, which use freight exchange platforms for national and transnational purposes. Based on these outcomes, the potential issues of the limited use of freight exchange platforms in the Balkan area have been recorded during the interview process and are presented below.

In total, carriers and clients have pointed out that they are eager to use freight exchange platforms on a daily basis for national and transnational routes. However, both freight user groups mentioned that in order for these initiatives to be successful, they have to be implemented from both user groups and relevant national authorities due to the legal aspects of cargo shipping procedures. Ceteris paribus, carriers have indicated several internal and external barriers that discourage them from exploiting digital freight transport initiatives such as: (i) limited hardware and software infrastructure for the carrier drivers; (ii) the fact that the majority of carrier drivers are mature (40+ years old), thus they need specific training programs to learn how to use these digital tools properly; and (iii) that the majority of these digital freight exchange tools are not available in mobile devices with a user-friendly interface. Therefore, these issues have to be tackled during future freight exchange platforms development.

6 Conclusions

To sum up, an extensive review on the recent freight exchange platforms has been conducted in a global level, in the EU area and more specifically, in three countries of the Balkan region (Albania, Bulgaria, and Greece). The information has been acquired from interviews with the representative Chambers of the key freight transport stakeholders. The first step of the proposed scheme includes the identification of successful freight transport platforms at a global level. Then, the most promising of them are further examined in the participating countries, and they were categorized into five (5) main groups according to their offered services and digital tools (such as service plurality, operability in mobile devices, etc.).

This mapping sets the basis for the interviews from key national freight transport stakeholders in order to identify critical issues that should be considered for the creation of a new innovative freight exchange platform with a strategically targeted holistic approach which aims to enhance the sustainability of users by minimizing transportation costs and groupage (Less than Container Loads).

Last but not least, the proposed approach could be used to investigate and categorize the relevant freight transport initiatives for the rest of the Balkan area (i.e., Serbia, Romania, North Macedonia, etc.) and a comparative analysis among all Balkan region countries could be achieved in an attempt to better identify additional obstacles that discourage freight transport stakeholders from adopting digital tools and services.

Acknowledgments. This research has been co-funded by the European Union and national funds of the participating countries through the Interreg Balkan – Mediterranean program "**SCOPE - Shared freight transport services connecting shipper and carrier operations**" (MIS: 5048545).

References

1. Markets and Markets Webpage: Freight Management System Market by Solution (Freight Tracking & Monitoring, Cargo Routing & Scheduling, Security, EDI, TMS, Order Management), End-user (3PLs, Forwarders, Brokers, Shippers), Transportation Mode, and Region – Global Forecast to 2023. https://www.marketsandmarkets.com/Market-Reports/freight-management-system-market-214631371.html. Accessed 15 May 2020
2. European Commission: Transport in the European Union: Current Trends and Issues. European Commission, Directorate-General Mobility and Transport, Brussels (2019). https://ec.europa.eu/transport/sites/transport/files/2019-transport-in-the-eu-current-trends-and-issues.pdf
3. Organisation for Economic Competitiveness in South East Europe: A Policy Outlook 2018, Competitiveness and Private Sector Development. OECD Library, Paris (2018). http://doi.org/10.1787/9789264298576-en
4. World Bank Webpage: Logistic Performance Index Dataset (database). http://lpi.worldbank.org. Accessed 23 Mar 2020
5. World Bank Webpage: World Development Indicators (database). http://databank.worldbank.org/data/reports.aspx?source=world-developmentindicators. Accessed 23 Mar 2020
6. TRANSMOGUL Homepage. https://www.f6s.com/transmogul. Accessed 23 Mar 2020
7. URSUSForwarding Webpage: Freight transport to/from Albania. https://www.ursusforwarding.com/eiropa/albania.html. Accessed 23 Mar 2020
8. Speditor.NET Homepage. http://www.freightmail.com. Accessed 23 Mar 2020
9. Trans.eu Homepage. https://www.trans.eu. Accessed 23 Mar 2020
10. TIMOCOM Homepage. https://www.TIMOCOM.bg/ or https://www.TIMOCOM.gr. Accessed 23 Mar 2020
11. Nestcargo Homepage. http://nestcargo.com/el/. Accessed 23 Mar 2020
12. Freights Homepage. https://www.freights.online/platform. Accessed 23 Mar 2020

Check for updates

Benchmarking Analysis of Road Safety Levels for an Extensive and Representative Dataset of European Cities

Katerina Folla[1], Paraskevas Nikolaou[2], Loukas Dimitriou[2](✉), and George Yannis[1]

[1] National Technical University of Athens, Iroon Polytechniou 9, 157 73 Zografou, Athens, Greece

[2] Laboratory for Transport Engineering, Department of Civil and Environmental Engineering, University of Cyprus, 75 Kallipoleos Street, P.O. Box 20537, 1678 Nicosia, Cyprus

lucdimit@ucy.ac.cy

Abstract. According to European urban sustainable planning guidelines, road safety corresponds to one of the most important elements of cities' performance. Several methods have been developed over the years for supporting policy-making, towards the improvement of road safety levels mainly at the national level. In this study, a methodological framework is developed, extending the macro-level (national) analysis and focusing at a higher spatial resolution, that of urban regions. The methodological approach is based on benchmarking analysis, able to suitably rank alternative cases/regions with distinctive characteristics within a multivariate comparative framework and on the investigation of the components that affect their ranking. In particular, an extensive and representative dataset from 101 European regions is collected and analyzed, incorporating their socio-economic, demographic and road infrastructure characteristics. Then, Data Envelopment Analysis (suitably adapted to road fatalities framework) has been developed, evaluating the urban regions' road safety performance over a period of 9 years. The resulted region ranking is further examined by using Tobit regression models for identifying the components that appear to affect their performance in different extents providing a valuable guiding 'tool' for experience/knowledge-transfer and policy-making. The datasets and the results are presented and discussed in detail, such as they will be useful not only for demonstration purposes but also they will be suitable as a benchmark for researchers and practitioners.

Keywords: Benchmarking analysis · Sustainable urban mobility · Road traffic fatalities · Stochastic frontier · Data Envelopment Analysis

1 Introduction

Road traffic injuries constitute a major public health problem that requires concerted efforts for effective prevention. Worldwide, it is estimated that 1.35 million people are killed in road accidents annually, with road traffic injuries being the leading cause of

E. G. Nathanail et al. (Eds.): CSUM2020, AISC 1278, pp. 1066–1075, 2021.
https://doi.org/10.1007/978-3-030-61075-3_102

death for children and young people aged between 5 and 29 years old [1]. The European Commission (EC) has committed to improve the safety of the European road network. On that purpose, EC has adopted a Road Safety Programme which aims to halve the number of road deaths by 2020, compared to the 2010 level. This target followed an earlier target set in 2001 to cut road fatalities by 50% compared to 2001, which was almost achieved [2].

In the European Union (EU), in 2018, there were around 25,100 fatalities in road accidents in the EU-28, recording a decrease of 21% compared to 2010 [3]. With an average of 49 road fatalities per one million inhabitants, Europe is the safest continent in the world, however, EU seems to be far from reaching the target of halving the number of road deaths by 2020 [1, 3]. Additionally, despite the significant progress in road safety improvement in the European Union, there are still significant variations between the European countries in terms of road safety performance, which may be attributed to the different socioeconomic environments, cultures and behavior characteristics, modal shares, etc.

At the regional level, the disparities may be even larger in the EU. An analysis has shown that despite the constant decrease of road fatalities in the EU since 1991, significant disparities are obvious among the regions of the EU, highlighting the different conditions affecting fatality rates in road transport at regional level [4].

Within this context, more and more countries are taking action to improve their road safety situation, while there is also an urgent need not only for the countries but also for the regions to work together more closely, to identify the common problems and improve their road safety performance by learning lessons from each other. Consequently, international benchmarking of the road safety performance of various regions could be a useful tool for national policymakers as well as international organizations working on road safety.

The objective of the current study is to evaluate the road safety performance of EU urban regions, taking into account the evolution of road safety, transport and economic characteristics over the decade 2008–2016. On that purpose, a representative sample has been created with data for 101 regions (at NUTS 2 level) from 13 European countries over the same period. Data concerning the vehicle fleet, road infrastructure and economic conditions of these regions were selected mainly from the EU CARE database and Eurostat. In this study, a methodological framework is developed, extending the macro-level (regional) analysis and focusing at a higher spatial resolution, that of urban regions. More specifically, suitably adapted to road fatalities' Data Envelopment Analysis was developed, evaluating the regions' road safety performance and identifying the best and under-performing regions. Then the road safety performance (efficiency score) of the region was concerned for measuring the effect that the different explanatory variables have on it and in this way shedding light on the factors that particular focus must be given for improving the road safety performance of the EU regions.

The remaining of this paper is organized as follows. Section 2 presents the results of previous work related to the current study. Section 3 describes the dataset used for the current analysis. Section 4 describes the methodology followed for the purposes of the current paper. Section 5 presents the results of the applied methodology. The paper ends concluding remarks in Sect. 6.

2 Literature Review

This section reviews the existing studies conducted on the assessment and benchmarking of road safety performance using various methodological techniques.

In order to comparatively evaluate the road safety performance of various countries or regions, risk indicators defined as the ratio of road safety outcomes (mainly fatalities) and measures of exposure (e.g. vehicle-kilometers traveled, population, registered vehicles, etc.) are often used. However, in the vast majority of the cases, the benchmarking results of a group of countries or regions are not consistent, with the ranking position of a country or a region being different depending on the risk indicator used. Within this context, numerous road safety indicators have been developed for international or interregional comparisons and monitoring of road safety progress. For instance, *Al-Haji* [5] proposed a road safety development index (RSDI) based on three domains of road safety, the outcomes (fatality rates), the user (road user behavior) and the system (vehicles, roads, socioeconomic conditions, enforcement and organization level). In the SUNflower next study [6], three different types of performance indicators were distinguished, i.e. road safety performance indicators, implementation performance indicators and policy performance indicators which were integrated into an overall road safety index. *Yannis et al.* [7] developed an SPI for the evaluation of the safety level of the road network to be used as a benchmark for cross-region comparisons.

Additionally, Data Envelopment Analysis (DEA) has been used in various studies, as a performance measurement technique. *Shen et al.* [8] developed a model in order to evaluate the road safety performance of the 27 EU countries and assess whether the road safety outcomes registered in a country correspond to their level of exposure. *Shen et al.* [9] developed a Malmquist productivity index (MPI) in order to evaluate the road safety performance of the EU countries in the period 2001–2010 taking also into account efficiency and technical changes over this period. In 2018, *Nikolaou & Dimitriou* [10], analyzed the road safety performance of EU-23 countries over a decade (2005–2014) taking into account their socio-economic and demographic background in order to support the decision-making process by using short-term and long-term targets.

At the regional level, there are few studies having implemented DEA for benchmarking road safety performance, all referring to regions within the borders of a country. More specifically, *Alper et al.* [11] estimated the relative efficiency of 197 local municipalities in traffic safety in Israel during 2004–2009, using DEA. The inputs reflected the resources allocated to the local municipalities and outputs concerned road accidents, while safety performance indicators were also used as intermediate variables. Moreover, the mortality rate and fatality rate were aggregated into a composite indicator through a multiple-layer DEA composite indicator model, aiming to identify the optimum combination of indicators' weights for 27 Brazilian states [12]. Finally, a double frontier DEA cross efficiency method was used to evaluate the road safety performance of 31 Iranian provinces based on safety management data and the number of fatalities in 2016 [13].

For analyzing and identifying the effect of different components on the regions' road safety performance a suitable regression model was developed, namely, Tobit.

The combination of the DEA method and Tobit regression has also been developed in several researches. For instance, *Tasnim and Afzal* [14] used Tobit regression model for observing what macro factors affect the efficiency of the knowledge spillover theory of entrepreneurship. For identifying the efficiency of 59 countries DEA method was developed. Another study used DEA for measuring the efficiency of 30 university science parks and Tobit regression model for analyzing the impact of possible influential factors [15].

3 Data Description

As shown in Sect. 2, numerous studies are investigating the assessment of road safety performance at the international level by developing indexes/indicators in order to rank the countries based on their road safety performance. This study aims to develop a methodological framework, extending the macro-level (regional) analysis and focusing at a higher spatial resolution, that of urban regions. In order to achieve adequate and meaningful results during comparisons, similar regions have to be included in benchmarking analyses, which have a similar level of development, motorization and a similar type of transport system [5]. On that purpose, an extensive and representative dataset from 101 European regions (NUTS 2 level) was collected and analyzed, incorporating their socio-economic, demographic and road infrastructure characteristics.

In the current study, the EU NUTS 2 subdivisions have been taken into account. "The nomenclature of territorial units for statistics, (NUTS) is a geographical nomenclature subdividing the economic territory of the European Union (EU) into regions at three different levels (NUTS 1, 2 and 3 respectively, moving from larger to smaller territorial units)" [16]. "The NUTS classification is based on the following main principles: (a) the NUTS regulation defines minimum and maximum population thresholds for the size of individual NUTS regions and (b) NUTS favors administrative divisions and thus, if available, administrative structures are used for the different NUTS levels" [4]. The population size of the NUTS 2 regions ranges between 800.000 inhabitants and 3.000.000 inhabitants [4].

The importance of spatial analyses has been demonstrated in the literature, while studies have shown that smaller regions may present significant similarities in road safety outcomes across the EU, without being affected by the presence of national borders [17, 18]. *Eksler* [19] has shown that the mortality ratios of EU NUTS 2 regions show greater variations compared to those estimated at national level, which is attributed to their demographic structure and population dynamics. More specifically, population density has been found to have a significant influence on the number of road accident fatalities, with densely populated regions showing better road safety performance, which could be attributed to lower travelling speeds, modal share variations, more developed infrastructure and differences in vehicle fleet patterns [20].

Thus, data for 101 NUTS 2 regions covering 13 European countries for the period 2008–2016 were collected. Countries for which data were available are depicted in Fig. 1. In detail, Fig. 1 shows all NUTS 2 regions included in the database. A geographical variation of fatality rates is obvious, with higher fatality rates being presented in the Southern and Eastern European countries and the lowest fatality rates in the

Northwestern countries. At a regional level, regions in Bulgaria and Romania have the worst road safety performance in terms of fatality rates (more than 100 fatalities per million population), while regions in Sweden, Austria, and Germany present the best performance (less than 25 fatalities per population). Accident fatality rates, however, vary at a large scale among the regions within the borders of a country, indicating the need for investigating road safety problems at a regional level.

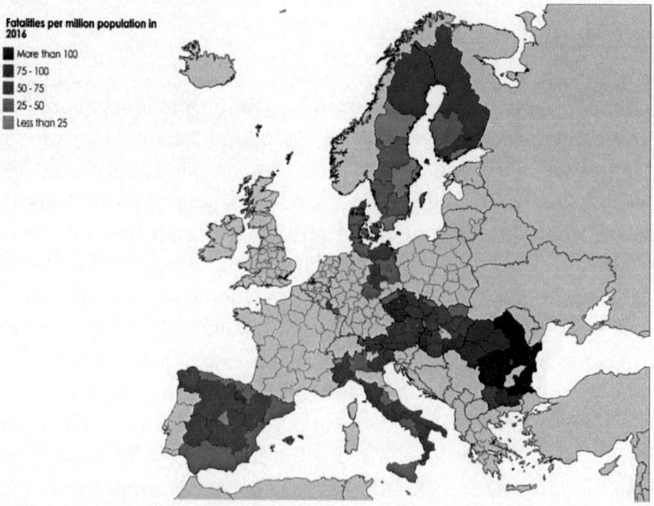

Fig. 1. Fatalities per million population by NUTS 2 region in the EU, 2017.

Therefore, time-series fatality data were retrieved from the CARE database, the EU database with disaggregate road accident data. Additionally, data on population and the Gross Domestic Product (GDP) for these regions were collected from Eurostat. Finally, data related to the vehicle fleet (in total and by type of vehicle) and to the available transport infrastructure by NUTS 2 region were collected from Eurostat. Transport infrastructure related data concern the length of the motorway network as well as of the remaining road network.

4 Methodology

This study develops two procedure concepts of benchmarking analysis (DEA) together with Tobit regression. The concept of DEA was conducted for measuring the efficiency of the regions' road safety performance considering their socio-economic, demographic and road infrastructure context.

As adapted from *Charnes et al.* [21] output-oriented DEA measures the efficiency scores for different Decision Making Units (DMUs) based on the concept of maximizing the product (i.e., output/s) keeping constant the input data. However, the concept developed in this study is that road fatalities (output) must be minimized and not maximizing as the original formation of the DEA method. Therefore, the suitable adapted to road safety DEA method is presented below:

$$\begin{aligned} & min \sum_{i=1}^{p} w_i y_{i,s} \\ & \text{Subject to} \sum_{i=1}^{p} w_i y_{i,s} \geq 1, s = 1, \ldots, n \\ & w_i \geq 0, i = 1, \ldots, p \end{aligned} \quad (1)$$

where, $y_{i,s}$ is the ith indicator of the sth DMU (region), w_i is the weight attributed to indicator $y_{i,s}$, n is the total number of DMUs (i.e., 101) and p is the total number of indicators. The best-performing regions (in terms of road safety) will be presented with an efficiency equal to one and the under-performing regions (in the same terms) will be presented with efficiency scores below to one.

The concept of Tobit was employed for analyzing what determinants affect the efficiency of the region's road safety performance. The reason for developing Tobit and not a classic regression model was due to the capability of Tobit of analyzing censored data and thus Tobit is also known as the censored regression model. The mathematical formation of Tobit model is based on *Tobin* [22] work and can be seen below:

$$\begin{aligned} y_t &= X_t\beta + u_t \; if \; X_t\beta + u_t > 0 \\ &= 0X_t + u_t \; if \; X_t\beta + u_t \leq 0 \\ & t = 1, 2, \ldots, N. \end{aligned} \quad (2)$$

where, N is the number of observations (i.e., 101), y_t is the dependent variable (efficiency scores), X_t is a vector of independent variables, β is a vector of unknown coefficients, and u_t is an independently distributed error term assumed to be normal with zero mean and constant variance σ^2.

Therefore this study followed the above methodological framework for identifying the road safety performance of the 101 EU regions in each time instant and analyzed the measured efficiencies for estimating the effect that each component has on them. The next section presents all the results from the DEA-Tobit analysis.

5 Results

This section presents the results form the developed methodological framework. The first implementation was the development of the DEA method which was suitable adapted to the road safety framework. In particular, road fatalities were concerned as output in DEA and the socio-economic, demographic and road infrastructure characteristics of the regions were concerned as inputs. The goal of this DEA, as mentioned above, was the minimization of road fatalities taking as constant the input characteristics. Figure 2 presents the outcome of this implementation. As can be seen 11 out of the 101 regions appeared to best-perform out of the years and thus policymakers should take a closer look at what they are doing in terms of road safety in order to adopt strategies and policies that are helping them to create and maintain their road safety performance.

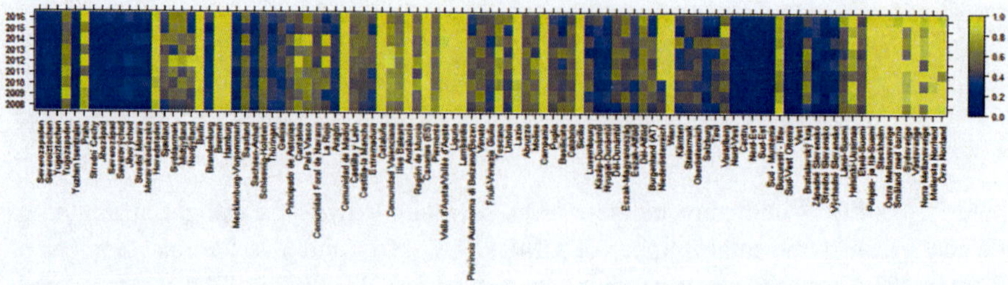

Fig. 2. Road safety performance of the 101 EU regions throughout the years (2008–2016).

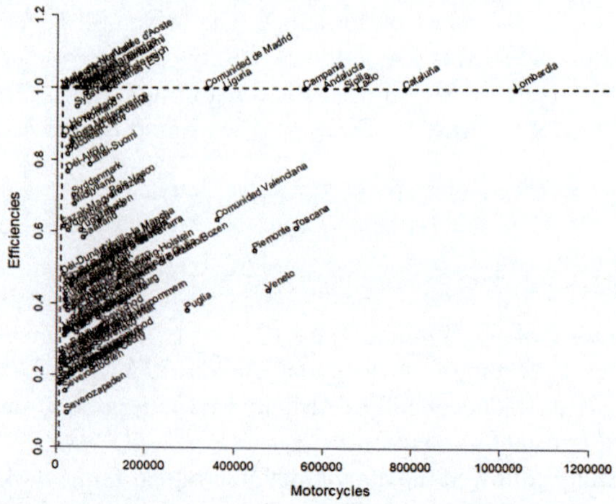

Fig. 3. DEA frontier of the regions based on motorcycles.

Overall from the DEA analysis, we were able to identify under and best-performing regions in terms of road safety. This identification is very important to local authorities to observe the overall "picture" of their road safety performance throughout the years and adopt the strategies that best-performing regions (benchmarks) are following. But the question raised here is: Which counties are considered as benchmarks?

In order to answer this question, we must clarify that for every different explanatory variable to road fatalities create a different effect on their road safety performance and thus for the different socio-economic, demographic and road infrastructure characteristics we have different benchmark regions. For instance, Fig. 3 presents the DEA frontier concerning the explanatory variable "Motorcycles". Taking examples from this figure, an under-performing region should focus on the benchmarking regions (e.g. Catluna, Lazio, etc.), which are best-performing, and adopt the strategies that might they follow especially on the context of motorcycles (laws, prohibitions, etc.).

As was discussed in the previous section for analyzing the efficiency scores of the regions' road safety performance is important incorporating this information as the dependent variable and measuring the effect that the explanatory variables have on their performance (efficiency). In this case, the Tobit regression model was considered more than appropriate. Table 1 presents the outcome of the Tobit model for each time instant.

Table 1. Results from Tobit regression models.

Variables/Year	2008	2009	2010	2011	2012	2013	2014	2015	2016
Intercept	2.537e-01*** (5.05e-02)	2.75e-01*** (4.94e-02)	3.52e-01*** (4.93e-02)	2.38e-01*** (4.83e-02)	3.49e-01*** (5.13e-02)	2.39e-01*** (5.03e-02)	3.25e-01*** (4.97e-02)	1.36e-01** (5.23e-02)	2.11e-01*** (4.66e-02)
Population	-1.490e-07*** (3.65e-08)	-1.53e-07*** (3.62e-08)	-1.53e-07*** (3.67e-08)	-1.36e-07*** (3.67e-08)	-1.88e-07*** (4.49e-08)	-2.06e-07*** (3.95e-08)	-1.92e-07*** (4.59e-08)	-1.65e-07*** (4.86e-08)	-1.56e-07*** (4.57e-08)
Vehicles	-1.866e-06* (9.44e-07)	-2.078e-06* (9.23e-07)	-2.32e-06* (9.23e-07)	5.84e-07* (2.42e-07)	-	-	-	-	-1.72e-06* (8.51e-07)
Lorries	2.451e-06* (9.94e-07)	2.70e-06** (9.75e-07)	3.17e-06** (9.73e-07)	-	1.19e-06*** (2.89e-07)	1.10e-06*** (2.89e-07)	1.14e-06*** (2.87e-07)	1.01e-06*** (2.97e-07)	2.59e-06** (8.93e-07)
Motorcycles	9.095e-07* (3.87e-07)	8.340e-07* (3.69e-07)	8.41e-07* (3.59e-07)	1.19e-06*** (3.43e-07)	8.12e-07* (3.62e-07)	3.98e-07. (2.30e-07)	8.33e-07* (3.44e-07)	9.08e-07** (3.51e-07)	6.42e-07* (3.25e-07)
Passenger Cars	1.790e-06. (10.00e-07)	1.991e-06* (9.79e-07)	2.21e-06* (9.82e-07)	-8.43e-07** (2.88e-07)	-1.99e-07. (1.17e-07)	-	-1.86e-07. (1.13e-07)	-2.50e-07* (1.15e-07)	1.57e-06. (9.00e-07)
Buses	4.52e-05** (1.41e-05)	4.54e-05** (1.38e-05)	3.29e-05* (1.43e-05)	5.369e-05*** (1.418e-05)	3.96e-05** (1.50e-05)	3.977e-05** (1.403e-05)	3.90e-05** (1.44e-05)	5.89e-05*** (1.47e-05)	5.05e-05*** (1.39e-05)
Motorway Density	4.75e-03*** (8.26e-04)	4.53e-03*** (8.09e-04)	4.22e-03*** (7.94e-04)	4.888e-03*** (7.987e-04)	4.49e-03*** (8.56e-04)	4.718e-03*** (8.365e-04)	4.375e-03*** (8.14e-04)	4.79e-03*** (8.47e-04)	4.48e-03*** (7.28e-04)
Other Roads	1.02e-05*** (2.18e-06)	1.22e-05*** (2.13e-06)	7.36e-06*** (2.10e-06)	1.234e-05*** (2.029e-06)	9.34e-06*** (2.11e-06)	1.085e-05*** (2.031e-06)	9.40e-06*** (2.01e-06)	1.42e-05*** (2.10e-06)	1.23e-05*** (1.90e-06)
GDP	3.11e-06** (1.03e-06)	3.59e-06*** (1.05e-06)	3.76e-06*** (9.54e-07)	3.445e-06*** (8.982e-07)	4.14e-06*** (9.48e-07)	3.118e-06*** (8.086e-07)	4.31e-06*** (8.76e-07)	4.31e-06*** (8.77e-07)	4.31e-06*** (7.66e-07)
Log-Lik.	28.21	30.61	31.11	32.08	24.26	26.92	28.15	24.88	32.49
AIC	-34.42	-39.23	-40.22	-44.16	-28.52	-35.83	-36.29	-29.76	-42.98

Note: Parenthesis denotes the standard error of the variables
-: denotes the non-statistically variables that were omitted from the model

As it appeared, regions with a high population record more fatal accidents and therefore their efficiency in road safety performance is dropping. Additionally, the other explanatory variables that seem to have a negative relation to the regions' efficiency is the number of vehicles, except from the year 2011 where the particular variable appeared to have a positive relationship with regions' efficiency and the passengers cars which is negatively correlated with efficiency in 2011, 2012, 2014 and 2015. The other variables appeared with a positive relationship with the regions' road safety performance. Special focus should also be given to the explanatory variable "GDP" which shows that regions with high GDP have also a good efficiency fact which means that they probably spent money on improving their road safety performance. In overall, the results obtained from Tobit regression models can be used from policymakers for supporting their work on focusing to particular factors (explanatory variables) that seem to affect the regions' road safety performance in either way (positively or negatively) not only for improving the performance of under-performing regions but also to maintain the good performance of best-performing regions.

Comparing the models based on the fit index Akaike Information Criterion (AIC) it seems that the Tobit model that has the best fit is the Tobit model of 2011 which we should pay close attention to for interpretation.

6 Conclusions

The road safety performance of regions and countries play a major role, especially in the European Health and Safety agenda. For decreasing or even eliminating the number of roads fatalities are important to identify the factors that either decrease or increase this number and find a "cure" to this "illness". However, before this step is essential to identify which regions/countries best and worst performed throughout the years.

This study has developed a straight forward methodological framework for analyzing and identifying the road safety performance of 101 EU regions (NUTS2) over a 9 years' period taking into consideration their socio-economic, demographic and road infrastructure context.

In detail, a benchmarking analysis, namely DEA, was first implemented for ranking and identifying best-performing and under-performing regions. DEA analysis was suitable adapted to road safety's framework. From this implementation, we are able to assist the underperforming regions by suggesting policies, laws, and enforcements that best-performing regions are following. For this reason benchmarking regions have been determined (Fig. 3).

The following procedure followed in this paper was the measurement of the effect that these explanatory variables have on the regions' road safety performance (efficiency) by developing Tobit regression models. Therefore, the outcome form Tobit models revealed the effect of the explanatory variables on the regions' efficiency. In this way, we are able to support the address of policymakers for improving the region's road safety performance. Special focus should be given from local authorities for improving their Public transportations (buses) because it can be seen that these factors show the highest positive relationship with the regions' efficiency.

References

1. WHO: Global status report on road safety 2018. World Health Organization, Geneva (2018)
2. ETSC: Ranking EU progress on road safety – 10th road safety performance index report. European Transport Safety Council, Brussels (2016)
3. European Commission Press Release. https://ec.europa.eu/commission/presscorner/detail/en/MEMO_19_1990. Accessed 22 July 2020
4. Eurostat: Eurostat regional yearbook. 2018 edn. Publications Office of the European Union, Luxembourg (2018)

5. Al-Haji, G.: Road Safety Development Index (RSDI): Theory, Philosophy and Practice (Ph. D. dissertation). Department of Science and Technology, Campus Norrköping, Linköping University, Norrköping (2007)
6. Wegman, F., Commandeur, J., Doveh, E., Eksler, V., Gitelman, V., Hakkert, S., Lynam, D., Oppe, S.: SUNflowerNext: Towards a Composite Road Safety Performance Index. SWOV, Leidschendam (2008)
7. Yannis, G., Weijermars, W., Gitelman, V., Vis, M., Chaziris, A., Papadimitriou, E., Azevedo, C.L.: Road safety performance indicators for the interurban road network. Accid. Anal. Prev. **60**, 384–395 (2013)
8. Shen, Y., Hermans, E., Bao, Q., Brijs, T., Wets, G., Wang, W.: International benchmarking of road safety: state of the art. Transp. Res. Part C **50**, 37–50 (2015)
9. Shen, Y., Hermans, E., Bao, Q., Brijs, T., Wets, G.: Road safety development in Europe: a decade of changes (2001–2010). Accid. Anal. Prev. **60**, 85–94 (2013)
10. Nikolaou, P., Dimitriou, L.: Evaluation of road safety policies performance across Europe: results from benchmark analysis for a decade. Transp. Res. Part A **116**, 232–246 (2018)
11. Alper, D., Sinuany-Stern, Z., Shinar, D.: Evaluating the efficiency of local municipalities in providing traffic safety using the data envelopment analysis. Accid. Anal. Prev. **78**, 39–50 (2015)
12. Bastos, J.T., Shen, Y., Hermans, E., Brijs, T., Wets, G., Ferraz, A.C.P.: Traffic fatality indicators in Brazil: state diagnosis based on data envelopment analysis research. Accid. Anal. Prev. **81**, 61–73 (2015)
13. Ganji, S.R.S., Rassafi, A., Xu, D.: A double frontier DEA cross efficiency method aggregated by evidential reasoning approach for measuring road safety performance. Measurement **136**, 668–688 (2019)
14. Tasnim, N., Afzal, M.: An empirical investigation of country level efficiency and national systems of entrepreneurship using data envelopment analysis (DEA) and the TOBIT model. J. Global Entrep. Res. **8**(1), 37 (2018)
15. Wu, W., Xie, F., Zhao, L.M..: A study on the operation of Chinese university science park based on DEA-Tobit model (2010). https://doi.org/10.1109/ICIEEM.2010.5646090
16. Eurostat. https://ec.europa.eu/eurostat/statistics-explained/index.php/Glossary:Nomenclature_of_territorial_units_for_statistics_%28NUTS%29. Accessed 22 July 2020
17. Eksler, V.: Exploring spatial structure behind the road mortality of regions in Europe. Appl. Spat. Anal. **1**, 133–150 (2008)
18. Papadimitriou, E., Ekshler, V., Yannis, G., Lassarre, S.: Modelling the spatial variation of road safety in Greece. In: Institution of Civil Engineers - Transport Proceedings, vol. 166, no. 1, pp. 49–58. ICE Publishing, Scotland (2013)
19. Eksler, V.: Road mortality in Europe: how sensitive is it to demographic structure and population dynamics? IATSS Res. **31**(1), 80–88 (2006)
20. Eksler, V., Lassarre, S., Thomas, I.: Regional analysis of road mortality in Europe. Public Health **122**(9), 826–837 (2008)
21. Charnes, A., Cooper, W., Rhodes, E.: Measuring the efficiency of decision making units. Eur. J. Oper. Res. **2**(6), 429–444 (1978)
22. Tobin, J.: Estimation of relationships for limited dependent variables. Econometrica **26**(1), 24 (1958)

A Cloud-Based Big Data Architecture for an Intelligent Green Truck

Nikos Dimokas[1(✉)], Dimitris Margaritis[1], Manuel Gaetani[2], and Alfredo Favenza[2]

[1] Centre for Research and Technology Hellas/Hellenic Institute of Transport, 6th km Charilaou – Thermi, 57001 Thessaloniki, Greece
dimokas@gmail.com

[2] LINKS Foundation, Via Pier Carlo Boggio 6, 10138 Turin, Italy

Abstract. The rapid growth of cloud computing in combination with the advances in the communication boosted the green and smart mobility. Smart mobility aims to reduce congestion and foster greener, cheaper and with less delay transportation. Many of the current options to reduce fuel consumption and CO_2 emissions have worked on light-duty vehicles but are not feasible for heavy duty vehicles. Reduction of emissions and consumables (e.g. urea) without sacrificing on emission standards is an important challenge for heavy-duty vehicles. The paper introduces a cloud-based system architecture exploiting the potentials of big data analytics to deliver an on-demand route optimization service reducing NOx emissions of heavy-duty vehicles. The system utilizes the information provided by the navigation systems, big data analytics such as predictive traffic and weather conditions, road topography and road network and information about vehicle payload, vehicle configuration and transport mission to develop a strategy for the best route and the best velocity profile. The performance evaluation attested the efficiency and effectiveness of the system, since the cumulative engine-out NOx has been decreased more than 13%.

Keywords: System architecture · Cloud computing · Data warehouse · Intelligent transport system · Green vehicle

1 Introduction

Heavy-Duty Vehicles (HDVs) carry out the major part of freight transport as reported by the U.S. Department of Energy [1] and the Department of Transportation [2] and are responsible for more than 15% of the fuel consumption in the transportation sector. The automotive industry has made a substantial effort in recent years to develop powertrain technologies that improve the fuel consumption efficiency of HDVs. However, due to increasing road freight traffic, the projections indicate that total HDVs energy use and CO_2 emissions are expected to remain stable at the cur-rent level over the long term, if no policy action is taken. This is clearly incompatible with the goal of reducing greenhouse gas emissions from transport by around 60% below 1990 levels by 2050.

The fuel consumption can vary for the same route when different speed and gear applied [3]. Recent advances in information and communication technologies and

E. G. Nathanail et al. (Eds.): CSUM2020, AISC 1278, pp. 1076–1085, 2021.
https://doi.org/10.1007/978-3-030-61075-3_103

especially the cloud computing [4], big data analytics and vehicle-to-infrastructure (V2I) communication may be used to gather real-time information from external services and the HDV in order to optimize the fuel consumption and minimize the CO_2 emissions.

The proposed system architecture tries to bring together the most advanced technologies from powertrain control and intelligent transport systems in order to achieve a global optimum for consumption of fuel as well as other energy sources and consumables while achieving Euro VI emission standards for heavy duty road haulage. Moreover, the system architecture contains a global optimizer, which consists of a set of dynamic, intelligent control and prediction components designed for effective powertrain management, utilizing the environment data related to the transport mission, road topography, weather and road conditions and surrounding vehicles.

The system architecture is a flexible and scalable cloud computing architecture involving leading big data management components and predictive and simulation modules. The cloud-based system architecture communicates with on-board control systems components, developed on the truck, interacting with all on-board sensors (e.g., cameras, load, temperature), actuators and already deployed control units. The primary goal of the system architecture is to generate best routes, speed profiles and optimizing the truck's emissions.

A major component of the system architecture is the data storage (big data) component that has been developed on a cloud system. This component receives all data generated by the truck and engine auxiliary systems, powertrain and engine control systems. These data are integrated with specific mission data, navigation data and external information system such as traffic, weather forecast, road topography and road network and information about transport mission (vehicle payload, vehicle configuration etc.) to implement big data analytics and intelligent - predictive model-based optimization algorithms.

The rest of the article is organized as follows: In Sect. 2 we describe the relevant work, while the Sect. 3 presents the system architecture and Sect. 4 presents the data warehouse approach. Section 5 presents the results and finally, Sect. 6 concludes the article.

2 Relevant Work

Today's advances [5] in vehicular telematics and advances in GIS (Geographic Information System), GPS (Global Positioning Systems), ITS (Intelligent Transportation Systems), V2V (Vehicle to Vehicle) communication, and VII (Vehicle Infrastructure Integration) create opportunities for predicting a vehicle's trip information with details such as the road type, traffic, weather condition, and the driver style [14, 15].

There are various vehicle technologies, which focus on reducing the energy consumption and emissions of ground vehicles by using these information sources and by predictive planning of the vehicle's motion for a certain prediction horizon [16, 17]. The optimization of power-train control and calibration according to real-world conditions is one of them. However, powertrain control systems receive limited or no input from the present or future driving environment, traffic and weather conditions.

Driver support information systems [18] are getting more and more popular based on modern technologies and frequently updated databases. Existing driver information systems usually provide fuel efficient driving recommendations (acceleration and gear shifting score, cruise control usage etc.) to drivers based on their overall driving history

A number of research projects have investigated the potential of a better managed engine and driver performance by proving information based on historical and real-time processed data. CO-GISTICS [6] was one of the first European projects fully dedicated to the deployment of cooperative intelligent transport systems (C-ITS) focused on logistics. The project partly aimed at the reduction of fuel consumption and the equivalent CO_2 emissions. A Decision Support System (DSS), which would support decisions of the actors involved, in order to optimize and improve the system performances and the customer satisfaction, has been developed. Operational decisions were made on a daily basis or in real-time. Similarly, to CO-GISTICS, the REDUCTION EU project [7] has developed methodologies for providing predictive analytics based on advanced data mining technology in order to meet the objectives of fleet management, such as decision making for driver-adaption, eco-routing, and consequently CO_2 emissions control and improved fuel economy.

Another eco-driving related initiative was the eCoMove. The project aimed to develop solutions to reduce inefficiencies responsible for energy waste in road transport, which are related to vehicle, driver behavior, and traffic management [8]. This was done by utilizing V2I communication in the form of eco-messages where traffic participants share data on their destinations, traffic situation data and fuel consumption. Furthermore, a digital map enhanced with slope, historical speed profile, energy consumption data, and traffic data is used to form a so called eco-Cooperative Horizon model to calculate a fuel-efficient driving strategy. In addition, a post-trip analysis is provided to the drivers for improving their driving characteristics.

A very recent and on-going research initiative from US is the so called "Co-Optimized Delivery Vehicles" [9]. It is aimed at technologies to improve the fuel efficiency of delivery vehicles through real-time powertrain optimization using two-way vehicle-to-cloud connectivity. A predictive cloud-based approach (physics-aware spatiotemporal data analytics, PSDA) is used to control the hybrid powertrain using efficient computational techniques combined with physics models for computational efficiency. Additionally, powertrain calibration is optimized and downloaded to the vehicle using V2C connectivity in real-time during a delivery route, compensating for exogenous parameter changes or unpredicted driver behavior.

The proposed architecture presents the creation of a global optimizer which consists of a set of dynamic, intelligent control and prediction components designed for effective powertrain management, utilizing the environment data related to the transport mission, road topography, weather and road conditions and surrounding vehicles.

3 System Architecture

The system architecture consists of several software modules distributed in two separate optimizers that work in a coordinated fashion. The first one is the cloud computing optimizer while the second is the on-board optimizer. This logical subdivision is reflected in the design of two separated but strongly inter-connected systems, the cloud computing system and the on-board system, respectively operating in a Platform-as-a-Service (PaaS) [10, 11] cloud computing environment and on the truck.

The system architecture (Fig. 1) presents a high-level representation of the two main subsystems and their modules. On the left side, the external data sources are presented. The data that are sent to the cloud computing system are the maps and topographic data, including basic geographical maps and ADAS information. Additionally, the weather forecast data, including short- and long-term information about climate condition on specific areas, and the traffic forecast data, including information on real-time traffic condition and incidents. Finally, mission data are sent from the fleet management company, including information about the departure and destination for a given mission and the truck's characteristics. Maps, weather and traffic data are received by the cloud computing system through specific data service interfaces.

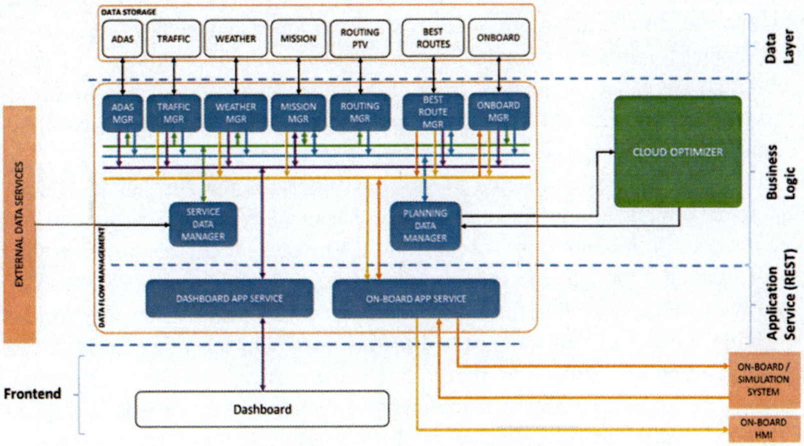

Fig. 1. System architecture

In the central part of the diagram, the cloud computing system implements the following functionalities:

- Data storage, including all database entities to store real-time and historical data from the external data services, mission related data, data from the on-board system and results of the cloud optimization process.
- Data flow management architecture, including data fetching, data transformation and data feeding components operating on raw data and post-processed data storages components.
- Cloud optimizer, including vehicle models and eco-route-planning/route-prediction component.
- Mission dashboard, including a frontend for the fleet management company to upload the mission data and monitor status of the missions.

The cloud computing system consists of several software components (managers). The service data manager is responsible to fetch periodically the data from the external data services and send them to the data storage in order to guarantee that enriched data for each mission are stored in the data storage. The communication between the external data services and the service data manager is based on the standard REST architecture. The ADAS manager is the component responsible to store and retrieve ADAS data to/from the data storage. The traffic manager is the component responsible to store and retrieve traffic data to/from the data storage. In the same way the weather, mission, routing, on-board manager are the components responsible to store and retrieve the weather, mission, routing and on-board data to/from the data storage.

Additionally, to the service data manager, the planning data manager accomplishes the functionality to fetch and fuse data stored in the data lake, prepare data and send it to the cloud optimizer for the computation of the best route related to a specific mission. After the computation is finished, the outputs (best route and velocity profile) are returned to the planning data manager, to store the results of the optimization process using the functionality provided by the best route manager.

The dashboard application service is responsible for the management of the missions submitted by a fleet specialist, through a web dashboard. The on-board application service is responsible for the management of the communication between the on-board components (system and HMI) and the cloud system.

The cloud optimizer is connected to the "data tier" and the "presentation tier" via the planning data manager and performs two key tasks, namely prediction of traffic and environment conditions and route optimization. On one hand, the traffic prediction component monitors, models and foresees the traffic and road conditions ahead of the truck, and feed the predicted information into the route optimization component for route selection with minimum energy to be consumed. On the other hand, the route optimization component calculates the global vehicle speed set-point trajectory and provides it to the traffic prediction component so that local vehicle speed trajectories can be updated continuously for the optimal route. The main outputs from the cloud Optimizer are the local vehicle speed trajectories and the best route.

Finally, the on-board control system is located on the truck and implements specific modules like Predictive Cruise Control Module (PCCM), including the Enhanced Predictive Cruise Control (ePCC) and the Power Demand Estimator components. Additionally, the Rapid Control Prototyping Module (RCPM), including the Energy Flow Coordinator, the Emission Estimator, the Emission Coordinator, the Emission Compensator and the Fuel Consumption Optimizer components are parts of the on-board system. Moreover, the Engine Control Unit (ECU), including the Engine Operating Mode Coordinator component, the Sensors Fusion Module, including all the components to fuse the data from the different sensors placed on the truck (e.g., radar, cameras, location, etc.) and finally the HMI module, including a visual interface for the driver to get relevant output from the system and provide input about the route environment. The on-board control system interacts with the truck actuators and sensors through a bidirectional communication channel. The on-board control system and the cloud computing system cooperate transferring information through a bidirectional channel like 4G to allow the overall system to perform the real-time optimization.

The whole system architecture is based on the Message Queuing Telemetry Transport (MQTT) standard publish-subscribe-based messaging protocol except from the communication with the external services where the REST architecture is used.

4 Data Storage Architecture

The planning data manager is implemented using a star-schema approach [12]. This approach has been chosen because it supports the integration of heterogeneous data source well. The architecture (Fig. 2) supports a number of different data sources. Further, a data warehouse approach is good at handling large datasets while still being easy to understand and query for non-experts such as traffic planners. These queries can both be complicated historical analysis queries or predictive queries.

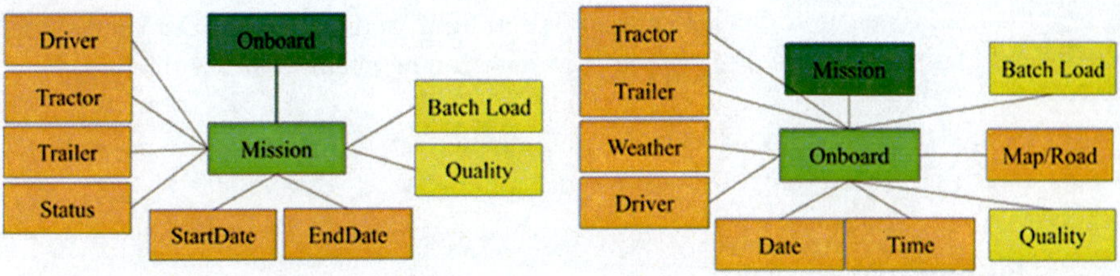

Fig. 2. Mission schema design (left side) and onboard schema design (right side).

4.1 Mission Data

The data warehouse design for the mission data is presented in the figure below. All rectangles indicate database tables and all lines are primary-key/foreign key constraints. The green tables are fact tables, the orange tables are dimension tables, and the yellow tables are utility dimension used by the planning data manager to keep track of where data is loaded from and what is the quality of the data.

The table Mission is the fact table [12] and one or more rows are stored for each mission. The Driver dimension contains information about drivers, e.g., age, nationality and years of experience. The Tractor dimension contains information about the tractor, e.g., make, model, number of axels, and engine size. The Trailer dimension contains information concerning each trailer, e.g., length, maximum load and number of axels. The Status dimension keeps track of the status of the mission, e.g., "prepared", "on-route" or "finished". The StartDate and EndDate dimension contains temporal information; for example, to find a mission that starts on a Saturday. The Batch Load dimension contains the details from where information was loaded. This makes it easy to find exactly the data that was loaded on a particular day. The Quality dimension keeps track of the quality of each row in the Mission fact table, e.g., if the mission data was altered during the Extract-Transform-Load (ETL) phase due to quality control policies set up by the trucking company.

4.2 The Onboard Data

The onboard data refers to the GPS and CAN-bus data that is collected from the trucks and sent to the cloud system (the planning data manager) to be integrated with the other data sources (Fig. 2).

The onboard schema design follows the same principles as for the mission schema. Each row with GPS/CAN bus data is stored in the Onboard fact table. To make it easy to query this data in a context, e.g., find all rows related to snow or rain, each row is annotated with information from the other data sources. This annotation is again done via the dimension tables. Please note that there are overlaps between the dimensions between the Mission and Onboard fact tables. Such shared dimensions make it possible to reuse code and better integrate data. The shared dimensions are Driver, Tractor and Trailer. The Batch Load and Quality dimensions are unique to each fact table.

The Weather dimension contains information about the weather station closest to each GPS row stored in the fact table. The Map/Road dimension contains information, e.g., on the road type (motorway, city-road) speed limit, and slope. The Date and Time dimensions make it efficient and simple for users to query temporal information, e.g., all rows on weekdays between 10:00 and 11:15.

The connection between Mission and Onboard fact tables makes it possible to associate each GPS/CAN bus row with a mission and vice versa.

4.3 Point Information

A mission is planned by specifying a number of points that the truck must pass, e.g., to delivery of pickup goods. Further, many of the data sources integrated in cloud use points to indicate where information is valid, e.g., traffic accidents or weather information. To handle all points' information in a uniform manner, all data sources that provide point-like information is stored in the single fact table (Fig. 3).

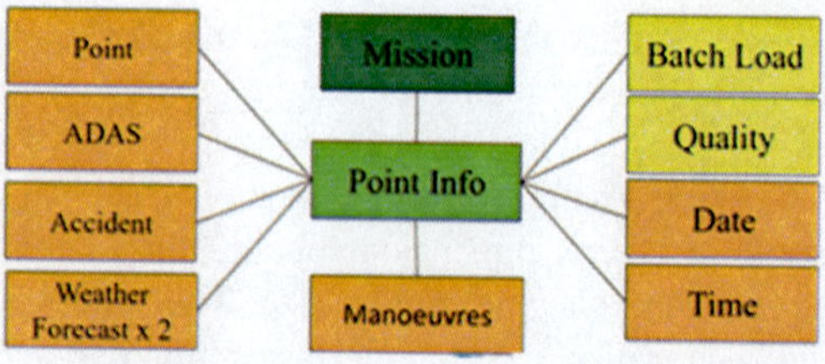

Fig. 3. Point info schema design.

The fact table is called Point Info. The shared dimensions are Date and Time. The new dimensions are Point that contains the latitude, longitude, and altitude. Further, well-known or often used point can be named. The ADAS dimension contains information, e.g., on slope and curvature. The Accident dimension contains information on accidents, e.g., duration and changes in speed. The Weather Forecast x 2 contains information on both a short-term and long-term weather. The Maneuvers dimension contains information such as "use right lane". The Point Info and Mission fact tables are connected to be able to query which points a mission passes and vice versa.

5 Performance Evaluation

The performance evaluation of the whole system has been organized and conducted by Ford Otosan. Two 2019 model year Ford F-Max tractors with identical hardware have been used for testing purposes. Both vehicles equipped with a 500 PS diesel engine and a 12-speed automated manual transmission. Both vehicles have a gross vehicle weight rating of 42 tons. The test was conducted on July 3rd and July 4th, 2019 between two Ford Otosan locations: Sancaktepe, İstanbul (departure) and İnönü, Eskişehir (arrival). The same mission was conducted twice on these dates to achieve consistent test results in the presence of traffic conditions. The test took approximately 5 h in one direction.

5.1 Test Results

The whole route has been divided in three segments. The three segments have been selected according to the longest trips during which the optiTruck control system was operational in an uninterrupted manner. Due to traffic conditions, the demo driver was occasionally forced to apply brakes or accelerate while disabling the optiTruck control system completely. When the traffic conditions were suitable, the demo driver has re-engaged the optiTruck control system during the mission.

The complete route was approximately 260 km, and some segments had to be discarded from the analysis since the driver did not activate the optiTruck control system in these road segments. Therefore, the numerical results are provided only for the selected two road segments (Table 1).

Table 1. Test summary.

Segment	Vehicle	Duration [sec]	Total Distance [km]	Mean Speed [kph]	Cumulative Engine Out NOx flow [kg. s/h]	Total Vehicle Weight [tons]
Segment 1	Demo	3939.0	49.4	45.0	0.32881	38.9
	Ref	2997.5	49.6	59.5	0.39847	38.0
Segment 2	Demo	1633.5	22.8	50.1	0.41927	35.6
	Ref	1638.6	22.9	50.4	0.48598	35.3

The results attested the efficiency and effectiveness of the system. The fuel consumption in the two segments has been similar with the usage of the system (there was a slight improvement). However, the cumulative engine-out NOx has been decreased more than 13% in each case. More precisely, in the first segment there was a reduction of 17.5% and in the second 13.7%. Figure 4 presents the cumulative engine-out NOx for the first segment. The whole distance was around 49 km. The reduction of the cumulative engine-out NOx increases as the distance grows.

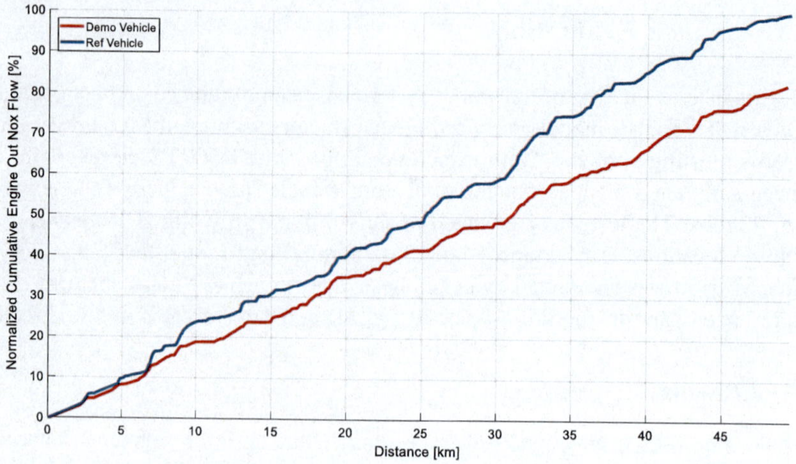

Fig. 4. Segment 1 cumulative engine-out NOx flow.

6 Conclusions

This paper presents a cloud-based architecture exploiting the potentials of big data analytics to deliver an on-demand route optimization service reducing emissions of heavy-duty vehicles. In particular, this highly scalable system is able – by fusing heterogeneous sources of data - to determine in real-time the best route and the optimal velocity profile that the truck has to maintain in order to minimize engine emissions.

The capacity to deal with high velocity of data streams coming from the running vehicles, enable the system to continuously update best-route and speed-profile adapting the route to fast changing traffic conditions in urban and suburban scenarios.

The results attested the significant reduction of the cumulative engine-out NOx. Future work will consider a detailed analysis of the performance of the system in a complex, fast changing roads scenarios with several trucks running at the same time.

Acknowledgements. Research supported by optiTruck project [13] that is funded by the European Commission under the Grant Agreement 713788.

References

1. Davis, S.C., Diegel, S.W., Boundy, R.G.: Transportation Energy Data Book. U.S. Department of Energy, Washington, DC, USA (2013)
2. U.S. Department of Transportation: Commodity Flow Survey United States, Technical report EC12TCF-US (2014)

3. Saltsman, S.: Impacts of connectivity and automation on vehicle operations. In: Global Symposium Proceedings on Connected Vehicle and Infrastructure, University of Michigan (2014)
4. Dikaiakos, M., Katsaros, D., Mehra, P., Pallis, G., Vakali, A.: Cloud computing: distributed internet computing for IT and scientific research. IEEE Internet Comput. **13**(5), 10–13 (2009)
5. Jiménez, F.: Intelligent Vehicles: Enabling Technologies and Future Developments. Butterworth-Heinemann, Oxford (2018)
6. Fanti, M.P., Iacobellis, G., Nolich, M., Rusich, A., Ukovich, W.: Cooperative logistics for sustainable mobility of goods. In: 1st Workshop Proceedings on the State of the Art and Challenges of Research Efforts, Bari, Italy (2014)
7. REDUCTION European Union Research Project. http://www.reduction-project.eu/. Accessed 23 Jan 2020
8. eCoMove European Union Research Project. http://www.ecomove-project.eu/. Accessed 23 Jan 2020
9. Vehicle Powertrain and Routing Co-Optimization. https://vpro.umn.edu/. Accessed 23 Jan 2020
10. Dhuldhule, P.A., Lakshmi, J., Nandy, S.K.: High performance computing cloud – a platform-as-a-service perspective. In: Proceedings of the International Conference on Cloud Computing and Big Data (CCBD), Shanghai, China, pp. 21–28. IEEE (2015)
11. Boniface, M., Nasser, B., Papay, J, Kyriazis, D.: Platform-as-a-service architecture for real-time quality of service management in clouds. In: 5th International Conference Proceedings on Internet and Web Applications and Services (ICIW), Barcelona, Spain, pp. 155–160 (2010)
12. Kimball, R., Ross, M.: The Data Warehouse Toolkit. Wiley, Hoboken (2013)
13. optiTruck European Union Research Project. http://optitruck.eu/. Accessed 23 Jan 2020
14. Mahmassani, H.S., Dong, J., Kim, J., Chen, R.B., Park, B.: Incorporating Weather Impacts in Traffic Estimation and Prediction Systems, US Department of Transport, USA (2009)
15. Work, D.B., Tossavainen, O.P., Blandin, S., Bayen, A.M., Iwuchukwu, T., Tracton, K.: An ensemble Kalman filtering approach to highway traffic estimation using GPS enabled mobile devices. In: 47th IEEE Conference on Decision and Control (CDC), pp. 5062–5068 (2008)
16. Vlassenroot, S., Willem, J., De Mol, J., Marchau, V., Brookhuis, K., Allaert, G.: Speed management through vehicle measures, intelligent transport systems and intelligent speed assistance (2005)
17. Kohut, N.J., Hedrick, J., Borrelli, F.: Integrating traffic data and model predictive control to improve fuel economy. In: 12th International IEEE Conference on Intelligent Transportation Systems (ITSC), pp. 155–160 (2009)
18. Li, W., Wu, G., Yao, D., Zhang, Y., Barth, M.J.: Dynamic en-route eco-navigation: strategy design, implementation and evaluation. In: 21st International IEEE Conference on Intelligent Transportation Systems (ITSC), Maui, Hawaii, USA, pp. 1888–1893 (2018)

Connecting Cruise Lines with Local Supply Chains for Enhancing Customer Experience: A Platform Application in Greece

Eleftherios Sdoukopoulos[1,2](✉), Vasiliki-Maria Perra[1], Maria Boile[1,2], Leonidas Efthymiou[3], Evi Dekoulou[3], and Yianna Orphanidou[3]

[1] Hellenic Institute of Transport, Centre for Research and Technology Hellas, 52 Egialias Str., 15125 Marousi, Athens, Greece
sdouk@certh.gr
[2] Department of Maritime Studies, University of Piraeus, M. Karaoli & 80 A. Dimitriou Str., 18534 Piraeus, Greece
[3] University of Nicosia Research Foundation, Metochiou 66, Engomi, Nicosia, Cyprus

Abstract. The Blue Economy constitutes a large and fast growing part of the European Economy. In 2017, it accounted for nearly € 74.3 billion in gross operating profit, and an annual turnover of approximately € 658 billion. Among all blue economic sectors, cruise tourism is one of the most important ones, with the Mediterranean region still accounting for the second most popular cruise destination in the world, after the Caribbean. The contribution of the cruise sector to the local economies of most ports of call however, proves to be quite limited, with negative externalities often surpassing derived benefits. Promoting and integrating 'traditional' products into cruise supply chains can be a promising solution addressing this situation. It may also provide passengers with an enhanced local-tasting experience that may make their journey even more memorable. Focusing on Greece and its numerous ports integrated into various cruise itineraries, this paper explores cruise network characteristics and identifies local products that can be promoted at different ports of call for boosting customer experience. Through a novel e-marketplace that is being outlined, cruise companies will be able to directly connect with local producers and suppliers and place their orders. Seamless information may also provide benefits with regard to improving transport and logistics processes.

Keywords: Cruise supply chains · Cruise destination · Port-city · E-marketplace · Blue economy

1 Introduction

The Blue Economy in Europe constitutes a rapidly increasing part of the economy of several coastal and insular EU Member States. In 2017, the established blue economic sectors (i.e. coastal tourism, maritime transport, ports, warehousing and construction of water projects, shipbuilding and repair, extraction and commercialization of marine

E. G. Nathanail et al. (Eds.): CSUM2020, AISC 1278, pp. 1086–1096, 2021.
https://doi.org/10.1007/978-3-030-61075-3_104

living sources and marine extraction of minerals, oil and gas) generated almost € 658 billion of turnover and € 180 billion of Gross Value Added (GVA) [1].

The cruise sector accounts for one of the cornerstones of the Blue Economy worldwide but also in Europe, contributing in the economy of the latter a record of € 47.86 billion in 2017, which represents a 16.9% increase compared to 2015 [2]. This upward trend is being supported by three key factors: (i) Europe serves as the world's 2nd biggest cruise passenger market (6.96 million European citizens embarked on a cruise in 2017) (ii) the Mediterranean region remains the world's 2nd most popular cruise destination, after the Caribbean, holding a share of 17% of global deployment, and (iii) European shipyards hold a leading position in the building of new cruise vessels, with cruise lines spending almost € 5.6 billion on new orders in 2017 [2, 3].

Within the Mediterranean region, besides home-porting that is being mostly concentrated in Italy (e.g. Venice) and Spain (e.g. Barcelona, Palma), Greece holds a leading position in cruise itineraries, with two or three Greek destinations on average being included in the route planning of cruises that span across the whole region or its Eastern part [4]. The resulting economic benefits for the local communities however, prove to be quite limited with negative externalities (e.g. congestion impacts) often surpassing them. In fact, although Greece ranks as the 3rd most attractive cruise destination in Europe after Italy and Spain (4.09 million passenger visits), it holds just the 9th position in terms of total economic impacts [2]. To this end, relevant targeted actions should be taken for boosting the local economy at the ports of call. These actions need to be well aligned with the different trends that are currently impacting the cruise sector, which may be summarized to the following: (a) greater accessibility of almost all income levels to cruising, which may further widen passengers' preferences and needs; (b) cruise passengers going beyond sightseeing, looking for transformational experiences that will make their journey more memorable; (c) cruise vessels transforming into smart cities, ensuring on-board internet connectivity and providing highly personalized experiences through smart-tech devices (e.g. key-chains, bracelets, etc.) and applications; (d) generation Z (i.e. birth years starting from mid to late 1990s) set to become the largest cruise generation outpacing Millennials. This generation proves to be more interested in the appeal of multiple destinations and new experiences; (e) increasing demand of off-peak cruises, for passengers looking to escape the cold in their home town or seeking to gain unique experiences during the winter; and (f) increase of solo travelers, since cruising often facilitates to connect with other passengers and form community bonds [5].

It is evident that cruising increasingly accounts for a highly experience-oriented product with greater diversity with respect to destinations reached and services provided. The cruise industry therefore, is currently heavily investing on improving both its on-board and off-board services [3, 5]. With regard to the latter, cooperation with local authorities and (business) communities is a prerequisite so that local landmarks, historical monuments, products, activities, excursions, etc. can be better promoted so that cruise destinations enhance their touristic attractiveness and increase their product demand (i.e. the cruise itinerary).

Within the aforementioned context, and considering local characteristics and the dynamics of key cruise destinations in Greece, emphasis is being provided within this paper on efficiently integrating local producers and suppliers into cruise supply chains, as a promising measure for boosting local economies and providing an additional 'tasting' experience to passengers at each destination. To this end, the rest of the paper is structured as follows: Sect. 2 provides an overview of the current status of the cruise sector in Greece and maps the existing and planned network, highlighting key destinations and routes; Sect. 3 outlines the structure of cruise supply chains and provides insights on existing practices but also possible opportunities to be exploited; Sect. 4 briefly describes the proposed solution (i.e. an e-marketplace) that can bring together cruise companies with local suppliers and producers at each cruise destination. More specifically, the relevant objectives of this technological solution are set and the methodology that will be followed for its development is being outlined. Section 5 concludes the paper by pointing out the potential benefits to be derived for the different stakeholders involved, and setting the next steps to be taken for platform's development, pilot-testing, and ultimately its commercialization and full-scale implementation.

2 Current Status of the Cruise Sector in Greece

As previously noted, although home-porting is very limited in Greece (with the port of Piraeus being the most important case), 28 cruise lines operating within the Mediterranean region included, in 2019, at least one Greek destination within their itineraries. According to recent data (i.e. 2017), the total economic impact (i.e. direct, indirect and induced) of the cruise sector in Greece amounted to € 913 million generating a total of 10.721 jobs, with 59% of the latter being related to the transportation and trade sector and 11% to manufacturing. Direct expenditures were estimated to approximately € 546 million and were mainly related to cruise operations and secondarily to expenditures of both passengers and crew at the ports of call [2]. The latter, which in 2017 ranged between € 150 and € 156 per cruise passenger, mainly concerned tours/excursions, food and beverages and merchandise [6].

Among the extended Greek port network, 56 ports are contributing, to a greater or lesser extent, to the aforementioned economic impact. Ten (10) of them (Fig. 1) can be considered as primary, accommodating more than 100.000 passengers per annum, with the rest 46 forming a secondary network of cruise ports. The large majority of them is being located in South Aegean, considering the high touristic attractiveness of the Cyclades and Dodecanese islands as well as of Crete, while others in the hinterland part of Greece are often linked to key touristic attractions (e.g. Acropolis, Olympia, etc.) and other local landmarks of historic and cultural value or natural beauty.

In 2019, 3.899 cruise calls and 5.537.500 cruise passenger movements were recorded in the aforementioned ports. This presents an increase of 14.3% and 15.6% respectively compared to 2018 (quite equal were also the numbers in 2017), when the lowest traffic over the last five years was recorded [7]. Estimates for 2020 were optimistic expecting to overpass last year's performance, but given the severe global impact of Covid-19 on tourism, a large deviation from initial expectations is almost certain [8].

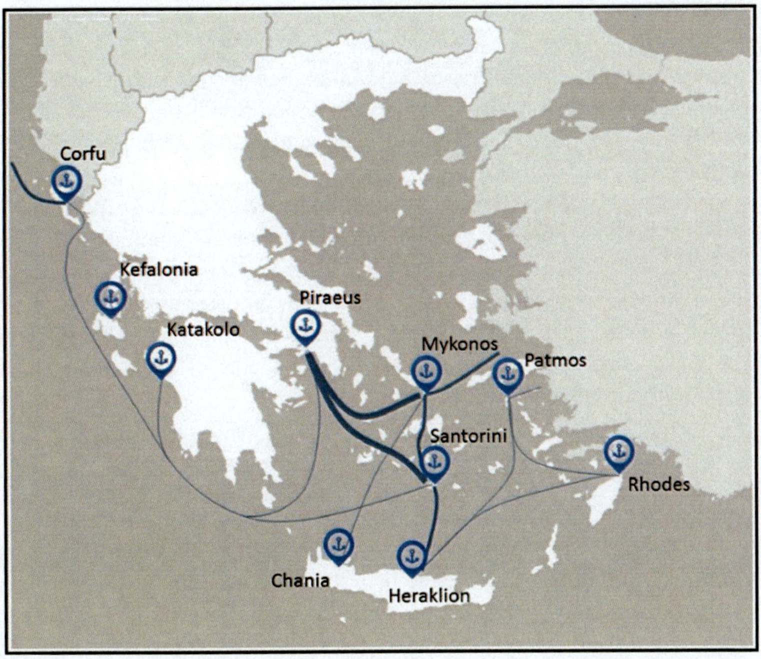

Fig. 1. Primary cruise ports in Greece, and key routes (itinerary legs) followed [authors]

Over the last five years, almost 92.7% of total passengers' movements were concentrated on the ten primary cruise ports of Greece, with the ports of Piraeus, Santorini, Mykonos and Corfu accounting for more than 500.000 passengers' movements per year. Relatively high was also the number of cruise vessels calling at these ports (80.7%), with the top 4 noted above serving more than 400 calls every year [7]. A mapping of the cruise network in Greece (April to December 2019), identified 277 unique cruise legs (i.e. from foreign to Greek port, between Greek ports and from Greek to foreign port), which were included in 829 different itineraries provided by the 28 cruise companies that were providing services within the whole Mediterranean region and/or at least its Eastern part. Based on a thorough review of data available from cruise lines and travel agents, the cruise network was to be expanded in 2020 with 333 cruise legs being identified as part of 891 itineraries. However, due to Covid-19, multiple cancellations or schedule changes have been made and it remains to be seen what will be the final impact on the sector once this health crisis is over [9].

For 2019, the majority of cruise services calling at Greek ports were Mediterranean ones (94.6%), while a few world cruises (5.4%) and national ones (e.g. in South Aegean Sea) were also identified. Mainly the national, as well as a few East Med cruises, used four Greek ports as home-ports (i.e. Piraeus, Corfu, Heraklion and Chania). These are to be reduced in 2020, since no cruise was found to be departing from the port of Chania this year. To this end, unfortunately, the list of "former home-ports" in Greece (i.e. Rhodes, Thessaloniki and Lavrio) continues to grow.

Greek cruise ports are mostly connected to destinations in the Adriatic Sea (i.e. North Italy, Croatia, Slovenia, Montenegro and Albania), East (i.e. West Turkey) and West Med (i.e. West Italy, Spain, France and Malta). Fewer connections were identified with destinations in Western Ionian Sea (i.e. Sicily and South Italy), East South Med (Cyprus and Israel), Red Sea (i.e. Egypt, Jordan and United Arab Emirates) and the Black Sea (Bulgaria and North Turkey).

With regard to intra-connectivity, connections between 42 cruise ports in Greece were identified. The ones concentrating the greatest traffic are among the ten primary ports, but several connections with ports of the secondary network are also identified. The port of Mykonos, for example, is connected with 20 other cruise ports in Greece. Apart from the primary ones (Fig. 1), several other smaller ports are being visited by cruise vessels after departing or before calling at the port of Mykonos (e.g. Milos, Hydra, Nafplio, Gythio, Kalamata, Rethymno, Kos).

3 The Cruise Supply Chain

Given the increasing size of cruise vessels, with the largest one today (i.e. the 'Symphony of the Seas' of Royal Caribbean) being able to accommodate approximately 6.600 passengers, served by a crew of nearly 2.200 members [10], it is evident that in order to successfully meet customer demand, the complexity and service requirements of modern cruise supply chains have considerably increased. To put this into better perspective, on a typical seven-day itinerary, nearly 60.000 eggs, 4.400 kg of chicken, 9.000 kg of potatoes and 315 kg of ice-cream, to mention a few, are required [11]. To add to this complexity, itinerary changes on a seasonal basis for meeting market demand, present another pressure on the structure of cruise supply chains. Efficient and well-coordinated supply chains are therefore a prerequisite for ensuring that a variety of choices and services of high quality are being provided to cruise passengers, and to this end relevant departments exist in all cruise companies, staffed with highly-skilled experts [12]. Besides food and beverages, the provision and quality of which largely affect the level of service provided to passengers, emphasis is also given to the management of all other necessary supplies (e.g. fuel, maintenance equipment, etc.).

Figure 2 below provides a graphical presentation of the planning levels of the cruise supply chain, highlighting the main processes undertaken and the different actors involved. Interesting insights on the factors that influence the decision of cruise companies with regard to supply chain structure, and particularly on the drivers for outsourcing are also provided by Diakomihalis and Stefanidaki [14]. Based on a series of interviews with industry experts, the authors classified these factors into the following six categories: (a) *cruise product*, including the characteristics of each itinerary (i.e. home-port, number of ports of call and time spent at each port) and the product classification as such (i.e. mass, premium, etc.), (b) *characteristics of the cruise company*, i.e. specialization and whether it is a parent or affiliated company since this can also affect the supply chain strategy, (c) *geographical area of operation,* including location of headquarters, proximity to distribution centers, and deployment of fleet activities in specific geographical areas, (d) *passengers' profile and preferences*, including demographic information (e.g. age, nationality, income), nutritional habits, etc., (e) *costs*, considering all different factors that are shaping the total cost of each cruise, and (f) *other factors influencing the orientation of the supply chain strategy* (e.g. regulatory frameworks, trust and know-how and experience of suppliers, etc.).

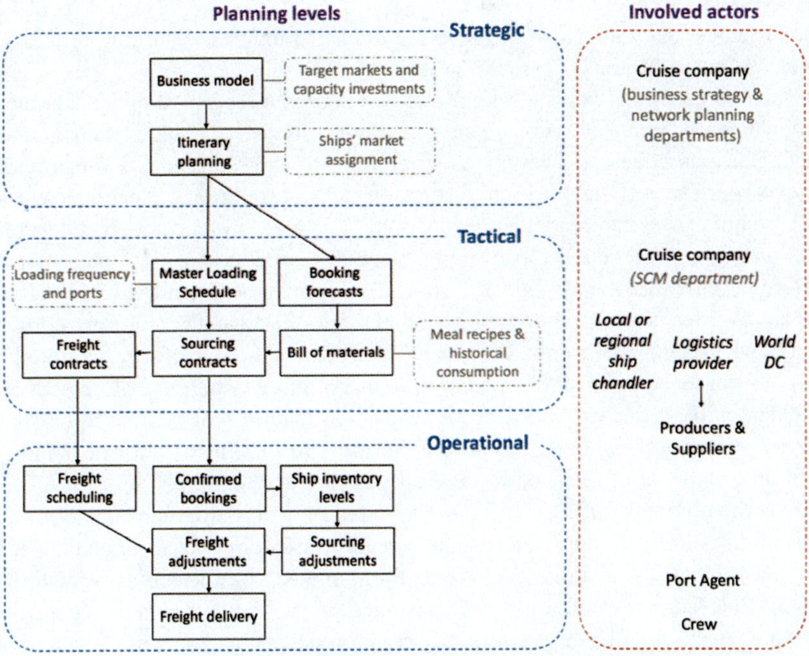

Fig. 2. Cruise supply chain: planning levels and involved actors [adapted from 12 & 13]

In most cases today, the Supply Chain Management departments of cruise companies collaborate with major logistics providers, so that required products of desired quality are supplied to the companies' cruise vessels in a reliable and cost-effective manner. To meet demand, logistics providers have established extended networks of local producers and suppliers in various countries, and considering their productivity rate, capacity and inventory, they orchestrate the necessary transport and consolidation processes so that all products are delivered to cruise vessels on time. Such a process usually takes place at home-ports, while for longer itineraries certain ports of call are also exploited. The necessary products arrive to the ports of call beforehand, often via containers, and are then loaded on cruise vessels upon their arrival.

3.1 An Emerging Opportunity for Cruise Destinations

The aforementioned practices coupled with the main trends impacting the sector, and particularly the search of passengers for new experiences, highlight a promising opportunity for cruise destinations, which once exploited can result into greater economic benefits being generated for the local community in each port of call. More specifically, and focusing on Greece, given the uniqueness of several local products at each destination (e.g. olive oil, cheese, beverages, traditional sweets, honey, etc.) with specific varieties not found elsewhere making their quality incomparable, their targeted promotion to cruise passengers and their integration to cruise supply chains can result in a twofold benefit. Cruise passengers will be able to add a new 'tasting' experience at each cruise destination of Greece, while local producers and suppliers can benefit from greater revenues and enjoy a wider visibility of their products, enlarging in that way their future sales potential.

In order to accommodate such a process, local producers and suppliers should be able to efficiently connect with cruise companies, so that the latter can order the products their passengers are willing to purchase or use these products for providing passengers, during their stay at each destination, a local traditional meal that will add another memory from that particular destination. Of course, all relevant procedures (i.e. connection of local producers and suppliers with cruise companies, placement of orders, order processing and delivery, etc.) should be done seamlessly, so that any delays are avoided and the short time windows that are available for delivering the products and loading them onboard can be met. This can be performed through an online platform that can have the form of an e-marketplace, where the aforementioned stakeholders will be the key users, but relevant information will be able to be shared with other related stakeholders (i.e. port authorities) so that, as mentioned before, all relevant procedures are facilitated to the best possible extent.

This is the objective of the 'NAYS – Exploitation of maritime information for the interoperability of cruise supply chains and services' research project, funded under the Interreg V-A Greece – Cyprus 2014–2020 Programme. More specifically, within its framework, the current conditions and dynamics of the cruise sector in Greece and Cyprus are analysed, and a network of local producers and suppliers and cruise companies is built for (a) collecting their views and perspectives on how the proposed platform can be better structured, what functionalities should it include and what other added-value services should it provide, (b) informing and steering the platform's development providing feedback at different stages of the process, (c) pilot-testing the platform and indicating any problems to be resolved, commenting on its usability (e.g. inserting information, placing an order, making a transaction, etc.) and user friendliness, and (d) providing recommendations on its commercialization strategy so that its uptake and exploitation in the future is ensured.

4 A Platform Solution for Integrating Local Products into Cruise Supply Chains

As indicated above, the first step for developing the proposed platform is to set-up a network of local producers and suppliers and cruise companies (i.e. prospective users of the platform) and analyze their needs, better guiding in that way the development of the platform's high-level architecture. To this end, a series of regional stakeholder workshops were held in Greece (in Rhodes, Chios and Chania) and Cyprus (in Nicosia) between May and June 2019, with 10–12 representatives from the aforementioned stakeholder groups participating on average. Participants were first informed of the project idea and the activities to be undertaken, as well as of the different stages when their feedback will be requested. The platform to be developed was positively evaluated by the participants, though certain prerequisites to be met were expressed.

More specifically, cruise companies were open to evaluate the potential of providing their passengers with this additional experience and multiple ways were proposed for doing so (e.g. introduce a traditional dish or meal at each or selected destinations, provide passengers with a list of local products that they can order before arrival and receive them at each cruise destination, etc.). However, they stressed out that the platform should ensure that (a) high-quality products are only offered (e.g. by requiring producers and suppliers submitting relevant standards and certificates) (b) details on every product are available (e.g. origin, weight, packaging, etc.), and (c) orders are processed in a timely manner, so that all products are delivered on time and without any deviations in terms of number or volume. To this end, as stressed out above, relevant information should be also shared, through the platform, with port authorities so that the delivery process is efficiently planned and executed. Interoperability with other systems, such as the National Maritime Single Windows which are still in development and pilot-testing phase in Greece and Cyprus respectively, should be also foreseen, so that the orders of passengers, for example, can be delivered in very short time after the arrival of the cruise vessel at the port.

Emphasis on time windows was also provided by local producers and suppliers, so that once they receive an order, they have enough time to prepare and deliver the relevant products, minimizing in that way any risk of experiencing delays. Both local producers and suppliers as well as cruise companies also pointed out that the platform should allow to evaluate the users that they partnered with, for highlighting any underperformance (e.g. low quality of products, delays in delivery, etc.), problems, etc. so that other platform users are also informed, in a similar way with what happens with the rating of hotels or apartments at booking sites.

Building upon these first insights described above, and for collecting more detailed feedback on the platform's structure and functions, a questionnaire survey was conducted during Jan-Feb 2020 for the two stakeholder groups targeted. Each questionnaire contained a set of 14 multiple choice questions, requesting the respondents' views on how key functions should be developed and performed. Overall, 56 responses were received, 43 from local producers and suppliers (LPS) and the remaining 13 from cruise companies (CC). Table 1 below summarizes the main findings of the survey.

Table 1. Main findings of the questionnaire survey

Function	Main findings *(R-number of respondents)*
Users' registration	Each user having a central account was preferred than setting-up multiple ones (e.g. of different departments) connected to one profile (43.9% of LPS & 53.9% of CC). Unlike CC, LPS were quite open to both options (36.6%)
Users' profiling	Inserting manually all information required for setting a user profile was favored against developing an automatic function that draws information from the users' internal systems (45% of LPS-*R40* & 54.5% of CC-*R11*)

(*continued*)

Table 1. (*continued*)

Function	Main findings *(R-number of respondents)*
Products	The platform should mostly facilitate the supply of food and beverages (41% and 43.6% respectively of LPS-*R39* & 100% of CC-*R8*). For each product, an analytical description of its characteristics should be provided, along with the available volume, packaging details and place of origin (order of importance of LPS-*R36* & CC-*R7*). Available products should be categorized per type and port of call (85.7% & 42.9% respectively of CC-*R7*), while a free-text search function should also be offered
Order placing and processing	It was commonly suggested that CC should place their orders and LPS can provide, if interested, their economic offer (57.1% of CC-*R7* & 50% of LPS-*R34*). CC will then examine all offers received and select their LPS. Relevant certificates/standards will be provided by the LPS to the platform and will be checked by the administrator. However, CC will be able to request any additional certificates, if deemed necessary. Before proceeding to processing the order, a direct communication between the CC and the LPS should be facilitated (71.4% of CC-*R7* & 79.4% of LSP-*R34*). LPS also indicated (55.8% of LPS-R34) that they would need at least 7 days for being able to process and deliver each order on time
Evaluation of users	The platform should require users to evaluate the ones that they have partnered with (82.3% of LPS-*R34* & 100% of CC-*R7*). They should be able to evaluate certain aspects (64.3% of LPS-*R28* & 57.1% of CC-*R7*), while it will also be beneficial to provide an overall score (e.g. 1 to 5 – poor to excellence). Both of this information should be visible to all users, while under-performing users (i.e. below 2) should receive a warning or penalty

5 Conclusions and Future Steps

Taking the aforementioned information as a basis, the next step for the development of the proposed platform is the setting of its high-level architecture. According to the latter, specific components will start to be developed following the SCRUM framework. More specifically, the development process will be broken down into consecutive sprints of short duration. At the end of each sprint, the developed component will be presented to the stakeholder group supporting the development process, so that valuable feedback and comments are received, which should be addressed in the next sprint. This process can ensure that the final product is fully aligned with the requirements of its prospective users, and thus its future uptake and exploitation potential, as well as anticipated benefits, increase.

The stakeholders already engaged in the consultation activities will support the platform's development process with the vision that its final version can be easily used for providing added value not only to the main stakeholders involved but also to others who will benefit to a lesser extent (i.e. port authorities and the local community). Once fully developed, the platform will be pilot-tested in selected ports of call so that any problems encountered are successfully addressed. Added value services that may be incorporated within the platform will be outlined. Finally, a thorough analysis will be conducted for setting-up the platform's business strategy and marketing plan, choosing also the most appropriate governance model to be followed.

Acknowledgement. Research for this paper was carried out within the framework of the 'NAYS – Exploitation of maritime information for the interoperability of cruise supply chains and services' research project, which is part of the "Interreg V-A Greece-Cyprus 2014–2020 Programme" that is being co-funded by the European Union and National Funds of Greece and Cyprus.

References

1. European Commission: The EU Blue Economy Report 2019. Publications Office of the European Union. Luxembourg (2019)
2. Cruise Lines International Association (CLIA): Contribution of Cruise Tourism to the Economies of Europe 2017 (2018). https://es.cruiseexperts.org/media/2971/2017-europe-economic-impact-report.pdf. Accessed 04 March 2020
3. Cruise Lines International Association (CLIA): 2020 State of the Cruise Industry report (2019). https://cruising.org/news-and-research/press-room/2019/December/clia-releases-20-state-of-the-cruise-industry-outlook-report. Accessed 04 March 2020
4. Pallis, T., Vaggelas, G., Arapi, C., Kladaki, P., Papachristou, E.: The prospects of development of cruise home-porting in Greece. DiaNEOsis Research and Policy Institute (2017)
5. Cruise Lines International Association (CLIA): 2019 Cruise Trends & Industry Outlook (2018). https://cruising.org/news-and-research/-/media/CLIA/Research/CLIA%202019%20State%20of%20the%20Industry.pdf. Accessed 05 March 2020
6. Greek Tourism Confederation Institute (INSETE): Study on Nautical Tourism (yachting, cruise and short sea shipping) (2018). http://www.insete.gr/Portals/0/meletes-INSETE/06/2018_Nautical_Tourism_Economic_Impact.pdf. Accessed 06 March 2020
7. Hellenic Ports Association (ELIME): Greek Cruise. https://www.elime.gr/krouaziera. Accessed 09 March 2020
8. Greek Travel Pages (GTP): Cruise Ship and Passenger Arrivals to Greece Rise in 2019 (2020). https://news.gtp.gr/2020/01/08/cruise-ship-and-passenger-arrivals-to-greece-rise-in-2019/. Accessed 09 March 2020
9. Cruise Critic: Coronavirus: Updated Cruise Ship Policies and Cancellations (2020). https://www.cruisecritic.co.uk/news/5016/. Accessed 13 March 2020
10. Ship Technology: Symphony of the Seas Cruise Ship (2020). https://www.ship-technology.com/projects/symphony-seas-cruise-ship/. Accessed 13 March 2020
11. CNN Travel: Feeding the largest cruise ships in the world (2019). https://edition.cnn.com/travel/article/cruise-ships-food-supplies/index.html. Accessed 13 March 2020

12. Véronneau, S., Roy, J.: Global service supply chains: an empirical study of current practices & challenges of a cruise line corporation. Tourism Manag. **30**, 128–139 (2009)
13. Tsourakis, C.: An Analysis of Cruise Supply Chain Management in European Region: The Outsourcing Decision Making of Food and Beverage Department. MSc Thesis, Erasmus University Rotterdam (2012)
14. Diakomihalis, M., Stefanidaki, E.: Cruise ship supply chain: a field study on outsourcing decisions. Int. J. Decis. Sci. Risk Manag. **3**, 369–383 (2011)

Investigating the Prospect of Adopting Artificial Intelligence Techniques from Transport Operators in Greece

Aristomenis Kopsacheilis(✉), Anastasia Nikolaidou, Georgios Georgiadis, Ioannis Politis, and Panagiotis Papaioannou

Transport Engineering Laboratory, Department of Civil Engineering, Faculty of Engineering, Aristotle University of Thessaloniki, GR-54124 Thessaloniki, Greece
kopsacheilis@civil.auth.gr

Abstract. The bloom of computer engineering and data science unveils new potential for several business sectors, regarding the enhancement of their operational functions. Transport stakeholders can benefit from computer science advances, such as Artificial Intelligence (AI) or Data Mining (DM) techniques, which can contribute to the utilization of existing data and maximize their productive efficiency. Reduced operating costs, decreased travel times, as well as faster freight delivery times are among several gains that can be achieved through the adoption of such techniques. Analysing transport related social media content could also provide useful insight into customers' needs and preferences based on their social media behaviour. However, the aforementioned computer science fields are relatively new and require advanced hardware as well as necessary expertise. Thus, the take – up from transport operators is not significant yet. In this paper, we provide an overview of the currently applied AI and DM techniques in corporate decision making, planning, and management practiced by various transport operators worldwide. Our literature review highlights original findings and case studies from selected freight and passenger transport operators in Greece. To further investigate their activities, a questionnaire survey was conducted to examine the existence of a systematic data collection system, the processes of data mining and analysis applied (if any) and most importantly the fields where the information derived from data analysis is exploited. Survey findings reveal the relative maturity of Greek transport operators for applying AI and DM techniques.

Keywords: Artificial intelligence maturity · Data mining · Technological equipment · Organizational readiness · Transport operators

1 Introduction

Artificial intelligence (AI) is the simulation of human intelligence processes by machines, especially computer systems and works by identifying patterns in large datasets through the application of complex algorithms [1]. AI applies logic-based

E. G. Nathanail et al. (Eds.): CSUM2020, AISC 1278, pp. 1097–1106, 2021.
https://doi.org/10.1007/978-3-030-61075-3_105

techniques and advanced analysis procedures, such as Machine Learning (ML), in order to interpret events, decide and choose appropriate actions [2]. Data Mining (DM), being an essential part of the AI techniques, is the process of finding anomalies, patterns and correlations within large data sets to predict outcomes. The predictive algorithms, as the result of DM, are the base for the AI application [3].

In 1956, John McCarthy was the first to adopt the term 'Artificial Intelligence' [4]. Even though agencies and enterprises quickly realised the added value that AI techniques could bring to their activities, technological limitations did not allow them to fully employ these methods. However, recent advances in the field of computer science and technology have contributed to the automation of many activities that previously required human effort. Over the past sixty years there have been advances in search algorithms, ML algorithms, and integrating statistical analysis into understanding the world at large. According to PwC's Global Artificial Intelligence [5], AI could contribute up to $15.7 trillion to global GDP in 2030. Between 2018 and 2019, organizations that have deployed AI grew from 4% to 14%, according to Gartner's 2019 Chief Information Officer (CIO) Agenda survey [6].

The AI applications in transport have been developing and implemented in a variety of ways. In 2017, the global market for transport-related AI technologies reached $1.2 to $1.4 billion, based on estimates from global research firms. According to projections the market could grow from $3.1 to $3.5 billion by 2023. The rapid growth of this market is driven by its many benefits [7]. Transport can benefit from AI through increased efficiency, enhanced pedestrian and driver safety, and lower costs.

With regard to the Greek reality and the implementation of AI by Greek companies, recent studies revealed that AI can act as a new factor for Greek economic growth over the next years. In fact, AI could contribute to a significant increase in the country's GDP (up to $195 billion) until 2035 [8]. Although Greek executives realize the value of AI, they seem hesitant in investing in it [8]. There are several deterrents in the application of AI. The limited personnel expertise regarding the implementation of AI, the outdated technological equipment and the inadequate quality of the available datasets are among the major factors for the low adoption rate of AI. To date, few Greek enterprises (3%) have implemented AI programs, while the corresponding global share is 45%. However, 23% of businesses are in early stages of investment [8]. Specific statistics for AI adoption and utilization in the Greek transport sector or any other specific sector are not available so far.

The objective of this research is to investigate the use of AI in the transport sector in Greece. An ad-hoc questionnaire survey was conducted to collect and analyse relevant data from a sample of Greek transport actors, which represent basic transport business domains, in order to investigate their readiness and maturity on AI deployment.

The remainder of the paper is structured as follows. The next section reviews basic AI applications on prominent transport sectors along with their benefits. The questionnaire survey on AI use in Greek transport sector, the sample of participants in the survey and the method for analysing the collected data, are explained in Sect. 3. Results are presented and discussed in Sect. 4. Finally, conclusions are summarised in Sect. 5.

2 Literature Review

Transport problems are often characterized by a wide variety of factors that cannot always be evaluated using conventional methods. For instance, the influence of users' behaviour to the creation of a transport problem (e.g. congestion, environmental impacts, poor accessibility etc.) is sometimes difficult to specify. Such restrictions can be overcome through the application of AI, which can form predictive models, even though the relationship between all factors is not evident [1]. This characteristic, along with the fact that many problems impose high computational requirements, as well as require the accounting of both qualitative and quantitative variables, has made AI a favourable option among researchers and operators when addressing a transport problem [4]. In transport, AI is systematically applied to various sectors among which are road safety, public transport, traffic management and logistics.

Regarding public transport systems the main goal when using AI is to enhance the user experience through an improved level of service, as well as to reduce operational costs and maximize efficiency. One way to increase service quality is to improve the provision of information, subsequently increasing the utility of involved means. Users state that the provision of accurate information could increase their tendency to use public transport [9]. Since, the accuracy of the information provided is influenced by many factors, such as congestion, weather, emergency events and accidents, there is a need for the application of models that can minimise the projection error. Attempts have been made to forecast bus stop arrival time through the exploitation of the GPS vehicle trace, using an Artificial Neural Network, with promising results [10]. Other applications of AI in the field of public transport, include the use of Ant Colony Optimization algorithms to optimize routing. Study results indicated a significant decrease in travel time and required transfers [11].

AI can also be used for enhancing ridesharing experience. Many ridesharing operators are reportedly exploiting AI methods, to optimize their routing strategies and therefore reduce waiting times and increase their ridership [12]. Furthermore, research has shown that general benefits, such as the increase of the daily income per driver, as well as a decrease in congestion levels can be achieved, through the application of deep learning algorithms, such as multi – agent reinforcement learning [13]. An optimization of travel demand has also been achieved through deep learning algorithms, which give operators the ability to forecast ride requests spatially and temporally and therefore allocate their fleet, accordingly, leading to cost savings. The algorithm employed, is using passenger pickup and drop-off historic data as input and through the exploitation of Convolutional Neural Network or a Long Short-Term memory algorithm, predicts the future demand in several time periods [14]. AI methods are also used for the identification of illegal ridesharing activity, thus helping city authorities and operators to deter such phenomena and protect customers [15].

Logistics operators benefit from the induction of AI techniques to their operational procedures. Studies suggest that the introduction of AI to logistics can reduce operating costs by 5% and increase profits by $25bn in a decade [16]. The main benefit that arises from using AI in logistics refers to the optimization of delivery routes and therefore the limitation of cost. Operators claim that for every mile a route is shortened in a day, the operator can moderate expenses by approximately $50 m annually [16]. Research has indicated that the use of Ant Colony Optimization can improve rescheduling of logistics services, especially in dynamic optimization procedures [17]. Furthermore, attempts to forecast travel time of vehicles have been made, using algorithms that exploit historic data and create spatiotemporal patterns, which are used for the optimal routing of logistics fleets [18].

Companies that are active on the recently introduced sector of shared mobility, which among others includes bike-sharing and e-scooters, deploy AI applications as well. A key characteristic of a shared system is its ability to serve users' demands at all times in various places inside the network, something that requires an efficient system rebalancing process. In dockless systems the rebalancing problem is more significant, since the variance of spatial data, regarding bicycle or e-scooter position, is considerably high. Therefore, operators have been using AI and ML algorithms, in an effort to overcome this obstacle and improve system efficiency. Research in China has shown that the application of deep learning algorithms with the exploitation of spatiotemporal historic data from bike-sharing operators, can determine an optimized rebalancing solution [19]. Additionally, demand projections can contribute to an increased level of service in a busy bike-sharing network. The evaluation of a ML algorithm, with historic data retrieved from the bike-sharing network of Washington D.C., indicated that accurate demand predictions can be achieved on a 24-h timeframe [20].

The European Commission (EC), realising the benefits that arise from the introduction of AI techniques into the operational procedures of organisations and enterprises across multiple economic sectors, is planning to increase the take-up of AI in the next decade. Specifically, EC is focusing its efforts to increase investments in the field, to €20 billion by the end of 2020, while the aim for the coming decade is to increase investments by the same amount annually [21]. Among the EC's investment targets, the focus regarding the transport sector is set on the introduction of more innovation, testing and experimentation centres. Furthermore, action is to be taken towards increasing AI accessibility especially to Small and Medium Enterprises (SMEs) [21]. However, these EC targets require increased capacity by organisations and enterprises. Currently, data indicate that companies do not have the required expertise on the field, since only a small margin of SMEs have achieved a high digitization level, while one third of employees is not familiar with basic computer skills [21].

3 AI Use by Transport Operators in Greece

In line with this study's objective, a questionnaire survey was accordingly designed and realised in order to capture the penetration of AI in Greek transport organisations. In total it consisted of 28 questions that aimed to capture organizations' understanding of AI, specify their respective degree of readiness to deploy the associated technologies and applications and highlight discouraging factors that influence their decision to introduce AI to their operational procedures. The participating sample comprised of six (6) transport actors, which are based in Thessaloniki, Greece's second largest city, operating on various services and included: (a) a freight logistics company (b) a public transport authority (c) a public transport operator (bus modes only) (d) an association of taxi operators (e) a micromobility operator (shared bicycles) and (f) a parking management company. The micromobility operator and the logistics company operate internationally, while the other actors operate locally.

The questionnaire survey was conducted in March 2020. The questionnaire form was distributed to the selected actors and was completed according to the views and insight of the IT personnel along with the most relevant directors of these organisations. The survey's questions (Table 1) aim at highlighting the maturity of each operator against three (3) core pillars which describe the critical foundations for applying AI techniques [22]. Therefore, the survey is divided in three (3) corresponding sections, i.e.: (a) the utilisation and organisational structures regarding Data collection/Mining methods (DM) (b) the assessment of Organizational Readiness (OR), regarding the deployment of AI techniques and (c) the evaluation of actors' Technological Equipment (TE). In order to analyse the maturity of each operator, we constructed an AI maturity index. The factors, which determine the maturity index, were based on existing literature and on the particularities of Greek transport operators and were estimated through the survey's questions (Table 1).

Table 1. Measurement of artificial intelligence (AI) maturity index.

Category	Question	Answers' points	Maximum score
DM	Utilisation of automated data collection methods	None: 0 Low: 0.25 Medium: 0.5 High: 0.75 Very high: 1	7
	Data update frequency		
	Utilisation of social media for data collection		
OR	Contribution of AI to organization's goals		
	Improvement of organizational activities through the exploitation of AI		
TE	Technological equipment update frequency		
	Personnel's familiarization with programming languages		
DM	Utilisation of cloud services for storing data	No: 0 Yes: 1	5
	Utilisation of data filtering procedures		
	Employment of specialised personnel for data collection procedures		
	Employment of specialised personnel for data analysis procedures		
OR	Use of AI to meet organization's goals		

Each question is answered using a single – choice answer type, i.e. a "yes/no" or a 5 – point answering scale. To calculate the AI maturity indices, each question was assigned with a maximum score of 1. In questions, which used a 5 – point scale, the score was allocated accordingly to each option, while in polar questions, the value of 1 was appointed to the affirmative option.

In general, a maturity index consists of a series of levels that describe the relative maturity of an organisation or enterprise regarding a certain feature or domain. The lowest level of the scale indicates the total absence of maturity and can characterize organisations that have low experience. On the other hand, the top level indicates total maturity [23].

In this study, the AI maturity appraisal scale we designed is presented in Table 2. It consists of four (4) stages of maturity, ranging from "AI Beginner" to "AI Mature". AI Beginners have not taken serious steps towards adjusting their operational procedures to assist the take – up of AI. Operators classified as AI Intermediates recognize the importance of AI and the added value that it can bring to their organisations and are preparing for the introduction of such techniques. The AI advanced category indicates operators that have completed the necessary preparations towards adopting AI, while at the same time have limited experience with applying AI methods. Operators on the AI mature cluster have introduced AI techniques and are systematically applying them to their operational procedures.

Table 2. Artificial intelligence (AI) maturity appraisal scale.

Score	0 to 3	3 to 6	6 to 9	9 to 12
Rating	AI Beginner	AI Intermediate	AI Advanced	AI Mature
Description	Has taken limited action towards adopting AI	Recognizes the benefit of AI Is preparing for the application of AI	Is sufficiently prepared for adopting AI Limited experience with AI	Is systematically applying AI techniques

4 Results and Discussion

Figure 1 illustrates the maturity level of the transport operators that took part in the survey. The circle at the centre of every graph indicates the maturity score of each operator, while the 3 concentric rings around the circle indicate the level of maturity of the operator in terms of the three core pillars mentioned in the previous section.

Almost all the actors investigated in this research could be classified as "AI Intermediate", with the exception of the micromobility operator which is classified as "AI Advanced". According to the survey data, the actors that were found to be "AI Intermediates" recognize the benefit of AI techniques and applications and are sufficiently prepared in terms of strategy, organizational set-up and data availability in order to move forward and implement AI technology and solutions in defined operational scenarios. They are ready to take the next step by making the necessary investments. They mostly collect data concerning the geographic location and monitoring of their fleet. Such data and information are usually used by the respective actors to record and monitor demand as well as optimize system management. Despite the fact, that all

"AI intermediate" actors display relatively high scores regarding DM and in less degree OR, their main weakness is the lack of the appropriate TE. Especially, for the public transport authority the TE score is remarkably low, because the organization does not possess technological equipment, which means that it relies on external partners in order to fulfil its relevant needs.

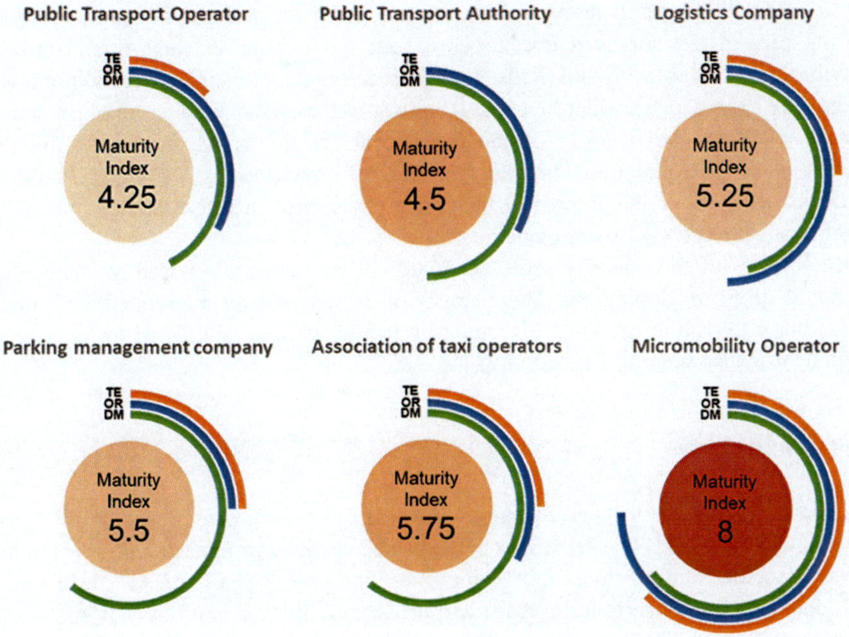

Fig. 1. Survey results: maturity index of selected Greek transport actors.

The public transport operator achieved the lowest AI maturity score (4.25) among the participating transport actors. Overall, the readiness percentage of the operator in question was low on all three pillars, with the TE one being the weakest. Among the contributing factors for this result, are the absence of an organized data management structure, which comes in contradiction with the real – time data collection mechanisms that the operator employs. Furthermore, the operator's personnel are characterized by a relatively low technological capacity, which comes as a deterrent to the further introduction of AI applications in the organization.

The micromobility operator, on the other hand, is classified as "AI Advanced" with an AI maturity score of 8. This operator is sufficiently prepared for adopting and moving forward with AI techniques but there are still certain gaps and limitations regarding its strategy, data collection/management capabilities and technological resources. The micromobility operator presents the highest OR in our sample, as it already applies AI techniques such as machine learning and neural network applications. The operator collects data through a plethora of automated mechanisms, such as automated vehicle location sensors, fleet monitoring through the media access control (MAC) address or Bluetooth etc. The datasets gathered are mainly exploited for fleet monitoring, economic information regarding the use of service and incident monitoring. Contrary to the other 5 operators that took part in the survey, the micromobility operator appears to have the required technological capacity for the exploitation of AI techniques, since its equipment is frequently updated, and its personnel is familiar with the necessary code programming languages.

Regarding the benefits deriving from the use of AI, according to the survey results, the majority of the surveyed executives expect AI to improve their organizational activities in different levels and fields. Transport actors direct their AI attention in areas where they expect to see wider benefits. Resource utilization and reduction of operating costs is cited by the majority of the respondents as the greatest benefit from the application of AI techniques. The transformation of organizations' service and product portfolio, as well as the improved customer experience are also among the main benefits of AI expressed by respondents.

Regardless of the maturity score achieved, all respondents referred to issues that prevent them from adopting AI. The majority of the respondents mentioned the limited accessibility to data as the most discouraging factor, while the outdated technological equipment is also a serious deterrent.

5 Conclusions

In the last decade the utilisation of AI in various business sectors, in transport alike, has significantly increased. The versatility of AI makes it a great tool for practitioners in the transport sector. While autonomous vehicles are the best example of AI use in transport, operators and experts in logistics and urban mobility are continuously including AI in their arsenal. The increase of AI use in transport is directly linked to the wide benefits that derive from its exploitation. Cost savings, increased road safety, improved travel and delivery times, as well as improved environmental conditions are some of the gains from the application of AI techniques in transport.

In this paper we evaluate the maturity level of selected transport operators in Greece, regarding their readiness in terms of applying AI techniques in their operational procedures. The evaluation was conducted through a questionnaire survey, which examined several characteristics, in three main categories, data collection/management, organizational readiness and technological equipment.

Results of the survey indicate a relatively low maturity level for the majority of the operators in the sample. It seems that the actors' operating location does not influence their maturity index results. More in detail, 5 of the 6 respondents were categorised in the "AI Intermediate" cluster, while one operator was assessed as "AI Advanced". The two organisations from the public transport sector achieved the two lowest scores, while the micromobility operator was the most mature in terms of adopting AI.

Results also showed that the majority of the examined Greek operators do not exploit AI technologies, such as those discussed in the literature review section. Furthermore, since their maturity index is relatively low, they are not equally prepared to their EU counterparts who have already incorporated such technologies.

The survey gave also insight in the factors that influence the take up of AI, with the shortage of data and the up-to-date technological equipment being the most important parameters. Reduced operating costs as well as the improvement regarding resource exploitation were highlighted as the main benefits from the introduction of AI applications. The main deterrents towards applying AI techniques in transport in Greece, are linked with the limited familiarization of personnel with the concept and the potential of AI, as well as with the outdated technological equipment. Furthermore, the limitations regarding data accessibility which are encountered by many stakeholders in Greece, not only active in the transport sector, form a barrier in the adoption of AI, since accurate results require large and detailed datasets.

The organisation of training courses on the benefits of AI and its prerequisites, can work as catalyst in the take – up of AI in the public and possibly private sector, since it will allow managers to comprehend the potential of AI to their organisation and increase the employees' capacity regarding the application of AI. Additionally, the incentivisation of companies regarding the frequent update of technological equipment can assist the introduction of AI. Finally, the development of a national open access database repository, which will concentrate data deriving from the research and industrial activities of several organisations or companies in Greece, can contribute to the uptake of AI by practitioners.

The paper's derived conclusions can prove as the foundation of future research for the determination of a more thorough AI maturity appraisal tool, as well as assist policy makers into focusing their efforts on specific sectors in order to increase the take – up of AI on transport operators.

Acknowledgements. The authors would like to acknowledge the six (6) Greek transport operators and their representatives who responded to our request and participated in the questionnaire survey. Their interviews significantly contributed to the completion of this study.

References

1. Patterson, D.: Introduction to Artificial Intelligence and Expert Systems. Prentice-Hall Inc., USA (1990)
2. Nilsson, N.J.: Principles of Artificial Intelligence. Morgan Kaufmann, Burlington (1982). https://doi.org/10.1007/978-3-662-09438-9
3. OLeary, D.E.: Artificial intelligence and big data. IEEE Intell. Syst. **28**(2), 96–99 (2013)
4. Sadek, A.W.: Artificial intelligence in transportation information for application. Washington, D.C. (2007)
5. Sizing the prize - PwC's Global Artificial Intelligence Study: Exploiting the AI Revolution | Printemps numérique. https://www.printempsnumerique.ca/en/studies/study/sizing-the-prize-pwcs-global-artificial-intelligence-study-exploiting-the-ai-revolution/. Accessed 21 July 2020
6. Gartner Survey of More Than 3,000 CIOs Reveals That Enterprises Are Entering the Third Era of IT. https://www.gartner.com/en/newsroom/press-releases/2018–10-16-gartner-survey-of-more-than-3000-cios-reveals-that-enterprises-are-entering-the-third-era-of-it. Accessed 21 July 2020

7. Conde, M.L., Twinn, I.: How Artificial Intelligence is Making Transport Safer, Cleaner, More Reliable and Efficient in Emerging Markets. Washington, D.C. (2019)
8. Greece: With an AI to the Future | Accenture. https://www.accenture.com/gr-en/insights/digital/greece-an-ai-future. Accessed 21 July 2020
9. Brebbia, C.A., Longhurst, J., Marco, E., Booth, C.: Sustainable Development and Planning IX. WIT Press, Ashurst (2017)
10. Gurmu, Z.K., Fan, W.D.: Artificial neural network travel time prediction model for buses using only GPS data. J. Public Transp. **17**(2), 3 (2014). https://doi.org/10.5038/2375-0901.17.2.3
11. Yang, Z., Yu, B., Cheng, C.: A parallel ant colony algorithm for bus network optimization. Comput. Aided Civ. Infrastruct. Eng. **22**(1), 44–55 (2007). https://doi.org/10.1111/j.1467-8667.2006.00469
12. Uber Might Be The First AI-First Company, Which Is Why They Don't Even Think About It Anymore. https://www.forbes.com/sites/johnkoetsier/2018/08/22/uber-might-be-the-first-ai-first-company-which-is-why-they-dont-even-think-about-it-anymore/#65254ae35b62. Accessed 21 July 2020
13. Li, M., Yang, Y., Wang, C., Qin, Z., Gong, Z., Wu, G., Jiao, Y., Wang, J., Ye, J.: Efficient ridesharing order dispatching with mean field multi-agent reinforcement learning. In: The Web Conference 2019 - Proceedings of the World Wide Web Conference, WWW 2019 (2019). https://doi.org/10.1145/3308558.3313433
14. Yao, H., Wu, F., Ke, J., Tang, X., Jia, Y., Lu, S., Gong, P., Li, Z., Ye, J., Chuxing, D.: Deep multi-view spatial-temporal network for taxi demand prediction. In: 32nd AAAI Conference on Artificial Intelligence, AAAI 2018 (2018)
15. Wang, L., Geng, X., Ma, X., Zhang, D., Yang, Q.: Ridesharing car detection by transfer learning. Artif. Intell. **273**, 1–18 (2019). https://doi.org/10.1016/j.artint.2018.12.008
16. In algorithms we trust - How AI is spreading throughout the supply chain | Special report | The Economist. https://www.economist.com/special-report/2018/03/28/how-ai-is-spreading-throughout-the-supply-chain. Accessed 21 July 2020
17. Silva, C.A., Sousa, J.M.C., Runkler, T.A.: Rescheduling and optimization of logistic processes using GA and ACO. Eng. Appl. Artif. Intell. **21**(3), 343–352 (2008). https://doi.org/10.1016/j.engappai.2007.08.006
18. Wen, R., Yan, W., Zhang, A.N.: Adaptive spatio-temporal mining for route planning and travel time estimation. In: Proceedings - 2017 IEEE International Conference on Big Data, Big Data 2017 (2017). https://doi.org/10.1109/BigData.2017.8258311
19. Pan, L., Cai, Q., Fang, Z., Tang, P., Huang, L.: A deep reinforcement learning framework for rebalancing dockless bike sharing systems. In: Proceedings of the AAAI Conference on Artificial Intelligence, vol. 33 (2019). https://doi.org/10.1609/aaai.v33i01.33011393
20. Giot, R., Cherrier, R.: Predicting bikeshare system usage up to one day ahead. In: IEEE SSCI 2014: 2014 IEEE Symposium Series on Computational Intelligence - CIVTS 2014: 2014 IEEE Symposium on Computational Intelligence in Vehicles and Transportation Systems, Proceedings (2015). https://doi.org/10.1109/CIVTS.2014.7009473
21. Commission staff working document, Liability for emerging digital technologies, Brussels (2018)
22. Pringle, T., Zoller, E.: How to achieve AI maturity and why it matters an AI maturity assessment model and road map for CSPs (2018)
23. Becker, J., Knackstedt, R., Pöppelbuß, J.: Developing Maturity Models for IT Management. Business & Information Systems Engineering, vol. 1 (2009). https://doi.org/10.1007/s12599-009-0044-5

Clustering of Urban Road Paths; Identifying the Optimal Set of Linear and Nonlinear Clustering Features

Glykeria Myrovali[1(✉)], Theodoros Karakasidis[2], Maria Morfoulaki[1], and Georgia Ayfantopoulou[1]

[1] Hellenic Institute of Transport/Centre for Research and Technology Hellas (HIT/CERTH), 6th Km Charilaou-Thermi Road, 57001 Thessaloniki, Greece
{myrovali,marmor,gea}@certh.gr

[2] Laboratory of Hydromechanics and Environmental Engineering, Department of Civil Engineering, University of Thessaly, 38334 Volos, Greece
thkarak@uth.gr

Abstract. Urban traffic is undoubtedly a dynamic phenomenon presenting variations over both time and space, that in the majority of cases are the result of a mixture of, either well known (i.e. weather, seasonality) or not easily predictable (i.e. events, accidents) external factors. Identification of similarities in the performance of different urban road paths under different traffic states (different travel demand conditions) is the main subject of the current paper. Floating taxi travel time data (timeseries per road path) collected in the framework of Thessaloniki Smart Mobility Living Lab (initiated and operated by CERTH/HIT) consist the basic input for the hierarchical clustering that is applied. Clustering applies upon different combinations of road paths' features (data points of travel time timeseries, descriptive statistics and mutual information of timeseries). The comparison of the clustering results based on average weekdays travel times per road path (from a six months period) with the respective results of a typical and an atypical day adds on the interpretability of underlying relations among paths under different states. The analysis reveals that resulting clusters can be a building block for the spatiotemporal understanding of urban traffic. Furthermore, it is shown that adding as clustering feature the criterion of mutual information of timeseries, therefore taking into account also non-linear dependences of the different road paths, the clustering interpretability is differentiated.

Keywords: Urban traffic state · Timeseries · Clustering · Mutual information criterion

1 Introduction

1.1 Urban Road Traffic; an Undoubtedly Dynamic System

Urban road traffic is undoubtedly a dynamic, therefore complex, system which becomes even more complex with the increase of city size and of the number of intersections (nodes) of the road network. The gradual (structured or even unstructured)

E. G. Nathanail et al. (Eds.): CSUM2020, AISC 1278, pp. 1107–1116, 2021.
https://doi.org/10.1007/978-3-030-61075-3_106

cities' evolution has increased the complexity of urban structures and functionalities, part of which is the urban transport network and urban road traffic. Road network structure (topology), socio-economic factors, weather effects, traffic management policies and mobility options, purpose of daily trips, mode choice, trips distribution during the day and drivers' behavior and choices are some of the aspects which contribute in system's complexity. The multiplicity of decisions (personal decisions of many discrete individuals, which are nevertheless strongly related) adds also in the complexity of urban traffic [1–8].

Urban traffic, being a non-static phenomenon, is subject to a series of variations related to both space and time. The former refers to the different traffic patterns per case (road path, sub-area, flow direction) while the latter relates to traffic variations over different time-periods:

- Short term variations caused by signaling (delays)
- Daytime variation- peak and non-peak hours
- Variation between days - weekdays, Saturday and Sunday
- Long-term variations, fluctuations over the years due to demographic, economic and geographical changes [9].

Under normal situations (recurrent and known patterns), when no special event has occurred, urban road systems operate in the expected way. When this balance is disturbed, with demand to exceed capacity ('*the maximum number of vehicles that can reasonably be expected to be served in the given time period*' [10]) or with the appearance of a special event, congestion phenomena are noticed [11, 12] while ex ante flexible and quickly adaptable traffic management schemes should have been mobilized. Spatiotemporal dynamics understanding is a main challenge for transport systems management. In this paper, complementing other spatiotemporal analysis, clustering analysis is estimated to be able to help identify operation scenarios that can support transportation system management. The key goal of the paper is not to compare the effectiveness of different clustering methodologies but to compare resulting clusters at different traffic states.

1.2 Aim and Structure

During the last decades, the large penetration of Intelligent Transport Systems in daily city's operations has changed the traditional way of collecting and analyzing data for transport planning and traffic management; large databases of traffic data have been built and are updated at real time. Floating car data, Bluetooth readers with application on mobility, cameras/radars and traffic related data mined from social media are jointly offering a broader picture of the traffic situation for the entire road network [13, 14].

Taking advantage of the existence of such a large traffic databases (time series) in the framework of Thessaloniki Smart Mobility Living Lab (https://smartmlab.imet.gr/), the current paper aims at presenting two practical approaches in clustering urban road paths; clustering based on point values and clustering based on global features extracted from the time series. The scope per research orientation defines the effectiveness of the different approaches while the interpretability of the clusters (injection of prior knowledge under an intuitive way) acts as a measure of effectiveness and accuracy.

The remainder of the paper is organized as follows. Section 2 presents the reference area where analysis took place, the main urban road paths and the timeseries used for the clustering of the road paths. Section 3 presents in brief the methodological approach while results are presented in Sect. 4. Concluding the paper, Sect. 5 highlights key advantages and opportunities of the proposed approach.

2 The Reference Research Area

2.1 Thessaloniki's Urban Road Network

The reference area for the current work consists of basic road paths of the wider area of Thessaloniki, a medium size Greek city with a population of around 800.000 inhabitants (2011). High dependence of private vehicles use (50%) is characterizing the city transport (SUMP of Thessaloniki, www.svakthess.imet.gr). The selection of road paths follows wider research scopes (availability of different traffic datasets and already existing geo-reference from CERTH/HIT research activity [15–22]). The examined paths (with coding, direction, location) are presented in Fig. 1.

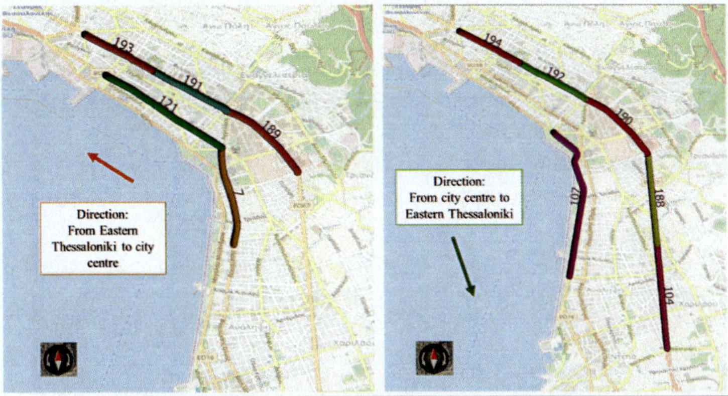

	path7	path104	path107	path121	path188	path190	path191	path192	path193	path194
Matlab coding	*1*	*2*	*3*	*4*	*5*	*6*	*7*	*8*	*9*	*10*
Path length (km)	1.3	1.6	2.3	1.5	1.4	1.1	0.95	0.95	0.85	0.85
Direction	west	east	east	west	east	east	west	east	west	east
Number of lanes	Bus lane + 2 operational		Bus lane + 4 operational	Bus lane + 2 operational						
Subpart of greater axis*	V. Olgas	K. Karamanli	M. Alexandrou	Tsimiski	K. Karamanli	Egnatia	Egnatia	Egnatia	Egnatia	Egnatia
Daily Average Flow / Passenger Car Unitσ (PCU)	35.000	35.000	45.000		35.000 – 40.000					

* subparts oflarger horizontal road axis are highlighted in the same color

Fig. 1. Examined urban road paths, Thessaloniki (GR), and respective coding for the analysis in MATLAB software.

Figure 1 contains also information on the operational characteristics of the paths; as a general comment we can say that all road paths examined are parts of the three large horizontal road axis serving a large percentage of daily trips. With the exception of Path 107 all the other paths are of similar length (1–1.6 km) while consist of one bus lane plus two more operational lanes.

2.2 The Traffic Data

After verifying the relatively high accuracy and validity of the information contained in traffic data collected via 1200 floating taxi (GDPR respecting) in the study area, travel time taxi GPS based timeseries were used for the current analysis. Travel time timeseries per road path in 15 min intervals (reference period: first semester of 2017, hereinafter A2017) consisted the basis of the analysis– 13.966.878 timestamps were used for calculating average travel times based on CERTH/HIT methodology [20, 21].

3 The Hierarchical Clustering Approach

3.1 Methodological Steps

Data retrieval from taxi floating data and travel time data per road path were the input data received from CERTH/HIT – additional elaboration of the data took place for the current work:

1a. Travel time timeseries per examined road path in 15-min intervals for a long period were extracted, interpolation took place for completing the timeseries where necessary (10560 data points for A2017).
1b. Travel time timeseries of specific days in 15-min intervals (one typical – expected traffic variations and one atypical, day with extreme weather conditions) were elicited.
1c. Data aggregation for calculating average travel times per road path in 15-min intervals took place developing the average travel time timeseries;
2. Timeseries features were also calculated for the average travel time timeseries (1c).

Finally, clustering applied and comparison among the resulting clusters based on point data and on timeseries feature took place.

3.2 Hierarchical Clustering

Time series clustering remains a challenging issue partially attributed to its high sensitivity to input data. In its generic notion, cluster analysis or clustering is the grouping of objects that present similarities. The objects are placed into groups, based on specific measures and similarity criteria according to the following principles:

- In each created group, the close observations resemble each other in the characteristics examined
- The objects belonging in the same group differ significantly from those of the rest groups.

The ultimate scope of clusters analysis is to draw useful conclusions about the timeseries while the efficiency of the methods used is related to the type and size of the timeseries but also to the use of similarity criteria (ex-ante detailed definition of research scope i.e. describing objects or describing relations?).

In the current paper, hierarchical clustering is applied given its principle advantage of not requiring a priori knowledge of the number of clusters and its visualization strength (dendograms generation) [23]. The agglomerative bottom-up clustering approach is exploited – starting with each observation as a cluster, the two closest clusters are joined into one while this repetitive process ends when there is only one cluster with all the data [24]. For a more detailed description on the hierarchical clustering, the interested reader is redirected to [25–27].

3.3 The Input Features for Clustering Trials

Following the distinction among ‘shape level’ and ‘structure level’ clustering of Wang et al. (2006), the current paper presents a clustering of urban road paths (timeseries) based on point values as well as on their structural characteristics. Structural characteristics in the current analysis include distribution characteristics and underlying relations: max, min, mean, skewness, kurtosis, autocorrelation (linear measure) and mutual information (non linear measure) [28].

4 Results

4.1 Long Length Timeseries vs Timeseries of a Typical and an Atypical Day

Aiming to identify similarities at the performance (travel time variations within the day) of road paths under typical (averaged travel time data of a long period) network operation and under non recurrent events and to explore whether abstract similar patterns among road paths exist, hierarchical clustering applied at different scenarios:

i. Scen.A1: Raw travel time timeseries (x-axis: 15 min intervals) of a large period, namely of one semester (weekdays) – 150 days, 96 values per day.
ii. Scen.A2: Raw travel time timeseries of a typical day (within the same semester, a day with typical and expected traffic conditions) – one day, 96 values.
iii. Scen.A3: Raw travel time timeseries of an atypical day (within the same semester, a day with extreme weather conditions) – one day, 96 values.

MATLAB software was used for applying clustering, dendrograms generation and cophenetic correlation coefficient calculation. Among the similarities/distances tested (Euclidean distance, squared Euclidean distance, single and centroid linkage), Euclidean distance (y-axis) presented the highest correlation coefficient for all three scenarios (varying between 0.6 and 0.7). Figure 2 shows the results of the hierarchical clustering procedure per scenario.

Running again the clustering methodology with average timeseries on weekdays of A2017 the clusters are the same as in A1 with slightly differentiated distances.

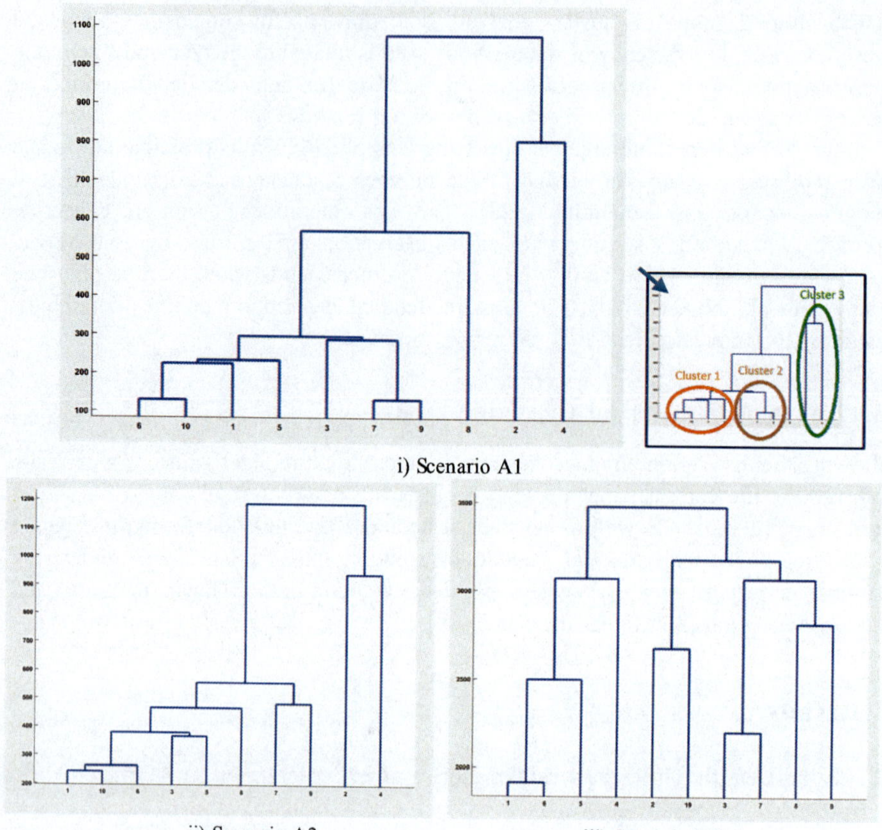

i) Scenario A1

ii) Scenario A2

iii) Scenario A3

Fig. 2. Hierarchical clusters dendrograms per scenario Ax.

The dendrograms of scenarios 1 and 2 presents high similarity (branches) fact that can be used as an indication that objects (road paths) similarity at a typical weekday is high enough to high dimensionality clustering such as travel time timeseries of a large period (approach incorporating the notion of averaging). The prior knowledge for the reference paths under expected traffic state verifies the 3 obvious clusters created; Cluster 1 consists of paths 190 and 194 that are paths of same length, consisting part of a greater axis crossing the city center (direction to the east) and from path 7 that has similar length and operational characteristics (1 bus lane + 2 lanes, average daily flows of similar levels) and serves the opposite direction, Cluster 2 consists also of the opposite direction paths to 190 and 194, namely of 191 and 193 (direction to the West) and 107 with direction to East, Cluster 3 consists of the last legs (with the opposite direction) of the two large horizontal axis that are examined. Path 192 seems to differentiate from the other paths, which is indeed expected (it serves also turnover trips around the city center) – correlation analysis undertaken as an extension to the above analysis, verifies also the above understanding and interpretation.

From the other side, when clustering road paths using travel times based on raw data of an atypical day (extreme weather conditions with severe congestion in the whole city) the clusters derived are, as expected, totally different from these of the first two scenarios. Distances are higher while also similarities change significantly; the developed Clusters are difficult to be interpreted given also the breakdown that was noticed the specific date in the city (snowcovered roads, widespread traffic jam).

4.2 Raw Data Clustering vs Clustering Based on Linear and a Combination of Non-linear Timeseries Characteristics

Serving the second goal of the current work, namely to compare the clustering results based on raw data and on features in time domain extracted from the raw datasets, hierarchical clustering was conducted based on distribution related features – max, min, mean, skewness, kurtosis – (linear measure of) autocorrelation and (non linear measure of) mutual information of the timeseries. Figure 3 shows the results of the structural characteristics based clustering (following a standardization procedure due to the existence of big differentiations in the values - subtracting of the mean of the attribute values and dividing by the standard deviation).

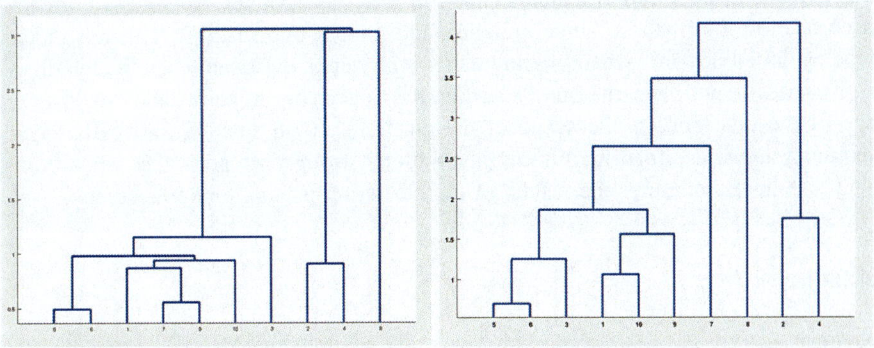

i) Scenario B1 – without mutual information   ii) Scenario B2 – including mutual information

Fig. 3. Hierarchical clustering dendrograms based on structural characteristics.

The resulting clusters in the two scenarios B1 and B2 are differentiated in a some degree; distances among road paths and positioning among clusters are not exactly the same which shows a relevant impact of taking into consideration also non linear features while conducting clustering of objects belonging to a dynamic system. Although the clusters do not differ to such a large degree, the morphology of the B2 dendrogram seems more clear – it is however an indication which should be further investigated however given the fact that traffic is a dynamic phenomenon it is well noted that including non linear features can enhance results interpretability.

5 Conclusions and Discussion

Many studies have dealt with clustering traffic data variables, however in the framework of the current work we focused on exploring the efficiency of dimensionality reduction of large datasets rather than selecting the most efficient clustering algorithm. In line with previous research results [26–29], the current paper showed that using dimensionality reduction methods can provide high accuracy clustering results with a very low computational cost; road paths clustering based on long length travel time timeseries is similarly enough with the clustering based on average travel time timeseries (averaging the 110 weekdays of A2017). Furthermore, the results of the clustering of average travel time timeseries of A2017 do not differ much from the clustering of road paths in a typical weekday. From the other side and as expected, the road paths clustering is totally different under unusual circumstances (a day with extreme weather conditions). Correlation and causality analysis [22] shows also the different underlying relations between timeseries under different traffic states, therefore a two steps analysis joining the results of correlation-causality and clustering is estimated to add in the interpretability of outputs –clusters can be only a part of the total necessary input for a decision support system. From the other side and when concept-based clustering, i.e. relationships interpretation is the scope of the research, clustering can be based on characteristics as distribution features (max, min & mean values, skewness, kurtosis) and correlation information (autocorrelation and mutual information). This advantage can be exploited not only when missing data in timeseries exist and when the length of the timeseries is not the same but also when the underlying relations and not just the shape of the timeseries is the scope of the clustering. And since we are talking for dynamic systems, as urban road traffic, the authors strongly supports that when combining non linear measures the results of the clustering present higher accuracy.

References

1. Barthelemy, M.: Spatial networks. Phys. Rep. **499**, 1–101 (2011)
2. Strano, E., Nicosia, V., Latora, V., Porta, S., Barthelemy, M.: Elementary processes governing the evolution of road networks. Sci. Rep. **2**, 296 (2012)
3. Van Ommeren, J., Rietveld, P., Nijkamp, P.: Job mobility, residential mobility and commuting: a theoretical analysis using search theory. Ann. Reg. Sci. **34**(2), 213–232 (2000)
4. Polyzos, S.: Urban Development, p. 541. Kritiki Publications, Athens (2015)
5. Polyzos, S., Tsiotas, D., Minetos, D.: Determining the driving factors of commuting: an empirical analysis from greece. J. Eng. Sci. Technol. Rev. **6**(3), 46–55 (2013)
6. Chowell, G., Hyman, J.M., Eubank, S., Castillo-Chavez, C.: Scaling laws for the movement of people between locations in a large city. Phys. Rev. **68**, 066102 (2003)
7. Kwon, J., Mauch, M., Varaiya, P.: Components of congestion: delay from incidents, special events, lane closures, weather, potential ramp metering gain, and excess demand. Transp. Res. Rec. J. Transp. Res. Board **1959**(1), 84–91 (2006)
8. Wen, T.H., Chin, W.C.B., Lai, P.C.: Understanding the topological characteristics and flow complexity of urban traffic congestion. Phys. A Stat. Mech. Appl. **473**, 166–177 (2017)
9. Weijermars, W.A.: Analysis of urban traffic patterns using clustering (2007)

10. Roess, R.P., McShane, W.R., Prassas, E.S.: Traffic Engineering, 2nd edn. Pretence Hall, USA (1998)
11. Rehborn, H., Klenov, S.L., Palmer, J.: An empirical study of common traffic congestion features based on traffic data measured in the USA, the UK, and Germany. Phys. A Stat. Mech. Appl. **390**(23–24), 4466–4485 (2011)
12. Wright, C., Roberg, P.: The conceptual structure of traffic jams. Transp. Policy **5**(1), 23–35 (1998)
13. Leduc, G.: Road traffic data: collection methods and applications. Working Pap. Energ. Transp. Clim. Change **1**(55), 1–55 (2008)
14. Efthymiou, D., Antoniou, C.: Use of social media for transport data collection. Procedia Soc. Behav. Sci. **48**, 775–785 (2012)
15. Salanova Grau, J.M., Toumpalidis, I., Chaniotakis, E., Karanikolas, N., Aifadopoulou, G.: Correlation between digital and physical world, case study in Thessaloniki. J. Location Based Serv. **11**(2), 118–132 (2018)
16. Salanova Grau, J.M., Mitsakis, E., Tzenos, P., Stamos, I., Selmi, L., Aifadopoulou, G.: Multisource data framework for road traffic state estimation. J. Adv. Transp. (2018)
17. Aifadopoulou, G., Salanova Grau, J.M., Tzenos, P., Stamos, I., Mitsakis, E.: Big and open data supporting sustainable mobility in smart cities – the case of Thessaloniki. In: Proceedings of 4th Conference on Sustainable Urban Mobility (CSUM2018), pp. 24–25. May, Skiathos Island, Greece (2019)
18. Mitsakis, E., Chrysohoou, E., Salanova Grau, J.M., Iordanopoulos, P., Aifadopoulou, G.: The sensor location problem: Methodological approach and application. Transport **32**(2), 113–119 (2017)
19. Mitsakis, E., Salanova Grau, J.M., Chrysohoou, E., Aifadopoulou, G.: A robust method for real time estimation of travel times for dense urban road networks using point-to-point detectors. Transport **30**(3), 1648–4142 (2015)
20. Stamos, I., Salanova Grau, J.M., Mitsakis, E., Aifadopoulou, G.: Modeling effects of precipitation on vehicle speed: floating car data approach. Transp. Res. Rec. J. Transp. Res. Board **2551**(1), 100–110 (2015)
21. Salanova Grau, J.M., Maciejewski, M., Bischoff, J., Estrada, M., Tzenos, P., Stamos, I.: Use of probe data generated by taxis. Big Data for Regional Science. Routledge Advances in Regional Economics, Science and Policy. Taylor & Francis Group, Abingdon (2017)
22. Myrovali, G., Karakasidis, T., Charakopoulos, A., Tzenos, P., Morfoulaki, M., Aifadopoulou, G.: Exploiting the knowledge of dynamics, correlations and causalities in the performance of different road paths for enhancing urban transport management. In: Freitas, P., Dargam, F., Moreno, J. (eds) Decision Support Systems IX: Main Developments and Future Trends. EmC-ICDSST 2019. Lecture Notes in Business Information Processing, vol 348. Springer, Cham (2019)
23. Keogh, E., Lin, J., Truppel, W.: Clustering of time series subsequences is meaningless: implications for past and future research. In: Proceedings of the 3rd IEEE International Conference on Data Mining, Melbourne, FL, USA, pp. 115–122 (2003)
24. An introduction to data science, Dr. Saed Sayad. https://www.saedsayad.com/clustering_hierarchical.htm. Accessed 06 2020
25. Steinbach, M., Karypis, G., Kumar, V.: A comparison of document clustering techniques. KDD Workshop Text Min. **400**(1), 525–526 (2000)
26. Kraskov, A., Stögbauer, H., Andrzejak, R.G., Grassberger, P.: Hierarchical clustering using mutual information. EPL Europhys. Lett. **70**(2), 278 (2005)
27. Kojadinovic, I.: Agglomerative hierarchical clustering of continuous variables based on mutual information. Comput. Stat. Data Anal. **46**(2), 269–294 (2004)

28. Wang, X., Smith-Miles, K., Hyndman, R.: Characteristic-based clustering for time series data. Data Min. Knowl. Discov. **13**, 335–364 (2006)
29. Steinbach, M., Ertoz, L., Kumar, V.: Challenges of clustering high dimensional data. In L.T. Wille, editor, New Vistas in Statistical Physics - Applications in Econophysics, Bioinformatics, and Pattern Recognition. Springer-Verlag (2003)

Exploring the Big Data Usage in Transport Modelling

Danai Tzika-Kostopoulou(✉) and Eftihia Nathanail

Department of Civil Engineering, University of Thessaly,
Pedion Areos, 38334 Volos, Greece
danaitzika@gmail.com

Abstract. Continuous growth of information and the increasing volume of data with high coverage in space and time open up new possibilities in the field of transport planning. In recent years, there is great research interest in how big data can be applied to the modelling and planning of transport systems. A literature survey of existing methodologies and applications of big dada in transportation is a useful tool for identifying strengths and capabilities for big data exploitation in different fields of application. The main objective of this paper is to provide a comprehensive overview of big data usage in transport planning, focusing on travel demand modelling. More specifically, the paper aims to examine whether analyzing big data can facilitate transport planning and to summarize the relative scientific discussion. Three big data sources have been examined: smart cards, mobile phones and social networks. Existing theories and studies are presented and classified according to the source of data, the methodology, the extracted transportation features and the validation of results. In the course of the review, the different big data sources are further analyzed regarding their special applications in the transport planning field. Finally, the paper concludes by presenting the barriers and gaps in the existing approaches as well as new research challenges.

Keywords: Big data · Travel demand modelling · Transport planning

1 Introduction

Predicting travel demand and understanding human mobility patterns is central to transport planning. The dominant methods to collect the appropriate information are household surveys and travel diaries. These "traditional" methods face specific difficulties especially due to time and budget constraints, while the main advantage is that they can examine mobility patterns taking into account sociodemographic characteristics [1]. On the other hand, the rapid rise of technology and the high advances in data acquisition provide the opportunity to track various data sets that could open up new possibilities in the field of transport planning and complement the existing approaches.

Currently, there is increasing research interest in how big data can be applied in various sectors of transportation. Such data can be derived from mobile phones, public transport automated fare collection systems, social networks, GPS devices, sensors, cameras, or even bank notes [2] and credit cards [3]. Several studies have attempted to

E. G. Nathanail et al. (Eds.): CSUM2020, AISC 1278, pp. 1117–1126, 2021.
https://doi.org/10.1007/978-3-030-61075-3_107

extract crucial attributes for transport planning and modelling by utilizing big data. Human mobility patterns [4, 5], trip identification [6, 7], activity inference [8] and mode detection have been explored using massive and mainly passive generated data.

The main objective of this paper is to showcase a comprehensive overview of big data usage in transport planning, focusing on travel demand modelling. It is attempted to summarize the existing approaches and to underline their challenges and limitations. This paper considers three big data sources (smart cards, mobile phones and social media) that have been widely utilized in transportation research and aims to answer the following questions: 1. Which are the current trends of big data usage in transport planning and what field do they serve; 2. How are they used in transport planning and what methods have been implemented; 3. To what extent are these methods evaluated and validated; 4. How examined big data collaborate with other data sources; 5. What are the most common extracted attributes; 6. What are the unresolved issues and the potentials that could be further explored.

The structure of the paper is as follows. Section 2 presents the process defining the eligible studies. Section 3 reviews the chosen relevant researches that have been conducted over the years and presents a classification of them. The topics are divided according to the big data source and the results of the analysis are presented. Finally, the findings are summarized and some limitations and research perspectives are further discussed.

2 Literature Search Strategy

The first step of this study was to define the reviewed literature. The main criteria were the relevance to the paper's purposes, the representativeness and the accuracy. To achieve this, several steps of selection with increasingly strict criteria were followed. At the beginning, key words were searched in google scholar platform. There were two categories of specific terms and results should contain at least one keyword of each group: 1. big data, passive data, mobile phone, call detail record (CDR), social media, Twitter, Foursquare (owing to their frequency of appearance), automatic fare collection systems, smart card, 2. transport planning, transport modelling, transportation, travel demand, public transport, mobility, travel. Following that, a big paper dataset was collected. In order to perceive the included studies as eligible, they had to satisfy the following criteria: published by scientific journals and conference proceedings, original research, related to transport planning, use data derived from mobile phone, social media or smart cards. After a first screening, a big set was rejected. Afterwards, by screening the abstracts, studies which either mainly dealt with real time applications and transport management or focused on other data sources were excluded.

Lastly, through cross reading, studies were selected based on their research aim, applied methodology and validation. After this last screening, a set of 42 papers remained for further analysis. The references cited by these papers were checked using google scholar. The same check was applied to these references. The whole process was repeated several times. In total 68 papers matched all the criteria.

Taking into account the limited space of the current paper, 33 studies are presented, which were considered as fundamental, representative or contemporary, and it is attempted to highlight the evolution in the researched field through the years (from 2005 till today). As the main objective of the paper is to investigate the progress of big data usage in travel demand modelling, studies attempting to serve this field are overemphasized.

3 Overview of Big Data Sources and Classification

There are various approaches about the definition and special features of big data [9–11]. In this paper three sources of passive generated data (smart cards, mobile devices and social networks), which are referred to in relevant literature as big data [12–16], are studied. While several research works have been conducted on using such data in the transportation field, the research purpose, the flavor of the data (different accuracy, granularity, aggregation level) and the implemented methodology vary a lot among these works. Therefore, it is attempted to summarize these works in a cohesive way and classify them according to their basic characteristics.

Initially, the studies were classified by their data source. Then, each one was classified according to the following criteria: publication year, special data features, supplementary dataset, transport attributes, methodological approach and validation. Finally, overall results about three data sources are presented, concerning the field they serve, in order to capture the current trends and to identify the dynamics and weaknesses of such datasets.

3.1 Automatic Fare Collection System and Smart Card

Smart card automatic fare collection systems (SC-AFC) are increasingly applied in transportation systems worldwide. Smart cards constitute a useful tool for many applications, such as identification, localization and payment, since they can store and process a wealth of information [17]. Smart cards usually store data such as card id, transportation mode, boarding times and locations and in some cases alighting times and locations. Although the main objective of such systems is the fare collection, their stored information could be exploited for other applications rather than their initial purpose. In the context of transport planning smart card data has been mostly utilized in travel pattern analysis, public transport modelling and performance assessment.

As Table 1 shows, only few studies deal with all means of public transport, while most of them focus on bus trips. Since smart cards provide boarding/alighting information depending on the type of card, most studies focus on extracting crucial attributes for transport modelling. The basic aim is individual trip reconstruction. Trip-chaining algorithm with several distance and time constraints is the most frequent implemented methodology. Another important task is to distinguish transfers from the activity location. The most common approach is by time-based rules. More recent studies attempted to extract the trip purpose and analyze activity patterns from smart card data using rule based, machine learning and probabilistic methods. Vehicle GPS or GIS data is a common supplementary dataset in the above studies. Finally, most studies lacked validation or simply compared their proposed methods with travel diary or survey data. Overall, smart cards contain large volumes of individual trip data, which

Table 1. Classification of studies regarding smart card data.

Authors	Year	Mode	Other data	Transport attribute	Methodology	Validation
Bagchi et al. [18]	2005	Bus	–	Travel patterns	Rule based	–
Trepanier et al. [19]	2007	Bus	–	Trip destination (OD estimation)	Trip-Chaining/Users classification	–
Chu et al. [20]	2008	Bus	AVL data	Transfer patterns	Rule based	–
Chakirov et al. [21]	2012	All modes	Land uses	Trip/Activity detection	Rule based/Probabilistic method (multinomial logit model)	V
Devillaine [22]	2012	Bus-metro	Land uses	Trip, Alighting stops/Activity detection	Trip-Chaining/Rule based classification	–
Yuan et al. [23]	2013	Bus	–	Trip, Stops/primary activity detection	Machine learning (Conditional random fields)	V
Han et al. [24]	2016	Bus-metro	Land use	Trip purpose/Activity patterns	Unsupervised machine-learning (continuous hidden Markov model)	V
Zou et al. [25]	2018	Subway	–	Home location/Trip purpose	Center point based algorithm/Rule based	V
Zhao et al. [26]	2018	Rail	AVL Data	Trip prediction	Bayesian n-gram model	V
Zhang et al. [27]	2018	Subway	–	Group Travel behavior	Rule based	–
Alsger et al. [28]	2018	Bus, ferry, train	Land uses, weather information, survey	Trip purpose	Rule based modelling (spatial, temporal & trip frequency)	V
Liu et al. [29]	2019	Bus, train, ferry	–	Journeys/fare policy and ridership correlation	Rule based/Spatial statistical analysis	–

cover extended periods, while the basic disadvantage is that they provide information only about specific public transport modes.

3.2 Mobile Phones

Mobile phone data, deriving from communication usage (event-driven data) and updates of the network (network-driven data), often contain time and location data. Since mobile phones have become one of the necessities in people's everyday life, they have emerged as the most frequently used data source [5]. In many cases, mobile phone data have been utilized by researchers in order to estimate OD matrices, to identify trip purposes or to examine human mobility patterns.

Table 2. Classification of studies regarding mobile phones data.

Authors	Year	Data type	Other data	Transport attribute	Methodology	Validation
Gonzalez et al. [4]	2008	Event-driven	–	Mobility patterns	Statistical analysis	–
Song et al. [30]	2010	Event-driven	–	Mobility patterns	Statistical analysis	V
Wang et al. [7]	2010	Event-driven	–	Trip identification/Travel mode	Spatiotemporal clustering/k-means clustering	V
Schlaich et al. [31]	2010	Network driven	Geodata	Trip identification/Travel mode	Rule based (temporal based – duration)	–
Isaacman et al. [32]	2011	Event-driven	ZIP codes	Trip identification/Important locations	Frequency based/Probabilistic approach	V
Iqbal et al. [6]	2014	Event-driven	Traffic counts	OD matrices	Frequency based, Scaling method	V
Toole et al. [33]	2015	Event-driven	Transport network, census	OD matrices/Activity/Trip assignment	Spatiotemporal clustering/Frequency based/(ITA) algorithm	V
Alexander et al. [8]	2015	Event-driven	census	OD trips by purpose and time	Spatiotemporal clustering/Frequency based	V
Horn et al. [34]	2017	Network driven	Geo data	Trip identification/Travel mode	Georeferencing/Rule based	–
Lee et al. [35]	2018	Event-driven	Census	Accessibility to public transit	Huff model-based FCA method	V
Bassolas et al. [36]	2019	Event-driven	–	Activity diaries	Frequency based, Rule based, Multi-Agent Simulation	–

As shown in Table 2, the majority of the researchers used event-driven and especially CDR data. Two of the presented studies examine mobility patterns mainly by statistical analysis and one examines the accessibility to the transport network. The rest of them focus on extracting necessary attributes for travel demand modelling and the initial challenge is to reconstruct individual trips. Some researchers extract OD matrices by a trip-based approach utilizing several rules. Later studies attempt to reconstruct individual trips by identifying stay locations, implementing mostly spatiotemporal clustering and frequency-based methods. As it is shown in Table 3, four researchers detect the trip purpose using mainly time frequency rules. The mode transport is detected in three studies by clustering and rule-based methods. In terms of validation, the majority compares the results with census and travel survey data. An important advantage of mobile phones data is that they can offer large samples and a broad spatial and temporal coverage. Lack of social-economic information and their noisy and sparse measurements are considered as their main disadvantages.

3.3 Social Networks

Another source of massive data that is gaining attention is social media, such as Facebook, Twitter and Foursquare. In literature, there are various efforts of mining transport information by analyzing text, hash-tag(s) or check-in data provided by social networks. The majority of them focuses on real time traffic management and on detecting incidents or anomalies in the network. However, there are relevant studies that attempt to utilize such data for planning purposes by analyzing aggregate or individual mobility patterns, deriving activity locations, estimating OD matrices and examining commuters' opinions regarding the transport system.

Table 3. Classification of studies regarding social media data.

Authors	Year	Data type	Other data	Transport attribute	Methodology	Validation
Cheng et al. [37]	2011	Check-in (Fr, Fb)	–	Mobility patterns	Statistics	–
Collins et al. [38]	2013	Text (T)	–	Riders opinion	Machine learning, Sentiment analysis	–
Hasan et al. [39]	2014	Check-in (Fr)	–	Individual activity patterns	Probabilistic topic model	–
Chaniotakis et al. [40]	2015	Geo-tagged (T)	–	Leisure activities (Trip purpose)	Classification/Statistics	V
Yang et al. [41]	2015	Check-in (Fr)	Land uses	OD patterns	Gravity model–based method/Hierarchical clustering	V
Lee et al. [42]	2016	Geo-tagged (T)	Census Land uses	OD matrices	Latent class analysis/Tobit regression model	V

(*continued*)

Table 3. (*continued*)

Authors	Year	Data type	Other data	Transport attribute	Methodology	Validation
Zhang et al. [43]	2017	Geo-tagged (T)	–	Travel behavior	x-means clustering	V
Huang et al. [44]	2017	Check-in (T)	Land uses-Census	Temporal patterns	Statistical analysis	–
Cheng et al. [45]	2018	Geo-tagged (T)	Land uses-Census	OD trips	Machine learning (random forest)	V
Shem-Tov et al. [46]	2020	Geo-tagged (G)	–	OD matrices (emergency)	Spatiotemporal algorithm (rule based)	–

T: Twitter, Fr: Foursquare, Fb: Facebook, G: Gowalla

Table 3 demonstrates that among social media Twitter and Foursquare, which mainly provides location-based information, are the most common in extracting transport-related data from users' posts. Almost all studies use check-in and geo-tagged data, while just one uses text mining techniques in order to obtain more information. Among the reviewed studies, the majority examines mobility pattens and travel behavior. There are some recent efforts on estimating OD matrices and trip purposes, showing that social media can be a helpful supplement for travel demand estimation and activity-based modelling. In such cases, supplementary data such as census data and land uses are often necessary. The main advantages of social media are their free access with low cost and that they contain many qualitive data, while the main disadvantage is the lack of transport information between two posts.

4 Conclusions

This paper analyzed three big data sources and attempts to summarize the existing approaches concerning their usage in the transport sector. In terms of transportation planning, the main focus of big data applications is on the following categories: travel demand modeling, travel patterns and assessing public transport. In the examined studies, trip identification was revealed as the most frequent extracted transport attribute. However, only few studies attempted to extract all the necessary attributes for travel demand estimation, while only one attempted to utilize big data in an agent-based microsimulation tool.

Many of the reviewed studies developed simple rule-based methods. Advanced models, machine learning techniques and probabilistic approaches are needed to leverage the potential of big data. Moreover, only few studies managed to combine various data sources in order to extract transport attributes. Combining different data sources could reveal new aspects of human mobility and improve the accuracy of the proposed methodologies. Finally, due to ground truth data scarcity, validation of the applied methods was sporadically implemented or the results were simply compared with aggregate data.

There is still much to be investigated in regards to transport planning and big data sources. To structure a step by step methodology for transportation applications is a difficult exercise due to the diversity of big data features. Therefore, a main challenge is to unveil big data dynamics and to examine at what extent they can change the traditional approaches.

References

1. Huang, H., Cheng, Y., Weibel, R.: Transport mode detection based on mobile phone network data: a systematic review. Transp. Res. Part C **101**, 297–312 (2019)
2. Brockmann, D., Hufnagel, L., Geisel, T.: The scaling laws of human travel. Nature **439**(7075), 462–465 (2006)
3. de Montjoye, Y.A., Hidalgo, C.A., Verleysen, M., Blondel, V.D.: Unique in the crowd: the privacy bounds of human mobility. Sci. Rep. **3**, 1376 (2013)
4. Gonzalez, M.C., Hidalgo, C.A., Barabasi, A.-L.: Understanding individual human mobility patterns- nature. Nature **453**, 779–782 (2008)
5. Calabrese, F., Diao, M., Lorenzo, G.D., Ferreira, J., Ratti, C.: Understanding individual mobility patterns from urban sensing data: a mobile phone trace example. Transp. Res. Part C **26**, 301–313 (2013)
6. Iqbal, M.S., Choudhury, C.F., Wang, P., González, M.C.: Development of origin-destination matrices using mobile phone call data. Transp. Res. Part C **40**, 63–74 (2014)
7. Wang, H., Calabrese, F., Lorenzo, G.D., Ratti, C.: Transportation mode inference from anonymized and aggregated mobile phone call detail records. In: 13th International Conference on Intelligent Transportation Systems, NJ, pp. 318–323. IEEE (2010)
8. Alexander, L., Jiang, S., Murga, M., González, M.C.: Origin-destination trips by purpose and time of day inferred from mobile phone data. Transp. Res. Part C **58**, 240–250 (2015)
9. De Mauro, A., Greco, M., Grimaldi, M.: A formal definition of big data based on its essential features. Libr. Rev. **65**(3), 122–135 (2016)
10. Emani, C.K., Cullot, N., Nicolle, C.: Understandable big data: a survey. Comput. Sci. Rev. **17**, 70–81 (2015)
11. Batty, M.: Big data, smart cities and city planning. Dialogues Hum. Geogr. **3**(3), 274–279 (2013)
12. Milne, D., Walter, D.: Big data and understanding change in the context of planning transport systems. J. Transp. Geogr. **76**, 235–244 (2019)
13. Chen, C., Ma, J., Susilo, Y., Liu, Y., Wang, M.: The promises of big data and small data for travel behavior (aka human mobility) analysis. Transp. Res. Part C **68**, 285–299 (2016)
14. Chandrasekar, P.: Big data and transport modeling: opportunities and challenges. Int. J. Appl. Eng. Res. **10**(17), 38038–38044 (2015)
15. Rashidi, H.T., Abbasi, A., Maghrebi, M., Hasan, S., Waller, S.T.: Exploring the capacity of social media data for modelling travel behaviour: opportunities and challenges. Transp. Res. Part C **75**, 197–211 (2017)
16. Schmöcker, J.D., Kurauchi, F., Shimamoto, H.: An overview on opportunities and challenges of smart card data analysis. In: Kurauchi, F., Schmöcker, J.D. (eds.) Public Transport Planning with Smart Card Data, pp. 1–9. CRC Press, Boca Raton (2017)
17. Pelletier, M.-P., Trepanier, M., Morecy, C.: Smart card use in public transit: a literature review. Transp. Res. Part C **19**, 557–568 (2011)

18. Bagchi, M., White, P.R.: The potential of public transport smart card data. Transp. Policy **12**, 464–474 (2005)
19. Trepanier, M., Chapleau, R., Tranchant, N.: Individual trip destination estimation in transit smart card automated fare collection system. J. Intell. Transp. Syst. Technol. Plann. Oper. **11**(1), 1–15 (2007)
20. Chu, K.K., Chapleau, R.: Enriching archived smart card transaction data for transit demand modeling. Transp. Res. Rec. J. Transp. Res. Board **2063**, 63–72 (2008)
21. Chakirov, A., Erath, A.: Activity identification and primary location modelling based on smart card payment data for public transport. Eidgenössische Technische Hochschule Zürich, IVT, Institute for Transport Planning and Systems, Zurich, Switzerland (2012)
22. Devillaine, F., Munizaga, M., Trépanier, M.: Detection of activities of public transport users by analyzing smart card data. Transp. Res. Rec. J. Transp. Res. Board **2276**, 48–55 (2012)
23. Yuan, N.J., Wang, Y., Zhang, F., Xie, X., Sun, G.: Reconstructing individual mobility from smart card transactions: a space alignment approach. In: 13th International Conference on Data Mining (ICDM), NJ, pp. 877–886. IEEE (2013)
24. Han, G., Sohn, K.: Activity imputation for trip-chains elicited from smart-card data using a continuous hidden Markov model. Transp. Res. Part B **83**, 121–135 (2016)
25. Zou, Q., Yao, X., Zhao, P., Wei, H., Ren, H.: Detecting home location and trip purposes for cardholders by mining smart card transaction data in Beijing subway. Transportation **45**, 919–944 (2018)
26. Zhao, Z., Koutsopoulos, H.N., Zhao, J.: Individual mobility prediction using transit smart card data. Transp. Res. Part C **89**, 19–34 (2018)
27. Zhang, Y., Martens, K., Long, Y.: Revealing group travel behavior patterns with public transit smart card data. Travel Behav. Soc. **10**, 42–52 (2018)
28. Alsger, A., Tavassoli, A., Mesbah, M., Ferreira, L., Hickman, M.: Public transport trip purpose inference using smart card fare data. Transp. Res. Part C **87**, 123–137 (2018)
29. Liu, Y., Wang, S., Xie, B.: Evaluating the effects of public transport fare policy change together with built and non-built environment features on ridership: the case in South East Queensland. Aust. Transp. Policy **76**, 78–89 (2019)
30. Song, S., Qu, Z., Blumm, N., Barabasi, A.L.: Limits of predictability in human mobility. Science **327**, 1018–1021 (2010)
31. Schlaich, J., Otterstätter, T., Friedrich, M.: Generating trajectories from mobile phone data. In: TRB 89th Annual Meeting Compendium of Papers, no. 10-0374. Transportation Research Board, Washington, DC, USA (2010)
32. Isaacman, S., Becker, R., Cáceres, R., Kobourov, S., Martonosi, M., Rowland, J., Varshavsky, A.: Identifying important places in people's lives from cellular network data. In: Lyons, K., Hightower, J., Huang, E.M. (eds.) Pervasive Computing 2011, LNCS, vol. 6696, pp. 133–151. Springer, Berlin (2011)
33. Toole, J.L., Colak, S., Sturt, B., Alexander, L.P., Evsukoff, A., González, M.C.: The path most traveled: travel demand estimation using big data resources. Transp. Res. Part C **58**, 162–177 (2015)
34. Horn, C., Gursch, H., Kern, R., Cik, M.: QZTool-automatically generated origin-destination matrices from cell phone trajectories. In: Stanton, N.A., Landry, S., Di Bucchianico, G., Vallicelli, A. (eds.) Advances in Human Aspects of Transportation, Advances in Intelligent Systems and Computing, vol. 484, pp. 823–833. Springer, Switzerland (2017)
35. Lee, W.K., Sohn, S.Y., Heo, J.: Utilizing mobile phone based floating population data to measure the spatial accessibility to public transit. Appl. Geogr. **92**, 123–130 (2018)
36. Bassolas, A., Ramascoa, J., Herranz, R., Cantu-Ros, G.O.: Mobile phone records to feed activity-based travel demand models: MATSim for studying a cordon toll policy in Barcelona. Transp. Res. Part A **121**, 56–74 (2019)

37. Cheng, Z., Caverlee, J., Lee, K., Sui, D.Z.: Exploring millions of footprints in location sharing services. In: 5th International Conference on Weblogs and Social Media, AAAI, Menlo Park, CA, pp. 81–88 (2011)
38. Collins, C., Hasan, S., Ukkusuri, S.V.: A novel transit rider satisfaction metric: rider sentiments measured from online social media data. J. Public Transp. **16**(2), 21–45 (2013)
39. Hasan, S., Ukkusuri, S.V.: Urban activity pattern classification using topic models from online geo-location data. Transp. Res. Part C **44**, 363–381 (2014)
40. Chaniotakis, E., Antoniou, C., Mitsakis, E.: Data for leisure travel demand from social networking services. In: 4th hEART Symposium. European Association Research in Transportation - hEART, DTU, Denmark (2015)
41. Yang, F., Jin, J.P., Cheng, Y., Zhang, J., Ran, R.: Origin-destination estimation for non-commuting trips using location-based social networking data. Int. J. Sustain. Transp. **9**(8), 551–564 (2015)
42. Lee, J., Gao, S., Goulias, K.: Can twitter data be used to validate travel demand models? In: 95th Annual Transportation Research Board Meeting, TRB, Washington, DC, USA, pp. 1–27 (2016)
43. Zhang, Z., Qing Heb, Q., Zhu, S.: Potentials of using social media to infer the longitudinal travel behavior: a sequential model-based clustering method. Transp. Res. Part C **85**, 396–414 (2017)
44. Huang, A., Gallegos, L., Lerman, K.: Travel analytics: understanding how destination choice and business clusters are connected based on social media data. Transp. Res. Part C **77**, 245–256 (2017)
45. Cheng, Z., Jian, S. Maghrebi, M., Rashidi, T.H., Waller, S.T.: Is social media an appropriate data source to improve travel demand estimation models? In: TRB 97th Annual Meeting Compendium of Papers, no. 18-06774. Transportation Research Board, Washington, DC, USA (2018)
46. Shem-Tov, B.I., Bekhor, S.: Extracting travel demand for emergency situations using location-based social network data. Transp. Res. Procedia **45**, 111–118 (2020)

Modelling Emerging Transport Solutions for Urban Mobility

Future Scenarios for Mobility Innovations and Their Impacts in Cities and Transport Models

Javier Burrieza Galán[1], Rita Rodríguez Vázquez[1], Oliva G. Cantú Ros[1], Georgia Ayfantopoulou[2], Josep Maria Salanova Grau[2], Maria Konstantinidou[2](✉), Rodric Frederix[3], and Péter Pápics[3]

[1] Nommon Solutions and Technologies, Madrid, Spain
{javier.burrieza, rita.rodriguez, oliva.garcia-cantu}@nommon.es
[2] Hellenic Institute of Transport- Centre for Research and Technology Hellas, Thessaloniki, Greece
{gea, jose, mariakon}@certh.gr
[3] Transport and Mobility Leuven, Leuven, Belgium
{rodric.frederix, peter.papics}@tmleuven.be

Abstract. Cities should try to combine the various conflicting interests in the field of urban mobility aiming to extract the maximum benefits of the different mobility options. Although this task concerns mainly the current situation, cities should also prioritize future mobility measures. The purpose of this paper is to explore how emerging mobility concepts would evolve under future mobility scenarios related to carsharing services, micromobility services, Demand Responsive Transport (DRT) services, Connected Autonomous Vehicles (CAVs), Urban Air Mobility and Mobility-as-a-Service. Two diverging or complementary scenarios have been associated to each concept expected to transform future urban mobility. A Delphi poll has identified the future scenarios related to data sources and transport modelling and assessed the effects that each mobility innovation may have on urban transportation. The impacts of these emerging mobility solutions on transport planning tools and techniques are also investigated and prioritized. Finally, data sharing between operators and policy-makers and the lack of skills among transport planners are rated by the Delphi poll as the most important gaps in terms of transport data sources and barriers that hinder the modelling of the new urban mobility options respectively.

Keywords: Urban mobility · Alternative futures · Impacts · Transport planning tools

1 Introduction

The disruptive technological developments and the societal changes operate as accelerators for the urban mobility transformation changes. The rapid development of ICTs is facilitating the operation of a wide range of new mobility alternatives grouped under the concept of smart mobility with uncertain impacts for the future of sustainable

E. G. Nathanail et al. (Eds.): CSUM2020, AISC 1278, pp. 1129–1138, 2021.
https://doi.org/10.1007/978-3-030-61075-3_108

mobility. The combination of emerging mobility services will boost multimodality, contributing to the promotion of the public transport. A cleaner, cheaper, safer and more efficient transport system with improved accessibility in low demand areas will be the first step towards the reduction of car ownership and road fatalities.

Regarding the success of carsharing systems, the population density and the collaboration between private and public stakeholders have been identified as critical factors [1–3]. The urban areas ensure high demand for operators and targeting to both public and private interests is positively evaluated [3]. In addition, a city with infrastructure for electric vehicle is more likely to provide a successful electric vehicle based carsharing system [2].

Similarly to all privately-led shared mobility services, it is not yet clear if micromobility companies will be profitable in the future, since they are based on their competence to raise funds [4, 5]. Although there are still gaps in the regulatory framework some cities are taking a leading role collaborating with the operators [6]. Beside the stakeholders' cooperation, agreements on data sharing [6–8] and regulations on the public space management [5–7] seem to affect the success of these systems.

In general, DRT are highly subsidized by authorities since they are low profit services [9, 10]. However, there are some Public Transport Authorities that evaluate that the DRT systems contribute positively to their supply networks. Among the success factors are the coordination between the itineraries of the DRT and the scheduled public transport [9, 11], the minimisation of costs and empty trips [11]. Moreover, the strong stakeholders' engagement in providing a sufficient fleet [11] and the customised communication strategies are crucial to the increased attractiveness of the service.

In CAVs field there are already vehicles that have incorporate some automate functionalities but the achievement of higher-level production depends not only on vehicle technical advancements but also on cooperative ITS infrastructure. Although there are some programmes for supporting testing and implementation of CAV technologies in many European countries, the EU is aiming to launch large scale pilots in order to assess the maturity of these technologies [12]. The prerequisites for the successful implementation of CAV can be clustered in two groups: user acceptance and management. It is not yet clear how the users will be convinced to trust a totally automated vehicle [13] while in terms of management, the public-private partnership seems to be a necessary condition [14].

Regarding the UAM there aren't currently commercial services fully responding to the basic operational requirements of these services, but there are some precursors across the world such as Voom services in São Paulo and Mexico City [15] and Uber helicopter services in New York [16]. There are also some relevant initiatives regarding UAM to be launched in the following years such as the case of Paris where a service between Charles de Gaulle airport and the city centre will be provided by Airbus and RATP for 2024 Olympic Games [17]. Furthermore, an initiative related to UAM has been launched by the European Innovation Partnership on Smart Cities and Communities (EIP-SCC). The initiative is supported by 43 members and includes a manifesto aiming to foster UAM demonstration projects.

Finally, regarding MaaS there are two main conditions for their successful implementation based on the experiences and the related literature [18, 19]: (i) the ability of a MaaS operator to deploy its services in a city or region; (ii) users' willingness to change their mobility behavior. MaaS schemes require not only the availability of a wide range of transport modes but also the willingness of operators to share their real-time data with third parties and to sell their services via e-ticketing [20]. Cost savings, user friendly applications, freedom of modal choice and vehicle and high customization levels are reported as fields of MaaS added value [21].

The present paper aims at the identification of the future scenarios for mobility innovations and their impacts in cities and transport models. After the review of the basic concepts behind each of the mobility innovation trends and the detection of the key factors or the obstacles for their successful integration in sustainable mobility, the future scenarios are described in Sect. 2. The results of the assessment of their impacts on the urban mobility resulted from the Delphi poll are outlined in Sect. 3. Finally, a summary of the main concluding remarks is provided in Sect. 4.

2 Future Mobility Scenarios

Two diverging or complementary scenarios for each of the emerging mobility concepts were developed as they are described below. The generation of the future scenarios was based on the literature review of roadmaps, white papers, vision papers and policy documents related to mobility. Moreover, the identification of the success factors and the gaps described in Sect. 1 contributed to the final formulation of the scenarios.

2.1 Alternative Futures for Carsharing Services

In the first medium term horizon scenario the electric carsharing operates as an additional transport mode in the cities. The implementation of stricter urban vehicle access regulations and parking management policies in the metropolitan areas imposes limitations in the use of the private cars. The transport sector is characterized by multimodality, combining mass and individual transport. This situation combined with the high acquisition cost of the electric cars benefits the integration of sharing schemes in the transport sector increasing its modal share up to 20–25%. Electric vehicle infrastructure exists in certain urban areas in order to serves the increasing use of electric vehicles sharing schemes. Electric carsharing is combined with public transport and the new mobility services offer more flexibility and better quality of combined transport options.

The second, long-term time horizon scenario, proposes a holistic housing solution which integrates aspects such as mobility, housing, energy distribution and ICT networks. In this scenario the electric sharing vehicle schemes are included in the modern collective housing policies. The different urban electricity needs throughout the day are balanced through smart energy grids. Thus, the holistic housing-mobility approach leads to an easily accessible community-based electric vehicle scheme with low parking requirements. The large-scale implementation of this approach contributes to the homogeneous modal share of electric vehicle solutions.

2.2 Alternative Futures for Micromobility Services

Micromobility is one of the most recent trends among shared mobility services. The first scenario describes a transport landscape where other transport modes are substituted by micromobility in short distance trips. The regulatory framework for micromobility has been defined in most cities and dedicated lanes for e-scooters and similar vehicles are provided to road users. Thus, they are used for daily short distance trips instead of car and taxi rides since car ownership appears a decline especially in young people who prefer new transport modes based on pay-per-use rather than a car purchase. In addition, since all of the competitors in the field are subjected to the same regulation, new business schemes will emerge based on the results of the pilot projects and the market's needs,

In the long-term scenario, micromobility is a part of a longer combined trip providing most flexible, efficient, and sustainable transport options. It is combined with public mass transport and following a strict regulatory framework it is completely integrated in the transport system. The mobility packages provided to travelers through MaaS platforms include the micromobility option mainly as a mode to reach or leave the transition stations or combined with the private car in Park&Ride solutions. The design is more user friendly for vulnerable users such as disabled or aged people.

2.3 Alternative Futures for Demand Responsive Transport Services

The new intelligent applications facilitated the diffusion of digital content among the travelers and allow the operators to improve the quality of the services (i.e. less waiting time, more accessibility) reducing the operational costs by 2030. On-demand transportation emerges as a supplementary alternative to traditional public transport services combined with the backbone public transport network in first/last mile. Citizens of remote areas and users of special categories such as ageing population or vulnerable users, being dissatisfied of the quality of the traditional bus and railway services, are attracted by the on-demand public services. User visibility allows the efficient demand management providing to travelers a variety of alternatives options of different prices and connectivity to other transport modes.

The on-demand transport system of the first scenario expands to freight and goods delivery by 2050. Both passengers and cargo are transported by the same vehicle in an efficient and safe way. More than one passenger and parcel are allowed in the same vehicle and thus the capacity of the vehicles is completely exploited. The consolidation of the goods flows increases the efficiency of the last mile deliveries since the unnecessary/empty vehicle movements in urban areas are avoided. Innovative logistics business models which integrate the use of public transport infrastructure for both passenger and goods delivery are developed thanks to shared data.

2.4 Alternative Futures for Connected Autonomous Vehicles

The use of Connected Autonomous Vehicles for passenger delivery will bring disruptive changes in the current transport system. In the short-term scenario, privately owned CAVs Level 4 have been already introduced into the car fleet by 2030.

Conventional vehicles are limited to highways and motorways. This means that the car-share fleets are used mainly for short-distance trips within cities. Due to high cost of the purchase and the maintenance, the CAVs' ownership is limited to the wealthier part of society. The intercity CAV traffic is prohibited in the inner cities, due to the lack of parking spaces and therefore, the creation of P+R parkings and the connection with the public transport is required. Since the majority of CAVs are electric, the level of the total emissions from cars will be reduced.

After the availability of Level 4 CAVs, Level 5 vehicles will also enter the market by 2050. Completely autonomous vehicles without human interaction over any kind of traffic situation will be available. All transport modes are connected thanks to the advanced information sharing systems and multimodal trips provided through MaaS platforms are promoted. As a result, private car use is decreased and a boost in the public transport use is observed. Fewer parking spaces are required and the majority of them are converted to hubs for mobility services. C-ITS equipment has been installed over various kinds of traffic infrastructure, when not only V2V but V2I connectivity is considered. Full automation provides more safety and convenience compared to the current fixed public transport lines.

2.5 Alternative Futures for Urban Air Mobility

Urban Air Mobility (UAM) is based on highly automated and efficient air vehicles for the transportation of passenger and cargo within metropolitan areas. In the short-term scenario, the willingness of manufacturers and other stakeholders to operate UAM services contributes to the advancement of regulatory framework for this mode but the high requirements of coordination among stakeholders hinders the progress in this field. Sustainable production levels have been achieved by the manufacturers but car production schemes are still at an early stage. Although European aviation authorities have leaded some demonstration projects in certain metropolitan areas, there are still issues that have not been solved. Thus, the adoption of a holistic standard system across Europe is not yet the case. Except of some large cities that have planned to use heliports and create some additional vertiports in suburban areas, the majority of the cities have a lack of related infrastructure. UAM is still an exclusive mode and it is not included in the public transportation system of the city.

In the long-term scenario, profitable business models for UAM are developed. Cooperation among different public and private stakeholders is achieved and the technological advances contribute to high production volumes. Standards for urban air traffic management have been set by the EU aviation authorities after intensive funding of research activities and demonstration projects. Large cities are funded for the deployment of UAM infrastructure. Vertiport network that covers airports, transport hubs and important nodes in suburban areas have been developed and UAM services have been incorporated in the mobility packages of MaaS aggregators. Thus, UAM is gradually become part of the cities' transport system.

2.6 Alternative Futures for Mobility-as-a-Service

The introduction of MaaS in the transport system is expected to result in a shift from private towards multimodal and sustainable modes. In the first scenario MaaS are limited in large cities at a local level offering to citizens an alternative for their daily trips instead of private car use. Individual providers are responsible for the organisation and pricing of the traffic services. The MaaS provider just plan trips and provide a limited number of fixed package purchasing simply bundled individual tickets of the specific transport providers. There is no flexibility in the pricing and route options and the same options are addressed to everyone.

In the second scenario, an EU-level network of MaaS services is created beyond city boundaries reaching areas that would be impossible to be served with traditional public transport modes. MaaS stakeholder group comprises of cooperating stakeholders from mobility providers to local and regional governments, who are all strongly engaged in the planning, operation and maintenance of transport services. There are no more bundles of individual tickets but single mobility offers from point A to B irrespective of the modes used. Individual mobility patterns and attributes are considered in the offerings of a MaaS app to design a package and a price that fits a given person's needs. Public transport services are still the key element of MaaS services (trains, trams, and metro lines).

3 Impacts

A Delphi poll is conducted in MOMENTUM project, assessing the impacts of emerging mobility solutions in urban transportation and the planning techniques and tools.

3.1 Future Impacts in Cities

In the 1st Round, the factors and impacts of emerging mobility solutions for cities are identified by the participants of the Delphi poll through open-ended questions. The collected answers were prioritized in the 2nd Round. Among the current factors underlying the emergence of the mobility solutions, the collaboration between public and private stakeholders, the societal benefits as well as the improvement of mobility services' quality for the users are mentioned. In terms of current impacts in European cities, 25% of the panel answered that there are no current impacts but they were aware about potential future impacts such as modal shift from public transport services and car ownership decrease.

In addition, the participants share their opinions about the factors for the successful implementation of the presented mobility options. The deployment of Urban Vehicle Access Regulations (UVARs) as well as parking regulations were the most common answers followed by services reliability and affordability and the cooperation between operators and cities towards the deployment of the systems. Interestingly, the answers for the future impacts were by far more positive, with few exceptions. In the 2nd

Round, these impacts were grouped in terms of importance in an adverse and a positive group as it is shown in Fig. 1 and Fig. 2 respectively.

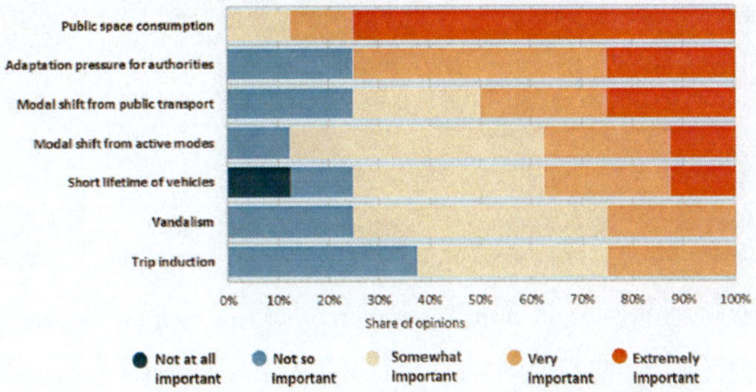

Fig. 1. Importance of new mobility options' adverse impacts

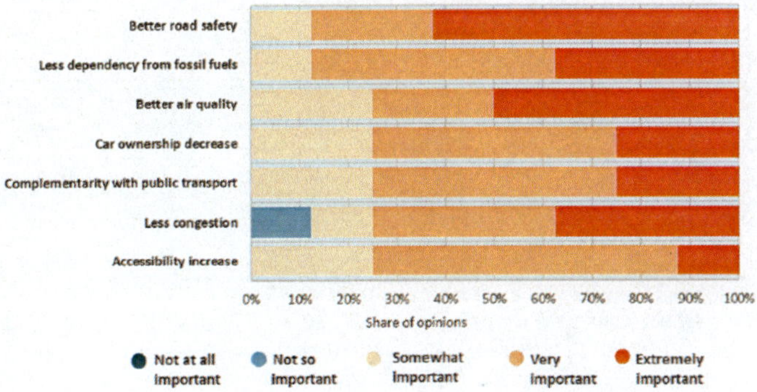

Fig. 2. Importance of new mobility options' benefits

3.2 Future Impacts in Transport Models

The Delphi poll also discussed about the effects of the emerging mobility solutions in transport planning tools and techniques. Participants believe that the shared mobility, micromobility and MaaS already represent a challenge for these techniques. The DRT, shared mobility systems and micromobility are expected to be included in transport planning techniques between 2020 and 2030. On the contrary, the majority of the panel considers that the inclusion of CAV and UAM in the transport planning techniques will not happen before 2030 and a percentage of 10–20% considers that this will not happen. The research gaps and challenges among transport data sources and modelling tools identified by the Delphi poll are presented in Fig. 3 and Fig. 4.

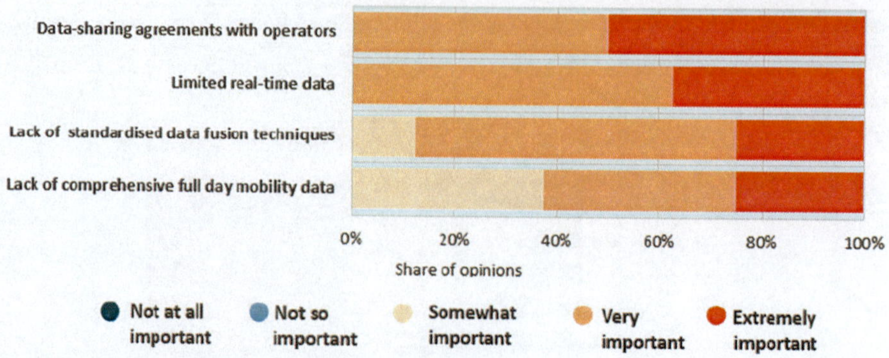

Fig. 3. Importance of gaps in transport data sources for new mobility options' modelling

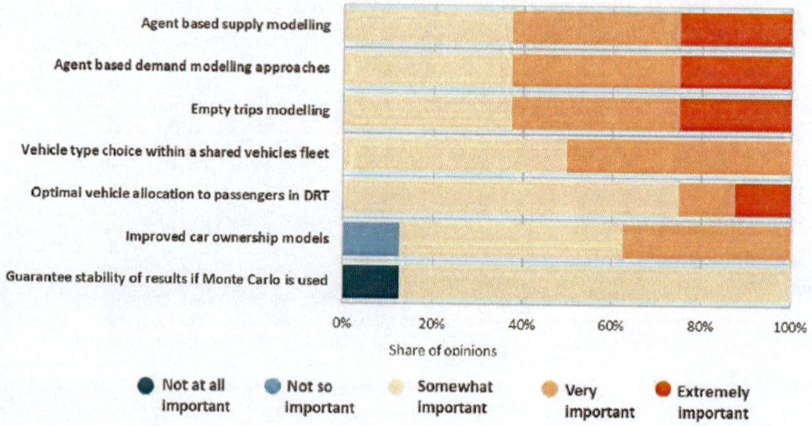

Fig. 4. Importance of transport modelling gaps for new mobility options' modelling

In terms of the barriers that appear in planning cycle during the modeling process of the new mobility options, the lack of skills among transport planners, the lack of resources among transportation departments and the limited cooperation between transport and urban planning are mentioned by the experts. Finally, the panel outlined that indicators related to emission levels, safety, accessibility, modal split, car ownership, public space consumption, energy consumption changes and general comprehensive indicators related to urban life quality will have crucial role in the future transport planning.

4 Conclusions

The present paper captures the future mobility scenarios of the emerging mobility concepts related to carsharing services, micromobility services, Demand Responsive Transport services, Connected Autonomous Vehicles, Urban Air Mobility and Mobility-as-a-Service. The identification through the Delphi poll of the more challenging factors for the successful implementation of these services and the effects that each mobility innovation may have on urban transportation and the transport models could be a valuable input for the stakeholders, from the planners to providers and local and regional governments, who are all participate in the planning, design and operation of transport services.

Acknowledgments. Research supported by the European project MOMENTUM-Modelling Emerging Transport Solutions for Urban Mobility funded from the European Union's Horizon 2020 research and innovation programme under grant agreement No 815069.

References

1. Arndt, W.H., Drews, F., Hertel, M., Langer, V., Wiedenhöft, E.: Topic guide: integration of shared mobility approaches in sustainable urban mobility. Rupprecht Consult (2019)
2. Seign, R., Bogenberger, KP.: Rescriptions for the successful diffusion of carsharing with electric vehicles. In: Conference on Future Automotive Technology (2013)
3. Terrien, C., Maniak, R., Chen, B., Shaheen, S.: Good practices for advancing urban mobility innovation: a case study of one-way carsharing. Res. Transp. Bus. Manage. **20**, 20–32 (2016)
4. Eckhardt, G.M., Houston, M.B., Jiang, B., Lamberton, C., Rindfleisch, A., Zervas, G.: Marketing in the sharing economy. J. Mark. **83**(5), 5–27 (2019)
5. Crozet, Y., Santos, G., Coldefy, J.: Shared mobility and MaaS - The regulatory challenges of urban mobility. Centre on Regulation in Europe. https://www.cerre.eu/publications/shared-mobility-and-maas-regulatory-challenges-urban-mobility. Accessed 12 March 2020
6. Zarif, R., Pankratz, D.M., Kelman, B.: Small is beautiful: Making micromobility work for citizens, cities, and service providers. https://www2.deloitte.com/us/en/insights/focus/future-of-mobility/micro-mobility-is-the-future-of-urban-transportation.html. Accessed 15 Febrauary 2020
7. Machado, P.: SUMP and Shared Mobility in Lisbon: Presented at the 6th European Conference on Sustainable Urban Mobility Plans, Gröningen, The Netherlands. https://www.eltis.org/sump. Accessed 18 Febrauary 2020
8. Clewlow, R.: The Micro-Mobility Revolution. Populus. https://medium.com/populus-ai/the-micro-mobility-revolution-95e396db3754. Accessed 25 March 2020
9. Mageean, J., Nelson, J.D.: The evaluation of demand responsive transport services in Europe. J. Transp. Geogr. **11**(4), 255–270 (2003)
10. Ryley, T.J., Stanley, P.A., Enoch, M.P., Zanni, A.M., Quddus, M.: Investigating the contribution of demand responsive transport to a sustainable local public transport system. Res. Transp. Econ. **48**, 364–372 (2014)
11. Enoch, M., Potter, S., Parkhurst, G., Smith, M.: Why do demand responsive transport systems fail. Presented at the Transportation Research Board 85th Annual Meeting, Washington DC, United States (2006)

12. European Commission: STRIA Roadmap on Connected and Automated Transport- Road Rail and Waterborne (2019)
13. Banerjee, S.S., Jha, S., Cyriac, J., Kalbarczyk, Z.T., Iyer, R.K.: Hands off the wheel in autonomous vehicles. A systems perspective on over a million miles of field data. In: 2018 48th Annual IEEE/IFIP International Conference on Dependable Systems and Networks (DSN), pp. 586–597. IEEE (2018)
14. National League of Cities: Autonomous Vehicle Pilots Across America (2018)
15. Lascara, B., Spencer, T., DeGarmo, M., Lacher, A., Maroney, D., Guterres, M.: Urban Air Mobility Landscape Report. MITRE (2018)
16. Porter, J.: The Uber for helicopters is now Uber. The Verge. https://www.theverge.com/2019/6/6/18655126/uber-copter-new-york-city-jfk-airport-elevate-helicopter-flying-taxi-price. Accessed 23 September 2019
17. UAS Vision: Airbus Studies Urban Air Mobility for 2024 Olympic Games in Paris. UAS Vision. https://www.uasvision.com/2019/06/24/airbus-studies-urban-air-mobility-for-2024-olympic-games-in-paris. Accessed 24 January 2020
18. Audouin, M.U.J.: Towards Mobility-as-a-Service: a cross-case analysis of public authorities' roles in the development of ICT-supported integrated mobility schemes. EPFL (2019)
19. Durand, A., Harms, L., Harms, L., Hoogendoorn-Lanser, S., Zijlstra, T.: Mobility-as-a-Service and Changes in Travel Preferences and Travel Behaviour: A Literature Review. KiM Netherlands Institute for Transport Policy Analysis, Netherlands (2018)
20. Li, Y., Voege, T.: Mobility as a service (MaaS): challenges of implementation and policy required. J. Transp. Technol. **7**(02), 95–106 (2017)
21. Harms, L., Durand, A., Hoogendoorn-Lanser, S., Zijlstra, T.: Exploring Mobility-as-a-Service: insights from literature and focus group meetings. Netherlands Institute for Transport Policy Analysis (KiM) (2018)

Cataloging and Assessing City-scale Mobility Data

Georgia Ayfantopoulou[1(✉)], Javier Burrieza Galán[2], Antonio Masegosa[3,4], Josep Maria Salanova Grau[1], Neofytos Boufidis[1], Ignacio Martín Martínez[2], Pablo Fernandez-Muga[3], and Oliva Cantú Ros[2]

[1] Centre for Research and Technology Hellas, Hellenic Institute of Transport, 57001 Thessaloniki, Greece
[2] Nommon Solutions and Technologies, 28006 Madrid, Spain
[3] DeustoTech, Faculty of Engineering, University of Deusto, 48007 Bilbao, Spain
[4] IKERBASQUE, Basque Foundation for Science, 48011 Bilbao, Spain

Abstract. In the era of data-driven decision making, the under-utilization of available data sources prevents organizations and corporations from unlocking their full potential and might even threaten their existence. On a city level, public authorities typically have access to numerous heterogeneous data sources, which are either generated by proprietary infrastructure or by collaborating local stakeholders. However, the adaptation to modern trends and adoption of new tools and methodologies by a local authority or even corporation can be overly slow, given the exceedingly complicated nature and sheer size of implementation. To this end, the authors propose an efficient, well-structured methodology towards city-wide data analytics and data-driven decision support systems for the transport sector. The main focus is on the identification and cataloguing of mobility-relevant data sources along with both qualitative and quantitative metadata. Those metadata indicators are used for assessing the quality and appropriateness of data within a) the general context of mobility and b) in relation to specific tasks and objectives. Further, a case study for the city of Thessaloniki is presented, where all the available mobility-relevant data sources have been organized, cataloged and described according to the aforementioned methodology.

Keywords: Mobility data · Data quality · Data catalogue

1 Introduction

The last five years, data have been referred to as the oil of current times [1], and with good reason. Tech giants with huge global penetration like Google, Amazon and Facebook are currently some of the most valuable firms. The main raw material that has enabled their growth is data.

Besides private companies, public authorities and decision makers are gradually adopting strategies to shift their operations and leverage multiple available sources of data. Data-driven studies and applications have recently been paving the way for

E. G. Nathanail et al. (Eds.): CSUM2020, AISC 1278, pp. 1139–1148, 2021.
https://doi.org/10.1007/978-3-030-61075-3_109

ever-smarter "smart cities", with several of those applications lying within the Transport domain and being considered highly relevant, if not vital, towards improving urban mobility. Several such applications have already been implemented, as an example [2] proposes a tool for the calculation of urban crowd flows utilizing heterogeneous big data while [3–5] go even further, attempting to forecast traffic flows using deep learning and other machine learning algorithms. [6] presents a more generic smart city application of big data, with focus on all data life circle from their generation to all processing steps and up to decision making. A big data analytics platform for public transportation information is designed in [7]. The automated tools presented in [8, 9] can be trained to detect traffic congestion and abnormal phenomena, while studies [10–14] focus on studying the utilization of taxi systems.

Emerging transport and mobility applications like vehicle sharing, Demand Responsive Transport (DRT), Mobility-as-a-Service (MaaS) and Connected Autonomous Vehicles (CAV) are built upon data processing frameworks. Cases like Uber, which exposed the unpreparedness of decision makers, reveal the need for a revision of policy making procedures to manage technological advancements.

A prerequisite to harnessing data is access, which in many cases is prohibited or limited. To this end, there is an obvious need to create unified data repositories in the geographical scope of each city, which host as many data sources as possible. In the case of sensitive or proprietary data, the repository might only contain samples of the data accompanied by metadata. In this way, a documentation of the existing data sources is available for all interested parties.

2 Methodology

In this section the authors propose a methodology for the implementation of a city-scale mobility data repository and the quality assessment of data sources it includes.

2.1 Data Catalogue

The repository is created to serve as a content aggregator unifying several sources under a common schema. Its completeness is critical, a thorough investigation should be implemented to detect all existing mobility data sources in a city. This procedure is burdened with hurdles and limitations, mainly posed by either inaccessible or non-existent data. Sources might be inaccessible due to the nature of data (sensitive, proprietary etc.). In the case of sensitive data, it is advised that the owners provide access to processed, pseudo-anonymized data which cannot be linked to a data object or a physical entity. The option to provide with aggregated, non-personal data should also be investigated.

The data repository may facilitate a large number of different sources and it is advised to categorize the sources in order to ensure an organized structure. Further, different applications require specific categories of input data and the interested parties can take advantage of this categorization, which serves as generic metadata.

All city-scale mobility data sources are segmented in five main categories, namely Transport Supply (TS), Transport Demand (TD), Maps and Cartography (MC),

Sociodemographic (SD) and Travel Times (TT). Several subcategories are found within the five main categories, as show in Table 1.

Table 1. Urban mobility related data sources.

Main Categories	Subcategories
Transport supply data sources	Public Transport Schedules and Lines, Transport Network, Taxi Service Car/Moto-Sharing, Bike-sharing MaaS, Parking
Transport demand data sources	Public Transport Smart Card, Other Public Transport Demand Data, Bike-sharing, Cycling, Pedestrian, Mobility Surveys, Telecom, MaaS, Car/Moto-Sharing, Traffic, Taxi Service, Social Media, Parking
Maps & Cartography data sources	Land Use, Weather, Social, Cultural or Sportive Events, Points of Interest
Socio-demographic data sources	Census, Income statistics, Tourism statistics, Car Ownership, Labour/Unemployment statistics, House price statistics, Other socio-demographic statistics
Travel times data sources	Travel Times

2.2 Data Quality Assessment

The output of data-driven applications heavily depends on the input data. Irregularities, inaccuracies and bias are only some possible flaws of data which may destroy or limit the quality of results. The data quality assessment is a pre-processing step considered highly important, as it further enriches data with information on the applicability of a data source to a given task.

The quality of a data source can be assessed from numerous points of view. In this section, a methodology is proposed to assess the quality and therefore usability and value of data sources with respect to mobility applications and studies. It is based on a factsheet that should be answered for each available data source and is organized in seven sections. It is considered that the questions should be investigated and answered by one or more individuals who participate in the design of the data recording infrastructure.

The factsheet's first section contains general metadata, the data source name, a link to the source in case it exists and the date of last factsheet update. The second section is a brief description with information on current applications of those data, further potential uses, classification of dataset as original or generated/derived by processing another dataset. Further, in this section it is stated whether and how this data is used in combination with another dataset, and lastly a comment on the reliability of the dataset (as low, medium, high or unknown) is requested, indicating the extent to which one can trust the data, from the point of view of its accuracy and precision.

The third section is related to availability, thus detailed information about the data owner and readiness to use is provided to interested parties. The access conditions are strictly specified (for instance open access or a non-disclosure agreement), followed by the way and technology for access (for instance any internet connection, a VPN connection, whether and what kind of authentication is required, API call limit or data volume exchange limit). The last questions concern the update frequency of the data, their future availability and the terms of use, where the sensitivity of data is clarified along with details on the sensitive, if any, data and how those should be processed to comply to the applicable data protection and privacy laws, e.g. General Data Protection Regulation (GDPR) in the case of European Union.

The fourth section consists of two sets of questions, the first focuses on the format and the second on the resolution of data. Regarding the format, the information specifies the file format of the data (for example.txt or.csv file), the ability to choose format when accessing the data, a link to the format documentation in case it exists, the approximate size of dataset and any proprietary tool or software required to access data. The data resolution is also highly critical for applications in the urban mobility domain; therefore, the factsheet facilitates all relevant information, including the temporal and geographical scope and granularity of the data. Further, any filtration of aggregation applied on the dataset is documented.

In the following factsheet section, general comments and other relevant information are requested as free text. The sixth section contains detailed descriptions of the information contained in the data source. In the last section, a short data sample is requested.

After completing the factsheet for each source, the final evaluation is implemented, and four quality indicators are evaluated concerning the reliability, resolution, availability-accessibility and relevance-usability. The indicators are evaluated in a three-value ordinal scale as low, medium or high. It has to be clarified that the last indicator is project specific, meaning that it is evaluated with respect to a specific application or project, therefore it is considered outside the scope of this paper and only the first three indicators will be evaluated. In all cases, some commends may be noted to underline or mention issues related to the data.

As a final step to complete the quality assessment, exploratory data analysis and calculation of descriptive statistics are encouraged for all sources. Each dataset contains certain information and can be processed and visualized in custom ways. Therefore, universally applicable, strict guidelines to complete this procedure do not exist. Since this study is focused on mobility data, researchers are encouraged to consider map visualizations for spatiotemporal data and aggregated time series graphs for transactional datasets. Extensive sources with huge volume of data will require intensive computational effort. In such cases it is advised to utilize subsets with sensible amount of data, preferably for the most recent period of time possible.

3 Case Study

3.1 Thessaloniki Data Catalogue

In this section the mobility data catalogue created for Thessaloniki is presented. It contains nineteen sets of data which span across the five categories defined in Sect. 2.

The main enabler of this extensive data collection is the Thessaloniki Smart Mobility Living Lab "ThessM@LL", a lab of the Hellenic Institute of Transport (HIT) which belongs to the Centre for Research and Technology Hellas. ThessM@LL was created in 2008 and is a member of the European Network of Living Labs (ENoLL). One of the lab's targets is to operate as a content aggregator, gathering multi-domain mobility data for Thessaloniki to enable the implementation and development of Transport domain studies and applications.

ThessM@LL owns and maintains technologically advanced equipment for automated mobility data recording. For instance, it monitors a network of 43 Bluetooth detectors located in key crossroads of the city, which sense Bluetooth-enabled devices in their vicinity. Further, the lab has created an extensive data repository from previously completed projects including transport network data, points of interest for the city, sociodemographic data, survey data etc. Lastly, ThessM@LL has cooperated with numerous public and private sector organizations in the framework of several projects and has created a local ecosystem in which entities collaborate and exchange data and information. As an outcome, the lab has real time access to data recorded by infrastructure owned by external entities, for instance data from traffic counters in the city maintained by public authorities, floating car data recorded from a taxi fleet of approximately half the city's taxis; more than 1.000 vehicles in total, and data from the city's bike sharing system.

The mobility data catalogue for Thessaloniki contains all related data sources available at ThessM@LL. Using the categorization defined in Sect. 2, the available data sources in the catalogue for Thessaloniki are presented in Table 2.

Table 2. Urban mobility related data sources identified in Thessaloniki use case.

Main Categories	Subcategories
Transport supply data sources	Public Transport Schedules and Lines, Transport Network, Taxi Service Car/Moto-Sharing, Bike-sharing
Transport demand data sources	Bike-sharing, Mobility Surveys, Car/Moto-Sharing, Traffic, Taxi Service, Social Media
Maps & Cartography data sources	Land Use, Weather, Points of Interest
Socio-demographic data sources	Census, Income statistics, Car Ownership, Labour/Unemployment statistics,
Travel times data sources	Travel Times

3.2 Thessaloniki Data Quality Assessment

In this section we present the results after the application of data quality assessment on the data catalogue for the city of Thessaloniki. The grouping in five main categories is also adopted this section.

Regarding the Transport supply data sources, all five available subcategories are generally assessed as high quality in all three categories (reliability, resolution, availability-accessibility). A comment has been noted for the taxi service supply source, to underline that the source is generated by one of several taxi fleets operating in the city. However, this fleet has approximately half the city's taxis and operates in all the region offering all available booking procedures (telephone, mobile app, traditional taxi stops etc.) therefore no bias is identified. Another comment focuses on the temporal scope of the Car/Moto sharing, as the database has been recorded since February 2020 with available for rental e-scooters records.

The Transport demand data sources are also evaluated as high quality, while several comments have been noted. The recency of the Car/Moto sharing demand data is also noted. The main concern for the traffic data subcategory is that the floating vehicles recording the data are taxis and not private vehicles, which may cause some particularities as professional drivers may behave differently. Lastly, the social media data present biases towards social media users which are probably not representative of the population. This assumption cannot be investigated statistically as the data are non-personal, aggregated counts of visitors per venue.

Maps & Cartography data sources for Thessaloniki are of high quality. In the comments it is mentioned that the last update for the data subcategories "land use" and "points of interest" is unknown.

The Socio-demographic data sources available in Thessaloniki are considered of medium quality. A common issue of the census, car ownership and labor/unemployment statistics data sources is that their last update occurred in 2011. The income data are available from a survey and not from tax income statistics.

The travel times for Thessaloniki are evaluated as high-quality data, however, it is noted that the times are only available for driving mode and not for other modes, for instance pedestrians or bicycle users.

Following the qualitative assessment, exploratory data analysis was performed along with the calculation of certain descriptive statistics to provide deeper insights on the quality of data sources. Several data sources for Thessaloniki, for instance the taxi generated floating car data, have a large temporal scope and high recording frequency, resulting in huge amounts of data. For the analysis of those sources a data subset was used containing all records during the autumn of 2019.

For the bike-sharing demand data source, the demand variability timeseries and duration distribution histogram were plotted (Fig. 1).

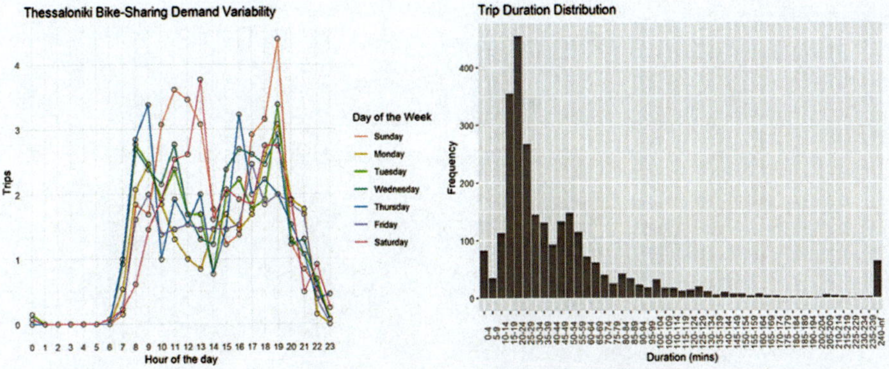

Fig. 1. Bike-sharing demand temporal variability and trip duration distribution.

For the traffic data source, the validity is visually evaluated by investigating the distributions of longitude and latitude coordinates (Fig. 2).

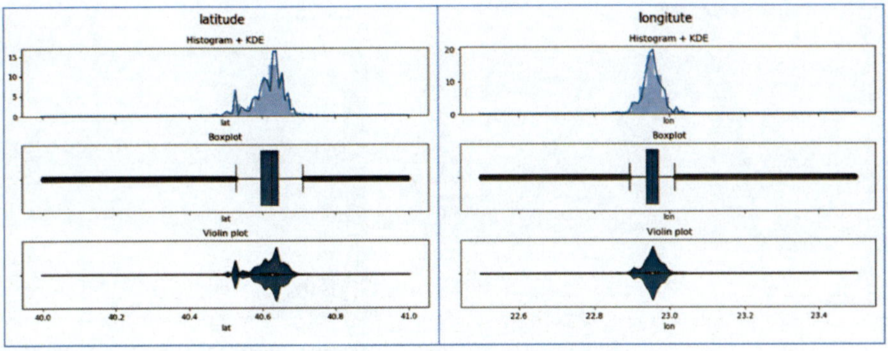

Fig. 2. Validity analysis of traffic data.

The speeds reference the Open Street Maps network and therefore can provide map visualizations (Fig. 3), where the colors used for streets indicate the traffic flow speeds.

The taxi service demand exploratory data analysis may contain graphs about demand variability in time and trip distance distribution (Fig. 4).

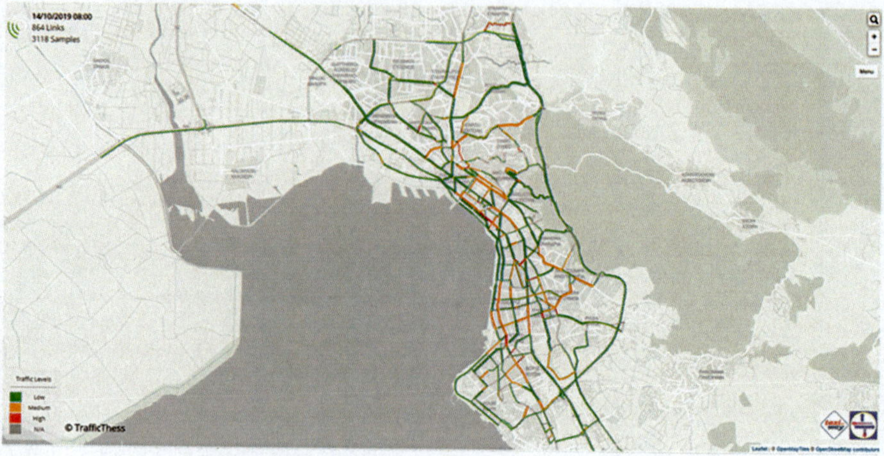

Fig. 3. Map visualization of congestion levels utilizing traffic data.

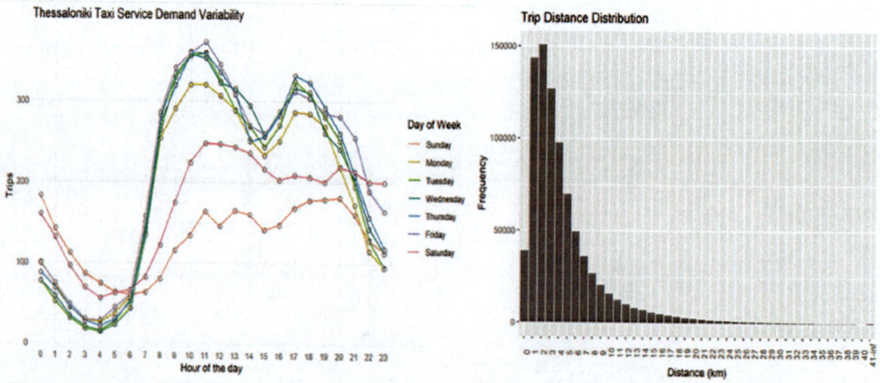

Fig. 4. Taxi service demand temporal variability and trip distance distribution.

The proposed exploratory data analysis for the social media data contains a time-series check-in events graph (Fig. 5).

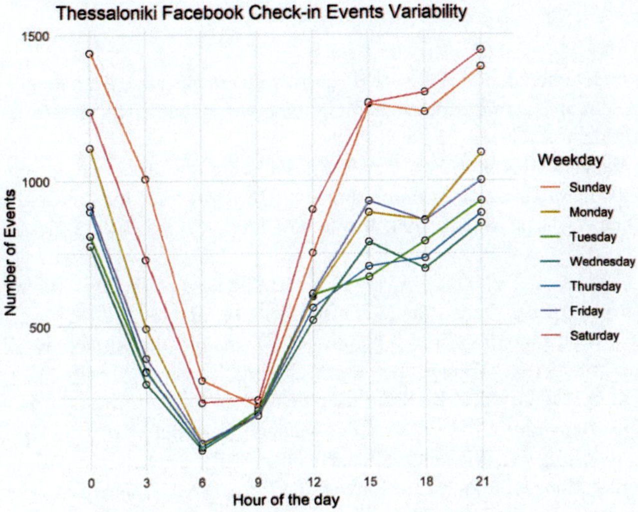

Fig. 5. Facebook check-in events temporal variability

4 Conclusions

This study presents a holistic mobility data catalogue for the city of Thessaloniki, containing nineteen sources and databases retrieved by privately-owned equipment or through collaborations with public and private organizations. Further, the paper contains a multi-perspective quality assessment methodology and its results after the application on the available data sources.

In general, the data catalogue for Thessaloniki is evaluated as high quality in all three aspects of reliability, resolution, and availability-accessibility. ThessM@LL, the lab that hosts this data repository, has achieved this quality standards by maintaining and upgrading its infrastructure both in terms of software and hardware, but also by leveraging collaborations with the most important mobility providers or controllers of the city. The high quality of data reliably enables their application on several projects and services; many of those have already been successfully implemented by HIT.

However, some weaknesses have also been identified. They stem from external data sources and specifically from census data. In Greece, the census is conducted every ten years, therefore the data may be outdated. However, this issue is detrimental only to a certain degree, as those data categories are mainly aggregated, holistic population statistics which are not changing rapidly. As a result, using the information generated up to some years ago is not expected to limit the potential of any application.

Acknowledgements. This study was realized within the framework of MOMENTUM project, an EU Horizon 2020 programme, funded under grant agreement No. 815069.

References

1. Regulating the internet giants - The world's most valuable resource is no longer oil, but data. https://www.economist.com/leaders/2017/05/06/the-worlds-most-valuable-resource-is-no-longer-oil-but-data. Accessed 05 April 2020
2. Tang, B., Yang, C., Xiang, L., Zeng, J.: Deriving real-time city crowd flows by heterogeneous big urban data. In: Proceedings - 2018 IEEE International Conference on Big Data, Big Data 2018, pp. 3485–3494. Institute of Electrical and Electronics Engineers Inc. (2019)
3. Lv, Y., Duan, Y., Kang, W., Li, Z., Wang, F.Y.: Traffic flow prediction with big data: a deep learning approach. IEEE Trans. Intell. Transp. Syst. **16**, 865–873 (2015)
4. Antoniou, C., Koutsopoulos, H.N., Yannis, G.: Dynamic data-driven local traffic state estimation and prediction. Transp. Res. Part C Emerg. Technol. **34**, 89–107 (2013)
5. Fu, G., Han, G.Q., Lu, F., Xu, Z.X.: Short-term traffic flow forecasting model based on support vector machine regression. J. S. China Univ. Technol. (Nat. Sci.) **41**(9), 71–76 (2013)
6. Suarez, P., Anderson, W., Mahal, V., Lakshmanan, T.R.: Impacts of flooding and climate change on urban transportation: a systemwide performance assessment of the boston metro area. Transp. Res. Part D Transp. Environ. **10**, 231–244 (2005)
7. Alic, A.S., Almeida, J., Aloisio, G., Andrade, N., Antunes, N., Ardagna, D., Badia, R.M., Basso, T., Blanquer, I., Braz, T., Brito, A., Elia, D., Fiore, S., Guedes, D., Lattuada, M., Lezzi, D., Maciel, M., Meira, W., Mestre, D., Moraes, R., Morais, F., Pires, C.E., Kozievitch, N.P., dos Santos, W., Silva, P., Vieira, M.: BIGSEA: a big data analytics platform for public transportation information. Futur. Gener. Comput. Syst. **96**, 243–269 (2019)
8. Guo, Y., Zhang, J., Zhang, Y.: A method of traffic congestion state detection based on mobile big data. In: 2017 IEEE 2nd International Conference on Big Data Analysis, ICBDA 2017, pp. 489–493. Institute of Electrical and Electronics Engineers Inc. (2017)
9. Adoni, W.Y.H., Nahhal, T., Aghezzaf, B., Elbyed, A.: The mapreduce-based approach to improve vehicle controls on big traffic events. In: 2017 International Colloquium on Logistics and Supply Chain Management: Competitiveness and Innovation in Automobile and Aeronautics Industries, LOGISTIQUA 2017, pp. 1–6. Institute of Electrical and Electronics Engineers Inc. (2017)
10. Salanova, J.M., Romeu, M.E., Amat, C.: Aggregated modeling of urban taxi services. Procedia Soc. Behav. Sci. **160**, 352–361 (2014)
11. Meenakshi, S., Senthilkumar, R.: Efficient taxi dispatching system in distributed environment. In: IEEE International Conference on Information, Communication, Instrumentation and Control, ICICIC 2017, pp. 1–6. Institute of Electrical and Electronics Engineers Inc. (2018)
12. Howard, A.J., Lee, T., Mahar, S., Intrevado, P., Myung-Kyung Woodbridge, D.: Distributed data analytics framework for smart transportation. In: 2018 IEEE 20th International Conference on High Performance Computing and Communications, IEEE 16th International Conference on Smart City, IEEE 4th International Conference on Data Science and Systems (HPCC/SmartCity/DSS), pp. 1374–1380 (2018)
13. Xie, J., Luo, J.: Construction for the city taxi trajectory data analysis system by Hadoop platform. In: 2017 IEEE 2nd International Conference on Big Data Analysis, ICBDA 2017, pp. 527–531. Institute of Electrical and Electronics Engineers Inc. (2017)
14. Shang, Z., Li, G., Bao, Z.: DITA: distributed in-memory trajectory analytics. In: Proceedings of the ACM SIGMOD International Conference on Management of Data, pp. 725–740. Association for Computing Machinery, New York, New York, USA (2018)

Integrating Modelling in Urban Policy Cycle and Decision Making

Georgia Ayfantopoulou, Maria Natalia Konstantinidou(✉), Maria Chatziathanasiou, and Josep Maria Salanova Grau

Hellenic Institute of Transport- Centre for Research and Technology Hellas, Thessaloniki, Greece
{gea, nconstantinidou, mariacha, jose}@certh.gr

Abstract. Urban mobility is subject to deep changes, imposed by technological advancements and new concepts of operation in transport and serious challenges on urban space are expected to be introduced. Transport modelling should follow the urban mobility trends to connect the new arrays of demands and the new services they will provide, guiding future mobility. The integration of urban mobility trends into mobility planning, through the 8 principles and the 4 phases of Sustainable Urban Mobility Plans (SUMPs) to support policy goals, is essential, as well. Modelling the complexity and uncertainty enforced by the urban mobility trends necessitate a move from common modelling techniques? How to accomplish a transition in mobility planning shifting from conventional and static planning approaches? Are current transport modelling approaches able to integrate urban mobility trends, to fulfil SUMP's implementation? Traditional planning and modelling methods cannot face the challenges of urban mobility trends integration. Long-term planning, collaborative and participatory planning processes, co-creation of scenarios for the future and measure packages using specialized tools on the one hand, multi-sectoral and dynamic modelling approaches, the Big Data potentials on the other hand, are cases of on-going research for transport planning and modelling new requirements respectively, identified through an extensive literature review. The integration of transport modelling requirements, in the urban policy cycle is anticipated to guarantee the reliability of the modelled scenarios, ensure the accuracy of modeling outcomes and offer guidance on how to accomplish the integration of data-driven methods in participatory and collaborative planning approaches.

Keywords: Transport modelling · SUMP · Decision making

1 Introduction

Approximately 50 years ago, a publication related to futurology of transport, highlighted a variety of elements, that would stimulate and change transport. It indicated that electrical energy would determine transport systems and future vehicles would be powered by electricity, the car driver would be able to dial the destination into a selector and do not drive until journey's end. Public transport vehicles would be scheduled to travel to predetermined destinations transporting 4 to 12 passengers, without drivers. Public transport stations would be frequently spaced, and public

E. G. Nathanail et al. (Eds.): CSUM2020, AISC 1278, pp. 1149–1158, 2021.
https://doi.org/10.1007/978-3-030-61075-3_110

transport systems would be able to be joined to urban structures (residencies, building) as well as cross under or over the current road network. A moving footway, like a horizontal escalator, would enable pedestrians to be accelerated to the speed of 10 miles per hour. At the end of this research it is stated that *"It remains to be seen whether these systems ever become part of normal life"* [1].

Announcements about the future can, subsequently, fall in or out since future is uncertain and in this lies the complexity for planners and policymakers. The process of planning for the future, is exceeding forecasting, it is targeting to develop a finer future [2]. Deep changes arising in the way urban mobility is progressing directly associated with technological trends in urban mobility and trends refer to the operation of transport systems, for instance Mobility as a Service (MaaS), Connected and Automated Vehicles (CAVs), Automated Vehicles (AVs), Drones, Big Data and more urban mobility trends.

It is understood that urban mobility trends are enforcing alternative transport planning and modelling approaches. Current transport planning and modelling approaches cannot fulfil urban mobility trends implementation in the SUMP policy cycle. Under the uncertainty and complicacy that describes urban mobility trends, in terms of impacts, regulatory framework and ownership models, transport planning should follow the urban mobility trends, in order to provide useful guidance on future mobility and to fill the gap correspond to the array of demands urban mobility trends will induce.

This paper lays on the MOMENTUM EU project, an ongoing European funded project to apprehend the consequences of the upcoming mobility options on the urban mobility ecosystem, to support metropolitan areas in the mission of creating the correct planning and policy approach and to take advantage of the full potential of these evolving urban mobility trends by developing the context for creating innovative data analysis methods, models as well as tools to support the planning process.

Primarily, an extensive literature review of the current transport planning and modelling approaches is attempting, in parallel with the analysis of the existing methodology of transport modelling integration into the SUMP policy cycle. Afterwards, the necessity for new requirements for transport modelling is investigated, as a result of the emergence of urban mobility trends. In the end, future transport planning requirements are specified and in combination with the previous analysis related to future transport modelling requirements, there is an effort to reveal key conditions for modelling integration in the SUMP policy cycle.

2 Outline of the Sustainable Urban Mobility Planning in Europe

Taking into consideration that local and regional authorities are the responsible policy making operators, in terms of transport planning, strategy structure, regulations and plans, there are additionally European guidelines providing the planning methodology, through a policy document. A Sustainable Urban Mobility Plan (SUMP) is a strategic plan outlined by the Urban Mobility Package (2013) from the European Commission (EC). The concept of this strategic plan has been revised and supplemented with numerous guides as well as briefings as regards particular topics (e.g. Mobility as a Service, electrification, sustainable urban logistics plan). The most recent revision of the SUMP guidelines [3] issued in October 2019.

A SUMP is targeting to enhance the living conditions in an urban space, fulfilling the mobility and accessibility requirements of individuals. It reflects the principles of integration, participation, and evaluation. Additionally, a SUMP is supported by 8 principles. The principles of the SUMP planning document [3] are demonstrated below:

- Plan for sustainable mobility in the 'functional city'
- Cooperate across institutional boundaries
- Involve citizens and stakeholders
- Assess current and future performance
- Define a long-term vision and a clear implementation plan
- Develop all transport modes in an integrated manner
- Arrange for monitoring and evaluation
- Assure quality.

The decision-making approach expressed by a SUMP contains 4 phases represented through a cycle, an urban policy cycle (Fig. 1). A SUMP, apart from the 4 phases, consists of 12 steps corresponding to the phases and all steps involve specific activities (32 activities in overall). The 4 phases indicated by the SUMP planning document [3] are "Preparation and analysis", "Strategy development", "Measure planning", "Implementation and monitoring".

Fig. 1. Sump policy cycle [3]

The phase "Preparation and analysis" is setting the basis for the planning process. The human, institutional, financial resources are explored and working structures are developing. In parallel, the outline of the mobility conditions of the study area is being shaped e.g. traffic flow analysis, mode shares. The specification of requirements in terms of planning timeline, functional urban area, legal requirements, and external support is a crucial process at this phase. The problems and opportunities analysis mark the completion of this phase. Reaching a common awareness of the problems and opportunities among relevant actors (e.g. decision makers, stakeholders, citizens) encourage and strengthen the levels of collaboration among them.

The phase "Strategy development" defines the cooperative formulation of scenarios, vision, goals, objectives, and indicators at a strategic level among citizens and stakeholders. The decisions of this phase offer a stable context for the sequent phase.

The phase "Measure planning" indicates the transition from the strategic to the operational phase. Measure packages described comprehensively with budget and risk appraisal are important components of this phase, but not without the acceptance of citizens, stakeholders, and political representatives.

The phase "Implementation and monitoring" indicates the implementation of the determined measures and their monitoring and assessment. Additionally, this final phase indicates the end of the SUMP cycle and in parallel the beginning of a new process, signifying the circular and continuous structure of the process suggested by the SUMP policy cycle.

The mobility planning process represented by the SUMP cycle is normally developed with the use of transport modelling or alternative tools used for policy-making. The selection of the modeling technique and the specific tool that will assist the SUMP planning process is undertaken from the responsible bodies, such as municipalities and transport planners. Factors influencing this decision refer to the acquirement of a former transport model, availability of data and budget constraints.

3 The Role of Transport Models in Mobility Planning

Transport modelling can reinforce a SUMP in several stages. Modelling can be a useful technique when developing the strategy of a SUMP, formulating the scenarios of potential futures as well as when predicting the impacts of measures. The outputs of transport models, complemented with economic analysis are graphic representations describing the outcome, the impact and the evaluation of a strategy and measure planning of the SUMP process, setting breeding ground for political discussion and public involvement [3, 4].

There are many different structural issues that should be considered to formulate a transport model. Basic structural issues refer to the level of planning detail (macroscopic, mesoscopic, microscopic), the level of facts certainty (deterministic, stochastic), the time perception (dynamic, static), the level of data detail (aggregation, disaggregation), the spatial aggregation (individuals, households), the operational capabilities (uni-modal, 4step model, LUTI, Activity-Based Model) and more. Additionally, the input variables determine the output variables, if the transport model has a national level of detail the output variables are expressed in terms of total mileage, total number of cars etc. The decision about the transport modelling structure is part of the early stages of the SUMP planning process. The choice of the most appropriate model structure needs to take into consideration monetary, time resources and the planning needs of the specific area.

The study of the literature demonstrated several approaches of transport modelling integration into the SUMP cycle. In these cases, the unravelling of the SUMP cycle phases along with the development of a transport model is aiming at providing a correlation between the model's competences and the different phases, steps, and activities of the SUMP. Transport modelling follows the iterative process of the SUMP policy cycle. Some key findings concerning the integration of transport models in the SUMP policy cycle, are related to the choice of the type of the transport model, the scope of the transport model in agreement with the scope of the SUMP, the phases and the activities of SUMP policy cycle in which the transport model is involved and also the time horizon of the transport model in consistency with the SUMP time horizons.

The approach of Okraszewska R. et al. (2008) indicates that since the primary stages of a SUMP, it should be identified if a transport model is required within the SUMP context or if an existing model will be updated. The initial phase of the SUMP can be related to the shaping of a meso-macro scopic model. At the second phase of the SUMP the transport model can be used as a supportive tool to represent future scenarios and the effects of the planned scenarios. A transport model can contain several levels of detail to assess the alternative scenarios. The basic layer can be the macro scopic model, which can function as an autonomous model accompanied by a meso and a micro scopic model. As a result of its various levels of detail, the transport model can offer insights at different levels of detail e.g. scenario analysis, effect of measures on modal split, assessment of traffic conditions, evaluation of detailed changes to traffic organization. At the third phase of the SUMP, the indicators for monitoring and evaluation are identified and then calculated by the model. The process of evaluation is applied for determining factors and indicators, in order to monitor and assess the measures. At the final phase of the SUMP, decision-makers adopt the SUMP and the city is committed to implement it. After the SUMP implementation, monitor and evaluation is taking place regularly by using the transport model, as well [5].

Another case refers to the integration of a LUTI model (Land Use Transport Interaction) into the SUMP policy cycle, introduced by Gavanas N. et al. (2016). At the initial phase of this integration process the standards of the LUTI model are setting e.g. the model's capabilities with respect to the requirements of a study area and the development of strategic scenarios. At the second phase, the model can be modified based on the goals agreed, aiming at providing thorough predictions of the impacts resulting from the decided measures for enhancing urban mobility and the land use system. After formulating the scenarios according to a synthesis of transport connected measures, the model can be a useful tool for appraising the consequences of the selected measures on the land uses and the decision making process of the SUMP by recognizing the most efficient measures.

In the third phase of the SUMP the monitoring of the selected indicators can provide with data the LUTI model, therefore it can be modified accordingly. Furthermore, it is capable of delivering precise estimates to examine the progress throughout the implementation of the SUMP. In parallel, these estimations can be exploited as contributions to the LUTI model, in order to revise it. Next, the revised model can be used to re-evaluate the target-based scenarios employing actual data and conclude on the improvement of the implementation of the SUMP as well as the success of its objectives as regards the urban space. The purpose of the final phase is the complete validation of the LUTI model to the SUMP's last milestone and to perform the essential modifications for the following SUMP. The model should be in line with the entire SUMP cycle [6]. An additional case, consistent with the integration framework of Gavanas N. et al. (2016) proposed by Pozoukidou G. et al. (2015, 2017) [7, 8].

The approach of Karon G et al. (2017) indicates that a transport model, integrated into a SUMP, should be applied to evaluate the results of the implementation of specific actions, enhancing sustainable mobility. A transport model integrated into a SUMP can be a 4-step trip-based travel demand model, consisting of the sub-models: trip generation (production and attraction), trip distribution, mode choice as well as distribution of traffic flow on the transport network. It is necessary to consider, the actions enhancing sustainable mobility and the outcome of their implementation. The actions enhancing sustainable mobility should be represented in the transport network of the model and the outcomes of actions enhancing sustainable mobility should be represented in the model's functions. Moreover, it is necessary to consider the SUMP implementation, in which specific actions are implemented in various time periods. Therefore, the assessment of the SUMP requires to prepare transport models for specified time horizons, for which the planned results of actions adopted in the plan are assumed to occur [4].

According to the SUMP guidelines (2019), the integration of the most suitable transport model requires assessing aspects, referring to addressing the issue, the level of decision-making, the desired accuracy of predictions, the detail of the expected results and the availability of data [3]:

Another approach [9], describing the transport modelling integration into the policy cycle indicates that a transport model can provide demand data for scenario analysis, designing and dimensioning new infrastructure or operational service. A transport model can assist identifying the effect of a transport plan on transport flows, demonstrating the way demand varies in relation to new infrastructure establishment and the conditions resulting from new infrastructure implementation. Additionally, a transport model can enhance the understanding of the way transport conditions will transform in response to changes in the number of inhabitants and employees, economic activity, car ownership and growth patterns and provide with standards of network capacity and necessity for further capacity.

4 Are Urban Mobility Trends Suggesting Different Transport Planning and Modelling Approaches?

Deep changes arising in the way urban mobility is progressing accelerated by advances in technological trends in urban mobility as well as trends refer to the operation of transport systems, for instance Mobility as a Service (MaaS), Demand Responsive Transport (DRT), Car-sharing, Moto-sharing, Bike-sharing, Micro mobility, Automated Vehicles (AVs), Drones and more. The SUMP policy cycle should take into consideration the upcoming revolution in urban mobility and shape a proactive approach to employ the positive aspects and minimize the unintended impacts. The transport models are expected to support decision-makers in exploring possible directions for urban mobility in the context of sustainability and similarly to meet the requirements of the urban mobility trends.

Urban mobility trends are accompanied with multiple impact scenarios, positive and negative ones. On the one hand, urban mobility trends, such as MaaS, are aligned with sustainability objectives, fact demonstrated by the connection among MaaS, public transport and active modes [10] and other urban mobility trends is possible to produce databases with travel data, which can be used for mobility planning and management [11]. On the other hand, urban mobility trends are accompanied with lack of a regulatory context with regards to business and ownership models [10]. Deciding not to structure or control their implementation, there is a risk to be introduced in urban areas, without the opportunity of planning.

Another example refers to Automation, which is considered to have a positive effect on public transport, in terms of efficiency, because it can provide to dynamic systems and it can reduce operation costs. Moreover, automated cars can improve the level of traffic safety, if the technologies are credible, facilitating the implementation of traffic calming measures. However, private CAVs involve threats related to sustainability. More detailed, if automation is integrated in private cars at accessible values, it is possible to boost the attractiveness of private cars and benefit the shift from public transport and active mobility to private car. Likewise, the urban space can be affected because it can enable urban sprawl conditions [12].

As it appears, a failure to manage the upcoming mobility changes effectively, will result in adverse effects on the urban environment and the society. These potential impact scenarios may lead to policy dilemmas that are going to be undertaken by transport and urban planners, transport modelers, relevant stakeholders, and citizens.

5 Modelling Integration in Urban Policy Cycle

Future mobility schemes should be prepared to exploit the opportunities of urban transport trends and reduce the unplanned effects of changes in the way individuals perform their trips. Current transport planning and modelling approaches are not able to integrate urban transport trends if they cannot face their challenges and impacts. A comprehensive literature review shed light on key transport planning and modelling requirements for the integration of the urban transport trends in the SUMP policy cycle.

During the primary phase of the SUMP policy cycle, where the setting up of the working structures, the planning framework and the analysis of the mobility situation are taking place the gathering of citizens and relevant actors is essential. A collaborative policy approach can lead towards high levels of participation among the actors involved. A collaborative policy approach requires dialogue with a variety of ideas and views to address problems and opportunities [12]. Citizens are directly affected by the urban transport trends and public or private actors related to urban transport trends, have different objectives, possibly conflicting, so their participation is crucial. Discussion and co-creation can cultivate trust among them [13].

Furthermore during the primary phase of the SUMP policy cycle, the use of a quantitative technique, such as transport modeling is valuable to review and evaluate the performance of the current situation, provide evidence for discussion for citizens and actors, enhancing the cooperative process, while delivering baseline indicators - i.e. modal split, route choice, volume to capacity ratios. To the extent of uncertainty imposed by urban mobility trends as regards the interaction of the transport sector with other sectors, transport modelling can be multi-sectoral/cross-sectoral, which means to consider other related sectors e.g. environment, energy, economy and consult national/international studies of economies, in order to identify and investigate future impacts, while perceiving travelling in time-dependent terms [12, 14, 15]. Examples of interaction areas include: interactions with the energy sector as electric vehicles are emerging, interactions with the public health sector from air pollution and active travel, interactions with society in terms of equity issues emerging from transport interventions.

During the second phase of the SUMP policy cycle, where the mobility strategy is developing, citizens and relevant actors supported by transport modelling outputs, can create the future scenarios, vision, set targets and indicators with longer time horizons, as a consequence of urban mobility trends uncertainty. The discussion of future scenarios between citizens and relevant actors can be evidence-based, using the model outputs and though cautious explanation of the modelling outputs to the citizens. The transport model can provide useful insights for the identification of indicators for all objectives and quantifiable targets.

The objectives of the second phase of the SUMP should balance among modal share development of environmentally friendly mobility alternatives, multimodality, decrease of private car trips, improved accessibility for individuals, users engagement in socially responsible behaviour within the community, enhancement of air quality and health of citizens and sustainability [13]. Transport models to satisfy the strategy development of the SUMP policy cycle, should elaborate on the aspect of multi-modality in urban mobility and should be capable of analysing private vehicles and public transport interactions. Additionally, dynamic integrated modelling with the consideration of several routing scenarios (e.g. a taxi or a car cruising for passengers or parking respectively) is developing by transport modelling software providers and the traffic dynamics of vehicles in a road network for promoting sustainability and protection of the environment, are under the microscope [12].

During the third phase of the SUMP policy cycle, policymakers are possible to face challenges with regards to a shift from collective modes to individualised modes or a shift from sustainable to motorised modes. Measure planning should take into consideration, the aforementioned undesirable effects and the protection of a level playing field amongst transport operators [13]. Transport models can contribute to the assessment of packages of measures and forecast the impacts of measures on the transport system. Furthermore, public authorities might plan the transition from the traditional mobility management towards a multimodal mobility management by exploiting the data exchange mechanisms scheme can offer.

Innovative procurement could be considered to guarantee that the decided solutions favour the transport users and do not create bottlenecks or monopolies. In addition, public procurement can be used to set requirements for data ownership or sharing or the management of open Application Programming Interfaces (APIs). [13] In the ultimate phase, transport models can advance monitoring and evaluation, for the strategic as well as measure level indicators that are estimated through the model, in order to assess the successes and failures and finally for measure adaptation and new indicators' definition. It seems possible that enhancements e.g. e-ticketing, mobile data, social media data, in Big Data availability that urban mobility trends can offer, could feed transport models and allow post-SUMP evaluation to take place in a consistent and accurate way [16]. Big Data can reinforce the link between evaluation and better use of evidence.

6 Conclusions

Future mobility is characterized by increased levels of uncertainty induced by the impacts of the urban mobility trends. Potential unintended effects of public authorities' weakness to develop policies to prevent a shift from collective modes to individual-based modes or a shift from sustainable to motorised modes, effective decision-making or unsuccessful adoption of quality standards for the new mobility providers, may result in reduced accessibility, spatial proximity, environmental issues, inequity issues and more adverse scenario impacts. It is obvious, that the urban mobility trends are imposing alternative transport planning and modelling requirements.

The formation and implementation of innovative mobility planning and transport modelling requirements is an important issue under research. Remarkable transport planning requirements indicate long-term planning, collaborative/participatory policy approach among decision makers, citizens and relevant stakeholders, scenario-based planning to tackle uncertainty, co-creation of future scenarios and vision among the involved actors, data ownership or management issues solving as well as regulation formulation for ownership and business models. Multi-sectoral or in some cases cross-sectoral modelling methodology, implementation of dynamic instead of static modelling approach and Big Data use are cases of on-going research lines with regards to transport modeling.

The integration of new transport modelling requirements, in the urban policy cycle is anticipated to guarantee the relativity of the modelled scenarios, ensure the confidence level of modeling outcomes and offer valuable insights on how to achieve a reliable integration of data driven methods in participatory and collaborative planning approaches. The advancement of this transport planning process with the integration of transport modelling is a key requirement to facilitate the collaborative policy approach through evidence, such as the visualization of the outcomes of future scenarios or the impacts of the urban mobility trends.

References

1. Harris, P.: Futurology of transport. Police J. **46**(3), 280–284 (1973)
2. Myers, D., Kitsuse, A.: Constructing the future in planning: a survey of theories and tools. J. Plann. Educ. Res. **29**(3), 221–231 (2000)
3. European Platform on Sustainable Urban Mobility Plans: Guidelines for developing and implementing a sustainable urban Mobility Plan. Second edition. https://www.eltis.org/mobility-plans/sump-guidelines. Accessed 24 July 2020

4. Karon, G., Krawczyk, G., Urbanek, A., Mikulski, J.: Sustainable urban mobility planning (SUMP) at subregional area level with the use of transportation model. Arch. Transp. Syst. Telemat. **10**(2), 19–26 (2017)
5. Okraszewska, R., Romanowska, A., Wołek, M., Oskarbski, J., Birr, K., Jamroz, K.: Integration of a multilevel transport system model into sustainable urban mobility planning. Sustainability **10**(2), 479 (2018). https://doi.org/10.3390/su10020479
6. Gavanas, N., Pozoukidou, G., Verani, E.: Integration of LUTI models into sustainable urban mobility plans (SUMPS). Eur. J. Environ. Sci. **6**, 11–17 (2016). https://doi.org/10.14712/23361964.2016.3
7. Pozoukidou, G., Gavanas, N., Verani, E.: Land use models and sustainable urban mobility plans: an integrative approach for strategic planning. Int. J. Sustain. Dev. Plann. **12**(5), 867–882 (2017). https://doi.org/10.2495/SDP-V12-N5-867-882
8. Pozoukidou, G., Gavanas, N., Verani, E.: Potential contribution of LUTI models in contemporary strategic planning for urban mobility: a case study of the metropolitan area of Thessaloniki. WIT Trans. Built Environ. **2**, 843–854 (2015)
9. Joint Assistance in Supporting Projects in European Regions: The Use of Transport Models in Transport Planning and Project Appraisal. JASPERS Appraisal Guidance. http://www.jaspersnetwork.org/display/HOME/Homepage. Accessed 24 July 2020
10. Arthur, D.: Little Future Lab: The Future of Mobility 3.0 - Reinventing mobility in the era of disruption and creativity. https://www.adlittle.com/en/insights/viewpoints/future-mobility-30. Accessed 24 July 2020
11. POLIS: Mobility as a Service: Implications for urban and regional Transport. https://www.polisnetwork.eu/wp-content/uploads/2017/12/polis-maas-discussion-paper-2017-final_.pdf. Accessed 24 July 2020
12. Burrieza, J., Djukic, T., Aifadopoulou, G., Salanova Grau, J.M., Masegosa, D.A., Rojo, M. J., Ripa, F., Frederix, R., Pápics, P., Himpe, W., Antoniou, C., Narayanan, S., Bousse, Y., Fabianski, C.: New Mobility Options and Urban Mobility: Challenges and Opportunities for Transport Planning and Modelling. https://h2020-momentum.eu/wp-content/uploads/2020/01/MOMENTUM-D2.1-New-Mobility-Options-and-Urban-Mobility.pdf. Accessed 24 July 2020
13. European Platform on Sustainable Urban Mobility Plans: Mobility as a Service (MAAS) and Sustainable Urban Mobility Planning. Practitioner Briefing. https://www.eltis.org/mobility-plans/topic-guides-and-practitioner-briefings. Accessed 24 July 2020
14. Blainey, S.P., Preston, J.M.: Predict or prophesy? Issues and trade-offs in modelling long-term transport infrastructure demand and capacity. Transp. Policy **74**, 165–173 (2019). https://doi.org/10.1016/j.tranpol.2018.12.001
15. Ben-Akiva, M., Bottom, J., Gao, S., Koutsopoulos, H.N., Wen, Y.: Towards disaggregate dynamic travel forecasting models. Tsinghua Sci. Technol. **12**(2), 115–130 (2007). https://doi.org/10.1016/S10070214(07)70019-6
16. Department for Transport: Appraisal and Modelling Strategy Consultation, Consultation Responses. Responses 41–61. https://assets.publishing.service.gov.uk/government/uploads/system/uploads/attachment_data/file/793025/responses-41-61-final.pdf. Accessed 24 July 2020

Author Index

E. G. Nathanail et al. (Eds.): CSUM2020, AISC 1278, pp. 1159–1162, 2021.
https://doi.org/10.1007/978-3-030-61075-3